T0221755

Modern Condensed Matter Physics

Modern Condensed Matter Physics brings together the most important advances in the field from recent decades. It provides instructors teaching graduate-level condensed matter courses with a comprehensive and in-depth textbook that will prepare graduate students for research or further study alongside reading more advanced and specialized books and research literature in the field.

This textbook covers the basics of crystalline solids as well as analogous optical lattices and photonic crystals, while discussing cutting-edge topics such as disordered systems, mesoscopic systems, many-body systems, quantum magnetism, Bose–Einstein condensates, quantum entanglement, and superconducting quantum bits.

Students are provided with the appropriate mathematical background to understand the topological concepts that have been permeating the field, together with numerous physical examples ranging from the fractional quantum Hall effect to topological insulators, the toric code, and Majorana fermions. Exercises, commentary boxes, and appendices afford guidance and feedback for beginners and experts alike.

Steven M. Girvin received his BS in 1971 from Bates College and his PhD in 1977 from Princeton University. He joined the Yale faculty in 2001, where he is Eugene Higgins Professor of Physics and Professor of Applied Physics. From 2007 to 2017 he served as Deputy Provost for Research. His research interests focus on theoretical condensed matter physics, quantum optics, and quantum computation; he is co-developer of the circuit QED paradigm for quantum computation.

Honors: Fellow of the American Physical Society, the American Association for the Advancement of Science, and the American Academy of Arts and Sciences; Foreign Member of the Royal Swedish Academy of Sciences, Member of the US National Academy of Sciences; Oliver E. Buckley Prize of the American Physical Society (2007); Honorary doctorate, Chalmers University of Technology (2017); Conde Award for Teaching Excellence (2003).

Kun Yang received his BS in 1989 from Fudan University and his PhD in 1994 from Indiana University. In 1999 he joined the faculty of Florida State University, where he is now McKenzie Professor of Physics. His research focuses on many-particle physics in condensed matter and trapped-cold-atom systems.

Honors: Fellow of the American Physical Society and the American Association for the Advancement of Science; Alfred Sloan Research Fellowship (1999); Outstanding Young Researcher Award, Overseas Chinese Physics Association (2003).

Modern Condensed Matter Physics

STEVEN M. GIRVIN
Yale University, Connecticut

KUN YANG
Florida State University

CAMBRIDGE
UNIVERSITY PRESS

Shaftesbury Road, Cambridge CB2 8EA, United Kingdom

One Liberty Plaza, 20th Floor, New York, NY 10006, USA

477 Williamstown Road, Port Melbourne, VIC 3207, Australia

314–321, 3rd Floor, Plot 3, Splendor Forum, Jasola District Centre, New Delhi – 110025, India

103 Penang Road, #05–06/07, Visioncrest Commercial, Singapore 238467

Cambridge University Press is part of Cambridge University Press & Assessment,
a department of the University of Cambridge.

We share the University's mission to contribute to society through the pursuit of
education, learning and research at the highest international levels of excellence.

www.cambridge.org
Information on this title: www.cambridge.org/9781107137394

DOI: 10.1017/9781316480649

© Steven M. Girvin and Kun Yang 2019

This publication is in copyright. Subject to statutory exception and to the provisions
of relevant collective licensing agreements, no reproduction of any part may take
place without the written permission of Cambridge University Press & Assessment.

First published 2019 (version 2, August 2023)

A catalogue record for this publication is available from the British Library

Library of Congress Cataloging-in-Publication data
Names: Girvin, Steven M., author. | Yang, Kun, 1967– author.
Title: Modern condensed matter physics / Steven M. Girvin (Yale University,
Connecticut), Kun Yang (Florida State University).
Description: Cambridge ; New York, NY : Cambridge University Press, [2019]
Identifiers: LCCN 2018027181 | ISBN 9781107137394
Subjects: LCSH: Condensed matter. | Electronic structure. | Atomic structure.
Classification: LCC QC173.454 .G57 2019 | DDC 530.4/1–dc23
LC record available at https://lccn.loc.gov/2018027181

ISBN   978-1-107-13739-4   Hardback

Additional resources for this publication at www.cambridge.org/Girvin&Yang

Cambridge University Press & Assessment has no responsibility for the persistence
or accuracy of URLs for external or third-party internet websites referred to in this
publication and does not guarantee that any content on such websites is, or will
remain, accurate or appropriate.

Brief contents

Preface   *page* xvii
Acknowledgments   xix

1   **Overview of Condensed Matter Physics**   1

2   **Spatial Structure**   9

3   **Lattices and Symmetries**   20

4   **Neutron Scattering**   44

5   **Dynamics of Lattice Vibrations**   64

6   **Quantum Theory of Harmonic Crystals**   78

7   **Electronic Structure of Crystals**   98

8   **Semiclassical Transport Theory**   164

9   **Semiconductors**   198

10 **Non-local Transport in Mesoscopic Systems**   222

11 **Anderson Localization**   252

12 **Integer Quantum Hall Effect**   301

13 **Topology and Berry Phase**   331

14 **Topological Insulators and Semimetals**   362

15 **Interacting Electrons**   376

16 **Fractional Quantum Hall Effect**   430

17   Magnetism   480

18   Bose–Einstein Condensation and Superfluidity   531

19   Superconductivity: Basic Phenomena and Phenomenological Theories   549

20   Microscopic Theory of Superconductivity   592

Appendix A.   Linear-Response Theory   632

Appendix B.   The Poisson Summation Formula   640

Appendix C.   Tunneling and Scanning Tunneling Microscopy   642

Appendix D.   Brief Primer on Topology   647

Appendix E.   Scattering Matrices, Unitarity, and Reciprocity   657

Appendix F.   Quantum Entanglement in Condensed Matter Physics   659

Appendix G.   Linear Response and Noise in Electrical Circuits   665

Appendix H.   Functional Differentiation   673

Appendix I.   Low-Energy Effective Hamiltonians   675

Appendix J.   Introduction to Second Quantization   680

References   685
Index   692

Contents

Preface   *page* xvii
Acknowledgments   xix

1 Overview of Condensed Matter Physics   1

1.1   Definition of Condensed Matter and Goals of Condensed Matter Physics   1
1.2   Classification (or Phases) of Condensed Matter Systems   3
    1.2.1   Atomic Spatial Structures   4
    1.2.2   Electronic Structures or Properties   4
    1.2.3   Symmetries   5
    1.2.4   Beyond Symmetries   6
1.3   Theoretical Descriptions of Condensed Matter Phases   6
1.4   Experimental Probes of Condensed Matter Systems   8

2 Spatial Structure   9

2.1   Probing the Structure   9
2.2   Semiclassical Theory of X-Ray Scattering   10
2.3   Quantum Theory of Electron–Photon Interaction and X-Ray Scattering   13
2.4   X-Ray Scattering from a Condensed Matter System   15
2.5   Relationship of $S(\vec{q})$ and Spatial Correlations   16
2.6   Liquid State versus Crystal State   17

3 Lattices and Symmetries   20

3.1   The Crystal as a Broken-Symmetry State   20
3.2   Bravais Lattices and Lattices with Bases   24
    3.2.1   Bravais Lattices   24
    3.2.2   Lattices with Bases   26
    3.2.3   Lattice Symmetries in Addition to Translation   29
3.3   Reciprocal Lattices   30
3.4   X-Ray Scattering from Crystals   34
3.5   Effects of Lattice Fluctuations on X-Ray Scattering   38
3.6   Notes and Further Reading   41

4 Neutron Scattering   44

4.1   Introduction to Neutron Scattering   44
4.2   Inelastic Neutron Scattering   46

4.3 Dynamical Structure Factor and f-Sum Rule 50
    4.3.1 Classical Harmonic Oscillator 54
    4.3.2 Quantum Harmonic Oscillator 56
4.4 Single-Mode Approximation and Superfluid ^{4}He 60

5 Dynamics of Lattice Vibrations 64
5.1 Elasticity and Sound Modes in Continuous Media 64
5.2 Adiabatic Approximation and Harmonic Expansion of Atomic Potential 68
5.3 Classical Dynamics of Lattice Vibrations 71

6 Quantum Theory of Harmonic Crystals 78
6.1 Heat Capacity 78
6.2 Canonical Quantization of Lattice Vibrations 83
6.3 Quantum Dynamical Structure Factor 88
6.4 Debye–Waller Factor and Stability of Crystalline Order 91
6.5 Mössbauer Effect 93

7 Electronic Structure of Crystals 98
7.1 Drude Theory of Electron Conduction in Metals 98
7.2 Independent Electron Model 104
7.3 Bloch's Theorem 105
    7.3.1 Band Gaps and Bragg Reflection 114
    7.3.2 Van Hove Singularities 115
    7.3.3 Velocity of Bloch Electrons 116
7.4 Tight-Binding Method 117
    7.4.1 Bonds vs. Bands 122
    7.4.2 Wannier Functions 122
    7.4.3 Continuum Limit of Tight-Binding Hamiltonians 124
    7.4.4 Limitations of the Tight-Binding Model 126
    7.4.5 s–d Hybridization in Transition Metals 129
7.5 Graphene Band Structure 133
7.6 Polyacetylene and the Su–Schrieffer–Heeger Model 138
    7.6.1 Dirac electrons in 1D and the Peierls instability 138
    7.6.2 Ground-State Degeneracy and Solitons 142
    7.6.3 Zero Modes Bound to Solitons 144
    7.6.4 Quantum Numbers of Soliton States and Spin–Charge Separation 147
7.7 Thermodynamic Properties of Bloch Electrons 148
    7.7.1 Specific Heat 149
    7.7.2 Magnetic Susceptibility 150
7.8 Spin–Orbit Coupling and Band Structure 153
7.9 Photonic Crystals 156
7.10 Optical Lattices 159
    7.10.1 Oscillator Model of Atomic Polarizability 160
    7.10.2 Quantum Effects in Optical Lattices 162

8 Semiclassical Transport Theory 164
8.1 Review of Semiclassical Wave Packets 164

8.2	Semiclassical Wave-Packet Dynamics in Bloch Bands	165
	8.2.1 Derivation of Bloch Electron Equations of Motion	169
	8.2.2 Zener Tunneling (or Interband Transitions)	171
8.3	Holes	171
8.4	Uniform Magnetic Fields	173
8.5	Quantum Oscillations	176
8.6	Semiclassical $\vec{E} \times \vec{B}$ Drift	179
8.7	The Boltzmann Equation	181
8.8	Boltzmann Transport	186
	8.8.1 Einstein Relation	191
8.9	Thermal Transport and Thermoelectric Effects	193

9 Semiconductors — **198**

9.1	Homogeneous Bulk Semiconductors	198
9.2	Impurity Levels	204
9.3	Optical Processes in Semiconductors	207
	9.3.1 Angle-Resolved Photoemission Spectroscopy	210
9.4	The p–n Junction	212
	9.4.1 Light-Emitting Diodes and Solar Cells	215
9.5	Other Devices	216
	9.5.1 Metal–Oxide–Semiconductor Field-Effect Transistors (MOSFETs)	216
	9.5.2 Heterostructures	217
	9.5.3 Quantum Point Contact, Wire and Dot	220
9.6	Notes and Further Reading	221

10 Non-local Transport in Mesoscopic Systems — **222**

10.1	Introduction to Transport of Electron Waves	222
10.2	Landauer Formula and Conductance Quantization	225
10.3	Multi-terminal Devices	231
10.4	Universal Conductance Fluctuations	233
	10.4.1 Transmission Eigenvalues	238
	10.4.2 UCF Fingerprints	240
10.5	Noise in Mesoscopic Systems	242
	10.5.1 Quantum Shot Noise	245
10.6	Dephasing	248

11 Anderson Localization — **252**

11.1	Absence of Diffusion in Certain Random Lattices	253
11.2	Classical Diffusion	256
11.3	Semiclassical Diffusion	258
	11.3.1 Review of Scattering from a Single Impurity	258
	11.3.2 Scattering from Many Impurities	262
	11.3.3 Multiple Scattering and Classical Diffusion	265
11.4	Quantum Corrections to Diffusion	267
	11.4.1 Real-Space Picture	268
	11.4.2 Enhanced Backscattering	269

11.5		Weak Localization in 2D	271
	11.5.1	Magnetic Fields and Spin–Orbit Coupling	273
11.6		Strong Localization in 1D	275
11.7		Localization and Metal–Insulator Transition in 3D	277
11.8		Scaling Theory of Localization and the Metal–Insulator Transition	279
	11.8.1	Thouless Picture of Conductance	279
	11.8.2	Persistent Currents in Disordered Mesoscopic Rings	282
	11.8.3	Scaling Theory	283
	11.8.4	Scaling Hypothesis and Universality	284
11.9		Scaling and Transport at Finite Temperature	287
	11.9.1	Mobility Gap and Activated Transport	291
	11.9.2	Variable-Range Hopping	292
11.10		Anderson Model	294
11.11		Many-Body Localization	297

12 Integer Quantum Hall Effect   301

12.1		Hall-Effect Transport in High Magnetic Fields	301
12.2		Why 2D Is Important	304
12.3		Why Disorder and Localization Are Important	305
12.4		Classical and Semiclassical Dynamics	306
	12.4.1	Classical Dynamics	306
	12.4.2	Semiclassical Approximation	308
12.5		Quantum Dynamics in Strong B Fields	309
12.6		IQHE Edge States	315
12.7		Semiclassical Percolation Picture of the IQHE	318
12.8		Anomalous Integer Quantum Hall Sequence in Graphene	321
12.9		Magnetic Translation Invariance and Magnetic Bloch Bands	324
	12.9.1	Simple Landau Gauge Example	327
12.10		Quantization of the Hall Conductance in Magnetic Bloch Bands	329

13 Topology and Berry Phase   331

13.1		Adiabatic Evolution and the Geometry of Hilbert Space	331
13.2		Berry Phase and the Aharonov–Bohm Effect	336
13.3		Spin-1/2 Berry Phase	339
	13.3.1	Spin–Orbit Coupling and Suppression of Weak Localization	343
13.4		Berry Curvature of Bloch Bands and Anomalous Velocity	344
	13.4.1	Anomalous Velocity	345
13.5		Topological Quantization of Hall Conductance of Magnetic Bloch Bands	348
	13.5.1	Wannier Functions of Topologically Non-trivial Bands	351
	13.5.2	Band Crossing and Change of Band Topology	352
	13.5.3	Relation Between the Chern Number and Chiral Edge States: Bulk–Edge Correspondence	353
13.6		An Example of Bands Carrying Non-zero Chern Numbers: Haldane Model	356
13.7		Thouless Charge Pump and Electric Polarization	358
	13.7.1	Modern Theory of Electric Polarization	360

14 Topological Insulators and Semimetals 362

14.1 Kane–Mele Model 362
14.2 $\mathbb{Z}_2$ Characterization of Topological Insulators 364
14.3 Massless Dirac Surface/Interface States 368
14.4 Weyl Semimetals 371
  14.4.1 Fermi Arcs on the Surface 372
  14.4.2 Chiral Anomaly 373
14.5 Notes and Further Reading 375

15 Interacting Electrons 376

15.1 Hartree Approximation 376
15.2 Hartree–Fock Approximation 378
  15.2.1 Koopmans' Theorem 381
15.3 Hartree–Fock Approximation for the 3D Electron Gas 382
  15.3.1 Total Exchange Energy of the 3DEG in the
      Hartree–Fock Approximation 384
15.4 Density Functional Theory 385
15.5 Kohn–Sham Single-Particle Equations 387
15.6 Local-Density Approximation 389
15.7 Density–Density Response Function and Static Screening 391
  15.7.1 Thomas–Fermi Approximation 394
  15.7.2 Lindhard Approximation 394
15.8 Dynamical Screening and Random-Phase Approximation 396
15.9 Plasma Oscillation and Plasmon Dispersion 397
  15.9.1 Plasma Frequency and Plasmon Dispersion from the RPA 397
  15.9.2 Plasma Frequency from Classical Dynamics 398
  15.9.3 Plasma Frequency and Plasmon Dispersion from
      the Single-Mode Approximation 399
15.10 Dielectric Function and Optical Properties 400
  15.10.1 Dielectric Function and AC Conductivity 400
  15.10.2 Optical Measurements of Dielectric Function 401
15.11 Landau's Fermi-Liquid Theory 402
  15.11.1 Elementary Excitations of a Free Fermi Gas 402
  15.11.2 Adiabaticity and Elementary Excitations of an Interacting Fermi Gas 404
  15.11.3 Fermi-Liquid Parameters 407
15.12 Predictions of Fermi-Liquid Theory 409
  15.12.1 Heat Capacity 409
  15.12.2 Compressibility 410
  15.12.3 Spin Susceptibility 411
  15.12.4 Collective Modes, Dynamical and Transport Properties 411
15.13 Instabilities of Fermi Liquids 412
  15.13.1 Ferromagnetic Instability 412
  15.13.2 Pomeranchuk Instabilities 413
  15.13.3 Pairing Instability 414
  15.13.4 Charge and Spin Density-Wave Instabilities 418
  15.13.5 One Dimension 419
  15.13.6 Two-Dimensional Electron Gas at High Magnetic Field 420

15.14		Infrared Singularities in Fermi Liquids	420
	15.14.1	Perfect Screening and the Friedel Sum Rule	420
	15.14.2	Orthogonality Catastrophe	422
	15.14.3	Magnetic Impurities in Metals: The Kondo Problem	423
15.15		Summary and Outlook	429

16   Fractional Quantum Hall Effect   **430**

16.1		Landau Levels Revisited	431
16.2		One-Body Basis States in Symmetric Gauge	433
16.3		Two-Body Problem and Haldane Pseudopotentials	435
16.4		The $\nu = 1$ Many-Body State and Plasma Analogy	438
	16.4.1	Electron and Hole Excitations at $\nu = 1$	441
16.5		Laughlin's Wave Function	442
16.6		Quasiparticle and Quasihole Excitations of Laughlin States	446
16.7		Fractional Statistics of Laughlin Quasiparticles	452
	16.7.1	Possibility of Fractional Statistics in 2D	452
	16.7.2	Physical Model of Anyons	455
	16.7.3	Statistics Angle of Laughlin Quasiholes	457
16.8		Collective Excitations	460
16.9		Bosonization and Fractional Quantum Hall Edge States	463
	16.9.1	Shot-Noise Measurement of Fractional Quasiparticle Charge	467
16.10		Composite Fermions and Hierarchy States	469
	16.10.1	Another Take on Laughlin's Wave Function	469
	16.10.2	Jain Sequences	470
16.11		General Formalism of Electron Dynamics Confined to a Single Landau Level	470
	16.11.1	Finite-Size Geometries	474
16.12		Relation between Fractional Statistics and Topological Degeneracy	476
16.13		Notes and Further Reading	478

17   Magnetism   **480**

17.1		Basics	480
17.2		Classical Theory of Magnetism	481
17.3		Quantum Theory of Magnetism of Individual Atoms	481
	17.3.1	Quantum Diamagnetism	482
	17.3.2	Quantum Paramagnetism	485
	17.3.3	Quantum Spin	486
17.4		The Hubbard Model and Mott Insulators	486
17.5		Magnetically Ordered States and Spin-Wave Excitations	491
	17.5.1	Ferromagnets	491
	17.5.2	Antiferromagnets	495
17.6		One Dimension	499
	17.6.1	Lieb–Schultz–Mattis Theorem	501
	17.6.2	Spin-1/2 Chains	502
	17.6.3	Spin-1 Chains, Haldane Gap, and String Order	506
	17.6.4	Matrix Product and Tensor Network States	510

17.7 Valence-Bond-Solid and Spin-Liquid States in 2D and Higher Dimensions 513

    17.7.1 $\mathbb{Z}_2$ Topological Order in Resonating Valence-Bond Spin Liquid 519

17.8 An Exactly Solvable Model of $\mathbb{Z}_2$ Spin Liquid: Kitaev's Toric Code 521

    17.8.1 Toric Code as Quantum Memory 525

17.9 Landau Diamagnetism 528

18 Bose–Einstein Condensation and Superfluidity 531

18.1 Non-interacting Bosons and Bose–Einstein Condensation 531

    18.1.1 Off-Diagonal Long-Range Order 534

    18.1.2 Finite Temperature and Effects of Trapping Potential 535

    18.1.3 Experimental Observation of Bose–Einstein Condensation 536

18.2 Weakly Interacting Bosons and Bogoliubov Theory 539

18.3 Stability of Condensate and Superfluidity 542

18.4 Bose–Einstein Condensation of Exciton-Polaritons: Quantum Fluids of Light 545

19 Superconductivity: Basic Phenomena and Phenomenological Theories 549

19.1 Thermodynamics 549

    19.1.1 Type-I Superconductors 550

    19.1.2 Type-II Superconductors 552

19.2 Electrodynamics 553

19.3 Meissner Kernel 556

19.4 The Free-Energy Functional 558

19.5 Ginzburg–Landau Theory 559

19.6 Type-II Superconductors 566

    19.6.1 Abrikosov Vortex Lattice 568

    19.6.2 Isolated Vortices 569

19.7 Why Do Superconductors Superconduct? 573

19.8 Comparison between Superconductivity and Superfluidity 576

19.9 Josephson Effect 579

    19.9.1 Superconducting Quantum Interference Devices (SQUIDS) 585

19.10 Flux-Flow Resistance in Superconductors 587

19.11 Superconducting Quantum Bits 587

20 Microscopic Theory of Superconductivity 592

20.1 Origin of Attractive Interaction 592

20.2 BCS Reduced Hamiltonian and Mean-Field Solution 594

    20.2.1 Condensation Energy 598

    20.2.2 Elementary Excitations 599

    20.2.3 Finite-Temperature Properties 602

20.3 Microscopic Derivation of Josephson Coupling 603

20.4 Electromagnetic Response of Superconductors 606

20.5 BCS–BEC Crossover 609

20.6 Real-Space Formulation and the Bogoliubov–de Gennes Equation 611

20.7 Kitaev's p-Wave Superconducting Chain and Topological Superconductors 614

20.8 Unconventional Superconductors 617
    20.8.1 General Solution of Cooper Problem 617
    20.8.2 General Structure of Pairing Order Parameter 619
    20.8.3 Fulde–Ferrell–Larkin–Ovchinnikov States 620
20.9 High-Temperature Cuprate Superconductors 621
    20.9.1 Antiferromagnetism in the Parent Compound 622
    20.9.2 Effects of Doping 624
    20.9.3 Nature of the Superconducting State 624
    20.9.4 Why d-Wave? 627

Appendix A. Linear-Response Theory 632

A.1 Static Response 632
A.2 Dynamical Response 634
A.3 Causality, Spectral Densities, and Kramers–Kronig Relations 636

Appendix B. The Poisson Summation Formula 640

Appendix C. Tunneling and Scanning Tunneling Microscopy 642

C.1 A Simple Example 642
C.2 Tunnel Junction 643
C.3 Scanning Tunneling Microscopy 645

Appendix D. Brief Primer on Topology 647

D.1 Introduction 647
D.2 Homeomorphism 648
D.3 Homotopy 648
D.4 Fundamental Group 650
D.5 Gauss–Bonnet Theorem 651
D.6 Topological Defects 654

Appendix E. Scattering Matrices, Unitarity, and Reciprocity 657

Appendix F. Quantum Entanglement in Condensed Matter Physics 659

F.1 Reduced Density Matrix 659
F.2 Schmidt and Singular-Value Decompositions 661
F.3 Entanglement Entropy Scaling Laws 662
F.4 Other Measures of Entanglement 663
F.5 Closing Remarks 664

Appendix G. Linear Response and Noise in Electrical Circuits 665

G.1 Classical Thermal Noise in a Resistor 665
G.2 Linear Response of Electrical Circuits 668
G.3 Hamiltonian Description of Electrical Circuits 670
    G.3.1 Hamiltonian for Josephson Junction Circuits 672

Appendix H. Functional Differentiation 673

Appendix I.   Low-Energy Effective Hamiltonians   675

I.1   Effective Tunneling Hamiltonian   675
I.2   Antiferromagnetism in the Hubbard Model   677
I.3   Summary   679

Appendix J.   Introduction to Second Quantization   680

J.1   Second Quantization   680
J.2   Majorana Representation of Fermion Operators   683

References   685
Index   692

Preface

This textbook is intended for both introductory and more advanced graduate-level courses in condensed matter physics and as a pedagogical reference for researchers in the field. This modern textbook provides graduate students with a comprehensive and accessible route from fundamental concepts to modern topics, language, and methods in the rapidly advancing field of quantum condensed matter physics.

The field has progressed and expanded dramatically since the publication four decades ago of the classic text by Ashcroft and Mermin [1], and its name has changed from Solid State Physics to Condensed Matter Physics, reflecting this expansion. The field of inquiry is vast and is typically divided into two halves. The first, often called "soft matter," covers the classical statistical physics of liquid crystals, glassy materials, polymers, and certain biological systems and materials. This area is nicely addressed in the textbook of Chaikin and Lubensky [2]. The second area, often called "hard matter" or "quantum matter," primarily covers the quantum physics of electrons in solids but these days also includes correlated quantum states of ultra-cold atomic gases and even photons. While a number of good textbooks [3–5] address various aspects of hard matter, the present text offers broader and more in-depth coverage of the field and provides physical intuition through many deep phenomenological descriptions, in addition to introducing the required mathematical background.

The present text is aimed primarily at graduate students and researchers in quantum condensed matter physics and provides encyclopedic coverage of this very dynamic field. While sharing a similar starting point with Ashcroft and Mermin, we have attempted to cover the aforementioned new developments in considerably greater depth and detail, while providing an overarching perspective on unifying concepts and methodologies. Chapters 1–9 cover traditional introductory concepts, but we have made considerable effort to provide a modern perspective on them. The later chapters introduce modern developments both in theory and in experiment. Among the new topics are coherent transport in mesoscopic systems, Anderson and many-body localization in disordered systems, the integer and fractional quantum Hall effects, Berry phases and the topology of Bloch bands, topological insulators and semimetals, instabilities of Fermi liquids, modern aspects of quantum magnetism (e.g. spinons, the Haldane gap, spin liquids, and the toric code), quantum entanglement, Bose–Einstein condensation, a pedagogical introduction to the phenomenology of superfluidity and superconductivity, superconducting quantum bits (qubits), and finally a modern review of BCS theory that includes unconventional pairing, high-temperature superconductivity, topological superconductors, and majorana fermions. We have also attempted to make contact with other fields, in particular ultra-cold atomic gases, photonic crystals, and quantum information science, emphasizing the unifying principles among different branches of physics. For this reason the text should also be of interest to students and practitioners outside condensed matter physics.

The text is intended to be accessible and useful to experimentalists and theorists alike, providing an introduction both to the phenomenology and to the underlying theoretical description. In particular, we provide the mathematical background needed to understand the topological aspects of condensed matter systems. We also provide a gentle and accessible introduction to scaling and renormalization group methods with applications to Anderson localization, the Kondo problem, and the modern approach to the BCS problem. The text assumes prior knowledge of quantum mechanics and statistical physics at the level of typical first-year graduate courses. Undergraduate preparation in condensed matter physics at the level of Kittel [6] would be useful but is not essential. We make extensive use of harmonic oscillator ladder operators but almost completely avoid second quantization for fermions until Chapter 17. In addition, we provide a pedagogical appendix for the reader to review second quantization.

Recent decades have seen the application of advanced methods from quantum field theory which provide effective descriptions of "universal" features of strongly correlated many-body quantum systems, usually in the long-wavelength and low-energy limit [7–15]. The present text provides a pedagogical gateway to courses on these advanced methods by introducing and using the language of many-body theory and quantum field theory where appropriate.

This book has evolved over several decades from course notes for graduate condensed matter physics taught by the authors at Indiana University and Florida State University, respectively. The content exceeds the amount of material which can be covered in a one-year course but naturally divides into an introductory portion (Chapters 1–10) which can be covered in the first semester. For the second semester, the instructor can cover Chapters 11–15 and then select from the remaining chapters which cover the fractional quantum Hall effect, magnetism, superfluidity, and superconductivity.

Acknowledgments

We are grateful to many people for kindly taking the time to provide feedback on the manuscript as it was developing. We would particularly like to acknowledge Jason Alicea, Collin Broholm, Jack Harris, Alexander Seidel, A. Douglas Stone, and Peng Xiong. SMG thanks KY for carrying the bulk of the writing load during the decade that SMG was serving as deputy provost for research at Yale. SMG also thanks Diane Girvin for extensive assistance with proofreading. KY would like to thank the students who took his course PHZ5491-5492 at Florida State University for their comments, in particular Shiuan-Fan Liou, Mohammad Pouranvari and Yuhui Zhang who have also helped draw many figures. He is also grateful to Li Chen for help proofreading several chapters. Over the years our research in condensed matter and quantum information theory has been supported by the NSF, DOE, ARO, the Keck Foundation, Yale University, Florida State University and the National High Magnetic Field Laboratory. This work was begun at Indiana University.

We are grateful to the staff of Cambridge University Press and especially to our editor Simon Capelin whose patient encouragement over the span of two decades helped us reach the finish line.

Finally, we are most grateful for the infinite patience of our families over the many years that this project was underway.

1 Overview of Condensed Matter Physics

Matter that we encounter in our everyday life comes in various forms: air is gaseous, water (between 0 °C and 100 °C under ambient pressure) is a liquid, while ice and various kinds of metals and minerals are crystalline solids. We also encounter other familiar forms of matter from our daily experience, including glasses and **liquid crystals**, which do not fall into the categories of gas, liquid, or solid/crystal. More exotic forms of matter exist under extreme conditions, like very low (all the way to almost absolute zero) or very high temperatures, extremely high pressures, very far from equilibrium, etc. Roughly speaking, "condensed matter physics" studies physical properties of matter in "condensed" forms (where the density is high enough that interaction among the constituent particles is crucial to these properties). In the rest of this chapter we will attempt to give a more precise (but far from unique) definition of condensed matter physics, and discuss how to classify, theoretically describe, and experimentally probe various forms or phases of condensed matter. In this book we will deal exclusively with condensed matter systems that are made of atoms or molecules and, in particular, the electrons that come with them (though on occasion we will study collective excitations in which photons play an important role). On the other hand, the methodology developed here and many specific results apply to more exotic condensed matter systems, like neutron stars and quark–gluon plasmas that are best described in terms of quarks, gluons, or nucleons.

1.1 Definition of Condensed Matter and Goals of Condensed Matter Physics

Matter which surrounds us is made of huge numbers (of order 10^{23}) of atoms or molecules, which have a characteristic size of 10^{-10} m or 1 Å. In the gaseous form, the typical interparticle distance is much larger than this characteristic size, thus the particles interact weakly (except when occasional interparticle collisions occur) and retain their integrity; in particular, electrons are attached to individual atoms or molecules. As a result, the physical properties of such gaseous matter are usually dictated by the properties of *individual* atoms or molecules, and we do *not* refer to the gaseous matter as condensed matter in most cases.[1]

---

[1] There are exceptions to this. For example, at very low temperatures, collective behavior that involves large numbers of atoms or molecules (such as Bose–Einstein condensation) may occur. Interactions may be weak in some sense, but not on the scale of the extremely low temperature. In such cases, the physical properties are dominated by the collective behavior, and we call such gaseous matter condensed matter as well, in the spirit of following discussions.

For matter in liquid/fluid or solid forms, on the other hand, the constituent atoms are in sufficiently close proximity that the distance between them is comparable to the size of individual atoms. As a result, these atoms interact with each other strongly. In many cases, some of the electrons (mostly the outer-shell ones) which were attached to individual atoms may be able to move throughout the system. (In a metal the electrons can move more or less freely. In an insulator they can move only by trading places.[2]) We take these interactions among the atoms (and their electrons) as the defining characteristic of a condensed matter system.

This characteristic of condensed matter, namely important interactions among the atoms (and possibly the loss of integrity of individual atoms due to detachment of outer-shell electrons), leads to a fundamental difference from gaseous matter, in that many properties of fluids and solids differ qualitatively from the properties of an aggregate of isolated atoms. When the atoms are in close proximity, low-energy states of the system have strongly correlated atomic positions. Owing to the motion of electrons throughout the system, low-energy states may then also have strongly correlated electronic positions. Low-energy excitations of the system usually involve subtle changes in the atomic or electronic degrees of freedom and have an energy scale much smaller than the binding energy scale for isolated atoms. Many physical properties of a system at a temperature T depend on those excitations which have an energy less than the thermal energy $k_B T$. The Boltzmann constant $k_B \approx 8.167 \times 10^{-5}\,\mathrm{eV\,K^{-1}}$ so that the thermal energy at room temperature is $\sim 2.35 \times 10^{-2}\,\mathrm{eV}$. The binding energy of an isolated hydrogen atom, the Rydberg $\sim 13.6\,\mathrm{eV}$, represents the vastly larger chemical energy scale. For this reason, many physical properties of condensed matter systems, from the absolute zero of temperature to temperatures many times higher than room temperature, reflect the possibilities for small rearrangement of correlations among the huge numbers of atomic degrees of freedom. Thus, the low-energy/temperature properties of condensed matter systems are *emergent*; that is, these are collective properties that make the "whole greater than the sum of the parts." These collective properties (examples of which include the rigidity of solids, superfluidity, superconductivity, and the various quantum Hall effects) emerge through such subtle correlated motions involving very large number of particles, and are associated with various quantum or classical phase transitions in the thermodynamic limit. Understanding these subtle correlations and how the physical properties of condensed matter systems depend on them is the business of condensed matter physics.

An isolated atom consists of a small positively charged nucleus composed of protons and neutrons and a surrounding "cloud" of negatively charged electrons. The size of the nucleus is given by the length scales of nuclear physics (that is, the length scale of the strong force) $\sim 10^{-15}\,\mathrm{m}$ to $10^{-14}\,\mathrm{m}$. Furthermore, the energy scale needed to excite the nucleus out of its ground state is roughly six orders of magnitude larger than typical chemical energy scales. Hence, it is almost always the case in condensed matter physics that the nucleus can be taken to be an inert point particle with only translational degrees of freedom. (The orientation of the spin of the nucleus is occasionally important.) The familiar picture of an atom has electrons occupying a set of atomic bound states. Electrons in the most strongly bound states are most likely to be close to the nucleus, while the most weakly bound states are most likely to be found far from the nucleus. The size of the electron cloud is given by the length scales of atomic physics $\sim 10^{-10}\,\mathrm{m}$ to $10^{-9}\,\mathrm{m}$. Of course, this picture of the atom is only approximate, since it is based on the quantum mechanics of a single electron moving in an average potential of all the other electrons. In fact the positions of electrons in an atom are correlated. It is now possible to solve the many-electron quantum mechanics problem for an atom with extremely high accuracy. (That is a story we will not allow ourselves to be distracted by here.) In a condensed matter system it is often the case that electrons in states which evolve from some of the more weakly

---

[2] Since all electrons of the same spin are indistinguishable we have to be a little careful in what we mean by "moving by trading places."

bound atomic orbitals move relatively freely through the entire system. The presence of these *itinerant* electrons is very important. For example, they allow electric currents to flow through the system with relative ease. Correlations among itinerant electrons are very intricate and, unlike the atomic case, we frequently do not yet understand them adequately. This is one of the frontiers of condensed matter physics.

It is usually appropriate to think of a condensed matter system as consisting of ion cores and (possibly) itinerant valence electrons. The ion cores are composed of the nucleus and those atomic orbitals which are still tightly bound to an individual atom in the condensed matter system. Itinerant electrons are not present in every system and, as we will understand better later, the distinction between ion-core electrons and itinerant electrons can become fuzzy.

We are interested in understanding how the macroscopic physical properties of condensed matter systems depend on their microscopic underpinning. We are interested in thermodynamic properties like the specific heat and the magnetic susceptibility and in structural properties, i.e. how the ion cores are distributed within the system. We will learn how a system carries electric and heat currents, and about many other interesting properties. All of these macroscopic properties reflect the quantum and statistical mechanics of the ion cores and the itinerant electrons.

Condensed matter physics is no longer content with understanding on a fundamental level the properties of those condensed matter systems which nature readily provides. Present frontiers are often found in attempts to understand the properties of artificial materials which frequently have unusual and sometimes unexplained properties. Often artificial materials are designed to have desired physical properties. Occasionally, new knowledge makes it possible to fabricate materials whose properties are extremely useful. Probably the most spectacular example of this, at present, is the engineered electronic material used in semiconductor devices (e.g. the microwave amplifiers in every cell phone). The hope of condensed matter physicists is that there will be many more equally spectacular examples in the future.

Condensed matter physics is such a vast, diverse, and rapidly growing field that it is impossible to do justice to all of it in a single volume. The field is roughly divided into two halves: "$\hbar = 0$" ("soft" condensed matter) and "$\hbar = 1$" (electronic/quantum). This book focuses mostly on the electronic/quantum subfield and attempts to include some of the numerous important scientific advances made since the time of the publication of the classic text by Ashcroft and Mermin [1]. Readers interested in the "$\hbar = 0$" side of things should consult the text by Chaikin and Lubensky [2].

We have attempted to keep the level of discussion accessible to beginning graduate students in physics. More advanced sections that can be skipped on first reading are marked with Knuth's "dangerous bend" TEX symbols and .

1.2 Classification (or Phases) of Condensed Matter Systems

Every system is different in some way.[3] It is, however, neither possible nor necessary to study every condensed matter system in nature, or every theoretical model one can write down. We would like, instead, to group different systems with *qualitatively* similar properties together and study their common and robust (often called "universal") properties. In order to do this, we need to classify condensed matter systems into various classes or phases. This type of analysis shows us, for example, that (rather remarkably) the thermodynamic critical point of a liquid/vapor system has the same

---

[3] Following Tolstoy's famous quote from *Anna Karenina*, "Happy families are all alike; every unhappy family is unhappy in its own way," this suggests that every system is an unhappy family of atoms.

universal properties as the magnetic critical point of the three-dimensional (3D) Ising model of classical magnetism.

Owing to the vast domain of condensed matter physics, there is not a single complete scheme to classify all condensed matter systems. In the following we discuss several different but overlapping schemes; which one is more appropriate depends on the specific class of systems being studied, and the specific properties of interest.

1.2.1 Atomic Spatial Structures

Since our condensed matter systems are made of atoms, we can classify them in terms of the spatial structures (or patterns of the positions) of the atoms, or more precisely, their ion cores or atomic nuclei. The two most familiar forms of condensed matter, namely solid and liquid, have their most significant difference precisely in this respect: in a solid the atoms form periodic arrays and have long-range positional order, while in a liquid the atomic positions do not have any such global pattern or long-range order. This difference in spatial structure also leads to the biggest difference in their physical properties: a solid has rigidity but cannot flow, because once the position of a single atom is fixed, so are those of all other atoms. A liquid, on the other hand, can flow easily and has no rigidity because there is no fixed pattern for the atomic positions. One immediate consequence is that a liquid can take any shape dictated by the container in which it is placed, while the shape of a piece of solid does not change as easily. We have learned in thermal physics that solids and liquids are different phases of matter, with a thermodynamic phase boundary separating them. Frequently, these phase boundaries are first-order ones, meaning that there is a finite discontinuity in the internal energy and the entropy in crossing the boundary.

1.2.2 Electronic Structures or Properties

Many physical properties of condensed matter systems, especially those important for applications, are dictated by the behavior of the electrons within them. We often classify solids (and sometimes even liquids) into insulators, semiconductors, metals, and superconductors. Such classifications, of course, are based on the ability of the system to transport electric charge. Electric current in condensed matter systems is almost always carried by electrons (nuclei, while charged objects as well, are too heavy to make a significant contribution, though there do exist materials which are fast-ion conductors of interest in the construction of batteries); in general, electric transport properties are dominated by electronic structure.

Many condensed matter systems are magnetic, and we can classify them in terms of their magnetic properties as paramagnets, diamagnets, ferromagnets, antiferromagnets, etc. Such magnetic properties are also dominated by electrons, through the magnetic moments from their spins and orbital motions. Very often the electric transport and magnetic properties are closely related. For example, non-magnetic metals tend to be **paramagnets**, while non-magnetic insulators tend to be **diamagnets**; **ferromagnets** tend to be metals while **antiferromagnets** are often insulators of a specific type, known as **Mott insulators**. Superconductors, on the other hand, are also "perfect diamagnets" in a very precise sense that we will discuss later.[4] We thus refer to electric transport and magnetic properties collectively as "electronic properties."

---

[4] A paramagnet has no magnetic polarization unless an external field is applied, in which case the magnetization is proportional to (and reinforces) the applied field. A diamagnet is similar, but its response opposes an applied field. Superconductors can act as perfect diamagnets in some regimes, completely resisting the penetration of any applied field into the sample by generating an exactly opposing field.

1.2.3 Symmetries

Many condensed matter phases can be characterized by their symmetry properties. A symmetry is a transformation which leaves the Hamiltonian H invariant. Typical examples include translation (in a system with no external potential), inversion about a special point ($\vec{r} \rightarrow -\vec{r}$ for a certain choice of origin), reflection about a particular plane, and time-reversal. A specific phase, either thermal (i.e. classical, at finite temperature T) or quantum ($T = 0$), may break some of the symmetries of the Hamiltonian. That is, the phase may have less symmetry than the Hamiltonian that describes the system. For example, the space translation and rotation symmetries are (spontaneously) broken by a crystalline solid, while they are respected by a liquid. A liquid (at least when averaged over a period of time) is completely uniform and translation-invariant. A solid, however, is not. In a perfect crystal, the atoms sit in particular places and once you know the position of a few of them, you can predict the positions of all the rest (modulo small thermal and quantum fluctuations), no matter how far away. If we translate the crystal by some arbitrary amount, the atoms end up in different places but (absent an external potential) the energy is left invariant. The initial position among all these degenerate possibilities is an accident of the history. This is what we mean when we say that the symmetry breaking is "spontaneous."

Ferromagnets and antiferromagnets both break spin rotation symmetry and time-reversal symmetry, while both of these symmetries are respected in paramagnets and diamagnets. An antiferromagnet has two sublattices of opposite spin and thus further breaks lattice **translation symmetry**, which is *unbroken* in a ferromagnet.

In a phase that breaks one or more symmetries, the pattern of symmetry breaking can be characterized by the so-called "**order parameters.**" An order parameter is a measurable physical quantity, which transforms non-trivially under a symmetry transformation that leaves the Hamiltonian invariant. To understand the meaning of this definition, consider a ferromagnet. Here the order parameter is the magnetization. Magnetization changes sign under time reversal (i.e. "transforms non-trivially under a symmetry transformation that leaves the Hamiltonian invariant"). Hence, if the order parameter is non-zero, the system has less symmetry than its Hamiltonian. Furthermore, we can deduce from the symmetry of the Hamiltonian that states with opposite values of the order parameter are degenerate in energy.

In a crystal the order parameter is the (Fourier transform of the) atomic density. The Fourier transform of the density will have sharp peaks at wave vectors commensurate with the lattice spacing. Under translation of the crystal by some arbitrary small amount, the phases of the (complex) Fourier amplitudes will change, but not their magnitudes. These are thus order parameters associated with broken translation symmetry. The order parameters are zero in the corresponding symmetry-unbroken phases, namely liquid phases (which are translation-symmetric). Similarly, paramagnet/diamagnet phases have zero magnetic order parameters since they are non-magnetic in the absence of an applied field, i.e. when the Hamiltonian is time-reversal symmetric.

Superfluids (which are in some ways closely related to superconductors yet different from them in a fundamental way because they are charge-neutral) have a complex order parameter with continuously variable phase. They break the continuous symmetry corresponding to particle number conservation.[5] This is a subtler and less familiar form of symmetry breaking, but its mathematical structure (including that of its order parameter) turns out to be identical to that of a ferromagnet whose magnetization vector is restricted to a plane (sometimes referred to as an "XY" or "easy-plane" magnet).

---

[5] In physics there is a one-to-one correspondence between conservation laws and *continuous* symmetries; in the context of field theory, this is known as Noether's theorem.

Symmetry is a unifying concept in physics, and is of paramount importance. It is therefore not surprising that it appears both in the atomic crystal structure and in the electronic structure classification schemes.

1.2.4 Beyond Symmetries

It was the great Soviet physicist Lev Landau who first advocated using symmetry as a classification scheme for phases. The Landau scheme has been tremendously successful, both for classifying phases and for developing phenomenological models of symmetry breaking in thermodynamic phase transitions. Owing to the success and the great influence of Landau and his followers, it was felt for many years that we could classify *all* condensed matter phases using symmetry. In recent years, however, especially since the discovery of the fractional quantum Hall effect in 1982, examples of *distinctive* phases with the *same* symmetry have been accumulating. Such phases obviously do *not* fit into the Landau symmetry scheme. In the following we discuss an elementary example, which the reader should be familiar with from her/his undergraduate-level solid state or modern physics courses.

The simplest types of metal and insulator are the so-called band metal and band insulator, formed by filling electron bands of a (perfect) crystal with non-interacting electrons (an idealization). Metals and insulators have the same symmetry properties, but are obviously different phases. So what is the (qualitative) difference between them? The difference, of course, lies in the fact that all bands are either completely filled or empty in a band insulator, while there is at least one partially filled band in a metal, resulting in one or more Fermi surface[6] sheet(s). This is a *topological* difference, as the number of Fermi surface sheets is a topological invariant (quantum number) that does *not* change (at least not continuously) when the shape or *geometry* of a Fermi surface is varied due to the band structure. Furthermore, it can be shown that when the number of Fermi surface sheets changes, the system must undergo a quantum phase transition (known as a Lifshitz transition) at which the ground-state energy or its derivatives become singular. However unlike the solid–liquid transition, there is no change in the symmetry properties of the system in a Lifshitz transition.

With the discovery of the fractional quantum Hall effect, physicists started to realize that many phases with excitation gaps separating their ground states from excited states have non-trivial topological properties. They are thus termed *topological phases* and are said to possess *topological order* [15]. All quantum Hall phases, as well as the recently discovered topological insulators, are examples of topological phases, which will be discussed in this book. Perhaps the most familiar but also under-appreciated example is an ordinary superconductor. In contrast to common wisdom and unlike superfluids (which are uncharged), there is no spontaneous symmetry breaking in a superconductor [16]. The topological nature of the superconducting phase is reflected in the fact that its ground-state degeneracy is dependent on the topology of the space in which it is placed, but *nothing else*. As we will learn, robust and topology-dependent ground-state degeneracy is one of the most commonly used methods to probe and characterize topological phases.

1.3 Theoretical Descriptions of Condensed Matter Phases

In atomic physics one typically attempts to give a full microscopic description of the atom or molecule by solving Schrödinger's equation to obtain the (few-electron) wave functions. Except for the simplest

---

[6] Because electrons are fermions, they obey the Pauli exclusion principle and each must occupy a different quantum state. As we will learn later, the ground state for non-interacting electrons has all states occupied up to the so-called Fermi energy, above which no states are occupied. The set of states having energy equal to the Fermi energy defines a surface with one or more sheets in momentum space.

case of a hydrogen-like atom with one electron, analytic exact solution is not possible, and numerical solutions become out of reach very quickly with increasing electron number. In an interacting many-electron condensed matter system, microscopic descriptions based on electron wave functions are still widely used. Such descriptions are of course approximate (except for very few highly idealized model systems), often based on some mean-field type of approximations or variational principles. Such descriptions are highly effective when the approximate or variational wave function is simple yet captures the most important correlations of the phase that the system is in. The most famous examples of this are the Bardeen–Cooper–Schrieffer (BCS) wave function for superconductors and the Laughlin wave function for fractional quantum Hall liquids.

Very often, even approximate microscopic descriptions are beyond reach. Fortunately, to understand physical properties that are probed experimentally and important for applications, we often need only understand how a condensed matter system responds to an external perturbation at low frequency or energy (compared with microscopic or atomic energy scales, typically eV), and/or at long wavelength (again compared with the atomic scale, 1 Å). Most microscopic degrees of freedom do *not* make significant contributions to the response in this limit. We thus need only focus on the low-energy, long-wavelength degrees of freedom that dominate such responses. Also fortunate is the fact that, very often, the physics simplifies significantly in the low-energy limit, rendering an accurate description in terms of these (often heavily "renormalized" or "effective") degrees of freedom possible. Such simplification can often be understood theoretically in terms of a **renormalization group** analysis.[7] A theoretical description of this type goes under the name of a "low-energy effective theory," and we will encounter several examples. Condensed matter systems in the same phase share the same low-energy effective theory (but possibly with different parameters), while different phases can be characterized by different low-energy effective theories. These concepts will become clearer as we study particular examples in later chapters.

Band metals again serve as a simple example of the discussion above. In the idealization of non-interacting electrons, the ground state is simply the single Slater determinant in which all electron levels below the Fermi energy are occupied and all above are empty, and excitations are created by moving one or more electrons from below the Fermi energy (or inside the Fermi surface) to above (or outside). It is already clear in this simple case that low-energy excitations involve electron states near the Fermi surface(s) *only*. Remarkably, this remains true even when electron–electron interactions are turned on (as long as no phase transition is caused by the interaction), and in some sense electrons near the Fermi surface continue to behave as if they do *not* interact with each other. It is thus possible to describe the low-energy/temperature properties of a metal in terms of the electrons that live near the Fermi surface (often called Landau quasiparticles), and such a low-energy effective theory is known as Landau's Fermi-liquid theory (yes, the same Landau, but symmetry or its breaking plays no role here!).

In symmetry-breaking phases, the low-energy degrees of freedom are often those giving rise to the order parameter. This is particularly true when a continuous symmetry is broken spontaneously, as guaranteed by the **Goldstone theorem**, which states that there is a *gapless* collective mode (known as a "**Goldstone mode**") associated with each **spontaneously broken continuous symmetry**. The gapless mode is nothing but the long-wavelength fluctuation of the order parameter. As a result of this, the corresponding low-energy effective theory often takes the form of a field theory (either quantum or classical, depending on whether we are at zero or finite temperature), in which the field is nothing but the local order parameter. Such low-energy effective field theories go under the name of Ginzburg–Landau theories (yes, the same Landau again).

---

[7] This is an analysis that relates the microscopic theory describing the system at short length and time scales to an effective theory (possibly having different effective masses and interactions for the particles) that describes the physics at longer length and time scales.

Ginzburg–Landau theories can also be used to describe topological phases. Actually, the birth place of (the original) Ginzburg–Landau theory is superconductivity, which is a topological phase. With proper generalizations they can also be used to describe the simplest types of quantum Hall phases. These types of Ginzburg–Landau theory are also written in terms of a local order parameter field. One might wonder how one can have an order parameter without a broken symmetry. The difference here is that in these topological phases the order parameters couple to gauge fields and are therefore *not* (by themselves) gauge-invariant; as a result, the presence of an order parameter does *not* signal symmetry breaking. Our understanding of topological phases is far from complete, and it is not clear what the low-energy effective theories are for some of the known examples of topological phases. On the other hand, many well-understood topological phases (in particular quantum Hall phases) can be well described by topological quantum field theories. This discussion is doubtless mysterious to the reader at the beginning, but the issues will be clarified as we proceed.

1.4 Experimental Probes of Condensed Matter Systems

Most experimental probes of condensed matter systems are based on linear response. As explained in Appendix A, one weakly perturbs the system, and measures how the system responds to the perturbation. In the linear response regime, the response is proportional to the perturbation, and one thus measures the ratio between the two, known as the response function (at the frequency and wavelength of the perturbation). For example, electric current is the response to a (perturbing) electric field or voltage drop, and the ratio between them is the conductance that one measures in a transport experiment, while magnetization is the response to an external magnetic field, and the ratio between them is the magnetic susceptibility measured in a magnetic measurement. In many cases, the frequency of the probing perturbation is low, and the wavelength is long compared with the characteristic microscopic scales of the system, and that is why we need only focus on the low-energy and long-wavelength properties of the system.

Different probes are used to probe different properties of the system. Atomic spatial structures are most often probed using X-ray scattering, while neutron scattering can probe both atomic lattice vibrations and magnetic excitations. Thermodynamic measurements (like specific heat) probe contributions from all degrees of freedom. Electronic contributions, on the other hand, dominate electric responses like conductivity. We will discuss these and many other experimental probes in later chapters.

2 Spatial Structure

The spatial arrangement of the nuclei or ion cores in a condensed matter system is often called its "structure." As we will see later, the structure of real matter can in principle be predicted theoretically by solving the Schrödinger equation and finding the structure which minimizes the total ground-state energy of the electrons and nuclei. In general, however (and certainly historically), it is determined experimentally and the language we will use to describe the structure is thus closely tied to experiment. Although the invention of modern probe techniques such as scanning tunneling microscopy has allowed direct imaging of the spatial positions of individual atoms, structure is more generally measured by elastic scattering experiments in which photons, neutrons, or electrons are scattered and the ion cores or nuclei in the system act as scattering centers. These scattering experiments play a very important role in experimental condensed matter physics and have been advanced to an extremely sophisticated level. The principal ideas, however, are quite simple, and we will introduce them here.

2.1 Probing the Structure

In order to measure the detailed microscopic structure of condensed matter systems we need a probe with a wavelength comparable to the typical distance between nuclei. For photons of energy E, the wavelength is $\lambda = hc/E$ so that

$$\lambda \sim \frac{12.4 \text{ Å}}{E/\text{keV}}, \tag{2.1}$$

where keV is kilo-electron-volts. For neutrons, the de Broglie wavelength is $\lambda = h/\sqrt{2ME}$, thus

$$\lambda \sim \frac{0.28}{(E/\text{eV})^{1/2}} \text{ Å}, \tag{2.2}$$

where M is the neutron mass.[1] For electrons the wavelength is longer by a factor of the square root of the neutron-to-electron mass ratio:

$$\lambda \sim \frac{12 \text{ Å}}{(E/\text{eV})^{1/2}}. \tag{2.3}$$

---

[1] Here we assume the neutron kinetic energy is small compared with its rest energy and is thus in the non-relativistic limit; as we will see shortly, this is appropriate. The same is true for electrons.

It turns out that for typical condensed matter systems with an average distance between nuclei of ~ 1–10 Å we would need to use X-ray photons in the keV energy range, electrons in the eV range, or thermal neutrons, like those that come out of the thermal moderator of a nuclear reactor. Low-energy electrons have a very short mean free path in solids since the electron charge interacts so strongly with the charges in the solid. It turns out that it is possible to use high-energy electrons, provided that one can study their deflections through small angles (small momentum transfers). X-rays are weakly scattering and penetrate samples easily. The **Born approximation** for weak scattering is generally applicable, which is convenient both mathematically and for physical interpretation of the scattering intensity. X-rays are conveniently generated by accelerating electrons into a metal target. The wavelength can be tuned by adjusting the accelerating voltage and by choice of target material. Extremely intense X-ray beams of widely tunable energy are also generated at **synchrotron radiation** facilities.

Neutrons are *very* weakly interacting (which means that samples smaller than tens of grams in mass are difficult to work with), and can more easily provide information on magnetic structures of the system than X-rays (although the latter is also possible). X-ray diffraction is the most common technique, so we shall discuss it first, and then discuss neutron scattering at length a little later. It turns out that neutrons have advantages in inelastic scattering experiments that can tell us something about the excitations of condensed matter systems. This will be discussed in detail in Chapter 4. In the remainder of this chapter we will focus on X-ray scattering [17].

2.2 Semiclassical Theory of X-Ray Scattering

Let us begin by considering classical X-ray scattering by a free, classical, non-relativistic electron (**Thomson scattering**). We can ignore scattering by the nuclei since they are too massive to be much perturbed (i.e. accelerated) by the electromagnetic field. The electron is assumed to have speed $v/c \ll 1$, so it couples primarily to the electric field $\vec{\epsilon}$ (rather than the magnetic field) of the X-rays (recall that in the Bohr model of hydrogen $v/c = \alpha \sim 1/137$ in the ground state). Hence we neglect the **Lorentz force** and Newton's law gives[2]

$$\delta\ddot{\vec{r}} = \frac{-e}{m_e}\vec{\epsilon}\,(\vec{r} + \delta\vec{r}, t), \tag{2.4}$$

where $\vec{r}$ is the equilibrium position of the electron, and $\delta\vec{r}$ is the deviation caused by external force. Now consider a plane wave incident on the electron,

$$\vec{\epsilon}\,(\vec{r}, t) = \vec{E}_{in}e^{i\vec{k}\cdot\vec{r}}e^{-i\omega t}. \tag{2.5}$$

The oscillatory electric force induces a harmonically oscillating dipole:

$$\vec{p}(t) = -e\,\delta\vec{r}(t) = -\frac{e^2}{m_e\omega^2}\vec{E}_{in}e^{i\vec{k}\cdot\vec{r}}e^{-i\omega t}, \tag{2.6}$$

where we have assumed that $\vec{k}\cdot\delta\vec{r} \ll 1$ (or small oscillation), so the position dependence of the phase of electric field is negligible. The electric field radiated by this oscillating dipole at position $\vec{R}$ is[3]

$$\vec{\epsilon}_a = \frac{e^2}{m_e c^2}[\hat{n}\times(\hat{n}\times\vec{E}_{in})]e^{i\vec{k}\cdot\vec{r}}e^{-i\omega t}\frac{e^{ik|\vec{R}-\vec{r}|}}{|\vec{R}-\vec{r}|}, \tag{2.7}$$

---

[2] We have also neglected the electric force due to the nuclei and other electrons in the system. This is justifiable because, at its equilibrium position, the total force from them on the electron would be zero; because the X-ray frequency is much higher than the characteristic oscillation frequency of an electron in an atom (of order $eV/\hbar$), the perturbation due to the X-ray field does not drive the electron far from its equilibrium position.

[3] See, for example, Jackson [18] or Zangwill [19] for a derivation of this expression.

where $\hat{n} = (\vec{R} - \vec{r})/|\vec{R} - \vec{r}| \approx \vec{R}/|\vec{R}|$ is the unit vector along the direction of radiation, $e^2/m_e c^2 \equiv$ $r_c = \alpha^2 a_B$ is the "**classical radius of the electron**" and $k \equiv \omega/c$ (we assume that the frequency of the radiated wave is the same as that of the incident wave, i.e. we neglect the small inelasticity of the scattering that is present quantum mechanically due to the **Compton effect**). The factor of $|\vec{R} - \vec{r}|^{-1}$ gives the usual fall off of amplitude for spherical waves due to energy conservation. The factor of r_c cancels out the length units on the right hand side of Eq. (2.7), and it will turn out that the scattering cross section for photons is proportional to r_c^2.

If we have more than one electron, then waves scattered from each electron will interfere. If the detector is far away at position $\vec{R}_D$, then in the denominator in Eq. (2.7) we can make the replacement

$$|\vec{R} - \vec{r}| \rightarrow R_D. \tag{2.8}$$

This, however, is too crude an approximation in the exponent (which determines the phase and hence the interference). Instead we must write

$$k|\vec{R} - \vec{r}| = k\sqrt{R_D^2 - 2\vec{r} \cdot \vec{R}_D + r^2}$$
$$\approx k R_D \left[1 - \frac{\vec{r} \cdot \vec{R}_D}{R_D^2} + \mathcal{O}\left(\frac{r^2}{R_D^2}\right)\right]. \tag{2.9}$$

See Fig. 2.1(a) for illustration. The direction to the detector determines the final state momentum of the X-ray:

$$\vec{k}' \equiv \vec{k} + \vec{q} = k\frac{\vec{R}_D}{R_D} = k\hat{n}, \tag{2.10}$$

since, if the target is far away, all the scattered beams reaching the detector are essentially parallel. Hence we obtain

$$\frac{e^{ik|\vec{R}-\vec{r}|}}{|\vec{R} - \vec{r}|} \approx \frac{e^{ikR_D}}{R_D} e^{-i(\vec{k}+\vec{q})\cdot\vec{r}}, \tag{2.11}$$

where $\hbar\vec{q}$ is the momentum taken from the crystal and then transferred to the X-ray beam. Thus

$$\vec{\epsilon}_a \approx \frac{e^2}{m_e c^2} \frac{e^{ikR_D}}{R_D} \left[\hat{n} \times (\hat{n} \times \vec{E}_{in})\right] e^{-i\omega t} e^{-i\vec{q}\cdot\vec{r}}. \tag{2.12}$$

The factor $e^{-i\vec{q}\cdot\vec{r}}$ clearly removes momentum $\hbar\vec{q}$ from the electron wave function in the quantum version of this calculation. This factor arises here because, as the electron moves from the origin to $\vec{r}$, the phase of the driving force due to the incident wave changes, as does the phase of the scattered wave since the distance to the detector changes. It is the sensitivity of the *phase* of the scattered wave to the position of the electron that allows the spatial structure of atoms (through the electrons they carry) to be detected in X-ray scattering.

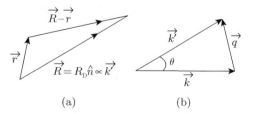

(a)   (b)

Figure 2.1 Geometrical illustrations of X-ray scattering.

Our result is readily generalized to the case of Z electrons (in an atom say) by replacing $e^{-i\vec{q}\cdot\vec{r}}$ in Eq. (2.12) by

$$f(\vec{q}\,) \equiv \sum_{j=1}^{Z} e^{-i\vec{q}\cdot\vec{r}_j}, \tag{2.13}$$

where f is known as the **atomic form factor** and can be written as

$$f(\vec{q}\,) = \int d^3\vec{r}\; e^{-i\vec{q}\cdot\vec{r}} \rho(\vec{r}), \tag{2.14}$$

where $\rho(\vec{r})$ is the electron density:

$$\rho(\vec{r}) \equiv \sum_{j=1}^{Z} \delta^3(\vec{r} - \vec{r}_j). \tag{2.15}$$

To approximate the effect of quantum mechanics we can replace $\rho(\vec{r})$ in (2.14) by its quantum expectation value $\langle\rho(\vec{r})\rangle$. We see that, at this level of approximation, the X-ray **scattering amplitude** directly measures the Fourier transform of the charge density. (One slight complication in the analysis of experimental data is that typically it is possible only to measure the intensity, which is proportional to $|f|^2$.) A simple interpretation of the form factor f is that it is the dimensionless ratio of the scattering amplitude produced by the full collection of electrons to that by a single electron:

$$f = \frac{\text{amplitude scattered by } N \text{ electrons}}{\text{amplitude scattered by 1 electron}}. \tag{2.16}$$

Notice from Eq. (2.14) that in the limit $q \to 0$ we have

$$f(0) = Z. \tag{2.17}$$

This result is interpreted as saying that the spatial resolution $\sim 2\pi/q$ is very poor for small q, and the atom looks as if it is a single charge $-Ze$ oscillating coherently in the applied electric field. The intensity scales with $|f(0)|^2 = Z^2$ in this case. On the other hand, for $q \to \infty$ we expect $f(q) \to 0$ since the (expectation value of the) charge density is smooth below some length scale. Hence we typically expect something like the result shown in Fig. 2.2. The width q_0 is inversely proportional to the atomic diameter.

For elastic scattering, q is related to the scattering angle θ by

$$q = 2k \sin(\theta/2) \tag{2.18}$$

as shown in Fig. 2.1(b), and so increases with angle and with photon energy. It should now be clear why we need to have $k \sim 1\,\text{Å}^{-1}$ for the scattering experiment to be effective: We know the interesting scatterings are for $q \sim 1\,\text{Å}^{-1}$. For $k \ll 1\,\text{Å}^{-1}$, such values of q are inaccessible as clear from the

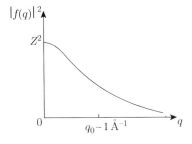

Figure 2.2 Typical atomic structure factor.

equation above. For $k \gg 1\,\text{Å}^{-1}$, scattering is restricted to very small scattering angle θ and thus requires extremely high angular resolution.

2.3 Quantum Theory of Electron–Photon Interaction and X-Ray Scattering

Our semiclassical result is valid in the limit that the X-ray energy $\hbar\omega$ is much greater than E_B, the atomic binding energy of the electrons. In this limit the electrons look (approximately) like free electrons. Our approximation is particularly poor if $\hbar\omega$ is near the energy of an allowed atomic transition for the electrons. To see why this is so, we will present a rough sketch of the fully quantum-mechanical treatment of the interaction between electrons and radiation field, and apply it to the elastic X-ray scattering problem. Readers unfamiliar with the quantization of the electromagnetic field may wish to consult textbooks on quantum mechanics that cover this topic (e.g. those by Schiff [20] and Merzbacher [21]), or study the section covering the closely analogous problem of quantization of lattice vibrations in Chapter 6 before proceeding.

The quantum Hamiltonian is

$$H = H_e + H_N + H_{EM}, \tag{2.19}$$

where H_N is the nuclear Hamiltonian, H_{EM} is the Hamiltonian of the free electromagnetic field (photons), and the electronic Hamiltonian is

$$H_e = \sum_{j=1}^{Z} \frac{1}{2m_e}\left[\vec{p}_j + \frac{e}{c}\vec{A}(\vec{r}_j) \right]^2 + V, \tag{2.20}$$

where V is the potential due to the nuclei and the electron–electron interactions, and $\vec{A}(\vec{r}_j)$ is the vector potential of the external radiation field evaluated at the position of the jth electron. We have chosen a gauge in which the scalar potential is zero, and $\vec{A}(\vec{r}, t)$ determines both the electric and the magnetic field:

$$\vec{B} = \nabla \times \vec{A}(\vec{r}, t); \qquad \vec{E} = -\frac{1}{c}\frac{\partial}{\partial t}\vec{A}(\vec{r}, t). \tag{2.21}$$

This is a convenient gauge that we will use repeatedly later in the book. We can ignore the coupling between the radiation field and the nuclei because $M_N \gg m$. Let us start with the $\vec{A}^2$ term in H_e:

$$\Delta H_2 = \sum_{j=1}^{Z} \frac{e^2}{2m_e c^2}[\vec{A}(\vec{r}_j)]^2. \tag{2.22}$$

This looks as if it involves only the free photon field, but actually the electronic charge density enters since $\vec{A}$ is evaluated at the position of each electron. In fact, we may rewrite this as

$$\Delta H_2 = \frac{e^2}{2m_e c^2} \int d^3r \, \rho(\vec{r})[\vec{A}(\vec{r})]^2, \tag{2.23}$$

where $\rho(\vec{r})$ is the electron density operator introduced in Eq. (2.15). Now we are interested in the piece of $\vec{A}^2$ which destroys a photon of wave vector $\vec{k}$ and creates a photon of wave vector $\vec{k}' = \vec{k} + \vec{q}$. Quantization of the electromagnetic field yields (in a normalization volume L^3)[4]

---

[4] See, e.g., Schiff [20], Eq. (56.35), p. 519. There is a small difference in the normalization factor of the photon-creation/annihilation operators between that of Schiff and that used here. Ours corresponds to the standard commutation relation for creation/annihilation operators: $[a_{\vec{p}\lambda}, a^{\dagger}_{\vec{p}'\lambda'}] = \delta_{\vec{p}\vec{p}'}\delta_{\lambda\lambda'}$.

$$\vec{A}(\vec{r}) = \sum_{\vec{p},\lambda} \hat{\epsilon}_{\vec{p}\lambda} \, \wedge_p \, (a^{\dagger}_{\vec{p}\lambda} + a_{-\vec{p}\lambda})e^{-i\vec{p}\cdot\vec{r}}, \tag{2.24}$$

where $\wedge_p \equiv (hc/L^3 p)^{1/2}$ and $a^{\dagger}_{\vec{p}\lambda}$ is the photon-creation operator for polarization mode λ (with polarization vector $\hat{\epsilon}_{\vec{p}\lambda} \perp \vec{p}$) and wave vector $\vec{p}$. To gain some intuition about the expression in Eq. (2.24), notice that the last factor on the RHS removes momentum $\hbar\vec{p}$ from the electron. The term in parentheses containing the annihilation and creation operators adds momentum $\hbar\vec{p}$ to the electromagnetic field by either creating a photon with that momentum or destroying a photon with the opposite momentum. Hence the overall momentum is conserved.

The portion of ΔH_2 relevant to photon scattering from mode $\vec{k}\lambda$ to mode $\vec{k}+\vec{q}, \lambda'$ is

$$\frac{e^2}{m_e c^2} \wedge_{|\vec{k}+\vec{q}|} \wedge_k \left(\hat{\epsilon}_{\vec{k}\lambda} \cdot \hat{\epsilon}_{\vec{k}'\lambda'} \right) \int d^3\vec{r} \; \rho(\vec{r})e^{-i\vec{q}\cdot\vec{r}}a^{\dagger}_{\vec{k}+\vec{q},\lambda'}a_{\vec{k}\lambda}. \tag{2.25}$$

The matrix element of ΔH_2 between the initial state

$$|i\rangle = \underbrace{|\phi\rangle}_{\text{electron state}} \quad \underbrace{a^{\dagger}_{\vec{k}\lambda}|0\rangle}_{\text{photon state}} \tag{2.26}$$

and the final state

$$|f\rangle = |\phi\rangle a^{\dagger}_{\vec{k}+\vec{q},\lambda'}|0\rangle \tag{2.27}$$

(where we assume elastic scattering which leaves the electronic state unchanged and gives the final photon the same energy as the initial) is

$$\langle f|\Delta H_2|i\rangle = \frac{e^2}{m_e c^2} f(\vec{q}) \, (\wedge_k)^2 \, \hat{\epsilon}_{\vec{k}\lambda} \cdot \hat{\epsilon}_{\vec{k}'\lambda'}, \tag{2.28}$$

where we have used the fact that $|\vec{k}+\vec{q}| = |\vec{k}|$ for elastic scattering and defined

$$f(\vec{q}) \equiv \int d^3\vec{r} \; e^{-i\vec{q}\cdot\vec{r}} \langle\phi|\rho(\vec{r})|\phi\rangle. \tag{2.29}$$

This is precisely of the same form as we found semiclassically. In the semiclassical picture the incident electric field accelerates the electrons and they then radiate. In this quantum perturbation calculation it appears that the electrons are always in the stationary state $|\phi\rangle$, so it is hard to see how they can be accelerating. Indeed, the average momentum is (and remains) zero:

$$\langle\phi| \sum_j \vec{p}_j |\phi\rangle = \vec{0}. \tag{2.30}$$

However, we must recall that, in the presence of a vector potential, the velocity is not $\vec{p}_j/m_e$ but rather $(\vec{p}_j + (e/c)\vec{A}_j)/m_e$.[5] Since $\vec{A}$ is time-dependent, we have acceleration without any (canonical) momentum change.

Still, we know that the X-ray changes its momentum state and must impart a total momentum $-\hbar\vec{q}$ to the electrons. Equation (2.29) tells us that we can impart this momentum to state $|\phi\rangle$ and still end up in the same state! Because the electron cloud around an atom has a relatively well-defined position, it has uncertain momentum, and giving it a kick does not automatically send the cloud into an orthogonal quantum state.[6] If the electrons remain in the same state, the momentum must be transferred to the nucleus via the strong electron–nucleus Coulomb interaction. The massive nucleus is essentially a

---

[5] We will encounter this point repeatedly later in the book.

[6] Though the fact that $f(q)$ becomes small for large q tells us that the probability of the system remaining in the ground state becomes smaller and smaller as the strength of the kick increases.

"momentum sink," as we will discuss in Section 6.5 when we study "recoiless" momentum transfer in the **Mössbauer effect**.

Exercise 2.1.
(i) Using Eq. (2.12), compute the power radiated away by the oscillating dipole within the semiclassical approximation.
(ii) Show, using Fermi's Golden Rule and the matrix element in Eq. (2.28), that the quantum rate of production of scattered photons is exactly consistent with the semiclassical expression for the radiated power.

The quantum process we have just described produces photon scattering at first order in perturbation theory in $A_j^2 \equiv |\vec{A}(\vec{r}_j)|^2$. There is another process which can do the same, namely second-order perturbation theory in the perturbation term in Eq. (2.20) which is linear in the vector potential:

$$\Delta H_1 \equiv \frac{e}{2m_e c} \sum_{j=1}^{Z} \left[\vec{p}_j \cdot \vec{A}(\vec{r}_j) + \vec{A}(\vec{r}_j) \cdot \vec{p}_j \right]. \tag{2.31}$$

The physical interpretation of the second-order perturbation theory amplitude given below in Eq. (2.32) is the following. In the first step, ΔH_1 causes the incoming photon to be absorbed and an electronic excitation to be created. In the second step, ΔH_1 acts again and the electron falls back to its original state (say), and the scattered photon is created. In the limit of extremely high X-ray photon energies where atomic absorption is small (and the perturbation-theory energy denominator is large), this process is negligible. However, if the X-ray energy happens to be close to an atomic transition, then energy will be nearly conserved in the intermediate state. This means that the energy denominator nearly vanishes:

$$\frac{e^2}{2m_e^2 c^2} \sum_{n=0}^{\infty} \langle f|\vec{p} \cdot \vec{A}|n\rangle \frac{1}{(\hbar\omega + E_0) - E_n} \langle n|\vec{p} \cdot \vec{A}|i\rangle, \tag{2.32}$$

where E_0 is the initial electron energy and E_n is the energy of the "virtual" excited state n. At these special values of $\hbar\omega$ one has resonances which greatly enhance this contribution to the scattering amplitude. This effect can be used (with a tunable X-ray source such as a synchrotron) to help locate specific types of atoms within the crystal structure. We will, however, ignore it in the remaining discussion of elastic X-ray scattering.

In the discussion above the first step action of ΔH_1 is a *virtual* process, and energy is *not* conserved in the corresponding *intermediate* state. There are, of course, real physical processes induced by a first-order action of ΔH_1, that correspond to absorption or emission of a photon by an electron (cf. Eq. (2.24); $\vec{A}$ either creates or annihilates a photon). Energy is conserved in such processes, with the change of electronic energy equal to that of the photon being absorbed or emitted. Such processes play a dominant role in the optical properties of condensed matter systems. These will be discussed at length in later chapters.

2.4 X-Ray Scattering from a Condensed Matter System

What happens if we now have a condensed matter system consisting of a large number of atoms? The theory we have derived for individual atoms still applies if we replace the electron density by the total electron density of the system. We first discuss X-ray scattering from condensed matter systems by making an approximation, which is often very good. We assume that the total electron density is approximately the sum of contributions from individual isolated atoms:

$$\rho(\vec{r}) = \sum_{i=1}^{N} \rho_a(\vec{r} - \vec{R}_i), \tag{2.33}$$

where i is the atom index, $\vec{R}_i$ is the position of atom i, N is the number of atoms, and ρ_a is the atomic electron density (in the condensed matter environment). This approximation can be quite poor for the valence electrons (which take part in chemical bonding) but will be quite accurate for systems composed of large atoms where most of the electrons are in tightly bound core levels. Let $F(\vec{q})$ be the form factor for the entire system:

$$F(\vec{q}) = \int d^3\vec{r} \, e^{-i\vec{q}\cdot\vec{r}} \sum_{i=1}^{N} \rho_a(\vec{r} - \vec{R}_i) \tag{2.34}$$

$$= \sum_{i=1}^{N} \int d^3\vec{r}' e^{-i\vec{q}\cdot(\vec{r}'+\vec{R}_i)} \rho_a(\vec{r}') \tag{2.35}$$

$$= \left(\sum_{i=1}^{N} e^{-i\vec{q}\cdot\vec{R}_i} \right) \left(\int d^3\vec{r}' e^{-i\vec{q}\cdot\vec{r}'} \rho_a(\vec{r}') \right). \tag{2.36}$$

We thus see

$$F(\vec{q}) = f(\vec{q}) W(\vec{q}), \tag{2.37}$$

where

$$W(\vec{q}) \equiv \sum_{i=1}^{N} e^{-i\vec{q}\cdot\vec{R}_i}. \tag{2.38}$$

Here $W(\vec{q})$ can be thought of as the "**crystal form factor**" which describes how the N atoms are distributed in the system just as $f(\vec{q})$ describes how charge is distributed in an individual atom. (We assume here for simplicity that all the atoms are identical.) We will see in Chapter 3 how to compute W.

Assuming that the atomic form factor is known, at least approximately, measurements of the elastic X-ray scattering intensity as a function of angle, which in the Born approximation is proportional to $|F(\vec{q})|^2$, can be used to determine

$$S(\vec{q}) \equiv \frac{1}{N} \langle\langle |W(\vec{q})|^2 \rangle\rangle, \tag{2.39}$$

where $\langle\langle \cdot \rangle\rangle$ stands for the thermal average. This is appropriate as the scattering experiment takes a time much longer than the microscopic relaxation time, and the time-average result is equivalent to the thermal average.[7] $S(\vec{q})$ is known as the **static structure factor** for the nuclear positions. The dependence of $S(\vec{q})$ on the wave vector $\vec{q}$ can be dramatically different depending on the spatial arrangements of the atoms in the condensed matter system. In the following sections we discuss how $S(\vec{q})$ is related to these spatial arrangements and the special properties of $S(\vec{q})$ when the atoms are arranged in a crystal.

2.5 Relationship of $S(\vec{q})$ and Spatial Correlations

The discussion in this section is quite general and applies to spatial correlations of any system of classical or quantum particles. It introduces some common definitions. We have in mind here atoms in

---

[7] Except in glassy and other out-of-equilibrium states, where the relaxation time becomes very long.

a condensed matter system whose locations are described by a probability distribution determined, in general, by the rules of quantum statistical mechanics. We will learn more later about how we could, at least in principle, develop a theory for that distribution function. For the moment, though, we are interested only in what X-ray scattering, by measuring the static structure function, tells us experimentally about that distribution function. The essential point is that, in the Born approximation, the scattering amplitude is a sum of terms corresponding to scattering from different atoms. The intensity is the square of the amplitude and hence contains pairwise interference terms which depend on the distance and relative orientation between pairs of atoms in the sample. We thus learn something in X-ray scattering (in the Born approximation or single-scattering limit) about the probability distribution for separations of pairs of particles.

We define a quantity known as the two-point distribution function:

$$n^{(2)}(\vec{r}, \vec{r}') \equiv \left\langle\left\langle \sum_{i \neq j} \delta(\vec{r} - \vec{r}_i)\delta(\vec{r}' - \vec{r}_j) \right\rangle\right\rangle. \tag{2.40}$$

Using the definition of the static structure factor in Eq. (2.39) and using invariance under a constant shift of all particles (and the assumption based on translational invariance that $n^{(2)}(\vec{r}, \vec{r}')$ depends only on $\vec{r}' - \vec{r}$) we find that

$$S(\vec{q}) = 1 + n \int d^3\vec{r}\, e^{i\vec{q}\cdot\vec{r}} g(\vec{r}), \tag{2.41}$$

where n is the mean particle density and $g(\vec{r}) \equiv n^{(2)}(\vec{r}, \vec{0})/n^2$. $g(\vec{r})$ is usually known as the **pair distribution function**. If the particles are arranged randomly, $g(\vec{r}) \equiv 1$. In an isotropic liquid $g(\vec{r})$ depends only on $r = |\vec{r}|$, and is often referred to as the radial distribution function. We should expect that $g(r) \to 1$ at large r. It is then useful to define

$$h(r) \equiv g(r) - 1 \tag{2.42}$$

as the pair correlation function since it is non-zero only when the particles are correlated, i.e. their positions are not random. We thus have

$$S(q) = N\delta_{\vec{q},0} + 1 + \tilde{h}(q), \tag{2.43}$$

where

$$\tilde{h}(q) \equiv n \int d^3\vec{r}\, e^{i\vec{q}\cdot\vec{r}} h(r). \tag{2.44}$$

Because $h(r)$ vanishes for large r, $\tilde{h}(q)$ is well-behaved at small wave vectors. This language is also used to describe correlations in the positions of itinerant electrons.

Exercise 2.2. Prove Eq. (2.41).

2.6 Liquid State versus Crystal State

The reason why the pair correlation function appears in the expression for the X-ray scattering intensity is that the X-ray scattering amplitude is proportional to the Fourier transform of the atomic density given in Eq. (2.36). The intensity is proportional to the square of this quantity and contains interference terms between scattering from one atom and some other atom. Whether this interference is constructive or destructive depends on the relative separation of the two atoms. This is how the two-body correlation function enters the calculation.

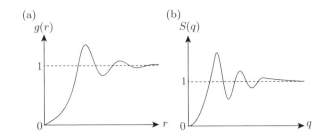

Figure 2.3 Typical pair correlation function (a) and structure factor (b) of a liquid.

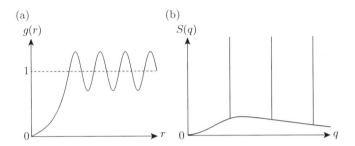

Figure 2.4 Typical pair correlation function (a) and structure factor (b) of an (idealized) 1D crystal.

Our task now is to understand what two-body correlations we would expect in various phases of matter such as crystalline solids, amorphous glasses, and liquids. At sufficiently high temperatures, all condensed matter systems are in the liquid or gaseous phase, with a pair correlation function and corresponding structure factor exemplified by that of Fig. 2.3. The system contains (varying degrees of) correlation at short distance, typically out to distances a few times larger than the scale of the inter-atomic separation, but the correlations die off quickly (usually in an exponential fashion) at larger distances.

At sufficiently low temperatures, the equilibrium state of the majority of materials is an ordered crystalline lattice made of an array of perfectly repeating units consisting of one or more atoms or molecules. Such a regular array optimizes the energy by putting nearest-neighbor atoms at suitable distances for chemical bonding and giving the atoms molecule-like environments. This leads to *long-range* correlation in the positions of the atoms, as illustrated in Fig. 2.4(a); it is also reflected as delta-function peaks in the structure factor (see Fig. 2.4(b)). Such long-range correlation, often referred to as **long-range order**, leads to *spontaneously broken symmetry*. In the case of a crystal, the presence of long-range positional order makes it possible to pin down the (average) positions of *all* atoms in the crystal by a small perturbation that pins the positions of a handful of atoms. This spontaneously breaks the translation symmetry of the system, as spatial homogeneity would have suggested that an atom should have the same probability of appearing anywhere in space.[8] The broken-symmetry nature of the crystalline states will be discussed in great detail in the next chapter. It should be noted, however, that we have available no rigorous mathematical theorems proving that any given class of Hamiltonians always has a crystalline ground state.

---

[8] Formally, spontaneously broken symmetry is signalled by a *finite* response to an *infinitesimal* perturbation that breaks the symmetry. Long-range correlation makes such a *singular* response possible, and pinning the position and orientation of a crystal by pinning positions of several atoms is an example of that. In 1D, pinning a single atom pins the entire crystal (though, as we shall see, thermal and quantum fluctuations actually destroy crystalline order in 1D); in higher dimensions this is not quite enough due to the rotational degrees of freedom of the crystal.

In 1984 the physics world was stunned by the unexpected discovery of a new class of equilibrium structures called **quasicrystals** [22]. These structures have X-ray diffraction patterns indicating long-range positional order (as in ordinary crystals) and yet they have axes with five-fold rotational symmetry – something which is known rigorously to be impossible for a true (periodic) crystal.

Some systems undergo a transition from a liquid state at high temperature to an amorphous glassy state (rather than an ordered crystalline state) at low temperatures. In some cases this happens because, no matter how slowly the system is cooled, it always falls out of equilibrium. Thus the true ground state might be a crystal, but the free-energy barrier that must be overcome to move from the liquid to the crystal state is too large to be traversed in a reasonable time period. It is also possible that in some cases the glassy state is a true equilibrium thermodynamic phase. This has been a long-standing and largely unresolved problem.

Other materials undergo transitions to "liquid crystal" states, which are peculiar states (of great technological importance for flat-panel displays) in which the molecules have various kinds of orientational and positional long-range order (as in a solid) and yet the substance is still a fluid. These "soft matter" phases are extensively discussed in Chaikin and Lubensky [2].

Finally, we note that one material, liquid helium, has the unique property of remaining fluid at atmospheric pressure all the way down to theoretical absolute zero temperature. As will be discussed in Section 4.4, the remarkable collective properties of this substance are controlled by the large quantum zero-point fluctuations due to the low mass of the helium atoms.

In the next chapter we will focus on the structure of ideal crystals and see in more detail how this can be probed with X-ray scattering.

3 Lattices and Symmetries

Many solids exist in the form of *crystals*. The shapes, colors, and sheer beauty of naturally occurring crystals have fascinated humans from the earliest times to the present day. It was recognized, both by the ancient Greeks and by scientists struggling in the nineteenth century to come to grips with the concept of atoms, that the angles of facets on the surface of crystals could be explained by the existence of atoms and their simple geometric packing into periodic and mathematically beautiful structures that have well-defined cleavage planes.

Crystals are the thermodynamic equilibrium state of many elements and compounds at room temperature and pressure. Yet, the existence of natural mineral crystals seems quite miraculous because this requires special non-equilibrium geological processes in the earth to concentrate particular elements in one place and arrange the atoms so perfectly. Without these non-equilibrium processes that somehow managed to put lots of copper atoms in northern Michigan, iron atoms in North Dakota, tin atoms in Britain, gold atoms in California, carbon atoms (diamonds) in South Africa, and lithium atoms in Chile, human civilization would be vastly different than it is today.

Laboratory synthesis of nearly perfect crystals is an art, some would even say a black art. Billions of dollars have been spent perfecting the synthesis of silicon crystals that are nearly atomically perfect across their 300 mm diameter so that wafers can be sliced from them and enter a 10 billion dollar processing plant to make computer chips. It is now possible to make diamond crystals which are easily distinguishable from naturally occuring ones because they are much more nearly perfect.

In this chapter we will develop the mathematical framework needed for studying crystal lattices and the periodic arrangements of atoms in them. This will provide us with the tools needed in the subsequent three chapters to understand scattering of neutrons used to probe crystal structures and study the classical and quantum propagation of vibrations through crystals. This will lead us into Chapter 7, where we will learn how electron waves propagate through the periodic potential formed by the atoms in a crystal. From this we will begin to understand the electronic bonds and bands which give energetic stability to the crystalline structure that generates them.

3.1 The Crystal as a Broken-Symmetry State

A crystal is a special state of **spontaneously broken translation symmetry**. What does this mean? In the simplest sense it means that, if we know the positions of a few atoms, we can predict the positions of all the others. To understand this more precisely, consider a system of point particles with an interaction that depends only on the magnitude of the separation of pairs. The Hamiltonian is

$$H = \frac{1}{2M} \sum_{j=1}^{N} \vec{p}_j^2 + \sum_{i<j} \upsilon(|\vec{r}_i - \vec{r}_j|). \tag{3.1}$$

This Hamiltonian is translationally invariant (or has Tr-symmetry). That is, if

$$\{\vec{r}_j \longrightarrow \vec{r}_j + \vec{\delta}, \forall j\} \tag{3.2}$$

then H is unchanged. Additional symmetries of H include rotation through any arbitrary angle (R),

$$\vec{r}_j \rightarrow \overleftrightarrow{R}\, \vec{r}_j, \tag{3.3}$$

and **space inversion** (or parity P) *about any point* $\vec{R}$,[1]

$$\vec{r}_j - \vec{R} \rightarrow -(\vec{r}_j - \vec{R}). \tag{3.4}$$

In a uniform liquid, the time-averaged density of particles $\langle \rho(\vec{R}) \rangle$ is a constant (independent of $\vec{R}$), and is symmetric under translation Tr, rotation R, and parity P, whereas in a crystal $\langle \rho(\vec{R}) \rangle$ becomes non-constant but periodic, thus breaking Tr, R, and P. The density is still invariant under certain *discrete translations* (by a distance equal to the spatial period), but not under general translations. It may also remain invariant under certain discrete rotations along specific axes, space inversion about specific inversion centers, or reflection about certain mirror planes.[2] Hence the state of the crystal has *less* symmetry than its Hamiltonian. This phenomenon is known as *spontaneously broken symmetry*. Strictly speaking, spontaneous symmetry breaking can occur only in the thermodynamic limit of infinite system size. For a finite system the finite thermal (or quantum zero-point) kinetic energy Q in the center-of-mass degree of freedom leads to a finite random velocity fluctuation $\sim (2Q/M_{\text{tot}})^{1/2}$ which makes the time average of the density independent of position; for thermal fluctuations in 3D we expect $Q = \frac{3}{2}k_B T$ from the equipartition theorem. Thus the fluctuation vanishes only in the thermodynamic limit where the total mass $M_{\text{tot}} \rightarrow \infty$.

Crystallization occurs (classically) because the particles arrange themselves into special positions to minimize their potential energy. With few exceptions, quantum condensed matter systems form crystalline solids at zero temperature ($T = 0$) because this minimizes the total energy.[3] There does not exist a rigorous proof that a crystal is the optimum arrangement to minimize $\langle H \rangle$ in Eq. (3.1), but experiments show that this is often true (quasicrystals being one exception). Classically, we need only find the configuration that minimizes the potential energy V. Helium is the only material with low enough atomic mass and weak enough interaction that quantum zero-point energy is so important that it leaves the system as a liquid at $T = 0$ and ambient pressure. The situation changes as the temperature T increases, as the goal at finite T is to minimize the free energy $F = E - TS$, not just the energy E; the higher T is, the more important the entropy S becomes. On the other hand, crystalline ordering limits the configurational space available to the constituent particles, and thus extracts an entropy cost. As a result, as T increases the full Tr, R, and P symmetries are restored

---

[1] In the presence of translation symmetry that is *not* spontaneously broken we often choose $\vec{R} = 0$ without loss of generality. When translation symmetry is broken, $\vec{R}$, which is known as the inversion center in this context, needs to be specified.

[2] The collection of all such remaining symmetry transformations forms a group known as the space group of the crystal. It is a subgroup of the direct product of full translation, rotation, and inversion groups. The space group can be used to classify all possible crystalline states. This is a typical example of classifying states or phases of matter in terms of their symmetry properties.

[3] Classically, the kinetic energy is neglibile at low enough temperatures. Quantum mechanically, the zero-point motion of the atoms always gives finite kinetic energy, which contributes to the relative stability of different possible structures that the crystal can assume. At finite temperatures, the thermal fluctuations of the positions of the atoms contribute an entropy term, and we need to minimize the thermodynamic free energy to find the correct structure. The entropic contributions are often small, but are occasionally significant.

through one or a sequence of melting transitions, and at high temperature the system will always be in the liquid phase with no ordering.

The degree to which a symmetry is broken is quantified by a measurable quantity called the order parameter, which is zero in the high-temperature disordered phase and non-zero in the low-temperature ordered phase. The order parameter transforms non-trivially under a symmetry operation, thus characterizing *how* (through its structure) and *how much* (through its magnitude) the symmetry is broken. Perhaps the most familiar example is the magnetization $\vec{M}$, which is the order parameter of a ferromagnet that breaks (spin) rotation symmetry. $\vec{M}$ transforms as a vector under rotation R,[4] thus R is broken when $\vec{M} \neq 0$, and its magnitude $|\vec{M}|$ quantifies the degree to which R is broken.

Can we find an *order parameter* which measures the broken translation symmetry? Consider the Fourier transform of the density

$$\langle \tilde{\rho}(\vec{k}) \rangle \equiv \frac{1}{\Omega} \int d^3 \vec{R} \langle \rho(\vec{R}) \rangle e^{-i\vec{k}\cdot\vec{R}}, \tag{3.5}$$

where Ω is the volume of the system. For a uniform liquid, constant density

$$\langle \rho(\vec{R}) \rangle = \bar{\rho} \tag{3.6}$$

implies that

$$\langle \tilde{\rho}(\vec{k}) \rangle = 0, \tag{3.7}$$

except for $\vec{k} = 0$, where $\langle \tilde{\rho}(\vec{0}) \rangle = \bar{\rho}$.

For a crystal, the situation is different. If we pick $\vec{k}$ to match the density modulations, $\langle \tilde{\rho}(\vec{k}) \rangle$ will be non-zero. Here is a simple 1D example.[5] Suppose that the density looks like that shown in Fig. 3.1 as the solid begins to form. For example, we might have

$$\langle \rho(R) \rangle = \bar{\rho} + b \cos\left(\frac{2\pi}{a} R - \theta\right), \tag{3.8}$$

where a is the lattice constant, b measures the strength of the ordering, and θ is a phase that determines where the density peaks. We see the *discrete* translation symmetry of Eq. (3.8) under

$$R \rightarrow R + na, \tag{3.9}$$

where n is any integer.

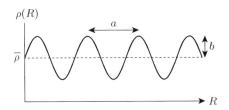

Figure 3.1 Density profile of a 1D crystal with harmonic density modulation.

---

[4] In other systems that break R (e.g. a nematic liquid crystal), the order parameter transforms as a higher-rank tensor under rotation. A vector is a rank-1 tensor. A nematic has a rank-2 tensor order parameter (similar to a moment of inertia tensor) which can be visualized as an arrow with no head, thus distinguishing a nematic from a ferromagnet in *how* R is broken. Using the structure of the order parameter of a broken-symmetry state is another example of classifying phases in terms of symmetry. It is actually closely related to using the residual symmetry to classify broken symmetry as discussed earlier. Coming back to the ferromagnet example, the full 3D rotation symmetry is characterized by the group SO(3). In the presence of a non-zero magnetization $\vec{M}$ the state is invariant under rotations along the direction of $\vec{M}$ *only*; such rotations form a (much smaller) rotation group, O(2). Specifying this specific residual symmetry is equivalent to specifying the direction of $\vec{M}$.

[5] Actually 1D solids exist only at zero temperature for classical particles, and 1D quantum solids do not exist even at zero temperature, but that is another story, which will be discussed in Chapter 6.

Rather than working with b as a measure of the ordering, it will prove convenient to consider the Fourier transform of the density

$$\tilde{\rho}(k) = \lim_{L \to \infty} L^{-1} \int_{-L/2}^{L/2} dR \left[\bar{\rho} + \frac{b}{2} \left(e^{2\pi i R/a} e^{-i\theta} + e^{-2\pi i R/a} e^{+i\theta} \right) \right] e^{-ikR}, \tag{3.10}$$

which yields

$$\tilde{\rho}(k) = \bar{\rho} \delta_{k,0} + \frac{b}{2} \left\{ e^{-i\theta} \delta_{k,2\pi/a} + e^{i\theta} \delta_{k,-2\pi/a} \right\}. \tag{3.11}$$

Thus $\tilde{\rho}(k)$ at $k = \pm 2\pi/a$ is an order parameter. Notice that the complex phase of $\tilde{\rho}(k)$ is arbitrary since the energy is independent of θ. Indeed, under translations θ, and hence the phase of $\tilde{\rho}(k)$, shift.[6] The fact that the order parameter is not invariant under translations (even though the Hamiltonian is) is characteristic of spontaneously broken (translation) symmetry. We see that the order parameter is characterized by three numbers. (i) The magnitude b, which is analogous to $|\vec{M}|$ of the ferromagnet; neither of them changes under the corresponding symmetry transformations. (ii) The phase angle θ, which is the quantity that *transforms non-trivially* under Tr; this is analogous to the *direction* of $\vec{M}$, which transforms non-trivially under R. In fact, if $\vec{M}$ is restricted to a plane (as in the so-called planar or XY ferromagnet) the two have identical mathematical structure. (iii) The wave vector k. This is characteristic of broken Tr; it does not appear in a ferromagnet (because $k = 0$ for $\vec{M}$), but does appear in antiferromagnets and other magnetic phases with spatially dependent magnetization in which the magnetic order parameters also break Tr symmetry.

Now suppose that we lower the temperature and the density modulation increases and sharpens as shown in Fig. 3.2. We can no longer use just a simple cosine wave to represent the particle density. The most general form that satisfies the discrete translation symmetry

$$\rho(R + na) = \rho(R) \tag{3.12}$$

is

$$\rho(R) = \sum_{m=-\infty}^{\infty} \tilde{\rho}_m e^{i \frac{2\pi m}{a} R}. \tag{3.13}$$

Note that $\rho(R)$ being real $\Rightarrow \tilde{\rho}_{-m} = \tilde{\rho}_m^*$.

We will return to this Fourier representation of the density shortly, but first let us consider the limit of (classical) particles at zero temperature, where the positions become infinitely sharply defined:

$$\rho(R) = \sum_{j=-\infty}^{\infty} \delta(R - ja - \theta a/2\pi). \tag{3.14}$$

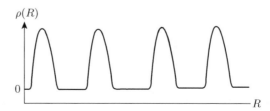

Figure 3.2 Density profile of a 1D crystal with generic density modulation.

---

[6] Remember that the symmetry breaking is *spontaneous*. The translation symmetry of H guarantees that all values of θ give the same energy. Its actual value is determined by the history of the crystal formation, or some infinitesimal random perturbation that breaks Tr symmetry but is *not* included in H.

The particles will sit exactly on the points of a periodic *lattice*. Notice that the set of wave vectors $\{2\pi m/a\}$ in Eq. (3.13) also constitute a lattice known as the **reciprocal lattice**. It should now be clear from this 1D toy model that it is useful to consider the translation symmetries of crystals both in real space and in wave-vector space through the use of Fourier series.

3.2 Bravais Lattices and Lattices with Bases

3.2.1 Bravais Lattices

Let us formalize things a bit by defining a **Bravais lattice**, which is the simplest mathematical object with the desired discrete translation symmetries. A Bravais lattice (in 3D) is a set of points of the form[7]

$$\vec{R}_{\vec{n}} = n_1\vec{a}_1 + n_2\vec{a}_2 + n_3\vec{a}_3, \tag{3.15}$$

where the vector $\vec{n} \equiv (n_1, n_2, n_3)$ is a triple of integers (of any sign) and the three vectors $\vec{a}_1, \vec{a}_2, \vec{a}_3$ (which must not be coplanar) are called the *primitive lattice vectors*.[8] Notice that a Bravais lattice has the property of being closed under addition and subtraction: if $\vec{R}_{\vec{n}}$ and $\vec{R}_{\vec{m}}$ are elements of the lattice, then

$$\vec{R}_{\vec{m}\pm\vec{n}} \equiv \vec{R}_{\vec{m}} \pm \vec{R}_{\vec{n}} \tag{3.16}$$

is also an element of the lattice. This means a Bravais lattice is *invariant* under a translation by a vector of the form (3.15); as a result, these vectors are called lattice vectors.[9] It is clear from the definition that every site on a Bravais lattice is precisely equivalent,[10] and that the lattice is necessarily infinite in extent. This idealization is a useful and appropriate starting point in considering bulk properties of crystalline material.[11]

Common examples in 3D are the

- simple cubic (SC) lattice: $\vec{a}_1 = a\hat{x}, \vec{a}_2 = a\hat{y}, \vec{a}_3 = a\hat{z}$;
- face-centered cubic (FCC) lattice: $\vec{a}_1 = (a/2)(\hat{y} + \hat{z}), \vec{a}_2 = (a/2)(\hat{z} + \hat{x}), \vec{a}_3 = (a/2)(\hat{x} + \hat{y})$;
- body-centered cubic (BCC) lattice: $\vec{a}_1 = (a/2)(\hat{y} + \hat{z} - \hat{x}), \vec{a}_2 = (a/2)(\hat{z} + \hat{x} - \hat{y}), \vec{a}_3 = (a/2)(\hat{x} + \hat{y} - \hat{z})$.

These are shown in Figs. 3.3(a)–(c). The FCC and BCC lattices can be obtained by decorating the SC lattice: starting with the SC lattice, adding lattice points at the center of every single cube gives rise to the BCC lattice, while adding lattice points at the center of every single square (faces of the cubes)

---

[7] We have chosen our coordinate system so that one of the lattice points lies at the origin.

[8] In general, d linearly independent primitive lattice vectors are needed to specify a d-dimensional Bravais lattice, in a way that is a straightforward generalization of Eq. (3.15). The set of primitive lattice vectors uniquely defines a Bravais lattice. For a given Bravais lattice, however, the choice of the primitive lattice vectors is *not* unique. The reader should convince her/himself that, starting from one specific set of primitive lattice vectors, there exist infinitely many linear combinations of them that give rise to the same Bravais lattice (except in 1D, where there is only one primitive lattice vector).

[9] While Eq. (3.16) follows from (3.15) trivially, it can also be shown that Eq. (3.15) follows from (3.16) (by constructing the primitive lattice vectors from the $\vec{R}$s that are close to the origin) as well. As a result, Eqs. (3.15) and (3.16) are *equivalent*, and both can be used as the definition of a Bravais lattice.

[10] Because any one of them can be moved to the origin by a translation of a lattice vector (3.15), which is a symmetry *respected* by the crystalline state. As we will see shortly, this special property of Bravais lattices does not hold for general lattices.

[11] Real crystals are, of course, *finite* in extent and have boundaries, whose presence *breaks* lattice translation symmetry, but we can ignore their effects when studying bulk properties.

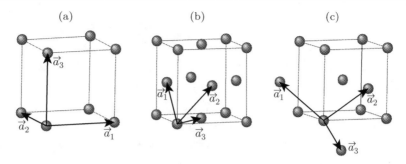

Figure 3.3 Simple (a), face-centered (b), and body-centered (c) cubic lattices and corresponding primitive lattice vectors.

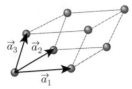

Figure 3.4 A primitive unit cell spanned by primitive unit vectors. While eight lattice sites are shown, each of them is shared by eight different primitive unit cells; thus there is a one-to-one correspondence between a lattice site and a primitive unit cell.

results in the FCC lattice. From this viewpoint one might wonder whether the added lattice points are equivalent to the original ones. The answer is, of course, yes, as all lattice points can be obtained using the (new) corresponding primitive lattice vectors.

Every site on a Bravais lattice has the same number of nearest neighbors. This important number is called the **coordination number** of the lattice. For the SC lattice the coordination number is $z = 6$, for the FCC lattice $z = 12$, and for the BCC lattice $z = 8$. Among Bravais lattices formed by simple elements,[12] FCC and BCC lattices are much more common than SC lattices, precisely because of their larger coordination number: as discussed earlier, the energetic reason behind crystal formation is minimization of potential energy, and a larger coordination number allows more (nearest-neighbor) pairs of atoms to be placed at the inter-atomic potential minimum.

The discussion above focuses on lattice *points*. An equivalent way to view/describe a Bravais lattice is to specify its **primitive unit cell** with some fixed shape which, when translated by all possible lattice vectors, precisely fills space without overlaps or voids.[13] This is perhaps a more physical viewpoint as atoms and molecules that form a crystal are *not* point objects. In particular electron wave functions spread throughout the entire space; how they spread out within a primitive unit cell in a way that respects lattice translation symmetry is one of the most important questions in condensed matter physics, which will be discussed in great detail in Chapter 7. Just like for primitive lattice vectors, the choice of primitive unit cell is not unique. One choice that is often used is the parallelepiped naturally defined by the primitive lattice vectors $\vec{a}_1$, $\vec{a}_2$, $\vec{a}_3$ as shown in Fig. 3.4. Regardless of the shape of the primitive unit cell, they must have the same volume of $|\vec{a}_1 \cdot (\vec{a}_2 \times \vec{a}_3)|$.

There is a particular type of primitive unit cell called the **Wigner–Seitz cell** which is uniquely defined as that volume which is closer to one particular lattice site (say the origin) than to any other. If $\vec{r}$ is in the Wigner–Seitz cell then

---

[12] As will be discussed in more detail later, compounds involving multiple elements cannot form Bravais lattices, and form lattices with bases instead.

[13] This implies that there is a one-to-one correspondence between a lattice site and a primitive unit cell.

$$|\vec{r}| < |\vec{r} - \vec{R}_{\vec{n}}| \tag{3.17}$$

$$\Rightarrow \quad r^2 < r^2 - 2\vec{r} \cdot \vec{R}_{\vec{n}} + \vec{R}_{\vec{n}} \cdot \vec{R}_{\vec{n}} \tag{3.18}$$

$$\Rightarrow \quad \left(\vec{r} - \frac{\vec{R}_{\vec{n}}}{2}\right) \cdot \vec{R}_n < 0. \tag{3.19}$$

This last relation tells us that the boundaries of the Wigner–Seitz cells are planes bisecting lattice vectors. This gives us a simple recipe for constructing the cell graphically and also shows us that the Wigner–Seitz cell has the full **point symmetry** of the Bravais lattice. The point symmetry is the set of symmetry transformations of the crystal (rotations, reflections, etc.) which leave a given point in the crystal invariant.

As an example, consider the 2D square lattice ($|\vec{a}_1| = |\vec{a}_2|$ with a 90° angle between $\vec{a}_1$ and $\vec{a}_2$; shown in Fig. 3.5). In this case the Wigner–Seitz cell happens to be related to the primitive cell by a simple translation (as shown in Fig. 3.5).

This is not always true, however. Consider, for example, the 2D triangular lattice ($|\vec{a}_1| = |\vec{a}_2|$ with a 60° angle between $\vec{a}_1$ and $\vec{a}_2$; also known as the hexagonal lattice). The primitive cell is a parallelogram, but the Wigner–Seitz cell is a hexagon (as shown in Fig. 3.6).

Exercise 3.1.
(a) Assume that a solid is made of hard spheres that are touching. Find the volume fraction occupied by the spheres for the SC, BCC, and FCC lattices.
(b) Now consider a 2D solid that is made of hard circular disks that are touching. Find the area fraction occupied by the disks for the square and triangular lattices.

3.2.2 Lattices with Bases

Now that we have defined the Bravais lattice, let us relate this mathematical object to physical crystals: *every crystal has a corresponding Bravais lattice*. The simplest crystals consist of identical atoms

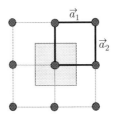

Figure 3.5 Square lattice and corresponding Wigner–Seitz cell (shaded region). The primitive unit cell corresponding to $\vec{a}_1$ and $\vec{a}_2$ is bounded by thick lines (and is unshaded).

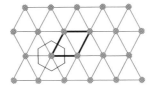

Figure 3.6 Triangular lattice and corresponding hexagonal Wigner–Seitz cell. The primitive unit cell is bounded by thick lines. In addition to repeating the unit cells, one may also construct a triangular lattice by tiling together equilateral triangles, with two different orientations (say, one pointing up and the other pointing down). These triangles can be viewed as the smallest building block of the lattice, and are called plaquettes.

located at the sites of a Bravais lattice. In general, however, there may be more than one atom (or molecule) per unit cell. That is, a general lattice is constructed by attaching a set of m inequivalent points to sites of a Bravais lattice:

$$\vec{R}_{\vec{n},s} \equiv \vec{R}_{\vec{n}} + \vec{\tau}_s; \quad s = 1, 2, \ldots, m. \tag{3.20}$$

These are lattices with bases, which are invariant under translation by the underlying Bravais lattice vectors defined in (3.15), but not necessarily so when translated by a vector connecting two arbitrary lattice sites. A simple 2D example is the honeycomb lattice, as shown in Fig. 3.7; note the vector $\vec{\tau}$ connecting the two neighboring atoms sitting on A and B sites is *not* a primitive translation vector. This is the lattice structure of **graphene**, which is a single sheet of carbon atoms. The underlying Bravais lattice is a triangular lattice in this case; thus the honeycomb lattice can be viewed as a triangular lattice with a two-point basis, or two interpenetrating triangular lattices formed by A and B sites (or "sublattices").

Like graphene, diamond, silicon, and germanium all have two atoms per unit cell. Because these are pure elemental crystals, the two atoms happen to be identical. The underlying Bravais lattice is FCC, and $\vec{\tau}$ is along the diagonal direction of the cube, with one-quarter the length of the diagonal: $\vec{\tau} = (a/4)(\hat{x} + \hat{y} + \hat{z})$. This lattice structure is often referred to as the diamond lattice; see Fig. 3.8.

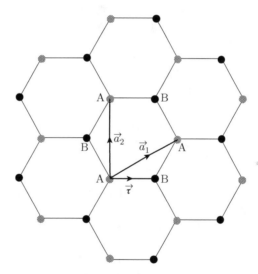

Figure 3.7 A honeycomb lattice has two sites per unit cell.

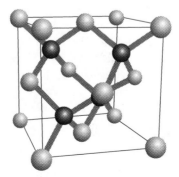

Figure 3.8 Diamond lattice. Dark and light dots represent sites of two sublattices. Neighboring sites (of opposite sublattices) are connected by rods. The zincblende lattice has the same structure, but with two different types of atoms occupying the two sublattices.

The diamond lattice can be viewed as the 3D generalization of the honeycomb lattice, in the following sense. The triangular lattice has coordination number $z = 6$, the largest possible in 2D. It can be constructed by tiling equilateral triangles (with two different orientations, say one pointing up and the other pointing down; see Fig. 3.6) with lattice sites sitting at the vertices. In the honeycomb lattice a second sublattice (say, B) is introduced, so that a B site sits at the center of each A triangle of one orientation (say, those pointing up). Similarly, each A site sits at the center of a B triangle. As a result, the honeycomb lattice has coordination number $z = 3$. While not easy to draw on paper, an FCC lattice ($z = 12$, the largest possible in 3D) can be constructed by tiling together regular tetrahedra, again with two different orientations. In the diamond lattice a second sublattice (say, B) is introduced, so that a B site sits at the center of each tetrahedron with one particular orientation. As a result, the diamond lattice has coordination number $z = 4$ (see Fig. 3.8).

One might wonder why carbon atoms choose to form such a (somewhat) complicated lattice with a relatively low coordination number (recall that we just mentioned earlier that FCC and BCC lattices are common because of their large coordination numbers). The answer lies in the chemistry. Carbon is a valence-4 atom. By forming the diamond lattice, each carbon atom uses one of its valence electrons to form a valence bond with each of its four nearest neighbors. Thus diamond is a typical example of a covalent crystal, which can be viewed as a huge molecule formed through these valence bonds. In covalent crystals, the valence electrons are localized in the nearest-neighbor bonds, instead of on individual atoms. Graphene is also made of carbon. In this case three of the four valence electrons of each carbon atom form valence bonds with the three neighboring atoms, and the remaining electron can wander around the entire honeycomb lattice. This is why graphene is metallic, and has many other fascinating properties that we will learn about later.

Other important 3D examples include sodium chloride, in which the Na^+ and Cl^- ions fill a simple cubic lattice with neighboring sites occupied by opposite charges. Because there are two types of ions, each unit cell contains one Na^+ and one Cl^- ion, and the underlying Bravais lattice is actually FCC. Sodium chloride is a typical example of an ionic crystal, in which the underlying particles that form the crystal are ions instead of neutral atoms, and they are bound together by the attractive Coulomb force between opposite charges. The zincblende lattice, which got its name from zinc sulfide (ZnS), can be viewed as the diamond lattice with two types of atoms sitting on two opposite sublattices. Other important binary (i.e. two-element) compounds that form the zincblende lattice include gallium arsenide (GaAs) and aluminum arsenide (AlAs), which we will encounter later in this book. It should be clear by now that compounds cannot form Bravais lattices, as each unit cell has to contain more than one atom (or actually ions in the case of sodium chloride).

It is often convenient to describe a Bravais lattice using a so-called **conventional unit cell**, which contains more than one lattice site but has a simpler geometrical shape. The volume of a conventional unit cell is an integer multiple of that of the primitive unit cell, with the integer n being equal to the number of lattice sites included in the cell. Just like the primitive unit cell, one can fill the entire space without overlap by repeating the conventional unit cell. The most commonly used conventional cell is the simple cubic cell in the cases of FCC and BCC lattices. Such a description is equivalent to treating the Bravais lattice *as if* it were a lattice with an n-point basis.

Exercise 3.2. Find the number of lattice sites in each conventional unit cell for FCC and BCC lattices.

Exercise 3.3.
 (i) Show that the three primitive lattice vectors of the FCC lattice, as shown in Fig. 3.3(b), span a regular tetrahedron.
 (ii) Show that, in the diamond lattice, another lattice site (on the opposite sublattice) is placed at the geometrical center of this tetrahedron.

Exercise 3.4.
 (i) Show that Eq. (3.16) follows from Eq. (3.15).
 (ii) Show that Eqs. (3.15) and (3.16) are actually equivalent, by showing that Eq. (3.15) also follows from Eq. (3.16). To do that, you need to show that, with the closure condition of Eq. (3.16), all lattice vectors can be written in the form of Eq. (3.15). This may require constructing (a particular set of) primitive lattice vectors. If you find doing this in 3D is too challenging, you may want to try 1D and 2D first.

3.2.3 Lattice Symmetries in Addition to Translation

Thus far our discussion has focused on the (remaining) discrete lattice translation symmetry of crystals, after the full translation symmetry has been broken spontaneously. There are also other spatial symmetries respected by crystals, which we discuss briefly here. We list these symmetries below, along with some simple examples. More examples can be found in the exercises. We remind the reader that a symmetry here means an *operation* that maps the crystalline lattice to itself.

1. **Rotation** through angles $2\pi/n$, with n being an integer, along certain axes (in 3D; in 2D an axis reduces to a single point). These axes are referred to as n-fold rotation axes. Examples: each lattice site of the square lattice is a four-fold rotation axis, and so is the center of each square.
2. **Reflection** about a plane (in 3D; in 2D a plane reduces to a line). Examples: each line that connects a lattice point to its nearest- or next-nearest-neighbor is a mirror line in the square or triangular lattice.
3. **Inversion** about a special point, known as an inversion center. This is specific to 3D; in 2D an inversion is equivalent to a π rotation about the inversion center, and thus is not an independent symmetry operation. But, because of its importance, people sometimes still specify whether a 2D crystal has an inversion center or not. Crystals with at least one inversion center are called **centrosymmetric**; otherwise they are called non-centrosymmetric. Example: each lattice site of a Bravais lattice is an inversion center, thus all Bravais lattices are centrosymmetric.
4. Combinations of all of the above, along with translation (not necessarily by a lattice vector).

By specifying the symmetries above (or lack thereof), crystals have been classified into many different categories (in addition to Bravais and non-Bravais lattices). The full classification is a highly technical subject that we will not delve into here. Instead, we will discuss specific examples that are physically important in this book.

Thus far we have assumed that the lattice sites are point-like objects, which transform trivially (or do not transform at all) under the symmetry operations above. In reality these objects may transform *nontrivially*. One example is a lattice formed by molecules with a non-spherical shape, whose orientation changes under rotation, inversion, and mirror reflection. In such cases the symmetry will be further reduced.

Exercise 3.5. Show that, for a Bravais lattice, any mid-point between two lattice sites $(\vec{R}_{\vec{m}} + \vec{R}_{\vec{n}})/2$ is an inversion center. This includes all lattice sites, and many more points.

Exercise 3.6. The triangular and honeycomb lattices each have a six-fold and a three-fold rotation axis; find where they are. For the honeycomb lattice, assume it is made of the same types of atoms in both A and B sublattices like in graphene. What happens to the six-fold rotation symmetry (of the honeycomb lattice) if the atoms are different for the A and B sublattices, like in hexagonal boron nitride, in which one sublattice is occupied by boron and the other by nitrogen?

Exercise 3.7. Find all the mirror lines of the square lattice.

Exercise 3.8. Show that the diamond lattice is centrosymmetric by finding an inversion center. Show that what you have just found is not an inversion center of the zincblende lattice, which, in fact, is non-centrosymmetric.

Exercise 3.9. ⬦ Show that five-fold rotation symmetry is inconsistent with lattice translation symmetry in 2D. Since 3D lattices can be formed by stacking 2D lattices, this conclusion holds in 3D as well.

3.3 Reciprocal Lattices

We saw earlier that Fourier-series expansions were a natural way to describe functions (such as atom or electron density) that have the periodicity of a lattice. Let us return now to this point.

A function $\rho(\vec{R})$ that has the translation symmetry of the lattice satisfies

$$\rho(\vec{R}) = \rho(\vec{R} + \vec{R}_{\vec{n}}), \tag{3.21}$$

with $\vec{R}_{\vec{n}}$ a lattice vector defined in Eq. (3.15). One way to construct such a periodic function from an *arbitrary* function f is

$$\rho(\vec{R}) = \sum_{n_1, n_2, n_3} f(\vec{R} - n_1 \vec{a}_1 - n_2 \vec{a}_2 - n_3 \vec{a}_3), \tag{3.22}$$

where the sums run over all integers. For example, f might be the electron density of the *basis atoms* in one unit cell. We can make a Fourier representation of ρ by writing

$$\rho(\vec{R}) = \sum_{\{\vec{G}\}} \tilde{\rho}_{\vec{G}} e^{i\vec{G} \cdot \vec{R}}, \tag{3.23}$$

where the set of wave vectors $\{\vec{G}\}$ is determined from the periodicity condition (3.21) by the requirement

$$e^{i\vec{G} \cdot (\vec{R} + n_1 \vec{a}_1 + n_2 \vec{a}_2 + n_3 \vec{a}_3)} = e^{i\vec{G} \cdot \vec{R}} \tag{3.24}$$

for every n_1, n_2, n_3. This in turn means that

$$\vec{G} \cdot \vec{a}_j = 2\pi m_j, \qquad j = 1, 2, 3, \tag{3.25}$$

where m_j is an integer (possibly zero). It turns out that the set $\{\vec{G}\}$ forms a Bravais lattice in *wave-vector space* known as the *reciprocal lattice*. The easiest way to see this is by noticing that, if

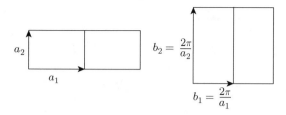

Figure 3.9 (a) A rectangular lattice with primitive lattice vectors $a_1\hat{x}$ and $a_2\hat{y}$. (b) Its reciprocal lattice.

$\vec{G}_1 \in \{\vec{G}\}$ and $\vec{G}_2 \in \{\vec{G}\}$, we must have $\vec{G}_1 \pm \vec{G}_2 \in \{\vec{G}\}$; from the closure condition of (3.16) we conclude that $\{\vec{G}\}$ forms a Bravais lattice. More physically, however, we will construct the primitive lattice vectors of the reciprocal lattice from those of the original (or direct) lattice below.

As a simple example, consider a 2D rectangular lattice shown in Fig. 3.9(a). We can define two basis vectors of the reciprocal lattice:

$$\vec{b}_1 = \frac{2\pi}{a_1}\hat{x}, \tag{3.26a}$$

$$\vec{b}_2 = \frac{2\pi}{a_2}\hat{y}. \tag{3.26b}$$

We see that

$$\begin{aligned}
\vec{b}_1 \cdot \vec{a}_1 &= 2\pi, \\
\vec{b}_1 \cdot \vec{a}_2 &= 0, \\
\vec{b}_2 \cdot \vec{a}_1 &= 0, \\
\vec{b}_2 \cdot \vec{a}_2 &= 2\pi.
\end{aligned} \tag{3.27}$$

Hence $\{G\}$ is the set of all vectors of the form

$$\vec{G} = m_1\vec{b}_1 + m_2\vec{b}_2, \tag{3.28}$$

since these (and only these) satisfy Eq. (3.25). Using Eq. (3.26a) and (3.26b) we see that the reciprocal lattice to the 2D rectangular lattice is as shown in Fig. 3.9(b). In this particular case, the reciprocal lattice happens to also be a rectangular lattice, but if $a_1 > a_2$ then $b_1 < b_2$. Often the reciprocal lattice is *not* of the same lattice type as the real-space lattice. In general for 2D we have

$$\vec{b}_i = \epsilon_{ij}(\hat{z} \times \vec{a}_j), \tag{3.29}$$

where $\hat{z}$ is a unit vector along the normal direction of the lattice plane, and $\epsilon_{ij} = -\epsilon_{ji}$ is the usual totally anti-symmetric Levi-Civita tensor (with two indices here, so that $\epsilon_{12} = -\epsilon_{21} = 1$, and $\epsilon_{11} = \epsilon_{22} = 0$). Also we are using the usual Einstein summation convention (in which repeated indices are summed over) here and throughout this book, unless noted otherwise. As a result

$$\vec{a}_m \cdot \vec{b}_j = 2\pi \delta_{mj}. \tag{3.30}$$

The Wigner–Seitz cell of the reciprocal space is known as the **first Brillouin zone**, made of points in reciprocal space that are closer to the origin than any other reciprocal lattice points. This object will be of central importance in our future study of the electronic states in a crystal and also the vibrational states of lattices. Occasionally we may find it useful to study the nth Brillouin zone, which consists of the points in reciprocal space that have the origin as their nth closest reciprocal lattice point.

Having seen a special case in 2D let us now make a general construction of the reciprocal lattice for any 3D lattice. Let

$$\omega \equiv \vec{a}_1 \cdot (\vec{a}_2 \times \vec{a}_3). \tag{3.31}$$

We see that $|\omega|$ is the volume of the primitive unit cell. Now define

$$\vec{b}_1 = \frac{2\pi}{\omega} \vec{a}_2 \times \vec{a}_3, \tag{3.32a}$$

$$\vec{b}_2 = \frac{2\pi}{\omega} \vec{a}_3 \times \vec{a}_1, \tag{3.32b}$$

$$\vec{b}_3 = \frac{2\pi}{\omega} \vec{a}_1 \times \vec{a}_2. \tag{3.32c}$$

These three equations can be written in the compact form

$$\vec{b}_j = \frac{1}{2} \frac{2\pi}{\omega} \epsilon_{jk\ell} (\vec{a}_k \times \vec{a}_\ell), \tag{3.33}$$

where $\epsilon_{jk\ell}$ is again the totally anti-symmetric Levi-Civita tensor, now with three indices.[14] It is clear from the construction that

$$\vec{a}_m \cdot \vec{b}_j = 2\pi \delta_{mj}, \tag{3.34}$$

and hence we have successfully found a set of primitive vectors for the reciprocal lattice.

It is clear from the above that $\vec{b}_1$ is perpendicular to the plane spanned by $\vec{a}_2$ and $\vec{a}_3$, etc. A very useful way to think about a 3D crystal is to view it as a (periodic) stacking of 2D lattices made of planes of atoms.[15] It turns out that the vector normal to the plane is always proportional to one of the reciprocal lattice vectors. One way to see this is to observe that it is always possible to choose a new set of primitive lattice vectors, with two of them being the primitive lattice vectors of the 2D lattice of interest. Then the construction of primitive lattice vectors of the reciprocal lattice above will automatically include one that is perpendicular to this plane. As a result, lattice planes can be conveniently labeled by this reciprocal lattice vector, as we discuss below.

Consider the reciprocal lattice vector

$$\vec{G}_{KLM} \equiv K\vec{b}_1 + L\vec{b}_2 + M\vec{b}_3. \tag{3.35}$$

Recall (from analytic geometry) that a plane can be characterized by a point in the plane and a vector normal to the plane. Hence the equation

$$\vec{G}_{KLM} \cdot \vec{r} = 0 \tag{3.36}$$

defines a plane perpendicular to $\vec{G}_{KLM}$ and passing through the origin.[16] Some subset of the crystal atoms will lie in this plane. Now, since $\vec{G} \cdot \vec{R}_j = 2\pi m_j$ for *all* lattice points, there exist equivalent (i.e. parallel) planes of atoms obeying

$$\vec{G}_{KLM} \cdot \vec{r} = 0, \pm 2\pi, \pm 4\pi, \ldots. \tag{3.37}$$

These equivalent planes are labeled by their **Miller indices**, KLM.[17] The Miller indices can be found directly in real space from the reciprocals of the intercepts of the axes and the plane. For example, if

$$\vec{R}_{n_1 n_2 n_3} = n_1 \vec{a}_1 + n_2 \vec{a}_2 + n_3 \vec{a}_3 \tag{3.38}$$

is a set of lattice points in the plane, we have

$$\vec{G}_{KLM} \cdot \vec{R}_{n_1 n_2 n_3} = (Kn_1 + Ln_2 + Mn_3)2\pi = 2\pi I, \tag{3.39}$$

---

[14] $\epsilon_{jk\ell} = +1$ if $jk\ell$ is a cyclic permutation of 123, -1 if $jk\ell$ is a non-cyclic permutation, and otherwise vanishes.

[15] This is particularly true when one thinks about surfaces of crystals (facets) or about reflections of X-rays or other particles from crystals.

[16] Again we choose our coordinate system so that there is a lattice point at the origin.

[17] The definition assumes that K, L, M have no common factors; i.e. G_{KLM} is the shortest reciprocal lattice vector perpendicular to the plane.

where I is a fixed integer (assumed non-zero) for a specific single plane. Setting n_2 and n_3 to zero, we can solve for $\bar{n}_1$, the value of n_1 at the intercept between the plane and the ray defined by $\vec{b}_1$:

$$\bar{n}_1 = \frac{I}{K}. \tag{3.40}$$

Similarly the other two intercepts are

$$\bar{n}_2 = \frac{I}{L}, \tag{3.41}$$

$$\bar{n}_3 = \frac{I}{M}, \tag{3.42}$$

and thus

$$(K, L, M) = \left(\frac{I}{\bar{n}_1}, \frac{I}{\bar{n}_2}, \frac{I}{\bar{n}_3} \right) \tag{3.43}$$

as claimed.

A 2D example is shown in Fig. 3.10.[18] The three upper lines (sloping down to the right) have intercepts (in units of a_1 and a_2) of (1, 1), (2, 2), and (3, 3). The reciprocals of these intercepts are (1, 1), (1/2, 1/2), and (1/3, 1/3). Multiplying each by an appropriate integer to obtain integer coefficients, we see that all three have Miller indices which are (1, 1). These lines (or 1D "planes") are in fact perpendicular to the reciprocal lattice vector

$$\vec{G}_{11} = \vec{b}_1 + \vec{b}_2. \tag{3.44}$$

The lower two curves have intercepts $(\infty, -1)$ and $(\infty, -2)$, respectively. Performing the same procedure, we see that the Miller indices for these planes are $(0, \bar{1})$, or equivalently $(0, 1)$. Here $\bar{1}$ means -1 in the standard notation.

Crystals tend to grow (and cleave) along faces with Miller indices which are small integers. It was known to the ancients even before atoms had been proven to exist that the angles formed by crystal facets were consistent with the idea that crystals are made up of tiny periodically repeating structural units (such as spheres).

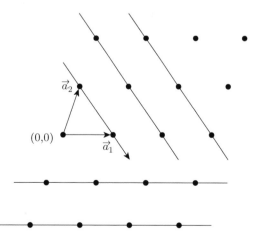

Figure 3.10 1D lattice lines in a 2D lattice.

---

[18] For 2D lattices the reciprocal lattice vectors are perpendicular to the lattice lines.

Exercise 3.10. Construct the reciprocal lattice of a 1D lattice with lattice constant a, and all (not just the first) of the Brillouin zones.

Exercise 3.11. Show that the reciprocal lattice of an FCC lattice is a BCC lattice (in reciprocal space), and vice versa. Determine the corresponding lattice constants of the reciprocal lattices, in terms of those of the direct lattices.

Exercise 3.12. Starting with a Bravais lattice in any dimension called A, show that the reciprocal lattice of A's reciprocal lattice is the original lattice A.

Exercise 3.13. As discussed in the text, it is very useful to view a 3D lattice as a periodic stacking of 2D lattice planes. Similarly, one can also view a 2D lattice as a stacking of 1D lattice lines (or chains). Here we explore how such stacking can be done in the most efficient manner as measured by the volume (area) fraction occupied by the hard spheres (disks) introduced in Exercise 3.1, and known as close packing. (a) Consider a 1D lattice made of a straight line of hard circular disks touching each other. Show that one obtains a triangular lattice by stacking such 1D chains in a manner that maximizes the area occupation fraction. (Hint: this means that the distance between neighboring chains is minimized.) Thus the triangular lattice is the only close-packed 2D lattice. (b) Now consider a 2D triangular lattice of hard spheres that are touching, and then imagine stacking such layers to form a 3D lattice. Show that, because the unit cell of the triangular lattice contains two triangles, there are two ways to stack the second layer on top of the first layer. Pick one of the two options, and call the first layer configuration A, and the second layer configuration B. Show that by interlaying A and B together one obtains a honeycomb lattice. (c) Now add the third layer. Show that one of the two options is to return to configuration A. By repeating the ABABAB... pattern in stacking one obtains the hexagonal close-packed (HCP) lattice. Show that the HCP lattice is a non-Braivais lattice with a two-point basis. Find the primitive lattice vectors of the underlying Bravais lattice. (d) Now take the other option for the third layer, and call its (2D) configuration C. Show that by repeating the ABCABCABC... pattern one obtains the FCC lattice. (e) Show that there exist infinitely many close-packed lattice structures in 3D, and that FCC and HCP are only two of them.

3.4 X-Ray Scattering from Crystals

We saw previously that, in the Born approximation, the X-ray scattering amplitude is proportional to the Fourier transform of the electron density at a wave vector $\vec{q}$ determined by the scattering angle: $\vec{k}_{\text{final}} = \vec{k}_{\text{initial}} + \vec{q}$. With our new knowledge about the reciprocal lattice we can see that it will play a central role in X-ray scattering structural determinations. If we assume zero temperature and classical atoms so that quantum (zero-point) and thermal fluctuations can be neglected, the electron density clearly has the periodicity of the lattice. Suppose we write it in the form

$$\rho(\vec{r}) = \sum_i \rho_{\text{a}}(\vec{r} - \vec{R}_i), \tag{3.45}$$

where the sum is over sites of the underlying Bravais lattice and ρ_a is the atomic electron density (in the crystalline environment). Let $F(\vec{q})$ be the X-ray form factor for the crystal,

$$F(\vec{q}) = \int d^3\vec{r}\, e^{-i\vec{q}\cdot\vec{r}} \sum_i \rho_a(\vec{r} - \vec{R}_i). \tag{3.46}$$

It is easily seen that

$$F(\vec{q}) = W(\vec{q})f(\vec{q}), \tag{3.47}$$

where

$$W(\vec{q}) \equiv \sum_i e^{-i\vec{q}\cdot\vec{R}_i} \tag{3.48}$$

is the lattice form factor introduced in Chapter 2. As we show below, in a periodic crystal W has the property that, if $\vec{q}$ is not an element of the reciprocal lattice, then W vanishes. If $\vec{q} = \vec{G}$ then $e^{-i\vec{G}\cdot\vec{R}_i} = 1$, which implies that W diverges. We can see this by writing the lattice sum explicitly,

$$W(\vec{q}) = \sum_{\ell,m,n} e^{-i\vec{q}\cdot(\ell\vec{a}_1 + m\vec{a}_2 + n\vec{a}_3)} = N \sum_{\{\vec{G}\}} \delta_{\vec{q},\vec{G}}, \tag{3.49}$$

or using the Poisson summation formula (see Appendix B) to obtain

$$W(\vec{q}) = \sum_{K,L,M} (2\pi)^3 \delta(2\pi K - \vec{q}\cdot\vec{a}_1)\delta(2\pi L - \vec{q}\cdot\vec{a}_2)\delta(2\pi M - \vec{q}\cdot\vec{a}_3)$$

$$= \frac{1}{|\omega|} \sum_{\{\vec{G}\}} (2\pi)^3 \delta^3(\vec{q} - \vec{G}), \tag{3.50}$$

where $\omega = \vec{a}_1 \cdot (\vec{a}_2 \times \vec{a}_3)$. The factor $|\omega|^{-1}$ is "obvious" for an orthorhombic lattice where all the lattice vectors are mutually orthogonal and $\omega = a_1 a_2 a_3$. With a little thought one can see that it is correct in the general case as well by noting that the triple product defining ω is closely related to the determinant of the Jacobian matrix needed in evaluating the integrals involving the expression in Eq. (3.50). We can verify the equivalence of these two results using the equivalence $\sum_{\vec{q}} \Leftrightarrow (L/2\pi)^3 \int d^3\vec{q}$, where L^3 is the normalization volume (see Box 3.1 on periodic boundary conditions for more discussion on this). Summing Eq. (3.49) over $\vec{q}$ in a small region near the origin yields

$$\sum_{\vec{q}}' W(\vec{q}) = N. \tag{3.51}$$

Performing the corresponding integral in Eq. (3.50) yields the same result,

$$\left(\frac{L}{2\pi}\right)^3 \int' d^3\vec{q}\, \frac{1}{|\omega|}(2\pi)^3 \delta^3(\vec{q}) = \frac{L^3}{|\omega|} = N. \tag{3.52}$$

The physical interpretation of this result is the following. At those scattering angles where the wave-vector transfer lies on the reciprocal lattice, the interference of the scattering from different atoms (or planes of atoms) is constructive. In all other directions it is destructive and there will be no scattering.[19]

We have finally arrived at the result

$$F(\vec{q}) = f(\vec{q})\frac{(2\pi)^3}{|\omega|} \sum_{\{\vec{G}\}} \delta^3(\vec{q} - \vec{G}), \tag{3.53}$$

---

[19] This result will be somewhat modified in the presence of fluctuations but still remains essentially correct; see the next section.

which tells us that interference scattering from the periodic repetition of the atoms in the crystal selects out only special allowed values of the wave vector and that the weight of the delta function is simply the form factor of an individual atom. The requirement $\vec{q} = \vec{G}$ is called the **Laue diffraction condition**. The intense diffraction peaks at $\vec{q} = \vec{G}$ are known as **Bragg peaks**.

Box 3.1. Periodic Boundary Conditions

Most of condensed matter physics is about the bulk properties of systems which are much longer than a typical atomic length scale in every direction. Local physical properties are usually only influenced within a few lattice constants of the surface. Understanding local properties in the vicinity of the surface is an important subfield of condensed matter physics. In order to understand the bulk properties of a system, however, it is not necessary to describe the surface realistically. It turns out to be mathematically convenient to specify the finite size of a system by imposing periodic boundary conditions. We can think of the system as a box with sides of length L_x, L_y, and L_z in the x, y, and z directions. We require that any physical property be identical on opposite faces of the system. If $f(\vec{r})$ is some local physical property of the system and $\vec{r}$ is on a "surface" of the system, $f(\vec{r} + L_\gamma \hat{\gamma}) = f(\vec{r})$ for $\gamma = x, y, z$. (These boundary conditions are not realistic, but they are convenient.) Adopting a common convention for Fourier transforms, we can expand $f(\vec{r}) = (1/L_x L_y L_z) \sum_{\vec{k}} f(\vec{k}) \exp(i\vec{k} \cdot \vec{r})$, where $f(\vec{k}) = \int_V d^3 r \, \exp(-i\vec{k} \cdot \vec{r}) f(\vec{r})$, $V = L_x L_y L_z$ is the volume of the system, and the sum over $\vec{k}$ is over the discrete set

$$\vec{k} = 2\pi \left(\frac{n_x}{L_x}, \frac{n_y}{L_y}, \frac{n_z}{L_z} \right),$$

where n_x, n_y, and n_z are integers. If $f(\vec{k})$ is smooth, then the sum over $\vec{k}$ can be replaced by an integral

$$\frac{1}{V} \sum_{\vec{k}} \longrightarrow \int \frac{d\vec{k}}{(2\pi)^3}$$

in the thermodynamic limit, and we recover (one convention for) the continuum Fourier transform. It is common, however, that $f(\vec{k})$ is not smooth, particularly near $\vec{k} = 0$, and mathematical manipulations sometimes are clearer if periodic boundary conditions for a finite system and discrete wave vectors are used until the end of the calculation, where the thermodynamic limit can usually be taken without ambiguity.

A very useful way to understand (or interpret) the Laue condition is in terms of conservation of crystal momentum (also called lattice momentum). We know that momentum conservation is a consequence of translation symmetry. A crystal breaks translation symmetry and thus momentum is *not* conserved. However, the residue (discrete) lattice translation symmetry dictates that the momentum of an X-ray photon can change only by a reciprocal vector times $\hbar$, or momentum is *conserved modulo* $\hbar\vec{G}$. This is known as conservation of crystal momentum, because, as we will see in later chapters, for particles or waves propagating in a crystal, wave vectors that differ by $\vec{G}$ are in a sense *equivalent*, as a result of which we need only use wave vectors whose domain is the first Brillouin zone (or any unit cell of the reciprocal space). Such a wave vector with reduced domain, after being multiplied by $\hbar$, is known as a crystal momentum. Conservation of crystal momentum is the consequence of (discrete) lattice translation symmetry, another example of symmetry leading to a conservation law.

The following figures help us understand the significance of Eq. (3.53). We take a 1D example for simplicity. The atomic form factor typically looks as shown in Fig. 3.11(a). As shown in Fig. 3.11(b), the lattice form factor is a uniform sequence of delta functions with spacing $b_1 = 2\pi/a$. The product

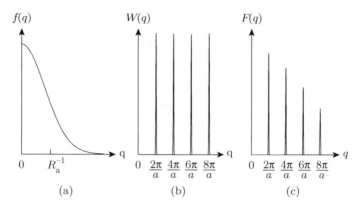

Figure 3.11 (a) A typical atomic form factor, where R_a is the size of an atom; (b) the lattice form factor of a 1D crystal; and (c) the total X-ray scattering form factor of a 1D crystal.

of the two gives $F(q)$ as shown in Fig. 3.11(c). Notice that, since $2\pi/R_a > 2\pi/a$, $f(q)$ decays on a scale greater than b_1 – typically a few times b_1.

We have been implicitly ignoring the possibility of a multi-atom basis in the crystal. If there are m atoms in the unit cell then $f(\vec{q})$ is replaced by

$$f(\vec{q}) = \sum_{s=1}^{m} e^{-i\vec{q}\cdot\vec{\tau}_s} f_s(\vec{q}), \tag{3.54}$$

where f_s is the atomic form factor of the sth atom located at position $\vec{\tau}_s$ within the unit cell.[20]

In the case of a multi-atom basis, it sometimes happens that, for certain values of $\vec{G}$, $f(\vec{G})$ vanishes due to destructive interference. This occurs for instance in diamond and silicon, which have two identical atoms in the unit cell (so that $f_1 = f_2$). These so-called "forbidden reflections" can cause confusion in structural determinations.

Now that we have established that the only possible X-ray reflections correspond to momentum transfers of $\hbar\vec{G}$, we must see under what conditions such momentum transfers are allowed by kinematics, i.e. are consistent with energy conservation. Since we are making the Born approximation, we ignore any changes in the X-ray dispersion relation induced by the presence of the crystal.[21] If the initial wave vector is $\vec{k}$ and the final wave vector is $\vec{k} + \vec{G}$, energy conservation (assuming elastic scattering) requires

$$|\vec{k} + \vec{G}|^2 = |\vec{k}|^2 \tag{3.55}$$

or

$$G^2 + 2\vec{k} \cdot \vec{G} = 0. \tag{3.56}$$

This means $\vec{k}$ must live on a Brillouin-zone boundary, which bisects a reciprocal lattice vector.

The so-called "Ewald construction" gives a graphical method for determining the allowed reflections. One simply draws the reciprocal lattice and places on it the incoming wave vector $\vec{k}$ with the *tip* of the vector at a reciprocal lattice point as shown in Fig. 3.12. A circle of radius k centered at the *base* of the vector $\vec{k}$ gives the locus of final wave vectors of the correct energy. If it happens that the circle intersects any lattice points, then those points correspond to allowed reflections, since there exists a

---

[20] $\vec{\tau}_s$ is measured from some arbitrary origin chosen for the unit cell.

[21] The so-called "dynamical diffraction theory" deals with the sometimes important effects of the tiny deviations from unity of the index of refraction at X-ray frequencies.

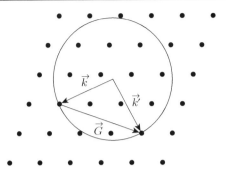

Figure 3.12 Illustration of the Ewald construction. The tip of the incoming wave vector $\vec{k}$ is placed on a reciprocal lattice point. A circle of radius $|\vec{k}|$ is then drawn, centered on the base of the vector $\vec{k}$. If that circle intersects any other points on the reciprocal lattice, then elastic scattering from $\vec{k}$ to $\vec{k}'$ is allowed because there exists a reciprocal lattice vector $\vec{G}$ such that $\vec{k}' = \vec{k} + \vec{G}$.

$\vec{G}$ such that $|\vec{k} + \vec{G}| = |\vec{k}|$. In general, there are no such points. One has to adjust the magnitude and direction of $\vec{k}$ to achieve a reflection. One can do this by rotating the sample relative to the X-ray beam, by using a powder sample consisting of randomly oriented crystallites, or by using non-monochromatic X-rays. In the latter case, a crystalline sample can be used as an X-ray *monochromator* since it acts like a diffraction grating in selecting out special "colors" (energies). Because the grating is 3D, nearby colors are not diffracted at all. This is very different from the behavior of an ordinary planar diffraction grating, which simply disperses different colors through different angles.

Exercise 3.14. Show that scattering from the identical atoms on a diamond lattice produces perfect destructive interference for reciprocal lattice vectors $\vec{G}_{KLM}$ if $K + L + M$ is twice an odd number (which is not the same thing as an even number!). Here the first basis vector for the reciprocal lattice is chosen to be $\vec{b}_1 = (2\pi/a)(\hat{y} + \hat{z} - \hat{x})$ and the remaining two are generated by cyclic permutations of the three terms in the second parenthesis. Here a is the lattice constant of the conventional cubic cell.

3.5 Effects of Lattice Fluctuations on X-Ray Scattering

Up until this point we have been assuming that the atoms are precisely located on the points of a perfect lattice. Using this assumption we found infinitely sharp diffraction beams characterized by $\vec{q} = \vec{G}$. Even neglecting lattice *defects*, we have been ignoring the effects of:

(1) finite temperature;
(2) quantum zero-point motion;
(3) dynamics (time-dependence).

There was a great debate early in the twentieth century about whether or not X-ray diffraction could ever be observed at finite temperatures (quantum fluctuations were unknown at the time). It was argued that at room temperature all the atoms would be randomly displaced by at least a small amount from their nominal lattice positions and this would totally destroy the diffraction pattern. This argument turns out to be incorrect, however, as we show in the following classical argument. (The full quantum derivation will be given later.)

Let us assume that thermal fluctuations cause

$$\vec{r}_j = \vec{R}_j + \vec{u}_j, \tag{3.57}$$

where $\vec{u}_j$ is the displacement of the jth atom from its nominal lattice position $\vec{R}_j$. Let us further assume that the time-dependence of $\vec{u}_j$ can be neglected. This is typically a good approximation for X-ray scattering, where the energy resolution $\Delta\epsilon$ is so poor that the atomic vibration periods are much longer than the corresponding temporal resolution given by the Heisenberg uncertainty relation, $\hbar/\Delta\epsilon$.

In the presence of fluctuations, the lattice structure factor becomes

$$W(\vec{q}) = \sum_i e^{-i\vec{q}\cdot(\vec{R}_i + \vec{u}_i)}. \tag{3.58}$$

The scattered intensity is proportional to the square of this quantity:

$$I_L(\vec{q}) = NS(\vec{q}) = \langle\langle |W(\vec{q})|^2 \rangle\rangle = \sum_{i,j} e^{i\vec{q}\cdot(\vec{R}_i - \vec{R}_j)}\langle\langle e^{i\vec{q}\cdot(\vec{u}_i - \vec{u}_j)} \rangle\rangle, \tag{3.59}$$

where $\langle\langle\cdot\rangle\rangle$ indicates the thermal expectation value.[22]

Thermal averaged quantities are perfectly translation-invariant (i.e. have the symmetry of the lattice), so we can take $\vec{R}_i = \vec{0}$ and multiply by a factor of N:

$$I_L(\vec{q}) = N \sum_j e^{-i\vec{q}\cdot\vec{R}_j}\langle\langle e^{i\vec{q}\cdot(\vec{u}_0 - \vec{u}_j)} \rangle\rangle. \tag{3.60}$$

Expanding the exponential yields

$$1 + i\vec{q}\cdot\langle\langle \vec{u}_0 - \vec{u}_j \rangle\rangle - \frac{1}{2}\langle\langle [\vec{q}\cdot(\vec{u}_0 - \vec{u}_j)]^2 \rangle\rangle + \cdots. \tag{3.61}$$

For a harmonic crystal (atoms connected by Hooke's Law springs), the distribution function for the $\vec{u}$s is a Gaussian (with zero mean),[23] in which case this series simply resums to an exponential,

$$e^{-\frac{1}{2}\langle\langle [\vec{q}\cdot(\vec{u}_0 - \vec{u}_j)]^2 \rangle\rangle} = e^{-2\Gamma(\vec{q}) + \langle\langle [\vec{q}\cdot\vec{u}_j][\vec{q}\cdot\vec{u}_0] \rangle\rangle}, \tag{3.62}$$

where

$$\Gamma(\vec{q}) \equiv \frac{1}{2}\langle\langle (\vec{q}\cdot\vec{u}_j)^2 \rangle\rangle \tag{3.63}$$

is independent of the site index j due to lattice translation symmetry. We will consider the computation of such thermal averages in detail in the following chapters. A simple one-variable example illustrating the above can be found in Exercise 3.15 (see also Exercise 3.16). For the moment let us assume simply that the fluctuations in $\vec{u}_0$ and $\vec{u}_j$ will be uncorrelated (statistically independent). While not accurate for small $|\vec{R}_j|$, this assumption gives us the mathematical simplification

$$\langle\langle [\vec{q}\cdot\vec{u}_0][\vec{q}\cdot\vec{u}_j] \rangle\rangle = 2\Gamma(\vec{q})\delta_{0j}. \tag{3.64}$$

Thus

$$\langle\langle e^{i\vec{q}\cdot(\vec{u}_0 - \vec{u}_j)} \rangle\rangle = e^{-2\Gamma(\vec{q})} + (1 - e^{-2\Gamma(\vec{q})})\delta_{0j} \tag{3.65}$$

and

$$I_L(\vec{q}) = NS(\vec{q}) = Ne^{-2\Gamma(\vec{q})} \sum_j e^{-i\vec{q}\cdot\vec{R}_j} + N(1 - e^{-2\Gamma(\vec{q})}). \tag{3.66}$$

---

[22] We again emphasize that this is a purely classical calculation.

[23] This is obvious for classical harmonic crystals, where the distribution function is simply proportional to $e^{-V(\{\vec{u}_j\})/k_B T}$ with V expanded to second order in $\vec{u}$, but the following is correct for quantum harmonic crystals as well, basically because harmonic oscillator wave functions are Gaussian-like. See the following chapters.

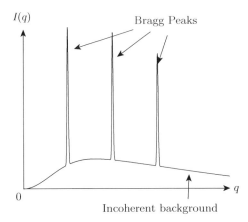

Figure 3.13 The X-ray scattering intensity of a crystal in the presence of lattice fluctuations, in which Bragg peaks coexist with a smooth background.

Restoring the atomic form factor and using Eqs. (3.49) and (3.50), we have

$$I(\vec{q}\,) = N\frac{(2\pi)^3}{|\omega|}e^{-2\Gamma(\vec{q})}|f(q)|^2 \sum_{\{\vec{G}\}} \delta^3(\vec{q} - \vec{G})$$

$$+ N(1 - e^{-2\Gamma(\vec{q})})|f(q)|^2. \tag{3.67}$$

The first term is exactly the same as before except that the sharp diffraction peaks are reduced in intensity by a factor of $e^{-2\Gamma(\vec{q})}$ known as the **Debye–Waller factor**. As shown in Fig. 3.13, the loss of intensity in the sharp diffraction peaks goes into a smooth weak "incoherent" background represented by the second term. Thus we have the somewhat surprising result that the diffracted beams only lose intensity rather than broaden when random disorder is introduced by thermal fluctuations. Notice that if there were no interference effects the scattered intensity from the N atoms would be $N|f(q)|^2$, which is essentially what the second incoherent term represents (but with reduced amplitude).

The Debye–Waller factor has the same effect on the scattering strength at the Bragg peaks as smoothing out the atomic charge density by convolving it (in real space) with a Gaussian (since the Fourier transform of a Gaussian is another Gaussian, and products in $\vec{k}$ space are convolutions in real space). The width of this Gaussian smearing is precisely the root-mean-square (rms) thermal position fluctuation. This smoothing does not, however, produce any incoherent scattering in between the Bragg peaks.

Our derivation of the Debye–Waller factor was based on the **harmonic approximation** that the fluctuations of $\vec{u}_j$ are Gaussian distributed as they are for a harmonic oscillator coordinate. Clearly, when the temperature is raised too high the atomic displacements become so large that this approximation fails. Above the melting temperature the atoms move freely and are not associated with any particular lattice site. The scattering intensity is then simply (see Chapter 2)

$$I(\vec{q}\,) = |f(q)|^2 \sum_{i,j} \langle\!\langle e^{i\vec{q}\cdot(\vec{r}_i - \vec{r}_j)} \rangle\!\rangle = NS(q)|f(q)|^2, \tag{3.68}$$

$I(q)$ has no sharp delta function peaks, but typically has smooth peaks near the positions of the smallest $|\vec{G}|$s in the corresponding crystal; see Fig. 3.14.

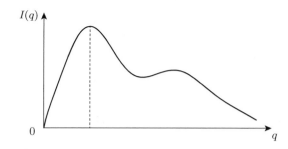

Figure 3.14 A schematic illustration of the X-ray scattering intensity of a liquid. The broad peaks are remnants of the Bragg peaks in the corresponding crystal and reflect the short-range solid-like correlations of the atomic positions in the liquid above the melting point of the crystal.

Exercise 3.15. Consider a variable x with range $(-\infty, +\infty)$ and a Gaussian distribution function $p(x) = (1/\sqrt{2\pi}\sigma)e^{-x^2/(2\sigma^2)}$. (i) Check that $p(x)$ is properly normalized. (ii) Calculate $\langle\langle x^{2n}\rangle\rangle$ for all positive integer ns. (iii) Show that $\langle\langle e^{-i\alpha x}\rangle\rangle = e^{-\alpha^2\langle\langle x^2\rangle\rangle/2}$. (iv) Expand $e^{-i\alpha x}$ as a Taylor series in x, calculate the average for every single term, resum the series, and show that you recover the result obtained in (iii).

Exercise 3.16. Consider a general probability distribution function of an arbitrary number of variables, and an observable A that depends on these variables in an arbitrary way. We may always write

$$\langle\langle e^A\rangle\rangle = e^C, \tag{3.69}$$

where C is given by the so-called cumulant expansion

$$C \equiv \sum_{n=1}^{\infty} \frac{1}{n!} c_n. \tag{3.70}$$

Without loss of generality, we may subtract a constant from A to impose the constraint that $\langle\langle A\rangle\rangle = 0$. Under this constraint show that the lowest few cumulants obey

$$c_1 = 0, \tag{3.71}$$

$$c_2 = \langle\langle A^2\rangle\rangle, \tag{3.72}$$

$$c_3 = \langle\langle A^3\rangle\rangle, \tag{3.73}$$

$$c_4 = \langle\langle A^4\rangle\rangle - 3\langle\langle A^2\rangle\rangle^2. \tag{3.74}$$

If the distribution of A is close to Gaussian then we expect the cumulant series to converge rapidly.

3.6 Notes and Further Reading

In this chapter we have focused on crystalline solids built from periodic arrays of atoms or molecules. Such periodic ordering breaks translation and rotation symmetries, and is a distinct feature not found in liquids, in which there is no such symmetry breaking. There also exist phases of matter that are *intermediate* between crystals and liquids. These are liquid crystals, including nematics that break rotation symmetry but preserve full translation symmetry, and smectics that break translation symmetry along certain direction(s). Another intriguing state of matter is the quasicrystal discovered by

D. Shechtman *et al.* [22] in 1984. In a quasicrystal, the atoms are ordered in a specific but aperiodic pattern. Surprisingly, such ordering gives rise to sharp Bragg peaks in the X-ray and electron scattering, but with a pattern and, in particular, rotation symmetries that are impossible from periodic crystals. In the original discovery, Shechtman *et al.* found in the electron scattering pattern an axis with a five-fold rotational symmetry (i.e., rotations around this axis by angle $2\pi/5$ leave the diffraction pattern invariant; see Fig. 3.15). Such rotation symmetry is known to be incompatible with periodic lattices (see Exercise 3.9). To learn more about liquid crystals and quasicrystals, refer to Chapter 2 of the book by Chaikin and Lubensky [2].

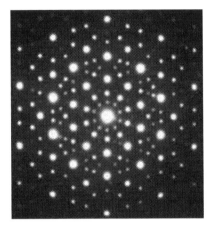

Figure 3.15 The electron scattering intensity of an AlMn compound that forms a quasicrytal. Note the five- and ten-fold rotation symmetries which are forbidden in a periodic crystal. The mechanism of (high-energy) electron scattering is very similar to that of X-ray scattering discussed earlier in this book. Figure adapted with permission from [22]. Copyright by the American Physical Society.

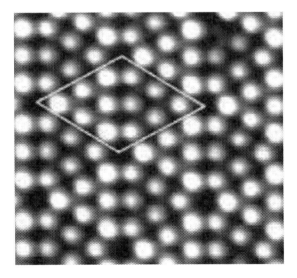

Figure 3.16 A real-space image of a silicon 111 surface obtained from a scanning tunneling microscope. The surface structure differs from that of the bulk due to surface reconstruction. The "7 × 7" reconstructed surface unit cell is indicated. The atoms visible on the surface are considerably further apart from each other than the atoms in the bulk, so it is difficult to ascertain from this image just how large the new unit cell is relative to the unreconstructed surface. Figure adapted with permission from [23]. Copyright by the American Physical Society.

As mentioned earlier, (periodic) crystals can be classified in terms of the remaining symmetries they possess, including (discrete) translation, rotation, mirror reflection, inversion, and their combinations. A good starting point to learn about this topic is Chapter 7 of the book by Ashcroft and Mermin [1]. Their book and the book by Marder [3] also contain discussions of lattice structures of many different materials (along with nice pictures!). In the present chapter we have only touched upon a handful of specific materials that we will encounter in later chapters.

Our discussions in this chapter have focused on ideal crystals, in which the *average* positions of the atoms form a perfect lattice. Real crystals almost always contain various kinds of defects, where the atoms are misplaced in various ways. A detailed discussion on defects in crystals can be found in Chapter 30 of the book by Ashcroft and Mermin [1]. Real crystals are also accompanied by surfaces. As we will discuss in great detail later, there is extremely interesting physics associated with surfaces, as well as interfaces separating two different crystals. Here we mention only that nowadays there exist high-resolution imaging techniques that allow direct probing of atomic as well as electronic structure at the surface. See Fig. 3.16 for an example of an image obtained using a **scanning tunneling microscope (STM)** (see Appendix C for an introduction). Such *real-space* probes are complementary to the momentum-space probes based on X-ray and other scattering experiments discussed in this and previous chapters. It should be kept in mind, though, that the crystal structure at the surface may differ from that in the bulk, due to **surface reconstruction**. In fact Fig. 3.16 is an example of such surface reconstruction.

4 Neutron Scattering

We have learned that scattering experiments are very useful in probing spatial structures. Thus far we have focused on elastic scattering, which involves momentum transfer from the probing particle/wave to our system, but no energy exchange between them.[1] Inelastic scattering, on the other hand, involves energy exchange as well as momentum exchange, and thus tells us not simply the static structure, but rather the dynamics/excitation spectrum of the system. In fact, elastic scattering can be viewed as a very special case of inelastic scattering where the energy transfer happens to be zero. Thus studying inelastic scattering will deepen our understanding of elastic scattering.

In this chapter we will learn why neutrons are excellent probes of crystalline solids and their excitations. We will develop the mathematics needed to understand inelastic scattering and to describe the dynamical correlation functions of atomic positions (and spins) that inelastic scattering can measure.

4.1 Introduction to Neutron Scattering

Radioactive decay produces high-energy neutrons. Here we will be concerned with thermal neutrons (i.e. neutrons that have diffused around inside a solid long enough to come to thermal equilibrium with it). These low-energy neutrons can be made mono-energetic by using a crystal as a **3D Bragg diffraction grating** (to select a single momentum, much as in X-ray scattering). This mono-energetic beam can then be scattered from a sample and the resulting energy *and* momentum transfer analyzed very accurately [24]. Hence one can measure the spectrum of lattice and electronic excitations in a solid much more precisely than with X-rays. Crudely speaking, since the speed of thermal neutrons is comparable to the speed of atoms vibrating in the sample, we cannot make the approximation of neglecting atomic motion during the scattering event (as we did for the case of X-ray scattering).[2]

---

[1] In principle, momentum transfer will always be accompanied by *some* energy exchange. In practice, however, such energy exchange may be too small to resolve within the energy/frequency resolution of the detector, and the scattering is thus effectively elastic. This point will be discussed further.

[2] Another way to understand this is to note that, for a characteristic wavelength of 1 Å, the energy of a neutron is about 0.1 eV, which matches the lattice vibration or phonon energy nicely. On the other hand, for X-rays and electrons the corresponding energies are of order 1 keV and 10 eV respectively, which are much higher than phonon or other characteristic excitation energies in condensed matter systems (except for plasmons), and thus less effective probes for them.

More precisely, the very good energy resolution available with neutrons means that the neutron wave packets are long and take many vibrational periods to pass by any given atom. As a result, motion of the atom (either classical or quantum mechanical) has significant effects on the scattering process. Hence we must develop a theory of the *dynamical correlations* being probed by the neutrons. This is quite different from the X-ray scattering, which takes a "snapshot" of the instantaneous configuration of the atoms, and is not sensitive to the dynamics. The source of the difference lies in the energy and, in particular, the energy *resolution* of the different probes, as we will discuss in more detail later.

Neutrons, like X-rays, interact only weakly, and hence have great penetration depths. They therefore undergo bulk diffraction but will give an appreciable signal only if large enough samples can be obtained. A microscopic crystallite will not suffice! Neutrons, being electrically neutral, interact primarily with the nuclei (via the strong force) and to a much lesser extent with the magnetic moments of the electrons. If a neutron experiences a magnetic interaction which causes its spin to flip, the scattering is called *incoherent*, because diffraction does *not* occur in this case. The reason is the following: in principle, we can look back in the sample and see which nucleus or electron had its spin flipped by the neutron. Thus we know where in the crystal the neutron scattered. The situation is very much like what happens in the famous two-slit diffraction experiment. When detectors are inserted to see which slit the particle went through, the interference pattern is destroyed. Crystal diffraction comes from not knowing which atom does the scattering so that we have a coherent superposition of all possibilities,

$$F(\vec{q}\,) = f(\vec{q}\,) \sum_j e^{-i\vec{q}\cdot\vec{r}_j}. \tag{4.1}$$

Spin-flip scattering (generally[3]) just gives a smooth background beneath the coherent diffraction peaks.

We will focus on the process in which the neutron scatters coherently but inelastically from the lattice. One might think that this would be an incoherent process since one could, in principle, look into the sample and see which atom received kinetic energy from the collision. This is not possible, however, because the atomic vibration immediately spreads out as a coherent wave packet of collective lattice vibrations called **phonons**. In contrast, nuclear spin excitations remain localized for a long time and give rise to the incoherent scattering described above. Electron spin excitations can propagate freely as collective spin waves known as **magnons**, and coherent neutron scattering (that does involve spin flips) is an excellent way to study these magnetic excitations.

You might wonder how long an excitation has to stay localized for it to cause incoherent scattering. The answer is that it has to be localized for a long time compared with the duration of the scattering event, which in turn is defined not by the energy of the probe particle, but rather by the energy *resolution* of the experiment (through the uncertainty principle).

To make this point clear, imagine that the neutron scatters from wave vector $\vec{p}$ to wave vector $\vec{p}\,' = \vec{p} + \vec{q}$, emitting an excitation with wave vector $-\vec{q}$ and energy $\hbar\omega(-\vec{q})$. With poor energy resolution, any value of $\hbar\omega$ could have occurred and hence any value of $\vec{q}$. The experiment is over before the excitation has time to propagate anywhere. On the other hand, with good energy resolution, $\hbar\omega(\vec{q})$ and hence $|\vec{q}|$ is sharply defined (and hence so is the scattering direction). The experiment takes so long that one cannot tell where the excitation was created – it has propagated through the sample. The excitation has sharp $\vec{q}$ and uncertain position.

Perhaps the simplest way to express all this is to note that, with good energy resolution, the neutron wave packet is highly extended and pieces of it scatter from a variety of points in the crystal at a

---

[3] Unless (as we discuss later) the energy resolution is *much* higher, so that the system becomes sensitive to propagation of the flipped (electron) spin through the lattice in the form of a spin wave. Exercise 4.2 deals with such situations.

variety of times. These different scattered waves sample the crystal's configuration at different times, and their interference gives a measure of the dynamical fluctuations in the atomic positions.

Let us now try to turn all these nice words into practical equations.

4.2 Inelastic Neutron Scattering

The total Hamiltonian describing the system being studied and the neutron used to probe it is

$$H_{\text{tot}} = H_{\text{n}} + H + H_{\text{int}}, \tag{4.2}$$

where H is the Hamiltonian of the system, $H_{\text{n}} = P_{\text{n}}^2/2M_{\text{n}}$ is the neutron kinetic energy, and H_{int} describes the interaction (to be specified later) between the neutron and the system.

Let the initial state of the neutron plus crystal system be

$$\Psi_i(\vec{R}, \{\vec{r}\}) = \psi_{\vec{p}}(\vec{R}) \Phi_i(\{\vec{r}\}), \tag{4.3}$$

where $\vec{R}$ is the neutron coordinate and $\{\vec{r}\}$ stands for $(\vec{r}_1, \vec{r}_2, \ldots, \vec{r}_N)$, the set of nuclear coordinates (atom positions). For simplicity we will focus mostly on the case of lattice vibrations and hence ignore the nuclear and electronic spin degrees of freedom; our results, however, are quite general (and apply to liquid states), and can be generalized to spin-dependent interactions straightforwardly (see Exercise 4.2). We assume that the only effect of the electrons is to produce chemical bonds, which give the effective Hamiltonian for the atoms a potential-energy term $V(\vec{r}_1, \vec{r}_2, \ldots, \vec{r}_N)$. We otherwise ignore the electrons insofar as degrees of freedom are concerned. We take the initial state of the neutron to be a plane wave,

$$\psi_{\vec{p}}(\vec{R}) = \frac{1}{\sqrt{L^3}} e^{i\vec{p}\cdot\vec{R}}, \tag{4.4}$$

and the initial energy to be the neutron kinetic energy plus the initial crystal energy:

$$\mathcal{E}_i = \frac{\hbar^2 p^2}{2M_{\text{n}}} + E_i. \tag{4.5}$$

After the scattering, the final state is

$$\Psi_f = \psi_{\vec{p}'}(\vec{R}) \Phi_f(\{\vec{r}\}), \tag{4.6}$$

with energy

$$\mathcal{E}_f = \frac{\hbar^2 p'^2}{2M_{\text{n}}} + E_f. \tag{4.7}$$

We note that the $E_{i,f}$ are eigenvalues of the system (or crystal) Hamiltonian H.

The neutron–nucleus interaction (via the strong force) is so short-ranged relative to the scale of the neutron de Broglie wavelength that we can closely approximate it by a delta function,

$$H_{\text{int}} \approx \alpha \sum_{j=1}^{N} \delta^3(\vec{R} - \vec{r}_j), \tag{4.8}$$

where, for the case of a crystal,

$$\vec{r}_j = \vec{R}_j + \vec{u}_j \tag{4.9}$$

is the instantaneous atomic position and $\vec{R}_j$ is the nominal lattice site.

We will use the Born approximation by assuming only a single scattering occurs. The scattering rate from the initial state i to a specific final state f can be most simply computed using Fermi's Golden Rule,

$$\Gamma_{if} = \frac{2\pi}{\hbar}\delta(\mathcal{E}_f - \mathcal{E}_i)|M_{if}|^2 = \frac{2\pi}{\hbar}\delta(E_f - E_i - \hbar\omega)|M_{if}|^2, \tag{4.10}$$

where $M_{if} \equiv \langle\Psi_f|H_{\text{int}}|\Psi_i\rangle$ is the matrix element of the neutron interaction with the crystal, while $\hbar\omega$ is the energy of the excitation produced and hence the energy lost by the neutron:

$$\hbar\omega = \frac{\hbar^2}{2M_n}(p^2 - p'^2). \tag{4.11}$$

Note that at finite temperatures $\hbar\omega$ could be negative due to absorption of energy by the neutron from some thermal excitation in the sample. In general we do not actually know the initial microscopic state of the crystal, and so we should ensemble average physical observables (such as the scattering rate) over all possible initial states. Thus the neutron scattering rate from $\vec{p}$ to $\vec{p}\,'$ is

$$\Gamma_{\vec{p}\vec{p}'} \equiv \frac{1}{Z}\sum_{i,f}e^{-\beta E_i}\Gamma_{if}, \tag{4.12}$$

where $Z = \sum_i e^{-\beta E_i}$ is the partition function, and $\beta = 1/k_BT$. Notice the distinction: Fermi's Golden Rule tells us to *sum* over all possible final states (with the neutron being in state $\vec{p}\,'$, there is no restriction on the state of the system), while statistical mechanics tells us to *average* over all possible initial states. The matrix element is

$$M_{if} = \frac{1}{L^3}\int d^3\vec{R}\, e^{-i\vec{p}\,'\cdot\vec{R}}\langle\Phi_f|\alpha\sum_j\delta^3(\vec{R} - \vec{r}_j)|\Phi_i\rangle e^{i\vec{p}\cdot\vec{R}} \tag{4.13}$$

$$= \frac{1}{L^3}\langle\Phi_f|\alpha\sum_{j=1}^N e^{-i\vec{q}\cdot\vec{r}_j}|\Phi_i\rangle. \tag{4.14}$$

Note the operator whose matrix element we need above is precisely what gives rise to the lattice form factor that we encountered in Chapter 2. The scattering rate can now be written as

$$\Gamma_{\vec{p}\vec{p}'} = \frac{2\pi\alpha^2}{\hbar L^6}\sum_{i,f}\delta(E_f - E_i - \hbar\omega)\frac{e^{-\beta E_i}}{Z}\sum_{j,k=1}^N\langle\Phi_i|e^{+i\vec{q}\cdot\vec{r}_j}|\Phi_f\rangle\langle\Phi_f|e^{-i\vec{q}\cdot\vec{r}_k}|\Phi_i\rangle. \tag{4.15}$$

If we were not troubled by the presence of the Dirac delta function, we could use the **completeness relation**

$$\sum_f|\Phi_f\rangle\langle\Phi_f| = \mathbb{1} \tag{4.16}$$

to simplify the expression for Γ. Here is how to get around the problem of the delta function. Write

$$\delta(E_f - E_i - \hbar\omega) = \frac{1}{2\pi\hbar}\int_{-\infty}^{\infty}dt\, e^{-i(E_f - E_i)t/\hbar}e^{+i\omega t} \tag{4.17}$$

and move this inside the matrix element:

$$\Gamma_{\vec{p}\vec{p}'} = \frac{\alpha^2}{\hbar^2 L^6}\int_{-\infty}^{\infty}dt\, e^{+i\omega t}\sum_{i,f}\frac{1}{Z}e^{-\beta E_i}$$

$$\times \sum_{j,k=1}^N\langle\Phi_i|e^{iHt/\hbar}e^{+i\vec{q}\cdot\vec{r}_j}e^{-iHt/\hbar}|\Phi_f\rangle\langle\Phi_f|e^{-i\vec{q}\cdot\vec{r}_k}|\Phi_i\rangle, \tag{4.18}$$

where we have used

$$H|\Phi_f\rangle = E_f|\Phi_f\rangle, \tag{4.19}$$

$$H|\Phi_i\rangle = E_i|\Phi_i\rangle. \tag{4.20}$$

Using the completeness relation, we have

$$\Gamma_{\vec{p}\vec{p}'} = \frac{\alpha^2}{\hbar^2 L^6} \int_{-\infty}^{\infty} dt\, e^{i\omega t} \frac{1}{Z} \sum_i e^{-\beta E_i} \sum_{j,k=1}^{N} \langle \Phi_i | e^{i\vec{q}\cdot\vec{r}_j(t)} e^{-i\vec{q}\cdot\vec{r}_k} | \Phi_i \rangle, \tag{4.21}$$

where $\vec{r}_j(t)$ is the Heisenberg representation of the position of the jth atom, $\vec{r}_j$:

$$\vec{r}_j(t) \equiv e^{iHt/\hbar} \vec{r}_j e^{-iHt/\hbar}. \tag{4.22}$$

Let us define the equilibrium **dynamical structure factor**:

$$S(\vec{q}, \omega) \equiv \frac{1}{N} \int_{-\infty}^{\infty} dt\, e^{i\omega t} \sum_{j,k=1}^{N} \langle\langle e^{i\vec{q}\cdot\vec{r}_j(t)} e^{-i\vec{q}\cdot\vec{r}_k(0)} \rangle\rangle, \tag{4.23}$$

where now $\langle\langle \cdot \rangle\rangle$ means ensemble average over all possible initial states $|\Phi_i\rangle$ with appropriate Boltzmann weighting factors. The scattering rate can now be expressed entirely in terms of *equilibrium correlations*:

$$\Gamma_{\vec{p}\vec{p}'} = \frac{N}{\hbar^2 L^6} \alpha^2 S(\vec{q}, \omega). \tag{4.24}$$

Now that we have the total scattering rate we can obtain the differential scattering cross section $d^2\sigma/d\Omega\, dE$. When first learning about scattering in quantum mechanics, one typically studies *elastic* scattering and finds that the rate of scattering of particles into the detector is

$$\frac{dn}{dt} = \mathcal{J} \frac{d\sigma}{d\Omega}\, d\Omega \tag{4.25}$$

where $\mathcal{J}$ is the incident flux, $d\sigma/d\Omega$ is the differential cross section and $d\Omega$ is the (small) solid angle subtended by the detector when viewed from the target. In the case of inelastic scattering we are assuming that our detector accepts particles in an energy window of width dE centered on some energy E, so that

$$\frac{dn}{dt} = \mathcal{J} \frac{d^2\sigma}{d\Omega\, dE}\, d\Omega\, dE. \tag{4.26}$$

The incident flux is the product of the neutron velocity and density (taken to be L^{-3}),

$$\mathcal{J} = \frac{\hbar p}{M_n L^3}, \tag{4.27}$$

so the rate of scattering into a small volume of phase space centered on final neutron momentum $\vec{p}'$ is

$$\Gamma_{\vec{p}\vec{p}'}\, d^3\vec{p}' \left(\frac{L}{2\pi}\right)^3 = \frac{\hbar p}{M_n L^3} \frac{d^2\sigma}{d\Omega\, dE}\, d\Omega\, dE, \tag{4.28}$$

where $(L/2\pi)^3$ is the **density of states in wave-vector space** and $d\Omega$ is the differential solid angle. Now, since $E = \hbar^2 p'^2/2M_n$ we have

$$d^3\vec{p}' = p'^2\, dp'\, d\Omega = \frac{M_n}{\hbar^2} p'\, dE\, d\Omega, \tag{4.29}$$

so we can write

$$\frac{\hbar p}{M_n L^3} \frac{d^2\sigma}{d\Omega\, dE} = \left(\frac{L}{2\pi}\right)^3 \frac{M_n}{\hbar^2} p' \Gamma_{\vec{p}\vec{p}'}; \tag{4.30}$$

$$\frac{d^2\sigma}{d\Omega\,dE} = \frac{L^6}{(2\pi)^3}\frac{M_n^2}{\hbar^3}\frac{p'}{p}\Gamma_{\vec{p}\vec{p}'}.$$ (4.31)

Using Eq. (4.24) yields

$$\frac{d^2\sigma}{d\Omega\,dE} = \frac{N\alpha^2}{(2\pi)^3\hbar^2}\frac{M_n^2}{\hbar^3}\frac{p'}{p}S(\vec{q},\omega).$$ (4.32)

The **scattering length** a for the nuclear potential is defined by the behavior of the scattering phase shift δ at low energies (small incident wave vector k),

$$\lim_{k\to 0} k\cot(\delta_k) = -\frac{1}{a},$$ (4.33)

and is related to the total (low-energy) cross section for scattering (from one atom) by

$$\sigma_0 = 4\pi a^2,$$ (4.34)

and for the delta function potential is given in the Born approximation by

$$a = \frac{M_n}{2\pi\hbar^2}\alpha,$$ (4.35)

$$\frac{d^2\sigma}{d\Omega\,dE} = \frac{Na^2}{2\pi\hbar}\left(\frac{p'}{p}\right)S(\vec{q},\omega),$$ (4.36)

$$\frac{d^2\sigma}{d\Omega\,dE} = \left(\frac{\sigma_0}{4\pi}\right)N\left(\frac{p'}{p}\right)\left[\frac{1}{2\pi\hbar}S(\vec{q},\omega)\right].$$ (4.37)

This is our final result. The factor $\sigma_0/4\pi$ is the differential cross section for *elastic* scattering from a single (fixed) atom, the factor of N represents the fact that there is more signal as the target increases in size, and (p'/p) is a phase-space factor that represents the reduced outgoing flux due to the slowing of the neutron when energy is lost to the target.

Exercise 4.1. We have seen a lot of similarities between inelastic neutron scattering and elastic X-ray scattering; for example, the latter involves the static structure factor while the former involves the dynamical structure factor. One notable difference is that the latter also involves the atomic form factor, but the former does not. Identify the origin of this difference.

Exercise 4.2. This exercise illustrates how one can use a spin-polarized neutron source and a spin-resolving detector to probe magnetic (or spin-flip) excitations in a system. In the text we assumed a spin-independent interaction between the neutron and nuclei, and neglected the fact that neutrons are spin-1/2 particles (since their spins do not change in the scattering process). There exist, however, magnetic (or spin-dependent) interactions between neutrons and (spinful) nuclei, as well as electrons. Consider the following spin-dependent interaction between the neutron and nuclei:

$$H'_{\text{int}} = \alpha'\sum_{j=1}^{N}\delta^3(\vec{R}-\vec{r}_j)\vec{S}_n\cdot\vec{S}_j,$$ (4.38)

where $\vec{S}_n$ and $\vec{S}_j$ are the spin operators of the neutron, and jth nucleus, respectively. Further assume that the incoming neutrons are fully spin-polarized in the up direction, and focus on the scattered neutrons whose spins are flipped to the down direction. Show that the rate or cross section for such spin-flip scattering measures the spin dynamical structure factor $S^{+-}(\vec{q},\omega)$, which replaces the usual (density) dynamical structure factor $S(\vec{q},\omega)$ in the expressions for the rate or cross section. Find the expression for $S^{+-}(\vec{q},\omega)$.

4.3 Dynamical Structure Factor and f-Sum Rule

The dynamical structure factor $S(\vec{q}, \omega)$ has two interpretations. First it can be viewed as a measure of the dynamical correlations in the ground (or, at finite T, the equilibrium) state. Namely, it is the mean square value of the density fluctuations of wave vector $\vec{q}$ and frequency ω. Secondly, as we shall see below, $(1/2\pi\hbar)S(\vec{q}, \omega)$ is the density of excited states at wave vector $\vec{q}$ *created by the density operator*. We have

$$S(\vec{q}, \omega) = \frac{1}{N} \int_{-\infty}^{\infty} dt \, e^{i\omega t} \sum_{j,k=1}^{N} \langle\langle e^{i\vec{q}\cdot\vec{r}_j(t)} e^{-i\vec{q}\cdot\vec{r}_k(0)} \rangle\rangle, \tag{4.39}$$

which is simply the Fourier transform of the density–density correlation, since

$$\hat{\rho}_{\vec{q}}(t) = \int d^3\vec{r} \, e^{-i\vec{q}\cdot\vec{r}} \sum_{j=1}^{N} \delta^3[\vec{r} - \vec{r}_j(t)] = \int d^3\vec{r} \, e^{-i\vec{q}\cdot\vec{r}} \hat{\rho}(\vec{r}, t). \tag{4.40}$$

Thus

$$S(\vec{q}, \omega) = \int_{-\infty}^{+\infty} dt \, e^{i\omega t} \Pi(\vec{q}, t), \tag{4.41}$$

where

$$\Pi(\vec{q}, t) \equiv \frac{1}{N} \langle\langle \hat{\rho}_{-\vec{q}}(t) \hat{\rho}_{+\vec{q}}(0) \rangle\rangle. \tag{4.42}$$

We can interpret the expression as telling us that the density waves in a crystal act as moving diffraction gratings that scatter and Doppler shift the neutron.

Notice from the definition of the density operator that the effect of $\rho_{+\vec{q}}$ is to boost the momentum of one of the particles (we do not know which one) by $-\hbar\vec{q}$, corresponding to the fact that the neutron has acquired momentum $+\hbar\vec{q}$. We can thus also think of $S(\vec{q}, \omega)$ as the spectral density of excited states that couple to the initial state through the density operator:

$$S(\vec{q}, \omega) = \frac{1}{Z} \sum_i e^{-\beta E_i} S_i(\vec{q}, \omega), \tag{4.43}$$

$$S_i(\vec{q}, \omega) = \frac{1}{N} \sum_f |\langle i|\rho_{\vec{q}}|f\rangle|^2 \, 2\pi \delta[\omega - (E_f - E_i)/\hbar]. \tag{4.44}$$

Notice that, if the dynamical structure factor is integrated over all frequencies, we obtain the static (i.e. $t = 0$) structure factor which we looked at in X-ray scattering:

$$S(\vec{q}) = \int_{-\infty}^{\infty} \frac{d\omega}{2\pi} S(\vec{q}, \omega) = \Pi(\vec{q}, t=0) = \frac{1}{N} \sum_{j,k} \langle\langle e^{i\vec{q}\cdot\vec{r}_j} e^{-i\vec{q}\cdot\vec{r}_k} \rangle\rangle. \tag{4.45}$$

The static structure factor is thus the zeroth frequency moment of the dynamic structure factor. The first frequency moment is related to the average energy of density excitations at wave vector $\vec{q}$ and is called the **oscillator strength** (generally called $f(\vec{q})$ but not to be confused with the atomic form factors discussed previously):

$$f(\vec{q}) \equiv \int_{-\infty}^{\infty} \frac{d\omega}{2\pi} \hbar\omega S(\vec{q}, \omega). \tag{4.46}$$

We can derive a general sum rule for the oscillator strength by the following manipulations:

$$f(\vec{q}) = \frac{1}{Z} \sum_i e^{-\beta E_i} f_i(\vec{q}), \tag{4.47}$$

where

$$
\begin{aligned}
f_i(\vec{q}) &\equiv \int_{-\infty}^{\infty} \frac{d\omega}{2\pi} \hbar\omega S_i(\vec{q}, \omega) \\
&= \int_{-\infty}^{\infty} \frac{d\omega}{2\pi} \hbar\omega \frac{1}{N} \sum_f \langle i|\rho_{-\vec{q}}|f\rangle 2\pi \delta[\omega - (E_f - E_i)/\hbar]\langle f|\rho_{+\vec{q}}|i\rangle \\
&= \frac{1}{N} \sum_f \langle i|(\rho_{-\vec{q}} E_f - E_i \rho_{-\vec{q}})|f\rangle \langle f|\rho_{+\vec{q}}|i\rangle \\
&= -\frac{1}{N} \sum_f \langle i|[H, \rho_{-\vec{q}}]|f\rangle \langle f|\rho_{+\vec{q}}|i\rangle = -\frac{1}{N} \langle i|[H, \rho_{-\vec{q}}]\rho_{+\vec{q}}|i\rangle.
\end{aligned} \tag{4.48}
$$

Upon ensemble averaging we have

$$
f(\vec{q}) = -\frac{1}{N} \langle\!\langle [H, \rho_{-\vec{q}}]\rho_{+\vec{q}} \rangle\!\rangle. \tag{4.49}
$$

One way to view this result is to note that, from the Heisenberg equation of motion

$$
\dot{\rho}_{-\vec{q}} = \frac{i}{\hbar}[H, \rho_{-\vec{q}}], \tag{4.50}
$$

the commutator is equivalent to taking a time derivative. Upon Fourier transformation, this time derivative produces the factor of ω in Eq. (4.46).

We could as well have put this in a different form:

$$
f_i(\vec{q}) = \frac{1}{N} \sum_f \langle i|\rho_{-\vec{q}}|f\rangle \langle f|[H, \rho_{+\vec{q}}]|i\rangle = \frac{1}{N} \langle i|\rho_{-\vec{q}}[H, \rho_{+\vec{q}}]|i\rangle; \tag{4.51}
$$

$$
f(\vec{q}) = \frac{1}{N} \langle\!\langle \rho_{-\vec{q}}[H, \rho_{+\vec{q}}] \rangle\!\rangle. \tag{4.52}
$$

Assuming inversion symmetry so that $f(-\vec{q}) = f(\vec{q})$, we can combine these two forms to obtain the double commutator,

$$
f(\vec{q}) = \frac{1}{2N} \langle\!\langle [\rho_{-\vec{q}}, [H, \rho_{+\vec{q}}]] \rangle\!\rangle. \tag{4.53}
$$

Now a remarkable simplification occurs. This double commutator turns out to be independent of the interactions among the particles! To see this, write the Hamiltonian in terms of the atomic coordinates as

$$
H = T + V(\vec{r}_1, \ldots, \vec{r}_N), \tag{4.54}
$$

$$
T \equiv \frac{1}{2M} \sum_{j=1}^{N} P_j^2, \tag{4.55}
$$

where for simplicity we assume that all the atoms have the same mass M, and

$$
\rho_{\vec{q}} \equiv \sum_{j=1}^{N} e^{-i\vec{q}\cdot\vec{r}_j}. \tag{4.56}
$$

Now note that, because all the position operators commute with each other,

$$
[V, \rho_{\vec{q}}] = 0. \tag{4.57}
$$

We have

$$
[\vec{P}_j \cdot \vec{P}_j, \rho_{+\vec{q}}] = \vec{P}_j \cdot [\vec{P}_j, \rho_{+\vec{q}}] + [\vec{P}_j, \rho_{+\vec{q}}] \cdot \vec{P}_j, \tag{4.58}
$$

with

$$[\vec{P}_j, \rho_{+\vec{q}}] = -\hbar\vec{q}\, e^{-i\vec{q}\cdot\vec{r}_j}. \tag{4.59}$$

Thus

$$[P_j^2, \rho_{+\vec{q}}] = -\{\vec{P}_j \cdot \hbar\vec{q}e^{-i\vec{q}\cdot\vec{r}_j} + e^{-i\vec{q}\cdot\vec{r}_j}\hbar\vec{q}\cdot\vec{P}_j\} \tag{4.60}$$

and

$$[\rho_{-\vec{q}}, [P_j^2, \rho_{+\vec{q}}]] = 2\hbar^2 q^2. \tag{4.61}$$

Putting this in Eq. (4.53), we finally arrive at the **oscillator strength sum rule** (or "f-sum rule")[4]

$$f(\vec{q}) = \frac{\hbar^2 q^2}{2M}. \tag{4.62}$$

This is nothing more than the kinetic energy that a single (stationary) atom would acquire from an impulse $\hbar\vec{q}$. The average energy of density excitations (modes that couple to the equilibrium state through $\rho_{\vec{q}}$) is simply given by

$$\Delta(\vec{q}) \equiv \frac{f(\vec{q})}{S(\vec{q})} = \frac{\int_{-\infty}^{\infty}(d\omega/2\pi)\hbar\omega S(\vec{q},\omega)}{\int_{-\infty}^{\infty}(d\omega/2\pi)S(\vec{q},\omega)}. \tag{4.63}$$

Using the f-sum rule we obtain

$$\Delta(\vec{q}) = \frac{\hbar^2 q^2}{2M}\frac{1}{S(\vec{q})}, \tag{4.64}$$

which is known as the **Bijl–Feynman formula**. It tells us that the average energy of density-wave excitations at wave vector $\vec{q}$ is the corresponding single-particle excitation energy renormalized by a factor $1/S(\vec{q})$ which describes static correlations in the equilibrium state. We now discuss the physical meaning of this remarkable result.

To gain some intuition on the f-sum rule and the Bijl–Feynman formula, let us apply them to the special case of a neutron (or any other probing particle) scattering off a *single* isolated atom. In this case we simply have $S(\vec{q}) = 1$. If the initial state of the atom is the ground state with momentum $\vec{k} = 0$, then the final state momentum is $-\vec{q}$ and energy $\hbar^2 q^2/2M$, in agreement with the Bijl–Feynman formula (4.64) (because the initial state energy is zero). We note that the final state energy is nothing but the familiar recoil energy in a collision; due to its presence there is almost always energy transfer accompanying any momentum transfer. Now consider the more generic case where the initial state of the atom has a non-zero momentum $\vec{k}$. In this case the energy transfer

$$\Delta\epsilon(\vec{q}) = \frac{\hbar^2}{2M}(|\vec{k}-\vec{q}|^2 - k^2) = \frac{\hbar^2}{2M}(q^2 - 2\vec{k}\cdot\vec{q}) \tag{4.65}$$

is not a constant. However, as long as the distribution of $\vec{k}$ is invariant under inversion $\vec{k} \to -\vec{k}$, we find

$$\langle\langle\Delta\epsilon(\vec{q})\rangle\rangle = \frac{\hbar^2 q^2}{2M}, \tag{4.66}$$

again in agreement with Eq. (4.64).

---

[4] A note about terminology: $f(\vec{q})$ is an integral over ω, and integration is equivalent to summation. In fact, we derived this sum rule by turning the integral over ω to a summation over all final states produced by the density operator. In general, sum rules usually refer to equalities involving integrals.

Applying the Bijl–Feynman formula (4.64) to a perfectly ordered crystal, we note from Eqs. (3.49) and (2.39) that, for wave vectors equal to a reciprocal lattice vector, the static structure factor diverges as $S(\vec{G}) = N$ (neglecting the Debye–Waller corrections) and the energy transfer

$$\Delta(\vec{G}) = \frac{\hbar^2 G^2}{2(NM)},\qquad(4.67)$$

while strictly speaking not zero, is negligibly small in the thermodynamic limit. This happens because the *entire* crystal (with mass NM) absorbs the momentum transfer as if it were a rigid body! The scattering is hence "recoilless"[5] and hence elastic. This tells us that scattering at $\vec{q} = \vec{G}$ is predominantly elastic, as we had learned previously. Fluctuations that lead to the Debye–Waller factor do not alter this conclusion.

For $\vec{q} \neq \vec{G}$, and in the presence of fluctuations, we have $S(\vec{q}) \sim O(1)$ (see Eq. (3.66), second term), and Eq. (4.64) tells us that the energy transfer is comparable to that of scattering off a single atom, and we essentially *always* have inelastic scattering in which excitations are created or destroyed by the scattering process.

Mathematically, the fact that the oscillator strength is independent of the particle interactions follows rigorously from the fact that the position operators all commute with each other. Physically, however, this result seems somewhat counter-intuitive. After all, if we strengthen the chemical bonds, all of the characteristic vibrational energies will go up. We must remember, however, that $S(\vec{q}, \omega)$ is the density of excitations at momentum $\hbar\vec{q}$ and energy $\hbar\omega$ *weighted by the square of the matrix element of the density operator* that produces the excitation. Evidently the matrix element must be reduced by just the right amount to compensate for the increased excitation energy, thereby leaving the oscillator strength invariant. That this is sensible can be seen from Eq. (4.64) for the average excitation energy. If this increases, it can only be because $S(\vec{q}) = (1/N)\langle\!\langle|\rho_{+\vec{q}}|^2\rangle\!\rangle$ has decreased. Clearly, however, if we stiffen the bonds connecting the atoms, the thermal and quantum density fluctuations will be reduced and $S(\vec{q})$ will decrease as required. Similarly, the density matrix elements entering $S(\vec{q}, \omega)$ will be correspondingly reduced by just the right amount to preserve the oscillator strength.

As the final step in our development of a physical interpretation of the dynamical structure factor, let us return to our comparison of scattering with and without good energy resolution. We have already seen from our analysis of the Debye–Waller factor that the static structure factor tells us that there will be intense scattering at the Bragg peak positions $\vec{q} = \vec{G}$. Thermal and quantum fluctuations will reduce the intensity of the peaks but *not* broaden them. The intensity removed from the Bragg peaks is redistributed into a smooth, nearly featureless, background which we previously labeled as "incoherent."

We now can see that, since the static structure factor is the integral over frequency of $S(\vec{q}, \omega)$, it must be that this "incoherent" background is really the result of *inelastic* scattering, which is detected because of the lack of energy resolution. It is very important to understand that the word *static* means equal time ($t = 0$, all possible ω), while *elastic* means $\omega = 0$ (all possible t). They are *not* equivalent. This is a common source of linguistic confusion because we tend to think of elastic scattering as arising from static (i.e. fixed, non-recoiling) scatterers.[6]

---

[5] This is similar to the recoilless emission and absorption of γ-rays in the Mössbauer effect to be discussed in Chapter 6.

[6] So what was "wrong" with the discussion of elastic X-ray scattering in Chapter 2, especially that of Section 2.3, which is also based on a Golden Rule calculation similar to what is done here? The problem is that there the nuclear or ion positions are assumed to be fixed, and do not change during the scattering process. The Golden Rule applies to *eigenstates* of the Hamiltonian H; but a state with fixed nuclear positions *cannot* be a true eigenstate of H, thus, strictly speaking, the Golden Rule does not apply (even if a quantum-mechanical average is performed at the end of the calculation). On the other hand, such a state has a characteristic energy *uncertainty* of order typical phonon energies, which is much smaller than the X-ray photon energy resolution; within this resolution, such states can be viewed as energy eigenstates. This is how the calculations in Section 2.3 can be justified.

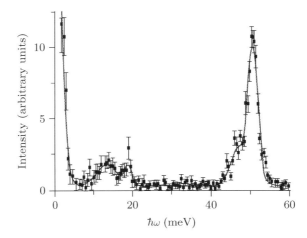

Figure 4.1 Phonon energy measured by inelastic neutron scattering. An energy scan of neutron intensity is shown at a fixed momentum transfer for uranium nitride. A pronounced peak at 50 meV reveals the optical phonon energy at this momentum. Figure adapted from [25] by permission from Springer Nature.

Equation (4.64) tells us that, if we have good energy resolution in our detector and set it for elastic scattering, then $S(\vec{q}, \omega = 0)$ will be large only on the Bragg peaks $\vec{q} = \vec{G}$, where the recoil energy deposited in the crystal is negligible. In order to have scattering at $\vec{q} \neq \vec{G}$ we must produce (or absorb) a density wave excitation which necessarily has non-zero energy. The density wave acts like a moving diffraction grating that imparts energy and momentum (not equal to $\hbar\vec{G}$) to the neutron. When we study the quantum theory of lattice vibrations, we will see that density wave excitations at wave vector $\vec{q}$ often have sharply defined energy $\hbar\omega(\vec{q})$ corresponding to phonon excitations. Thus, if we set our detector orientation so that we detect neutrons with momentum transfer $-\hbar\vec{q}$, an energy scan will reveal not a smooth incoherent background, but rather a sharp peak at an energy loss (or gain) equal to the phonon energy, as shown in Fig. 4.1. Conversely, if we set the detector for fixed energy loss $\hbar\Omega$, and scan in angle, the signal will be sharply peaked at the corresponding momentum. If we add up the signals from different energy losses $\hbar\Omega$, the sum of all these sharp peaks yields the smooth featureless background shown in Fig. 3.13.

Now that we have established some formal properties of $S(\vec{q}, \omega)$ and the density–density correlation function, we want to learn how to compute these quantities for realistic models of solids. This will lead us into the study of quantized lattice vibrations – phonons, which we will develop in subsequent chapters. As an introduction here we will study a single harmonic oscillator, first classically and then quantum mechanically.

4.3.1 Classical Harmonic Oscillator

It is useful as a warm-up exercise to consider a *single classical particle* in a 1D harmonic oscillator potential. The Hamiltonian is

$$H = \frac{P^2}{2M} + \frac{1}{2}M\Omega^2 X^2 \tag{4.68}$$

and we would like to compute

$$\Pi(q, t) \equiv \langle\langle e^{iqX(t)} e^{-iqX(0)} \rangle\rangle. \tag{4.69}$$

The system is classical, so the time evolution is entirely determined by the initial values of position $X(0) \equiv X_0$ and momentum $P(0) \equiv P_0$. We have the exact solution of the classical equations of motion,

$$X(t) = X_0 \cos(\Omega t) + \frac{P_0}{M\Omega} \sin(\Omega t), \tag{4.70}$$

which is easily seen to obey the initial conditions. Thus

$$\Pi(q,t) = \left\langle\!\!\left\langle e^{iq[\cos(\Omega t)-1]X_0} e^{i\frac{q}{M\Omega}\sin(\Omega t)P_0} \right\rangle\!\!\right\rangle. \tag{4.71}$$

Now, in a classical system X_0 and P_0 are *independent* random variables given by the Boltzmann probability distribution

$$P(X_0, P_0) = \frac{1}{Z} e^{-\beta\left(\frac{P_0^2}{2M} + \frac{1}{2}M\Omega^2 X_0^2\right)}, \tag{4.72}$$

where

$$Z \equiv \frac{1}{h} \int dX_0 \int dP_0 \, e^{-\beta\left(\frac{P_0^2}{2M} + \frac{1}{2}M\Omega^2 X_0^2\right)}. \tag{4.73}$$

(Recall that Z is dimensionless, so we need a factor of $1/h$, which is the "area" of a quantum state in phase space. The necessity of this factor was one of the important mysteries of classical physics which was later resolved by the discovery of quantum mechanics.)

Let us return to our original definition of $\Pi(q,t)$ in Eq. (4.69) and make an expansion in powers of q:

$$\begin{aligned}
\Pi(q,t) = \; & 1 + iq[\langle\!\langle X(t)\rangle\!\rangle - \langle\!\langle X(0)\rangle\!\rangle] \\
& - \frac{1}{2} q^2 [\langle\!\langle X(t)^2\rangle\!\rangle + \langle\!\langle X(0)^2\rangle\!\rangle] \\
& + q^2 [\langle\!\langle X(t)\,X(0)\rangle\!\rangle] + \cdots.
\end{aligned} \tag{4.74}$$

From parity and time translation symmetries we have

$$\langle\!\langle X(t)\rangle\!\rangle = \langle\!\langle X(0)\rangle\!\rangle = 0 \tag{4.75}$$

and

$$\langle\!\langle X(t)^2\rangle\!\rangle = \langle\!\langle X(0)^2\rangle\!\rangle, \tag{4.76}$$

so that

$$\Pi(q,t) = 1 + q^2[\langle\!\langle X(t)X(0)\rangle\!\rangle - \langle\!\langle X(0)X(0)\rangle\!\rangle] + \cdots. \tag{4.77}$$

Now it turns out that, for Gaussian-distributed random variables, this type of series always *exactly* sums into a simple exponential

$$\Pi(q,t) = e^{q^2(\langle\!\langle X(t)X(0)\rangle\!\rangle - \langle\!\langle X(0)X(0)\rangle\!\rangle)}. \tag{4.78}$$

The fact that we need only examine terms up to second order follows from the fact that a Gaussian distribution is entirely characterized by its mean and its variance (see Exercises 3.15 and 3.16). Using the solution of the equations of motion (4.70), we obtain

$$\langle\!\langle X(t)X(0)\rangle\!\rangle = \langle\!\langle X_0^2\rangle\!\rangle \cos(\Omega t) + \langle\!\langle X_0 P_0\rangle\!\rangle \frac{\sin(\Omega t)}{M\Omega}. \tag{4.79}$$

Since in a classical system in equilibrium position and velocity are totally uncorrelated random variables, namely

$$\langle\!\langle X_0 P_0\rangle\!\rangle = 0, \tag{4.80}$$

we have

$$\Pi(q,t) = e^{-q^2\langle\!\langle X_0^2\rangle\!\rangle(1-\cos(\Omega t))}. \tag{4.81}$$

Exercise 4.3. Calculate $S(q, \omega)$ for this classical problem using the result of Eq. (4.81) and an appropriate analog of Eq. (4.23) for its definition.

4.3.2 Quantum Harmonic Oscillator

Let us now treat the same problem of a single harmonic oscillator quantum mechanically rather than classically. We have

$$\Pi(q, t) = \langle\!\langle e^{iq\hat{X}(t)} e^{-iq\hat{X}(0)} \rangle\!\rangle, \tag{4.82}$$

where

$$\hat{X}(t) \equiv e^{iHt/\hbar} X e^{-iHt/\hbar} \tag{4.83}$$

is the Heisenberg representation of the position operator which obeys the equations of motion

$$\frac{d\hat{X}(t)}{dt} = \frac{i}{\hbar}[H, \hat{X}(t)] = \frac{1}{M}\hat{P}(t), \tag{4.84}$$

$$\frac{d\hat{P}(t)}{dt} = \frac{i}{\hbar}[H, \hat{P}(t)] = -M\Omega^2 \hat{X}(t). \tag{4.85}$$

These are of course identical to the classical equations of motion and their solution is formally identical:

$$\hat{X}(t) = X \cos(\Omega t) + \frac{1}{M\Omega} \sin(\Omega t) P, \tag{4.86}$$

$$\hat{P}(t) = P \cos(\Omega t) - M\Omega \sin(\Omega t) X, \tag{4.87}$$

where X and P are the Schrödinger representation operators (i.e. $\hat{X}(0)$ and $\hat{P}(0)$). These solutions are readily verified by direct substitution into the equations of motion.

The dynamical correlation function becomes

$$\Pi(q, t) = \langle\!\langle e^{iq[X \cos(\Omega t) + \frac{1}{M\Omega} \sin(\Omega t) P]} e^{-iqX} \rangle\!\rangle. \tag{4.88}$$

Now, as in the classical case, we expand this to second order in q. As it turns out, (a quantum version of) the cumulant expansion works even for quantum operators (see Exercises 3.15 and 3.16, and Box 4.1)! One hint that this is so comes from the fact that the harmonic oscillator ground state has a Gaussian wave function. Hence X and P act in a sense like random Gaussian-distributed classical variables. The complications due to the non-commutativity of X and P miraculously do not ruin the convergence of the cumulant expansion. More precisely, by expressing X and P in terms of the ladder operators a and $a^\dagger$, and using the result in Box 4.1, we obtain

$$\Pi(q, t) = e^{q^2 g(t)}, \tag{4.89}$$

where

$$\begin{aligned}
g(t) \equiv &\langle\!\langle X^2 \rangle\!\rangle \cos(\Omega t) + \frac{1}{M\Omega} \sin(\Omega t)\langle\!\langle PX \rangle\!\rangle \\
&- \frac{1}{2}\left\{ \langle\!\langle X^2 \rangle\!\rangle \cos^2(\Omega t) + \frac{1}{M^2\Omega^2} \sin^2(\Omega t)\langle\!\langle P^2 \rangle\!\rangle \right. \\
&\left. + \langle\!\langle X^2 \rangle\!\rangle + \frac{1}{M\Omega} \cos(\Omega t) \sin(\Omega t)\langle\!\langle XP + PX \rangle\!\rangle \right\}.
\end{aligned} \tag{4.90}$$

Now, using the expressions for P and X in terms of creation and annihilation operators

$$X = X_{\text{ZPF}}(a + a^\dagger),\tag{4.91}$$

$$P = \frac{-i\hbar}{2X_{\text{ZPF}}}(a - a^\dagger),\tag{4.92}$$

where the zero-point position uncertainty is $X_{\text{ZPF}} \equiv \sqrt{\hbar/2M\Omega}$, we have (in contrast to the classical case where $\langle PX \rangle = 0$) the paradoxical result[7]

$$\langle\langle PX \rangle\rangle = \frac{-i\hbar}{2}\langle\langle[a, a^\dagger]\rangle\rangle = \frac{-i\hbar}{2},\tag{4.93}$$

$$\langle\langle XP \rangle\rangle = \frac{-i\hbar}{2}\langle\langle[a^\dagger, a]\rangle\rangle = \frac{+i\hbar}{2}.\tag{4.94}$$

Furthermore, from the quantum version of the virial theorem for a harmonic oscillator, we have

$$\langle\langle T \rangle\rangle = \langle\langle V \rangle\rangle,\tag{4.95}$$

or equivalently, directly from Eqs. (4.91) and (4.92), we have

$$\frac{1}{M^2\Omega^2}\langle\langle P^2 \rangle\rangle = \langle\langle X^2 \rangle\rangle,\tag{4.96}$$

so that

$$g(t) = \left[\langle\langle X^2 \rangle\rangle(\cos(\Omega t) - 1) - i X_{\text{ZPF}}^2 \sin(\Omega t)\right].\tag{4.97}$$

Now in the high-temperature limit where $k_B T \gg \hbar\Omega$ and therefore $\langle\langle X^2 \rangle\rangle$ is much greater than X_{ZPF}^2, thermal fluctuations dominate over quantum fluctuations and we obtain the previous classical expression for $\Pi(q, t)$. Since in the classical limit $\Pi(q, t)$ is symmetric in t, the spectral density is symmetric in ω:

$$S(q, \omega) = S(q, -\omega).\tag{4.98}$$

This tells us that a neutron scattering from a classical oscillator can gain or lose energy equally well.

In contrast, at zero temperature we have

$$\langle 0|X^2|0\rangle = X_{\text{ZPF}}^2\langle 0|(a + a^\dagger)^2|0\rangle = X_{\text{ZPF}}^2,\tag{4.99}$$

so that

$$\Pi(q, t) = e^{-q^2 X_{\text{ZPF}}^2} e^{q^2 X_{\text{ZPF}}^2 e^{-i\Omega t}}.\tag{4.100}$$

Expanding the second exponential in a power series, we immediately see that this has Fourier components only at *positive* frequencies:

$$S(q, \omega) = e^{-q^2 X_{\text{ZPF}}^2} \sum_{n=0}^{\infty} \frac{\left(q^2 X_{\text{ZPF}}^2\right)^n}{n!} 2\pi\delta(\omega - n\Omega).\tag{4.101}$$

The physical interpretation of this result is simply that an oscillator in its quantum ground state can only absorb energy, not emit it. A very useful observation is to note that the number of quanta which are excited, n, is Poisson distributed:

$$P_n = e^{-\lambda}\frac{\lambda^n}{n!},\tag{4.102}$$

---

[7] Note that the product of two Hermitian operators is not itself a Hermitian operator if the two do not commute. It is this fact which allows the expectation value to be imaginary, which it must be in order to be consistent with the canonical commutation relation $[X, P] = i\hbar$.

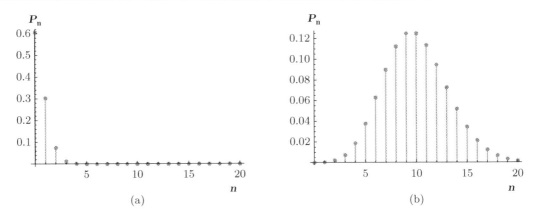

Figure 4.2 The Poisson distribution $P_n = e^{-\lambda}\lambda^n/n!$ for $\lambda = 0.5$ (a) and $\lambda = 10$ (b).

where $\lambda \equiv q^2 X_{ZPF}^2$ is a measure of the momentum transfer. The harder the neutron hits the oscillator, the more likely it is that n will be large. The average number excited is

$$\bar{n} = \sum_{n=0}^{\infty} n P_n = \lambda, \tag{4.103}$$

and hence the average energy transfer is

$$\Delta = \bar{n}\hbar\Omega = \frac{\hbar^2 q^2}{2M}, \tag{4.104}$$

in agreement with the f-sum rule.

If $\lambda < 1$, P_n is monotone decreasing as shown in Fig. 4.2(a). However, if $\lambda \gg 1$, P_n is approximately a Gaussian centered on $\bar{n} = \lambda$ and having variance $\sigma^2 = \bar{n}$ as shown in Fig. 4.2(b).

> **Exercise 4.4.** For $\lambda \gg 1$, find a semiclassical expression for the distribution $P_E(E)$ of final state excitation energies and show that this is consistent with Eq. (4.102). Hint: the initial state momentum P_0 is Gaussian distributed. Find the energy absorbed by a free particle of mass M when momentum $\hbar q$ is transferred to it. Give a physical argument explaining why the harmonic potential can be neglected at this level of approximation.

What happens in the intermediate regime at temperatures greater than zero but not large compared with $\hbar\Omega/k_B$? In this case $\langle\langle X^2 \rangle\rangle$ exceeds X_{ZPF}^2, and Eq. (4.97) acquires negative frequency components. Because $g(t)$ must be exponentiated in Eq. (4.89) and then Fourier transformed in Eq. (4.41) to find $S(q,\omega)$, it is not completely obvious how the negative-frequency components of $g(t)$ will enter $S(q,\omega)$. However, directly from Eqs. (4.43) and (4.44) it is possible to show that

$$S(\vec{q}, -\omega) = e^{-\beta\hbar\omega} S(\vec{q}, \omega). \tag{4.105}$$

Thus we see that the spectral weight at negative frequencies is thermally activated. For $T = 0$ there is no spectral weight at $\omega < 0$, while in the classical limit $k_B T \gg \hbar\omega$, $S(\vec{q}, \omega)$ becomes symmetric in ω as discussed above. These points are extensively discussed in Appendix A and in [26], where it is shown that the spectral density at positive frequencies corresponds to the ability of the system to absorb energy from the neutron, while that at negative frequencies corresponds to the ability to emit energy to the neutron.

> **Exercise 4.5.** Derive the result in Eq. (4.105).

Linear response theory (see Appendix A) tells us that the susceptibility to a perturbation that couples to the density is given by an expression similar to Eq. (4.41):

$$\chi(\vec{q}, \omega) = \int_{-\infty}^{\infty} dt \; e^{i(\omega + i\eta)t} \Pi_R(\vec{q}, t), \tag{4.106}$$

where η is a small positive convergence factor and Π_R is the **retarded** **correlation function** for the density fluctuations:

$$\Pi_R(\vec{q}, t) \equiv -\frac{i}{\hbar} \theta(t) \langle\!\langle [\hat{\rho}_{-\vec{q}}(t), \hat{\rho}_{\vec{q}}(0)] \rangle\!\rangle, \tag{4.107}$$

where the step function enforces **causality**. Again working directly from the definition of the thermal average, it is straightforward to show that

$$-2 \operatorname{Im} \chi(\vec{q}, \omega) = S(\vec{q}, \omega) - S(\vec{q}, -\omega) = (1 - e^{-\beta\hbar\omega}) S(\vec{q}, \omega). \tag{4.108}$$

This result is known as the *fluctuation–dissipation* theorem because it relates the fluctuation spectrum $S(\vec{q}, \omega)$ to the dissipative (imaginary) part of the response function. The difference of the two spectral densities enters because one corresponds to energy gain and the other to energy loss, and hence they contribute oppositely to the dissipation [26].

Notice that, from Eqs. (4.105) and (4.108), it follows that the imaginary part of χ is anti-symmetric in frequency:

$$\operatorname{Im} \chi(\vec{q}, -\omega) = -\operatorname{Im} \chi(\vec{q}, +\omega). \tag{4.109}$$

As discussed in Appendix A, this is related to the fact that the (real-space) density change associated with a perturbation must be real.

Box 4.1. Expectation Values of Certain Operators for 1D Harmonic Oscillators

We want to consider $\langle\!\langle e^{\bar{u}a^\dagger} e^{-ua} \rangle\!\rangle$, where u is any complex number and $\bar{u}$ is its conjugate. First note that

$$\langle n | e^{+\bar{u}a^\dagger} e^{-ua} | n \rangle = \sum_{k=0}^{n} \frac{|u|^k (-1)^k}{(k!)^2} \frac{n!}{(n-k)!} = L_n(|u|^2), \tag{4.110}$$

where $L_n(x)$ is a Laguerre polynomial. This leads (in thermal equilibrium) to

$$\langle\!\langle e^{\bar{u}a^\dagger} e^{-ua} \rangle\!\rangle = \frac{\sum_{n=0}^{\infty} (e^{-\beta\Omega})^n L_n(|u|^2)}{(1 - e^{-\beta\Omega})^{-1}}, \tag{4.111}$$

where Ω is the oscillator frequency (and we have set $\hbar = 1$). Now we recall the generating function of Laguerre polynomials:

$$\sum_{n=0}^{\infty} z^n L_n(|u|^2) = (1-z)^{-1} e^{|u|^2 z/(z-1)}. \tag{4.112}$$

Let

$$z = e^{-\beta\Omega} \Rightarrow \frac{z}{1-z} = \frac{e^{-\beta\Omega}}{1 - e^{-\beta\Omega}} = \frac{1}{e^{\beta\Omega} - 1} = \langle n \rangle \tag{4.113}$$

$$\Rightarrow \langle\!\langle e^{\bar{u}a^\dagger} e^{-ua} \rangle\!\rangle = e^{-|u|^2 \langle n \rangle}. \tag{4.114}$$

We could have guessed this immediately if we had replaced a and $a^\dagger$ by $\sqrt{n}$ and ignored the fact that they are operators. It turns out that this is exactly true. Equivalently we could have used the cumulant expansion.

Exercise 4.6. Using the linear response expressions in Appendix A, derive Eq. (4.106).

4.4 Single-Mode Approximation and Superfluid ^{4}He

Now that we understand the oscillator strength sum rule (f-sum rule), we can develop an extremely powerful approximation for the dynamical structure factor $S(\vec{q}, \omega)$, known as the **single-mode approximation** (SMA). It often happens that, for small $\vec{q}$, density oscillations (such as sound waves) are natural collective modes. In this case it is often a good approximation to assume that $\rho_{\vec{q}}$ acting on the ground state produces a state

$$|\vec{q}\rangle \equiv \rho_{\vec{q}} |0\rangle, \tag{4.115}$$

which is very nearly an eigenstate of the Hamiltonian, having some eigenvalue $\hbar\Omega(\vec{q})$. This means that the sum over excited eigenstates in Eq. (4.44) is dominated by a single term $|f\rangle = |\vec{q}\rangle$ and the ground state dynamical structure factor has the form

$$S(\vec{q}, \omega) = 2\pi S(\vec{q})\delta[\omega - \Omega(\vec{q})]. \tag{4.116}$$

Notice that the factor multiplying the delta function must be the static structure factor in order to satisfy Eq. (4.45). With this simple form, however, we can easily evaluate Eq. (4.63) to find

$$\epsilon(\vec{q}) = \hbar\Omega(\vec{q}) = \frac{\hbar^2 q^2}{2M} \frac{1}{S(\vec{q})}. \tag{4.117}$$

That is, since there is only a single mode that saturates the oscillator strength sum, its energy must equal the average energy. Thus (within this approximation) knowledge of the static structure factor tells something us about a dynamical quantity – the collective mode energy.

Perhaps the first successful application of the SMA is Feynman's theory of the collective mode spectrum of superfluid ^{4}He, shown in Fig. 4.3.[8] At long wavelengths (small q) we have a linearly dispersing ($\epsilon(q) \propto q$) collective phonon mode, while there is a pronounced minimum at $q_0 \sim 1/R$ (where R is the average spacing between He atoms) around which the excitations are known as rotons. Feynman used Eq. (4.115) as a *variational* approximation to the wave function of these modes, with (4.117) giving the corresponding variational energies, which turn out to agree with the experimentally determined spectrum very well, at least qualitatively; see Fig. 4.4.

To understand why the SMA works well for superfluid ^{4}He, let us start, by way of introduction, with the simple harmonic oscillator. The ground state is of the form

$$\psi_0(x) \sim e^{-\alpha x^2}. \tag{4.118}$$

Suppose we did not know the excited state and tried to make a variational ansatz for it. Normally we think of the variational method as applying only to ground states. However, it is not hard to see that the first excited state energy is given by

$$\epsilon_1 = \min\left\{\frac{\langle\psi|H|\psi\rangle}{\langle\psi|\psi\rangle}\right\}, \tag{4.119}$$

provided that we do the minimization over the set of states ψ which are constrained to be orthogonal to the ground state ψ_0. One simple way to produce a variational state which is automatically orthogonal to the ground state is to change the parity by multiplying by the first power of the coordinate:

---

[8] The phenomenology of superfluidity and its relation to the collective mode spectrum of ^{4}He liquid are discussed in standard textbooks on statistical mechanics. See, e.g., [28]. They will also be discussed in Chapter 18.

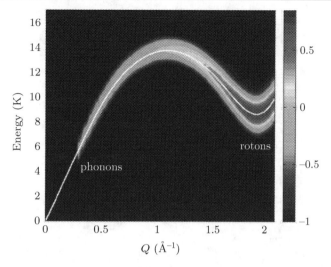

Figure 4.3 Collective mode spectrum of superfluid ^{4}He, revealed by inelastic neutron scattering. Figure adapted with permission from [27]. Copyright by the American Physical Society.

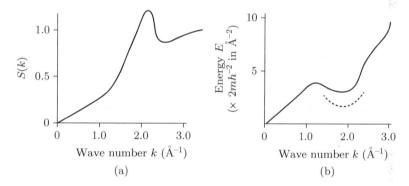

Figure 4.4 Feynman's theory of collective mode spectrum of superfluid ^{4}He, based on single-mode approximation. (a) Static structure factor of ^{4}He. (b) Collective mode spectrum based on the single-mode approximation. The dashed curve is a fit to specific heat data. Figure adapted with permission from [29]. Copyright by the American Physical Society.

$$\psi_1(x) \sim x e^{-\alpha x^2}. \tag{4.120}$$

Variation with respect to α of course leads (in this special case) to the *exact* first excited state.

With this background, let us now consider the case of phonons in superfluid ^{4}He. Feynman argued that, because of the Bose statistics of the particles, there are no low-lying single-particle excitations. This is in stark contrast to a Fermi gas, which has a high density of low-lying excitations around the Fermi surface. Feynman argued that the only low-lying excitations in ^{4}He are collective density oscillations that are well described by a family of variational wave functions (that has no adjustable parameters) of the form (4.115), which, in terms of wave functions is

$$\psi_{\vec{q}} = \frac{1}{\sqrt{N}} \rho_{\vec{q}} \Phi_0(\vec{r}_1, \dots, \vec{r}_N) \tag{4.121}$$

where $\Phi_0(\vec{r}_1, \dots, \vec{r}_N)$ is the exact (but often unknown) ground state and

$$\rho_{\vec{q}} \equiv \sum_{j=1}^{N} e^{-i\vec{q}\cdot\vec{r}_j} \tag{4.122}$$

is the Fourier transform of the density. The physical picture behind this is that at long wavelengths the fluid acts like an elastic continuum and $\rho_{\vec{q}}$ can be treated as a generalized oscillator normal-mode coordinate. In this sense Eq. (4.121) is then analogous to Eq. (4.120). To see that $\psi_{\vec{q}}$ is orthogonal to the ground state we simply note that

$$\langle \Phi_0 | \psi_{\vec{q}} \rangle = \frac{1}{\sqrt{N}} \langle \Phi_0 | \rho_{\vec{q}} | \Phi_0 \rangle$$
$$= \frac{1}{\sqrt{N}} \int d^3 R \, e^{-i\vec{q} \cdot \vec{R}} \langle \Phi_0 | \rho(\vec{R}) | \Phi_0 \rangle, \qquad (4.123)$$

where

$$\rho(\vec{R}) \equiv \sum_{j=1}^{N} \delta^3(\vec{r}_j - \vec{R}) \qquad (4.124)$$

is the density operator. If Φ_0 describes a translationally invariant liquid ground state then the Fourier transform of the mean density vanishes for $q \neq 0$. Another way to see this orthogonality is if Φ_0 is a momentum eigenstate (with eigenvalue zero), then $\psi_{\vec{q}}$ is another momentum eigenstate with a *different* eigenvalue $(-\hbar\vec{q})$.

There are several reasons why $\psi_{\vec{q}}$ is a good variational wave function, especially for small q. First, it contains the ground state as a factor. Hence it contains all the special correlations built into the ground state to make sure that the particles avoid close approaches to each other, without paying a high price in kinetic energy. Secondly, $\psi_{\vec{q}}$ builds in the features we expect on physical grounds for a density wave. To see this, consider evaluating $\psi_{\vec{q}}$ for a configuration of the particles like that shown in Fig. 4.5(a) which has a density modulation at wave vector $\vec{q}$. This is not a configuration that maximizes $|\Phi_0|^2$, but, as long as the density modulation is not too large and the particles avoid close approaches, $|\Phi_0|^2$ will not fall too far below its maximum value. More importantly, $|\rho_{\vec{q}}|^2$ will be much larger than it would for a more nearly uniform distribution of positions. As a result, $|\psi_{\vec{q}}|^2$ will be large and this will be a likely configuration of the particles in the excited state. For a configuration like that in Fig. 4.5(b), the phase of $\rho_{\vec{q}}$ will shift, but $|\psi_{\vec{q}}|^2$ will have the same magnitude. This is analogous to the parity change in the harmonic oscillator example. Because all different phases of the density wave are equally likely, $\psi_{\vec{q}}$ has a mean density which is uniform (translationally invariant).

Having justified Eq. (4.115) as a good approximation for the collective mode, we expect Eq. (4.117) to be a good (variational) approximation for the collective mode spectrum of superfluid ^{4}He. At long wavelengths $S(\vec{q}) \sim q$ (see Fig. 4.4(a)), giving rise to the linear dispersion of the phonons. At shorter

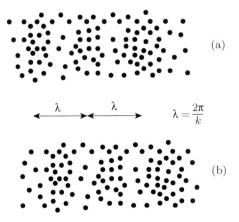

Figure 4.5 (a) Configuration of particles in which the Fourier transform of the density at wave vector k is non-zero. (b) The Fourier amplitude will have a similar magnitude for this configuration but a different phase.

wavelengths, there is a peak in $S(q)$ at the position near where the first Bragg peak would be in the solid. This leads to the famous "roton minimum" in $\epsilon(q)$ (see Fig. 4.4(b)). In the ultimate short-wavelength limit $q \to \infty$, we have $S(\vec{q}) \to 1$ and thus

$$\epsilon(q) \to \frac{\hbar^2 q^2}{2M}. \tag{4.125}$$

This is nothing but the *single-particle* excitation energy at very high momentum.

Historically, the form of the spectrum of superfluid ^{4}He, especially the presence of a linear phonon mode at long wavelength and a roton minimum, was first suggested by Landau on the basis of his analysis of specific-heat data. However, to fully understand how ^{4}He can flow without dissipation, we need to invoke the fact that the superfluid state is a state that spontaneously breaks the U(1) symmetry associated with particle number conservation. We will explain this point in Chapter 18 and Section 19.7. Here, we close this section by pointing out in this context that the linear phonon mode is nothing but the Goldstone mode associated with the broken symmetry, while the presence of the roton minimum reflects the tendency of the atoms to crystallize at the corresponding wave vector.[9] Thus the most important features of the collective mode reflect the symmetry that is broken, or what the system intends to (but does not yet) break!

Exercise 4.7.

(i) Consider N non-interacting bosons in vacuum. Write down the ground state wave function, and show that Eq. (4.115) gives rise to a set of *exact* excited states. Find the corresponding excitation energies.

(ii) Calculate the static structure factor $S(q)$ of the ground state of (i), and show that the single-mode approximation (4.117) matches the exact spectrum you found in (i) perfectly.

Exercise 4.8.

(i) Consider a gas of non-interacting fermions of mass m and finite density in vacuum, whose ground state is a Fermi sea with Fermi wave vector k_F. Show that the expression for the density-wave state in Eq. (4.115) does not give rise to exact excited states. This means that the density-wave state can only be written as a superposition of excited states with different excitation energies.

(ii) Find the range of excitation energies. (Hint: study Fig. 15.1(b).)

---

[9] In fact, ^{4}He does crystallize at sufficiently high pressure. In the crystalline solid, $S(\vec{G})$ diverges and the roton minimum drops all the way down to zero as discussed earlier.

5 Dynamics of Lattice Vibrations

Now that we have warmed up in the previous chapter with a study of the simple 1D harmonic oscillator, let us attempt a more realistic model of the lattice vibrations in a 3D crystal. We begin here with a classical analysis and later, in Chapter 6, we will study the underlying quanta of sound waves known as phonons. These lattice vibrations carry energy and contribute signicantly to the heat capacity and the thermal conductivity of many solids. Particularly in insulators, where the electronic contributions are nearly negligible, lattice vibrations dominate the thermodynamic properties.

It may seem complicated to have to deal with the enormous number of motional degrees of freedom ($3N$ in a crystal of N atoms), but we will see in this chapter that the translation symmetry of the crystalline lattice allows vast simplification of the description of the dynamics of the particles.

To get started, we will first ignore the discrete atomic structure of the crystal and treat it as an **elastic continuum** with complete translation symmetry. After studying elasticity in the continuum, we will move on to a more realistic approximation that takes into account the detailed microscopic structure of the atomic lattice. This is will lead to significant differences in the collective mode properties at short wavelengths, where the modes are sensitive to the underlying discrete atomic structure of the solid.

5.1 Elasticity and Sound Modes in Continuous Media

Elastic distortions of the crystal displace the atoms from their nominal equilibrium positions. The displacement vector $\vec{u}_j$ was defined in Eq. (4.9). The elastic continuum approximation is appropriate if $\vec{u}_j = \vec{u}(\vec{R}_j)$ varies *slowly* with $\vec{R}_j$, in which case we can ignore the discrete nature of $\vec{R}_j$, and treat it as a *continuous* variable $\vec{r}$. In this continuum limit $\vec{u}_j = \vec{u}(\vec{R}_j)$ is replaced by a **displacement field** $\vec{u}(\vec{r})$ describing the elastic deformations.[1] Microscopically $\vec{u}(\vec{r})$ is the displacement vector of atoms with equilibrium positions $\vec{R}_j \approx \vec{r}$. Then for small fluctuations the kinetic energy is

$$T = \frac{1}{2}\rho_0 \int d^3\vec{r}\,|\dot{\vec{u}}(\vec{r})|^2, \tag{5.1}$$

---

[1] The continuum limit is often referred to as the field-theory limit, in which degrees of freedom (fields) depend on the continuous spatial coordinates. Upon quantizing these degrees of freedom, we obtain a quantum field theory. The reader should compare the elasticity theory discussed here with both the scalar and the vector field theories in a standard relativistic quantum field theory course, and identify their similarities and differences.

where ρ_0 is the mass density (in the undistorted configuration). By translation symmetry, the potential energy is invariant if $\vec{u}(\vec{r})$ is a constant. Hence the potential energy can only depend on gradients of $\vec{u}$. Naively one would therefore write

$$V = \frac{1}{2} \int d^3\vec{r} \{\Gamma_{\mathrm{L}} (\nabla \cdot \vec{u})^2 + \Gamma_{\mathrm{T}} |\nabla \times \vec{u}|^2\}. \tag{5.2}$$

These two terms seem to be the simplest translation- and rotation-invariant scalars that one can write down. The first represents longitudinal distortions and the second represents the cost of transverse (shear) distortions. In the former the density fluctuates since the displacements have a non-zero divergence, $\nabla \cdot \vec{u} \neq 0$ (see Exercise 5.2).

Unfortunately, the expression in Eq. (5.2) is not quite correct. It is fine for the case of distortions that satisfy periodic boundary conditions (and can thus be expanded as linear combinations of plane waves), but is not correct in the general case. Consider a uniform infinitesimal rotation $\delta\vec{\Omega}$ applied to the crystal which costs *no* elastic energy. Then the displacement field obeys

$$\vec{u}(\vec{r}) = \delta\vec{\Omega} \times \vec{r} \Leftrightarrow u^\mu = \epsilon^{\mu\nu\gamma} \delta\Omega^\nu r^\gamma, \tag{5.3}$$

$$\partial_\gamma u^\mu = \epsilon^{\mu\nu\gamma} \delta\Omega^\nu, \tag{5.4}$$

or equivalently

$$\delta\Omega^\nu = \frac{1}{2} \epsilon^{\nu\gamma\mu} \partial_\gamma u^\mu = \frac{1}{4} \epsilon^{\nu\gamma\mu} (\partial_\gamma u^\mu - \partial_\mu u^\gamma). \tag{5.5}$$

Here we use Greek letters for spatial (*not* space-time!) indices of $\vec{u}$s (Roman letters are reserved for lattice site indices), with *no* distinction between upper and lower indices, and the Einstein summation convention is being used. This notation and convention will be used throughout the book.

In order for it to be independent of the global rotation angle, the elastic energy must be independent of the anti-symmetric part of $\partial_\gamma u^\mu$ and depend only on the so-called **strain tensor**, which is *symmetric* under exchange of μ and ν:

$$u_{\mu\nu} \equiv \frac{1}{2} (\partial_\mu u_\nu + \partial_\nu u_\mu). \tag{5.6}$$

Physically $u_{\mu\nu}$ describes the local *distortion* (but not rotation) of an object, and is the source of elastic energy cost and internal stress. For an *isotropic* elastic continuum[2] the potential energy can then be written in the harmonic approximation in terms of the two different scalars that can be formed from the strain tensor:[3]

$$V = \frac{1}{2} \int d^3\vec{r} (\lambda u_{\gamma\gamma} u_{\nu\nu} + 2\mu u_{\gamma\nu} u_{\gamma\nu}). \tag{5.7}$$

A little algebra shows that the so-called **Lamé coefficients** λ and μ are related to the stiffnesses for longitudinal and transverse waves in Eq. (5.2) by (see Exercise 5.3)

$$\Gamma_{\mathrm{L}} = \lambda + 2\mu \tag{5.8}$$

and

$$\Gamma_{\mathrm{T}} = \mu. \tag{5.9}$$

---

[2] A crystal is, of course, *not* isotropic at the atomic scale, as crystalline order breaks rotation symmetry (more on this point later). However, in polycrystalline solids, glasses, and polymers etc., the system is effectively isotropic beyond a certain microscopic scale.

[3] Recall that, to obtain scalars from tensors, all indices must be contracted. The two terms in Eq. (5.7) correspond to the only two possible ways of making such contractions for a term quadratic in $u_{\mu\nu}$. Owing to rotation symmetry, only such scalars are allowed in the Lagrangian of a completely isotropic system. This is similar to the fact that only Lorentz scalars are allowed in the Lagrangian of relativistic field theories that respect Lorentz invariance. As a result, we have only two independent elasticity constants in an isotropic elastic medium.

μ is also known as the **shear modulus**, which is a measure of the elastic energy cost of shear distortion, in which the shape of an element in the elastic medium is distorted without changing its volume (or mass density). In such a classification the complementary quantity is the **bulk modulus** B, which parameterizes the energy cost of uniform expansion or compression without changing the shape of an element. B is nothing but the inverse uniform **compressibility** of the medium, and is related to the Lamé coefficients as follows:

$$B = -V \left(\frac{\partial P}{\partial V} \right)_T = \lambda + 2\mu/d, \qquad (5.10)$$

where P is pressure, V is volume, and d is the dimensionality. It is clear that $\Gamma_T = \mu$ because a transverse distortion does not induce a change of local density, while Γ_L must involve both B and μ since a longitudinal distortion induces both volume and shape changes of an element.[4]

Suppose for simplicity that we have a time-dependent distortion of the form

$$\vec{u}(\vec{r}, t) = a(t) \hat{\epsilon}_{\vec{q}\lambda} \cos(\vec{q} \cdot \vec{r}), \qquad (5.11)$$

where $\hat{\epsilon}$ is the polarization vector and $\hat{q} \cdot \hat{\epsilon}_{\vec{q}\lambda} = 1$ for a longitudinal mode and $|\hat{q} \times \hat{\epsilon}_{\vec{q}\lambda}| = 1$ for a transverse mode. The Lagrangian becomes

$$\mathcal{L} = T - V$$
$$= \frac{1}{2} \frac{L^3}{2} \rho_0 \dot{a}^2(t) - \frac{1}{2} \frac{L^3}{2} \{ (\vec{q} \cdot \hat{\epsilon}_{\vec{q}\lambda})^2 \Gamma_L + (\vec{q} \times \hat{\epsilon}_{\vec{q}\lambda})^2 \Gamma_T \} a^2(t), \qquad (5.12)$$

where L^3 is the volume of the system and we have used the fact that the cos-squared term averages to $\frac{1}{2}$. This looks like the Lagrangian for a simple harmonic oscillator of mass $m \equiv \frac{1}{2}\rho_0 L^3$ and spring constant $K_{L(T)} \equiv \frac{1}{2} L^3 q^2 \Gamma_{L(T)}$ for the longitudinal (transverse) case. The mode frequency is therefore

$$\omega_{L(T)}^2 = \frac{K_{L(T)}}{m} = q^2 \frac{\Gamma_{L(T)}}{\rho_0}. \qquad (5.13)$$

Thus the **speeds of sound** for the two types of linearly dispersing modes are

$$v_{L(T)} = \left[\frac{\Gamma_{L(T)}}{\rho_0} \right]^{1/2}. \qquad (5.14)$$

In 3D we have one longitudinal mode, and two independent transverse modes with the same speed. ($\Gamma_L > \Gamma_T$, so the longitudinal sound velocity is always larger than the transverse sound velocity.) In d dimensions there is one longitudinal mode and there are $d - 1$ transverse modes.[5]

---

[4] While we had 3D in mind, the discussion of the strain tensor and its elastic energy cost apply to arbitrary dimensions. Formally, combining the (d-dimensional) vectors ∂_μ and u_ν, one obtains a rank-2 tensor $\partial_\mu u_\nu$ with d^2 components. It is natural to decompose this into an anti-symmetric and a symmetric tensor, with $d(d-1)/2$ and $d(d+1)/2$ components, respectively. Recall that, in d dimensions, there are $d(d-1)/2$ rotational degrees of freedom, and the anti-symmetric components of $\partial_\mu u_\nu$ correspond precisely to them. One can further decompose the symmetric components, which form the strain tensor $u_{\mu\nu}$, into a *traceless* tensor and a pure *scalar*: $u_{\mu\nu} = [u_{\mu\nu} - (1/d)u_{\gamma\gamma}\delta_{\mu\nu}] + (1/d)u_{\gamma\gamma}\delta_{\mu\nu}$; the former is shear distortion and the latter corresponds to uniform expansion/compression. The shear and bulk moduli couple to them, respectively. Coming back to 3D, we have thus decomposed the nine-component tensor $\partial_\mu u_\nu$ into a scalar (one component), an anti-symmetric tensor (three components), and a traceless symmetric tensor (five components). It is *not* a coincidence that this counting as well as the symmetry properties under exchange are identical to the problem of adding two angular momenta, each with quantum number $l = 1$, and decomposing the result into $l_{\text{tot}} = 0, 1, 2$, with degeneracy 1, 3, and 5, respectively. The reader is encouraged to figure out why. Rotation invariance means that terms entering the energy functional must be scalars. Since at quadratic level the scalar and traceless rank-2 tensor components of the strain tensor can only combine with themselves to form scalars, we have two independent scalars at the quadratic level, and thus two independent elastic constants.

[5] In contrast, a liquid has only the longitudinal mode (sound mode), but *no* transverse modes. This is because a liquid has a finite bulk modulus, but zero shear modulus, as there is no restoring force for shape change.

The elastic continuum approximation is generally good at long wavelengths for Bravais lattices. We will find that at shorter wavelengths (comparable to the lattice spacing) the discrete non-continuum nature of the lattice will modify the (linear) dispersion relation.

A second failing of the elastic continuum model is the following. We have seen that only longitudinal sound waves have density fluctuations associated with them. This would seem to mean that neutron scattering experiments could not detect transverse modes. This is false, however, since, as we shall see, a transverse mode at wave vector $\vec{q}$ can have density fluctuations at $\vec{q} + \vec{G}$ due to the discrete translation symmetry of the lattice.[6]

A further complication is that, except in high-symmetry directions, the modes of a lattice are actually a mixture of longitudinal and transverse. This is because, due to the broken rotation symmetry, Eq. (5.7) needs to be generalized to

$$V = \frac{1}{2} \int d^3\vec{r} \, C_{\alpha\beta\gamma\delta} u_{\alpha\beta} u_{\gamma\delta}, \tag{5.15}$$

where $C_{\alpha\beta\gamma\delta}$ is a set of 21 (in 3D) elastic constants (instead of $6 \times 6 = 36$ due to the constraint $C_{\alpha\beta\gamma\delta} = C_{\gamma\delta\alpha\beta}$), instead of just two elastic constants for the isotropic medium.[7] Residual (discrete) rotation and other symmetries further reduce the number of independent components of $C_{\alpha\beta\gamma\delta}$; for cubic symmetry (the highest lattice symmetry in 3D) the number is down to three. Using Eq. (5.15) instead of Eq. (5.7) in the Lagrangian to derive and solve for the equations of motion will yield $\vec{u}(\vec{r}, t)$ neither parallel nor perpendicular to the wave vector $\vec{q}$ in general.

A third failing of the simple continuum model is found in non-Bravais lattices. Because these lattices have more than one atom in their basis, there exist additional **"optical" phonon** modes in which these atoms vibrate out of phase with each other. These modes are often at relatively high frequencies even in the limit of long wavelengths. To see why this is allowed, consider the case of two atoms per unit cell described in a generalized elastic continuum picture by two distinct displacement fields, $\vec{u}_1(\vec{r})$ and $\vec{u}_2(\vec{r})$. An allowed scalar in the potential energy is

$$V_{12} = \frac{1}{2} K \int d^3\vec{r} (\vec{u}_1 - \vec{u}_2)^2. \tag{5.16}$$

Because of the relative minus sign such a term is translation-invariant without requiring any derivatives. Hence the analog of Eq. (5.13) will have no powers of q^2 in it, and the mode frequency will not vanish at long wavelengths. These modes are called "optical" because, in ionic compounds, fluctuations in $(\vec{u}_1 - \vec{u}_2)$ create electrical polarization, which couples to electromagnetic waves.

Exercise 5.1. Derive Eq. (5.1) from the kinetic energy of discrete atoms.

Exercise 5.2. Show that, to lowest order in displacement $\vec{u}(\vec{r})$, the local density of a continuous medium becomes $\rho(\vec{r}) = \rho_0[1 - \nabla \cdot \vec{u}(\vec{r})]$, where ρ_0 is the density without distortion.

Exercise 5.3. Verify Eqs. (5.8) and (5.9), using the facts that under periodic boundary condition $\vec{u}(\vec{r})$ may be decomposed into Fourier components, and different Fourier components do not mix in Eq. (5.7) (verify it!).

---

[6] The essential point is that, for a transverse mode, the fact that $\vec{q} \cdot \vec{u}_{\vec{q}} = 0$ does not imply that $(\vec{q} + \vec{G}) \cdot \vec{u}_{\vec{q}} = 0$.

[7] The fact that elasticity constants C carry indices is an immediate indication of broken rotation symmetry: the energy cost of a distortion depends on the directions of the distortion!

> **Exercise 5.4.** Consider a 2D elastic medium. Find the number of independent shear distortions, and show how they distort an element with the shape of a square.

> **Exercise 5.5.** Find the number of independent components of $C_{\alpha\beta\gamma\delta}$ in Eq. (5.15), in d dimensions.

5.2 Adiabatic Approximation and Harmonic Expansion of Atomic Potential

Let us turn now to a more detailed consideration of the dynamics of vibrations in a discrete lattice system. Let us assume a Bravais lattice (no basis) and an *effective* atomic Hamiltonian

$$H = \sum_j \frac{P_j^2}{2M} + U(\vec{r}_1, \ldots, \vec{r}_N), \tag{5.17}$$

where j labels an atom and $\vec{r}_j = \vec{R}_j + \vec{u}_j$ as usual. We do not write down any part of the Hamiltonian involving the electron coordinates. We assume, however, that the nuclear motion is slow so that the relatively light electrons are able to follow the nuclear motion *adiabatically*. That is, we assume that, for any configuration of the nuclei, the electrons are in the ground state wave function they would have if the nuclei were rigidly held in that configuration and were not moving. The approximation works because the nuclei are much more massive than the electrons. In the following we illustrate how to arrive at Eq. (5.17) starting from the full Hamiltonian of the system that includes the electronic degrees of freedom.

If $\psi_{[\vec{r}_j]}(\vec{\xi}_1, \ldots, \vec{\xi}_M)$ is the adiabatic electronic state, it obeys the Schrödinger equation

$$H_e \psi_{[\vec{r}_j]}(\vec{\xi}_1, \ldots, \vec{\xi}_M) = E_{[\vec{r}_j]} \psi_{[\vec{r}_j]}(\vec{\xi}_1, \ldots, \vec{\xi}_M), \tag{5.18}$$

where H_e is the Hamiltonian for the electrons assuming the nuclei are fixed at $[\vec{r}_j]$, and the $\vec{\xi}$s are electron coordinates including spin. The lowest eigenvalue depends on the $[\vec{r}_j]$, which here are not dynamical degrees of freedom but merely parameters in the Hamiltonian insofar as the electrons are concerned (in the adiabatic approximation). With modern electronic structure methods it is possible (though not necessarily easy) to numerically compute $E_{[\vec{r}_j]}$ and its dependence on the nuclear coordinates.

The full wave function is taken to be a simple product of nuclear and electronic wave functions:

$$\Phi(\vec{r}_1, \ldots, \vec{r}_N; \vec{\xi}_1, \ldots, \vec{\xi}_M) = \phi(\vec{r}_1, \ldots, \vec{r}_N)\psi_{[\vec{r}_j]}(\vec{\xi}_1, \ldots, \vec{\xi}_M). \tag{5.19}$$

The full Hamiltonian is

$$\tilde{\mathcal{H}} = H_n + H_e, \tag{5.20}$$

where $H_n = T_n + V_{nn}$ is the Hamiltonian for the nuclei which contains the kinetic energy

$$T_n = \sum_{j=1}^N \frac{P_j^2}{2M}, \tag{5.21}$$

and the pair-wise nucleus–nucleus interaction term V_{nn}. If we neglect the terms in the Schrödinger equation where the nuclear kinetic energy T_n acts on the electronic wave function $\psi_{[\vec{r}_j]}$ (justified by the much bigger mass that suppresses this term, compared with the electron kinetic energy), we find a separate Schrödinger equation for the nuclear wave function,

$$(T_n + V_{nn} + E_{[\vec{r}_j]})\phi(\vec{r}_1, \ldots, \vec{r}_N) = \epsilon\phi(\vec{r}_1, \ldots, \vec{r}_N). \tag{5.22}$$

This simply means that the effective potential energy seen by the nuclei contains a term which is the total electronic energy for that nuclear configuration plus the nucleus–nucleus Coulomb repulsion:

$$U(\vec{r}_1, \ldots, \vec{r}_N) = V_{nn} + E_{[\vec{r}_j]}. \tag{5.23}$$

This is the potential which enters the *effective Hamiltonian* for the nuclei, Eq. (5.17). This is an excellent approximation in insulators, where the excited electronic states that we are neglecting lie above a large gap. It often is not a bad approximation even in metals, but it is on weaker ground there. The effects can sometimes be profound, leading, for example, to superconducting pairing of the electrons at low temperatures caused by exchange of virtual phonon excitations.

It is important to note that, in Eq. (5.23), the original nucleus–nucleus interaction V_{nn} is repulsive and thus a *positive* term. The electronic energy term $E_{[\vec{r}_j]}$, on the other hand, is negative as it is dominated by the *attractive* interaction between nuclei and electrons. It is the electrons that give rise to the attraction between atoms and hold the crystal together (see Exercise 5.6 for a simple calculation illustrating this point). The same physics is, of course, what binds atoms together to form molecules. In this sense a crystal can be viewed as a huge molecule, in which atoms are arranged periodically.[8]

In purely ionic insulators the effective potential U may be well approximated by a sum of pairwise central interactions since the electrons are in closed shells and are largely inert:

$$U = \sum_{i<j} V(|\vec{r}_i - \vec{r}_j|). \tag{5.24}$$

In general U is much more complex than this and contains three-body and higher terms.[9] In order to avoid hopeless complexities, we are forced to adopt the *harmonic approximation*. We assume that the fluctuations $\vec{u}_j$ are small so that we can expand U in a Taylor series and truncate it at second order:

$$U(\vec{r}_1, \ldots, \vec{r}_N) = U(\vec{R}_1, \ldots, \vec{R}_N) + \sum_{j=1}^{N} \vec{u}_j \cdot \nabla_j U \Big|_{\{\vec{r}=\vec{R}\}}$$
$$+ \frac{1}{2} \sum_{j,k} (\vec{u}_j \cdot \nabla_j)(\vec{u}_k \cdot \nabla_k) U \Big|_{\{\vec{r}=\vec{R}\}} + \cdots. \tag{5.25}$$

The linear term necessarily vanishes because each atom experiences no net force at the equilibrium position. We can ignore $U(\vec{R}_1, \ldots, \vec{R}_N) = U_{eq}$ since it is a constant. Hence, in the harmonic approximation

---

[8] For this reason crystals can be classified in terms of the type of chemical bonding that holds them together, including ionic crystals (example: sodium chloride with ionic bonds between Na^+ and Cl^-), covalent crystals (example: diamond with covalent bonds between neighboring carbon atoms), molecular crystals (bonded by the van der Waals force), hydrogen-bonded crystals, etc. It is worth noting that such a classification is somewhat loose, with no sharp distinction between different classes. One class of examples consists of III–V and II–VI semiconductors (to be discussed in Chapter 9), which have both ionic and covalent character. Metals are often said to be bonded by metallic bonds, which have no counterpart in molecules. Metallic bonding is best understood in terms of band theory, which will be discussed at length in Chapter 7.

[9] These terms arise because the electronic energy depends in a non-trivial way on the nuclear positions. For example (thinking in terms of chemical bonds), there can be bond-bending forces which depend on the position of triples of ions.

$$U - U_{\text{eq}} = \frac{1}{2} \sum_{j,k} \vec{u}_j \cdot \overleftrightarrow{D}_{jk} \cdot \vec{u}_k = \frac{1}{2} \sum_{j,k} u_j^\mu D_{jk}^{\mu\nu} u_k^\nu, \tag{5.26}$$

where the **elastic tensor** is

$$\overleftrightarrow{D}_{jk} = \nabla_j \nabla_k U; \tag{5.27}$$

$$D_{jk}^{\mu\nu} = \partial_{r_j^\mu} \partial_{r_k^\nu} U. \tag{5.28}$$

Notice the important fact that this contains only pair-wise interactions among the atoms even if the full U contains n-body effective interactions. It is fortunate that the harmonic approximation is usually quite accurate, at least for temperatures well below the melting point. This comes about because the large nuclear masses tend to limit the size of the quantum zero-point motion of the atoms. Compounds containing hydrogen bonds can be exceptions to this rule, as is helium, which remains liquid all the way down to absolute zero (at ambient pressure).

The Hamiltonian has thus been simplified to a set of coupled harmonic oscillators:

$$H_{\text{L}} = \sum_j \frac{P_j^2}{2M} + \frac{1}{2} \sum_{j,k} \vec{u}_j \cdot \overleftrightarrow{D}_{jk} \cdot \vec{u}_k. \tag{5.29}$$

For a Bravais lattice the elastic tensor has certain fundamental symmetries.

1. $D_{jk}^{\mu\nu}$ can depend on j and k only through the quantity $(\vec{R}_j - \vec{R}_k)$ and hence has the same discrete translation symmetries as the crystal.

2. $D_{jk}^{\mu\nu} = D_{kj}^{\nu\mu}$ since

$$\frac{\partial^2 U}{\partial u_j^\mu \, \partial u_k^\nu} = \frac{\partial^2 U}{\partial u_k^\nu \, \partial u_j^\mu}. \tag{5.30}$$

3. $D_{jk}^{\mu\nu} = D_{kj}^{\mu\nu}$ since every Bravais lattice has inversion symmetry (which takes $\vec{u} \to -\vec{u}$ and $\vec{R} \to -\vec{R}$).

4. $\sum_k D_{jk}^{\mu\nu} = 0$ since uniform translation of the lattice ($\vec{u}_j = \vec{d} \;\; \forall j$) must give zero energy,

$$0 = \frac{1}{2} \sum_{jk} \vec{d} \cdot \overleftrightarrow{D}_{jk} \cdot \vec{d} = \frac{N}{2} \left(\sum_k D_{jk}^{\mu\nu} \right) d^\mu d^\nu \tag{5.31}$$

independently of $\vec{d}$.

These symmetries will help us solve for the normal modes of the lattice.

Exercise 5.6. This exercise illustrates how electron(s) contribute to the effective potential between atoms/nuclei, and how the adiabatic (also known as the **Born–Oppenheimer**) approximation works. Consider a 1D "molecule" made of two nuclei with masses M_1 and M_2, respectively, and an electron with mass $m \ll M_{1,2}$. The potential-energy term of the total Hamiltonian reads

$$V(X_1, X_2, x) = -\frac{1}{2} K_1 (X_1 - X_2)^2 + \frac{1}{2} K_2 [(x - X_1)^2 + (x - X_2)^2],$$

where $X_{1,2}$ and x represent the coordinates of the nuclei and the electron, respectively. The first and second terms represent repulsion between the nuclei and attraction between the electron and the nuclei.

(i) Use the adiabatic approximation to eliminate the electron degree of freedom and obtain the effective potential $U(X_1, X_2)$ between the nuclei, and find the condition that K_1 and K_2 must satisfy in order for the molecule to be stable (or bound). It should be clear that the molecule is really bound by the presence of the electron. It is easy to generalize the above to the many nuclei and many electrons that form a crystal.

(ii) Assuming the stability condition is indeed satisfied, find the oscillation frequency of this molecule (within the adiabatic approximation).

(iii) This three-body Hamiltonian can actually be solved exactly because it is nothing but coupled harmonic oscillators. Find the corresponding frequency in the exact solution, and show that it is identical to what you found in (ii), for the special case of $M_1 = M_2$. For $M_1 \neq M_2$, show that the exact frequency approaches what you found in (ii) in the limit $m \ll M_{1,2}$. Give a physical explanation for why the adiabatic approximation is exact (in this toy model) when $M_1 = M_2$.

5.3 Classical Dynamics of Lattice Vibrations

We are ultimately interested in a quantum theory of the lattice vibrations. However, we know that, for a harmonic system, the quantum eigenvalues are simply $\hbar$ times the classical oscillation frequencies. It is thus useful to first perform a classical normal mode analysis of H_L defined in Eq. (5.29). The classical equation of motion is

$$M\ddot{\vec{u}}_j = -\nabla_j U = -\overset{\leftrightarrow}{D}_{jk} \cdot \vec{u}_k. \tag{5.32}$$

Let us assume a plane-wave solution of the form

$$\vec{u}_j = e^{-i\omega t} e^{i\vec{q}\cdot\vec{R}_j} \hat{\epsilon}(\vec{q})\tilde{u}(\vec{q}), \tag{5.33}$$

where $\hat{\epsilon}$ is a unit vector denoting the polarization direction of the vibrational mode and $\tilde{u}$ is an (arbitrary) amplitude. Substituting this ansatz into the equation of motion, we obtain

$$-M\omega^2 \hat{\epsilon}(\vec{q}) = -\sum_k \overset{\leftrightarrow}{D}_{jk} \cdot \hat{\epsilon}(\vec{q}) e^{i\vec{q}\cdot(\vec{R}_k - \vec{R}_j)}. \tag{5.34}$$

Let us define the Fourier transform of $\overset{\leftrightarrow}{D}$:

$$\overset{\leftrightarrow}{\mathcal{D}}(\vec{q}) \equiv \sum_k \overset{\leftrightarrow}{D}_{jk} e^{i\vec{q}\cdot(\vec{R}_k - \vec{R}_j)}. \tag{5.35}$$

This is independent of j by translation symmetry.

The equation of motion now reduces to the eigenvalue problem

$$M\omega^2 \hat{\epsilon}(\vec{q}) = \overset{\leftrightarrow}{\mathcal{D}}(\vec{q}) \cdot \hat{\epsilon}(\vec{q}). \tag{5.36}$$

From the symmetry of $\overset{\leftrightarrow}{D}$ we see that its Fourier transform obeys $\overset{\leftrightarrow}{\mathcal{D}}(\vec{0}) = \overset{\leftrightarrow}{0}$, and hence $\omega(\vec{q})$ must vanish in the limit $\vec{q} \to \vec{0}$. This is a nice illustration of *Goldstone's theorem*, which states that such a gapless mode must occur in a translationally invariant Hamiltonian. The essential physics is that uniform translations of all atoms cost exactly zero energy. In a very-long-wavelength oscillation, the displacements are locally uniform and hence the potential energy cost is very low and the restoring force is small. The same physics appeared in the elastic continuum model considered at the beginning of the chapter, where we found it was impossible to write down a translationally invariant energy which did not involve two gradients of the displacement field.

Figure 5.1 Illustration of a chain of identical atoms coupled to each other by springs.

As a simple example where we can compute the elastic tensor, let us consider a 1D chain of atoms coupled harmonically to their near neighbors, as in Fig. 5.1. The potential energy is

$$U = \frac{1}{2}\lambda \sum_n (x_{n+1} - x_n - a)^2, \tag{5.37}$$

where a is the lattice constant. Using $R_{n+1} - R_n = a$, and $x_n = R_n + u_n$, we obtain

$$U = \frac{1}{2}\lambda \sum_n (u_{n+1} - u_n)^2, \tag{5.38}$$

where the u_n are now scalar displacements, and

$$\frac{\partial U}{\partial u_j} = \lambda(2u_j - u_{j-1} - u_{j+1}). \tag{5.39}$$

The elastic "tensor" is now a scalar:

$$D_{jk} = \frac{\partial^2 U}{\partial u_k \, \partial u_j} = \lambda(2\delta_{jk} - \delta_{k,j-1} - \delta_{k,j+1}), \tag{5.40}$$

and its Fourier transform is

$$\mathcal{D}(q) = 2\lambda[1 - \cos(qa)]. \tag{5.41}$$

The eigenvalue equation (which now has no polarization vectors) becomes

$$M\omega^2 = \mathcal{D}(q) \tag{5.42}$$

and gives the solution

$$\omega(q) = \sqrt{\frac{\lambda}{M}} \, [2(1 - \cos(qa))]^{1/2} = 2\sqrt{\frac{\lambda}{M}} \left| \sin\left(\frac{qa}{2}\right) \right|, \tag{5.43}$$

which vanishes linearly for small q (see Fig. 5.2) with a slope $\partial\omega/\partial q = \sqrt{\lambda a^2/M}$ which determines the **speed of sound**. The linear mode dispersion predicted by continuum elastic medium theory is violated when the wavelength becomes comparable to the lattice constant a, reflecting the importance of lattice effects at this scale. The mode frequency appears to be periodic in q but the solutions outside the Brillouin zone of the reciprocal lattice turn out to be redundant. To see this note that in d dimensions the spatial phase factor in

$$\vec{u}_j = e^{-i\omega t} e^{i\vec{q}\cdot\vec{R}_j} \hat{\epsilon}(\vec{q})\tilde{u}(\vec{q}) \tag{5.44}$$

is invariant under $\vec{q} \to \vec{q} + \vec{G}$. There are no solutions for $\vec{q}$ outside the Brillouin zone which are linearly independent of solutions inside. *Hence the collective modes can be labeled by wave vectors $\vec{q}$ inside the first Brillouin zone only.*

How many independent vibration modes are there? In our 1D example the Brillouin zone is the interval $(-\pi/a, +\pi/a]$ and

$$N_{\text{modes}} = \int_{-\pi/a}^{\pi/a} dq \left(\frac{L}{2\pi}\right) = \frac{L}{a} = N_{\text{atoms}}, \tag{5.45}$$

so that we have one mode for each atom in the lattice. There is one degree of freedom (coordinate) per atom and so each vibrational mode is equivalent to one degree of freedom, as it should be.

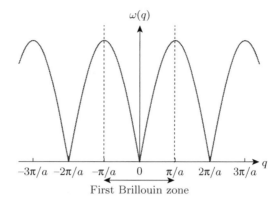

Figure 5.2 The phonon dispersion curve of a 1D Bravais crystal.

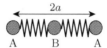

Figure 5.3 A unit cell with size $2a$ containing two atoms.

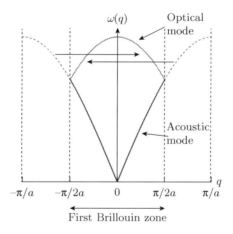

Figure 5.4 Illustration of Brillouin-zone folding due to doubling of the unit cell in real space. Two segments of the mode in the original (extended) first Brillouin zone are shifted (or "folded back") into the reduced first Brillouin zone, to form the optical branch.

What would happen if we arbitrarily pretended that our uniform spring system had a unit cell which was twice as long ($2a$ instead of a)? Then the unit cell would contain a basis consisting of not one but two atoms, as shown in Fig. 5.3. There would thus be two degrees of freedom per unit cell. On the other hand, the Brillouin zone would now be only *half as big*. Instead of a diagram looking like Fig. 5.2, we would have two branches of vibration modes, as in Fig. 5.4. Of course, the physical modes have not changed at all. The dispersion curve we had before has simply been "folded back" (actually shifted by $G = 2\pi/2a = \pi/a$) into a smaller zone. The lower branch is called acoustic because the two atoms in the unit cell tend to move in phase. The upper branch is called optical because the two atoms in the unit cell move nearly out of phase. If, instead of being identical as in this example, the two atoms had opposite charge, then the optical mode would produce a time-dependent electrical polarization of the crystal which would couple to light waves (hence the name optical mode).

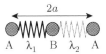

Figure 5.5 Atoms in a doubled unit cell oscillating out of phase in the long-wavelength optical mode.

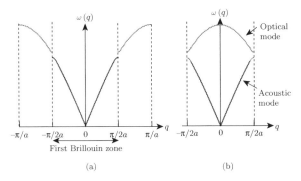

Figure 5.6 The same as in Fig. 5.1, but with different spring constants on even and odd links; thus the unit cell must have two atoms.

Figure 5.7 Vibration mode spectra of a 1D chain with a two-point basis, in the extended (a) and reduced (b) zone schemes. The lower branch is the acoustic mode and the upper branch is the optical mode. Note the gap at the zone boundary.

To see that the optical branch has the basis atoms moving out of phase, we simply note that the optical mode near $q = 0$ in the "**reduced zone scheme**" corresponds in the "**extended zone scheme**" to a mode near $q = \pi/a$. The eigenvector is

$$u_j = e^{iqR_j} = e^{i\frac{\pi}{a}R_j}. \tag{5.46}$$

For atoms inside the unit cell $R_{j+1} - R_j = a$, so $u_{j+1} = -u_j$. For the reduced zone this mode is at $q = 0 = \pi/a - G$ because the first atoms in all the unit cells are in phase (since $\pi/a \times 2a = 2\pi$). See Fig. 5.5.

Of course, near $q = \pi/2a$ the acoustic and optical modes become indistinguishable and hence degenerate. The total number of acoustic modes is

$$N_{\text{acoustic}} = N_{\text{cell}} = \frac{L}{2\pi} \int_{-\pi/2a}^{\pi/2a} dq = \frac{L}{2a} = \frac{N_{\text{atoms}}}{2}, \tag{5.47}$$

but there is an equal number of optical modes, so the total is still N_{atoms} as it was before.

Suppose now we really did have a doubled unit cell because the spring constants alternate, see Fig. 5.6. Then it turns out that the degeneracy of the modes at $q = \pi/2a$ is lifted (due to the reduced symmetry of the Hamiltonian) and the spectrum looks like Fig. 5.7(b) in the reduced zone scheme or like Fig. 5.7(a) in the extended zone scheme.

Let us now derive this result by assuming an eigenvector of the plane-wave form:

$$u_j = \epsilon_u e^{iqR_j} e^{-i\omega t}, \tag{5.48}$$

$$v_j = \epsilon_v e^{iqR_j} e^{-i\omega t}, \tag{5.49}$$

where u_j represents the displacement of the first atom in the jth unit cell and v_j represents the displacement of the second atom in the jth unit cell. The coefficients ϵ_u and ϵ_v represent the amplitude and phase of the motion of the two basic atoms. Also note that, since the unit cell has doubled, $R_j = 2aj$ is now the location of the jth unit cell, not the jth atom. The potential energy is now

$$U = \frac{1}{2} \sum_j \{\lambda_2 (u_j - v_j)^2 + \lambda_1 (u_j - v_{j-1})^2\}, \tag{5.50}$$

so the equations of motion are (assuming equal masses for the two atoms):

$$M\ddot{u}_j = -\frac{\partial U}{\partial u_j} = -\lambda_2 (u_j - v_j) - \lambda_1 (u_j - v_{j-1}); \tag{5.51}$$

$$M\ddot{v}_j = -\frac{\partial U}{\partial v_j} = -\lambda_2 (v_j - u_j) - \lambda_1 (v_j - u_{j+1}). \tag{5.52}$$

Substituting the plane-wave solution gives the eigenmode equations:

$$\left[-M\omega^2 + (\lambda_1 + \lambda_2)\right]\epsilon_u - [\lambda_2 + \lambda_1 e^{-i2qa}]\epsilon_v = 0; \tag{5.53}$$

$$\left[-M\omega^2 + (\lambda_1 + \lambda_2)\right]\epsilon_v - [\lambda_2 + \lambda_1 e^{+i2qa}]\epsilon_u = 0. \tag{5.54}$$

These equations will have consistent solutions only for values of ω at which the determinant of the coefficient matrix vanishes:

$$\begin{vmatrix} (-M\omega^2 + \lambda_1 + \lambda_2) & -(\lambda_2 + \lambda_1 e^{-i2qa}) \\ -(\lambda_2 + \lambda_1 e^{+i2qa}) & (-M\omega^2 + \lambda_1 + \lambda_2) \end{vmatrix} = 0; \tag{5.55}$$

$$\left[+M\omega^2 - (\lambda_1 + \lambda_2)\right]^2 = |\lambda_2 + \lambda_1 e^{-i2qa}|^2; \tag{5.56}$$

$$\omega_{\pm}^2 = \frac{1}{M}\{(\lambda_1 + \lambda_2) \pm |\lambda_2 + \lambda_1 e^{-i2qa}|\}. \tag{5.57}$$

The two (positive) ω solutions of this equation correspond to the acoustic and optical branches:

$$\omega_{\pm}^2 = \frac{1}{M}\left\{(\lambda_1 + \lambda_2) \pm \sqrt{(\lambda_2 + \lambda_1 \cos(2qa))^2 + \lambda_1^2 \sin^2(2qa)}\right\} \tag{5.58}$$

$$= \frac{1}{M}\left\{(\lambda_1 + \lambda_2) \pm \sqrt{\lambda_2^2 + \lambda_1^2 + 2\lambda_1\lambda_2 \cos(2qa)}\right\}. \tag{5.59}$$

If $\lambda_1 = \lambda_2 = \lambda$ then

$$\omega_{\pm}^2 = \frac{\lambda}{M}\{2 \pm \sqrt{2 + 2\cos(2qa)}\}. \tag{5.60}$$

At the zone center and zone boundary, we have

$$\omega_-(q = 0) = 0, \tag{5.61}$$

$$\omega_-(q = \pi/2a) = \sqrt{\frac{2\lambda}{M}}, \tag{5.62}$$

$$\omega_+(q = 0) = 2\sqrt{\frac{\lambda}{M}}, \tag{5.63}$$

$$\omega_+(q = \pi/2a) = \sqrt{\frac{2\lambda}{M}}. \tag{5.64}$$

We see that the two modes are degenerate at the zone boundary, which is to be expected when $\lambda_1 = \lambda_2$, because we are simply reproducing the original dispersion curve of Eq. (5.43) in a reduced zone scheme.

If $\lambda_1 \neq \lambda_2$, then the degeneracy is lifted and

$$\omega_\pm^2(q = \pi/2a) = \frac{1}{M}\{(\lambda_1 + \lambda_2) \pm |\lambda_2 - \lambda_1|\}. \tag{5.65}$$

Notice that the group velocity vanishes at the zone boundary, as can be seen from the dispersion curves in Fig. 5.7. If $\lambda \equiv (\lambda_1 + \lambda_2)/2$ and $\delta\lambda \equiv |\lambda_2 - \lambda_1|$ then

$$\omega_\pm(q = \pi/2a) = \sqrt{\frac{2\lambda}{M}}\left\{1 \pm \frac{\delta\lambda}{2\lambda}\right\}^{1/2}, \tag{5.66}$$

so that there is a finite gap, which for $\delta\lambda/\lambda \ll 1$ obeys

$$\Delta\omega \equiv \omega_+(q = \pi/2a) - \omega_-(q = \pi/2a) \approx \sqrt{\frac{2\lambda}{M}}\left(\frac{\delta\lambda}{2\lambda}\right). \tag{5.67}$$

In 3D there are three directions for displacing the atoms and hence three independent polarization modes for each $\vec{q}$ in a Bravais lattice: $\epsilon_\alpha(\vec{q})$, $\alpha = 1, 2, 3$, obtained from solving Eq. (5.36).[10] The total number of modes is thus

$$N_{\text{modes}} = 3 \int_{\vec{q} \in \text{BZ}} d^3q \left(\frac{L}{2\pi}\right)^3 = 3\left(\frac{L}{2\pi}\right)^3 \Omega_{\text{BZ}}, \tag{5.68}$$

with Ω_{BZ} being the volume in reciprocal space of the Brillouin zone:

$$\Omega_{\text{BZ}} = |\vec{G}_1 \cdot (\vec{G}_2 \times \vec{G}_3)| = \frac{(2\pi)^3}{\Omega_{\text{cell}}}, \tag{5.69}$$

where Ω_{cell} is the volume of the real-space primitive unit cell. Hence

$$N_{\text{modes}} = 3\frac{L^3}{\Omega_{\text{cell}}} = 3N_{\text{atoms}}. \tag{5.70}$$

The presence of three branches of gapless (**acoustic phonon**) modes follows from the fact that the crystalline state spontaneously breaks translation symmetry, and obeys the corresponding Goldstone theorem. In 3D there are three generators of translation, all of which transform the crystalline state *non-trivially*. In general we expect d branches of gapless (acoustic phonon) modes in a d-dimensional crystal.[11]

If there is an N_b-atom basis in the crystal then $N_{\text{modes}} = 3N_bN_{\text{cells}} = 3N_{\text{atoms}}$, with $3N_{\text{cells}}$ of them forming the three acoustic branches whose frequency vanishes linearly in the long-wavelength limit, and the remaining modes forming optical branches.

Exercise 5.7. Consider a 2D square lattice made of atoms with mass M and interaction energy

$$U = \frac{1}{2a^2}\sum_{(ij)}\lambda_{(ij)}[(\vec{R}_i - \vec{R}_j) \cdot (\vec{u}_i - \vec{u}_j)]^2,$$

---

[10] Recall that for an isotropic continuous medium there is one longitudinal mode ($\hat{\epsilon} \times \vec{q} = \vec{0}$) and there are two transverse modes ($\hat{\epsilon} \cdot \vec{q} = 0$). However, for lattices, these designations are not precise since (except for special symmetry directions of $\vec{q}$) the polarization of a mode is never purely longitudinal or transverse.

[11] It is interesting to contrast this with superfluid ^{4}He, which does *not* break translation symmetry but supports a single branch of a longitudinal phonon mode, which we understood in Section 5.1 as due to the presence of bulk modulus but no shear modulus in a liquid. We also noted in Section 4.4 that this gapless phonon mode is the Goldstone mode associated with a broken U(1) symmetry. This symmetry has only one generator, resulting in one Goldstone mode, regardless of dimensionality. It is quite common in physics that the same result can be understood from multiple viewpoints. It should also be clear by now that, despite their differences, using the term phonon to describe collective modes both in liquid and in crystalline states is appropriate, as they are atom density waves in both cases.

where a is the lattice constant, (ij) represent pairs of atoms, and $\lambda_{(ij)}$ equals λ_1 and λ_2 for nearest- and next-nearest neighbors, respectively, and is zero otherwise. Find the frequency and polarization vectors of the normal modes along the directions of nearest-neighbor bonds.

Exercise 5.8. Consider a 1D chain of atoms as depicted in Fig. 5.3, made of two types of atoms with mass M_A and M_B for the atoms on A and B sites, respectively. The only interaction is between nearest-neighbor atoms, of the form of Eq. (5.37). Find the lattice wave spectrum.

6 Quantum Theory of Harmonic Crystals

In our study of the classical theory of the harmonic crystal in Chapter 5, we learned how to compute the classical normal-mode frequency $\omega_\lambda(\vec{q})$ for wave vector $\vec{q}$ and polarization λ. The quantum energy for this normal mode is simply $\left(n + \frac{1}{2}\right)\hbar\omega_\lambda(\vec{q})$, where $n = 0, 1, 2, \ldots$ is the excitation level of the oscillator. Since these levels are evenly spaced we can view them as consisting of the zero-point energy of the vacuum, $\frac{1}{2}\hbar\omega_\lambda(\vec{q})$, plus the total energy of n non-interacting "particles," each carrying energy $\hbar\omega_\lambda(\vec{q})$. By analogy with the quanta of the electromagnetic field, photons (whose name comes from the same Greek root for light as in the words photograph and photosynthesis), these quantized sound-wave excitations are known as *phonons* (based on the same Greek root for sound as in the words microphone, megaphone, telephone, and phonograph).[1] If there are $n_{\vec{q}\lambda}$ phonons in plane-wave state $\vec{q}$, they contribute $n_{\vec{q}\lambda}\hbar\omega_\lambda(\vec{q})$ to the energy above the ground state. Because $n_{\vec{q}\lambda}$ can be any non-negative integer, not just zero and one, it must be that phonons (like photons) are bosons. We will see shortly that the operators that create and destroy phonons do indeed obey Bose commutation relations.

Thus a quantum crystal can absorb or emit vibration energy only in discrete lumps (quanta). At high temperatures this discreteness is not very noticeable, since even the highest-frequency phonon modes typically have energies $\hbar\omega$ not much larger than thermal energies at room temperature. At low temperatures, however, the quantized nature of the excitations becomes very important and can dramatically alter various physical properties such as the specific heat and thermal conductivity. As we will learn in this chapter, quantum effects will also strongly modify the dynamical structure factor $S(\vec{q}, \omega)$ at low temperatures.

6.1 Heat Capacity

In classical physics the equipartition theorem tells us that the average energy stored in a degree of freedom which enters the Hamiltonian quadratically is $\frac{1}{2}k_{\mathrm{B}}T$. Thus, for a harmonic oscillator,

$$\bar{\epsilon} = \left\langle\!\!\left\langle \frac{1}{2m}p^2 + \frac{1}{2}m\omega^2 x^2 \right\rangle\!\!\right\rangle = k_{\mathrm{B}}T. \tag{6.1}$$

Note that this is true independently of the mass and the spring constant for the oscillator.

---

[1] As we will see later in this chapter, the analogy between phonons and photons goes very far.

We have learned from our normal-mode analysis that a crystal with N atoms has $3N$ coordinates and $3N$ momenta which are contained in the $3N$ harmonic oscillator normal modes of the system. Thus classically the total thermal vibration energy stored in the crystal is

$$\bar{E} = 3N k_B T, \tag{6.2}$$

and we immediately obtain the law of Dulong and Petit for the heat capacity (specific heat times the volume) at constant volume:

$$C_V = \frac{\partial \bar{E}}{\partial T} = 3N k_B. \tag{6.3}$$

This is a very general result which is independent of all microscopic details and assumes only the validity of the harmonic approximation. It is found to work very well above some characteristic temperature T^*, which is typically 200–500 K in many crystalline materials. Below this temperature, however, it fails dramatically and severely overestimates the actual heat capacity, which is found to decrease rapidly toward zero as $T \to 0$.

Actually, classical thermodynamics already gives us a hint that there is something seriously wrong with the law of Dulong and Petit. Recall that the change in entropy upon going from temperature T_1 to temperature T_2 is given by

$$S(T_2) - S(T_1) = \int_{T_1}^{T_2} \frac{dT}{T} C_V(T). \tag{6.4}$$

If C_V were a constant as given by Dulong and Petit, then the entropy would diverge logarithmically toward minus infinity as $T_1 \to 0$, thus violating the rules $S(T) \geq 0$ and $S(0) = 0$.

The failure of the law of Dulong and Petit at low temperatures was a nagging problem in the pre-quantum era and eventually helped lead to the downfall of classical physics. Quantum physics has a very simple but radical solution to the problem: for temperatures $k_B T \ll \hbar\omega$ the heat capacity of an oscillator becomes exponentially small because of the finite excitation gap in the discrete spectrum of the oscillator. Quantum effects cause high-frequency modes to "freeze out" at low temperatures. The very same mechanism eliminates the "ultraviolet catastrophe" in blackbody radiation. For a given temperature, modes at very high frequencies (i.e. in the "ultraviolet") are not thermally excited.

We can make a very rough estimate of the typical energy quantum. We use the fact that the squares of the phonon frequencies are given by the spring constants of the corresponding harmonic oscillators (remember there is an effective harmonic oscillator for each vibration or phonon mode), also known as force constants of the system in this context. A force constant is the second derivative of energy with respect to distance. Using the Rydberg $e^2/2a_B$ as the typical energy scale, for electronic energies and the **Bohr radius** a_B as the typical length scale, we find e^2/a_B^3 is the typical scale for force constants. It follows that the square of the energy of a typical phonon is $\hbar^2\omega^2 \sim \left(\hbar^2/M a_B^2 \right) \left(e^2/a_B \right)$. Here M is the nuclear mass. If M were the electron mass m, the right-hand side here would just be on the order of a Rydberg squared. It follows that the phonon energy scale is smaller than a Rydberg by a factor, very roughly, of $\sqrt{m/M}$. For a typical atomic mass number of about 50 the reduction factor is between 100 and 1000, so that $\hbar\omega \sim 10$ meV to 100 meV, which is comparable to the thermal energy at 100 K to 1000 K. This very rough argument does actually give a reasonable estimate of the range of typical phonon energies which occur in most condensed matter systems. Systems with strong chemical bonds will tend to have larger force constants and be toward or sometimes exceed the upper end of this range. Systems composed of heavy nuclei will tend to be toward the low end or sometimes below the bottom end of this range. The main point we want to make here is that, in most materials, the typical phonon energies are in the same range as the thermal energy at room temperature. This fact is important in explaining the heat capacity of condensed matter systems.

It was Einstein who first realized the connection between the problem of blackbody radiation and the heat capacity of solids. He made a very simple model, now known as the **Einstein model**, which captures the essential physics. Einstein assumed that the normal modes of the solid all had the same frequency Ω so that the heat capacity would be given by

$$C_V = 3Nc_\Omega(T), \tag{6.5}$$

where $c_\Omega(T)$ is the heat capacity of a quantum oscillator of frequency Ω. This is easily found from the partition function,

$$Z = \sum_{n=0}^{\infty} e^{-\beta\left(n+\frac{1}{2}\right)\hbar\Omega} = \frac{e^{-\beta\frac{1}{2}\hbar\Omega}}{1 - e^{-\beta\hbar\Omega}}, \tag{6.6}$$

which yields, for the average energy $\bar{\epsilon} = -\partial \ln Z/\partial\beta$,

$$\bar{\epsilon} = \left(\bar{n} + \frac{1}{2}\right)\hbar\Omega = \left[n_B(\beta\hbar\Omega) + \frac{1}{2}\right]\hbar\Omega, \tag{6.7}$$

where $\bar{n}$ is the mean excitation number and is given by the Bose–Einstein function

$$n_B(\beta\hbar\Omega) = \frac{1}{e^{\beta\hbar\Omega} - 1}. \tag{6.8}$$

The heat capacity is thus

$$c_\Omega(T) = \frac{\partial\bar{\epsilon}}{\partial T} = k_B(\beta\hbar\Omega)^2 \frac{e^{-\beta\hbar\Omega}}{(1 - e^{-\beta\hbar\Omega})^2}. \tag{6.9}$$

This vanishes exponentially for $k_B T = 1/\beta \ll \hbar\Omega$ and goes to the classical value $c_\Omega(T) \to k_B$ at high temperatures as shown in Fig. 6.1.

The Einstein model is not bad for the optical modes of a crystal since they tend to have frequencies which are approximately constant, independently of the wave vector. Acoustic modes, however, have a frequency which vanishes linearly at small wave vectors. Thus, no matter how low the temperature

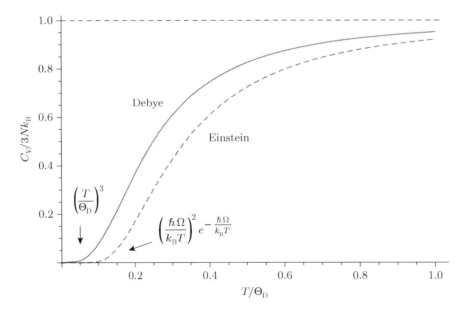

Figure 6.1 The phonon contribution to the heat capacity per atom within the Debye (solid line) and Einstein (dotted line) approximations. Θ_D is the **Debye temperature**, and the oscillator frequency of the Einstein model is chosen to be $\Omega = k_B\Theta_D/\hbar$.

is, some of the modes are still contributing classically to the heat capacity. **Peter Debye** made an improved model based on an elastic continuum picture in which the dispersion is exactly linear:

$$\omega = v_s q. \tag{6.10}$$

As a first approximation he ignored the optical modes and assumed that the speed of sound, v_s, was the same for all three acoustic modes (one longitudinal and two transverse). In order to get the correct number of degrees of freedom ($3N$ coordinates), he introduced an ultraviolet cutoff k_D on q so that

$$3\frac{4\pi}{3}k_D^3\left(\frac{L}{2\pi}\right)^3 = 3N. \tag{6.11}$$

That is, the first Brillouin zone was approximated by a sphere having the same volume. This determines the radius of the sphere k_D.

The maximum phonon frequency in the Debye model is $\omega_D = v_s k_D$. Hence there is a characteristic "Debye temperature" (typically on the scale of room temperature)

$$\Theta_D \equiv \frac{\hbar v_s k_D}{k_B} \tag{6.12}$$

above which the classical results become accurate. As the temperature is lowered below Θ_D, more and more of the modes are frozen out and the heat capacity decreases. This decrease is, however, not exponentially fast as in the Einstein model.

As a first crude approximation to analyze this situation, let us assume $T \ll \Theta_D$ and take modes with $\hbar\omega < k_B T$ to be purely classical and modes with $\hbar\omega > k_B T$ to be completely quantum and hence frozen out. Then the effective number of classical modes relative to the total is

$$f \equiv \frac{(4\pi/3)[\tilde{q}(T)]^3}{(4\pi/3)k_D^3}, \tag{6.13}$$

where $\tilde{q}$ is defined by $\hbar\tilde{q}v_s = k_B T$. Thus the fraction of active modes is

$$f = \left(\frac{T}{\Theta_D}\right)^3 \tag{6.14}$$

and the low-temperature heat capacity is given by

$$C_V(T) \sim 3Nk_B\left(\frac{T}{\Theta_D}\right)^3. \tag{6.15}$$

We cannot expect the prefactor to be accurate at this level of approximation, but the exponent of three in this power-law form agrees very nicely with experiment for a wide variety of materials.[2]

It is not much more difficult to solve the Debye model exactly. The specific heat contributed by the three acoustic phonon modes is simply

$$C_V = 3\left(\frac{L}{2\pi}\right)^3 \int' d^3\vec{q}\; c_{\Omega(q)}(T), \tag{6.16}$$

where the prime on the integral indicates that it is cut off at $q = k_D$, and c_Ω is the quantum oscillator heat capacity evaluated at $\Omega = v_s q$. Our previous crude approximation was simply a step function $c_{\Omega(q)} = k_B \theta(k_B T/\hbar v_s - q)$. The exact expression is

$$C_V = 3\left(\frac{L}{2\pi}\right)^3 \int_0^{k_D} 4\pi q^2\, dq\; k_B(\beta\hbar v_s q)^2 \frac{e^{-\beta\hbar v_s q}}{\left(1 - e^{-\beta\hbar v_s q}\right)^2}. \tag{6.17}$$

---

[2] In metals the electrons make an additional contribution that is linear in T. We will learn more about this later.

Changing variables to

$$x \equiv \beta \hbar v_s q \tag{6.18}$$

and using $\beta \hbar v_s k_D = \Theta_D / T$ yields

$$C_V = 3 N k_B \left(\frac{T}{\Theta_D} \right)^3 Y\left(\frac{\Theta_D}{T} \right), \tag{6.19}$$

where

$$
\begin{aligned}
Y\left(\frac{\Theta_D}{T} \right) &\equiv 3 \int_0^{\Theta_D/T} dx\, x^4 \frac{e^{-x}}{(1-e^{-x})^2} \\
&= -3 \int_0^{\Theta_D/T} dx\, x^4 \frac{d}{dx} \frac{1}{1-e^{-x}} \\
&= -3 \left\{ \frac{(\Theta_D/T)^4}{1-e^{-\Theta_D/T}} - 4 \int_0^{\Theta_D/T} dx\, x^3 \frac{1}{1-e^{-x}} \right\} \\
&= -3 \left\{ \left(\frac{\Theta_D}{T} \right)^4 \left(\frac{1}{1-e^{-\Theta_D/T}} - 1 \right) \right. \\
&\qquad \left. - 4 \int_0^{\Theta_D/T} dx\, x^3 \left(\frac{1}{1-e^{-x}} - 1 \right) \right\}.
\end{aligned}
\tag{6.20}
$$

The resultant full temperature dependence of C_V is plotted in Fig. 6.1 and contrasted with that of the Einstein model. At asymptotically low temperatures the first term vanishes exponentially and the second can be approximated by

$$
\begin{aligned}
Y(\infty) &= 12 \int_0^\infty dx\, x^3 \sum_{n=1}^\infty e^{-nx} \\
&= 72 \sum_{n=1}^\infty \frac{1}{n^4} = \frac{4}{5} \pi^4 \sim 77.93.
\end{aligned}
\tag{6.21}
$$

Our previous crude approximation was equivalent to replacing $Y(\infty)$ by unity. Hence the prefactor was not well approximated in our previous simple estimate. The qualitative feature that the heat capacity vanishes at T^3 was correctly captured, as can be seen from the exact expression in Eq. (6.19).

In a system where the speed of sound depends on polarization of the mode or is anisotropic, care is required if we wish to choose Θ_D to reproduce the exact low-temperature heat capacity. Since $\Theta_D = \hbar v_s k_D / k_B$ in the isotropic model and since the heat capacity is proportional to Θ_D^{-3}, the appropriate generalization to the anisotropic case is to average the inverse third power of the speed:

$$\Theta_D = \frac{\hbar k_D}{k_B} \langle v_s^{-3} \rangle^{-1/3}, \tag{6.22}$$

where

$$\langle v_s^{-3} \rangle \equiv \frac{1}{3} \sum_{\lambda=1}^3 \frac{1}{4\pi} \int d^2\hat{q} \left[\frac{1}{v_{s\lambda}(\hat{q})} \right]^3. \tag{6.23}$$

Here $\hat{q}$ is a vector on the unit sphere indicating the propagation direction and $\lambda = 1, 2, 3$ refers to the three polarization modes of acoustic phonons.

Note that, if there are optical phonons present, they will make only an exponentially small contribution at low temperatures and hence can be neglected in this analysis.

Finally, notice that thermally excited acoustic phonons behave very much like thermally excited blackbody radiation. The speed of light is simply replaced by the speed of sound. Recalling that the

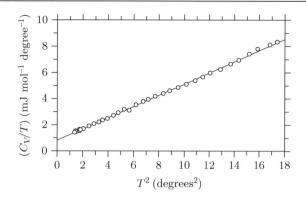

Figure 6.2 The low-temperature specific heat of gold, plotted in a way (see the text) that allows easy separation of the phonon ($\propto T^3$) and electron ($\propto T$) contributions. Figure adapted with permission from [30]. Copyright by the American Physical Society.

energy density of blackbody radiation scales as T^4, we see immediately that blackbody radiation has a T^3 heat capacity just as the phonons do.[3]

In insulators, the heat capacity is dominated by phonons. In metals there is an additional small contribution due to low-energy electronic excitations which vanishes only linearly with T. This is negligible except at low temperatures $T \ll \Theta_D$, where it eventually dominates. The electronic part can be isolated by plotting C_V/T vs. T^2 for $T \ll \Theta_D$. The zero-temperature limit of C_V/T gives the electronic contribution while the slope gives a measure of the phonon contribution and hence Θ_D. See Fig. 6.2 for an example.

Exercise 6.1. Find the density of phonon states per unit volume per unit energy for the Debye model in d dimensions. Show that it has the form $\rho(\epsilon) \sim \epsilon^{d-1} \theta(k_B \Theta_D - \epsilon)$. Determine the corresponding T dependence of C_V at low T for general d. It is interesting to note that in 1D the phonon and electron contributions to C_V have the same T dependence, despite their difference in statistics. This indicates that one can establish a one-to-one mapping of the low-energy states between bosonic and fermionic systems in 1D, which is the basis of **bosonization** or **fermionization**. Further discussion of this can be found in the description of edge states in the fractional quantum Hall effect presented in Section 16.9.

6.2 Canonical Quantization of Lattice Vibrations

To get a better idea of what a phonon is, consider a set of decoupled or independent oscillators in 1D,[4] as shown in Fig. 6.3, with Hamiltonian

$$H = \sum_i \frac{P_i^2}{2M} + \frac{1}{2}\lambda u_i^2, \tag{6.24}$$

where u_i is the coordinate of the ith oscillator located at R_i. Each oscillator has the *same* frequency $\Omega = \sqrt{\lambda/M}$. This corresponds precisely to the Einstein model for lattice vibrations.

---

[3] The main difference is that phonons have a longitudinal mode which photons do not have. More on this point later.

[4] We choose 1D mainly to avoid notational cluttering related to polarization, which is not essential for the main ideas to be discussed.

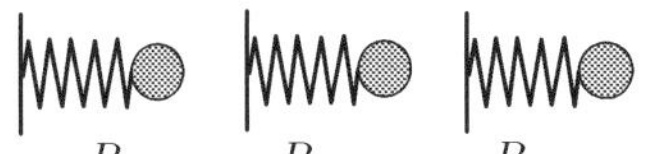

Figure 6.3 A set of decoupled harmonic oscillators.

Suppose that all the oscillators are in their ground states. We call this the vacuum or zero-phonon state, and denote it $|0\rangle$. Now suppose we raise the jth oscillator into its first excited state using the appropriate raising operator

$$|1_j\rangle \equiv a_j^\dagger |0\rangle, \tag{6.25}$$

where

$$a_j^\dagger = \sqrt{\frac{M\Omega}{2\hbar}} u_j - i \sqrt{\frac{1}{2M\hbar\Omega}} P_j \tag{6.26}$$

is the phonon creation operator for site j. Clearly we have

$$[a_i, a_j] = 0; \qquad [a_i, a_j^\dagger] = \delta_{ij}. \tag{6.27}$$

Thus the as satisfy *bosonic* commutation relations. We say there is one phonon "particle" in the system and it is located at $\vec{R}_j$. Thus the oscillator raising operator at $a_j^\dagger$ is now viewed as a particle (phonon) creation operator. For this simple Hamiltonian $|1_j\rangle$ happens to be an exact eigenstate. The phonon cannot travel anywhere because the oscillators are uncoupled. Obviously multiple phonon states (with phonons on the same or different sites) can be constructed straightforwardly.

While it is unnecessary for such decoupled oscillators, we can advance a complete description of these phonon modes in *momentum* space, by introducing the corresponding phonon creation operators through Fourier transformation:[5]

$$a^\dagger(k) = \frac{1}{\sqrt{N}} \sum_{j=1}^{N} e^{ikR_j} a_j^\dagger \tag{6.28}$$

$$= \frac{1}{\sqrt{N}} \sum_{j=1}^{N} e^{ikR_j} \left(\sqrt{\frac{M\Omega}{2\hbar}} u_j - i \sqrt{\frac{1}{2M\hbar\Omega}} P_j \right), \tag{6.29}$$

which satisfies commutation relations similar to Eqs. (6.27) (the reader is urged to verify this):

$$[a(k), a(k')] = 0; \qquad [a(k), a^\dagger(k')] = \delta_{k,k'}. \tag{6.30}$$

The oscillator raising temperature $a^\dagger(k)$ creates a phonon at wave vector k, whose *crystal momentum* is $\hbar k$. The inverse Fourier transformation is simply

$$a_j^\dagger = \frac{1}{\sqrt{N}} \sum_{k \in 1BZ} e^{-ikR_j} a^\dagger(k). \tag{6.31}$$

Notice that the wave-vector sum is restricted to the first Brillouin zone (1BZ), as is appropriate for crystal momentum. Henceforth this will be implicit throughout. The Hamiltonian (6.24) is diagonal in terms of both a_j and $a(k)$:

$$H = \sum_j \hbar\Omega[a_j^\dagger a_j + 1/2] = \sum_k \hbar\Omega[a^\dagger(k)a(k) + 1/2]. \tag{6.32}$$

---

[5] The sum does *not* create multiple phonons! It simply tells us that there is an amplitude of finding the particle at different locations.

We know classically that, if we excite one atom in a crystal, the vibrational energy (wave packet) will spread throughout the system due to coupling among the oscillators. The true normal modes are the plane-wave-like excitations we studied in the previous chapter. The spreading of the wave packets is controlled by the dispersion $\omega(k)$, which is *not* independent of k. The group velocity is $\partial \omega / \partial k$, which quantum mechanically becomes the speed of the phonon particles. Thus in the generic situation it is impossible to construct *localized* phonon eigenmodes as in Eqs. (6.25) and (6.26).[6] As a result, phonon eigenmodes and their corresponding operators exist in momentum space only, in terms of which we expect the lattice Hamiltonian (within the harmonic approximation) to take the form

$$\mathcal{H} = \sum_k \hbar \omega(k)[a^\dagger(k)a(k) + 1/2]. \tag{6.33}$$

Amazingly, the presence of dispersion alters Eq. (6.29) only in a minimal way, namely replacing the constant mode frequency Ω with the dispersion $\omega(k)$ (which does not alter the commutation relations in Eqs. (6.30); again the reader should verify this):

$$a^\dagger(k) = \frac{1}{\sqrt{N}} \sum_j e^{ikR_j} \left[\sqrt{\frac{M\omega(k)}{2\hbar}} u_j - i \sqrt{\frac{1}{2M\hbar\omega(k)}} P_j \right]. \tag{6.34}$$

The easiest way to convince oneself that this is the case is to substitute (6.34) into (6.33) to find

$$\mathcal{H} = \sum_j \frac{P_j^2}{2M} + \frac{1}{2} \sum_{jl} D_{jl} u_j u_l, \tag{6.35}$$

where one needs to use the 1D version of Eq. (5.36) in which the polarization vector reduces to a pure number (and cancels out on the two sides).

For readers who feel the justification of Eq. (6.34) above is too slick, we offer a more systematic derivation below. First we take advantage of the lattice translation symmetry of the problem by going to momentum space, and introduce the (lattice) Fourier transform of u and that of P,

$$u(k) = \frac{1}{\sqrt{N}} \sum_j e^{ikR_j} u_j = u^\dagger(-k), \tag{6.36}$$

$$P(k) = \frac{1}{\sqrt{N}} \sum_j e^{ikR_j} P_j = P^\dagger(-k), \tag{6.37}$$

which satisfy the (non-trivial) commutation relations

$$[u(k), P(k')] = i\hbar \delta_{k,-k'}. \tag{6.38}$$

We now express the Hamiltonian (6.35) in terms of $u(k)$ and $P(k)$:

$$\mathcal{H} = \sum_k \left[\frac{1}{2M} P(k)P(-k) + \frac{1}{2} D(k)u(k)u(-k) \right] \tag{6.39}$$

$$= \sum_k \left[\frac{1}{2M} P(k)P(-k) + \frac{1}{2} M\omega^2(k)u(k)u(-k) \right]. \tag{6.40}$$

We see the above looks (almost) like a set of decoupled harmonic oscillators (one for each k), except modes with k and $-k$ are still coupled. Such decoupling is a powerful consequence of translation symmetry, which we will take advantage of again and again in this book. In principle, we can further decouple the k and $-k$ modes by making symmetric and anti-symmetric combinations of $u(k)$ and

---

[6] It is possible for the decoupled oscillator case because the phonon mode has *no* dispersion: $\omega(k) = \Omega$.

$u(-k)$ etc. However, the problem is now sufficiently simple that we are ready to write down the appropriate creation operator by making an analogy to the case of a single harmonic oscillator:

$$a^{\dagger}(k) = \sqrt{\frac{M\omega(k)}{2\hbar}}u(k) - i\sqrt{\frac{1}{2M\hbar\omega(k)}}P(k), \tag{6.41}$$

which is equivalent to Eq. (6.34).

The generalization of this to the plane-wave phonon creation/annihilation operators in a 3D Bravais lattice crystal with atoms of mass M is rendered only slightly more complicated by the presence of polarization vectors (which is *necessary* due to the *vector* nature of $\vec{u}$ and $\vec{P}$):

$$a_{\alpha}^{\dagger}(\vec{k}) \equiv \frac{1}{\sqrt{N}} \sum_{j} e^{+i\vec{k}\cdot\vec{R}_{j}} \hat{\epsilon}_{\alpha}(\vec{k})$$

$$\cdot \left[\left(\frac{M\omega_{\alpha}(\vec{k})}{2\hbar} \right)^{1/2} \vec{u}_{j} - i \left(\frac{1}{2M\hbar\omega_{\alpha}(\vec{k})} \right)^{1/2} \vec{P}_{j} \right], \tag{6.42}$$

where $\alpha = 1, 2, 3$ labels the three polarization modes corresponding to the three degrees of freedom per atom, and $\hat{\epsilon}_{\alpha}$ is the corresponding polarization vector obtained from solving Eq. (5.36).

We can derive the commutation relations for the phonon operators by starting from

$$[u_{j}^{\mu}, P_{\ell}^{\nu}] = i\hbar\delta_{j\ell}\delta_{\mu\nu}, \tag{6.43}$$

which, once combined with the orthonormal nature of $\hat{\epsilon}_{\alpha}$, yields (independently of the form of $\omega_{\alpha}(\vec{k})$)

$$\left[a_{\alpha}(\vec{k}), a_{\beta}(\vec{q}) \right] = 0, \qquad \left[a_{\alpha}(\vec{k}), a_{\beta}^{\dagger}(\vec{q}) \right] = \delta_{\alpha\beta}\delta_{\vec{k},\vec{q}}. \tag{6.44}$$

Thus the phonon creation and destruction operators obey the canonical commutation relations for those of bosons (i.e. each normal mode corresponds to a state that bosons can occupy).

It is straightforward to invert Eq. (6.42) to express the coordinate $\vec{u}_{j}$ in terms of the phonon operators:

$$\vec{u}_{j} = \frac{1}{\sqrt{N}} \sum_{\vec{k},\alpha} \left(\frac{\hbar}{2M\omega_{\alpha}(\vec{k})} \right)^{1/2} \hat{\epsilon}_{\alpha}(\vec{k}) e^{i\vec{k}\cdot\vec{R}_{j}} \left[a_{\alpha}(\vec{k}) + a_{\alpha}^{\dagger}(-\vec{k}) \right]. \tag{6.45}$$

Note carefully the momentum arguments of the phonon operators. To decrease the momentum of the phonon system by $\hbar\vec{k}$, one can either *destroy* a phonon of wave vector $\vec{k}$ or *create* a phonon of wave vector $-\vec{k}$. You will find it instructive to work through the algebra to see how this combination arises. The algebra is simplified by noting that the polarization vectors form a resolution of the identity:

$$\sum_{\alpha} \hat{\epsilon}_{\alpha}(\vec{k}) \left(\hat{\epsilon}_{\alpha}(\vec{k}) \cdot \vec{u}_{\ell} \right) = \vec{u}_{\ell}. \tag{6.46}$$

It is instructive to note the similarity between Eqs. (6.45) and (2.24). This is not an accident, but is instead rooted in the fact that $\vec{A}$ and $\vec{u}$ are both *vector* operators (or fields). In fact, due to their striking similarity, photons and optical phonons can convert into each other when propagating in a crystal. In such cases the collective normal modes are coherent superpositions of photons and optical phonons known as **polaritons**[7] or **phonon-polaritons** to distinguish them from the **exciton-polariton** collective modes which will be discussed in Chapter 9.

---

[7] As noted before, the biggest difference between photons and phonons lies in the fact that photons cannot have longitudinal polarization, as a result of which there are only two independent photon modes for each $\vec{k}$, while there are three for phonons. This is because the longitudinal component of $\vec{A}$ is a pure gauge and thus does not correspond to a physical degree of freedom. Look for more discussions on this point in the superconductivity chapters later in the book.

The momentum of the jth particle is similarly expressed as[8]

$$\vec{P}_j = \frac{-i}{\sqrt{N}} \sum_{\vec{k},\alpha} \left(\frac{\hbar M \omega_\alpha(\vec{k})}{2} \right)^{1/2} \hat{\epsilon}_\alpha(\vec{k}) e^{i\vec{k}\cdot\vec{R}_j} \left[a_\alpha(\vec{k}) - a_\alpha^\dagger(-\vec{k}) \right]. \tag{6.47}$$

Recall that (in the harmonic approximation) the Hamiltonian is

$$\mathcal{H} = \sum_j \frac{\vec{P}_j^2}{2M} + \frac{1}{2} \sum_{jk} \vec{u}_j \cdot \overset{\leftrightarrow}{D}_{jk} \cdot \vec{u}_k. \tag{6.48}$$

Using $\hat{\epsilon}_\alpha(\vec{k}) = \hat{\epsilon}_\alpha(-\vec{k})$, the kinetic energy may be written as

$$T = -\frac{1}{4} \sum_{\vec{k},\alpha} \hbar\omega_\alpha(\vec{k}) \left[a_\alpha(\vec{k}) - a_\alpha^\dagger(-\vec{k}) \right]\left[a_\alpha(-\vec{k}) - a_\alpha^\dagger(\vec{k}) \right]. \tag{6.49}$$

Likewise the potential energy can be expressed in terms of the Fourier transform of the dynamical matrix using

$$M\omega_\alpha^2(\vec{k}) = \hat{\epsilon}_\alpha(\vec{k}) \cdot \overset{\leftrightarrow}{D}(\vec{k}) \cdot \hat{\epsilon}_\alpha(\vec{k}), \tag{6.50}$$

$$V = \frac{1}{4} \sum_{\vec{k},\alpha} \hbar\omega_\alpha(\vec{k}) \left[a_\alpha(\vec{k}) + a_\alpha^\dagger(-\vec{k}) \right]\left[a_\alpha(-\vec{k}) + a_\alpha^\dagger(\vec{k}) \right]. \tag{6.51}$$

Separately these are complicated, but together they form a very simple Hamiltonian,

$$\mathcal{H} = \sum_{\vec{k},\alpha} \hbar\omega_\alpha(\vec{k}) \left\{ a_\alpha^\dagger(\vec{k}) a_\alpha(\vec{k}) + \frac{1}{2} \right\}. \tag{6.52}$$

Note that the **zero-point energy** makes a contribution of

$$\Delta\epsilon = \frac{1}{N} \sum_{\vec{k},\alpha} \frac{1}{2} \hbar\omega_\alpha(\vec{k}) \tag{6.53}$$

to the binding energy per particle of the crystal. The total zero-point energy scales linearly with N as one would expect.

We end this section with the following comments. (i) While a crystal is made of atoms with *strong* and complicated interactions (see Eq. (5.17)), its elementary excitations are *weakly* interacting phonons.[9] While it is very difficult to solve the actual ground state of Eq. (5.17), to understand the low-energy/temperature properties of the crystal we need only work with an *effective* Hamiltonian of the form of Eq. (5.29), which is much simpler than Eq. (5.17) and (at this level) non-interacting.[10] (ii) While details of phonon excitations (number of branches, dispersion of each branch, shape of the 1BZ etc.) depend on specifics of the (already-simplified) effective Hamiltonians, they *all* include

[8] Up to a proportionality constant, the conjugate momentum of vector potential $\vec{A}$ is electric field $\vec{E}$, which is proportional to the time derivative of $\vec{A}$ in the gauge where the scalar potential is zero, and one has a similar expression for it in the quantum theory for radiation or photons.

[9] Phonons are non-interacting within the harmonic approximation we have been using. Higher-order terms in the expansion of $U(\vec{r}_1, \ldots, \vec{r}_N)$ in Eq. (5.17) give rise to interaction among phonons, but such interactions get weaker and weaker at lower and lower energy/temperature, as anharmonicity nearly vanishes in this limit. This phenomenon is quite common in condensed matter physics. Another example we will encounter is a Fermi liquid in which elementary excitations of a strongly interacting fermion system are weakly interacting quasiparticles. See Chapter 15.

[10] In fact Eq. (5.17) is itself an effective Hamiltonian, in which the electronic degrees of freedom are frozen into their instantaneous ground state (or "integrated out" in field-theory jargon). In physics we often deal with a hierarchy of effective Hamiltonians, and choose the appropriate level depending on the energy scale of interest.

d branches of acoustic phonons with linear dispersions at long wavelength (and hence low energy), differing *only* in the speeds of phonons from one material to another. This is an example of *universality* that emerges in the low-energy limit, which allows the use of even simpler effective models (applicable at even lower energy/temperature, namely the Debye model in the present case) that capture the *universal* low-energy/temperature properties of *all* systems in the same universality class (e.g. T^3 behavior of the heat capacity of *all* 3D crystals). It is such universality that allows us to understand the most robust physical properties of many condensed matter systems without having to study them one-by-one. The classification of condensed matter systems, discussed generally (and somewhat abstractly) in Chapter 1, is equivalent to identifying these universality classes.

This is a good time for the reader to revisit Section 2.3, and compare the quantum theories of photons and phonons.

> **Exercise 6.2.** Verify the commutation relations of Eq. (6.30).

> **Exercise 6.3.** Verify the commutation relations of Eq. (6.38).

> **Exercise 6.4.** Verify the commutation relations of Eq. (6.44).

6.3 Quantum Dynamical Structure Factor

Let us return now to the calculation of the dynamical structure factor which we considered previously in the quantum case only for a single oscillator. We have

$$\Pi(\vec{q}, t) = \frac{1}{N} \sum_{j,k} e^{i\vec{q} \cdot (\vec{R}_j - \vec{R}_k)} \langle\!\langle e^{i\vec{q} \cdot \vec{u}_j(t)} e^{-i\vec{q} \cdot \vec{u}_k(0)} \rangle\!\rangle, \tag{6.54}$$

where $\vec{u}_j(t)$ is the jth displacement in the Heisenberg representation. As we noted earlier, the cumulant resummation is exact, even in the quantum case. Hence we have

$$\Pi(\vec{q}, t) = \frac{1}{N} \sum_{j,k} e^{i\vec{q} \cdot (\vec{R}_j - \vec{R}_k)} e^{-2\Gamma(\vec{q})} e^{f_{jk}(t)}, \tag{6.55}$$

where as usual the Debye–Waller factor $e^{-2\Gamma(\vec{q})}$ contains

$$\Gamma(\vec{q}) \equiv \frac{1}{2} \langle\!\langle (\vec{q} \cdot \vec{u}_j)^2 \rangle\!\rangle, \tag{6.56}$$

and

$$f_{jk}(t) \equiv \langle\!\langle [\vec{q} \cdot \vec{u}_j(t)][\vec{q} \cdot \vec{u}_k(0)] \rangle\!\rangle. \tag{6.57}$$

This is a simple generalization of our previous result for a single oscillator. We note that one of the contributions to Π is from correlations of $\vec{u}_k$ at time zero with $\vec{u}_j$ at time t. Such correlations occur because a disturbance at site k can propagate to site j in a finite time t. We will see shortly that the easiest way to compute this is to simply express $\vec{u}_j$ and $\vec{u}_k$ in terms of the normal-mode creation and annihilation operators.

Recall that when we carried out this procedure for the single oscillator, we found that at zero temperature

$$f(t) = q^2 X_{\text{ZPF}}^2 e^{-i\Omega t}. \tag{6.58}$$

Hence (at $T = 0$)

$$e^{f(t)} = \sum_{m=0}^{\infty} \frac{\left(q^2 X_{\mathrm{ZPF}}^2\right)^m}{m!} e^{-im\Omega t} \tag{6.59}$$

contains only *positive*-frequency components corresponding to the creation of m quanta. The same thing occurs in the present case except that a variety of different frequencies $\omega_\alpha(\vec{k})$ will occur. We will therefore refer to the mth term in the expansion of $e^{f(t)}$ as the m-phonon contribution to the dynamical structure factor. Of course, at finite temperature some of the m phonons may be *absorbed* rather than emitted, thereby lowering the energy loss (possibly even making it negative). However, we will still refer to the mth term as the m-phonon term.

The dynamical structure factor is given by

$$S(\vec{q}, \omega) = \int_{-\infty}^{\infty} dt\, e^{i\omega t}\, \Pi(\vec{q}, t). \tag{6.60}$$

The expansion in terms of the number of phonons involved yields

$$S(\vec{q}, \omega) = \sum_{m=0}^{\infty} S_m(\vec{q}, \omega), \tag{6.61}$$

where

$$S_m(\vec{q}, \omega) \equiv \frac{e^{-2\Gamma(\vec{q})}}{Nm!} \sum_{j,k} e^{i\vec{q}\cdot(\vec{R}_j - \vec{R}_k)} \int_{-\infty}^{\infty} dt\, e^{i\omega t} \langle\!\langle [\vec{q} \cdot \vec{u}_j(t)][\vec{q} \cdot \vec{u}_k(0)] \rangle\!\rangle^m. \tag{6.62}$$

The zero-phonon piece is

$$S_0(\vec{q}, \omega) = e^{-2\Gamma(\vec{q})} \frac{1}{N} \sum_{j,k} e^{i\vec{q}\cdot(\vec{R}_j - \vec{R}_k)} 2\pi\delta(\omega) \tag{6.63}$$

$$= e^{-2\Gamma(\vec{q})} 2\pi\delta(\omega) \frac{1}{\Omega_{\mathrm{cell}}} \sum_{\{\vec{G}\}} \delta^3(\vec{q} - \vec{G}). \tag{6.64}$$

This clearly gives only elastic scattering. The sharp Bragg peaks have their intensity reduced by the Debye–Waller factor, but they are not broadened.

The one-phonon term is

$$S_1(\vec{q}, \omega) = \frac{e^{-2\Gamma(\vec{q})}}{N} \sum_{j,k} e^{i\vec{q}\cdot(\vec{R}_j - \vec{R}_k)} \int_{-\infty}^{\infty} dt\, e^{i\omega t} \langle\!\langle [\vec{q} \cdot \vec{u}_j(t)][\vec{q} \cdot \vec{u}_k(0)] \rangle\!\rangle. \tag{6.65}$$

Expressing the displacement operators in terms of phonon operators, we have

$$S_1(\vec{q}, \omega) = e^{-2\Gamma(\vec{q})} \int_{-\infty}^{\infty} dt\, e^{i\omega t} \sum_{\alpha,\alpha'} \frac{\hbar}{2M} \left[\frac{1}{\omega_\alpha(\vec{q})\omega_{\alpha'}(\vec{q})} \right]^{1/2}$$
$$\times [\vec{q} \cdot \hat{\epsilon}_\alpha(\vec{q})][\vec{q} \cdot \hat{\epsilon}_{\alpha'}(\vec{q})] \times \left\langle\!\!\left\langle \left[a_\alpha(-\vec{q}, t) + a_\alpha^\dagger(\vec{q}, t) \right] \left[a_{\alpha'}(\vec{q}, 0) + a_{\alpha'}^\dagger(-\vec{q}, 0) \right] \right\rangle\!\!\right\rangle. \tag{6.66}$$

Now for a harmonic system we have the simple result

$$a_\alpha(\vec{q}, t) \equiv e^{iHt/\hbar} a_\alpha(\vec{q}) e^{-iHt/\hbar} = e^{-i\omega_\alpha(\vec{q})t} a_\alpha(\vec{q}). \tag{6.67}$$

Using this and the fact that only terms which conserve phonon occupations have non-zero thermal averages, we obtain (assuming inversion symmetry so that $\omega_\alpha(\vec{q}) = \omega_\alpha(-\vec{q})$) and $\hat{\epsilon}_\alpha(\vec{q}) = \hat{\epsilon}_\alpha(-\vec{q})$)

$$S_1(\vec{q}, \omega) = e^{-2\Gamma(\vec{q})} \sum_\alpha \frac{2\pi\hbar}{2M\omega_\alpha(\vec{q})} \left[\vec{q} \cdot \hat{\epsilon}_\alpha(\vec{q})\right]^2$$
$$\times \{\delta[\omega + \omega_\alpha(\vec{q})]\langle\langle a_\alpha^\dagger(\vec{q})a_\alpha(\vec{q})\rangle\rangle$$
$$+ \delta[\omega - \omega_\alpha(\vec{q})]\langle\langle a_\alpha(\vec{q})a_\alpha^\dagger(\vec{q})\rangle\rangle\}. \tag{6.68}$$

The phonon occupation number is the Bose factor (since the phonon number is not conserved, the chemical potential is zero):

$$n_\alpha(\vec{q}) \equiv \langle\langle a_\alpha^\dagger(\vec{q})a_\alpha(\vec{q})\rangle\rangle = \frac{1}{e^{\beta\hbar\omega_\alpha(\vec{q})} - 1}. \tag{6.69}$$

Thus we finally obtain

$$S_1(\vec{q}, \omega) = e^{-2\Gamma(\vec{q})} \sum_\alpha \frac{2\pi\hbar}{2M\omega_\alpha(\vec{q})} \left[\vec{q} \cdot \hat{\epsilon}_\alpha(\vec{q})\right]^2$$
$$\times \{n_\alpha(\vec{q})\delta[\omega + \omega_\alpha(\vec{q})] + [n_\alpha(\vec{q}) + 1]\delta[\omega - \omega_\alpha(\vec{q})]\}. \tag{6.70}$$

The first term in curly brackets describes *absorption* of a phonon by the probe particle (a neutron, say) and is proportional to the number of phonons present in the sample. The second term in curly brackets describes *stimulated emission* of a phonon by the probe particle. This can occur even at $T = 0$ when no phonons are present in equilibrium. Note that the coefficient in front of the term in curly brackets is dimensionless. Ignoring geometric factors, it is basically

$$\frac{\hbar^2 q^2}{2M} \frac{1}{\hbar\omega_\alpha(\vec{q})}. \tag{6.71}$$

Also note that in the classical limit $n_\alpha(\vec{q}) \gg 1$, and $S(\vec{q}, \omega)$ becomes symmetric in ω as we discussed earlier.

In reality, due to finite detector resolution and imperfections in the lattice, the delta-function peaks in Eq. (6.70) are broadened; but they are often sufficiently pronounced to be unambiguously associated with a single-phonon process. See Fig. 4.1 for an example.

One interesting point about $S_1(\vec{q}, \omega)$, the single-phonon contribution to the dynamical structure factor, is that it contains a coefficient $\left[\vec{q} \cdot \hat{\epsilon}_\alpha(\vec{q})\right]^2$. Thus it would seem that neutrons can only excite longitudinal phonons. This turns out not to be the case however. To see this, simply add an arbitrary non-zero reciprocal lattice vector to $\vec{q}$. We saw earlier that the atomic displacements are unaffected by this:

$$\tilde{u}_\alpha(\vec{q} + \vec{G}) = \frac{1}{N} \sum_j e^{-i(\vec{q}+\vec{G})\cdot\vec{R}_j} \hat{\epsilon}_\alpha(\vec{q} + \vec{G}) \cdot \vec{u}_j$$
$$= \frac{1}{N} \sum_j e^{-i\vec{q}\cdot\vec{R}_j} \hat{\epsilon}_\alpha(\vec{q}) \cdot \vec{u}_j = \tilde{u}_\alpha(\vec{q}). \tag{6.72}$$

(We define $\hat{\epsilon}_\alpha(\vec{q} + \vec{G}) = \hat{\epsilon}_\alpha(\vec{q})$.) Hence the phonons live only inside the first Brillouin zone. However, by adding $\vec{G}$ to $\vec{q}$ we have effectively allowed the neutron to simultaneously Bragg scatter (by $\vec{G}$) and emit a phonon of wave vector $\vec{q}$. This means that the factor $\left[(\vec{q} + \vec{G}) \cdot \hat{\epsilon}(\vec{q})\right]^2$ which enters here no longer constrains $\hat{\epsilon}$ to be parallel to $\vec{q}$. Hence, transverse modes can be excited by neutron scattering.

Scattering events characterized by $\vec{G} \neq \vec{0}$ are called **umklapp** events. Because of the existence of umklapp processes, the law of conservation of momentum becomes

$$-\Delta\vec{P}_{\text{neutron}} = \hbar\left(\vec{q}_{\text{phonon}} + \vec{G}\right), \tag{6.73}$$

where $\vec{G}$ is any reciprocal lattice vector. This is another example of conservation of crystal momentum that we encountered in Chapter 3, now in an inelastic scattering event. It is important to understand that $\hbar\vec{q}_{\text{phonon}}$ is *not* a true momentum carried by the phonon. This is due to the fact that the crystal has discrete rather than continuous translation symmetry. This allows the momentum change of the neutron to be divided between the phonon and the center-of-mass motion of the crystal.

Multi-phonon contributions to the dynamical structure factor can be laboriously computed by taking higher terms in the series. However, it is more instructive to imagine a scenario in which the detector of scattered particles (X-ray photons, say) has poor energy resolution and all multi-phonon processes are accepted equally. Then the total scattering rate is given by the static structure factor

$$
\begin{aligned}
S(\vec{q}) &\equiv \int_{-\infty}^{\infty} \frac{d\omega}{2\pi} S(\vec{q}, \omega) = \Pi(\vec{q}, 0) \\
&= \frac{1}{N} \sum_{j,k} e^{i\vec{q}\cdot(\vec{R}_j - \vec{R}_k)} \langle\!\langle e^{i\vec{q}\cdot\vec{u}_j} e^{-i\vec{q}\cdot\vec{u}_k} \rangle\!\rangle \\
&= e^{-2\Gamma(\vec{q})} \frac{1}{N} \sum_{j,k} e^{i\vec{q}\cdot(\vec{R}_j - \vec{R}_k)} e^{\langle\!\langle |\vec{q}\cdot\vec{u}_j| |\vec{q}\cdot\vec{u}_k| \rangle\!\rangle},
\end{aligned}
\tag{6.74}
$$

which is equivalent to the expression we obtained earlier in Eqs. (3.60)–(3.63).

6.4 Debye–Waller Factor and Stability of Crystalline Order

It is interesting to evaluate the Debye–Waller factor for crystals in different dimensions. For a d-dimensional crystal

$$
e^{-2\Gamma(\vec{q})} = e^{-\langle\!\langle |\vec{q}\cdot\vec{u}_j|^2 \rangle\!\rangle} \approx e^{-\frac{q^2}{Nd} \sum_j \langle\!\langle \vec{u}_j \cdot \vec{u}_j \rangle\!\rangle},
\tag{6.75}
$$

where in the last step we assumed that the fluctuations of $\vec{u}$ are isotropic. This is typically a good approximation at low temperature, where fluctuations are dominated by long-wavelength phonons and are less sensitive to microscopic details. Expressing the displacements in terms of phonon operators yields

$$
\begin{aligned}
2\Gamma(\vec{q}) &\approx \frac{q^2}{Nd} \sum_{\vec{k},\alpha} \frac{\hbar}{2M\omega_\alpha(\vec{k})} \left\langle\!\left\langle \left[a_\alpha(\vec{k}) + a_\alpha^\dagger(-\vec{k}) \right] \left[a_\alpha(-\vec{k}) + a_\alpha^\dagger(\vec{k}) \right] \right\rangle\!\right\rangle \\
&= \frac{1}{Nd} \frac{\hbar^2 q^2}{2M} \sum_{\vec{k},\alpha} \frac{1}{\hbar\omega_\alpha(\vec{k})} \left[2n_\alpha(\vec{k}) + 1 \right].
\end{aligned}
\tag{6.76}
$$

Let us first consider the quantum limit $T = 0$, so that $n_\alpha(\vec{k}) = 0$. For simplicity, we consider only acoustic phonons and approximate the dispersion by

$$
\omega_\alpha(\vec{k}) = v_s k.
\tag{6.77}
$$

As a further simplification, we use the Debye model and take the Brillouin zone to be a d-dimensional sphere of radius k_D. Recall that we can fix k_D by the requirement that the Brillouin zone contain precisely N points. For example, in a 3D cubic lattice with lattice constant a,

$$
\frac{4\pi}{3} k_D^3 \left(\frac{L}{2\pi} \right)^3 = N = \frac{L^3}{a^3}
\tag{6.78}
$$

$$
\Rightarrow k_D = \left(\frac{2\pi}{a} \right) \left(\frac{3}{4\pi} \right)^{1/3}.
\tag{6.79}
$$

With these simplifications we have

$$2\Gamma(\vec{q}) \approx \frac{1}{d} \frac{\hbar^2 q^2}{2M} \int_{1BZ} d^d\vec{k} \frac{1}{N} \left(\frac{L}{2\pi}\right)^d \frac{1}{\hbar v_s k} \tag{6.80}$$

$$\approx \frac{1}{d} \frac{\hbar^2 q^2}{2M} \left(\frac{a}{2\pi}\right)^d \left(\frac{1}{\hbar v_s}\right) A_d \int_0^{k_D} dk \frac{k^{d-1}}{k}, \tag{6.81}$$

where A_d is the value of the angular integrations (the surface area of a d-dimensional unit sphere: $A_3 = 4\pi$, $A_2 = 2\pi$, $A_1 = 2$). For $d > 1$ we have

$$\int_0^{k_D} dk \, k^{d-2} = \left(\frac{1}{d-1}\right) k_D^{d-1} \tag{6.82}$$

and $\Gamma(\vec{q})$ is finite. Hence quantum **zero-point fluctuations** do *not* destroy sharp Bragg diffraction peaks. Note, however, that there is a logarithmic divergence in the integral for $d = 1$. For a system of finite length L, the smallest meaningful phonon wave vector is $k = 2\pi/L$ ($k = 0$ corresponding to center-of-mass translation). Substituting this as an "infrared cutoff" we find

$$\int_{2\pi/L}^{k_D} dk \, k^{-1} = \ln(k_D L/2\pi) \tag{6.83}$$

and hence

$$e^{-2\Gamma(q)} \approx e^{-2\frac{\hbar^2 q^2}{2M}\left(\frac{a}{\hbar v_s}\right)\ln(k_D L/2\pi)} = \left(\frac{L k_D}{2\pi}\right)^{-\frac{\hbar^2 q^2}{M}\left(\frac{a}{\hbar v_s}\right)}. \tag{6.84}$$

Thus the Bragg peak is killed off by the infrared divergence. It turns out that, instead of a sequence of delta functions

$$S(q) \sim \frac{1}{a} \sum_{\{G\}} \delta(q - G), \tag{6.85}$$

one obtains power-law divergences in $S(q)$ near the reciprocal lattice vectors:

$$S(G + \delta q) \sim (\delta q)^{-\gamma}. \tag{6.86}$$

Thus, even at $T = 0$, 1D solids do not exist. Zero-point motion in a sense "melts" them.

At finite temperatures the infrared divergences are even worse. For small k we have

$$2n_\alpha(\vec{k}) + 1 \sim \frac{2}{\beta \hbar v_s k}, \tag{6.87}$$

so the logarithmic divergence now occurs in one higher dimension. For $d = 2$ we have

$$\int_0^{k_D} dk \, k^{d-1} \frac{1}{\hbar v_s k} \frac{2}{\beta \hbar v_s k} \sim \frac{2}{\beta(\hbar v_s)^2} \ln(k_D L/2\pi). \tag{6.88}$$

Thus, at arbitrarily small but finite temperatures, 2D solids inevitably are melted by thermal fluctuations,[11] in a manner very similar to the way quantum fluctuations melt a 1D crystal discussed above. We thus find that crystalline order is stable against both quantum (at $T = 0$) and thermal (for finite T) fluctuations only at or above three dimensions.

The notion that a quantum system ($T = 0$) has properties similar to a classical system ($T > 0$) in one higher dimension is deep and pervasive. It can be shown to be true for many (but not all) models, on quite general grounds using path integration to represent the quantum partition function in

---

[11] Actually there remains some order which can disappear in a subtle sequence of phase transitions. See, e.g., Section 9.5 of the book by Chaikin and Lubensky [2].

d dimensions as a classical problem in $d + 1$ dimensions, in which thermal fluctuations of the latter correspond to quantum fluctuations of the former.[12]

Exercise 6.5. Normally the Debye–Waller factor vanishes in 1D and 2D at finite temperature, and in 1D at zero temperature as well. The situation can be different for cases with long-range coupling between atoms, which can result in non-linear phonon dispersion of the form $\omega(k) \sim k^\alpha$ at long wavelength, with $\alpha < 1$.[a]

   (i) Show that the Debye–Waller factor does not vanish for a 2D lattice at finite T, as long as $0 < \alpha < 1$. Show that the same is true for a 1D lattice at $T = 0$.

  (ii) Find the range of α within which the Debye–Waller factor does not vanish for the 1D lattice at finite T.

 (iii) Now consider a more exotic phonon dispersion at long wavelength of the form $\omega(k) \sim k|\ln k|^\gamma$. Find the conditions γ must satisfy in order for the Debye–Waller factor not to vanish for a 2D lattice at finite T, or a 1D lattice at $T = 0$, respectively.

[a] The reader is encouraged to take a look at Problem 22.1 in the book of Ashcroft and Mermin [1], which provides some background information on the present problem.

6.5 Mössbauer Effect

The Mössbauer effect is a remarkable manifestation of the quantum nature of phonons. Radioactive ^{57}Co beta decays to an excited state of ^{57}Fe. This excited state of the iron nucleus decays by emitting a gamma-ray photon of energy $\hbar\Omega \approx 14.4$ keV. The lifetime of the ^{57}Fe excited state is so long ($\tau \sim 0.1\,\mu$s) that the energy uncertainty of the gamma-ray photon is only[13] $\Delta\epsilon \sim 1.6 \times 10^{-8}$ eV [31]. This photon can strike a ^{57}Fe sample and undergo resonant absorption. However, upon thinking about these two processes we realize that we have neglected the recoil energy which arises because the gamma ray carries finite momentum,

$$\hbar k = \frac{\hbar\Omega}{c}. \tag{6.89}$$

The total energy E of the excited ^{57}Fe goes into recoil plus the gamma-ray photon energy:

$$E = \frac{\hbar^2 k^2}{2M} + \hbar\Omega = \frac{\hbar^2\Omega^2}{2Mc^2} + \hbar\Omega. \tag{6.90}$$

Assuming the recoil energy is small (since $Mc^2 \gg 14$ keV), we have

$$\hbar\Omega \approx E\left(1 - \frac{E}{2Mc^2}\right); \tag{6.91}$$

$$E - \hbar\Omega \approx 2 \times 10^{-3} \text{ eV}. \tag{6.92}$$

Thus the recoil energy is much greater than the gamma-ray linewidth (energy uncertainty). Furthermore the recoil price is paid both for emission and for absorption. Hence there is no possibility that

---

[12] In some cases the quantum correlations in a d-dimensional system behave like those of a classical system in $d + z$ dimensions. The so-called **dynamical critical exponent** z is often 1 but can also be 2 or larger. For detailed discussions, see the book on quantum phase transitions by Sachdev [12].

[13] This energy width is about an order of magnitude larger than the theoretical limit set by the Heisenberg energy–time uncertainty relation and is due to various small perturbations on the nuclei and additional tiny effects like the second-order Doppler shift (relativistic time dilation) associated with the thermal vibrations of the nuclei.

this process can occur, since energy will not be conserved. Nevertheless, when Rudolph Mössbauer performed the experiment, he found that resonant absorption does indeed occur! How can this be?

The answer lies in the fact that the ^{57}Fe atoms in the emitter and absorber are embedded in a crystal lattice. Some fraction of the time the recoil momentum does not produce any phonon excitations – it excites only the center-of-mass motion of the entire crystal. This costs negligible energy and hence gives "recoilless" emission and absorption. Thus recoil effects can be quite different in quantum systems than in their classical counterparts.

Let us make an oversimplified model of the gamma-ray emission. Momentum $\hbar \vec{k}$ is suddenly given to a particular atom j.[14] This impulse process can be represented by the lattice vibrational wave function

$$\psi_{\text{after}}\{\vec{r}\} = e^{i\vec{k} \cdot \vec{r}_j} \psi_{\text{before}}\{\vec{r}\}. \tag{6.93}$$

What is the energy transfer? It is *not* a fixed number, because ψ_{after} is *not* an eigenstate of $\mathcal{H}$. Before computing the full probability distribution for the energy transfer, let us carry out a warm-up exercise to gain some intuition about the Mössbauer effect. Consider a simple harmonic oscillator in 1D. The ground state wave function is a Gaussian:

$$\psi_{\text{before}}(x) = \left(\frac{1}{2\pi x_{\text{ZPF}}^2}\right)^{\frac{1}{4}} e^{-\frac{1}{4}\frac{x^2}{x_{\text{ZPF}}^2}}. \tag{6.94}$$

The scale x_{ZPF} of the zero-point position fluctuations sets the scale for the momentum uncertainty. Because of the momentum uncertainty, it is possible to deliver a finite impulse $\hbar k$ to the oscillator and still be in the ground state! The probability of remaining in the ground state is given by

$$P_0 = |\langle \psi_{\text{after}} | \psi_{\text{before}} \rangle|^2 = |\langle \psi | e^{ikx} | \psi \rangle|^2 = e^{-\frac{1}{2}k^2 x_{\text{ZPF}}^2}. \tag{6.95}$$

In a 3D crystal, the initial vibration associated with the sudden impulse spreads through the lattice because ψ_{after} is not a stationary state. The probability distribution of the energy transfer $\hbar \omega$ is simply given by

$$\begin{aligned}
P(\vec{k}, \omega) &= \sum_n 2\pi \langle\langle |\langle n | \psi_{\text{after}} \rangle|^2 \delta[\omega - (E_n - E_i)/\hbar]\rangle\rangle \\
&= \sum_n 2\pi \langle\langle |\langle n | e^{i\vec{k} \cdot \vec{r}_j} | i \rangle|^2 \delta[\omega - (E_n - E_i)/\hbar]\rangle\rangle \\
&\equiv \int_{-\infty}^{\infty} dt \, e^{i\omega t} \left\langle\!\!\left\langle e^{-i\vec{k} \cdot \vec{r}_j(t)} e^{i\vec{k} \cdot \vec{r}_j(0)} \right\rangle\!\!\right\rangle.
\end{aligned} \tag{6.96}$$

In the above $\langle\langle \cdot \rangle\rangle$ represents thermal averaging over the initial state i, the summation is over all eigenstates of the system, and the last step follows from manipulations similar to those that lead to the dynamical structure factor in Section 4.2. Notice, however, that $P(\vec{k}, \omega)$ is *not* the dynamical structure factor we have been studying, since we know precisely which atom suffered the recoil. There is *no* interference among processes involving different atoms. The energy-transfer distribution is properly normalized, as will be shown in Eq. (6.101).

For a harmonic crystal we can make the cumulant expansion as usual and obtain a result closely analogous to that found for the dynamical structure factor $S(\vec{k}, \omega)$:

$$P(\vec{k}, \omega) = e^{-2\Gamma(\vec{k})} \int_{-\infty}^{\infty} dt \, e^{i\omega t} e^{f(t)}, \tag{6.97}$$

---

[14] We can tell which one because of the change in internal state of the nucleus associated with the absorption of the gamma ray.

where

$$f(t) = \left\langle\!\!\left\langle [\vec{k} \cdot \vec{u}_j(t)]\,[\vec{k} \cdot \vec{u}_j(0)] \right\rangle\!\!\right\rangle \tag{6.98}$$

and $e^{-2\Gamma(\vec{k})}$ is the Debye–Waller factor defined in Eq. (6.75). We can expand this in a series of contributions involving m phonons just as we did for the dynamical structure factor. The zero-phonon piece is what gives recoilless emission.[15] This is simply given by replacing $e^{f(t)}$ by the zeroth term in its Taylor-series expansion

$$e^{f(t)} \to 1, \tag{6.99}$$

so that

$$P_0(k, \omega) = e^{-2\Gamma(\vec{k})} 2\pi \delta(\omega). \tag{6.100}$$

For a three-dimensional crystal at zero temperature, the full $P(k, \omega)$ looks like that schematically illustrated in Fig. 6.4. There is a delta function (with finite weight) at zero frequency representing recoilless emission and then an inelastic background that rises from zero (initially) linearly with frequency. As long as the Debye–Waller factor is finite, there will be a finite probability of recoilless emission and absorption.

We have the following sum rule:

$$\int_{-\infty}^{\infty} \frac{d\omega}{2\pi} P(\vec{k}, \omega) = e^{-2\Gamma(\vec{k})} e^{f(t=0)} = 1, \tag{6.101}$$

which simply shows that the energy distribution is normalized. Notice that, because there are no interference terms involving scattering from different atoms, the static structure factor does not enter this result as it did in the normalization of the dynamic structure factor. We also have

$$\int_{-\infty}^{\infty} \frac{d\omega}{2\pi} \hbar\omega P(\vec{k}, \omega) = \left.\frac{df}{dt}\right|_{t=0} = \frac{\hbar^2 k^2}{2M}, \tag{6.102}$$

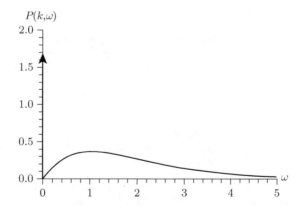

Figure 6.4 Schematic illustration of the Mössbauer spectral density $P(k, \omega)$ describing the probability of recoil energy loss $\hbar\omega$ for a fixed value of the momentum $\hbar k$ transferred to a single atom in a 3D crystal (at $T = 0$). In addition to a delta-function peak describing recoilless emission, there is a smooth background that is linear in frequency at low frequencies. See Exercise 6.6 for further discussion.

---

[15] In principle, at finite temperature there could be contributions from the two-phonon (and higher) piece in which one phonon is emitted and the other absorbed. However this only gives a smooth background and does not contribute to the sharp delta function in the distribution at exactly zero energy shown in Eq. (6.100).

which is simply the f-sum rule we derived earlier. It tells us that the mean recoil energy is the same as for a ^{57}Fe atom in free space (i.e. the classical result). The crystal has no effect on the mean recoil energy – only the shape of the energy distribution is affected.

It is very instructive to compare $P(\vec{k}, \omega)$ in Eq. (6.97) with the dynamical structure factor $S(\vec{k}, \omega)$ using Eqs. (6.60) and (6.55). They are very similar in form, but $P(\vec{k}, \omega)$ involves only a single atom (since we can tell from examining the nuclei which one absorbed the gamma ray). $S(\vec{k}, \omega)$ contains the physics of the interference due to scattering from different sites. The zero-phonon piece of $S_0(\vec{k}, \omega)$ given in Eq. (6.64) is the same as that of $P_0(\vec{k}, \omega)$ in Eq. (6.100) except that the allowed momentum transfers are restricted to the reciprocal lattice due to the interference. This is not true for $P_0(\vec{k}, \omega)$.

The Mössbauer effect is such a sensitive probe of small frequency shifts that it was used successfully to detect the gravitational red-shift of a photon climbing the height of a building [31], only three years after its discovery! This is but one example of the many impacts of condensed matter physics on other branches of science.

Exercise 6.6. Using the expression for the atomic displacements in terms of phonon operators given in Eq. (6.45), derive the form of $P(k, \omega)$ for low frequencies schematically illustrated in Fig. 6.4. To simplify the calculation, assume an isotropic system in which (at long wavelengths) longitudinal and transverse phonons have the same sound velocity, v_s. In this special case one can make the replacement $\sum_\alpha (\vec{q} \cdot \hat{\epsilon}_\alpha)(\vec{q} \cdot \hat{\epsilon}_\alpha) = q^2$. The integrals will be greatly simplified if, instead of assuming a sharp cutoff at a Debye wave vector k_D, one recognizes the "ultraviolet cutoff" via insertion of a factor e^{-k/k_D} to regulate the integrals. This approximation will not affect the low-frequency part of the spectral density. After deriving the zero-phonon and one-phonon contributions to $P(k, \omega)$, show that the two-phonon and higher contributions contribute only higher powers of ω in the Taylor-series expansion of the smooth background and hence do not need to be evaluated if one is interested only in very small energy transfers.

Exercise 6.7. The Mössbauer effect can be used to measure the entire recoil distribution $P(\vec{k}, \omega)$ shown in Fig. 6.4, not just the recoilless emission and absorption probability. This is done by moving the absorber at varying velocities relative to the emitter and taking advantage of the Doppler shift of the gamma-ray energy as seen in the frame of the absorber. Assuming recoiless emission, find the velocity that the absorber must have in order to absorb a gamma ray when the absorption-event recoil produces a phonon of energy 1 meV.

Exercise 6.8. General relativity predicts that the frequency of a clock varies with height above the earth's surface according to $\Delta v/v_0 = gh/c^2$, where g is the local acceleration due to gravity and the height h is assumed to be small with respect to the radius of the earth (so that g can be treated as a constant). Thus a gamma ray emitted at height zero will be red-shifted relative to the absorption frequency of an absorber at height h by frequency Δv. This can be compensated for by the Doppler effect, by giving the absorber a downward velocity v. Find v as a function of the height of the absorber relative to the emitter. What is its numerical value for a height of 20 m? At a height of 20 m, how large is the gravitational red-shift compared with the linewidth of ^{57}Co, $\Delta\epsilon \sim 1.6 \times 10^{-8}$ eV? (Now you know why this was a difficult experiment!). Find the vertical velocity (including the sign!) of the source relative to the absorber such that the first-order Doppler shift cancels out the gravitational shift. Estimate the degree of temperature stability required in order that the second-order (relativistic time-dilation) shift is not a significant source of error in this experiment. (Hint: for simplicity use the classical equipartion theorem to obtain the nuclear kinetic energy.)

Exercise 6.9. Suppose that a 14.4 keV gamma ray could be trapped in a box with reflecting walls. (Why would this not be possible in practice?). Find the increase in mass of the box when the gamma ray is trapped inside. Compute the change in work done in lifting the box by a height h near the earth's surface. Describe quantitatively and qualitatively how this is related to the frequency shift of the gamma ray in the Pound–Rebka experiment [31].

7 Electronic Structure of Crystals

Now that we have understood the basic ideas of how to determine the atomic structure of crystals using X-ray and neutron scattering, we are ready to turn to the properties of electrons in crystals. This may seem like a completely new topic, but in fact many of the essential concepts we learned previously such as

- lattice symmetry
- reciprocal lattice
- Brillouin zone
- conservation of crystal momentum (momentum modulo $\hbar\vec{G}$)

will be directly applicable to the problem of solving the Schrödinger equation for the wave function of an electron moving in the presence of the potential produced by the periodic array of atoms. Indeed, the problem of scattering of electron waves from the periodic array of atoms is very closely related to the problem of the scattering of X-ray and neutron waves from crystals. There is, however, one crucial difference – because of the Coulomb interaction the electrons scatter very strongly from the periodic lattice potential, and we typically cannot use perturbation theory (the Born approximation) as we did in studying neutron and X-ray scattering.

The energy spectrum of electrons in a crystal controls all of its essential properties: whether it is an insulator or a metal, whether it is transparent or opaque, what color it is, and so forth. Before we begin our study of these properties, it is worthwhile to review a little of the history.

7.1 Drude Theory of Electron Conduction in Metals

In 1853 Wiedemann and Franz made the observation that good electrical conductors are also generally good thermal conductors. For example, stainless steel, a poor electrical conductor (due to the fact that it is a disordered alloy) is also a poor thermal conductor. The latter property is prized both by low-temperature experimentalists and by the makers of pots and pans. Silver, on the other hand, is an excellent electrical conductor, and if you have ever held a silver spoon in a bowl of hot soup, you will know that it is also an excellent thermal conductor.

The **"Wiedemann–Franz law"** states that κ/σ, the ratio of thermal to electrical conductivity, is approximately the same for a wide variety of metals. (See Section 8.9 for more details.) In 1881 Lorentz noted that this ratio is temperature-dependent, but the **"Lorentz number"**

$$\frac{\kappa}{\sigma T} \sim (2\text{--}3) \times 10^{-8} \text{ W}\,\Omega/\text{K}^2 \tag{7.1}$$

is roughly independent of temperature.

Subsequently, **Matthiessen** noted that at low temperature the resistivity of a metal containing impurities could be expressed to a good approximation in the form

$$\frac{1}{\sigma(T)} = \frac{1}{\sigma_i} + \frac{1}{\sigma_0(T)}, \tag{7.2}$$

where $1/\sigma_i$ is proportional to the impurity density and $\sigma_0(T)$ is the conductivity of the pure metal and diverges for $T \to 0$. We can interpret this rule as saying that the temperature-independent impurity resitivity and the temperature-dependent resistivity are (to a good approximation) simply additive.

In 1879 Hall discovered that a current in a wire in the presence of a magnetic field induces an electric field at right angles to the current and the field. The magnitude (and sometimes even the sign) of the effect varied from metal to metal.

The next major development came in 1897 when J. J. Thompson discovered the electron. He used cathode ray tubes to measure the charge-to-mass ratio of the electron by deflecting the beam in a magnetic field. He also postulated the "plum pudding" model, which described atoms as consisting of electrons moving in a smooth blob of positively charged "pudding."

In 1900 P. Drude developed the **"Drude model"** of electron transport in metals. It was understood that electrons could somehow be "detached" from atoms, since cathode ray tube experiments had shown that electrons could be "boiled off" a heated cathode. Drude assumed that, in metals, at least some of the electrons detached themselves from the atoms and were free to move throughout the material. Drude applied the newly developed "kinetic theory" of gases to these electrons. If each atom were to donate one electron, the spacing between particles in the electron "gas" would be only $\sim 2\,\text{Å}$ or so. Despite this, Drude audaciously assumed he could treat the electrons as a dilute, nearly ideal gas, and assumed that one could ignore the Coulomb interactions among the particles.[1]

According to the kinetic theory of gases, the equation of motion for the mean particle velocity is

$$\langle \dot{\vec{v}} \rangle = -\frac{e}{m_e}\vec{E} - \frac{1}{\tau}\langle \vec{v} \rangle, \tag{7.3}$$

where $-e/m$ is the charge-to-mass ratio, $\vec{E}$ is the local electric field (resulting from, and often but not always equal to, the externally applied electric field), and τ is the characteristic viscous damping time due to collisions with the lattice of atoms.[2]

In steady state, where $\langle \dot{\vec{v}} \rangle = \vec{0}$, the **drift velocity** is linear in the electric field:

$$\langle \vec{v} \rangle = -\frac{e\tau}{m_e}\vec{E}. \tag{7.4}$$

If the number density is n, the current density is

$$\vec{J} = -ne\langle \vec{v} \rangle = \frac{ne^2\tau}{m_e}\vec{E}. \tag{7.5}$$

---

[1] Naturally, since quantum mechanics had not yet been invented, Drude was also ignoring a lot of things he did not even dream of, such as the wave nature of the electron, the Pauli exclusion principle, etc.

[2] This equation takes the form of Newton's law, and the first term on the RHS obviously comes from the force due to the electric field $\vec{E}$. The second term is due to collisions and requires some explanation. The assumption here is that for each electron a collision occurs with probability dt/τ in an infinitesimal time interval dt, after which its velocity is randomized so that the average velocity is zero. As a result, the loss rate of $\langle \vec{v} \rangle$ due to collisions is $(1/\tau)\langle \vec{v} \rangle$. Collisions of the electrons with each other are less important since conservation of center-of-mass momentum implies conservation of total current in electron–electron collisions, and is not included here. (This argument fails in the quantum theory because scattering of the electron waves off the periodic lattice potential can destroy **Galilean invariance**.) More detailed and microscopic discussion of collisions will be presented in Chapter 8. This is a good time to attempt Exercise 7.3.

Thus we obtain the famous Drude result for the conductivity

$$\sigma = \frac{ne^2\tau}{m_e}. \tag{7.6}$$

Exercise 7.1. Solve Eq. (7.3) for the case of an oscillatory electric field at frequency ω

$$\vec{E} = \vec{E}_0 e^{j\omega t}, \tag{7.7}$$

where we use the electrical engineering convention with $j = -i$ rather than the quantum mechanical convention for the time dependence.[a] From this, derive the formula for the ac Drude conductivity

$$\sigma(\omega) = \frac{ne^2\tau}{m_e} \frac{1}{1 + j\omega\tau}. \tag{7.8}$$

(i) Give a physical explanation for why τ drops out in the limit of high frequencies, and explain the significance of the fact that the conductivity becomes purely imaginary.

(ii) Compute the complex admittance (the inverse of the impedance) of an inductor L and resistor R in series and show that it has an identical form to the Drude conductivity. Give a physical explanation for why the inertial mass of the electron acts like an inductance. The effective inductance associated with the electron mass is called the "kinetic inductance."

[a] One frequent point of confusion is that in electrodynamics it is common practice to define the standard time dependence of quantities such as currents to be $\exp(+j\omega t)$, whereas in quantum mechanics it is standard to use $\exp(-i\omega t)$. In writing the Drude conductivity here we are using the electrical engineering convention rather than the quantum mechanics convention. The conversion from one to the other is easily made via the mnemonic $j = -i$. (Electrical engineers reserve the symbol i for current.)

Box 7.1. Electromagnetic Units

We will generally use cgs units for electrical quantities except where explicitly noted otherwise. This is purely for notational simplicity so that we do not have to carry the factors of $4\pi\epsilon_0$ etc. that appear in SI units. In order to obtain results for transport coefficients in SI units it is useful to note that the e^2 in cgs expressions can be eliminated in a variety of ways. In addition to the straightforward replacement $e^2 \longrightarrow e^2/(4\pi\epsilon_0)$, we have, for example,

$$\frac{e^2}{2a_B} \approx 13.6 \text{ eV},$$

where $a_B \sim 0.53$ Å is the Bohr radius and the volt (V) on the right-hand side is an SI volt and e on the right-hand side is the SI electron charge $\sim 1.6 \times 10^{-19}$ coulombs.

Another useful trick is to note that in cgs units $e^2 = (e^2/\hbar c)\hbar c$ and use the fact that the dimensionless **fine-structure constant** is $(e^2/\hbar c)_{cgs} = (e^2/4\pi\epsilon_0\hbar c)_{SI} \sim 1/137.035999$.

Finally, note that in cgs units conductivities (in 3D) have units of velocity. Maxwell's impedance of the vacuum, $4\pi/c$ in the cgs system, is $\sqrt{\mu_0/\epsilon_0} \sim 376.730313461$ ohms, which is exactly defined in the SI system. The connection of this to the von Klitzing constant giving the quantum of resistance is discussed in Box 10.1.

One can estimate the "mean free time" τ from the typical particle speed $\bar{v} \equiv \left(\langle \vec{v}^2 \rangle\right)^{1/2}$ and the "mean free path" ℓ:

$$\tau = \frac{\ell}{\bar{v}}. \tag{7.9}$$

Classically the mean free path for a particle traveling through a random array of scatterers is given by

$$\ell^{-1} = \nu\sigma, \tag{7.10}$$

where ν is the density of scatterers and σ is their cross section (not to be confused with the conductivity!). We have to be careful when considering transport properties, however, because small-angle scattering events do not degrade the current as effectively as large-angle events. It turns out that this can be remedied by defining a transport mean free path (essentially the distance the particle typically travels before it has scattered through a large angle $\sim 90°$):

$$\ell_{TR}^{-1} = \nu \int d\Omega (1 - \cos\theta) \frac{d\sigma}{d\Omega}, \tag{7.11}$$

where θ is the scattering angle and $d\Omega = \sin\theta \, d\theta \, d\varphi$ is the differential solid angle; one should replace ℓ with ℓ_{TR} when calculating transport properties. Notice that, if the differential cross section is constant (isotropic scattering), $\ell_{TR} = \ell$. If, however, $d\sigma/d\Omega$ is peaked in the forward direction, $\ell_{TR} \gg \ell$. Typically the scattering is roughly isotropic, and ℓ_{TR} and ℓ have the same order of magnitude.[3]

To a classical physicist thinking about electrons ricocheting among the atoms it was "obvious" that $\ell_{TR} \sim \ell \sim a$, where a is the lattice constant. Likewise, not knowing about Fermi statistics, it was natural to assume the standard result from the kinetic theory of gases,

$$\frac{1}{2}m_e \bar{v}^2 = \frac{3}{2}k_B T. \tag{7.12}$$

Using the room-temperature value

$$\bar{v} \sim 10^7 \text{ cm/s} \tag{7.13}$$

and $\ell \sim 10^{-7}$ cm gives $\tau \sim 10^{-14}$ s. Substituting this into Eq. (7.6) (and assuming each atom in the crystal contributes of order one electron) gives

$$\rho \equiv \frac{1}{\sigma} \sim 10^{-6} \left(\frac{T}{T_{room}} \right)^{1/2} \text{ ohm} \cdot \text{cm}. \tag{7.14}$$

Surprisingly (and purely coincidentally!), this is the right order of magnitude for typical metals at room temperature, although in actuality the exponent of $1/2$ is often closer to 1 at not-too-low temperatures.

One can make similar arguments about the thermal conductivity, since electrons carry energy (and entropy) as well as charge. The heat current $\vec{J}_Q$ carried by the electrons is related to the temperature gradient by

$$\vec{J}_Q = -\kappa_e \, \nabla T. \tag{7.15}$$

Classical kinetic theory predicts

$$\kappa_e = \frac{2}{3} \tau \bar{v}^2 c_e, \tag{7.16}$$

where c_e is the classical expression for the electronic specific heat (per unit volume), $c_e = \frac{3}{2}nk_B$. Eliminating $\bar{v}^2$ yields

$$\kappa_e = \frac{3nk_B^2 T}{m_e} \tau. \tag{7.17}$$

---

[3] An interesting exception occurs in modulation-doped quantum wells (which we will encounter in Chapter 9) where the electrons are scattered by remote donor ions. In this case ℓ_{TR} can be as large as $\sim 10^2 \ell$. The difference between ℓ_{TR} and ℓ will be further elaborated in Chapter 8.

Using the Drude conductivity yields for the Lorentz number[4]

$$L \equiv \frac{\kappa_e}{\sigma T} = 3 \left(\frac{k_B}{e} \right)^2 \sim 2.2 \times 10^{-8} \left(\frac{V}{K} \right)^2, \tag{7.18}$$

which again is of the right order of magnitude.[5]

Finally, since the resistivity depends on the scattering *rate*, and since one expects that the total scattering rate in the presence of impurities is the sum of the scattering rate in the pure system plus the rate due to impurity scattering, Matthiessen's rule follows naturally.

All is not well, however! We have already seen that the temperature dependence of the conductivity is not correct. In addition we face the same problem about the specific heat of the electrons as we did previously for the phonons. Classically one expects the law of Dulong and Petit to apply to the electrons as well as the ion cores. Thus there should be a contribution of $\frac{3}{2}k_B$ per electron, which will be comparable to that of the lattice. Even at temperatures well above Θ_D, where the law of Dulong and Petit works for the lattice, the electronic contribution to the specific heat in metals is orders of magnitude smaller than expected classically. Despite this, the classical model predicts the right order of magnitude for the Lorentz number.

Another particularly strong indication that something is wrong with the classical model is found in the Hall effect. This effect, discovered in 1879, is the appearance of a transverse electric field in a conductor carrying current perpendicular to an applied magnetic induction. This transverse field arises from the Lorentz force

$$\vec{F}_L = \frac{q}{c} \vec{v} \times \vec{B} \tag{7.19}$$

which acts on electrons (of charge $q = -e$) moving in a magnetic field. Because the force is at right angles to the velocity, it does no net work on the electrons, but does change the direction of their motion. In the presence of a magnetic field, Eq. (7.3) becomes

$$\langle \dot{\vec{v}} \rangle = -\frac{e}{m_e} \vec{E} - \frac{e}{m_e c} \langle \vec{v} \rangle \times \vec{B} - \frac{1}{\tau} \langle \vec{v} \rangle. \tag{7.20}$$

In steady state the LHS vanishes and for $\vec{B} = B\hat{z}$ we obtain the **resistivity tensor** that relates the electric field to the current density $\vec{J}$,

$$E_\mu = \rho_{\mu\nu} J_\nu, \tag{7.21}$$

with

$$\rho = \begin{pmatrix} \sigma^{-1} & +B/nec & 0 \\ -B/nec & \sigma^{-1} & 0 \\ 0 & 0 & \sigma^{-1} \end{pmatrix}, \tag{7.22}$$

where the diagonal elements are simply the inverse of the usual Drude conductivity and the off-diagonal component $\rho_{yx} = -\rho_{xy} = -B/nec$ is called the Hall resistivity. If the boundary conditions on the conductor force the current to flow in the x direction, there is the usual electric field in the same direction, but in addition a transverse electric field $E_y = \rho_{yx} J_x$ will appear at right angles to the current. A simple kinematic argument requires that this electric field balance the Lorentz force on the moving electrons.

---

[4] We note that the units used here of (volts/kelvin)^2 are equivalent to the watts-ohms/kelvin^2 used in Eq. (7.1).

[5] Note that this classical result is rigorously independent of temperature if τ is a constant that is independent of energy.

In the classical case, the Hall resistivity is linear in the magnetic field, and it is convenient to define the so-called **Hall coefficient**[6] by

$$r_{\rm H} \equiv \frac{\rho_{yx}}{B_z} = -\frac{1}{nec}, \tag{7.23}$$

where n is the electron density. (In SI units $r_{\rm H} = -1/ne$.) The Hall effect is thus handy for determining the carrier density in a conductor and the sign of the charge of those carriers. Measurements show that Eq. (7.23) does indeed predict the correct order of magnitude of the observed Hall electric field, but in some cases (such as aluminum), it actually gets the sign wrong! This was completely baffling to classical physicists.

Another interesting feature of the Drude-model resistivity tensor in Eq. (7.22) is that the diagonal elements are unaffected by the magnetic field. (See, however, Exercise 7.2.) This lack of so-called **"magnetoresistance"** (dependence of the diagonal elements of ρ on B) is an artifact of the classical treatment of transport within the Drude model. In Chapters 10, 11, and 12, we will study quantum transport and see striking examples of experimental data exhibiting strong magnetoresistance. Quantum transport involves interference of amplitudes corresponding to a particle taking more than one path to the same final state. The vector potential associated with the magnetic field alters the relative phases of these interfering amplitudes via the **Aharonov–Bohm effect**, which will be discussed in Section 13.2.

Exercise 7.2. Invert the resistivity tensor in Eq. (7.22) to obtain the conductivity tensor relating electric field to current via

$$J_\mu = \sigma_{\mu\nu} E_\nu. \tag{7.24}$$

Show that, even though there is no magnetoresistance, the diagonal elements of the conductivity *do depend* on the magnetic field.

One of the great triumphs of quantum mechanics was the resolution of all these paradoxes. It turns out, as we shall see, that **Fermi–Dirac statistics** introduces a very high quantum energy scale (the Fermi energy $\epsilon_F \sim 1$–3 eV, corresponding to a temperature $T_F \sim (1$–$3) \times 10^4$ K). This is analogous to the Debye temperature in the sense that quantum effects freeze out electronic contributions to the specific heat below this scale.

Furthermore, quantum mechanics shows that Drude's classical picture of electrons scattering off the ion cores is completely wrong. The electrons are waves which can travel coherently through the periodic lattice *without* scattering (in a certain sense which will be explained shortly). Thus the mean free path is much larger than Drude realized. On the other hand, the Fermi energy makes the typical electron speed much larger, and so the mean free time $\tau = \ell/\bar{v}$ accidentally comes out close to the value Drude estimated (for room temperature)! We will also learn later about the theory of "holes," which will explain via Fermi–Dirac statistics why some materials behave as if their charge carriers were positive and hence have Hall coefficients with the "wrong" sign.

As we will see later in this chapter, and in particular in Chapter 8, the Drude formula (Eq. (7.6)) often still holds in the correct quantum mechanical description of electric transport, but *every* parameter that enters it needs to be modified or re-interpreted. As discussed above, the scattering time τ is determined *not* by electrons scattering off the periodic array of ions, but by scattering due to *deviations* from perfect periodicity. The electron mass m_e should be replaced by an *effective* mass which can deviate from the bare electron mass by a significant amount and is determined by the

---

[6] It is common practice to use the symbol $R_{\rm H}$ for the Hall coefficient, but, in discussing the quantum Hall effect in 2D systems, we use $r_{\rm H}$ and reserve $R_{\rm H}$ for the Hall resistance, which will be introduced shortly.

spectrum of electron eigenstates in the crystal (known as the band structure). The electron density n, which is actually *not* very well defined in the Drude model, becomes a well-defined quantity that is either the density of electrons in the so-called conduction band, or the density of holes in the so-called valence band. And even the universal constant e requires re-interpretation in the latter case; the holes behave like positively charged particles. While such a sign change does not affect the Drude formula because it involves e^2, the Hall resistivity *is* sensitive to this sign change, as discussed above. Perhaps most fundamentally, the correct quantum-mechanical theory is capable of predicting when a crystal is actually an *insulator* (which many materials are) that has zero conductivity at $T = 0$. This corresponds to having $n = 0$ in the Drude formula, but the Drude theory is not able to explain when and why this happens.

We now start our development of this quantum-mechanical theory of electron motion in a periodic crystal.

Exercise 7.3. Assume the probability that an electron experiences a collision between time t and $t + dt$ is dt/τ. (i) Let t_c be the time between two subsequent collisions. Show that the probability distribution function for t_c is $p(t_c) = (1/\tau)e^{-t_c/\tau}$. (ii) Calculate $\langle t_c \rangle$ and $\langle t_c^2 \rangle$. (iii) Let n be the number of collisions experienced by the electron in a time interval of length t (say, between t' and $t' + t$). Show that the probability distribution for n is the Poisson distribution $P(n) = e^{-\bar{n}}\bar{n}^n/n!$, where $\bar{n}$ is its average. Find the value of $\bar{n}$.

7.2 Independent Electron Model

We will be interested in solving the Schrödinger equation for electrons in a condensed matter system with some fixed set of positions of the nuclei. In particular, when the system is in the crystalline state the nuclei are located on lattice sites $\{\vec{R}_I\}$. Neglecting for the moment the spin degree of freedom of the electrons, the Schrödinger equation for the nth eigenfunction is

$$\mathcal{H}\Phi_n(\vec{r}_1, \ldots, \vec{r}_N) = E_n \Phi_n(\vec{r}_1, \ldots, \vec{r}_N), \tag{7.25}$$

where the Hamiltonian is

$$\mathcal{H} = \sum_{i=1}^{N} \left(\frac{-\hbar^2 \nabla_i^2}{2m_e} - \sum_I \frac{e^2 Z}{|\vec{r}_i - \vec{R}_I|} \right) + \frac{1}{2} \sum_{i \neq j}^{N} \frac{e^2}{|\vec{r}_i - \vec{r}_j|}. \tag{7.26}$$

The presence of the (last) electron–electron Coulomb interaction term couples the degrees of freedom of different electrons and makes this many-body Schrödinger equation difficult to solve. Fortunately, it is frequently (but not always!) a good approximation to replace $\mathcal{H}$ by an effective independent-particle Hamiltonian

$$\mathcal{H} = \sum_{i=1}^{N} H_i = \sum_{i=1}^{N} \left(\frac{-\hbar^2 \nabla_i^2}{2m_e} + v_{\text{eff}}(\vec{r}_i) \right). \tag{7.27}$$

There are (at least) two different approaches leading to effective independent-particle Hamiltonians of this form. We will discuss these approaches and their limitations in Chapter 15. It is clear that $v_{\text{eff}}(\vec{r})$ is some sort of effective potential which includes the interaction of an electron with the nuclei and represents in some average way the interaction of an electron with all other electrons. The most naive approximation to a condensed matter system is one where $v_{\text{eff}}(\vec{r})$ is replaced by a constant, which we can take to be the zero of energy. This leaves us with the so-called free-electron-gas model. In this model the single-particle states are plane waves, which can be labeled by a wave vector,

$$\langle \vec{r} \, | \vec{k} \rangle = \frac{1}{\sqrt{L^3}} \exp\left(i \vec{k} \cdot \vec{r}\right), \tag{7.28}$$

where L^3 is the normalization volume. These plane waves have eigenenergy

$$\epsilon_{\vec{k}} = \frac{\hbar^2 k^2}{2 m_{\mathrm{e}}} \tag{7.29}$$

and are eigenstates of the velocity operator

$$\vec{V} = \frac{d\vec{r}}{dt} = \frac{i}{\hbar}[H, \vec{r}] = \frac{i}{2 m_{\mathrm{e}} \hbar}[\vec{p} \cdot \vec{p}, \vec{r}] = \frac{i}{m_{\mathrm{e}} \hbar} \vec{p}(-i\hbar) = \frac{\vec{p}}{m_{\mathrm{e}}} = \frac{-i\hbar \nabla}{m_{\mathrm{e}}} \tag{7.30}$$

with eigenvalue $\hbar \vec{k}/m_{\mathrm{e}}$. Because the velocity is a constant of motion, the electrical resistance is strictly zero, since a current does not decay in the free-electron model.

The next-simplest model (and one which we will now study extensively), sometimes called the independent-particle model, uses

$$v_{\mathrm{eff}}(\vec{r}) = V(\vec{r}), \tag{7.31}$$

where V is a periodic potential with the symmetries of the crystalline lattice:

$$V(\vec{r} + \vec{R}_j) = V(\vec{r}), \tag{7.32}$$

where $\vec{R}_j$ is a lattice vector of the crystal. Amazingly, as we will see shortly, one is able to make many concrete statements about the electronic structure that are based on this symmetry property *alone*, without additional information on the detailed form of $V(\vec{r})$. Recall that in Drude's classical model it was scattering from the periodic potential that produced a very short mean free path (on the order of the lattice spacing) and hence was responsible for the finite electrical resistance. We now need to learn how to solve this problem quantum mechanically. We will find from a powerful result known as **Bloch's theorem** that quantum coherence makes the mean free path infinite. Hence the electrical resistance is actually zero in this model for metals with perfectly ordered crystalline lattices. In the end, we will find that Drude's transport formulas are still very useful, provided that we compute the scattering time τ (due to impurities or phonons which disturb the perfect crystalline order) quantum mechanically rather than classically.

7.3 Bloch's Theorem

Bloch's theorem is a powerful tool in the study of *independent* electrons moving in a periodic potential. It is of such importance that electrons moving in a periodic potential are often referred to as "Bloch electrons." We learned in our study of elastic X-ray and neutron scattering that the periodicity of the structure enforces a conservation law,

$$\Delta \vec{k} = \vec{G}, \tag{7.33}$$

where $\Delta \vec{k}$ is the change of wave vector of the probe particle and $\vec{G}$ is an element of the reciprocal lattice. We frequently treat X-ray and neutron scattering in the Born approximation, but we cannot do this for low-energy electrons because the potential is too strong. There is a lot of multiple scattering and, as we shall see, a special sort of coherence develops, which effectively gives the electrons an infinite mean free path in a perfect crystal, a situation which is quite impossible classically.

Let us begin by finding the solutions to the Schrödinger equation for a single particle in a periodic potential with Hamiltonian

$$H = \frac{\vec{p}^{\,2}}{2m} + V(\vec{r}). \tag{7.34}$$

Now $V(\vec{r})$ (and hence H) has the translation symmetry of the lattice:

$$V(\vec{r} + \vec{a}_j) = V(\vec{r}), \tag{7.35}$$

where $\vec{a}_j$ is one of the primitive vectors of the lattice. For example, V might be a sum of atomic potentials

$$V(\vec{r}) = \sum_{\{\vec{R}_j\}} v(\vec{r} - \vec{R}_j), \tag{7.36}$$

where $\vec{R}_j$ is a lattice site. In general (because of chemical bonding and other effects), v will *not* be identical to the free-atom potential; this is just an example that has the right translation symmetry.

We would like to take advantage of the lattice translation symmetry, Eq. (7.35), in the solution of H. Let us start with the simplest possible case, $V(\vec{r}) = $ constant. In this case H is invariant under *any* translation. Formally, this is the consequence of the fact that in this case

$$[H, T_{\vec{l}}] = 0 \tag{7.37}$$

for any vector $\vec{l}$, where

$$T_{\vec{l}} = e^{i\vec{p}\cdot\vec{l}/\hbar} \tag{7.38}$$

is the (unitary) translation operator (see Exercise 7.4) that translates the system by $\vec{l}$. This can be seen from how $T_{\vec{l}}$ transforms a wave function,

$$T_{\vec{l}} \psi(\vec{r}) = \psi(\vec{r} + \vec{l}), \tag{7.39}$$

and how it transforms an operator,

$$T_{\vec{l}} F(\vec{r}, \vec{p}) T_{\vec{l}}^{\dagger} = T_{\vec{l}} F(\vec{r}, \vec{p}) T_{-\vec{l}} = F(\vec{r} + \vec{l}, \vec{p}). \tag{7.40}$$

Exercise 7.4. Prove Eq. (7.39) by expanding the exponential in $T_{\vec{l}}$ in a power series and comparing the Taylor-series expansion of the RHS of (7.39).

Since $\vec{l}$ is arbitrary in Eq. (7.37), we must have

$$[H, \vec{p}] = 0, \tag{7.41}$$

namely H commutes with *every* component of $\vec{p}$, which are the *generators* of translation. Because different components of $\vec{p}$ commute with each other, eigenstates of H may be *chosen* to be eigenstates of $\vec{p}$, which are plane waves. They are also eigenstates of any $T_{\vec{l}}$, which also commute among themselves.

Box 7.2. Symmetries in Quantum Mechanics

A quantum system is said to have a certain symmetry (translation, rotation, spatial inversion, particle–hole, gauge, time-reversal, etc.) if carrying out the symmetry operation leaves the Hamiltonian H invariant. Most symmetry operations (time-reversal being one exception, see Section 7.8) are effected via unitary transformations obeying $UHU^{\dagger} = H$. Just because the Hamiltonian is left invariant by the transformation does *not* mean that its eigenstates are invariant. For example, the free-particle Hamiltonian is invariant under translation by an arbitrary distance, but its plane-wave eigenstates are *not*. Under translation $\vec{r} \rightarrow \vec{r} + \vec{a}$, the wave function gets multiplied by a phase,

$$e^{i\vec{k}\cdot\vec{r}} \rightarrow e^{i\vec{k}\cdot\vec{a}} e^{i\vec{k}\cdot\vec{r}}. \tag{7.42}$$

This is a characteristic feature of symmetries in quantum mechanics. As another example, consider reflection (spatial inversion) symmetry $P x P^\dagger = -x$ in a 1D problem. Because $P^2 = I$, $P P^\dagger = I$ implies $P^\dagger = P$. If the potential has even parity, $P V(x) P = V(x)$, then $[H, P] = 0$, and H and P can be simultaneously diagonalized. Because $P^2 = I$, the simultaneous eigenstates must obey $P|\psi\rangle = \pm|\psi\rangle$. Thus we see in the case of this discrete symmetry that the phase in the factor multiplying the transformed wave function can only be 0 or π.

The group theory of symmetries in quantum mechanics is a vast topic and the reader is referred to the many existing books devoted specifically to the topic. See, for example, Ref. [32].

When $V(\vec{r})$ is not a constant, we no longer have Eq. (7.41), and Eq. (7.37) is not true for general $\vec{l}$. However, with Eq. (7.35) we have

$$[H, T_j] = 0, \tag{7.43}$$

where

$$T_j = T_{\vec{a}_j} \tag{7.44}$$

is one of the primitive translation operators. Furthermore, the translation operators all commute with each other:[7] $[T_j, T_k] = 0$ for all j, k. Hence H and $\{T_j\}$ can all be simultaneously diagonalized. That is, we can choose our energy eigenstates ψ to be eigenstates of $\{T_j\}$:

$$
\begin{aligned}
T_j \, \psi &= \lambda_j \psi; \\
\psi(\vec{r} + \vec{a}_j) &= \lambda_j \psi(\vec{r}); \\
\psi(\vec{r} + n\vec{a}_j) &= T_j^n \psi = \lambda_j^n \psi.
\end{aligned}
\tag{7.45}
$$

If we now invoke periodic boundary conditions

$$\psi(\vec{r} + M\vec{a}_j) = \psi(\vec{r}), \tag{7.46}$$

where M is a large integer, then we must have

$$\lambda_j^M = 1. \tag{7.47}$$

This means that

$$\lambda_j = e^{i\vec{k}\cdot\vec{a}_j}, \tag{7.48}$$

where $\vec{k}$ is one of the usual wave vectors allowed in a periodic system,

$$\vec{k} = \frac{\vec{G}}{M}, \tag{7.49}$$

and $\vec{G}$ is a member of the reciprocal lattice. The reader should verify that Eq. (7.47) is satisfied.

It follows that the most general solution of the periodic Schrödinger equation is

$$\psi_{\vec{k}}(\vec{r}) = e^{i\vec{k}\cdot\vec{r}} u_{\vec{k}}(\vec{r}), \tag{7.50}$$

where $u_{\vec{k}}(\vec{r})$ is a periodic function having the translation symmetry of the lattice:

$$u_{\vec{k}}(\vec{r} + \vec{a}_j) = u_{\vec{k}}(\vec{r}), \qquad j = 1, 2, 3. \tag{7.51}$$

It is straightforward to see that Eq. (7.50) is the most general form consistent with Eqs. (7.45) and (7.48). This is known as **Bloch's theorem**.

---

[7] Complications arise if a magnetic field is present and the appropriately generalized *magnetic* translation operators do not necessarily commute. This will be discussed in Chapter 12.

The simplest example we could have is the case

$$u_{\vec{k}}(\vec{r}) = 1;$$ (7.52)

$$\psi_{\vec{k}}(\vec{r}) = e^{i\vec{k}\cdot\vec{r}}.$$ (7.53)

This, of course, is the answer for the case $V = $ constant. This "empty lattice" model is often a useful check on one's calculations.

For the more general case, we can draw on our experience from Chapter 3 in writing functions with the periodicity of the lattice using the reciprocal lattice vectors:

$$u_{\vec{k}}(\vec{r}) = \sum_{\{\vec{G}\}} a_{\vec{G}}(\vec{k})e^{i\vec{G}\cdot\vec{r}}.$$ (7.54)

Thus Bloch's theorem tells us that the general solution is of the form

$$\psi_{\vec{k}}(\vec{r}) = \sum_{\{\vec{G}\}} a_{\vec{G}}(\vec{k})e^{i(\vec{k}+\vec{G})\cdot\vec{r}}.$$ (7.55)

This should look familiar from our earlier discussion of conservation of crystal momentum, and it is clear that $\hbar\vec{k}$, being a good quantum number and thus a conserved quantity, is the electron's *crystal momentum*. The electron originally at $\vec{k}$ keeps getting scattered so that its momentum is uncertain modulo the $\vec{G}$ vectors. As with X-ray photons or neutrons propagating in a crystal, it is tempting to view $\hbar\vec{k}$ as the electron momentum, but it is *not*, as the above argument makes clear. However, it turns out that we can often ignore the distinction between momentum and crystal momentum, as we shall see later. In particular, as we will discuss in Chapter 8, it is possible to make wave packets out of Bloch waves, and their group velocity is given by the usual expression (see Section 7.3.3 for the derivation)

$$\vec{v}_{\vec{k}} = \frac{1}{\hbar}\nabla_{\vec{k}}\epsilon(\vec{k}).$$ (7.56)

These packets travel coherently through the lattice and, if there is no disorder, the mean free path will be infinite. This is in stark contrast to the results of the classical Drude picture, where the electrons scatter chaotically off each of the ions and have a very short mean free path.

Without loss of generality, we can restrict the wave vector $\vec{k}$ to the first Brillouin zone. To see this, write

$$\psi_{\vec{k}}(\vec{r}) = \sum_{\{\vec{G}\}} b_{\vec{G}}(\vec{k})e^{i(\vec{k}+\vec{G}'+\vec{G})\cdot\vec{r}},$$ (7.57)

where $b_{\vec{G}}(\vec{k}) \equiv a_{\vec{G}+\vec{G}'}(\vec{k})$. Now we can always choose $\vec{G}'$ so that $\vec{k}' \equiv \vec{k} + \vec{G}'$ lies in the first Brillouin zone. Thus we can always write the wave functions in the form

$$\psi_{n\vec{k}'}(\vec{r}) = e^{i\vec{k}'\cdot\vec{r}}u_{n\vec{k}'}(\vec{r}),$$ (7.58)

where $\vec{k}'$ is restricted to the first Brillouin zone, and use the integer label n, known as the band index, to make up for this restriction.

It turns out that Eq. (7.58) is the most general form for the eigenfunctions of a periodic potential. It tells us that the energy eigenstates are labeled by a lattice wave vector $\vec{k}$ defined within the first Brillouin zone, and an integer-valued band index n. The situation is similar to that for the eigenstates of a central potential, which are labeled by three integer-valued quantum numbers, nlm. There the presence of the angular momentum quantum numbers l and m is guaranteed by *rotation* symmetry, similarly to how the $\vec{k}$ quantum number here is guaranteed by the lattice translation symmetry; these quantum numbers dictate which representation of the symmetry group the eigenstate(s) form. However, these symmetry quantum numbers do not uniquely determine the state (except in very special

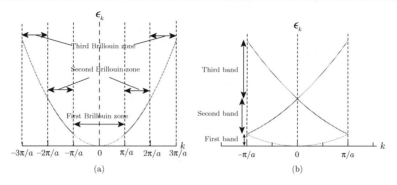

Figure 7.1 Electron dispersion of a 1D empty lattice, in the extended-zone scheme (a) and in the reduced-zone scheme (b).

cases); in general one needs an additional energy quantum number n. (In atoms n is known as the "principal" quantum number. In solids it is the band index.) Eigenstates that differ *only* in n have the same symmetry properties, but different energies.

As an illustrative example, let us consider a 1D empty lattice model. We assume a lattice constant a, but take the periodic potential to vanish. The eigenfunctions are, of course,

$$\psi_k(x) = e^{ikx}, \tag{7.59}$$

with eigenvalue $\epsilon_k \equiv \hbar^2 k^2/2m$. Unlike the case of phonons, it is meaningful for electron de Broglie waves to have wave vectors outside the first Brillouin zone. In the so-called "extended zone" scheme we simply plot the energy in the usual way, as shown in Fig. 7.1(a).

In the so-called "reduced-zone" scheme we map all wave vectors back into the first Brillouin zone by shifting them by integer multiples of $G_1 = 2\pi/a$, resulting in the energy-band diagram shown in Fig. 7.1(b). The energies of the various bands are given by:

$$\epsilon_{0k} = \frac{\hbar^2}{2m} k^2; \tag{7.60}$$

$$\epsilon_{1k} = \frac{\hbar^2}{2m} \left[k - \text{sgn}(k)G_1 \right]^2; \tag{7.61}$$

$$\epsilon_{nk} = \frac{\hbar^2}{2m} \left[k - \text{sgn}(k)nG_1 \right]^2. \tag{7.62}$$

Except for the fact that there are an infinite number of bands, this empty lattice band structure is reminiscent of the phonon dispersion curves we studied for the uniform lattice, where we chose an artificially doubled unit cell which hid some of the translation symmetry. (Recall Fig. 5.4.) In both cases there are degeneracies between different bands at the zone boundary or zone center because of the extra translation symmetry. We saw for the phonons that, if this symmetry is removed by choosing alternating masses and/or spring constants, the degeneracy is lifted. We will see shortly that, if the electrons see a periodic potential rather than an empty lattice, their degeneracies are also lifted.

Knowing the general form of the solution for Bloch waves from Eq. (7.58) and using

$$\vec{p}\,\psi_{n\vec{k}} = e^{i\vec{k}\cdot\vec{r}}\left(\vec{p} + \hbar\vec{k}\right)u_{n\vec{k}}, \tag{7.63}$$

the Schrödinger equation can be rewritten as

$$h_{\vec{k}} u_{n\vec{k}}(\vec{r}) = \epsilon_{n\vec{k}} u_{n\vec{k}}(\vec{r}), \tag{7.64}$$

where

$$h_{\vec{k}} \equiv \frac{1}{2m}\left(\vec{p} + \hbar\vec{k}\right)^2 + V(\vec{r}).\tag{7.65}$$

Now notice that the equation need only be solved within a single unit cell because of the periodicity of $u_{n\vec{k}}$. This vast simplification makes clear the true power of Bloch's theorem. We know on general grounds that for a given $\vec{k}$ the solutions will be *discrete* (labeled by n) because of the finite size of the unit cell. We have essentially reduced the problem to a particle in a box.[8]

Let us return to our 1D empty-lattice example. The Schrödinger equation reduces to

$$\frac{1}{2m}(p + \hbar k)^2 u_{nk}(x) = \epsilon_{nk} u_{nk}(x),\tag{7.66}$$

with periodic boundary condition

$$u_{nk}\left(-\frac{a}{2}\right) = u_{nk}\left(+\frac{a}{2}\right); \qquad u'_{nk}\left(-\frac{a}{2}\right) = u'_{nk}\left(+\frac{a}{2}\right).\tag{7.67}$$

Choosing

$$u_{nk}(x) = \sqrt{\frac{1}{a}} e^{-inG_1 \operatorname{sgn}(k)x}\tag{7.68}$$

satisfies the boundary condition and recovers our previous result. Note that the normalization corresponds to that of a unit cell. The factor $\operatorname{sgn}(k)$ is included purely for later convenience.

Let us now do a non-empty lattice. Consider the case of a weak periodic potential of the form

$$V(x) = v\cos(G_1 x) = \frac{v}{2}(e^{iG_1 x} + e^{-iG_1 x}).\tag{7.69}$$

Clearly this potential can scatter an electron at wave vector k to $k \pm G_1$. Of particular interest is the zone-boundary case $k_\pm = \pm\frac{1}{2}G_1$. The states ψ_{k_+} and ψ_{k_-} are degenerate but mixed by this perturbation:

$$\langle \psi_{k_+} | V | \psi_{k_-} \rangle = \frac{v}{2}.\tag{7.70}$$

States with $k = \pm\frac{3}{2}G_1$ are also mixed in, but they are far away in energy so we can ignore them if v is sufficiently small.

In the reduced-zone scheme

$$\psi_{k_+} = \psi_{n=0,k=+\frac{G_1}{2}},\tag{7.71}$$

$$\psi_{k_-} = \psi_{n=1,k=+\frac{G_1}{2}},\tag{7.72}$$

so the mixing is between states at the *same* k (note that ks that differ by G_1 are *equivalent*) but in *different* bands. Within degenerate perturbation theory we need to diagonalize the Hamiltonian in the degenerate subspace:

$$\begin{pmatrix} \langle n=0, k=\frac{G_1}{2} | H | n=0, k=\frac{G_1}{2} \rangle & \langle n=0, k=\frac{G_1}{2} | H | n=1, k=\frac{G_1}{2} \rangle \\ \langle n=1, k=\frac{G_1}{2} | H | n=0, k=\frac{G_1}{2} \rangle & \langle n=1, k=\frac{G_1}{2} | H | n=1, k=\frac{G_1}{2} \rangle \end{pmatrix}$$
$$= \begin{pmatrix} \hbar^2 G_1^2/8m & v/2 \\ v/2 & \hbar^2 G_1^2/8m \end{pmatrix}.\tag{7.73}$$

---

[8] This simplification is analogous to reducing a 3D Schrödinger equation to a 1D Schrödinger equation by using the rotation symmetry of a spherical potential.

This has eigenvalues $\epsilon_\pm = h^2 G_1^2/8m \pm v/2$ corresponding to eigenvectors

$$\chi_\pm = \frac{1}{\sqrt{2}} \begin{pmatrix} 1 \\ \pm 1 \end{pmatrix}. \tag{7.74}$$

This means that the states are

$$|\psi_\pm\rangle = \frac{1}{\sqrt{2}} \left\{ \left| n = 0, k = \frac{G_1}{2} \right\rangle \pm \left| n = 1, k = \frac{G_1}{2} \right\rangle \right\}, \tag{7.75}$$

which yields (properly normalized in a unit cell)

$$\psi_+(x) = \sqrt{\frac{2}{a}} \cos\left(\frac{G_1 x}{2}\right), \tag{7.76}$$

$$\psi_-(x) = i \sqrt{\frac{2}{a}} \sin\left(\frac{G_1 x}{2}\right). \tag{7.77}$$

These are, of course, standing waves. As shown in Fig. 7.2, the antinodes of ψ_+ are at the peaks of the potential (if $v > 0$), and so the eigenvalue is positive. The antinodes of ψ_- are in the valleys, and so it has the lower energy. In the language of chemistry, we can think of these as "anti-bonding" and "bonding" orbitals (more on this later).

As a result of the lifting of the degeneracy, the band structure looks like that shown in Fig. 7.3. To explore this further, let us examine the energy levels slightly away from the zone boundary at

$$k = \frac{1}{2} G_1 - \delta k. \tag{7.78}$$

The unperturbed states are no longer exactly degenerate, but they are still close together, so it is appropriate to continue to use degenerate perturbation theory. The subspace Hamiltonian is now

$$h = \begin{pmatrix} (\hbar^2/2m)(G_1/2 - \delta k)^2 & v/2 \\ v/2 & (\hbar^2/2m)(-G_1/2 - \delta k)^2 \end{pmatrix}. \tag{7.79}$$

Diagonalization of this gives the energy dispersion of bands $n = 0$ and $n = 1$ near the zone boundary, as shown in Fig. 7.3.

The band gaps produced by the lattice potential can have a profound impact on the transport properties of a material. If the chemical potential lies in a band gap [i.e. all states in bands below the gap (known as valence bands) are filled at zero temperature and states in bands above the gap (known as conduction bands) are empty], then the system is an insulator at $T = 0$. The Pauli principle requires

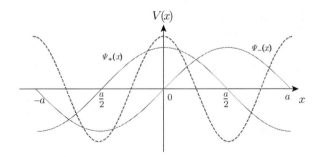

Figure 7.2 Degenerate perturbation solutions of wave functions with the wave vector at the zone boundary (ψ_+ and ψ_-, solid lines), in the presence of a cosine potential (dashed line).

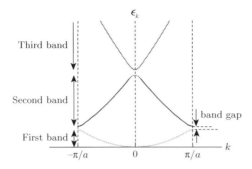

Figure 7.3 Band structure in the presence of a weak 1D periodic potential.

that an electron can absorb energy only by moving across the gap. Infinitesimal electric fields (at low frequencies) cannot accomplish this, and the system is effectively inert.[9]

As a result of these considerations, we are naturally led to ask how many electrons it takes to fill all the states in a band. In one dimension, we have (ignoring spin)

$$N_{\text{states}} = \int_{-\frac{G_1}{2}}^{+\frac{G_1}{2}} \left(\frac{L}{2\pi} \right) dk = \frac{L}{2\pi} \frac{2\pi}{a} = \frac{L}{a} = N_{\text{cells}}. \tag{7.80}$$

Thus the number of states in a single band is equal to the number of unit cells. In general dimensions

$$N_{\text{states}} = \left(\frac{L}{2\pi} \right)^d \int_{\text{BZ}} d^d \vec{k} = N_{\text{cells}}, \tag{7.81}$$

since the volume of the Brillouin zone is $(2\pi)^d/|\omega|$, where $|\omega|$ is the volume of the real-space unit cell. Thus the **density of states** is one per cell per band per spin.

Recall that the number of phonon states per polarization mode is also N_{cells}. The total number of modes (including polarizations) is, however, finite: d times the number of atoms per unit cell. We again emphasize that the electrons have an infinite number of bands going up to arbitrarily large energies.

In a Bravais lattice (one atom/cell) and taking into account the two spin states for electrons, it takes $2N_{\text{cells}} = 2N_{\text{atoms}}$ electrons to fill a band. Thus we have the following prediction of the independent-electron model: odd valence elements that form Bravais lattices are metals, because in this case one expects the highest band populated by electrons to be *half-filled*. The alkali element atoms all have odd valence and are all good metals. By the same argument, one would expect even-valence elements that form Bravais lattices to be insulators. Si, Ge, and C (diamond form) all satisfy this rule.[10] But there are many exceptions to this expectation, due to overlaps in band dispersions as illustrated in Fig. 7.4. In portions of the Brillouin zone the states in the nth band are actually higher in energy than states (in another part of the zone) in the $(n + 1)$th band. As a result, electrons are transferred from band n to band $n + 1$, leaving two partially filled bands. Zn is divalent, but is a good metal because of large interband transfer of electrons. Other examples of good metals formed by divalent elements include alkaline earth metals, formed by elements on the second column of the periodic table. When the band overlap is small, resulting in a low density of charge carriers, the system is known as a **semimetal**, and Bi is a good example of that.

---

[9] We emphasize that all of this is still assuming the independent-electron model.

[10] In fact these elements do *not* form Bravais lattices. It is a straightforward exercise to show that the conclusion that these even-valence elements are insulators remains valid, regardless of the actual lattice structure.

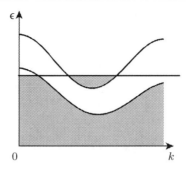

Figure 7.4 Schematic band structure of a semimetal, in which band dispersions overlap in energy, and the Fermi energy crosses more than one band. Shaded regions represent states fully occupied by electrons at zero temperature. The lower band has a small "hole pocket" that is unoccupied and the upper band has a small "electron pocket" that is occupied.

Box 7.3. Floquet Theory of Driven Systems

(N.B. $\hbar = 1$ in this discussion.)

We have learned from Bloch's theorem that the solution of the Schrödinger equation for electrons moving in a crystal is vastly simplified by recognizing the discrete translation symmetry of the spatially periodic crystal potential. **Floquet's theorem** gives an analogous result for Hamiltonians that are periodic in *time*. For example, periodic external magnetic field drives are frequently applied to quantum spin systems (see Chapter 17), and periodic laser fields and microwave fields are frequently applied to cold atoms in optical lattices (see Section 7.10). The purpose of these drives is to induce novel effective Hamiltonians (without excessively heating up the system). These effective Hamiltonians have the advantage that they are tunable *in situ* (by changing the drives), and they can place the system into different non-trivial topological phases of the types described in Chapter 13. For example, periodic drives have been successfully used to create strong pseudomagnetic fields seen by cold atoms in optical lattices, even though the atoms are charge neutral!

Suppose that a time-dependent Hamiltonian is invariant under discrete time translation by "distance" T

$$H(t + T) = H(t).\tag{7.82}$$

Then, by analogy with the quasi-momentum k in Bloch's theorem, we can define a quasi-energy ω such that the state evolution obeys

$$|\Psi(t + T)\rangle = e^{-i\omega T}|\Psi(t)\rangle, \quad |\Psi(t)\rangle = e^{-i\omega t}|u_{\omega,n}(t)\rangle,\tag{7.83}$$

where $-\pi/T \le \omega \le \pi/T$ lives in the first "Brillouin zone," and the periodic part of the state $|u_{\omega,n}(t)\rangle = |u_{\omega,n}(t + T)\rangle$ carries the quasi-energy label and a "band index," n, in analogy with Eq. (7.58).

Just as Bloch's theorem tells us that momentum is conserved modulo reciprocal lattice vectors $\vec{G}$, Floquet's theorem tells us that energy is conserved modulo $\Omega = 2\pi/T$. For the case of a single, sinusoidal driving term in the Hamiltonian, Ω is its frequency and the energy non-conservation comes from absorbing and emitting quanta of the drive frequency. For the case of multiple drive frequencies (which are rationally related), the "unit cell size," T, is determined by the shortest interval at which all the drives are back in phase with each other (after being in phase at some arbitrary time zero).

Another good example of a semimetal is graphene, where there are *no* overlapping bands at the Fermi energy, but the valence and conduction band are *degenerate* at special points in the Brillouin zone, as a result of which there is *no* band gap. We discuss the band structure of graphene in Section 7.5.

> **Exercise 7.5.** Consider an empty lattice ($V = 0$) in 2D, but pretend we have a square lattice with lattice constant a. Find the energy range over which the first and second bands overlap.

7.3.1 Band Gaps and Bragg Reflection

We have seen from our study of a periodic potential in 1D that scattering between two degenerate states can create a band gap. Let us now consider the general case in arbitrary dimension with a (weak) periodic potential,

$$V(\vec{r}) = \sum_{\{\vec{G}\}} V_{\vec{G}} e^{i\vec{G}\cdot\vec{r}}. \tag{7.84}$$

This potential can scatter an electron from $\vec{k}$ to $(\vec{k} + \vec{G})$. These two unperturbed states will be degenerate, provided that

$$\frac{\hbar^2}{2m_e}|\vec{k}|^2 = \frac{\hbar^2}{2m_e}|\vec{k} + \vec{G}|^2. \tag{7.85}$$

This, however, is identical to the Laue condition for X-ray diffraction,

$$|\vec{k}| = |\vec{k} + \vec{G}|. \tag{7.86}$$

Hence we can use the Ewald construction just as before, and whenever $\vec{k}$ satisfies the Laue condition we can expect a band gap[11] (within the simple degenerate perturbation theory approximation we have been using).

Consider the wave vectors shown in Fig. 7.5. The dashed line is a Bragg plane which is the perpendicular bisector of $\vec{G}$. Any point on this plane defines a pair of vectors which satisfy the Laue condition. Hence we expect that a contour map of the energy-band dispersion $\epsilon_n(\vec{k})$ in the extended-zone scheme to qualitatively resemble that shown in Fig. 7.6. The iso-energy contours bulge out toward the Bragg planes because the potential energy is negative in that region (for the $n = 0$ band) and must be compensated for by larger kinetic energy. The discontinuities at the Bragg planes are the origin of the band gaps. It is possible to prove that lines of constant energy intersect the Bragg planes at right angles, as the following construction will show.

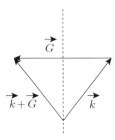

Figure 7.5 Illustration of the Laue condition, applied here to the scattering of electrons by the lattice potential.

---

[11] This is even true for the propagation of X-rays in crystals, although the band gaps are typically very small. This is the subject of the so-called "dynamical diffraction theory." See Section 7.9 for related discussion.

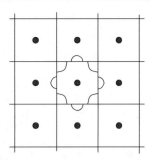

Figure 7.6 Equal-energy contours of Bloch bands, in the extended-zone scheme. The lines meet the Bragg planes at right angles and are generically discontinuous there.

In degenerate perturbation theory the wave function is of the form

$$\psi_{\vec{k}} = a e^{i\vec{k}\cdot\vec{r}} + b e^{i(\vec{k}+\vec{G})\cdot\vec{r}}. \tag{7.87}$$

The 2×2 matrix form of the Schrödinger equation is

$$h \begin{pmatrix} a \\ b \end{pmatrix} = E \begin{pmatrix} a \\ b \end{pmatrix}, \tag{7.88}$$

where

$$h \equiv \begin{pmatrix} \epsilon_{\vec{k}} & V_{\vec{G}}^* \\ V_{\vec{G}} & \epsilon_{\vec{k}+\vec{G}} \end{pmatrix}. \tag{7.89}$$

The two eigenvalues are

$$E_\pm = \epsilon_+ \pm \sqrt{\epsilon_-^2 + |V_{\vec{G}}|^2}, \tag{7.90}$$

where

$$\epsilon_\pm \equiv \frac{1}{2}(\epsilon_{\vec{k}} \pm \epsilon_{\vec{k}+\vec{G}}). \tag{7.91}$$

The gradient of the energy eigenvalue is

$$\nabla_{\vec{k}} E_\pm = \nabla_{\vec{k}}\epsilon_+ \pm \left(\nabla_{\vec{k}}\epsilon_-\right) \frac{\epsilon_-}{\sqrt{\epsilon_-^2 + |V_{\vec{G}}|^2}}. \tag{7.92}$$

Using $\nabla_{\vec{k}}\,\epsilon_+ = (\hbar^2/m_e)\left(\vec{k} + \frac{1}{2}\vec{G}\right)$ and the fact that $\epsilon_- = 0$ on the Bragg plane, we obtain

$$\nabla_{\vec{k}} E_\pm = \frac{\hbar^2}{m_e}\left(\vec{k} + \frac{1}{2}\vec{G}\right). \tag{7.93}$$

As can be seen from Fig. 7.5, this gradient vector lies in the Bragg plane and hence has no component perpendicular to it. The constant-energy lines must therefore meet the plane at right angles.

7.3.2 Van Hove Singularities

The periodic potential distorts the energy dispersion away from its free-electron value. It is true on quite general grounds that these distortions inevitably lead to singularities in the electronic density of states known as **van Hove singularities**. We know that we can label all the states by $(\vec{k}, n)$, where $\vec{k}$ lies in the first Brillouin zone. We will find it convenient, however, to let $\vec{k}$ run outside the first zone. A 1D example is shown in Fig. 7.7. Then we can label the (redundant) solutions in such a way that $\epsilon_{n\vec{k}}$ is a smooth periodic bounded function in reciprocal space. It must therefore be the case that $\epsilon_{n\vec{k}}$ has smooth local maxima and minima within the first zone at which $\nabla_{\vec{k}}\epsilon_{n\vec{k}} = \vec{0}$. This topological fact

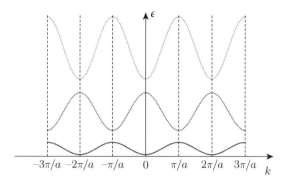

Figure 7.7 The 1D band structure plotted in the extended-zone scheme. Van Hove singularities are associated with extrema of the dispersion curves.

guarantees the existence of singularities in the density of states, since near each extremum there lies a large number of states with nearly the same energy.

Consider the (single-spin) density of states per unit volume for band n:

$$\rho_n(E) = \int_{\mathrm{BZ}} \frac{d^3\vec{k}}{(2\pi)^3} \delta(E - \epsilon_{n\vec{k}}). \tag{7.94}$$

Consider now the surface $S(E)$ in k space having $\epsilon_{n\vec{k}} = E$. We can define a local coordinate $\vec{k}_\perp$ perpendicular to this surface. The energy eigenvalue near the surface is given by

$$\epsilon_{n\vec{k}} = E + \vec{k}_\perp \cdot \nabla_{\vec{k}} \epsilon_{n\vec{k}}, \tag{7.95}$$

where the gradient is evaluated on the surface (i.e. at $\vec{k}_\perp = \vec{0}$). The density of states therefore is

$$\rho_n(E) = \int_{S(E)} \frac{dS}{(2\pi)^3} \int dk_\perp \, \delta(-\vec{k}_\perp \cdot \nabla_{\vec{k}} \epsilon_{n\vec{k}}). \tag{7.96}$$

Using the fact that the energy gradient is necessarily parallel to $\vec{k}_\perp$ we have

$$\rho_n(E) = \int_{S(E)} \frac{dS}{(2\pi)^3} \frac{1}{|\nabla_{\vec{k}} \epsilon_{n\vec{k}}|}, \tag{7.97}$$

which clearly shows that extrema in the energy will lead to singularities.

7.3.3 Velocity of Bloch Electrons

As another example of how band structure determines the physical properties of Bloch electrons, and that crystal momentum is *not* momentum, we consider the relation between the velocity of a Bloch electron and its crystal momentum $\vec{k}$, and prove Eq. (7.56). Our starting points are Eqs. (7.64) and (7.65). Treating $\vec{k}$ as a set of *parameters* of $h_{\vec{k}}$ and using the **Hellmann–Feynman theorem** (see Exercise 7.6), which states that

$$\frac{\partial E(\lambda)}{\partial \lambda} = \langle n(\lambda)| \frac{\partial H(\lambda)}{\partial \lambda} |n(\lambda)\rangle, \tag{7.98}$$

where $H(\lambda)$ is a λ-dependent Hamiltonian, $|n(\lambda)\rangle$ is one of its eigenstates and $E(\lambda)$ is the corresponding eigenvalue, we have

$$\frac{1}{\hbar} \nabla_{\vec{k}} \epsilon(\vec{k}) = \frac{1}{\hbar} \langle u_{n\vec{k}}| \frac{\partial h_{\vec{k}}}{\partial \vec{k}} |u_{n\vec{k}}\rangle = \langle u_{n\vec{k}}| \frac{\vec{p} + \hbar\vec{k}}{m_\mathrm{e}} |u_{n\vec{k}}\rangle$$

$$= \langle \psi_{n\vec{k}}| \frac{\vec{p}}{m_\mathrm{e}} |\psi_{n\vec{k}}\rangle; \tag{7.99}$$

thus Eq. (7.56) follows. Note that here the lattice momentum $\vec{k}$ is viewed as a *parameter* that lives in the first Brillouin zone, which specifies a member in a *family* of Hamiltonians, $\{h_{\vec{k}}\}$, $\vec{k} \in$ 1BZ. This is a viewpoint we will take again and again in this book.

Exercise 7.6. Prove the Hellmann–Feynman theorem, Eq. (7.98). Hint: take advantage of the fact that the norm of the eigenfunction is fixed at unity, which implies that $(\partial/\partial\lambda)\langle n(\lambda)|n(\lambda)\rangle = 0$.

Exercise 7.7. A 1D lattice has lattice constant a and a band with energy dispersion

$$\epsilon_k = -t\cos(ka).$$

Find the density of states and identify the nature of the van Hove singularities.

Exercise 7.8. Now consider the 2D version of Exercise 7.7, where the dispersion is $\epsilon_{\vec{k}} = -t(\cos(k_x a) + \cos(k_y a))$. Identify the energy where there is a van Hove singularity, and determine the leading energy dependence of the density of states near the van Hove singularity.

Exercise 7.9. Calculate the velocity of electrons with the band dispersion in Exercise 7.7, and identify the regions in momentum space where the velocity is parallel and anti-parallel to the momentum, respectively.

7.4 Tight-Binding Method

So far we have been studying nearly free electrons that are only weakly perturbed by a periodic potential. Generically we found broad bands separated by narrow gaps, as shown in Fig. 7.8(a). Now we will look at the opposite limit of strong periodic potentials which produce narrow bands separated by large gaps, as shown in Fig. 7.8(b).

One way to achieve this situation is to imagine a crystal made up of very widely separated atoms. To a good approximation, the atomic orbitals remain undistorted. The only change is that an electron in one atom has a small amplitude for tunneling to states on the neighboring atoms (see Appendix C

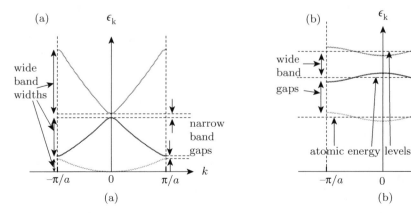

Figure 7.8 Band structures with (a) large band widths and small band gaps, and (b) large band gaps and small band widths. The latter can often be well described using the tight-binding method, which starts from atomic electron levels.

for an introduction). This tunneling (also referred to as hopping in this context), as we shall soon see, broadens the sharp atomic energy levels into narrow bands.

To describe this type of band structure we will now introduce the so-called "**tight-binding model**." Let $|j\rangle$ be the relevant state on the jth atom having wave function

$$\langle \vec{r}\,|j\rangle \equiv \varphi_0(\vec{r} - \vec{R}_j) \tag{7.100}$$

centered on the atomic position $\vec{R}_j$. If we restrict our attention to this single orbital on each atom, then Bloch's theorem *uniquely* constrains the form of the eigenfunction in a Bravais lattice with lattice wave vector $\vec{q}$ to be

$$|\psi_{\vec{q}}\rangle \equiv \frac{1}{\sqrt{N}} \sum_{j=1}^{N} e^{i\vec{q}\cdot\vec{R}_j} |j\rangle. \tag{7.101}$$

To see why this is so, consider

$$\langle \vec{r}|\psi_{\vec{q}}\rangle \equiv \psi_{\vec{q}}(\vec{r}) = \frac{1}{\sqrt{N}} \sum_{j} e^{i\vec{q}\cdot\vec{R}_j} \varphi_0(\vec{r} - \vec{R}_j) \tag{7.102}$$

$$= e^{i\vec{q}\cdot\vec{r}} u_{\vec{q}}(\vec{r}), \tag{7.103}$$

where

$$u_{\vec{q}}(\vec{r}) \equiv \frac{1}{\sqrt{N}} \sum_{j=1}^{N} e^{-i\vec{q}\cdot(\vec{r}-\vec{R}_j)} \varphi_0(\vec{r} - \vec{R}_j). \tag{7.104}$$

Now clearly $u_{\vec{q}}(\vec{r} + \vec{a}) = u_{\vec{q}}(\vec{r})$ for any lattice vector $\vec{a}$. Hence Bloch's theorem is satisfied. There is no other linear combination of these orbitals that satisfies the requirements of translation symmetry (with the same lattice wave vector). Hence, within the approximation of neglecting mixing between different atomic orbitals, there is only a single state for each $\vec{q}$, and it must be an energy eigenstate with eigenvalue (within this approximation) given by

$$\epsilon_{\vec{q}} = \frac{\langle \psi_{\vec{q}}|H|\psi_{\vec{q}}\rangle}{\langle \psi_{\vec{q}}|\psi_{\vec{q}}\rangle}. \tag{7.105}$$

Let us first evaluate the denominator in this expression. Recall that for a single atom (in the independent-electron approximation) each atomic orbital $\varphi_n(\vec{r} - \vec{R}_j)$ is orthogonal to the others:

$$\int d^3\vec{r}\ \varphi_n^*(\vec{r} - \vec{R}_j)\varphi_m(\vec{r} - \vec{R}_j) = \delta_{nm}, \tag{7.106}$$

because they are all eigenstates of a Hermitian Hamiltonian operator, each with a different eigenvalue (ignoring degeneracies). This does *not* imply, however, that orbitals on *different* atoms are orthogonal. In particular,

$$\langle j|j'\rangle \equiv \int d^3\vec{r}\ \varphi_0^*(\vec{r} - \vec{R}_j)\varphi_0(\vec{r} - \vec{R}_{j'}) \tag{7.107}$$

is generically non-zero (though of course exponentially small if the atoms are widely separated). Thus the Bloch states are not normalized to unity:

$$\langle \psi_{\vec{q}}|\psi_{\vec{q}}\rangle = \frac{1}{N} \sum_{j,\ell} e^{-i\vec{q}\cdot(\vec{R}_j-\vec{R}_\ell)}\langle j|\ell\rangle$$

$$= \sum_{\ell} e^{i\vec{q}\cdot\vec{R}_\ell}\langle j = 0|\ell\rangle \equiv \eta(\vec{q}). \tag{7.108}$$

Despite the non-orthogonality of the atomic orbitals, the Bloch states *are* automatically orthogonal:

$$\langle \psi_{\vec{q}} | \psi_{\vec{q}'} \rangle = \eta(\vec{q}) \delta_{\vec{q}, \vec{q}'}. \tag{7.109}$$

This is because the Bloch states are eigenvectors of the unitary (not Hermitian) translation operators:

$$T_{\vec{a}} \psi_{\vec{q}} = e^{i\vec{q}\cdot\vec{a}} \psi_{\vec{q}}, \tag{7.110}$$

$$\langle \psi_{\vec{q}} | T_{\vec{a}} | \psi_{\vec{q}'} \rangle = e^{i\vec{q}'\cdot\vec{a}} \langle \psi_{\vec{q}} | \psi_{\vec{q}'} \rangle. \tag{7.111}$$

We can also evaluate this by means of

$$\langle \psi_{\vec{q}} | T_{\vec{a}} | \psi_{\vec{q}'} \rangle = \left(\langle \psi_{\vec{q}'} | T_{\vec{a}}^{\dagger} | \psi_{\vec{q}} \rangle \right)^{*} \tag{7.112}$$

and using $T_{\vec{a}}^{\dagger} = T_{\vec{a}}^{-1} = T_{-\vec{a}}$ to obtain

$$\left(\langle \psi_{\vec{q}'} | \psi_{\vec{q}} \rangle e^{-i\vec{q}\cdot\vec{a}} \right)^{*} = e^{i\vec{q}\cdot\vec{a}} \langle \psi_{\vec{q}} | \psi_{\vec{q}'} \rangle. \tag{7.113}$$

Equating these two results yields

$$(e^{i\vec{q}'\cdot\vec{a}} - e^{i\vec{q}\cdot\vec{a}}) \langle \psi_{\vec{q}} | \psi_{\vec{q}'} \rangle = 0. \tag{7.114}$$

The term in parentheses vanishes if $\vec{q}' = \vec{q} + \vec{G}$, where $\vec{G}$ is a reciprocal lattice vector. However, since $\vec{q}$ and $\vec{q}'$ are restricted to the first Brillouin zone, the term in parentheses vanishes only for $\vec{q} = \vec{q}'$. Hence the states must be orthogonal for $\vec{q} \neq \vec{q}'$.

To evaluate the matrix element of the numerator, write the Hamiltonian in the form

$$H = \frac{\vec{p}^{\,2}}{2m} + \sum_{\ell} v(\vec{r} - \vec{R}_{\ell}), \tag{7.115}$$

$$H|j\rangle = \left[\frac{\vec{p}^{\,2}}{2m} + v(\vec{r} - \vec{R}_{j}) + \Delta v_{j} \right] |j\rangle, \tag{7.116}$$

where

$$\Delta v_{j} \equiv \sum_{\ell \neq j} v(\vec{r} - \vec{R}_{\ell}). \tag{7.117}$$

Here v is the individual atomic potential, unmodified by the presence of the other atoms. Hence

$$H|j\rangle = \varepsilon_{0}|j\rangle + \Delta v_{j}|j\rangle, \tag{7.118}$$

where ε_{0} is the atomic energy eigenvalue. Thus

$$H|\psi_{\vec{q}}\rangle = \varepsilon_{0}|\psi_{\vec{q}}\rangle + \frac{1}{\sqrt{N}} \sum_{j} e^{i\vec{q}\cdot\vec{R}_{j}} \,\Delta v_{j}|j\rangle \tag{7.119}$$

and

$$\langle \psi_{\vec{q}} | H | \psi_{\vec{q}} \rangle = \varepsilon_{0}\eta(\vec{q}) + \Lambda(\vec{q}), \tag{7.120}$$

where

$$\Lambda(\vec{q}) \equiv \frac{1}{N} \sum_{j,k} e^{i\vec{q}\cdot(\vec{R}_{j}-\vec{R}_{k})} \langle k|\Delta v_{j}|j\rangle. \tag{7.121}$$

Thus we obtain for the energy eigenvalue

$$\epsilon_{\vec{q}} = \varepsilon_{0} + \frac{\Lambda(\vec{q})}{\eta(\vec{q})}. \tag{7.122}$$

For the case of widely separated atoms we anticipate that $\Lambda(\vec{q})$ will be very small and $\eta(\vec{q})$ will be close to unity. We can parameterize $\Lambda(\vec{q})$ by assuming that

$$\langle j | \Delta v_j | j \rangle = \Delta\varepsilon \tag{7.123}$$

and

$$\langle k | \Delta v_j | j \rangle = t_0 \tag{7.124}$$

if k is a nearest neighbor of j. The parameter t_0 is often referred to as the transfer integral or hopping matrix element, as it is what allows the electron to hop from one atom to another (without it the electron would be bound to a single atom forever). We assume the remaining matrix elements vanish. Then we have

$$\Lambda(\vec{q}) = \Delta\varepsilon + t_0 z \gamma(\vec{q}), \tag{7.125}$$

where z is the coordination number of the lattice (the number of nearest-neighbors of each atom), and

$$\gamma(\vec{q}) \equiv \frac{1}{z} \sum_{\vec{\delta}} e^{-i\vec{q}\cdot\vec{\delta}}, \tag{7.126}$$

where $\vec{\delta}$ is summed over the set of vectors connecting a lattice site to its nearest neighbors.

Similarly, we can assume

$$\langle j | \ell \rangle = \begin{cases} 1 & \text{if } j = \ell \\ \zeta & \text{if } j \text{ and } \ell \text{ are nearest neighbors} \\ 0 & \text{otherwise.} \end{cases} \tag{7.127}$$

Then

$$\eta(\vec{q}) = 1 + z\zeta\gamma(\vec{q}), \tag{7.128}$$

and the energy is

$$\epsilon_{\vec{q}} = \varepsilon_0 + \frac{\Delta\varepsilon + t_0 z \gamma(\vec{q})}{1 + z\zeta\gamma(\vec{q})}. \tag{7.129}$$

Since all this is approximate anyway, one often does not bother to compute t_0 and ζ but rather treats them as phenomenological constants. Often one ignores ζ and just uses the following simple form:

$$\epsilon_{\vec{q}} = A + z t_0 \gamma(\vec{q}), \tag{7.130}$$

which yields a band width $W = 2z|t_0|$ (since $-1 < \gamma < +1$).

The tight-binding dispersion in 1D would thus be (if $t_0 < 0$, which is often the case)

$$\epsilon_q = -\frac{W}{2} \cos(qa), \tag{7.131}$$

as shown in Fig. 7.9 (where it is assumed that $A = 0$). When the atoms are widely separated, t_0 will be very small, and the band will be nearly flat. This means that a wave packet

$$\phi(\vec{r}, t) = \int_{\text{BZ}} d^d\vec{q} \; \psi_{\vec{q}}(\vec{r}) e^{-i\epsilon_{\vec{q}} t/\hbar} \tag{7.132}$$

that is localized near the origin will nearly be a stationary eigenstate of H and hence remain localized near the origin for a long time. The physical interpretation of this lack of dispersion is simply that it takes a long time $t \sim \hbar/W$ for the electron to tunnel between atomic sites if they are widely separated.

This simple example shows that the dispersion of tight-binding bands is often far from the simple parabola appropriate to free electrons. We can, however, define an **effective mass** m^* in terms of the

Table 7.1 Comparison of the band properties between nearly free electrons in a weak periodic potential and the tight-binding limit of a strong periodic potential

	Nearly free electrons	Tight-binding limit
Starting point	Plane waves	Atomic Orbitals
Band width	Large	Small
Origin of band width or band dispersion	Kinetic energy ("obvious")	Hopping (subtle)
Band gap	Small	Large
Origin of band gap	Bragg scattering from periodic potential (subtle)	Spacing between atomic energy levels ("obvious")

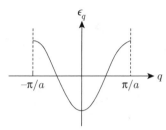

Figure 7.9 Dispersion of a 1D tight-binding band.

band curvature near the minimum. In the simple 1D example discussed above, the energy near the minimum is

$$\epsilon_q - \epsilon_0 \approx \frac{W}{4} q^2 a^2 = \frac{\hbar^2}{2m^*} q^2, \tag{7.133}$$

from which we can derive

$$m^* = \frac{2\hbar^2}{W a^2}. \tag{7.134}$$

For weak tunneling, W is small and m^* is large.

Before getting into discussions of more specific topics in the context of the tight-binding method, we compare in Table 7.1 the difference in size and origin of the band width and band gap between nearly free electrons and the tight-binding limit of a periodic potential. It is easy to understand the band dispersion, but harder to understand the band gap in the former, while the origin of the band gap is more obvious in the latter, but the presence of the band width/dispersion comes from the subtler effect of hopping.

Exercise 7.10. Consider a 1D tight-binding model with long-range hopping, such that the band dispersion is $\epsilon_k = \sum_{n=1}^{\infty} t_n \cos(kn)$, where t_n is the hopping matrix element between nth-nearest neighbors, and we have set the lattice constant $a = 1$ for simplicity. For carefully chosen t_n the band dispersion can be made exactly quadratic: $\epsilon_k = k^2$ for the entire first Brillouin zone. Find the t_n that accomplish this. (Hint: the band dispersion takes the form of a Fourier series.)

Exercise 7.11. Even if we are allowed to keep only one orbital for each atom, lattice translation symmetry does not completely determine the Bloch wave functions if we have a non-Bravais lattice. The following example illustrates this point. Consider a 1D crystal in the tight-binding

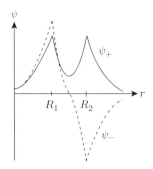

Figure 7.10 Illustration of bonding and anti-bonding electron wave functions of a diatomic molecule.

approximation (using a single atomic S-orbital only), and (for simplicity) treat the orbitals on different atoms as being orthogonal. (i) Suppose that the crystal dimerizes so that the nearest-neighbor hopping matrix elements alternate between two different values t_1 and t_2, corresponding to spacings between neighboring atoms of a_1 and a_2, respectively. Compute the complete band structure exactly (i.e. find all the eigenfunctions and eigenvalues of H). (Note: the eigenfunctions will have two components, one for each of the two atomic orbitals per unit cell.) (ii) Now consider the special case of no dimerization, so that $a_1 = a_2 = a$ and $t_1 = t_2 = t_0$, and show that in this case the band structure is equivalent to Eq. (7.131), but expressed here in a reduced-zone scheme. Note the similarity between the situation here and that of the phonon spectrum discussed in Section 5.3.

7.4.1 Bonds vs. Bands

We are familiar from chemistry with the concept of bonding and anti-bonding molecular orbitals made up of linear combinations of atomic orbitals (LCAOs). For instance, consider two hydrogen atoms forming a hydrogen molecule. The appropriate (unnormalized) diatomic molecular orbitals are approximately

$$\psi_\pm(\vec{r}) = \frac{1}{\sqrt{2}}\left\{\phi_{1S}(\vec{r} - \vec{R}_1) \pm \phi_{1S}(\vec{r} - \vec{R}_2)\right\}, \tag{7.135}$$

which are illustrated in Fig. 7.10. ψ_+ is the bonding orbital because it is nodeless and hence has lower kinetic energy. The potential energy is also lower since there is no node and the electron density is more concentrated in the region in between, but close to, the two nuclei.

Notice that the $+$ and $-$ phase factors are the same ones that occur in the 1D tight-binding model at $k = 0$ (zone center) and $k = \pm\pi/a$ (zone boundary). Hence we can roughly connect the bandwidth W to the chemistry by

$$W \sim z(\epsilon_{AB} - \epsilon_B), \tag{7.136}$$

where ϵ_B and ϵ_{AB} are the energies of the bonding and anti-bonding orbitals, respectively. For the case of one electron per atom, the band will be half full (due to spin), and chemical bonding occurs because electrons fill only the lower (bonding) half of the band. The six-atom benzene ring is a classic example. The chemist's "resonating bonds" are closely related to the condensed matter physicist's Bloch states.

7.4.2 Wannier Functions

We saw earlier that Bloch states with different wave vectors are automatically orthogonal even if the underlying atomic orbitals are not. One might inquire whether it is possible to use the Bloch states to work backwards and derive local orbitals that are actually orthogonal. This was accomplished by Gregory Wannier, after whom these orbitals are named. The **Wannier function** for band n centered on the jth unit cell is given by

$$|\chi_{nj}\rangle \equiv \frac{1}{\sqrt{N}} \sum_{\vec{q}} e^{-i\vec{q}\cdot\vec{R}_j} |\Psi_{n\vec{q}}\rangle, \tag{7.137}$$

where $|\Psi_{n\vec{q}}\rangle$ is the Bloch state for the nth band:

$$|\Psi_{n\vec{q}}\rangle = \frac{1}{\sqrt{\eta(\vec{q})}} \frac{1}{\sqrt{N}} \sum_{\ell} e^{i\vec{q}\cdot\vec{R}_\ell} |n\ell\rangle, \tag{7.138}$$

and $\eta(\vec{q})$ is the normalization factor defined in Eq. (7.108), which is needed because the atomic orbitals on different sites are not orthogonal.

Notice that, if the atomic orbitals *were* orthogonal, we would have $\eta(\vec{q}) = 1$, and Eq. (7.137) would simply reduce to

$$|\chi_{nj}\rangle = |nj\rangle. \tag{7.139}$$

If the orbitals overlap slightly then

$$|\chi_{nj}\rangle = \sum_{\ell} b_n(\vec{R}_\ell - \vec{R}_j)|n\ell\rangle, \tag{7.140}$$

where b_n is an oscillating function strongly peaked at sites ℓ close to site j. In the case of strong overlap $b_n(\vec{R}_\ell - \vec{R}_j)$ becomes long-ranged and the whole tight-binding model becomes doubtful. However, even if the tight-binding approximation breaks down and the Bloch states $|\Psi_{n\vec{q}}\rangle$ need to be obtained by other means, one can still use Eq. (7.137) to generate a set of localized orbitals (also known as Wannier functions), which may not look like atomic orbitals at all, but are nevertheless the same on each lattice site (or in each unit cell):

$$\langle \vec{r}|\chi_{nj}\rangle = f_n(\vec{r} - \vec{R}_j); \tag{7.141}$$

that is, f_n is independent of j.

In interacting systems, local Wannier functions are useful because the electronic correlations that need to be built into the many-body wave function tend to be local in space.

One subtlety of the Wannier functions is that Eq. (7.137) does *not* define them uniquely. The reason is that the Bloch states $|\Psi_{n\vec{q}}\rangle$ are well defined up to an arbitrary $\vec{q}$-dependent phase; that is, $e^{i\theta(\vec{q})}|\Psi_{n\vec{q}}\rangle$ is an equally legitimate set of Bloch states, where $\theta(\vec{q})$ is an arbitrary function of $\vec{q}$. Changing $\theta(\vec{q})$, however, leads to non-trivial changes of $|\chi_{nj}\rangle$ and $f_n(\vec{r})$. One can take advantage of this freedom of Bloch states to "optimize" the Wannier function, by choosing the phase factor that minimizes its spatial spreading (maximally localizes it). It can be shown that for ordinary bands one can always construct Wannier functions that are exponentially localized, just like atomic orbitals. However, as we will discuss in Chapter 13, bands can have non-trivial topological structures and carry non-trivial topological quantum numbers. For such topologically non-trivial bands it is *impossible* to construct exponentially localized Wannier functions. This reflects the fact that the physics associated with such bands is intrinsically *non-local* and cannot be captured by starting from a highly localized set of basis states.

We end this subsection with several exercises illustrating important properties of Wannier functions, which will be used repeatedly in later sections and chapters.

Exercise 7.12. From the definition of the Wannier functions, prove (i) their orthonormality

$$\langle \chi_{n'j'}|\chi_{nj}\rangle = \delta_{nn'}\delta_{jj'}; \tag{7.142}$$

and (ii) that the original normalized Bloch states can be reconstructed from them as

$$|\Psi_{n\vec{q}}\rangle = \frac{1}{\sqrt{N}} \sum_{\ell} e^{i\vec{q}\cdot\vec{R}_{\ell}} |\chi_{n\ell}\rangle. \tag{7.143}$$

Exercise 7.13. The electron Hamiltonian can be formally expressed in terms of the band dispersions and Bloch states (including corresponding bras) as

$$H = \sum_{n} H_n, \tag{7.144}$$

where the (sub)Hamiltonian corresponding to band n is

$$H_n = \sum_{\vec{q}} \epsilon_{n\vec{q}} |\Psi_{n\vec{q}}\rangle\langle\Psi_{n\vec{q}}|. \tag{7.145}$$

Show that H_n may be expressed in terms of the Wannier functions (including corresponding bras) as

$$H_n = \sum_{jj'} t_n(\vec{R}_j - \vec{R}_{j'}) |\chi_{nj}\rangle\langle\chi_{nj'}|, \tag{7.146}$$

where

$$t_n(\vec{R}) = \frac{1}{N} \sum_{\vec{q}} \epsilon_{n\vec{q}} \, e^{i\vec{q}\cdot\vec{R}} \tag{7.147}$$

is the lattice Fourier transform of $\epsilon_{n\vec{q}}$. Thus the Hamiltonian can always be cast in a tight-binding form for all the bands in terms of Wannier functions (as local orbitals), although the orbitals may not be well localized, and hopping over arbitrarily large distances may be involved.

7.4.3 Continuum Limit of Tight-Binding Hamiltonians

The tight-binding approximation (for a single band) greatly simplifies the problem by reducing the Hilbert space to one atomic orbital or Wannier function per atom. As a result, instead of working with the full Bloch wave function that depends on the continuous variable $\vec{r}$, we need only work with the amplitude of the state on atom j, which we will call Φ_j. The full Bloch wave function is constructed from Φ_j and the orthonormal local orbital $|j\rangle$ (we neglect the orbital or band index to simplify notation in this subsection):

$$|\Psi_{\vec{q}}\rangle = \sum_j \Phi_j |j\rangle. \tag{7.148}$$

For a Bravais lattice, the discrete translation symmetry dictates that

$$\Phi_j = \frac{1}{\sqrt{N}} e^{i\vec{q}\cdot\vec{R}_j}. \tag{7.149}$$

Later on we will encounter situations in which lattice translation symmetry is not sufficient to determine Φ_j, and it needs to be solved from the appropriate Hamiltonian. We note that in many cases the (coarse-grained) physical properties of the electron states depend mostly on Φ_j but not sensitively on the details of the orbital wave function $\varphi(\vec{r})$, so Φ_j is often the focus of the study and viewed as the wave function itself.

We now consider the Schrödinger equation that Φ_j satisfies. For simplicity we consider a 1D lattice with nearest-neighbor tunneling or hopping only (see Appendix C and Exercise 7.13):

$$H = \sum_j |t|(-|j\rangle\langle j+1| + 2|j\rangle\langle j| - |j+1\rangle\langle j|). \tag{7.150}$$

Note that the second term in the parentheses simply adds a constant term to the energy such that the band minimum is at $\epsilon = 0$, and the minus signs in front of the first and third terms ensure that the band minimum is located at the center of the first Brillouin zone.[12] Plugging Eq. (7.148) into the Schrödinger equation $H|\Psi\rangle = \epsilon|\Psi\rangle$ yields

$$|t|(-\Phi_{j-1} + 2\Phi_j - \Phi_{j+1}) = \epsilon\Phi_j. \tag{7.151}$$

One immediately notices that the LHS of Eq. (7.151) looks like the discrete lattice version of the second derivative, and the equation may be replaced by

$$-\frac{\hbar^2}{2m^*}\frac{d^2}{dx^2}\Phi(x) = \epsilon\Phi(x), \tag{7.152}$$

if Φ_j varies slowly with j, which allows us to replace it with $\Phi(x = ja)$ in the same spirit as that of replacing lattice displacement $\vec{u}_j$ with $\vec{u}(\vec{r})$ in Chapter 5. Here a is the lattice spacing and the effective mass is

$$m^* = \frac{\hbar^2}{2|t|a^2}, \tag{7.153}$$

consistent with Eq. (7.134).

The fact that Φ_j varies slowly with j means that the state has small q, or is a linear combination of states with small q. We learned earlier that the dispersion of such states is well approximated by the effective mass approximation. Here we showed that its wave function $\Phi_j = \Phi(x = ja)$ follows a continuum Schrödinger equation with the same effective mass. We note that the *full* wave function is (essentially) the product of Φ and the orbital wave function φ, which varies much faster. From Eq. (7.148) we have (now going to general dimension)

$$\Psi_{\vec{q}}(\vec{r}) = \sum_j \Phi_j \varphi(\vec{r} - \vec{R}_j), \tag{7.154}$$

which has the continuum limit

$$\Psi_{\vec{q}}(\vec{r}) \approx \Phi_{\vec{q}}(\vec{r})\psi_{\vec{q}}(\vec{r}), \tag{7.155}$$

with (for a Bravais lattice) $\Phi_{\vec{q}}(\vec{r}) \approx e^{i\vec{q}\cdot\vec{r}}$ and $\psi_{\vec{q}}$ being the spatially periodic function

$$\psi_{\vec{q}} \equiv \sum_j e^{-i\vec{q}\cdot(\vec{r}-\vec{R}_j)}\varphi(\vec{r} - \vec{R}_j). \tag{7.156}$$

For this reason $\Phi(\vec{r})$ is also referred to as the **envelope function** of the full wave function.

The usefulness of Eq. (7.152), which is the continuum limit of the original tight-binding equation Eq. (7.151), lies in two aspects. (i) Conceptually this allows us to apply known properties of the continuum Schrödinger equation to Bloch electrons. In fact, this is the reason why Bloch electrons behave in many circumstances as if the periodic potential were not there. Later in this chapter we will also encounter situations in which the lattice Schrödinger equation reduces to the continuum **Dirac equation**, and insight about the latter is indispensable to the understanding of the physics. (ii) Practically, we often need to solve the electron wave function in the presence of a periodic potential *and* additional perturbations, usually coming from a slowly varying external electromagnetic field. Taking the continuum limit allows us to divide and conquer the effects of a periodic potential and (slowly varying) external perturbations, namely we first reduce the lattice Hamiltonian to Eq. (7.152)

---

[12] Note that we can always choose the phase of the orbitals on one sublattice of a bipartite lattice system to make the sign of the hopping be positive or negative (or even complex). Equivalently, the wave vector $\vec{q}$ is not a physical observable because it is gauge-dependent.

or its analog, and then add on top of it the smooth external perturbations. We will encounter examples of this strategy in later chapters.

Exercise 7.14. Derive Eq. (7.151).

Exercise 7.15. In the text we showed that a continuum limit naturally emerges for states near the band bottom, where Φ_j varies slowly with j. For states near the band maximum (at the zone boundary $K = \pi/a$) this is not the case for Φ_j. However, it is possible to construct a slowly varying variable from Φ_j, by factoring out a j-dependent factor that varies with the wave vector K. Identify this variable and drive the continuum Schrödinger equation it satisfies from Eq. (7.151). Hint: it will be useful to shift the band energy by a constant so that the band maximum is at $\epsilon = 0$.

7.4.4 Limitations of the Tight-Binding Model

So far we have considered the case of a single atomic orbital. Neglecting the other orbitals allowed us to uniquely determine the Bloch energy eigenfunctions from consideration of translation symmetry alone. This idealization is an oversimplification insofar as real solids are concerned. As shown in Fig. 7.11, the width of the separate atomic bands increases exponentially with decreasing atomic spacing. When the bands begin to overlap, it becomes essential to include hybridization (coherent mixing). If this mixing brings in too many different orbitals, the method begins to be inappropriate. The most general wave function including hybridization has the form

$$\psi_{m\vec{q}} = e^{i\vec{q}\cdot\vec{r}} u_{m\vec{q}}(\vec{r}), \tag{7.157}$$

with

$$u_{m\vec{q}}(\vec{r}) = \sum_{n=0}^{\infty} a_{mn\vec{q}} \sum_j e^{-i\vec{q}\cdot\vec{R}_j} \varphi_n(\vec{r} - \vec{R}_j). \tag{7.158}$$

Strictly speaking, the sum over the complete set of atomic orbitals φ_n must include the continuum scattering states as well. We will ignore this complication. In situations where the scattering states are important, the tight-binding approximation is inappropriate anyway.

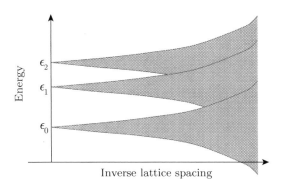

Figure 7.11 Schematic illustration of the evolution of discrete atomic energy levels into ever broader bands as the atoms in a crystal are brought closer together. Initially each band is clearly identifiable with a single atomic state, but eventually the bands begin to overlap and band mixing can no longer be neglected.

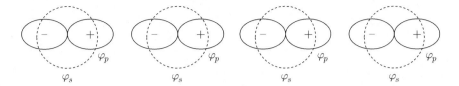

Figure 7.12 The 1D tight-binding model with two atomic orbitals per site. The s orbital is denoted φ_s and the p_x orbital is denoted φ_p. Note that the near-neighbor p–p hopping integral will (probably) have the opposite sign to that of the near-neighbor s–s hopping matrix element. The near-neighbor s–p coupling will alternate in sign. Inversion symmetry implies that the on-site s–p coupling will vanish just as it does in the isolated atom.

Now translation symmetry alone is no longer sufficient to uniquely determine the wave function. The hybridization coefficients $a_{mn\vec{q}}$ that describe the mixing of the mth and nth bands (or, more precisely, the amplitude of the contribution of the nth atomic orbital to the mth band) must be found by diagonalizing the Hamiltonian matrix. Translation symmetry is still a powerful aid, however. If we consider the mixing of M bands, the Hamiltonian is only $M \times M$ (for each $\vec{q}$) and not of dimension $M N_{\text{atom}}$. The special case we studied before was simply a 1×1 matrix.

As a simple illustration, consider Fig. 7.12, which shows a line of atoms with two orbitals per site: an s orbital $\langle \vec{r} \, | \varphi_{sj} \rangle = \varphi_s(\vec{r} - \vec{R}_j)$ with atomic energy level ϵ_s and a p_x orbital $\langle \vec{r} \, | \varphi_{pj} \rangle = \varphi_p(\vec{r} - \vec{R}_j)$ with energy ϵ_p.[13] The s-wave Bloch state is

$$|\Phi_{s\vec{k}}\rangle = \frac{1}{\sqrt{N}} \sum_j e^{i\vec{k}\cdot\vec{R}_j} |\varphi_{sj}\rangle \tag{7.159}$$

and the p_x Bloch state is

$$|\Phi_{p\vec{k}}\rangle = \frac{1}{\sqrt{N}} \sum_j e^{i\vec{k}\cdot\vec{R}_j} |\varphi_{pj}\rangle. \tag{7.160}$$

Here we have ignored the non-orthogonality correction to the normalization (i.e. taken $\eta(\vec{k}) = 1$). The tight-binding eigenvectors are of the form

$$|\psi_{\vec{k}}\rangle = a_{\vec{k}} |\Phi_{s\vec{k}}\rangle + b_{\vec{k}} |\Phi_{p\vec{k}}\rangle. \tag{7.161}$$

The band-mixing amplitudes are found by solving (for each $\vec{k}$) the 2×2 matrix Schrödinger equation

$$h_{\vec{k}} \begin{pmatrix} a_{\vec{k}} \\ b_{\vec{k}} \end{pmatrix} = \epsilon_{\vec{k}} \begin{pmatrix} a_{\vec{k}} \\ b_{\vec{k}} \end{pmatrix}, \tag{7.162}$$

where

$$h_{\vec{k}} \equiv \begin{pmatrix} \langle \Phi_{s\vec{k}} | H | \Phi_{s\vec{k}} \rangle & \langle \Phi_{s\vec{k}} | H | \Phi_{p\vec{k}} \rangle \\ \langle \Phi_{p\vec{k}} | H | \Phi_{s\vec{k}} \rangle & \langle \Phi_{p\vec{k}} | H | \Phi_{p\vec{k}} \rangle \end{pmatrix}. \tag{7.163}$$

Let us now turn to the problem of evaluating these matrix elements.

We assume the following non-zero matrix elements of H:

$$\langle \varphi_{sj} | H | \varphi_{sj} \rangle = \tilde{\epsilon}_s, \tag{7.164}$$

$$\langle \varphi_{sj} | H | \varphi_{sj+1} \rangle = \langle \varphi_{sj+1} | H | \varphi_{sj} \rangle = t_{ss}, \tag{7.165}$$

$$\langle \varphi_{pj} | H | \varphi_{pj} \rangle = \tilde{\epsilon}_p, \tag{7.166}$$

---

[13] The p orbitals come in trios. In atomic physics we often label them using the L_z quantum number, $m = 0, \pm 1$. In condensed matter physics it is often convenient to use linear combinations of them that stretch along the $\hat{x}, \hat{y},$ and $\hat{z}$ directions, labeled $p_x, p_y,$ and p_z orbitals. In this toy example we consider only the p_x orbital.

$$\langle \varphi_{pj}|H|\varphi_{pj+1}\rangle = \langle \varphi_{pj+1}|H|\varphi_{pj}\rangle = t_{pp}, \tag{7.167}$$

$$\langle \varphi_{pj}|H|\varphi_{sj+1}\rangle = \langle \varphi_{sj+1}|H|\varphi_{pj}\rangle = t_{sp}, \tag{7.168}$$

$$\langle \varphi_{pj+1}|H|\varphi_{sj}\rangle = \langle \varphi_{sj}|H|\varphi_{pj+1}\rangle = t_{ps}. \tag{7.169}$$

Note that, from the symmetry of the p orbitals, $t_{ps} = -t_{sp}$, and we also anticipate that t_{pp} will (most likely) have its sign opposite that of t_{ss}. Note that parity (inversion) symmetry causes

$$\langle \varphi_{pj}|H|\varphi_{sj}\rangle = 0. \tag{7.170}$$

Using these matrix elements, it is straightforward to obtain

$$\langle \Phi_{s\vec{k}}|H|\Phi_{s\vec{k}}\rangle = \tilde{\epsilon}_s + 2t_{ss}\gamma(\vec{k}), \tag{7.171}$$

$$\langle \Phi_{p\vec{k}}|H|\Phi_{p\vec{k}}\rangle = \tilde{\epsilon}_p + 2t_{pp}\gamma(\vec{k}), \tag{7.172}$$

$$\langle \Phi_{s\vec{k}}|H|\Phi_{p\vec{k}}\rangle = 2it_{ps}\tilde{\gamma}(\vec{k}), \tag{7.173}$$

$$\langle \Phi_{p\vec{k}}|H|\Phi_{s\vec{k}}\rangle = -2it_{ps}\tilde{\gamma}(\vec{k}), \tag{7.174}$$

where

$$\gamma(\vec{k}) = \cos(ka) \tag{7.175}$$

$$\tilde{\gamma}(\vec{k}) = \sin(ka). \tag{7.176}$$

Notice that, since both the p orbital and the wave vector $\vec{k}$ are odd under parity, parity symmetry implies

$$\langle \Phi_{p\vec{k}}|H|\Phi_{s\vec{k}}\rangle = -\langle \Phi_{p(-\vec{k})}|H|\Phi_{s(-\vec{k})}\rangle. \tag{7.177}$$

Hence the simple parity rule which prevents mixing of s and p orbitals in the isolated atom is destroyed by the presence of the wave vector $\vec{k}$ in the solid and the s and p bands are allowed to mix if $\vec{k} \neq \vec{0}$. However, at the zone center or boundary the symmetry is restored and $\langle \Phi_{p\vec{k}}|H|\Phi_{s\vec{k}}\rangle = 0$ for $k = 0, \pi/a$. That is, the mixing vanishes by parity symmetry as it does in the isolated atom.[14]

> **Exercise 7.16.** Use parity argument to show that $\langle \Phi_{p\vec{k}}|H|\Phi_{s\vec{k}}\rangle = 0$ for $k = 0, \pi/a$ (i.e. at zone center and boundary).

The 2×2 block Hamiltonian becomes

$$h_{\vec{k}} = \begin{pmatrix} \tilde{\epsilon}_s + 2t_{ss}\gamma(\vec{k}) & 2it_{ps}\tilde{\gamma}(\vec{k}) \\ -2it_{ps}\tilde{\gamma}(\vec{k}) & \tilde{\epsilon}_p + 2t_{pp}\gamma(\vec{k}) \end{pmatrix}. \tag{7.178}$$

Diagonalization of this gives the two band-energy eigenvalues and eigenfunctions for each $\vec{k}$. For simplicity consider the special case $\tilde{\epsilon}_s = \tilde{\epsilon}_p = 0$, $t_{ss} = -t_{pp} = 1/2$, and $t_{ps} = \lambda/2$. Then

$$h_{\vec{k}} = \begin{pmatrix} \cos(ka) & i\lambda \sin(ka) \\ -i\lambda \sin(ka) & -\cos(ka) \end{pmatrix}, \tag{7.179}$$

which has eigenvalues

$$\epsilon_{\pm}(\vec{k}) = \pm\sqrt{\lambda^2 + (1 - \lambda^2)\cos^2(ka)}. \tag{7.180}$$

---

[14] This is more obvious for $k = 0$ than for $k = \pi/a$. To understand the latter, we need to invoke the fact that inversion takes π/a to $-\pi/a$, but recall that $\pm\pi/a$ are effectively the *same* point in reciprocal space since they differ by a reciprocal lattice vector. In 1D the zone boundary is a single point. In higher dimensions some special points on the zone boundary are invariant under inversion. More on this in Chapter 14.

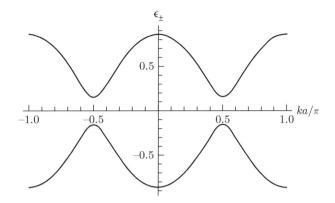

Figure 7.13 Plot of the energy dispersion for the 1D two-orbital model given in Eq. (7.180) for $\lambda = 0.15$.

Notice that, if $\lambda = 1$, the bands are perfectly flat (dispersionless). We can understand this physically by considering the hopping of a particle from $|\varphi_{sj}\rangle$ to $|\varphi_{sj+2}\rangle$ by way of the intermediate states on site $j+1$. There are two separate "paths," one going through $|\varphi_{sj+1}\rangle$ and one through $|\varphi_{pj+1}\rangle$. The matrix elements are such that these two paths suffer perfect destructive interference, and the particle is unable to move. Zero group velocity is the same as a perfectly dispersionless band. The band dispersion for the case $\lambda = 0.15$ is plotted in Fig. 7.13. We label the upper band "s" and the lower band "p" because at the zone center and zone boundary the upper band has eigenvector $\begin{pmatrix} 1 \\ 0 \end{pmatrix}$ and the lower band has eigenvector $\begin{pmatrix} 0 \\ 1 \end{pmatrix}$ (since $h_{\vec{k}}$ is diagonal at these points). Of course, at any finite wave vector each band is a mixture of s and p.

What if the parity were opposite for a band at the zone center and zone boundary? It turns out that this is an extremely interesting case where the band has non-trivial topology. We will return to this case in Chapter 14.

Exercise 7.17. A certain 1D solid aligned in the x direction consists of "S" atoms and "P" atoms in an infinite alternating sequence SPSPSP.... The S atoms have a single atomic s-orbital energy ϵ_s. The P atoms have p_x, p_y, and p_z orbitals of energy ϵ_p. In a nearest-neighbor tight-binding model, the transfer integrals from an S-atom to the P-atom on its right are J_x, J_y, J_z.
(a) What are the corresponding transfer integrals to the P-atom on the left?
(b) Which of the Js are zero by symmetry?
(c) Compute the band structure and sketch the energy levels.
(d) Why is there a total of four bands, and why are some of them dispersionless?
(e) Suppose the P atom contains a spin–orbit coupling of the form $\Delta H = \Gamma \vec{L} \cdot \vec{S}$, where $\vec{L}$ is the orbital angular momentum and $\vec{S}$ is the spin operator. What are the energy levels and eigenstates of an isolated P atom? Describe how you would solve for the tight-binding band structure in this case. Write down all the transfer matrix elements, but do not bother to solve the Hamiltonian.

7.4.5 *s–d* Hybridization in Transition Metals

The transition metals (which live in the middle part of the periodic table) have both s and d valence electrons. The $3d$ transition series runs from Sc ($3d^1 4s^2$) to Ni ($3d^8 4s^2$). The $4d$ series runs from

Y ($4d^1 5s^2$) to Pd ($4d^{10} 5s^0$), and the $5d$ (lanthanide) series runs from La ($5d^1 6s^2$) to Pt ($5d^9 6s^1$). The electronic band structure of these materials is dominated by s–d hybridization, and the electronic properties vary in systematic and interesting ways across the series.

In a free atom for which the Hamiltonian is spherically symmetric, the single-particle orbitals in the independent-electron approximation have the usual form

$$\psi_{n\ell m}(\vec{r}) = R_{n\ell}(r) Y_\ell^m(\theta, \varphi). \tag{7.181}$$

For d-waves ($\ell = 2$) the spherical harmonics have the form

$$Y_2^0 = \sqrt{\frac{5}{4\pi}} \left(\frac{3}{2} \cos^2 \theta - \frac{1}{2} \right) \tag{7.182}$$

$$Y_2^{\pm 1} = \mp \sqrt{\frac{15}{8\pi}} \sin\theta \, \cos\theta \, e^{\pm i\varphi} \tag{7.183}$$

$$Y_2^{\pm 2} = \frac{1}{4} \sqrt{\frac{15}{2\pi}} \sin^2\theta e^{\pm 2i\varphi}. \tag{7.184}$$

In a solid, the spherical symmetry is not present because each atom is perturbed by the presence of its neighbors. The transition metals either have cubic symmetry (either BCC or FCC) or are hexagonal close-packed (HCP). The HCP structure differs from FCC only at the second-neighbor distance and beyond.[15] Hence it is a good approximation to take the potential surrounding each atom to have cubic symmetry. The resulting perturbation has two major effects:

(1) s–d mixing (because of the lack of continuous rotation symmetry);
(2) Y_ℓ^m mixing of different m states.

There is no mixing of p states into the s and d because parity is still a good symmetry in a cubic environment.[16] We will focus here on the second effect, in which the $2\ell + 1 = 5$ different angular momentum substates (which are degenerate in the isolated atom) are mixed together, but the total angular momentum $\ell = 2$ does not change.

We need to find the appropriate linear combinations of the Y_2^m spherical harmonics which yield cubic symmetry. A good starting point is writing

$$\psi_{n2\beta}(\vec{r}) = \left(\frac{15}{4\pi} \right)^{1/2} R_{n2}(r) \frac{1}{r^2} \Phi_\beta, \tag{7.185}$$

where the Φ_β are the following $2\ell + 1 = 5$ states:

$$\Phi_{xy} = xy, \tag{7.186}$$

$$\Phi_{xz} = xz, \tag{7.187}$$

$$\Phi_{yz} = yz, \tag{7.188}$$

$$\Phi_{x^2-y^2} = \frac{1}{2}(x^2 - y^2), \tag{7.189}$$

$$\Phi_{z^2} = \frac{1}{2\sqrt{3}}(3z^2 - r^2). \tag{7.190}$$

---

[15] See Exercise 3.13, or Chapter 4 of Ref. [1].

[16] We have to be careful with this statement. We have already seen that, when we find the Bloch wave solutions of the Schrödinger equation, the wave vector $\vec{k}$ is odd under parity. It turns out that this means that p states are strictly absent only at the Γ point (the zone center), where $\vec{k} = 0$, and other special points in reciprocal space that are invariant under parity transformation (see Exercise 7.16).

The first three are known as t_{2g} orbitals, while the latter two are known as e_g orbitals. These two groups of orbitals do *not* mix under cubic point-group transformations. As a simple example, consider a rotation of $\pi/2$ about the z axis. This symmetry operation leaves a (suitably oriented) cube invariant and maps $x \to y, y \to -x, z \to z$. Thus, under this operation

$$\Phi_{xy} \to -\Phi_{xy}, \tag{7.191}$$

$$\Phi_{xz} \to \Phi_{yz}, \tag{7.192}$$

$$\Phi_{yz} \to -\Phi_{xz}, \tag{7.193}$$

and the remaining two states remain invariant.

Because the Y_ℓ^m states are mixed together by the crystal potential (often called the **crystal field**), we say that the angular momentum is *quenched*. In the free atom, the d-wave states Y_2^m are degenerate. In the crystalline environment, this degeneracy is lifted and the states are separated in energy by what is referred to as the **crystal field splitting**. For a single atom in the presence of a cubic symmetric potential, states within the t_{2g} and e_g groups remain degenerate; see Exercise 7.18 for an example illustrating this. In the presence of additional symmetry breaking, including generic wave vectors in Bloch bands, the remaining degeneracies are lifted, but the splitting between the t_{2g} and e_g groups is typically larger than that within each group.

The main features of the electronic structure of the transition metals can be crudely described by treating the s electrons as *free* and treating the d orbitals in the *tight-binding* approximation. This works (to an extent) because the $3d$ orbitals mostly lie inside the $4s$ (and the $4d$ inside the $5s$). The s orbitals overlap too strongly to be described by the tight-binding picture. The d band states can be written as

$$\psi_{\vec{k}}(\vec{r}) = e^{i\vec{k}\cdot\vec{r}} \, \Phi_{\vec{k}}(\vec{r}), \tag{7.194}$$

where

$$\Phi_{\vec{k}}(\vec{r}) = \sum_j e^{-i\vec{k}\cdot(\vec{r}-\vec{R}_j)} \sum_{\beta=1}^{5} f_\beta(\vec{k})\Phi_\beta(\vec{r}-\vec{R}_j) \tag{7.195}$$

and β refers to the labels on the cubic orbitals Φ_{xy}, Φ_{yz}, etc. More abstractly, we can represent the state as a column vector of the five amplitudes:

$$|\vec{k}\rangle = \begin{pmatrix} f_{xy} \\ f_{xz} \\ f_{yz} \\ f_{x^2-y^2} \\ f_{z^2} \end{pmatrix}. \tag{7.196}$$

These are eigenvectors of a 5×5 block Hamiltonian which can be found by a generalization of the method outlined in the earlier discussion. The density of states in the resulting band structure typically looks something like that shown qualitatively in Fig. 7.14. The five peaks are the remnants of the five crystal-field-split atomic orbitals. If the lattice constant is decreased, the bandwidth will increase. Each band contains $2(2\ell + 1) = 10$ states (including the two-fold spin degeneracy) per atom, and the total band width is not terribly large (~ 5 eV), so the density of states is very large compared with that in the s (really the s–p) band.

We can use these facts to develop a qualitative picture of the bonding of the transition metals and understand the different pieces of physics that determine the equilibrium lattice spacing. An illustration of the systematic trends in the **cohesive energy**[17] of the transition metals is shown in

---

[17] The cohesive energy is basically the binding energy of the crystal. The more negative it is, the lower the crystal's energy compared with that of individual isolated atoms. For a definition and further discussion, see Section 15.6.

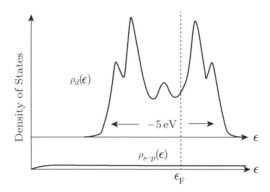

Figure 7.14 Schematic illustration of the qualitative features of the density of states in a transition metal. The d atomic orbitals are crystal-field split and then broadened into overlapping bands. The s orbital electrons live mostly outside the d orbitals, so in the solid they form a much wider $s-p$ band with a correspondingly lower density of states. The location of the Fermi level moves from left to right (at least relative to the d bands) as the atomic number (and hence the number of filled atomic d orbitals) increases across the series.

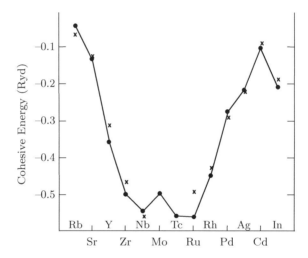

Figure 7.15 Systematic trend of the cohesive energy across the $4d$ transition-metal series from atomic number 39 (Y) to 48 (Cd). The cohesive energy is largest in the middle of the series because the d band is half filled there and only bonding states are occupied. The solid dots are theoretical band-structure calculations and the crosses are experimental values. Figure adapted with permission from [33]. Copyright by the American Physical Society.

Fig. 7.15. The cohesive energy is largest in the middle of the series because the d band is half filled there, and only bonding states are occupied. If it is less than half filled, fewer bonding states are occupied. If it is more than half filled, anti-bonding states are also occupied. Either way, the energy is less favorable.

Notice, however, that, as long as the d band is not completely empty or completely full, an increase in the total band width always gives stronger bonding and lower energy. Thus the d bonds would prefer to see a *smaller* value of the lattice spacing. This tendency is opposed, however, by the $s-p$ electrons. They are light and moving more or less freely, forming a nearly free Fermi gas. Reduction of the lattice constant would increase the density of this gas and hence raise the Fermi energy, which scales as

$$\epsilon_F \sim \frac{\hbar^2}{2m^*} a^{-2}. \tag{7.197}$$

The equilibrium lattice constant a is controlled by the competition between the attractive force of the d bonds and the repulsive Fermi pressure of the $s–p$ electrons.

Note from Fig. 7.14 that the strong d-band bonding makes the cohesive energies in the middle of the transition series very large. This is why these metals are so refactory (able to withstand high temperatures). Tungsten (W), for example, has a cohesive energy of nearly 9 eV per atom. This is one of the reasons why it makes a good light-bulb filament.

Exercise 7.18. We know that eigenfunctions of a single-electron Hamiltonian with a rotationally invariant potential take the form of Eq. (7.181), with states of fixed n and l but different m being degenerate. Now consider adding a perturbing potential of the form

$$\Delta V(x, y, z) = \lambda(x^4 + y^4 + z^4), \qquad (7.198)$$

which breaks full rotation symmetry but respects cubic symmetry. That is, the symmetry of the system has been reduced from spherical to cubic by ΔV. Treat ΔV to first order perturbatively for the set of degenerate states with $l = 2$ and any n (but assume there is no accidental degeneracy for different ns), and show that the five-fold degeneracy is split into two levels, with three- and two-fold degeneracies, respectively, and the angular parts of the states in each group take the forms of those of the t_{2g} and e_g orbitals, or any linear combinations within each group.

7.5 Graphene Band Structure

The two most common forms of elemental carbon in nature are diamond and graphite. The diamond lattice (see Fig. 3.8) is an FCC lattice with a two-point basis, while graphite is a layered system formed by stacking layers of carbon atoms forming honeycomb lattices (see Fig. 3.7). A single atomic layer of graphite, called graphene, can be isolated and is found to exhibit fascinating physical properties. In this section we use the tight-binding method to study the electronic properties of graphene.

As usual, we start with the atomic orbitals of a single carbon atom, whose electronic configuration is $1s^2 2s^2 2p^2$. Obviously, the $1s$ orbital has very low energy (i.e. large binding energy) and forms deep and completely filled $1s$ band(s) regardless of the crystalline structure formed by the carbon atoms. These deep bands are irrelevant to the low-energy physics. On the other hand, the behavior of the four electrons in the $2s$ and $2p$ orbitals depends on the crystalline structure, and determines the low-energy properties of the system. First of all, the single $2s$ and three $2p$ orbitals are very close in energy to begin with. Secondly, the rotation symmetry is broken by the crystalline environment, and these orbitals are thus mixed by the crystal field. As a result, these four orbitals (often referred to as sp^3 or simply sp orbitals) need to be treated on an equal footing, and the resultant bands are referred to as sp^3 bands.

We start by considering diamond. In this case we have two atoms per unit cell, thus eight sp^3 orbitals per unit cell and eight sp^3 bands. The coordination number of the diamond lattice is $z = 4$. In fact, one can view an atom and its four nearest neighbors in a diamond lattice as forming a tetrahedron, with one atom at the center and the four neighbors sitting on the four corners (see Exercise 3.3). Since each atom has four sp^3 orbitals, it can pair one orbital with the corresponding orbital on each neighbor, to form bonding and anti-bonding orbitals. As a result, the eight sp^3 bands naturally group into four (lower-energy) bonding bands and four (higher-energy) anti-bonding bands, with a large gap separating them. The four sp^3 electrons per atom (and thus eight per unit cell) fully occupy the four bonding bands and leave the anti-bonding bands empty.[18] As a result, diamond is an extremely good band insulator.

We now turn our discussion to graphene. Here again we have two atoms per unit cell, but the main difference is that the coordination number of the honeycomb lattice is $z = 3$. We thus expect, by going through similar arguments as for diamond above, that there will be three fully occupied bonding bands and three empty anti-bonding bands at high energy (these are called σ bands). Locally one can understand them by mixing the s orbital with p_x and p_y orbitals to form orbitals that extend along the directions of nearest-neighbor bonds, known as σ orbitals. The bonding and anti-bonding combinations of these σ orbitals have energies far below and far above the Fermi energy, respectively. This leaves us with one electron, and one orbital per atom that does not pair up with neighboring atoms, and it is this orbital that dominates the low-energy physics of graphene. This remaining orbital is basically the p_z orbital that sticks out of the graphene plane and is called the π orbital. In the following we will treat the bands formed by the π orbitals using the tight-binding method and ignore their coupling with other orbitals. For simplicity, we will also ignore the non-zero overlaps between π orbital wave functions on different carbon atoms. Alternatively, we can also formulate the problem in terms of the (orthonormal) Wannier orbitals that correspond to the π bands.

Using $|i\rangle$ to represent the π orbital of the ith atom and setting its energy to be zero, the tight-binding Hamiltonian takes the form

$$
\begin{aligned}
H &= -t \sum_{\langle ij \rangle} (|i\rangle\langle j| + |j\rangle\langle i|) \\
&= -t \sum_{i \in \mathrm{A}, \vec{\delta}} (|i\rangle\langle i + \vec{\delta}| + \mathrm{h.c.}),
\end{aligned}
\tag{7.199}
$$

where $t \approx 2.8\,\mathrm{eV}$ is the nearest-neighbor hopping matrix element, $\langle ij \rangle$ represent a nearest-neighbor pair (thus i and j must be on opposite sublattices), and $\vec{\delta}$ is a vector connecting an A site to one of its three nearest neighbors (see Fig. 7.16). Using the lattice constant $a = |\vec{\delta}| \approx 1.42\,\text{Å}$ as the unit of length, we have $\vec{\delta}_1 = (1, 0)$ and $\vec{\delta}_{2,3} = (-1/2, \pm\sqrt{3}/2)$, and the primitive lattice vectors of the underlying triangular Bravais lattice are $\vec{a}_{1,2} = (3/2, \pm\sqrt{3}/2)$. We neglect hopping between more distant neighbours for simplicity. To take advantage of lattice translation symmetry, we would like to

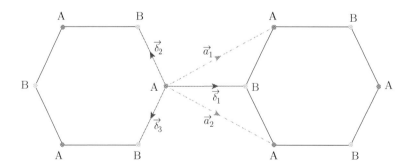

Figure 7.16 Graphene's honeycomb lattice structure. The honeycomb lattice is a triangular lattice with a two-atom basis. The primitive lattice vectors of the underlying triangular lattice are given by the dashed lines.

---

[18] For this reason diamond is a typical example of a covalent-bonded solid, as the (full) occupation of these bonding bands can be viewed in real space as two neighboring carbon atoms forming a covalent bond by each contributing one electron to occupy the bonding orbital. Hence, the whole lattice can be viewed as a huge molecule bound together by these covalent chemical bonds. This also explains why carbon chooses the diamond lattice with coordination number four, instead of more close-packed lattices.

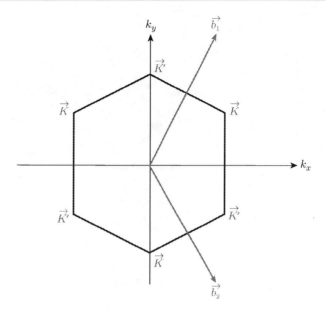

Figure 7.17 The first Brillouin zone of graphene's underlying triangular lattice. $\vec{b}_1$ and $\vec{b}_2$ are primitive unit vectors of the reciprocal space. $\vec{K}$ and $\vec{K}'$ are the two inequivalent Brillouin zone corners that become Dirac points of the spectrum.

construct eigenstates of lattice momentum, as we did in Section 7.4. Since we have two atoms per unit cell, there are two such states for each momentum $\vec{q}$ (one for each sublattice):

$$|\vec{q}\rangle_{A/B} = \frac{1}{\sqrt{N}} \sum_{j \in A/B}^{N} e^{i\vec{q}\cdot\vec{R}_j} |j\rangle. \tag{7.200}$$

Here N is the number of unit cells, and $\vec{q}$ is restricted to the first Brillouin zone, which is a hexagon (see Fig. 7.17). Of particular importance to our discussion later are the zone corners; while a hexagon has six corners, only two of them, namely $\vec{K} = (2\pi/3, 2\pi/3\sqrt{3})$ and $\vec{K}' = -\vec{K}$, are inequivalent, as the others differ from one of these two by a reciprocal lattice vector.

If we had only one atom per unit cell, these momentum eigenstates would automatically diagonalize the Hamiltonian. Here this is not the case, but instead they bring the Hamiltonian (7.199) into block diagonal form, in a way analogous to the calculation of the classical lattice vibration spectrum for a 1D non-Bravais lattice with a two-point basis in Section 5.3:

$$H = -t \sum_{\vec{q}} (|\vec{q}\rangle_A, |\vec{q}\rangle_B) h_{\vec{q}} (\langle\vec{q}|_A, \langle\vec{q}|_B)^T, \tag{7.201}$$

where $h_{\vec{q}}$ is a 2×2 Hermitian matrix,

$$h_{\vec{q}} = \begin{pmatrix} 0 & f(\vec{q}) \\ f^*(\vec{q}) & 0 \end{pmatrix}, \tag{7.202}$$

in which

$$f(\vec{q}) = \sum_{\vec{\delta}} e^{i\vec{q}\cdot\vec{\delta}} = \sum_{i=1}^{3} e^{i\vec{q}\cdot\vec{\delta}_i}. \tag{7.203}$$

The reader is urged to verify the above. Here we only point out the similarity between Eqs. (7.203) and (7.126), and comment that here $f(\vec{q})$ shows up as an off-diagonal matrix element in $h_{\vec{q}}$ because it

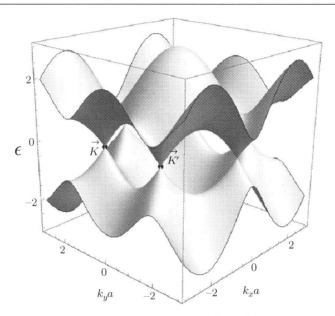

Figure 7.18 Graphene's band structure. Note the Dirac points at $\vec{K}$ and $\vec{K}'$. The other degeneracy points are not independent because they are connected by reciprocal lattice vectors to either $\vec{K}$ or $\vec{K}'$.

corresponds to the coupling between sites on opposite sublattices in the original Hamiltonian (7.199), while $\gamma(\vec{k})$ of Eq. (7.126) enters the eigenenergy directly because it is (part of) a diagonal matrix element of a 1×1 matrix!

Diagonalizing $h_{\vec{q}}$ yields the spectra of graphene's conduction and valence bands:

$$\epsilon_{\pm}(\vec{q}) = \pm t |f(\vec{q})|, \tag{7.204}$$

as plotted in Fig. 7.18. The Fermi energy is precisely at $\epsilon_F = 0$ (in the absence of extrinsic doping due to charge transfer from a nearby substrate or from adsorbed impurity atoms; see Chapter 9 for more extensive discussions of effects of impurity doping). The most notable feature of this graphene band structure is that $\epsilon_{\pm}(\vec{K}) = \epsilon_{\pm}(\vec{K}') = 0 = \epsilon_F$, namely the conduction and valence bands are degenerate at the two inequivalent Brillouin-zone corners, and there is thus *no* band gap. As a result, graphene is (effectively) a semimetal, in sharp contrast to diamond, which is a band insulator with a large band gap. Expanding $\epsilon_{\pm}(\vec{q})$ near $\vec{K}$ we find

$$\epsilon_{\pm}(\vec{K} + \vec{k}) \approx \pm v_F |\vec{k}|, \tag{7.205}$$

where $v_F = \frac{3}{2}ta \approx 10^7$ m/s is the (quite large) graphene Fermi velocity. The same result is obtained near $\vec{K}'$. Thus the dispersions near these degenerate points are linear and resemble those of massless Dirac fermions, except that the speed of light is replaced by v_F. As a result, $\vec{K}$ and $\vec{K}'$ are often referred to as Dirac points in graphene.

The analogy to Dirac fermions actually goes beyond the linear dispersion. As we show below, the states near $\vec{K}$ and $\vec{K}'$ satisfy Dirac-like equations. To do that we need to fix a deficiency of $h_{\vec{q}}$, namely it is *not* periodic in reciprocal space: $h_{\vec{q}} \neq h_{\vec{q}+\vec{G}}$. The origin of this deficiency lies in the fact that $\vec{\delta}$ is not a lattice vector of the underlying (triangular) Bravais lattice, as a result of which $f(\vec{q}) \neq f(\vec{q}+\vec{G})$. Fortunately, this can be easily fixed by a gauge transformation, namely we introduce an additional ($\vec{q}$-dependent) phase to $|\vec{q}\rangle_B$ and work instead with

$$|\vec{q}\rangle_{\tilde{B}} = i e^{-i\vec{q}\cdot\vec{\delta}_1} |\vec{q}\rangle_B, \tag{7.206}$$

resulting in a change of $h_{\vec{q}}$ to

$$\tilde{h}_{\vec{q}} = -t \begin{pmatrix} 0 & \tilde{f}(\vec{q}) \\ \tilde{f}^*(\vec{q}) & 0 \end{pmatrix}, \tag{7.207}$$

with

$$\tilde{f}(\vec{q}) = i e^{-i\vec{q}\cdot\vec{\delta}_1} f(\vec{q}) = i[1 + e^{i\vec{q}\cdot(\vec{\delta}_2-\vec{\delta}_1)} + e^{i\vec{q}\cdot(\vec{\delta}_3-\vec{\delta}_1)}]. \tag{7.208}$$

It is easy to see that $\tilde{f}(\vec{q}) = \tilde{f}(\vec{q}+\vec{G})$, because $\vec{\delta}_2 - \vec{\delta}_1 = -\vec{a}_2$ and $\vec{\delta}_3 - \vec{\delta}_1 = -\vec{a}_1$ are both lattice vectors. Clearly $\tilde{h}_{\vec{q}}$ and $h_{\vec{q}}$ describe the same physics. Writing $\vec{q} = \vec{K} + \vec{k}$ and expanding $\tilde{f}(\vec{q})$ to linear order in $\vec{k}$, we obtain

$$\tilde{f}(\vec{q}) \approx \nabla_{\vec{q}} \tilde{f}(\vec{q})|_{\vec{q}=\vec{K}} \cdot \vec{k} = -\frac{3}{2}(k_x - ik_y), \tag{7.209}$$

thus

$$\tilde{h}_{\vec{q}\approx\vec{K}} \approx v_{\mathrm{F}} \begin{pmatrix} 0 & k_x - ik_y \\ k_x + ik_y & 0 \end{pmatrix} \tag{7.210}$$

$$= v_{\mathrm{F}}(\sigma_x k_x + \sigma_y k_y) = v_{\mathrm{F}}\vec{\sigma}\cdot\vec{k}, \tag{7.211}$$

with σ_x, σ_y the first two Pauli matrices, and $\vec{\sigma} = (\sigma_x, \sigma_y)$. Introducing a "momentum" operator $\vec{p}$ whose eigenvalue is $\vec{k}$ with $|\vec{k}| \ll |\vec{K}|$, we can express the above as an *effective* continuous Hamiltonian describing states near $\vec{K}$ (see Exercise 7.20):

$$H_{\vec{K}} = v_{\mathrm{F}}(\sigma_x p_x + \sigma_y p_y) = v_{\mathrm{F}}\vec{\sigma}\cdot\vec{p}, \tag{7.212}$$

which takes the form of a 2D massless Dirac Hamiltonian.[19] Here $\vec{p}$ should be understood as the operator of lattice momentum *measured from* $\vec{K}$. We can perform a similar analysis near $\vec{K}'$, and arrive at another effective continuous Hamiltonian there:

$$H_{\vec{K}'} = v_{\mathrm{F}}(-\sigma_x p'_x + \sigma_y p'_y). \tag{7.213}$$

In the above $\vec{p}\,'$ is understood as the operator of lattice momentum *measured from* $\vec{K}'$. The two effective Hamiltonians $H_{\vec{K}}$ and $H_{\vec{K}'}$ can be written in a unified manner by introducing a valley degree-of-freedom index, $\tau_z = \pm 1$ for $\vec{K}$ and $\vec{K}'$, respectively, in terms of which we have

$$H = v_{\mathrm{F}}(\tau_z \sigma_x p_x + \sigma_y p_y). \tag{7.214}$$

Note that the biggest difference between the Dirac equation and the Schrödinger equation is that the former has negative-energy solutions. In the present context (and other condensed matter contexts), these solutions have a very clear physical meaning, namely they represent states of the valence band! While in graphene there is no mass term, a straightforward generalization of the model leads to a mass term, which is proportional to the band gap (see Exercise 7.19). Using a condensed matter analogy, we may actually view the vacuum as an electronic band insulator with a huge band gap of $2m_ec^2 \approx 1\,\mathrm{MeV}$ (the cost of producing an electron–positron pair), and the positive- and negative-energy states are nothing but the conduction and valence band states! More fascinating is the fact that there can be *different* types of Dirac mass, and they correspond to different topologies for the bands involved. We will encounter such examples in Chapters 13 and 14.

---

[19] Recall that in 3D one needs four mutually anti-commuting matrices to construct the Dirac Hamiltonian (three for the three components of momentum and one for the mass term), and these are 4×4 matrices. In 2D we need only three such matrices, and the 2×2 Pauli matrices suffice. For massless Dirac fermions we use only two of them, but the third matrix is needed when there is a mass term (see Exercise 7.19).

Exercise 7.19. We discussed graphene's band structure using the tight-binding model on a honeycomb lattice, with nearest-neighbor hopping t. Now consider the same model, but add a site-dependent energy for the local orbital, that has the value $+V$ for all A sublattice sites and $-V$ for all B sublattice sites. (i) Calculate the band dispersions in this case, and in particular the band gap. (ii) Show that, when $V \neq 0$, a mass term is added to the Dirac Hamiltonians (7.212) and (7.213), and calculate this mass. Find its relation to the band gap. We note that the Hamiltonian of this exercise describes a single sheet of boron nitride (chemical formula BN), which forms a honeycomb lattice with boron and nitrogen atoms occupying A and B sublattices, respectively. Obviously the energies are different on these two sublattices for π-electrons, resulting in a large band gap.

Exercise 7.20. Use procedures similar to those used in Section 7.4.3 and in particular Exercise 7.15 to show that properly constructed envelope wave functions for states near $\vec{K}$ and $\vec{K}'$ satisfy corresponding Dirac equations. Note that the envelope wave function contains two components, one for each sublattice. They must vary slowly from one unit cell to another, but can change rapidly between neighboring sites (which belong to opposite sublattices); this is the reason why we need to keep two components.

7.6 Polyacetylene and the Su–Schrieffer–Heeger Model

7.6.1 Dirac electrons in 1D and the Peierls instability

In Section 7.5 we learned that electronic states in graphene near the Fermi energy are described by the massless Dirac equation. A band gap, which corresponds to a mass term in the Dirac equation, can be induced by breaking the symmetry between the A and B sublattices (see Exercise 7.19). As it turns out, the appearance of the Dirac equation as an effective theory for electronic states at low energy is quite common in condensed matter physics. The simplest example is actually a half-filled 1D tight-binding band, with dispersion relation given by Eq. (7.131). To see how the Dirac equation emerges, let us pretend the unit cell contains two atoms and has (a doubled) size $2a$, with a reduced 1BZ of $-\pi/2a < q \le \pi/2a$, and band dispersion illustrated in Fig. 7.19.[20] The *single* Dirac point is at $k_0 = \pi/2a$ (because k_0 and $-k_0$ are equivalent on the reduced-zone scheme!) with $\epsilon_{k_0} = 0$, near which the spectrum reads (in units where $\hbar = 1$)

$$\epsilon_\pm(k_0 + k) = \pm v_F |k|. \tag{7.215}$$

At half filling we have Fermi energy $\epsilon_F = 0$, and the lower (valence) band is completely filled, forming the "Dirac sea," while the upper (conduction) band is empty. The absence of a gap separating these two bands is because they are really parts of a *single* band (see Exercise 7.11 for more discussion on this point).

The tight-binding Hamiltonian written in momentum space and in the doubled unit cell basis takes the form

$$h_q = \begin{pmatrix} 0 & -2t\cos(qa) \\ -2t\cos(qa) & 0 \end{pmatrix} \approx \begin{pmatrix} 0 & v_F k \\ v_F k & 0 \end{pmatrix} = v_F k \sigma_x, \tag{7.216}$$

---

[20] Note the similarity to Fig. 5.4! The reader should ponder the origin of this similarity.

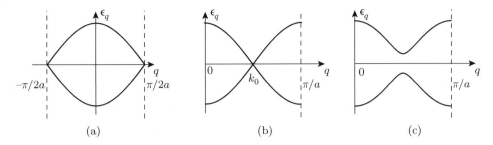

Figure 7.19 (a) Dispersion of a 1D tight-binding band plotted in a reduced (halved)-zone scheme. (b) The same dispersion plotted in a shifted zone with $0 < q \leq \pi/a$ so that the Dirac point at $k_0 = \pi/2a$ appears at the center of this range. (c) Dispersion with a band gap, corresponding to that of a massive Dirac fermion near $k_0 = \pi/2a$.

where $k = q - k_0$ is the deviation of lattice momentum from the Dirac point, and $v_F = 2ta$ is the Fermi velocity playing the role of the speed of light here. By the same arguments as in Section 7.5, we can write down the corresponding low-energy effective Hamiltonian for this system as

$$H = v_F \sigma_x p, \tag{7.217}$$

where p is the (1D) momentum operator. This is the 1D Dirac Hamiltonian for massless particles.

Since in 1D we need only one of the three Pauli matrices (σ_x) for the kinetic energy of the Dirac particle, we can actually have two different types of mass terms, $m_z \sigma_z$ and $m_y \sigma_y$. The origin of m_z is easy to see and similar to that of graphene; this is a diagonal term corresponding to (staggered) on-site energy m_z for odd sites and $-m_z$ for even sites (see Exercise 7.19). Obviously the asymmetry between even and odd sites reduces the lattice translation symmetry, and doubles the unit cell to size $2a$; a band gap of $2|m_z|$ gets opened up.

What could give rise to a mass term of the form $m_y \sigma_y$? Since σ_y is purely off-diagonal, it can only come from hopping between even and odd sites, including nearest neighbors. It turns out that dimerization (as we encountered in Exercise 7.11) leads to such a term. To illustrate this point, we write the hopping matrix elements on odd and even bonds as

$$t_{1,2} = t \pm \delta t, \tag{7.218}$$

in terms of which the momentum-space Hamiltonian takes the form

$$
\begin{aligned}
h_q &= \begin{pmatrix} 0 & -2t\cos(qa) - 2i\,\delta t \sin(qa) \\ -2t\cos(qa) + 2i\,\delta t \sin(qa) & 0 \end{pmatrix} \\
&\approx \begin{pmatrix} 0 & v_F k - 2i\,\delta t \\ v_F k + 2i\,\delta t & 0 \end{pmatrix} = v_F k \sigma_x + 2\,\delta t\,\sigma_y,
\end{aligned}
\tag{7.219}
$$

where we have again expanded q near the Dirac point k_0 as $q = k_0 + k$, and used the fact that $\sin(k_0 a) = 1$. We thus find the mass $m_y = 2\,\delta t$.

In the presence of one or both mass terms, the spectrum becomes

$$\epsilon_{\pm}(k) = \pm\sqrt{v_F^2 k^2 + m^2}, \tag{7.220}$$

with

$$m^2 = m_y^2 + m_z^2. \tag{7.221}$$

Compared with the massless case of Eq. (7.215), we find the valence band is pushed down in energy while the conduction band energy goes up. Thus, for the half-filled lattice, the electronic energy goes down as the valence band is fully occupied while the conduction band is empty. Let us now calculate this energy gain:

$$\Delta E = \sum_k [\epsilon_-(k, m) - \epsilon_-(k, m = 0)] \tag{7.222}$$

$$\approx \frac{L}{2\pi} \int_{-\frac{\pi}{2a}}^{\frac{\pi}{2a}} \left[-\sqrt{v_F^2 k^2 + m^2} + v_F|k| \right] dk$$

$$\approx -\frac{m^2 L}{4\pi v_F} \ln \left(\frac{v_F}{|m|a} \right). \tag{7.223}$$

We note that in the above we used Dirac spectra for the entire (reduced) 1BZ, even though this is accurate only near the Dirac point. This is a valid approximation for $|m| \ll v_F a$, or when the mass gap is small compared with the band width, since the energy difference between the massive and massless cases vanishes quickly as one moves away from the Dirac point. The expression above thus gives the correct leading dependence on m, with corrections of order $O(m^2)$.[21]

The discussion above suggests that the system can lower its electronic energy by having a Dirac mass in the electronic spectrum. One way to do that is to spontaneously dimerize the lattice, which leads to an m_y mass term. On the other hand, this requires a lattice distortion, which would lead to an elastic energy cost. To study the interplay and competition between the electronic and elastic energy changes, we need to consider them simultaneously. The following simple Hamiltonian known as the **Su–Schrieffer–Heeger (SSH) model** treats the electronic and lattice degrees of freedom on equal footing:

$$H = \sum_j [-t(u_{j+1} - u_j)|j\rangle\langle j + 1| + \text{h.c.}] + \frac{\lambda}{2} \sum_j (u_{j+1} - u_j)^2. \tag{7.224}$$

The first term represents hopping whose matrix element is a function of the change of nearest-neighbor distance due to lattice distortion,[22] while the second term is the elasticity energy cost due to such distortion. For small distortions we may approximate t by Taylor-series expansion as

$$t(u_{j+1} - u_j) \approx t - \frac{\alpha}{2}(u_{j+1} - u_j), \tag{7.225}$$

where α is a constant. For lattice dimerization we have

$$u_j = u(-1)^j, \tag{7.226}$$

resulting in

$$\delta t = \alpha u \tag{7.227}$$

and an elasticity energy cost of $U = 2\lambda N u^2$. Combining these with the electron energy gain of Eq. (7.223), we find the total energy change to be

$$\Delta E + \Delta U \approx -\frac{(\alpha u)^2 L}{\pi v_F} \ln \left(\frac{v_F}{\alpha a |u|} \right) + 2\lambda N u^2. \tag{7.228}$$

Owing to the logarithmic factor, the first term dominates at small u and the system can always lower its total energy by spontaneous dimerization, an effect known as the **Peierls instability**.

---

[21] This approximation is in the same spirit as using the Debye model (instead of the actual phonon dispersion) to calculate the phonon contribution to the specific heat and the Debye–Waller factor in Chapter 6. Such approximations capture the leading physical effect in the simplest possible setting.

[22] Since u is the phonon degree of freedom, the u-dependence of t represents electron–phonon interaction! Here we treat u as a *classical* variable.

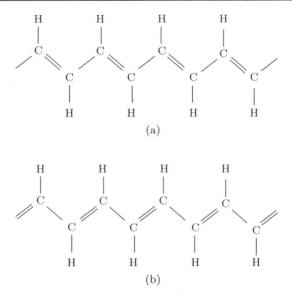

Figure 7.20 The atomic configuration of polyacetylene. Solid lines correspond to chemical bonds. Panels (a) and (b) correspond to the two possible dimerization patterns, A and B, in which double bonds occur on alternating links which have been shortened by the lattice distortions.

Quasi-1D systems are quite common in condensed matter. One class of such systems consists of polymers, which are huge chain-like molecules with carbon atoms forming the backbone, coupled together by relatively weak inter-chain forces. Electrons can move much more easily along the chain than they can by hopping between chains, and their motion can be considered effectively 1D. A particularly important example is **polyacetylene**, which is made of chains of carbon and hydrogen atoms with chemical formula $(CH)_x$ $(x \rightarrow \infty)$. Among the various spatial arrangements of carbon and hydrogen atoms, the most stable is $trans\text{-}(CH)_x$, shown in Fig. 7.20. Each unit cell contains two carbons and two hydrogens, and thus ten valence electrons (two from hydrogens and eight from carbons). In fact, the chemical environment of carbon here is quite similar to that in graphene, namely it has three nearest neighbors; as a result its three σ electrons form valence bonds with the three neighbors (or completely fill the corresponding four bonding bands along with the electrons from hydrogens), while the remaining π electron is free to move along the chain. If the carbon–carbon distance (and thus hopping matrix element) were a constant, the π electrons would see an *effective Bravais* lattice, and would half fill the corresponding tight-binding band. The Peierls instability dictates that the carbon chain spontaneously dimerizes, resulting in two *degenerate* ground state configurations, as shown in Figs. 7.20(a) and (b), respectively. The combined effects of π electron hopping and lattice dimerization in polyacetylene are properly described by the SSH model. Since a band gap is opened up at the Fermi energy, the system is an insulator, in agreement with experiments.

Polyacetylene can be turned into a conductor by doping it with impurities that introduce extra charge carriers into the system (see Chapter 9 for more extensive discussions of effects of impurity doping). One major puzzle is that, at low doping, these charge carriers appear to carry charge *only*, but no spin, as shown in Fig. 7.21: the conductivity increases very rapidly with impurity concentration, but the **magnetic susceptibility** (defined in Section 7.7.2) remains essentially zero until the impurity density becomes sufficiently large. This puzzle was resolved by Su, Schrieffer, and Heeger, who showed that charge carriers in polyacetylene at low impurity concentration are not simply electrons in the conduction band, but rather **charged solitons** that carry *no* spin. We turn now to an investigation of this novel phenomenon.

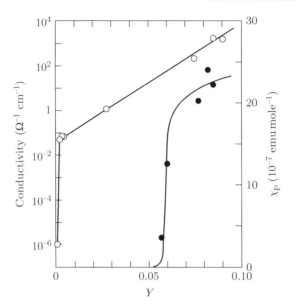

Figure 7.21 Conductivity and magnetic susceptibility of polyacetylene as functions of dopant concentration Y. At low doping the conductivity (hollow circles) increases quickly with dopant concentration, while the magnetic susceptibility (black dots) remains essentially zero, indicating that the charge carriers introduced by doping carry *no* spin. Figure reprinted with permission from [34]. Copyright by the American Physical Society.

7.6.2 Ground-State Degeneracy and Solitons

As discussed above, there is a two-fold ground state degeneracy due to dimerization, corresponding to either configuration A or B in Fig. 7.20. In an excited state, we can have A and B domains coexisting; in this case there is a domain wall (called a soliton) separating them. We now study the atomic configuration of a soliton and its energy cost, using the SSH model. To this end, we need an energy functional that describes the energetics associated with dimerization and lattice distortion in general.

As discussed in Chapter 5, a natural starting point for such an energy functional is continuum elasticity theory, which requires that the lattice displacement u varies slowly over the scale of the lattice spacing. This is *not* the case here, as u changes sign from one lattice site to the next in the dimerized ground state. We can, however, introduce a staggered displacement

$$\phi_j = (-1)^j u_j, \tag{7.229}$$

which will turn out to be a slowly varying variable.[23] In particular,

$$\phi_j = \pm u_0 \tag{7.230}$$

is a *constant* for the A and B ground states, respectively. We expect the energy functional (which includes contributions from both electronic and lattice energies) to take the form

$$E[\phi(x)] = \int dx \left\{ \frac{A}{2} \left[\frac{d\phi(x)}{dx} \right]^2 - B\phi^2(x) + C\phi^4(x) + \cdots \right\}, \tag{7.231}$$

---

[23] A similar procedure was used in Exercise 7.15. If the reader has not attempted that exercise yet, this is the time to do it.

where all coefficients are positive, and terms of higher order either in ϕ itself or in its derivative are neglected. There is a very important difference between Eq. (7.231) and the elastic energy functionals in Chapter 5, namely there are terms depending on ϕ itself, *in addition to* those that depend on $d\phi(x)/dx$. This is because a constant ϕ does *not* correspond to a uniform displacement of the whole lattice. The negative coefficient of the ϕ^2 term ensures that the ground state has non-zero ϕ, and the symmetry

$$E[\phi(x)] = E[-\phi(x)] \tag{7.232}$$

leads to the two-fold ground state degeneracy

$$\phi = \pm u_0 = \pm\sqrt{\frac{B}{2C}}. \tag{7.233}$$

The reader has probably recognized that Eq. (7.231) takes the same form as that of the Landau theory of ferromagnets in the Ising universality class, and ϕ is the order parameter whose presence signals the symmetry breaking.

In addition to the ground state configurations, there are also other stable configurations of $\phi(x)$ that are local minima of energy; these configurations satisfy the extremum condition

$$\frac{\delta E[\phi(x)]}{\delta\phi(x)} = 0 \Rightarrow A\frac{d^2\phi(x)}{dx^2} = -2B\phi(x) + 4C\phi^3(x). \tag{7.234}$$

This is nothing but the Lagrange equation of a particle moving in a 1D potential, if we identify x as time, ϕ as the position of the particle, and A as the particle's mass. The potential seen by this particle is[24]

$$\tilde{V}(\phi) = B\phi^2 - C\phi^4, \tag{7.235}$$

as shown in Fig. 7.22. The two ground states, A and B, correspond to the "particle" sitting on the two maxima of the potential at $\phi = \pm u_0$.

Another type of solution of Eq. (7.234) corresponds to the "particle" starting at A, rolling down the hill, and eventually ending asymptotically at B. This means the system is in the A configuration for $x \to -\infty$ and the B configuration for $x \to \infty$, and there is a soliton separating these two domains. An anti-soliton corresponds to the opposite situation, in which the "particle" rolls down from B and ends up at A. See Fig. 7.22 for illustrations. The actual soliton/anti-soliton solutions take the form

$$\phi(x) = \pm u_0 \tanh[(x - x_0)/\xi], \tag{7.236}$$

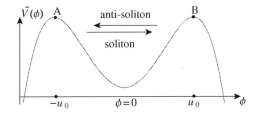

Figure 7.22 Effective potential of Eq. (7.235). The two ground states correspond to the "particle" sitting on one of the two maxima of the potential at $\phi = \pm u_0$. A soliton corresponds to the "particle" rolling down from one maximum and gradually coming to a halt at the other, and the anti-soliton corresponds to the reverse. In this picture the spatial position x plays the role of time.

---

[24] Note the change of sign from the corresponding terms in Eq. (7.231)! This is because the Lagrangian is kinetic energy (corresponding to $(A/2)\left[d\phi(x)/dx\right]^2$ here) *minus* the potential energy.

where x_0 is the center of the soliton/anti-soliton, and

$$\xi = \sqrt{A/B} \tag{7.237}$$

is its width. (Thinking of x as "time," x_0 is when the event occurs and ξ is its duration.) For polyacetylene $\xi \approx 7a \gg a$, thus justifying a continuum treatment.

7.6.3 Zero Modes Bound to Solitons

In the presence of a soliton or anti-soliton, we have an x-dependent ϕ, which gives rise to an x-dependent Dirac mass $m_y(x) = \alpha \phi(x)$ that changes sign at x_0. As we show below, this leads to a **mid-gap bound state** localized near x_0, at energy precisely $\epsilon = 0$. The Dirac equation (in which we set $\hbar = v_F = 1$ to simplify things)

$$\begin{pmatrix} 0 & -i\,d/dx - im(x) \\ -i\,d/dx + im(x) & 0 \end{pmatrix} \begin{pmatrix} \psi_1(x) \\ \psi_2(x) \end{pmatrix} = \epsilon \begin{pmatrix} \psi_1(x) \\ \psi_2(x) \end{pmatrix} \tag{7.238}$$

reduces to two decoupled first-order equations for $\epsilon = 0$:

$$\frac{d\psi_1(x)}{dx} = m(x)\psi_1(x), \tag{7.239}$$

$$\frac{d\psi_2(x)}{dx} = -m(x)\psi_2(x), \tag{7.240}$$

the solutions of which are

$$\psi_1(x) = c_1 e^{+\int_0^x m(x')dx'}, \tag{7.241}$$

$$\psi_2(x) = c_2 e^{-\int_0^x m(x')dx'}. \tag{7.242}$$

For a soliton we need to set $c_1 = 0$ as otherwise $|\psi_1| \to \infty$ as $x \to \pm\infty$. Similarly, for an anti-soliton we need to set $c_2 = 0$. It is also clear from these solutions that there are no normalizable **zero modes** unless $m(x)$ changes sign between $x = -\infty$ and $x = +\infty$. Let us define the "topological charge" for the lattice solitons,

$$Q = \frac{1}{2}\{\text{sign}[m(\infty)] - \text{sign}[m(-\infty)]\}. \tag{7.243}$$

Let $N_{1(2)}$ be the number of (normalizable) $\psi_{1(2)}$ zero-mode solutions. There is an important relation between these and the topological charge

$$N_2 - N_1 = Q. \tag{7.244}$$

This relation between the number of zero modes and the topological charge is a simple example of what is known as the "**Atiyah–Singer index theorem.**"[25]

One might wonder why the bound state is precisely at zero energy, and how robust this zero mode is – for example, is this still true for the original lattice model (without taking the continuum limit)? As it turns out, the answer is yes, and the zero energy of a *single* bound state is guaranteed by a special property of the Hamiltonian. Let us start with the Dirac Hamiltonian with only the (possibly x-dependent) m_y mass term:

$$H_D = \sigma_x p + m_y(x)\sigma_y. \tag{7.245}$$

Since H_D involves terms proportional to σ_x and σ_y but *not* σ_z, we have the anti-commutator

$$\{H_D, \sigma_z\} = 0. \tag{7.246}$$

---

[25] See, e.g., Ref. [35].

Recall that, when the Hamiltonian commutes with an operator O, O is the generator of a symmetry, with all of its familiar consequences. Now we have a situation in which the Hamiltonian *anti*-commutes with an operator C. (In this particular case $C = \sigma_z$.) For reasons that will become clear soon, this can be viewed as a special type of symmetry, which we call **particle–hole symmetry** (sometimes referred to as **chiral symmetry** in the literature). Some consequences of this symmetry are similar to, but different from, ordinary symmetry. One of them is the following: if $|\psi\rangle$ is an eigenstate of H with eigenvalue ϵ, then $C|\psi\rangle$ is an eigenstate of H with eigenvalue $-\epsilon$. (The reader should verify this! See Exercise 7.21.) As a result, eigenstates of Hamiltonians with such particle–hole symmetry must come in pairs with opposite eigenvalues (hence the name particle–hole symmetry), *except for those with exactly $\epsilon = 0$*. Thus a single (or unpaired) bound state *must* have $\epsilon = 0$. We also note that the particle–hole symmetry (7.246) is not specific to the (continuum limit) Dirac Hamiltonian. We can define

$$\Sigma_z = \sum_j (-1)^j |j\rangle\langle j|, \tag{7.247}$$

which anti-commutes with the hopping Hamiltonian (with nearest-neighbor hopping only). Thus the bound state being at exactly zero energy is also true for the original lattice Hamiltonian.

Exercise 7.21. If $\{H, C\} = 0$ and $H|\psi\rangle = \epsilon|\psi\rangle$, prove that

$$H|\psi'\rangle = -\epsilon|\psi'\rangle \tag{7.248}$$

and (provided that $\epsilon \neq 0$)

$$\langle\psi|\psi'\rangle = 0, \tag{7.249}$$

where $|\psi'\rangle \equiv C|\psi\rangle$.

Exercise 7.22. Show that the operator defined in Eq. (7.247) anti-commutes with the hopping Hamiltonian, as long as hopping exists between even and odd sites only.

As an example which is the opposite of the continuum limit, consider the extreme case of dimerization such that the hopping on weaker bonds is zero. As shown in Fig. 7.23, there is a lattice site s at the boundary separating the A and B configurations (i.e. a soliton) *decoupled* from the rest of the lattice, and $|s\rangle$ is precisely the zero mode.

We have seen that the presence of a soliton or anti-soliton gives rise to a new mid-gap bound state or zero mode. The total number of electron states of the system, however, must be fixed; for example, in polyacetylene the number of electrons (in the relevant bands) is simply equal to the number of carbon atoms. Thus the appearance of this new state implies that the number of states must be reduced elsewhere. Owing to the particle–hole symmetry of the model, it must be that the number of states is reduced by *one-half* from each of the conduction band and the valence band. But this sounds bizarre; what is the meaning of half a state? Of course, the actual number of states in a given energy range must be an integer, and the resolution of this paradox depends on the specific setting of the system, in particular the *boundary condition* of the system.

Figure 7.23 A soliton and its zero mode in the extremely dimerized limit where hopping on the weaker bond is zero. Solid lines represent bonds with non-zero hoppings. A soliton separating the A and B configurations resides on site s.

First consider the case of periodic boundary conditions (equivalent to arranging the atoms to form a ring), and assume an even number of atoms, which means the system contains an integer number of unit cells. In this case a *single* soliton configuration is impossible because the periodic boundary condition forces the topological charge Q to vanish. Hence only soliton–anti-soliton *pairs* can be created. This is most easily understood by considering the extreme dimerization case shown in Fig. 7.23. Since there is one zero mode associated with the soliton and another associated with the anti-soliton, the number of states removed from the conduction band and the valence band is $\frac{1}{2} \times 2 = 1$. Generalization to multiple soliton–anti-soliton pairs is straightforward.[26]

Now consider the more interesting case of a periodic boundary condition but with an *odd* number of atoms. A similar consideration finds that there should be at least one soliton or anti-soliton, and their total number must be odd. Now count the number of states the conduction or valence band *would* have if there were no solitons. Since the number of atoms is odd, the unit cell number and thus number of states per band would be half of an integer! Of course, this is impossible, which is another reason why there must be an odd number of solitons/anti-solitons, so that the extra half state can be absorbed.[27]

Periodic boundary conditions are a very useful mathematical construction, convenient for understanding many conceptual issues. In reality, atoms are usually not arranged in rings, and systems frequently have open boundaries instead. This gives rise to the possibility of boundary or **edge states**. Consider, for example, a generic single-orbital-per-atom tight-binding system with open boundary conditions, particle–hole symmetry, and an odd number of atoms. Since the total number of electron states is odd, and since particle–hole symmetry guarantees that states with non-zero energy ϵ always have a partner at $-\epsilon$, there must exist one state with exactly zero energy. If there is no dimerization, then this state is just an extended standing wave in the middle of the band. However, if the system is dimerized, then there is a gap (surrounding zero energy) in the bulk and the zero mode must live on one of the boundaries (and decay exponentially into the bulk). It is easy to see from the case of extreme dimerization shown in Fig. 7.23 that, for the case of an open chain with an odd number of atoms, the isolated site s must be located at one end of the chain. In this case, the zero mode is strongly confined to that end of the chain. As the dimerization becomes weaker, the zero mode decays more and more slowly into the bulk, eventually reaching all the way to the other end of the chain, thus making contact with the standing-wave mode discussed above for the limit of no dimerization. For the open chain, the appearance of a soliton and its zero mode is always accompanied by a corresponding change in the number of zero modes at the boundaries; see Exercise 7.24. This fact resolves the paradox that each soliton takes half a state from the upper and lower bands, and yet the number of states in each band must be an integer.

We used to think that, in the thermodynamic limit, "details" like boundary conditions and the specific number of particles in the system (say, this number being even or odd) were unimportant. Here we see that this is *not* the case. The reason is that the bulk has some rigid structure (here it is dimerization), which is *topological* in nature. Such topological structure puts rigid constraints on the number of topological defects (solitons) that depend in detail on boundary conditions etc. Very often the *bulk* topological structure forces the appearance of low-energy modes at the *edge*, as we see here. We will encounter many more examples of these later in this book (e.g. in the quantum Hall effect).

---

[26] We note that, when the distance between a soliton and its neighboring anti-soliton(s) is large but finite, there can be a small overlap (and hence hybridization) between the bound state solutions, and their energy will differ from zero by an exponentially small amount. Thus $N_1 = N_2 = 0$, but this is not inconsistent with the index theorem in Eq. (7.244), since $Q = 0$.

[27] In contrast to the previous example, now we have $|Q| = 1$ and there is a solution at *exactly* zero energy even if multiple soliton–anti-soliton pairs are present.

Table 7.2 Charge and spin quantum numbers of a soliton, depending on the number of electrons occupying the zero mode

Occupation number of zero mode	Charge	Spin
0	$+|e|$	0
1	0	1/2
2	$-|e|$	0

Exercise 7.23. This exercise illustrates that the zero mode bound to the soliton requires protection from particle–hole symmetry. Consider a staggered potential term of the form

$$H' = m_z \Sigma_z, \tag{7.250}$$

with Σ_z defined in Eq. (7.247). (i) Show that, once H' is included in the Hamiltonian, it no longer anti-commutes with Σ_z, thus the particle–hole symmetry is broken. (ii) Show that the presence of H' does *not* alter the zero-mode wave function, but shifts its energy away from zero. (iii) Show that the energy shifts in opposite direction for zero modes bound to solitons and those bound to anti-solitons. (iv) Show that sufficiently large $|m_z|$ moves the zero mode into either the conduction or the valence band, thus making it "disappear." (v) Discuss what happens when the system is a ring with periodic boundary conditions but the ring contains an odd number of atoms and a single domain wall. What is the resolution of the ambiguity regarding which way the zero mode should move?

7.6.4 Quantum Numbers of Soliton States and Spin–Charge Separation

The soliton zero mode may be empty, occupied by one electron, or occupied by two electrons with opposite spin. Each soliton costs a certain finite amount of elastic/electronic energy to create. However, if the gap between the valence and conduction bands is sufficiently large, the cheapest way[28] to accommodate extra (positive or negative) charge from doping is to create a finite density of solitons. What are the charge and spin quantum numbers associated with these states? Naively one would think they are the same as the corresponding quantum numbers of electrons. This is *not* the case, as we need to calculate the *difference* between these quantum numbers and those of the ground state *without* the soliton. Let us start with the state with the zero mode empty. Since the (completely filled) valence band has lost half a state (that would have been occupied by spin-up and -down electrons), this state has a missing electron compared with the ground state and thus charge $+|e|$, but *no* spin. Similarly, the state with the zero mode fully occupied has charge $-|e|$, and no spin, while the state with the zero mode occupied by one electron has *no* charge but spin-1/2. These results are summarized in Table 7.2. Again the easiest way to convince oneself of the validity of these (seemingly strange) results is to inspect the extreme dimerization model of Fig. 7.23, and calculate the *difference* of (average) charge and spin on site s with and without the soliton.

---

[28] Assume the chemical potential sits in the center of the band gap. To bring in an extra electron from the chemical potential either one has to promote it up above the gap to the conduction band or one has to pay the price of creating a soliton lattice distortion which creates a state in the middle of the band (i.e. at the chemical potential). For a large enough band gap, the soliton is always cheaper to create. See additional discussion further below.

We see here the fascinating phenomenon of **spin–charge separation**, namely the elementary excitations of the system carry either the spin *or* the charge quantum number of an electron, but not both. This is an example of quantum number fractionalization, namely elementary excitations of the system carrying a fraction of the quantum number(s) of the fundamental constituent particles. We will see more examples of this in Chapter 16. In particular, we will gain an understanding of how fractional quantum numbers can be sharply defined despite being subject to strong fluctuations (as long as there is an excitation gap which pushes these fluctuations up to high frequencies).

Coming back to polyacetylene, we find (as noted earlier) that there are two ways to accommodate the added electronic charge due to doping. The first is to simply put electrons in the conduction band. Because (from particle–hole symmetry) the chemical potential is at the mid-point of the gap, this costs energy $|m_y|$, or half the band gap due to dimerization. The second is to create solitons, and put electrons into the zero modes. The energy cost here is that of the soliton energy, calculable using the energy functional of Eq. (7.231), at least in the continuum limit. As it turns out, the energy cost in polyacetylene is lower for the latter. So the added charges in polyacetylene show up as *spinless* charged solitons, whose motion gives rise to conduction. Since they carry no spin, they do not contribute to the magnetic response, as indicated in Fig. 7.21.

> **Exercise 7.24.** Consider the extremely dimerized chain of Fig. 7.23 with open boundary conditions. Show that, while there can be zero mode(s) at the boundaries, a change in the number of solitons in the bulk is always accompanied by a change in the number of the boundary zero mode(s).

>
>
> **Exercise 7.25.** We discussed the Peierls instability of a half-filled band in 1D. Now consider a 1D band that is one-third filled. (i) Is there a similar instability? If so, justify your answer, and specify the pattern of lattice distortions. (ii) Specify the types of solitons, and the charge and spin quantum numbers of the soliton states.

7.7 Thermodynamic Properties of Bloch Electrons

We learned earlier in this chapter about Bloch's theorem, which tells us that within the independent electron approximation we can label the energy eigenvalues $\epsilon_{n\vec{k}}$ of a crystal by a band index n and a wave vector $\vec{k}$ lying inside the first Brillouin zone. The density of states per unit volume per spin is therefore

$$\rho(\epsilon) \equiv \sum_n \rho_n(\epsilon) = \sum_n \int_{\text{BZ}} \frac{d^3\vec{k}}{(2\pi)^3} \delta(\epsilon - \epsilon_{n\vec{k}}). \tag{7.251}$$

A crude estimate of the contribution of the nth band to ρ is

$$\rho_n \sim \frac{1}{|\omega| W_n}, \tag{7.252}$$

where $|\omega|$ is the unit-cell volume and W_n is the band width (typically of order 1–10 eV).

As we shall see below, the density of states plays a central role in determining the thermodynamic properties of the electrons in a crystal. In simple band insulators there is a gap in the energy spectrum between the highest occupied and lowest unoccupied Bloch band states. Hence the density of states at the Fermi level vanishes and thermal excitations of the electrons across the gap are exponentially suppressed. (Recall that room temperature corresponds to $k_B T \sim 25$ meV, while band gaps are often ~ 0.5–$5\,\text{eV}$.) In a metal there is a finite density of states available for thermal excitation around the Fermi level. The electrons therefore contribute significantly to the thermodynamics at low temperatures. We focus below on the case of metals.

7.7.1 Specific Heat

The specific heat is defined as the rate of change of the internal energy per unit volume $\mathcal{E}$ of a system with temperature:

$$c_V \equiv \left.\frac{\partial \mathcal{E}}{\partial T}\right|_V. \tag{7.253}$$

This is an example of a response function, which measures the rate of change of some measurable property of the system as we change some externally controllable parameter of the system. What does measuring the specific heat tell us about a system of Bloch electrons? Recall from elementary statistical mechanics that for non-interacting fermions the total energy density of the system is

$$\mathcal{E} = \frac{2}{L^d} \sum_{\vec{k},n} \epsilon_{n,\vec{k}} f^\circ(\epsilon_{n,\vec{k}}), \tag{7.254}$$

where

$$f^\circ(\epsilon) = \frac{1}{e^{\beta(\epsilon-\mu)} + 1} \tag{7.255}$$

is the Fermi–Dirac distribution function in which $\beta = 1/k_B T$ is the inverse temperature, and the factor of 2 accounts for the electron spin. The quantity μ is the chemical potential, which depends on the temperature T and possibly other parameters (such as the external magnetic field). The phrase Fermi energy is commonly used interchangeably with μ at low temperature, but we will make a distinction here because the T-dependence of $\mu(T)$ is important, and define the Fermi energy to be

$$\epsilon_F = \mu(T = 0) = \mu_0. \tag{7.256}$$

We can express the energy density in terms of the density of states:

$$\mathcal{E} = 2 \int_{-\infty}^{\infty} d\epsilon\, \rho(\epsilon) f^\circ(\epsilon) \cdot \epsilon. \tag{7.257}$$

The same is true for any property of the system which is expressible as a sum over states of some function which depends only on the energy of the state. A very useful low-temperature expansion of the type of integral of Eq. (7.257) is known as the Sommerfeld expansion (see Exercise 7.26). This uses the fact that an integral of the form

$$I = \int_{-\infty}^{\infty} d\epsilon\, H(\epsilon) f^\circ(\epsilon) \tag{7.258}$$

has the low-temperature expansion

$$I = \left\{\int_{-\infty}^{\mu} d\epsilon\, H(\epsilon)\right\} + \frac{(\pi k_B T)^2}{6} H'(\mu) + \frac{7(\pi k_B T)^4}{360} H'''(\mu) + \cdots. \tag{7.259}$$

Applying this to the internal energy density, we have

$$\mathcal{E} = \left\{2 \int_{-\infty}^{\mu} d\epsilon\, \epsilon \rho(\epsilon)\right\} + \frac{(\pi k_B T)^2}{3}\left[\rho(\mu) + \mu\rho'(\mu)\right] + \cdots. \tag{7.260}$$

Note that this expansion is actually an expansion in $k_B T/W$, where W is the "band width" which we could define as the typical energy interval which contains one Bloch eigenstate per unit cell.

For typical crystals (or at least in circumstances where the independent electron approximation works!) $W \sim 1$ eV so that $W/k_B \sim 10^4$ K. At room temperature (or even well above room temperature) these low-temperature expansions therefore converge rapidly. An exception occurs when a van Hove singularity in the density of states lies within an energy range $\sim k_B T$ of μ. In that case, these low-temperature expansions will fail because ρ is then such a strong function of energy.

The internal energy above was evaluated in the grand-canonical ensemble as a function of temperature T and chemical potential μ. For quantum systems of indistinguishable particles it turns out to be simplest to do calculations in this ensemble. However, the physical situation for electrons in a crystal is that the total number of electrons per unit volume, n, is held fixed (via charge neutrality maintained by the long-range Coulomb force), and so the chemical potential must be a function of temperature:

$$n = \left\{ 2 \int_{-\infty}^{\mu(T)} d\epsilon \rho(\epsilon) \right\} + \frac{(\pi k_B T)^2}{6} \rho'(\mu) + \cdots \tag{7.261}$$

$$= n + 2[\mu(T) - \mu_0]\rho(\mu) + \frac{(\pi k_B T)^2}{3} \rho'(\mu) + \cdots$$

$$\Rightarrow \mu(T) = \mu_0 - \frac{(\pi k_B T)^2}{6} \frac{\rho'(\mu)}{\rho(\mu)} + \text{higher powers of } T. \tag{7.262}$$

At fixed n, we therefore have, from Eq. (7.260),

$$\mathcal{E} = \mathcal{E}_0 + 2[\mu(T) - \mu_0] [\mu_0 \rho(\mu_0)] + \frac{(\pi k_B T)^2}{3} \left[\rho(\mu_0) + \mu_0 \rho'(\mu_0) \right] + \cdots$$

$$\approx \mathcal{E}_0 + \frac{(\pi k_B T)^2}{3} \rho(\mu_0), \tag{7.263}$$

where $\mathcal{E}_0$ and μ_0 are the zero-temperature values of $\mathcal{E}$ and μ, and we have dropped all terms of higher order than T^2. Miraculously, the dependence on $\rho'(\mu_0)$ drops out! This yields for the specific heat

$$c_V = \frac{\pi^2 k_B^2 T}{3} 2\rho(\mu_0) \equiv \gamma T, \tag{7.264}$$

where the linear T coefficient

$$\gamma = \frac{2\pi^2}{3} k_B^2 \rho(\mu_0) \tag{7.265}$$

depends on the density of states at the Fermi energy $\rho(\mu_0)$ *only*.

When we studied the lattice specific heat we found that the internal energy of acoustic phonons scaled as T^{d+1} in d dimensions. For degenerate (i.e. $T \ll T_{\text{Fermi}}$) fermions the internal energy scales as T^2 in any dimension. This is because of the Pauli exclusion principle. Only fermions within $\sim k_B T$ of the Fermi energy can be thermally excited, and each carries thermal energy $\sim k_B T$, yielding $\mathcal{E} - \mathcal{E}_0 \sim T^2$. Their contribution to the internal energy and thus c_V dominates over that from phonons at sufficiently low T, for $d > 1$.

One can extract the electronic contribution to the specific heat by plotting $c_V(T)/T$ vs. T^2 (see Fig. 6.2). Extrapolating this to zero temperature yields the coefficient γ and is a measure of the density of states at the Fermi level (in the independent-electron approximation). For interacting electrons, the quantity extracted from experiment via this method is referred to as the **thermodynamic density of states**.

7.7.2 Magnetic Susceptibility

Both the orbital and the spin degrees of freedom of an electron couple to an external magnetic field, and both contribute to the magnetization of the electron system. It is often the case that the spin and orbital moments can be measured separately or that the spin moment is much larger than the orbital moment. We will assume that is the case here and calculate the response to a magnetic field which couples only to the spins. In the presence of a magnetic field, the spin doublet associated with each Bloch orbital is split by the **Zeeman energy**. The intrinsic magnetic moment $\vec{\mu}$ of an electron is related to its spin angular momentum $\vec{S}$ by

$$\vec{\mu} = -\frac{1}{\hbar} g \mu_B \vec{S} = -\frac{g}{2} \mu_B \vec{\sigma}, \tag{7.266}$$

where the **Bohr magneton** is $\mu_B \equiv e\hbar/2m_ec \approx 0.579 \times 10^{-8}$ eV/G ~ 0.5 kelvin/tesla, and the Landé g factor obeys $g \approx 2$. Note the minus sign in the above equation indicating that the magnetic moment of the electron is aligned opposite to the spin direction because of the negative charge. If **spin–orbit coupling**[29] and the effect of the magnetic field on the kinetic energy operator can both be neglected then

$$H = -\frac{\hbar^2 \nabla^2}{2m} + V(\vec{r}) + \frac{g}{2} \mu_B \vec{B} \cdot \vec{\sigma}. \tag{7.267}$$

The eigenstates are then simple products of spatial orbitals and spinors, and the eigenvalues within this approximation are

$$\epsilon_{n,\vec{k},\sigma} = \epsilon_{n,\vec{k}} \pm \frac{g}{2} \mu_B B \sigma, \tag{7.268}$$

where $\sigma = \pm 1$. The high-energy state has its spin aligned with the field and the low-energy state has its spin oppositely directed. We can calculate the total energy as a function of the field. The combined density of states (for both spins) in a field is simply related to the single-spin density of states in the absence of a field:

$$\rho_B(\epsilon) = [\rho(\epsilon + \mu_B B) + \rho(\epsilon - \mu_B B)]. \tag{7.269}$$

Because we expect $\mu_B B \ll \epsilon_F$ we can make an expansion to order $(\mu_B B/\epsilon_F)^2$ so that at $T = 0$

$$
\begin{aligned}
\mathcal{E} &= \int_{-\infty}^{\mu} d\epsilon \, \epsilon \rho_B(\epsilon) \\
&= \int_{-\infty}^{\mu+\mu_B B} d\epsilon (\epsilon - \mu_B B)\rho(\epsilon) + \int_{-\infty}^{\mu-\mu_B B} d\epsilon (\epsilon + \mu_B B)\rho(\epsilon) \\
&= \mathcal{E}(B = 0, \mu) + \int_{\mu}^{\mu+\mu_B B} d\epsilon \, \epsilon \rho(\epsilon) + \int_{\mu}^{\mu-\mu_B B} d\epsilon \, \epsilon \rho(\epsilon) \\
&\quad - (\mu_B B) \int_{\mu-\mu_B B}^{\mu+\mu_B B} d\epsilon \, \rho(\epsilon) \\
&= \mathcal{E}(B = 0, \mu) + \left\{ \left[\rho(\mu) + \mu\rho'(\mu)\right](\mu_B B)^2 - 2(\mu_B B)^2 \rho(\mu) \right\} \\
&= \mathcal{E}(B = 0, \mu) + \mu\rho'(\mu)(\mu_B B)^2 - (\mu_B B)^2 \rho(\mu). \tag{7.270}
\end{aligned}
$$

If the density of states is not constant then the particle number will shift with B if μ is held fixed:

$$
\begin{aligned}
n &= n(B = 0, \mu) + \rho'(\mu)(\mu_B B)^2 \\
&= n(B = 0, \mu_0) + 2\rho(\mu_0)(\mu - \mu_0) + \rho'(\mu_0)(\mu_B B)^2. \tag{7.271}
\end{aligned}
$$

If n is held fixed, the chemical potential shifts accordingly:

$$\mu - \mu_0 = -\frac{1}{2}\frac{\rho'(\mu_0)}{\rho(\mu_0)}(\mu_B B)^2. \tag{7.272}$$

[29] Spin–orbit coupling can in some cases significantly alter the effective value of g, even changing its sign.

Thus we can write

$$\mathcal{E}(B, n) = \mathcal{E}(B = 0, n) - (\mu_B B)^2 \rho(\mu_0), \tag{7.273}$$

in which $\rho'(\mu_0)$ again drops out, and the magnetization (magnetic moment per unit volume) becomes (we neglect the distinction between B and H here)

$$M \equiv -\frac{\partial \mathcal{E}(B, n)}{\partial B} = \mu_B^2 2\rho(\mu_0) B. \tag{7.274}$$

The spin magnetic susceptibility is therefore

$$\chi \equiv \frac{\partial M}{\partial B} = \mu_B^2 2\rho(\mu_0). \tag{7.275}$$

We can think of the equilibrium spin polarization as resulting from a balance between the Zeeman energy reduction from flipping minority spins into majority spins and the band energy cost associated with the polarization. The meaning of the above result is simply that the band energy cost to polarize the electron gas is lower if the density of states is higher. Hence the magnetic susceptibility increases with increasing density of states. See Exercie 7.29.

We have seen that the electronic specific heat in a crystal is linear in T and proportional to the density of states at the Fermi energy. The **spin susceptibility** is also proportional to $\rho(\mu_0)$. These are the only material parameters which enter the expressions for χ and $\gamma \equiv \lim_{T \to 0} c_V(T)/T$. The ratio

$$\frac{\chi/\mu_B^2}{3c_V/(\pi^2 k_B^2 T)} = 1 \tag{7.276}$$

within the independent-electron approximation. This ratio is often called the "**Wilson ratio**" after Ken Wilson, who considered it in the context of the Kondo problem (involving a local spin moment coupled to a Fermi sea; see Section 15.14.3 for further discussion). Even if the band structure is such that the density of states is significantly modified from the free-electron value, the Wilson ratio remains unity for the independent-electron model. Deviations of the Wilson ratio from unity are a signal that interaction corrections to the independent-electron approximation are important in a material; this will be discussed in Chapter 15.

Exercise 7.26. This exercise illustrates how the Sommerfeld expansion works. The key observation is that the Fermi–Dirac distribution function, Eq. (7.255), deviates from the step function $\theta(\mu - \epsilon)$ only for $|\mu - \epsilon| \lesssim k_B T$.

(i) Define $\Delta f^\circ(\epsilon) = f^\circ(\epsilon) - \theta(\mu - \epsilon)$. Show that $\Delta f^\circ(\epsilon)$ is an odd function of $\epsilon - \mu$.

(ii) Now consider the integral

$$I = \int_{-\infty}^{\infty} d\epsilon\, H(\epsilon) f^\circ(\epsilon) = \int_{-\infty}^{\infty} d\epsilon\, H(\epsilon)[\theta(\mu - \epsilon) + \Delta f^\circ(\epsilon)]$$

$$= \int_{-\infty}^{\mu} d\epsilon\, H(\epsilon) + \int_{-\infty}^{\infty} d\epsilon\, H(\epsilon) \Delta f^\circ(\epsilon). \tag{7.277}$$

Expand $H(\epsilon)$ around $\epsilon = \mu$ in the last term, and use (i) to show that

$$I = \int_{-\infty}^{\mu} d\epsilon\, H(\epsilon) + \sum_{n=1}^{\infty} \frac{2(k_B T)^{2n}}{(2n - 1)!} H^{(2n-1)}(\mu) I_{2n-1}, \tag{7.278}$$

where $H^{(m)}$ is the mth derivative of H, and the constants

$$I_{2n-1} = \int_{0}^{\infty} \frac{x^{2n-1}\, dx}{e^x + 1} \tag{7.279}$$

can be expressed in terms of the Riemann zeta function and Bernoulli numbers, which are known explicitly. See Appendix C of the book by Ashcroft and Mermin [1] for more discussion. Note that the derivation here is slightly different (and simpler), resulting in different but equivalent expressions. Equation (7.259) follows from the results above.

Exercise 7.27. Develop a similar low-temperature expansion for the integral

$$I = \int_{-\infty}^{\infty} d\epsilon \, H(\epsilon) \frac{\partial f^{\circ}(\epsilon)}{\partial \epsilon}. \tag{7.280}$$

Exercise 7.28. Derive Eqs. (7.271) and (7.273), following steps similar to those leading to Eq. (7.270).

Exercise 7.29. Compute both the Zeeman energy and the kinetic energy of a free Fermi gas as a function of spin polarization at zero temperature. By minimizing the total energy, find the magnetization as a function of B and reproduce the result derived above for the susceptibility in terms of the density of states.

Exercise 7.30.
(i) Show that, in the presence of particle–hole symmetry, $\rho(\epsilon) = \rho(-\epsilon)$.
(ii) Show that in this case $\mu(T) = 0$ for any T, if $\mu(T = 0) = 0$.

Exercise 7.31. Calculate graphene's specific heat at low T with $B = 0$, and its magnetization at low B with $T = 0$. Note that the density of states is zero at the Fermi energy, and thus the corresponding formulas in this section are not directly applicable. On the other hand, the fact that graphene's band structure has particle–hole symmetry, namely $\rho(\epsilon) = \rho(-\epsilon)$ (please prove), is useful.

7.8 Spin–Orbit Coupling and Band Structure

In our discussion of band structure thus far we have largely ignored the spin degree of freedom. We are allowed to do that because we have so far been dealing with *spin-independent* Hamiltonians, in which electron spin is *conserved*. Once we have obtained the (spatial) eigenfunction $\psi_{n\vec{k}}(\vec{r})$, we can simply multiply it by the ($\vec{r}$-independent) spin wave function to obtain the full wave functions for the up- and down-spin states, using the fact that S_z is a good quantum number:

$$\psi_{n\vec{k}}(\vec{r}) \begin{pmatrix} 1 \\ 0 \end{pmatrix} \quad \text{and} \quad \psi_{n\vec{k}}(\vec{r}) \begin{pmatrix} 0 \\ 1 \end{pmatrix}. \tag{7.281}$$

The only physical effect of the spin degree of freedom is to contribute a two-fold spin degeneracy for each Bloch state, which results in the doubling of the number of electrons each Bloch band can accommodate. This is, of course, crucial for the determination of whether a crystal is a metal or an insulator. Alternatively, one can also say that the spatial Bloch wave functions $\psi_{n\vec{k}}(\vec{r})$ really describe *two different* bands, one for spin-up and one for spin-down electrons. This latter viewpoint is more easily generalized to the cases with spin-dependent Hamiltonians, a task to which we now turn.

The simple factorization of the spatial and spin wave functions in Eq. (7.281) is an artifact of neglecting the spin-dependence in the crystal Hamiltonian, which is *always* present due to the inevitable spin–orbit coupling:

$$H_{SO} = \frac{[\nabla V(\vec{r}) \times \vec{p}\,] \cdot \vec{S}}{2m_e^2 c^2}, \tag{7.282}$$

where $\vec{S}$ is the spin operator of the electron. H_{SO} is obtained by taking the low-velocity limit of the Dirac equation (that results in the Schrödinger equation), and is thus a relativistic effect. Physically it can be understood as the Zeeman coupling between the magnetic moment carried by electron spin $\vec{\mu} = -(g/\hbar)\mu_B \vec{S}$ and the effective magnetic field seen by the electron in the reference frame in which it is at rest, while the other charges move around it. For a single atom in a spherically symmetric environment, spin–orbit coupling manifests itself as a term of the form $\lambda \vec{L} \cdot \vec{S}$ in the Hamiltonian. Such a term can be re-expressed in terms of the total angular momentum $\vec{J} = \vec{L} + \vec{S}$ of the electron. The situation is more complex in a solid where spherical symmetry is lacking and electrons are not associated with individual atoms.

Except for core electrons of heavy elements, we have electron speed $v \ll c$, and the relativistic spin–orbit effect is weak and thus often neglected in studies of band structures. Qualitatively, however, its presence invalidates the factorization of Eq. (7.281) because $[S_z, H_{SO}] \neq 0$, and the (spin-dependent) Bloch wave function should be written in a more general two-component form:

$$\psi_{n\vec{k}}(\vec{r}, \sigma) = e^{i\vec{k}\cdot\vec{r}} \begin{pmatrix} u_{n\vec{k}}^{\uparrow}(\vec{r}) \\ u_{n\vec{k}}^{\downarrow}(\vec{r}) \end{pmatrix}. \tag{7.283}$$

Here $\sigma = \uparrow, \downarrow$, and $u_{n\vec{k}}^{\uparrow}(\vec{r})$ and $u_{n\vec{k}}^{\downarrow}(\vec{r})$ are *different* periodic spatial wave functions. Clearly S_z is *not* a good quantum number in $\psi_{n\vec{k}}(\vec{r}, \sigma)$. If we return to the case without spin–orbit coupling so that S_z is conserved, the band index n should be understood to include the spin index. As a result, bands always come in degenerate pairs, even if we keep only one spatial orbital per atom in a Bravais lattice. In general, this degeneracy is lifted in the presence of spin–orbit coupling. As we will discuss below, the degeneracy is preserved if (and only if) both time-reversal and inversion symmetry are present.

While spin–orbit coupling (7.282) breaks spin rotation symmetry, it is invariant under time-reversal transformation. This is because under time-reversal transformation Θ we have

$$\vec{r} \to \Theta \vec{r} \Theta^{-1} = \vec{r}, \tag{7.284}$$

$$\vec{p} \to \Theta \vec{p} \Theta^{-1} = -\vec{p}, \tag{7.285}$$

$$\vec{S} \to \Theta \vec{S} \Theta^{-1} = -\vec{S}, \tag{7.286}$$

resulting in

$$\Theta H \Theta^{-1} = H \quad \text{or} \quad [\Theta, H] = 0. \tag{7.287}$$

The time-reversal symmetry has major implications for the spectra of spin-1/2 particles such as electrons. The most important is the **Kramers degeneracy**, which states that all energy levels come in degenerate pairs. This is because, if $|\psi\rangle$ is an eigenstate of H with eigenvalue ϵ, then so is $|\psi'\rangle = \Theta|\psi\rangle$. But this does not prove the degeneracy as it is possible that

$$|\psi'\rangle = \Theta|\psi\rangle = \alpha|\psi\rangle, \tag{7.288}$$

with α being a constant, namely $|\psi'\rangle$ and $|\psi\rangle$ are the *same* state. Now let us prove that this is impossible for spin-1/2 particles. In this case we have[30]

$$\Theta = \sigma_y K, \tag{7.289}$$

where K is the complex conjugation operator (that turns the wave function it acts on into its complex conjugate). Since σ_y is purely imaginary, we find

$$\Theta^2 = \sigma_y K \sigma_y K = -\sigma_y^2 K^2 = -1. \tag{7.290}$$

If Eq. (7.288) were true, we would have $\Theta^2|\psi\rangle = |\alpha|^2|\psi\rangle$ or $|\alpha|^2 = -1$, which is impossible. Thus $|\psi'\rangle$ and $|\psi\rangle$ cannot be the same state, and thus form a degenerate pair.[31]

We now study how the Kramers degeneracy is manifested in band structure. Hitting the Bloch state (7.283) with Θ yields a different Bloch state (of a different band) with the same energy:

$$\Theta\psi_{n\vec{k}}(\vec{r}, \sigma) = \Theta e^{i\vec{k}\cdot\vec{r}} \begin{pmatrix} u_{n\vec{k}}^{\uparrow}(\vec{r}) \\ u_{n\vec{k}}^{\downarrow}(\vec{r}) \end{pmatrix} = i e^{-i\vec{k}\cdot\vec{r}} \begin{pmatrix} -u_{n\vec{k}}^{\downarrow*}(\vec{r}) \\ u_{n\vec{k}}^{\uparrow*}(\vec{r}) \end{pmatrix} \tag{7.291}$$

$$= \psi_{n',-\vec{k}}(\vec{r}, \sigma). \tag{7.292}$$

The new state $\psi_{n',-\vec{k}}(\vec{r}, \sigma)$ has lattice momentum $-\vec{k}$ because the $e^{i\vec{k}\cdot\vec{r}}$ factor in $\psi_{n\vec{k}}(\vec{r}, \sigma)$ is turned into $e^{-i\vec{k}\cdot\vec{r}}$ by the complex conjugation operator K.[32] We thus find that n and n' form a Kramers pair of bands, with

$$\epsilon_n(\vec{k}) = \epsilon_{n'}(-\vec{k}). \tag{7.293}$$

However, this does not imply that these two band dispersions are identical, as the above is not the same as $\epsilon_n(\vec{k}) = \epsilon_{n'}(\vec{k})$.

We do get degenerate bands in the presence of parity (P) or inversion symmetry, under which

$$\vec{r} \to P\vec{r}P = -\vec{r}, \tag{7.294}$$

$$\vec{p} \to P\vec{p}P = -\vec{p}, \tag{7.295}$$

$$\vec{S} \to P\vec{S}P = \vec{S}. \tag{7.296}$$

So, as long as there exists a center of inversion (choice of coordinate origin) such that $V(\vec{r}) = V(-\vec{r})$, parity is a symmetry of the Hamiltonian, including spin–orbit coupling. Under parity transformation we have $\vec{k} \to -\vec{k}$, thus $\epsilon_n(\vec{k}) = \epsilon_n(-\vec{k})$. Combining this with Eq. (7.293), we find that the two Kramers bands are degenerate:

$$\epsilon_n(\vec{k}) = \epsilon_{n'}(\vec{k}), \tag{7.297}$$

as in the case of spin-independent bands.[33] But here this is the combined effect of time-reversal and parity symmetries – the lack of either will break this degeneracy.

---

[30] See, e.g., Ref. [36], Section 4.4.

[31] It can be shown that $\Theta^2 = \pm 1$ for integer- and half-integer-spin particles, respectively. Thus Kramers degeneracy exists for all half-integer spin particles in the presence of time-reversal symmetry, but *not* for integer-spin particles. This is the first example of fascinating differences between integer and half-integer spins due to their quantum nature, with profound consequences to be explored later (see Chapters 14 and 17).

[32] Because of the presence of K, Θ is an *anti*-linear operator, namely $\Theta(a|\psi_1\rangle + b|\psi_2\rangle) = a^*\Theta|\psi_1\rangle + b^*\Theta|\psi_2\rangle \neq a\Theta|\psi_1\rangle + b\Theta|\psi_2\rangle$. This is very different from the linear operators that we deal with almost exclusively in quantum mechanics.

[33] Another way to see Eq. (7.297) is to hit the Bloch wave function (7.283) with ΘP, which results in a degenerate state with the *same* $\vec{k}$.

In the presence of the degeneracy (7.297), we can pair the two degenerate bands up and introduce a set of **pseudospin** operators for each $\vec{k}$, with $\vec{S}(\vec{k}) = (\hbar/2)\vec{\sigma}(\vec{k})$ (this is always possible for any two-level system):

$$\sigma_1(\vec{k}) = |\vec{k}, n\rangle\langle\vec{k}, n'| + |\vec{k}, n'\rangle\langle\vec{k}, n|, \tag{7.298}$$

$$\sigma_2(\vec{k}) = -i|\vec{k}, n\rangle\langle\vec{k}, n'| + i|\vec{k}, n'\rangle\langle\vec{k}, n|, \tag{7.299}$$

$$\sigma_3(\vec{k}) = |\vec{k}, n\rangle\langle\vec{k}, n| - |\vec{k}, n'\rangle\langle\vec{k}, n'|. \tag{7.300}$$

More importantly, however, it is always possible to choose the basis in such a way that

$$P|\vec{k}, n\rangle = |-\vec{k}, n\rangle, \tag{7.301}$$

$$P|\vec{k}, n'\rangle = |-\vec{k}, n'\rangle, \tag{7.302}$$

$$\Theta|\vec{k}, n\rangle = |-\vec{k}, n'\rangle, \tag{7.303}$$

$$\Theta|\vec{k}, n'\rangle = -|-\vec{k}, n\rangle. \tag{7.304}$$

With the conditions above, our pseudospin operator $\vec{S}(\vec{k})$ transforms in exactly the same way as the original spin operator $\vec{S}$ under P and Θ, and can in practice be treated as a spin operator for essentially all purposes. Since it is closely related to, but not exactly the same as, the original spin operator, some modifications may be necessary in its usage. One important example is that, when considering its coupling to a magnetic field, the Landé g-factor for the pseudospin is often significantly modified from the free-electron value of 2 (sometimes even changing its sign!), due to the spin–orbit coupling.[34]

Exercise 7.32. Show that, for spinless particles, time-reversal symmetry guarantees for each band that $\epsilon_n(\vec{k}) = \epsilon_n(-\vec{k})$, even for crystals without an inversion center. (We remind the reader that in a crystal with inversion symmetry one can choose a point as the origin of space such that $V(\vec{r}) = V(-\vec{r})$. Crystals without an inversion center are called non-centrosymmetric.) Show that this conclusion is valid for electrons as well, if spin–orbit coupling can be neglected. Show as a result that $\vec{v}(\vec{k} = 0) = \nabla_{\vec{k}}\epsilon_n(\vec{k})\big|_{\vec{k}=0} = \vec{0}$.

Exercise 7.33. Show that, under the same conditions as those of Exercise 7.32, we also have $\epsilon_n(\vec{G}/2 + \vec{k}) = \epsilon_n(\vec{G}/2 - \vec{k})$, as a result of which $\vec{v}(\vec{k} = \vec{G}/2) = \vec{0}$. Here $\vec{G}$ is a reciprocal lattice vector, and $\vec{G}/2$ is thus a point on the Brillouin zone boundary. This may lead to a van Hove singularity at $\epsilon_n(\vec{G}/2)$.

7.9 Photonic Crystals

Bloch's theorem is a central result not only for this chapter, but for the entire book. Thus far we have been discussing its implications for electron waves propagating in a periodic crystal of atoms, and we will continue this discussion throughout the remainder of this book. In later chapters the emphasis will be on how electron motion is affected when *deviations* from perfect periodicity are present, either due to external perturbations or due to lattice defects. Before diving into these discussions, however, we would like to emphasize that Bloch's theorem is very general. It applies not only to electron waves

---

[34] In general g becomes a rank-2 tensor in this case, and the spin magnetization need not be parallel to the magnetic field.

in a crystal formed by atoms, but also to any waves that satisfy wave equations with discrete lattice periodicity such as elastic waves in a mechanical medium and light waves in a dielectric medium. In this section we discuss Maxwell's equations for the propagation of light in the presence of a non-uniform but periodic dielectric medium, known as a **photonic crystal**. We will see that propagation of electromagnetic waves (or equivalently the photons) is strongly modified in photonic crystals, very much like electron motion in a crystal. In the next section we will turn things around and study the motion of matter waves (atoms) in periodic potentials formed by standing waves of light.

We start with Maxwell's equations in the presence of a medium:

$$\nabla \cdot \vec{B}(\vec{r}, t) = 0; \qquad \nabla \cdot \vec{D}(\vec{r}, t) = 4\pi \rho(\vec{r}, t); \tag{7.305}$$

$$\nabla \times \vec{E}(\vec{r}, t) + \frac{1}{c} \frac{\partial \vec{B}(\vec{r}, t)}{\partial t} = 0; \tag{7.306}$$

$$\nabla \times \vec{H}(\vec{r}, t) - \frac{1}{c} \frac{\partial \vec{D}(\vec{r}, t)}{\partial t} = \frac{4\pi}{c} \vec{J}(\vec{r}, t). \tag{7.307}$$

Here $\rho(\vec{r}, t)$ and $\vec{J}(\vec{r}, t)$ are, respectively, the density and current density of free charges (i.e. mobile as in a metal instead of bound as in an insulating dielectric), which will be set to zero from now on.

These equations need to be supplemented by two additional equations describing the matter:

$$\vec{B} = \mu \vec{H}; \qquad \vec{D} = \epsilon \vec{E}. \tag{7.308}$$

For non-magnetic materials the magnetic permeability μ is very close to 1 (in cgs units), and will be set to unity from now on. The dielectric constant ϵ is non-uniform and periodic in a photonic crystal, and, as we will see shortly, plays a role similar to the periodic potential in the Schrödinger equation.[35]

We look for eigensolutions of the form

$$\vec{B}(\vec{r}, t) = \vec{B}(\vec{r}) e^{-i\omega t}; \qquad \vec{E}(\vec{r}, t) = \vec{E}(\vec{r}) e^{-i\omega t}. \tag{7.309}$$

Plugging these into Maxwell's equations, and eliminating $\vec{E}$ in favor of $\vec{B}$, we arrive at the so-called *master equation* that plays a role similar to the static (or time-independent) Schrödinger equation:

$$\nabla \times \left[\frac{1}{\epsilon(\vec{r})} \nabla \times \vec{B}(\vec{r}) \right] = \frac{\omega^2}{c^2} \vec{B}(\vec{r}), \tag{7.310}$$

in which

$$\hat{\Xi} = \nabla \times \left[\frac{1}{\epsilon(\vec{r})} \nabla \times \right] \tag{7.311}$$

plays the role of the Hamiltonian (and is carefully ordered to be Hermitian!), and ω^2/c^2 corresponds to the eigenenergy. The solution of Eq. (7.310) must also satisfy the transverse (no monopoles) condition $\nabla \cdot \vec{B}(\vec{r}) = 0$. We chose to eliminate $\vec{E}$ in favor of $\vec{B}$ because it is easier to obtain an explicitly Hermitian Hamiltonian [37]. Exercises 7.34 and 7.35 address this issue.

Clearly the symmetries of Ξ are identical to those of $\epsilon(\vec{r})$. In a photonic crystal we have

$$\epsilon(\vec{r}) = \epsilon(\vec{r} + \vec{R}), \tag{7.312}$$

where $\vec{R}$ is a lattice vector. In this case Bloch's theorem applies to Eq. (7.310), whose solutions form *photonic* bands. Everything we learned in this chapter applies to the photonic bands (except that photons are bosons and the Pauli exclusion principle does *not* apply). In particular, the presence of photonic band gaps means that photons or electromagnetic waves *cannot* propagate inside the photonic crystal, if their frequency falls within the band gaps. Also the propagation velocity is

---

[35] In general ϵ depends on frequency, can be non-local (i.e. it can depend on the wave vector), and can have both real and imaginary parts. We ignore all these complications here. In particular, by neglecting the imaginary part of ϵ we assume that the medium causes no loss or dissipation. In this case the index of refraction is $n = \sqrt{\epsilon}$.

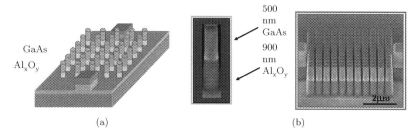

GaAs
Al$_x$O$_y$

500
nm
GaAs

900
nm
Al$_x$O$_y$

2μm

(a)   (b)

Figure 7.24 (a) Schematic illustration of a 2D photonic crystal formed by dielectric rods made of GaAs. (b) An actual image obtained from a scanning electron microscope. Figure (a) adapted with permission from [38]. Figure (b) adapted from S. Assefa (2004), "The Development of Novel /Passive and Active Photonic-Crystal Devices." http://hdl.handle.net/1721.1/18053 with the permission of Massachusetts Institute of Technology.

determined by the band dispersion, instead of simply c/n. Photonic crystals, which are artificial[36] macroscopic structures, provide the possibility of controlling the behavior of photons in ways that are not possible otherwise [37]. An illustrative example is shown in Fig. 7.24.

A particularly simple and commonly used photonic crystal is the **distributed Bragg reflector (DBR)**. This is a "**superlattice**"[37] consisting of alternating layers of two materials A and B with differing indices of refraction, n_A and n_B. Such superlattices can be created by sputtering many alternating (amorphous) layers onto a glass substrate, and can be used to create high-reflectivity mirrors at some particular free-space wavelength λ. If the two layers have respectively optical thicknesses $n_A d_A$ and $n_B d_B$ close to $\lambda/4$, there will be a band gap which will give the light an exponentially decaying solution inside the material and cause the reflectivity R to be extremely high. High-quality commercial Bragg mirrors used in optical **Fabry–Pérot cavities** can achieve reflectivities of $R \sim 99.999\%$. Bragg mirrors play an essential role in the semiconductor laser diodes that are ubiquitous in modern technology. They will also make an appearance when we discuss experiments creating **Bose–Einstein condensates** of polaritons in Chapter 18.

Exercise 7.34. Let us define an inner product between two magnetic field configurations $\vec{B}_1(\vec{r})$ and $\vec{B}_2(\vec{r})$ to be

$$(\vec{B}_1, \vec{B}_2) = \int d^3r \left[\vec{B}_1^*(\vec{r}) \cdot \vec{B}_2(\vec{r}) \right]. \tag{7.313}$$

Show that Ξ defined in Eq. (7.311) is Hermitian in the sense that

$$(\vec{B}_1, \Xi\vec{B}_2) = (\Xi\vec{B}_1, \vec{B}_2). \tag{7.314}$$

Exercise 7.35. Instead of eliminating $\vec{E}$ in favor of $\vec{B}$, we can also do the opposite and arrive at an equation for $\vec{E}(\vec{r})$ that is similar to (7.310). (i) Find this equation. (ii) Show that it is not possible to interpret this equation as an eigenequation of a Hermitian operator.

[36] Naturally occurring minerals such as opals and naturally occurring structures in bird feathers and insects also act as photonic crystals which produce pleasing colorations and other interesting optical effects.

[37] See Chapter 9 for a discussion of semiconductor superlattices which can modify the propagation of electrons. Because the wavelength of light is much larger than atomic scales, the fact that sputtered layers are amorphous on the atomic scale is not very important. However, semiconductor superlattices for electrons need to be crystalline (and hence created with atomic beam epitaxy) since electrons have wavelengths on the scale the atomic spacing.

Exercise 7.36. Consider a distributed Bragg reflector consisting of alternating layers of index of refraction $n_A = 1.5$ and $n_B = 2.0$, with thickness $d_A = 0.2\,\mu$m and $d_B = 0.15\,\mu$m. Find the range of wavelengths reflected by the first stop band (band gap).

7.10 Optical Lattices

Interactions between photons/electromagnetic waves and matter play a major role in condensed matter as well as atomic, molecular, and optical (AMO) physics. In Section 7.9 we learned that (condensed) matter can be used to engineer band structures for photons. In this section we discuss the opposite possibility, namely using laser light to set up periodic potentials for atoms or molecules, and engineer their band structures. (Remember that quantum mechanics tells us that atoms propagate as waves too!) Such engineered periodic potentials for atoms and molecules are known as **optical lattices**.

Let us start with the simplest possible optical lattice, namely a 1D lattice formed by two counter-propagating laser beams that result in a standing electromagnetic wave (see Fig. 7.25(a) for an illustration), whose electric field (which we approximate as classical) takes the form

$$\vec{E}(\vec{r}, t) = \vec{E}(\vec{r})\cos(\omega t) = E_0\,\hat{e}\cos(\vec{k} \cdot \vec{r})\cos(\omega t), \qquad (7.315)$$

with $\omega = c|\vec{k}|$. As we learned in Chapter 2, the dominant interaction between an atom and an electromagnetic wave is through the coupling between the atom's electrons and the electric field; in particular, an oscillating electric field induces an oscillating dipole moment with the same frequency:

$$\vec{d}(\omega) = \alpha(\omega)\vec{E}(\omega), \qquad (7.316)$$

where $\alpha(\omega)$ is the dipole susceptibility. For oscillating fields, the susceptibility is generally complex, so some care is required in the interpretation. Since the electric field $\vec{E}(t)$ is real, we must have $\vec{E}(-\omega) = \vec{E}^*(+\omega)$, a condition which is indeed satisfied by Eq. (7.315). In order for $\vec{d}(t)$ to be real, we must require the same condition on the **polarizability**, $\alpha(-\omega) = \alpha^*(+\omega)$. Thus Eqs. (7.315) and (7.316) combine to yield the following time-dependent polarization for an atom located at position $\vec{r}$:

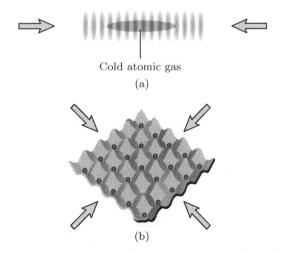

Cold atomic gas

(a)

(b)

Figure 7.25 (a) A standing wave formed by two laser beams propagating in opposite directions, resulting in a 1D optical lattice for atoms trapped in it. The standing wave can also be created between two mirrors (not shown) of an optical cavity. (b) A 2D optical lattice formed by four laser beams. 3D optical lattices can be formed using more laser beams. Figure adapted with permission from [39].

$$\vec{d}(t) = \frac{1}{2}\vec{E}(\vec{r})[\alpha(+\omega)e^{-i\omega t} + \alpha(-\omega)e^{+i\omega t}] \tag{7.317}$$

$$= \vec{E}(\vec{r})[\operatorname{Re}\alpha(+\omega)\cos(\omega t) + \operatorname{Im}\alpha(+\omega)\sin(\omega t)]. \tag{7.318}$$

The real part of the susceptibility yields the component of the dipole moment that is oscillating in phase with the electric field and the imaginary part describes the component that is oscillating 90° out of phase with the electric field. Only the former stores any energy. (As will be discussed below, the imaginary part of the susceptibility corresponds to dissipation.) The energy of a permanent dipole in an electric field is $U = -\vec{d} \cdot \vec{E}$. However, for dipole moments that are induced by the field the correct expression is $U = -\frac{1}{2}\vec{d} \cdot \vec{E}$. The potential energy stored in the oscillating dipole averaged over one optical cycle is thus

$$V(\vec{r}) = -\frac{1}{4}\operatorname{Re}\alpha(\omega)|\vec{E}(\vec{r})|^2 = -v_0\left[\frac{1 + \cos(2\vec{k}\cdot\vec{r})}{2}\right], \tag{7.319}$$

where $v_0 = \frac{1}{4}\operatorname{Re}\alpha(\omega)E_0^2$, and the additional factor of $1/2$ comes from averaging over time. This plays the role of a periodic potential seen by the atom whose period, $\lambda/2$, is *one-half* of the optical wavelength. One can, obviously, control the lattice structure and intensity by using multiple laser beams and controlling their intensities, wavelengths, and relative phases. For example, one can generate a 3D simple cubic lattice using three pairs of counter-propagating beams along three orthogonal directions. Figure 7.25(b) schematically illustrates two pairs of laser beams being used to generate a 2D lattice structure.

Because the atomic nucleus is much more massive than the electron, the atom's position does not oscillate significantly at optical frequencies, and we are justified in replacing the electric field squared by its time average over an optical cycle to obtain a static potential. We see that, if $\operatorname{Re}\alpha(+\omega)$ is positive, the atom is attracted to regions of strong electric field (i.e. the anti-nodes of the standing wave). Conversely, if $\operatorname{Re}\alpha(+\omega)$ is negative, the atom is attracted to the nodes of the electric field.[38]

As noted above, the imaginary part of the polarizability corresponds to dissipation. To see this, consider the time-averaged power supplied by the electric field to the atom:

$$\langle P \rangle = \langle \dot{\vec{d}} \cdot \vec{E} \rangle$$
$$= -\omega\vec{E}(\vec{r})\langle[\operatorname{Re}\alpha(+\omega)\sin(\omega t) - \operatorname{Im}\alpha(+\omega)\cos(\omega t)] \cdot \vec{E}(\vec{r})\cos(\omega t)\rangle$$
$$= +\frac{\omega}{2}\operatorname{Im}\alpha(+\omega)|\vec{E}(\vec{r})|^2, \tag{7.320}$$

which is positive provided that $\operatorname{Im}\alpha(+\omega)$ is positive. At the classical level, $\operatorname{Im}\alpha(+\omega)$ represents *radiation damping*. The laser field accelerates the electrons, causing them to radiate electric field in a dipole pattern as shown in Eq. (2.7) and thus lose energy at a rate proportional to the square of their acceleration [18, 19].

7.10.1 Oscillator Model of Atomic Polarizability

In Section 2.2 we considered a classical picture of the polarizability of an atom at X-ray frequencies, frequencies so high that it is not unreasonable to approximate the electrons in the atom as free and unperturbed by each other or the atomic nucleus. Here we will consider a simple harmonic oscillator model of atomic polarizability which captures the essential physics when optical fields are nearly resonant with the transition frequency between the atomic ground state and some excited state. Atoms

---

[38] In certain precision experiments with lattices of atoms, it can be advantageous to have the atoms sitting on lattice sites that are nodal points because the atoms are then less perturbed by the lasers used to create the optical lattice potential.

are obviously not harmonic oscillators, but, if they are driven only weakly, they will have only a small probability of being in a low-lying excited state and a negligible probability of being in any higher excited states. The same is true for a damped harmonic oscillator. Furthermore, if in the full quantum theory we compute the polarization of the atom with perturbation theory to lowest order in the field, the polarization is linear in the electric field. For a harmonic oscillator, the polarization response is exactly linear.

We will take the oscillator frequency to match the excitation frequency of interest in the atom[39] and fit the oscillator damping parameter to the spontaneous-emission lifetime of the excited state. Taking the electric field to be $\vec{E} = \hat{x}E(t)$, we will assume a 1D harmonic oscillator with mass m_e, charge q, and spring constant k, obeying Newton's equation of motion

$$m_e \ddot{x} = -kx - \gamma \dot{x} + q E(t). \tag{7.321}$$

By Fourier transforming and solving for the relation between polarization and electric field, we obtain the exact polarizability at the drive frequency ω,

$$\alpha(\omega) = \frac{q^2}{m_e} \frac{1}{\omega_0^2 - \omega^2 - i\omega\gamma/m_e}, \tag{7.322}$$

where $\omega_0 = \sqrt{k/m_e}$ is the resonance frequency of the oscillator. We see that $\alpha(-\omega) = \alpha^*(+\omega)$, as required in order for the polarization to be real.

We also see that the polarization response is large near $\pm\omega_0$. It turns out that, because of the small size of the fine-structure constant, $\gamma \ll \omega_0$, so the damping rate is weak relative to the frequency. Thus, in the region $\omega \sim +\omega_0$ we can approximate the polarizability as

$$\alpha(\omega) \approx \frac{q^2}{m_e} \frac{1}{(\omega_0 - \omega)2\omega_0 - i\omega_0\gamma/m_e} \approx \frac{q^2}{2m_e\omega_0} \frac{1}{\omega_0 - \omega - i\kappa/2}, \tag{7.323}$$

where $\kappa \equiv \gamma/m_e$ and $1/\kappa$ is the excited state spontaneous-emission lifetime. Equivalently, Eq. (7.323) tells us that $\kappa/2$ is the amplitude damping rate for the oscillator and, since the energy scales with the square of the amplitude, κ is the energy damping rate of the oscillator. The energy of the undriven oscillator obeys

$$\frac{d\mathcal{E}}{dt} = -\kappa\mathcal{E}. \tag{7.324}$$

Figure 7.26 illustrates the resonant polarization response of a driven, damped harmonic oscillator. The plot shows the dimensionless response function $\kappa g(\omega) \equiv \kappa/(\omega_0 - \omega - i\kappa/2)$. The imaginary part of $g(\omega)$ is a Lorentzian with full width at half-maximum of κ. For drive frequencies below the resonance ("red detuning"), the real part of the polarization response function is positive, indicating that the oscillator polarization is in phase with the driving electric field, and so the atom would be attracted to the anti-nodes of the driving field. For drive frequencies above the resonance ("blue detuning"), the atom would be attracted to the nodes of the driving field. For large detuning from resonance, $|\omega_0 - \omega| \gg \kappa$, the imaginary part of the response falls off as $\kappa^2/2(\omega_0 - \omega)^2$, while the real part falls off much more slowly as $\kappa/(\omega_0 - \omega)$. Thus the experimentalist can take advantage of the relatively large polarizability over a wide frequency range near resonance without paying the price of strong dissipation very close to the resonance. This dissipation introduces heating and decoherence into the motion of the atoms in the optical potential for reasons that will become clear when we discuss the quantum theory of these processes.

---

[39] We assume there is only a single optically allowed transition that is close to resonance with the laser frequency.

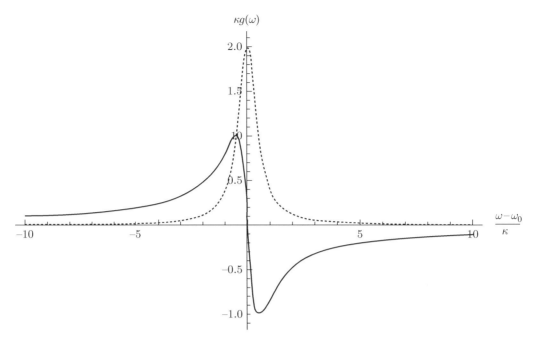

Figure 7.26 Resonant polarization response of a damped harmonic oscillator driven by an electric field of frequency ω near the resonance frequency $+\omega_0$. The solid (dashed) lines show the real (imaginary) parts of the dimensionless response function $\kappa g(\omega)$. Here κ is the energy damping rate of the oscillator.

7.10.2 Quantum Effects in Optical Lattices

Our entire discussion has so far treated both the laser field and the atom motion in the optical potential classically. The quantum theory for X-ray scattering was briefly described in Section 2.3. Here we will provide a qualitiative discussion of quantum effects that were neglected in our classical treatment above. Let us begin by treating the atom as a quantum-mechanical wave moving in the optical lattice which we treat as a fixed, conservative potential given by Eq. (7.319). Suppose that the matter wave is in a plane-wave state

$$\Psi(\vec{r}) = e^{i\vec{K}\cdot\vec{r}}. \tag{7.325}$$

The potential-energy term in the Schrödinger equation acting upon this wave function has a constant part, $-v_0/2$, and a spatially modulated part that changes the momentum of the particle to $\hbar(\vec{K} \pm 2\vec{k})$ when the particle scatters from the potential:

$$V(\vec{r})\Psi(\vec{r}) = -\frac{v_0}{4}\left[2e^{i\vec{K}\cdot\vec{r}} + e^{i(\vec{K}+2\vec{k})\cdot\vec{r}} + e^{i(\vec{K}-2\vec{k})\cdot\vec{r}}\right]. \tag{7.326}$$

How do we understand this from the point of view of the laser light being a quantum field composed of photons? We should think that the two counter-propagating laser fields have a macroscopically large number of photons N in two states with momentum $\pm\hbar k$. When the detuning is large, $|\omega_0 - \omega| \gg \kappa$, energy is not conserved when the atom absorbs a photon, and so the absorption is "virtual" and must be quickly followed by emission of a photon. If the atom virtually absorbs a photon of momentum $\hbar\vec{k}$, it can re-emit the photon in any direction. However, because of the macroscopic occupation of the modes with momentum $\pm\hbar\vec{k}$, there is a strong tendency for stimulated emission into those two modes. If the photon emission is stimulated into $+\hbar\vec{k}$, the net momentum transfer is zero. This accounts for the first term in Eq. (7.326). If the photon emission is stimulated into $-\hbar\vec{k}$, then the net momentum transfer from the photon field to the atom field is $+2\hbar\vec{k}$. Conversely, if the atom virtually absorbs a photon of momentum $-\hbar \vec{k}$, the stimulated emission into $-\hbar \vec{k}$ gives zero momentum transfer, but the stimulated emission into $+\hbar \vec{k}$ gives a momentum transfer to the atom field of $-2\hbar \vec{k}$. These four processes account for the three terms in Eq. (7.326), and the fact that the zero-momentum-transfer term is twice as large as the others.

From the energy–time uncertainty principle, the duration of the virtual intermediate excited state of the atom is of order the inverse detuning. When this is large, the duration is so short that there is little probability of spontaneous emission of a photon into one of the (large number of) modes that is not macroscopically occupied. In this limit the classical theory of the optical potential becomes essentially exact. However, as the detuning becomes small and approaches the line width, the duration of the intermediate state begins to approach the spontaneous-emission lifetime, and spontaneous emission of photons in random directions becomes important. Such processes give random momentum recoil kicks to the atom and heat up its motion in the periodic potential. They also introduce decoherence in the motion since the environment is effectively "observing" the position of the atom through its spontaneous fluorescence emission. These heating and decoherence effects can be avoided by detuning the laser beams from the atomic resonance, at a small price in increased laser intensity to maintain the same strength of the optical lattice potential.

Atomic physicists have been able to load both bosonic and fermionic atoms into optical lattices and see band insulators, interaction-induced insulators, metals, superfluids, and magnetism dynamics. Optical lattices are a rich playground for "quantum simulations" of condensed matter models of interacting particles, such as the Hubbard model which we will introduce in Section 17.4.

8 Semiclassical Transport Theory

Bloch's theorem tells us that the energy eigenfunctions for a crystal in the independent-electron model are plane waves modulated by a periodic function

$$\psi_{n\vec{k}}(\vec{r}) = e^{i\vec{k}\cdot\vec{r}}\, u_{n\vec{k}}(\vec{r}). \tag{8.1}$$

The wave function is extended throughout the crystal, so it is inappropriate to use Drude's picture that electrons are classical particles which collide randomly with the atoms. Instead, quantum interference allows the electron waves to propagate coherently throughout the crystal, giving the electrons an infinite mean free path (in the absence of impurities, disorder, or lattice distortions by phonons).

It turns out, however, that it is possible to make a semiclassical version of the Drude model in which Bloch-state wave packets play the role of the particles. Drude's formula for the electrical conductivity will still hold, but the mean free time τ (due to scattering from disordered impurities) must be computed quantum mechanically.

We will begin with a review of quantum wave-packet dynamics in free space and then discuss what happens when the wave packets are made of Bloch waves moving through the periodic potential of the lattice. In some ways the dynamics will be surprisingly similar, but certain features will be quite different.

8.1 Review of Semiclassical Wave Packets

As a warm-up, let us remind ourselves how to construct a **Gaussian wave packet** in *free* space, where energy eigenfunctions are simple plane waves. A wave packet centered at $\vec{R}$ and carrying average momentum $\vec{K}$ takes the form (without worrying about normalization)

$$\phi_{\vec{R},\vec{K}}(\vec{r}) = e^{i\vec{K}\cdot(\vec{r}-\vec{R})} e^{-\frac{|\vec{r}-\vec{R}|^2}{2\Delta r^2}} \sim \int_{-\infty}^{+\infty} \frac{d^3\vec{k}}{(2\pi)^3}\, e^{-\frac{|\vec{k}-\vec{K}|^2}{2(\Delta k)^2}}\, e^{i\vec{k}\cdot\vec{r}} e^{-i\vec{k}\cdot\vec{R}}. \tag{8.2}$$

Here Δr and Δk are the widths (in each of the spatial directions) of the wave packet in real space and momentum space, respectively. We have $\Delta r\,\Delta k = 1$ and thus the uncertainty relation bound is saturated. We can think of the wave packet as describing a classical particle having position $\vec{R}$, momentum $\hbar\vec{K}$, and energy $\epsilon(\vec{K}) = \hbar^2 K^2/2m_e$. The velocity of propagation of this particle is the wave-packet group velocity

$$\vec{v}(\vec{K}) \equiv \frac{1}{\hbar} \left. \nabla_{\vec{k}} \epsilon(\vec{k}) \right|_{\vec{k}=\vec{K}} = \frac{\hbar \vec{K}}{m_e}. \tag{8.3}$$

In the presence of a smooth (on the scale of the de Broglie wavelength) external potential $U(\vec{r})$, we have **Ehrenfest's theorem**:

$$\hbar \frac{d\vec{K}}{dt} = m_e \frac{d^2 \vec{R}}{dt^2} = -\nabla U(\vec{R}) = \vec{F}, \tag{8.4}$$

which is the quantum-mechanical analog of Newton's second law.

In the next section we discuss how to generalize this way of constructing wave packets to Bloch electrons, as well as their equations of motion.

8.2 Semiclassical Wave-Packet Dynamics in Bloch Bands

When an electron moves in a weak and slowly varying (in both space and time) external electromagnetic field *in addition* to the periodic lattice potential, we expect its motion to be (primarily) confined to a given Bloch band.[1] It is thus natural to construct wave packets using Bloch states with a fixed band index n. The most obvious modification to the free-space case is to replace the plane-wave factor $e^{i\vec{k}\cdot\vec{r}}$ in Eq. (8.2) by the Bloch wave function $\psi_{n\vec{k}}(\vec{r})$:

$$\phi^n_{\vec{R},\vec{K}}(\vec{r}) \equiv \int_{1\mathrm{BZ}} \frac{d^3\vec{k}}{(2\pi)^3} e^{-\frac{|\vec{k}-\vec{K}|^2}{2(\Delta k)^2}} e^{-i\vec{k}\cdot\vec{R}} \psi_{n\vec{k}}(\vec{r}). \tag{8.5}$$

Accompanying this modification is the fact that the wave vector $\vec{k}$ is restricted to the first Brillouin zone, instead of being unlimited as in Eq. (8.2). This means that Δk is meaningful only if it is small compared with the linear size of the first Brillouin zone. As a result, the real-space size $\Delta r \sim 1/\Delta k$ must be large compared with the lattice spacing. Thus a semiclassical wave-packet description of Bloch electrons assumes that there is a hierarchy of three distinct length scales

$$L_1 \ll L_2 \ll L_3, \tag{8.6}$$

where $L_1 = a$ is the lattice constant, $L_2 \sim \Delta r$ is the wave-packet size, and L_3 is the scale which characterizes the slow spatial variation of the applied external fields.

There is an important subtlety with the construction in Eq. (8.5). As we discussed in Section 7.4.2, the Bloch wave function $\psi_{n\vec{k}}(\vec{r})$, or more precisely its periodic part $u_{n\vec{k}}(\vec{r})$, carries an arbitrary $\vec{k}$-dependent phase factor; any change of this phase factor modifies the wave packet in Eq. (8.5). Thus we need a (k-space) "gauge-fixing" condition[2] to fix $u_{n\vec{k}}(\vec{r})$, such that Eq. (8.5) is uniquely defined (up to an *overall* phase). It can be shown that, in the presence of both time-reversal and inversion symmetries, one can choose to have[3]

$$\int d^3\vec{r}\, u^*_{n\vec{k}}(\vec{r})[\nabla_{\vec{k}} u_{n\vec{k}}(\vec{r})] = \vec{0}. \tag{8.7}$$

---

[1] This is because wave functions in different bands generally differ significantly in their rapid atomic-scale oscillations and also generally differ significantly in energy. Hence the transition-matrix elements for slowly varying perturbations are small.

[2] Even in the absence of a lattice potential, we need this condition for plane-wave solutions which have the form $\psi_{\vec{k}} = u_{\vec{k}} e^{i\vec{k}\cdot\vec{r}}$, with $u_{\vec{k}}(\vec{r}) = e^{i\theta_{\vec{k}}}$. We are so used to automatically taking $\theta_{\vec{k}} = 0$ that we forget that this default is actually a gauge fixing.

[3] See Section 13.4 for proof.

With this condition the phase of $u_{n\vec{k}}(\vec{r})$ is uniquely determined once we fix the phase at one point in $\vec{k}$ space. Furthermore, with the condition (8.7) the wave packet (8.5) is indeed centered at $\vec{R}$.

We want to emphasize that the derivation of the semiclassical equations of motion that we present in this chapter is the traditional one. In Section 13.4 we will learn that this derivation actually misses an "anomalous velocity" term which is important when either time-reversal symmetry or inversion symmetry is absent, which leads to the presence of **Berry curvature**. For the rest of this chapter, however, we will assume the presence of both time-reversal and inversion symmetries, and use the gauge-fixing condition (8.7). The necessary modifications when either of the two conditions is violated will be discussed in Section 13.4.

If the wave-vector uncertainty Δk is small, then $L_2 \sim 1/\Delta k \gg L_1 = a$. If we make the approximation of assuming that $u_{n\vec{k}}(\vec{r})$ varies slowly with $\vec{k}$ on the scale of Δk (which is guaranteed by Eq. (8.7)), we can pull it out of the integral above to obtain

$$
\begin{aligned}
\phi^n_{\vec{R},\vec{K}}(\vec{r}) &\approx u_{n\vec{K}}(\vec{r}) \int \frac{d^3\vec{k}}{(2\pi)^3} e^{-\frac{|\vec{k}-\vec{K}|^2}{2(\Delta k)^2}} e^{+i\vec{k}\cdot(\vec{r}-\vec{R})} \\
&\sim u_{n\vec{K}}(\vec{r}) e^{-\frac{(\Delta k)^2}{2}|\vec{r}-\vec{R}|^2} e^{+i\vec{K}\cdot(\vec{r}-\vec{R})} \\
&= \psi_{n\vec{K}}(\vec{r}) e^{-\frac{(\Delta k)^2}{2}|\vec{r}-\vec{R}|^2} e^{-i\vec{K}\cdot\vec{R}}.
\end{aligned}
\tag{8.8}
$$

We see that the Gaussian in Eq. (8.2) becomes the *envelope* function of the Bloch electron wave packet, which is multiplied by the rapidly oscillating (on the scale of the lattice spacing a) Bloch wave function. Since $L_2 \gg L_1$, the wave vector $\vec{K}$ is well-defined and simultaneously, since $L_2 \ll L_3$, the position $\vec{R}$ is also well-defined. Hence we can think of the wave packet as describing a (semi)classical particle having position $\vec{R}$, (crystal) momentum $\hbar\vec{K}$, and energy $\epsilon_{n\vec{K}}$.

As discussed in Section 7.3.3, the velocity of propagation of this particle is the wave-packet group velocity

$$
\vec{v}_n(\vec{K}) \equiv \frac{d\vec{R}}{dt} = \frac{1}{\hbar}\nabla_{\vec{k}}\epsilon_n(\vec{k})\Big|_{\vec{k}=\vec{K}}.
\tag{8.9}
$$

When the effective-mass approximation is valid, the velocity reduces to a form similar to that for a free electron:

$$
\vec{v}_n(\vec{K}) = \frac{\hbar\vec{K}}{m^*},
\tag{8.10}
$$

where m_e is replaced by an effective mass m^*, and $\vec{K}$ is measured from the band minimum. In general, however, even the direction of $\vec{v}$ is modified so that it is no longer parallel to $\vec{K}$. Recall that $\vec{v}_n(\vec{K})$ must be perpendicular to the surface of constant energy passing through the point $\vec{K}$. This in turn can be strongly distorted by the periodic crystalline potential, especially near the zone boundaries. Such distortions can have important consequences, as we shall see later.

We now turn to the question of how to include the effect on the dynamics of a slowly varying external potential $U(\vec{R})$. Because we are taking both position and wave vector to be well-defined, it seems reasonable to write the semiclassical Hamiltonian as

$$
H = \epsilon_n(\vec{K}) + U(\vec{R}).
\tag{8.11}
$$

This should be a constant of the motion so that

$$
\frac{dH}{dt} = 0 = \frac{d\vec{K}}{dt} \cdot \nabla_{\vec{K}}\epsilon_n(\vec{K}) + \frac{d\vec{R}}{dt} \cdot \nabla U(\vec{R}).
\tag{8.12}
$$

Using the expression for the group velocity,

$$
\frac{d\vec{R}}{dt} = \vec{v}_n(\vec{K}) = \frac{1}{\hbar}\nabla_{\vec{K}}\epsilon_n(\vec{K}),
\tag{8.13}
$$

we deduce that

$$\hbar \frac{d\vec{K}}{dt} = -\nabla U(\vec{R}) = -e\vec{E}(\vec{R}), \qquad (8.14)$$

which looks like Hamilton's classical equation of motion.

Generalizing this result to include a weak smoothly varying (in space and time) electromagnetic field yields

$$\hbar \frac{d\vec{K}}{dt} = -e\left[\vec{E}(\vec{R}, t) + \frac{1}{c}\vec{v}(\vec{K}) \times \vec{B}(\vec{R}, t)\right] = \vec{F}, \qquad (8.15)$$

where $\vec{F}$ is the external electromagnetic force. Remember that $\vec{E}$ excludes the rapidly varying microscopic electric fields within each unit cell. These have already been taken into account in solving for the rapidly oscillating part of the wave function $u_{n\vec{K}}$. We must also remember that, because the wave function does contain a rapidly oscillating part, $\hbar\vec{K}$ is not really a true momentum.

While Eq. (8.15) looks simple and appealing (and is effectively the Bloch-wave version of Ehrenfest's theorem), its derivation from first principles has proven quite complicated. In Section 8.2.1 we present a heuristic derivation. There we also discuss subtle issues like how to properly define $\vec{K}$ in a gauge-invariant manner, when a vector potential is present.

Now assume $\vec{K}$ is near a band minimum (at $\vec{k}_0$), and further assume one can use the (isotropic) effective mass approximation:

$$\epsilon_n(\vec{K}) \approx \epsilon_n(\vec{k}_0) + \frac{\hbar^2|\vec{K} - \vec{k}_0|^2}{2m^*}. \qquad (8.16)$$

In this case we have

$$\frac{d^2\vec{R}}{dt^2} = \frac{\hbar}{m^*}\frac{d\vec{K}}{dt} = \frac{\vec{F}}{m^*}, \qquad (8.17)$$

which takes the same form as Eq. (8.4) or Newton's second law, as long as we use the effective mass m^* instead of the bare mass m_e. Here we made the simplifying assumption of an isotropic (or constant) effective mass; generalization to an anisotropic effective mass tensor is straightforward. We should note, though, that the effective-mass approximation breaks down once the wave packet has moved far away from the band minimum (in momentum space) due to the external force.

We now come to a quite subtle point. What happens when the wave vector $\vec{K}(t)$ that solves Eq. (8.15) drifts over to the zone boundary? We do not expect that weak, smooth external forces can induce an interband transition across the excitation gap to the next band. Hence the electron must somehow remain in the same band. What happens is that the electron is Bragg reflected back to the opposite side of the Brillouin zone, where the drift motion starts all over again. This seems like an unphysical and discontinuous process, but in fact everything is continuous. Indeed, the wave function at a point on the zone boundary is *identical* to the wave function at the corresponding point on the opposite side of the Brillouin zone (because in effect these two points are to be identified as the same point).[4]

As a simple example, consider a 1D band structure like that shown in Fig. 7.3, in which the electron is subject to a constant electric field E. Note that the group velocity vanishes at both ends of the first Brillouin zone, so there is no discontinuity in the velocity observable as a result of the Bragg reflection. This is clear in the extended-zone picture, where $K(t) = K_0 - eEt/\hbar$ increases smoothly without bound, but the velocity is periodic in time since the band energy is periodic in K.

---

[4] There is a subtlety on this point in the presence of a magnetic field which we will revisit in Section 13.5 when we study the topological aspects of the quantum Hall effect. The wave functions on opposite sides of the Brillouin zone can differ in phase due to properties of the magnetic translation group.

These periodic oscillations in velocity lead to periodic oscillations in position known as Bloch oscillations. In practice, such oscillations are extremely difficult to observe for electrons in a naturally occuring crystal because of scattering from impurities and phonons. This "friction" destroys the quantum coherence, causes large sudden changes in $\vec{K}$ and ultimately controls the rate at which the electrons drift "downhill" in potential energy. On the other hand, Bloch oscillations have been observed in artificially synthesized crystals, including electrons in semiconductor super-lattices (to be discussed in Chapter 9) [40], cold atoms moving in a rigid and impurity-free optical lattice [41], and even photons in a photonic crystal [42]. There is even a sense in which the ac Josephson effect (to be discussed in Section 19.9) corresponds to highly coherent Bloch oscillations.

It is important to appreciate that the finite band gaps play a crucial role in confining the electrons to a single band. Clearly, if there were no periodic crystalline potential, an external electric field would cause the electron velocity to increase without bound as illustrated in Fig. 8.1(a). This is contrasted with Fig. 8.1(b), where band gaps suppress interband transitions. If the electron remains in one band, its kinetic energy is bounded and it cannot simply accelerate downhill. The electron must oscillate in real space so that its potential energy also remains bounded (and able to oscillate out of phase with the kinetic energy to keep the total energy fixed).

A rough criterion for when the band gap Δ prevents interband transitions, which is discussed in some detail in Section 8.2.2, is

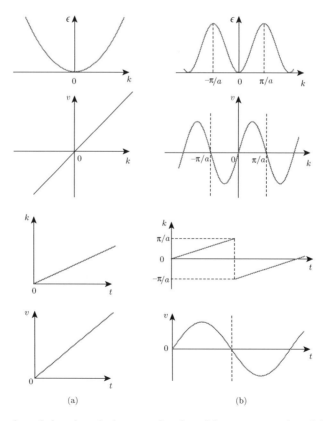

(a)   (b)

Figure 8.1 The dispersion relation, the velocity v as a function of the wave vector k, and the time-dependences of k and v when a constant electric field is present, in the absence (a) and presence (b) of a 1D periodic potential. Here a is the lattice spacing.

$$eEa \ll \frac{\Delta^2}{\epsilon_F}, \tag{8.18}$$

$$\hbar\omega_c = \hbar\frac{eB}{m_e c} \ll \frac{\Delta^2}{\epsilon_F}. \tag{8.19}$$

Here ϵ_F is the electron Fermi energy measured from the bottom of the conduction band, and ω_c is the cyclotron frequency. The process of interband jumping is known as Zener tunneling or electric (or magnetic) breakdown.

A major observable consequence of the inability of weak external fields to cause interband transitions is that filled bands are *inert*. The Pauli principle prohibits *intra*band transitions within a filled band and hence there are no low-lying excitations at all. In the semiclassical picture, each electron is drifting across the Brillouin zone and then is Bragg reflected to start the process all over again. However, since every possible state is occupied, no observables are changing with time; i.e. there is no observable effect of the external electric field. See, however, Section 13.5 for a discussion of the quantized Hall conductance which can occur in filled bands.

8.2.1 Derivation of Bloch Electron Equations of Motion

Recall that the lattice momentum $\vec{k}$ is introduced through the eigenvalues of the lattice translation operator, Eq. (7.48). It is better known, however, as the wave vector of the plane-wave factor of the Bloch wave function, in Eq. (7.50). We note that Eq. (7.50) is *specific* to the eigenstates of the Hamiltonian Eq. (7.34); on the other hand, both the wave function and the Hamiltonian are *gauge-dependent*. More specifically, we are allowed to make the following *local* gauge transformation:

$$\psi \to e^{i\varphi(\vec{r}, t)}\psi, \qquad \vec{p} \to \vec{p} - \hbar\nabla\varphi, \tag{8.20}$$

under which the Hamiltonian and lattice translation operators become

$$\mathcal{H} = \frac{(\vec{p} - \hbar\nabla\varphi)^2}{2m_e} + V(\vec{r}); \tag{8.21}$$

$$T_j = e^{i(\vec{p} - \hbar\nabla\varphi)\cdot\vec{a}_j}. \tag{8.22}$$

Clearly Eq. (7.50) is no longer valid with a generic choice of φ, but Eq. (7.48) still holds, as long as we use the φ (or gauge)-dependent definition of T_j above. As a result, $\vec{k}$ (defined using the eigenvalue of the gauge-invariant translation operator) is *gauge-independent or -invariant*. Obviously, in the absence of a magnetic field, we can stick to Eq. (7.34) for the Hamiltonian, and avoid the nuisance of gauge-dependence. But, as we will see soon, in the presence of a magnetic field the issue of the gauge is unavoidable (due to the presence of the vector potential), and one has to carefully distinguish between gauge-dependent and gauge-invariant quantities. Only the latter correspond to physical observables.

Now consider the lattice Hamiltonian in the presence of (possibly time-dependent) external electromagnetic potentials:

$$\mathcal{H} = \frac{[\vec{p} + (e/c)\vec{A}(\vec{r}, t)]^2}{2m_e} + V(\vec{r}) - e\phi(\vec{r}, t), \tag{8.23}$$

for which the electromagnetic fields are

$$\vec{B} = \nabla \times \vec{A}; \qquad \vec{E} = -\nabla\phi - \frac{1}{c}\frac{\partial\vec{A}}{\partial t}. \tag{8.24}$$

For a *generic* state subject to this Hamiltonian, we can no longer define a lattice wave vector $\vec{k}$. However, for a (normalized) wave packet $|\vec{K}, \vec{R}\rangle$ whose spatial spread is small compared with the scale of variation for $\vec{B}$ and $\vec{E}$, it is still possible to define its lattice wave vector $\vec{K}$ *approximately*, with accuracy growing with the size of the wave packet, in the following manner. We first define its position

$$\vec{R} = \langle \vec{K}, \vec{R} | \hat{\vec{r}} | \vec{K}, \vec{R} \rangle. \tag{8.25}$$

Next we expand $\vec{A}$ around $\vec{R}$ to linear order:

$$A_i(\vec{r}) \approx A_i(\vec{R}) + \partial_j A_i|_{\vec{R}} \, \delta r_j$$
$$= A_i(\vec{R}) + \frac{1}{2} \eta_{ij} \, \delta r_j + \frac{1}{2} f_{ij} \, \delta r_j, \tag{8.26}$$

where $\delta \vec{r} = \vec{r} - \vec{R}$ and repeated indices are summed over (as usual), and we have defined symmetric and anti-symmetric tensors:

$$\eta_{ij} \equiv \partial_j A_i + \partial_i A_j, \tag{8.27}$$
$$f_{ij} \equiv \partial_j A_i - \partial_i A_j = -\epsilon_{ijl} B_l. \tag{8.28}$$

Now define a vector field $\vec{\xi}(\vec{r})$ by

$$\xi_i(\vec{r}) = A_i(\vec{r}) + \frac{1}{2} \eta_{ij} \, \delta r_j. \tag{8.29}$$

From the symmetry $\eta_{ij} = \eta_{ji}$ it is straightforward to show that $\nabla \times \vec{\xi} = \vec{0}$. Hence there exists a scalar field $\Phi_{\vec{R}}(\vec{r})$ such that

$$\vec{\xi}(\vec{r}) = -\frac{\hbar c}{e} \nabla \Phi_{\vec{R}}(\vec{r}). \tag{8.30}$$

We thus arrive at

$$\vec{A}(\vec{r}) \approx -(\hbar c/e) \nabla \Phi_{\vec{R}}(\vec{r}) - \frac{1}{2} \delta \vec{r} \times \vec{B}(\vec{R}). \tag{8.31}$$

The first term above is obviously a pure gauge. Higher-order terms in $\delta \vec{r}$ are negligible for slowly varying $\vec{B}$. The lattice momentum $\vec{K}$ can be determined through

$$e^{i \vec{K} \cdot \vec{a}_j} \approx \langle \vec{K}, \vec{R} | T_j | \vec{K}, \vec{R} \rangle = \langle \vec{K}, \vec{R} | e^{i(\vec{p} - \hbar \nabla \Phi_{\vec{R}}) \cdot \vec{a}_j} | \vec{K}, \vec{R} \rangle. \tag{8.32}$$

Alternatively, one can also *construct* $|\vec{K}, \vec{R}\rangle$ in the presence of the (gauge part of the) vector potential, in a manner similar to Eq. (8.8):

$$\phi_{\vec{R}, \vec{K}}^n(\vec{r}) = e^{i \Phi_{\vec{R}}(\vec{r})} \psi_{n\vec{K}}(\vec{r}) e^{-\frac{(\Delta k)^2}{2} |\vec{r} - \vec{R}|^2} e^{-i \vec{K} \cdot \vec{R}}. \tag{8.33}$$

Taking a derivative with respect to t on both sides of Eq. (8.32), we obtain

$$i e^{i \vec{K} \cdot \vec{a}_j} \left(\frac{d\vec{K}}{dt} \cdot \vec{a}_j \right) \approx \langle \vec{K}, \vec{R} | \frac{dT_j}{dt} | \vec{K}, \vec{R} \rangle$$
$$= \langle \vec{K}, \vec{R} | \frac{\partial T_j}{\partial t} | \vec{K}, \vec{R} \rangle + \frac{1}{i\hbar} \langle \vec{K}, \vec{R} | [T_j, \mathcal{H}] | \vec{K}, \vec{R} \rangle$$
$$\approx -i e^{i \vec{K} \cdot \vec{a}_j} \left[\frac{d\nabla \Phi_{\vec{R}}}{dt} \right.$$
$$\left. + \frac{e}{\hbar} \left(\nabla \phi + \langle \vec{K}, \vec{R} | \frac{\vec{p} + (e/c)\vec{A}}{m_e c} | \vec{K}, \vec{R} \rangle \times \vec{B}(\vec{R}) \right) \right] \cdot \vec{a}_j$$
$$\approx -i e^{i \vec{K} \cdot \vec{a}_j} \frac{e}{\hbar} \left[\vec{E}(\vec{R}) + \frac{1}{c} \frac{d\vec{R}}{dt} \times \vec{B}(\vec{R}) \right] \cdot \vec{a}_j, \tag{8.34}$$

from which Eq. (8.15) follows. In the derivation above we used the fact that $\vec{A}(\vec{R}) \approx -(\hbar c/e) \nabla \Phi_{\vec{R}}$, and a gauge-covariant version of Eq. (7.99), which also leads to Eq. (8.9).

8.2.2 Zener Tunneling (or Interband Transitions)

We consider the motion of a Bloch electron subject to a static and uniform electric field $\vec{E}$. One way to model the electric field is by adding a static external electric potential term $e\vec{E} \cdot \vec{r}$ to the Hamiltonian, which breaks the lattice translation symmetry. Alternatively, we could also encode $\vec{E}$ through a spatially uniform but time-dependent vector potential:

$$\vec{A}(t) = -ct\vec{E}, \tag{8.35}$$

which respects lattice translation symmetry. The price we pay is that now the electron sees a time-dependent Hamiltonian:

$$\mathcal{H} = \frac{(\vec{p} - et\vec{E})^2}{2m_e} + V(\vec{r}), \tag{8.36}$$

which takes the same form as Eq. (7.65), once we identify (the time-dependent parameter) $\vec{k}(t)$ as $-et\vec{E}/\hbar$. This immediately implies Eq. (8.14) for the case of a uniform external electric field.

The assumption that the electron stays in a given band thus corresponds to an *adiabatic* evolution of the electron state as $\vec{k}$ varies with t. Adiabaticity holds as long as $\vec{k}$ varies slowly with t, which corresponds to weak field. Zener tunneling, or interband transition, corresponds to breakdown of adiabaticity. The amplitude of finding the electron in band $j \neq n$ at later time is of order[5]

$$a_j \sim \frac{\hbar}{\Delta_{jn}^2} \langle j | \frac{\partial \mathcal{H}}{\partial t} | n \rangle, \tag{8.37}$$

where Δ_{jn} is the energy difference between states in the jth and nth bands (with the *same* lattice momentum). Clearly, suppression of Zener tunneling implies that $|a_j| \ll 1$, and we expect $|a_j|$ to be largest for neighboring bands. Noting that $\partial \mathcal{H}/\partial t = -e\vec{E} \cdot \vec{v}$, where $\vec{v}$ is the electron velocity *operator*, we may approximate the absolute value of the matrix element above by $eE\epsilon_F/(\hbar k_F) \sim eE\epsilon_F a/\hbar$, where we used the fact that electrons at the Fermi energy ϵ_F typically have the highest velocity of order $\epsilon_F/(\hbar k_F)$, and in metals we typically have $k_F \sim 1/a$ (a is the lattice constant here). From these the condition Eq. (8.18) follows.

Exercise 8.1. Consider a Hamiltonian of the form $H = \epsilon_n((1/\hbar)[\vec{p} + (e/c)\vec{A}(\vec{R})]) + U(\vec{R})$, where $\epsilon_n(\vec{k})$ is the band dispersion and $\vec{R}$ and $\vec{p}$ are operators satisfying the canonical commutation relation $[R_i, p_j] = i\hbar_{ij}$. Use the quantum-mechanical Heisenberg equations of motion to derive Eqs. (8.9) and (8.15), by identifying $\hbar\vec{K}$ as the expectation value of $\vec{p} + (e/c)\vec{A}(\vec{R})$.

8.3 Holes

The idea that quantum mechanics causes a filled band to be inert has several fundamental consequences. First and foremost, it explains why in the Drude model it is correct to focus only on the valence electrons and ignore all the rest which fill the low-lying bands associated with the atomic core orbitals.

Secondly, we are led naturally to the concept of "holes." Consider a band with all of its states occupied except for one. Rather than specifying a huge list of occupied states, it is easier to specify the one state that is unoccupied. We can think of the system as containing only a single object called

---

[5] See, e.g., Eq. (35.27) of Schiff [20]. The amplitude oscillates with t.

a hole. The particular state of the hole is specified by the band index n and the wave vector of the missing electron.

What are the properties of this hole? Because a filled band plus the ion-core background is charge-neutral, a hole must have positive charge since it corresponds to a missing electron. It is one of the great triumphs of the quantum theory that it can explain how it is that the charge carriers in certain materials appear to be positively charged particles.

The total crystal momentum and current associated with a filled band are zero. Since the crystal momentum contributed by an electron in state $\vec{k}$ is $\hbar\vec{k}$, removing this electron (thus creating a hole) results in a state with total crystal momentum $-\hbar\vec{k}$. It is thus natural to assign a wave vector

$$\vec{k}' = -\vec{k} \tag{8.38}$$

for this hole. Similarly, measuring energy from that of the completely filled band, this hole carries energy

$$\epsilon_\mathrm{h}(\vec{k}') = -\epsilon(\vec{k}) = -\epsilon(-\vec{k}'), \tag{8.39}$$

where $\epsilon(\vec{k})$ is the band dispersion.[6]

We are now in a position to derive the equations of motion of a hole wave packet with position $\vec{R}'$ and wave vector $\vec{K}'$, namely a state in which an electron is *removed* from a wave packet with $\vec{R} = \vec{R}'$ and wave vector $\vec{K} = -\vec{K}'$, in an otherwise completely filled band. First of all, we expect the hole velocity to be that of the wave packet (see Eq. (8.9)):

$$\vec{v}_\mathrm{h}(\vec{K}') \equiv \frac{d\vec{R}'}{dt} = \frac{d\vec{R}}{dt} = \frac{1}{\hbar} \left.\nabla_{\vec{k}}\,\epsilon(\vec{k})\right|_{\vec{k}=\vec{K}=-\vec{K}'} = \frac{1}{\hbar} \left.\nabla_{\vec{k}'}\epsilon_\mathrm{h}(\vec{k}')\right|_{\vec{k}'=\vec{K}'}. \tag{8.40}$$

The last equality above is reassuring as we obtain the same result by using the hole-dispersion relation of Eq. (8.39).

Secondly, it follows from Eqs. (8.38) and (8.15) that

$$\hbar\frac{d\vec{K}'}{dt} = -\hbar\frac{d\vec{K}}{dt} = +e\left[\vec{E}(\vec{R}',t) + \frac{1}{c}\vec{v}_\mathrm{h}(\vec{K}') \times \vec{B}(\vec{R},t)\right] = \vec{F}', \tag{8.41}$$

namely the "force" $\vec{F}'$ felt by the hole is that of the force exerted by the external electromagnetic field on a particle with charge $+e$. We thus find that the hole carries charge $+e$, which is consistent with our intuition that a hole is equivalent to a missing electron.

We should emphasize at this point that we can choose to describe a partially filled band either in terms of electrons or in terms of holes; these two viewpoints are related by a particle–hole transformation and are mathematically equivalent. The *physical* difference between the motion of an electron and that of a hole is revealed by the following observation. Electrons tend to occupy low-energy states of a band; as a result *unoccupied states* tend to be near the *top* of a band. An equivalent way of saying this is that a hole near the *top* of a band is a state of *lower* energy than one with the hole near the bottom of the band. This is, of course, the reverse of what it is for electrons, and is reflected in the hole dispersion Eq. (8.39). In equilibrium we therefore expect the holes to congregate near the top of the band (if there are not too many holes). Suppose that the band has a parabolic *maximum* of the form

$$\epsilon(\vec{k}) \approx \epsilon_0 - \frac{\hbar^2}{2m^*}|\vec{k}|^2, \tag{8.42}$$

with $m^* > 0$, which is an effective-mass approximation of the band structure near $\vec{k}_0$ with *negative* mass, and we have chosen the momentum of the maximum to be the origin of momentum space.[7] The electron velocity is

---

[6] To simplify the notation we suppress the band index n in this section, with the understanding that we always discuss hole(s) in a specific band.

[7] This choice has no effect on the most important result to follow; see Exercise 8.2.

$$\vec{v}(\vec{k}) = \frac{1}{\hbar}\nabla_{\vec{k}}\epsilon(\vec{k}) = -\frac{\hbar\vec{k}}{m^*}, \tag{8.43}$$

which is in the *opposite* direction of the crystal momentum. A hole with the *same* velocity, on the other hand, has

$$\vec{v}_{\mathrm{h}}(\vec{k}') = -\frac{\hbar\vec{k}}{m^*} = \frac{\hbar\vec{k}'}{m^*}, \tag{8.44}$$

which is *parallel* to its crystal momentum, an intuitively more satisfying situation. More importantly, the acceleration for either an electron or a hole with a wave packet constructed from states near this band maximum is

$$\frac{d^2\vec{R}}{dt^2} = \frac{d}{dt}\vec{v} = \frac{+e}{m^*}\left[\vec{E}(\vec{R},t) + \frac{1}{c}\vec{v}\times\vec{B}(\vec{R},t)\right]. \tag{8.45}$$

The acceleration is thus in the direction of the electromagnetic force, as would be expected for a *positively* charged particle with *positive* mass, $m^* > 0$. In this interpretation we have traded the minus sign of the electron effective mass for the sign change of the charge in the equation of motion, when we switch to a description in terms of hole(s). As a result, we can think of the motion of a hole as that of a *positively* charged particle, with *positive* (effective) mass, that obeys Newton's laws.

These results give a very natural understanding of the otherwise mysterious fact that some materials (e.g. aluminum) have a *positive* Hall coefficient, on which we will have more to say in Section 8.6. Again generalization to the case of anisotropic mass (tensor) is straightforward.

Exercise 8.2. Consider the band dispersion near one of its maxima of the form

$$\epsilon_n(\vec{k}) \approx \epsilon_0 - \frac{\hbar^2}{2m^*}|\vec{k}-\vec{k}_0|^2. \tag{8.46}$$

Use this expression to derive Eq. (8.45) for a wave packet constructed using Bloch states $\psi_n(\vec{k})$ with $\vec{k}\approx\vec{k}_0$, and explain why $\vec{k}_0$ does not enter.

8.4 Uniform Magnetic Fields

The semiclassical equation of motion in the presence of the Lorentz force is[8]

$$\hbar\dot{\vec{k}} = -\frac{e}{c}\vec{v}(\vec{k})\times\vec{B}; \tag{8.47}$$

$$\dot{\vec{k}} = -\frac{e}{\hbar^2 c}\nabla_{\vec{k}}\epsilon_n(\vec{k})\times\vec{B}. \tag{8.48}$$

We see that

$$\frac{d\epsilon_n(\vec{k})}{dt} = \dot{\vec{k}}\cdot[\nabla_{\vec{k}}\,\epsilon_n(\vec{k})] = 0, \tag{8.49}$$

so $\epsilon_n(\vec{k})$ is a constant of the motion. Because the Lorentz force is perpendicular to the velocity, it can do no work on the particle.[9]

Consider a 2D system with iso-energy contours for some band as shown in Fig. 8.2(a). The electronic motion in reciprocal space will be along these contours. Suppose that the band minimum is

---

[8] To simplify the notation, from now on we use $\vec{k}$ to represent both the Bloch wave vector and its average in a wave packet, or, equivalently, we no longer distinguish between $\vec{k}$ and $\vec{K}$. Similarly, we no longer distinguish between $\vec{r}$ and $\vec{R}$.

[9] How then can an electric motor possibly work? Think through this carefully.

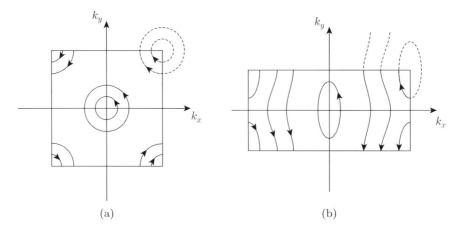

Figure 8.2 Momentum-space orbits of Bloch wave packets subject to a uniform magnetic field perpendicular to a 2D lattice. The motion is along lines of constant energy in the band structure. In (a) all orbits are closed (in the extended-zone scheme) and in (b) some are closed and others are open.

at the zone center and there is a maximum at the zone corners. Then, for $\vec{B} = +|B|\hat{z}$, the motion will be counterclockwise on orbits near the zone center and clockwise for the orbits encircling (in the extended-zone scheme) the zone corners (all of which correspond to the *same* point in the first Brillouin zone). These separate motions are referred to as **electron orbits** and **hole orbits**, respectively. They are distinguished by the fact that the area of an electron orbit *increases* with energy while that of a hole orbit *decreases*.

In an anisotropic system the iso-energy contours may look like those shown in Fig. 8.2(b). We see that some of the orbits are closed but others are open and unbounded (in the extended-zone scheme).[10] As we will see shortly, the topology of the orbits can have a significant effect on such properties as the magnetoresistance.

So far we have considered only 2D examples and ignored motion along the direction of the field $\vec{B}$. Examination of the equation of motion shows that

$$\dot{\vec{k}} \cdot \vec{B} = 0, \tag{8.50}$$

so that $k_\parallel \equiv \vec{k} \cdot \hat{B}$ is a constant of the motion.

Now that we understand the motion of the wave vector through reciprocal space, we can ask about the corresponding trajectory in real space. First, let us consider the components of the motion perpendicular to the field. We have

$$\hat{B} \times \hbar\dot{\vec{k}} = -\frac{e}{c}\hat{B} \times \left(\vec{v}_n(\vec{k}) \times \vec{B}\right)$$

$$= -\frac{eB}{c}\left\{\vec{v}_n(\vec{k}) - \hat{B}\left[\hat{B} \cdot \vec{v}_n(\vec{k})\right]\right\}$$

$$= -\frac{eB}{c}\vec{v}_\perp, \tag{8.51}$$

---

[10] In the reduced-zone scheme the first Brillouin zone is actually a (d-dimensional) torus due to the periodicity, and all orbits are closed. However, there are still qualitative differences among them: those that are closed in the extended-zone scheme are contractible (meaning they can shrink to a point continuously), while those that are open in the extended-zone scheme wrap around the torus, and thus are not contractible. As described in Appendix D, such differences are topological in nature, and rooted in the fact that the torus is not simply connected (or, equivalently, that it has a non-trivial **fundamental group** [35, 43, 44].

where $\hat{B}$ is the unit vector parallel to $\vec{B}$, and $\vec{v}_\perp$ represents components of $\vec{v}$ perpendicular to $\vec{B}$. Integrating the velocity over time, we obtain the position

$$\vec{r}_\perp(t) - \vec{r}_\perp(0) = -\frac{\hbar c}{eB} \hat{B} \times \left[\vec{k}(t) - \vec{k}(0) \right]. \tag{8.52}$$

This tells us that the $\vec{r}_\perp$ orbit is simply a scaled version of the $\vec{k}_\perp$ orbit but rotated by 90° about the $\hat{B}$ axis. If, for example, the $\vec{k}_\perp$ orbit is a circle, then the real-space orbit is also a circle, just as it would be for a particle in free space. For the open orbits shown in Fig. 8.2(b), the real-space orbit is a meandering line that moves primarily left (right) for the left (right) open orbit shown.

Real-space motion along the $\hat{B}$ direction can be complicated despite the fact that $k_\parallel$ is a constant of the motion:

$$z(t) - z(0) = \int_0^t d\tau \frac{1}{\hbar} \frac{\partial}{\partial k_z} \epsilon_n[k_x(\tau), k_y(\tau), k_z]. \tag{8.53}$$

Only for the case of ordinary parabolic dispersion $\epsilon \sim k^2$ is $v_\parallel$ constant when $k_\parallel$ is constant. In general, $v_\parallel$ oscillates as $\vec{k}_\perp$ moves through its orbit.

For the case of a closed orbit, the period is given by

$$T = t_2 - t_1 = \int_{t_1}^{t_2} dt = \oint \frac{|d\vec{k}|}{|\dot{\vec{k}}|} \tag{8.54}$$

$$= \oint \frac{|d\vec{k}|}{\left|(\nabla_{\vec{k}} \epsilon_n)_\perp\right|} \frac{\hbar^2 c}{eB}, \tag{8.55}$$

where t_1 and t_2 are the starting and ending times of one period. We can give a geometrical interpretation to this expression by considering two orbits separated by a small energy $\Delta\epsilon$ as shown in Fig. 8.3. Let $\vec{\Delta}(\vec{k})$ be the local perpendicular connecting the two curves. Then

$$\Delta\epsilon = \nabla_{\vec{k}} \epsilon_n \cdot \vec{\Delta}(\vec{k}) = \left|(\nabla_{\vec{k}} \epsilon_n)_\perp\right| \left|\vec{\Delta}(\vec{k})\right|. \tag{8.56}$$

Thus we have

$$T = \frac{\hbar^2 c}{eB} \oint |d\vec{k}| \frac{\left|\vec{\Delta}(\vec{k})\right|}{\Delta\epsilon} = \frac{\hbar^2 c}{eB} \frac{A(\epsilon + \Delta\epsilon) - A(\epsilon)}{\Delta\epsilon} = \frac{\hbar^2 c}{eB} \frac{\partial A}{\partial \epsilon}\bigg|_{k_z}, \tag{8.57}$$

where $A(\epsilon)$ is the area enclosed in the iso-energy orbit in reciprocal space. For the special case of parabolic dispersion with effective mass m^*,

$$\epsilon = \frac{\hbar^2 k^2}{2m^*}, \tag{8.58}$$

we have

$$A = \pi(k_x^2 + k_y^2) = \frac{2\pi m^*}{\hbar^2} \epsilon - \pi k_z^2; \tag{8.59}$$

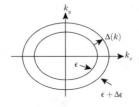

Figure 8.3 Two closely spaced equal-energy lines with the same k_z.

$$\left.\frac{\partial A}{\partial \epsilon}\right|_{k_z} = \frac{2\pi m^*}{\hbar^2}, \tag{8.60}$$

thus

$$T = \frac{2\pi}{\omega_c}, \tag{8.61}$$

where

$$\omega_c \equiv \frac{eB}{m^*c} \tag{8.62}$$

is the **cyclotron frequency**. In this special case all particles orbit at the same frequency independently of their energy. The fast particles go on larger-diameter orbits in real space and hence return at the same time as the slower particles.[11]

This circular motion is observable in a **cyclotron resonance** experiment as a peak in the dissipative conductivity at frequency $\omega = \omega_c$. Measurement of ω_c yields the cyclotron effective mass m^*, which is not necessarily the same as the mass deduced from the specific heat (either because of peculiarities of the band structure or because of interaction effects). For non-simple band structures, ω_c will be different for different orbits and the resonance will be washed out except for frequencies of certain *extremal orbits* at values of k_z for which $\partial A/\partial \epsilon|_{k_z}$ goes through a maximum or a minimum (so that nearby orbits all have nearly the same frequency). Tilting the B field will sample different extremal orbits and allows one to map out the geometry of the Fermi surface, in a way similar to what will be discussed in the following section.

Exercise 8.3. In 2D the period $T(\epsilon)$ for periodic orbits at energy ϵ in a magnetic field is a direct measure of the density of states $\rho(\epsilon)$ at energy ϵ. Express $\rho(\epsilon)$ in terms of $T(\epsilon)$, the magnetic field B, and universal constants.

Exercise 8.4. Consider an anisotropic 2D band structure $\epsilon(\vec{k}) = -2t_x \cos k_x - 2t_y \cos k_y$, with $t_x > t_y > 0$. (i) Find the energy ranges for electron-like closed orbits, open orbits, and hole-like closed orbits when a perpendicular magnetic field is applied. (ii) Show that there are van Hove singularities in the density of states at the energies separating neighboring energy ranges. This means that the period $T(\epsilon)$ diverges as ϵ approaches these energies.

8.5 Quantum Oscillations

As one learns in any course on quantum mechanics, and as we will further discuss in great detail in Chapter 12, quantum motion of charged particles in a uniform magnetic field is exactly soluble, resulting in discrete **Landau levels**. The same conclusion holds for Bloch electrons subject to a uniform magnetic field when the effective-mass approximation holds. Exact solution of the Schrödinger equation for a generic band structure is no longer possible when a magnetic field is present, but applying the semiclassical **Bohr–Sommerfeld quantization** rule to the periodic motion again results in Landau-level-like discrete energy levels. Such quantization of the electron energy by the magnetic field alters the electron density of states (DOS) significantly, resulting in oscillations (as a function

---

[11] The fact that the orbits are isochronous in this case is related to the fact that there is a harmonic oscillator Hamiltonian hidden here. This will be discussed in Chapter 12. Note that $\partial A/\partial \epsilon$ changes sign for the hole orbits, indicating that the motion is reversed.

of the magnetic field) of any physical properties of a metal that are sensitive to the DOS at the Fermi level. These phenomena, collectively known as quantum oscillations, have become a powerful tool to probe Fermi surfaces.

We start with the 2D case for its simplicity. As we learned in Section 8.4, closed orbits in momentum space correspond to periodic electron motion in real space when a magnetic field is present, with periodicity $T(\epsilon, B)$ depending on the (band) energy ϵ and magnetic field B. The Bohr–Sommerfeld quantization rule then dictates that the energy of the electron gets *quantized* into discrete levels near ϵ, with spacing

$$\Delta\epsilon = \frac{2\pi\hbar}{T} = \frac{2\pi eB}{\hbar c}\left(\frac{\partial A(\epsilon)}{\partial \epsilon}\right)^{-1}, \tag{8.63}$$

resulting in

$$\left(\frac{\partial A(\epsilon)}{\partial \epsilon}\right)\Delta\epsilon = \Delta A = \frac{2\pi eB}{\hbar c}. \tag{8.64}$$

This leads to a sequence of areas in reciprocal space corresponding to the sequence of discrete energies ϵ_n,

$$A_n = \frac{2\pi eB}{\hbar c}(n + \nu), \tag{8.65}$$

where $n = 0, 1, 2, \ldots$ and ν is a fraction. For non-relativistic particles we normally set $\nu = 1/2$ to account for zero-point motion of the particle in the ground state, resulting in $\epsilon_n = (n + 1/2)\hbar\omega_c$ when the effective-mass approximation is appropriate. These are precisely the Landau-level energies. For massless Dirac particles ν turns out to be zero due to a Berry-phase contribution that precisely cancels out the zero-point motion factor (more on this point later).

Now consider a metal with Fermi energy ϵ_F, with corresponding Fermi surface area $A(\epsilon_F)$ (assuming that the $\epsilon(\vec{k}) = \epsilon_F$ line is closed). When $A(\epsilon_F)$ matches A_n for a particular n, the DOS at ϵ_F has a pronounced peak. This happens in a periodic fashion as B varies. More precisely, from Eq. (8.65) we find that, when

$$\Delta n = 1 = \frac{\hbar c}{2\pi e}A(\epsilon_F)\Delta\left(\frac{1}{B}\right), \tag{8.66}$$

a new peak appears, thus the DOS oscillates periodically as $(1/B)$ varies, with period

$$\Delta\left(\frac{1}{B}\right) = \frac{2\pi e}{\hbar c}\frac{1}{A(\epsilon_F)}. \tag{8.67}$$

In the (somewhat unusual) situation of multiple Fermi surfaces, there is one period for each of them, and the total DOS oscillates with multiple periods superimposed on each other.

The situation gets more complicated in 3D, where, in addition to ϵ, another conserved quantity, $k_\parallel = \vec{k} \cdot \hat{B}$, needs to be specified for an orbit in reciprocal space, and the period for a closed orbit $T(\epsilon, k_\parallel)$ depends on both of them. Applying the quantization rule (8.63) results in a family of energy levels depending on $k_\parallel$ *continuously*. In this case it is useful to view the 3D system as a one-parameter family of 2D systems, parameterized by $k_\parallel$. To determine the DOS at ϵ_F, we need to sum up (or integrate over $k_\parallel$) the contributions from these 2D systems. Since $A(\epsilon_F, k_\parallel)$ also depends on $k_\parallel$ continuously, one might worry that the integration over $k_\parallel$ will wash out the periodic dependence on $(1/B)$, because each of them has a different period according to Eq. (8.67). Fortunately this is not the case. The periods corresponding to the so-called extremal orbits on the Fermi surface will survive the integration over $k_\parallel$:

$$\Delta\left(\frac{1}{B}\right) = \frac{2\pi eB}{\hbar c}\frac{1}{A_e(\epsilon_F)}, \tag{8.68}$$

where the extremal area $A_e(\epsilon_F)$ corresponds to $A(\epsilon_F, k_\parallel)$ with a value of $k_\parallel$ such that

$$\frac{\partial A(\epsilon_F, k_\parallel)}{\partial k_\parallel} = 0. \tag{8.69}$$

This is because the period (8.68) is stationary with respect to change of $k_\parallel$.[12] Again, in the presence of multiple Fermi surfaces and/or multiple extremal areas on a single Fermi-surface sheet, each of them makes an oscillatory contribution to the DOS with period determined by Eq. (8.68), with the total DOS oscillating with multiple periods superimposed on each other.

As we learned in Section 7.7 and elsewhere, many physical properties depend sensitively on the DOS at the Fermi energy. All of them will therefore oscillate with B in principle. Historically such oscillations were first observed in magnetization and conductivity, known as **de Haas–van Alphen** and **Shubnikov–de Haas oscillations**, respectively. Nowadays other oscillations are routinely observed in high-quality crystals, including oscillations of the specific heat, Hall conductivity, and even mechanical properties. See Figs. 8.4(a) and 8.4(b) for examples.

By varying the direction of the magnetic field and determining the oscillation period(s) (usually through fast Fourier transformation), one can extract $A_e(\epsilon_F)(s)$ as functions of direction (see Fig. 8.4(c) for an example). This allows one to reconstruct the Fermi surface(s) of a metal. Quantum

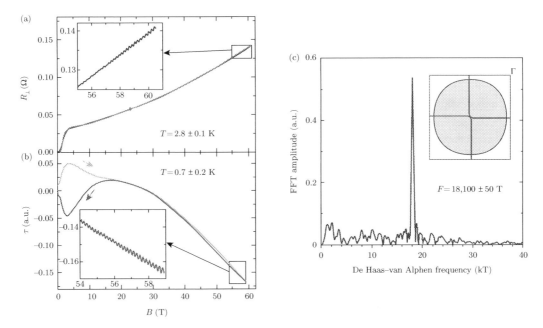

Figure 8.4 Quantum oscillation data of over-doped $Tl_2Ba_2CuO_{6+\delta}$ (Tl2201), which is a member of the cuprate high-temperature superconductor family (to be discussed in Section 20.9). (a) The magnetic field dependence of the resistance perpendicular to the copper–oxygen plane. (b) The magnetization measured via torque due to the fact that the magnetic moment and the magnetic field are not parallel. The oscillatory parts are amplified and shown in the insets. (c) The result of fast Fourier transformation (FFT) of the oscillatory magnetization (the de Haas–van Alphen effect) as a function of $1/B$, revealing a single frequency of $F \approx 18,100\,T$, indicating that there is a single (2D) Fermi surface whose area is consistent with other measures of the Fermi surface (as shown in the inset). Figure reprinted with permission from [45].

---

[12] This is analogous to the reason why the classical trajectory survives path integration in the semiclassical limit of quantum mechanics, because its contribution to the action is stationary with respect to variation of the path.

oscillation is now the most widely used method to measure Fermi surfaces. For more details, see Ref. [46].

8.6 Semiclassical $\vec{E} \times \vec{B}$ Drift

Let us now examine what happens when electrons simultaneously see both an electric field and a magnetic field. First, consider a free electron moving in a circular orbit in a magnetic field $B\hat{z}$. Now imagine jumping into a new frame of reference moving with velocity $\vec{v}_f = -v\hat{x}$ relative to the laboratory. In this frame a uniform drift velocity

$$\vec{v}_d = -\vec{v}_f = +v\hat{x} \tag{8.70}$$

is superposed on the circular motion.

What happens to the electromagnetic field under this change of frames? To lowest order in v/c the Lorentz transformation yields

$$\vec{B}_{\text{new}} = \vec{B}_{\text{old}}, \tag{8.71}$$

$$\vec{E}_{\text{new}} = \frac{1}{c}\vec{v}_f \times \vec{B}_{\text{old}} = -\frac{\vec{v}_d}{c} \times \vec{B} = \frac{v}{c}B\hat{y}. \tag{8.72}$$

The change of frame generates both an electric field (Faraday induction) and a net drift velocity for the electrons. Hence we are entitled to conclude that, in the presence of crossed $\vec{E}$ and $\vec{B}$ fields, charged particles will have an average drift velocity which satisfies

$$\vec{E} = -\frac{\vec{v}_d}{c} \times \vec{B} \tag{8.73}$$

or

$$\frac{\vec{v}_d}{c} = \frac{\vec{E} \times \vec{B}}{B^2} \tag{8.74}$$

independently of the sign of the charge.[13]

If we have a uniform density n of electrons, the drift current density is

$$\vec{J} = -ne\vec{v}_d = -\frac{nec}{B}\vec{E} \times \hat{B}. \tag{8.75}$$

We can identify the Hall conductivity from

$$J_x = \sigma_{xy}E_y; \tag{8.76}$$

$$\sigma_{xy} = -\frac{nec}{B}. \tag{8.77}$$

Because the current is perpendicular to the electric field, the diagonal components of the conductivity tensor vanish:

$$\sigma_{xx} = \sigma_{yy} = 0. \tag{8.78}$$

We have intentionally chosen to derive this via the Lorentz transformation in order to emphasize that (for free electrons at least) the Hall effect is purely kinematical. The Hall coefficient depends only on the density of charge, not on any dynamical properties such as the particle mass (or even interactions and correlations among the particles).

In the presence of a periodic crystal potential we are not free to make a change of frame of reference, since this would cause the electrons and the background lattice to move together. We want to

---

[13] Of course, if the charge is zero the original motion is linear (not circular) and continues unaffected by the fields.

study the situation where the electrons are moving relative to the lattice. To do this, we return to the semiclassical equation of motion

$$\hbar \dot{\vec{k}} = -\frac{e}{c} \vec{v}_n(\vec{k}) \times \vec{B} - e\vec{E}. \tag{8.79}$$

As long as $\vec{E} \perp \vec{B}$, $k_\parallel \equiv \vec{k} \cdot \hat{B}$ remains a constant of motion. Just as before, we want to solve for the component of velocity perpendicular to $\vec{B}$:

$$\hat{B} \times \hbar \dot{\vec{k}} = -\frac{e}{c} \hat{B} \times [\vec{v}_n(\vec{k}) \times \vec{B}] - e\hat{B} \times \vec{E}, \tag{8.80}$$

$$\frac{\hbar c}{eB} \hat{B} \times \dot{\vec{k}} = -\vec{v}_\perp + \vec{v}_{\mathrm{d}}, \tag{8.81}$$

where $\vec{v}_{\mathrm{d}}$ is the drift velocity defined in Eq. (8.74). Thus we obtain

$$\vec{v}_\perp = -\frac{\hbar c}{eB} \hat{B} \times \dot{\vec{k}} + \vec{v}_{\mathrm{d}}. \tag{8.82}$$

This is identical to Eq. (8.51) except for the addition of the drift term. The time evolution of the position then obeys

$$\vec{r}_\perp(t) - \vec{r}_\perp(0) = \vec{v}_{\mathrm{d}}t - \frac{\hbar c}{eB} \hat{B} \times \left(\vec{k}(t) - \vec{k}(0) \right). \tag{8.83}$$

To proceed further we have to solve for the trajectory in reciprocal space using the semiclassical equation of motion, Eq. (8.79) rewritten as

$$\hbar \dot{\vec{k}} = -\frac{e}{\hbar c} \nabla_{\vec{k}} \epsilon_n(\vec{k}) \times \vec{B} - e\vec{E} \tag{8.84}$$

$$= -\frac{e}{\hbar c} \nabla_{\vec{k}} \bar{\epsilon}(\vec{k}) \times \vec{B}, \tag{8.85}$$

where

$$\bar{\epsilon}(\vec{k}) \equiv \epsilon_n(\vec{k}) - \hbar\vec{k} \cdot \vec{v}_{\mathrm{d}}. \tag{8.86}$$

We can view $\bar{\epsilon}$ as the "Doppler-shifted" energy in the comoving drift frame of reference. The particle trajectory is now along lines of constant $\bar{\epsilon}$ instead of ϵ_n. In the linear response regime of small $\vec{E}$, $\vec{v}_{\mathrm{d}}$ is small, so *closed* trajectories along constant $\bar{\epsilon}$ and ϵ_n are almost identical. However, this is *not* true for open trajectories, along which $\vec{k}$ wanders without bound, and the $\hbar\vec{k} \cdot \vec{v}_{\mathrm{d}}$ terms grow without bound no matter how small $\vec{v}_{\mathrm{d}}$ is.

As noted earlier, the solutions to Eq. (8.85) may either be open or closed orbits. If all the electrons live on closed orbits then $\vec{k}(t) - \vec{k}(0)$ is a bounded oscillatory function and makes zero contribution to the dc current deduced from Eq. (8.83). (It does, however, contribute to the high-frequency cyclotron resonance.) Thus, for the case where only closed orbits are occupied by electrons, the dc current obeys Eq. (8.75) and we obtain the free-electron result for the conductivity given in Eqs. (8.77) and (8.78), *independently of the details of the band structure*.

In general a band will contain both open and closed orbits. As we shall see, the open orbits complicate the analysis and can significantly alter the transport. It is often the case that orbits at energies near the bottom of the band are closed (electron) orbits, while orbits at energies near the top of the band are closed (hole) orbits. (Recall that the names electron and hole here simply refer to the sign of the energy derivative of the k-space orbit area.) The open orbits typically occur at intermediate energies. Thus, if a band has a sufficiently large number of electrons in it, both open and closed orbits will be occupied.

A special situation arises if *all* of the open orbits are occupied. Then we can view the system as an inert filled band plus *holes* occupying only closed orbits, or, more precisely, a filled band with electrons *removed* from closed orbits. In this case the Hall conductivity is the *opposite* of what the contributions from these closed orbits would be if they were occupied by electrons:

$$\sigma_{xy} = +\frac{n_{\mathrm{h}}ec}{B}, \tag{8.87}$$

where n_{h} is the density of holes, because a completely filled band has zero Hall conductivity.[14] Notice that in this case the Hall conductivity has the *"wrong" sign*, which is of course naturally interpreted as the result of the motion of positively charged holes. Thus quantum mechanics and the Pauli exclusion principle can explain the anomalies in the sign of the Hall effect which plagued classical physics. Indeed, the explanation of the anomalous sign of the Hall coefficient in aluminum was one of the early triumphs of the quantum theory applied to solids.

If only part of the open orbits are occupied, then we cannot avoid having to deal with them because (in the extended-zone scheme) $\vec{k}(t) - \vec{k}(0)$ increases without bound in Eq. (8.83). Consider as an example the open orbits shown in Fig. 8.2(b) and the limit of very large magnetic field ($B > 1\,T = 10^4$ G). Equation (8.79) tells us that $\dot{\vec{k}}$ becomes large at high fields, so the time it takes the particle to traverse the Brillouin zone becomes small. Thus we can approximate the velocity of the particles on this orbit by the average velocity along the orbit $\vec{V} \equiv \left\langle v_n(\vec{k}) \right\rangle$. The drift term is proportional to B^{-1} and also becomes negligible at high fields. If the left and right open orbits shown in Fig. 8.2(b) are equally occupied then their currents will cancel each other out and there will be no net current. In practice what happens is that states whose velocities are such that the electrons are absorbing energy from the electric field become more heavily occupied than states losing energy to the electric field.[15] The precise details of the population asymmetry depend on collision rates for the particles with each other and with impurities and phonons. The upshot is that the net current has the form

$$\vec{J} = \sigma_0 \hat{V}(\hat{V} \cdot \vec{E}) + \overset{\leftrightarrow}{\sigma}_1 \cdot \vec{E}, \tag{8.88}$$

where $\overset{\leftrightarrow}{\sigma}_1$ is related to the drift term and vanishes for $B \to \infty$, and σ_0 is controlled primarily by the details of the collisions and the population asymmetry.

In a wire, the current is forced to be parallel to the axis of the wire. If this direction happens not to be parallel to $\hat{V}$ then an enormous electric field (nearly perpendicular to $\hat{V}$) will be required to drive a current through the wire. This implies a "magnetoresistance" which diverges as $B \to \infty$. For a closed-orbit system the magnetoresistance either saturates at a finite value or increases relatively slowly with B.

8.7 The Boltzmann Equation

Now that we understand something about band structure and the dynamics of individual Bloch wave packets we can consider collisions: electron–electron, electron–phonon, and electron–impurity.

We will continue with our semiclassical picture in which external fields are slowly varying so that there is a large separation in lengths and, as discussed previously, we can view both $\vec{k}$ and $\vec{r}$ as simultaneously well-defined.

The basic quantity of interest will be the dimensionless distribution function $f(\vec{r}, \vec{k}, t)$, which is the probability that the state $\vec{k}$ (in band n) is occupied at time t by an electron located near $\vec{r}$. More precisely, $f(\vec{r}, \vec{k}, t)d^3r\, d^3\vec{k}/(2\pi)^3$ is the number of electrons inside the (infinitesimal) phase-space

[14] Except for certain topologically non-trivial bands; see Chapter 13.

[15] This result is derived in Eq. (8.131).

volume $d^3\vec{r}\,d^3\vec{k}$ near $(\vec{r},\vec{k})$. We implicitly assume that we are dealing with a single band n, and for simplicity we will ignore spin except where explicitly noted.

The distribution function is useful because it tells us everything we need to know (semiclassically) to compute observables such as the number density,

$$n(\vec{r},t) = \frac{N_\Omega}{\Omega} = \frac{1}{\Omega}\sum_{\vec{k}} f(\vec{r},\vec{k},t) = \frac{1}{\Omega}\frac{\Omega}{(2\pi)^3}\int d^3\vec{k}\ f(\vec{r},\vec{k},t)$$

$$= \int \frac{d^3\vec{k}}{(2\pi)^3} f(\vec{r},\vec{k},t), \tag{8.89}$$

where Ω is a real-space volume near $\vec{r}$ and N_Ω is the number of electrons in it. Ω needs to be sufficiently small that intensive quantities like $n(\vec{r},t)$ can be taken as a constant, and sufficiently large that the allowed values of $\vec{k}$ are dense enough that the summation over $\vec{k}$ may be replaced by an integral. The actual value of Ω is unimportant because it drops out of the final expression; this will be the case in calculations of all physical quantities. Similarly, the electrical current density is

$$\vec{J}(\vec{r},t) \equiv -e\int \frac{d^3\vec{k}}{(2\pi)^3} \vec{v}_n(\vec{k}) f(\vec{r},\vec{k},t). \tag{8.90}$$

Collision times are typically very short (10^{-8}–10^{-14} s) and, in the absence of external electric fields and temperature gradients, the electrons rapidly come to thermal equilibrium. The distribution function is then simply the (equilibrium) Fermi–Dirac distribution

$$f(\vec{r},\vec{k},t) = f_{\vec{k}}^\circ \equiv \frac{1}{e^{\beta\left(\epsilon_n(\vec{k})-\mu\right)}+1}. \tag{8.91}$$

In the presence of perturbations, the distribution can be driven out of equilibrium, which will lead to electrical and thermal transport currents. The size of these currents depends on how far out of equilibrium the system falls. This in turn depends on the strength of the perturbations and on how quickly the system can return to equilibrium when the driving force is turned off. The longer the equilibration time, the more easily the system can be driven out of equilibrium.

In the absence of collisions, a particle found at time t at $\vec{r}(t), \vec{k}(t)$ can only have come from

$$\vec{r}(t-dt) = \vec{r}(t) - \vec{v}(\vec{k})dt, \tag{8.92}$$

$$\vec{k}(t-dt) = \vec{k}(t) - \frac{1}{\hbar}\vec{F}(\vec{r},\vec{k},t)dt, \tag{8.93}$$

where dt is an infinitesimal time interval and $\vec{F}$ is the externally applied electric and magnetic force. The distribution function therefore obeys

$$f(\vec{r},\vec{k},t)d^3\vec{r}(t)d^3\vec{k}(t) = f\left(\vec{r}-\vec{v}(\vec{k})dt, \vec{k}-\frac{1}{\hbar}\vec{F}(\vec{r},\vec{k},t)dt, t-dt\right)$$

$$\times d^3\vec{r}(t-dt)d^3\vec{k}(t-dt), \tag{8.94}$$

or equivalently (using **Liouville's theorem**[16] on conservation of the phase-space volume $d^3\vec{r}(t)d^3\vec{k}(t) = d^3\vec{r}(t-dt)d^3\vec{k}(t-dt)$ and making a lowest-order Taylor-series expansion of f)

$$\frac{\partial f}{\partial t} + \vec{v}(\vec{k})\cdot\nabla_{\vec{r}}f + \frac{1}{\hbar}\vec{F}\cdot\nabla_{\vec{k}}f = 0. \tag{8.95}$$

---

[16] See, e.g., p. 64 of Ref. [28]. Liouville's theorem applies here because the semiclassical equations of motion can be cast in the form of Hamilton's equations; see Exercise 8.1.

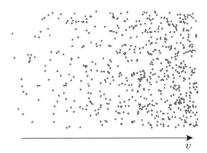

Figure 8.5 Illustration of a particular term in the Boltzmann equation. If there is a density gradient of particles from left to right and the particles are drifting in the same direction, the density at a fixed point in space will decrease with time.

This equation simply describes the (smooth) flow of particles in phase space. As a simple example, consider the swarm of particles shown in Fig. 8.5. Notice that their density increases from left to right and that the particles are moving to the right. At some later time the swarm will have moved to the right, reducing the density at the point $\vec{r}$. This illustrates the contribution

$$\frac{\partial f}{\partial t} = -\vec{v}(\vec{k}) \cdot \nabla_{\vec{r}} f \qquad (8.96)$$

to the decrease of f with time. Similarly, external forces cause the particles to drift in momentum space. This accounts for the second term in Eq. (8.95).

So far we have been ignoring the possibility of collisions. Their introduction requires some discussion. For definiteness, let us consider electron–impurity scattering. An impurity potential could in principle be represented by the force term $\vec{F}$ in Eq. (8.95). This is not appropriate, however, because the impurity potential characteristic length scale is the lattice constant. An electron scattering from an impurity can easily change its momentum by an amount comparable to the size of the Brillouin zone (i.e. comparable to the reciprocal lattice spacing). Equation (8.95) is designed to describe the smooth continuous flow of particles (actually wave packets whose spatial extent is large compared with the lattice constant) in their semiclassical phase space. The effect of a collision is to suddenly move a particle to a completely new part of reciprocal space. This invalidates the conditions needed for Liouville's theorem to apply. In order to describe this short-length-scale physics we must introduce a separate collision term to Eq. (8.95) to account for its impact on f:

$$\frac{\partial f}{\partial t} + \vec{v}(\vec{k}) \cdot \nabla_{\vec{r}} f + \frac{1}{\hbar} \vec{F} \cdot \nabla_{\vec{k}} f = \left(\frac{\partial f}{\partial t} \right)_{\text{coll}}. \qquad (8.97)$$

This is the celebrated **Boltzmann equation** developed in the nineteenth century for the (classical) kinetic theory of gases (but here being used semiclassically). Given the distribution $f(\vec{r}, \vec{k}, t_0)$ at some initial time t_0, we can use the Boltzmann equation to find the distribution at any later time.

The detailed form of the collision term depends on the specifics of the source of collision and parameters like the electron–electron interaction strength, the impurity differential scattering cross section, etc. We will see how to calculate this shortly. As an introduction, however, it is useful to consider a phenomenological model for the collision term known as the relaxation-time approximation. We know that without collisions the microstate occupation numbers cannot change (ignoring slow drifts due to external fields) and the system cannot equilibrate. The main effect of the collision term is to control how rapidly an initial non-equilibrium distribution eventually decays back to the final equilibrium distribution. Hence a useful phenomenological ansatz is

$$\left(\frac{\partial f}{\partial t} \right)_{\text{coll}} = -\frac{1}{\tau} \left(f(\vec{r}, \vec{k}, t) - f_{\vec{k}}^{\circ} \right), \qquad (8.98)$$

where τ is a single characteristic relaxation time for the distribution. For a spatially homogeneous system with no external forces the Boltzmann equation becomes

$$\frac{\partial f}{\partial t} = \left(\frac{\partial f}{\partial t}\right)_{\text{coll}}. \tag{8.99}$$

Defining

$$\delta f = f(\vec{r}, \vec{k}, t) - f_{\vec{k}}^{\circ}, \tag{8.100}$$

we then have

$$\left(\frac{\partial \delta f}{\partial t}\right)_{\text{coll}} = -\frac{1}{\tau}\delta f, \tag{8.101}$$

which has the solution

$$\delta f(\vec{r}, \vec{k}, t) = \delta f(\vec{r}, \vec{k}, 0)e^{-t/\tau}. \tag{8.102}$$

Thus all deviations of the momentum distribution away from equilibrium die out exponentially in a single characteristic time, τ. Clearly this is an oversimplification,[17] but it has the essential behavior correct.

Coming back to a microscopic description, the collisions are described by a scattering rate from the initial state $\vec{k}$ to the final state $\vec{k}'$, $W_{\vec{k}\vec{k}'}$, which is usually computed using the Born approximation and Fermi's Golden Rule. The total rate of scattering *out* of state $\vec{k}$ is

$$\frac{1}{\tau_{\text{out}}(\vec{k})} = \sum_{\vec{k}'} W_{\vec{k}\vec{k}'}[1 - f_{\vec{k}'}]f_{\vec{k}} = \Omega \int \frac{d^3\vec{k}'}{(2\pi)^3} W_{\vec{k}\vec{k}'}[1 - f_{\vec{k}'}]f_{\vec{k}}, \tag{8.103}$$

where Ω is the normalization volume introduced at the beginning of this section (whose precise value is unimportant and will drop out) for the Bloch wave function, and the factor $f_{\vec{k}}$ is present because there can be no scattering out if $\vec{k}$ is unoccupied and the $1 - f_{\vec{k}'}$ factor accounts for the Pauli blocking of possible final states $\vec{k}'$. Similarly the rate of scattering *into* state $\vec{k}$ is

$$\frac{1}{\tau_{\text{in}}(\vec{k})} = \Omega \int \frac{d^3\vec{k}'}{(2\pi)^3} W_{\vec{k}'\vec{k}}[1 - f_{\vec{k}}]f_{\vec{k}'}. \tag{8.104}$$

Thus the collision term becomes

$$\left(\frac{\partial f}{\partial t}\right)_{\text{coll}} = -\Omega \int \frac{d^3\vec{k}'}{(2\pi)^3} \{W_{\vec{k}\vec{k}'}[1 - f_{\vec{k}'}]f_{\vec{k}} - W_{\vec{k}'\vec{k}}[1 - f_{\vec{k}}]f_{\vec{k}'}\}. \tag{8.105}$$

We emphasize that in the semiclassical picture we are working with *rates* and *probabilities*, not quantum *amplitudes* and *interference*. If there are two routes to the same final state, we generally add the separate rates, not the amplitudes. Ignoring interference in this manner is, of course, not always justifiable.

At this point it is worth commenting that, within the Born approximation or Golden Rule calculation, the scattering amplitude $M_{\vec{k}'\vec{k}}$ due to a *single* scatterer like a single phonon or a single impurity is inversely proportional to the sample volume: $M_{\vec{k}'\vec{k}} \propto 1/\Omega$. The simplest example for this is perhaps scattering of plane waves by a single delta-function scatterer $U(\vec{r}) = U_i\delta(\vec{r} - \vec{R}_i)$, located at position $\vec{R}_i$. In this case we have

---

[17] In general there can be more than one characteristic time scale, and the relaxation rate may be non-linear in the deviation from equilibrium. Another complication is that elastic scattering from static impurities can relax anisotropies in the momentum distribution but it *cannot* change the energy distribution. Finally if the density is inhomogeneous, the simplest version of the relaxation approximation violates local particle number conservation because it allows the density fluctuations to relax without the necessity of diffusion.

$$M_{\vec{k}'\vec{k}} = \langle \vec{k}'|U(\vec{r})|\vec{k}\rangle = (U_i/\Omega)e^{i(\vec{k}-\vec{k}')\cdot\vec{R}_i}. \tag{8.106}$$

As a result, its contribution to the scattering rate

$$W_{\vec{k}'\vec{k}} = \frac{2\pi}{\hbar}|M_{\vec{k}'\vec{k}}|^2\delta(\epsilon_{\vec{k}'} - \epsilon_{\vec{k}}) \tag{8.107}$$

is proportional to $1/\Omega^2$. Once we sum up the contributions from all the scatterers inside the volume, we have $W \propto 1/\Omega$, assuming that the number of the scatterers is proportional to Ω. As a result, the *total* scattering rates like those in Eqs. (8.103) and (8.104) are *independent* of Ω, but are proportional to the *concentration* of the scatterers. But, as mentioned earlier, in quantum mechanics one should add up *amplitudes* like $M_{\vec{k}'\vec{k}}$ from different scatterers, then square the sum to obtain the correct scattering rate. By summing the scattering rates from different scatterers instead, we have neglected interference effects, which can be dramatic under certain circumstances. This will be elaborated in Chapter 11.

As a simple example, we will consider below a situation with constant spatial density and elastic scattering from static impurities. In this case the elastic impurity scattering rate $W_{\vec{k}\vec{k}'}$ is non-zero only for states of equal energy. Further assuming isotropic band dispersion as well as isotropic (or s-wave) scattering for simplicity yields

$$W_{\vec{k}\vec{k}'} = W_0\delta\left(\epsilon_n(\vec{k}) - \epsilon_n(\vec{k}')\right), \tag{8.108}$$

so that Eq. (8.105) simplifies to

$$\left(\frac{\partial f}{\partial t}\right)_{\text{coll}} = -W_0\Omega \int \frac{d^3\vec{k}'}{(2\pi)^3}\delta\left(\epsilon_n(\vec{k}) - \epsilon_n(\vec{k}')\right)(f_{\vec{k}} - f_{\vec{k}'}) \tag{8.109}$$

$$= -\frac{1}{\tau}\int \frac{d\Omega'}{4\pi}(f_{\vec{k}} - f_{\vec{k}'}), \tag{8.110}$$

where $1/\tau = W_0\Omega\rho$, ρ is the density of states (at energy $\epsilon_n(\vec{k})$), and the integral is over the solid angle Ω' of the different possible orientations of the vectors $\vec{k}'$ obeying $|\vec{k}'| = |\vec{k}|$. We have assumed here that the surfaces of constant energy are spherical (or the band dispersion is isotropic) so that this last condition is equivalent to $\epsilon_n(\vec{k}') = \epsilon_n(\vec{k})$.

Suppose we now make the further assumption that

$$\int \frac{d\Omega'}{4\pi} f_{\vec{k}'} = f_{\vec{k}}^\circ; \tag{8.111}$$

that is, we assume that the *directionally averaged* probability of a particle having a certain energy $\epsilon_n(\vec{k}')$ is the same as in equilibrium. Then the only possible deviations from equilibrium are that the momentum distribution is not *isotropic*. Using this assumption, Eq. (8.110) yields

$$\left(\frac{\partial f}{\partial t}\right)_{\text{coll}} = -\frac{1}{\tau}(f_{\vec{k}} - f_{\vec{k}}^\circ). \tag{8.112}$$

Thus, within this highly restrictive set of assumptions, the relaxation-time approximation becomes *exact*. Indeed, the distribution at each wave vector relaxes *independently* of all the others. This is, of course, true only under the conditions of completely isotropic dispersion and scattering as assumed above.

Exercise 8.5. We know that under equilibrium conditions the collision term in the Boltzmann equation, Eq. (8.105), must vanish. Detailed balance requires something more stringent: not only does the integral on the RHS vanish, but the integrand itself vanishes. Show that this is true if

$$\frac{W_{\vec{k}\vec{k}'}}{W_{\vec{k}'\vec{k}}} = e^{-\beta\left[\epsilon_n(\vec{k}') - \epsilon_n(\vec{k})\right]}. \tag{8.113}$$

This is known as the **detailed-balance condition**. Show that this condition is satisfied for elastic scattering. (Strictly speaking, this is true only in the presence of time-reversal symmetry so that the scattering is reciprocal.)

Exercise 8.6. This exercise illustrates how the electron–phonon interaction leads to inelastic scattering, and why Eq. (8.113) is satisfied. Suppose $\epsilon_{\vec{k}} - \epsilon_{\vec{k}'} = \hbar\omega > 0$. Then, to scatter from $\vec{k}$ to $\vec{k}'$, the electron needs to emit a phonon with the energy difference. A phonon is created in this process, and the matrix element must therefore involve an electronic contribution $M_{\vec{k}\vec{k}'}$ (similar to that in elastic scattering), and another factor due to phonon creation. (i) Show that the scattering rate for this process takes the form

$$W_{\vec{k}\vec{k}'} = \frac{2\pi}{\hbar} |M_{\vec{k}\vec{k}'}|^2 \left[n_{\mathrm{B}}(\hbar\omega_{\mathrm{ph}}) + 1 \right] \delta(\hbar\omega - \hbar\omega_{\mathrm{ph}}), \tag{8.114}$$

where $n_{\mathrm{B}}(\hbar\omega_{\mathrm{ph}})$ is the phonon occupation before the scattering. (ii) Show that the rate for the reverse process, in which a phonon is absorbed, is

$$W_{\vec{k}'\vec{k}} = \frac{2\pi}{\hbar} |M_{\vec{k}'\vec{k}}|^2 n_{\mathrm{B}}(\hbar\omega_{\mathrm{ph}}) \delta(\hbar\omega - \hbar\omega_{\mathrm{ph}}). \tag{8.115}$$

(iii) Now show that Eq. (8.113) is satisfied if $|M_{\vec{k}'\vec{k}}|^2 = |M_{\vec{k}\vec{k}'}|^2$ and we are allowed to replace $n_{\mathrm{B}}(\hbar\omega_{\mathrm{ph}})$ by its equilibrium thermal average. Note that the difference between $W_{\vec{k}\vec{k}'}$ and $W_{\vec{k}'\vec{k}}$ lies in the contribution to the former that comes from spontaneous emission of the phonon. The physics here is very similar to photon emission and absorption by an atom. Historically Einstein was able to write down the spontaneous photon emission rate (his "A" coefficient) in terms of the stimulated photon absorption/emission rate (his "B" coefficient, calculable without invoking the quantum nature of photons), by insisting on detailed balance.

The reader may wish to refer to the discussion of detailed balance in Appendix A when attempting this exercise.

8.8 Boltzmann Transport

Now that we have introduced the Boltzmann equation and understand something about the collision term, we are in a position to compute transport properties. While in principle it is possible to compute the full non-linear response to external fields, we will limit ourselves to the usual linear-response transport coefficients. This allows us to simplify the equations by assuming that weak external fields produce only weak deviations of $f_{\vec{k}}$ from $f_{\vec{k}}^{\circ}$.

It is convenient to parameterize the deviations from equilibrium through a function $\phi_{\vec{k}}$ via

$$f_{\vec{k}} = f^{\circ}\left(\epsilon_n(\vec{k}) - \phi_{\vec{k}}\right) \approx f_{\vec{k}}^{\circ} - \frac{\partial f^{\circ}}{\partial \epsilon}\phi_{\vec{k}}. \tag{8.116}$$

This is always possible as both $f_{\vec{k}}$ and f° range between 0 and 1, and turns out to be a convenient way to describe a shifted Fermi sea that is carrying current as illustrated in Fig. 8.6 by having

$$\phi_{\vec{k}} = \epsilon_n(\vec{k}) - \epsilon_n(\vec{k} - \vec{k}_0) \approx \hbar\vec{k}_0 \cdot \vec{v}_n(\vec{k}), \tag{8.117}$$

f° f

Figure 8.6 A shifted Fermi sea of electrons due to an external electric field. The electric field pushes the electrons in the opposite direction in momentum space and drives the distribution function out of equilibrium. The effect of the field is balanced by scattering (as described by the collision term in the Boltzmann equation), resulting in a steady state in which the whole Fermi sea is shifted. The size of this shift in the illustration is highly exaggerated relative to typical experimental situations in the linear-response regime.

where $\vec{k}_0$ is the shift. Substituting this into Eq. (8.97) and keeping only terms up to first order in $\vec{E}$ and $\phi_{\vec{k}}$ yields the *linearized* **Boltzmann equation**[18]

$$-\frac{\partial f^\circ}{\partial \epsilon}\left\{\frac{\partial \phi_{\vec{k}}}{\partial t} + \vec{v}(\vec{k}) \cdot \nabla_{\vec{r}}\phi_{\vec{k}}\right\} + \frac{1}{\hbar}\vec{F} \cdot \nabla_{\vec{k}} f_{\vec{k}}^\circ = \left(\frac{\partial f}{\partial t}\right)_{\text{coll}}. \tag{8.118}$$

Note that the last term on the LHS involves the *equilibrium* distribution $f_{\vec{k}}^\circ$ instead of $f_{\vec{k}}$, because the $\vec{F}$ factor is already linear in the perturbation. We anticipate that the collision term will also be a linear functional of $\phi_{\vec{k}}$ since it vanishes for the equilibrium distribution. Using

$$\frac{1}{\hbar}\nabla_{\vec{k}} f_{\vec{k}}^\circ = \frac{1}{\hbar}(\nabla_{\vec{k}}\epsilon_{n\vec{k}})\frac{\partial f^\circ}{\partial \epsilon} = \vec{v}(\vec{k})\frac{\partial f^\circ}{\partial \epsilon}, \tag{8.119}$$

Eq. (8.118) finally becomes

$$-\frac{\partial f^\circ}{\partial \epsilon}\left\{\frac{\partial \phi_{\vec{k}}}{\partial t} + \vec{v}(\vec{k}) \cdot \nabla_{\vec{r}}\phi_{\vec{k}} - \vec{F} \cdot \vec{v}(\vec{k})\right\} = \left(\frac{\partial f}{\partial t}\right)_{\text{coll}}. \tag{8.120}$$

To proceed further we assume that a weak constant electric field $\vec{E}$ produces a spatially homogeneous (but not necessarily isotropic) momentum distribution in steady state. Using $\partial f/\partial t = 0$ (steady-state condition) and $\nabla_{\vec{r}} f = 0$ (homogeneous-state condition), Eq. (8.118) becomes

$$-e\frac{\partial f^\circ}{\partial \epsilon}\left[\vec{E} \cdot \vec{v}(\vec{k})\right] = \left(\frac{\partial f}{\partial t}\right)_{\text{coll}}. \tag{8.121}$$

In order to evaluate the collision term, we assume elastic impurity scattering (but not necessarily isotropic s-wave scattering as before). Using $W_{\vec{k}\vec{k}'} = W_{\vec{k}'\vec{k}}$ (valid for elastic scattering in the presence of time-reversal symmetry) in Eq. (8.105) yields (to lowest order in $\phi_{\vec{k}}$)

$$\left(\frac{\partial f}{\partial t}\right)_{\text{coll}} = -\Omega \int \frac{d^3\vec{k}'}{(2\pi)^3} W_{\vec{k}\vec{k}'}(f_{\vec{k}} - f_{\vec{k}'})$$

$$= \Omega \int \frac{d^3\vec{k}'}{(2\pi)^3}\frac{\partial f^\circ}{\partial \epsilon} W_{\vec{k}\vec{k}'}(\phi_{\vec{k}} - \phi_{\vec{k}'}). \tag{8.122}$$

This is an integral equation for $\phi_{\vec{k}}$. To solve it we need to make some guesses on the form of the functional dependence of $\phi_{\vec{k}}$ on $\vec{k}$, so that the integral on the RHS can be evaluated. Assuming again

---

[18] The equilibrium distribution function f° may be viewed as either a function of $\vec{k}$ or a function of $\epsilon = \epsilon(\vec{k})$. It is convenient to switch between them, depending on whether one needs to take a derivative with respect to $\vec{k}$ or ϵ. Later on we will also encounter derivatives of f° with respect to temperature T and chemical potential μ, which are parameters of f°.

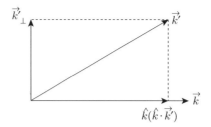

Figure 8.7 Decomposition of $\vec{k}'$ into components parallel and perpendicular to $\vec{k}$.

that the band dispersion is isotropic, and $W_{\vec{k}\vec{k}'}$ depends only on the relative angle between $\vec{k}$ and $\vec{k}'$, the only preferred direction in the problem is that selected by the external field. Furthermore, we expect $\phi_{\vec{k}}$ to vanish for $|\vec{E}| = 0$, and we are making the linear-response approximation. Hence it seems reasonable to make the following ansatz for the form of $\phi_{\vec{k}}$:

$$\phi_{\vec{k}} = (\vec{E} \cdot \hat{k})\Lambda(\epsilon_{n\vec{k}}). \tag{8.123}$$

This is a reasonable guess, as in linear response we must have $\phi_{\vec{k}} \propto \vec{E}$; but $\vec{E}$ is a vector and the only way to turn it into a scalar is to contract it with another vector, and the only vector that $\phi_{\vec{k}}$ depends on is (the direction of) $\vec{k}$. It will turn out that this form is *exact* within the linear-response regime. With this ansatz the collision term becomes

$$\left(\frac{\partial f}{\partial t}\right)_{\text{coll}} = \frac{\partial f^{\circ}}{\partial \epsilon}\Lambda(\epsilon_{n\vec{k}})\vec{E} \cdot \left[\Omega \int \frac{d^3\vec{k}'}{(2\pi)^3} W_{\vec{k}\vec{k}'}(\hat{k} - \hat{k}')\right]. \tag{8.124}$$

Now $W_{\vec{k}\vec{k}'}$ can depend on $\vec{k}'$ only through the angle between $\vec{k}$ and $\vec{k}'$ because the medium is assumed to be isotropic and the length of $\vec{k}'$ is fixed (since the scattering is elastic).

It is convenient to resolve $\vec{k}'$ into components parallel and perpendicular to $\vec{k}$:

$$\vec{k}' = \hat{k}(\hat{k} \cdot \vec{k}') + \vec{k}'_{\perp} \tag{8.125}$$

as illustrated in Fig. 8.7. Assuming that the scattering is azimuthally symmetric about the incident direction $\hat{k}$, we have

$$\int \frac{d^3\vec{k}'}{(2\pi)^3} W_{\vec{k}\vec{k}'}\vec{k}'_{\perp} = \vec{0}. \tag{8.126}$$

Hence Eq. (8.124) becomes

$$\left(\frac{\partial f}{\partial t}\right)_{\text{coll}} = \frac{\partial f^{\circ}}{\partial \epsilon}\Lambda(\epsilon_{n\vec{k}})(\vec{E} \cdot \hat{k})\left[\Omega \int \frac{d^3\vec{k}'}{(2\pi)^3} W_{\vec{k}\vec{k}'}(1 - \hat{k} \cdot \hat{k}')\right]. \tag{8.127}$$

The single-particle (i.e. ignoring Pauli blocking) scattering rate is

$$\frac{1}{\tau} \equiv \sum_{\vec{k}'} W_{\vec{k}\vec{k}'} = \Omega \int \frac{d^3\vec{k}'}{(2\pi)^3} W_{\vec{k}\vec{k}'}. \tag{8.128}$$

This, however, is *not* what appears in Eq. (8.127). Instead we have

$$\frac{1}{\tau_{\text{TR}}(\epsilon)} \equiv \Omega \int \frac{d^3\vec{k}'}{(2\pi)^3} W_{\vec{k}\vec{k}'}(1 - \cos\theta_{\vec{k}\vec{k}'}), \tag{8.129}$$

where τ_{TR} (which is in general energy-dependent) is called the **"transport lifetime."** The essential physics represented by the $(1 - \cos\theta)$ factor is that small-angle scattering is ineffective at relaxing the momentum distribution, since the momentum does not change very much with each collision. Conversely, scattering angles near π are doubly effective because they reverse the particle momentum.

In typical metallic alloys, τ_{TR} is larger than, but of the same order of magnitude as, τ. This is because the range of the impurity potential is of order the lattice spacing and hence of order the Fermi wavelength. This means that the partial wave expansion for the impurity scattering phase shifts converges at least moderately rapidly. Thus the scattering amplitude contains a significant s-wave contribution.[19] In contrast, modulation-doped semiconductor **quantum wells** (see Chapter 9) can have τ_{TR} orders of magnitude larger than τ. This is because most of the scattering is due to the very smooth potentials of remote donor ions separated from the quantum well by a thick (~ 0.1–1 μm) spacer layer.

Equation (8.127) can now be written as

$$\left(\frac{\partial f}{\partial t}\right)_{\text{coll}} = \frac{\partial f^\circ}{\partial \epsilon} \Lambda(\epsilon_{n\vec{k}})(\vec{E} \cdot \hat{k}) \frac{1}{\tau_{TR}(\epsilon)} = \frac{1}{\tau_{TR}(\epsilon)} \frac{\partial f^\circ}{\partial \epsilon} \phi_{\vec{k}}. \tag{8.130}$$

Using this in Eq. (8.121), we find

$$\phi_{\vec{k}} = -e\left[\vec{E} \cdot \vec{v}(\vec{k})\right]\tau_{TR}(\epsilon). \tag{8.131}$$

This has units of energy, and indeed has the form anticipated in Eq. (8.117), which tells us that states with velocities such that the particle is absorbing energy from the external field have an increased occupancy in the non-equilibrium state.[20] Also important is the fact that it is consistent with the ansatz of Eq. (8.123) for an isotropic system with $\vec{v}(\vec{k}) \parallel \hat{k}$, and thus our solution is self-consistent.

Now that we have found the non-equilibrium distribution function, it is straightforward to compute the current density flowing in response to the applied field:

$$\vec{J} = -2e \int \frac{d^3\vec{k}}{(2\pi)^3} \vec{v}(\vec{k}) f_{\vec{k}}, \tag{8.132}$$

where the factor of 2 accounts for the spin degree of freedom. Using the fact that the current vanishes in the equilibrium state, Eqs. (8.131) and (8.116) yield

$$\vec{J} = -2e \int \frac{d^3\vec{k}}{(2\pi)^3} \vec{v}(\vec{k}) e[\vec{E} \cdot \vec{v}(\vec{k})]\tau_{TR}(\epsilon)\frac{\partial f^\circ}{\partial \epsilon}. \tag{8.133}$$

The conductivity defined by

$$\vec{J} = \sigma \vec{E} \tag{8.134}$$

is thus (for an isotropic system)

$$\sigma = 2e^2 \int \frac{d^3\vec{k}}{(2\pi)^3} \tau_{TR}(\epsilon) \left(-\frac{\partial f^\circ}{\partial \epsilon}\right) \frac{\left|\vec{v}(\vec{k})\right|^2}{3}. \tag{8.135}$$

The derivative of the Fermi function is sharply peaked at the Fermi energy and hence this integral is essentially a 2D integral over the Fermi surface. For a parabolic band with effective mass m^*,

$$\sigma = 2e^2 \tau_{TR}(\epsilon_F) \int_0^\infty d\epsilon\, \rho(\epsilon)\left(-\frac{\partial f^\circ}{\partial \epsilon}\right)\frac{2}{3}\frac{\epsilon}{m^*}, \tag{8.136}$$

---

[19] If only the s partial wave contributes to scattering, then the $\cos\theta_{\vec{k}\vec{k}'}$ term in Eq. (8.129) averages to zero after integration, and we would have $\tau = \tau_{TR}$.

[20] The electric field is doing work on the electron system in steady state, but our lowest-order perturbation-theory solution to the Boltzmann equation with elastic scattering does not contain the physics of the energy transfer from the electrons to the phonon heat bath. More on this shortly.

where $\rho(\epsilon)$ is the density of states per spin. Integration by parts shows that

$$2 \int_0^\infty d\epsilon \, \rho(\epsilon) \epsilon \left(-\frac{\partial f^\circ}{\partial \epsilon} \right) = \frac{3}{2} n, \tag{8.137}$$

where

$$n \equiv 2 \int_0^\infty d\epsilon \, \rho(\epsilon) f^\circ(\epsilon) \tag{8.138}$$

is the density. Thus we arrive at the final expression

$$\sigma = \frac{ne^2 \tau_{\mathrm{TR}}(\epsilon_{\mathrm{F}})}{m^*}, \tag{8.139}$$

which is precisely the Drude result except that m_{e} has been replaced by m^* and $\tau_{\mathrm{TR}}(\epsilon_{\mathrm{F}})$ has to be computed quantum mechanically from Eq. (8.129).

While the Drude formula above seems to suggest that σ is proportional to the density n, which is given by the entire Fermi *sea*, it should be clear from Eqs. (8.133) and (8.135) that, at low temperature (much lower than the Fermi temperature), conduction is dominated by electrons near the Fermi *surface*. This is manifested in Eq. (8.139) by the fact that the quantity playing the role of the scattering time, $\tau_{\mathrm{TR}}(\epsilon_{\mathrm{F}})$, is the transport scattering time of electrons *at the Fermi energy*. For anisotropic systems and other systems where the simple effective-mass approximation cannot be used, Eq. (8.135) cannot be brought into the Drude form, and the integration over the Fermi surface needs to be evaluated with the specific band structure.

To emphasize the fact that the conductivity is a property of the Fermi surface, here we present an alternative evaluation of Eq. (8.135), in which τ_{TR} represents $\tau_{\mathrm{TR}}(\epsilon_{\mathrm{F}})$:

$$\sigma = 2e^2 \rho(\epsilon_{\mathrm{F}}) \frac{1}{3} v_{\mathrm{F}}^2 \tau_{\mathrm{TR}}. \tag{8.140}$$

Defining the transport mean free path

$$\ell \equiv v_{\mathrm{F}} \tau_{\mathrm{TR}}, \tag{8.141}$$

we have

$$\sigma = 2e^2 \rho(\epsilon_{\mathrm{F}}) \frac{1}{3} \frac{\ell^2}{\tau_{\mathrm{TR}}}. \tag{8.142}$$

If we view the particles as undergoing a random walk of step length ℓ and time τ_{TR} per step, then the **diffusion constant** in 3D is

$$D \equiv \frac{1}{3} \frac{\ell^2}{\tau_{\mathrm{TR}}}. \tag{8.143}$$

Hence we obtain the **Einstein relation** (see Section 8.8.1)

$$\sigma = e^2 \frac{dn}{d\mu} D, \tag{8.144}$$

where $dn/d\mu \equiv 2\rho(\epsilon_{\mathrm{F}})$ is the combined density of states (for both spins) *at the Fermi level*. Thus the fundamental difference between a band metal and band insulator lies in the fact that $dn/d\mu > 0$ for the former, while $dn/d\mu = 0$ for the latter.[21]

We must now address a quite subtle point underlying our assumption of a linear response to the external electric field. We have assumed linear response is valid and our equations self-consistently

---

[21] In Chapter 11 we will encounter another type of insulator known as an Anderson insulator, in which $dn/d\mu > 0$, but the diffusion constant D vanishes due to disorder-induced localization.

tell us that the current density is indeed linear in the applied electric field. We found that the conductivity is inversely proportional to the rate (in the absence of an applied field) at which the momentum distribution (and hence the current) relaxes back to equilibrium due to elastic scattering from impurities. This result agrees well with experiment, yet a little thought shows us that something is wrong. If current flows in the direction of the applied electric field, the external field is doing work on the electrons at rate

$$P = \int d^3\vec{r} \, \vec{J} \cdot \vec{E} = \int d^3\vec{r} \, \sigma |\vec{E}|^2. \tag{8.145}$$

However, if we only have elastic scattering, the energy cannot relax and rises without limit so that our assumption of steady state is invalidated. The solution to this conundrum is the following. The power added to the system is of course very small when the electric field is weak. Hence even an inelastic scattering rate (due to electron–phonon coupling) which is much smaller than the elastic scattering rate will still be able to cool the electrons and prevent the energy runaway. The inelastic scattering can be weak enough that its effect on the conductivity is negligible and yet (for weak enough applied field) the inelastic scattering rate is large enough to keep the temperature from rising significantly. Of course, we need a way for the heat transferred to the phonons to be carried away into a substrate or other thermal bath. The reason why we do *not* need to specify the details of inelastic scattering in our linear-response calculation is that its effect (which gives rise to dissipation) is of order $|\vec{E}|^2$, as seen in Eq. (8.145). Note that it could occur in some systems that the inelastic scattering is very strong and must therefore be taken into account in the calculation of the conductivity (especially if the elastic scattering due to disorder is very weak). It can also occur that the rate of inelastic scattering vanishes as a high power of temperature at low temperatures, $1/\tau_{\text{inelastic}} \sim T^p$. For sufficiently large exponent p, even weak electric fields can cause a singular temperature rise [47].

So far we have considered only linear response at zero frequency within a semiclassical transport approximation. At very high frequencies we must use a fully quantum theory because we must recognize that the individual quanta of the applied electric field carry significant energy $\hbar\omega$. For high enough frequency, this energy may be large enough to create phonons or cause electrons to jump across a band gap into a different band. Such quantum effects arise when we study the optical conductivity (see Section 15.10.2). For the case of lower frequencies (e.g. RF, microwave, and, in some cases, far infrared) in situations where the transport lifetime is the only relevant time scale against which to measure the frequency, a simple extension of our semiclassical results is possible, and we obtain the Drude model result for the ac conductivity given in Exercise 7.1.

Exercise 8.7. Derive Eq. (8.137) which is valid for a parabolic band by integrating by parts. Hint: it will be useful to recall that $\rho(\epsilon) = A\epsilon^{1/2}$. If you use $\partial \ln \rho / \partial \epsilon = 1/2\epsilon$ you will not have to bother to evaluate the quantity A.

8.8.1 Einstein Relation

Earlier in this section we considered a thermodynamically large conductor that is capable of supporting a non-equilibrium but *steady* current. In practice this requires that it be connected to leads, which play the roles of source(s) and drain(s) for the current, such that the current does not lead to charge build up or redistribution in the conductor. The crucial roles of leads in transport will be discussed in much greater detail in Chapter 10. In this subsection we consider an *isolated* wire of *finite* length with an applied electric field or, equivalently, potential gradient, as illustrated in Fig. 8.8. Obviously there can be no net current in such an *isolated* system in equilibrium.

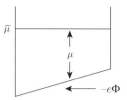

Figure 8.8 Illustration of the Einstein relation. $\tilde{\mu}$ is the global or electrochemical potential, which is a constant in an equilibrium state. The local chemical potential μ (which depends on the local electric potential Φ) determines the local electron density.

The electrical current driven by the electric field[22] is

$$\vec{J}_E = \sigma(-\nabla\Phi). \tag{8.146}$$

However, since the wire is finite, charge will build up at one end, and there will also be a density gradient that produces an equal and opposite diffusion current

$$\vec{J}_D = -eD(-\nabla n). \tag{8.147}$$

This equation defines the diffusion constant D. Using

$$\nabla n = \frac{dn}{d\mu} e \nabla\Phi \tag{8.148}$$

and the equilibrium condition

$$\vec{J}_E + \vec{J}_D = \vec{0}, \tag{8.149}$$

we obtain Einstein's result

$$\sigma = e^2 D \frac{dn}{d\mu}. \tag{8.150}$$

This relation is exact, even in the presence of electron–electron interactions.

In the equilibrium state considered here, the *global* chemical potential $\tilde{\mu}$ must be a constant as illustrated in Fig. 8.8. The local density of electrons, however, is determined by the *local* chemical potential μ, namely the global chemical potential measured from the local electron potential energy. We thus find

$$\tilde{\mu} \equiv -e\Phi + \mu. \tag{8.151}$$

Because of its dependence on the local electric potential, $\tilde{\mu}$ is also known as the **electrochemical potential**. Contrary to a common misunderstanding, what is measured in a **voltage** measurement (say, by a voltmeter) is the difference of electrochemical potentials between two contacts, *not* their electric potential difference. Related to this is the fact that an electric current is driven by the *combination* of electric field and local chemical potential difference, not just the former. The conductivity determines the current associated with the electric field, and the diffusion coefficient and density of states determine the response to a chemical potential gradient. As we have seen, the Einstein relation connects these two, and this allows us to express the current in a wire in terms of the voltage drop along the wire via Ohm's law,

$$I = \frac{V}{R}, \tag{8.152}$$

---

[22] It is crucial to understand that this is not necessarily the externally applied electric field. The conductivity is defined to be the current response to the net electric field inside the sample.

independently of whether it is a diffusion current or an electric-field-induced current or a combination of the two. (Here the conductance $G = 1/R$ is determined by the conductivity and the geometry.) As we will see in Chapter 10, theoretically it is often convenient to consider the current as being driven by voltages determined by chemical potential differences rather than electric fields.[23]

8.9 Thermal Transport and Thermoelectric Effects

Because the Boltzmann transport picture is applicable to non-equilibrium situations, we can use it to study systems where there are temperature gradients and to compute the thermal conductivity and the thermoelectric power.

The expression for the heat current density carried by the electrons is

$$\vec{J}^Q = 2 \int \frac{d^3\vec{k}}{(2\pi)^3} \left[\epsilon(\vec{k}) - \mu \right] \vec{v}(\vec{k}) f_{\vec{k}}. \tag{8.153}$$

This is similar to the expression for the number current but is weighted by the energy of the particle relative to the chemical potential. Recall that the chemical potential is the energy cost of adiabatically adding a particle to the system, and adiabatic means without heat flow. Hence it makes sense to say that the contribution to the heat current should be weighted by $\left[\epsilon(\vec{k}) - \mu \right]$.

Let us now consider the linear response to weak gradients in the temperature T, chemical potential μ, and electrostatic potential Φ:

$$\vec{J} = L^{11}\vec{\Sigma} + L^{12}(-\nabla T), \tag{8.154}$$

$$\vec{J}^Q = L^{21}\vec{\Sigma} + L^{22}(-\nabla T), \tag{8.155}$$

where

$$\vec{\Sigma} \equiv \vec{E} + \frac{1}{e}\nabla\mu = -\left(\frac{1}{-e}\right)\nabla(-e\Phi + \mu). \tag{8.156}$$

As discussed in Section 8.8.1, the combination $\tilde{\mu} \equiv -e\Phi + \mu$ is known as the electrochemical potential.[24] We will see shortly why this particular combination appears. In particular, we are used to thinking of the electrostatic potential as the voltage. However, it is very important to understand that the voltage at some point in a circuit is actually related to the full electrochemical potential via $\tilde{\mu} = -eV$. Hence $\vec{\Sigma} = -\nabla V$ is related to the voltage gradient.

As before, it is convenient to parameterize the deviations from equilibrium by the expression[25] in Eq. (8.116). In steady state, the Boltzmann equation, Eq. (8.97), becomes (to lowest order in temperature gradients and fields)

$$\vec{v}(\vec{k}) \cdot \nabla_{\vec{r}} f^\circ + \frac{1}{\hbar}\vec{F} \cdot \nabla_{\vec{k}} f^\circ = \left(\frac{\partial f}{\partial t}\right)_{\text{coll}}. \tag{8.157}$$

---

[23] Since electrons are charged, a local chemical potential difference inevitably induces an electric field through the charge density gradient it generates. But in the idealization of non-interacting electrons one can in principle separate these two.

[24] In this context μ is also referred to as the *local* chemical potential to remove any possible ambiguity in terminology.

[25] In this case we need to allow $\phi_{\vec{k}}$ to be $\vec{r}$-dependent as the system is no longer homogeneous. This does not change the manipulations in the Boltzmann equation as all the summations or integrals are performed in momentum space. We will thus keep the possible $\vec{r}$-dependence of $\phi_{\vec{k}}$ implicit below.

Previously we dropped the spatial gradient term, but now we cannot because $f°$ (which describes the *local* equilibrium) depends on T and μ, both of which vary with position:

$$\nabla_{\vec{r}} f° = \left(\frac{\partial f°}{\partial T} \right) \nabla T + \left(\frac{\partial f°}{\partial \mu} \right) \nabla \mu. \tag{8.158}$$

We now take advantage of the fact that

$$\frac{\partial f°}{\partial \mu} = -\left(\frac{\partial f°}{\partial \epsilon} \right) \tag{8.159}$$

and

$$\frac{\partial f°}{\partial T} = \frac{\left[\epsilon(\vec{k}) - \mu \right]}{T} \left(-\frac{\partial f°}{\partial \epsilon} \right) \tag{8.160}$$

to write

$$\nabla_{\vec{r}} f° = -\left(\frac{\partial f°}{\partial \epsilon} \right) \left[\frac{\epsilon(\vec{k}) - \mu}{T} \nabla T + \nabla \mu \right]. \tag{8.161}$$

Using Eq. (8.119) the Boltzmann equation becomes

$$-\left(\frac{\partial f°}{\partial \epsilon} \right) \vec{v}(\vec{k}) \cdot \left\{ \frac{\epsilon(\vec{k}) - \mu}{T} \nabla T + \nabla \mu + e\vec{E} \right\} = \left(\frac{\partial f}{\partial t} \right)_{\text{coll}}, \tag{8.162}$$

which is of the same form of Eq. (8.121) if we identify the terms inside the curly brackets with $e\vec{E}$, and can thus be solved the same way. This leads to the solution

$$\phi_{\vec{k}} = \tau_{\text{TR}}(\epsilon_{\vec{k}}) \vec{v}(\vec{k}) \cdot \left\{ -e\vec{\Sigma} + \frac{\epsilon(\vec{k}) - \mu}{T} (-\nabla T) \right\}, \tag{8.163}$$

which is similar to Eq. (8.131) if we have only elastic impurity scattering.

Now that we know the deviation from (local) equilibrium, we can compute the electrical and thermal current densities:

$$\vec{J} = -2e \int \frac{d^3\vec{k}}{(2\pi)^3} [f_{\vec{k}} - f_{\vec{k}}°] \vec{v}(\vec{k}), \tag{8.164}$$

$$\vec{J}^Q = 2 \int \frac{d^3\vec{k}}{(2\pi)^3} [f_{\vec{k}} - f_{\vec{k}}°] \left(\epsilon(\vec{k}) - \mu \right) \vec{v}(\vec{k}). \tag{8.165}$$

We again assume the system is isotropic and define

$$\sigma(\epsilon) \equiv 2e^2 \tau_{\text{TR}}(\epsilon) \int \frac{d^3\vec{k}}{(2\pi)^3} \delta\left(\epsilon - \epsilon(\vec{k}) \right) \frac{1}{3} \left[v(\vec{k}) \right]^2. \tag{8.166}$$

Physically this is the conductivity for $\epsilon_{\text{F}} = \epsilon$ and $T = 0$ (in which case $\partial f°/\partial \epsilon$ becomes a delta function; see Eq. (8.135)).

It is useful to now define

$$\mathcal{L}^{(\alpha)} \equiv \int d\epsilon \left(-\frac{\partial f°}{\partial \epsilon} \right) (\epsilon - \mu)^\alpha \sigma(\epsilon). \tag{8.167}$$

Using Eq. (8.163) it is then straightforward to derive

$$\vec{J} = \mathcal{L}^{(0)} \vec{\Sigma} + \mathcal{L}^{(1)} \left(\frac{1}{-eT} \right) (-\nabla T), \tag{8.168}$$

$$\vec{J}^Q = \left(\frac{1}{-e} \right) \mathcal{L}^{(1)} \vec{\Sigma} + \frac{1}{(-e)^2} \mathcal{L}^{(2)} \frac{1}{T} (-\nabla T), \tag{8.169}$$

so that the linear response coefficients are

$$L^{11} = \mathcal{L}^{(0)}, \tag{8.170a}$$

$$L^{12} = \frac{1}{-eT}\mathcal{L}^{(1)}, \tag{8.170b}$$

$$L^{21} = \frac{1}{-e}\mathcal{L}^{(1)}, \tag{8.170c}$$

$$L^{22} = \frac{1}{e^2 T}\mathcal{L}^{(2)}. \tag{8.170d}$$

The result

$$L^{21} = TL^{12} \tag{8.171}$$

is true quite generally and is one of the equalities among the linear-response coefficients collectively known as **Onsager relations**. See Appendix A for further discussion of linear-response theory and the Onsager relations.

At low temperatures ($k_B T \ll \epsilon_F$) we can use the Sommerfeld expansion (see Exercise 7.26 and in particular Exercise 7.27) for $(-\partial f^\circ / \partial \epsilon)$ to obtain

$$L^{11} = \sigma, \tag{8.172a}$$

$$L^{21} = TL^{12} = -\frac{\pi^2}{3e}(k_B T)^2 \left(\frac{\partial \sigma}{\partial \epsilon}\right)_{\epsilon=\mu}, \tag{8.172b}$$

$$L^{22} = \frac{\pi^2}{3}\frac{k_B^2 T}{e^2}\sigma(\mu). \tag{8.172c}$$

Now that we have the linear-response coefficients, we can relate them to the standard transport experiments with which we are familiar. Electrical conduction is measured with zero temperature gradient. Hence $L^{11} = \sigma(\mu)$ is the conductivity.

Thermal conductivity is measured under conditions of zero electrical current flow. From Eq. (8.154), $\vec{J} = \vec{0}$ implies

$$\vec{\Sigma} = \frac{L^{12}}{L^{11}}\nabla T \tag{8.173}$$

and

$$\vec{J}^Q = -\kappa_e \nabla T, \tag{8.174}$$

where the electronic contribution to the thermal conductivity κ_e is obtained by substituting Eq. (8.173) into Eq. (8.155):

$$\kappa_e = L^{22} - \frac{L^{21}L^{12}}{L^{11}}. \tag{8.175}$$

The second term in this expression vanishes like $\sim T^3$ in a metal and is generally negligible with respect to the first term, which vanishes only linearly. This is then consistent with the Wiedemann–Franz law and predicts for the so-called Lorenz number

$$\frac{\kappa_e}{\sigma T} \approx \frac{L^{22}}{\sigma T} = \frac{\pi^3}{3}\left(\frac{k_B}{e}\right)^2, \tag{8.176}$$

which is numerically quite close to what was obtained in Drude theory; see Eq. (7.18).

Phonons can also make a contribution to the thermal conductivity which we are ignoring here, as it vanishes with higher powers of T at low temperature (just like the specific heat). The Wiedemann–Franz law states that in a metal, where the thermal conductivity is dominated by the electrons, the thermal conductivity is proportional to the electrical conductivity and the Lorenz number is independent of temperature.

Another standard transport (related) quantity is the thermoelectric power (also known as the **thermopower**) Q, which measures the voltage gradient produced by a temperature gradient:

$$\vec{\Sigma} = -\nabla V = Q \, \nabla T, \tag{8.177}$$

under the condition of zero electric current $\vec{J} = \vec{0}$. It is readily verified that

$$Q = \frac{L^{12}}{L^{11}} = -\frac{\pi^2}{3}\left(\frac{k_B}{e}\right)k_B T\left(\frac{1}{\sigma}\frac{\partial \sigma}{\partial \epsilon}\right). \tag{8.178}$$

The second equality above is known as the **Mott formula** and is valid only for the case of elastic impurity scattering in a metal such as we are considering here. To estimate the typical size of Q we note that $k_B/e \approx 86\,\mu\text{V/K}$ and $(1/\sigma)\partial\sigma/\partial\epsilon \sim 1/\epsilon_F$, so at room temperature

$$|Q| \sim \frac{\pi^2}{3}86\,\mu\text{V/K} \times \frac{k_B T}{\epsilon_F} \sim 1\,\mu\text{V/K}. \tag{8.179}$$

Hence the effect is rather small (though still useful for making thermocouple temperature sensors).

One unusual aspect of the thermopower (compared with the conductivity and other longitudinal transport coefficients) is that it is sensitive to the *sign* of the charge carrier, as can be seen from Eq. (8.178). In this regard it is similar to the Hall conductivity. If there is perfect particle–hole symmetry about the Fermi level then Q vanishes. In interacting systems with inelastic scattering, Q is very sensitive to the details of the scattering and is often difficult to compute reliably.

We have been explicitly ignoring the phonons throughout this discussion. It sometimes happens that a temperature gradient or an electrical current drives the phonons significantly out of equilibrium. If the predominant electron scattering is via phonons rather than impurities, this can lead to an additional contribution known as the **phonon-drag** thermopower. Very pure lithium crystals exhibit an anomaly in their thermopower at low temperatures due to this effect.

Exercise 8.8. Use the Drude formula $\sigma = ne^2\tau/m_e$ in the Mott formula (8.178), and assume that the scattering time τ has no energy dependence. In this case the only dependence of σ on the (Fermi) energy ϵ comes from the electron density n. Show in this case that the thermopower at low temperatures obeys

$$Q = -\frac{S}{N_e e}, \tag{8.180}$$

where N_e is the number of conduction electrons and S is the entropy they carry. Miraculously, the scattering time τ does not appear in this expression!

Exercise 8.9. The thermopower is a combination of the thermoelectric coefficients L^{12} and L^{11}. They are not well-defined (or diverge) in the absence of scattering (i.e. in the clean limit). The thermopower, on the other hand, is well-defined and finite in the clean limit, and easily calculable by combining thermodynamic and hydrodynamic relations. The key observation is that the electron liquid must be in mechanical equilibrium when no current is allowed to flow.

(i) Show that this requires $\nabla P = ne\,\nabla\Phi$, where P is the pressure of the electron liquid, n is its local number density, and Φ is the local electric potential.

Since P is an intensive quantity, it can only depend on other intensive quantities, in this case, T and μ. We thus have

$$\nabla P = \left(\frac{\partial P}{\partial \mu}\right)_T \nabla \mu + \left(\frac{\partial P}{\partial T}\right)_\mu \nabla T. \tag{8.181}$$

(ii) Use appropriate Maxwell relations between thermodynamic quantities to show that $(\partial P/\partial \mu)_{T,V} = (\partial N_e/\partial V)_{T,\mu} = N_e/V = n$ and $(\partial P/\partial T)_{\mu,V} = (\partial S/\partial V)_{T,\mu} = S/V$. The last steps follow from the extensivity of S, N_e, and V, which are proportional to each other when the intensive quantities μ and T are fixed.

(iii) Now show that the thermopower in this clean limit is given by Eq. (8.180).

This exercise illustrates that the thermopower is not an ordinary transport coefficient, but more like a thermodynamic/hydrodynamic quantity. It is a measure of the entropy carried by mobile electrons, divided by their number. The derivation above does not assume non-interacting electrons, because electron–electron interactions do not alter the thermodynamic and hydrodynamic relations used. On the other hand, being in the clean limit is important, as impurity scattering can give rise to additional forces (effectively frictional forces) that alter the force equilibration relation. If the mean free path is a strong function of the carrier energy, this effect can also modify the thermopower.

9 Semiconductors

Thus far we have focused our discussion of electron dynamics and transport properties on metals, where, due to the presence of partially filled band(s), we have a high density of free carriers even at zero temperature ($T = 0$), and transport coefficients are mainly determined by impurity and other scattering processes. For example the T-dependence of conductivity, *à la* Drude

$$\sigma = \frac{ne^2\tau}{m_e},$$

mainly comes from the T-dependence of the scattering time τ (since the carrier density n is essentially T-independent). Since the scattering rate $1/\tau$ increases with increasing T (as there are more sources of inelastic scattering such as phonons), σ decreases with increasing T in a metal.[1]

In a band insulator, all bands are either empty or full in the ground state, as a result of which $n = 0$ at $T = 0$, thus $\sigma = 0$ as well. On the other hand, there is always some conduction at finite T, due to thermally activated carriers (namely some electrons get thermally excited from valence bands to conduction bands). As a result, σ *increases* with increasing T in an insulator.[2] Quantitatively the number of activated carriers is very sensitive to the size of the band gap. A very important class of materials called semiconductors are band insulators for which the band gap is not too large. As a result, there is substantial conduction at room temperature due to thermally activated carriers. Empirically, semiconductors are often identified by the property that the (easily measurable) electrical conductivity increases with temperature. More importantly, it is relatively easy to introduce charge carriers into these materials through chemical doping and other methods, thus one can control or even design their physical properties both for fundamental research and for industrial applications. These important materials are the focus of this chapter.

9.1 Homogeneous Bulk Semiconductors

Strictly speaking, semiconductors are band insulators. In practice there is measurable conduction at room temperature if the band gap Δ is less than approximately $2\,\text{eV}$. (Room temperature corresponds

---

[1] There are exceptions to this rule, all believed to be related to inelastic scattering weakening the effects of **Anderson localization** (see Chapter 11).

[2] This statement applies to band insulators and other types of insulators alike, including the Anderson and Mott insulators that we will encounter later in the book. There is growing interest in the possibility of so-called many-body localized states in which σ may remain zero at finite T. For a discussion see Section 11.11.

to a thermal energy scale of approximately $0.025\,\mathrm{eV}$.) The most important semiconductors are the group-IV elements (Si, $\Delta \simeq 1.12$ eV; Ge, $\Delta \simeq 0.67$ eV), binary compounds of group-III and group-V elements (so-called III–V semiconductors), and compounds of group-II and group-VI elements (II–VI semiconductors). See the following tables for the relevant part of the periodic table for semiconductors, and some examples of III–V and II–VI semiconductors.

II	III	IV	V	VI
	Al	Si	P	
Zn	Ga	Ge	As	Se
Cd	In	Sn	Sb	Te
Hg				

III–V	II–VI
AlSb, GaSb,	HgTe,
InAs, GaP,	ZnTe,
GaAs, AlAs,	CdTe
InSb	

Like diamond, Si, and Ge in group IV, these compound semicondutors are (almost[3]) purely covalently bonded, but their band gaps are considerably smaller.[4] The binary compounds are covalent but the natural trend is that III–VI compounds are slightly ionic, and II–VI compounds are still more ionic because of the charge transfer between the elements. Interestingly, the III–V and II–VI compounds in the Sn row are low-band-gap semiconductors with important optical technology applications because the low band gap permits infrared photons to excite electrons across the gap.

The various III–V and II–VI compounds (in all rows) usually occur in the zincblende crystal structure, which is the same as the diamond structure (see Fig. 3.8) formed by Si and Ge, except that the two types of atoms separately occupy the two sublattices. In the following we focus on these structures exclusively, and do not distinguish among them unless truly necessary.

The physical properties of semiconductors, like those of all materials, are dominated by the bands closest to the Fermi energy, namely the conduction band (right above the Fermi energy) and valence band(s) (right below the Fermi energy). Which atomic orbitals are responsible for these bands? All the relevant atoms have valence electrons occupying ns and np orbitals, with $n = 3, 4, 5, 6$. The s orbital has lower energy than the three p orbitals (see Fig. 9.1(a)), and the latter is further split by spin–orbit coupling. Each unit cell contains two atoms, and these orbitals can form bonding and anti-bonding

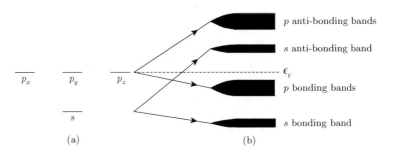

Figure 9.1 Illustration of atomic orbitals forming conduction and valence bands in semiconductors. (a) Valence electrons occupy some of the s and p orbitals in the atoms that form semiconductors. (b) These orbitals form bonding and anti-bonding bands in semiconductors.

---

[3] Because the two elements in the structure are not identical, III–V and II–VI compounds are slightly ionic. This is why, for example, GaAs is piezoelectric.

[4] The trend of decreasing band gap from Si to Ge continues to Sn. This group IV element is peculiar in that it has several possible crystal structures. In one of them, so-called "grey tin," it has a band gap which is almost exactly zero. See Ref. [48].

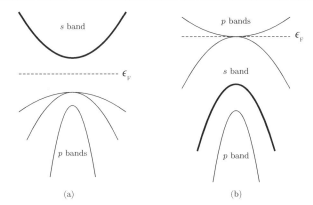

Figure 9.2 (a) Illustration of the GaAs band structure near the Brillouin zone center Γ. The conduction band (thick line) is an s band, while the three valence bands closest to the Fermi energy ϵ_F are all p bands. (b) Illustration of the HgTe band structure near the Brillouin zone center Γ. Note the inversion between s and p bands compared with the case of GaAs (which has the usual ordering).

combinations; altogether they form eight bands (four bonding and four anti-bonding). Each unit cell also contains eight valence electrons from the two atoms in it, and as a result four of these bands are occupied while the other four are empty.

Which of these bands are closest to the Fermi energy ϵ_F? Obviously the s bonding band and p anti-bonding bands will have lowest and highest energies, respectively, and thus lie farthest from ϵ_F. In most cases the p bonding bands are pushed below the s anti-bonding band, and ϵ_F is between these bands (see Fig. 9.1(b)). As a result, there is a single (low-energy) conduction band made of s anti-bonding orbitals, but three nearly degenerate valence bands near ϵ_F made of p bonding orbitals.[5] As a result, in most semiconductors the electron band structure and dynamics are simpler than those for holes, because for the latter one needs to deal with spin–orbit coupling and must specify which band a hole of interest is in (or which "hole" one is talking about; see below). We note that there are also cases in which the ordering of bands deviates from that in Fig. 9.1 in parts of the Brillouin zone; in such cases we say there is "band inversion." Inversion between conduction and valence bands leads to extremely interesting topological physics that we will discuss in Chapter 14.

The typical band structure in the diamond/zincblende structure has the maxima of the valence bands at the center of the Brillouin zone, which is called the Γ point in standard band-structure notation. The minimum of the conduction band may be at Γ, in which case the material is called a **direct-gap** semiconductor.[6] A good example of this is gallium arsenide (GaAs), whose band structure near Γ is illustrated in Fig. 9.2(a). The s-orbital conduction band is separated from the p-orbital valence bands by a band gap $\Delta = 1.4\,\text{eV}$. It is interesting to compare this with mercury telluride (HgTe), Fig. 9.2(b). In that case the s band is pushed *below* two of the p bands near Γ and becomes (part of the) valence band, resulting in band inversion. Thus, if one insists on using the energy difference between the s band and p bands as the definition of the band gap Δ, then one would conclude that HgTe has a "negative" band gap! In the meantime, two of the three p bands remain degenerate at Γ, which is also where ϵ_F lies. As a result, HgTe is actually a semimetal. As we will discuss in Chapter 14, one can lift this degeneracy, open a gap at ϵ_F, and turn HgTe into a **topological insulator** by cleverly manipulating its structure.

The origin of the splitting of the p bands and the band inversion in HgTe is spin–orbit coupling (SOC). As discussed in Section 7.8, within a single atom, SOC takes the form

---

[5] In fact, they may be exactly degenerate at certain points in the Brillouin zone.
[6] Otherwise it is called an indirect-gap semiconductor. More on this shortly.

$$H_{\text{SO}} = \xi(r)\vec{L} \cdot \vec{S}, \qquad (9.1)$$

$$\xi(r) = \frac{1}{2m_e^2 c^2 r} \frac{dV(r)}{dr}, \qquad (9.2)$$

where $\vec{L}$ and $\vec{S}$ are the orbital angular momentum and spin operators, respectively. For the s orbital we have orbital angular momentum quantum number $l = 0$, thus SOC has no effect on its energy. But for p orbitals $l = 1$, and H_{SO} splits the six (including the two-fold spin degeneracy) degenerate p orbitals into two groups, with $j = 3/2$ (four-fold degenerate, higher energy) and $j = 1/2$ (two-fold degenerate, lower energy). This is the origin of the splitting of the p bands seen in Fig. 9.1(a) at the Γ point, and the (remaining) degeneracy of the two higher p bands which actually form a $j = 3/2$ manifold.[7] This degeneracy is further split as one moves away from the Γ point, where the (rotational) symmetry is lowered. The different dispersions result in two different effective masses. As a result, these two (energetically most important) valence bands are often referred to as "light-hole" and "heavy-hole" bands. We know that the heavier an element is, the stronger the SOC is.[8] Mercury is so heavy that its presence pushes the $j = 3/2$ p bands above the s band at the Γ point, resulting in the band inversion shown in Fig. 9.2(b).

The conduction band minima can also be away from Γ, in which case we have an **indirect-gap** semiconductor. For this case, there are several conduction band minima that are related to each other through crystalline rotation and/or inversion symmetries. Each minimum is also referred to as a "**valley**," and the number of these valleys is called the **valley degeneracy**. Two of the most important examples are as follows.

Silicon There are six global minima occurring at six symmetry equivalent points along the (100) and equivalent directions. Thus the valley degeneracy is six. $\Delta \simeq 1.12$ eV.

Germanium The conduction band minima are at the zone boundary in the (111) and symmetry-related directions. The valley degeneracy is only four (because pairs of zone boundary points are effectively the same point). $\Delta \simeq 0.67$ eV.

For most properties of semiconductors at room temperature and below, only the states near the minima of the conduction band or maxima in the valence band are relevant, so it is useful to expand the dispersion relations around a band extremum and parameterize the quadratic terms by an (inverse) effective mass tensor:

$$\epsilon(\vec{k}) = \epsilon_{\text{c}} + \frac{\hbar^2}{2} \sum_{\alpha\beta} k_\alpha (M_{\text{c}}^{-1})_{\alpha\beta} k_\beta \quad \leftarrow \text{Conduction band},$$

$$\epsilon(\vec{k}) = \epsilon_{\text{v}} - \frac{\hbar^2}{2} \sum_{\alpha\beta} k_\alpha (M_{\text{v}}^{-1})_{\alpha\beta} k_\beta \quad \leftarrow \text{Valence band},$$

$$\epsilon_{\text{c}} = \epsilon_{\text{v}} + \Delta,$$

where ϵ_{c} and ϵ_{v} are the conduction and valence band edges, respectively, and the wave vectors $(\vec{k})$ are measured from the extremum. Since M^{-1} is real and symmetric, it can always be diagonalized by an orthogonal transformation, i.e. we can always choose a coordinate system where the effective mass tensor is diagonal. Owing to the high symmetry at the Γ point, M reduces to a scalar number and is simply called the effective mass. For extrema away from Γ one needs to specify a direction for each eigenvalue of M.

---

[7] Here a two-fold "spin" degeneracy is factored out as is normally done when spin is conserved, even though with SOC spin is not a good quantum number anymore.

[8] SOC is a relativistic (Dirac equation) effect. In elements with large atomic number, the electrons move very rapidly when they are near the large positive charge of the nucleus, resulting in stronger relativistic effects.

In the absence of impurities, the density of thermally activated electrons in the conduction band, $n_c(T)$, must match the density of thermally activated holes in the valence band, $p_v(T)$:

$$n_c(T) = p_v(T). \tag{9.3}$$

At very low temperature $k_B T \ll \Delta$, we expect[9]

$$n_c(T) \sim e^{-(\epsilon_c - \mu)/k_B T}, \tag{9.4}$$

$$p_v(T) \sim e^{-(\mu - \epsilon_v)/k_B T}, \tag{9.5}$$

where μ is the chemical potential that approaches ϵ_F as $T \to 0$. Here we have neglected T-dependent coefficients (prefactors multiplying the exponentials) that are sensitive to the details of band structures but depend on T only in a power-law fashion. We thus find as $T \to 0$

$$\mu = (\epsilon_c + \epsilon_v)/2, \tag{9.6}$$

namely $\mu(T \to 0) = \epsilon_F$ is right in the middle of the band gap, and

$$n_c(T) = p_v(T) \sim e^{-\frac{\Delta}{2k_B T}}. \tag{9.7}$$

The temperature dependence of the conductivity is dominated by that of the carrier density, and thus also has an exponential form:

$$\sigma \sim \sigma_0 e^{-\frac{\Delta}{2k_B T}} = \sigma_0 e^{-\frac{E_a}{k_B T}}, \tag{9.8}$$

where

$$E_a = \Delta/2 \tag{9.9}$$

is known as the **activation energy**, and such an exponential dependence on (inverse) temperature is known as activated behavior or the **Arrhenius law**.

The activation energy being equal to only one-half of the energy gap is consistent with the "**law of mass action**," which is familiar from chemistry. See Box 9.1.

Box 9.1. Law of Mass Action

Suppose that a semiconductor is in equilibrium with blackbody radiation at temperature T. Further suppose that the radiation photons with energy near the band gap Δ create particle–hole pairs at rate $\omega_+ e^{-\beta \Delta}$ (with $\beta = 1/k_B T$). Suppose also that there is a two-body recombination process in which particles and holes annihilate. We expect this "reaction rate" to be proportional to the product of the electron and hole densities: $R = \omega_- n_c p_h = \omega_- n_c^2$ (since $p_h = n_c$). The kinetic equation for the density of electrons is therefore

$$\dot{n}_c = \omega_+ e^{-\beta \Delta} - \omega_- n_c^2. \tag{9.10}$$

Setting the time derivative to zero gives a steady-state electron density with activation energy equal to $\Delta/2$ consistent with Eq. (9.7):

$$n_c(T) = \sqrt{\frac{\omega_+}{\omega_-}} e^{-\beta \frac{\Delta}{2}}. \tag{9.11}$$

---

[9] In the low-carrier-density limit we are interested in, the difference between the Boltzmann distribution and the Fermi–Dirac distribution is negligible. This statement applies both to electrons and to holes, and we may treat holes as real particles with their own thermal distribution function. See Exercise 9.1 for further discussions on these.

While having the same density, electrons and holes do not necessarily contribute the same to the electrical current in transport, because they have different mobilities. The **mobility** α is defined as the ratio between the drift speed of a charge carrier and the electric field that drives the charge motion,

$$v = \alpha E, \tag{9.12}$$

and is a measure of how mobile the carrier is. In most cases we have $\alpha_e > \alpha_h$, as electrons tend to have smaller effective mass and move more quickly.

The thermopower of a semiconductor can be much larger than that for a metal. While electrons and holes give rise to contributions of *opposite* sign to the thermopower (due to their opposite charge), the electrons in the conduction band have much higher mobility than the holes in the valence band (there is no particle–hole symmetry). As a result, the majority of the current is carried by the electrons and (when $|\nabla T| = 0$)

$$\frac{J^Q}{J} \sim \frac{\epsilon_c - \mu}{e} \sim \frac{1}{2}\frac{\Delta}{e}, \tag{9.13}$$

where ϵ_c is the conduction band energy minimum and Δ is the band gap. When $|\nabla T| = 0$,

$$\vec{J} = L^{11}\vec{\Sigma} = \sigma\vec{\Sigma}, \tag{9.14a}$$

$$\vec{J}^Q = L^{21}\vec{\Sigma}, \tag{9.14b}$$

so that

$$\frac{J^Q}{J} = \frac{L^{21}}{L^{11}} = T\frac{L^{12}}{L^{11}} = TQ, \tag{9.15}$$

and the thermopower is given by

$$Q \sim -\frac{1}{2}\left(\frac{k_B}{e}\right)\left(\frac{\Delta}{k_B T}\right), \tag{9.16}$$

which can be large since Δ can be on the order of an electron volt. This effect has commercial applications in thermoelectric cooling devices.

Exercise 9.1. (i) Show that, in the limit $\epsilon - \mu \gg k_B T$, the Fermi–Dirac distribution function Eq. (7.255) reduces to the Boltzmann distribution: $f^\circ(\epsilon) \approx e^{-(\epsilon-\mu)/k_B T} \ll 1$. (ii) Now consider the opposite limit of $\mu - \epsilon \gg k_B T$. Now $f^\circ(\epsilon) \to 1$. Show that $1 - f^\circ(\epsilon) \approx e^{(\epsilon-\mu)/k_B T} \ll 1$. In this case $1 - f^\circ(\epsilon)$ is naturally interpreted as the occupation number of a hole state, with $-\epsilon$ being its energy and $-\mu$ being the chemical potential of the hole. These are all consistent with the discussions of Chapter 8, where the concept of holes is introduced.

Exercise 9.2. (i) Assuming that the conduction and valence bands have isotropic but different effective masses, and no valley degeneracy, determine the T-dependent coefficients in front of the exponentials which we omitted form Eqs. (9.4) and (9.5) in the limit $T \to 0$, and show that Eq. (9.6) is valid despite the difference between conduction and valence bands. (ii) Generalize (i) to the case of anisotropic effective mass and to multi-valley cases (including the possibility that the valley degeneracy may be different for electrons and holes), and show that Eq. (9.6) is still valid. In both parts it would be useful to first calculate the density of states near the conduction band bottom and valence band top, in terms of the eigenvalues of the effective mass tensors.

9.2 Impurity Levels

Because of their technological importance, much effort was devoted in the twentieth century to techniques for the production of nearly perfect crystals of semiconductors, particularly Si. In practice, however, there always exist some chemical impurities in any crystal, and these can have dramatic effects in semiconductors – some of which are highly useful. In particular, impurities (often called **dopants**) can introduce extra charge carriers that go predominantly into either the conduction band (as electrons) or the valence band (as holes), thus dramatically increasing the conductivity. In this case we have an "**extrinsic**" semiconductor, in contrast to a pure undoped semiconductor (which is called "**intrinsic**"), where the only charge carriers are thermally excited electrons and holes that come in equal density. For technological purposes (to be described later) one often introduces impurities intentionally and in an extremely well-controlled manner, so that the semiconductors acquire various desired properties. To understand this physics, we need to study the electronic states that come with the impurities.

Let us start with a specific example. Consider an elemental semiconductor like Si or Ge, made of group-IV atoms. Now consider a group-V impurity, say P, that substitutes for one of the host atoms. The biggest effect of this impurity is that it contributes an extra electron that *cannot* be accommodated in the (covalently bonded states of the) valence band(s). As a result, we expect the electron to go to the conduction band, where it can contribute to transport, and for this reason such an impurity is called a donor (since it donates an electron). This, however, does not happen at $T = 0$. This is because the donor atom also has an ion (nucleus plus core electrons) with one extra (positive) charge compared with the host atom, as a result of which this extra electron is bound to the donor in the ground state. But, as we will see shortly, it is much easier to excite this bound electron into the conduction band (than it is to excite electrons from valence bands into the conduction band), because the binding energy ϵ_b is much smaller than the band gap Δ. Let us find out what ϵ_b is.

If this extra electron were living in vacuum and were bound to a unit nuclear charge, then we would simply have a hydrogen-atom problem and would find $\epsilon_b = 1$ Ryd. $= m_e e^4/(2\hbar^2) \approx 13.6\,\text{eV}$, and the binding length a_b would be the Bohr radius $a_B = \hbar^2/(m_e e^2) \approx 0.53\,\text{Å}$, where m_e is the electron mass in vacuum. The host semiconductor, however, has two important effects on the quantum-mechanical motion of this electron, both of which tend to make ϵ_b smaller and a_b bigger. The first is the large dielectric constant ϵ of the semiconductor – which is typically between 10 and 20 because of the relatively small band gap.[10] This reduces the strength of the Coulomb attraction, by replacing e^2 with e^2/ϵ. The other effect is band structure. We have already discussed the idea that, since the valence bands are already full, the states available to this extra electron are those in the conduction band.[11] If this electron lives near the band bottom (an approximation that is valid if the binding length is much larger than the lattice constant), then the effective-mass approximation would be appropriate, in which case we need to replace m_e by the effective mass m^* in the expressions for ϵ_b and a_b, which further

---

[10] A heuristic way to understand the relation between the dielectric constant ϵ and the band gap Δ is that, if the latter were zero, we would have a metal which exhibits perfect screening ($\epsilon \to \infty$). On the other hand, vacuum (with $\epsilon = 1$) may be viewed as a band insulator with $\Delta = 2m_e c^2$ (namely electron–positron creation energy; a positron is nothing but a hole in a very deep valence band!). Extrapolating between these two points tells us that the smaller Δ is, the bigger ϵ becomes. The underlying physics is that the dielectric constant describes screening of charges via the polarization of the material due to the virtual excitation of particle–hole pairs across the excitation gap.

[11] This does not necessarily mean that the electron has to be in an unbound or scattering state. By making linear combinations of the conduction band states the system can form bound states, just like bound states in vacuum are formed by linear combinations of plane waves. Another way to understand this is to view the bound state as a *stationary* wave packet formed by states in the conduction band.

reduces ϵ_b while increasing a_b, since in semiconductors we generally have $m^* \ll m_e$. Combining these two effects, we find[12]

$$\epsilon_b = \frac{m^*}{m_e} \frac{1}{\epsilon^2} \, \text{Ryd} \sim 0.01 \, \text{eV} \ll \Delta, \tag{9.17}$$

$$a_b = \frac{m_e}{m^*} \epsilon a_B \gg a_B. \tag{9.18}$$

The results, namely that ϵ_b is small compared with both the band gap and the band width, and a_b is much larger than the lattice spacing, justify *a posteriori* the usage of the effective-mass approximation and the continuum description that comes with it. The fact that the effective Rydberg constant (donor binding energy) is so small means that many donors will be ionized even at room temperature.

The existence of such a bound state shows up as a delta-function peak in the electron density of states *inside* the band gap.[13] This donor level $\epsilon_d = \epsilon_c - \epsilon_b$ is much closer to the conduction band edge than the valence band edge; see Fig. 9.3(a) for an illustration.[14] As $T \to 0$, the chemical potential $\mu = \epsilon_F$ is between ϵ_d and ϵ_c, thus much closer to the conduction band than the valence band. As T increases in the presence of a finite density of donors, μ will decrease but stay close to ϵ_d, until

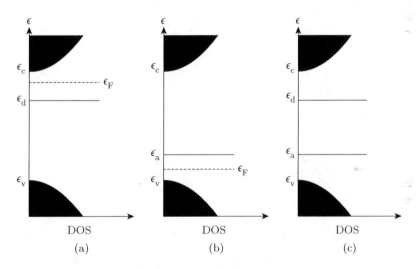

Figure 9.3 Illustration of the electronic density of states (DOS) and impurity levels in semiconductors. (a) An n-type semiconductor with a donor level at ϵ_d. (b) A p-type semiconductor with an acceptor level at ϵ_a. (c) A compensated semiconductor with both donor and acceptor levels.

[12] Appropriate modification for anisotropic effective mass is needed for these expressions.

[13] In quantum mechanics one is used to associating bound states with attractive potentials and with negative-energy eigenvalues. In general, bound states give rise to discrete energy spectra, while continuum spectra are associated with scattering states. This means in a crystal that the bound state energy must be in between bands, i.e. inside a band gap. The mid-gap bound states we encountered in Section 7.6 are also examples of this. Another somewhat counter-intuitive fact is that a repulsive impurity potential can also give rise to bound states. This is the case for an acceptor impurity, and will be discussed later.

[14] In this case, the bound state is (mostly) formed by states in the conduction band. As a result, there is one fewer scattering state in the conduction band, while the numbers of states in other bands are not affected. The acceptor level near the top of the valence band (to be discussed below) is mostly formed by states in the valence band, as a result of which there is one fewer scattering state in the valence band. On the other hand, if a bound state is deep inside a band gap, it must involve states from multiple bands. Again the mid-gap bound states we encountered in Section 7.6 are good examples of this.

most of the electrons bound to the donors have been thermally excited into the conduction band. At this point very few holes exist in the valence band, thus the (mobile) charge carriers in the system are dominantly electrons. As a result, we call a semiconductor with dominantly donor impurities an n-type semiconductor (n standing for negative charge carrier).

Now consider having a group-III impurity (say Al), which has one fewer valence electron than the host atoms. This creates a hole, which is bound to the impurity called an acceptor (because it takes an electron from the system, thus creating a hole, in contrast to what a donor does). This shows up as an acceptor level right above the valence band edge, and the difference between them is the binding energy of the hole. See Fig. 9.3(b) for an illustration. When there is a finite density of acceptors in the semiconductor, the chemical potential stays much closer to the valence band than the conduction band, thus holes dominate and the semiconductor is called p-type (for positive charge carrier).

In general, if an impurity atom has higher valence than the host atom it substitutes, it acts as a donor, whereas if it has lower valence it acts as an acceptor. When both donors and acceptors are present, then we have both donor and acceptor levels present, a situation illustrated in Fig. 9.3(c). In this case we say that the semiconductor is compensated. The properties of a semiconductor are quite sensitive to whether it is compensated or not, and to the degree of compensation (namely how well the donor and acceptor densities match).

Exercise 9.3. (i) Assume that the conduction and valence bands have isotropic but different effective masses, and no valley degeneracy. There is a donor level with binding energy $\epsilon_b \ll \Delta$, and the density of donors is n_d. Calculate the chemical potential as a function of T at low temperature ($k_B T \ll \Delta$), and determine the temperature scale T^* above which μ moves away from the donor level ϵ_d toward the center of the band gap. (ii) For $T < T^*$, calculate the electron and hole densities $n_c(T)$ and $p_v(T)$, and compare them with the intrinsic case at the same temperature. In particular, compare the product $n_c(T) p_v(T)$ in the two cases. The "law of mass action" states that the product should be independent of the position of the chemical potential and hence the same in the two cases.

Exercise 9.4. In the text we showed that a donor gives rise to a bound state whose energy is slightly below the conduction band edge, by mapping the Hamiltonian to a hydrogen-atom-like problem. Use similar arguments to show that an acceptor gives rise to a bound state whose energy is slightly *above* the valence band edge, and relate the energy difference between the acceptor level and the valence band edge to the hole effective mass (assumed to be isotropic) and dielectric constant of the host semiconductor.

Exercise 9.5. A bound state below the conduction band edge caused by the presence of the donor potential is made up of a linear combination of conduction band states. The purpose of this exercise is to show that the total number of states above the bottom of the conduction band is reduced by one for each bound state, so that the grand total number of states in the band remains the same as it is in the absence of the donor potential. (We neglect mixing between bands.) As a simple illustration of the basic physics, consider a 1D attractive delta-function potential in free space,

$$H = -\frac{\hbar^2}{2m}\frac{d^2}{dx^2} - \lambda\delta(x), \tag{9.19}$$

where $\lambda > 0$ is a constant.

 (i) Find the (single) bound state $\psi^{\mathrm{b}}(x)$ and its binding energy.

 (ii) Find the positive-energy scattering states $\psi_k^{\mathrm{e,o}}(x)$, where (e,o) stand, respectively, for even and odd spatial parity.

(iii) The completeness relation states that

$$\psi^{\mathrm{b}}(x)\psi^{\mathrm{b}}(x') + \sum_k \left\{\psi_k^{\mathrm{o}}(x)\psi_k^{\mathrm{o}}(x') + \psi_k^{\mathrm{e}}(x)\psi_k^{\mathrm{e}}(x')\right\} = \delta(x - x'). \tag{9.20}$$

Pretend that you don't know the bound state wave function, only the scattering states. Compute the scattering state contribution to the completeness relation and use this to derive the bound state wave function. Explicitly show that, for a $\lambda < 0$ where there is no bound state, the scattering states are actually complete.

(iv) Using the scattering phase shifts (of the even and odd parity states), θ_k, count the change in the total number of states in the continuum due to the impurity potential. Show that this change exactly compensates for the formation of the bound state, so the total number of states remains the same. Hint: use the method described in Section 15.14.1 to count the change in the total number of scattering states. Note that odd-parity states are unaffected by the delta-function potential.

9.3 Optical Processes in Semiconductors

In earlier chapters we discussed how to probe electronic properties of solids by measuring the dc conductivity, which is the electrons' response to a static electric field. Another widely used method is studying the systems' optical properties, which probe the electrons' response to high-frequency electromagnetic waves shining on them. This is the focus of this section. While we phrase our discussion in the context of semiconductors, many of the ideas apply to metals as well.

In Section 2.3 we developed the quantum theory of the electromagnetic coupling of electrons (the electron–photon interaction). We found that, through a first-order process involving the ΔH_1 term in Eq. (2.31), an electron can absorb or emit a photon, and change its state. Such a single-photon absorption/emission process must conserve both energy and crystal momentum. We note that, for optical photon absorption, the electronic energy change is typically of order $1\,\mathrm{eV}$ or less, and the corresponding photon momentum $|\Delta\epsilon|/c$ is much smaller than the characteristic lattice momentum of order $\hbar/\text{Å}$ and thus negligible (because the speed of light is so high).

Assuming the crystal is in the ground state, it can only absorb photons. In the following we discuss processes that involve a single-photon absorption.

- Interband transitions. As illustrated in Fig. 9.4, an electron can absorb a photon, and jump from the valence band to the conduction band. Obviously the photon energy has to match or exceed the band gap Δ for energy conservation to be satisfied. Thus, in principle, an absorption experiment can be used to measure Δ. But, as we will see shortly, there are several complications with this idea. The simplest case is that of a direct-band-gap semiconductor, in which the lowest photon energy needed is indeed Δ, which excites an electron from the top of the valence band to the bottom of the conduction band, as illustrated in Fig. 9.4(a). Since the photon momentum is (essentially) zero, momentum

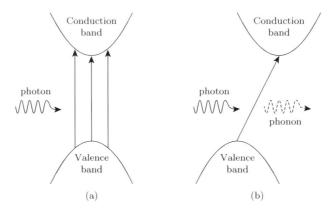

Figure 9.4 (a) Interband transition process in a direct-band-gap semiconductor, in which an electron absorbs a single photon. (b) Interband transition process in a semiconductor with an indirect band gap, in which a phonon is emitted or absorbed in addition to the single-photon absorption.

conservation is also satisfied. For an indirect-band-gap semiconductor, the corresponding process does *not* conserve lattice momentum. As a result, a phonon must be emitted or absorbed (the latter is possible at finite temperature) to compensate for the difference in lattice momenta of the electronic states at valence band top and conduction band bottom.[15] This is illustrated in Fig. 9.4(b). Since one more particle is involved, the photon energy threshold is no longer the same as Δ, and the matrix element tends to be much smaller for such a process.

- Exciton generation. In the discussion above, we neglected the interaction between the electron excited into the conduction band and the hole it leaves behind. In reality the electron and the hole attract each other due to the opposite charges they carry. Such attraction has no effect on the energy of the final state if the electron and hole are not bound together. However, there do exist bound states of the electron–hole pair, whose energies are lower than Δ. Using considerations similar to the discussions on impurity states in the previous section, we find the exciton binding energy (which is the difference between Δ and the energy of the excitonic state) and binding length to be

$$\epsilon_{\rm b} = \frac{\mu^*}{m_{\rm e}} \frac{1}{\epsilon^2} \, {\rm Ryd}, \tag{9.21}$$

$$a_{\rm b} = \frac{m_{\rm e}}{\mu^*} \epsilon a_{\rm B}, \tag{9.22}$$

where

$$\mu^* = \frac{m_{\rm e}^* m_{\rm h}^*}{m_{\rm e}^* + m_{\rm h}^*} \tag{9.23}$$

is the reduced mass of the electron–hole pair. As noted previously, the effect of the background dielectric constant ϵ of the material is to significantly reduce the binding energy scale relative to the Rydberg constant and increase the binding length scale relative to the lattice constant, thereby justifying the use of the effective-mass approximation.

In an ideal situation we expect to see contributions from interband transitions and excitons in an optical absorption experiment, as illustrated schematically in Fig. 9.5. In reality, however, due to the presence of impurities, the involvement of phonons, and many other complications, optical absorption data is not always straightforward to interpret.

---

[15] Since the speed of sound (or phonon velocity) is much smaller than the speed of light, a phonon typically has much lower energy than a photon but much higher (lattice) momentum!

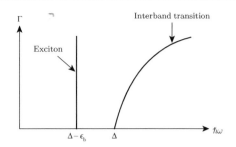

Figure 9.5 Schematic illustration of the optical absorption rate Γ as a function of the photon energy $\hbar\omega$. The delta-function-like peak is due to excitons (which can come in different energies; here we assume a single energy for simplicity), and the continuum starts at band gap Δ. The exciton Rydberg series contains an infinite number of hydrogen-like bound states below the band gap that affect the absorption spectrum near Δ, but are not shown here.

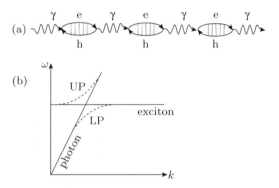

Figure 9.6 (a) coherent conversion of a photon (γ) into an electron–hole pair (e–h) bound into an exciton via repeated Coulomb interaction (dotted lines). (b) The dispersion relation $\omega = (c/n)k$ for a photon crosses the dispersion relation for the exciton. Because the speed of light c/n is so high, the two curves cross at very small wave vectors, and the variation of the exciton energy with the center-of-mass momentum $\hbar k$ is negligible on the scale of the figure. Hence the parabolic dispersion of the exciton energy appears to be flat. The dotted lines indicate the avoided crossing of the energy levels as the photon and exciton hybridize to form the upper (UP) and lower (LP) polariton branches. Away from this crossing region, the photon and exciton have well-defined and separate identities.

In addition, the absorption spectrum in Fig. 9.5 is overly simplified because, in fact, excitons are collective modes in crystals that can propagate coherently. Since they can be created from photons and the reverse of that process, namely annihilation of an electron–hole pair to generate a photon, is also allowed, the exciton and photon modes can convert back and forth into each other as illustrated in Fig. 9.6(a). The resulting collective mode is a coherent superposition of the photon and exciton known as an **exciton-polariton**,[16] and has the dispersion relation illustrated in Fig. 9.6(b). Exciton-polaritons are effectively bosons and, in special circumstances, can undergo Bose–Einstein condensation, as we will discuss in Chapter 18.

- Cyclotron resonance. As was mentioned in Chapter 8 and will be further discussed in great detail in Chapter 12, a uniform magnetic field re-organizes electronic states into Landau levels. Transitions between Landau levels induced by photons show up in optical absorption experiments as resonances whose energy matches that of the Landau level spacing. See Fig. 9.7 for an example. These are the

---

[16] Note that this collective mode should be distinguished from the phonon-polariton we encountered in Chapter 6.

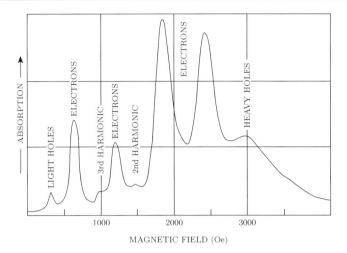

Figure 9.7 A typical optical absorption spectrum of germanium as a function of the magnetic field strength, for a fixed infrared light frequency and generic field direction. We see one resonance peak for light holes and another for heavy holes, and four electron resonance peaks from the four minima of the electron band. Traces of higher-harmonic contributions are also visible. Figure adapted with permission from [49]. Copyright (1956) by the American Physical Society.

cyclotron resonances discussed from a semiclassical viewpoint in Chapter 8. There we already saw that the resonance frequency tells us about the (isotropic) effective mass of electrons. In semiconductors, the effective mass is often anisotropic. In these cases, the resonance frequency depends not only on the strength of the magnetic field, but also on its direction. A detailed study of the direction dependence of the resonance frequency allows one to determine the entire effective-mass tensor.

Exercise 9.6. Consider the conduction band of a semiconductor, with a single minimum at the Brillouin zone center. Choose the coordinate system such that the effective-mass tensor is diagonal, with eigenvalues m_x, m_y, and m_z, respectively. Find the cyclotron frequency with a magnetic field B applied along the x, y, and z directions, respectively.

9.3.1 Angle-Resolved Photoemission Spectroscopy

In the optical processes discussed thus far, one can measure the energy *difference* between the initial and final states of the electron (through the photon energy that is absorbed), but neither of these two states is separately and directly probed. As a result, we do not know the *absolute* energy of either of them. In an **angle-resolved photoemission spectroscopy (ARPES)** measurement, an electron is actually knocked out of the crystal after absorbing a photon, and the outgoing electron's momentum (and sometimes its spin) is carefully measured. See Fig. 9.8(a) for an illustration. As a result, we know not only the energy and momentum of the photon absorbed, but also those of the final state of the electron.[17] From conservation of energy and momentum, we can deduce the initial state the electron was in before absorbing the photon. Such detailed spectroscopic information allows one to directly probe the crystal's band structure, as we discuss below.

---

[17] The energy and momentum resolution of ARPES has dramatically improved in recent decades. One of the motivations for the push toward higher resolution was the discovery of high-temperature superconductors and the desire to directly observe the d-wave structure of the superconducting gap function (see Section 20.9.3).

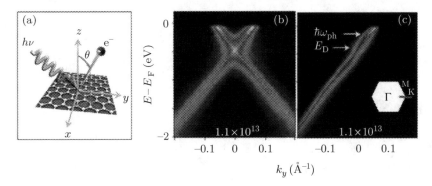

Figure 9.8 (a) Schematic illustration of angle-resolved photoemission spectroscopy (ARPES). A high-energy photon with energy $h\nu$ hits a crystal (in this case graphene) and gets absorbed by an electron, which then leaves the crystal. The photoelectron's momentum vector is carefully measured. Graphene's band dispersion perpendicular (b) and parallel (c) to the horizontal bar near the K point in the inset of panel (c), revealed through an intensity (represented by grayscale) plot of ARPES. Both scans go through the K point (which is one of the Dirac points, whose energy is indicated by E_D in panel (c)). The linear dispersion is clearly visible. Deviation from linearity near the phonon frequency $\hbar\omega_{ph}$ (as indicated in panel (c)) may be due to the electron–phonon interaction. Figure adapted with permission from [50]. Copyright (2014) by the American Physical Society.

The photoelectron's energy is

$$E = \frac{\hbar^2}{2m_e}(K_\perp^2 + \vec{K}_\parallel^2) = h\nu + \epsilon(k_\perp, \vec{k}_\parallel), \tag{9.24}$$

where $h\nu$ is the photon energy, $\vec{K}$ is the momentum of the electron after absorbing the photon (and escaping the crystal), $\vec{k}$ is the (crystal) momentum of the electron in its initial state before absorbing the photon, and the indices $\perp$ and $\parallel$ denote their components perpendicular and parallel to the crystal's surface, respectively. In an ARPES experiment $h\nu$ is controlled and $\vec{K}$ is carefully measured. If the surface is sufficiently clean and flat, then the parallel components of lattice momentum are conserved due to the corresponding (reduced) lattice translation symmetry:

$$\vec{K}_\parallel = \vec{k}_\parallel + \vec{G}, \tag{9.25}$$

where $\vec{G}$ is any one of the 2D reciprocal lattice vectors corresponding to the 2D lattice translation symmetry in the presence of the surface. We then find that Eq. (9.24) tells us something about the band dispersion:

$$\epsilon(k_\perp, \vec{k}_\parallel) = \frac{\hbar^2}{2m_e}[K_\perp^2 + (\vec{k}_\parallel + \vec{G})^2] - h\nu. \tag{9.26}$$

Since the presence of the surface breaks (lattice) translation symmetry in the direction perpendicular to the surface, $k_\perp$ cannot be deduced from the final state, and some approximation on it is needed in order to obtain the band dispersion. Fortunately, in a 2D system like graphene, $k_\perp$ does not exist and Eq. (9.26) allows complete determination of the band structure, at least for the states that are occupied by electrons. Another important example of 2D band structure determined by ARPES is the surface states of topological insulators, which we will encounter later.

As discussed above, ARPES probes the part of bands that are occupied by electrons. The inverse of it, anti-ARPES or inverse photoemission, probes the empty part of the band structure. In this process an electron with controlled energy and momentum is injected into the crystal, enters an empty state in the crystal, and emits a photon that is detected.[18] Thus the combination of ARPES and anti-ARPES in principle allows complete determination of the band structure. They also reveal physics related to electron–electron and electron–phonon interactions. For reasons discussed above, they are ideal for measurements of 2D band structures, including those of 2D crystals or surface states (bands) of 3D crystals. They are also widely used to measure 3D band structures, although care must be taken to ensure that what is seen are intrinsic bulk (instead of surface) properties of the crystals.

9.4 The p–n Junction

What we have considered thus far are *homogeneous* semiconductors, either pure or doped with *uniform* dopant type and density. In this and the following sections we will study devices made of different types of semiconductors (with differences in host material and/or dopant type/density), with sophisticated structures (often involving other materials) designed to achieve specific functionality both for fundamental research and for technological applications. Different materials have different chemical potentials in isolation. When they are grown together or put in contact with each other, charge redistribution often occurs, so that the electrochemical potential[19] is uniform and the whole system is in equilibrium. Such charge redistribution can have a significant effect on the local environment of the charge carriers (electrons or holes). This is behind much of the physics responsible for the functionality of many semiconductor-based devices. The simplest example is the p–n junction (**semiconductor diode**), to which we now turn.

We have seen that the chemical potential and carrier type/densities in a semiconductor can be altered by doping, i.e. by adding appropriate impurities. Quite dilute concentrations of impurities are sufficient to have a large effect. More specifically, in an n-type semiconductor, the chemical potential is near the donor level or conduction band edge, whereas in a p-type semiconductor, the chemical potential is near the acceptor level or valence band edge, in an appropriate temperature range. When we put two such semiconductors together (with the same host material, often grown in a single process but with different dopants introduced in different regions at different stages of growth), we have a p–n junction, and charge motion across the junction is necessary to reach equilibration, leading to interesting physical effects.

The simplest example is an abrupt p–n junction, for which

$$N_{\mathrm{d}}(x) = \begin{cases} N_{\mathrm{d}} & x > 0 \\ 0 & x < 0, \end{cases} \tag{9.27}$$

$$N_{\mathrm{a}}(x) = \begin{cases} 0 & x > 0 \\ N_{\mathrm{a}} & x < 0, \end{cases} \tag{9.28}$$

where N_{d} and N_{a} are the donor and acceptor densities, respectively. For $x < 0$ we have a p-type semiconductor, for $x > 0$ we have an n-type semiconductor, and the interface or junction is at $x = 0$. If there were no charge redistribution across the interface the (electro)chemical potential would be lower for $x < 0$ (p-side) and higher for $x > 0$ (n-side); see Fig. 9.9(a). This leads to electrons moving from the n-side to the p-side, resulting in a change of the (local) electric potential, until the electrochemical potential is constant throughout the system; see Fig. 9.9(b). Since we are interested in the physics of devices at sufficiently low temperature, we use electrochemical potential and Fermi energy (ϵ_{F}) interchangeably in this chapter.

---

[18] In fact Fig. 9.8(a) could also illustrate this process, if one understands that the electron is moving into the crystal and the photon is coming out and detected.

[19] The electrochemical potential defined in Eq. (8.151) is the sum of the *local* chemical potential (say, measured from the local conduction band edge) and the local electric potential energy of the electrons.

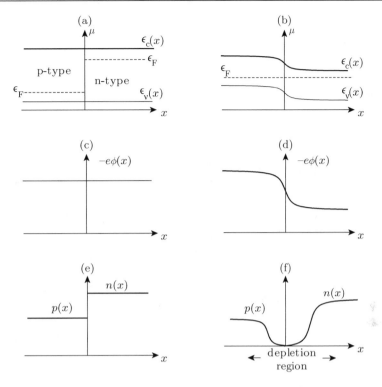

Figure 9.9 Illustration of a p–n junction before (left panels) and after (right panels) equilibration. Top row: chemical potential as a function of position; middle row: electrostatic potential energy as a function of position; bottom row: (number) density of free carriers (holes or electrons) as a function of position. Note that the acceptor density for $x < 0$ happens to be somewhat lower than the donor density for $x > 0$.

The charge redistribution has a number of physical effects, two of which are particularly important. The first is the band bending (namely the spatial dependence of the band edge) visible in Fig. 9.9(b), due to the non-uniform electric potential developed in response to charge redistribution (see Fig. 9.9(d)).[20] This effect will be discussed in greater detail in the following sections. The second effect is the electron motion from the n-side to the p-side, resulting in *annihilation* of electrons and holes near the interface, and the formation of a **depletion layer** (see Fig. 9.9(f)) in which the (unbound) carrier densities are significantly reduced from those far away from the interface. As a result, the local resistivity is much higher in the depletion layer, and the resistance of the whole device is often dominated by this region. Also, the total net charge density (from the combination of charged dopant ions and the reduced number of carriers) is non-zero in this region due to the reduction of the carrier density. Thus the variation in the electric potential is also concentrated in this transition region (compare Figs. 9.9(b) and (d)). These two quantities are related through Poisson's equation:

$$\frac{d^2\Phi(x)}{dx^2} = -4\pi\rho(x)/\epsilon, \qquad (9.29)$$

where ϵ is the dielectric constant and

$$\rho(x) = e[\delta p_v(x) - \delta n_c(x)] \qquad (9.30)$$

[20] Here we assume that the electric potential varies smoothly on the scale of the lattice spacing, so that bands are well-defined locally. This is the same condition as that for the validity of semiclassical approximations in Chapter 8.

is the *deviation* of the local carrier charge density from that of isolated p- or n-semiconductors. The bigger its magnitude, the lower the local carrier density and conductivity. While detailed microscopic analysis is needed to solve for $\Phi(x)$, we know the difference between $e\Phi(x \to \infty)$ and $e\Phi(x \to -\infty)$ must match the chemical potential difference between the n-type and p-type semiconductors *before* equilibration:

$$e\,\Delta\Phi = e\Phi(\infty) - e\Phi(-\infty) = \mu_n - \mu_p = e\,\Delta\Phi_{\text{equ}} \approx \epsilon_c - \epsilon_v = \Delta > 0. \qquad (9.31)$$

The equation above defines $\Delta\Phi_{\text{equ}}$, the electric potential difference at equilibrium (where the electrochemical potential is constant throughout the sample).

The most important functionality of a p–n junction is rectification, namely the resistance for *forward bias* (higher voltage on the p-side) is lower than for *reverse bias* (higher voltage on the n-side). Qualitatively this is quite easy to understand by extending the analysis above to the *non-equilibrium* situation with an applied voltage difference. The electric potential difference is now

$$e\,\Delta\Phi = e\,\Delta\Phi_{\text{equ}} - eV, \qquad (9.32)$$

which is smaller for positive V and larger for negative V. This means that the transition region for Φ, which is also the depletion region for charge carriers, is narrower for positive V and wider for negative V; see Fig. 9.10. In terms of carriers, for positive bias, holes from the p-doped region and electrons from the n-doped region both move *toward* the junction and thus populate (and narrow) the depletion layer, reducing its resistance. For negative bias the exact opposite happens: holes from the p-doped region and electrons from the n-doped region both move *away* from the junction and thus further deplete (and widen) the depletion layer, increasing its resistance. See Fig. 9.10 for an illustration. Since the depletion region dominates the resistance, rectification follows.

The p–n junction is a basic building block for many important electronic devices. In addition to their use in semiconductor diodes, which are simply p–n junctions with two leads connected to them (making them two-terminal devices, see Fig. 9.11(a)), p–n junctions are also the most important components in a **bipolar transistor**, which is basically a p–n–p or n–p–n junction connected to three leads (thus a three-terminal device, see Fig. 9.11(b)). The middle region (called the base) has a very narrow width, comparable to that of the depletion layer of a single p–n junction. If the base were not connected to an external terminal, then the two p–n junctions connected in series would always have opposite biases and very little current could flow through. The terminal connected to the base can

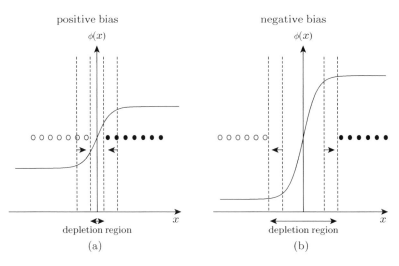

Figure 9.10 The electric potential profile of a p–n junction, with positive (a) and negative (b) voltage biases.

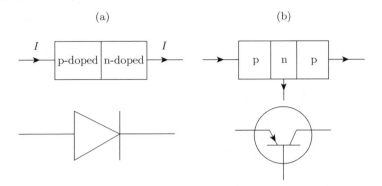

Figure 9.11 (a) A semiconductor diode and its corresponding circuit diagram. (b) A bipolar transistor of type p–n–p (as opposed to n–p–n) and its corresponding circuit diagram.

inject current into and out of it, as well as control its voltage. It turns out that, because the base is very thin and excess carriers can diffuse through it, the current-carrying capability of the depletion layer is very sensitive to the voltage biases and carrier densities. As a result, the current going though the transistor is very sensitive to small currents injected into the base through the terminal connected to it. Transistors can be used as electric switches and amplifiers of electric signals, a topic which will be discussed further below when we discuss another type of semiconductor-based transistor. In the following we discuss two electro-optical devices, which are also based on p–n junctions.

9.4.1 Light-Emitting Diodes and Solar Cells

Semiconductor diodes and transistors are based on the rectification and other electric transport properties of p–n junctions. It turns out they also have very special optical properties that can be used to build electro-optical devices that turn electric energy/signal into optical energy/signal, and vice versa.

For over a century since Edison's original invention, most light bulbs have been built on the same physical mechanism, namely using electric energy to heat up the filament, and harness the visible light from its blackbody radiation. This electricity → heat → visible-light process is very inefficient, as visible light only makes up a very narrow window in the blackbody radiation spectrum (even for optimal temperature). Obviously one could improve the efficiency by turning electric energy directly into visible light. One class of such devices is light-emitting diodes (LEDs), as illustrated in Fig. 9.12(a). When a p–n junction is positively biased, holes from the p-doped region and electrons from the n-doped region both move into the junction region. Some of them annihilate each other and emit a photon as a result.[21] The photon energy is very close to the band gap Δ. Since for semiconductors we typically have $\Delta \sim 1$–$2\,\mathrm{eV}$, it falls within or is very close to the visible-light range. By choosing a semiconductor with an appropriate band gap, one can control the color of the light emitted, which is also important in many devices. In addition to having an appropriate Δ, another strong preference is using direct-band-gap semiconductors. This is because, in an indirect-band-gap semiconductor, the electron–hole annihilation process must involve phonon emission or absorption in order to have conservation of both energy and momentum, and this reduces the efficiency (rate) of this process.

Historically the first LED, actually emitting in the infrared, was based on a GaAs p–n junction, and was built in the 1950s. Later red, yellow, and green LEDs were built using semiconductors with successively larger band gaps. Blue LEDs, needed to cover the entire visible-light range (so that

---

[21] This is the inverse of the single-photon absorption process illustrated in Fig. 9.4.

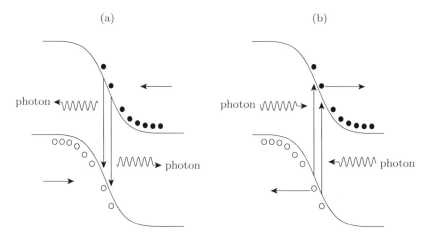

Figure 9.12 Illustrations of the physical mechanisms behind (a) light-emitting diodes (LEDs) and (b) solar cells.

LED-based lamps can emit white light, for example), pose a special challenge. The bigger band gap required strongly limits the semiconductor materials that can be used. A natural candidate is GaN, whose band gap is about 3.4 eV at room temperature, which is perfect for this purpose. The difficulty, however, lies in doping it with acceptors to make the material p-type. The breakthrough in this endeavor came in the 1990s, and the three scientists (Akasaki, Amano, and Nakamura) responsible for it were awarded the Nobel Prize in Physics in 2014.

LEDs turn electric energy into light efficiently. Solar cells (or photovoltaic devices) do the exact opposite: turning light (usually from the sun, but not necessarily visible light) into electric energy. The basic physical process is the reverse of that in LEDs, and is illustrated in Fig. 9.12(b): light (or photons) generates electron–hole pairs, the electrons and holes are pulled in opposite directions by the electric field that exists in the junction (see Fig. 9.10, and note that the (negative) gradient of the electric potential is the electric field), resulting in photo-electric current or voltage, depending on whether the circuit is closed or open. Many considerations related to those for LEDs also enter the design of the most efficient solar cells.

It should be clear that different types of photo-electric detectors or sensors, that turn electric signals into photo-signals and vice versa, can be built on the same principles.

9.5 Other Devices

9.5.1 Metal–Oxide–Semiconductor Field-Effect Transistors (MOSFETs)

We have seen that variation of the electric potential in a semiconductor leads to spatial variation of band edges, or band bending. In the case of a p–n junction, the chemical potential (or Fermi energy) remains inside the band gap throughout the system. Creating stronger band bending through the application of large bias voltages can pull the conduction band edge *below* the Fermi energy in certain regions, leading to a *degenerate* electron gas[22] that is *confined* in space. This opens the door for basic science research on electron systems in reduced dimensionality, as well

[22] Degenerate means that the density is high enough that the Pauli principle is important: the Fermi energy is larger than the thermal energy. For semiconductor inversion layers, it may be necessary to go to cryogenic temperatures to achieve strong degeneracy.

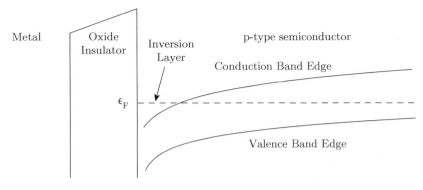

Figure 9.13 An illustration of a metal–oxide–semiconductor field-effect transistor (MOSFET).

as numerous electronic devices based on this piece of physics. One of the most important examples is the **metal–oxide–semiconductor field-effect transistor (MOSFET)**, which is illustrated in Fig. 9.13.

Consider p-type Si, whose Fermi energy is inside the band gap but close to the valence band edge. It terminates on an oxide insulator (e.g. SiO_2) with a very large band gap, that electrons cannot penetrate into. Behind this oxide barrier is a metallic gate on which one can apply a voltage bias. When a positive voltage is applied to the metallic gate, the bands bend downward as one approaches the semiconductor/oxide interface. If the bias is large enough, the conduction band edge is pulled *below* the Fermi energy within a thin layer known as the inversion layer (because the carriers are now electrons instead of holes). We thus obtain a degenerate electron gas that is confined to this (2D) inversion layer. Because of the finite thickness of the inversion layer, electron motion in the direction perpendicular to the interface is confined and forms discrete "particle in a box" quantum levels (known as **electric subbands**), while the motion in the plane parallel to the interface is "free" in the sense that the solutions of the Schrödinger equation form extended Bloch waves. Very often, only the lowest electric subband level is below the Fermi energy and populated (at very low temperature). Because of the quantum excitation gap for motion perpendicular to the interface, in this limit electron motion in the conduction band is completely frozen along this direction, and confined to the interface. We thus have a genuine **2D electron gas (2DEG)**, even though the inversion layer is *not* atomically thin (as it is in graphene).[23] Obviously the conductivity of the system is very sensitive to whether this inversion occurs or not, which can be controlled by the bias voltage or electric field. This property can be used to build devices that switch conduction on and off depending on a voltage signal, thus the name **field-effect transistor (FET)**.

The MOSFET is one of the most important devices for large integrated circuits such as those used as microprocessors and semiconductor memories [51]. Our focus in this book, however, will be on the fascinating fundamental science of 2DEGs realized in this and other systems. A good example is the integer quantum Hall effect to be discussed in Chapter 12, which was first discovered by von Klitzing in a Si MOSFET.

9.5.2 Heterostructures

The devices we have discussed thus far involve a single (host) semiconductor. More sophisticated devices, known as heterostructures, involve different types of semiconductors. GaAs and AlAs are

---

[23] Because of the excitation gap between electric subbands, the dynamics is truly 2D. The only effect of the finite extent in the third dimension is to slightly modify the interactions by softening the Coulomb repulsion between electrons at short distances.

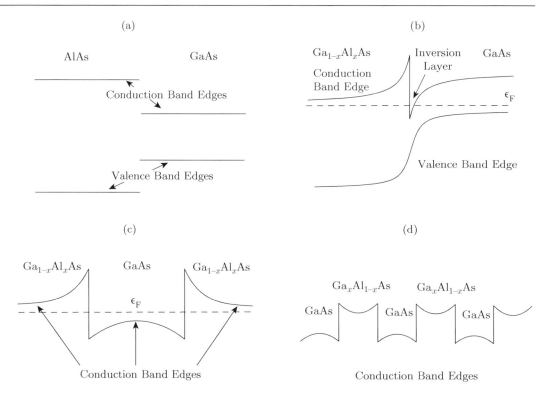

Figure 9.14 Heterostructures. (a) A GaAs/AlAs heterojunction without band bending. (b) The formation of an inversion layer in a GaAs/Ga$_{1-x}$Al$_x$As heterojunction due to band bending. (c) A Ga$_{1-x}$Al$_x$As/GaAs/Ga$_{1-x}$Al$_x$As quantum well. (d) A superlattice formed by periodically repeating Ga$_{1-x}$Al$_x$As/GaAs/Ga$_{1-x}$Al$_x$As quantum wells.

both III–V semiconductors with zincblende structure, with almost identical lattice constants (5.63 Å and 5.62 Å, respectively). It is thus relatively easy [using **molecular beam epitaxy (MBE)**] to grow one material on top of the other with a sharp and flat interface.[24] On the other hand, they have a big difference in band gap (1.42 eV vs. 2.16 eV). Thus, in the absence of band bending, there will be a step-function jump of the conduction band edge across the interface; see Fig. 9.14(a). A heterostructure with a single interface is also called a heterojunction. Very often the alloy Ga$_{1-x}$Al$_x$As is used instead of AlAs. Here the band gap interpolates between those of GaAs ($x = 0$) and AlAs ($x = 1$), even though, strictly speaking, there is no lattice translation symmetry in an alloy, and bands are not well-defined.

When the semiconductors are doped, charge redistribution occurs due to mismatch of chemical potentials, in a way similar to what happens in a p–n junction. This leads to band bending and the creation of an inversion layer; see Fig. 9.14(b). We thus get a 2DEG confined to the junction area (on the lower-band-gap side), *without* applying a gate voltage. Another commonly used structure to confine a 2DEG is the quantum well illustrated in Fig. 9.14(c), where a lower-band-gap semiconductor (say, GaAs) is sandwiched between two higher-band-gap semiconductors (say, Ga$_{1-x}$Al$_x$As). With the Fermi energy between the two conduction band edges, electrons are confined inside the well. In both

---

[24] If two materials have a big difference in lattice structure and/or constant (known as lattice mismatch), there will be large strain at the interface, with (generally) undesirable consequences.

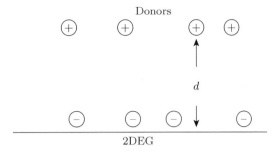

Figure 9.15 Illustration of modulation doping. The dopant ions (assumed to be positively charged, as is the case for donors) are placed at a distance d away from the 2D electron gas (2DEG). The larger d is, the more weakly electrons in the 2DEG are scattered by these dopants.

cases electron motion in the perpendicular direction has discretely quantized energy levels. Again, in most cases only the lowest level is populated.[25]

As is clear by now, doping plays a pivotal role in the functionality of semiconductor-based structures and devices. On the other hand, the ionized donors and acceptors that produce the mobile carriers are also one of the dominant sources of carrier scattering at low temperature, limiting carrier mobility and other related physical properties of the system. One widely used method to minimize the scattering effect due to charged impurity ions is **modulation doping**, namely putting the dopants relatively far away from the 2DEG confined to the heterojunction or quantum well. This is illustrated in Fig. 9.15, and further elaborated in Exercise 9.7. The fractional quantum Hall effect (to be discussed in Chapter 16), one of the most fascinating of all many-electron phenomena, was first discovered in a modulation-doped GaAs/Ga$_{1-x}$Al$_x$As heterojunction, and is still being actively studied in various semiconductor heterostructures. The high mobility of the 2DEG is crucial for preventing destabilization of such collective many-electron states by the random disorder potential.

High-electron-mobility field-effect transistors (**HEMTs**) constructed from modulation-doped semiconductor heterostructures are widely used in the low-noise high-frequency amplifiers commonly found in cell phones and are also much prized by radio astronomers. They are invaluable in the measurement circuits that read out the states of superconducting qubits (see Section 19.11) in quantum computers.

Repeating the quantum wells periodically results in a so-called superlattice (see Fig. 9.14(d)), in which periodicity is restored in the perpendicular direction and Bloch's theorem again applies. This allows one to *design* the band structure and do fascinating physics with it. For example, so-called quantum cascade lasers emit light when electrons make interband transitions within such "designer band structures."

Exercise 9.7. Electrons in a 2DEG interact with charged donors (with charge $+|e|$) via the 3D Coulomb potential. (i) Determine the effective 2D potential between an electron and a donor, as a function of the distance between the electron and the projection of the donor to the plane of the 2DEG (assume the 2DEG has no thickness; see Fig. 9.15 for an illustration). (ii) Calculate the Fourier transform of this effective 2D potential. (iii) Calculate the ratio between ordinary and transport scattering rates, as a function of $k_F d$, where k_F is the electron Fermi wave vector and d

---

[25] Sometimes the lowest few excited levels in the perpendicular direction are populated, and we need to specify which level the electron is in. Thus, in addition to the 2D Bloch momentum, we need a discrete index n similar to a Bloch band index to specify the electronic state. This is often referred to as the **electric subband index**, to acknowledge the fact that all these states come from the same 3D conduction band, due to quantum confinement.

is the distance between the donor and the 2DEG, using an isotropic effective-mass approximation for the conduction band. There are no correlations between the positions of the donors (other than that they have the same d). (iv) Assuming an equal 2D density of electrons and donors, calculate the electron mobility as a function of k_F and d. Obtain its numerical value for $d = 1000$ Å, electron effective mass $m^* = 0.07m_e$ (where m_e is the electron mass in vacuum), dielectric constant $\epsilon = 13$, and electron density $n_e = 10^{11}$ cm^{-2}. These numbers are typical for a 2DEG in a modulation-doped GaAs/Ga$_{1-x}$Al$_x$As heterojunction or quantum well.

9.5.3 Quantum Point Contact, Wire and Dot

In earlier subsections we have seen how to confine electron motion to an inversion layer, through the band bending effect of either an applied gate voltage or band engineering using heterostructures. As a natural next step, researchers have successfully combined these two methods to further confine the electron motion to form more complicated structures. One starts with a 2DEG formed in a heterostructure, and then patterns metallic gates with suitable size and shape on the surface of the device. Since the 2DEG is typically only about 1000 Å beneath the surface, the electrons are very sensitive to the

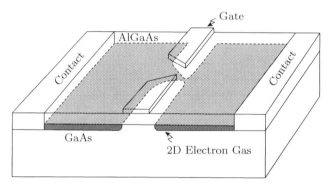

Figure 9.16 A gate-controlled quantum point contact. Figure adapted from [52] with the permission of the American Institute of Physics.

Figure 9.17 A gate-confined quantum dot. Lighter regions are metallic gates or electrodes. The four interior electrodes confine electrons within the dot in the buried quantum-well layer. The outer electrodes serve as contacts for electrons to tunnel into or out of the dot. Figure adapted with permission from [53].

gate voltage applied. Thus a modest gate voltage can deplete the electrons underneath the gate, and effectively create a barrier to confine the electron motion. For example, one can use gates to separate a 2DEG into two regions, and allow electrons to move from one region to another only through a small constriction known as a **quantum point contact**. See Fig. 9.16 for an illustration. One can also confine electron motion to 1D wires or even zero-dimensional regions, known as **quantum dots**. An example of a gate-confined quantum dot is shown in Fig. 9.17.

9.6 Notes and Further Reading

Compared with other chapters, the discussion in the present chapter is more descriptive and qualitative, and less mathematical and quantitative. This is particularly true for the sections on devices. For a more quantitative theory of the p–n junction, the interested reader can consult the book by Ashcroft and Mermin [1]. Semiconductor devices comprise a huge and very active field, and the present chapter can provide only a very brief introduction. There are many excellent monographs on this topic, and Ref. [51] is one example.

10 Non-local Transport in Mesoscopic Systems

In our earlier study of semiclassical Boltzmann transport theory we dealt with *probabilities* and *rates* rather than quantum *amplitudes*. The end result is essentially a microscopic justification of Ohm's law, namely the ability of a conductor to carry current is parameterized by a *local* quantity, the conductivity. Owing to this (assumed) locality, when we have two conductors connected in parallel (as in Fig. 10.1), an electron has to pass through *either* conductor 1 *or* conductor 2; we thus simply sum up the *probabilities* of it passing through each individual conductor, as a result of which the total **conductance** is the sum of the conductances of the individual conductors:

$$G = G_1 + G_2. \tag{10.1}$$

In a fully quantum-mechanical description of electron motion and transport, however, electrons are waves, which can go through *both* conductors; as a result we should sum up the *amplitudes* of each electron passing through each individual conductor. The probability is the square of the sum of the amplitudes rather than the sum of the squares. This results in *interference* contributions to the conductance that are *not* captured in the Boltzmann transport theory.

Our main goal in this chapter is to present a purely quantum-mechanical formulation of transport theory which is completely different from the semiclassical Boltzmann picture. In this picture a system whose resistance is to be measured is viewed as a target which scatters the electron waves incident upon it. As we shall see, the conductance is then related to the transmission coefficient for the scatterer.

10.1 Introduction to Transport of Electron Waves

Quantum interference of electron waves is dramatically manifested in the Aharonov–Bohm effect (see Section 13.2), which has been observed experimentally through electron transport in metals as shown in Fig. 10.2. The device being used is a small metallic ring with two arms; each arm is like an individual conductor connected to the other arm in parallel. A magnetic field introduces a magnetic flux through the ring, which modulates the relative *phase* of the quantum amplitudes of the electron going through the two arms. This controls the interference and results in an oscillatory dependence of the conductance of the ring on this flux, with period

$$\Phi_0 = \frac{hc}{e}, \tag{10.2}$$

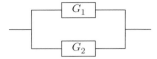

Figure 10.1 Two conductors connected in parallel. According to Ohm's law, the total conductance is the sum of individual conductances, $G = G_1 + G_2$, but the relation can fail for fully coherent quantum transport of electron waves.

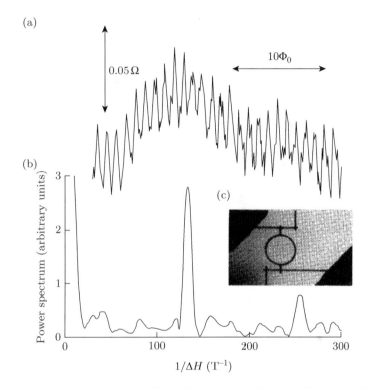

Figure 10.2 Electron micrograph of a mesoscopic metal ring (c), and the observed Aharonov–Bohm oscillation in resistance as a function of the magnetic flux going through the ring (a). (b) The Fourier transform of the data in (a), from which the oscillation period can be read out. Figure adapted with permission from [54]. Copyright (1985) by the American Physical Society.

the **flux quantum**. Such an oscillation clearly invalidates Eq. (10.1), as the magnetic field or flux through the ring has negligible effect on the conductances of the individual arms. In Section 10.4 we will also discuss an example in which, when two resistors are connected in series, the total resistance in general is *not* the sum of the resistances of individual resistors.

The scattering theory of transport (largely associated with the names of R. Landauer and M. Büttiker) is especially useful in the study of **mesoscopic** devices.[1] With modern lithographic techniques (as used, for example, in integrated circuits for computer chips) it is possible to construct objects whose size is smaller than ℓ_φ, the characteristic length between inelastic collisions (which destroy **phase coherence**; see later), and the small metallic ring of Fig. 10.2 is an example of that. The value of ℓ_φ depends on the type of system and increases with decreasing temperature, but typically is less than $\sim 1\,\mu$m except at temperatures well below 1 K.

---

[1] Mesoscopic means larger than microscopic atomic length scales (~ 0.1–1 nm), but smaller than the macroscopic scales at which semiclassical transport applies. Typically it refers to micron length scales.

If the system size is less than ℓ_φ then it is completely inappropriate to use a classical picture in which scattering rates from separate impurities are added independently. It turns out that the transmission coefficient depends crucially on the interference effects controlled by the precise positions of all the impurities. Moving even a single impurity a distance of order the lattice spacing (more precisely, moving it by a significant fraction of the Fermi wavelength) can substantially affect the transport. This effect, which goes by the name of "**universal conductance fluctuations**" (UCF) has been observed experimentally (in fact, the slow and non-periodic dependence of the background resistance on magnetic field in Fig. 10.2 is a manifestation of UCF), and we will discuss it in some detail in Section 10.4.

A very simple but dramatic example of the non-local nature of transport, to be discussed in the next section, is that a narrow mesoscopic wire containing no impurities at all can still exhibit a non-zero resistance. Remarkably, the conductance of such a wire is quantized in multiples of the fundamental quantum unit e^2/h, *independently* of the length of the wire; thus conductivity (which is typically a local[2] property of a conductor!) loses its meaning here.

The above are examples of the profound effect of phase coherence in mesoscopic systems, namely that electrical conduction ceases to be a local property. The Aharonov–Bohm effect discussed earlier is perhaps the most dramatic manifestation of this non-locality. The overall conductance depends on the total geometry, and it makes no sense to speak of a local conductivity relating the local current to the local electric field. In fact the Aharonov–Bohm effect manifests itself even in *isolated rings* (without external leads so that one cannot measure the conductance) in the form of the so-called **persistent current**. We will discuss persistent currents again in Chapter 11 for the case when disorder is present, but the basic physics is very simple and is illustrated in Exercise 10.1.

As we will learn in Section 10.6, what destroys phase coherence and the corresponding interference effects in electron transport is inelastic scattering. In our earlier study of (semiclassical) Boltzmann transport theory, we noted that elastic impurity scattering can relax the momentum distribution but is incapable of relaxing the energy distribution. Implicit in our calculation of the Drude conductivity was an assumption that *weak* inelastic scattering maintains the energy distribution in (local) equilibrium but has no effect on the momentum relaxation. This can be justified in the linear-response approximation, where the applied electric field is infinitesimal and we keep effects to linear order in it. Inelastic scattering, which results in an energy dissipation rate that is of order the electric field squared, does not show up *explicitly* in the linear-response regime if its strength is much weaker than that of elastic scattering. But, in general, inelastic scattering cannot be neglected, even in Boltzmann transport theory. As we discuss in Section 10.6, the existence of inelastic scattering is also important for washing out quantum interference effects and thereby justifying the use of semiclassical transport theory at length scales beyond ℓ_φ. We will discuss the precise mechanism of **dephasing**, namely how inelastic scattering destroys phase coherence, in Section 10.6.

After this brief introduction to coherent transport in mesoscopic structures, let us now explore the various novel effects in more detail.

Exercise 10.1. To better understand persistent currents and the Aharonov–Bohm effect, consider a toy model of a quantum rotor consisting of an electron that is free to rotate around the z axis at a fixed radius R so that it has a moment of inertia $I = mR^2$. A magnetic flux tube containing flux $\Phi/\Phi_0 = \beta$ is placed on the z axis so that the Hamiltonian is

---

[2] Typically the current at a point is sensitive only to the electric field within a short distance of that point. The relevant characteristic distance scales are the Fermi wavelength (essentially the particle spacing) and the mean free path. Here the mean free path is infinite, implying non-locality, yet the conductance is finite.

$$H = \frac{\hbar^2}{2I}\left(-i\frac{\partial}{\partial\theta} + \frac{eR}{\hbar c}A_\theta\right)^2,$$

where θ is the angular coordinate.

(i) Show that choosing a gauge in which the azimuthal component of the vector potential is a constant yields

$$\frac{eR}{\hbar c}A_\theta = \beta$$

independently of the value of R.

(ii) Find the eigenfunctions of H, and show that they are completely independent of the flux in the flux tube.

(iii) Show that the eigenvalues of H do change with the flux, but that the spectrum is periodic in β with period 1.

(iv) Calculate the ground state current as a function of β, and show that it is proportional to $dE_0/d\beta$, where E_0 is the ground state energy.

(v) Show that, for small Φ, the magnetic flux generated by the current tends to reduce Φ (if its effect is included); thus the response is diamagnetic.

10.2 Landauer Formula and Conductance Quantization

Consider a narrow wire connected to two reservoirs as shown in Fig. 10.3. We imagine that the wire has perfectly uniform cross-sectional shape and size and is free of any disorder and inelastic scattering. We can think of this wire as a waveguide that transmits electron waves from one reservoir to the other. It will turn out to be important for our purposes that the width W of the wire be comparable to the electron de Broglie wavelength. In terms of the Fermi wave vector, this means that $k_F W$ should not be too large. This is difficult to achieve in an ordinary metal since the wire would have to be only a few atoms in diameter. With modern lithographic techniques it is possible to construct semiconductor wires containing a 2D electron gas (2DEG) of rather low density and correspondingly large Fermi wavelength (~ 30 nm) that fits our criterion.

As a model of this situation, let us take the wire to run in the x direction and take the 2DEG to be confined by hard walls at $y = 0$ and $y = W$. In the independent-electron approximation the eigenstates of the Hamiltonian are given by[3]

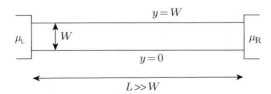

Figure 10.3 A semiconductor 2DEG quantum wire with length L and width W, connected to two leads with chemical potentials μ_L and μ_R, respectively.

---

[3] Here we ignore the periodic part of the Bloch wave function, which is irrelevant to our analysis, and keep only the envelope function.

$$\psi_{nk}(x, y) = \sqrt{\frac{2}{LW}} \, e^{ikx} \, \sin\left(\frac{n\pi y}{W}\right); \quad n = 1, 2, 3, \ldots, \tag{10.3}$$

with corresponding eigenvalues

$$\epsilon_{nk} = \frac{\hbar^2 k^2}{2m^*} + \frac{\hbar^2 \pi^2}{2m^* W^2} n^2, \tag{10.4}$$

where m^* is the semiconductor band effective mass, and n is referred to as the subband index.[4] This solution is valid in the interior of the wire. We assume that the connection to the reservoir is made by gradually and smoothly increasing the wire width. In this limit the interior wave function connects "adiabatically" to the reservoir and the wave is transmitted from the first reservoir into the wire and then into the second reservoir without reflection.[5]

We assume that the length of the wire L obeys $L \ll \ell_\varphi$ and that the size of the reservoirs is much greater than ℓ_φ. Hence, after an electron has entered the reservoir, it stays there for a long time and comes to local thermal equilibrium.

We can now ask ourselves what the resistance of this wire is. Naively it would seem to be resistanceless since there is no backscattering and the electron waves are transmitted perfectly. This picture is incorrect, however, as we shall now see.

Consider the situation where the right reservoir is held at chemical potential μ_R and the left reservoir has its chemical potential elevated to $\mu_L > \mu_R$. It will turn out that the non-equilibrium current in the wire will be finite, and hence, if the reservoirs are sufficiently large, the small "leak" from one to the other will not disrupt the internal equilibrium of the reservoirs. Hence μ_L and μ_R are each well-defined.

All of the electrons in the wire that are right moving must have come from the reservoir on the left (since there are no reflections). Thus the distribution function for right movers ($k > 0$) in the nth transverse mode is $f^\circ(\epsilon_{nk} - \mu_L)$. Likewise the distribution function for left movers ($k < 0$) is $f^\circ(\epsilon_{nk} - \mu_R)$. At zero temperature, both left- and right-moving states are fully occupied up to μ_R. However, as illustrated in Fig. 10.4, in the interval $\mu_R < \epsilon < \mu_L$, only right-moving states are occupied. It is these states which carry the net current.

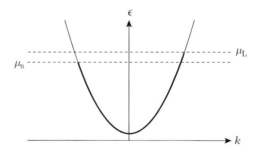

Figure 10.4 Electron dispersion and occupation at zero temperature. The thickened part of the dispersion represents occupied states.

---

[4] Obviously n is very much like a band index, but it arises from the spatial confinement of electron motion in the $\hat{y}$ and $\hat{z}$ directions (without such confinement the electrons would be in a single 3D conduction band). We thus use the term subband to distinguish the present case. The same term is used to describe spatially confined holes in a 3D valence band.

[5] If W changes rapidly (i.e. on the scale of k_F^{-1}), then the waves would be partially reflected from the entrance and the exit connecting the wire to the reservoir. This would complicate the analysis somewhat. In many cases, the confinement potential that determines the shape of the conducting channel is set electrostatically by voltages on (somewhat) remote gates, and so the shape automatically varies smoothly on the scale of k_F^{-1}.

The net current is (including a factor of 2 for spin)

$$I = 2 \sum_{n=1}^{\infty} \int_{-\infty}^{\infty} dk \left(\frac{L}{2\pi} \right) \left(\frac{-ev_{nk}}{L} \right) \tag{10.5}$$

$$\times \left[f^{\circ}(\epsilon_{nk} - \mu_{\mathrm{L}})\theta(k) + f^{\circ}(\epsilon_{nk} - \mu_{\mathrm{R}})\theta(-k) \right]. \tag{10.6}$$

The factor $L/2\pi$ is the density of states in k space. The factor $(-ev_{nk}/L)$ is the charge density[6] times the velocity. Hence the wire length L will drop out. Using $\epsilon_{nk} = \bar{\epsilon}_k + \gamma n^2$, where $\gamma \equiv \hbar^2 \pi^2 / 2m^* W^2$ and $\bar{\epsilon}_k \equiv \hbar^2 k^2 / 2m^*$, we see that the velocity

$$v_{nk} = \frac{1}{\hbar} \frac{d\epsilon_{nk}}{dk} = \frac{1}{\hbar} \frac{d\bar{\epsilon}_k}{dk} \tag{10.7}$$

depends only on k and is independent of n. Thus

$$I = -\frac{2e}{h} \int_0^{\infty} dk \frac{d\bar{\epsilon}_k}{dk} \sum_{n=1}^{\infty} \left[f^{\circ}(\bar{\epsilon}_k + \gamma n^2 - \mu_{\mathrm{L}}) - f^{\circ}(\bar{\epsilon}_k + \gamma n^2 - \mu_{\mathrm{R}}) \right]$$

$$= -\frac{2e}{h} \int_0^{\infty} d\bar{\epsilon} \sum_{n=1}^{\infty} \left[f^{\circ}(\bar{\epsilon} + \gamma n^2 - \mu_{\mathrm{L}}) - f^{\circ}(\bar{\epsilon} + \gamma n^2 - \mu_{\mathrm{R}}) \right]. \tag{10.8}$$

In the limit $T \to 0$ the Fermi distribution function reduces to the step function $f^{\circ}(x) = \theta(x)$, which takes values 0 and 1 only. After a change of integration variable from $\bar{\epsilon}$ to the total energy ϵ, we have

$$I = -\frac{2e}{h} \sum_{n=1}^{\infty} \int_{\gamma n^2}^{\infty} d\varepsilon [\theta(\mu_{\mathrm{L}} - \varepsilon) - \theta(\mu_{\mathrm{R}} - \varepsilon)]. \tag{10.9}$$

It is clear that the integrand (and thus the integral) in Eq. (10.9) is zero for $\mu_{\mathrm{R}} \approx \mu_{\mathrm{L}} \approx \mu < \gamma n^2$, as both θ functions are equal to zero. Physically this is simply because $\epsilon_{nk} > \mu$ for any k, and, as a result, all states with such n are empty. The branches of states with such ns are referred to as closed channels. On the other hand, for $\mu > \gamma n^2$ and $\mu_{\mathrm{R}} < \epsilon < \mu_{\mathrm{L}}$ (assuming $\mu_{\mathrm{R}} < \mu_{\mathrm{L}}$ for the moment, but the result is valid for $\mu_{\mathrm{R}} > \epsilon > \mu_{\mathrm{L}}$ as well), the integrand is one, otherwise the integrand is zero. We thus find

$$I = -\frac{2e}{h}(\mu_{\mathrm{L}} - \mu_{\mathrm{R}})N, \tag{10.10}$$

where

$$N \equiv \sum_{n=1}^{\infty} \theta(\mu - \gamma n^2) \tag{10.11}$$

is the number of the so-called **open channels**, namely those that are occupied by electrons for some range of k.[7]

---

[6] The 1D electron *number* density (associated with a given mode) is $\int_0^W dy |\psi_{nk}(x, y)|^2 = 1/L$. Another way to understand the $(-ev_{nk}/L)$ factor is to note that L/v_{nk} is the time it takes an electron occupying the state ψ_{nk} to move through the wire, thus $(-ev_{nk}/L)$ is its contribution to the net current.

[7] We are assuming $\mu_{\mathrm{L}} - \mu_{\mathrm{R}}$ is small enough that N is the same for μ_{R} and μ_{L}. This is, of course, guaranteed to be the case in the linear-response regime, where $\mu_{\mathrm{L}} - \mu_{\mathrm{R}} \to 0$.

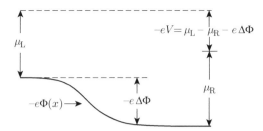

Figure 10.5 A generic situation of current flowing in a wire, with differences both in electric potential (Φ) and in local chemical potential (μ) between the two leads. The voltage V measures the electrochemical potential difference, $\mu_L - \mu_R - e\,\Delta\Phi$.

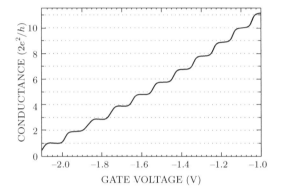

Figure 10.6 Conductance as a function of gate voltage in a quantum point contact, showing Landauer quantization. A quantum point contact can be viewed as a quantum wire with very small length, in which the Landauer formula still applies. The gate voltage controls the number of open channels. Figure adapted with permission from [55]. Copyright (1991) by the American Physical Society.

Now the voltage drop is the electrochemical potential difference[8]

$$-eV = (\mu_L - \mu_R), \tag{10.12}$$

and so

$$I = N\,\frac{2e^2}{h}\,V, \tag{10.13}$$

and the conductance is quantized at

$$G = N\frac{2e^2}{h} \sim 2\frac{N}{25{,}812.807\,\Omega}. \tag{10.14}$$

This conductance quantization has been observed experimentally as shown in Fig. 10.6. The quantum unit of conductance, e^2/h, and the corresponding resistance unit h/e^2, are extremely important combinations of fundamental constants that appear repeatedly later in this chapter, and in following chapters on Anderson localization and quantum Hall effects. More discussion of the resistance quantum can be found in Box 10.1.

---

[8] In this model there is *no* electric field or electric potential difference, and the difference between the chemical potentials of the two leads, which drives the current in the wire, is the electrochemical potential difference. One can also consider the more complicated situation in which both chemical-potential and electric-potential differences exist (see Fig. 10.5), but the result will not change, as was made clear in the general discussion in Section 8.8.1. See Exercise 10.3 for an explicit analysis for the present case.

While we arrived at Eq. (10.14) by using a parabolic free-electron-like dispersion, it should be clear from the derivation itself that this result is actually *independent* of the details of the band dispersion, as its effects cancel out in the two places where the dispersion relation enters, namely the electron (group) velocity, v_{nk}, and the density of states, which is *inversely* proportional to v_{nk}. Thus the conductance depends on the number of open channels N *only*.

The fact that a perfect wire with no scattering has a finite resistance (which is independent of the length of the wire) reflects the fact that the wire's current-carrying ability is *finite* even in the absence of scattering.[9] Since there is a finite resistance, there must be finite dissipation. This actually occurs solely in the reservoirs. The particles with the higher chemical potential entering the reservoir thermalize with it by emitting phonons, thus dissipating their excess energy as heat in that reservoir. Because of this the dissipation can be attributed to a "contact" or "injection" resistance even though (paradoxically) the transmission from the wire to the reservoir is reflectionless, and the magnitude of the resistance is determined by properties of the wire (the channel number, as determined by the width of the wire). Thus the wire resistance is truly a *global* (in contrast to *local*) property of the whole device.

Suppose now that there is an impurity in the middle of the wire which reflects waves in the nth subband with quantum amplitude r_n and transmits the waves with amplitude t_n. We evaluate these amplitudes for waves at the Fermi energy and assume their values are not strongly energy-dependent.[10] It is possible that the scattering will cause the subband index n to change. We will temporarily ignore this complication.

In steady state, the current is divergenceless and hence must be the same at every point along the wire. Picking an arbitrary point somewhere to the left of the impurity, we find that Eq. (10.9) is modified to

$$I = -\frac{2e}{h} \sum_{n=1}^{\infty} \int_{\gamma n^2}^{\infty} d\bar{\epsilon} [(1 - |r_n|^2)\theta(\mu_L - \bar{\epsilon}) - |t_n|^2 \theta(\mu_R - \bar{\epsilon})]. \tag{10.15}$$

Using $1 - |r_n|^2 = |t_n|^2$, we see that states with energies below μ_R still make no contribution to the net current. In the interval $\mu_R < \epsilon < \mu_L$ there are now some left movers because of reflection from the impurity. The net current is

$$I = -\frac{2e}{h}(\mu_L - \mu_R)T, \tag{10.16}$$

with total transmission probability (summed over all open channels)

$$T \equiv \sum_{n=1}^{\infty} |t_n|^2 \theta(\mu_R - \gamma n^2) = N\bar{T}, \tag{10.17}$$

where $\bar{T}$ is the mean transmission probability per open channel and N is the number of open channels. Thus the conductance is given by

$$G = \frac{2e^2}{h}T. \tag{10.18}$$

This is known as the **Landauer formula.**

---

[9] A loose analogy to this is the fact that a freeway with no traffic lights can still only accommodate a finite flow of traffic, which is (roughly) proportional to the number of lanes it has, which is similar to the number of channels in the wire. The length of the freeway, however, plays no role in determining its traffic-carrying capability.

[10] As we have seen, at $T = 0$ the net current is carried by the electrons whose energies are between μ_L and μ_R. Thus, in the linear-response limit of $\mu_L \to \mu_R = \epsilon_F$, transport properties are completely determined by the electronic wave functions at the Fermi energy. For sufficiently low temperature and small $|\mu_L - \mu_R|$ this remains true, assuming that the transmission probability does not change with energy too rapidly.

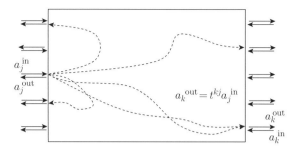

Figure 10.7 Schematic illustration of the transmission of electron waves from input port j on the left to various output ports on the left and right sides of a disordered sample. The transmission matrix t^{kj} gives the transmission amplitude for a wave entering at port j on the left and exiting at port k on the right. Notice that there can be multiple paths whose amplitudes add (and interfere) to produce the net transmission amplitude t^{kj}. The port (mode) labels j and k index the modes in some specified basis. They could, for example, label the different subbands n, n' in the disorder-free electrodes bringing the waves into the disordered region. Note that, in general, the number of ports on the left need not equal the number of ports on the right. The scattering matrix (S **matrix**) for this type of problem is discussed in greater detail in Appendix E.

Our formulation has not been completely general, since we assumed that the impurity perturbation is diagonal in the subband index. If there are N open channels then the transmission coefficient t is actually an $N \times N$ transmission matrix,[11] with matrix elements $t^{nn'}$ being the transmission amplitude into the outgoing channel n for an incoming electron in channel n', as illustrated in Fig. 10.7. We should therefore use

$$T = \mathrm{Tr}(t^\dagger t) = \sum_{nn'} |t^{nn'}|^2 = \sum_{nn'} T^{nn'} \tag{10.19}$$

in Eq. (10.18), resulting in

$$G = \frac{2e^2}{h} \mathrm{Tr}(t^\dagger t) = \frac{2e^2}{h} \sum_{nn'} T^{nn'}. \tag{10.20}$$

Here

$$T^{nn'} = |t^{nn'}|^2 \tag{10.21}$$

is the probability that an incoming electron in channel n' gets transmitted to channel n. In the limit of a large number of open channels (a wide wire) and for random potential disorder, $t^{nn'}$ becomes a large matrix with random (but correlated) entries. We will explore the impact of this randomness on the statistical properties of the conductance shortly. First, however, we will generalize the Landauer formula to the case of a mesoscopic system with more than two terminals (each of which has some number of open channels).

Exercise 10.2. Consider a clean quantum wire with a single (possibly) open channel, in which the electron dispersion is $\epsilon(k) = \hbar^2 k^2 / 2m$, and μ is the chemical potential. At zero temperature, the two-terminal (Landauer) conductance of the wire is $G(\mu) = (2e^2/h)\theta(\mu)$; thus there is a quantum phase transition at $\mu = 0$ where G jumps. At finite temperature T (not to be confused with the transmission coefficient), the transition is no longer sharp.

---

[11] For simplicity, we are assuming here that the number of open channels in each lead is N. In general, the numbers for different leads do not need to be equal.

(i) Calculate $G(\mu, T)$.

(ii) Show that $G(\mu, T)$ has a scaling form, namely $G(\mu, T) = \tilde{G}(\mu/k_B T)$, and that, at the zero-temperature transition point ($\mu = 0$), $G(\mu = 0, T) = e^2/h$ is independent of T (and thus "universal").

Exercise 10.3. Consider the situation in Fig. 10.5, where the distribution of the left and right movers entering the wire is controlled by the local chemical potentials of the right and left leads, μ_R and μ_L, respectively. Calculate the current in the wire, and show that Eq. (10.13) is valid, upon identifying the voltage V with the electrochemical potential difference between the two leads via $-eV = \mu_L - \mu_R - e\,\Delta\Phi$. Hint: note that some of the left movers with low kinetic energy will bounce back due to the potential barrier.

Exercise 10.4. In Fig. 10.3 we have a wire formed by restricting the motion of a 2D electron gas to a strip of width W. (i) For large W, show that the open channel number $N \simeq k_F W/\pi$, where k_F is the electron Fermi wave vector, satisfying $(\hbar k_F)^2/(2m^*) = \epsilon_F$. (ii) In the above we assumed that the motion of the electron along the z direction is completely frozen. Now consider a more generic 3D wire aligned along the x direction, with a uniform cross-sectional area A, whose linear dimension is much larger than $1/k_F$. Give an estimate of N without making an assumption about the particular shape of the cross section.

Box 10.1. Quantum of Resistance

In the nineteenth century, Maxwell discovered that the impedance of free space is (in cgs units) $4\pi/c$. In SI units this corresponds to $Z_0 \equiv \sqrt{\mu_0/\epsilon_0} = \mu_0 c = 1/\epsilon_0 c \sim 376.730313461$ ohms, a quantity which is *exactly* defined because both μ_0 and c are defined quantities in SI units. This impedance appears physically in electrodynamics as E/H, the ratio (in SI units) of the electric field strength to the magnetic field strength for plane waves in free space. In the early part of the twentieth century, Sommerfeld discovered that in quantum electrodynamics the dimensionless coupling constant between electrons and photons is the fine-structure constant (expression in SI units)

$$\alpha \equiv \frac{e^2}{4\pi\epsilon_0 \hbar c} \approx \frac{1}{137.035999139(31)}, \tag{10.22}$$

where the figures in parentheses give the uncertainty in the last two digits (2014 CODATA recommended value). This tells us immediately that the combination of fundamental constants h/e^2 has units of electrical resistance and an SI value of (2014 CODATA recommended value)

$$\frac{h}{e^2} = \frac{Z_0}{2\alpha} \approx 25{,}812.8074555(59) \text{ ohms.} \tag{10.23}$$

10.3 Multi-terminal Devices

M. Büttiker generalized the Landauer formula to **multi-terminal configurations** such as the one illustrated in Fig. 10.8. Each of the four ports is assumed to be connected adiabatically to independent

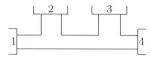

Figure 10.8 A four-terminal mesoscopic device. Each of the terminals is adiabatically connected to an independent reservoir with its own local chemical potential.

reservoirs. The generalization of the Landauer formula (cf. Eq. (10.16)) for the net current entering reservoir j is[12]

$$I_j = -\frac{2e}{h} \sum_k (T_{jk}\mu_k - T_{kj}\mu_j), \qquad (10.24)$$

where

$$T_{jk} = \sum_{n,n'=1}^{N} |t_{jk}^{nn'}|^2 \qquad (10.25)$$

is the analog of Eq. (10.19). $t_{jk}^{nn'}$ is the transmission amplitude from mode n' of terminal k to mode n of terminal j. T_{jk} is thus the trace over open channels of the probability that an electron leaving reservoir k is transmitted to reservoir j. T is a real matrix and is also symmetric except in so-called "non-reciprocal" systems (i.e. lacking time-reversal symmetry due to, e.g., an external magnetic field B), where it may be that $T_{jk} \neq T_{kj}$. The reader is directed to Appendix E for a general discussion of scattering, reciprocity, and unitarity.

Suppose that all reservoirs have the same chemical potential μ. Then the net current must vanish at every port:

$$I_j = -\frac{2e}{h} \sum_k (T_{jk} - T_{kj})\mu = 0, \qquad (10.26)$$

which gives us the sum rule

$$\sum_k T_{jk} = \sum_k T_{kj}. \qquad (10.27)$$

Using Eq. (10.27) in Eq. (10.24) yields

$$I_j = -\frac{2e}{h} \sum_k T_{jk}(\mu_k - \mu_j). \qquad (10.28)$$

With this result in hand we can now introduce the crucial idea of a **voltage probe**. This is defined as a reservoir whose electrochemical potential has been adjusted (or is "floating" in electronics parlance) so that the net current entering it is precisely zero. If, in the example shown in Fig. 10.8, ports 2 and 3 are chosen to be voltage probes, then their respective voltages (relative to some arbitrary ground) are

$$-eV_2 = \mu_2, \quad -eV_3 = \mu_3, \qquad (10.29)$$

and their voltage difference (which is independent of the choice of ground) is $-eV_{23} = \mu_2 - \mu_3$. Choosing μ_2 and μ_3 to make the currents entering ports 2 and 3 vanish is the quantum analog of having a high input impedance in a classical voltmeter.[13]

---

[12] Again, for simplicity, we assume that there is no potential gradient, so the local chemical potential is the same as the electrochemical potential.

[13] In a circuit made of macroscopic or classical devices, an ideal voltmeter has nearly infinite input impedance so that it draws negligible current from the circuit and thus does not perturb the circuit (or the resulting voltage readings).

From Eq. (10.28), the chemical potential in a voltage probe is given by the weighted average

$$\mu_j = \frac{\sum_{k \neq j} T_{jk} \mu_k}{\sum_{k \neq j} T_{jk}}. \tag{10.30}$$

This formula gives us a way of understanding what a voltage probe measures when it is in contact with a system that does not have a well-defined local thermodynamic equilibrium.

If, in the example in Fig. 10.8, there is a finite current flowing from port 1 to port 4, making ports 2 and 3 voltage probes constitutes a so-called "**four-terminal measurement**." Importantly, if there are no scatterers in the region between ports 2 and 3 and if the ports are perfectly left/right symmetric, one can show that V_{23} vanishes, unlike the two-terminal result obtained previously. In general, however, because we are dealing with quantum scattering in an asymmetric environment, the value (and even the sign!) of V_{23} can have a non-trivial dependence on the global geometry.

More generally speaking, adding contacts to the system, even in the form of voltage probes (which supposedly do not affect the circuit classically), inevitably changes the system and results of measurements involving other contacts. This is yet another example of the fact that in a quantum world any measurement inevitably affects the objects being measured.

10.4 Universal Conductance Fluctuations

Throughout this chapter we have been discussing mesoscopic samples at low temperatures where it is possible for the phase coherence length ℓ_φ to exceed the sample size. The sample is then in a special regime of coherent transport. For a disordered sample, the details of the wave interference pattern that control the probability of electron transmission through the sample depend on the precise location of all the impurities, much as a **laser speckle pattern** depends on the details of the frosted glass through which the light is passing. We turn now to a deeper examination of Eqs. (10.20) and (10.21) to understand the statistical properties of the conductance over the ensemble of possible random disorder realizations.

The big-picture idea here is that, in systems that are smaller in size than the **phase coherence length**, transport coefficients are not "self-averaging." That is, even though the sample may contain a huge number of randomly located impurities, there is no "central limit theorem" type of averaging effect that guarantees that the properties of any given sample are well represented by an ensemble average over all possible disorder realizations. This is in sharp contrast to the case of macroscopic samples, which can be viewed as a large network of classical resistors, each representing a segment of the sample on the scale of the phase coherence length. While their resistance values are determined by quantum interference, these resistors are effectively classical because there are no wave interference effects between them. As a result, the transport properties are in fact self-averaging and the observed fluctuations in transport properties at macroscopic length scales[14] are very small.

In the mesoscopic regime, it is interesting to ask the following question: how much does the conductance change if a single impurity moves slightly to a new position? It turns out that the conductance change can be many orders of magnitude larger than one might have imagined. Conductance fluctuations can be observed when impurities hop to new locations or when a change in the magnetic field or electron density (and hence the Fermi wave vector) causes the wave interference pattern to change. Remarkably, these random fluctuations of the conductance have a universal scale on the order of e^2/h in all dimensions and are known as universal conductance fluctuations (UCFs). In the following we

---

[14] That is, length scales vastly larger than the microscopic phase coherence length.

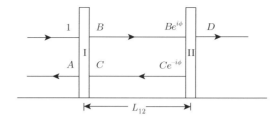

Figure 10.9 A quantum wire with a single conduction channel and two impurities (I and II). A, B, C, and D are the amplitudes of waves propagating in various regions and directions. The incoming wave from the left has amplitude 1 at impurity I. There is no incoming wave from the right.

illustrate this explicitly by considering a single-channel 1D wire with two impurities,[15] and then give more general arguments for 2D and 3D.

Consider a wire with two impurities, as illustrated in Fig. 10.9. Suppose we know *everything* about the individual impurities, including the transmission and reflection amplitudes for both directions of incidence. Does that uniquely determine the conductance G of the wire? The answer is *no*. As we shall see below, because of interference effects G also depends on L_{12}, the separation between the impurities. As a result, when one of the impurities moves, the conductance fluctuates. To calculate G we need to obtain the overall transmission probability T by solving the scattering problem when both impurities are present. Assume that an electron is incident from the left with Fermi wave vector k_F and amplitude 1. It hits the first impurity and then either reflects or is transmitted and hits the second impurity. If the electron ultimately transmits to the right then it either does so immediately or bounces a number of times inside the "Fabry–Pérot cavity"[16] formed by the two impurities and then eventually moves right with amplitude D; we have

$$T = |D|^2. \tag{10.31}$$

To determine D we need equations involving D as well as the amplitudes of plane waves between the two impurities that travel in both directions, B and C, to the right of the first impurity (see Fig. 10.9). They satisfy the following relations:

$$B = t_1 + C r_1', \tag{10.32}$$

$$C e^{-i\phi} = B e^{i\phi} r_2, \tag{10.33}$$

$$D = B e^{i\phi} t_2, \tag{10.34}$$

where the primed variables are the transmission and reflection amplitudes for right incidence (which have the same magnitude as their left-incidence counterparts but possibly different phases), and

$$\phi = k_F L_{12} \tag{10.35}$$

is the phase change of the plane wave between the two impurities.[17] Solving these equations yields

$$D = \frac{t_1 t_2 e^{i\phi}}{1 - r_2 r_1' e^{2i\phi}} \tag{10.36}$$

---

[15] The discussion below of the two-impurity case follows that in Imry's book [56].

[16] In optics, a Fabry–Pérot cavity is an optical resonator consisting of two (usually curved) mirrors separated by a distance L. If the mirrors have high reflectivity, the transmission through the resonator has a series of sharp peaks with a frequency spacing $\Delta f = c/2L$ known as the free spectral range (FSR). The frequency width of the resonances is inversely proportional to the finesse F, the mean number of round trips the photons make in bouncing between the mirrors before leaking out due to the small but non-zero transmission coefficient.

[17] Here again we use the fact that what matters is electrons at the Fermi energy, whose wave vector is k_F. We assume that the bias voltage is small enough that it does not affect the Fermi wave vector.

and thus

$$T = \frac{T_1 T_2}{1 + R_1 R_2 - 2\sqrt{R_1 R_2} \cos\theta}, \tag{10.37}$$

with

$$\theta = 2\phi + \arg(r_2 r_1'). \tag{10.38}$$

We immediately see that G depends not only on the transmission/reflection probabilities of the individual impurities, but also the distance between them; in particular, unless $R_1 R_2 \ll 1$ or $T_1 T_2 \ll 1$, the *change* of $G = (2e^2/h)T$ is of order e^2/h, for a change of L_{12} of order $1/k_F$. A similar change in G is induced by a change of k_F of order $1/L_{12}$.[18] It is clear that such non-locality and the corresponding fluctuation in G is due to the interference among waves generated by multiple reflections between the two impurities, whose phases depend on L_{12}.

Simple as it is, this two-impurity model actually goes a long way in illustrating some of the dramatic and often counter-intuitive effects of phase coherence and quantum interference on transport. To illustrate this, let us consider the resistance of the wire instead of the conductance. As we discussed earlier, the resistance of a *clean* wire is $h/2e^2$; this should be understood as the resistance of the contacts. To single out the contribution of an impurity to the resistance, it is natural (at least in some semiclassical sense) to subtract $h/2e^2$ from the total resistance of the wire, since the contacts and the impurity are connected in series. We thus find, using $h/2e^2$ as the resistance unit, the (dimensionless) resistance (or impedance) due to a single impurity to be[19]

$$Z = \frac{1}{T} - 1 = \frac{R}{T}. \tag{10.39}$$

Now for the two-impurity case we have

$$Z = \frac{R}{T} = \frac{R_1 + R_2 - 2\sqrt{R_1 R_2} \cos\theta}{T_1 T_2}, \tag{10.40}$$

which also fluctuates, and in general

$$Z \neq Z_1 + Z_2 = R_1/T_1 + R_2/T_2, \tag{10.41}$$

meaning that quantum resistances do *not* add up when connected in series! This remains true even if we smear out the fluctuations by averaging over θ:

$$\langle Z \rangle = \frac{R}{T} = \frac{R_1 + R_2 - 2\sqrt{R_1 R_2}\langle\cos\theta\rangle}{T_1 T_2} = \frac{R_1 + R_2}{T_1 T_2} > Z_1 + Z_2, \tag{10.42}$$

namely, *on average*, the resistance of two impurities is *bigger* than the sum of the resistances of individual impurities!

Exercise 10.5. Compute the average T in Eq. (10.37) to obtain the average *conductance* and show that it is *different* from the inverse of the average resistance (including contact resistances). This is another hallmark of mesoscopic transport.

This is nothing but the precursor of Anderson localization (to be discussed in Chapter 11) induced by quantum interference, which renders all electronic states localized in 1D disordered systems, and

---

[18] This is *in addition* to the changes of t_1 and t_2 which depend on k_F; such changes are negligible when $\delta k_F \sim 1/L_{12}$ is very small.

[19] Here we use Z for the impedance, since r and R represent the reflection amplitude and probability, respectively.

makes these systems insulating in the thermodynamic limit. For this special two-impurity case, however, we can easily arrange their configuration (which changes ϕ and thus θ) to have the opposite effect, namely $Z < Z_1 + Z_2$. An extreme example is the case with $R_1 = R_2$, and we choose L_{12} such that $\theta = 2n\pi$, in which case we have $Z = 0$. The resistances not only fail to add up, but they actually cancel each other out! This is, of course, the familiar phenomenon of resonant transmission through a Fabry–Pérot cavity, where constructive interference enhances transmission, while destructive interference suppresses reflection completely. However as illustrated by Eq. (10.42), on average, quantum interference tends to enhance backscattering and thus suppress transmission, which is the origin of Anderson localization, which will be discussed in Chapter 11.

One can also argue for UCFs more generally (and in arbitrary dimensions) in the following way. The Landauer expression for the dimensionless conductance from Eq. (10.20) is[20]

$$g = \frac{h}{2e^2} G = \sum_{i,j} |t^{ij}|^2. \tag{10.43}$$

We will here consider the semiclassical limit, where the disorder is sufficiently weak that the mean free path ℓ is much larger than the Fermi wavelength ($k_F \ell \gg 1$), but much smaller than the system size L. We further assume that Anderson localization effects (to be discussed in Chapter 11) are negligible on the scale of L (i.e., if the states are localized in the thermodynamic limit, the localization length ξ obeys $\xi > L$). In short, we are dealing with a good metal in the mesoscopic regime ($\ell_\varphi \gg L$). In this case, the transmission amplitude t^{ji} from the ith entrance channel to the jth exit channel is, in the semiclassical **Feynman path integral** formulation,[21] a sum over contributions from a huge number M of different paths, each with a random complex amplitude α_k. Equivalently, a wave packet arriving in entrance mode i diffuses through the sample, scattering randomly off the impurities and splitting up into multiple wave packets, M of which arrive at exit port j with random amplitudes and phases. Thus

$$t^{ji} \sim \sum_{k=1}^{M} \alpha_k, \tag{10.44}$$

and, by the central limit theorem, we expect t^{ji} to be a complex, Gaussian-distributed variable.

We can find the variance of t^{ji} from the following argument. In the limit $k_F \ell \gg 1$, we can apply the Drude formula of Eq. (8.139) to obtain in d dimensions (neglecting dimensionless factors associated with the detailed shape of the sample, e.g. the aspect ratio)

$$g = \frac{h}{2e^2} G \sim \frac{h}{2e^2} L^{d-2} \sigma = \pi \left(\frac{n}{k_F^d} \right) (k_F L)^{d-1} \left(\frac{\ell}{L} \right) \sim N \left(\frac{\ell}{L} \right), \tag{10.45}$$

where $N \sim (k_F L)^{d-1}$ is the number of open channels. (For $d = 3$, N is essentially the number of Fermi wavelengths squared that fit in the cross-sectional area of the sample. See Exercise 10.4, where the reader is asked to find a more accurate expression including prefactors.)

From Eq. (10.43) we see that

$$\bar{g} = N^2 \langle |t^{ji}|^2 \rangle, \tag{10.46}$$

[20] To be clear, i and j are input and output channel indices, not measurement port indices.

[21] In the simple example of Fig. 10.9, each additional bounce the electron experiences between the two impurities before transmitting corresponds to a different path that contributes to t. As we will see repeatedly later in the book, the Feynman path-integral formulation, while not always as easy to make as quantitatively rigorous as solving Schrödinger's equation, often reveals *qualitative* aspects of the physics of interest more straightforwardly, and is thus indispensable for developing physical insights.

where the angular brackets indicate the ensemble average over the disorder. From this we deduce

$$\langle T^{ji} \rangle = \langle |t^{ji}|^2 \rangle = \frac{1}{N}\left(\frac{\ell}{L}\right). \tag{10.47}$$

With this result in hand, let us now turn to a calculation of the variance of the dimensionless conductance, which by definition is

$$\langle (\delta g)^2 \rangle = \sum_{jklm} \{ \langle T^{jk} T^{lm} \rangle - \langle T^{jk} \rangle \langle T^{lm} \rangle \}. \tag{10.48}$$

Continuing with our assumption (which turns out to be good) that t^{jk} is a complex Gaussian-distributed variable, let us see what happens if we further assume that all of the amplitudes are statistically uncorrelated (except for the fact that $T^{jk} = T^{kj}$ if reciprocity holds; see Appendix E). A standard property of such correlators of Gaussian variables (known as **Wick's theorem** in the quantum case) is $\langle t^* t \, t^* t \rangle = 2\langle t^* t \rangle^2$ (assuming $\langle t \rangle = 0$). From this it follows that

$$\langle (\delta g)^2 \rangle = \sum_{jklm} \bar{T}^2 \{ \delta_{jk}\delta_{lm} + \delta_{jl}\delta_{km} \} = 2N^2 \bar{T}^2 = 2\left(\frac{\ell}{L}\right)^2 \ll 1. \tag{10.49}$$

This simple model result predicts that the variance of the dimensionless conductance is many orders of magnitude smaller than unity. However, this is in strong disagreement with experimental observations that the variance is of order unity.

The origin of this large and striking discrepancy is a seemingly small and subtle violation of our assumption that the transmission amplitudes are statistically independent [57]. Numerics, as well as sophisticated diagrammatic calculations, show that correlations exist between T^{jk} and T^{lm}. The size of these correlations depends on the relationships among the indices j, k, l, m. However, for our purposes, the most important correlation is a residual "long-range" positive correlation that is independent of all four indices,

$$\langle T^{jk} T^{lm} \rangle = \left[\{ \delta_{jk}\delta_{lm} + \delta_{jl}\delta_{km} \} + \frac{D_3}{\bar{g}^2} \right] \bar{T}^2, \tag{10.50}$$

where D_3 is a constant of order unity. Using Eq. (10.46), we see that the correction term is very small:

$$\frac{D_3}{\bar{g}^2} \bar{T}^2 = \frac{D_3}{N^4}. \tag{10.51}$$

This seems negligible until we realize that the summation in Eq. (10.49) contains N^4 terms. This tiny correction thus dominates, and we end up with a variance of order unity,

$$\langle (\delta g)^2 \rangle \approx D_3 \sim 1. \tag{10.52}$$

This result for the universal conductance fluctuations is quite remarkable. Despite the fact that the average dimensionless conductance depends on the mean free path ℓ, the variance of the same quantity is universal[22] and independent of ℓ.

The fact that the residual correlations are positive and independent of the mode indices tells us that, if one transmission probability is larger (or smaller) than expected, then (on average) so are *all* of the others! What is the physical origin of these additional "long-range" correlations? We can develop a qualitative picture by examining Fig. 10.10, which illustrates that the waves contributing to T^{kj} and T^{ml} can cross each other and hence suffer the same impurity scattering (shown as the gray regions in the figure). Because they see the same disorder, the two transmission coefficients become statistically

---

[22] It does depend on the dimension d and on the symmetry of the Hamiltonian under time-reversal, and depends weakly on the shape (aspect ratio) of the sample.

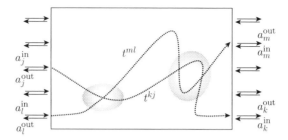

Figure 10.10 Illustration of semiclassical trajectories contributing to $|t^{kj}|^2$ and $|t^{ml}|^2$. The shaded gray areas indicate places where the trajectories pass through the same regions of the disorder potential. This induces correlations between the two transmission probabilities. If the shaded regions have a particular disorder realization that increases or decreases the transmission probability, then both $|t^{kj}|^2$ and $|t^{ml}|^2$ will increase or decrease together. This produces the extra correlations described by the D_3 term in Eq. (10.50).

correlated.[23] The actual calculation of the correlation coefficient D_3 in Eq. (10.50) is quite technical. See [57] for details along with a discussion of the application of these ideas to laser speckle patterns.

10.4.1 Transmission Eigenvalues

As discussed in Appendix E, if the number of open channels on the left side of the sample in Fig. 10.7 is N, and the number of open channels on the right side is M, then the S matrix describing scattering from inputs to outputs is an $(N + M) \times (N + M)$ unitary matrix which, if the system is reciprocal, is also symmetric. The matrix t of transmission amplitudes connecting the N incoming channels on the left with the M outgoing channels on the right is an $M \times N$ rectangular matrix (i.e. one having M rows and N columns) whose elements are a subset of the elements of S. For various purposes, including the calculation of **shot noise** that will be carried out in Section 10.5, it is useful to explore the concept of transmission eigenvalues. Because t is rectangular, we cannot simply diagonalize it. We have to carry out a **singular-value decomposition** (SVD) as described in Appendix F:

$$t = V\tau U. \qquad (10.53)$$

Here τ is an $M \times N$ matrix whose elements are zero except on the diagonal, where the entries are the "singular values" of t. V is an $M \times M$ unitary matrix which acts on the M output channels to create a new output mode basis, and U is an $N \times N$ unitary matrix which acts on the N input channels to create a new input mode basis.

The physical meaning of the singular value τ_j is that, if we prepare an incoming wave on the left in the jth (new) basis mode, it will transmit with amplitude τ_j into the jth (new) basis mode on the right, and not into any other modes on the right. (Note that there is enough freedom in the two unitaries to guarantee that we can, without loss of generality, choose all the τ_j to be real and positive.) We will primarily be interested in the transmission probability $(\tau_j)^2$ rather than the transmission amplitude. Notice also that the number of singular values is $\min\{M, N\}$. Therefore the number of non-zero transmission coefficients cannot exceed $\min\{M, N\}$.

With these mathematical preliminaries completed, let us consider the $N \times N$ Hermitian matrix

$$\mathcal{T} = t^\dagger t. \qquad (10.54)$$

---

[23] The following analogy used by some practitioners in the field may be useful: in driving a long distance, drivers from two different towns who are headed to two different destinations are likely at some point to end up taking the same freeway part of the way. They therefore see the same road conditions.

Recall from Eq. (10.19) that the trace of this matrix gives the dimensionless conductance (per spin) of the sample. In terms of the SVD of t we can write

$$\mathcal{T} = U^\dagger \tau V^\dagger V \tau U = U^\dagger \tau^2 U. \tag{10.55}$$

We see therefore that $U\mathcal{T}U^\dagger$ diagonalizes the Hermitian matrix $\mathcal{T}$ and that its eigenvalues are the transmission probabilities in the new mode basis for the incoming channels on the left side of the system. $\mathrm{Tr}\,\mathcal{T}$ is simply the sum of the transmission probabilities for the new basis modes. Notice that the unitary V has dropped out of consideration. This is because we do not care what particular superposition of outgoing modes on the right the transmitted wave happens to go into, only that it transmits.

Exercise 10.6. Consider the problem of finding the transmission amplitudes from the M modes on the right to the N modes on the left in Fig. 10.7. Use the results of Appendix E to find what matrix you need to diagonalize instead of $t^\dagger t$. Show that U drops out and only V is important in the diagonalization. Do not assume reciprocity.

The fluctuations in $\mathrm{Tr}\,\mathcal{T}$ give us the UCF we discussed above. It is interesting to ask for more information about the statistical properties of $\mathcal{T}$ beyond simply the fluctuations in its trace. In particular, as we will see when we study the noise properties of mesoscopic systems, it would be very useful to know the probability distribution of the transmission coefficients (eigenvalues of $\mathcal{T}$). There is a well-developed theory of large Hermitian random matrices whose entries are independently drawn from a real or complex Gaussian distribution. Unfortunately this theory is not directly applicable to matrices of the form given in Eq. (10.54) because the entries of $\mathcal{T}$ are not Gaussian-distributed and, as we saw above, are not statistically uncorrelated with each other. It turns out that the probability of finding an eigenvalue $T = (\tau_j)^2$ of $\mathcal{T}$ is given for large L/ℓ by [58, 59]

$$P(T) \approx \frac{\ell}{2L} \frac{1}{T\sqrt{1-T}}, \tag{10.56}$$

in the interval $\delta < T < 1$, where $\delta = e^{-2L/\ell}$. $P(T) = 0$ outside this interval. Notice that the mean transmission coefficient is given by

$$\bar{T} = \int_0^1 dT\, T\, P(T) \approx \frac{\ell}{L}, \tag{10.57}$$

in agreement with the result in Eq. (10.45).

This distribution is bimodal, with large weight at small values of T but also some weight for T approaching unity. This peculiar distribution is a result of the subtle statistical properties of the correlations among the elements of $\mathcal{T}$ discussed above. Most eigenchannels are highly reflecting, but some are nearly perfectly transmitting. By random chance some "lucky" transmission coefficients can approach their upper bound of unity. Others can be extremely small. In the most "unlucky" case, the forward-moving amplitude of a wave propagating through a diffusive medium could be continually damped by backscattering from the disorder, but the backscattered waves never re-scatter into the forward direction. This leads to a transmission probability [56, 58, 59] that is exponentially small in the system size, $T \sim e^{-2L/\ell}$. Hence the transmission coefficients for different channels can lie between these extremes. As we will see later in Exercise 10.11, the bimodal nature of the distribution leads to a reduction in the shot noise associated with current flowing through such a mesoscopic diffusive conductor.

10.4.2 UCF Fingerprints

Let us return now to the question of how much the conductance fluctuates when we move a single impurity. In the earlier example of two impurities in a single-channel 1D wire, we saw a change of conductance $\delta G \sim e^2/h$ when a single impurity is moved by a distance $\sim 1/k_F$. To get a hint of how big δG is in higher dimensions due to motion of a single impurity, consider the limit $k_F\ell \gg 1$, where ℓ is the mean free path, and we can use a semiclassical random-walk picture [60]. The typical number of steps (each of length ℓ) in a random walk that travels a distance L (the linear size of the sample) obeys $N \sim (L/\ell)^2$. Dividing this by the number of impurities $(L/\ell)^d$ gives the average number of visits to each impurity: $\sim (L/\ell)^{2-d}$. Thus, in 2D, every impurity site is visited approximately once, no matter how large the sample is. The situation is thus very similar to that of 1D, and moving a single impurity will have a significant effect on the conductance. In 3D the probability of visiting a given site falls off inversely with L, or the phases of a fraction of order ℓ/L of the paths are affected by the motion of a single impurity. Using the analysis above, we find of order $\delta g \sim \sqrt{\ell/L}$ in 3D. Thus, in 3D, moving a single impurity does not cause significant relative fluctuations in the conductance. One would have to move many impurities in order to create a fluctuation in g of order unity (which, as noted above, is the natural scale for UCFs in any dimension). Typically we do not have experimental control over the precise positions of impurities in a disordered sample. Occasionally, however, some impurities can thermally hop between two (say) quasi-stable positions. Slow thermal fluctuations of such so-called **"two-level systems"** in disordered conductors can lead to low-frequency noise in the conductance [60]. As we will discuss in Section 10.5, this type of mechanism is implicated in the so-called "$1/f$ noise" universally seen in the parameters of many physical systems. See also Exercise 10.8.

It should be clear from the discussions above that an external magnetic field can also induce UCFs, as its flux modifies the relative phases of interfering paths, via the Aharonov–Bohm effect (see Section 13.2). More quantitatively, we expect the conductance to change by $O(e^2/h)$ when the magnetic flux going through the sample changes by one flux quantum. In fact, in experiments it is *not* easy to move (or control) the positions of impurities. Instead one varies the (relative) phases of different paths by applying and changing the magnetic field, or varying the Fermi wave vector (which is controllable in 2DEGs in semiconductors through an applied gate voltage). UCFs are manifested in the random but completely reproducible dependence of the conductance on such variables. Examples are shown in Fig. 10.11 for the magnetic field dependence, and Fig. 10.12 for the gate-voltage dependence.

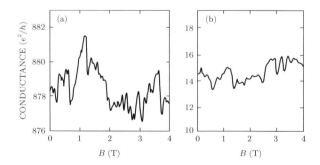

Figure 10.11 Universal conductance fluctuation (UCF) manifested in the magnetic field (B) dependence of the conductance of mesoscopic conductors. (a) The slowly varying part of the conductance as a function of B of the data in Fig. 10.2, after filtering out the rapid periodic oscillations due to the Aharonov–Bohm effect. The remaining slow fluctuations are due to the magnetic flux going into the interior of the wires, instead of that going into the hole of the ring which gives rise to Aharonov–Bohm oscillation. (b) The conductance as a function of B in a mesoscopic silicon MOSFET device. Despite the big difference between these two systems and their average conductances, the size of the fluctuation is of order e^2/h in both cases. Figure adapted from [61], original data source [62], with permission. Copyright (1986) by the American Physical Society.

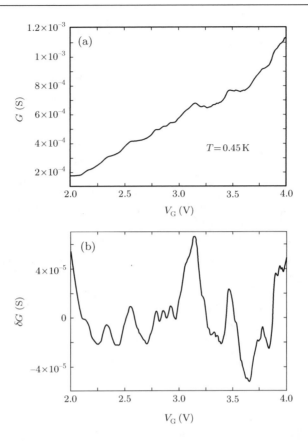

Figure 10.12 (a) The gate-voltage dependence of the conductance and (b) its fluctuating part after background subtraction in a mesoscopic silicon MOSFET device. The background subtraction is necessary since there is an overall dependence on the gate voltage in the average conductance. Note that the unit of conductance is the siemens (S), and $e^2/h \approx 1/25{,}812.807\,\Omega \approx 3.87 \times 10^{-5}$ S. Thus the fluctuation in conductance is of order e^2/h. Figure adapted with permission from [63]. Copyright (1988) by the American Physical Society.

In fact, such a dependence is specific to the particular impurity configuration of the sample, and can be viewed as a unique "fingerprint" of the specific sample.

Coming back to the case of 1D with *many* impurities, each impurity is visited many times, and the fluctuations in the conductance are very broadly distributed. The arguments given above do not really apply since they assumed the semiclassical diffusing wave model with no strong localization. As we will see in Chapter 11, in 1D the localization effects are strong (as already hinted in the two-impurity case studied above), and the localization length is on the same scale as the mean free path. This makes the fluctuations in the conductance so strong that the *logarithm* of the conductance is Gaussian-distributed. With fluctuations as wild as this, ensemble averaging is not very meaningful. Average and typical results are often very different. One can estimate the typical value of some quantity by averaging its logarithm and exponentiating the result:

$$g_{\text{typ}} \equiv e^{\langle \ln g \rangle}. \tag{10.58}$$

This procedure prevents the extreme tails in the distribution from dominating the result, which is what happens when the ordinary ensemble average is computed. In a sense this procedure finds the "average order of magnitude" of the conductance and then exponentiates it to find the typical value. Consider as an illustration a situation in which g has probability $1 - \epsilon$ of being 2.3 and probability ϵ of being 10^{11}. If $\epsilon = 10^{-9}$ then a sampling will almost always yield the value $g = 2.3$. Very rarely, however, it will yield $g = 10^{11}$. Thus

$$\langle g \rangle = (1 - 10^{-9})2.3 + 10^{-9}10^{+11} = 102.3, \tag{10.59}$$

while

$$\langle \ln g \rangle = (\ln 2.3)(1 - 10^{-9}) + 11(\ln 10)10^{-9} \simeq \ln 2.3, \tag{10.60}$$

so that

$$g_{\text{typ}} \simeq 2.3. \tag{10.61}$$

Thus g_{typ} is a much better representation of what an experiment will typically see since it is not dominated by **rare fluctuation events**.

Exercise 10.7. Consider a quantum wire with a single open channel, in which the Fermi wave vector is k_{F}. The electron effective mass is m.

 (i) There is a single delta-function impurity in the wire with potential $V(x) = V_0\delta(x)$. Calculate the two-terminal (Landauer) conductance of the wire.

 (ii) Now consider two identical delta-function impurities: $V(x) = V_0[\delta(x - a/2) + \delta(x + a/2)]$. Calculate the two-terminal (Landauer) conductance of the wire, and show that it oscillates with a, the distance between the two scatterers. This sensitivity to the relative configuration of the two impurities (as well as to k_{F}) is a manifestation of the kind of interference physics that leads to universal conductance fluctuations.

10.5 Noise in Mesoscopic Systems

The conductance G relates the *average* current going through a conductor to the voltage drop: $\langle I \rangle = GV$. However, for fixed V the instantaneous current I due to the motion of the individual discrete electrons actually is a function of time t that fluctuates around $\langle I \rangle$; see Fig. 10.13 for an illustration. Such fluctuations are known as current noise. Similarly, if we current bias a sample such that the current is fixed, there will be fluctuation or noise in the voltage drop that one measures. It turns out there is considerable information encoded in the noise spectrum (to be defined below); as Rolf Landauer put it, "noise is the signal!" For the rest of the section we consider a voltage-biased sample, namely the voltage is fixed, while the current is noisy.

Before discussing the various sources of noise, which is the main purpose of this section, let us first introduce the **noise spectrum** (or **spectral density**), which is used to quantitatively analyze the noise:

$$S_I(\omega) = \langle |\delta I(\omega)|^2 \rangle, \tag{10.62}$$

where

$$\delta I(\omega) = \lim_{\tau \to \infty} \frac{1}{\sqrt{\tau}} \int_{t'}^{t'+\tau} dt \, \delta I(t)e^{i\omega t}, \tag{10.63}$$

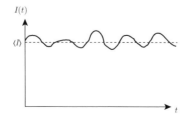

Figure 10.13 Current fluctuation noise.

$\delta I(t) = I(t) - \langle I \rangle$ is the (instantaneous) fluctuation of the current, and $\langle \cdot \rangle$ represents averaging over initial time t' of the period with length τ within which the Fourier transformation of $\delta I(t)$ is performed. We assume that the interval τ is much larger than the **autocorrelation time** of the noise and that averaging over t' is equivalent to ensemble averaging over all possible realizations of the noise. It is then easy to show the result of the **Wiener–Khinchin theorem**, namely that $S_I(\omega)$ is the Fourier transform of the correlation function of $\delta I(t)$:

$$S_I(\omega) = \int_{-\infty}^{\infty} dt \langle \delta I(t') \delta I(t' + t) \rangle e^{i\omega t}. \tag{10.64}$$

A proof of the Wiener–Khinchin theorem can be found in Appendix G in the closely related context of voltage noise. For a thorough discussion of noise in quantum systems, the reader is directed to the review of Clerk *et al.* [26].

Loosely speaking, there are three different sources of noise, which we discuss in turn.

- Thermal or **Johnson–Nyquist noise**. This is equilibrium noise, namely the noise when the system is in thermodynamic equilibrium, and thus $\langle I \rangle = 0$. It originates from the thermal fluctuations of the microstate of the system; for non-interacting electrons, this is simply due to the equilibrium statistical fluctuations in the occupation number of single-electron eigenstates (which can carry current). Obviously, such fluctuations (and corresponding noise) vanishes in the zero-temperature limit $T \to 0$. In fact, as explained in Appendices A and G, the zero-frequency limit of the thermal noise is related to the conductance via the **fluctuation–dissipation theorem** [26]:

$$\lim_{\omega \to 0} S_I(\omega) = 2k_B T G. \tag{10.65}$$

Here S_I describes the fluctuations, and the conductance G is the *dissipative* response function. It is clear that the bigger G is, the easier it is to have current flowing through the sample, thus the bigger the current noise. The noise grows linearly with temperature T because only states within an energy window of order $k_B T$ of the Fermi energy are partially occupied on average and hence have fluctuating occupancies. This physics is further illustrated in Exercise 10.10. It is important to note that, because this is equilibrium noise that does not rely on quantum interference effects, it survives in macroscopic samples whose size greatly exceeds the electron dephasing length beyond which interference effects are destroyed (see Section 10.6).

In order to have current fluctuations/noise one needs to have a closed circuit. For an open circuit, the current is identically zero but one can instead measure voltage fluctuations/noise, which is closely related. This is discussed in Appendix G. In particular the reader should convince herself/himself that Eq. (10.65) is equivalent to Eq. (G.24).

We now turn the discussion to non-equilibrium noise sources, namely those that would induce noise only when $\langle I \rangle \neq 0$. Such noise can survive in the zero-temperature limit, and can thus be distinguished from thermal noise.

- $1/f$ Noise. Here f stands for frequency; obviously such noise grows at low frequency. This is caused by the change of G with time, due to, e.g., motion of impurities, which can lead to significant fluctuations in G as discussed earlier. For fixed voltage V, this leads to fluctuations in current: $\delta I = \delta G V$, and thus the noise power scales with the square of the current,

$$S_I \propto V^2 \propto \langle I \rangle^2. \tag{10.66}$$

The frequency dependence of this type of noise is generally believed to arise from the broad distribution of time scales over which the motion of the impurities occurs. This can occur, for example, if the motion of the impurities is thermally activated across barriers and the barriers have a distribution of heights which is approximately flat. The basic physics is illustrated in Exercise 10.8. It should be noted that the $1/f$ functional form of the noise spectrum is only semi-empirical; the

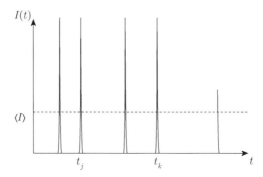

Figure 10.14 Classical shot noise. For classical particles, the contribution to the current is a sequence of delta-functions at the time they move through the detector.

specific dependence may vary from system to system, and any power-law spectrum of the form $S \sim 1/f^{\theta}$ is generically referred to as "$1/f$."

• Shot noise, also known as Poisson or Schottky noise. This is the most fundamental source of non-equilibrium noise, reflecting the discreteness of electron (or any other particle responsible for electric current) charge. We will discuss this in some detail for the rest of the section.

If electrons were classical particles, whose positions and momenta can (in principle) be determined with arbitrary precision, the instantaneous current going through a cross section of a sample would be a sequence of delta-functions peaked at the time when an electron moves through this cross section (see Fig. 10.14 for an illustration):

$$I(t) = e \sum_j \delta(t - t_j). \tag{10.67}$$

Thus the noise spectrum is

$$S_I(\omega) = \langle |\delta I(\omega)|^2 \rangle = \frac{e^2}{\tau} \sum_{j,k} \langle e^{i\omega(t_j - t_k)} \rangle = \frac{e^2 \langle N \rangle}{\tau} = e \langle I \rangle, \tag{10.68}$$

where $\langle N \rangle$ is the average number of electrons going through the sample during period τ. In the above we assumed that the temporal locations of the delta-function peaks are *uncorrelated*, and thus the average of the phase factors in the brackets is zero unless $j = k$. We find that this (classical) shot noise has the following distinctive features: (i) it is frequency-independent; (ii) its magnitude is proportional to $\langle I \rangle$ (not $\langle I \rangle^2$ as was the case for $1/f$ noise); and (iii) the proportionality constant is simply the charge of the particle.[24] One can use (i) and (ii) to distinguish shot noise from other types of noise such as $1/f$ noise (one needs to go to sufficiently high frequency so that shot noise dominates). Schottky, who was the first to discuss shot noise, proposed using property (iii) to measure the electron charge. Caution is required, however, even in the classical regime because Coulomb repulsion induces correlations among the electrons that suppress charge fluctuations at low frequencies. As we will see below, for non-interacting but quantum-mechanical electrons, their wave nature combined with Pauli statistics gives rise to corrections to Eq. (10.68) which can be significant. Finally we note that shot noise has been used to probe the **fractional charge** of certain quasiparticles in fractional quantum Hall systems [64, 65].

---

[24] In the literature one often sees $S_I(\omega) = 2e\langle I \rangle$ instead; this factor-of-2 discrepancy comes from the practice (common in engineering) of adding the noise spectrum at ω and that at $-\omega$, which are the same (classically). See the discussion in Appendices A and G and in Clerk *et al.* [26].

Exercise 10.8. The following is a simple model for $1/f$ noise. Consider a single impurity atom that can take two possible positions, 1 and 2, and jumps back and forth between them randomly with time constant τ, namely the probability that it jumps during the time interval between t and $t + dt$ is dt/τ (recall the similarity to the collision probability in the Drude model). For these two positions the conductance takes values G_1 and G_2, respectively, resulting in current $I_1 = G_1 V$ and $I_2 = G_2 V$, respectively, when the voltage is fixed to be V. As a result, there is current noise. (i) Draw a typical plot of current as a function of time, cf. Fig. 10.13. Such time dependence gives rise to what is known as **random telegraph noise**. (ii) Find the average current. (iii) Find the average number of jumps $\bar{n}(t)$ between t' and $t' + t$. (iv) Use the fact that the probability for the number of jumps n follows the Poisson distribution $P(n) = e^{-\bar{n}} \bar{n}^n / n!$ (cf. Exercise 7.3) to show that

$$\langle \delta I(t') \delta I(t' + t) \rangle = \frac{(I_1 - I_2)^2}{4} e^{-2|t|/\tau}. \tag{10.69}$$

Clearly the correlation time of the noise is $\tau/2$ in this case. (v) Calculate the noise spectrum $S_I(\omega)$ from the above.

In reality, there are many impurities in the system, each with a different time constant. (vi) Show that, if the probability distribution function of the time constant $p(\tau) \propto 1/\tau$ over a wide range of τ, we have $S_I(\omega) \propto 1/\omega$ over a wide range of ω, namely we have $1/f$ noise.

(vii) Assume for simplicity that the two positions for each impurity are degenerate in energy and the jumps between the positions are thermally activated. Show that, for a flat distribution of energy barrier heights, $p(\tau) \propto 1/\tau$ over a wide range of τ.

10.5.1 Quantum Shot Noise

In quantum mechanics, electron motion is described by propagating waves. It is *not* possible to determine the electron position at a given time in a current measurement. In fact, in the absence of scattering, electron energy eigenstates are plane waves which are also eigenstates of momentum and thus current; there is *no* uncertainty (and thus noise) in the outcome of a current measurement on electrons occupying such eigenstates.[25] In this case the only source of noise would be thermal fluctuations in the electron occupation number in such states. Such thermal fluctuations disappear at zero temperature, which is what we will focus on for the rest of the section. Thus quantum current is much "quieter" than classical current.

The situation changes in the presence of impurity scattering. In this case, electron energy eigenstates (stationary states) are no longer current eigenstates, as a result of which there is uncertainty (and noise) in the outcome of a current measurement. In the following we analyze the noise spectrum of a single-channel quantum wire with a single impurity first, and then generalize it to the multi-channel case.

Assuming the frequency-independence of the shot noise carries over to the quantum case (at least for low frequencies), we can simplify the calculation by taking (with some care [66–68]) the $\omega \to 0$ limit of Eq. (10.62):

$$S_I(\omega \to 0) = \frac{e^2}{\tau} (\langle N_\mathrm{T}^2 \rangle - \langle N_\mathrm{T} \rangle^2), \tag{10.70}$$

---

[25] There is a subtlety here: momentum and position cannot be measured simultaneously, so in measuring the momentum (the current) we need to have a spatially extended region in which to measure it if we are not to severely perturb the system via measurement back action.

where N_T is the number of electrons transmitted through the wire during the measuring period τ. Let us first consider the equilibrium case $\mu_L = \mu_R$ (see Fig. 10.4), and assume that the system is in the ground state at zero temperature. In this case, the current is exactly zero on either side of the impurity, since the left-moving and right-moving states are either both occupied or both empty. The net number of electrons transmitted is $N_T = 0$ for any period, and there is no noise. Now consider the out-of-equilibrium situation $\mu_L > \mu_R$. In this case, the difference $\Delta\mu = \mu_L - \mu_R$ completely determines the *additional* electron current flowing into the wire from the left lead, and the corresponding electron number N in a given period. If there is perfect transmission $T = 1$ (or $R = 0$), there is no uncertainty in $N_T = N$. In a semiclassical picture, we can envision the current being carried by electrons occupying wave packets formed by right-moving plane-wave states with energy $\mu_L > \epsilon > \mu_R$; every single wave packet is occupied by an electron and propagates through the wire, thus the electron flow is completely regular and noiseless.

In the presence of an impurity each wave packet has a probability T of transmitting and a probability R of being reflected and returning to the left lead. The stochastic nature of this process[26] gives rise to fluctuations in N_T and noise in the current, which we now calculate. For N wave packets, we have

$$\langle N_T \rangle = \sum_{j=0}^{N} j C_N^j T^j R^{N-j} = T \frac{\partial}{\partial T} \sum_{j=0}^{N} C_N^j T^j R^{N-j}$$
$$= T \frac{\partial}{\partial T} (T + R)^N = NT(T + R)^{N-1} = NT, \qquad (10.71)$$

where the standard combinatorial factor is

$$C_N^j = \binom{N}{j} = \frac{N!}{(N-j)! \, j!}.$$

In the manipulations above, we have treated T and R as *independent* variables, and only used the fact $T + R = 1$ in the last step to obtain the obvious result. Similar manipulations can be used to calculate $\langle N_T^2 \rangle$:

$$\langle N_T^2 \rangle = \sum_{j=0}^{N} j^2 C_N^j T^j R^{N-j} = T \frac{\partial}{\partial T} \sum_{j=0}^{N} j C_N^j T^j R^{N-j}$$
$$= T \frac{\partial}{\partial T} [NT(T + R)^{N-1}] = (NT)^2 + NT(1 - T). \qquad (10.72)$$

We thus find the quantum shot-noise spectrum for a single-channel quantum wire takes the form

$$S_I(\omega) = \frac{e^2}{\tau} NT(1 - T) = e\langle I \rangle(1 - T) = \frac{2e^2}{h} eVT(1 - T), \qquad (10.73)$$

which reduces to the classical shot-noise spectrum (10.68) in the limit $T \ll 1$. This is not surprising, since in this limit the transmitted wave packets are few and far between; their size is thus much smaller than the typical distance between them, and they can be treated as (classical) point particles. Conversely, if all the channels are perfectly transmitting (i.e. there are no impurities), then we recover the noiseless quantum result for an ideal channel. The partition noise is maximized when $T = 1/2$, where the randomness associated with "choosing" whether to transmit or reflect is maximized.

---

[26] Note that, from the point of view of electron waves, the transmitted and reflected wave amplitudes are deterministic solutions of the Schrödinger equation. However, when we actually measure what happens in a scattering event, each electron is either transmitted or reflected. It cannot be both transmitted and reflected. Thus the act of measurement induces wave-function collapse that leads to so-called **partition noise** for the electrons (viewed as particles).

The multi-channel generalization of Eq. (10.73) is obvious:

$$S_I(\omega) = \frac{2e^2}{h} eV \sum_n T_n(1 - T_n),$$ (10.74)

where n is the channel index (we assume there is no mixing between different channels, or, equivalently, work in the basis in which the transmission matrix t_{ij} is diagonal). Again, if *all* the transmission probabilities satisfy $T_n \ll 1$, we would obtain classical shot noise, Eq. (10.68).

As we discussed in Section 10.4, in a randomly disordered mesoscopic conductor, *most* channels have very low transmission probabilities. But there exist a number of *effective channels*, whose transmission probabilities are close to one; conduction is dominated by these channels. As a result, the quantum shot noise is reduced from the classical value. Detailed calculation based on the probability distribution for the transmission coefficients finds that the shot noise is reduced by a factor of 3 from that of Eq. (10.68). See Exercise 10.11 for more details.

Exercise 10.9. Generalization of Landauer formula to finite temperature (T): Consider a quantum wire with a single open channel, whose transmission probability is $|t(\epsilon)|^2$, where ϵ is the energy of an incident electron (in this exercise T represents the temperature, so we use $|t(\epsilon)|^2$ to represent the transmission probability). Show that the conductance obeys

$$G(T) = \frac{2e^2}{hk_B T} \int_0^\infty |t(\epsilon)|^2 f^\circ(\epsilon)[1 - f^\circ(\epsilon)]d\epsilon,$$ (10.75)

where $f^\circ(\epsilon)$ is the Fermi–Dirac distribution function. (ii) Consider a particular single-electron state with energy ϵ. The probability that it is occupied is $f^\circ(\epsilon)$, and as a result the average occupation is $\langle n \rangle = f^\circ(\epsilon)$. Calculate the fluctuation of the occupation, $\langle n^2 \rangle - \langle n \rangle^2$. At this point you should see the relation between the fluctuation of occupation and the conductance calculated above.

Exercise 10.10. This exercise concerns the current fluctuations of a wire in equilibrium. Consider a quantum wire with a single open channel, with no impurity and in equilibrium, so the left and right movers have the same temperature T and chemical potential μ. Prove Eq. (10.65) in this case by calculating the current noise spectrum $\lim_{\omega \to 0} S_I(\omega)$, using a method similar to the derivation of Eq. (10.70). Discuss how this generalizes to multi-channel cases with impurities. Hint: you can assume that electrons come from both left and right in a sequence of wave packets as we discussed in semiclassical transport theory in Chapter 8, with the occupation probability given by $f^\circ(\epsilon)$. The fluctuation of the number of electrons passing a given point is due to the fluctuation in occupation of these wave packets.

Exercise 10.11. Consider a disordered conductor with a huge number of open channels. In this case the transmission probability T_n forms an essentially continuous distribution, with a probability density function that turns out to be given by Eq. (10.56). Note that this distribution peaks at small T and $T = 1$, with the latter dominating both the conductance and the shot noise. Show that for this distribution the shot noise is $S_I(\omega) = Fe\langle I \rangle$, where the so-called Fano factor $F = \langle T(1 - T) \rangle / \langle T \rangle = 1/3$.

10.6 Dephasing

The novel and often counter-intuitive behavior of electron transport in the mesoscopic regime is due to the wave nature of electron motion, and the resultant quantum interference effects that occur when such waves maintain phase coherence. However, as we discuss in this section, due to ubiquitous interactions between the electron and its environment, the electron can become **entangled** with the environment, and such entanglement (see Appendix F) may destroy phase coherence and the effects due to interference. In the more traditional picture in which the environment is treated classically, we say that the phase of the electron wave function gets randomized when the electron undergoes inelastic scattering processes, and quantum interference effects are eventually washed out by such dephasing processes at sufficiently long time or length scales, at least when the temperature is not zero. Beyond such time and length scales (known as the phase-breaking time and length, respectively), the electron motion is incoherent, and it behaves much like a classical particle.

When an electron scatters elastically from an impurity, no recoil energy is imparted to the impurity. This will naturally be the case, provided that the impurities are infinitely massive. If the impurities have finite mass then the average energy lost to recoil will be finite, although there will still be a finite probability of recoilless scattering via the Mössbauer effect (recall the discussion in Chapter 6). The recoil energy will go into the creation of one or more phonons. Of course, the electron need not scatter from an impurity in order to produce phonons. The perturbing potential produced by a phonon vibration allows the electron to scatter even in the absence of impurities. At finite temperatures, this scattering can take the form of either emission or absorption of phonons, while at zero temperature only emission is possible, provided that the electron has sufficient energy.[27]

Before discussing the full quantum treatment of inelastic scattering, it is useful to develop a picture in which the impurity motion is treated classically. First consider a wave which bounces off a moving mirror. The reflected wave will be Doppler-shifted by the motion of the mirror. If this mirror is part of an interferometer then clearly the interference fringes will oscillate in time and be washed out when time-averaged. More generally, the Hamiltonian that describes electron motion needs to include the impurity (or environmental) degree of freedom (even if it is treated classically):

$$H = H_0(x) + H_{\text{env}}(y) + H_{\text{int}}(x, y), \tag{10.76}$$

where x and y are the electron and impurity coordinates, respectively, while H_0, H_{env}, and H_{int}, respectively, describe the electron motion, the impurity motion, and the interactions between them. The time-dependence of y means that the electron sees a time-dependent Hamiltonian $H_0(x) + H_{\text{int}}[x, y(t)]$. Clearly, fluctuations of y will affect and randomize the phase of the electron wave function $\psi(x)$.[28] More quantitatively, if ϕ is the phase, we have

$$\langle e^{i\phi} \rangle \approx e^{-\langle \delta\phi^2 \rangle/2} \to 0 \tag{10.77}$$

once we have

$$\delta\phi \approx \frac{1}{\hbar} \int_o^{\tau_\phi} \delta H_{\text{int}} \, dt \sim O(1), \tag{10.78}$$

---

[27] As noted before, phonons and photons are quite similar in their properties, and the phonon absorption/emission mechanism is also quite similar to that for photons described in Section 9.3. In the zero-temperature limit there are no phonons present, and the electrons occupy only states with the lowest possible energies that are allowed by the Pauli exclusion principle. As a result, electrons can neither absorb nor emit phonons (or any other excitations of the crystal), and thus dephasing is not possible.

[28] Such fluctuations also give rise to noise (say, of the $1/f$ type) as discussed earlier.

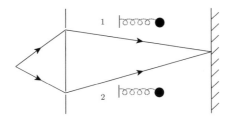

Figure 10.15 An electron propagates through two slits, with the two associated paths possibly interfering with each other. There are two harmonic oscillators, each coupled to one of the two paths. These constitute **"which-path" detectors** and can therefore affect the interference pattern of the waves.

where δH_{int} is the fluctuation of H_{int} due to the fluctuation of y (and we treat both x and y as classical variables here), and τ_ϕ is the phase-breaking time defined through the relation above. More commonly used is its inverse $1/\tau_\phi$, known as the phase-breaking or dephasing rate.

A more careful (and conceptually more correct!) treatment of dephasing of the electrons by the environment takes into account the quantum nature of the environmental degrees of freedom such as the phonons. As a toy model, consider a two-slit interference experiment in which the environment "records" which slit the electron passes through; see Fig. 10.15. That is, imagine that, if the electron passes through the first slit, it scatters inelastically and promotes some local oscillator degree of freedom from its ground state to its first excited state. Similarly, if the electron passes through the second slit, it excites another oscillator. By examining the state of the oscillators, we can deduce which slit the electron passed through and hence there ought not to be an interference pattern. Interference occurs when there are two amplitudes or paths leading to the same final state for the environment. In the absence of the oscillators, the final state of the electron has the form

$$|\psi\rangle = |\psi_1\rangle + |\psi_2\rangle = \int dx [\psi_1(x)|x\rangle + \psi_2(x)|x\rangle] \tag{10.79}$$

on the surface of the detector screen, where $\psi_1(x)$ and $\psi_2(x)$ are the amplitudes from path 1 and path 2, respectively.[29] This leads to the probability density of detecting the electron at X:

$$\begin{aligned} p(X) = |\psi_1(X) + \psi_2(X)|^2 &= |\psi_1(X)|^2 + |\psi_2(X)|^2 \\ &+ 2\text{Re}[\psi_1^*(X)\psi_2(X)], \end{aligned} \tag{10.80}$$

where the last term represents the interference. In the presence of the oscillators, the electron and oscillator states (which are the relevant degrees of freedom of the environment) are entangled:

$$|\Psi\rangle = \int dx [\psi_1(x)|x\rangle \otimes |1, 0\rangle + \psi_2(x)|x\rangle \otimes |0, 1\rangle]. \tag{10.81}$$

The probability density of detecting the electron at X without specifying the state of the environment is

$$\langle \Psi | \delta(X - x) | \Psi \rangle = \sum_{n_1, n_2} \langle \Psi | (|n_1, n_2\rangle \delta(X - x) \langle n_1, n_2|) | \Psi \rangle \tag{10.82}$$

$$= \text{Tr}\, [\rho_e \delta(X - x)] = |\psi_1(X)|^2 + |\psi_2(X)|^2, \tag{10.83}$$

---

[29] Equivalently, but more accurately, in the absence of interaction with the environment, the state of the *entire* system (including the electron and its environment) can be written as $|\Psi\rangle = |\psi\rangle \otimes |\text{env}\rangle = \{\int dx [\psi_1(x)|x\rangle + \psi_2(x)|x\rangle]\} \otimes |\text{env}\rangle$, where $|\text{env}\rangle$ is the state of the environment. We do not need to specify $|\text{env}\rangle$ insofar as the interference is concerned, because $|\Psi\rangle$ is a direct product between the electron state and the environment state, thus there is *no entanglement* between them.

where the summation in the first line traces out the environmental degrees of freedom (in this case the possible states of the oscillators), and

$$\rho_e = \mathrm{Tr}_{\mathrm{env}}|\Psi\rangle\langle\Psi| = \sum_{n_1, n_2} \langle n_1, n_2|\Psi\rangle\langle\Psi|n_1, n_2\rangle = |\psi_1\rangle\langle\psi_1| + |\psi_2\rangle\langle\psi_2| \qquad (10.84)$$

is the *reduced* density matrix of the electron (see Appendix F) obtained by tracing out the environmental degrees of freedom in the density matrix of the entire system (made of the electron and its environment), $|\Psi\rangle\langle\Psi|$. Because in this case the environmental states left behind by the two possible electron paths are orthogonal, ρ_e is an *incoherent* mixture of $|\psi_1\rangle$ and $|\psi_2\rangle$, and the interference terms drop out.[30]

It can happen that the two final states of the environment, $|E_1\rangle$ and $|E_2\rangle$, are *not* orthogonal. Suppose for example, that the two oscillators described above are in fact coupled together. Then it may happen that $\langle E_2|E_1\rangle = \alpha \neq 0$, which means that we can no longer be certain which slit the electron passed through. This will at least partially restore the interference pattern, because the "which-path" information has been partially **erased** via the coupling of the two oscillators. Quantitatively, we find in this case

$$\rho_e = |\psi_1\rangle\langle\psi_1| + |\psi_2\rangle\langle\psi_2| + \alpha|\psi_1\rangle\langle\psi_2| + \alpha^*|\psi_2\rangle\langle\psi_1|, \qquad (10.85)$$

and

$$p(X) = |\psi_1(X)|^2 + |\psi_2(X)|^2 + 2\,\mathrm{Re}[\alpha\psi_1(X)\psi_2^*(X)]. \qquad (10.86)$$

Comparing this with Eq. (10.80), it is clear the magnitude of the interference fringe oscillations ("fringe contrast") due to the third term is reduced by a factor of $|\alpha|$. (If α is complex, the fringes are also shifted spatially.)

In a more realistic model involving electrons in a disordered metal scattering inelastically from phonons, the same effect as that found in our toy model can occur. If an electron emits a phonon, either on one path or on another, the interference effect is not necessarily destroyed. If the wavelength of the phonon is larger than the spatial separation of the two emission points, it will be difficult to tell which path the phonon was emitted on. Semiclassically we can say that a long-wavelength phonon causes the local potential to oscillate up and down. However, if the wavelength is long enough, this looks like a spatially uniform potential which modifies the phase associated with all paths equally (see Eq. (10.78)), but does not induce a random phase *difference* between the different paths. Hence the interference is not lost. If the wavelength is short enough, then the two possible phonon states are distinguishable and orthogonal, and so the interference is destroyed.

It should become clear from the discussion above that how much interference there is depends on how *entangled* the electron is with its environment. The more entanglement there is, the less interference remains. The information about entanglement is fully encoded in the electron **reduced density matrix** ρ_e, if the entire system is in a pure state. There are various ways to quantify the amount of entanglement. One popular quantity to study is the entanglement entropy, which is the von Neumann entropy of ρ_e:

---

[30] The reader probably encountered density matrices in a course on statistical mechanics (see Ref. [28]), where the concept is introduced to acknowledge the fact that we do *not* know precisely which state the system is in, and have to introduce an ensemble of systems with an appropriate probability distribution for their states. Here the reduced density matrix plays a role identical to the density matrix in statistical mechanics, except that the source of the lack of precise knowledge about the system (the electron in the present case) is different: it is due to the entanglement between the system of interest and its environment. In fact, in the example discussed here, we actually do have complete knowledge of the state for the *combination* of the system and environment, but not of a part of it, if that part is entangled with the other part! We will encounter other examples of such and related counter-intuitive aspects of entanglement later. See Appendix F for further discussion of entanglement and the entanglement entropy.

$$S_E = -\text{Tr}[\rho_e \ln \rho_e].\qquad(10.87)$$

Interference is lost once S_E reaches $O(1)$. Thus the dephasing rate can also be understood as the rate of entanglement (entropy) generation between the electron and its environment.[31] We note that entanglement and the entanglement entropy, concepts of central importance in quantum information theory, are finding wider and wider application in condensed matter and quantum many-body physics. For a more detailed discussion, see Appendix F.

Exercise 10.12. Consider the electron reduced density matrix ρ_e in Eq. (10.85). Assume $|\psi_1\rangle$ and $|\psi_2\rangle$ have the same norm and are orthogonal to each other, and ρ_e is properly normalized. Calculate the entanglement entropy between the electron and its environment as a function of α.

---

[31] On the other hand, it has nothing to do with *how much* energy is transferred between an electron and its environment. In principle one can have entanglement and thus dephasing *without* energy transfer. This would be the case if the oscillator or phonon frequency were identically zero! In reality, however, dephasing is almost always accompanied by some energy transfer, but the point we want to emphasize here is that it is not related to the amount of energy transfer.

11 Anderson Localization

In Chapter 8 we encountered the Einstein relation for the conductivity:

$$\sigma = e^2 \frac{dn}{d\mu} D, \tag{11.1}$$

where D is the diffusion constant, $dn/d\mu$ is the combined (including both spins) density of states (DOS) per unit volume at the Fermi level for non-interacting electrons at zero temperature; in the presence of interactions and/or finite temperature $dn/d\mu$ is (essentially) the isothermal compressibility (see Eq. (15.162)). By definition, an insulator has $\sigma = 0$, which requires either $dn/d\mu = 0$ or $D = 0$. We already see that (except for very unusual cases) we can only have true insulators at $T = 0$, as finite temperature would generically give arise to non-zero D and $dn/d\mu$.

Thus far, the only insulators we have encountered are band insulators, in which $dn/d\mu = 0$ because the Fermi energy lies inside a band gap, and as a result the DOS is zero. These are systems like diamond or sodium chloride which have a gap between the highest fully occupied band (the valence band) and the lowest unoccupied band (the conduction band). The density of states is zero in this gap, and the system cannot respond (dissipatively) to weak electric fields. It can polarize (i.e. it has a dielectric constant greater than unity), but it cannot carry current at zero temperature where there are no thermally excited carriers. Hence the origin of band-insulator physics lies in simple single-electron physics and the formation of (filled) bands due to the periodic crystalline structure. Later in this book we will encounter Mott insulators, in which $dn/d\mu = 0$ due to strong electron–electron interactions, even though, in the absence of interactions, the system would have *partially* filled bands and thus be metallic. In both of these cases we can have $D > 0$, namely charge carriers introduced into the system are mobile, so that an initial density perturbation will diffuse away. However, in both cases one has to overcome a finite excitation gap to introduce charge carriers or produce such a density perturbation, and at zero temperature (only) we have $dn/d\mu = 0$.

We are now going to encounter a new type of insulator called the **Anderson insulator**, in which $dn/d\mu > 0$, but $D = 0$ due to the fact that electron eigenstates at the Fermi level have been *localized* by the random impurity potential, so that the charges can no longer diffuse in the long-time or large-distance limit. These systems are metals in the absence of disorder and they have a finite density of states at the Fermi level. As disorder (due, say, to alloying) is increased, the conductivity falls and finally goes to zero at some critical value of the disorder, at which point a **metal–insulator phase**

transition occurs. The density of states at the Fermi energy remains finite[1] and yet the system cannot carry dc current. As we will discuss in detail, the origin of this is quantum interference, whose effects have already shown up in our discussion of mesoscopic transport in Chapter 10, but are completely neglected in the semiclassical Boltzmann transport theory.

11.1 Absence of Diffusion in Certain Random Lattices

The semiclassical Boltzmann transport theory which we have studied up to now leads to the classical Drude result for the conductivity,

$$\sigma = \frac{ne^2 \tau_{TR}}{m_e}. \tag{11.2}$$

The only difference is that the transport scattering time τ_{TR} is computed quantum-mechanically. Classically, electrons scatter chaotically from the individual atoms in the crystal, giving a mean free path which is very short – namely on the scale of the lattice constant. Quantum-mechanically, Bloch's theorem tells us that the electrons do not scatter from the periodic crystal potential. Stated another way, for a perfectly periodic potential, quantum interference among the waves scattered from individual atoms is constructive and, while the waves are distorted (into Bloch waves), they remain plane-wave like and continue to propagate normally. Thus, in a crystal, the waves effectively scatter only from *deviations* from the perfect periodic potential due to the presence of impurities or lattice distortions associated with phonons.

Within the semiclassical Boltzmann approach, the scattering rate of a quantum wave packet is computed for each individual impurity, and the total scattering rate is simply the sum of the rates for scattering from the individual impurities. Because the semiclassical model deals only with rates and probabilities, it completely neglects quantum interference effects that occur when quantum probability amplitudes are correctly treated. The Drude formula in Eq. (11.2) would suggest the system is always metallic as long as the carrier density n is not zero, since τ_{TR} is always finite. However, we will find that there are important interference effects from scattering off different impurities, but, because the impurities are randomly located, accounting for these effects can be quite subtle. Inclusion of Coulomb interactions among the electrons in the presence of disorder can lead to even more subtle effects.

It turns out that, in strongly disordered metals, especially in lower dimensions, quantum interference effects can become so strong (and destructive rather than constructive) that they cause the system to become insulating. This phenomenon is known as **Anderson localization**. In 1958, P. W. Anderson [69] wrote an insightful paper entitled "On the absence of diffusion in certain random lattices." It was motivated by the problem of spin-diffusion in magnetic insulators being studied by magnetic resonance. Sir Nevill Mott then applied Anderson's idea to electrons moving in disordered metals and realized that, under some conditions, they too would fail to diffuse and would become insulating. What happens is that the electronic wave functions become "localized." They decay exponentially at large distances from some peak position and so do not extend across the sample and cannot carry current from one side of the sample to the other.

Note that the density of states typically remains finite and roughly constant through the Anderson localization transition. This means that, for any given eigenvalue, there are states arbitrarily close

---

[1] That is, the thermodynamic density of states is finite. If we increase the chemical potential μ, and wait for the system to re-equilibrate, electrons will flow into the system and occupy new levels whose energies lie in the interval between the old and new values of the chemical potential. The equilibration time is formally infinite because the states are localized and the conductivity is zero. To achieve equilibrium in practice, it may be necessary to raise the temperature of the system to temporarily turn on the conductivity.

by in energy. However, states that are close by in energy tend to be far away spatially (beyond the localization length) so that the matrix elements of the current operator between states of similar energy vanish. States that are nearby spatially are far away in energy. Hence it is impossible for the system to carry a dc current.

Because the Born approximation used to compute τ_{TR} treats the scattering only in first-order perturbation theory, it would seem to be valid only for low densities of weakly scattering impurities. Indeed, for sufficiently strong disorder the perturbation series fails to converge and a quantum phase transition to an insulating state occurs. Remarkably, in 1D and 2D, the radius of convergence of the series actually vanishes and all states are localized, no matter how weak the disorder is! In 2D, it turns out that the states are barely localized and the effect can be difficult to detect except at cryogenic temperatures. We shall explore this "**weak localization**" effect in some detail.

Even if we neglect the Coulomb interactions among the electrons, in order to fully describe quantum transport we will need to deal with the following Hamiltonian:

$$H = H_0 + V(\vec{r}) + H_\varphi, \qquad (11.3)$$

where H_0 is the Bloch Hamiltonian that includes kinetic energy and *periodic* potential, $V(\vec{r})$ is the *random* potential due to impurities, and H_φ describes the interaction between the particle and its environment (which we encountered in Chapter 10). The latter gives rise to the dephasing effects of inelastic scattering (e.g. from phonons) which ultimately destroy the subtle interference effects on long time scales. We are used to the idea of finding the energy eigenvalues and corresponding eigenfunctions of a Hamiltonian and deducing physical properties of a quantum system from these. This is impossible here, *even if* we ignore the H_φ term. If we had a single impurity we could solve the scattering problem. Or, if we had an infinite number of impurities but they were laid out on a regular lattice, we could use Bloch's theorem to take advantage of the discrete translation symmetry and make the problem tractable. You might think that, for weak impurity scattering, we could simply do perturbation theory. This is also not correct, because even for weak disorder, where the scattering mean free path may be much larger than the spacing between impurities, the mean free path is typically still very much smaller than the macroscopic system size. Thus multiple scattering is inevitable as the electron travels from one end of the sample to the other during transport. If we think of a classical random walk whose step size is the mean free path ℓ, the typical number of steps needed to walk from the center to the edge of a sample of size L is $N \sim (L/\ell)^2$, which grows without bound as the sample size increases. Obviously, including the H_φ term makes the problem even more intractable as this would require either solution of a time-dependent problem for the electron (if the environment is treated classically), or working in a much bigger Hilbert space that includes environmental degrees of freedom when they are treated quantum mechanically.

We will see below that there is a sense in which we can use perturbation theory to estimate the mean free path when the disorder is weak, but we are forced to go to infinite order in perturbation theory to properly understand diffusive motion of the electrons. Rather than attempting to work with formally exact expressions for the eigenfunctions and eigenvalues of the random Hamiltonian, we will see that it is best to take an alternate approach. In our study of non-local coherent transport in mesoscopic systems in Chapter 10 we took a view of transport as a scattering problem in which the entire sample was the scatterer. We found that the conductance of the sample was related to the transmission probability for electron waves impinging on the sample. Here we will find that **quantum diffusion** and transport in macroscopic samples is best studied using the Feynman path-integral formulation, by examining the quantum evolution of an electron wave packet which begins centered at some point in the random potential and spreads out over time, exploring its surroundings (by taking many different interfering paths). As we will see, the effects of inelastic scattering and dephasing due to the H_φ term in H can also be taken into account in such an approach, at least semi-quantitatively.

It will also be useful to think about the corresponding picture in momentum space. An initial plane (or Bloch) wave state with (crystal) momentum $\hbar \vec{k}$ is an eigenstate of the kinetic energy (or the periodic lattice Hamiltonian) but of course not an eigenstate of the full disorder Hamiltonian.[2] It will thus not be a stationary state, but will rather begin to develop amplitudes at other momenta and undergo a kind of diffusion around the Fermi surface in momentum space.

With some effort we will be able to connect the evolution of an electron wave packet under the Schrödinger equation to a picture of semiclassical diffusion in which the particle undergoes a classical random walk whose step size (the mean free path) is computed from the disorder potential using quantum mechanics. The semiclassical description of transport in terms of mean free paths deals with probabilities rather than quantum amplitudes and so misses crucial interference effects between different scattering events that are present in the full quantum theory. As we will see, these interference effects produce what are called "quantum corrections" to the classical transport coefficients. These quantum corrections are especially important at low temperatures, where the time and distance between inelastic scattering events which scramble the quantum phase are very large. We will find that it is useful to imagine a macroscopic sample as being made up of a collection of small coherent samples whose dimensions are given by the phase coherence length. Below that scale the non-local coherent transport picture of Chapter 10 applies, but at larger scales, these systems can be combined together like classical incoherent resistors to yield the full transport properties of the macroscopic system.

As noted above, in 1D and 2D, the quantum corrections to zero-temperature transport are enormous even for weak disorder, and the diffusion actually comes to a halt. This also occurs in 3D if the disorder is strong enough. A wave packet initially located at some point in the sample will never leave that vicinity, but will rather remain trapped within a fixed distance called the **Anderson localization length**. We can always formally express the initial wave packet as a superposition of exact energy eigenstates of the Hamiltonian. In the absence of disorder, the energy eigenstates are plane waves extending throughout the sample. As discussed above, the random potential causes non-perturbative changes in the eigenstates, and these changes drive the transport from ballistic to quantum diffusion. The Anderson localization phenomenon occurs when the nature of these exact eigenstates changes from being extended (more or less uniformly) throughout the sample to being spatially localized. These localized states should not be thought of as bound states that are trapped in regions where the random potential happens to be especially attractive. Unlike bound states in atomic physics, which occur below the bottom of the continuum, these localized states are in the midst of the continuum. There are other localized states at essentially the same energy throughout the sample but, in the absence of inelastic scattering, the particle cannot travel from one to the other.

A convenient device for characterizing whether states in a d-dimensional system are localized or extended is the so-called **inverse participation ratio**. Consider the integral

$$P_q = \int d^d r \, |\Psi(\vec{r})|^{2q}, \tag{11.4}$$

where q is a parameter. For $q = 1$ we have $P_1 = 1$ from the usual normalization condition on any eigenfunction. For $q > 1$ we expect from simple dimensional analysis that P_q will scale with the system size L and localization length ξ according to

$$P_q \sim \begin{cases} \xi^{-d(q-1)} & \text{for localized states (independently of } L); \\ L^{-d(q-1)} & \text{for extended states.} \end{cases} \tag{11.5}$$

---

[2] Unless explicitly noted, from now on we will ignore the difference between real and crystal momentum, kinetic energy and band energy etc, as they are not important for the discussion of effects of the random potential. We can thus ignore the presence of a periodic potential and corresponding band structure, unless noted otherwise.

In 3D, states at the Fermi energy are extended for weak disorder and localized for strong disorder. These two phases are separated by a **quantum critical point** at which the localization length diverges and we obtain "fractal" wave functions obeying

$$P_q \sim L^{-D_q(q-1)}. \tag{11.6}$$

The density distribution $\rho(\vec{r}) = |\Psi(\vec{r})|^2$ is a fractal object with effective spatial dimension $D_q < d$. It is in fact a "multi-fractal" in which the effective dimension D_q depends on the value of the parameter q which is chosen for the analysis.

11.2 Classical Diffusion

We have already studied the very important concept of diffusion. In a metallic conductor, the electrons have a finite diffusion constant. We learned from the Einstein relation that the diffusion constant is proportional to the conductivity of the metal. Classically, we visualize diffusion as occurring when particles undergo random walks. Thus, for example, electrons in a disordered metal treated as bouncing chaotically among the impurities would undergo a random walk. In quantum mechanics, however, electrons are waves which evolve in time according to the deterministic Schrödinger equation. It requires careful thought to develop a physical picture that correctly connects the evolution of quantum wave functions with the classical notion of diffusion.

Classical mechanics is also deterministic in its time evolution, although chaotic motion can appear to be random because of the extreme sensitivity of the dynamics to small changes in the initial conditions. When the motion is chaotic, it is often useful and appropriate to take a statistical approach based on a physical picture in which the dynamics is assumed to randomly and uniformly explore all the available phase space. This motion, while random, is still subject to the constraints of overall conservation of energy, momentum, and particle number. In the presence of fixed random scatterers, momentum conservation is of course lost.

Diffusion is fundamentally a consequence of particle number conservation in the absence of momentum conservation.[3] Suppose that a density wave is induced in a classical gas of particles subject to diffusion. This initial density inhomogeneity will disappear exponentially with some characteristic time $\tau(q)$ that depends on the (magnitude of the) wave vector $\vec{q}$ of the disturbance,

$$\rho_{\vec{q}}(t) = \rho_{\vec{q}}(0)e^{-t/\tau(q)}, \tag{11.7}$$

as diffusion returns the system to its homogeneous equilibrium state.

A density fluctuation at zero wave vector is simply a change in the total number of particles. Because of the conservation law obeyed by the dynamics, such a fluctuation cannot relax and $\tau(q = 0)$ must be infinity. Hence, for small but non-zero wave vectors, we expect (by continuity) the relaxation time to be very large. In order for diffusion to relax the density fluctuation, particles must undergo random walks that carry them from high-density regions to low-density regions located half a wavelength

---

[3] There are some tricky issues here. We are used to thinking about diffusion of atoms in an ordinary gas. This, however, is so-called *tracer diffusion* in which we follow the motion of an individual particle that has been "tagged." The momentum of this individual particle is of course not conserved; only the total momentum of the system is conserved. Hence this satisfies the conditions for diffusion. If we look not at an individual atom, but at the total density of atoms, there is no diffusion. A wave in the total density oscillates as a sound wave rather than diffusing away. This is intimately related to the overall conservation of momentum. If the gas is contained in a porous medium that acts as a momentum sink which causes the sound modes to be heavily overdamped at long wavelengths, then, in addition to tracer diffusion, one will observe collective diffusion of the total density. In general, the collective diffusion and tracer diffusion coefficients need not be the same.

away. For a random walk, the time required scales with the square of the distance. Hence it turns out that the relaxation rate must vanish at small wave vectors as

$$\frac{1}{\tau(q)} = Dq^2, \tag{11.8}$$

where D is the diffusion constant.

To derive this result more quantitatively, recall that the diffusion constant is defined to be the proportionality constant in **Fick's law**[4] relating the particle current to the density gradient (under the assumption that the relationship is linear):

$$\vec{J} = -D\,\nabla\rho. \tag{11.9}$$

Invoking particle number conservation (here it is again!), we have the **continuity equation**

$$\nabla \cdot \vec{J} + \frac{\partial\rho}{\partial t} = 0. \tag{11.10}$$

Combining these yields the diffusion equation

$$\frac{\partial\rho}{\partial t} = D\,\nabla^2\rho. \tag{11.11}$$

Fourier transformation in space yields

$$\frac{\partial\rho_{\vec{q}}}{\partial t} = -Dq^2\rho_{\vec{q}}. \tag{11.12}$$

The solution to this equation is given by Eq. (11.7), with the relaxation time given by Eq. (11.8) as claimed above.

Finally, we may relate all these to random walks by assuming that the particles undergo a random walk with characteristic space and time steps which are microscopic compared with the scales of interest. Thus, on the scales of physical interest, the path taken by any particle will consist of a very large number of steps. The central limit theorem tells us that the sum of a large number of random variables is Gaussian distributed and has a variance (width squared) proportional to the number of steps. Thus a particle starting at the origin in d dimensions will have a probability distribution which is a Gaussian whose width squared increases linearly with time (since the number of steps in the random walk increases linearly with time):

$$P(\vec{r}, t) = (2\pi\alpha t)^{-d/2} e^{-\frac{r^2}{2\alpha t}}. \tag{11.13}$$

The parameter α depends on the microscopic details of the step size and rate. We can relate it to the macroscopic diffusion constant by Fourier transformation:

$$\tilde{P}(\vec{q}, t) = e^{-\frac{1}{2}\alpha q^2 t}. \tag{11.14}$$

Comparing this with Eqs. (11.7) and (11.8) allows us to identify $D = \alpha/2$, so that the real-space time evolution in Eq. (11.13) becomes

$$P(\vec{r}, t) = (4\pi Dt)^{-d/2} e^{-\frac{r^2}{4Dt}}. \tag{11.15}$$

---

[4] Fick's law is based on the assumption that there is a linear relation between the mean particle current and the gradient of particle density. The particle current does not need to result from any force on the particles (due for example to the particles repelling each other). Imagine that we have non-interacting particles independently undergoing random walks. If there are more particles on the right than there are on the left, it is more likely for particles to travel from right to left simply because there are more particles available to do that than the reverse. In some sense, entropy increase is the driving "force."

We can make contact with the microscopic physics by the following semiclassical argument. We assume that, since only electrons near the Fermi surface contribute to the conductivity, we may treat the diffusing particles as moving at fixed speed v_F between collisions. Taking the collisions to be Poisson distributed in time, the probability distribution of times between collisions which randomize the velocity is (see Exercise 7.3)

$$p(\tau') = \frac{1}{\tau_{TR}} e^{-\tau'/\tau_{TR}}, \tag{11.16}$$

where τ_{TR} is the transport lifetime defined previously. In a time interval t the mean number of steps in the random walk is t/τ_{TR}. Hence the mean square distance walked in time t is

$$\langle r^2 \rangle = \Delta^2 t/\tau_{TR}, \tag{11.17}$$

where the mean square step length in the random walk is given by (see Exercise 7.3)

$$\Delta^2 = v_F^2 \langle (\tau')^2 \rangle = 2v_F^2 \tau_{TR}^2 = 2\ell^2. \tag{11.18}$$

The tricky factor of 2 (which Drude himself actually missed) arises from the fact that the mean square free time is twice the square of the mean free time:

$$\langle (\tau')^2 \rangle \equiv \int_0^\infty d\tau' \, p(\tau')(\tau')^2 = 2\tau_{TR}^2. \tag{11.19}$$

Combining Eqs. (11.17) and (11.18) and using $\langle r^2 \rangle = 2Ddt$ from Eq. (11.15), we finally arrive at the result

$$D = \frac{1}{d} \frac{\ell^2}{\tau_{TR}} = \frac{1}{d} v_F^2 \tau_{TR}. \tag{11.20}$$

Notice that this last equality tells us that the diffusion constant is linear in the mean free time. This is of course consistent with the fact that the Drude conductivity is also linear in this same quantity.

> **Exercise 11.1.** Substitute the second equality in Eq. (11.20) into the Einstein relation and show that it yields precisely the Drude expression for the conductivity.

11.3 Semiclassical Diffusion

11.3.1 Review of Scattering from a Single Impurity

Before we begin considering the subtleties of the transport theory of quantum waves interacting with randomly located impurities, let us briefly review the theory of a quantum particle scattering from a single target. This has already been covered in part in Chapter 8, but we review it here in more detail to better understand how to treat the random locations of the impurities. We ignore Coulomb interactions among the electrons as well as interaction with the environment, and assume a simple single-particle Hamiltonian of the form

$$H = H_0 + V(\vec{r}), \tag{11.21}$$

where $H_0 = p^2/2m$, and $V(\vec{r})$ represents the potential energy of interaction felt by the particle in the vicinity of the target, which we will assume results from a single impurity atom centered at the origin $\vec{r} = 0$.[5] In a typical scattering experiment a collimated beam of electrons (in our case) with a given

---

[5] For Bloch electrons we should also include the periodic potential in H_0, and $V(\vec{r})$ represents the *deviation* from such periodicity due to a single impurity. Including the periodic potential complicates the analysis but does not change the basic physics. For simplicity, we ignore the periodic potential in the rest of the chapter unless noted otherwise.

fixed energy is fired at the target. The goal of the experiment is to detect particles that are scattered by collision with the target through different angles and from that determine the nature of the target potential $V(\vec{r})$. The total rate Γ of particles scattering out of the original beam direction is given by

$$\Gamma = I_0 \sigma, \tag{11.22}$$

where I_0 is the beam intensity (in units of particle number per unit area per unit time passing through a plane perpendicular to the beam direction), and σ is called the scattering cross section and has units of area. If the target is a hard sphere, then (classically) σ is literally the cross section of that sphere. In general, however, $V(\vec{r})$ is a smooth potential and the geometric size of the target is only approximately defined. The quantum scattering cross section may or may not be close to the characteristic geometric size and may well vary (possibly sharply) with the beam energy at which the experiment is performed. This is because quantum mechanically the beam particles are waves, and interference effects can cause the cross section to vary as the wavelength (beam energy) is varied. For example, while the physical size of a sodium atom is on the $1\,\text{Å}$ scale, the actual cross section for scattering photons of yellow light is orders of magnitude larger and is on the scale of the square of the wavelength of the light. This is because of a resonance phenomenon that occurs when the photon energy is near the transition energy to the first electronic excited state of the atom.

In order to learn more about the scattering target from the experiment, it is useful to measure more than just the total scattering cross section. We saw in Chapter 4 a discussion of inelastic neutron scattering. Here we will focus on the simpler case of elastic scattering. Suppose that the detector has many separate segments set to collect particles that have been scattered through different polar angles θ and azimuthal angles φ (measured relative to the original beam axis). The rate of detection in each segment of the detector will be

$$d\Gamma(\theta, \varphi) = I_0 \frac{d\sigma(\theta, \varphi)}{d\Omega} d\Omega, \tag{11.23}$$

where $d\sigma(\theta, \varphi)/d\Omega$ is the differential cross section (with units of area/per unit solid angle) and $d\Omega$ is the small solid angle subtended by the detector segment. The total cross section is simply the differential cross section integrated over all possible scattering angles:

$$\sigma = \int d\Omega \frac{d\sigma(\theta, \varphi)}{d\Omega}. \tag{11.24}$$

Classically, most of the particles in a wide beam miss the target completely and continue straight ahead. Those particles that happen to have a small impact parameter (minimum distance from the origin) will be deflected by the force associated with the potential $V(\vec{r})$. The angle of this deflection (and hence the differential cross section) is readily computed for each value of the impact parameter using Newton's law. Quantum mechanically, we must solve the Schrödinger equation in order to determine the differential cross section. In the absence of the impurity potential, the eigenfunctions of H_0 are simple plane waves,

$$\psi_{\vec{k}}^{(0)}(\vec{r}) = \frac{1}{\sqrt{\Omega}} e^{i\vec{k}\cdot\vec{r}}, \tag{11.25}$$

where Ω is the sample volume (not to be confused with solid angle!). In the presence of $V(\vec{r})$ the eigenfunctions of H become

$$\psi_{\vec{k}}^{(+)}(\vec{r}) = \frac{1}{\sqrt{\Omega}} \left[e^{i\vec{k}\cdot\vec{r}} + \delta\psi_{\vec{k}}^{(+)}(\vec{r}) \right], \tag{11.26}$$

where for 3D in the limit $r = |\vec{r}| \to \infty$

$$\delta\psi_{\vec{k}}^{(+)}(\vec{r}) \to \frac{e^{ikr}}{r} f(\vec{k}, \vec{k}'), \tag{11.27}$$

where $\vec{k}' = k\hat{r}$, and the superscript $(+)$ indicates that we look for eigenfunctions of H in which the correction term $\delta\psi_{\vec{k}}^{(+)}(\vec{r})$ takes the form of the outgoing wave above. Then

$$\frac{d\sigma(\theta, \varphi)}{d\Omega} = |f(\vec{k}, \vec{k}')|^2. \tag{11.28}$$

For spherically symmetric potentials, conservation of angular momentum can be helpful. Particles in the incoming plane wave have well-defined momentum but completely uncertain impact parameter and hence uncertain angular momentum. We can actually write the plane wave as a coherent superposition of different angular momentum eigenstates and then solve the scattering problem for each angular momentum channel separately. The solution for the outgoing waves scattered by the target is entirely encapsulated in the **scattering phase shift** δ_ℓ in each angular momentum channel. For the case of a potential whose range is much shorter than the de Broglie wavelength of the beam particles, the scattering will be dominated by the s-wave ($\ell = 0$) channel because slow particles with large angular momentum must have a large impact parameter and hence miss the target. Only δ_0 will be non-zero, and it turns out that scattering in the s-wave channel is isotropic, so the differential cross section is independent of the scattering angle. Conversely, for a longer-range potential, many different angular momentum channels will have non-zero scattering phase shifts, and the differential scattering cross section will vary significantly with angle. This angular dependence arises from interference among different angular momentum channels.

For the case of a weak scattering potential, one can use low-order perturbation theory (the Born approximation) to compute the effect of the scatterer. This is in practice the only systematic method to solve scattering problems in the absence of rotation symmetry (and the partial-wave method is not applicable). To start, we re-write the Schrödinger equation (note that $\epsilon = \hbar^2 k^2 / 2m_e$)

$$H|\psi_{\vec{k}}^{(+)}\rangle = \epsilon|\psi_{\vec{k}}^{(+)}\rangle = \frac{\hbar^2 k^2}{2m_e}|\psi_{\vec{k}}^{(+)}\rangle \tag{11.29}$$

in the form of

$$(\nabla^2 + k^2)\delta\psi_{\vec{k}}^{(+)}(\vec{r}) = \frac{2m_e}{\hbar^2}\psi_{\vec{k}}^{(+)}(\vec{r}), \tag{11.30}$$

in which we used the fact that $(\nabla^2 + k^2)e^{i\vec{k}\cdot\vec{r}} = 0$. Using the fact that in 3D the solution to the equation

$$(\nabla^2 + k^2)G^{(+)}(\vec{r}, \vec{r}') = -4\pi\delta(\vec{r} - \vec{r}') \tag{11.31}$$

with the desired large-distance asymptotic behavior is[6]

$$G^{(+)}(\vec{r}, \vec{r}') = \frac{e^{ik|\vec{r}-\vec{r}'|}}{|\vec{r} - \vec{r}'|}, \tag{11.32}$$

we obtain

$$\delta\psi_{\vec{k}}^{(+)}(\vec{r}) = -\frac{m_e}{2\pi\hbar^2}\int d^3r'\, V(\vec{r}')\psi_{\vec{k}}^{(+)}(\vec{r}')\frac{e^{ik|\vec{r}-\vec{r}'|}}{|\vec{r} - \vec{r}'|}, \tag{11.33}$$

from which one reads out (by taking the $r \to \infty$ limit) the scattering amplitude

$$f(\vec{k}, \vec{k}') = -\frac{m_e}{2\pi\hbar^2}\int d^3r'\, e^{-i\vec{k}'\cdot\vec{r}'}V(\vec{r}')\psi_{\vec{k}}^{(+)}(\vec{r}') = -\frac{m_e\Omega}{2\pi\hbar^2}\langle\vec{k}'|\hat{V}|\psi_{\vec{k}}^{(+)}\rangle. \tag{11.34}$$

---

[6] For the special case of $k = 0$ this is simply the familiar Green function of the Poisson equation, which gives us the Coulomb potential of a point charge. The extra factor $e^{ik|\vec{r}-\vec{r}'|}$ compensates for the k^2 term on the LHS. Note that we could also have $e^{-ik|\vec{r}-\vec{r}'|}$, but that would give the asymptotic behavior at large r that corresponds to inverse scattering instead of scattering.

While this is only a formal result as it involves the unknown scattering state $|\psi_{\vec{k}}^{(+)}\rangle$, replacing it with the unperturbed plane wave state $|\vec{k}\rangle$ immediately yields the familiar first Born approximation.

As noted in Chapter 8, rather than using the full machinery of scattering theory, one can jump directly to the answer by using Fermi's Golden Rule to compute the rate of transitions from the incoming momentum $\vec{k}$ to the final momentum $\vec{k}' = \vec{k} + \vec{q}$, from which the differential cross section can also be extracted. The matrix element for scattering between the two plane-wave states is

$$M_{\vec{k},\vec{k}+\vec{q}} = \langle \vec{k} + \vec{q}|V|\vec{k}\rangle = \frac{1}{\Omega}v[\vec{q}], \tag{11.35}$$

where $v[\vec{q}]$ is the Fourier transform of the impurity potential:

$$v[\vec{q}] = \int d^d r \, e^{-i\vec{q}\cdot\vec{r}} V(\vec{r}). \tag{11.36}$$

This is precisely what enters the scattering amplitude of Eq. (11.34), if $|\psi_{\vec{k}}^{(+)}\rangle$ is replaced by $|\vec{k}\rangle$. It is a good exercise to show explicitly that Fermi's Golden Rule and the first Born approximation are equivalent when applied to scattering problems.

Going to higher orders in the Born series requires knowledge about corrections to the wave function, $\delta\psi_{\vec{k}}^{(+)}(\vec{r})$. This can be obtained through the following manipulations. Just as for Eq. (11.30), we re-write Eq. (11.29) as

$$(\epsilon - H_0)|\psi_{\vec{k}}^{(+)}\rangle = \hat{V}|\psi_{\vec{k}}^{(+)}\rangle, \tag{11.37}$$

and perform a formal operator inversion to obtain the Lippmann–Schwinger equation

$$|\psi_{\vec{k}}^{(+)}\rangle = |\vec{k}\rangle + \hat{G}\hat{V}|\psi_{\vec{k}}^{(+)}\rangle, \tag{11.38}$$

where

$$\hat{G} = \frac{1}{\epsilon - H_0 + i0^+} \tag{11.39}$$

is the **resolvent operator**. Note the positive infinitesimal imaginary part is necessary to ensure that $\hat{G}$ is well-defined, and $|\psi_{\vec{k}}^{(+)}\rangle$ has the appropriate asymptotic behavior of Eq. (11.27). Equation (11.38) can be solved through iteration to yield the perturbative series

$$|\psi_{\vec{k}}^{(+)}\rangle = \sum_{n=0}^{\infty}(\hat{G}\hat{V})^n|\vec{k}\rangle, \tag{11.40}$$

from which we obtain the $(n + 1)$th-order Born approximation for the scattering amplitude:

$$\begin{aligned}
f^{(n+1)}(\vec{k}, \vec{k}') &= -\frac{m_e}{2\pi\hbar^2}\langle \vec{k}'|\hat{V}\hat{G}\hat{V}\hat{G}\cdots\hat{G}\hat{V}|\vec{k}\rangle \\
&= -\frac{m_e}{2\pi\hbar^2}\sum_{\vec{k}_1,\ldots,\vec{k}_n} M_{\vec{k},\vec{k}'}(\vec{k}_1,\ldots,\vec{k}_n),
\end{aligned} \tag{11.41}$$

where

$$M_{\vec{k},\vec{k}'}(\vec{k}_1,\ldots,\vec{k}_n) = v(\vec{k}_1 - \vec{k})G(\vec{k}_1)v(\vec{k}_2 - \vec{k}_1)G(\vec{k}_2)\cdots G(\vec{k}_n)v(\vec{k}' - \vec{k}_n) \tag{11.42}$$

and

$$G(\vec{k}_j) = \langle \vec{k}_j|\hat{G}|\vec{k}_j\rangle = \frac{1}{\epsilon - \hbar^2 k_j^2/(2m_e) + i0^+}. \tag{11.43}$$

Note that the amplitude of Eq. (11.42) has the familiar form of products of matrix elements of $\hat{V}$ (in momentum basis) and energy denominators, and that the energy of the intermediate state $\hbar^2 k_j^2/2m_e$

does *not* need to be equal to ϵ. $G(\vec{q})$ is usually referred to as the Green function, whose Fourier transform is Eq. (11.32), with proper normalization.

A natural interpretation of Eq. (11.42) is that, instead of being scattered directly from $\vec{k}$ to $\vec{k}'$ via the matrix element $M_{\vec{k},\vec{k}'}$ defined in Eq. (11.35), the electron is scattered $n + 1$ times, and goes through a sequence of n *intermediate* states, $\vec{k}_1, \vec{k}_2, \ldots, \vec{k}_n$. Such a sequence may be viewed as a *path* that the electron has taken in momentum space. Since it stays in these intermediate states for only a finite amount of time (of order τ when there is a finite density of impurities, as we discuss next), their energies do *not* need to be equal to the initial energy of the electron, but instead have uncertainties of order h/τ.

It should be emphasized that the Born series (11.40) is an expansion in powers of V. If we have a strong scattering potential, such an expansion may not converge (and actually, for a singular potential with a hard core, cannot even be defined). However, as long as we are dealing with a *single* impurity with a finite-range potential, the *exact* scattering state can always be written in the form of Eq. (11.26) with the asymptotic behavior of Eq. (11.27), namely the correction to the plane-wave component due to V carries negligible weight in the thermodynamic limit. As a result, the scattering cross section is finite, and the probability that an *individual* electron initially in the plane-wave state gets scattered by this impurity is actually infinitesimal (the reason why there is a finite signal in a scattering experiment is because a finite *flux* of particles is sent in). This is very different in the case of a finite density of impurities, to which we now turn.

11.3.2 Scattering from Many Impurities

We now consider what happens when we have many impurities that are *randomly* distributed in the sample, with finite average density. For mathematical convenience, it is useful to imagine that we have a large *ensemble* of samples containing all possible realizations of the impurity configurations. As we saw in our study of mesoscopic systems, quantum interference effects can cause the transport properties of a sample to depend on the precise locations of all the impurities. In general, however, for macroscopic samples whose sizes are much larger than the phase coherence length, we expect that, while the transport properties depend on global properties such as the mean density of impurities, they should not depend on the microscopic details of the impurity positions, because what is measured will be the *average* of many mesoscopic subsystems that make up the macroscopic sample. It is mathematically convenient in this case to compute the ensemble-averaged transport properties, which will be a good representation of (almost) any particular element of the ensemble.

We assume that the disorder realizations are completely random and unaffected by whether or not the resulting energy of the electron gas is favorable or unfavorable. We refer to this situation as **quenched disorder**. It can be achieved by heating a sample to high temperatures at which the impurities become mobile and then suddenly cooling ("quenching") the sample, which freezes the impurities in place. The opposite extreme is **annealed disorder**, in which the sample is cooled slowly, allowing the system to remain in thermal equilibrium. In this case the impurity positions are less random because configurations that lower the overall energy of the electrons are favored.

We will model the quenched-disorder case by assuming that the impurity potential has a random Gaussian distribution defined by its zero mean,[7]

$$\overline{V(\vec{r})} = 0, \tag{11.44}$$

---

[7] Obviously a non-zero mean only results in a constant shift of potential which has no effect. The assumption of a Gaussian distribution allows one to reduce multi-point correlation functions of the potential to products of two-point correlation functions as introduced below. This is not crucial to the discussion for the rest of this section.

and a two-point correlation function which is both translation- and rotation-invariant,

$$\overline{V(\vec{r})V(\vec{r}\,')} = F(|\vec{r}-\vec{r}\,'|). \tag{11.45}$$

Here F is a peaked function characterized by a height V_0^2 and width λ. This pseudo-translation/rotation symmetry is one of the mathematical advantages of considering the entire ensemble. The random realization of disorder in each sample (strongly) breaks translation symmetry, but the statistical properties of the disorder are translation-invariant. For further mathematical simplicity it is sometimes convenient to approximate F as a Dirac delta function. In d spatial dimensions we then have translationally invariant local correlations of the form

$$F(\vec{r}-\vec{r}\,') = \eta^2 \delta^d(\vec{r}-\vec{r}\,'), \tag{11.46}$$

where $\eta^2 = V_0^2 \lambda^d$. This special case is called the **Gaussian white-noise distribution**.[8]

Of course, it is easy to solve the Schrödinger equation for the ensemble-averaged potential (since it vanishes), but that is not what we need. We are asked to solve the Schrödinger equation for each random potential in the ensemble, use the resulting random wave functions to compute the electronic transport properties, and only then ensemble average those physically observable properties. This is *much* harder! Furthermore, we are interested in the transport properties at very long times, namely times much longer than the mean free time between scatterings. Thus low-order perturbation theory in powers of the impurity potential is completely inadequate. Nevertheless, this is where we will begin, by inspecting the scattering matrix element in Eq. (11.36).

We can think of the integral in (11.36) as being a sum of a large number ($\sim \Omega/\lambda^d$) of random terms (from many different impurities), each of which is statistically independent. Hence we expect $v[\vec{q}]$ to be Gaussian distributed even if $V(\vec{r})$ is not, thanks to the central limit theorem. For our model assumptions we have

$$\overline{v[\vec{q}]} = 0, \tag{11.47}$$

$$\overline{v[\vec{q}]v^*[\vec{q}\,']} = \int d^d r \int d^d r' e^{i\vec{q}\cdot\vec{r}} e^{-i\vec{q}\,'\cdot\vec{r}\,'} \overline{V(\vec{r})V(\vec{r}\,')}$$

$$= \Omega F[\vec{q}] \delta_{\vec{q},\vec{q}\,'}, \tag{11.48}$$

where δ is the Kronecker delta and $F[\vec{q}]$ is the Fourier transform of $F(\vec{r})$ and (for isotropic correlations) depends only on the magnitude of $\vec{q}$, not on the direction. This simple result comes from the pseudo-translation invariance and tells us that all the Fourier components are statistically independent except for the constraint (which often causes confusion) associated with $V(\vec{r})$ being real:

$$v[-\vec{q}] = v^*[\vec{q}]. \tag{11.49}$$

If the random potential is smooth on some length scale λ which is large compared with the Fermi wavelength, then $F[\vec{q}]$ will be strongly varying and decrease rapidly for $q\lambda \gg 1$. For example, if F is a Gaussian

$$F(\vec{r}) = V_0^2 e^{-\frac{r^2}{2\sigma^2}}, \tag{11.50}$$

then the Fourier transform is also a Gaussian,

$$F[q] = V_0^2 (2\pi\sigma^2)^{d/2} e^{-\frac{1}{2}q^2\sigma^2}, \tag{11.51}$$

which rolls off rapidly with q for large correlation length σ. In the opposite limit of white-noise disorder (small σ), Eq. (11.46) yields

---

[8] The reader probably recognizes by now that the way to characterize the disorder potential here is quite similar to that for current noise in Section 10.5. In fact, the disorder potential can be viewed as potential noise (or fluctuation) in space (instead of time).

$$\overline{v[\vec{q}\,]v^*[\vec{q}\,']} = \eta^2 \Omega \delta_{\vec{q},\vec{q}\,'}, \tag{11.52}$$

and the power spectrum of the disorder is independent of q.[9]

Now that we have the statistical properties of the random potential we can study how electrons scatter from the potential fluctuations. Unlike the case of a single impurity discussed earlier, here it is no longer meaningful to talk about the scattering cross section (either total or differential), as one expects it to scale linearly with the number of impurities, which diverges in the thermodynamic limit. The meaningful quantity here is the scattering rate, which, as we discussed in Chapter 8, is finite in the presence of a finite density of scatterers. We will start by filling in some of the microscopic details left out in Chapter 8. Let us imagine that an electron starts in plane-wave state $\vec{k}$. We would like to know the rate at which it scatters out of this state. Fermi's Golden Rule (which is lowest-order perturbation theory and therefore completely equivalent to the first Born approximation) yields for the ensemble-averaged scattering rate[10]

$$\frac{1}{\tau} = \frac{2\pi}{\hbar}\frac{\Omega}{(2\pi)^d}\int d^d q\,\overline{|M_{\vec{k},\vec{k}+\vec{q}}|^2}\delta(\epsilon_{\vec{k}+\vec{q}} - \epsilon_{\vec{k}}) \tag{11.53}$$

$$= \frac{2\pi}{\hbar}\int \frac{d^d q}{(2\pi)^d}F(q)\delta(\epsilon_{\vec{k}+\vec{q}} - \epsilon_{\vec{k}}). \tag{11.54}$$

This first-Born-approximation expression for the quantum scattering rate of course does not contain multiple-scattering effects. Notice, however, that because the quantum scattering rate contains the two-point impurity correlator $F(q)$, this expression *does* include pair-wise quantum interference between scattering off two different impurities into the same final state. These interference effects can be important if the impurity positions are correlated. For example, $F(q)$ might have a peak in the vicinity of $q = 2k_\mathrm{F}$ causing there to be enhanced backscattering across the Fermi surface. However, in the limit of white-noise disorder (or, equivalently, isotropic s-wave point scatterers with no positional correlations) F is independent of momentum transfer, and we can simplify this expression to

$$\frac{1}{\tau} = \frac{2\pi}{\hbar}F(0)\rho(\epsilon_{\vec{k}}), \tag{11.55}$$

where $\rho(\epsilon_{\vec{k}})$ is the density of states (per unit energy per unit volume per spin) at energy $\epsilon_{\vec{k}}$:

$$\rho(\epsilon_{\vec{k}}) = \int \frac{d^d q}{(2\pi)^d}\delta(\epsilon_{\vec{k}+\vec{q}} - \epsilon_{\vec{k}}). \tag{11.56}$$

As noted previously in Chapter 8, for smooth disorder potentials, the scattering is dominated by small momentum transfers, which are less effective at relaxing the current. In this case the transport scattering rate can be considerably smaller than the total scattering rate:

$$\frac{1}{\tau_{\mathrm{TR}}} = \frac{2\pi}{\hbar}\frac{\Omega}{(2\pi)^d}\int d^d q (1 - \cos\theta_{\vec{k}+\vec{q},\vec{k}})\overline{|M_{\vec{k},\vec{k}+\vec{q}}|^2}\delta(\epsilon_{\vec{k}+\vec{q}} - \epsilon_{\vec{k}}) \tag{11.57}$$

$$= \frac{2\pi}{\hbar}\int \frac{d^d q}{(2\pi)^d}\frac{q^2}{2k^2}F(q)\delta(\epsilon_{\vec{k}+\vec{q}} - \epsilon_{\vec{k}}). \tag{11.58}$$

---

[9] Here again we are using the noise terminology, by simply trading the frequency dependence of the noise spectrum for that of wave-vector dependence here. Note that, if we use the Gaussian representation of the delta function in Eq. (11.46), then we have $\eta^2 = V_0^2 (2\pi\sigma^2)^{d/2}$.

[10] Here we are treating this as a single-electron problem and hence ignoring all the Fermi occupation factors which appear in the corresponding expression in Chapter 8. This does not lead to any errors as long as the electrons are non-interacting, and the reader should convince herself/himself that this expression is equivalent to the corresponding ones there.

The corresponding transport mean free path is

$$\ell = v_F \tau_{TR}. \tag{11.59}$$

We will assume that the potential is sufficiently weak (or the impurities are sufficiently dilute) that the mean free path ℓ obeys $k_F \ell \gg 1$. That is, the electron travels many wavelengths between scattering events. Thus we can consider semiclassical wave packets that have both position and momentum that are relatively well-defined. In the semiclassical Boltzmann equation picture, we then extend the analysis of the dynamics to the long times that are needed to study diffusive transport by considering multiple scattering events described by scattering rates rather than quantum amplitudes. As we shall see, this approach neglects important quantum interference effects which can slow or even halt diffusion altogether via the phenomenon of Anderson localization.

> **Exercise 11.2.** Use energy conservation to derive the identity $(1 - \cos\theta_{\vec{k}+\vec{q},\vec{k}}) = q^2/2k^2$ used in Eq. (11.58).

11.3.3 Multiple Scattering and Classical Diffusion

So far we have invoked quantum mechanics in only the most elementary ways possible. We have taken the particle speed to be the quantum-mechanical Fermi speed, and we (in principle) have computed the transport lifetime using the impurity scattering matrix element and Fermi's Golden Rule. All of this seems reasonable, at least in the semiclassical limit $k_F \ell \gg 1$.[11]

However, we have completely neglected the fact that the electrons are really waves and subject to quantum interference. In addition, the fact that we have evaluated the classical random walks assuming that the particles undergo a very large number of collisions naturally implies that the simple Born approximation in the quantum treatment is completely inadequate. By definition, an electron wave impinging on a target whose thickness exceeds the mean free path cannot be described by the lowest-order Born approximation because the fraction of the incident amplitude which goes into the scattered wave is of order unity. We could attempt to extend the perturbation series to higher order, but this would be extremely tedious and is ultimately doomed to failure. Even if we had the strength to compute the first n terms, the result would become invalid at times exceeding approximately $n\tau$, since beyond that time more than n scattering events are likely to have occurred.

Our task now is to understand how to handle the strong multiple scattering of the quantum waves in a random system which classically exhibits diffusion. We need to understand the physics of quantum diffusion. As a first step toward this goal, we present rough arguments allowing us to "derive" classical diffusion from the propagation of quantum waves. Particle conservation, which is so essential to classical diffusion, makes itself felt in one-body quantum mechanics via the fact that the Hamiltonian is Hermitian and the time evolution is unitary, so that

$$\int d^d\vec{r}\, |\Psi|^2 = 1 \tag{11.60}$$

---

[11] The way to see that this corresponds to the semiclassical limit is to note that $1/k_F$ is the minimum size a wave packet whose energy is $\sim \epsilon_F$ can have. Thus $k_F \ell \gg 1$ means we can think of the electron as a wave packet with fixed position and momentum (subject to the uncertainty principle) when propagating between two scattering events. If one prefers to associate the (semi)classical limit with the limit $\hbar \to 0$, simply note that $k_F = mv_F/\hbar \to \infty$ in this limit. Thus an expansion in powers of $\hbar$ is equivalent to an expansion in powers of $1/(k_F\ell)$, which we will call the semiclassical expansion.

is an exact constant of the motion. Hence the Fourier transform of the density

$$\rho_{\vec{q}} \equiv \int d^d \vec{r} \, e^{-i\vec{q}\cdot\vec{r}} |\Psi|^2 \tag{11.61}$$

must necessarily change slowly with time in the limit of small wave vectors, just as it does classically.

We can make some progress by attempting to at least crudely estimate the contribution from arbitrary orders in the perturbation series. It will be useful to find the minimal approximation to the quantum mechanics which will yield classical diffusion at long times. Let us begin by considering the nth-order term in the perturbation series for the time evolution of a wave packet which is initially centered on some random location. We will assume that either the density of scattering centers is small or the scatterers are sufficiently weak that the condition $k_F \ell \gg 1$ is satisfied. Naively, this is the condition within the semiclassical Boltzmann picture in order for one to have a "good" metal. We will see shortly, however, that, in low dimensions, quantum corrections to this picture will be significant beyond some characteristic localization length scale ξ. In 2D $\xi \gg \ell$ if the disorder is weak, and in the large intermediate-scale regime $\ell \ll r \ll \xi$ we expect to find that quantum diffusion is very similar to classical diffusion. Our initial goal, then, should be to reproduce the results of classical diffusion from a semiclassical wave-packet picture. Only then can we begin to consider the quantum corrections to this semiclassical approximation.

Because the mean free path is large compared with the Fermi wavelength, we can choose the initial size of the wave packet to be small enough that it has a well-defined position on the scale of the mean free path. Without loss of generality we take it to be initially located at the origin. We would like to study how the wave packet evolves quantum mechanically. Conceptually, the easiest way to take the semiclassical limit is to formulate quantum mechanics using the Feynman path integral. Instead of thinking about the propagation of particles as waves using the Schrödinger equation, Feynman imagines that quantum particles are indeed particles, but the way to compute the amplitude for them to travel from one point to another is to assume that they can take all possible paths, not just the classical path. Each path is assigned a quantum amplitude (related to the classical action for that path) and the summation over all these paths leads to wave-like interference effects. The so-called Feynman path integral has the form

$$\Psi(\vec{r}, t) = \int \mathcal{D}\vec{x}(t') e^{iS[\vec{x}(t')]/\hbar} = \frac{1}{\mathcal{N}} \sum_{\text{paths}} e^{iS[\vec{x}(t')]/\hbar}, \tag{11.62}$$

where $\vec{x}(t')$ is a path (in space-time) satisfying $\vec{x}(0) = \vec{0}$ and $\vec{x}(t) = \vec{r}$, $S[\vec{x}(t')]$ is its classical action, and the integration measure of $\mathcal{D}\vec{x}(t)$ and the normalization $1/\mathcal{N}$ ensure that Eq. (11.60) is satisfied. Typically S (which enters Eq. (11.62) as a *phase* factor) varies quickly as the path $\vec{x}(t')$ varies, as a result of which contributions from most paths cancel out in the semiclassical limit $\hbar \to 0$. The exceptions are *classical* paths $\vec{x}_{cl}$, on which S is stationary (see Appendix H for a reminder about functional differentiation in the calculus of variations),

$$\left. \frac{\delta S}{\delta \vec{x}(t')} \right|_{\vec{x}_{cl}} = \vec{0}, \tag{11.63}$$

as a result of which S varies slowly around classical paths, and their contributions to the path integral do *not* suffer from the cancelation due to rapid phase variation mentioned above. Thus, in the semiclassical limit, we have

$$\Psi(\vec{r}, t) \approx \sum_{j=1}^{N(\vec{r},t)} a_j(\vec{r}, t), \tag{11.64}$$

where $N(\vec{r}, t)$ is the total number of classical paths which reach position $\vec{r}$ at time t, j is the index for a specific classical path, and

$$a_j(\vec{r}, t) = \frac{1}{\mathcal{N}} e^{iS[\vec{x}_j(t')]/\hbar} = \frac{1}{\mathcal{N}} e^{i\theta_j} \tag{11.65}$$

is its contribution to $\Psi(\vec{r}, t)$. In the absence of any potential, there will be only a single classical path that arrives at $\vec{r}$ at time t, namely the one in which the particle moves in a straight line with fixed velocity $\vec{r}/t$. In the presence of disorder, the chaotic scattering off the impurities will permit the existence of *many* possible paths, each with a different initial velocity vector.

We note that a collection of classical particles located at the origin at $t = 0$ but with completely random initial velocity directions will take all possible *classical* paths with equal probability. Furthermore, the density profile of the classical particles follows the diffusion equation as discussed previously. Hence we expect that the number of classical paths reaching $\vec{r}$ at time t is proportional to the solution of the diffusion equation

$$N(\vec{r}, t) \sim k_{\mathrm{F}}^{-d} P(\vec{r}, t), \tag{11.66}$$

where $P(\vec{r}, t)$ is the solution to the diffusion equation given in Eq. (11.15). The prefactor k_{F}^{-d}, which is needed to convert the Gaussian density into a dimensionless number, can be viewed as the fuzzy quantum "size" of the particle.

The probability of finding the particle at space-time point $(\vec{r}, t)$ is (in the semiclassical limit)

$$|\Psi|^2 = \sum_{j,k=1}^{N(\vec{r},t)} a_k^*(\vec{r}, t) a_j(\vec{r}, t) = (1/\mathcal{N}^2) \sum_{j,k=1}^{N(\vec{r},t)} e^{-i(\theta_k - \theta_j)}. \tag{11.67}$$

We see that the amplitude is the sum of a large number of random complex numbers, due to the difference of the action S from one classical path to another. Furthermore, these phase factors vary rapidly with position. The net result is a density which is a rapidly fluctuating function of position. This is nothing more than the electronic analog of a laser "speckle" pattern associated with randomly interfering waves. The fluctuations occur on the scale of the Fermi wavelength. For $k_{\mathrm{F}}\ell \gg 1$, this length scale is much smaller than the mean free path, which in turn is much smaller than the macroscopic length scales over which we are examining the diffusion.

The microscopic details of the "speckle" fluctuations are determined by the exact locations of all the impurities. In our study of transport in small mesoscopic systems in Chapter 10, we were interested in the effects of these fluctuations which produce sample-to-sample variations in the transport. Here, for the purpose of studying diffusion in macroscopic samples, we are not interested in these microscopic details, and we can safely coarse-grain (i.e. smooth out) the density, or, equivalently, ensemble average the density over the impurity positions, without affecting the diffusive behavior on long length scales. Carrying out this averaging causes all of the cross terms to vanish, leaving only N instead of N^2 terms, resulting in

$$\overline{|\Psi|^2} \approx \frac{N(\vec{r}, t)}{\mathcal{N}^2} = \frac{k_{\mathrm{F}}^{-d}}{\mathcal{N}^2} P(\vec{r}, t) = (4\pi Dt)^{-d/2} e^{-\frac{r^2}{4Dt}}, \tag{11.68}$$

where we have used Eq. (11.60) to fix $\mathcal{N} \approx k_{\mathrm{F}}^{d/2}$. This is in agreement with Eq. (11.15). Thus (by a series of somewhat shady arguments!) we have recovered classical diffusion, starting from the semiclassical propagation of quantum waves.

11.4 Quantum Corrections to Diffusion

It turns out that our arguments used above to recover classical diffusion from the quantum mechanics of waves in a disordered medium miss a subtle quantum interference effect. In a coherent system, the probability of a return to the origin is actually larger than that calculated above. Hence quantum

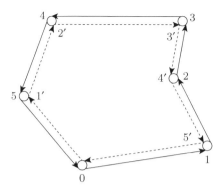

Figure 11.1 Paths that bring a particle back to the origin. For each such path (solid line), there is a *different* path (dashed line) that is its time-reversed version, in which the (space) loop is traversed in the opposite direction. These paths interfere constructively in the presence of time-reversal symmetry. Hence the quantum probability of a return to the origin is higher than the classical probability. Thus this quantum correction lowers the diffusion constant.

corrections reduce the diffusion constant below the value expected from semiclassical Boltzmann transport theory. Let us now see how this comes about in two different pictures, one in real space and one in momentum space.

11.4.1 Real-Space Picture

Consider the random path shown in Fig. 11.1 (solid line) which returns to the origin at time t: $\vec{x}(t' = 0) = \vec{x}(t' = t) = \vec{0}$. Among the large set of other paths which also return to the origin, there is one special path that deserves additional attention. This is the path which is the time-reverse of the original one: $\vec{x}_{\text{rev}}(t') = \vec{x}(t - t')$. As shown in Figure 11.1, this path visits the same set of impurities but in the reverse order. In a system which is time-reversal invariant (i.e. there are no magnetic fields or magnetic scatterers present), the quantum amplitude contributed to the wave function by the reverse path is not random (as we implicitly assumed previously) but in fact is precisely equal to the amplitude contributed by the forward path, because the actions for the two paths are identical: $S[\vec{x}(t')] = S[\vec{x}_{\text{rev}}(t')]$.

This correlation in the phases between certain paths causes *constructive* interference, which enhances the probability of a return to the origin. To see how this works, let us reconsider Eq. (11.64) in light of our new information that something special happens for $\vec{r} = \vec{0}$:

$$\Psi(\vec{0}, t) = \sum_{j=1}^{N(\vec{0},t)/2} 2a_j(\vec{0}, t). \tag{11.69}$$

Because of the phase correlations, we actually have only half as many random uncorrelated amplitudes as before, but now each amplitude is twice as large. Carrying out the same averaging procedure for the wave intensity, we find that the quantum probability of a return to the origin is precisely twice the classical return probability:

$$\overline{|\Psi(\vec{0}, t)|^2} = 2(4\pi Dt)^{-d/2}. \tag{11.70}$$

By conservation of probability, we see that an enhancement of the probability of returning to the origin implies a reduction in the probability of diffusion away from the origin. How does this effect come about in terms of quantum interference? This can be understood from Fig. 11.2, which shows a pair of paths ending away from the origin but containing an intermediate closed loop which is traversed

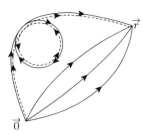

Figure 11.2 Paths with and without loops. Those with loops interfere constructively with a *different* path in which the loop is traversed in the opposite direction.

in opposite directions. It turns out that (roughly speaking) the increased probability of having closed loops reduces the probability for open segments and hence reduces the effective diffusion constant.

11.4.2 Enhanced Backscattering

The argument given above for the quantum corrections to the Boltzmann transport picture was framed in terms of time-reversed pairs of paths in real space. We found that constructive interference leads to an enhancement of the probability of the particle returning to its starting point. It is instructive to consider this effect in momentum space. Recall from our discussion of the difference between the scattering time and the transport scattering time that the scattering events which reverse the direction of the particle's motion are the ones which are most effective at relaxing the momentum distribution and hence the current. It turns out that, if we consider the perturbation series for a multiple-scattering process that takes a particle from $\vec{k}$ to the time-reversed state $-\vec{k}$, we will find pairs of time-reversed paths in momentum space that lead to the same final state. Their interference leads to an enhanced probability for backscattering which is the momentum-space analog of the enhanced probability for a return to the origin in real space.

The scattering rate from $\vec{k}$ to $\vec{k}'$ is given by the square of the total scattering amplitude obtained by summing over all possible intermediate states:[12]

$$|f(\vec{k}, \vec{k}')|^2 = \left| \sum_{n=0}^{\infty} f^{(n+1)}(\vec{k}, \vec{k}') \right|^2 = \left(\frac{m_e \Omega}{2\pi \hbar^2} \right)^2 \left| \sum_{\{\vec{k}_j\}} M_{\vec{k}, \vec{k}'}(\{\vec{k}_j\}) \right|^2. \tag{11.71}$$

Note that, because all the different intermediate states lead to the same final state, we must perform the summation before squaring, and so interference is obtained. The evaluation of the terms in the Born series for the scattering rate is complicated by the fact that each term contains a product of many different Fourier components $v(\vec{q})$ of the random potential. Each of these factors is a random complex number, and it is not trivial to work out the disorder ensemble average of such products.

We saw previously that closed-loop paths in real space could be traversed in two different directions and the resulting quantum amplitudes were identical because of time-reversal symmetry. The correlation (lack of statistical independence) in these random amplitudes causes an increase in the probability of the diffusing particle returning to its starting point. The momentum-space analog of this is an enhanced probability for backscattering from state $\vec{k}$ to the time-reversed state $-\vec{k}$. As we shall see, this enhanced backscattering increases the rate at which a pulse of current comes to a halt.

As an illustration of the source of this enhanced backscattering rate, consider the third-order perturbation-theory process shown in Fig. 11.3. This shows a particle scattering from state $\vec{k}$ to the time-reversed state $-\vec{k}$ via two possible routes:

[12] This summation over intermediate states may be viewed as a summation over paths in *momentum space*.

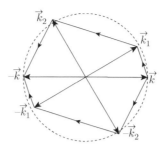

Figure 11.3 Illustration of two different backscattering processes that interfere constructively. Note that, in perturbation theory, the intermediate states are virtual and so do *not* need to have the same energy as the initial and final states.

$$\vec{k} \longrightarrow +\vec{k}_1 \longrightarrow +\vec{k}_2 \longrightarrow -\vec{k}; \tag{11.72}$$

$$\vec{k} \longrightarrow -\vec{k}_2 \longrightarrow -\vec{k}_1 \longrightarrow -\vec{k}. \tag{11.73}$$

The third-order perturbation-theory expression for the effective backscattering amplitude contributed by the upper path is

$$M_{\vec{k},-\vec{k}}(\vec{k}_1, \vec{k}_2) = v(\vec{k}_1 - \vec{k}) G(\vec{k}_1) v(\vec{k}_2 - \vec{k}_1) G(\vec{k}_2) v(-\vec{k} - \vec{k}_2), \tag{11.74}$$

while that for the lower path is

$$M_{\vec{k},-\vec{k}}(-\vec{k}_2, -\vec{k}_1) = v(-\vec{k}_2 - \vec{k}) G(-\vec{k}_2) v(\vec{k}_2 - \vec{k}_1) G(-\vec{k}_1) v(\vec{k}_1 - \vec{k}). \tag{11.75}$$

Because in a time-reversal-invariant system the energy eigenvalues $\epsilon(\vec{k})$ and $\epsilon(-\vec{k})$ are automatically degenerate, the Green function (i.e. the energy denominator) obeys $G(\vec{k}_j) = G(-\vec{k}_j)$. The matrix elements are also identical. Thus we have

$$M_{\vec{k},-\vec{k}}(\vec{k}_1, \vec{k}_2) = M_{\vec{k},-\vec{k}}(-\vec{k}_2, -\vec{k}_1). \tag{11.76}$$

This tells us that these amplitudes, corresponding to time-reversed paths in *momentum space*, automatically interfere constructively and enhance the backscattering rate from $\vec{k}$ to $-\vec{k}$. The same is true for all multiple scatterings at any order in perturbation theory.

A beautiful example of this enhanced backscattering is the **"glory"** effect. While traveling in an airplane which is flying above a smooth cloud layer, it is sometimes possible to see the shadow of the airplane cast on the clouds. Surrounding the shadow will be a bright glow known as the glory. This is the result of the enhanced scattering of the light at angles close to 180°. Light rays from the sun which just miss hitting the airplane scatter around inside the cloud and then return to your eye. If the airplane is well above the cloud layer, then the scattering angle for these rays will be close to 180°.

The same effect can sometimes be seen if you are standing on a mountain summit with friends. If the sun is behind you and you see your shadows cast on a fog layer below the summit, you will see a bright halo of light surrounding the shadow of your own head, but not those of your friends. Of course, each of your friends sees only themselves surrounded by glory, so don't think of yourself as being special!

Quantitative measurement of the light-scattering intensity for laser light scattered from colloidal suspensions shows that the intensity is essentially constant except for a small regime near 180° where the intensity is doubled by the constructive interference effect in backscattering. What is it that determines the angular width of this region of enhancement? The perturbation-theory calculation performed above can be repeated for the case where the final wave vector differs slightly from $-\vec{k}$. The results show that the angular width of the enhancement of the backscattered beam is given approximately by the dimensionless ratio of the wavelength of the light λ to the mean free path,

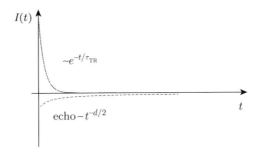

Figure 11.4 Current decay as a function of time. Solid line: exponential decay as expected from Drude theory. Dashed line: echo due to enhanced backscattering with a long power-law tail.

$$\Delta\theta \sim \frac{\lambda}{\ell}. \tag{11.77}$$

Thus the angular width can be viewed as the natural result of diffraction from a source of finite width ℓ.

For electrons, the enhanced backscattering reduces the conductivity. In the absence of quantum corrections to classical diffusion, a delta-function electric field pulse

$$\vec{E}(t) = \vec{E}_0 \delta(t) \tag{11.78}$$

applied to a disordered metal results in a current flow that jumps up to a finite value and then decays exponentially in time as the diffusing electrons gradually "forget" their initial direction of motion. This behavior, which occurs on the characteristic time scale τ_{TR}, is illustrated in Fig. 11.4 and, as noted in Exercise 11.3, can be derived from the frequency dependence of the simple Drude/Boltzmann form of the conductivity. It turns out that the quantum corrections to the diffusion create a long-time tail or "echo" of negative current due to the enhanced backscattering as illustrated schematically in Fig. 11.4. In d spatial dimensions, the exponent of the power-law decay is $d/2$ and is related to the classical return probability that we examined previously in Eq. (11.70). This is because only paths that contain closed loops that return to their starting point suffer the quantum interference correction. As we shall explore further in the next section, this means that, in low dimensions $d \leq 2$, the area under the echo pulse curve actually diverges, indicating that perturbation theory in the disorder strength is breaking down and signaling that all states are localized for $d \leq 2$.

Exercise 11.3. Use the frequency dependence of the Drude conductivity (see Exercise 7.1)

$$\sigma(\omega) = \frac{ne^2 \tau_{\mathrm{TR}}}{m_{\mathrm{e}}} \frac{1}{1 + j\omega\tau_{\mathrm{TR}}} \tag{11.79}$$

to show that the classical Drude part of the response to the electric field in Eq. (11.78) is given by

$$\vec{J}(t) = \vec{E}_0 \frac{ne^2}{m_{\mathrm{e}}} \theta(t) e^{-t/\tau_{\mathrm{TR}}}. \tag{11.80}$$

11.5 Weak Localization in 2D

Without doing any detailed calculations, we can see on physical grounds that, if the classical return probability is large, then quantum corrections to the diffusion (and hence the conductivity) will be large. As mentioned above, the strength of the negative-current "echo" is proportional to the excess return probability caused by the quantum interference (but this excess is precisely the classical return probability $P(\vec{0}, t)$ since the quantum effects simply double the probability).

The conductivity at frequency ω relates the current to the electric field via

$$\vec{J}(\omega) = \sigma(\omega)\vec{E}(\omega). \tag{11.81}$$

The Fourier transform of the delta-function electric field in Eq. (11.78) is simply a constant. Therefore the conductivity is proportional to the Fourier transform (in the electrical engineering notation) of the current produced by a delta-function electric field:

$$\sigma(\omega) = \frac{1}{E_0} \int_0^\infty dt\, e^{-j\omega t} J(t). \tag{11.82}$$

This means in particular that the quantum correction to the classical conductivity is proportional to the Fourier transform of the current "echo." For the case of the dc conductivity, the Fourier transform simply becomes the time integral. Detailed calculations show that the (negative) correction to the dc conductivity is

$$\delta\sigma = -\frac{e^2}{h} 4D \int_{t_0}^\infty dt\, P(\vec{0}, t) = -\frac{e^2}{h} 4D \int_{t_0}^\infty dt\, (4\pi D t)^{-d/2}, \tag{11.83}$$

where (since all our random-walk arguments are valid only for long times) t_0 is some short-time cutoff of order τ_{TR} (the point at which the motion starts to change from being ballistic to diffusive). We see that in 3D this integral converges to a finite value, but it diverges logarithmically in 2D and even more strongly in 1D.

The divergence of $\delta\sigma$ indicates that, in 1D and 2D, the conductivity is actually *zero* in the large-distance and long-time limit! In other words, the diffusive motion of electrons stops beyond a certain time scale, and they are localized within a corresponding length scale. This in turn indicates that all energy eigenstates are localized no matter how weak the disorder is. We thus find that the metallic state described by Drude theory is unstable against weak disorder, which drives the system into an Anderson insulating state.

In 2D the quantum corrections are relatively weak within a fairly wide range of time scales – the integral in Eq. (11.83) barely diverges, increasing only logarithmically with the upper cutoff. Thus, for "good" metal films with $k_F\ell \gg 1$, we have only the phenomenon of "weak localization." The states at the Fermi level (which are responsible for transport) are in fact exponentially localized, but only beyond an enormous exponentially large length scale

$$\xi \sim \ell e^{\frac{\pi}{2}(k_F\ell)}. \tag{11.84}$$

For weak disorder this length can vastly exceed the size of the sample (or, as we will discuss further below, the inelastic scattering length), making the localization too weak to detect. As a result, electron motion is diffusive and well described by semiclassical Boltzmann transport theory for length scales $\ell \ll L \ll \xi$.

To get an idea of the origin of this exponentially large localization length, consider following the time evolution of a wave packet initially localized near the center of a disordered square sample of dimension L embedded in an otherwise disorder-free infinite sample. Classically the characteristic time for a diffusing particle to escape from the disordered region obeys

$$t_{\text{esc}} = \lambda \frac{L^2}{D}, \tag{11.85}$$

where λ is a factor of order unity.

If we use t_{esc} as the upper cutoff in Eq. (11.83), we find that the quantum correction to the conductivity in 2D obeys

$$\delta\sigma = -\frac{e^2}{h}\frac{1}{\pi}\ln\left(\frac{t_{esc}}{t_0}\right). \tag{11.86}$$

Using Eq. (11.85) and taking

$$t_0 \sim \tau_{TR} \sim \frac{\ell^2}{D} \tag{11.87}$$

yields (ignoring a small constant term dependent on the precise value of λ)

$$\delta\sigma = -\frac{e^2}{h}\frac{2}{\pi}\ln\left(\frac{L}{\ell}\right). \tag{11.88}$$

In 2D the Drude conductivity can be expressed as

$$\sigma_0 = \frac{ne^2\tau_{TR}}{m_e} = \frac{e^2}{h}(k_F\ell). \tag{11.89}$$

The quantum corrections become comparable to this at scale

$$L \sim \ell e^{\frac{\pi}{2}k_F\ell}, \tag{11.90}$$

indicating that this is the localization length as claimed in Eq. (11.84). Beyond this scale our approximate treatment of quantum corrections breaks down since the localization is becoming strong. We again emphasize that, even though ℓ and $1/k_F$ are microscopic lengths, ξ can be enormous for weak disorder where $k_F\ell \gg 1$.

For a sample of finite size $L \ll \xi$, $\delta\sigma$ is only a small correction to σ_0. As we will discuss in detail in Section 11.9, even if we have an infinite sample, insofar as quantum interference effects are concerned, at finite temperature the sample consists of independent, incoherently connected, blocks which have an *effective* size of order the phase coherence length ℓ_φ. Even at relatively low temperatures, this effective size can easily be substantially smaller than ξ. In these situations the effects of localization physics are weak, hence the name weak localization. This also points to the need for experiments that directly probe the origin of the localization (and the dephasing) to reveal this fascinating but subtle physics, to which we now turn.

Exercise 11.4. Derive the second equality in Eq. (11.89).

11.5.1 Magnetic Fields and Spin–Orbit Coupling

The weak localization due to enhanced backscattering that we have been discussing depends crucially on having time-reversal invariance. If a magnetic field is present, particles traversing closed loops in opposite directions will acquire a difference in phase due to the Aharonov–Bohm effect; see Fig. 11.5. If the field is strong enough, this will occur on scales shorter than the phase coherence length and begin to destroy the weak localization. This will enhance the conductivity. The field scale at which this occurs can be as small as 0.1–1 Tesla and provides useful information to help calibrate the value of ℓ_φ as a function of temperature. Because weak localization is removed, the conductivity is enhanced. That is, there is a *negative* magnetoresistance.

Figure 11.6 shows experimental transport data illustrating the negative magnetoresistance of a 2D disordered system caused by the destruction of time-reversal symmetry as a magnetic field is applied. At weak fields and the lowest temperatures one sees a small region of positive magnetoresistance.

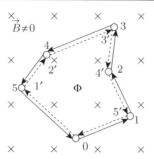

Figure 11.5 The same as Fig. 11.1, but now in the presence of a magnetic field which destroys time-reversal symmetry. There is now a phase difference in the contributions to the returning amplitude from these two paths: $\Delta\phi = 2\pi(2\Phi/\Phi_0)$, where Φ is the magnetic flux enclosed in the loop, $\Phi_0 = hc/e$ is the flux quantum, and the factor of 2 in front of Φ is due to the fact that the two paths enclose the flux in opposite directions; thus the flux *difference* between the two paths is 2Φ.

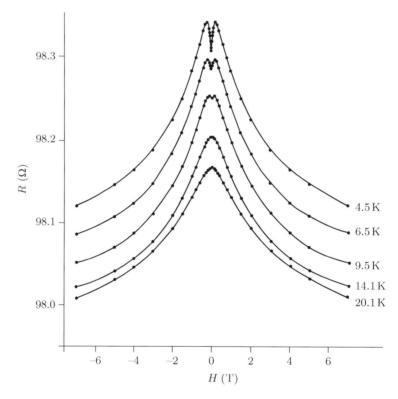

Figure 11.6 The resistance of a 2D disordered copper film as a function of the applied magnetic field for various temperatures. Notice that, at zero field, the resistance increases as the temperature is lowered. This is the result of weak localization and the fact that the phase coherence length increases as the temperature is lowered. The overall drop in resistance as the magnetic field is increased is due to the destruction of weak localization by the fact that the magnetic field breaks time-reversal symmetry. The small region of positive magnetoresistance at very weak fields and lowest temperatures is associated with the destruction of the spin–orbit-induced anti-localization. Figure reprinted from [70] with permission from Elsevier.

This turns out to be the result of spin–orbit scattering of the electrons from high-atomic-weight atoms (possibly impurities) such as gold in the material, which results in *anti*-localization, namely *suppressed* backscattering due to quantum interference. The physical origin of this is an additional Berry phase (see Chapter 13) due to the involvement of electron spin in the scattering processes. This

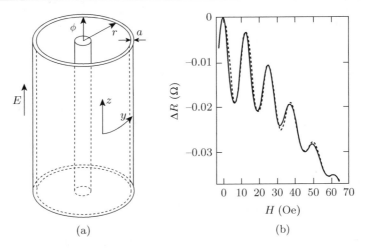

Figure 11.7 (a) A cylinder with a very small thickness a enclosing magnetic flux Φ, with current flowing along the direction of its axis (z) driven by electric field E. (b) The resistance of a lithium cylinder as a function of Φ (solid line), which shows oscillation with period $\Phi_0/2$ and agrees well with the theoretical prediction (dotted line). Figure adapted with permission from [71]. Copyright (1987) by the American Physical Society.

changes the backscattering interference from constructive to destructive. A more detailed discussion of this novel effect can be found in Section 13.3.1. A magnetic field also suppresses the anti-localization effects, and thus increases the resistance.

Perhaps a more dramatic manifestation of the origin of weak localization is the magnetoresistance of a cylinder-shaped conductor, with magnetic field applied along its axis (see Fig. 11.7(a)). If we neglect the flux that goes into the interior of the thin film, the only effect of the field is modulating the phase of electron paths that wrap around the cylinder, via the Aharonov–Bohm effect (see Section 13.2). Weak localization is due to the constructive interference between time-reversed pairs of paths returning to the origin, in the presence of time-reversal symmetry. Such paths, which are closed loops, fall into distinct topological classes labeled by a **winding number** n (see Appendix D). Time-reversed pairs of loops have opposite n, as they wrap around the cylinder in opposite directions. The flux going through the hole of the cylinder thus introduces an **Aharonov–Bohm phase** difference (see Chapter 13)

$$\Delta\phi_n = 2n(2\pi)\Phi/\Phi_0. \tag{11.91}$$

For $n = 0$ (topologically trivial loops) Φ has no effect. But for loops with $n \neq 0$ the constructive interference is suppressed by Φ, *unless* Φ is an integer multiple of $\Phi_0/2$. We thus expect the magnetoresistance to oscillate with Φ, with period $\Phi_0/2$. This is indeed seen experimentally; see Fig. 11.7(b).

Exercise 11.5. Discuss the origin of the difference in the periodicity (in units of the flux quantum Φ_0) of the magnetoresistance in Fig. 10.2 and Fig. 11.7.

11.6 Strong Localization in 1D

The analog of Eq. (11.89) for 1D is (a form which could have been guessed from simple dimensional analysis)

$$\sigma_0 = \frac{e^2}{h}4\ell. \tag{11.92}$$

Evaluation of Eq. (11.83) in 1D yields a much stronger divergence at long times and (when the value of λ in Eq. (11.85) is determined in detail):

$$\delta\sigma = -\frac{e^2}{h}2(L - \ell).$$ (11.93)

Thus we see that the corrections are *linear* in the length rather than logarithmic. This tells us that the quantum correction in Eq. (11.93) is comparable to the classical conductivity in Eq. (11.92) even at microscopic scales and that the localization length ξ is on the scale of the mean free path ℓ. Thus localization in 1D is always strong. Kinematically this is related to the fact that in 1D there are only two directions – forward or backward, thus essentially *all* scatterings are backscatterings, and an electron immediately senses the quantum interference enhancement of backscatterings when it travels a distance of order ℓ. This also means that, unlike in 2D, there is *no* range of length scales between ξ and ℓ for which the electron motion is diffusive. In other words, there is *no* quantum diffusion in 1D.

The quantum corrections are so strong in 1D that we cannot completely trust the approximate picture that we have used to estimate them. More careful calculations, however, confirm the picture that states in 1D are indeed strongly localized – the localization length ξ is in fact the same as the naive transport mean free path ℓ computed in the Born approximation. This means that the conductance of a 1D wire of length L has the form

$$G \sim \frac{2e^2}{h}e^{-L/\xi}.$$ (11.94)

The value of the prefactor has been chosen so that for $L/\xi \longrightarrow 0$ the result matches the quantized Landauer conductance of a clean wire (with the factor of 2 accounting for spin) which we encountered previously in Chapter 10.

Exercise 11.6. Using Eq. (11.94) and physical arguments, find an expression for the conductance of a 1D wire of length L in the temperature regime where the phase coherence length obeys $\xi \ll \ell_\varphi \ll L$. Assume that, for length scales larger than ℓ_φ, the classical Ohm law for resistors in series is valid. See Section 11.9 for further discussion of the concept of phase coherence length.

Thouless made the important observation that a wire does not have to be mathematically 1D in order to suffer this "strong localization" effect. Much as we saw when we studied 2DEGs in Section 9.5.1, quantization of the transverse motion can make a finite-diameter wire act dynamically like a 1D wire by introducing an excitation gap. For sufficiently large wires, this gap becomes very small. Nevertheless, Thouless found that these larger wires could still behave one-dimensionally insofar as their localization properties are concerned, provided that they were 1D with respect to diffusion. That is, one has to consider times much longer than the time required to diffuse a distance equal to the width so that the classical return probability behaves like that in a 1D system.[13]

Imagine that we start with a clean wire (electron "waveguide") of finite cross-sectional area A so that $N \sim k_F^2 A$ channels are open (see Exercise 10.4). Then the Landauer conductance is

$$G = N\frac{2e^2}{h}.$$ (11.95)

Introduction of impurities will cause inter-channel scattering and backscattering so that, as we found previously in our study of the Landauer–Büttiker multi-channel transport picture,

---

[13] We will see shortly that this means that the temperature must be very low so that the inelastic-scattering time (phase coherence time) is much larger than the time it takes electrons to diffuse over the width of the wire.

$$G = \frac{2e^2}{h} \operatorname{Tr}(t^\dagger t), \tag{11.96}$$

where t is the $N \times N$ transmission matrix. That is, t_{ij} is the amplitude for going from entrance channel i to exit channel j on the far side of the sample. Thouless showed that making the wire twice as long does not double the resistance as one might expect from the classical result of Ohm's law for series addition of resistors. Instead he found that for very long wires the resistance increases faster than expected and the conductance becomes exponentially small:

$$G \sim \frac{2e^2}{h} e^{-L/L_0}. \tag{11.97}$$

Here L_0 is the length of wire for which the Landauer result yields $G \sim 2e^2/h$, that is, the length for which $\operatorname{Tr}(t^\dagger t) \sim 1$. At this length scale the number of *effective* channels (introduced in Section 10.4) is down to 1, and the system starts to behave as a genuine 1D wire. Thus we identify L_0 with the localization length ξ. Unlike in the single-channel case, where $\xi \approx \ell$, we can have $\xi \gg \ell$ if N is large. This makes sense when we consider that a 2D sample (which we have already seen can have an exponentially large localization length) can be viewed as a multi-channel 1D sample, when it has a large but finite width.

11.7 Localization and Metal–Insulator Transition in 3D

We have seen that, since the time integral of the classical return probability converges for dimensions greater than two, the leading perturbative quantum corrections to the classical diffusion do not lead to localization, so the metallic state is *stable* against weak disorder. On the other hand, strong enough disorder can still drive the system into an Anderson insulator state, with a metal–insulator phase transition occurring at some finite value of the disorder. Because the phase transition occurs at finite disorder, it is not accessible via simple perturbative analysis. If, however, we assume that the metal–insulator phase transition is continuous, we can make some progress using renormalization-group and scaling-theory ideas.

One of the central features of a **continuous phase transition** is that there exists a characteristic length scale, the correlation length ξ, which diverges at the critical point. On the insulating side of the transition, it seems natural to define ξ as the localization length of the eigenfunctions at the Fermi energy. As the localization length grows larger and larger, the system is just barely insulating. On the other side of the transition, we expect it to be just barely conducting. We can identify the correlation length on the metallic side from a dimensional analysis of the conductivity. In 3D the resistivity has units of ohm-meters, and so the conductivity must have the form

$$\sigma = \frac{e^2}{h} \frac{1}{\xi}, \tag{11.98}$$

where ξ has units of length. We identify this as the correlation length. As we shall see in Section 11.8, physically this is the length scale above which Ohm's law starts to apply, and the conductivity becomes well-defined.[14] Since (by assumption) this length diverges at the critical point, it follows that the conductivity must rise continuously from zero as the metallic phase is entered.

---

[14] Namely the conductance of a cubic-shaped conductor with length L starts to scale with L in the form of $G(L) = \sigma L^{d-2} = \sigma L$. Equivalently ξ may be identified as the scale at which $G(L = \xi) \sim e^2/h$. It is clear that for $L < \xi$ Ohm's law does *not* apply, as the quantum nature of conduction implies that it must be non-local, as discussed in detail in Chapter 10.

Typically one expects that, for a continuous transition, the correlation length diverges as a power law in the tuning parameter that controls the distance to the transition. Thus, if δ is a measure of this distance (we will encounter an explicit example below), we expect

$$\xi \sim |\delta|^{-\nu}, \tag{11.99}$$

where ν is the so-called correlation length-exponent. Using this, we obtain a power-law form for the conductivity,

$$\sigma \sim |\delta|^{\nu}, \tag{11.100}$$

on the metallic side of the transition. On the insulating side of the transition, the fact that the eigenfunctions are almost extended implies that the static dielectric constant diverges. In Chapter 15 we will learn that the 3D Fourier transform of the $1/r$ Coulomb potential is $v(q) = 4\pi e^2/q^2$, and that in a clean metal the statically screened Coulomb interaction (as a function of the wave vector q) is given by

$$v_{\rm s}(q) = \frac{4\pi e^2}{q^2 + q_{\rm TF}^2}, \tag{11.101}$$

where $q_{\rm TF}^2 \equiv 4\pi e^2\, dn/d\mu$ is the square of the Thomas–Fermi screening wave vector (see Eq. (15.84)). Writing the screened interaction in terms of the bare Coulomb interaction and the (wave-vector-dependent) **dielectric function**

$$v_{\rm s}(q) = \frac{4\pi e^2}{q^2} \frac{1}{\epsilon(q)} \tag{11.102}$$

yields

$$\epsilon(q) = 1 + \frac{q_{\rm TF}^2}{q^2}. \tag{11.103}$$

In a disordered insulator the static screening will look metallic on length scales smaller than the localization length. Beyond that scale the system will be polarizable but cannot exhibit true metallic screening since the electrons are prevented from traveling distances greater than the localization length. Hence it is reasonable to expect that the dielectric function looks something like

$$\epsilon(q) \approx 1 + \frac{q_{\rm TF}^2}{q^2 + \xi^{-2}}. \tag{11.104}$$

Thus, as the transition is approached from the insulating side, the zero-wave-vector static dielectric constant diverges as

$$\epsilon(0) \sim (q_{\rm TF}\xi)^2 \sim |\delta|^{-2\nu}. \tag{11.105}$$

The physical origin of this divergence is the large polarizability that occurs when the eigenfunctions have large spatial extent. Crudely speaking, each localized electron which is perturbed by a weak applied electric field will have a dipole moment on the order of $p \sim e\xi$. The polarizability χ is proportional to $p^2 \propto \xi^2$, resulting in $\epsilon(0) \propto \xi^2$ when ξ becomes large. This divergence is needed to match smoothly the infinite static dielectric constant (found at $q = 0$) in the metal.

Figure 11.8 presents data showing the divergence of the dielectric susceptibility ($4\pi\chi = \epsilon - 1$) in a silicon sample as the metal–insulator transition is approached from the insulating side by increasing the density of phosphorus donors. Beyond the phase-transition point, characterized by a critical dopant density $n_{\rm c}$ (and the distance to criticality is $\delta = n - n_{\rm c}$), the conductivity rises continuously from zero as the system enters the metallic regime.[15] These results strongly suggest that the metal–insulator

---

[15] Note that the metallic phase corresponds to higher dopant density, where the system is *more* disordered. On the other hand, in this regime there is also a higher carrier density and Fermi energy, which tends to make the system more

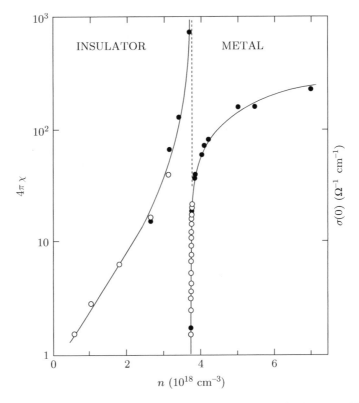

Figure 11.8 (left axis) Divergence of the dielectric susceptibility as the metal–insulator transition is approached from the insulating side. (right axis) Vanishing of the conductivity as the transition is approached from the metallic side. The tuning parameter that drives the phosphorus-doped silicon sample through the transition is the density of phosphorus donor atoms. Open and closed circles represent data points from different sources. See the original reference for details. Figure adapted with permission from [72]. Copyright (1982) by the American Physical Society.

transition is continuous. However, because Coulomb interactions can play an important role in such transitions, they are not necessarily well described by simple Anderson localization transitions of *non-interacting* electrons driven purely by disorder. This fact complicates the interpretation of the data.

11.8 Scaling Theory of Localization and the Metal–Insulator Transition

11.8.1 Thouless Picture of Conductance

In classical physics, and more specifically the Drude and (semiclassical) Boltzmann transport theories, electric transport is characterized by the conductivity σ, which is a *local* property of a conductor (the local value of the current is linearly proportional to the local value of the electric field). Starting from Chapter 10, however, we began to understand that *quantum* transport is intrinsically *non-local*; thus

---

metallic. See Fig. 11.15 and related discussions later in this chapter. Also playing a role is the fact that, as the dopant density increases, the distance between neighboring dopants decreases and the wave functions of their bound states start to overlap more and more strongly. Thus electrons bound to the donors can also tunnel more easily from one donor to another, and eventually delocalize.

electric transport is better characterized by the conductance G, which can have a highly non-trivial dependence on the linear *size* (or length scale) L of the conductor being studied. Such scale dependence reveals the nature of electronic states (e.g. whether they are localized or not) at the Fermi energy.

Assuming Ohm's law and the Einstein relation do apply (at a certain scale), we have the conductance of a cubic-shaped conductor in d dimensions (see Fig. 11.9(a)):

$$G(L) = \sigma L^{d-2} = e^2 D \frac{dn}{d\mu} L^{d-2} = e^2 \frac{D}{L^2} \frac{d(L^d n)}{d\mu} = \frac{2e^2}{\Delta t\, \Delta E}, \tag{11.106}$$

where

$$\Delta t = L^2/D \tag{11.107}$$

is the time it takes an electron to diffuse over a distance (of order) L, and

$$\Delta E = \left(\frac{d(L^d n/2)}{d\mu} \right)^{-1} = \left(\frac{dN(E)}{dE} \right)^{-1} \bigg|_{E_F} \tag{11.108}$$

is the average energy-level spacing at the Fermi energy of this *finite-size* conductor ($N(E)$ is number of states below energy E for each spin).

Thouless made the observation that, because the conductor is connected to leads (so that current can flow through it), its energy levels are *not* sharp; since an electron typically spends time Δt inside the conductor before diffusing out of it (compare Eq. (11.107) with Eq. (11.85)), from the Heisenberg uncertainty relation we expect a broadening of the energy level at the Fermi energy of order

$$E_T = h/\Delta t, \tag{11.109}$$

known as the **Thouless energy**. We thus have

$$G(L) = \frac{2e^2}{h} \frac{E_T}{\Delta E} = g(L) \frac{2e^2}{h}, \tag{11.110}$$

where

$$g(L) = \frac{E_T}{\Delta E} \tag{11.111}$$

is a *dimensionless* measure of the conductance (see Eq. (10.43)). In the Thouless picture, g is the ratio between the energy-level uncertainty and the energy-level spacing of the conductor at the Fermi level.

In theoretical (and, in particular, numerical) studies of conductors, we often consider *isolated* systems in order to simplify the physical setting. In an isolated and finite-size system, the energy levels are discrete and *sharp*. Is it possible to estimate E_T in such studies? Thouless further observed that leads are coupled to the conductor at its boundaries and can be viewed as a boundary effect; thus E_T is related to the sensitivity of the energy level to a *change* of the boundary condition. Thouless thus suggested estimating E_T as the change of energy for the state (labeled i) at the Fermi level when the boundary condition changes from periodic to antiperiodic along one direction (while sticking to a periodic boundary condition in the other directions):

$$E_T \approx |E_i(\phi = \pi) - E_i(\phi = 0)|, \tag{11.112}$$

where ϕ is the boundary-condition angle for the wave function in the $\hat{x}$ direction:

$$\psi(x + L) = e^{i\phi} \psi(x). \tag{11.113}$$

Using the typical value of the energy shift in Eq. (11.112) (evaluated at the Fermi level) as the *definition* of the Thouless energy E_T and substituting it into Eq. (11.111), we obtain

$$g_T(L) = [E_T/\Delta E]|_{E_F}. \tag{11.114}$$

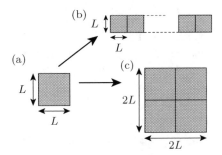

Figure 11.9 (a) A cubic-shaped conductor with linear size L. (b) Repeating *exactly* the same conductor along one direction to form a 1D wire with perfectly periodic potential. (c) Blocking together 2^d such conductors with the same macroscopic properties, but different microscopically, to form a cubic-shaped conductor with linear size $2L$.

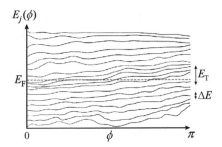

Figure 11.10 Schematic band structure for a quasi-1D metal constructed from periodically repeated blocks. The average level spacing is ΔE. For a good metal we expect that the Thouless energy (E_T) greatly exceeds the level spacing: $E_T \gg \Delta E$. If one follows a single level from the zone center to the zone boundary, the typical change in energy of that level is E_T. Since this greatly exceeds the level spacing and since (generically) the levels do not cross, it must be the case that the motion of nearby levels is correlated. The range of correlation covers g_T levels. Hence this is the typical number of levels in the band structure that cross the Fermi energy E_F and thus the Thouless number g_T determines the Landauer conductance. Note the boundary condition angle ϕ (defined in (11.113)) is related to the Bloch momentum via $k = \phi/L$, where L is the linear size of the block.

g_T is known as the **Thouless number** or **Thouless conductance**; it is expected to be close to (at least in order of magnitude), but not necessarily equal to, the actual dimensionless conductance g.

The (approximate) identification of g_T as g can be justified for the case of a "good" metal, meaning one for which g, $g_T \gg 1$, through the following consideration. Let us repeat this conductor *identically* along the $\hat{x}$ direction a large number of times to form a 1D wire (see Fig. 11.9(b)). This wire has a *periodic* potential with periodicity L; in other words the original conductor with size L has become a unit cell of the wire. For this wire we know from the Landauer formula that its conductance is

$$G(L) = \frac{2e^2}{h} N, \tag{11.115}$$

where N is the number of open channels. Notice that there is no backscattering in this wire because, by construction, the potential is *periodic*. The unit cell may be large, but Bloch's theorem for periodic potentials still applies. When we form such a wire by repetition, each energy level in the original conductor *broadens* to form a band,[16] and the band width is approximately equal to the level uncertainty in Eq. (11.112), as $\phi = 0, \pi$ correspond to bonding and anti-bonding states between neighboring blocks. A schematic band structure is shown in Fig. 11.10. Recalling Bloch's theorem, we see that

---

[16] Very much like each atomic energy level forms a band when atoms form a lattice; such broadening is precisely the source of the energy-level uncertainty discussed by Thouless.

$\phi = 0$ corresponds to the zone center and $\phi = \pi$ corresponds to the zone boundary in the band structure diagram. N is the number of bands that *intersect* the Fermi energy. Since the density of bands is the same as the density of states in a single block (i.e. the density of states per block per unit energy), $1/\Delta E$, we obtain (see Fig. 11.10), using the definition of Eq. (11.114),

$$N \approx [E_T/\Delta E]|_{E_F} = g_T(L). \tag{11.116}$$

On the other hand, for such a wire with no backscattering, G is *independent* of the length; this means that the smallest building block of the wire, the original block of conductor with length L, must have exactly the same G. Thus we have

$$g(L) = N \approx g_T(L). \tag{11.117}$$

11.8.2 Persistent Currents in Disordered Mesoscopic Rings

The model described above of an infinite 1D wire constructed by periodically repeating a single random segment of metal (as shown in Fig. 11.9(b)) would seem to be quite unrealistic. However, it turns out to be closely related to a realistic description of a disordered mesoscopic ring of metal. In Exercise 10.1 we considered the persistent currents that flow in a 1D (single-channel) ring enclosing magnetic flux and having zero disorder. Consider now the case of a quasi-1D disordered ring containing many transverse modes (multiple channels). The mathematics of this problem is identical to the band structure of a periodically repeated disordered block of metal (see Exercise 11.7). The band structure schematically illustrated in Fig. 11.10 shows the energy levels as a function of the wave vector in such a 1D wire.

Consider now a multi-channel wire in the shape of a ring so that there are periodic boundary conditions. One can show (see Exercise 11.7) that the energy levels of this ring vary as a function of magnetic flux through the ring in *exactly* the same way that the energy levels in the infinitely repeating wire vary with the wave vector (as shown in Fig. 11.10). (We neglect the small corrections to the geometry due to the bending of the wire into a ring by treating it as a linear wire segment of length L with periodic boundary conditions.) The range of wave vectors in the first Brillouin zone is $-\pi/L < k < +\pi/L$, while the range of flux corresponding to these wave vectors is $-\Phi_0/2 < \Phi < +\Phi_0/2$.

The quasi-random variations of the energy levels with flux through the ring constitute a unique "fingerprint" for the particular realization of disorder in a given ring. Since the derivative of the total energy[17] of the occupied levels with respect to the flux is the total persistent current flowing in the ring (see Exercise 10.1), one can measure this "fingerprint" by measuring the variation of the persistent current with the flux through the ring. These currents are tiny (on the scale of nanoamperes), but with modern measurement techniques they can be determined with high precision as illustrated in Fig. 11.11. Just as the conductance of a mesoscopic sample can be related to the Thouless energy, so can the typical size of the persistent current fluctuations which are given at zero temperature by the natural scale $E_T/\Phi_0 = eD/L^2 \sim ev_F\ell/L^2$, where ℓ is the mean free path for the electrons diffusing in the random potential. This diffusion current scale should be contrasted with the (much larger) ballistic current scale, ev_F/L, which is relevant to the disorder-free case as discussed in Exercise 10.1. At finite temperatures $k_B T > E_T$, the persistent current is strongly suppressed by thermal averaging over the different single-particle level occupancies. The presence of spin–orbit coupling, Coulomb interactions, and/or superconducting fluctuations can lead to interesting corrections even at zero temperature.

---

[17] Note that, for an isolated ring, the particle number N, rather than the chemical potential μ, is constant. Thus, in computing the current, the derivative of the total energy with respect to the flux must be taken at constant N rather than constant μ.

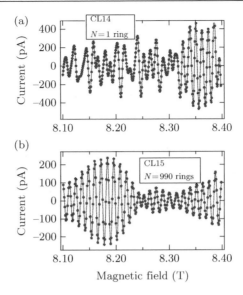

Figure 11.11 (a) Persistent current through a single mesoscopic ring as a function of the applied field. (b) The average persistent current per ring in an group of 990 rings. The rapid oscillation period corresponds to one flux quantum through the area enclosed by the ring. The slow oscillation scale is related to the finite cross-sectional area of the metal in the ring itself (and the magnetic flux that penetrates it). Note the similarity to Fig. 10.2. Figure reprinted from William Shanks, Ph.D. Thesis, Yale University, 2011. Copyright by W. E. Shanks. All rights reserved. See also [73].

Exercise 11.7. Consider an electron moving around a disordered ring with circumference L and area A. Let x be the coordinate along the ring. For simplicity neglect the size of the ring in directions perpendicular to $\hat{x}$. Clearly x and $x + L$ correspond to the same point, thus the electron wave function must satisfy the periodic boundary condition: $\psi(x) = \psi(x + L)$. The Schrödinger equation for eigenfunctions,

$$\left[\frac{p^2}{2m} + V(x)\right]\psi_n(x) = E_n\psi_n(x), \tag{11.118}$$

needs to be solved with this periodic boundary condition. Now consider adding a flux Φ through the ring. Show that $E_n(\Phi) = \epsilon_n(k)$, where $\epsilon_n(k)$ is the band dispersion of electron motion in the periodic potential $V(x + L) = V(x)$, and $k = (2\pi/L)\Phi/\Phi_0$, where Φ_0 is the flux quantum.

11.8.3 Scaling Theory

It is striking that, as we saw above, a random but periodically repeated block yields a 1D conductor. We turn now to the more realistic case where the blocks are independently random with no correlations among them in the microscopic configuration of the disorder potential. This will lead to completely different physics, namely localization.

In a scaling analysis, one attempts to relate physical properties of a system with one (sufficiently large) size to those of another, with as little *microscopic* input as possible. For the case of the size dependence of the conductance, let us put 2^d blocks of conductors together to form a single block with size $2L$ (see Fig. 11.9(c)). These 2^d blocks have the same *macroscopic* properties such as carrier densities, impurity densities/distribution, and conductance, but differ in microscopic details such as the precise impurity configuration. The question now is, what is the (dimensionless) conductance $g(2L)$? According to Eq. (11.111), we need to obtain detailed information about the energy-level spacing and uncertainty of the $2L$ block to answer this question. Thouless further observes that the eigenstates and energies of the $2L$ block should be closely related to those of the L blocks; in particular, if the blocks were decoupled, the Hamiltonian of the $2L$ block would simply be the sum of the Hamiltonians of the L blocks. Of course, these blocks are coupled together, and we (loosely) attribute such coupling to a term in the total Hamiltonian:[18]

$$H_{2L} = \sum_j H_L^j + H_T, \qquad (11.119)$$

where H_L^j is the Hamiltonian of the jth decoupled block, and H_T is the coupling between blocks. Now imagine calculating properties of the $2L$ block by treating H_T as a perturbation. One expects that a dimensionless quantity like $g(2L)$ can be expressed as a perturbation series involving ratios between the matrix elements of H_T and the energy differences of $\sum_j H_L^j$. The latter is of order $\Delta E(L)$. The former is actually responsible for the energy-level uncertainty of individual blocks, and thus should be of order $E_T(L)$. For small $g(L)$ we might expect the perturbation to be weak, and lowest-order perturbation theory might suffice. For large $g(L)$, we expect the scale of the coupling term $E_T(L)$ to be larger than the typical level spacing in the block $\Delta E(L)$, rendering perturbative analysis difficult. Nevertheless, one might expect to be able to express the dimensionless conductance at scale $2L$ in the form

$$g(2L) = F\left[\frac{E_T(L)}{\Delta E(L)}\right] = F[g(L)], \qquad (11.120)$$

in terms of some (unknown) scaling function F that relates the conductance of a piece of conductor with size L to that of a piece of size $2L$. If this is true, and if we know the form of the scaling function as well as the conductance for some given length scale, we can determine the conductance at *all* scales, *without* detailed knowledge of the microscopics of the system. Notice the key assumption here that the scaling function does not depend explicitly on the scale L or any microscopic details, but rather only on the dimensionless ratio of the two characteristic energy scales.

11.8.4 Scaling Hypothesis and Universality

In 1979 Abrahams *et al.* [74] (see also Wegner [75]) proposed a very bold hypothesis, namely that such a scaling function not only exists, but is also *universal*, in the sense that its form is *independent* of microscopic details such as the carrier concentration, band structure, impurity concentration, specific form of impurity potential etc. Instead it depends only on *qualitative* features of the system, including the dimensionality and, in particular, symmetry properties, including the presence or absence of time-reversal symmetry and spin–orbit couplings.

In theoretical analysis, it is often convenient to cast the scaling relation Eq. (11.120) in a differential form, in which the length scale L is changed continuously, and one focuses on how the conductance changes accordingly,

$$\frac{d\log g}{d\log(L/\ell)} = \beta(g), \qquad (11.121)$$

where the logarithms are introduced for convenience, and ℓ is some microscopic length scale (such as the mean free path) inserted to make the argument of the logarithm dimensionless. Clearly its actual value is irrelevant, and so is the base of the logarithm. In this continuum picture, the scaling hypothesis asserts that the scaling function $\beta(g)$ is a function of the conductance g only, at least for sufficiently large length scales ($L \gg \ell$); its specific form depends only on the few qualitative properties of the system mentioned above.

---

[18] One concrete example is the Anderson model (11.153). The coupling term would then be the near-neighbor hopping terms that correspond to hops connecting neighboring blocks.

Now let us check the validity of Eq. (11.121) in a couple of known cases.

- For the case of a good metal, $g \gg 1$, Ohm's law should be valid in certain regimes. In this case the conductivity σ is a fixed number that is independent of scale, and the (dimensionful) conductance obeys $G(L) = \sigma L^{d-2}$. This means that

$$\beta(g) = d - 2 \tag{11.122}$$

depends on the dimensionality d *only*. This is clearly consistent with the scaling hypothesis, and a fancy way of stating that Ohm's law tells us that, when you lump a number of conductors together (some in series and some in parallel), the total conductance depends only on the conductance of the original conductor, *not* on what it is made of. The dimensionality enters because it tells you how many conductors need to be connected in parallel when you double the linear size.
- In the opposite limit of an Anderson insulator, where states at the Fermi energy are localized, we expect, for large L, $g(L) = g_0 e^{-L/\xi}$, where $g_0 \sim O(1)$. Thus

$$\beta(g) = -L/\xi = \ln(g/g_0) \to \ln g; \tag{11.123}$$

again the scaling hypothesis is valid for sufficiently large L (such that $g \ll 1$).

Abrahams *et al.* further assumed that $\beta(g)$ is a continuous and *monotonic* function of g, that smoothly interpolates between the large- and small-g limiting behaviors discussed above. This is partially justified from perturbative treatment of disorder scattering (in the sense of semiclassical expansion in powers of $1/k_F\ell$) in the large-g (or weak-disorder) limit, which yields[19]

$$\beta(g) = d - 2 - a/g + O(1/g^2), \tag{11.124}$$

where a is a constant depending on the "universality class" of the system (e.g. whether there is spin–orbit scattering or not). For ordinary potential scattering, we have $a = 1/\pi > 0$. This *negative* correction to the Ohmic behavior of $\beta(g)$ is a result of the enhanced backscattering due to quantum interference discussed earlier, and is consistent with the assumption of a monotonic $\beta(g)$, which is sketched in Fig. 11.12 for $d = 1, 2, 3$.

The qualitative difference between 3D and 1D or 2D is reflected in the fact that, for 1D and 2D, we *always* have $\beta(g) < 0$. This means that, no matter how large g is initially, it decreases with

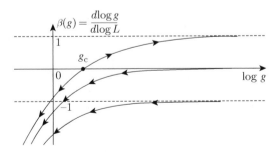

Figure 11.12 Scaling functions for (from bottom to top) dimensions $d = 1, 2, 3$. The arrows indicate the directions of flow of g under the scaling of Eq. (11.121). Notice that for $d = 1, 2$ the flow is always towards $g = 0$. In $d = 3$ there is an unstable fixed point at a critical value of $g = g_c$ where the scaling function passes through zero and changes sign.

---

[19] Note that we just saw that simple low-order perturbation theory in hopping between neighboring blocks makes sense only for $g \ll 1$. The perturbation theory used in the limit $g \gg 1$ is subtle and involves a resummation of an infinite number of particular terms in the naive perturbation theory. These terms capture the physics of the enhanced backscattering between time-reversed pairs of states $\vec{k}$ and $-\vec{k}$ discussed previously.

increasing L, and will eventually cross over to the insulating behavior of Eq. (11.123), indicating that, all states are localized as discussed earlier. An important difference between 2D and 1D is that in 2D $\beta(g) = -O(1/g) \ll 1$ for large g, thus this decrease is very slow initially, and the effect of localization is therefore *weak*. In contrast, in 1D $\beta(g) = -1 - O(1/g)$ is negative and *finite* to begin with. The reduction of g and flow to insulating behavior is very rapid, and hence we use the term "strong localization."

In 3D, on the other hand, $\beta(g)$ crosses zero at a critical value of $g = g_c$. For $g > g_c$, $\beta(g)$ increases with increasing L, and eventually reaches the Ohmic regime described by Eq. (11.122); this corresponds to the metallic phase. For $g < g_c$, $\beta(g)$ decreases with increasing L, and eventually reaches the insulating regime described by Eq. (11.123); this corresponds to the Anderson insulator phase in 3D.

The point $g = g_c$ is very special. Because $\beta(g = g_c) = 0$, g does *not* change with increasing L. This is thus a *fixed point* under scaling. It actually describes the *critical point* separating the metallic and insulating phases. At this metal–insulator transition, the characteristic length scale ξ is infinite. This is an *unstable* fixed point, in the sense that any *infinitesimal* deviation from g_c, namely $g = g_c + \delta g$, will drive g *away* from g_c under scaling (or increasing L), toward one of the *stable* phases, namely metal (with $g \to \infty$) or insulator (with $g \to 0$).[20]

The behavior of $\beta(g)$ *near* the critical point $g = g_c$ tells us about the system's properties near the transition. There we can linearize the scaling curve by approximating

$$\beta(g) \approx y(\log g - \log g_c) = y \log(g/g_c), \tag{11.125}$$

where y is the slope of the scaling curve at $g = g_c$. Assuming the initial value of g is $g_0 = g_c(1 + \delta)$ ($|\delta| \ll 1$ so the system starts very close to the phase boundary) at microscopic scale $L_0 = \ell$, and, integrating both sides of Eq. (11.121) with the approximation above, we obtain

$$\frac{\log[g(L)/g_c]}{\log(g_0/g_c)} = \left(\frac{L}{L_0}\right)^y. \tag{11.126}$$

This leads to a length scale ξ at which $|\log(g/g_c)| = O(1)$, thus

$$\xi \sim \ell \cdot |\delta|^{-(1/y)}. \tag{11.127}$$

Physically ξ is the length scale at which g is sufficiently far from g_c that the system is *clearly* in one of the two stable phases: for $g < g_c$ the system is insulating, so ξ is nothing but the localization length, while for $g > g_c$ the system is a metal, and beyond this scale Ohmic scaling (Eq. (11.122)) starts to dominate; we thus have the conductivity of the form Eq. (11.98). In both cases ξ is the *diverging* correlation length upon approaching the critical point (δ is a measure of the distance to criticality). On comparing this with Eq. (11.99), we find that the correlation length exponent obeys

$$\nu = 1/y. \tag{11.128}$$

Thus the scaling function not only tells us what the stable phases are, but also determines the details of the critical behavior near the (continuous) phase transition.

---

[20] In some sense $g = 0$ and ∞ are *stable* fixed points, in 3D, as g will be brought back to them under scaling if the deviation from them in g is not too big. Readers familiar with renormalization group (RG), and in particular its application to critical phenomena, must have recognized that the language we are using here is identical to that in the RG: g plays the role of a running coupling constant, Eq. (11.121) is an RG flow equation with $\beta(g)$ determining the flow; stable fixed points of $\beta(g)$ correspond to stable phases while unstable fixed points of $\beta(g)$ correspond to critical points. Note that under scale changes by a factor of λ the correlation length changes by a factor of ξ/λ. If the correlation length is infinite, however, it remains infinite under coarse-graining. This is why the critical point is an unstable fixed point. Away from the fixed point, ξ is finite and ξ/λ becomes smaller under coarse-graining, meaning that we have moved further away from the fixed point. Hence the critical fixed point is unstable.

Exercise 11.8.

(i) The scaling theory of localization states that

$$\beta(g) = \frac{d\log(g)}{d\log(L)} = d - 2 - \frac{a}{g}, \qquad (11.129)$$

where g is the dimensionless conductance, L is the system size, d is the dimensionality, and a is a known number of order 1. Let us assume $g = g_0$ for $L = l$. Calculate $g(L)$ for $L > l$, and $d = 1, 2, 3$.

(ii) Now focus on 2D and use the fact that $a = 1/\pi$ for potential scattering. At a microscopic length scale of the mean free path ℓ, the quantum interference effect has yet to kick in and we can use the Drude result of Eq. (11.89) for conductivity at this scale. Combine this with the results of part (i) to determine the localization length ξ. Hint: at the scale of ξ the conductance should approach zero very quickly.

11.9 Scaling and Transport at Finite Temperature

We have just studied how the conductance should scale with the system size. In practice, experiments done on 2D metal films with moderate disorder do reveal a reduced conductance, but it is never seen to vary with the sample size. Instead it is independent of the sample size, but varies with temperature. More generally, macroscopic conductors have a temperature-dependent resistivity or conductivity that is *independent of size*, while the conductance scales with the system size according to Ohmic scaling, $G \sim \sigma L^{d-2}$. Why is this? In order to answer this question, we need to understand the role of the inelastic scattering which inevitably occurs at finite temperatures.

Crudely speaking, inelastic scattering causes an electron to "forget" its phase after a phase coherence time τ_φ and hence destroys the possibility of quantum interference (a detailed discussion of the dephasing mechanism was presented in Section 10.6). Thus the "echo" of negative current (a quantum interference effect) is destroyed for times greater than τ_φ, and the quantum corrections to the Boltzmann conductivity are removed on length scales larger than the corresponding scale ℓ_φ, the so-called phase coherence length. This length can be as large as microns at low temperatures so that it vastly exceeds the elastic mean free path, and yet it is much smaller than the dimensions of any macroscopic sample. We should then view any macroscopic sample as made up of a large number of independent segments of size ℓ_φ. Because there are no quantum interference effects beyond this size, the scaling relation (11.121) stops working once we have $L \sim \ell_\varphi$, and we can view these segments as independent classical resistors connected together according to Ohm's law. The conductance of the segments $G(L = \ell_\varphi)$ is controlled by the size of the quantum corrections at that scale, and gives the observed conductivity of the macroscopic sample

$$\sigma = G(\ell_\varphi)\ell_\varphi^{2-d}. \qquad (11.130)$$

As the temperature is lowered, the scale increases, which makes the quantum corrections stronger. In 2D this lowers the conductance of each segment and causes the measured macroscopic conductivity to vary with temperature.

At low temperatures the inelastic scattering rate often assumes a power-law form,

$$\frac{1}{\tau_\varphi} = AT^p. \qquad (11.131)$$

The prefactor A and the exponent p depend on the nature of the inelastic scattering. Typically, however, Coulomb interactions among the electrons dominate, and in this case one can show that

$p = 1$. Phonons generally produce a larger exponent and therefore are less important at very low temperatures.

The effect of a finite phase coherence time is to impose an upper cutoff on the time integration in Eq. (11.83). In 2D this leads to

$$\delta\sigma = -\frac{e^2}{h}\frac{1}{\pi}\ln\left(\frac{\tau_\varphi}{t_0}\right), \tag{11.132}$$

which, using Eq. (11.131), leads to a logarithmic temperature dependence

$$\delta\sigma = -\frac{e^2}{h}\frac{p}{\pi}\ln\left(\frac{T_0}{T}\right), \tag{11.133}$$

where T_0 is the temperature scale associated with the short-time cutoff, t_0. Such a logarithmic temperature dependence of the conductivity has indeed been seen experimentally in thin films. Interestingly, it turns out that the effect of Coulomb interactions (which have so far been neglected in our discussion) in the presence of disorder can also produce a logarithmic temperature dependence in 2D which complicates the analysis. The details are rather technical and beyond the scope of the present discussion. We note here only that the magnetoresistance experiments described in Section 11.5.1 can help sort out the various contributions to the transport.

In 3D the metallic phase is stable, and the temperature dependence of the conductivity of a good metal is usually dominated by scattering rates (elastic and inelastic; recall Matthiesen's rule in Eq. (7.2)), instead of quantum interference. Of more interest (in the present context) is what happens near the metal–insulator (or localization) transition. This occurs at finite disorder and is not perturbatively accessible. Thus our arguments about the effect of inelastic-scattering-induced dephasing on the quantum echo cannot be used to determine the temperature dependence of the conductivity. We can, however, argue that if phase coherence is destroyed on length scales larger than ℓ_φ then Eq. (11.98) should be replaced by (see also Eq. (11.130))

$$\sigma = \frac{e^2}{h}\frac{1}{\ell_\varphi}. \tag{11.134}$$

This equation should be valid on either side of the transition sufficiently close to the critical point that the phase coherence length is shorter than the correlation length ξ.[21]

If the electron is only very weakly localized and can be viewed as undergoing semiclassical diffusion, then the phase coherence length will be related to the phase coherence time by

$$\ell_\varphi^2 = D\tau_\varphi. \tag{11.135}$$

It is very important to understand that the assumption of semiclassical diffusion underlying Eq. (11.135) is appropriate to the case of weak localization in 2D, but is quite inappropriate near the localization phase transition in 3D, where the disorder is strong and it turns out that the diffusion coefficient is strongly scale-dependent (as we will show below). In this case we need a different argument.

At a quantum critical point such as that for the 3D metal–insulator transition, one expects on general grounds that there is a power-law relationship between the characteristic length and time scales. That is, associated with the diverging correlation length one expects a diverging correlation time

$$\xi_\tau \sim \xi^z, \tag{11.136}$$

---

[21] The validity of this statement is perhaps clearer on the metallic side of the transition, but it should also be true on the insulating side because the system "can't tell" that it is on the insulating side if the dephasing length cuts off the scaling.

where z is the so-called **dynamical critical exponent**. Thus we predict that the phase coherence length is related to the phase coherence time by

$$\ell_\varphi \sim \tau_\varphi^{1/z} \qquad (11.137)$$

rather than by the simple diffusion relation given in Eq. (11.135), which is equivalent to $z = 2$. Furthermore, in an *interacting* system at its critical point, the only characteristic (low) energy scale is $k_B T$, so we expect[22]

$$\frac{h}{\tau_\varphi} \sim k_B T \qquad (11.138)$$

and thus

$$\ell_\varphi \sim T^{-1/z}. \qquad (11.139)$$

Thus we expect that, at a 3D interacting critical point, the conductivity will behave as

$$\sigma \sim T^{1/z}. \qquad (11.140)$$

This is illustrated in Fig. 11.13.

The value of the exponent z depends on the dimensionality and the types of interactions present. In the quantum Hall localization phase transition, which we will study later, there is good evidence that $z = 1$, probably due to the long-range Coulomb interaction. For the 3D metal–insulator transition there is good evidence that $z = 2$ despite the presence of Coulomb interactions. For a non-interacting system at its critical point, one can argue that the only characteristic time scale associated with the length scale L is related to the characteristic energy-level spacing ΔE at that scale:

$$\tau_L \sim \frac{h}{\Delta E}. \qquad (11.141)$$

Using Eq. (11.108) yields

$$\tau_L \sim h\frac{dn}{d\mu}L^d. \qquad (11.142)$$

Thus we deduce $z = d$ in d dimensions for a non-interacting system with a finite density of states.

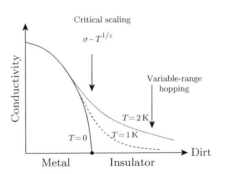

Figure 11.13 A cartoon of the temperature dependence of the conductivity near the metal–insulator transition in an interacting 3D system. At the critical point the conductivity obeys a power law in temperature. Deep in the insulating phase, variable-range hopping dominates. Deep in the metallic phase, phonons and electron–electron scattering hinder rather than help the conductivity and so give a weak positive temperature coefficient of resistivity.

---

[22] Strictly speaking, we mean here that the renormalization group fixed point Hamiltonian for the system contains interactions. The bare Hamiltonian may contain interactions which turn out to be irrelevant and flow away to zero under coarse graining.

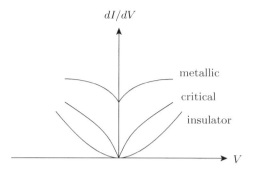

Figure 11.14 A cartoon of the differential tunneling conductance (which measures the tunneling density of states; see Appendix C) near the metal–insulator transition at zero temperature. In the metallic phase there is a power-law cusp known as a zero-bias anomaly. At the critical point the cusp goes all the way to zero conductance and may have a different exponent. In the insulator there is a pseudogap in the density of states. These effects are caused by enhancement of the Coulomb correlations in the presence of strong disorder. At finite temperatures, the singularities are rounded out for voltages obeying $eV < k_B T$.

On dimensional grounds we expect the diffusion constant at the critical point to scale like

$$D \sim \frac{L^2}{\tau_L} \sim L^{2-d} \tag{11.143}$$

for non-interacting electrons. Using the Einstein relation, this implies that in 3D the conductivity at the critical point vanishes like L^{-1}, in agreement with Eq. (11.98) as well as the scaling relation (11.121) which implies *conductance* is scale-invariant. Because the diffusion constant varies strongly with scale, we have to be careful in using it to relate lengths to times as we did for the 2D case in Eq. (11.135).

For an interacting system at its critical point, dimensional arguments suggest that the diffusion constant should scale like $D \sim L^2/\tau_L \sim L^{2-z}$ and that the **tunneling density of states** (see Appendix C for an introduction to this quantity, which is different from the thermodynamic density of states) should scale as[23] $\rho \sim L^{-d}\tau_L \sim L^{z-d}$. Thus, if interactions make $z < d$, the diffusivity is enhanced but the tunneling density of states is suppressed. (In actual experiments at finite temperatures one should replace L by ℓ_φ in these expressions.) The essential physics is that in disordered systems it takes a longer time for interacting electrons to diffuse away from each other than it does for them to fly ballistically away as they do in a clean metal. This makes it more difficult for the tunneling electron to enter the sample and end up in a low-energy state (in which it avoids close approach to other electrons). Such effects have been seen in tunneling into dirty metals. Related effects persist away from the critical point as well: on the metallic side of the transition there is a downward cusp in the tunneling density of states at the Fermi level. On the insulating side there is a **pseudogap** with a strongly suppressed tunneling density of states. These effects are illustrated in Fig. 11.14.

Exercise 11.9. We saw above that, on dimensional grounds, the tunneling density of states at the Fermi level should scale as $\rho \sim L^{-d}\tau_L$ in a system at its interacting critical point. We also noted that at finite temperature the system size L should be replaced by the phase-breaking length ℓ_φ. At zero temperature, but finite bias voltage for the tunneling measurement, L should be replaced by a length, L_V, associated with the bias voltage. (i) Given the value of the dynamical

---

[23] ρ has units of states per unit volume per unit energy. The factor L^d supplies a volume. The factor L^z supplies a time which is equivalent to an inverse energy.

critical exponent, z, find how L_V scales with the voltage. (ii) The tunneling density of states can be measured because it is proportional to the differential conductance dI/dV in the tunneling experiment. At an interacting critical point, the differential conductance vanishes as a power law for a small voltage,

$$\frac{dI}{dV} \sim V^\theta.$$

Use your result for L_V to predict the exponent θ that will be observed in experiment in a 3D sample. (Assume $z < d$ is known.)

So far we have been looking at the effect of finite temperature on the conductivity near the metal–insulator transition in 3D, or in the weakly localized regime in 2D. We now turn to an examination of the conductivity deep in the localized insulating phase.

11.9.1 Mobility Gap and Activated Transport

Imagine that a system is in the insulating phase because of strong disorder. This means that the chemical potential lies in the regime of localized states, as shown in Fig. 11.15. It may be, however, that carriers above a certain energy beyond the Fermi level are delocalized. This characteristic energy E^* is known as the **mobility edge**. The energy interval between the Fermi level and E^* is known as the **mobility gap**. It is not a true gap in the density of states such as occurs in a band insulator since the density of states remains finite in this region. However, this mobility gap has a similar effect on the temperature dependence of the conductivity to that of a band gap.

At finite T, electrons that have been thermally excited above the mobility gap will contribute to the conductivity. Hence the conductivity will have an activated form (or follow an Arrhenius law) just like in a clean semiconductor:

$$n \propto \sigma \sim \sigma_0 e^{-\beta(E^*-E_F)} = \sigma_0 e^{-\beta\Delta}, \tag{11.144}$$

where n is the density of electrons occupying extended states, and the activation energy is equal to the mobility gap:

$$E_a = \Delta = E^* - E_F. \tag{11.145}$$

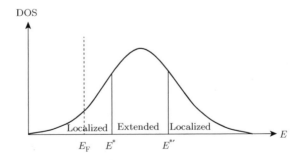

Figure 11.15 A sketch of the density of states (DOS) and mobility edges of the 3D Anderson model with $W < W_c$. Localized states occur near the band edges, while extended states exist near the band center (unless all states are localized by strong disorder, which is the case for $W > W_c$). The boundaries between extended and localized states are the mobility edges (E^* and $E^{*\prime}$ in the figure). The metal–insulator transition corresponds to the point where the Fermi energy (which can be tuned by changing the carrier density as in the example of Fig. 11.8) coincides with one of the mobility edges. When the Fermi energy E_F is in the localized region, the distance between E_F and the closest mobility edge is called the mobility gap. At finite temperatures, the mobility gap determines the thermal activation energy for the conductivity (the energy needed to excite an electron from the Fermi level into the extended states).

Comparing this with Eq. (9.8), we see that in a clean semiconductor E_a is half the band gap because the Fermi energy is in the middle of this gap.

In Fig. 11.15 we assume the Fermi energy is below the lower mobility gap, $E_F < E^*$. It is also possible that Fermi energy is above the upper mobility edge, $E_F > E^{*\prime}$. In this case it is still in the localized region of DOS, and the system is again an Anderson insulator. The mobility gap in this case is

$$\Delta = E_F - E^{*\prime}. \tag{11.146}$$

The easiest way to understand this is to perform a particle–hole transformation and view the conduction as due to hole motion. Then Δ is the energy that a hole needs to receive in order for it to become delocalized. In general Δ is the distance between E_F and the closest mobility edge.

We will shortly find out that there is an additional conduction mechanism called variable-range hopping in which particles hop incoherently among localized states. At extremely low temperatures in systems with large mobility gaps, this mechanism can dominate over activation to the mobility edge.

11.9.2 Variable-Range Hopping

In the previous subsection we considered one conduction mechanism in an Anderson insulator at finite temperature, namely electrons (or holes) get thermally activated above the mobility gap. This is very similar to the conduction mechanism in a clean band insulator, as well as in Mott insulators that (as we will see later) support a charge gap due to electron–electron interaction. It turns out there is another conduction mechanism, that is unique to Anderson insulators, to which we now turn.

Because the density of states for non-interacting electrons in a disordered material is generally non-zero at the Fermi energy, there will exist (empty) states which are arbitrarily close in energy to those occupied by electrons right below the Fermi level. However, in the localized phase, states which are nearby in energy tend to be separated spatially by large distances. If they were not, the system would conduct. In order to move through such a localized system, the electrons must change their energies. This can, for example, be done by emitting and absorbing phonons, provided that the temperature is non-zero. Thus, at sufficiently low temperatures, the motion will consist of a series of incoherent hops between localized states. There will be long pauses between the rare hopping events while the particle waits for a thermal fluctuation to drive it into the next hop.

Suppose that we consider hopping between two states ψ_1 and ψ_2 whose center positions are separated by a distance R (see Fig. 11.16). Mott argued that the typical matrix element squared for a hop of distance R will be of order $e^{-R/\xi}$ due to the exponentially small overlap[24] of the localized states (assuming that the phonon acts as a local perturbation).

Suppose without loss of generality that the energy splitting of the two states is $\delta\epsilon \equiv \epsilon_2 - \epsilon_1 > 0$. We will first assume that the lower energy state is occupied by an electron and that the upper one is empty. If $k_B T \ll \delta\epsilon$, the density of phonons of the required energy $\delta\epsilon$ will be very low. This will contribute a factor $n_B(\delta\epsilon) \sim e^{-\beta\delta\epsilon}$ so that the transition rate will be proportional to

$$\Gamma \sim e^{-\beta\,\delta\epsilon} e^{-R/\xi}. \tag{11.147}$$

---

[24] The two states are, strictly speaking, orthogonal, and have zero overlap since they are eigenstates of the Hamiltonian with different eigenvalues. Each state is exponentially small in the region of concentration of the other, but the exact orthogonality comes from different oscillations in the signs of the wave functions. The spatial variation of the perturbation associated with (say) a phonon makes the matrix element non-zero, but it is still expected to be exponentially small in R/ξ.

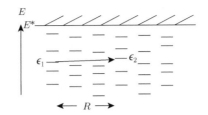

Figure 11.16 A sketch of an inelastic hopping process between two localized states. Above E^*, states are extended. Below E^*, states are localized; the bars represent their energies (vertical) as well as their spatial locations (horizontal).

At low temperatures the system will strongly prefer to hop between states with the smallest possible value of $\delta\epsilon$. On the other hand, in order to find a state to hop to for which $\delta\epsilon$ is small, the particle will have to look out to very large distances. To see why this is so, recall from Eq. (11.108) that the typical level spacing in a box of size R is $\delta\epsilon \sim ((dn/d\mu)R^d)^{-1}$. Thus small $\delta\epsilon$ typically requires searching a large volume $\sim R^d$ to find a state close by in energy.

Using the typical minimum level separation at scale R yields

$$\Gamma \sim \exp\left[-\frac{1}{k_B T (dn/d\mu) R^d} - \frac{R}{\xi} \right]. \tag{11.148}$$

This varies extremely strongly with distance and we may safely assume that the hopping rate is dominated by hops whose distance R minimizes the exponent:

$$\frac{\partial \ln \Gamma}{\partial R} = \frac{d R^{-(d+1)}}{k_B T \, dn/d\mu} - \frac{1}{\xi} = 0. \tag{11.149}$$

This yields

$$\frac{R}{\xi} \sim \left(\frac{T_0}{T} \right)^{\frac{1}{d+1}}, \tag{11.150}$$

where

$$T_0 \sim \left(\frac{1}{d} k_B \xi^d \frac{dn}{d\mu} \right)^{-1} \tag{11.151}$$

is an estimate of the characteristic energy scale associated with the localization length. This clearly demonstrates that for $T \ll T_0$ the typical hopping distance greatly exceeds the localization length and so our assumption is self-consistent.

Evaluating the hopping rate at the optimal distance R yields

$$\Gamma \sim e^{-a\left(\frac{T_0}{T} \right)^{\frac{1}{d+1}}}, \tag{11.152}$$

where a is a constant of order unity (whose precise value is beyond the ability of the present arguments to compute). As the temperature is lowered, the typical hop distance grows larger and larger yielding "stretched exponential" rather than simple Arrhenius behavior (i.e. the exponential of a fractional power of the inverse temperature rather than simply the inverse temperature). This stretched exponential is very small at low temperatures, but it still vanishes much more slowly than the Arrhenius law associated with activation to the mobility edge discussed previously. Eventually, variable-range hopping always dominates at asymptotically low temperatures.

The above derivation ignored the possibility that it was the upper state that was occupied. In this case the electron must emit rather than absorb a phonon, and the associated factor is much larger: $n_B(\delta\epsilon) + 1 \sim 1$. However, the probability that the electron will be found in the upper state is correspondingly lower, so this process makes a similar contribution to the overall rate and the essential features of the result are unchanged.

Efros and Shklovskii [76] showed that, if long-range Coulomb interactions are important in the localized phase, states that are nearby in energy must be pushed even further apart spatially in order to stabilize the ground state. This leads to a modified exponent in 3D of $1/2$ rather than $1/4$ in the stretched exponential. This effect also leads to a sharp dip in the tunneling density of states near the Fermi level, which varies as $|\epsilon - \epsilon_F|^{1/2}$. This "soft" gap is often referred to as the **Coulomb gap**. This soft gap occurs for any value of the Fermi energy and is not associated with a particular critical point of the type discussed in Exercise 11.9.

11.10 Anderson Model

Our discussions so far have mostly been about the continuum, and from the perspective of interference-enhanced backscattering by impurities. The starting point of such discussions is plane (or Bloch) waves, which are eigenstates of a disorder-free Hamiltonian. We have been considering how they are "perturbed" by impurities. In this section we offer a different perspective using a tight-binding lattice model, which is what Anderson did in his quite remarkable original paper [69]. We also briefly explore a modern question – can interacting systems at finite temperature exhibit "many-body localization?"

We begin our discussion by introducing the original model that Anderson considered. The **Anderson-model** Hamiltonian takes the form

$$\hat{H} = \hat{V} + \hat{T} = \sum_i V_i |i\rangle\langle i| - \sum_{ij} t_{ij} |i\rangle\langle j|, \qquad (11.153)$$

where i, j are lattice sites, and V_i is a random on-site potential energy which is drawn from a certain distribution function $P(V)$, a typical example of which is a uniform distribution in the range $-W/2 < V < W/2$, with no correlations between different sites. t_{ij} is the hopping-matrix element between sites i and j. For simplicity one usually considers only nearest-neighbor hopping, with t taken to be a constant, and neglects further-neighbor hoppings. Thus, for $W = 0$, Eq. (11.153) reduces to the familiar tight-binding model with nearest-neighbor hopping. As we learned in Chapter 7, in this limit we should take advantage of the lattice translation symmetry by forming Bloch states using plane-wave-like superpositions of different $|i\rangle$, which, in the case of a Bravais lattice, immediately yields eigenstates of $\hat{H} = \hat{T}$. For finite but small W, the $\hat{V}$ term breaks the lattice translation symmetry and introduces scattering between Bloch states with different wave vectors. We can (under appropriate conditions, including but not limited to $W \ll t$) treat such scattering processes perturbatively, which is what we have been (implicitly) doing thus far for the continuum case.

What we are interested in here, on the other hand, is localization physics driven by W. It is thus natural to consider the opposite limit of $W \gg t$. Let us start by considering $t = 0$. In this case we simply have

$$\hat{H}|i\rangle = \hat{V}|i\rangle = V_i|i\rangle, \qquad (11.154)$$

namely $|i\rangle$ is an exact eigenstate of $\hat{H}$ with eigenvalue $E = V_i$. Obviously all states are localized in this limit. This is, of course, an extreme example of Anderson localization: without t we simply have isolated atoms, whose eigenstates are simply atomic orbitals localized on individual atoms.

Let us now consider the more generic situation in which we have a non-zero $t \ll W$. In this case it seems quite natural to treat $\hat{T}$ as a perturbation of $\hat{V}$ in Eq. (11.153), and develop a perturbative series for eigenstates of $\hat{H}$ in powers of t. To do this we use the machinery reviewed in Section 11.3.1, in which we treated $\hat{V}$ as a perturbation of $\hat{T}$. In particular, the equivalent of Eq. (11.38) here is

$$|\psi_j\rangle = |j\rangle + \hat{G}\hat{T}|\psi_j\rangle, \qquad (11.155)$$

where $|\psi_j\rangle$ is the perturbed eigenstate of $\hat{H}$ that is connected to $|j\rangle$ in the limit $t \to 0$, and

$$\hat{G} = \frac{1}{E - \hat{V}} \tag{11.156}$$

is the appropriate resolvent operator here, with E being the appropriate eigenenergy. The corresponding Born-like series is thus

$$|\psi_j\rangle = |j\rangle + \sum_{j_1} \frac{t_{j_1 j}|j_1\rangle}{E - V_{j_1}} + \sum_{j_1 j_2} \frac{t_{j_2 j_1} t_{j_1 j}|j_2\rangle}{(E - V_{j_2})(E - V_{j_1})} + \cdots. \tag{11.157}$$

We thus find that $|\psi_j\rangle$ has amplitudes at $j' \neq j$, but such amplitudes are typically of order $(t/W)^l$, where l is the minimum number of lattice steps between j and j'. This is because such amplitudes show up only in the lth-order term in Eq. (11.157), in which the numerator is simply t^l while the denominator is *typically* of order W^l. One needs to be careful here as there is some probability that $|E - V_k|$ is very small for some site k that enters the series in Eq. (11.157), resulting in a large contribution. One may worry that the existence of such accidental resonant terms might destroy the convergence of the series. Fortunately, given some small parameter Λ, the condition $|E - V_k| < \Lambda$ is satisfied only with small probability that vanishes linearly with Λ, and (typically) the farther away from j one needs to go to find such resonant sites; this brings with it higher powers of t/W. Anderson [69] has an additional argument that, if two states are nearly degenerate, the coupling between them lifts the degeneracy, restoring the convergence of the series if the terms are summed appropriately. As a result, the probability that the series in Eq. (11.157) converges approaches unity for $|t/W|$ that is less than some critical value. When this is the case, eigenstates with energy E are localized. It turns out that this critical value is infinite for $d = 1, 2$ for any E, rendering all eigenstates localized in 1D and 2D. In 3D, the series in Eq. (11.157) *diverges* for $W < W_c \approx 16t$ and $E = 0$, indicating that states at $E = 0$ (the center of the Bloch band) become delocalized or extended at $W = W_c$. Further decreasing W renders eigenstates delocalized over an increasing range of energy. The localized and extended states are separated by so-called mobility edges; see Fig. 11.15 for an illustration. Note that the density of states (DOS) is continuous both for extended and for localized states, and also across the mobility edges; as a result, it cannot be used to distinguish these different cases. This is surprising, particularly if one makes the (somewhat loose) analogy between localized states and bound states, whose spectra are discrete and show up as delta-function peaks in the DOS (see, e.g., Fig. 9.3). On the other hand, one can define a *local* density of states (LDOS), which clearly distinguishes between localized and extended states. See Exercise 11.10.

In our discussions thus far, we have been considering tuning the system through the quantum phase transition between metal and insulator by changing the strength of the disorder. It is also possible to go through the transition by changing the density of carriers, as we have seen in Fig. 11.8. As the density is increased, the Fermi energy E_F goes up. This makes the potential energy of the disorder relatively less important and tends to increase the mean free path both in absolute terms and in terms of the dimensionless parameter $k_F \ell$. As discussed earlier, when E_F reaches a mobility edge, the metal–insulator transition occurs. When the Fermi level is below the mobility edge but the temperature is finite, electrons can be excited across the mobility gap, leading to thermally excited conduction obeying the Arrhenius law described in Eq. (11.144). This gives way to variable-range hopping at lower temperatures.

It is not really possible to change the carrier density in an elemental metal. However, in alloys like NbSi or phosphorus-doped silicon (P:Si) the alloying or doping does control the carrier density (see Fig. 11.8 for an example). In certain semiconductor structures that form 2D electron gases, it is possible to tune the carrier density electrostatically using voltages applied to nearby metallic gates.

We have been mostly ignoring particle interactions in our discussion. In practice, increasing the carrier density tends to reduce the importance of Coulomb interactions relative to the kinetic energy and also improves the screening of the disorder potential (to be discussed in detail in Chapter 15). Both of these effects also drive the system away from the insulating phase and toward the metallic phase. However, the presence of interactions poses some new conceptual questions in the context of localization physics, which we discuss in the following section. Here we simply point out that, in the presence of interactions, it is no longer possible to study localization by inspecting single-particle eigenstates and their characterization using quantities like inverse participation ratios. This is because a many-body eigenstate is, in general, an extremely complicated superposition of products of N_e single-particle states, where N_e is the number of electrons in the system, typically of order 10^{23}. While that is already a mind-bogglingly huge number, it is *not* the origin of the extreme complexity of a many-body wave function. The complexity comes from the fact that these N_e electrons can be placed in a (much) larger number of different single-electron orbitals N_o. The combinatorial factor $C_{N_o}^{N_e}$ can easily reach $10^{(10^{23})}$. Dealing with such levels of complexity is the ultimate challenge in condensed matter physics.

Exercise 11.10. Consider the Anderson model with N sites, whose normalized eigenstates are $|\psi_\alpha\rangle$ with eigenenergy E_α, $\alpha = 1, 2, \ldots, N$. Define the properly normalized density of states (DOS) as

$$\rho(E) = \frac{1}{N} \sum_{\alpha=1}^{N} \delta(E - E_\alpha). \tag{11.158}$$

(i) Show that $\rho(E)$ satisfies the normalization condition

$$\int_{-\infty}^{\infty} \rho(E) dE = 1. \tag{11.159}$$

Now introduce the local density of states (LDOS) at site i:

$$\rho_i(E) = \sum_{\alpha=1}^{N} \delta(E - E_\alpha) |\langle i | \psi_\alpha \rangle|^2. \tag{11.160}$$

(ii) Show that

$$\int_{-\infty}^{\infty} \rho_i(E) dE = 1. \tag{11.161}$$

(iii) Show that

$$\frac{1}{N} \sum_i \rho_i(E) = \rho(E), \tag{11.162}$$

namely the DOS is the spatial *average* of the LDOS.

Exercise 11.11.

(i) Consider the special limit of the Anderson model (11.153) with $V_j = 0$ for every site j (or, equivalently, $W = 0$ with W defined below Eq. (11.153)). Show in this case that the local density of states on site i obeys $\rho_i(E) = \rho(E)$ if we have a Bravais lattice with periodic boundary condition. Plot $\rho(E)$ vs. energy for the case of a 1D lattice.

(ii) Now consider the opposite limit of $t = 0$, with the distribution of V_i parameterized by $W > 0$. Find and plot $\rho(E)$ in this case.

(iii) For a specific instance of the disorder realization, find and plot $\rho_i(E)$ for a *specific* site i, and discuss the origin of the difference between $\rho_i(E)$ and $\rho(E)$ in this case, in the limit of large N. Assume $t = 0$.

(iv) Consider a more generic situation in which $W > 0$ and $t > 0$, with $t \ll W$ such that all states are localized. Construct a schematic plot for $\rho_i(E)$, and discuss its difference from what you plotted for part (i), in which case all of the states are extended.

11.11 Many-Body Localization

We turn now to the question of localization in interacting systems at finite temperature. In an interacting system, the many-body wave function cannot simply be given by listing a set of occupied single-particle orbitals which may or may not be spatially localized. Instead we have a complex superposition of (exponentially) many different sets of occupied states. Crudely speaking, we have to learn how to talk about localization in Hilbert space rather than in real space.

We learned previously that electrons coupled to a phonon bath at finite temperatures suffer interactions that cause a finite phase coherence length. Beyond that length, the quantum interference that leads to Anderson localization is shut off and localization fails. The modern question of **many-body localization (MBL)** is essentially the following. If we have an interacting system which is isolated (e.g. a cold atomic gas in an isolated optical lattice not coupled to an external bath that can cause dephasing), are the states at finite energy density (above the ground state) localized? Loosely speaking, do the inelastic-scattering processes associated with the interactions cause the system to act as its own (dephasing) bath? This is a complex question which at the time of writing is still under debate. Just as Anderson localization is very strong in 1D, it appears from numerical evidence that MBL is strong as well. It is quite possible that in higher dimensions MBL is not stable against these self-dephasing effects, but at the time of writing, this is not yet firmly established.

Let us begin the discussion by reminding ourselves that the reason why non-interacting systems are tractable, despite the huge numbers of electrons and available orbitals, is the equally large number of conservation laws, namely the occupation number of each single-particle eigenstate, n_s, where s is the label of the single-particle eigenstate, is a good quantum number. n_s can be either 0 or 1 for fermions, and any non-negative integer for bosons. As a result, a many-body eigenstate is labeled by a *specific* collection of these occupation numbers, $\{n_s\}$, and the complexity associated with the Hilbert-space combinatoric factor $C_{N_0}^{N_e} \sim 10^{(10^{23})}$ does not enter.

In a clean system s is the (lattice) momentum $\vec{k}$, while in the $t = 0$ limit of the Anderson model (11.153) it is simply lattice-site label j. We thus see the difference between the two extreme limits of extended and localized phases: the conserved quantities are particle numbers in plane waves (which spread out over the entire space) and local atomic orbitals, respectively. More generally, the conserved quantities are associated with *local* observables like n_l, where l is the label of a *localized* single-particle eigenstate, in the localized phase. As we illustrate below, this way of identifying localization can be generalized to systems with interactions as well.

To be concrete, let us return to the original Anderson model, and then generalize it to include interactions. Anderson's original motivation was to understand spin diffusion (which can be measured, for example, using NMR methods) in the presence of randomness. Different nuclei have differently sized spins, but let us assume the spin-1/2 case for simplicity. The same model can be applied to spinless particles (or, say, spin-polarized electrons) hopping on a lattice. In that context the degrees of freedom on each lattice site j can be represented as a spin-1/2 system, with the state $|\uparrow\rangle_j$ representing $|j\rangle$ being occupied by a particle, while $|\downarrow\rangle_j$ represents $|j\rangle$ being empty. In terms of spin operators Eq. (11.153) can be written as (up to a constant)

$$\hat{H} = \sum_i V_i S_i^z - \sum_{ij} t_{ij} S_i^+ S_j^-. \tag{11.163}$$

The reader should check that the Hamiltonians (11.153) and (11.163) are equivalent within the Hilbert space in which there is a single particle, or a single up-spin. Anderson was considering how this up-spin diffuses under the Hamiltonian (11.163), which is equivalent to how an electron diffuses under the Hamiltonian (11.153).

We now apply the Hamiltonian (11.163) to the entire spin Hilbert space, which can describe the spin-diffusion problem under general conditions (not simply restricted to the case of a single up-spin). It is no longer possible to map (11.163) to a non-interacting particle Hamiltonian.[25] We are thus facing an *interacting* system with many ($\sim 10^{23}$) degrees of freedom (spins in this case), or a genuine many-body problem. For reasons discussed above, such problems are very hard to solve.

To gain some insight into localization physics let us again consider the limit $t = 0$. In this case Eq. (11.163) reduces to a collection of *decoupled* spins in a site-dependent magnetic field V_j pointing along the z direction,

$$\hat{H}_0 = \sum_i V_i S_i^z, \tag{11.164}$$

and the many-spin eigenstates are simply the tensor product of single-spin eigenstates:

$$|\{s_j\}\rangle = |s_1 s_2 \cdots s_N\rangle, \tag{11.165}$$

where $s_j = \pm 1/2$ is the eigenvalue of S_j^z, and N is the number of lattice sites or spins. Again the reason why we can solve the (trivial) Hamiltonian (11.164) is due to the existence of an extensive set of conservation laws, namely each S^z is separately conserved. In fact, the members of the set $\{S_j^z\}$ form a complete set of conserved observables in the sense that they not only commute with the Hamiltonian (11.164) (and thus are conserved), but also commute with each other (thus they share a common set of eigenstates):

$$[\hat{H}_0, S_i^z] = [S_i^z, S_j^z] = 0, \tag{11.166}$$

and specifying their eigenvalues completely determines the state (11.165).

The Hamiltonian (11.164) is trivial because it describes *non-interacting* spins. Before going back to the full interacting spin Hamiltonian (11.163), let us first introduce a more special type of spin–spin interaction to (11.164), and consider

$$\hat{H}_0' = \sum_i V_i S_i^z + \sum_{ij} J_{ij} S_i^z S_j^z, \tag{11.167}$$

where the second term introduces interactions between different spins.[26] Because it involves S^z only, we still have

$$[\hat{H}_0', S_i^z] = [S_i^z, S_j^z] = 0, \tag{11.168}$$

and the eigenstates of $\hat{H}_0'$ are still those in Eq. (11.165), even though the eigenenergies now depend on $\{J_{ij}\}$ and $\{s_j\}$ in a more complicated way. We see that the solvability[27] of the *interacting* spin

---

[25] Hamiltonians that are quadratic in boson or fermion operators are exactly soluble. For spin-1/2 systems, the spins act like hard-core bosons – the spin operators commute on different sites, but S_i^+ cannot act twice on the same site without annihilating the state. Fermion operators have the same property, but the no-double-occupancy constraint is a natural and automatic consequence of the anti-commutation properties of the fermion operators.

[26] For "real" spins, this term is a natural consequence of exchange interactions. For the case of pseudospins representing whether or not a lattice site is occupied or empty, this term represents interactions between the particles.

[27] We have to be careful what we mean by solvability here. Each and every individual state of the form in Eq. (11.165) is an exact eigenstate of $\hat{H}_0'$. However, figuring out which one is the ground state is provably hard (complexity class

Hamiltonian (11.167) is due to the existence of the complete set of conserved observables $\{S_j^z\}$. Their role is very similar to that of the occupation numbers of single-particle states in a non-interacting-particle system. Obviously, because of this extensive set of conservation laws, spins do *not* diffuse under the *interacting* Hamiltonian (11.167), and as a result we have many-body localization.

Let us now finally return to Eq. (11.163) or its generic extension, in which we do *not* have an *a priori known* complete set of conserved observables like $\{S_j^z\}$. However, in the basis in which $\hat{H}$ is diagonal, we can always label its eigenstates by a set of binary numbers $|\{l_j\}\rangle$ similar to what we have in Eq. (11.165), with $j = 1, 2, \ldots, N$, and $l_j = \pm 1/2$. We can now introduce a set of N operators $\{L_j\}$ that commute with each other and $\hat{H}$, whose complete set of common eigenstates are $|\{l_j\}\rangle$. Hence the $\{L_j\}$ form a complete set of conserved obervables, similarly to $\{S_j^z\}$ for $\hat{H}_0'$. In fact, we can choose $L_j = S_j^z$ for $t = 0$.[28]

For $0 < t \ll W$, we expect there exist L_j that are smoothly (or perturbatively) connected to S_j^z, in a sense similar to $|\psi_j\rangle$ being perturbatively connected to $|j\rangle$ (see Eq, (11.157)):

$$L_i = f_i S_i^z + \sum_{a,b=x,y,z} \sum_{jk} f_{i;jk}^{ab} S_j^a S_k^b + \sum_{abc} \sum_{jkm} f_{i;jkm}^{abc} S_j^a S_k^b S_m^c + \cdots . \tag{11.169}$$

L_i is a *local* conserved observable, if such an expansion converges, which requires the expansion coefficients f to decay sufficiently rapidly as a function of distance between i and $j, k, m, \ldots$. It should be clear that we have many-body localization if we have a complete set of local conserved observables, in the same sense as having all single-particle eigenstates localized in a system with non-interacting particles [77, 78]. An immediate consequence is that spin-diffusion and other transport coefficients vanish, due to the constraints imposed by these conservation laws.

In the language of quantum information theory, if the expansion in Eq. (11.169) converges, it means that the interacting Hamiltonian can be diagonalized by a finite-depth "circuit" consisting of a small series of local unitary operations which "dress" each spin operator with additional nearby spin operators.[29] Whether or not the series converges is an even more subtle question than for single-particle Anderson localization. As already noted, the dimension of the Hilbert space is exponentially larger for the MBL problem, and estimating the distance (in Hilbert space rather than in real space) between nearly degenerate many-body states and the perturbation-theory matrix elements that connect them is not trivial. However, the basic physical picture is one of self-consistency: if MBL occurs, the system is frozen and unable to act as its own bath to dephase and mix states, and thus unable to delocalize them in Hilbert space. In this case the locally conserved operators $\{L_i\}$ are adiabatically connected to the bare spin operators (and have support only over other spin operators on nearby sites). If this fails, then self-consistently the system *is able* to act as its own bath and dephase local spin degrees of freedom by entangling them in high-order correlations spread across larger distances. This leads us to another very illuminating angle to reveal the unusual physics of MBL through the lens of entanglement and thermalization, which is discussed in Appendix F.

NP-complete) in general. Even the disorder-free Ising model in 3D has remained unsolved at finite temperature. We can write down the energy of every single microstate but cannot compute the macroscopic properties associated with the partition function (e.g. the free energy).

[28] Note that this choice is not unique. For a more detailed discussion on this, see Ref. [77]. This complication does not affect our discussion.

[29] When we study Landau Fermi-liquid theory in Chapter 15, we will see a similar idea – namely that (at least at low energies near the Fermi surface) the energy is approximately diagonal in the occupation numbers of Landau "quasiparticles," which are bare electrons in plane-wave states dressed by a small cloud of multi-particle–hole pairs. MBL is different, however, in that the conserved operators are localized in real space and exist even for high-energy many-body states.

Exercise 11.12. To help you understand Eq. (11.169), consider a two-site lattice with the Hamiltonian in Eq. (11.163). Assume that

$$\min|V_1 \pm V_2| \gg t,$$

where $t = t_{12} = t_{21}$. For $t = 0$, the Hamiltonian is diagonal in the S^z basis, and the four energy levels are widely separated. For the case of small but non-zero t, use the methods described in Appendix I to find a local quantum circuit (unitary transformation) which will perturbatively eliminate t to lowest order. With this in hand, transform the S^z operators

$$L_i = U S_i^z U^\dagger$$

and explicitly obtain an expression for the terms given schematically in Eq. (11.169). Appendix I focuses on low-energy states, but the methods readily generalize to the present problem, where we are considering all four energy eigenstates of this small lattice.

12 Integer Quantum Hall Effect

The quantum Hall effect (QHE) is one of the most remarkable condensed matter phenomena discovered in the second half of the twentieth century. It rivals superconductivity in its fundamental significance as a manifestation of quantum mechanics on macroscopic scales. Like superconductivity, there is dissipationless transport and it has fundamental implications for electrical metrology. Surprisingly, it provides a new method with which to measure the fine-structure constant with extreme precision in a dirty and disordered solid-state system. In this chapter we will study the integer version of the quantum Hall effect (IQHE), which takes place in a 2D electron gas (2DEG) in a high magnetic field. The universal values of the Hall resistance observed in transport were a total experimental surprise, but can now be understood by invoking disorder and single-particle quantum dynamics, but largely ignoring electron–electron interactions. In Chapter 16 we will study the fractional quantum Hall effect (FQHE), where the physics is that of strongly correlated interacting electrons, which bizarrely exhibit states described by fractional quantum numbers!

Furthermore, we will learn in Chapters 13 and 14 that the remarkable physics and universality of the transport properties is a manifestation of the appearance of topology in the mathematical description of these systems. The new theoretical ideas associated with the role of topology have implications in many different fields of physics beyond condensed matter.

We turn now to an introduction to the phenomenology of the IQHE.

12.1 Hall-Effect Transport in High Magnetic Fields

As noted in Chapter 7, the Lorentz force associated with an applied magnetic field causes a component of the electric field to develop which is perpendicular both to the current and to the magnetic field. The conductivity and resistivity therefore become tensors:

$$J_\mu = \sigma_{\mu\nu} E_\nu, \tag{12.1}$$

$$E_\mu = \rho_{\mu\nu} J_\nu. \tag{12.2}$$

For magnetic field $\vec{B} = B\hat{z}$ and current density $\vec{J} = J\hat{x}$, the requirement that the $\hat{y}$ component of the electric force cancels out the corresponding component of the Lorentz force (so that there is no current in the $\hat{y}$ direction) yields the standard classical expression for the Hall resistivity[1]

---

[1] The corresponding expression in SI units has no factor of c.

$$\rho_{yx} = \frac{B}{nqc}, \qquad (12.3)$$

where q is the charge of the current-carrying particles. This expression shows that ρ_{yx} grows linearly with the magnetic field B, with a coefficient that depends only on the carrier density n and charge q, but *not* on the scattering time or other measures of randomness. As a result, the Hall effect can be very useful for measuring the density of carriers and the sign of their charge (electron-like or hole-like), and is indeed widely used for this purpose in ordinary metals and doped semiconductors.

The basic experimental observation of the quantum Hall effect in a high-mobility 2DEG at very low temperature is dramatically different from the classical result given above; see Fig. 12.1. While following the linear field dependence of (12.3) at weak field, beyond several tesla one observes instead a series of so-called **quantum Hall plateaux** on which the Hall resistance R_H is quantized:[2]

$$R_H = \frac{V_y}{I_x} = \rho_{yx} = \frac{1}{\nu}\frac{h}{e^2}, \qquad (12.4)$$

where ν is a quantum number whose meaning will be made clear soon. This quantization is universal and independent of all microscopic details such as the type of semiconductor material, the purity of the sample, the precise value of the magnetic field on a specific plateau, and so forth. As a result, the effect is now used to maintain[3] the standard of electrical resistance by metrology laboratories

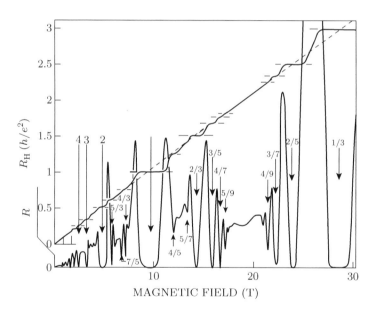

Figure 12.1 Transport data of a high-mobility two-dimensional electron gas (2DEG), showing the Hall resistance R_H and the (dissipative) longitudinal resistance R as functions of the magnetic field, at low temperature. Plateaux in R_H, and corresponding deep minima in R, are labeled by the quantum number ν. Integer and fractional values of ν correspond to the integer and fractional quantum Hall effects, respectively. Figure reprinted from [79] with permission from Elsevier.

---

[2] See the discussion further below about the fact that resistance and resistivity have the same units in 2D.

[3] Maintain does *not* mean *define*. The SI ohm is defined in terms of the kilogram, the second, and the speed of light (formerly the meter). It is best realized using the reactive impedance of a capacitor whose capacitance is computed from first principles. This is an extremely tedious procedure, and the QHE is a very convenient method for realizing a fixed, reproducible impedance to check for drifts of resistance standards. It does not, however, *define* the ohm. Rather it is a measure of h/e^2 in units of the SI ohm. At the time of writing a proposal is under active consideration to redefine the SI system of units in such a way that, like the speed of light, both Planck's constant and the electron charge will be

around the world. In addition, since the speed of light now has a defined value,[4] measurement of h/e^2 is equivalent to a measurement of the fine-structure constant of fundamental importance in quantum electrodynamics.

Associated with each of these plateaux is a dramatic decrease in the (dissipative) longitudinal resistance $R \propto \rho_{xx} \to 0$ which drops by as much as 13 orders of magnitude in the plateau regions, in sharp contrast to the Drude theory which, as we saw in Eq. (7.22), predicts a B-independent ρ_{xx} because its effect (the Lorentz force) is canceled out by the Hall electric field. Clearly the system is undergoing a sequence of phase transitions into highly idealized dissipationless states. As we will discuss later, a vanishingly small dissipation (ρ_{xx}) is crucial to the observability of the quantization of R_H.

In the so-called integer quantum Hall effect (IQHE) discovered by von Klitzing, Dorda, and Pepper in 1980 [80], the quantum number ν is a simple integer to within an experimental precision of about 10^{-11} and an absolute accuracy of about 10^{-9} (both being limited by our ability to do resistance metrology [81]).[5] We will discuss the origin of this precise quantization in this chapter, using a non-interacting-electron model.

In 1982, Tsui, Störmer, and Gossard [82] discovered that, in certain devices with reduced (but still non-zero) disorder, the quantum number ν could take on rational fractional values. This so-called fractional quantum Hall effect (FQHE) is the result of quite different underlying physics involving strong Coulomb interactions and correlations among the electrons. The particles condense into special quantum states whose excitations have the bizarre property of carrying *fractional* quantum numbers, including fractional charge and fractional statistics that are intermediate between ordinary Bose and Fermi statistics. The FQHE has proven to be a rich and surprising arena for the testing of our understanding of strongly correlated quantum systems. With a simple twist of a dial on her apparatus, the quantum Hall experimentalist can cause the electrons to condense into a bewildering array of new "vacua," each of which is described by a different topological quantum field theory. The novel "topological order" describing these phases is completely unprecedented and has revolutionized our understanding of the types of order that can occur in quantum fluids.

To take a first peek into the fascinating physics we are going to explore, and to get an idea about the physical meaning of the quantum number ν, we compare Eq. (12.4) with the classical expression for the Hall resistivity given in Eq. (12.3). Clearly the classical expression knows nothing about Planck's constant and thus quantum mechanics. Nevertheless, we can insist on using h/e^2, the quantum unit of resistance, in the expression above:

$$R_H = \frac{B}{nec} = \frac{h}{e^2}\frac{Be}{nhc} = \frac{h}{e^2}\frac{BA}{hc/e}\frac{1}{nA} = \frac{h}{e^2}\frac{\Phi}{\Phi_0}\frac{1}{N_e} = \frac{h}{e^2}\frac{N_\Phi}{N_e}, \qquad (12.5)$$

where A is the area of the system, $N_e = nA$ is the number of electrons, $\Phi = BA$ is the flux enclosed in the system,

$$\Phi_0 = \frac{hc}{e} \qquad (12.6)$$

---

exactly defined. This will have the benefit that the kilogram will no longer be defined in terms of the mass of a lump of platinum–iridium alloy that was cast in a foundry in 1879 and has been slowly losing weight every time it is cleaned.

[4] With the speed of light defined, the meter is no longer defined by an artifact (the meter bar) but rather by the distance light travels in a certain defined fraction of the second.

[5] As von Klitzing noted in his 1985 Nobel lecture, several experimentalists in addition to himself had seen flat Hall plateaus on many occasions over several years prior to his February 1980 measurement showing that the value of the Hall resistance on the plateaus was very precisely quantized in units of h/e^2 independently of the sample geometry and the precise value of the magnetic field. Any of these people could have made the discovery, but it appears that it was von Klitzing's early training in metrology that made the decisive difference.

is the flux quantum, and N_Φ is the number of flux quanta in the system. Comparing Eqs. (12.3) and (12.5), we can identify the quantum number

$$\nu = \frac{N_e}{N_\Phi}. \qquad (12.7)$$

as the number of electrons per flux quantum in the system. For reasons that will become apparent, this ratio is commonly referred to as the (Landau-level) filling factor. The observation of the QHE suggests that when the ratio between the electron and flux numbers lies close to certain special numbers, the 2DEG forms very special collective quantum ground states whose transport properties are robust against changes of system parameters such as the magnetic field. The purpose of this chapter and Chapter 16 is to reveal the nature of such states.

We begin with a brief description of why two-dimensionality is important to the universality of the result. We then give a review of the classical and semiclassical theories of the motion of charged particles in a magnetic field. Next, we consider the limit of low temperatures and strong fields where a full quantum treatment of the dynamics is required. After that we will be in a position to understand the importance of Anderson localization in the observed quantization, as well as quantum phase transitions in the IQHE. In Chapter 13 we will show that the Hall conductance is actually equal to a topological invariant known as the **first Chern number** of the filled (magnetic) Bloch bands, thus revealing the topological nature of its quantization. This further elucidates the robustness of the quantization against (non-universal) microscopic details. In Chapter 16 we will study the physics described by the novel wave function invented by Robert Laughlin [83] to describe the special condensed state of the electrons on the $\nu = 1/3$ plateau, as well as the physics behind the many other fractional quantum Hall plateaux visible in Fig. 12.1.

12.2 Why 2D Is Important

As we learned in our study of scaling in the localization transition, resistivity (which is what theorists calculate) and resistance (which is what experimentalists measure) for *classical* systems (in the shape of a hypercube) of size L are related by

$$R = \rho L^{(2-d)}. \qquad (12.8)$$

Two dimensions (2D) is therefore special since, in this case, the resistance of the sample is *scale-invariant*. Closely related to this is the fact that, in 2D, R and ρ have the *same* dimensionality as h/e^2. However, R is still *geometry-dependent*. This is illustrated in Fig. 12.2, where we have a rectangular-shaped sample of size $L \times W$. In this case a correction factor enters the expression for R:

$$R = \rho \frac{L}{W}. \qquad (12.9)$$

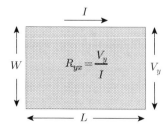

Figure 12.2 Illustration of scale- and geometry-independence of the Hall resistance for a rectangular-shaped sample.

This factor is the aspect ratio L/W, a dimensionless geometric property of the sample (describing its shape[6] but *not* its size).

The situation is very different in a Hall measurement. Assuming current flows in the horizontal (x) direction, we have

$$R_{yx} = V_y/I_x = E_y W/I_x = E_y/(I_x/W) = E_y/J_x = \rho_{yx}. \tag{12.10}$$

Thus the aspect ratio L/W (or any other geometric parameter) does *not* enter the relation between R_{yx} and ρ_{yx}. This turns out to be crucial to the universality of the result. In particular, it means that one does not have to measure the physical dimensions of the sample to one part in 10^{10} in order to obtain the Hall resistivity to that precision. Since the locations of the edges of the sample are not well-defined enough for one to even contemplate such a measurement, this is a very fortunate feature.

The above arguments are essentially based on dimensional analysis specific to 2D. It further turns out that, since the dissipation is nearly zero in the QHE states, even the shape of the sample and the precise location of the Hall voltage probes are almost completely irrelevant. (There is no voltage drop along the direction of the current so the sample edges are isopotentials; as a result mis-aligning the Hall voltage probes, which normally would pick up some *longitudinal* voltage drop if R were non-zero, has no consequence since $R \to 0$.) We will trace the lack of dissipation to a quantum effect that is also peculiar to 2D, namely the quenching of the kinetic energy by the strong magnetic field. Taken together, all of the factors we have mentioned above mean that there is essentially no need to distinguish between the Hall resistance and the Hall resistivity.

12.3 Why Disorder and Localization Are Important

Paradoxically, the extreme universality of the transport properties in the quantum Hall regime occurs because of, rather than in spite of, the random disorder and uncontrolled imperfections which the devices contain. Anderson localization in the presence of disorder plays an essential role in the quantization, but this localization is strongly modified from the usual case because of the strong magnetic field.

In our study of Anderson localization we found that in 2D all states are localized even for arbitrarily weak disorder. The essence of this weak localization effect was shown to be the logarithmic divergence of the time integral of the current "echo" associated with the quantum interference corrections to classical transport. We further saw that these quantum interference effects relied crucially on the existence of time-reversal symmetry. In the presence of a strong quantizing magnetic field (which breaks time-reversal symmetry), the localization properties of the disordered 2D electron gas are radically altered. We will shortly see that there exists a sequence of novel phase transitions, not between a metal and an insulator, but rather between two distinctly different integer quantum Hall states (here we view the insulating state as a special quantum Hall state with zero Hall conductance).

In the absence of any impurities, the 2DEG is translationally invariant and there does not exist any preferred frame of reference.[7] As a result, we can transform to a frame of reference moving with velocity $-\vec{v}$ relative to the lab frame. In this frame the electrons appear to be moving at velocity $+\vec{v}$ and carrying current density

$$\vec{J} = -ne\vec{v}, \tag{12.11}$$

---

[6] Practitioners in the field often describe the aspect ratio as the length of the sample in "squares," that is, the number of squares of side W that fit in the length L.

[7] This assumes that we can ignore the periodic potential of the crystal, which is of course fixed in the lab frame. Within the effective-mass approximation this potential modifies the mass but does not destroy the Galilean invariance, since the energy is still quadratic in the momentum.

where n is the areal density. In the lab frame, the electromagnetic fields are

$$\vec{E} = \vec{0}, \tag{12.12}$$

$$\vec{B} = B\hat{z}. \tag{12.13}$$

In the moving frame they are Lorentz-transformed (to lowest order in v/c) to

$$\vec{E} = -\frac{1}{c}\vec{v} \times \vec{B}, \tag{12.14}$$

$$\vec{B} = B\hat{z}. \tag{12.15}$$

This Lorentz transformation picture is precisely equivalent to the usual statement that an electric field must exist which just cancels out the Lorentz force $-(e/c)\vec{v} \times \vec{B}$ in order for the device to carry the current straight through without deflection. Thus we have

$$\vec{E} = \frac{B}{nec}\vec{J} \times \hat{B}. \tag{12.16}$$

The resistivity tensor is defined by Eq. (12.2), and from this we can make the identification

$$\overleftrightarrow{\rho} = \frac{B}{nec} \begin{pmatrix} 0 & +1 \\ -1 & 0 \end{pmatrix}. \tag{12.17}$$

The conductivity tensor from Eq. (12.1) is the matrix inverse of this, so that

$$\overleftrightarrow{\sigma} = \frac{nec}{B} \begin{pmatrix} 0 & -1 \\ +1 & 0 \end{pmatrix}. \tag{12.18}$$

Notice that, paradoxically, the system looks insulating since $\sigma_{xx} = 0$ and yet it also looks like a perfect conductor since $\rho_{xx} = 0$. In an ordinary insulator $\sigma_{xy} = 0$, so $\sigma_{xx} = 0$ implies $\rho_{xx} = \infty$. Here $\sigma_{xy} = -nec/B \neq 0$, so the inverse of $\overleftrightarrow{\sigma}$ always exists.

The argument given above relies only on Lorentz covariance. The only property of the 2DEG that entered was the density. The argument works equally well irrespective of whether the system is classical or quantum, and whether the electron state is liquid, vapor, or solid. It simply does not matter. Thus, in the absence of disorder, the Hall effect teaches us nothing about the system other than its density. We cannot even see that the kinetic energy is quantized into Landau levels. The Hall resistivity is simply a linear function of the magnetic field whose slope tells us about the electron density. In the quantum Hall regime we would therefore see none of the novel physics exhibited in Fig. 12.1 if we had a perfect sample, since disorder is *needed* in order to destroy translation invariance.

We have thus shown that the random impurity potential is a necessary condition for deviations from simple linear behavior to occur. Understanding how it leads to the IQHE is our task in the remainder of this chapter. A hint that Anderson localization must play a major role comes from the fact on the quantum Hall plateaux there is no dissipative response ($\sigma_{xx} \rightarrow 0$), suggesting that states at the Fermi level are localized (in the bulk).

12.4 Classical and Semiclassical Dynamics

12.4.1 Classical Dynamics

The classical equations of motion for a particle of charge $-e$ moving in 2D under the influence of the Lorentz force $-(e/c)\vec{v} \times \vec{B}$ caused by a magnetic field $\vec{B} = B\hat{z}$ are

$$m_e\ddot{x} = -\frac{eB}{c}\dot{y}; \qquad \tag{12.19}$$

$$m_e\ddot{y} = +\frac{eB}{c}\dot{x}.$$

The general solution of these equations corresponds to motion in a circle of arbitrary radius R:

$$\vec{r} = R[\cos(\omega_c t + \delta), \sin(\omega_c t + \delta)]. \tag{12.20}$$

Here δ is an arbitrary phase for the motion and

$$\omega_c \equiv \frac{eB}{m_e c} \tag{12.21}$$

is known as the classical **cyclotron frequency**. Notice that the period of the orbit is independent of the radius and that the tangential speed

$$v = R\omega_c \tag{12.22}$$

controls the radius. A fast particle travels in a large circle but returns to the starting point in the same length of time as a slow particle which (necessarily) travels in a small circle. The motion is thus *isochronous*, much like that of a harmonic oscillator, whose period is independent of the amplitude of the motion. This apparent analogy is not an accident, as we shall see when we study the Hamiltonian (which we will need for the full quantum solution).

Because of some subtleties involving distinctions between canonical and mechanical momentum in the presence of a magnetic field, it is worth reviewing the formal Lagrangian and Hamiltonian approaches to this problem. The above classical equations of motion follow from the Lagrangian

$$\mathcal{L} = \frac{1}{2} m_e \dot{x}^\mu \dot{x}^\mu - \frac{e}{c} \dot{x}^\mu A^\mu, \tag{12.23}$$

where $\mu = 1, 2$ refers to x and y, respectively, and $\vec{A}$ is the vector potential evaluated at the position of the particle. (We use the Einstein summation convention throughout this discussion.) Using

$$\frac{\delta \mathcal{L}}{\delta x^\nu} = -\frac{e}{c} \dot{x}^\mu \, \partial_\nu A^\mu \tag{12.24}$$

and

$$\frac{\delta \mathcal{L}}{\delta \dot{x}^\nu} = m_e \dot{x}^\nu - \frac{e}{c} A^\nu, \tag{12.25}$$

the Euler–Lagrange equation of motion becomes

$$m_e \ddot{x}^\nu = -\frac{e}{c} [\partial_\nu A^\mu - \partial_\mu A^\nu] \dot{x}^\mu. \tag{12.26}$$

Using

$$\vec{B} = \nabla \times \vec{A}, \tag{12.27}$$

or equivalently

$$B^\alpha = \epsilon^{\alpha\beta\gamma} \, \partial_\beta A^\gamma, \tag{12.28}$$

it is easy to show that Eq. (12.26) is equivalent to Eqs. (12.19).

Once we have the Lagrangian, we can deduce the canonical momentum

$$p^\mu \equiv \frac{\delta \mathcal{L}}{\delta \dot{x}^\mu} = m_e \dot{x}^\mu - \frac{e}{c} A^\mu, \tag{12.29}$$

and the Hamiltonian[8]

$$H[\vec{p}, \vec{x}] \equiv \dot{x}^\mu p^\mu - \mathcal{L}(\dot{\vec{x}}, \vec{x})$$
$$= \frac{1}{2m_e} \left(p^\mu + \frac{e}{c} A^\mu \right) \left(p^\mu + \frac{e}{c} A^\mu \right) = \frac{\Pi^\mu \Pi^\mu}{2m_e}. \tag{12.30}$$

---

[8] Recall that the Lagrangian is canonically a function of the positions and velocities, while the Hamiltonian is canonically a function of the positions and momenta.

The quantity

$$\Pi^{\mu} \equiv p^{\mu} + \frac{e}{c} A^{\mu} \tag{12.31}$$

is known as the *mechanical* momentum. Hamilton's equations of motion

$$\dot{x}^{\mu} = \frac{\partial H}{\partial p^{\mu}} = \frac{\Pi^{\mu}}{m_{\mathrm{e}}}, \tag{12.32}$$

$$\dot{p}^{\mu} = -\frac{\partial H}{\partial x^{\mu}} = -\frac{e}{m_{\mathrm{e}}c} \left(p^{\nu} + \frac{e}{c} A^{\nu} \right) \partial_{\mu} A^{\nu} \tag{12.33}$$

show that it is the mechanical momentum, not the canonical momentum, that is equal to the usual expression related to the velocity,

$$\Pi^{\mu} = m_{\mathrm{e}} \dot{x}^{\mu}. \tag{12.34}$$

Equation (12.33) can be recast as

$$\dot{\Pi}^{\mu} = -\frac{e}{m_{\mathrm{e}}c} \Pi^{\nu} [\partial_{\mu} A^{\nu} - \partial_{\nu} A^{\mu}]. \tag{12.35}$$

Equivalently, using Hamilton's equations of motion we can recover Newton's law for the Lorentz force given in Eq. (12.26) by simply taking a time derivative of $\dot{x}^{\mu}$ in Eq. (12.32).

The distinction between canonical and mechanical momentum can lead to confusion. For example, it is possible for the particle to have a finite velocity while having zero (canonical) momentum! Furthermore, the canonical momentum is dependent (as we will see later) on the choice of gauge for the vector potential and hence is not a physical observable. The mechanical momentum, being simply related to the velocity (and hence the current) is physically observable and gauge-invariant. The classical equations of motion involve only the curl of the vector potential, so the particular gauge choice is not very important at the classical level. We will therefore delay discussion of gauge choices until we study the full quantum solution, where the issue is unavoidable.

12.4.2 Semiclassical Approximation

Recall that, in the semiclassical approximation previously developed when we studied transport theory in Chapter 8, we considered wave packets $\Psi_{\vec{R}(t),\vec{K}(t)}(\vec{r}, t)$ made up of a linear superposition of Bloch waves. These packets are large on the scale of the de Broglie wavelength so that they have a well-defined central wave vector $\vec{K}(t)$, but small on the scale of everything else (external potentials, etc.) so that they simultaneously can be considered to have well-defined mean position $\vec{R}(t)$. (Note that $\vec{K}$ and $\vec{R}$ are *parameters* labeling the wave packet, not arguments of the wave function.) We then argued that the solution of the Schrödinger equation in this semiclassical limit gives a wave packet whose parameters $\vec{K}(t)$ and $\vec{R}(t)$ obey the appropriate analog of the classical Hamilton equations of motion,

$$\dot{R}^{\mu} = \frac{\partial \langle \Psi_{\vec{R},\vec{K}} | H | \Psi_{\vec{R},\vec{K}} \rangle}{\partial (\hbar K^{\mu})}; \tag{12.36}$$

$$\hbar \dot{K}^{\mu} = -\frac{\partial \langle \Psi_{\vec{R},\vec{K}} | H | \Psi_{\vec{R},\vec{K}} \rangle}{\partial R^{\mu}}. \tag{12.37}$$

Naturally this leads to the same circular motion of the wave packet at the classical cyclotron frequency discussed above. For weak fields and fast electrons, the radii of these circular orbits will be large compared with the size of the wave packets, and the semiclassical approximation will be valid. However, at strong fields, the approximation begins to break down because the orbits are too small and because $\hbar\omega_{\mathrm{c}}$ becomes a significant (large) energy. Thus we anticipate that the semiclassical regime requires that the cyclotron energy be much smaller than the Fermi energy, $\hbar\omega_{\mathrm{c}} \ll \epsilon_{\mathrm{F}}$.

We have already seen hints that the problem we are studying is really a harmonic oscillator problem. For the harmonic oscillator there is a characteristic energy scale $\hbar\omega$ (in this case $\hbar\omega_c$) and a characteristic length scale[9] ℓ for the zero-point fluctuations of the position in the ground state. The analogous quantity in this problem is the so-called **magnetic length**

$$\ell \equiv \sqrt{\frac{\hbar c}{eB}} \approx \frac{257\,\text{Å}}{\sqrt{B/\text{T}}}. \tag{12.38}$$

The physical interpretation of this length is that the area $2\pi\ell^2$ contains one quantum of magnetic flux $\Phi_0 = hc/e$. That is to say, the density of magnetic flux is

$$B = \frac{\Phi_0}{2\pi\ell^2}. \tag{12.39}$$

For the system to be in the semiclassical limit one then requires that the de Broglie wavelength be small on the scale of the magnetic length, or $k_F\ell \gg 1$. This condition turns out to be equivalent to $\hbar\omega_c \ll \epsilon_F$, so they are not separate constraints.

Exercise 12.1. Use the Bohr–Sommerfeld quantization condition that the orbit have a circumference containing an integral number of de Broglie wavelengths to find the allowed orbits of a 2D electron moving in a uniform magnetic field. Show that each successive orbit encloses precisely one additional quantum of flux in its interior. Hint: it is important to make the distinction between the canonical momentum (which controls the de Broglie wavelength) and the mechanical momentum (which controls the velocity). The calculation is simplified if one uses the symmetric gauge $\vec{A} = -\frac{1}{2}\vec{r} \times \vec{B}$, in which the vector potential is purely azimuthal and independent of the azimuthal angle.

12.5 Quantum Dynamics in Strong *B* Fields

Since we will be dealing with the Hamiltonian and the Schrödinger equation, our first order of business is to choose a gauge for the vector potential. One convenient choice is the so-called **Landau gauge**,

$$\vec{A}(\vec{r}) = -xB\hat{y}, \tag{12.40}$$

which obeys $\nabla \times \vec{A} = -B\hat{z}$ (we change the sign or direction of the B field simply to avoid some awkward minus signs here as well as in Chapter 16). In this gauge, the vector potential points in the y direction but varies only with the x position. Hence the system still has translation invariance in the y direction. Notice that the magnetic field is translationally invariant, but the Hamiltonian is not! This is one of many peculiarities of dealing with vector potentials. Translation leaves the curl of $\vec{A}$ invariant but not $\vec{A}$ itself. That is, a gauge change has occurred. The eigenvalues of the Hamiltonian are invariant under translation but the eigenfunctions are subject to a unitary transformation which depends on the gauge choice.

Exercise 12.2. Show for the Landau gauge that, even though the Hamiltonian is not invariant for translations in the x direction, the physics is still invariant, since the change in the Hamiltonian that occurs under translation is simply equivalent to a gauge change. Prove this for any arbitrary gauge, assuming only that the magnetic field is uniform.

---

[9] Not to be confused with the mean free path, for which we used the same symbol in previous chapters!

We will also ignore the periodic potential and associated band structure due to the lattice for the moment. This is justifiable as long as the effective-mass approximation is valid.[10] Possible band structure effects at very high magnetic fields will be discussed in Section 12.9. The Hamiltonian can thus be written in the Landau gauge as

$$H = \frac{1}{2m_{\mathrm{e}}} \left[p_x^2 + \left(p_y - \frac{eB}{c} x \right)^2 \right]. \tag{12.41}$$

Taking advantage of the translation symmetry in the y direction, let us attempt a separation of variables by writing the wave function in the form

$$\psi_k(x, y) = e^{iky} f_k(x). \tag{12.42}$$

This has the advantage that it is an eigenstate of p_y and hence we can make the replacement $p_y \longrightarrow \hbar k$ in the Hamiltonian. After separating variables, we have the effective 1D Schrödinger equation

$$h_k f_k(x) = \epsilon_k f_k(x), \tag{12.43}$$

where (note the similarity to Eq. (7.65))

$$h_k \equiv \frac{p_x^2}{2m_{\mathrm{e}}} + \frac{1}{2m_{\mathrm{e}}} \left(\hbar k - \frac{eB}{c} x \right)^2. \tag{12.44}$$

This is simply a 1D displaced harmonic oscillator

$$h_k = \frac{p_x^2}{2m_{\mathrm{e}}} + \frac{1}{2} m_{\mathrm{e}} \omega_{\mathrm{c}}^2 \left(x - k\ell^2 \right)^2, \tag{12.45}$$

whose frequency is the classical cyclotron frequency and whose central position (also known as the **guiding center**[11] position)

$$X_k = k\ell^2 \tag{12.46}$$

is (somewhat paradoxically) determined by the y momentum quantum number. Thus for each plane wave chosen for the y direction there will be an entire family of energy eigenvalues

$$\epsilon_{kn} = \left(n + \frac{1}{2} \right) \hbar \omega_{\mathrm{c}}, \tag{12.47}$$

which depend only on n but are completely independent of the y momentum $\hbar k$. The corresponding (unnormalized) eigenfunctions are

$$\psi_{nk}(\vec{r}) = e^{iky} H_n(x/\ell - k\ell) e^{-\frac{1}{2\ell^2}(x - k\ell^2)^2}, \tag{12.48}$$

where H_n is (as usual for harmonic oscillators) the nth Hermite polynomial (in this case displaced to the new central position X_k).

Exercise 12.3. Verify that Eq. (12.48) is in fact a solution of the Schrödinger equation as claimed.

---

[10] Provided that the lattice constant is much smaller than the magnetic length ℓ, it is a good approximation to simply use the band mass instead of the bare mass in Schrödinger's equation.

[11] Classically the guiding center is the center of the cyclotron motion of a charged particle in a magnetic field. What we have here is its quantum analog. In the Landau gauge used here, it is simply the X coordinate of the center of motion, the Y position being uncertain. We will later encounter in Eq. (16.8) a gauge-invariant formulation of the guiding center as a 2D vector position operator.

These harmonic oscillator levels are called **Landau levels**. Owing to the lack of dependence of the energy on k, the degeneracy of each level is enormous, as we will now show. We assume periodic boundary conditions in the y direction. Because of the vector potential, it is *impossible* to simultaneously have periodic boundary conditions in the x direction. However, since the basis wave functions are harmonic oscillator polynomials multiplied by strongly converging Gaussians, they rapidly vanish for positions away from the center position X_k. Let us suppose that the sample is rectangular with dimensions L_x, L_y and that the left-hand edge is at $x = 0$ and the right-hand edge is at $x = L_x$. Then the values of the wave vector k for which the basis state is substantially inside the sample run from $k = 0$ to $k = L_x/\ell^2$. It is clear that the states at the left edge and the right edge differ strongly in their k values, and hence periodic boundary conditions are impossible.[12]

The total number of states in *each* Landau level is then

$$N = \sum_k 1 = \frac{L_y}{2\pi} \int_0^{L_x/\ell^2} dk = \frac{L_x L_y}{2\pi \ell^2} = N_\Phi, \tag{12.49}$$

where

$$N_\Phi \equiv \frac{B L_x L_y}{\Phi_0} \tag{12.50}$$

is the number of flux quanta penetrating the sample. Thus there is one state per Landau level per flux quantum. As a result, when the filling factor ν of Eq. (12.7) takes an integer value, we have in the ground state exactly ν (lowest-energy) Landau levels completely filled, and all higher Landau levels empty.[13] Such a state is clearly special, as there is a gap for excitations *in the bulk*, and the 2DEG is *incompressible*, just like in band insulators.[14]

Notice that, even though the family of allowed wave vectors is only 1D, we find that the degeneracy of each Landau level is extensive in the 2D area. The reason for this is that the spacing between wave vectors allowed by the periodic boundary conditions $\Delta_k = 2\pi/L_y$ *decreases* with increasing L_y, while the range of allowed wave vectors $[0, L_x/\ell^2]$ *increases* with increasing L_x. The reader may also worry that for very large samples, the range of allowed values of k will be so large that it will fall outside the first Brillouin zone, forcing us to include band mixing and the periodic lattice potential beyond the effective-mass approximation. This is not true, however, since the canonical momentum is a gauge-dependent quantity. The value of k in any particular region of the sample can be made small by shifting the origin of the coordinate system to that region (thereby making a gauge transformation).

The width of the harmonic oscillator wave functions in the nth Landau level is of order $\sqrt{n}\ell$. This is microscopic compared with the system size, but note that the spacing between the centers

$$\Delta = \Delta_k \ell^2 = \frac{2\pi \ell^2}{L_y} \tag{12.51}$$

is vastly smaller (assuming that $L_y \gg \ell$). Thus the supports of the different basis states are strongly overlapping (but they are still orthogonal due to their different k quantum numbers).

---

[12] The best one can achieve is so-called quasi-periodic boundary conditions in which the phase difference between the left and right edges is zero at the bottom and rises linearly with height, reaching $2\pi N_\Phi \equiv L_x L_y/\ell^2$ at the top. The eigenfunctions with these boundary conditions are elliptic theta functions, which are linear combinations of the Gaussians discussed here. These issues will be discussed more generally and in a gauge-independent manner in Section 12.9.

[13] Here we neglected the fact that electrons carry spin, and each Landau-level orbital state can accommodate two electrons. This is because the magnetic field also lifts the spin degeneracy, and we choose to view different spin states with the same orbital Landau-level wave functions as *different* Landau levels.

[14] We will see very soon that the situation is quite different at the edge of the sample.

Exercise 12.4. Using the fact that the energy for the nth harmonic oscillator state is $(n + \frac{1}{2})\hbar\omega_c$, present a semiclassical argument explaining the result claimed above, namely that the width of the support of the wave function scales as $\sqrt{n}\ell$.

Exercise 12.5. Using the Landau gauge, construct a Gaussian wave packet in the lowest Landau level of the form

$$\Psi(x, y) = \int_{-\infty}^{+\infty} a_k e^{iky} e^{-\frac{1}{2\ell^2}(x - k\ell^2)^2} \, dk,$$

choosing a_k in such a way that the wave packet is localized as closely as possible around some point $\vec{R}$. What is the smallest-size wave packet that can be constructed without mixing in higher Landau levels?

Having now found the eigenfunctions for an electron in a strong magnetic field, we can relate them back to the semiclassical picture of wave packets undergoing circular cyclotron motion. Consider an initial semiclassical wave packet located at some position and having some specified *mechanical* momentum $\hbar\vec{K}$. In the semiclassical limit the mean energy of this packet will greatly exceed the cyclotron energy ($\hbar^2 K^2/2m \gg \hbar\omega_c$) and hence it will be made up of a linear combination of a large number of different Landau-level states centered around $\bar{n} = \hbar^2 K^2/2m\hbar\omega_c$:

$$\Psi(\vec{r}, t) = \frac{L_y}{2\pi} \sum_n \int dk \, a_n(k) \psi_{nk}(\vec{r}) e^{-i(n + \frac{1}{2})\omega_c t}. \tag{12.52}$$

Notice that, in an ordinary 2D problem at zero field, the complete set of plane-wave states would be labeled by a 2D continuous momentum label. Here we have one discrete label (the Landau-level index) and a 1D continuous label (the y wave vector). Thus the "sum" over the complete set of states is actually a combination of a summation and an integration.

The details of the initial position and momentum are controlled by the amplitudes $a_n(k)$. We can immediately see, however, that, since the energy levels are exactly evenly spaced, the motion is exactly periodic:

$$\Psi\left(\vec{r}, t + \frac{2\pi}{\omega_c}\right) = -\Psi(\vec{r}, t). \tag{12.53}$$

If one works through the details, one finds that the motion is indeed circular and corresponds to the expected semiclassical cyclotron orbit.

Exercise 12.6. Start with a Gaussian wave packet (not projected into any particular Landau level) with an initial position and velocity. Express it in terms of Landau-level eigenfunctions and show that its time evolution corresponds to a semiclassical circular cyclotron orbit.

For simplicity we will restrict the remainder of our discussion mostly to the lowest Landau level, where the (correctly normalized) eigenfunctions in the Landau gauge are (dropping the index $n = 0$ from now on)

$$\psi_k(\vec{r}) = \frac{1}{\sqrt{\pi^{1/2} L_y \ell}} e^{iky} e^{-\frac{1}{2\ell^2}(x - k\ell^2)^2} \tag{12.54}$$

and every state has the same energy eigenvalue $\epsilon_k = \frac{1}{2}\hbar\omega_c$; see Fig. 12.3(a). Many of the following results will be applicable to higher Landau levels as well, and we will point them out when appropriate.

We imagine that the magnetic field is high and hence the Landau-level splitting is very large compared with the scale of the disorder, so that we can ignore higher Landau levels. (There are some

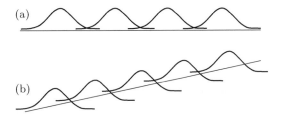

Figure 12.3 Illustration of lowest-Landau-level wave functions in the Landau gauge, which are plane waves along the $\hat{y}$ direction (into the paper) and Gaussian along the $\hat{x}$ direction (left to right), with different guiding center locations. (a) Without the electric field, all states are degenerate. (b) With the electric field along the $\hat{x}$ direction, the degeneracy is lifted.

subtleties here, to which we will return.) Because the states are all degenerate, any wave packet made up of any combination of the basis states will be a stationary state. The total current will be zero. We anticipate, however, from semiclassical considerations that there should be some remnant of the classical circular motion visible in the local current density. To see this, note that the current density in the kth basis state is

$$\langle \vec{J}(x, y) \rangle = -\frac{e}{m_e} \psi_k^*(x, y) \left[-i\hbar \nabla + \frac{e}{c} \vec{A}(x, y) \right] \psi_k(x, y), \tag{12.55}$$

and the y component of the current is

$$\langle I_y \rangle = -\frac{e}{m_e \pi^{1/2} \ell} \frac{1}{L_y} \int_{-\infty}^{\infty} dx \, e^{-\frac{1}{2\ell^2}(x - k\ell^2)^2} \left(\hbar k - \frac{eB}{c} x \right) e^{-\frac{1}{2\ell^2}(x - k\ell^2)^2}$$

$$= \frac{e\omega_c}{\pi^{1/2} \ell} \frac{1}{L_y} \int_{-\infty}^{\infty} dx \, e^{-\frac{1}{\ell^2}(x - k\ell^2)^2} (x - k\ell^2). \tag{12.56}$$

We see from the integrand that the current density is anti-symmetric about the peak of the Gaussian at $X_k = k\ell^2$ and hence the total current vanishes. This anti-symmetry is the remnant of the semiclassical circular motion, as the *local* current density points in opposite directions on the two sides of X_k. The reader should verify that exactly the same happens in all Landau levels.

One might be surprised by the vanishing current we found in Eq. (12.56), given the fact we are dealing with plane waves (along the y direction), which normally describe propagating states. This is one manifestation of the subtlety associated with the difference between canonical and mechanical momenta mentioned earlier. In Eq. (12.56) the contribution of the canonical momentum to the total current is canceled out *exactly* by the vector potential contribution to the mechanical momentum, yielding a null result. This could perhaps be anticipated from the observation that the energy of the Landau-level states is *independent* of the wave vector k, thus the group velocity is given by

$$v_y = \frac{1}{\hbar} \frac{\partial \epsilon_{kn}}{\partial k} = 0. \tag{12.57}$$

The lack of dependence of the energy on k can in turn be traced to the fact that k determines the guiding center position X_k, but, in the absence of a potential-energy term in the Hamiltonian (12.41), the energy is independent of the guiding center position. As we will see below, the situation changes completely when we add such a potential energy.

Let us now consider the case of a uniform electric field pointing in the x direction and giving rise to the potential energy

$$V(\vec{r}) = +eEx. \tag{12.58}$$

This still has translation symmetry in the y direction, so our Landau gauge choice is still the most convenient. Again, on separating variables we see that the solution is nearly the same as before, except that the displacement of the harmonic oscillator is slightly different. The Hamiltonian in Eq. (12.44) becomes

$$h_k = \frac{p_x^2}{2m_e} + \frac{1}{2}m_e\omega_c^2\left(x - k\ell^2\right)^2 + eEx. \tag{12.59}$$

Completing the square, we see that the oscillator is now centered at the new (and slightly shifted) position

$$X_k' = k\ell^2 - \frac{eE}{m_e\omega_c^2} = X_k - \frac{eE}{m_e\omega_c^2} \tag{12.60}$$

and the energy eigenvalue is now linearly dependent on the particle's peak position $X_k' \approx X_k$ (and therefore linear in the y momentum):

$$\epsilon_{kn} = \left(n + \frac{1}{2}\right)\hbar\omega_c + eEX_k' + \frac{1}{2}m_e\bar{v}^2, \tag{12.61}$$

where, as we will discuss further below,

$$\bar{v} \equiv c\frac{E}{B} \tag{12.62}$$

is the drift velocity. Because of the shift in the peak position of the wave function, the perfect anti-symmetry of the current distribution is destroyed and there is a net current

$$\langle I_y \rangle = -e\frac{\bar{v}}{L_y}, \tag{12.63}$$

showing that $\bar{v}$ is simply the usual $c\vec{E} \times \vec{B}/B^2$ drift velocity. Thus the time it takes the electron to travel through the system is $L_y/\bar{v}$. This result can be derived either by explicitly doing the integral for the current or by noting that the wave-packet group velocity is

$$\frac{1}{\hbar}\frac{\partial\epsilon_{kn}}{\partial k} = \frac{eE}{\hbar}\frac{\partial X_k'}{\partial k} = \frac{eE}{\hbar}\frac{\partial X_k}{\partial k} = \bar{v}, \tag{12.64}$$

independently of the values of k and the Landau-level index n (since the electric field is a constant in this case, giving rise to a strictly linear potential). Thus we have recovered the correct kinematics from our quantum solution.

We now have a clear understanding of the origin of the three terms on the RHS of Eq. (12.61). The first term is of course the original Landau level energy. The second term is the *potential* energy, which is determined by the peak (or guiding center) position. The last term is the *additional* kinetic energy associated with the non-zero drift velocity. Owing to the second term, the applied electric field "tilts" the Landau levels in the sense that their energy is now linear in position as illustrated in Fig. 12.3(b). This means that the degeneracy between states in the same Landau level gets lifted, but there are now degeneracies between states in *different* Landau levels, because the difference in Landau-level energy (the first term of Eq. (12.61)) can be compensated for by the difference in potential energy (the second term of Eq. (12.61)). Nevertheless, what we have found here are the exact eigenstates (i.e. the stationary states). It is not possible for an electron to decay into one of the other degenerate states, because they have different canonical momenta. If, however, disorder or phonons are available to break translation symmetry, then these decays become allowed, and dissipation can appear. The matrix elements for such processes are small if the electric field is weak, because the degenerate states are widely separated spatially due to the small tilt of the Landau levels.

Exercise 12.7. It is interesting to note that the exact eigenstates in the presence of the electric field can be viewed as displaced oscillator states in the original (zero E field) basis. In this basis the displaced states are linear combinations of all the Landau-level excited states of the same k. Use first-order perturbation theory to find the amount by which the $n = 1$ Landau level is mixed into the $n = 0$ state. Compare this with the exact amount of mixing computed using the exact displaced oscillator state. Show that the two results agree to first order in E. Because the displaced state is a linear combination of more than one Landau level, it can carry a finite current. Give an argument, derived from perturbation theory, for why the amount of this current is inversely proportional to the B field, but is independent of the mass of the particle. Note that this quantum perturbation-theory result is in agreement with Eq. (12.62). Hint: how does the mass affect the Landau-level energy spacing and the current operator?

12.6 IQHE Edge States

Now that we understand drift in a uniform electric field, we can consider the problem of electrons confined in a Hall bar of finite width by a non-uniform electric field. For simplicity, we will consider the situation where the potential $V(x)$ is smooth on the scale of the magnetic length, but this is not central to the discussion. If we assume that the system still has translation symmetry in the y direction, the solution to the Schrödinger equation must still be of the form

$$\psi(x, y) = \frac{1}{\sqrt{L_y}} e^{iky} f_k(x). \tag{12.65}$$

The function f_k will no longer be a simple harmonic wave function as we found in the case of the uniform electric field. However, we can anticipate that f_k will still be peaked near (but in general not precisely at) the point $X_k \equiv k\ell^2$. The eigenvalues ϵ_{kn} will no longer be linear in k but will still reflect the kinetic energy of the cyclotron motion plus the local potential energy $V(X_k)$ (plus small corrections analogous to the one in Eq. (12.61)). This is illustrated in Fig. 12.4.

We see that the group velocity

$$\vec{v}_k = \frac{1}{\hbar} \frac{\partial \epsilon_{kn}}{\partial k} \hat{y} \approx \frac{\ell^2}{\hbar} \frac{dV(x)}{dx} \bigg|_{x=X_k} \hat{y} \tag{12.66}$$

has the opposite sign on the two edges of the sample. This means that in the ground state there are edge currents of opposite sign flowing in the sample. The semiclassical interpretation of these currents is that they represent "skipping orbits" in which the circular cyclotron motion is interrupted by collisions with the walls at the edges as illustrated in Fig. 12.5.

One way to analyze the Hall effect in this system is quite analogous to the Landauer picture of transport in narrow wires that we discussed in Chapter 10. The edge states play the role of the left-

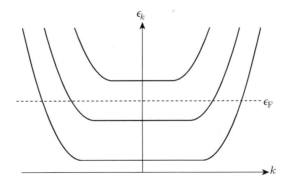

Figure 12.4 Landau-level dispersions in the presence of a confining potential, for a sample with Hall bar geometry.

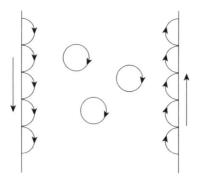

Figure 12.5 Cyclotron orbits of electrons, in the presence of a magnetic field pointing into the paper. These orbits are circles in the bulk; however, near the sample edges they cannot be completed due to collisions with the boundaries, thus the electron trajectories are a sequence of "skipping" orbits, resulting in non-zero (average) drift velocities along the edges. The drift velocities point in opposite directions in the opposite edges.

and right-moving states at the two Fermi points. Because momentum in a magnetic field corresponds to position, the edges are essentially real-space realizations of the Fermi surface. A Hall voltage drop across the sample in the x direction corresponds to a difference in electrochemical potential between the two edges. As in our analysis of the Landauer formulation of transport, we will choose to apply this in the form of a chemical-potential difference and ignore any changes in electrostatic potential.[15] This has the effect of increasing the number of electrons in skipping orbits on one edge of the sample and/or decreasing the number on the other edge. Previously the net current due to the two edges was zero, but now there is a net Hall current. To calculate this current we have to add up the group velocities of all the occupied states,

$$I = -\frac{e}{L_y} \int_{-\infty}^{+\infty} dk \, \frac{L_y}{2\pi} \frac{1}{\hbar} \frac{\partial \epsilon_k}{\partial k} n_k, \tag{12.67}$$

where for the moment we assume that, in the bulk, only a single Landau level is occupied and n_k is the probability that state k in that Landau level is occupied. Assuming zero temperature and noting that the integrand is a perfect derivative, we have

$$I = -\frac{e}{h} \int_{\mu_L}^{\mu_R} d\epsilon = -\frac{e}{h} [\mu_R - \mu_L]. \tag{12.68}$$

The definition of the Hall voltage drop is

$$(-e)V_H \equiv (-e)[V_R - V_L] = [\mu_R - \mu_L]. \tag{12.69}$$

Hence

$$I = \nu \frac{e^2}{h} V_H, \tag{12.70}$$

where we have now allowed for the possibility that ν different Landau levels are occupied in the bulk and hence there are ν separate edge channels contributing to the current. This is the analog of having ν "open" channels in the Landauer transport picture. In the Landauer picture for an ordinary wire, we are considering the longitudinal voltage drop (and computing G_{xx}), while here we have the Hall

---

[15] This has led to various confusions in the literature. If there is an electrostatic potential gradient then some of the net Hall current may be carried in the bulk rather than at the edges, but the final answer is the same. In any case, the essential part of the physics is that the only places where there are low-lying excitations are the edges.

voltage drop (and are computing $G_{xy} = \sigma_{xy}$). The analogy is quite precise, however, because we view the right and left movers as having distributions controlled by separate chemical potentials. It just happens in the QHE case that the right and left movers are physically separated in such a way that the voltage drop is transverse to the current. Using the above result and the fact that the current flows at right angles to the voltage drop, we have the desired results

$$\sigma_{xx} = 0, \tag{12.71}$$

$$\sigma_{xy} = \nu \frac{e^2}{h}, \tag{12.72}$$

with the quantum number ν being an integer.

So far we have been ignoring the possible effects of disorder. Recall that, for a single-channel 1D wire in the Landauer picture, a disordered region in the middle of the wire will reduce the conductivity to

$$I = \frac{e^2}{h} T, \tag{12.73}$$

where T is the probability that an electron will be transmitted through the disordered region. The reduction in transmitted current is due to *backscattering*. Remarkably, in the QHE case, the backscattering is essentially zero in very wide samples. To see this, note that, in the case of the Hall bar, scattering into a backward-moving state would require transfer of the electron from one edge of the sample to the other, since the edge states are spatially separated. For samples which are very wide compared with the magnetic length (more precisely, compared with the Anderson localization length), the matrix element for this is exponentially small (see Exercise 12.9). In short, there can be nothing but forward scattering. An incoming wave given by Eq. (12.65) can only be transmitted in the forward direction, with the outgoing wave at most suffering a simple phase shift δ_k as it continues forward:

$$\psi(x, y) = \frac{1}{\sqrt{L_y}} e^{i\delta_k} e^{iky} f_k(x). \tag{12.74}$$

This is because no other states of the same energy are available. If the disorder causes Landau-level mixing at the edges to occur (because the confining potential is relatively steep) then it is possible for an electron in one edge channel to scatter into another, but the current is still going in the same direction so that there is no reduction in the overall transmission probability. It is this *chiral* (unidirectional) nature of the edge states which is responsible for the fact that the Hall conductance is correctly quantized independently of the disorder.[16] This physics will be further illustrated in Exercise 12.9.

In fact, when the Fermi energy ϵ_F is between Landau levels, disorder not only does *not* cause backscattering that would destroy the quantization of σ_{xy}, but is actually responsible for the visibility of the quantized value over a finite range of magnetic field or electron density. That is, disorder is responsible for the finite width of the quantum Hall plateaux. This is because, without disorder, all states in the bulk have energies precisely at one of the Landau-level energies, and so does ϵ_F, unless the filling factor ν is *exactly* an integer, which requires fine-tuning. These are the special points

---

[16] Our discussion on the absence of backscattering here has been quantum mechanical. In fact, this conclusion holds for classical particles as well. Imagine distorting the shape of the boundary in Fig. 12.5, or adding some random potential near the boundary. This will distort the shape of the skipping orbitals, but will not change the fact that electrons near the right edge can only have overall upward motion, and electrons near the left edge can only have overall downward motion.

when ϵ_F jumps from one Landau level to the next one. Having edge states whose energies deviate away from the quantized Landau-level energies does not solve this problem (see Exercie 12.8). Disorder will broaden the Landau levels in the bulk and provide a reservoir of (localized) states which will allow the chemical potential to vary smoothly with density. These localized states will not contribute to the transport, so the Hall conductance will be quantized over a plateau of finite width in B (or density) as seen in Fig. 12.1. Thus obtaining the universal value of the quantized Hall conductance to a precision of 10^{-10} does not require fine-tuning the applied B field to a similar precision.

The localization of states in the bulk by disorder is an essential part of the physics of the quantum Hall effect, as we saw when we studied the role of translation invariance. We learned previously that, in zero magnetic field, all states are (weakly) localized in 2D. In the presence of a quantizing magnetic field, most states are actually strongly localized (as discussed in the following section). However, if all states were localized, then it would be impossible to have a quantum phase transition from one QHE plateau to the next. To understand how this works, it is convenient to work in a semiclassical percolation picture to be described below.

Exercise 12.8. Show that the number of edge states whose energies lie in the gap between two Landau levels scales with the length L of the sample, while the number of bulk states scales with the area. Use these facts to show that the range of magnetic field in which the chemical potential lies in between two Landau levels scales to zero in the thermodynamic limit. Hence, finite-width quantized Hall plateaux cannot occur in the absence of disorder that produces a reservoir of localized states in the bulk whose number is proportional to the area.

Exercise 12.9. We discussed the similarity between transport via edge states in the integer quantum Hall effect and the Landauer transport in 1D quantum wires. What is very different is the effect of impurity scattering. (i) First consider a 1D wire of length L, in the presence of a delta-function impurity potential: $V(x) = V_0 \delta(x)$. Show that the matrix element for backscattering is proportional to $1/L$. (ii) Now consider a Hall bar of length L_y and width L_x, in the presence of a single impurity: $V(x, y) = V_0 \delta(x - x_0)\delta(y - y_0)$. Show that the matrix element for backscattering is exponentially small as a function of L_x/ℓ for all edge channels (where ℓ is the magnetic length), regardless of the position of the impurity (x_0, y_0). Here we assume that the electron eigenstates are plane waves along the $\hat{y}$ direction and there is a confinement potential along the $\hat{x}$ direction.

12.7 Semiclassical Percolation Picture of the IQHE

Let us consider a smooth random potential caused, say, by ionized silicon donors remotely located away from the 2DEG in the GaAs semiconductor host. (See Fig. 9.15 and Exercise 9.7.) We take the magnetic field to be very large so that the magnetic length is small on the scale over which the potential varies. In addition, we ignore the Coulomb interactions among the electrons.

What is the nature of the eigenfunctions in this random potential? We have learned how to solve the problem exactly for the case of a constant electric field and know the general form of the solution when there is translation invariance in one direction; see Eq. (12.65). We found that the wave functions were plane waves running along lines of constant potential energy and having a width perpendicular to them which is very small and on the order of the magnetic length. The reason for this is the discreteness of the kinetic energy in a strong magnetic field. It is impossible for an electron stuck in a given

Landau level to continuously vary its kinetic energy. Hence energy conservation restricts its motion to regions of constant potential energy. In the limit of infinite magnetic field where Landau-level mixing is completely negligible, this confinement to lines of constant potential becomes exact.[17]

From this discussion it then seems very reasonable that, in the presence of a smooth random potential, with no particular translation symmetry, the eigenfunctions will have support lying close to contour lines of constant potential energy on the random potential surface. Thus, low-energy states will be found lying along contours in deep valleys in the potential landscape while high-energy states will be found encircling "mountain tops" in the landscape. Naturally these extreme states will be strongly localized about these extrema in the potential.

To understand the nature of states at intermediate energies, it is useful to imagine gradually filling a random landscape with water as illustrated in Fig. 12.6. In this analogy, sea level represents the chemical potential for the electrons. When only a small amount of water has been added, the water will fill the deepest valleys and form small lakes. As the sea level is increased, the lakes will grow larger and their shorelines will begin to take more complex shapes. At a certain critical value of the sea level, a phase transition will occur in which the shoreline percolates from one side of the system to the other. As the sea level is raised still further, the ocean will cover the majority of the land and only a few mountain tops will stick out above the water. The shoreline will no longer percolate but only surround the mountain tops.

As the sea level is raised still higher, a sequence of additional percolation transitions will occur as each successive Landau level passes under water. If Landau-level mixing is small and the disorder potential is symmetrically distributed about zero, then the critical value of the chemical potential for the nth percolation transition will occur near the center of the nth Landau level:

$$\mu_n^* = \left(n + \frac{1}{2}\right)\hbar\omega_{\mathrm{c}}. \tag{12.75}$$

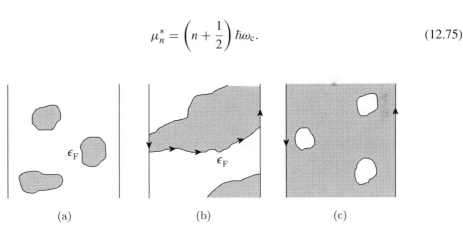

(a)   (b)   (c)

Figure 12.6 Illustration of filling a Landau level with electron liquid, as the (average) electron density or Fermi energy ϵ_{F} increases. (a) Low density, below the percolation threshold. (b) At the percolation threshold. (c) Above the percolation threshold.

---

[17] As the magnetic length goes to zero, all (reasonable) potentials begin to look smooth and locally linear, thus producing semiclassical $\vec{E} \times \vec{B}$ drift. We are led to the following somewhat paradoxical picture. The strong magnetic field should be viewed as putting the system in the quantum limit, in the sense that $\hbar\omega_{\mathrm{c}}$ is a very large energy (comparable to the zero-field Fermi energy, ϵ_{F}). At the same time (if one assumes the potential is smooth), one can argue that, since the magnetic length is small compared with the scale over which the random potential varies, the random potential begins to look locally linear and the system is in a semiclassical limit where small wave packets (on the scale of ℓ) follow classical $\vec{E} \times \vec{B}$ drift trajectories. The way to reconcile these statements is the following. In the strong-field limit, the electron motion can be decomposed into two parts: the very fast (frequency ω_{c}) cyclotron motion (with radius of order ℓ) around the guiding center (the center of the circles in Fig. 12.5), which is going to the extreme quantum limit, and the very slow $\vec{E} \times \vec{B}$ drift motion of the guiding center itself, which is semiclassical.

This percolation transition corresponds to the transition between quantized Hall plateaux. To see why, note that, when the sea level is below the percolation point, most of the sample is dry land. The electron liquid is therefore insulating. When the sea level is above the percolation point, most of the sample is covered with water. The electron liquid is therefore connected throughout the majority of the sample, and a quantized Hall current can be carried. Another way to see this is to note that, when the sea level is above the percolation point, the confining potential will make a shoreline along the full length of each edge of the sample. The edge states will then be populated in this case, and they carry current from one end of the sample to the other as we discussed in the previous section. In other words, we get a (new) open conduction channel.

We can also understand from this picture why the dissipative conductivity σ_{xx} has a sharp peak just as the plateau transition occurs. (Recall the data in Fig. 12.1). Away from the critical point the circumference of any particular patch of shoreline is finite. The period of the semiclassical orbit around this is finite and hence so is the quantum level spacing. Thus there are small energy gaps for excitation of states across these real-space Fermi levels. Adding an infinitesimal electric field will only weakly perturb these states due to the gap and the finiteness of the perturbing matrix element, which will be limited to values on the order of $\sim eED$, where D is the diameter of the orbit. If, however, the shoreline percolates from one end of the sample to the other, then the orbital period diverges and the gap vanishes. An infinitesimal electric field can then cause dissipation of energy.

Another way to see this is that, as the percolation transition is approached from above, the edge states on the two sides will begin taking detours deeper and deeper into the bulk. They will thus begin communicating with each other as the localization length diverges and the zig-zags of the shorelines bring the two edges close to each other. Thus, electrons in one edge state can be backscattered into the other edge states and ultimately reflected from the sample as illustrated in Fig. 12.7. Precisely at the percolation level, we can even have shorelines directly connecting the two edges through the bulk, as seen in Fig. 12.6(b). Below the percolation threshold the incoming edge mode is *always* connected to the mode on the other side, and the transmission coefficient through the sample is zero.

Because the random potential broadens out the Landau levels (spreading the density of states across some finite width), the quantized Hall plateaux will have finite width. As the chemical potential is varied in the regime of localized states in between the Landau-level peaks, only the occupancy of localized states is changing. Hence the transport properties remain constant until the next percolation transition occurs. It is important to have the disorder present in order to produce this finite density of states and to localize those states.

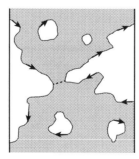

Figure 12.7 The electron distribution when the Fermi energy is above but very close to the percolation threshold. The grey area indicates the region occupied by electrons in the random potential. Backscattering between counter-propagating edge states becomes possible as they come close together in space, as indicated by the dashed line representing a quantum tunneling path connecting the two edges where they approach each other. The white regions in the interior are "mountain peaks" which are above "sea level" and therefore unoccupied by electrons. Their shorelines have finite length, and so local excitations have at least a small gap and cannot cause dissipation for weak electric fields at low frequencies.

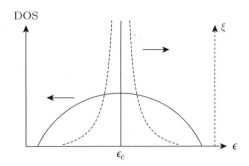

Figure 12.8 Schematic density of states (DOS; solid line) and localization length (ξ, dashed line) of a disorder-broadened Landau level, as a function of energy. ξ diverges at a single energy, ϵ_c.

It is known that, as the percolation point is approached in 2D, the characteristic size (diameter) of the shoreline orbits diverges like

$$\xi \sim |\delta|^{-4/3}, \tag{12.76}$$

where δ measures the deviation of the sea level from its critical value. The shoreline structure is not smooth, and in fact its circumference diverges with a larger exponent $C \sim |\delta|^{-7/3} \sim \xi^{7/4}$, showing that these are highly ramified fractal objects whose circumference scales as the 7/4th power of the diameter.

So far we have assumed that the magnetic length is essentially zero. That is, we have ignored the fact that the wave-function support extends a small distance transverse to the isopotential lines. If two different orbits with the same energy pass near each other but are classically disconnected, the particle can still tunnel between them if the magnetic length is finite. This quantum tunneling causes the localization length to diverge faster than the classical percolation model predicts. Numerical simulations find that the localization length diverges like [84, 85]

$$\xi \sim |\delta|^{-\nu}, \tag{12.77}$$

where the exponent ν (not to be confused with the Landau-level filling factor!) has a value close (but probably not exactly equal) to 7/3 rather than the 4/3 described above for classical percolation. It is believed that this exponent is the same, irrespective of the Landau-level index. The energy dependence of the density of states as well as the localization length is illustrated in Fig. 12.8. It is interesting to compare that with Fig. 11.15; loosely speaking, in Fig. 12.8 the finite range of extended states (in energy) of Fig. 11.15 shrinks to a single critical energy.

Experiments on the quantum critical behavior are quite difficult, but there is evidence, at least in selected samples which show good scaling, that ν is indeed close to 7/3. Why Coulomb interactions that are present in real samples do not spoil the agreement with the numerical simulations which is based on non-interacting electron models (as in our discussions in this chapter) is something of a mystery at the time of writing. For a discussion of these issues see Ref. [86].

12.8 Anomalous Integer Quantum Hall Sequence in Graphene

Thus far our discussion of the IQHE has mostly focused on the Landau levels of electrons whose wave functions satisfy Schrödinger's equation in the presence of a uniform magnetic field. On the other hand, as we learned in Section 7.5, the low-energy electronic structure of graphene is described by the Dirac equation. In this section we discuss the electronic states of Dirac electrons in the presence of a uniform magnetic field, and the corresponding integer quantum Hall sequence that turns out to be quite different from that for Schrödinger electrons.

The low-energy states of graphene live near the two Dirac points K and K', each described by a corresponding two-component wave function that satisfies the Dirac equation:[18]

$$H_{K,K'} = v_F(\pm \sigma_x p_x + \sigma_y p_y). \tag{12.78}$$

In the presence of a magnetic field we need to perform the standard minimum substitution of

$$\vec{p} \to \vec{\Pi} = \vec{p} + \frac{e}{c}\vec{A}(\vec{r}), \tag{12.79}$$

where $\vec{\Pi}$ is the mechanical momentum. In the following we focus on the K point; solutions near the K' point can be obtained in a similar manner.

In Section 12.5 we proceeded by choosing the Landau gauge to fix the explicit form of the vector potential $\vec{A}$. We can proceed in a similar manner here. But the physics (including the energy spectrum) is gauge-independent, and in the following we warm up by obtaining the Schrödinger-electron Landau-level spectrum in a gauge-independent manner. Before considering the Dirac Hamiltonian, let us re-examine the Schrödinger-electron Hamiltonian

$$H = \frac{1}{2m_e}\left(\vec{p} + \frac{e}{c}\vec{A}\right)^2 = \frac{1}{2m_e}\vec{\Pi}^2 = \frac{1}{2m_e}(\Pi_x^2 + \Pi_y^2). \tag{12.80}$$

Unlike the canonical momentum $\vec{p}$, the different components of $\vec{\Pi}$ do *not* commute:

$$[\Pi_x, \Pi_y] = -\frac{i\hbar e}{c}(\partial_x A_y - \partial_y A_x) = \frac{i\hbar e B}{c} = \frac{i\hbar^2}{\ell^2}, \tag{12.81}$$

where for convenience we again choose $\vec{B} = -B\hat{z}$. We see that the commutation relation between Π_x and Π_y is similar to that between x and p_x; thus Eq. (12.80) takes the form of a 1D harmonic oscillator Hamiltonian. Introducing ladder operators

$$a = \frac{\ell}{\sqrt{2}\hbar}(\Pi_x + i\Pi_y), \qquad a^\dagger = \frac{\ell}{\sqrt{2}\hbar}(\Pi_x - i\Pi_y), \tag{12.82}$$

we have $[a, a^\dagger] = 1$ and

$$H = \hbar\omega_c(a^\dagger a + 1/2). \tag{12.83}$$

We thus immediately obtain the Landau-level spectrum

$$\epsilon_n = \hbar\omega_c(n + 1/2), \tag{12.84}$$

without specifying the gauge.

Now consider the corresponding Dirac equation for point K:

$$H_K = v_F\vec{\sigma}\cdot\vec{\Pi} = v_F\begin{pmatrix} 0 & \Pi_x - i\Pi_y \\ \Pi_x + i\Pi_y & 0 \end{pmatrix} = \epsilon_D\begin{pmatrix} 0 & a^\dagger \\ a & 0 \end{pmatrix}, \tag{12.85}$$

where $\epsilon_D = \sqrt{2}\hbar v_F/\ell$ is an energy scale playing a role similar to $\hbar\omega_c$. To obtain the spectrum of (12.85), we employ the usual trick of solving the Dirac equation by considering the *square* of the Dirac Hamiltonian:

$$H_K^2 = \epsilon_D^2\begin{pmatrix} a^\dagger a & 0 \\ 0 & aa^\dagger \end{pmatrix} = \epsilon_D^2\begin{pmatrix} a^\dagger a & 0 \\ 0 & a^\dagger a + 1 \end{pmatrix}, \tag{12.86}$$

---

[18] Recall that the two components represent the amplitudes on the two sublattices, The two-component sublattice degree of freedom is often referred to as the pseudospin. The physical electron spin contributes a trivial factor of two; it will be ignored until the end of the discussion.

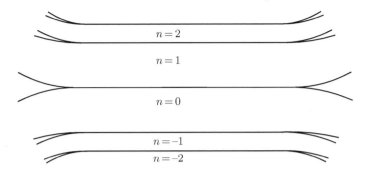

Figure 12.9 The Landau-level spectrum of graphene, including edge states.

which is of diagonal form, and the eigenvalues are $\epsilon_D^2|n|$ with $|n| = 0, 1, 2, \ldots$. Thus the spectrum of H_K is

$$\epsilon_n = \pm\epsilon_D\sqrt{|n|} = \text{sgn}(n)\epsilon_D\sqrt{|n|}, \tag{12.87}$$

with $n = 0, \pm1, \pm2, \ldots$, and $\text{sgn}(n)$ is the sign of n; see Fig. 12.9 for illustration.

We thus find that Dirac electrons in a uniform magnetic field also form Landau levels, but with some key differences from the Schrödinger Landau levels. (i) Unlike Schrödinger Landau levels, Dirac Landau levels are *not* equally spaced. The equal spacing in the Schrödinger case is simply a reflection of the constant density of states for $B = 0$ in 2D for parabolic dispersion. The Dirac dispersion gives a linearly rising density of states and hence the Landau-level spacing must decrease with energy. (ii) Dirac Landau levels come with both positive and negative energies. The former correspond to conduction band (or electron-like) states, while the latter correspond to valence band (or hole-like) states. (iii) Perhaps the most remarkable feature is the existence of a Landau level right at zero energy, $n = 0$, which is formed by conduction and valence band states with equal weight (as guaranteed by particle–hole symmetry of the underlying Hamiltonian). As we now discuss, the existence of the $n = 0$ Landau level, which is the hallmark of massless Dirac particles, gives rise to an "anomalous" sequence of integer quantum Hall states.

As we learned earlier, each electron Landau level, if fully occupied, contributes 1 (in units of e^2/h) to the Hall conductance σ_{xy}. Similarly, each hole Landau level, if fully occupied (by holes), contributes -1. At the neutral point we expect $\sigma_{xy} = 0$. On the other hand, this corresponds to a *half-filled* $n = 0$ Landau level! As a result, when the zeroth Landau level is fully occupied, if only the K point were present, we would expect $\sigma_{xy} = \frac{1}{2}e^2/h$, namely we have quantization at half integers (instead of integers) for σ_{xy}. The remedy, of course, is provided by the existence of the other Dirac point K', which contributes an identical set of Landau levels. Including contributions from both K and K' points, as well as the two-fold spin degeneracy, we expect the IQHE seqeunce in graphene to be[19]

$$\sigma_{xy} = 4\left(n + \frac{1}{2}\right)\frac{e^2}{h} = (\pm2, \pm6, \pm10, \ldots)\frac{e^2}{h}. \tag{12.88}$$

The experimental observation (see Fig. 12.10) of this sequence in 2005 clearly demonstrated the Dirac nature of electrons in graphene, and led to the 2010 Nobel prize in physics for graphene work. It is worth noting that the edge-state spectra of Dirac Landau levels are also different from

---

[19] Here, for simplicity, we have neglected the Zeeman splitting of electron spin states, since these splittings are nominally much smaller than the Landau-level spacing, and unimportant except in very-high-quality graphene samples. Interactions can in some cases lead to exchange enhancement of the spin splitting and the existence of quantum Hall ferromagnetism.

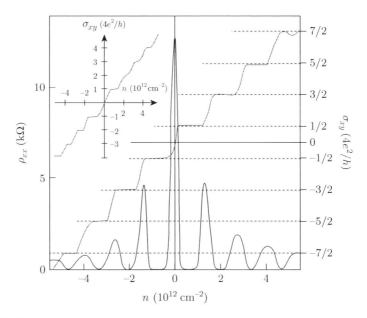

Figure 12.10 The Hall conductivity σ_{xy} and the longitudinal resistivity ρ_{xx} of graphene as a function of the carrier density at $B = 14$ T and $T = 4$ K. The inset shows the transport data of bilayer graphene, another fascinating system that we will not discuss here. Figure adapted from [87] by permission from Springer Nature.

their Schrödinger counterparts: For each Landau level with $n \neq 0$, and for each spin, there are two branches of edge modes at each edge (one from each of the two Dirac points), and dispersing in opposite ways for $n > 0$ (electron-like) and $n < 0$ (hole-like). The $n = 0$ Landau level also has two branches, one electron-like and one hole-like, reflecting the fact it is an equal superposition of conduction and valence band states. These are illustrated in Fig. 12.9. The integer quantum Hall sequence of Eq. (12.88) can also be understood using edge transport in terms of the Dirac Landau-level spectrum.

> **Exercise 12.10.** Use the Landau gauge to obtain the eigenstate wave functions of the Hamiltonian in Eq. (12.85), and do the same for the corresponding Hamiltonian of the K' point.

12.9 Magnetic Translation Invariance and Magnetic Bloch Bands

By now the reader should have become quite familiar with Landau-level states and wave functions, at least in the Landau gauge. It appears that these states behave (somewhat) like Bloch states in a Bloch band, as they all carry a band-like index n (the Landau-level index), and a momentum-like quantum number k; also we can use the band dispersion to calculate the electron velocity, as we did in Eq. (12.66), in a manner similar to Eq. (8.9) in Chapter 8. But the differences are also quite obvious: in Landau gauge we have a 1D k-space, whereas in a 2D system the Bloch wave vector should be a 2D vector defined in the first Brillouin zone. Most importantly, here k is a *gauge-dependent* quantity; its value and even *direction* depend on the choice of gauge.[20] Physics, on the other hand,

---

[20] In our earlier discussion we chose a specific Landau gauge in which $\vec{A}$ points in the $\hat{y}$ direction so k is the $\hat{y}$-direction momentum; but this choice is arbitrary in the absence of any potential, and we could have chosen $\vec{A}$ to be in any direction and the direction of k would change accordingly.

is gauge-independent. In the following we generalize Bloch's theorem in the presence of a uniform magnetic field (and eventually to include the possibility of non-uniform but periodic magnetic field), in a gauge-independent manner.

We start by reminding ourselves about the origin of Bloch's theorem. In the case of a periodic potential without a magnetic field, the Hamiltonian is invariant under a set of discrete lattice translations $T_{\vec{a}}$,

$$T_{\vec{a}} H T_{\vec{a}}^{\dagger} = H, \tag{12.89}$$

where $T_{\vec{a}} = e^{\frac{i}{\hbar} \vec{p} \cdot \vec{a}}$ and $\vec{a}$ is a lattice vector. Now, for an eigenstate of H with $H\psi = \epsilon\psi$, we have

$$T_{\vec{a}} H T_{\vec{a}}^{\dagger} T_{\vec{a}} \psi = H T_{\vec{a}} \psi = \epsilon T_{\vec{a}} \psi, \tag{12.90}$$

namely $T_{\vec{a}}\psi$ is also an eigenstate of H with the *same* eigenvalue. We thus expect that, under a translation which leaves the Hamiltonian invariant, the eigenfunction remains unchanged up to an overall phase factor[21]

$$T_{\vec{a}}\psi = \lambda_{\vec{a}}\psi = e^{i\vec{k}\cdot\vec{a}}\psi, \tag{12.91}$$

which *defines* the Bloch wave vector $\vec{k}$.

Now consider

$$H = \frac{1}{2m_{\mathrm{e}}} \left[\vec{p} + \frac{e}{c} \vec{A}(\vec{r}) \right]^2, \tag{12.92}$$

with $\nabla \times \vec{A}(\vec{r}) = \vec{B}$. Performing the same manipulation on the eigenstate ψ as in Eq. (12.90), we obtain

$$T_{\vec{a}} H T_{\vec{a}}^{\dagger} T_{\vec{a}} \psi = \frac{1}{2m_{\mathrm{e}}} \left[\vec{p} + \frac{e}{c} \vec{A}(\vec{r} + \vec{a}) \right]^2 T_{\vec{a}} \psi = \epsilon T_{\vec{a}} \psi. \tag{12.93}$$

We find $T_{\vec{a}}\psi$ is an eigenstate of a *different* Hamiltonian,

$$H' = \frac{1}{2m_{\mathrm{e}}} \left[\vec{p} + \frac{e}{c} \vec{A}(\vec{r} + \vec{a}) \right]^2, \tag{12.94}$$

that in general is *not* the same as the original Hamiltonian H. As a result, Bloch's theorem does *not* follow.

The origin of this is of course the $\vec{r}$ dependence of $\vec{A}$, which ruins the translation symmetry of H. On the other hand, for a uniform magnetic field, $\vec{A}(\vec{r})$ and its translated version, $\vec{A}(\vec{r} + \vec{a})$, correspond to the *same* magnetic field:

$$\nabla \times \vec{A}(\vec{r}) = \nabla \times \vec{A}(\vec{r} + \vec{a}) = \vec{B}. \tag{12.95}$$

This means that $\vec{A}(\vec{r})$ and $\vec{A}(\vec{r} + \vec{a})$ are related by a gauge transformation:

$$\vec{A}(\vec{r} + \vec{a}) = \vec{A}(\vec{r}) + \nabla f_{\vec{a}}(\vec{r}). \tag{12.96}$$

We can thus perform a *gauge transformation* to bring H' back to H:

$$e^{i\phi_{\vec{a}}(\vec{r})} H' e^{-i\phi_{\vec{a}}(\vec{r})} = H, \tag{12.97}$$

where

$$\phi_{\vec{a}}(\vec{r}) = \frac{e f_{\vec{a}}(\vec{r})}{\hbar c}. \tag{12.98}$$

---

[21] Here we assume for simplicity that the eigenstate is non-degenerate. More care has to be taken in the degenerate case.

This immediately suggests that we can combine translation and the corresponding gauge transformations to form a **magnetic translation operator**,

$$\tilde{T}_{\vec{a}} = e^{i\phi_{\vec{a}}(\vec{r})} T_{\vec{a}}, \tag{12.99}$$

which *commutes* with H:

$$[\tilde{T}_{\vec{a}}, H] = 0. \tag{12.100}$$

We have thus found the equivalent of translation symmetry, in the case of a uniform magnetic field.

There is, however, a fundamental difference between ordinary translation and magnetic translation. We know that $[T_{\vec{a}}, T_{\vec{b}}] = 0$ for arbitrary $\vec{a}$ and $\vec{b}$. This is *not* true for $\tilde{T}$, because it involves both position and momentum operators. As a result, it is impossible to have eigenstates of H that are also *simultaneous* eigenstates of all magnetic translation operators. On the other hand, for carefully chosen *different* vectors $\vec{a}_1$ and $\vec{a}_2$ that satisfy appropriate conditions, we can have

$$[\tilde{T}_{\vec{a}_1}, \tilde{T}_{\vec{a}_2}] = 0. \tag{12.101}$$

Let us find out what these conditions are. Equation (12.101) is equivalent to

$$\begin{aligned}
\tilde{T}_{\vec{a}_2} = \tilde{T}_{\vec{a}_1} \tilde{T}_{\vec{a}_2} \tilde{T}_{\vec{a}_1}^{\dagger} &= e^{i\phi_{\vec{a}_1}(\vec{r})} T_{\vec{a}_1} e^{i\phi_{\vec{a}_2}(\vec{r})} T_{\vec{a}_2} T_{\vec{a}_1}^{\dagger} e^{-i\phi_{\vec{a}_1}(\vec{r})} \\
&= e^{i\phi_{\vec{a}_1}(\vec{r}) + i\phi_{\vec{a}_2}(\vec{r}+\vec{a}_1) - i\phi_{\vec{a}_1}(\vec{r}+\vec{a}_2)} T_{\vec{a}_2} \\
&= e^{i\phi_{\vec{a}_1}(\vec{r}) + i\phi_{\vec{a}_2}(\vec{r}+\vec{a}_1) - i\phi_{\vec{a}_1}(\vec{r}+\vec{a}_2) - i\phi_{\vec{a}_2}(\vec{r})} \tilde{T}_{\vec{a}_2}.
\end{aligned} \tag{12.102}$$

Thus the condition we need $\vec{a}_1$ and $\vec{a}_2$ to satisfy is

$$[\phi_{\vec{a}_1}(\vec{r}) - \phi_{\vec{a}_1}(\vec{r}+\vec{a}_2)] - [\phi_{\vec{a}_2}(\vec{r}) - \phi_{\vec{a}_2}(\vec{r}+\vec{a}_1)] = 2\pi N \tag{12.103}$$

$$= \frac{e}{\hbar c} \oint_{\diamond} \vec{A}(\vec{r}) \cdot d\vec{r}$$

$$= \frac{e\Phi_{\diamond}}{\hbar c}$$

$$= \frac{2\pi \Phi_{\diamond}}{\Phi_0}. \tag{12.104}$$

Here N is an integer, $\diamond$ stands for the parallelogram spanned by $\vec{a}_1$ and $\vec{a}_2$, and $\Phi_{\diamond} = \vec{B} \cdot (\vec{a}_1 \times \vec{a}_2)$ is the flux it encloses. This phase is nothing more than the Aharonov–Bohm phase acquired in moving around the perimeter of the parallelogram. To arrive at Eq. (12.104) we combined Eqs. (12.96) and (12.98). Thus the condition for Eq. (12.101) is simply

$$\frac{\Phi_{\diamond}}{\Phi_0} = N, \tag{12.105}$$

namely the parallelogram spanned by $\vec{a}_1$ and $\vec{a}_2$ encloses an integer number of flux quanta N. For parallel $\vec{a}_1$ and $\vec{a}_2$, $N = 0$ and Eq. (12.105) is trivially satisfied. In this case $\tilde{T}_{\vec{a}_1}$ and $\tilde{T}_{\vec{a}_2}$ are *not* independent, and we only get a 1D k quantum number for ψ from the magnetic translation symmetry of H, as in the case we studied earlier using the Landau gauge. To obtain a 2D vector quantum number $\vec{k}$ for ψ, the minimum condition is $N = 1$, or the minimum *magnetic* unit cell must enclose one flux quantum. We can then combine Eqs. (12.100) and (12.101) to generalize Bloch's theorem and define *magnetic* Bloch bands and states that are simultaneous eigenstates of H, $\tilde{T}_{\vec{a}_1}$, and $\tilde{T}_{\vec{a}_2}$; clearly, in the present case they correspond to Landau levels and corresponding states.

Using the magnetic translation symmetry, we can obtain some of the key properties of Landau-level states, without actually solving H. One very special feature of the magnetic unit cell in this case is that the condition Eq. (12.105) determines only its area, *not* its shape or the actual lattice structure. Correspondingly, the reciprocal lattice structure is not determined either, other than the area of the Brillouin zone. On the other hand, we know that the band dispersion is a periodic function in the reciprocal lattice; the only function that is periodic in such a *continuously* varying family of lattice structures is a constant. We thus conclude that each of the magnetic Bloch bands must have a constant energy, namely they form degenerate Landau levels! Furthermore, since each magnetic unit cell contains one flux quantum, and the number of states in each band is equal to the number of unit cells, we come to the conclusion that the Landau-level degeneracy must be equal to the number of flux quanta enclosed in the system, as we found earlier.

The discussions above can be easily generalized to the case where a magnetic field coexists with a periodic potential, $V(\vec{r})$. In this case, in order to find $\tilde{T}_{\vec{a}_1}$ and $\tilde{T}_{\vec{a}_2}$ that commute with each other and both commute with

$$H = \frac{1}{2m_e}\left[\vec{p} + \frac{e}{c}\vec{A}(\vec{r})\right]^2 + V(\vec{r}), \tag{12.106}$$

we need $\vec{a}_1$ and $\vec{a}_2$ to be lattice vectors of $V(\vec{r})$ (but not necessarily primitive ones), in addition to satisfying Eq. (12.105). Thus, if the original unit cell of $V(\vec{r})$ contains a rational number p/q of flux quanta, with p and q mutually prime integers, we need to repeat the original unit cell q times to form a **magnetic unit cell**, which contains an integer number p of flux quanta. In this case each original band (in the absence of $\vec{B}$) is split into q magnetic subbands.[22] Equivalently, one can also say that each Landau level (in the absence of $V(\vec{r})$) is split into p magnetic subbands. The area of the (minimal) magnetic unit cell is determined by $\vec{B}$. However, in this case the lattice vectors are not arbitrary, since they must be consistent with the translation symmetry of $V(\vec{r})$. Hence the argument above showing that the Landau-level states are degenerate does not apply and, in general, these magnetic Bloch bands are *not* degenerate. It should also be clear that $\vec{B}(\vec{r})$ does *not* need to be a constant, but need only be periodic from one magnetic unit cell to another, in order for these conclusions to hold.

Owing to the (magnetic) translation symmetry, an eigenfunction of the Hamiltonian in Eq. (12.106) may be written as

$$\psi_{n\vec{k}}(\vec{r}) = e^{i\vec{k}\cdot\vec{r}}u_{n\vec{k}}(\vec{r}). \tag{12.107}$$

In the absence of a magnetic field (where we can choose the gauge $\vec{A} = \vec{0}$), Eq. (12.91) tells us that the Bloch function satisfies periodic boundary conditions

$$u_{n\vec{k}}(\vec{r} + \vec{a}_j) = u_{n\vec{k}}(\vec{r}), \tag{12.108}$$

for any lattice translation vector $\vec{a}_j$. In the presence of a magnetic field, the magnetic translation operator is changed and Eq. (12.99) tells us that the Bloch function satisfies *generalized* periodic boundary conditions:

$$\tilde{T}_{\vec{a}_j}u_{n\vec{k}}(\vec{r}) = e^{i\phi_{\vec{a}_j}(\vec{r})}u_{n\vec{k}}(\vec{r} + \vec{a}_j) = u_{n\vec{k}}(\vec{r}). \tag{12.109}$$

12.9.1 Simple Landau Gauge Example

To better understand all these somewhat formal mathematical results, it is useful to consider the specific case of a uniform magnetic field treated in the Landau gauge. Suppose that the periodic potential yields a rectangular unit cell of width a_x and height a_y. We will take the magnetic field strength to be such that the unit cell contains a single flux quantum. Thus

$$a_x a_y = 2\pi \ell^2. \tag{12.110}$$

---

[22] For an arbitrary magnetic flux through the original unit cell, it may be necessary in some cases to choose extremely large p and q in order to find an accurate rational approximation to the flux density. The mathematical details needed to treat the case of irrational flux density are ignored in the present discussion.

The corresponding reciprocal lattice unit cell is also rectangular, with sides

$$G_x = \frac{2\pi}{a_y} = \frac{a_x}{\ell^2}, \tag{12.111}$$

$$G_y = \frac{2\pi}{a_x} = \frac{a_y}{\ell^2}. \tag{12.112}$$

To make the example particularly easy, we will assume that the periodic potential simply vanishes. The Landau gauge solutions of the Schrödinger equation are therefore those given in Eq. (12.48). It is very hard to see how to reconcile this expression with the standard form of the Bloch-wave solution given in Eq. (12.107). First, the solutions in Eq. (12.48) are parameterized by a 1D wave vector k, whereas those in Eq. (12.107) involve a 2D wave vector $\vec{k}$. Secondly, as was noted earlier in the discussion below Eq. (12.48), those solutions are localized Gaussians in the x direction and do not at all obey the generalized periodic boundary condition given in Eq. (12.109).

We escape from this quandary by noting that the discrepancies between Eqs. (12.48) and (12.107) are simply an accident associated with the complete degeneracy of the Landau level when the periodic potential vanishes. Because of the degeneracy, linear combinations of different solutions in Eq. (12.48) are also solutions. By choosing the linear combinations correctly, we can put the solutions into the standard Bloch-wave form.

Consider the following (unnormalized) linear combination of Landau wave functions centered on different x positions:

$$\Psi_{nkq}(\vec{r}) = \sum_{m=-\infty}^{+\infty} e^{iqmG_x\ell^2} \psi_{n(k+mG_x)}(\vec{r})$$

$$= \sum_{m=-\infty}^{+\infty} e^{iqmG_x\ell^2} e^{i(k+mG_x)y} F_n(x - [k + mG_x]\ell^2), \tag{12.113}$$

where

$$F_n(z) \equiv H_n(z/\ell)e^{-z^2/2\ell^2}. \tag{12.114}$$

These new basis states (which are essentially Jacobi elliptic theta functions) depend on k only through $k \mod G_x$ and on q only through $q \mod G_y$, so we can without loss of generality employ the restrictions $-G_x/2 \leq k < +G_x/2$ and $-G_y/2 \leq q < +G_y/2$. The presence of the new parameter q compensates for the restriction on k and will be the route to obtaining a 2D wave vector.

Under the translation $x \to x + a_x$, the basis states transform according to

$$\Psi_{nkq} \to e^{iqa_x} e^{iG_x y} \Psi_{nkq}, \tag{12.115}$$

while, under the translation $y \to y + a_y$, the basis states transform according to

$$\Psi_{nkq} \to e^{ika_y} \Psi_{nkq}. \tag{12.116}$$

If we define a 2D wave vector $\vec{K} = (q, k)$, we can write this in the Bloch-wave form of Eq. (12.107),

$$\Psi_{nkq} = e^{i\vec{K}\cdot\vec{r}} u_{n\vec{K}}(\vec{r}), \tag{12.117}$$

where

$$u_{n\vec{K}}(\vec{r}) \equiv e^{-i\vec{K}\cdot\vec{r}} \Psi_{nkq} \tag{12.118}$$

obeys the generalized periodic boundary conditions

$$u_{n\vec{K}}(x, y + a_y) = u_{n\vec{K}}(x, y), \tag{12.119}$$

$$u_{n\vec{K}}(x + a_x, y) = e^{2\pi iy/a_y} u_{n\vec{K}}(x, y). \tag{12.120}$$

The final step in the derivation is to note that the Landau gauge vector potential given in Eq. (12.40) transforms under translation by one of the lattice vectors $\vec{a}_j$ according to

$$\vec{A}(\vec{r} + \vec{a}_j) = \vec{A}(x, y) + \nabla f_{\vec{a}_j}(x, y), \tag{12.121}$$

where

$$f_{\vec{a}_j}(x, y) = -(a_{jx}B)y, \tag{12.122}$$

and hence

$$\phi_{\vec{a}}(\vec{r}) = -\frac{a_{jx}y}{\ell^2}, \tag{12.123}$$

which, for the case $a_{jx} = a_x$, reduces to

$$\phi_{\vec{a}}(\vec{r}) = -2\pi \frac{y}{a_y}. \tag{12.124}$$

Thus Eq. (12.120) is equivalent to Eq. (12.109) as required.

Exercise 12.11. Suppose that the sample has dimensions $N_x a_x$ by $N_y a_y$. Count the number of allowed 2D wave vectors $\vec{K}$ for the Bloch waves and show that this matches the number of allowed 1D wave-vector states k, $N_\Phi = N_x N_y$, in agreement with Eq. (12.49).

Exercise 12.12. Consider an electron moving in the xy plane with a non-uniform but periodic magnetic field along the $\hat{z}$ direction: $\vec{B}(\vec{r}) = \hat{z}B(\vec{r})$, with $\vec{r}$ being a 2D vector. Its periodicity is specified by $B(\vec{r} + \vec{a}_j) = B(\vec{r})$, with $\{\vec{a}_j\}$ forming a Bravais lattice with primitive unit cell area A. The magnetic flux enclosed in such a unit cell is $p\Phi_0/q$, where Φ_0 is the flux quantum and p and q are mutually prime integers. Find the size of the magnetic unit cell, and corresponding magnetic translation operators similar to that of Eq. (12.99).

12.10 Quantization of the Hall Conductance in Magnetic Bloch Bands

In the previous section we learned that Landau levels are nothing but very special cases of magnetic Bloch bands, which in turn are generalizations of Bloch bands in the presence of magnetic fields. This naturally leads to the following conundrum. In Chapter 7 we were told that completely filled bands do *not* carry current, and systems with Fermi energies inside a band gap are insulators. Yet we find that completely filled Landau levels give rise to quantized contributions to the Hall conductance, and the system can carry a *dissipationless* Hall current (and behave almost like a superconductor!) when its Fermi energy lies in between Landau levels. So something must be missing when we simply identified systems with completely filled and completely empty bands as insulators. We learned in this chapter that the dissipationless Hall current is carried by edge states, but need to understand better why we have these edge states.

We would also like to ask whether the quantization of the Hall conductance (that we demonstrated for Landau levels) is robust for magnetic Bloch bands. The answer is yes, and, like the answers to the questions posed in the previous paragraph, has a deep topological origin which we will explore in Chapter 13. In a few words, the Hall conductance is a *topological* invariant of a Bloch band, which is *insensitive* to continuous changes of the band dispersion (as long as no band crossing is involved). The contribution of individual states in a band to the Hall conductance can be viewed (along with other characteristics) as a *geometric* property of the band. The overall Hall conductance of an isolated band (i.e. a band separated by an energy gap from other bands) is a topological invariant that does not depend on the particular geometry of the band. Furthermore, it is possible to show from purely topological arguments when gapless edge modes (such as the quantum Hall edge currents) will appear at the boundary between two different (topologically distinct) gapped regions. To fully appreciate these words, however, we need to take a detour and study some geometrical and topological features of quantum-mechanical systems in general, and then apply them to Bloch bands. This will be our task in Chapter 13, aided by the additional pedagogical material in Appendix D. In Chapter 14 we will extend these ideas to study the remarkable concepts of the topological insulator and topological semimetal.

13 Topology and Berry Phase

Geometry and topology have begun to play an increasingly important role in modern quantum physics. Geometry was invented by the ancients partly for the purpose of measuring distances and surveying the areas of farm fields, etc. Indeed, the very word means "measuring the earth." However, the formal Euclidean plane geometry that we first learn in high school explicitly eschews discussion of absolute distance. We are allowed to use a compass and a straight edge to construct geometric figures, but we are not allowed to use a ruler. In this sense, Euclidean plane geometry can thus be characterized as the study of invariants under linear scale changes and rotations. (An isoceles triangle made twice as large and then rotated through some angle is still an isoceles triangle with the same three internal angles, and two of the sides, though now twice as long, are still equal in length.) Topology is the study of invariants under more general arbitrary continuous transformations. (A coffee cup with one handle can be continuously deformed into a donut. An isoceles triangle can be continuously transformed into a circle.) There are far fewer invariants under arbitrary distortions than there are under linear scale changes, but, by the same token, those topological invariants are very robust and therefore quite special and fundamental. See Appendix D for a brief primer on the basic concepts of topology.

In recent years it has come to be recognized that the values of certain physical observables are universal and invariant under changes in the Hamiltonian. For example, in Chapter 12 we learned that the Hall conductance (or resistance) on a quantum Hall plateau is very precisely quantized, with a universal value completely independent of the details of the Hamiltonian (such as the particle mass, the strength of the magnetic field, and the particular realization of the random disorder potential). It is now understood [88] that this universality is the result of the fact that the physical observable is represented mathematically by a **topological invariant**. This so-called "topological protection" means that the quantity remains invariant even when the Hamiltonian undergoes changes (say, from one sample to the next, or the magnetic field is varied), unless the system goes through a quantum phase transition that results in a discontinuous change of the corresponding topological invariant.

We learn plane geometry in high school before we learn topology in college or graduate school. Hence we will start with a study of the geometry of Hilbert space in order to equip ourselves to move on to the study of topological invariants in quantum physics.

13.1 Adiabatic Evolution and the Geometry of Hilbert Space

To explore the geometry of Hilbert space we need to move around inside Hilbert space. We already have some notion of length and distance because we have a well-defined inner product between vectors. However, the fact that this is a complex vector space adds a new twist. Physical states are represented by rays,[1] not vectors in Hilbert space, and, as we shall see, this leads to interesting and novel features.

One nice way to move around in Hilbert space is to follow a particular eigenstate of the Hamiltonian as we slowly (adiabatically) vary the Hamiltonian. Such adiabatic processes can also be of great physical importance. For example, the naturally slow motion of the heavy nuclei in a diatomic molecule can be profitably viewed as a slow adiabatic variation of parameters in the Hamiltonian for the light (and speedy) electrons (see Exercise 5.6), which often have a significant excitation gap. It is this large excitation gap which prevents excitations of electrons into higher states as the nuclei slowly move.

Consider a Hamiltonian that depends on time through a set of parameters $\vec{\mathcal{R}} = (\mathcal{R}_1, \mathcal{R}_2, \ldots, \mathcal{R}_D)$:

$$H(t) = H[\vec{\mathcal{R}}(t)]. \tag{13.1}$$

$\vec{\mathcal{R}}$ can be viewed as a point or a vector in a D-dimensional parameter space, where D is the number of independent parameters, and not necessarily related to the spatial dimension of the physical system d (or even the number of particles in the system). For every $\vec{\mathcal{R}}$ we have an ($\vec{\mathcal{R}}$-dependent) set of orthonormal eigenstates of $H(\vec{\mathcal{R}})$:

$$H(\vec{\mathcal{R}})|n(\vec{\mathcal{R}})\rangle = \epsilon_n(\vec{\mathcal{R}})|n(\vec{\mathcal{R}})\rangle. \tag{13.2}$$

For simplicity, let us assume that the spectrum of H is discrete and non-degenerate everywhere in the $\vec{\mathcal{R}}$ space. Then the **adiabatic theorem** tells us that, if the system is initially in the nth eigenstate of H,

$$|\psi_n(t = 0)\rangle = |n[\vec{\mathcal{R}}(t = 0)]\rangle, \tag{13.3}$$

then it will evolve into the nth eigenstate of H at later times, as long as the time-variation of H is sufficiently slow:[2]

$$|\psi_n(t)\rangle = C_n(t)|n[\vec{\mathcal{R}}(t)]\rangle. \tag{13.4}$$

Thus a path in parameter space $\vec{\mathcal{R}}(t)$ defines (maps onto) a path in Hilbert space $|\psi_n(t)\rangle$.

Since time-evolution under H is a unitary transformation, the proportionality constant $C_n(t)$ has to be a pure phase.[3] For time-independent H, we simply have

$$C_n(t) = \exp\left[-\frac{i}{\hbar}\epsilon_n t\right]. \tag{13.5}$$

Generalizing to the case of H being time-dependent, we can write, without loss of generality,

$$C_n(t) = e^{i\gamma_n(t)} \exp\left[-\frac{i}{\hbar} \int_0^t dt' \, \epsilon_n(t')\right], \tag{13.6}$$

where the second factor above is the straightforward generalization of Eq. (13.5). We will see below that the phase $\gamma_n(t)$ in the first factor is not zero in general, and contains much physics. It accounts for a special aspect of the state evolution not captured by the second factor. Combining Eqs. (13.4) and (13.2) and plugging into the Schrödinger equation

$$i\hbar\frac{\partial}{\partial t}|\psi_n(t)\rangle = H[\vec{\mathcal{R}}(t)]|\psi_n(t)\rangle, \tag{13.7}$$

---

[1] A ray is the equivalence class of vectors all related to a given vector by multiplication by an overall complex phase $e^{i\theta}$ for any real θ. More generally, since overall normalization does not matter, a ray is defined as the equivalence class of vectors all related to a given vector by multiplication by any complex number λ.

[2] The time scale that defines "slow" is set by the (inverse of the) excitation gap between the nth state and neighboring states.

[3] Since $U^\dagger U = 1$, unitary transformations preserve the inner product between two states and that of a state with itself.

and taking the inner product with $\langle \psi_n(t) |$ on both sides, we obtain

$$\frac{\partial}{\partial t} \gamma_n(t) = i \langle n[\vec{\mathcal{R}}(t)] | \frac{\partial}{\partial t} | n[\vec{\mathcal{R}}(t)] \rangle. \tag{13.8}$$

Thus

$$\gamma_n(t) = i \int_0^t dt' \langle n[\vec{\mathcal{R}}(t')] | \frac{\partial}{\partial t'} | n[\vec{\mathcal{R}}(t')] \rangle \tag{13.9}$$

$$= \int_C \vec{\mathcal{A}}^n(\vec{\mathcal{R}}) \cdot d\vec{\mathcal{R}}, \tag{13.10}$$

where the so-called "**Berry connection**"[4]

$$\vec{\mathcal{A}}^n(\vec{\mathcal{R}}) = i \langle n(\vec{\mathcal{R}}) | \frac{\partial}{\partial \vec{\mathcal{R}}} | n(\vec{\mathcal{R}}) \rangle \tag{13.11}$$

is a (vector) function of $\vec{\mathcal{R}}$ but *not* an explicit function of t. Thus the additional phase γ_n depends only on the path C that $\vec{\mathcal{R}}$ traverses in the parameter space, *not* on its actual time-dependence. Thus γ_n is of *geometric* origin;[5] in contrast the phase $-(1/\hbar) \int_0^t dt' \, \epsilon_n(t')$ is called a *dynamical* phase because it does depend on the time-dependence of the control path in parameter space.

Exercise 13.1. Let us explore the fact that the adiabatic phase depends only on the path $\vec{\mathcal{R}}$ taken through parameter space and not on how that path is taken in time (as long as it is slow enough for the adiabatic approximation to be valid). Show that Eq. (13.9) exhibits "*time reparameterization invariance.*" That is, replace Eq. (13.9) by

$$\gamma_n(t) = i \int_0^t dt' \langle n[\vec{\mathcal{R}}(\tau)] | \frac{\partial}{\partial t'} | n[\vec{\mathcal{R}}(\tau)] \rangle, \tag{13.12}$$

where the modified time variable $\tau(t')$ is an arbitrary (smooth) function of t', subject only to the constraint that the starting and ending times (and therefore the end points of the trajectory) are left invariant, $\tau(0) = 0$, $\tau(t) = t$, and show that the result is completely independent of the functional form of $\tau(t')$. See Appendix D for further discussion.

Exercise 13.2. Using the fact that $\langle n(\vec{\mathcal{R}}) | n(\vec{\mathcal{R}}) \rangle = 1$ and hence $(\partial/\partial \vec{\mathcal{R}}) \langle n(\vec{\mathcal{R}}) | n(\vec{\mathcal{R}}) \rangle = 0$, show that $\vec{\mathcal{A}}^n(\vec{\mathcal{R}})$ and γ_n are purely real. The interpretation of this mathematical result is the following. In computing the Berry connection, we are looking for the part of the change in $|\psi\rangle$ with $\vec{\mathcal{R}}$ that is parallel to the Hilbert space vector $|\psi\rangle$ itself. Given that the length of the vector is fixed, the change must either be orthogonal to $|\psi\rangle$ or represent a change in the overall phase of $|\psi\rangle$. Only the latter contributes to the Berry connection, and the contribution is such that $\langle n(\vec{\mathcal{R}}) | (\partial/\partial \vec{\mathcal{R}}) | n(\vec{\mathcal{R}}) \rangle$ is purely imaginary.

---

[4] A word about our somewhat sloppy notation here. In the Dirac notation $\langle \psi | \mathcal{O} | \psi \rangle$ represents the expectation value of the operator $\mathcal{O}$, or equivalently, the inner product between $\mathcal{O} | \psi \rangle$ and $| \psi \rangle$. $\partial/\partial \vec{\mathcal{R}}$ is not a physical observable (represented by an operator acting on states in the Hilbert space). Hence the proper interpretation of the notation is that it should be taken to mean the inner product between $(\partial/\partial \vec{\mathcal{R}}) | \psi \rangle$ and $| \psi \rangle$.

[5] For a useful analogy from daily life, consider you are driving home from your workplace and watching the odometer of your car to measure the distance between these two locations. The final result does *not* depend on how fast you drive (which is a piece of dynamical information!), but does depend on the route you choose, since distance is a purely geometric property of the route (or path).

It is important to emphasize, however, that in general γ_n is *gauge-dependent*, and thus *not* a physical observable. This is because Eq. (13.2) does not determine the overall phase of $|n(\vec{\mathcal{R}})\rangle$; in particular, we can make an $\vec{\mathcal{R}}$-dependent phase change

$$|n(\vec{\mathcal{R}})\rangle \Rightarrow e^{i\zeta(\vec{\mathcal{R}})}|n(\vec{\mathcal{R}})\rangle, \tag{13.13}$$

which leads to

$$\vec{\mathcal{A}}^n(\vec{\mathcal{R}}) \Rightarrow \vec{\mathcal{A}}^n(\vec{\mathcal{R}}) - \partial_{\vec{\mathcal{R}}}\zeta(\vec{\mathcal{R}}) \tag{13.14}$$

and

$$\gamma_n \Rightarrow \gamma_n + \zeta[\vec{\mathcal{R}}(t=0)] - \zeta[\vec{\mathcal{R}}(t)]. \tag{13.15}$$

For this reason the potential physical effects of γ_n were ignored for many years. It was implicitly assumed that γ_n could always be "gauged away" and therefore had no physical content.

In 1984 Berry [89] (see also Refs. [90, 91] for more recent perspectives; parts of our discussions follow those of Ref. [91]) made the important discovery that, for adiabatic processes defined by motion over a closed loop in parameter space (i.e. ones for which $\vec{\mathcal{R}}(t=0) = \vec{\mathcal{R}}(t_{\text{final}})$, and thus H returns to its original value), the gauge-dependence of γ_n in Eq. (13.15) disappears. In these cases the line integral of the Berry connection around a closed loop in parameter space,

$$\gamma_n = \oint \vec{\mathcal{A}}^n(\vec{\mathcal{R}}) \cdot d\vec{\mathcal{R}}, \tag{13.16}$$

is *gauge-independent* and is known as the **Berry phase**. It is thus (in principle) observable and can have physical consequences if it turns out to be non-zero.

The reader may have recognized by now that the (gauge-dependent) **Berry connection** $\vec{\mathcal{A}}^n(\vec{\mathcal{R}})$ is reminiscent of the familiar vector potential of a magnetic field, while the Berry phase, being a loop integral of $\vec{\mathcal{A}}^n(\vec{\mathcal{R}})$, is like the (gauge-independent) magnetic flux passing through the loop (in parameter space, not "real" space). This analogy with electromagnetism immediately suggests the existence of a gauge-independent local field known as the **Berry curvature**:

$$\omega_{\mu\nu}^n(\vec{\mathcal{R}}) = \partial_{\mathcal{R}_\mu}\mathcal{A}_\nu^n(\vec{\mathcal{R}}) - \partial_{\mathcal{R}_\nu}\mathcal{A}_\mu^n(\vec{\mathcal{R}}), \tag{13.17}$$

which is an anti-symmetric rank-2 tensor in the parameter space (similar to the field tensor $F_{\mu\nu}$ in electromagnetism). Just as we do in electromagnetism, we can use Stokes' theorem to relate the line integral of the connection around the boundary of a region to the surface integral of the curvature over that region:

$$\gamma_n = \oint_{\partial S} \vec{\mathcal{A}}^n(\vec{\mathcal{R}}) \cdot d\vec{\mathcal{R}} = \frac{1}{2}\int_S d\mathcal{R}_\mu \wedge d\mathcal{R}_\nu \, \omega_{\mu\nu}^n(\vec{\mathcal{R}}), \tag{13.18}$$

where S stands for the area in parameter space enclosed by the loop ∂S, $d\mathcal{R}_\mu \wedge d\mathcal{R}_\nu = -d\mathcal{R}_\nu \wedge d\mathcal{R}_\mu$ stands for an infinitesimal (oriented) element of the surface,[6] and again we use the Einstein summation convention. The factor of 1/2 compensates for the fact that each element is summed over twice due to the summation convention. If the parameter space happens to be 3D, we can express the Berry curvature ω^n as a kind of "magnetic field" in terms of the curl of the connection vector,[7]

$$\vec{b}^n = \nabla_{\vec{\mathcal{R}}} \times \vec{\mathcal{A}}^n = i\langle \nabla_{\vec{\mathcal{R}}}n(\vec{\mathcal{R}})| \times |\nabla_{\vec{\mathcal{R}}}n(\vec{\mathcal{R}})\rangle, \tag{13.19}$$

---

[6] This "exterior" or "wedge" product notation is standard in differential geometry, see [35]. It represents the surface area of the parallelogram formed by the two vectors being multiplied. For the simplest case to visualize, namely a 2D surface embedded in 3D, it defines a surface normal vector given simply by the cross product of the two vectors.

[7] If ω is a rank-2 anti-symmetric tensor in D dimensions, it contains $D(D-1)/2$ independent components. For the special case of $D = 3$, there are three independent components, the same number as in a 3D vector. As a result, it can be represented by a vector in 3D via $b^\delta = \frac{1}{2}\epsilon^{\delta\mu\nu}\omega_{\mu\nu}$, where ϵ is the fully anti-symmetric tensor (Levi-Civita symbol). In this context $\vec{b}$ is often called a pseudovector, reflecting the fact that it is really a rank-2 anti-symmetric tensor.

and the analogy with the vector potential/magnetic field is complete:

$$\gamma_n = \int_S \vec{b}^n \cdot d\vec{S}. \tag{13.20}$$

Motivated by this analogy, the Berry phase is also occasionally referred to as the **Berry flux**. In the following we use $\omega_{\mu\nu}$ and $\vec{b}$ interchangeably to represent the Berry curvature when $D = 3$. Indeed, we will see some examples in which $\vec{b}$ actually is a physical magnetic field.

Exercise 13.3. Show from its definition in Eq. (13.17) that we can write the Berry curvature as

$$\omega_{\mu\nu}^n = i[\langle \partial_{\mathcal{R}_\mu} n(\vec{\mathcal{R}})|\partial_{\mathcal{R}_\nu} n(\vec{\mathcal{R}})\rangle - \langle \partial_{\mathcal{R}_\nu} n(\vec{\mathcal{R}})|\partial_{\mathcal{R}_\mu} n(\vec{\mathcal{R}})\rangle]. \tag{13.21}$$

In Eqs. (13.17) and (13.21), the Berry curvature $\omega_{\mu\nu}^n$ is expressed in terms of the *variation* of the eigenstate $|n(\vec{\mathcal{R}})\rangle$ with respect to the parameters $\vec{\mathcal{R}}$. In practice it is often convenient to express $\omega_{\mu\nu}^n$ in terms of quantities at fixed $\vec{\mathcal{R}}$. To this end we will show below that

$$\omega_{\mu\nu}^n(\vec{\mathcal{R}}) = i \sum_{n' \neq n} \frac{\langle n(\vec{\mathcal{R}})|(\partial H/\partial \mathcal{R}_\mu)|n'(\vec{\mathcal{R}})\rangle \langle n'(\vec{\mathcal{R}})|(\partial H/\partial \mathcal{R}_\nu)|n(\vec{\mathcal{R}})\rangle - \text{c.c.}}{[\epsilon_n(\vec{\mathcal{R}}) - \epsilon_{n'}(\vec{\mathcal{R}})]^2}. \tag{13.22}$$

Equation (13.22) is explicitly gauge-invariant and often more useful in numerical calculations of $\omega_{\mu\nu}^n$ because the numerically generated eigenvector $|n(\vec{\mathcal{R}})\rangle$ is gauge-dependent and may carry an $\vec{\mathcal{R}}$-dependent phase factor that (accidentally[8]) does *not* vary smoothly with $\vec{\mathcal{R}}$; this makes the derivatives in Eqs. (13.11) and (13.17) ill-defined, but does not affect the evaluation of Eq. (13.22).

To prove the equivalence of Eqs. (13.22) and (13.17), we take the derivative with respect to $\mathcal{R}_\nu$ on both sides of Eq. (13.2), which yields

$$\left(\frac{\partial H}{\partial \mathcal{R}_\nu}\right)|n(\vec{\mathcal{R}})\rangle = \left(\frac{\partial \epsilon_n}{\partial \mathcal{R}_\nu}\right)|n(\vec{\mathcal{R}})\rangle + (\epsilon_n - H)|\partial_{\mathcal{R}_\nu} n(\vec{\mathcal{R}})\rangle. \tag{13.23}$$

This leads to

$$\langle n'(\vec{\mathcal{R}})| \left(\frac{\partial H}{\partial \mathcal{R}_\nu}\right)|n(\vec{\mathcal{R}})\rangle = (\epsilon_n - \epsilon_{n'})\langle n'(\vec{\mathcal{R}})|\partial_{\mathcal{R}_\nu} n(\vec{\mathcal{R}})\rangle \tag{13.24}$$

for $n \neq n'$. Similarly,

$$\langle n(\vec{\mathcal{R}})| \left(\frac{\partial H}{\partial \mathcal{R}_\mu}\right)|n'(\vec{\mathcal{R}})\rangle = (\epsilon_n - \epsilon_{n'})\langle \partial_{\mathcal{R}_\mu} n(\vec{\mathcal{R}})|n'(\vec{\mathcal{R}})\rangle. \tag{13.25}$$

A good example of a pseudovector is the (real) magnetic field. One way to see that it is a pseudovector is to observe that, under spatial inversion that takes $\vec{r} \to -\vec{r}$, a true vector field transforms with odd parity as $\vec{A}(\vec{r}) \to -\vec{A}(-\vec{r})$, whereas a pseudovector field transforms with even parity as $\vec{B}(\vec{r}) \to +\vec{B}(-\vec{r})$. The reason why the magnetic field is even is that it is the curl of a true vector. The curl and the vector potential are both odd under spatial inversion, and their combination is even.

[8] For example, two different computer programs that find the eigenvectors of a given Hamiltonian matrix could produce results for (some of) the eigenvectors differing by a minus sign, or, in the case of complex eigenvectors, by an arbitrary overall phase that could be different for each eigenvector. Suppose, for instance, that the first program reports its eigenvectors using the arbitrary choice that the first entry in each eigenvector is real and positive. The second program might choose to make the last entry of the eigenvector real and positive. The two programs might produce identical results for some regions of parameter space and (suddenly) different results in another region (because in that region the eigenvector develops a sign difference between its first and last entries).

Combining these and using the resolution of the identity, $\sum_{n'} |n'\rangle\langle n'|$, shows that Eqs. (13.22) and (13.21) are equivalent. Equation (13.22) looks like a perturbation-theory result, but in fact it is formally exact.

13.2 Berry Phase and the Aharonov–Bohm Effect

A beautiful illustration of a non-trivial Berry connection can be found in the **Aharonov–Bohm effect**. Consider an infinitely long, very thin, tube of magnetic flux running along the z axis as shown in Fig. 13.1. Let there be a strong potential barrier which prevents charged particles from ever entering the thin region containing the magnetic field. Hence the particles never see the Lorentz force and classically the magnetic flux has no effect whatsoever on the motion. Quantum mechanically, however, there is an effect. Even though the magnetic field is zero outside the flux tube, the electromagnetic vector potential $\vec{A}(\vec{r})$ is not. The vector potential obeys $\nabla \times \vec{A} = \vec{0}$ everywhere the particle travels and hence, in any given local region, it can be gauged away. Nevertheless, there is a non-trivial global effect associated with the fact that, for any closed trajectory which winds around the flux tube, the line integral of the vector potential obeys

$$\oint d\vec{r} \cdot \vec{A}(\vec{r}) = n\Phi, \tag{13.26}$$

where Φ is the total flux inside the flux tube and n is the integer winding number of the trajectory around the tube. The winding number is a well-defined topological invariant (see Appendix D),

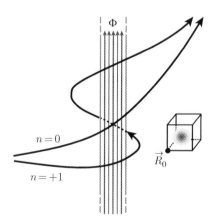

Figure 13.1 A tube of magnetic flux running along the z axis surrounded by a barrier which prevents charged particles from entering the region of the magnetic field. Because $\vec{B} = \nabla \times \vec{A} = \vec{0}$ outside the barrier, the particles see no Lorentz force and hence the classical motion, though affected by the presence of the barrier, is independent of the value of the magnetic flux. Quantum mechanically, however, the flux has a non-trivial effect on the interference between the complex amplitudes associated with topologically distinct paths (which cannot be deformed into each other continuously). The two paths shown here are topologically distinct, as their difference (to take the difference, simply reverse the arrow on one of them) forms a closed loop with winding number $n = +1$ (closed loops formed by pairs of topologically equivalent paths all have winding number $n = 0$). This allows one to assign a winding number for each individual path, once a specific path has been *defined* to have zero winding number as shown in the figure. Because they are of different length, they will have different corresponding quantum amplitudes. The phase difference between the amplitudes will also depend on the total flux Φ inside the flux tube, even though the Lorentz force is zero everywhere along both trajectories. The box in the lower right illustrates a method for adiabatically transporting a trapped electron along any desired trajectory. The orientation of the box is fixed, and the position $\vec{R}_0$ of one corner is slowly varied. The finite size of the box provides the excitation gap necessary for adiabatic transport.

because the 3D space has been punctured by the excluded region of the flux tube. Continuous deformation of the trajectory cannot change the winding number, because passing through the punctured region is not allowed.[9]

Even though locally there appears to be no effect on the quantum mechanics, globally there is, and charged particles scatter from the flux tube in a way that varies with the product of the charge q and the total flux Φ. One way to see this is to consider the Feynman path-integral approach, which assigns to any trajectory $\vec{r}(t)$ an amplitude

$$e^{\frac{i}{\hbar} S[\vec{r}\,]}, \tag{13.27}$$

where S is the classical action for that trajectory,

$$S[\vec{r}\,] = \int_0^t d\tau \, \mathcal{L}[\dot{\vec{r}}, \vec{r}\,]. \tag{13.28}$$

Recall that in the presence of a vector potential the Lagrangian undergoes a modification (that leads to the Lorentz force in the classical equations of motion)

$$\mathcal{L} \to \mathcal{L} + (-e)\vec{A}(\vec{r}) \cdot \dot{\vec{r}}. \tag{13.29}$$

Thus we see that the amplitude associated with any given trajectory acquires a vector-potential-dependent phase

$$e^{i\theta} = e^{\frac{i(-e)}{\hbar} \int d\vec{r} \cdot \vec{A}(\vec{r})}, \tag{13.30}$$

which for closed trajectories is topological (and gauge invariant) – it depends only on the winding number of the trajectory around the Aharonov–Bohm flux tube. Rather than solving the scattering problem, we will here follow Berry's example [89] and show that adiabatic transport of an electron around the flux tube yields a Berry phase equal to the Aharonov–Bohm phase,

$$\theta = \frac{i(-e)}{\hbar c} \oint d\vec{r} \cdot \vec{A}(\vec{r}) = -n2\pi \frac{\Phi}{\Phi_0}, \tag{13.31}$$

where Φ_0 is the flux quantum.

In order to do adiabatic transport of an electron (say) around the flux tube, we require that the Hamiltonian has an excitation gap. We can achieve this by trapping the electron inside a box with hard walls[10] as shown in Fig. 13.1. One corner of the box is located at position $\vec{R}_0$. The position of the electron is $\vec{r}$ and the Hamiltonian is

$$H = \frac{1}{2m_e}\left[\vec{p} + \frac{e}{c}\vec{A}(\vec{r})\right]^2 + V(\vec{r} - \vec{R}_0). \tag{13.32}$$

---

[9] Using topology jargon, the puncture due to the flux tube has changed 3D space from being simply connected to being multiply connected, with closed trajectories or loops forming different topological classes, each of which is characterized by an integer-valued winding number. Formally, this statement can be expressed by noting that the "fundamental group" [35, 43, 44] of the punctured space is $\pi_1 = \mathbb{Z}$, where $\mathbb{Z}$ is the group made of all integers with the group operation defined to be addition. The elements of the fundamental group are equivalence classes of trajectories labeled by their winding number and the group operation is composition of two trajectories. The equation $\pi_1 = \mathbb{Z}$ simply states the fact that, if we "glue" two loops together to form a new loop, its winding number is the sum of the winding numbers of the two original loops. This explains why the group operation is addition; the fact that the group is $\mathbb{Z}$ is explained (loosely speaking) by the fact that any loop with winding number n (necessarily an integer) can be smoothly deformed into any other loop with the same winding number, but not into any loop with a different winding number. See Appendix D for further details.

[10] Any other strongly confining potential such as a harmonic well will also work so long as there is a finite energy gap from the ground state to the first excited state. We also require that the confinement be strong enough that the wave function does not leak out into the vicinity of the flux tube.

We implicitly assume that the orientation of the box remains fixed as we transport it by varying $\vec{R}_0$ so that we can neglect the dependence of the potential on the orientation.[11] We will further assume that the particle in the box remains completely outside the flux tube at all times so that we may take $\nabla \times \vec{A} = \vec{0}$ and thus can choose an electromagnetic vector potential of the form

$$\vec{A}(\vec{r}) = \frac{\Phi_0}{2\pi} \nabla \chi(\vec{r}), \tag{13.33}$$

with[12]

$$\oint d\vec{r} \cdot \vec{A} = \frac{\Phi_0}{2\pi} \oint d\vec{r} \cdot \nabla \chi(\vec{r}) = n\Phi, \tag{13.34}$$

where n is the winding number of the closed loop around the flux tube containing flux Φ. An obvious (possible) gauge choice is having $\chi(\vec{r}) = (\Phi/\Phi_0)\varphi(\vec{r})$, with $\varphi(\vec{r})$ being the azimuthal angle of $\vec{r}$; however, we will not specify a particular gauge at this stage.

If the quantum ground state wave function is $\xi_0(\vec{r} - \vec{R}_0)$ when $\vec{A}$ vanishes everywhere, then, in the presence of the flux tube, the solution of the Schrödinger equation can be written as

$$\psi(\vec{r}) = e^{-i\chi_{\vec{R}_0}(\vec{r})} \xi_0(\vec{r} - \vec{R}_0), \tag{13.35}$$

with

$$\chi_{\vec{R}_0}(\vec{r}) \equiv \frac{2\pi}{\Phi_0} \int_{\vec{R}_0}^{\vec{r}} d\vec{r}' \cdot \vec{A}(\vec{r}'), \tag{13.36}$$

where the line integral is evaluated along some path lying inside the box. Because $\vec{A}$ has no curl anywhere inside the box, the integral is independent of the path over which it is evaluated and hence $\chi(\vec{r})$ is well-defined everywhere inside the box. (We do not require χ globally, only inside the box.)

As we have emphasized several times, we can multiply $\psi(\vec{r})$ by an arbitrary (global) phase factor $e^{i\theta}$ and still have a valid solution. Furthermore, the phase $\theta(\vec{R}_0)$ could be a function of the location in parameter space (but *not* a function of the real-space coordinate $\vec{r}$). Different choices of $\theta(\vec{R}_0)$ correspond to different gauge choices for the Berry connection in parameter space. Different choices of $\chi(\vec{r})$ correspond to different electromagnetic gauge choices. Let us choose the gauge for the Berry connection by requiring that $\psi(\vec{r}) = e^{i\theta(\vec{R}_0)} e^{-i\chi_{\vec{R}_0}(\vec{r})} \xi_0(\vec{r} - \vec{R}_0)$ be real and positive at some point $(\vec{R}_0 + \vec{\Delta})$ in the interior of the box (where $\xi_0(\vec{\Delta}) \neq 0$). Without loss of generality, we assume that $\xi_0(\vec{r} - \vec{R}_0)$ is everywhere real and that $\xi_0(\vec{\Delta})$ is positive. Then we have very simply $\theta(\vec{R}_0) = +\chi_{\vec{R}_0}(\vec{R}_0 + \vec{\Delta})$.

Using

$$\nabla_{\vec{R}_0} \theta(\vec{R}_0) = \nabla_{\vec{R}_0} \chi_{\vec{R}_0}(\vec{R}_0 + \vec{\Delta}) = \frac{2\pi}{\Phi_0} \left\{ \vec{A}(\vec{R}_0 + \vec{\Delta}) - \vec{A}(\vec{R}_0) \right\}, \tag{13.37}$$

we find that the Berry connection is then

$$\vec{\mathcal{A}}(\vec{R}_0) = i \langle \psi | \nabla_{\vec{R}_0} | \psi \rangle \tag{13.38}$$

$$= -\frac{2\pi}{\Phi_0} \vec{A}(\vec{R}_0 + \vec{\Delta}) + i \int d^3\vec{r}\, \xi_0(\vec{r} - \vec{R}_0) \nabla_{\vec{R}_0} \xi_0(\vec{r} - \vec{R}_0) \tag{13.39}$$

---

[11] It should be clear by now that, in this example, the parameters $\vec{\mathcal{R}}$ describing the Hamiltonian, H, are simply $\vec{R}_0$, the three real-space coordinates of (the corner of) the box.

[12] Note that, while $\vec{A}$ has no curl outside the excluded region, its line integral around a closed loop with winding number $n \neq 0$ is finite. There is no continuous function $\chi(\vec{r})$ which is globally well-defined (single-valued), but we will not require this.

$$= -\frac{2\pi}{\Phi_0} \vec{A}(\vec{R}_0 + \vec{\Delta}) + \frac{i}{2} \int d^3\vec{r} \, \nabla_{\vec{r}} \xi_0^2(\vec{r} - \vec{R}_0) \tag{13.40}$$

$$= -\frac{2\pi}{\Phi_0} \vec{A}(\vec{R}_0 + \vec{\Delta}), \tag{13.41}$$

where we have used the fact that the integrand in the second term in Eq. (13.40) is a total derivative so that the integral vanishes (because ξ_0 vanishes on the boundary of the box).

If we now slowly vary the parameter $\vec{R}_0$ in a loop such that $(\vec{R}_0 + \Delta)$ encircles the flux tube with winding number $n = +1$, we have

$$\gamma = \oint d\vec{R}_0 \cdot \vec{\mathcal{A}}(\vec{R}_0) = -\oint d\vec{R}_0 \cdot \frac{2\pi}{\Phi_0} \vec{A}(\vec{R}_0 + \vec{\Delta}) = -2\pi \frac{\Phi}{\Phi_0}. \tag{13.42}$$

Thus, the Berry phase is nothing but the Aharonov–Bohm phase in this simple case, and the origin of the Berry flux is the physical magnetic flux confined in the physical flux tube.[13]

13.3 Spin-1/2 Berry Phase

As another simple example, let us consider a spin-1/2 system, whose most general Hamiltonian (up to a constant which is irrelevant here[14]) takes the form

$$H = \vec{h} \cdot \vec{\sigma} = \sum_{j=1}^{3} h_j \sigma_j, \tag{13.43}$$

where σ_j is the jth Pauli matrix. Of course, the Hamiltonian for all two-level systems can be described this way. In this case the parameter space of $\vec{h}$ is the familiar 3D Euclidean space[15] E_3. As will soon be clear, it is convenient to write the position vector in parameter space, $\vec{h}$, using polar coordinates:

$$\vec{h} = (h_1, h_2, h_3) = h(\sin\theta \cos\varphi, \sin\theta \sin\varphi, \cos\theta). \tag{13.44}$$

The two eigenstates with energies $\pm h$ are

$$|-\rangle = \begin{pmatrix} \sin(\theta/2)e^{-i\varphi} \\ -\cos(\theta/2) \end{pmatrix}, \qquad |+\rangle = \begin{pmatrix} \cos(\theta/2)e^{-i\varphi} \\ \sin(\theta/2) \end{pmatrix}. \tag{13.45}$$

> **Exercise 13.4.** Prove that the expressions in Eq. (13.45) do indeed represent eigenstates obeying $H|\pm\rangle = \pm h|\pm\rangle$. Further show that the spin polarization is parallel (anti-parallel) to the magnetic field $\langle\pm|\vec{\sigma}|\pm\rangle = \pm\vec{h}/h$.

It is interesting to note that, while the eigenvalues depend on the magnitude of h, the eigenstates depend only on the orientation of the "quantization axis," which is determined by two of the three parameters of $\vec{h}$, namely θ and φ, that form the space of S^2 (the surface of the unit sphere). It is easy to calculate the Berry connection and curvature on the S^2 space, say, for the ground state:

---

[13] The minus sign in the expression for the Berry phase simply reflects the negative charge of the electron.

[14] Even if the constant is one of the parameters that varies as one moves around in parameter space, it still contributes only a dynamical phase, not a geometric phase.

[15] More precisely, it is a punctured version of E_3 in which the origin $\vec{h} = \vec{0}$ has been removed. This is necessary because the energy-level spacing (the excitation gap) $2|\vec{h}|$ vanishes at the origin.

$$\mathcal{A}_\theta^- = \langle -|i\, \partial_\theta|-\rangle = 0; \tag{13.46}$$

$$\mathcal{A}_\varphi^- = \langle -|i\, \partial_\varphi|-\rangle = \sin^2(\theta/2); \tag{13.47}$$

$$\omega_{\theta\varphi}^- = \partial_\theta \mathcal{A}_\varphi^- - \partial_\varphi \mathcal{A}_\theta^- = \frac{1}{2}\sin\theta. \tag{13.48}$$

Thus the Berry flux through a small element of S^2 is

$$\omega_{\theta\varphi}^-\, d\theta\, d\varphi = \frac{1}{2}\sin\theta\, d\theta\, d\varphi = \frac{1}{2}d\Omega, \tag{13.49}$$

where $d\Omega = \sin\theta\, d\theta\, d\varphi$ is the solid angle subtended by the element when viewed from the origin. As we will discuss further below in Eq. (13.55), this is a remarkable result which tells us that the "magnetic" field (Berry curvature) is a constant everywhere on the surface of a sphere in parameter space. The integral of the Berry curvature over the entire surface is thus

$$\int \omega_{\theta\varphi}^-\, d\theta\, d\varphi = \frac{1}{2}\int d\Omega = 2\pi, \tag{13.50}$$

and the Berry phase of a loop on S^2 is

$$\gamma_- = \frac{1}{2}\int_S d\Omega = \frac{1}{2}\Omega_S, \tag{13.51}$$

or one-half of the solid angle subtended by the trajectory, as viewed from the center of the sphere (and with a sign determined by the handedness of the trajectory).

It is important to understand that, in writing Eq. (13.45), we have implicitly made a particular gauge choice for the Berry connection. More generally, we could have written

$$|-\rangle = \begin{pmatrix} \sin(\theta/2)e^{-i\varphi} \\ -\cos(\theta/2) \end{pmatrix} e^{i\Theta_-(\theta,\varphi)} \tag{13.52}$$

$$|+\rangle = \begin{pmatrix} \cos(\theta/2)e^{-i\varphi} \\ \sin(\theta/2) \end{pmatrix} e^{i\Theta_+(\theta,\varphi)}, \tag{13.53}$$

where $\Theta_\pm(\theta,\varphi)$ are two distinct and arbitrary (smooth) functions of the position in parameter space. It is straightforward to show, however, that the Berry curvature remains unaffected by this gauge change in the Berry connection.

Exercise 13.5. Prove that the Berry curvature for adiabatic transport of the states in Eqs. (13.52) and (13.53) is independent of the gauge choice determined by the arbitrary (smooth) function $\Theta_\pm(\theta,\varphi)$.

It is easy to generalize our results for spin-1/2 to the general case of a spin s coupled to a Zeeman field.[16] One finds that the Berry phase and curvature is simply proportional to the size s of the spin:

$$\gamma_- = s\int_S d\Omega = s\Omega_S, \tag{13.54}$$

Exercise 13.6. Derive Eq. (13.54) for arbitrary spin angular momentum s. Hint: use the fact the angular momentum components are the generators of rotations and that they obey the commutation algebra $[S^x, S^y] = i\hbar S^z$ and cyclic permutations thereof.

---

[16] As a side note, we point out that the Zeeman coupling $\vec{h}\cdot\vec{S}$ is the most general possible Hamiltonian for spin-1/2. This is not true for higher spins.

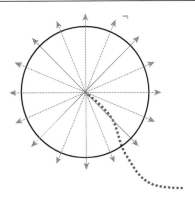

Figure 13.2 A monopole that radiates either Berry or magnetic flux (solid lines with arrows). Dirac proposed realizing a magnetic monopole in the form of a semi-infinite solenoid (dotted line) carrying a single quantum of magnetic flux hc/e. The total magnetic flux emanating from the surface of the sphere is hc/e, provided that we neglect the singular contribution from the infinitely narrow solenoid. This can be neglected because it produces an Aharonov–Bohm phase shift of $\pm 2\pi$ and has no observable physical effects. This **Dirac monopole** thus acts like a point magnetic charge with no string attached.

Going back to the Cartesian coordinate parameter space of $\vec{h}$, it is clear that the Berry curvature seen by the lower (adiabatic) energy eigenstate $|-\rangle$ is

$$\vec{b}_- = s\frac{\vec{h}}{h^3}, \tag{13.55}$$

which has the form of a magnetic field generated by a **magnetic monopole** with charge s located at the origin of parameter space; see Fig. 13.2. This magnetic field's magnitude is equal to s times the local **Gaussian curvature** of the surface of a sphere with radius h, and its direction is the local normal direction of this surface. Adiabatic evolution assumes the existence of an excitation gap, which in this model is $\epsilon_+ - \epsilon_- = 2|\vec{h}|$. We thus see that the gap vanishes at the origin of parameter space and we can think of the degeneracy at that point "radiating" Berry curvature $\vec{b}_-$ just as a magnetic monopole is a point source of magnetic field.

Exercise 13.7. Calculate the Berry curvature of the state $|+\rangle$, and show that it is the opposite of that of $|-\rangle$; thus the sum of the Berry curvatures of these two states is zero everywhere in parameter space.

The first observation of the spin Berry phase was made by the Pines group [92] using nuclear magnetic resonance techniques to drive the effective magnetic field through a small loop and then use a kind of interferometry to compare the phases acquired by the ground and excited states. The same experiment has also now been done using manipulation of a superconducting quantum bit acting as an artificial two-level atom [93], and the ability to implement "geometric phase gates" plays a prominent role in the nascent field of quantum information processing.

Our discussion would not be complete without pointing out that, for a closed loop on the sphere S^2, there is an ambiguity as to which part of the sphere is interior to the path and which part is exterior. As can be seen in Fig. 13.3, a small loop with right-hand[17] orientation enclosing region S can also be viewed as a left-hand-oriented loop enclosing a large region $\bar{S}$ complementary to S. Thus there are

---

[17] That is, if the fingers of your right hand curl in the direction of the motion around the loop, your thumb is pointing radially outward away from the surface.

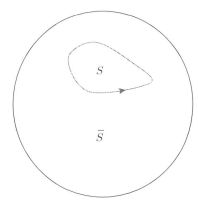

Figure 13.3 Illustration of an oriented path on the sphere S^2. The path can be viewed as having a right-handed orientation enclosing region S or a left-handed orientation enclosing the complementary region $\bar{S}$.

two different possible values for the Berry phase, $\gamma_- = s\Omega_S$ and $\bar{\gamma}_- = -s\Omega_{\bar{S}}$ (where we are using a convention in which the solid angle is always taken to be positive). In order for the Berry phase to be mathematically well-defined, we require that these two methods of calculation give the same answer:

$$\gamma_- = \bar{\gamma}_- \quad \mod (2\pi). \tag{13.56}$$

Using the geometric fact that the total solid angle of the sphere is 4π, we know that

$$\Omega_S + \Omega_{\bar{S}} = 4\pi, \tag{13.57}$$

and thus

$$\gamma_- - \bar{\gamma}_- = s(4\pi) = 0 \quad \mod (2\pi). \tag{13.58}$$

We thus see that our self-consistency condition in Eq. (13.56) is true, provided that the spin s is an integer multiple of $1/2$.

This remarkable topological result is a beautiful way to explain the quantization of spin in non-relativistic quantum mechanics, if we use Eq. (13.54) or Eq. (13.55) as the *definition* of the spin s. Because the Berry curvature in parameter space is like that of a magnetic monopole, the topological argument for the quantization of spin is closely related to the Dirac argument for the quantization of the magnetic charge of monopoles. One way to create an effective monopole, illustrated in Fig. 13.2, is to have a semi-infinitely long solenoid which abruptly terminates at the position where we want the monopole to be located. In order for the magnetic flux emanating from the end of the solenoid to act like the field from a point (magnetically charged) particle, the "Dirac string" (i.e. the solenoid) has to be undetectable. This will be the case provided that the flux inside the solenoid is an integer multiple of the flux quantum hc/e (or h/e in SI units) so that an electron circling the solenoid will pick up an Aharonov–Bohm phase of $\pm 2\pi$ which is equivalent to zero phase shift as if there were no string. (The product of the charge quantum e and the (SI) flux quantum h/e being Planck's constant guarantees that the Aharonov–Bohm phase is $h/\hbar = 2\pi$.) Dirac further argued, by reversing the logic, that quantization of electric charge is due to the presence of magnetic monopoles.[18]

---

[18] Although they are allowed in certain quantum field theories, there is no convincing evidence for the existence of magnetic monopoles at the time of writing.

13.3.1 Spin–Orbit Coupling and Suppression of Weak Localization

As an example of another physically observable effect of the Berry phase, let us consider how it affects electron backscattering and localization in the presence of a random disorder potential. As we discussed in Chapter 11, backscattering is enhanced by constructive quantum interference between different scattering processes that are related to each other by time-reversal transformation; see Fig. 11.1 and in particular Fig. 11.3. However, electron spin was not considered there. Neglecting spin would be appropriate if the spin state does *not* change during the scattering processes. This is true for pure potential scattering in the absence of spin–orbit coupling, but is no longer the case for pure potential scattering in the presence of spin–orbit coupling or for scattering due to spin-dependent forces. In the presence of spin–orbit coupling, the spin of the energy eigenstates depends on momentum. Hence each potential scattering that changes the momentum may be accompanied by a corresponding change in spin state. As a result, the electron spin precesses through *different* paths in spin space (or the unit sphere parameterized by θ, φ) for the different sequences of scattering processes illustrated in Figs. 11.3 and 11.1. For the case of scattering from a smoothly varying potential, the evolution of the spin can be treated as adiabatic motion on the Bloch sphere, and the phase *difference* between these two paths is precisely

$$\gamma = \gamma_{C_1} - \gamma_{C_2} = \gamma_C = \Omega_C/2, \tag{13.59}$$

where C_1 and C_2 represent these two paths in spin space (see Fig. 13.4 for an illustration), and $C = C_1 - C_2$ is a *closed loop* with C_1 and C_2 as its segments (with proper orientations). We thus find that spin–orbit coupling induces a non-zero phase difference among different backscattering processes, whose origin is nothing but the Berry phase associated with spin. Its effect is very similar to that of a magnetic field which induces an Aharonov–Bohm phase difference between time-reversal-related paths (as illustrated in Fig. 11.5), and thus *suppresses* weak-localization effects.

As an example of this, let us consider an extreme form of spin–orbit coupling, namely a Hamiltonian with Dirac (instead of Schrödinger) kinetic energy in 2D:

$$H = v_F \vec{\sigma} \cdot \vec{p} + V(\vec{r}), \tag{13.60}$$

where v_F is the Fermi velocity. This Hamiltonian describes electrons near the Dirac points in graphene (where, as described in Section 7.5, $\vec{\sigma}$ is a pseudospin corresponding to the A, B sublattice degree of freedom) or the surface states of a topological insulator (to be discussed in Chapter 14). For the case of the Hamiltonian in Eq. (13.60), the direction of spin (or pseudospin) in an energy eigenstate is always parallel or anti-parallel to the momentum and thus $\vec{k}$. As a result, for the two scattering

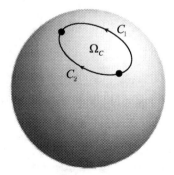

Figure 13.4 Illustration of the spin Berry phase difference between different paths (C_1 and C_2) that an electron spin traverses. This phase difference suppresses enhanced backscattering and weak localization, in the presence of spin–orbit coupling.

processes illustrated in Fig. 11.3, the loop C is precisely the equator and thus $\gamma_C = \pi$, resulting in a complete *cancelation* of backscattering! The resulting enhancement of diffusion is known as weak *anti-localization*. A magnetic field which introduces an additional phase difference between these two processes would suppress this anti-localization effect by ruining this perfect cancelation; this results in *positive* magnetoresistance, in contrast to what we saw in Chapter 11, where spin–orbit coupling was absent.

The Dirac equation calculation was presented here because of its simplicity – the spin direction in the plane-wave energy eigenstates is exactly parallel (or anti-parallel) to the momentum. This is a good description for the graphene (pseudospin) band structure. For more general Schrödinger bands with spin–orbit coupling, the quantitative computations are more complex, but the essential physics of the (real) spin Berry phase contribution to the backscattering amplitude is the same.

13.4 Berry Curvature of Bloch Bands and Anomalous Velocity

Owing to the (magnetic) translation symmetry, an eigenfunction of the Hamiltonian in Eq. (12.106) may be written as

$$\psi_{n\vec{k}}(\vec{r}) = e^{i\vec{k}\cdot\vec{r}} u_{n\vec{k}}(\vec{r}), \tag{13.61}$$

where $u_{n\vec{k}}(\vec{r})$ satisfies (generalized) periodic boundary conditions:

$$\tilde{T}_{\vec{a}_j} u_{n\vec{k}}(\vec{r}) = e^{i\phi_{\vec{a}_j}(\vec{r})} u_{n\vec{k}}(\vec{r} + \vec{a}_j) = u_{n\vec{k}}(\vec{r}). \tag{13.62}$$

We can solve for $u_{n\vec{k}}(\vec{r})$ and the corresponding band dispersion by combining Eq. (13.62) with the $\vec{k}$-dependent Schrödinger equation it satisfies:

$$h(\vec{k}) u_{n\vec{k}}(\vec{r}) = \epsilon_n(\vec{k}) u_{n\vec{k}}(\vec{r}), \tag{13.63}$$

where

$$h(\vec{k}) = \frac{1}{2m}\left[\vec{p} + \hbar\vec{k} + \frac{e}{c}\vec{A}(\vec{r})\right]^2 + V(\vec{r}). \tag{13.64}$$

In this formulation $\{\vec{k}\}$ becomes a set of *parameters* of the Hamiltonian $h(\vec{k})$. We can thus define the corresponding Berry curvature for each band (with band index n) in this $\vec{k}$-space. Because this space has the same number of components as the spatial dimension, the curvature tensor (in 3D[19]) can be expressed in a compact vector form (recall the discussion leading to Eq. (13.19)):

$$\vec{b}_n(\vec{k}) = i\langle\partial_{\vec{k}} u_{n\vec{k}}| \times |\partial_{\vec{k}} u_{n\vec{k}}\rangle. \tag{13.65}$$

Before getting into a description of the physical effects associated with (non-zero) $\vec{b}_n(\vec{k})$, we briefly comment on the geometry and topology of $\vec{k}$-space. We know that, as the lattice momentum/wave vector, $\vec{k}$ is defined inside the first Brillouin zone (1BZ). Naively, one would think the 1BZ is a segment of a line in 1D, a parallelogram or hexagon in 2D, and some kind of polygon in 3D. However, due to the periodic structure in reciprocal space, $\vec{k}$s that differ by a reciprocal lattice vector should be identified as the *same* point. As a consequence, opposite edges or faces of 1BZ should be identified with each other, and a 1BZ in d-dimensions corresponds to a d dimensional torus, $\mathcal{T}_d$. In 1D this is just a ring or circle. While their geometry depends on details of the lattice structure and the magnetic field, topologically all 1BZs with the same dimensionality are equivalent.

---

[19] For the case of 2D where $\vec{k}$ lies in the XY plane, $\vec{b}_n$ has only a $\hat{z}$ component and hence is effectively a scalar.

13.4.1 Anomalous Velocity

As we warned the reader in Section 8.2, our derivation of the semiclassical equations of motion for Bloch-band wave packets was not quite correct due to a subtle Berry phase effect in certain band structures. It is now time to address this point. We will show in this subsection that non-zero Berry curvature $\vec{b}_n(\vec{k})$ yields a correction to the (quantum-mechanical) average velocity of Bloch states, in the presence of a uniform or slowly varying (on the scale of a wave packet) external electric field.

Recall that we showed in Chapter 7 that

$$\langle \psi_{n\vec{k}} | \vec{v} | \psi_{n\vec{k}} \rangle = \langle u_{n\vec{k}} | \frac{1}{\hbar} \frac{\partial h(\vec{k})}{\partial \vec{k}} | u_{n\vec{k}} \rangle = \frac{1}{\hbar} \frac{\partial \epsilon_n(\vec{k})}{\partial \vec{k}}. \tag{13.66}$$

An external perturbing field *modifies* the state $|\psi_{n\vec{k}}\rangle$, which can lead to a correction to $\langle \vec{v} \rangle$. In the following, we consider the effect of a weak uniform electric field, and treat it using first-order perturbation theory in a manner very similar to Exercise 12.7. The perturbation is

$$H' = e\vec{E} \cdot \vec{r}, \tag{13.67}$$

and we have[20] (to first order in $\vec{E}$)

$$|\delta\psi_{n\vec{k}}\rangle = e\vec{E} \cdot \sideset{}{'}\sum_{n',\vec{k}'} \frac{|\psi_{n'\vec{k}'}\rangle \langle \psi_{n'\vec{k}'} | \vec{r} | \psi_{n\vec{k}} \rangle}{\epsilon_{n\vec{k}} - \epsilon_{n'\vec{k}'}} \tag{13.68}$$

$$= e\vec{E} \cdot \sideset{}{'}\sum_{n',\vec{k}'} \frac{|\psi_{n'\vec{k}'}\rangle \langle \psi_{n'\vec{k}'} | [\vec{r}, H] | \psi_{n\vec{k}} \rangle}{(\epsilon_{n\vec{k}} - \epsilon_{n'\vec{k}'})^2} \tag{13.69}$$

$$= i\hbar e\vec{E} \cdot \sideset{}{'}\sum_{n',\vec{k}'} \frac{|\psi_{n'\vec{k}'}\rangle \langle \psi_{n'\vec{k}'} | \vec{v} | \psi_{n\vec{k}} \rangle}{(\epsilon_{n\vec{k}} - \epsilon_{n'\vec{k}'})^2}. \tag{13.70}$$

In the calculation of the matrix element $\langle \psi_{n'\vec{k}'} | \vec{v} | \psi_{n\vec{k}} \rangle$, it is easy to see that the contributions from each unit cell (in real space) are identical except for a unit-cell-position ($\vec{R}$)-dependent phase factor $e^{i(\vec{k}-\vec{k}')\cdot\vec{R}}$. The matrix element thus vanishes unless[21] $\vec{k} = \vec{k}'$. This allows us to make the simplification

$$\langle \psi_{n'\vec{k}'} | \vec{v} | \psi_{n\vec{k}} \rangle = \langle u_{n'\vec{k}} | \frac{1}{\hbar} \frac{\partial h(\vec{k})}{\partial \vec{k}} | u_{n\vec{k}} \rangle \delta_{\vec{k}'\vec{k}} \tag{13.71}$$

and removes the sum over $\vec{k}'$ in Eq. (13.70), so that we are left with a single summation $n' \neq n$. From now on we focus on $|u_{n\vec{k}}\rangle$, and Eq. (13.70) reduces to

$$|\delta u_{n\vec{k}}\rangle = ie\vec{E} \cdot \sum_{n'\neq n} \frac{|u_{n'\vec{k}}\rangle \langle u_{n'\vec{k}} | (\partial h(\vec{k})/\partial \vec{k}) | u_{n\vec{k}} \rangle}{(\epsilon_{n\vec{k}} - \epsilon_{n'\vec{k}})^2}. \tag{13.72}$$

As a result, we find (to first order in $\vec{E}$)

$$\langle \vec{v} \rangle = \langle u_{n\vec{k}} | \frac{1}{\hbar} \frac{\partial h(\vec{k})}{\partial \vec{k}} | u_{n\vec{k}} \rangle + \langle \delta u_{n\vec{k}} | \frac{1}{\hbar} \frac{\partial h(\vec{k})}{\partial \vec{k}} | u_{n\vec{k}} \rangle + \langle u_{n\vec{k}} | \frac{1}{\hbar} \frac{\partial h(\vec{k})}{\partial \vec{k}} | \delta u_{n\vec{k}} \rangle$$

$$= \frac{1}{\hbar} \frac{\partial \epsilon_n(\vec{k})}{\partial \vec{k}} + \vec{v}_{\mathrm{a}}(n, \vec{k}), \tag{13.73}$$

---

[20] Experts will recognize that $\vec{r}$ is not, strictly speaking, a well-defined operator in a space of plane-wave-like states. We escape this mathematical subtlety (without proof) by transforming the matrix element to that of the velocity operator.

[21] Strictly speaking, we should write $\vec{k} = \vec{k}' + \vec{G}$ for any reciprocal lattice vector $\vec{G}$. However, we can safely restrict $\vec{k}$ and $\vec{k}'$ to the 1BZ because we are allowing for different band indices in our expressions.

where the "anomalous velocity," which is a correction to Eq. (13.66) due to electric-field-induced band mixing (i.e. interband coherence), is

$$\vec{v}_{\mathrm{a}}(n,\vec{k}) = \frac{ie}{\hbar} \sum_{n' \neq n} \frac{\langle u_{n\vec{k}} | (\partial h(\vec{k})/\partial \vec{k}) | u_{n'\vec{k}} \rangle [\vec{E} \cdot \langle u_{n'\vec{k}} | (\partial h(\vec{k})/\partial \vec{k}) | u_{n\vec{k}} \rangle] - \mathrm{c.c.}}{(\epsilon_{n\vec{k}} - \epsilon_{n'\vec{k}})^2} \qquad (13.74)$$

$$= \frac{ie}{\hbar} \vec{E} \times \left[\sum_{n' \neq n} \frac{\langle u_{n\vec{k}} | (\partial h(\vec{k})/\partial \vec{k}) | u_{n'\vec{k}} \rangle \times \langle u_{n'\vec{k}} | (\partial h(\vec{k})/\partial \vec{k}) | u_{n\vec{k}} \rangle}{(\epsilon_{n\vec{k}} - \epsilon_{n'\vec{k}})^2} \right] \qquad (13.75)$$

$$= \frac{e}{\hbar} \vec{E} \times \vec{b}_n(\vec{k}), \qquad (13.76)$$

where we used Eq. (13.22) in the last step.

Recognizing that a weak electric field $\vec{E}$ induces semiclassical time-evolution (or a "velocity" in momentum space) for the lattice momentum as in Eq. (8.14), we can write Eq. (13.76) in the following suggestive form:

$$\vec{v}_{\mathrm{a}}(n,\vec{k}) = -\left(\frac{d\vec{k}}{dt} \right) \times \vec{b}_n(\vec{k}). \qquad (13.77)$$

Notice that the RHS takes a form very similar to the Lorentz force, except that $d\vec{k}/dt$ is the velocity in *momentum* (instead of real) space, and the Berry curvature $\vec{b}_n(\vec{k})$ can be viewed as magnetic field (or flux density), also in momentum space.[22]

Combining the contribution from the anomalous velocity with the equations of motion derived in Chapter 8, we obtain the following corrected equations that are now highly symmetric:

$$\hbar \frac{d\vec{R}}{dt} = \nabla_{\vec{k}} \epsilon_n(\vec{k}) - \hbar \left(\frac{d\vec{k}}{dt} \right) \times \vec{b}_n(\vec{k}); \qquad (13.78)$$

$$\hbar \frac{d\vec{k}}{dt} = -\nabla_{\vec{R}} V(\vec{R}) - \frac{e}{c} \left(\frac{d\vec{R}}{dt} \right) \times \vec{B}(\vec{R}). \qquad (13.79)$$

As we see under the exchange of (wave-packet) position $\vec{R}$ and lattice momentum $\vec{k}$, the band energy $\epsilon_n(\vec{k})$ (which is the momentum-dependent energy) exchanges with the external potential $V(\vec{R})$ (which is the position-dependent energy), while the Berry curvature exchanges with the magnetic field. We would not have such a beautiful duality between real and momentum spaces without the anomalous velocity term.[23]

While Eq. (13.76) derived above is correct, strictly speaking, in the presence of the H' term (Eq. (13.67)), the Hamiltonian no longer respects (magnetic) lattice translation symmetry, thus its eigenstates can no longer be characterized by a lattice wave vector $\vec{k}$. Another problem with the derivation above is that H' in Eq. (13.67) is unbounded, and thus, strictly speaking, cannot be treated as a perturbation. To avoid these issues, in the following we use an additional time-dependent vector potential to model the electric field, as we did in Section 8.2.2:

$$\vec{E} = -\frac{1}{c} \frac{\partial \vec{A}'}{\partial t}; \qquad \vec{A}' = -c\vec{E}t. \qquad (13.80)$$

Since $\vec{A}'$ is independent of $\vec{r}$, lattice symmetry is respected. The price to pay is that we now need to deal with a time-dependent version of the Hamiltonian (13.64):

---

[22] Recall that a real magnetic field can be viewed as Berry curvature in real space, as we learned in Section 13.2.

[23] This is similar to the lack of electromagnetic duality without the displacement current term in Maxwell's equations.

$$h(\vec{k}, t) = h[\vec{k}(t)] = \frac{1}{2m}\left[\vec{p} + \hbar\vec{k}(t) + \frac{e}{c}\vec{A}(\vec{r})\right]^2 + V(\vec{r}), \tag{13.81}$$

where

$$\vec{k}(t) = \vec{k}_0 + \frac{e}{\hbar c}\vec{A}' = \vec{k}_0 - \frac{e\vec{E}t}{\hbar}, \tag{13.82}$$

from which we immediately obtain

$$\hbar\dot{\vec{k}} = -e\vec{E}, \tag{13.83}$$

which is a special case of Eq. (8.14). The weak-electric-field limit is thus the adiabatic limit, where $\vec{k}$ varies slowly with time. However, even in this limit there is a small admixture of states in different bands (see Eq. (8.37)) that is proportional to $\vec{E}$; this coherent admixture is responsible for the anomalous velocity, to which we now turn.

We now formulate the calculation of the anomalous velocity more generally in the context of adiabatic time-evolution and corresponding transport. As in Section 13.1, we consider a Hamiltonian $H(\vec{\mathcal{R}}) = H(\mathcal{R}_1, \mathcal{R}_2, \ldots)$ that depends on a set of parameters $(\mathcal{R}_1, \mathcal{R}_2, \ldots)$. We further introduce generalized velocity operators

$$V_\mu = \frac{\partial H(\vec{\mathcal{R}})}{\partial \mathcal{R}_\mu} = \partial_\mu H. \tag{13.84}$$

Now let $\vec{\mathcal{R}} = \vec{\mathcal{R}}(t)$ vary slowly in time, and consider the adiabatic time-evolution of a state $|\psi_n(t)\rangle$ that will remain very close, but *not* identical, to the nth (non-degenerate) instantaneous eigenstate of $H(\vec{\mathcal{R}})$, $|n[\vec{\mathcal{R}}(t)]\rangle$, whose eigenvalue is $\epsilon_n[\vec{\mathcal{R}}(t)]$. To single out the anomalous contribution to V_μ from the adiabatic evolution of $|\psi_n(t)\rangle$, we introduce

$$\tilde{H}(\vec{\mathcal{R}}) = H(\vec{\mathcal{R}}) - \epsilon_n(\vec{\mathcal{R}}), \tag{13.85}$$

thus

$$V_\mu = \frac{\partial\epsilon_n(\vec{\mathcal{R}})}{\partial\mathcal{R}_\mu} + \frac{\partial\tilde{H}(\vec{\mathcal{R}})}{\partial\mathcal{R}_\mu}, \tag{13.86}$$

and the last term above gives rise to the anomalous velocity:

$$V_\mu^a = \langle\psi_n(t)|\frac{\partial\tilde{H}(\vec{\mathcal{R}})}{\partial\mathcal{R}_\mu}|\psi_n(t)\rangle. \tag{13.87}$$

We now consider the time-evolution of $|\psi_n(t)\rangle$. After factoring out the dynamical phase factor,

$$|\psi_n(t)\rangle = |\tilde{\psi}_n[\vec{\mathcal{R}}(t)]\rangle \exp\left[-\frac{i}{\hbar}\int_0^t dt'\,\epsilon_n(t')\right], \tag{13.88}$$

the ket $|\tilde{\psi}_n[\vec{\mathcal{R}}(t)]\rangle$ satisfies a (reduced) Schrödinger equation

$$i\hbar\frac{\partial}{\partial t}|\tilde{\psi}_n[\vec{\mathcal{R}}(t)]\rangle = \tilde{H}[\vec{\mathcal{R}}(t)]|\tilde{\psi}_n[\vec{\mathcal{R}}(t)]\rangle. \tag{13.89}$$

Of course, using $|\tilde{\psi}_n[\vec{\mathcal{R}}(t)]\rangle$ (instead of $|\psi_n(t)\rangle$) in Eq. (13.87) yields the same result. Now, taking $\partial_\mu = \partial/\partial\mathcal{R}_\mu$ on both sides of Eq. (13.89), and leaving the t-dependence of $\vec{\mathcal{R}}$ implicit, we have

$$i\hbar\,\partial_\mu[|\partial_t\tilde{\psi}_n(\vec{\mathcal{R}})\rangle] = [\partial_\mu\tilde{H}(\vec{\mathcal{R}})]|\tilde{\psi}_n(\vec{\mathcal{R}})\rangle + \tilde{H}(\vec{\mathcal{R}})[|\partial_\mu\tilde{\psi}_n(\vec{\mathcal{R}})\rangle, \tag{13.90}$$

from which we obtain, using Eq. (13.87),

$$V_\mu^a = -\langle\tilde{\psi}_n(\vec{\mathcal{R}})|\tilde{H}|\partial_\mu\tilde{\psi}_n(\vec{\mathcal{R}})\rangle + i\hbar\langle\tilde{\psi}_n(\vec{\mathcal{R}})|\partial_\mu|\partial_t\tilde{\psi}_n(\vec{\mathcal{R}})\rangle \tag{13.91}$$

$$= i\hbar[\langle\partial_t\tilde{\psi}_n(\vec{\mathcal{R}})|\partial_\mu\tilde{\psi}_n(\vec{\mathcal{R}})\rangle - \langle\partial_\mu\tilde{\psi}_n(\vec{\mathcal{R}})|\partial_t\tilde{\psi}_n(\vec{\mathcal{R}})\rangle]. \tag{13.92}$$

In the last step we used the fact that

$$i\hbar\langle\tilde{\psi}_n|\partial_t\tilde{\psi}_n\rangle = \langle\tilde{\psi}_n|\tilde{H}|\tilde{\psi}_n\rangle = O[(\partial_t\mathcal{R})^2] \tag{13.93}$$

(see Eq. (8.37)) and is thus negligible in the adiabatic limit. Also, using the fact that in the adiabatic limit we have $|\tilde{\psi}_n\rangle \to |n(\vec{\mathcal{R}})\rangle$, and

$$\frac{\partial}{\partial t} = \sum_\nu \left(\frac{\partial\mathcal{R}_\nu}{\partial t}\right)\partial_\nu, \tag{13.94}$$

we obtain

$$V^a_\mu = i\hbar\sum_\nu [\langle\partial_\nu n(\vec{\mathcal{R}})|\partial_\mu n(\vec{\mathcal{R}})\rangle - \langle\partial_\mu n(\vec{\mathcal{R}})|\partial_\nu n(\vec{\mathcal{R}})\rangle](\partial_t\mathcal{R}_\nu) \tag{13.95}$$

$$= -\hbar\sum_\nu \left(\frac{\partial\mathcal{R}_\nu}{\partial t}\right)\omega_{\mu\nu}, \tag{13.96}$$

from which Eq. (13.76) follows as a special case where $\vec{\mathcal{R}} = \vec{k}$. We note that the derivation above was formulated entirely within a single band, and we did *not* need to invoke the existence of other bands explicitly.

Equation (13.76) helps establish some important symmetry properties of the Bloch-band Berry curvature $\vec{b}_n(\vec{k})$. Under time-reversal transformation, we have $\vec{v} \to -\vec{v}$, $\vec{k} \to -\vec{k}$, and $\vec{E} \to \vec{E}$. Hence, in the presence of **time-reversal symmetry**, we have

$$\vec{b}_n(\vec{k}) = -\vec{b}_n(-\vec{k}). \tag{13.97}$$

Under space inversion or parity transformation, we have $\vec{v} \to -\vec{v}$, $\vec{k} \to -\vec{k}$, and $\vec{E} \to -\vec{E}$. Thus, in the presence of inversion symmetry, we have

$$\vec{b}_n(\vec{k}) = \vec{b}_n(-\vec{k}). \tag{13.98}$$

Notice that, as we discussed earlier, the Berry curvature vector $\vec{b}$ has the same properties as an ordinary magnetic field $\vec{B}$. It is odd under time reversal and even under parity (i.e. it is a pseudovector). As a result, $\vec{b}_n(\vec{k}) = \vec{0}$ everywhere when both symmetries are present. On the other hand, in the absence of either symmetry we generically have non-zero $\vec{b}_n(\vec{k})$ and $\vec{v}_a(n,\vec{k})$, and the latter contributes to transport. Since the anomalous velocity is proportional to the vector (cross) product of the electric field and the Berry curvature, it is therefore *perpendicular* to both. As a result, it plays a particularly important role in the Hall effect and other related transport phenomena (e.g. the anomalous Hall effect in ferromagnets [94]).

13.5 Topological Quantization of Hall Conductance of Magnetic Bloch Bands

In the presence of a magnetic field, time-reversal symmetry is broken, thus $\vec{b}_n(\vec{k})$ and $\vec{v}_a(n,\vec{k})$ are (generically) non-zero; this leads to the Hall effect. For the case of a uniform magnetic field (and no periodic potential) discussed earlier in Chapter 12, $\vec{v}_a(n,\vec{k})$ is nothing but the $\vec{E} \times \vec{B}$ drift velocity we encountered back in Chapter 8. Equation (13.76) allows us to discuss the Hall effect and in particular, its quantization under more general settings. For a filled magnetic band in 2D, we have the Hall current density from electrons in that band:

$$\vec{J}_n = -\frac{e}{A}\sum_{\vec{k}\in 1BZ} \vec{v}_a(n,\vec{k}) = \sigma^n_{xy}\hat{z}\times\vec{E}, \tag{13.99}$$

where A is the area of the system. Thus

$$\sigma_{xy}^n = \frac{e^2}{\hbar A} \sum_{\vec{k} \in 1BZ} b_n(\vec{k}) = \frac{e^2}{h} \frac{1}{2\pi} \int_{\vec{k} \in 1BZ} b_n(\vec{k}) dk_x \, dk_y. \tag{13.100}$$

We note that in 2D $\vec{b}_n(\vec{k})$ always points in the $\hat{z}$ direction, and can thus be viewed as a scalar (with both signs). Equation (13.100) suggests that the Hall conductance of a filled band is proportional to the *total* Berry curvature of the band. We are going to show below that

$$\frac{1}{2\pi} \int_{\vec{k} \in 1BZ} b_n(\vec{k}) dk_x \, dk_y = C_n \tag{13.101}$$

is an integer known as the (first) **Chern number**, as long as the band is not degenerate with other bands for any $\vec{k}$. This leads to the quantization of the Hall conductance in units of e^2/h.

Before proving Eq. (13.101), let us discuss a related result in geometry, which is more familiar and in a simpler setting. This is the famous **Gauss–Bonnet theorem**, which states that the total curvature of a closed 2D surface M is quantized in units of 2π:

$$\frac{1}{2\pi} \int_M K \, dA = 2 - 2g_M, \tag{13.102}$$

where K is the local Gaussian curvature and g_M is the **genus** or "number of handles" of the surface M.[24] A few examples are illustrated in Fig. 13.5. A physicist's "proof" of Eq. (13.102) can be found in Appendix D.

Equation (13.102) is significant as it tells us the total curvature of a 2D surface is a quantized *topological invariant* of the surface. That is, smooth changes of its geometry, which change the curvature K *locally*, do *not* change g_M which is a *global* or *topological* property of the surface. Similarly, a smooth change of the band structure changes the local Berry curvature, but cannot change the (quantized) Chern number and the Hall conductance. Thus σ_{xy} is a global property of the magnetic band, and its quantization has a topological origin similar to the Gauss–Bonnet theorem.

As another warm-up, let us inspect the total Berry curvature of a closed (2D) surface M in the (3D) parameter space of the ground state of the spin-1/2 (or two-level) problem discussed in Section 13.3:

$$C_- = \frac{1}{2\pi} \int_M \vec{b} \cdot d\vec{S}. \tag{13.103}$$

From Gauss' theorem we find that C_- depends only on the Berry monopole charge enclosed in M: $C_- = 1$ of M encloses the origin (which is where the monopole is located, see Fig. 13.6(a)), and $C_- = 0$ otherwise (Fig. 13.6(b)). We thus find that C_-, which is analogous to the Chern number, takes two possible quantized values. Since C_- is *independent* of the shape or geometry of M, it is a topological invariant of the surface M. Note that the surfaces with $C_- = 0$ and $C_- = 1$ are topologically distinct in this case because, in order to connect the two smoothly, one needs to have

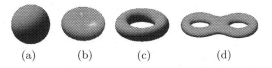

(a) (b) (c) (d)

Figure 13.5 A few examples of closed 2D surfaces. (a) A sphere. (b) A genus-zero closed surface, which is topologically equivalent to a sphere. (c) A torus, with genus one. (d) A genus-two torus.

---

[24] More precisely, the genus of a connected, orientable surface is the maximum number of cuts along non-intersecting closed simple curves that can be made without breaking the object into disconnected pieces.

surfaces that contain the origin (Fig. 13.6(c)). As discussed in Section 13.3, the origin $\vec{h} = 0$ is a singular point where the Hamiltonian and the gap between the two states vanish, which should be removed from the parameter space. As a result, surfaces of the type illustrated in Fig. 13.6(c) are not allowed, leading to two distinct topological classes of closed surfaces.[25]

Now let us prove Eq. (13.101), following Thouless *et al.* [88]. We write the left-hand side as

$$\frac{1}{2\pi} \int_{\vec{k} \in 1BZ} \omega_n(\vec{k}) dk_x \, dk_y = \frac{1}{2\pi} \oint \vec{\mathcal{A}}_n(\vec{k}) \cdot d\vec{k} \tag{13.104}$$

$$= \frac{i}{2\pi} \oint d\vec{k} \cdot \left[\int d^2\vec{r} \, u_{n\vec{k}}^*(\vec{r}) \nabla_{\vec{k}} \, u_{n\vec{k}}(\vec{r}) \right], \tag{13.105}$$

where the loop integral in $\vec{k}$ space is around the boundary of the first Brillouin zone (1BZ); see Fig. 13.7. Note that the opposite edges of the 1BZ differ from each other by a reciprocal lattice vector $\vec{G}$, and are thus *equivalent*. Naively one would think this equivalence means

$$\psi_{n\vec{k}}(\vec{r}) = \psi_{n,\vec{k}+\vec{G}}(\vec{r}) \tag{13.106}$$

and thus

$$u_{n\vec{k}}(\vec{r}) = e^{-i\vec{G}\cdot\vec{r}} u_{n,\vec{k}+\vec{G}}(\vec{r}). \tag{13.107}$$

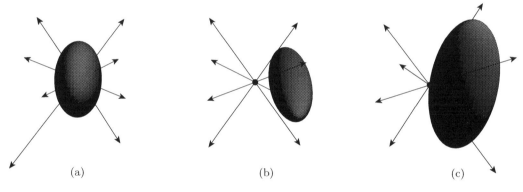

(a)   (b)   (c)

Figure 13.6 A closed surface in the Hamiltonian parameter space of a two-level system that: (a) encloses the origin; (b) does not enclose the origin; and (c) intersects the origin. The origin is a singular point where the level spacing vanishes. It is the source of the Berry flux (represented by arrows), very much like a magnetic monopole, which is the source of magnetic flux.

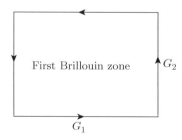

First Brillouin zone G_2

G_1

Figure 13.7 Illustration of integration of Berry connection around the boundary of the first Brillouin zone (1BZ). Notice that opposite edges of the 1BZ differ from each other by a reciprocal lattice vector and are thus equivalent.

[25] The reader is encouraged to ponder the similarity and difference between the situation here and that in the Aharonov–Bohm effect. Does the point singularity here lead to distinct topological classes of closed loops?

If that were the case we would find that the integral of Eq. (13.105) vanishes due to the cancelation of contributions from opposite edges of the 1BZ. However, in general, as with all quantum symmetries, we should only expect $\psi_{\vec{k}}(\vec{r})$ to return to itself under $\vec{k} \to \vec{k} + \vec{G}$ *up to a phase factor*; this implies

$$\psi_{n\vec{k}}(\vec{r}) = e^{-i\theta_n(\vec{k})}\tilde{\psi}_{n\vec{k}}(\vec{r}), \tag{13.108}$$

with $\tilde{\psi}_{n\vec{k}}(\vec{r}) = \tilde{\psi}_{n,\vec{k}+\vec{G}}(\vec{r})$ being a periodic function in reciprocal space. Obviously, if $\theta_n(\vec{k})$ is a constant, Eq. (13.105) yields zero. For non-trivial $\theta_n(\vec{k})$ we have an additional contribution due to the $\vec{k}$-dependence of θ_n:

$$\frac{i}{2\pi} \oint d\vec{k} \cdot \left[\int d^2\vec{r}\, u^*_{n\vec{k}}(\vec{r}) \nabla_{\vec{k}} u_{n\vec{k}}(\vec{r}) \right] = \frac{1}{2\pi} \oint d\vec{k} \cdot \nabla_{\vec{k}} \theta_n(\vec{k}) = C_n, \tag{13.109}$$

thus proving Eq. (13.101). In the above we used the fact that since $e^{-i\theta_n(\vec{k})}$ is single-valued, the *change* of $\theta_n(\vec{k})$ around a loop must be an integer multiple of 2π.[26] Bands having non-zero Chern numbers are topologically non-trivial, and are often referred to as **Chern bands** in the literature.

In the presence of time-reversal symmetry, Eq. (13.97) guarantees that $C_n = 0$ for all bands. However, as we will see in Chapter 14, this does *not* mean that all these bands are topologically trivial. Indeed, in the presence of spin–orbit coupling we can have topologically non-trivial bands because of the coupling of the spin degree of freedom to the momentum. As we will discuss in Chapter 14, these bands are characterized by a set of topological quantum numbers that can take only two possible values, 0 or 1. Such bands can exist both in 2D and in 3D crystals, and, as for the Chern-band case studied here, the gaps between bands with different topological quantum numbers generically close at the boundary of the sample; more on this point later.

13.5.1 Wannier Functions of Topologically Non-trivial Bands

In Section 7.4.2 we demonstrated that it is possible, using Bloch states of a Bloch band, to construct complete and orthonormal localized orbitals known as Wannier functions. This procedure can also be carried out for magnetic Bloch bands, including those carrying non-zero Chern numbers. Here we show that, while this is true, it is *not* possible to have exponentially localized Wannier functions for such topologically non-trivial bands. In the following we present a very simple argument due to Thouless [95]. Let us assume that such exponentially localized Wannier functions do exist. Then one can use them to reconstruct the Bloch states:

$$|\Psi_{n\vec{q}}\rangle = \frac{1}{\sqrt{N}} \sum_j e^{i\vec{q}\cdot\vec{R}_j} |\chi_{nj}\rangle, \tag{13.110}$$

where $|\chi_{nj}\rangle$ is the Wannier function at site j. Exponential localization guarantees the absolute convergence of the summation over j. This immediately implies $|\Psi_{n\vec{q}}\rangle = |\Psi_{n\vec{q}+\vec{G}}\rangle$, thus the Chern number must be zero, as discussed earlier. Thus, to have a non-zero Chern number, the summation over j above *cannot* be absolutely convergent, which implies that the Wannier function has a decay that is slower than $1/r^2$ at large distance r from its center.[27]

---

[26] The mathematical origin of this quantization is very similar to the quantization of vorticity of a vortex in a superfluid or superconductor (to be discussed in Chapter 19), except here the vorticity is in *momentum* space instead of in real space.

[27] Note that this discussion applies to 2D systems, which are the only ones that have band structures that can be labeled by the (first) Chern number.

Physically, this reflects the fact that such topologically non-trivial Bloch states are *intrinsically non-local*. We have already seen one aspect of such non-locality in *real space*, namely that the band spectra are very sensitive to the presence of edges or boundaries; the presence (or absence) of edges is a topological property of the system in which the electrons live. We also saw that these edge states are robust, *not* associated with any symmetries (or their breaking), and cannot be destroyed by disorder (unless it is so strong as to destroy the state in the bulk, for example, the quantum Hall effect itself). On the other hand, ultra-local Wannier functions cannot sense and thus be sensitive to the (change of) topology of the system. As we will see through more examples in later chapters, such sensitivity to system topology (e.g. dependence of the ground state degeneracy on the genus of the manifold in which the system lives) is a hallmark of topological phases.

13.5.2 Band Crossing and Change of Band Topology

Being an integer, the Chern number and thus the Hall conductance *cannot* change continuously with changes of band structure; the quantization is topologically protected. A change in Chern number can occur only when the band gap vanishes and two bands become degenerate at one point in the first Brillouin zone; see Fig. 13.8. This is also referred to as a band crossing. At such a degeneracy point there is ambiguity as to which state belongs to which band, and thus the local Berry curvature is undefined there.[28] This ambiguity or singularity is also reflected in Eq. (13.22), where the energy denominator vanishes. Typically such a degeneracy/crossing occurs on tuning one external parameter, and the Chern numbers of the two bands each change by ± 1 before and after the crossing, with their *sum* remaining unchanged. In the following we discuss such band crossings in some detail.

When two bands come close together in some region of the 1BZ we can ignore the presence of other bands that are far away in energy (their effects can be treated perturbatively if necessary), and reduce the band structure calculation to a two-level problem similar to that discussed in Section 13.3. We immediately see that, in order for the gap in a two-level system to vanish, all three components of the (effective) magnetic field $\vec{h}$ that couple to the pseudospin $\vec{\sigma}$ acting on the two-level Hilbert space (see Eq. (13.43)) must vanish. However, in 2D we have only two independent components of the lattice momentum $\vec{k}$, and thus generically we will *not* encounter a band crossing, unless we *fine-tune* one more parameter in the Hamiltonian, which we call m.[29] Thus we have $\vec{h} = \vec{h}(k_x, k_y, m)$. Without loss

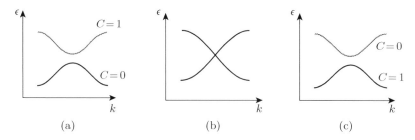

Figure 13.8 Change of Chern numbers through band degeneracy/crossing. (a) Before crossing. (b) Degeneracy or crossing between two bands occurs at one point in the first Brillouin zone. (c) After crossing, the Chern numbers of the two bands involved change (typically by ± 1), with their sum unchanged.

---

[28] The reader should note that the change of genus in 2D surfaces through a singularity as illustrated in Fig. D.1(b), where the local Gauss curvature is not well-defined at the point at which the two arms join, is similar to the band-crossing point in Fig. 13.8(b).

[29] As we will learn in Section 14.4, the situation is very different in 3D, where we can have such crossing *without* fine-tuning, leading to fascinating topological physics.

of generality, we may choose the crossing point to be at $k_x = k_y = m = 0$ by proper parameterization, with the minimal (non-zero) band gap at $k_x = k_y = 0$ for $m \neq 0$. By defining the direction of $\partial \vec{h}/\partial m$ to be the $\hat{z}$ direction in the pseudospin space and choosing its magnitude to be 1, and using a proper linear transformation in the (k_x, k_y) space (which alters its geometry but does not affect its topology), we can bring the two-level Hamiltonian (13.43) to the form

$$h(\vec{k}) = v(\pm k_x \sigma_x + k_y \sigma_y) + m \sigma_z, \tag{13.111}$$

or

$$h = v(\pm p_x \sigma_x + p_y \sigma_y) + m \sigma_z, \tag{13.112}$$

where v is a velocity scale.[30] We thus encounter the 2D Dirac Hamiltonian again, with a band crossing that occurs when the mass m is tuned to zero. As we learned in Section 13.3, the crossing point is a "magnetic" monopole carrying 2π Berry flux. Thus the Chern number of each band changes by ± 1 when the crossing occurs. Also, because the Berry flux is opposite for the two levels, the sum of their changes is zero. The change of band topology is accompanied by (or caused by!) a change of the sign of the Dirac mass, which is similar to what we learned in Section 7.6 for a 1D band.

The reader should note the similarity between the change of Chern number here and the change of total Berry flux through a 2D closed surface in the 3D parameter space of the spin-1/2 Hamiltonian illustrated in Fig. 13.6. Here the 1BZ for any fixed m (Fig. 13.8(b)) is a closed surface (a torus) in the 3D parameter space (k_x, k_y, m), and $m = 0$ corresponds to the special case that this surface contains the singular point $\vec{h} = 0$ of Fig. 13.6(c), where the total Berry flux through the surface (the Chern number in the present case) changes discontinuously.

13.5.3 Relation Between the Chern Number and Chiral Edge States: Bulk–Edge Correspondence

The facts discussed above about topological protection can be used to make very powerful and general arguments about the guaranteed existence of gapless modes at the edges of (free-electron) materials whose band structure contains gapped bands with non-trivial topological quantum numbers. Suppose that the Hamiltonian contains a parameter λ. Further suppose that when $\lambda < 0$ the band structure has the form shown in Fig. 13.8(a), and that, when $\lambda > 0$, the band structure has the form shown in Fig. 13.8(c). Now imagine an interface between two materials, one with $\lambda < 0$ and one with $\lambda > 0$. As a first approximation, imagine that the interface is not sharp but gradual and can be modeled by having λ vary smoothly in space from negative to positive. If the spatial variation is sufficiently slow, we can define a local band structure which varies in space. In this "adiabatic" interface limit it is clear that the gap has to close at the interface. This result remains generically robust even in the limit of a sharp interface. Furthermore, if, instead of an interface between two materials, we simply have a material with an edge (e.g. the second material is the vacuum), the same argument applies. The bulk bands are topologically non-trivial but the vacuum "bands" are topologically trivial. Hence the gap generically has to close at the edge.

---

[30] Without the proper linear transformation in the (k_x, k_y) space one would generically encounter a rank-two velocity tensor $v^{\mu\nu}$ describing how (k_x, k_y) are coupled to (σ_x, σ_y): $h = \sum_{\mu,\nu=x,y} v^{\mu\nu} k_\mu \sigma_\nu + m \sigma_z$, whose role is very similar to the rank-two effective-mass tensor in the effective-mass approximation. The transformation thus consists of first diagonalizing this velocity tensor (exactly the same as is often done to the effective mass tensor), and then performing a scale transformation to make the two velocity eigenvalues identical in magnitude, thus reducing the velocity tensor to a single number v. The two cases with the $\pm$ sign in front of σ_x correspond to the cases of the two velocity eigenvalues having the same or opposite signs, giving rise to two different chiralities of the Dirac fermions. The meaning of chirality will be discussed in connection with Eq. (13.120).

To prove the robustness of the gapless edge/interface modes, and understand their dispersion relations, we may model the electronic states near the edge/interafce using the Dirac Hamiltonian (13.112), with a *spatially dependent* mass that changes sign at the edge/interface. As we learned in Section 7.6 in 1D, this leads to a zero-energy state bound to the 0D edge/interface. Here the edge/interface is 1D, and the bound state develops a chiral dispersion along the edge/interface; see Exercise 13.8.

After these quite general arguments, let us return to the specific case of Landau levels as examples of topologically non-trivial bands with $C \neq 0$. When the potential $V(\vec{r}) = 0$, but in the presence of a uniform external magnetic field, we have seen that each Landau level is a (special type of) magnetic Bloch band, and each has Chern number $C = 1$. Correspondingly, they each carry one branch of gapless edge modes at the edge. When $V(\vec{r}) \neq 0$ we can (depending on the potential) have magnetic Bloch bands with non-zero Chern number $C \neq 1$. We will show below that, for each band that carries non-zero Chern number $C_n \neq 0$, there must be $|C_n|$ branches of chiral edge modes. It will turn out that the chirality (or direction of propagation) is determined by the sign of C_n.

To proceed, we consider a rectangular system with size $L_x \times L_y$, impose (magnetic) periodic boundary conditions along the $\hat{y}$ direction, and leave the boundary conditions along the $\hat{x}$ direction open. Then the geometry of the system is that of a cylinder (wrapped around the $\hat{y}$ direction), and contains two edges. See Fig. 13.9 for an illustration. To induce a Hall current along the $\hat{x}$ direction, we insert a *time-dependent* magnetic flux through the hole of the cylinder, which results in an electric field along the $\hat{y}$ direction through Faraday's law:

$$E_y = \frac{1}{cL_y} \frac{d\Phi(t)}{dt}. \tag{13.113}$$

The resultant *bulk* Hall current moves charge from one edge to the other. We consider a specific case in which the flux Φ increases slowly (and thus adiabatically) from 0 to one flux quantum Φ_0 over a long period T, and calculate the amount of charge transferred from one edge to another:

$$Q = \int_0^T dt \int_0^{L_y} dy\, J_x = \sigma_{xy}^n \int_0^T dt \int_0^{L_y} dy\, E_y = \sigma_{xy}^n \Phi_0/c = C_n e. \tag{13.114}$$

In the above we consider only the contribution from band n that is fully occupied in the bulk, and find that precisely C_n electrons have been transferred. We note that, after Φ has changed from 0 to one flux quantum Φ_0, the Hamiltonian of the 2DEG has returned to the original one (up to a gauge transformation[31]):

$$H(\Phi = 0) \cong H(\Phi = \Phi_0), \tag{13.115}$$

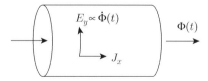

Figure 13.9 Illustration of the Laughlin and Halperin construction for flux insertion through a cylinder. Not shown is the large magnetic field pointing radially outward normal to the surface of the sample that induces the quantum Hall effect. See Refs. [96, 97].

---

[31] Again, the physics is that the flux tube produces a phase shift of 2π for a particle encircling it, and this is physically unobservable.

where $\cong$ stands for equivalence. The spectra of $H(\Phi = 0)$ and $H(\Phi = \Phi_0)$ are identical (cf. Exercise 10.1), and so are the corresponding eigenstates up to a gauge transformation. Thus the charge transfer we found in Eq. (13.114) must be the result of changes of occupation numbers of states localized at the edge. Furthermore, since the whole process is adiabatic, the states at $t = 0$ and $t = T$ must have an infinitesimal energy difference in the thermodynamic limit. This means that the states whose occupation numbers have changed must be at the Fermi energy, even though the Fermi energy lies in a band gap in the bulk. This immediately implies the existence of gapless edge states, whose spectra must look like that in Fig. 12.4 for $C_n = 1$. For general $C_n \neq 0$ there are $|C_n|$ branches of chiral edge modes, and for $C_n < 0$ the velocities of these edge modes are directed oppositely to the velocities of those with $C_n > 0$. For the whole system we need to add up the Chern numbers of all *occupied* bands (those below the Fermi energy in the bulk) to obtain the total Chern number:

$$C = \sum_{n \in \text{occupied}} C_n, \tag{13.116}$$

and precisely C electrons are transferred from one edge to the other during this adiabatic process. We thus have $|C|$ gapless chiral modes (that cross the Fermi energy) at the edge.[32]

The above argument shows that adiabatic addition of a flux quantum through the cylinder illustrated in Fig. 13.9 results in transfer of an integer number of electrons from one edge to the other. This implies that the Hall conductivity is quantized in integer multiples of e^2/h as discussed in Chapter 12 and confirms that this precise quantization is topologically protected. One obvious remark is that the reader should not forget that zero is an integer! The system might simply be insulating and have zero Hall conductivity. To distinguish this from the integer quantum Hall states with $C \neq 0$ but nevertheless having Fermi energy in a bulk band gap, the latter are also referred to as **Chern insulators** as they would be classified as band insulators in the transitional classification scheme of Chapter 7 (that does not take into account the topology of bands).

The other less obvious remark is that our entire discussion has been based on the idea that the occupation numbers of individual single-particle states are well-defined (at least in some Fermi-liquid sense; see Section 15.11). When we study the fractional quantum Hall effect in Chapter 16 we will see that strong electron–electron interactions can produce a strikingly different result, namely that the topological quantum numbers can be rational fractions! In fact, if we simply repeat the Laughlin–Halperin construction of threading a flux quantum through the cylinder, we find that it is not an electron which is transferred from one edge to the other, but rather a fractionally charged quasiparticle. We will further see that this is deeply connected to the fact that the ground state degeneracy of the interacting many-body system with (magnetic) periodic boundary conditions depends on the genus of the manifold on which the electrons move – a result which is clearly of topological origin.

Exercise 13.8. Consider a 2D, two-component Dirac Hamiltonian with an x-dependent mass:

$$H = v_{\text{F}}(\sigma_x p_x + \sigma_y p_y) + m(x)\sigma_z, \tag{13.117}$$

with $m(x \to \pm\infty) = \pm m_0$, with $m_0 > 0$. Using the translation-invariance along the y direction, energy eigenstates can be chosen to be plane waves along the y direction with wave vector k_y. (i) Show that for $k_y = 0$ there is a single zero-energy bound state localized near $x = 0$. Hint: the eigenvalue equation for this zero mode is closely related to those of the SSH model

---

[32] In principle we may also have C_+ forward-propagating modes and C_- backward-propagating modes, with $C_+ - C_- = C$. However, backscattering due to disorder would normally localize pairs of counter-propagating modes, leaving us with only the $|C|$ remaining (unpaired) chiral modes that are topologically protected.

discussed in Section 7.6. In particular, see Exercise 7.23. To reveal the similarity, it is useful to choose a representation of the Pauli matrices in which σ_y, instead of σ_z, is diagonal. This representation will be particularly useful for the next question. (ii) Now consider generic values of k_y, and show that this zero-energy bound state becomes a dispersive branch of states, which are precisely the edge/interface states between two integer quantum Hall states (or Chern insulators) whose Chern numbers differ by one. Find the edge-state dispersion relation. (iii) How does the dispersion/chirality change for $m_0 < 0$?

13.6 An Example of Bands Carrying Non-zero Chern Numbers: Haldane Model

As an example of bands carrying non-zero Chern numbers, we discuss a model introduced by Haldane in 1988 [98]. This is basically a tight-binding model on the honeycomb lattice. As we learned in Section 7.5, when only nearest-neighbor hopping t is included, the two bands are degenerate at two Dirac points K and K', and there is no band gap. In this case the system has both time-reversal Θ and inversion (or parity, $\mathcal{P}$) symmetries, so the Berry curvature is zero everywhere (except at K and K' where it is not defined). Time-reversal Θ symmetry also guarantees the Hall conductance is zero. To have well-defined Chern numbers we need to open up a band gap. To have non-zero Chern numbers and Hall conductance we need to break Θ symmetry.

The easiest way to open a band gap is to introduce an energy difference $2m$ between the A and B sublattices, which breaks $\mathcal{P}$ but not Θ (see Exercise 7.19). In this case the Hamiltonian in momentum space becomes

$$h(\vec{k}) = \begin{pmatrix} m & -tf(\vec{k}) \\ -tf^*(\vec{k}) & -m \end{pmatrix},$$ (13.118)

with band dispersions

$$\epsilon_\pm(\vec{k}) = \pm\sqrt{m^2 + t^2|f(\vec{k})|^2},$$ (13.119)

and a band gap of $2|m|$ is opened up at K and K'. In this case the Chern numbers are well-defined but zero for both bands, as guaranteed by Θ symmetry. It is instructive to understand how this comes about in detail, as the breaking of $\mathcal{P}$ allows non-zero local Berry curvature in the band, even though the overall Chern number vanishes.

For simplicity let us assume $|m| \ll |t|$ (the quantization of the Chern numbers is of course independent of the magnitude of $|m|$). In this case, the effect of m is important only near the Dirac points, where $|f(\vec{k})|$ is small. There $h(\vec{k})$ reduces to *massive* Dirac Hamiltonians (note the similarity to Eq. (13.112), except that here the chirality is opposite for K and K'):

$$H_{K,K'} = v_F(\pm\sigma_x p_x + \sigma_y p_y) + m\sigma_z,$$ (13.120)

where the plus sign applies to the K Dirac point and the minus sign applies to the K' Dirac point. Solving these Hamiltonians gives rise to the pseudospin configurations of the conduction band near K and K' as illustrated in Fig. 13.10 (assuming $m > 0$). We see that near K and K' the pseudospin wraps around the upper hemisphere, and each configuration gives rise to integrated Berry curvature of π in magnitude (recall the two-level system discussed in Section 13.3), but in opposite directions; in other words the Berry curvatures are of opposite sign near K and K'. Thus, as guaranteed by Θ symmetry, the total Chern number vanishes:

$$C_+ = \frac{1}{2} - \frac{1}{2} = 0. \tag{13.121}$$

Similarly $C_- = 0$.

In order to have non-zero Chern numbers, we need to have a situation where the Berry curvatures are of the same sign near K and K'. This would be accomplished if the mass were of opposite signs at K and K':

$$H'_{K,K'} = v_{\mathrm{F}}(\pm\sigma_x p_x + \sigma_y p_y) \pm m'\sigma_z, \tag{13.122}$$

or equivalently, in a notation that unifies K and K',

$$H' = v_{\mathrm{F}}(\tau_z\sigma_x p_x + \sigma_y p_y) + m'\tau_z\sigma_z, \tag{13.123}$$

where we have introduced another set of Pauli matrices $\vec{\tau}$ that acts on the two-level valley or Dirac point space, with $\tau_z = \pm 1$ for K and K', respectively.[33] Now let us discuss how to get such a mass term from the microscopic lattice Hamiltonian. First of all, notice that it is diagonal in the sublattice index. The simplest term of this kind would be an on-site energy, but that just gives the regular mass term in Eq. (13.120). The other possibility is a hopping term between the sites in the *same* sublattice, the simplest of which is next-nearest-neighbor hopping; see Fig. 13.11. Furthermore, to have non-zero Chern number and Hall conductance, such a term must break Θ symmetry; this is achieved by having

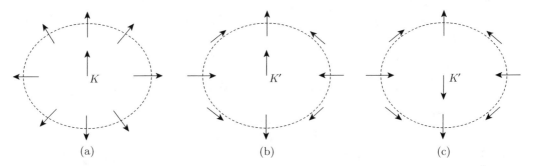

(a)   (b)   (c)

Figure 13.10 Pseudospin configurations of the conduction band near the Dirac points. (a) At the K point, in the presence of either regular mass or Haldane mass. The pseudospin is up at the K point and winds in the xy plane further away. (b) At the K' point, in the presence of regular mass. The pseudospin is up at the K' point but winds in the opposite direction around the equator further away. (c) At the K' point, in the presence of the Haldane mass, which causes the pseudospin to point down at the K' point.

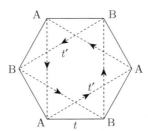

Figure 13.11 Next-nearest-neighbor hopping on the honeycomb lattice. The arrows indicate the direction of hopping, with hopping-matrix element t', which is *different* from the matrix element t'^* for hopping in the opposite direction.

---

[33] The $\vec{\tau}$ matrices are independent of $\vec{\sigma}$, which acts on the pseudospin or sublattice space.

an imaginary next-nearest-neighbor hopping $t' = i|t'|$. With complex hopping-matrix elements, the corresponding direction of hopping must be specified, as hopping over the same bond in the opposite direction has a matrix element which is the complex conjugate of the original one.

In the presence of t', we have

$$h(\vec{k}) = \begin{pmatrix} |t'|f'(\vec{k}) & -tf(\vec{k}) \\ -tf^*(\vec{k}) & -|t'|f'(\vec{k}) \end{pmatrix}, \tag{13.124}$$

with

$$f'(\vec{k}) = \sin(\vec{k} \cdot \vec{a}_1) - \sin(\vec{k} \cdot \vec{a}_2) - \sin[\vec{k} \cdot (\vec{a}_1 - \vec{a}_2)]. \tag{13.125}$$

For $|t'| \ll |t|$, the Hamiltonian reduces to Eq. (13.123), with

$$m' = -\frac{3\sqrt{3}}{2}|t'|. \tag{13.126}$$

The two bands are topologically non-trivial, with

$$C_\pm = \pm 1. \tag{13.127}$$

Exercise 13.9. Use Eqs. (13.120) and (13.123) to calculate the local Berry curvature near K and K', in the presence of regular and Haldane masses, respectively, and show that the Chern numbers of the two bands are zero for the former, and ± 1 for the latter.

13.7 Thouless Charge Pump and Electric Polarization

We started this chapter by considering adiabatic time evolution, and found the presence of Berry phases in such processes. Thus far our discussion of the importance of the Berry phase and Berry curvature has been limited to charge transport by static Hamiltonians, for example in the (quantization of the) Hall conductivity. We also found that, due to the presence of the anomalous velocity, which is closely related to the Berry curvature of a Bloch band, a completely filled band *can* transport Hall current. The Hall effect exists in 2D and in 3D. A natural question is, can a completely filled 1D band give rise to charge transport? In this section we show that the answer is yes, via what is known as the **Thouless charge pump** [99, 100], which is a *time-dependent* but adiabatic process.

Consider a Hamiltonian that depends on a single parameter λ in a periodic manner:

$$H(\lambda) = H(\lambda + 1). \tag{13.128}$$

Let λ be time-dependent:

$$\lambda(t) = t/\tau, \tag{13.129}$$

as a result of which H is time-dependent and periodic with period τ. Since the band of interest is completely filled, and we assume that it has a finite gap to other bands, we expect the time-evolution of the system to be adiabatic in the limit $\tau \to \infty$.

Now consider H to be a 1D periodic lattice Hamiltonian (generalization to higher dimensions is straightforward) with the Bloch wave vector k being a good quantum number. We then have a 2D parameter space for the k-dependent Hamiltonian $h(k, \lambda)$. Owing to the periodicity (13.128), the topology of this parameter space is a torus and identical to that of a 2D Brillouin zone. Owing to the time-dependence (13.129), a Bloch state picks up an anomalous velocity (see Eq. (13.96))

$$v_a^n(k) = -\omega^n(k, \lambda)/\tau, \tag{13.130}$$

where n is the band index, and ω is the (only component of) Berry curvature in the (k, λ) space. Now consider the amount of charge that moves through the system in a period τ, due to electrons completely filling band n:

$$Q = \int_0^\tau dt \langle J(t) \rangle = \frac{-e}{2\pi} \int_0^\tau dt \int_{1BZ} v_a^n(k) dk = e \int_0^1 d\lambda \int_{1BZ} \omega^n(k, \lambda) dk = eC_n, \quad (13.131)$$

where C_n is an integer-valued Chern number defined by the equation above. It should be clear that the origin of the quantization is identical to that of the Chern number of 2D bands, since here λ plays a role identical to that of the second momentum component. We thus find that a completely filled 1D band can also transport charge, provided that we have a suitable *time-dependent* Hamiltonian (or potential). For periodic time-dependence, such transport is quantized in an integer number of electrons per cycle. Such a system is known as a Thouless charge pump.

It is worth noting not only that the quantization of a Thouless charge pump has the same mathematical/topological origin as the quantization of the Hall conductivity, but also that they are physically equivalent to certain degree. This is because the Hall current along the x direction can be viewed as the result of pumping by a (linearly) time-dependent A_y, as seen in Eq. (13.80). Even though A_y is not periodic in time, the resultant Hamiltonian is periodic, modulo a so-called large gauge transformation; see Exercise 13.10.

Actual experiments realizing charge pumps can be quite different than the simple pedagogical example given above, and indeed do not necessarily follow the Thouless model of adiabatic pumping. **Electron "turnstiles"** consist of a set of voltage-biased metallic islands connected by tunnel junctions [101]. If the voltage biases are cyclically modulated, charge can be driven through the turnstile in a controlled manner using an external bias. These devices rely on strong Coulomb interactions to fix the number of electrons allowed through the system at one per cycle. However, there is irreversible relaxation of the electrons as they move downhill in energy through the device (generating heat along the way).

Reversible **electron pumps** were subsequently developed [102]. These pump charge through the device in a direction that depends only on the direction of motion around the closed loop in parameter space and not on the externally applied bias (i.e. they can pump electrons "uphill" as well as "downhill"), in contrast to the turnstile. Electron turnstiles and pumps are interesting for metrological purposes because they can convert the cycle frequency f to current $I = ef$ with accuracies of $\sim 10^{-8}$ [103]. However, the internal states of the mesoscopic normal-metal islands are not gapped, and so the pumping is only quasi-adiabatic – becoming closer and closer to adiabatic only in the limit of zero pumping frequency. As we will learn in Chapter 20, superconductivity can gap out all the low-energy excited states of a metallic island, and a superconducting charge pump should therefore easily be able to enter the adiabatic pumping regime. Unfortunately, such devices are plagued by "quasiparticle poisoning" caused by broken Cooper pairs.[34]

The quantum dots [104] discussed in Section 9.5.3 do not involve a 1D spatially periodic system. They do, however, involve more than one control parameter (gate voltage), a necessary (but not sufficient) condition if the system is to be driven through a closed loop in parameter space that contains (or may contain) a net Berry flux which will lead to charge pumping. Optical lattices can be controlled by varying the relative phase and amplitude of different laser beams. Because there is only a single state per lattice site (in a given band), these systems can come much closer to the truly adiabatic case envisioned by Thouless for ground-state pumping [105, 106].

---

[34] Cooper pairs and quasiparticles will be introduced in Chapter 20.

13.7.1 Modern Theory of Electric Polarization

Electric polarization plays an important role in the electrodynamics of solids and the electrostatics of ferroelectric materials. The modern theory of polarization resolves various computational difficulties and ambiguities by extending the picture of Thouless pumping to consider adiabatic time evolution over a finite time interval, resulting in a final Hamiltonian and ground state at t with $H(\lambda) = H(t/\tau) \neq H(0)$. As we demonstrate below, we can use the Berry curvature to calculate the *change* of *electric polarization* of the system. We again illustrate this in 1D, where things are particularly simple because polarization can be represented as a scalar (instead of a vector).

We know that the derivative (or gradient) of the local polarization $P(x)$ is related to the local charge density [18, 19]:

$$\partial_x P(x, t) = -\rho(x, t). \tag{13.132}$$

From the continuity equation

$$\partial_t \rho(x, t) = -\partial_x J(x, t) \tag{13.133}$$

we obtain

$$\partial_x [\partial_t P(x, t) - J(x, t)] = 0. \tag{13.134}$$

In a spatially periodic system we have[35]

$$\partial_t P(x, t) = J(x, t), \tag{13.135}$$

and integrating over t leads to

$$\Delta\langle P \rangle = \int_0^t dt' \langle J(t') \rangle = \sum_n \frac{-e}{2\pi} \int_0^t dt' \int_{1BZ} v_a^n(k) dk \tag{13.136}$$

$$= e \sum_n \int_0^\lambda d\lambda' \int_{1BZ} \omega^n(k, \lambda') dk. \tag{13.137}$$

In the above, the average is over x, or a unit cell if the system is strictly periodic, and the summation is over fully occupied bands. We thus find that the change of polarization P is determined by the Berry curvature $\omega^n(k, \lambda)$. It turns out that the absolute value of P is not well-defined due to (among other reasons) boundary charge contributions; only its difference between two states that are adiabatically connected to each other is well-defined and measurable. For further discussions of this subtle point see Ref. [91]. This modern formulation of the electric polarization is important in the study of the electrodynamics of solids and in such physical phenomena as ferroelectricity. It is also very helpful in removing ambiguities in the numerical computation of polarizations within band-structure calculations that study small system sizes using periodic boundary conditions.

Exercise 13.10. Consider a 2D system with finite size L_y along the y direction, and impose periodic boundary conditions on the wave function such that

$$\psi(x, y + L_y) = \psi(x, y). \tag{13.138}$$

This describes a physical situation illustrated in Fig. 13.9. Naively, one could "gauge away" a constant vector potential along the y direction via $A'_y = A_y + d\phi(y)/dy$. (i) Find a $\phi(y)$

---

[35] This neglects possible effects associated with charge build-up at the boundaries. In 3D the corresponding equation is $\nabla \cdot (\partial_t \vec{P} - \vec{J}) = 0$, which allows for an additional divergence-free contribution in the equation below. We do not discuss these subtleties here.

that makes $A'_y = 0$. (ii) Show that for generic values of A_y, after this gauge transformation the transformed wave function no longer satisfies the boundary condition (13.138). (iii) Find the special values of A_y such that, after the gauge transformation, the transformed wave function still satisfies the boundary condition (13.138), and find the corresponding values of the flux through the hole of the cylinder in Fig. 13.9. Such gauge transformations are often referred to as large gauge transformations. For these values of A_y the Hamiltonian is gauge-equivalent to that with $A_y = 0$. (iv) Discuss the relation between these results and those of Exercise 10.1.

14 Topological Insulators and Semimetals

In Chapter 13 we learned that the origin of the quantization of the Hall conductivity in an integer quantum Hall state is topology. More specifically, in units of e^2/h, the Hall conductivity of completely filled bands that are separated from empty bands by a gap is a topological quantum number. This Chern number can take any integer value, thus forming the set $\mathbb{Z}$. This is our first encounter with topological quantum numbers in this book. This understanding naturally leads to the following questions. (i) The Hall conductivity is non-zero only when the time-reversal symmetry is broken. Are there topological quantum numbers in the presence of time-reversal symmetry? (ii) Do topological quantum numbers always form the set $\mathbb{Z}$, or are there other possibilities? (iii) The integer quantum Hall effect is specific to 2D. Can we have topologically non-trivial states in other dimensions, especially 3D? (iv) The Chern number exists in insulators only, as it is well-defined only when a gap exists between occupied and unoccupied states. Do topological quantum numbers exist in metals where there is no such gap?

As we will learn in this chapter, the answer to all these questions is yes. Furthermore, we will explore the physical consequences of the presence of these new topological quantum numbers. As with integer quantum Hall states, the system supports topologically protected edge or surface states, when these topological quantum numbers take non-trivial values. These edge/surface states are particularly important for the transport properties of the system. In the case of 3D systems, the surface states can also be probed directly using angle-resolved photoemission spectroscopy (ARPES). Experimentally, it is the detection of these topologically protected edge/surface states that establishes the topologically non-trivial nature of a given system.

14.1 Kane–Mele Model

We discussed in the previous chapter two ways of opening up a gap in the graphene band spectrum, or giving the Dirac fermions masses. (i) The regular mass, which is the same at the two Dirac points; it comes from the sublattice potential difference, which breaks parity or inversion symmetry, but respects time-reversal symmetry. (ii) The Haldane mass, which has opposite signs at the two Dirac points; it can be generated by having imaginary next-nearest-neighbor hopping-matrix elements, which break time-reversal symmetry but respect inversion symmetry. These can be summarized by a four-component Dirac-like Hamiltonian

$$H = v_{\mathrm{F}}(\tau_z \sigma_x p_x + \sigma_y p_y) + m_{\mathrm{R}}\sigma_z + m_{\mathrm{H}}\tau_z \sigma_z, \tag{14.1}$$

where the σ and τ Pauli matrices act on the sublattice and valley (or K and K') spaces, respectively. Under inversion (or parity transformation, $\mathcal{P}$) we reverse the A and B sublattices as well as the K and K' points, thus $\sigma_z \to -\sigma_z$ and $\tau_z \to -\tau_z$. Under time-reversal transformation (Θ) we reverse the K and K' points, but the A and B sublattices remain the same, thus $\sigma_z \to \sigma_z$ and $\tau_z \to -\tau_z$. It is thus clear that the regular mass term (m_R) is even under Θ but odd under $\mathcal{P}$, while the Haldane mass term (m_H) is odd under Θ but even under $\mathcal{P}$. Then a natural question to ask is the following one: can one have a mass term that respects both $\mathcal{P}$ and Θ? Kane and Mele [107, 108] pointed out that the answer is yes, provided that one invokes spin–orbit coupling. The simplest possibility would be a mass term of the form

$$m_{KM}\tau_z\sigma_z S_z, \tag{14.2}$$

where S_z is the $\hat{z}$ component of the electron spin operator $\vec{S}$. Since $\vec{S}$ is even under $\mathcal{P}$ but odd under Θ, it is easy to see that this mass term is even under both $\mathcal{P}$ and Θ. The way to generate this **Kane–Mele mass term** is simply to have two copies of the Haldane model, one for spin-up and the other for spin-down electrons, with *opposite* next-nearest-neighbor hoppings. Obviously the charge Hall conductance (or total Chern number) is zero due to cancelation between the up- and down-spin electrons, as guaranteed by the presence of time-reversal symmetry. But there is a quantized *spin* Hall conductance when the chemical potential is in the band gap. The spin Hall conductance is the *difference* between the Hall conductances of up- and down-spin electrons, corresponding to the *spin* current response to a potential gradient. Clearly such a state, initially named a "**quantum spin Hall state**," is topologically non-trivial. It has two branches of edge modes for each edge, one for spin-up and another for spin-down, propagating in *opposite* directions. They are thus consistent with time-reversal symmetry and the overall system is *not* chiral, since, under the time-reversal transformation Θ, both the orbital motion and the spin direction change sign. The directions of the spin and velocity of the modes are tied together, and such modes are called helical. We thus find that there can be topologically non-trivial bands in the *presence* of time-reversal symmetry, characterized by gapless helical edge modes.

The model considered thus far is very special in that the $\hat{z}$ component of spin is conserved, allowing us to define a spin current and hence the spin Hall conductance and other spin transport coefficients. A generic spin–orbit coupling (which respects time-reversal symmetry), on the other hand, breaks spin rotation symmetry and spin conservation along all directions, thus forbidding a well-defined spin Hall conductance. A natural question then is whether or not the gapless helical edge modes are robust in this case (since up- and down-spin states are now mixed, which may open up a gap at their crossing, a possible scenario illustrated in Fig. 14.1). Kane and Mele showed that the answer is that these modes are indeed robust, as they are protected by time-reversal symmetry and the corresponding guaranteed **Kramers degeneracy**; they do *not* require protection from S_z conservation. This is best illustrated by considering a translationally invariant edge, say along the $\hat{x}$ direction, in which case all eigenstates (including bulk and edge states) are labeled by a quantum number k_x. Under Θ, $k_x \to -k_x$, we thus find that $k_x = 0$ and $k_x = \pi/a$ are invariant under Θ, and there must be a Kramers degeneracy for *all* states at these two points, which forbids the scenario of Fig. 14.1(b). Furthermore, any Θ-invariant random potential *cannot* localize these edge modes for the same reason, since states with k_x and $-k_x$ form Kramers pairs. This means that any backscattering matrix elements (if present) between them that would lift the Kramers degeneracy are forbidden and must vanish. We thus find that time-reversal symmetry protects the helical edge modes here, in a way similar to that in which chirality protects the edge states in the integer quantum Hall effect.[1] Since, in the absence of conservation of

---

[1] Recall that, in that case, the backscattering matrix elements vanish because the backward-moving mode is on the opposite side of the sample from the forward-moving mode.

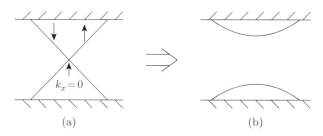

Figure 14.1 Illustration of the dispersion relation and robustness of helical edge modes in a 2D topological insulator. Shaded regions represent bulk states. (a) A pair of helical edge modes with up- and down-spin electrons propagating in opposite directions. They cross at $k_x = 0$. (b) A generic perturbation that mixes up- and down-spin states could open up a gap in the edge mode spectrum. This is *not* possible, however, when such a perturbation respects time-reversal symmetry, as the crossing at $k_x = 0$ is protected by Kramers degeneracy.

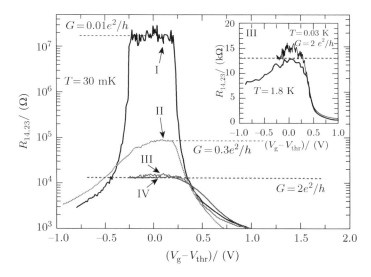

Figure 14.2 The longitudinal four-terminal resistance, $R_{14,23}$, of various normal (thickness $d = 5.5$ nm) (I) and inverted ($d = 7.3$ nm) (II, III, and IV) HgTe/CdTe quantum-well structures as a function of the gate voltage V_g, measured at zero magnetic field and temperature $T = 30$ mK. The device sizes are (20.0×13.3) μm^2 for devices I and II, (1.0×1.0) μm^2 for device III, and (1.0×0.5) μm^2 for device IV. The inset shows $R_{14,23}(V_g)$ for two samples from the same wafer, having the same device size (III) at $T = 30$ mK (upper curve) and $T = 1.8$ K (lower curve) on a linear scale. Figure reprinted with permission from [110].

(the $\hat{z}$ component of) spin, one can no longer define a spin Hall conductance, such topologically non-trivial states in the presence of time-reversal symmetry are now called (2D) **topological insulators**, characterized by their robust helical edge modes. These edge modes contribute to charge transport, and give rise to quantized contributions to the two-terminal conductance. As it turns out, spin–orbit coupling in graphene is too weak for such effects to be observable at accessible temperatures. On the other hand, such a quantized two-terminal conductance has been observed in HgTe/CdTe quantum-well structures, confirming earlier predictions by Bernevig, Hughes, and Zhang [109] that this material is a 2D topological insulator for appropriate well thickness; see Fig. 14.2.

14.2 $\mathbb{Z}_2$ Characterization of Topological Insulators

As discussed in Chapter 13, an integer quantum Hall or Chern insulator state is uniquely characterized by an integer ν, which is its Hall conductance in units of e^2/h. In terms of edge states, ν is the number

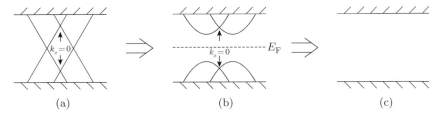

Figure 14.3 Illustration of the lack of robustness of an even number of Kramers pairs of edge modes in a 2D topological insulator. Shaded regions represent bulk states. (a) Two such pairs with four crossings; only the two crossings at $k_x = 0$ are protected by Kramers degeneracy. (b) Gaps open up at the other two crossings; no edge modes intersect the Fermi energy when it lies in this gap. (c) The edge modes disappear into bulk bands completely.

of chiral edge modes that intersect the Fermi energy at each edge.[2] The set of all integers is referred to as $\mathbb{Z}$; thus the set of all integer quantum Hall states corresponds to $\mathbb{Z}$. Owing to time-reversal symmetry, counter-propagating pairs of edge modes of a topological insulator form Kramers doublets; thus a natural guess is that topological insulators are characterized by the number of such pairs, and hence also correspond to the set $\mathbb{Z}$. This guess, however, is too naive. Let us assume that we have two pairs of such modes, as illustrated in Fig. 14.3(a). In this case we have four crossings between the counter-propagating modes, but only two of them, assumed in this case to be at $k_x = 0$, are protected by Kramers degeneracy;[3] the other two are not protected, and gaps may be opened there (see Fig. 14.3(b))! If the Fermi energy lies in this gap, then there is *no* edge mode intersecting the Fermi energy. In fact, in this case the edge modes may continuously move into the bulk bands, and disappear from the spectrum completely, as illustrated in Fig. 14.3(c). We thus find that the case with two pairs of edge modes is equivalent to there being no edge modes at all, and thus topologically trivial.

It is easy to convince oneself that cases with even numbers of pairs of edge modes are all equivalent and are topologically trivial, while cases with odd numbers of pairs of edge modes are all equivalent and are topologically non-trivial, guaranteeing at least one pair of robust edge states. These two classes correspond to the simplest non-trivial set $\mathbb{Z}_2$ (the set of integers modulo 2). Thus 2D topological insulators are characterized by a $\mathbb{Z}_2$ quantum number ν that takes two possible values, namely 0 and 1, or, equivalently, even and odd, corresponding to the trivial and non-trivial states, respectively.

Now two natural, and closely related questions are as follows. (i) How does a Bloch band (or, more precisely, a Kramers pair of bands) become topologically non-trivial in the presence of time-reversal symmetry? (ii) Given a band structure, how do we determine the $\mathbb{Z}_2$ topological quantum number ν? Before addressing these questions, it is useful to make an analogy with the topology of another type of band that is more familiar in daily life, namely a rubber band. The normal rubber band (if viewed as a 2D manifold, neglecting its thickness) has the same topology as an open-ended cylinder, see Fig. 14.4. We can change its topology by cutting it open, twisting one of the open edges once, and then gluing the two edges together, resulting in a Möbius band (often called a Möbius "strip"). The Möbius band is topologically distinct from the cylinder, as it has only *one* side (face); one can start from any point in the band, and return to the opposite side of the same point by traversing around the band.[4] One can, of

---

[2] In the most general case both forward- and backward-propagating edge modes can be present. In this case it is their *difference* that defines the integer quantum number: $\nu = \nu_+ - \nu_-$, where ν_+ and ν_- are the numbers of forward- and backward-propagating edge modes, respectively. In the presence of disorder forward- and backward-propagating modes can mutually localize due to backscattering, and it is only the net number ν that is robust.

[3] The other possibility allowed by time-reversal symmetry is that the protected crossing occurs at the zone boundary, $k_x = \pi/a$.

[4] In topology jargon, the Möbius band is the simplest example of a non-orientable manifold, while the cylinder, which has two faces, is orientable.

Figure 14.4 (a) A cylinder, which is topologically equivalent to an ordinary rubber band. (b) A Möbius band, obtained by cutting a cylinder open, twisting it once, and then gluing the edges together.

course, twist the rubber band as many times as one wants before gluing the open edges together. The reader should be able to convince herself/himself that all odd twists result in a band with one side as in a Möbius band, and even twists result in two sides as in a cylinder. In fact, all bands with even twists are topologically equivalent to a cylinder, while all bands with odd twists are topologically equivalent to a Möbius band. We thus find that the rubber bands fall into two topological classes, which are labeled by a $\mathbb{Z}_2$ "quantum number" (corresponding to an even or odd number of twists).[5]

We now come to the question (i) posed above. As it turns out, the way to make a Bloch band topologically non-trivial is also "twisting" a topologically trivial band an *odd* number of times, in a way that will be made precise below. Let us start with a trivial band. As we learned in Chapter 7, in the tight-binding limit one can construct a band from linear combinations of a *single* atomic orbital on different atoms, and the origin of the band gap is simply the gap between atomic orbitals. Obviously such a band is smoothly connected to decoupled atomic orbitals, and should be topologically trivial because decoupled atomic orbitals do not even know about the periodic crystalline structure the atoms form, and in particular the corresponding Brillouin zone! How do we characterize atomic orbitals? Owing to rotational symmetry, each orbital in an isolated atom is characterized by orbital and total angular momentum quantum numbers l and $j = |l \pm 1/2|$, and a parity quantum number $\mathcal{P} = (-1)^l$. The crystalline potential, on the other hand, breaks rotational symmetry, and thus states in a Bloch band no longer have well-defined l and j quantum numbers. However, for crystals with inversion symmetry, $\mathcal{P}$ can remain a good quantum number. This turns out to be a particularly convenient case for consideration of band topology, and we will therefore assume crystal inversion symmetry to be present and explore its consequences.

As with time-reversal transformation, under inversion or parity transformation $\mathcal{P}$ we have $\vec{k} \to -\vec{k}$, thus a generic Bloch state with lattice momentum $\vec{k}$ is not an eigenstate of $\mathcal{P}$. However, in d spatial dimensions there exists a set of 2^d special (zone center and boundary) wave vectors $\vec{k}_j$ that are equivalent to $-\vec{k}_j$. At these special $\vec{k}$s, Bloch states are also eigenstates of $\mathcal{P}$. In the simple tight-binding limit, they all have the same parity of the corresponding atomic orbital for a Bravais lattice. In this case the band is topologically trivial. Since the number of such $\vec{k}$s is even, the product of the parity eigenvalues is $+1$ (for this topologically trivial case):

$$\prod_{j=1}^{2^d} \delta_j = +1, \tag{14.3}$$

where j is an index labeling these special $\vec{k}$s, and δ_j is the parity eigenvalue of the corresponding Bloch state.

---

[5] Such rubber bands (or 2D manifolds, to be more precise) embedded in spaces of dimension $d > 3$ can be continuously untwisted into each other, as long as they are in the same topological class. This is not possible in our 3D world, which might lead one to think (incorrectly) that rubber bands with different numbers of twists are all topologically distinct. However, topological classification cares only about the *intrinsic* properties of an object, *not* about how it is (externally) embedded.

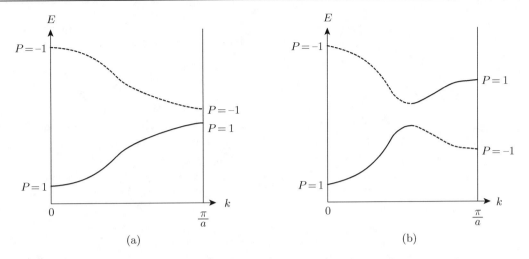

Figure 14.5 (a) Two neighboring 1D bands in the tight-binding limit, each with well-defined parities. (b) A band crossing occurred near the zone boundary $k = \pi/a$, resulting in two "twisted" bands with opposite parities at zone center and boundary. (Because wave vectors are odd under spatial inversion, band parity is not well-defined except at wave vectors obeying $-\vec{k} = +\vec{k} \mod \vec{G}$, where $\vec{G}$ is a reciprocal lattice vector. Thus parity can be well-defined only at the zone center and boundary.)

Exercise 14.1. For the cases of (i) a $d = 2$ square lattice and (ii) a $d = 2$ triangular lattice, find the 2^d special wave vectors that are equivalent under spatial inversion.

Being a topological quantum number, the $\mathbb{Z}_2$ index of a band (from now on we use a single band to represent the Kramers pair of bands when time-reversal symmetry is present) can change only when neighboring bands cross and a band gap closes; see Fig. 14.5 for an example in 1D. Such a crossing leads to "twisted" bands that are topologically non-trivial because they have opposite parities at the zone center and boundary. Obviously such bands *cannot* evolve smoothly to the (topologically trivial) atomic limit, where the parity must be the same for all states at $\mathcal{P}$-invariant wave vectors. Each time such a crossing occurs, the $\mathbb{Z}_2$ indices of the bands involved flip, and one can track the parity (meaning whether the number is even or odd) of such crossings by inspecting the product of parities (meaning eigenvalue of the parity operator!) of states at the $\mathcal{P}$-invariant wave vectors, leading us to the **Fu–Kane formula** [111]:

$$(-1)^{\nu} = \prod_{j=1}^{2^d} \delta_j, \tag{14.4}$$

where $\nu = 0, 1$ is the $\mathbb{Z}_2$ index of the band being considered. Note that the product of $(-1)^{\nu}$ over all bands is always $+1$. That is, the number of topologically non-trivial bands is always even.

The expression above cannot be applied to crystals *without* inversion symmetry, as parity is not a good quantum number anywhere in the Brillouin zone in such cases. However, it is still very useful, especially the way we arrived at it, at least in two ways. (i) One can imagine a path, in some parameter space, that can smoothly connect the band Hamiltonian to one *with* inversion symmetry. As long as the band gaps (for the band being considered) do not close along the way, they have the same $\mathbb{Z}_2$ index and Eq. (14.4) can be used. (ii) More physically, one can start from the atomic limit, and ask how many band crossings (between conduction and valence bands, or the ones right above and below the Fermi energy) must occur in order for one to arrive at the actual band structure; the parity of this number determines the $\mathbb{Z}_2$ index. As a typical example, the topologically non-trivial state realized in HgTe/CdTe quantum wells is induced by a crossing (discussed in Chapter 9) between an s band and a

p band.[6] One can also calculate the $\mathbb{Z}_2$ index directly from the band wave functions, similarly to what we did in calculating Chern numbers. Such a calculation is quite involved, and the interested reader may consult references listed in Section 14.5 for details.

Our discussions have so far not been careful in drawing distinctions between 2D and 3D systems. In 2D, the topologically non-trivial $\mathbb{Z}_2$ bands may be viewed as (somewhat) straightforward generalizations of Chern bands to the time-reversal-symmetric case. There is no Chern band (and no corresponding integer quantum Hall effect) in 3D, but so-called "**weak topological insulators**" can exist in 3D and are essentially just stacks of weakly coupled 2D topological insulators. So-called "**strong topological insulators**" in 3D have topologically non-trivial band structures which cannot be continuously deformed into stacks of weakly coupled 2D topological insulators. It turns out that, in 3D, full characterization of the band topology requires specification of four distinct $\mathbb{Z}_2$ quantum numbers. One of these is defined by Eq. (14.4) (which remains valid for 3D crystals with inversion symmetry), and it is this $\mathbb{Z}_2$ parameter that defines whether or not the system is a strong topological insulator. If the system is not a strong topological insulator, namely $\nu = 0$, it is the other three $\mathbb{Z}_2$ parameters which determine whether or not it is a weak topological insulator, and, if so, what the stacking directions of the underlying 2D topological insulator are.

14.3 Massless Dirac Surface/Interface States

We have encountered the two-component Dirac equation several times in this book already. It is often used to describe two neighboring bands and, in particular, their crossings. As we have seen in the last section, crossing of *Kramers pairs* of bands (in the presence of time-reversal symmetry) plays a crucial role in generating topologically non-trivial $\mathbb{Z}_2$ bands. Such a crossing involves four bands, and can be described by a four-component Dirac equation. This is particularly appropriate when inversion symmetry is also present, as a Kramers pair of bands will then be degenerate everywhere in the Brillouin zone (see Section 7.8). The (original) four-component **Dirac Hamiltonian**

$$H_{\mathrm{D}} = v_{\mathrm{F}}\vec{\alpha} \cdot \vec{p} + m\beta \tag{14.5}$$

was introduced by Dirac to describe electrons in a vacuum. (We have replaced the speed of light c by the Fermi velocity v_{F}.) Here $\vec{\alpha}$ and β are the 4×4 Dirac matrices satisfying the following anti-commutation relations (also known as the **Clifford algebra**):

$$\{\alpha_i, \alpha_j\} = 2\delta_{ij}, \qquad \{\alpha_i, \beta\} = 0, \qquad \beta^2 = 1, \tag{14.6}$$

where i and j are spatial indices that run from 1 to d (we consider the cases $d = 2$ and $d = 3$ on an equal footing here). A specific representation of these Dirac matrices (known as the **Weyl representation**, and particularly useful for our later discussions) is

$$\vec{\alpha} = \begin{pmatrix} \vec{\sigma} & 0 \\ 0 & -\vec{\sigma} \end{pmatrix}, \qquad \beta = \begin{pmatrix} 0 & I \\ I & 0 \end{pmatrix}, \tag{14.7}$$

where $\vec{\sigma}$ is the set of three Pauli matrices and I is the 2×2 identity matrix.[7]

---

[6] Bulk HgTe is actually a semimetal (see Fig. 9.2). Quantum confinement of the quantum-well structure (as discussed in Chapter 9) opens up a gap at the Fermi energy and turns it into an insulator. Depending on the width of the well, the ordering of bands at the Γ point changes, leading to topologically distinct 2D band structures.

[7] These are equivalent to the more familiar Pauli–Dirac representation in which

$$\vec{\alpha} = \begin{pmatrix} 0 & \vec{\sigma} \\ \vec{\sigma} & 0 \end{pmatrix} \quad \text{and} \quad \beta = \begin{pmatrix} I & 0 \\ 0 & -I \end{pmatrix}$$

via a unitary transformation. It is a good exercise for the reader to find the unitary transformation explicitly. For more details see, e.g., Ref. [112].

We wish to use Eq. (14.5) to describe two pairs of Kramers degenerate bands, and its content needs to be re-interpreted: the momentum $\vec{p}$ is measured from a specific point in the Brillouin zone corresponding to a (minimal) direct band gap. This point will be referred to as the Dirac point from now on. The absolute value of the mass $|m|$ is one-half of this minimal band gap; v_F is a velocity parameter determined not by the speed of light, as in the original Dirac equation, but rather by the band structure [essentially the rate at which (in the absence of the mass term) the gap opens up as one moves away from the degeneracy at the Dirac point]. The positive- and negative-energy states correspond to the conduction and valence band states, respectively. The band gap vanishes when $m = 0$ (and the Dirac particles become "massless"), and band crossing occurs at the point where the mass m changes *sign*. This is all quite similar to what happens when two bands cross, as discussed in Section 13.5.2.

In non-relativistic quantum mechanics, we know how the parity operator transforms a (one-component) wave function:

$$\mathcal{P}\psi(\vec{x}) = \psi(-\vec{x}). \tag{14.8}$$

For the Dirac particles with a four-component wave function, the transformation is slightly more complicated:[8]

$$\mathcal{P}\psi(\vec{x}) = \beta\psi(-\vec{x}), \tag{14.9}$$

namely we also need to transform the internal wave function using the mass operator β. For Bloch states described by the Dirac equation, it reflects the fact that there are internal degrees of freedom (in addition to the spatial plane-wave factor) that transform non-trivially under parity. At the Dirac point we have $\vec{k} = \vec{0}$, and the Hamiltonian reduces to

$$H_D = m\beta \tag{14.10}$$

and the parity operator $\mathcal{P}$ simply reduces to β (since the wave function has no $\vec{x}$ dependence). The band gap is $2|m|$ because the conduction and valence bands correspond, respectively, to the eigenvectors of β with eigenvalues ± 1 if the mass is positive and ∓ 1 if the mass is negative. Hence we find that the conduction and valence band states must have *opposite* parity at the Dirac point, and a band crossing results in an exchange of their parity eigenvalues, which is consistent with our discussions in the previous section. We thus find that a topology change of a band is triggered (again) by a sign change of a Dirac mass.

We now turn our discussion to the interface between two insulators with opposite $\mathbb{Z}_2$ topology, including the surface of a topological insulator, which can be viewed as the interface between a topological insulator and vacuum; the latter carries a trivial $\mathbb{Z}_2$ index by definition.[9] In light of the discussion above, there must be a sign change of a Dirac mass across such an interface. We learned in Section 7.6 that a zero-energy bound state (zero mode) appears at such an interface for the *two-component* Dirac equation. This result can be generalized to higher dimensions, and to the *four-component* Dirac equation. The easiest way to see this is to consider such an interface perpendicular to the $\hat{x}$ direction, so that the components of the momentum along the other directions are conserved. First consider the special case with $k_j = 0$ for $j > 1$. (That is, assume a solution which is translation-invariant along the plane of the interface and hence is an eigenstate of p_y and p_z with eigenvalue zero.) In this case Eq. (14.5) reduces to a four-component Dirac Hamiltonian in 1D (along the $\hat{x}$ direction), with an x-dependent mass term:

---

[8] See, e.g., page 24 of Ref. [112]. One way to see that the parity transformation must involve β is by noting that under $\mathcal{P}$ we have $\vec{p} \to -\vec{p}$; thus, in order for H_D to be invariant under $\mathcal{P}$, we must have $\alpha \to -\alpha$ and $\beta \to \beta$. We thus need a unitary operator that anti-commutes with α but commutes with β. β itself is essentially the only choice.

[9] In fact vacuum can be viewed as an insulator with a huge band gap that equals twice the electron rest energy, if we view the filled Dirac sea as valence band states.

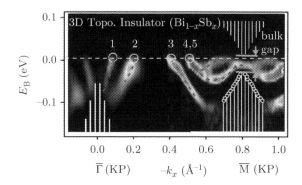

Figure 14.6 Photoemission data on surface states of a topological insulator. The data exhibits an odd number of surface bands, which cross the Fermi surface at the points numbered 1–5. Figure reprinted from [113], with permission from Springer Nature.

$$H = v_F \alpha_x p_x + m(x)\beta. \tag{14.11}$$

It is easy to show that there are *two* zero-energy bound states if $m(x)$ changes sign for $x \to \pm\infty$ (not surprisingly, when you double the number of components, you also double the number of solutions!). Using translation invariance along the interface, these solutions can be easily generalized to states with finite momentum parallel to the interface, with linear dispersion (and the velocity is simply v_F; see Exercise 14.3). These interface states represent *two-component* massless Dirac particles in $d - 1$ dimensions. For $d = 2$ they are nothing but the helical edge states discussed earlier. For $d = 3$ the direct observation of such Dirac surface states of many compounds confirmed earlier theoretical predictions that they are topological insulators on account of their (bulk) band structures; see Fig. 14.6 for an example.

As already mentioned in Section 14.1, the gapless nature of the surface/interface states acquires protection from time-reversal symmetry, Θ. In addition, the combination of time-reversal symmetry and the Dirac nature of these surface/interface states prevents them from being Anderson-localized; see Section 13.3.1. If Θ is broken either explicitly (say, by an external magnetic field) or spontaneously (by spontaneous ordering of magnetic moments carried by magnetic impurities), a gap may be opened at the surface/interface. See Exercise 14.4 for an example. Topological insulators are thus examples of a class of topological phases of matter whose gapless (or otherwise unusual) surface/interface states require the protection of certain symmetries. These are often referred to as **symmetry-protected topological (SPT) phases**.[10] Breaking the protecting symmetry in the bulk, on the other hand, may actually lead to new topological states. A particularly noteworthy example is the quantum anomalous Hall state realized in magnetic topological insulators that breaks time-reversal symmetry spontaneously, and was first observed in Ref. [115].

Exercise 14.2. Show that the α and β matrices of Eq. (14.7) satisfy the Clifford algebra (14.6).

Exercise 14.3. (a) Show that the Hamiltonian of Eq. (14.11) supports two zero-energy bound states. (b) Generalize the results of (a) to finite momentum along the interface, and obtain the dispersion of these interface bound states. Hint: note the similarity to Exercise 13.8.

---

[10] Perhaps the very first example of an SPT phase is the SSH model discussed in Section 7.6, which requires protection from particle–hole symmetry. See Exercise 7.23.

> **Exercise 14.4.** Repeat part (a) of Exercise 14.3, but add a Zeeman splitting term $\mu B \Sigma_x$ (which breaks time-reversal symmetry) to the Hamiltonian of Eq. (14.11), where
>
> $$\vec{\Sigma} = \begin{pmatrix} \vec{\sigma} & 0 \\ 0 & \vec{\sigma} \end{pmatrix}$$
>
> is the electron spin operator divided by $\hbar/2$ (both in the Weyl and in the Pauli–Dirac representation), and show that the bound states, if present, are no longer at zero energy.

14.4 Weyl Semimetals

In the absence of either time-reversal or inversion symmetry, we no longer have exactly degenerate bands, and crossings between two (but *not* four) bands are possible. As we learned in Section 13.5.2, in 2D this can occur only if one parameter is fine-tuned. Here we study what happens in 3D.

For reasons similar to those elucidated in Section 13.5.2, such a band crossing, if it does occur in 3D, can be described by the following Hamiltonian near the crossing point (chosen to be at $\vec{k} = 0$):[11]

$$h(\vec{k}) = \pm v(k_x \sigma_x + k_y \sigma_y + k_z \sigma_z) = v\vec{\sigma} \cdot \vec{k}, \tag{14.12}$$

or

$$h = \pm v\vec{\sigma} \cdot \vec{p}. \tag{14.13}$$

Note that the only difference between this **Weyl Hamiltonian** in 3D and the two-component massless Dirac Hamiltonian in Eq. (7.212) for 2D is the spatial dimension. Unlike in the 2D case, however, such a crossing is *stable* in 3D. The reason is that, in order to reach such a crossing, one needs to tune three independent parameters, which correspond precisely to the three components of the lattice momentum $\vec{k}$ in 3D. Perturbing the Hamiltonian (14.13), say by a "mass"-like term $m\sigma_z$, merely moves the location of the crossing point in momentum space, without eliminating it. The Hamiltonians of Eq. (14.13) were introduced by Weyl to describe massless particles with fixed chiralities (the $\pm$ signs correspond to $\pm$ chiralities) in high-energy physics (in which case $v = c$ is the speed of light).[12] For this reason the 3D band crossing described by the Hamiltonian (14.13) is also called a **Weyl point** or **Weyl node**, and, if the Fermi energy of the system coincides with Weyl points but has no other interception with bands, the system is called a **Weyl semimetal**. Gapless fermions described by the Weyl Hamiltonian (14.13) are also called **Weyl fermions**.

Another way to understand the stability of Weyl points is to recognize that they are "magnetic monopoles" in the 3D *momentum* space, due to the (quantized!) 2π Berry flux they carry (cf. Section 13.3). Like magnetic flux, Berry flux can only appear or disappear at these monopoles. Thus the stability of Weyl points is of topological nature. From this we also immediately understand that Weyl points must exist in pairs with opposite chiralities, and can disappear only when the members of such a pair annihilate each other.

---

[11] Similarly to what is discussed in Section 13.5.2, to bring the Hamiltonian to this form, in general one needs to make a proper choice of a coordinate system that includes re-scaling of velocity parameters that are in general anisotropic to begin with.

[12] It should be clear that the four-component massless Dirac Hamiltonian (14.5) comprises two decoupled Weyl Hamiltonians with opposite chirality. A non-zero mass term mixes the two Weyl Hamiltonians and destroys the conservation of chirality.

14.4.1 Fermi Arcs on the Surface

The Weyl semimetal is yet another state we have encountered that has distinct topological properties. Unlike integer quantum Hall states or topological insulators, however, there is *no* energy gap in the bulk. Does it also have topologically protected gapless surface states? One might guess the answer is no, since, in the absence of a bulk gap, any surface states ought to be able to leak into the bulk. However, as we are going to see in this section, there *are* indeed topologically protected gapless surface states, that form very unusual structures known as **Fermi arcs**.

To reveal the physics in the simplest possible setting, we assume there is a single pair of Weyl points with opposite chiralities,[13] and their momentum difference has a non-zero projection along the $\hat{z}$ direction. The two Weyl points are at $k_z = k_{\pm}$, respectively. See Fig. 14.7. A key observation is that, if we treat k_z as a *parameter*, then for each value (or "slice") of k_z there is a *2D band structure* that depends on (k_x, k_y). In particular, except for $k_z = k_{\pm}$, there is a *gap* in the 2D band structure at the Fermi energy. One can thus define a (k_z-dependent) Chern number of the system, $C(k_z)$. Being a topological quantum number, $C(k_z)$ can only (and does) change at the (2D) band-crossing points $k_z = k_{\pm}$. In fact, the Weyl Hamiltonian (14.13) takes precisely the form of Eq. (13.112), if m is identified as $v k_z$. This means that C must be *different* in the range $k_- < k_z < k_+$, and in the complement of that range, implying that C must be *non-zero* for some range of k_z. To proceed, we consider a particularly simple case in which $C = 1$ for $k_- < k_z < k_+$, and $C = 0$ otherwise. Furthermore, consider a sample that has finite cross section in the xy plane but is infinitely long and translationally invariant along the $\hat{z}$ direction, so that k_z is still a good quantum number. For reasons detailed in Section 13.5, we expect *gapless* edge states (that intersect the Fermi energy) for $k_- < k_z < k_+$, but *not* for other values of k_z. These states are bound to the surface, and, because they are topologically protected, their presence does *not* rely on symmetries, not even on the translation symmetry along the $\hat{z}$ direction which is being invoked here for ease of analysis.

Since the Fermi energy intersects an open segment in the 1D k_z-space, this intersection is called a Fermi arc. See Fig. 14.8 for an illustration for the simplest case illustrated in Fig. 14.7. Real Weyl semimetals typically have multiple pairs of Weyl points, and can thus have multiple Fermi arcs, depending on the surface orientation. In 2015 such Fermi arcs on the surface of TaAs were revealed using ARPES, marking the discovery of Weyl semimetals[116, 117], following earlier theoretical predictions [118–120].

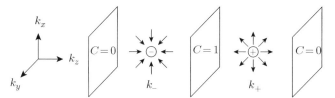

Figure 14.7 Illustration of Weyl points. The two Weyl points, at $k_z = k_{\pm}$, have positive and negative chirality, respectively. They are positive and negative monopoles of Berry flux in momentum space, as indicated by the arrows. For each fixed value of k_z, there is a 2D band structure that depends on (k_x, k_y); the parallelograms represent 2D slices of the 3D Brillouin zone for particular values of k_z, which are 2D Brillouin zones themselves. As long as $k_z \neq k_{\pm}$, the 2D bands have a gap at the Fermi energy, and the total Chern number C of the system is well defined. In the simplest case we have $C = 1$ for $k_- < k_z < k_+$ and $C = 0$ otherwise.

---

[13] It turns out this requires breaking of time-reversal symmetry. If inversion symmetry is broken but time-reversal symmetry is intact, there are at least two such pairs. This is because under time-reversal transformation a Weyl point at wave vector $\vec{K}$ transforms to a Weyl point at wave vector $-\vec{K}$ with the *same* chirality. The requirement that the net chirality be zero (as guaranteed by the fact that the total Berry flux going through the Brillouin zone is zero) thus requires that Weyl points must come in integer multiples of four in the presence of time-reversal symmetry.

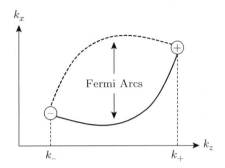

Figure 14.8 Illustration of Fermi arcs at the surfaces of a Weyl semimetal with only two Weyl points. Consider a crystal that is infinite in the $\hat{x}$ and $\hat{z}$ directions, and has a finite thickness d along the $\hat{y}$ direction. There are thus two infinitely large boundaries separated by d. Owing to the presence of lattice translation symmetries along the $\hat{x}$ and $\hat{z}$ directions all states (including bulk and boundary states) carry two good quantum numbers, k_x and k_z. For reasons discussed in the text, there is a single branch of topologically protected boundary states at each boundary that intersect the Fermi energy, for $k_- < k_z < k_+$. Such intersections form an arc at each boundary, illustrated with solid ($y = 0$ surface) and dashed ($y = d$ surface) lines. They connect the projections of the two Weyl points onto the $k_x k_z$ plane, illustrated by the circles. As one approaches these circles, the gap for the bulk states gets smaller and smaller, and hence the surface states penetrate deeper and deeper into the bulk. At the circles the surface states are completely mixed up with bulk states (or "disappear" into the bulk), allowing the arcs of opposite surfaces to be connected via the bulk.

14.4.2 Chiral Anomaly

As discussed earlier, when the mass term vanishes in the four-component Dirac Hamiltonian (14.5), it decomposes into two decoupled Weyl Hamiltonians with opposite chiralities. Naively one would then expect that in this case not only the total particle number N but also the numbers of particles with fixed chiralities, $N_{\pm}$, would be conserved. It was found, however, that this is *not* true when electromagnetic fields are present, a phenomenon known as the **chiral anomaly**.[14] In the condensed matter context of Weyl semimetals, the chiral anomaly is much easier to understand, as $N_{\pm}$ corresponds to the number of electrons (or holes) near the two Weyl nodes, which belong to the *same* (conduction or valence) band. Thus there is no fundamental reason why they need to be separately conserved. It is very interesting, however, to understand quantitatively how electromagnetic fields break the separate conservation of N_+ and N_-.

Let us start by considering a uniform magnetic field $\vec{B} = -B\hat{z}$. In this case k_z remains a good quantum number. In the meantime we expect electron motion in the xy plane will be re-organized into Landau levels. This is indeed the case. Let us first consider $k_z = k_+$. In this case the σ_z term in the Weyl Hamiltonian (14.13) vanishes (keep in mind that p_z or k_z is measured from k_+ there), and it simply reduces to a 2D Dirac Hamiltonian in a perpendicular magnetic field, which we encountered in Section 12.8. It is straightforward to generalize the solutions to generic k_z (see Exercise 14.5), and obtain the dispersions of the Landau levels as illustrated in Fig. 14.9(a). Particularly noteworthy is the dispersion of the $n = 0$ Landau level, which passes zero at $k_z = k_+$ and has a positive slope. A similar set of dispersion curves exists near the other Weyl node, except that there the $n = 0$ dispersion curve passes zero at $k_z = k_-$ and has a *negative* slope; see Fig. 14.9(b). The two $n = 0$ dispersion curves will meet and form a *1D* dispersion at two Fermi points $k_F = k_{\pm}$, as illustrated in Fig. 14.9(c).

---

[14] In quantum field theory, an anomaly refers to the phenomenon that a classical symmetry gets destroyed after quantization. For the chiral anomaly, the relevant symmetry is chiral symmetry, corresponding to *independent* gauge transformation for fermions with different chiralities.

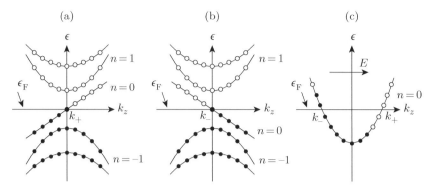

Figure 14.9 (a) The low-energy spectrum near the positive Weyl node at $k_z = k_+$, when a magnetic field $\vec{B} = -B\hat{z}$ is present. Different branches correspond to different Landau levels for motion in the xy plane. Note in particular the linear spectrum of the $n = 0$ Landau level. Filled circles represent occupied states and open circles represent empty ones. (b) The same as (a) but for the negative Weyl node at $k_z = k_-$. (c) In the presence of an additional electric field $\vec{E} = E\hat{z}$, electrons are pumped from the positive Weyl node to the negative Weyl node via the 1D dispersion connecting the $n = 0$ Landau level at the two Weyl nodes, giving rise to the chiral anomaly.

Now consider, in addition to $\vec{B} = -B\hat{z}$, that there is also an electric field $\vec{E} = E\hat{z}$. In this case k_z is no longer conserved, and evolves according to

$$\hbar \frac{dk_z}{dt} = -eE. \tag{14.14}$$

The Landau-level states with $n \neq 0$ are either fully occupied or empty, resulting in there being no redistribution of electrons. The situation is very different for the $n = 0$ Landau-level states whose dispersion and occupations are illustrated in Fig. 14.9(c), where the time-evolution above results in pumping of electrons from the $+$ Weyl node to the $-$ Weyl node. Let us calculate the pumping rate. We have

$$\frac{dN_\pm}{dt} = \pm N_\Phi \frac{L_z}{2\pi} \frac{dk_z}{dt} = \mp V \frac{e^2 E B}{4\pi^2 \hbar^2 c}, \tag{14.15}$$

where L_z is the length of the system along the $\hat{z}$ direction, $N_\Phi = A/(2\pi \ell^2)$ is the number of flux quanta enclosed in the cross section A of the system, which is also the Landau level degeneracy, and $V = AL_z$ is the volume of the system. The result above is a special case of the more general result obtained in field theory:

$$\frac{dn_\pm(\vec{r})}{dt} = \frac{e^2}{4\pi^2 \hbar^2 c} [\vec{E}(\vec{r}) \cdot \vec{B}(\vec{r})], \tag{14.16}$$

where $n_\pm(\vec{r})$ is the density of Weyl fermions with $\pm$ chiralities.

The chiral anomaly has observable effects in magneto-transport. As illustrated in Fig. 14.9(c), the magnetic field gives rise to $N_\Phi \propto B$ conduction channels along the direction of $\vec{B}$, which contribute $N_\Phi e^2/h$ to the conductance for current flowing in this direction, in the absence of scattering. In real materials which have scattering and other conduction channels (e.g. other bands that intersect the Fermi energy), these additional 1D conduction channels give rise to *negative* magnetoresistance for $\vec{E} \parallel \vec{B}$ [121, 122]: $\delta\rho \propto -B^2$. On the other hand, for $\vec{E} \perp \vec{B}$ one usually gets *positive* magnetoresistance. This difference is a manifestation of the chiral anomaly in transport.

Exercise 14.5.
(a) Consider the Weyl Hamiltonians (14.13) in the presence of a uniform magnetic field $\vec{B} = -B\hat{z}$, and consider the orbital but not the Zeeman effect of the field. Solve its spectrum.
(b) How does the Zeeman coupling affect the spectrum?

14.5 Notes and Further Reading

In addition to Weyl semimetals, there are also other semimetals in 3D that are of interest. As we learned in Section 14.3, when both inversion $\mathcal{P}$ and time-reversal Θ symmetries are present, topologically distinct band structures are separated by a point in parameter space in which the spectrum contains a massless Dirac point. If the Fermi energy coincides with this Dirac point, the system is called a Dirac semimetal. As discussed earlier, a Dirac point is made of a pair of Weyl points with opposite chiralities. Its stability requires additional symmetries (beyond $\mathcal{P}$ and Θ) to protect it. There do exist materials that support such stable Dirac points.

Dirac and Weyl points are point nodes, namely isolated points in momentum space where the energy gap between the conduction and valence bands vanishes. There can also be line nodes, namely 1D lines in the 3D momentum space where the conduction and valence bands become degenerate. Such line nodes again need extra symmetries to protect them, and there is intensive effort at present to look for semimetals in which the Fermi energy matches such line nodes.

Crystalline symmetries can also give rise to new topological insulators, that require their protection *in addition* to time-reversal symmetry. They are called topological crystalline insulators [123].

At the time of writing, topological insulators and semimetals are still a rapidly developing field of research. Several review articles and books that cover the topics in much more detail are available. A partial list of them includes the following: M. Z. Hasan and C. L. Kane, Topological insulators, Rev. Mod. Phys. **82**, 3045 (2010); Xiao-Liang Qi and Shou-Cheng Zhang, Topological insulators and superconductors, *Rev. Mod. Phys.* **83**, 1057 (2011); Shun-Qing Shen, *Topological Insulators*, Springer-Verlag, Berlin/Heidelberg (2012); B. A. Bernevig and Taylor L. Hughes, *Topological Insulators and Topological Superconductors*, Princeton University Press, Princeton (2013); N. P. Armitage, E. J. Mele, and A. Vishwanath, Weyl and Dirac semimetals in three dimensional solids, *Rev. Mod. Phys.* **90**, 15001 (2018).

15 Interacting Electrons

As we discussed previously in Chapter 7, a full quantum-mechanical description of electron motion requires solution of the many-electron Hamiltonian, Eq. (7.26), which includes the electron–electron Coulomb interaction. This is a formidably difficult problem. However, many electronic properties of condensed matter systems can be understood using an independent-electron approximation, in which each electron moves in an identical external potential composed of the electrostatic attraction to the (fixed) nuclei and the electrostatic repulsion of the average charge-density of the other electrons, of the form given in Eq. (7.27). In this chapter we will first discuss how this reduction from a many-body problem to a single-particle problem is carried out systematically for the ground state, then generalize it to study response functions and (certain) collective excitations of the system. We will provide a formal justification for using a single-particle-based picture to describe interacting many-particle systems (known as **Landau Fermi-liquid theory**), and eventually (and perhaps most importantly) discuss how such a single-particle description breaks down in certain cases.

15.1 Hartree Approximation

To make our discussion more general, we write Eq. (7.26) in a more general form:

$$\hat{H} = \hat{T} + \hat{V}_{\text{ext}} + \hat{U} = \sum_{i=1}^{N} \left(-\frac{\hbar^2 \nabla_i^2}{2m} + v_{\text{ext}}(\vec{r}_i) \right) + \frac{1}{2} \sum_{i \neq j}^{N} u(\vec{r}_i, \vec{r}_j), \tag{15.1}$$

where $\hat{T}$, $\hat{V}_{\text{ext}}$, and $\hat{U}$ represent the kinetic energy, the (one-body) external potential, and the two-body interaction term, respectively. For electrons in solids $v_{\text{ext}}(\vec{r})$ and $u(\vec{r}_i, \vec{r}_j)$ correspond to the potential due to nuclei and the electron–electron Coulomb interaction, respectively. In the rest of this section we do *not* assume any specific form or even symmetry for these terms.

In order to facilitate approximations later on, we express $\hat{U}$ in terms of the electron number density operator

$$\hat{n}(\vec{r}) = \sum_{i=1}^{N} \delta(\vec{r} - \vec{r}_i), \tag{15.2}$$

which results in

$$\hat{U} = \frac{1}{2} \int d\vec{r} \int d\vec{r}\,' u(\vec{r}, \vec{r}\,')\hat{n}(\vec{r})\hat{n}(\vec{r}\,') - \sum_i u(\vec{r}_i, \vec{r}_i). \tag{15.3}$$

Note that the last term above cancels out the self-interaction terms in the first term, which are not present in Eq. (15.1). For translationally invariant u, we have $u(\vec{r}_i, \vec{r}_i) = u(\vec{r}_i - \vec{r}_i) = \text{constant}$.[1] If u is *not* translationally invariant, this last term is a non-trivial single-particle term, which can nevertheless be absorbed into $\hat{V}_{\text{ext}}$. We will thus ignore it in the following.

To proceed, we separate $\hat{n}$ into its expectation value and the quantum fluctuations on top of that:

$$\hat{n}(\vec{r}) = n(\vec{r}) + \delta\hat{n}(\vec{r}), \tag{15.4}$$

with

$$n(\vec{r}) = \langle \hat{n}(\vec{r}) \rangle; \tag{15.5}$$
$$\delta\hat{n}(\vec{r}) = \hat{n}(\vec{r}) - n(\vec{r}). \tag{15.6}$$

In the above $\langle \hat{n}(\vec{r}) \rangle$ can be either the ground state expectation value of $\hat{n}(\vec{r})$, or an ensemble average. We focus mostly on the former. We can now re-write Eq. (15.3) as (ignoring the last term)

$$\hat{U} = \frac{1}{2} \int d\vec{r} \int d\vec{r}\,' u(\vec{r}, \vec{r}\,')[n(\vec{r}) + \delta\hat{n}(\vec{r})][n(\vec{r}\,') + \delta\hat{n}(\vec{r}\,')] \tag{15.7}$$

$$= \int d\vec{r}\, \hat{n}(\vec{r}) \left[\int d\vec{r}\,' u(\vec{r}, \vec{r}\,')n(\vec{r}\,') \right]$$

$$+ \frac{1}{2} \int d\vec{r} \int d\vec{r}\,' u(\vec{r}, \vec{r}\,')[\delta\hat{n}(\vec{r})\delta\hat{n}(\vec{r}\,') - n(\vec{r})n(\vec{r}\,')]. \tag{15.8}$$

In the spirit of the mean-field approximation, we ignore the last term above, which includes a constant and a term that is quadratic in $\delta\hat{n}$ (and thus small compared with other terms if the fluctuations are weak). The term that we do keep is linear in $\hat{n}$ and thus an (effective) one-body term, which *approximates* the effects of $\hat{U}$. Combining it with $\hat{V}$, we obtain the effective potential seen by the particles:

$$v_{\text{eff}}(\vec{r}) = v_{\text{ext}}(\vec{r}) + v_{\text{H}}(\vec{r}), \tag{15.9}$$

which includes the so-called **Hartree potential**

$$v_{\text{H}}(\vec{r}) = \int d\vec{r}\,' u(\vec{r}, \vec{r}\,')n(\vec{r}\,') \tag{15.10}$$

that needs to be determined *self-consistently*, because $n(\vec{r}\,')$ depends on the state of the system, which in turn depends on $v_{\text{eff}}(\vec{r})$. This is the **Hartree approximation**, which can be applied to any multi-particle system made of bosons, fermions, or distinguishable particles.

Coming back to our original problem of electrons moving in a crystal, we need to solve the following set of Hartree equations:

$$\left[-\frac{\hbar^2 \nabla^2}{2m} + v_{\text{eff}}(\vec{r}) \right] \varphi_\alpha(\vec{r}) = \epsilon_\alpha \varphi_\alpha(\vec{r}), \tag{15.11}$$

$$n(\vec{r}) = \sum_\alpha |\varphi_\alpha(\vec{r})|^2 \theta(\epsilon_{\text{F}} - \epsilon_\alpha), \tag{15.12}$$

$$v_{\text{eff}}(\vec{r}) = \sum_{\vec{R}} \frac{-Ze^2}{|\vec{r} - \vec{R}|} + e^2 \int d\vec{r}\,' \frac{n(\vec{r}\,')}{|\vec{r} - \vec{r}\,'|}, \tag{15.13}$$

---

[1] Note that for the Coulomb interaction this constant is infinity! This is a manifestation of the subtleties involving singular interactions. The Coulomb interaction is also long-ranged and needs to be handled carefully (especially when taking the thermodynamic limit).

where the $\varphi_\alpha(\vec{r})$ are normalized single-particle wave functions (which for simplicity we assume to be spin-independent, and the index α includes the spin index). These equations are also known as **self-consistent field** equations. When the nuclei are located on lattice sites, $v_{eff}(\vec{r})$ is periodic, and it is possible to solve these equations accurately by taking advantage of Bloch's theorem, provided the number of atoms per primitive cell of the crystal is not too large. At present, calculations with up to ≈ 100 atoms per primitive cell are often practical, and techniques now being developed sometimes make calculations with ≈ 1000 atoms per primitive cell practical. In practice, to self-consistently determine the single-particle potential, $v_{eff}(\vec{r})$, one typically makes an initial guess at $v_{eff}(\vec{r})$, finds all occupied solutions of the single-particle Schrödinger equation, recalculates $v_{eff}(\vec{r})$ and then iterates the procedure until Eq. (15.9) is satisfied with sufficient accuracy.

It is clear that the key approximation made here is neglecting the $\delta \hat{n}(\vec{r}) \delta \hat{n}(\vec{r}')$ term in Eq. (15.8). The justification is that, if the fluctuations $\delta \hat{n}$ are small (which is the basis of all mean-field-type approximations), then the quadratic term $\delta \hat{n}(\vec{r}) \delta \hat{n}(\vec{r}')$ must be even smaller. This would indeed be the case if the many-particle wave function were a simple product of the φ_αs. However, electrons are fermions, which means that their wave function must be anti-symmetric under exchange, and *cannot* take a product form. As we will see very soon, this anti-symmetry requirement gives rise to additional (so-called exchange) contributions to matrix elements for $\delta \hat{n}(\vec{r}) \delta \hat{n}(\vec{r}')$ which are not small, even when the corresponding matrix elements of $\delta \hat{n}$ are small. We thus need to do a better job in our approximation scheme to take into account these exchange contributions. There are two other single-particle-based approximations in common use that explicitly take into account the anti-symmetry requirement, which often give rise to better descriptions of the physics of a system, without adding any essential complications. We will discuss these approaches, namely the **Hartree–Fock approximation** and **density-functional theory**, in turn.[2]

15.2 Hartree–Fock Approximation

This approximation is widely used in atomic, molecular, condensed matter, and nuclear physics and captures a lot of important physics, although it is usually quantitatively and sometimes qualitatively in error. It improves on the Hartree approximation by taking into account the anti-symmetry requirement for many-fermion wave functions.

Unlike in the Hartree approximation where one starts with the Hamiltonian and tries to reduce it to a single-particle form, in the Hartree–Fock approximation one starts with many-electron *wave functions* which satisfy the anti-symmetry requirement. For non-interacting electrons, the N-electron eigenstates are Slater determinants:

$$\Psi(\vec{r}_1 S_1, \ldots, \vec{r}_N S_N) = \begin{vmatrix} \varphi_1(\vec{r}_1, S_1) & \varphi_1(\vec{r}_2, S_2) & \cdots & \varphi_1(\vec{r}_N, S_N) \\ \varphi_2(\vec{r}_1, S_1) & \varphi_2(\vec{r}_2, S_2) & \cdots & \varphi_2(\vec{r}_N, S_N) \\ \cdots & \cdots & \cdots & \cdots \\ \varphi_N(\vec{r}_1, S_1) & \varphi_N(\vec{r}_2, S_2) & \cdots & \varphi_N(\vec{r}_N, S_N) \end{vmatrix}$$

$$= \frac{1}{\sqrt{N!}} \sum_P (-1)^{\mathcal{P}} \varphi_{P_1}(\vec{r}_1, S_1) \cdots \varphi_{P_N}(\vec{r}_N, S_N), \tag{15.14}$$

where $\{\varphi_i\}$ is a set of orthonormal single-electron states, S is a spin label, $P = (P_1, P_2, \ldots, P_N)$ is a permutation of $(1, 2, \ldots, N)$, and $\mathcal{P}$ is its parity (even or odd).[3] The Hartree–Fock approximation

---

[2] Closely related approximation schemes with the same names exist for bosonic systems in which the wave function is symmetric under exchange. We do not discuss them here.

[3] Students unfamiliar with the permutation group should consult a text on linear algebra or group theory. There is a brief and clear discussion of the properties of permutation "operators" on p. 587 of the quantum mechanics text by Messiah [124].

for the ground state of a many-electron Hamiltonian is the single Slater determinant with the lowest (variational) energy. It is exact to leading order in a perturbative treatment of the interaction term in the Hamiltonian, and so the Hartree–Fock approximation also arises as a simple approximation in many-body perturbation theory. The Hamiltonian consists of one-body terms ($\hat{T}$ and $\hat{V}$) and an interaction term $\hat{U}$. In the following we calculate the expectation value of these terms in a single Slater determinant. Minimization of this energy with respect to variation of the single-particle wave functions gives rise to the Hartree–Fock single-particle equations.

It is easy to guess that the expectation value of the kinetic-energy term in the Hamiltonian,

$$\langle\Psi|\hat{T}|\Psi\rangle = \sum_{i=1}^{N}\langle\Psi|-\frac{\hbar^2\nabla_i^2}{2m}|\Psi\rangle, \tag{15.15}$$

is simply the sum of the kinetic energy of the occupied orbitals. It is nevertheless illuminating to show this explicitly. We begin by looking at the first term in this sum – the kinetic energy of particle "1." The sums on the right-hand side below are over the $N!$ permutations. Using the definition of the Slater determinant, and defining $\bar{\varphi}_i(\vec{r}, S)$ as the complex conjugate of $\varphi_i(\vec{r}, S)$, we find

$$\langle\Psi|-\frac{\hbar^2\nabla_1^2}{2m}|\Psi\rangle = \frac{1}{N!}\sum_{P,P'}(-1)^{P+P'}\langle\varphi_{P_2'}|\varphi_{P_2}\rangle\cdots\langle\varphi_{P_N'}|\varphi_{P_N}\rangle$$

$$\times\left[\int d\vec{r}_1\sum_{S_1',S_1}\bar{\varphi}_{P_1'}(\vec{r}_1,S_1')\delta_{S_1'S_1}\left(-\frac{\hbar^2\nabla_1^2}{2m}\right)\varphi_{P_1}(\vec{r}_1,S_1)\right] \tag{15.16}$$

$$= \frac{1}{N!}\sum_{P}\langle\varphi_{P_1}|-\frac{\hbar^2\nabla^2}{2m}|\varphi_{P_1}\rangle, \tag{15.17}$$

where we used the fact that since elements 2 through N of P and P' have to be identical (otherwise at least one of the overlaps above is zero due to the orthonormality of the orbitals), element 1 must be identical in P and P' as well. Thus there is a contribution to the RHS of Eq. (15.16) only if $P = P'$. Hence we need only perform a single summation over P, which can be further decomposed into a summation over all possible values of P_1, and the $(N-1)!$ permutations of the remaining indices which yields an overall factor of $(N-1)!$. We thus find

$$\langle\Psi|-\frac{\hbar^2\nabla_1^2}{2m}|\Psi\rangle = \frac{1}{N}\sum_{i=1}^{N}\langle\varphi_i|-\frac{\hbar^2\nabla^2}{2m}|\varphi_i\rangle. \tag{15.18}$$

The key point illustrated by this mathematical result is that the particles are (truly) indistinguishable and we cannot actually paint number labels on them (i.e. we don't know which particle is number "1"). If we insist on labeling the particles, we cannot tell which orbital is occupied by which particle.

After summing over the kinetic energy of all N particles (each with identical contribution), we obtain

$$\langle\Psi|T|\Psi\rangle = \sum_{i=1}^{N}\langle\varphi_i|-\frac{\hbar^2\nabla^2}{2m}|\varphi_i\rangle. \tag{15.19}$$

The calculation for the one-body potential term $\hat{V}$ is identical.

We can do a similar, but slightly more complicated calculation for the two-body interaction term $\hat{U}$. In this case for every term in $\hat{U}$, say $u(\vec{r}_1, \vec{r}_2)$, there are *two* possible values of P' given P, one for $P' = P$ as in the above, while the other corresponds to particles 1 and 2 *exchanging* their orbitals; in the latter case P' and P differ by exchanging a single pair and thus have *opposite* parity. We find[4]

---

[4] The reader is urged to verify this explicitly, during which process (s)he will begin to appreciate the added complexity in keeping track of the indices. (S)he is also encouraged to perform similar calculations for interactions among three

$$\langle\Psi|\hat{U}|\Psi\rangle = \frac{1}{2}\sum_{i,j}^{N}\left[\langle\varphi_i\varphi_j|u|\varphi_i\varphi_j\rangle - \langle\varphi_i\varphi_j|u|\varphi_j\varphi_i\rangle\right] = E_{\mathrm{H}} + E_{\mathrm{X}}, \tag{15.20}$$

where

$$\langle\varphi_i\varphi_j|u|\varphi_l\varphi_m\rangle = \sum_{SS'}\int d\vec{r}\int d\vec{r}\,'\bar{\varphi}_i(\vec{r}, S)\bar{\varphi}_j(\vec{r}\,', S')u(\vec{r}, \vec{r}\,')\varphi_l(\vec{r}, S)\varphi_m(\vec{r}\,', S'). \tag{15.21}$$

Note that the second term inside the brackets of Eq. (15.20) above, known as the **exchange integral** or exchange term, is *not* an expectation value. The minus sign in front of it comes from the opposite parity between P and P' mentioned above. The two terms in the brackets give rise to the Hartree (E_{H}) and exchange (E_{X}) contributions to the total energy.

The Hartree–Fock energy is a functional of the N one-particle wave functions, $\varphi_i(\vec{r})$ (with $i = 1, \ldots, N$), and their complex conjugate, $\bar{\varphi}_i(\vec{r})$.[5] We will specialize to the case of spin-independent interactions and no spin–orbit couplings. Then the one-body states have an orbital and a spin label, and in the second term on the right-hand side of Eq. (15.20) the exchange term gets contributions only from pairs of one-body states with the same spin label (see Exercise 15.2). To emphasize the importance of this fact we will make the spin index explicit below, while in earlier discussions it was hidden in i. This exchange term is responsible for magnetism in a great variety of systems, as we will see later.

Minimizing the energy expectation value with respect to $\bar{\varphi}_{i,\sigma}(\vec{r})$ leads to

$$\frac{\delta\langle\Psi|\hat{H}|\Psi\rangle}{\delta\bar{\varphi}_{i,\sigma}(\vec{r})} = \epsilon_{i,\sigma}\varphi_{i,\sigma}(\vec{r}), \tag{15.22}$$

where the LHS above is a *functional* derivative of the energy with respect to $\bar{\varphi}_{i,\sigma}(\vec{r})$,[6] while the RHS comes from the constraint

$$\frac{\delta\langle\Psi|\Psi\rangle}{\delta\bar{\varphi}_{i,\sigma}(\vec{r})} = 0, \tag{15.23}$$

and $\epsilon_{i,\sigma}$ is a Lagrange multiplier. We thus obtain the famous Hartree–Fock equations:

$$\left[\frac{-\hbar^2}{2m}\nabla^2 + v_{\mathrm{ext}}(\vec{r})\right]\varphi_{i,\sigma}(\vec{r}) + \sum_j{}'\int d\vec{r}\,'u(\vec{r}, \vec{r}\,')$$

$$\times\left[\sum_{\sigma'}{}'[\bar{\varphi}_{j,\sigma'}(\vec{r}\,')\varphi_{j,\sigma'}(\vec{r}\,')]\varphi_{i,\sigma}(\vec{r}) - [\bar{\varphi}_{j,\sigma}(\vec{r}\,')\varphi_{j,\sigma}(\vec{r})]\varphi_{i,\sigma}(\vec{r}\,')\right]$$

$$= \left[\frac{-\hbar^2}{2m}\nabla^2 + v_{\mathrm{ext}}(\vec{r}) + v_{\mathrm{H}}(\vec{r})\right]\varphi_{i,\sigma}(\vec{r}) + \int d\vec{r}\,'v_{\mathrm{X}\sigma}(\vec{r}, \vec{r}\,')\varphi_{i,\sigma}(\vec{r}\,')$$

$$= \epsilon_{i,\sigma}\varphi_{i,\sigma}(\vec{r}), \tag{15.24}$$

where $\sum'$ represents summation over *occupied* orbitals, and the *non-local* exchange potential is given by

and even more particles. These will clearly reveal the cumbersome bookkeeping that arises in dealing with single Slater determinants, which are actually the *simplest* many-body wave functions. Such clumsiness points to the need to use the so-called second quantized representation of many-body states, in which the exchange symmetry is built in upfront. See Appendix J for an introduction to second quantization, and compare the same calculation performed there.

[5] In the presence of spin or other internal degrees of freedom, $\varphi_i(\vec{r})$ has multiple components and can be written as a column vector. In such cases $\bar{\varphi}_i(\vec{r})$ should be understood as its Hermitian conjugate, which is a row vector.

[6] Readers not familiar with functional differentiation should consult Appendix H.

$$v_{X\sigma}(\vec{r}, \vec{r}\,') = -\sum_{j}{}' u(\vec{r}, \vec{r}\,')\bar{\varphi}_{j,\sigma}(\vec{r}\,')\varphi_{j,\sigma}(\vec{r}). \tag{15.25}$$

Just like the Hartree equations, these equations have to be solved self-consistently since the sum over j in the interaction term involves orbitals which are not known *a priori*. In the following we discuss some important *differences* between the Hartree–Fock and Hartree equations.

While Eq. (15.24) is formally a set of non-linear equations, we solve it by treating it as a set of linear eigenvalue equations for the $\varphi_{i,\sigma}(\vec{r})$, and then ensure self-consistency through iteration (which is assumed to converge), just as for the solution of Hartree equations. The difference between Eq. (15.24) and the corresponding Hartree equation is the presence of the second term in the second line, which comes from the exchange term in Eq. (15.20). This is a *non-local* term because it relates $\varphi_{i,\sigma}(\vec{r})$ to $\varphi_{i,\sigma}(\vec{r}\,')$, which resembles arbitrary distance hopping in a tight binding model.

Another way to interpret the exchange term is the following. The Hartree term comes from the Coulomb potential produced by the charge density of each occupied orbital $\bar{\varphi}_{j,\sigma'}(\vec{r}\,')\varphi_{j,\sigma'}(\vec{r}\,')$. The exchange term is in a sense the Coulomb potential produced by an effective (complex) charge density coming from the interference of two orbitals $\bar{\varphi}_{j,\sigma}(\vec{r}\,')\varphi_{i,\sigma}(\vec{r}\,')$. This new feature renders Eq. (15.24) more difficult to solve than the Hartree equations, which take the form of ordinary Schrödinger equations.

In the Hartree approximation, we started by making an approximation on the Hamiltonian, and then solved the resultant single-particle Hamiltonian of the form in Eq. (7.27). The eigenvalues ϵ_i, for either occupied or empty orbitals, naturally *seem* to have the interpretation of eigenenergies of the original Hamiltonian, namely the total energy of the system is approximated as $\sum_i \epsilon_i n_i$, where n_i is occupation number of orbital i. As discussed earlier the point of departure for the Hartree–Fock approximation is the *wave function* and the variational principle. The use of the single Slater determinant wave function and the variational principle makes the Hartree–Fock approximation better than the Hartree approximation in the description of the ground state wave function and energy. However, the parameter ϵ_i in Eq. (15.24) is a Lagrange multiplier, whose physical meaning is not immediately clear. In particular, it is not obvious how it is related to the energy of the system. In the following we will show that it can be interpreted as an excitation energy, but in a very limited sense.

Exercise 15.1. Pretend that electrons are *distinguishable* particles. In this case an appropriate single-particle approximation to the many-electron ground state wave function is a product wave function: $\Psi(\vec{r}_1 S_1, \ldots, \vec{r}_N S_N) = \prod_i^N \varphi_i(\vec{r}_i S_i)$. Use the variational principle to determine the equations that the $\varphi_i(\vec{r}_i S_i)$ satisfy. These equations are very similar, but not identical, to the Hartree equations we obtained in the previous section.

Exercise 15.2. Assume the single-electron states $|\varphi_i\rangle$ are eigenstates of electron spin S_z. Show that the exchange term in Eq. (15.20) vanishes unless $|\varphi_i\rangle$ and $|\varphi_j\rangle$ have the same S_z eigenvalue.

Exercise 15.3. Calculate $\frac{1}{2}\int d\vec{r} \int d\vec{r}\,' u(\vec{r}, \vec{r}\,')\langle\Psi|\delta\hat{n}(\vec{r})\delta\hat{n}(\vec{r}\,')|\Psi\rangle$, where $|\Psi\rangle$ is a single Slater determinant, and show it is equal to the second (exchange) term in Eq. (15.20).

15.2.1 Koopmans' Theorem

Assume $|\Psi\rangle$ is a single Slater determinant made of the set of N (occupied) orbitals $\{\varphi_j\}$, that satisfy the Hartree–Fock equation (15.24). We suppress spin indices here and below for simplicity. Now consider another single Slater determinant $|\Psi'\rangle$, in which a particle is removed from orbital $|\varphi_l\rangle$. We would like to calculate the energy difference between them:

$$\langle \Psi | \hat{H} | \Psi \rangle - \langle \Psi' | \hat{H} | \Psi' \rangle = \langle \varphi_l | T + v_{\text{ext}} | \varphi_l \rangle$$

$$+ \sum_{j \neq l}^{N} \left[\langle \varphi_l \varphi_j | u | \varphi_l \varphi_j \rangle - \langle \varphi_l \varphi_j | u | \varphi_j \varphi_l \rangle \right]$$

$$= \int d\vec{r}\, \bar{\varphi}_l(\vec{r}) \left[\frac{-\hbar^2}{2m} \nabla^2 + v_{\text{ext}}(\vec{r}) \right] \varphi_l(\vec{r})$$

$$+ \sum_{j=1}^{N} \int d\vec{r} \int d\vec{r}\,' u(\vec{r}, \vec{r}\,') \bar{\varphi}_l(\vec{r})$$

$$\times \{ [\bar{\varphi}_j(\vec{r}\,') \varphi_j(\vec{r}\,')] \varphi_l(\vec{r}) - [\bar{\varphi}_j(\vec{r}\,') \varphi_j(\vec{r})] \varphi_l(\vec{r}\,') \}$$

$$= \epsilon_l, \tag{15.26}$$

where in the last step we used the fact that $\varphi_l(\vec{r})$ satisfies Eq. (15.24). This result is known as **Koopmans' theorem** after its discoverer, Tjalling Koopmans.

Since $|\Psi\rangle$ is an approximation to the N-particle ground state, $|\Psi'\rangle$ can be viewed as a single-hole excitation. What is the energy of this excitation? Since $|\Psi\rangle$ and $|\Psi'\rangle$ have different particle numbers, we can in the case of a single atom or molecule view $-\epsilon_l$ as the ionization energy. To properly compare the two energies in a solid we need to use the grand canonical ensemble, and include a chemical potential term in the definition of the total energy (more on this in Section 15.11). For our purpose here it suffices to say that we need to measure the single-particle energy relative to the chemical potential or Fermi energy ϵ_F. Thus Eq. (15.26) suggests that the excitation energy of a hole in orbital φ_l is (approximately)

$$\epsilon_F - \epsilon_l > 0, \tag{15.27}$$

and ϵ_l does indeed have the meaning of a single-particle energy.

However, this interpretation breaks down once we start to remove *more* particles, say another particle in orbital $\varphi_{l'}$, from $|\Psi'\rangle$. In this case Eq. (15.26) no longer applies (to $|\Psi'\rangle$ and a new two-hole state that follows from it), because the orbitals forming $|\Psi'\rangle$ no longer satisfy Eq. (15.24), due to the removal of orbital φ_l from the summation. In particular,

$$\langle \Psi | \hat{H} | \Psi \rangle \neq \sum_{j=1}^{N} \epsilon_j. \tag{15.28}$$

Exercise 15.4. Solving the Hartree–Fock equation (15.24) yields not only the (self-consistent) occupied single-particle orbitals, but a complete set of single-particle wave functions, each with a corresponding eigenvalue ϵ_i. Assume $|\Psi\rangle$ is an N-particle single Slater determinant made of such self-consistent occupied orbitals. Now add one more electron to an originally unoccupied orbital, $|\varphi_l\rangle$, and call the resultant $(N + 1)$-particle single Slater determinant $|\Psi'\rangle$. Show that $\langle \Psi' | \hat{H} | \Psi' \rangle - \langle \Psi | \hat{H} | \Psi \rangle = \epsilon_l$.

15.3 Hartree–Fock Approximation for the 3D Electron Gas

As discussed earlier, in general the Hartree–Fock equation (15.24) is quite difficult to solve. However, as is usually the case in physics, solution becomes possible and quite straightforward when there is high symmetry. We thus consider the so-called **"jellium" model**, in which the external potential corresponds to a *uniform* neutralizing positive background. This jellium model is a convenient cartoon for the external potential from the nuclei which is present in all condensed matter systems. It is useful for model calculations, which are greatly simplified by the *full* translational invariance of the model, as compared with the (reduced) lattice translation symmetry of a crystalline solid. The full translational invariance dictates that simple plane waves are *exact* solutions of Eq. (15.24), just as for free-electron systems.[7] The Hartree–Fock ground state is thus the Fermi-sea state, in which all momentum states inside the spherical Fermi surface are occupied and all momentum states outside the spherical Fermi surface are empty. Perhaps surprisingly, the jellium model turns out to be a reasonably good approximation for a variety of elemental metals where the ion cores occupy a small fraction of the volume of the system.[8]

Since the system is neutral, the electron charge density balances the positive background charge *exactly*.[9] As a result, the external potential and the Hartree potential cancel out in Eq. (15.24), and $v_{X\sigma}(\vec{r}, \vec{r}\,') = v_{X\sigma}(\vec{r} - \vec{r}\,')$, ensuring the translational invariance of the equation and its plane-wave solutions. The eigenenergies are

$$\epsilon_{\mathrm{HF}}(\vec{k}) = \frac{\hbar^2 k^2}{2m} + \Sigma(k), \tag{15.29}$$

where $\Sigma(k)$ is the Fourier transform of $v_{X\sigma}(\vec{r} - \vec{r}\,')$. Using the fact that the Fourier transform of $u(\vec{r} - \vec{r}\,') = e^2/|\vec{r} - \vec{r}\,'|$ is $4\pi e^2/q^2$, we obtain

$$\Sigma(k) = -\frac{1}{L^3} \sum_{|\vec{k}'| < k_{\mathrm{F}}} \frac{4\pi e^2}{|\vec{k} - \vec{k}'|^2} = -\int \frac{d\vec{k}'}{(2\pi)^3} \theta(k_{\mathrm{F}} - k') \frac{4\pi e^2}{|\vec{k} - \vec{k}'|^2}$$

$$= \frac{-2e^2 k_{\mathrm{F}}}{\pi} \left[\frac{1}{2} + \frac{(1 - x^2)}{4x} \ln \left| \frac{1 + x}{1 - x} \right| \right], \tag{15.30}$$

where L^3 is the system volume, and $x = k/k_{\mathrm{F}}$. In a many-body approach, the quantity $\Sigma(k)$ appears as the Hartree–Fock approximation to the **self-energy** of the 3D electron gas (3DEG). Note that

$$\Sigma(k_{\mathrm{F}}) = \frac{-e^2 k_{\mathrm{F}}}{\pi}; \qquad \left. \frac{\partial \Sigma_{\mathrm{HF}}}{\partial k} \right|_{k=k_{\mathrm{F}}} = \infty. \tag{15.31}$$

The fact that the $\Sigma(k)$ has an infinite slope at k_{F} in the Hartree–Fock approximation means that the low-temperature limit of C/T (where C is the heat capacity) vanishes in the Hartree–Fock approximation, if we identify $\epsilon_{\mathrm{HF}}(\vec{k})$ as the (single-electron) excitation spectrum of the system. This result for the heat capacity is equivalent to the density of states vanishing at the Fermi level is *not* correct and occurs because the Hartree–Fock approximation does not incorporate screening of the Coulomb interaction between pairs of electrons due to the presence of the other electrons, an important piece of physics which we will come to a little later.

---

[7] The reader has probably recognized by now that the jellium model is very similar to the empty lattice model encountered in Chapter 7, where we discussed the band structure of non-interacting electrons.

[8] A deeper understanding of why this apparently crude approximation is sometimes reasonably accurate comes from so-called pseudopotential theory. See, for example, *Pseudopotentials in the Theory of Metals* by W. A. Harrison (Benjamin, New York, 1966). Modern pseudopotentials play an important role in contemporary electronic structure calculations. The essential idea is that one replaces the singular Coulomb potential between a given valence electron and a given ion by a smooth weak potential that produces the same scattering phase shift (despite missing some high-spatial-frequency oscillations of the wave function near the ionic cores).

[9] Strictly speaking, this is true in the *bulk*. Near the boundary of the system, generally speaking, this cancelation between the electron and background charge is not exact, which gives rise to fascinating surface physics. For a theoretical treatment of bulk physics only, we can impose periodic boundary conditions and remove the surfaces.

15.3.1 Total Exchange Energy of the 3DEG in the Hartree–Fock Approximation

Since the external potential and the Hartree potential cancel out in the jellium model, we have $E_H = 0$ and the only correction to the energy of the electron gas comes from the exchange contribution to the energy. The **exchange energy** of the electron gas is

$$E_X = -\frac{1}{2} \cdot \frac{2}{L^3} \sum_{|\vec{k}|, |\vec{k}'| < k_F} \frac{4\pi e^2}{|\vec{k} - \vec{k}'|^2} = \sum_{|\vec{k}| < k_F} \Sigma(k). \tag{15.32}$$

Note that the factor of 2 in $2/L^3$ accounts for the spin degeneracy, which is canceled out by the overall factor of $\frac{1}{2}$ in Eq. (15.20). So the above is only *one-half* of the sum of exchange contribution to the single-particle-energy-like quantity, $\epsilon_{HF}(k)$. We want to know how large the exchange energy is compared with the kinetic energy of the non-interacting electron ground state, since this gives us a measure of how important interactions are in the electron gas. Later we will be able to incorporate this result into a discussion of screening. We evaluate the exchange energy in Eq. (15.32) by using Eq. (15.30) for the electron self-energy to obtain

$$E_X = -\frac{e^2 k_F^4 L^3}{\pi^3} \cdot \int_0^1 \left[\frac{1}{2} + \frac{(1 - x^2)}{4x} \ln \left| \frac{1 + x}{1 - x} \right| \right] x^2 dx \tag{15.33}$$

which yields the following simple result

$$E_X = N \cdot \left(\frac{-3e^2 k_F}{4\pi} \right), \tag{15.34}$$

where N is the total electron number. The physical origin of this exchange energy is the fact that each electron is surrounded by an **"exchange hole"**–a small region with radius $\sim 1/k_F$ which the other electrons avoid because of the Pauli exclusion principle and the anti-symmetry requirement on the wave function. This keeps the electrons apart and lowers the total Coulomb interaction energy.

It is conventional in the electron-gas literature to express the density in terms of a **dimensionless quantity** r_s, rather than the Fermi wave vector. r_s is defined as the radius (in Bohr-radius units) of a sphere containing one electron:

$$n = \frac{3}{4\pi r_s^3 a_B^3} = \frac{k_F^3}{3\pi^2} \qquad \Rightarrow \qquad (k_F r_s a_B)^3 = \frac{9\pi}{4}. \tag{15.35}$$

Defining $\alpha = (4/9\pi)^{1/3} \approx 0.52106$, we find that the exchange energy per electron (in Rydbergs) is

$$\frac{E_X}{N} = -\left(\frac{e^2}{2a_B} \right) \left[\frac{1}{r_s} \cdot \frac{3}{2\pi\alpha} \right] \simeq -\frac{0.916}{r_s} \text{ Ryd.} \tag{15.36}$$

This result should be compared with the kinetic energy per electron. An elementary calculation gives

$$\frac{T}{N} \simeq \frac{2.21}{r_s^2} \text{ Ryd.} \tag{15.37}$$

Note that interactions have a relatively larger importance when r_s is larger, i.e. when the density is *lower*. This seems surprising at first, but is a consequence of the long-range interactions between electrons and the Pauli exclusion principle. The importance of interactions in the ground state is measured with respect to that of the kinetic energy, which is necessarily present because of the Pauli exclusion principle. Since each particle occupies a volume of radius $r_s a_B$, the Heisenberg uncertainty principle says that the typical momentum of each particle is $\sim \hbar/(r_s a_B)$. This is the physics behind the result for the kinetic energy per electron quoted above (see Exercise 15.6). If the interaction energy goes down with distance more slowly than $\sim 1/r^2$, interactions will be more important when the density is low and less important when the density is high. For electrons with Coulomb interactions, interactions are of lesser importance in high-density electron systems where the kinetic energy dominates.

> **Exercise 15.5.** Consider an electron gas in 1D, with density n, and let the electrons interact with each other through a delta-function interaction: $V(x_1, x_2) = V_0 \delta(x_1 - x_2)$. Use the Hartree–Fock approximation to calculate the kinetic energy and the interaction energy of the ground state per electron for the following two cases: (i) the electrons are spinless fermions; and (ii) the electrons are spin-1/2 fermions. What does this (approximate) result predict about the possibility of ferromagnetism when $V_0 > 0$?

> **Exercise 15.6.** Show that the (dimensionless) Hamiltonian of the jellium model can be written as
>
> $$H = \sum_i \left[-\nabla_i^2 + r_s \left(\frac{4}{9\pi} \right)^{1/3} \left(\sum_{j \neq i} \frac{1}{|\vec{r}_i - \vec{r}_j|} - \frac{2}{3\pi^2} \int \frac{d\vec{r}}{|\vec{r}|} \right) \right],$$
>
> by taking the energy and length units to be ϵ_F and $1/k_F$. This shows the relative importance of the Coulomb and kinetic terms is proportional to r_s.

15.4 Density Functional Theory

While already representing dramatic simplifications of the original interacting electron problem, solutions of the Hartree or Hartree–Fock equations can nevertheless still be computationally demanding. Furthermore, their reliability is often hard to assess.[10] There are two approaches to "improve" on Hartree or Hartree–Fock approximation, without sacrificing the conceptual simplicity of using single-particle language to describe interacting systems. One is an exact reformulation of the orginal problem known as density functional theory (DFT), to be discussed here, while the other is a phenomenological description known as **Landau's Fermi-liquid theory** (to be discussed in Section 15.11).

As long as we are dealing with electrons, the kinetic term $\hat{T}$ and the Coulomb term $\hat{U}$ are always the same in the Hamiltonian $\hat{H}$ (see Eq. (15.1)). Different systems (ranging from individual atoms and molecules to the most complex condensed matter system) differ only in the external potential term, $\hat{V}$. It would thus be highly desirable to "solve" the $\hat{T}$ and $\hat{U}$ terms under the most general setting first, and then tackle the specific problem by adding the $\hat{V}$ term corresponding to the system of interest.[11] Note that the $\hat{V}$ term can be expressed in terms of the density operator $\hat{n}(\vec{r})$,

$$\hat{V}_{\text{ext}} = \sum_{i=1}^{N} v_{\text{ext}}(\vec{r}_i) = \int d\vec{r} \, v_{\text{ext}}(\vec{r}) \hat{n}(\vec{r}), \tag{15.38}$$

and thus its expectation value is sensitive only to the density expectation value $n(\vec{r}) = \langle \hat{n}(\vec{r}) \rangle$:

$$\langle \hat{V}_{\text{ext}} \rangle = \int d\vec{r} \, v_{\text{ext}}(\vec{r}) n(\vec{r}). \tag{15.39}$$

This suggests trying to formulate the problem in terms of $n(\vec{r})$, instead of the wave function. This turns out to be possible, thanks to a couple of innocent-looking theorems proved by Hohenberg and Kohn in the 1960s [125]. We now state, discuss, and prove these theorems.

---

[10] As mentioned earlier, the Hartree–Fock approximation corresponds to (a set of) lowest-order (in the Coulomb interaction strength) perturbative corrections to the non-interacting electron problem. Unfortunately the Coulomb interaction is *not* weak in most cases.

[11] We neglect spin–orbit coupling here.

Theorem 1. *The ground-state of a many-electron system is a functional of the electron density.*

Saying that the many-body ground state is a functional of the ground state electron density is just a mathematical way of saying that two ground state wave functions must be identical if the expectation values of the density operator are identical for these wave functions. That is, there is a one-to-one correspondence between the ground state wave function and its density expectation value. The proof proceeds by *reductio ad absurdum*, i.e. by assuming that the statement is false and showing that this leads to a contradiction.

Proof: Let $|\Psi\rangle$ and $|\Psi'\rangle$ be two different many-body wave functions which are the ground states for different external potentials v_{ext} and v'_{ext}. Assume $\langle\Psi|\hat{n}(\vec{r})|\Psi\rangle = \langle\Psi'|\hat{n}(\vec{r})|\Psi'\rangle$. Then

$$E' = \langle\Psi'|H'|\Psi'\rangle < \langle\Psi|H'|\Psi\rangle \tag{15.40}$$

$$< \langle\Psi|H|\Psi\rangle + \langle\Psi|(H'-H)|\Psi\rangle \tag{15.41}$$

$$< E + \langle\Psi|(V'-V)|\Psi\rangle. \tag{15.42}$$

The inequality is introduced into the above relations by the variational theorem because $|\Psi\rangle$ is *not* the ground state of H'. Interchanging primed and unprimed quantities in the above relation leads to a similar inequality,

$$E < E' + \langle\Psi'|(V-V')|\Psi'\rangle, \tag{15.43}$$

and combining the two inequalities leads to

$$E' + E < E + E' + [\langle\Psi|(V'-V)|\Psi\rangle + \langle\Psi'|(V-V')|\Psi'\rangle]. \tag{15.44}$$

Using the fact that $|\Psi\rangle$ and $|\Psi'\rangle$ have the same density, the terms in the brackets cancel out, leaving us with the contradiction

$$E' + E < E + E'. \tag{15.45}$$

Therefore the assumption that the two states have the same density must be incorrect. The theorem follows. Note that this line of argument assumes that the ground state is non-degenerate.[12]

Theorem 2. *The ground state energy may be expressed as a functional of the density and this functional is minimized by the true ground state density.*

Proof: From the theorem above we know that, for any density $n(\vec{r})$ that corresponds to the ground state of some external potential, there is a corresponding ground state $|\Psi[n]\rangle$. We can thus define an **energy functional**

$$E[n] \equiv E_{\text{int}}[n] + \int d\vec{r}\; v_{\text{ext}}(\vec{r})n(\vec{r}), \tag{15.46}$$

$$E_{\text{int}}[n] \equiv \langle\Psi[n]|T+U|\Psi[n]\rangle. \tag{15.47}$$

Let $n(\vec{r})$ be the true ground state density for the external potential $v_{\text{ext}}(\vec{r})$. Let $n'(\vec{r})$ be some other density and $|\Psi'[n]\rangle$ be the corresponding ground state. Then

$$E[n'] = \langle\Psi[n']|T+U|\Psi[n']\rangle + \int d\vec{r}\; v_{\text{ext}}(\vec{r})n'(\vec{r}) \tag{15.48}$$

$$= \langle\Psi[n']|H|\Psi[n']\rangle > E[n]. \tag{15.49}$$

Again the inequality follows from the variational theorem because $|\Psi[n']\rangle$ is *not* the ground state of the Hamiltonian H. We thus find that $E[n]$ is minimized by the ground state density $n(\vec{r})$.

---

[12] Generalization to the case with ground state degeneracies is possible, but will not be discussed here.

In the above we *assumed* that $n(\vec{r}\,)$ can be realized as the ground state density of some external potential V. Such densities are said to be **V-representable**. It is easy to show, however, that there exist $n(\vec{r}\,)$ that are *not* V-representable. The simplest examples are densities of single-particle systems that vanish somewhere; since ground state wave functions are nodeless, such densities cannot be from ground states. Thus the domain of the functional defined in Eqs. (15.46) and (15.47) is restricted to V-representable densities. Except for the case with a single electron, this domain is actually not known; namely, we do not know all the densities that are V-representable. Levy and Lieb extended the domain of the energy functional by working with densities that correspond to *any* N-electron wave function $|\Psi\rangle$, not necessarily a ground state of some V. Such densities are said to be **N-representable**. They define the internal part of the energy functional as

$$E_{\text{int}}[n] = \min_{|\Psi\rangle \to n(\vec{r}\,)} \langle \Psi | T + U | \Psi \rangle, \qquad (15.50)$$

where $\min_{|\Psi\rangle \to n(\vec{r}\,)}$ means minimization with respect to all $|\Psi\rangle$s that give rise to the density $n(\vec{r}\,)$. Since a V-representable density must be N-representable but not vice versa, the energy functional defined this way has a much bigger domain. For the rest of the chapter we will assume the densities we encounter are N-representable, so that the corresponding energy functional exists.

The above two theorems, simple as they may look, are profound and highly counter-intuitive. They claim that all the complexity of the N-variable, complex ground state wave function $|\Psi\rangle$ (with N ranging from 1 to any *arbitrarily large* integer) is fully encoded in a single-variable real function $n(\vec{r}\,)$. But there is no free lunch; the underlying complexity must have been relegated to the complicated relation between $|\Psi[n]\rangle$ and $n(\vec{r}\,)$, which is *not* known except for $N = 1$, where $|\Psi[n]\rangle = \sqrt{n(\vec{r}\,)}$.[13] Note further that the intrinsic part of the functional in Eq. (15.46), $E_{\text{int}}[n]$, is *universal* for electronic systems. It has the same form for atomic hydrogen and crystalline $NbSe_2$ and everything in between. This great generality means that this functional must be quite complicated and subtle. As is often the case, something of such generality is of great conceptual importance, but useless in practice. In order to make practical use of the (unknown) density functional, we have to make (often rather simple and naive) approximations to it. An important step in making density functional theory useful for a wide variety of electronic systems was made by Kohn and Sham, who used a trick which allowed the kinetic-energy part of this functional, which is often the most important part, to be treated essentially exactly.

15.5 Kohn–Sham Single-Particle Equations

As discussed above, it is quite difficult to calculate $E_{\text{int}}[n]$ for electrons with Coulomb interactions. We can, however, consider the much simpler problem of *non-interacting* electrons, for which the Hohenberg–Kohn theorems also apply, but with a *different* intrinsic energy functional $E_{\text{int}}[n]$ which we do know how to calculate (in principle). Let $|\Psi_S[n]\rangle$ be the single Slater determinant which is the ground state for the case of non-interacting electrons. The total electron density for an N-electron system is (for the special case of a single Slater determinant) the sum over the contributions from the N occupied single-particle orbitals:

$$n(\vec{r}\,) = \sum_{i,\sigma} \underbrace{n_{i,\sigma}}_{\text{occupation numbers}} |\varphi_i(\vec{r}\,)|^2. \qquad (15.51)$$

---

[13] Here we have used the fact that the ground state wave function is real and nodeless, and can thus be chosen to be positive definite. Obviously such a simple relation does not exist for excited states. This can be used as a counter-example for attempts to formulate a similar density-based description for excited states.

The occupation numbers above have the value 1 or 0 and sum to N. If the electron system is not spin-polarized, each orbital is occupied twice. (We assume that there is no spin-polarization at present.) Since $U = 0$, $E_{\text{int}}[n]$ has contributions from T only and hence can be calculated straightforwardly:

$$T_{\text{S}}[n] = \langle \Psi_{\text{S}}[n] | T | \Psi_{\text{S}}[n] \rangle = \sum_{i,\sigma} n_{i,\sigma} \, \langle \varphi_i | - \frac{\hbar^2 \nabla^2}{2m} | \varphi_i \rangle. \tag{15.52}$$

This quantity is the internal energy functional for a fictional system of non-interacting electrons.

Kohn and Sham [126] separated the total energy functional for interacting electrons into pieces in the following way:

$$E_{\text{int}}[n] = \langle \Psi_{\text{S}}[n] | T | \Psi_{\text{S}}[n] \rangle + E_{\text{H}}[n] + E_{\text{xc}}[n]$$
$$\equiv T_{\text{S}}[n] + E_{\text{H}}[n] + E_{\text{xc}}[n], \tag{15.53}$$

where

$$E_{\text{H}}[n] = \frac{1}{2} \int d\vec{r} \int d\vec{r}\,' \, n(\vec{r}) \frac{e^2}{|\vec{r} - \vec{r}\,'|} n(\vec{r}\,') \tag{15.54}$$

is the Hartree contribution to the Coulomb interaction energy of $|\Psi_{\text{S}}[n]\rangle$, and $E_{\text{xc}}[n]$ is called the **exchange-correlation energy functional** and is defined by the above equation. Here T_{S} is the kinetic energy of a fictional non-interacting electron system with the same density $n(\vec{r})$ as the interacting system being studied. Note that the above definitions assume that every possible ground state density function which occurs for an interacting electron system also occurs for the (fictional) non-interacting electron system, presumably in the presence of an external potential which is in general different from that of the interacting electron system. Note that we have *not* made any approximation at this point.

The exchange-correlation energy functional above is defined with deceptive simplicity as the difference between the exact functional and the sum of $T_{\text{S}}[n]$ and $E_{\text{H}}[n]$. It is also common to further separate the exchange correlation energy into exchange ($E_{\text{X}}[n]$) and correlation ($E_{\text{c}}[n]$) contributions. The exchange energy functional is defined as the Hartree–Fock exchange energy functional, so that when the correlation contribution is neglected we recover the Hartree–Fock approximation. The correlation energy functional is then defined as the difference between the exact functional and the Hartree–Fock approximation to the exact functional. We will not perform such a separation here.

The possible charge densities are determined by the possible occupied orbitals, thus minimizing the total energy with respect to density is equivalent to minimization with respect to the single-particle orbitals subject to the constraint of orthonormality. Doing that gives rise to the widely used **Kohn–Sham single-particle equations** [126]:

$$\left[\frac{-\hbar^2 \nabla^2}{2m} + v_{\text{ext}}(\vec{r}) + v_{\text{H}}(\vec{r}) + v_{\text{xc}}(\vec{r}) \right] \varphi_i(\vec{r}) = \epsilon_i \varphi_i(\vec{r}), \tag{15.55}$$

where

$$v_{\text{H}}(\vec{r}) = e^2 \int d\vec{r}\,' \frac{n(\vec{r}\,')}{|\vec{r} - \vec{r}\,'|} \tag{15.56}$$

and v_{xc} is defined by the functional derivative (see Appendix H)

$$v_{\text{xc}}(\vec{r}) = \frac{\delta E_{\text{xc}}[n]}{\delta n(\vec{r})}. \tag{15.57}$$

The single-particle eigenvalues in these equations again originate from the Lagrange multipliers used to enforce the normalization constraint, just as in the Hartree–Fock approximation. For the Hartree and exchange-correlation energy terms the chain rule is used to take the functional derivative with respect to $\bar{\varphi}_i(\vec{r})$, so these terms are the product of the functional derivative with respect to the density and the functional derivative of the density with respect to $\bar{\varphi}_i(\vec{r})$ which is $\varphi_i(\vec{r})$. We note that the

Kohn–Sham equation (15.55) resembles the Hartree equation (15.11), and is *much simpler* than the Hartree–Fock equation (15.24), since there is no non-local potential term in it. Most amazingly, we still have not yet made any approximations!

These equations have proven to be tremendously useful in understanding the properties of many electronic systems. However, in order to use them we need to find a practical approximation for the exchange-correlation energy functional. We again emphasize that all of the complex correlated many-body physics of the interacting electrons is hidden in $v_{xc}(\vec{r})$, which is a functional of the density distribution in the ground state. The most commonly used approximation to this "infinitely complex" functional is discussed below.

15.6 Local-Density Approximation

In order to solve the Kohn–Sham equations it is necessary to make some approximation for the "exchange-correlation" energy functional $E_{xc}[n]$. This functional is not known exactly – indeed, it must be fantastically complex because it contains all of the underlying physics of every possible complex quantum ground state of every possible material! The most commonly used approximation is the **local-density approximation (LDA)** [126]

$$E_{xc}[n] \simeq \int d\vec{r}\, n(\vec{r}) \epsilon_{xc}(n(\vec{r})) \equiv E_{xc}^{LDA}[n], \tag{15.58}$$

where $\epsilon_{xc}(n)$ is the exchange-correlation energy per particle for the jellium model of a uniform (interacting) electron gas with density n ($n\epsilon_{xc}(n)$ is then the exchange-correlation energy per unit volume).[14] In the local-density approximation, the exchange-correlation energy density at each point in space is approximated by the exchange-correlation energy density of a uniform-density electron system with that density. It can be regarded as the first term in an expansion of the exchange-correlation energy in gradients of the electron density. In the local-density approximation

$$v_{xc}(\vec{r}) = \mu_{xc}(n(\vec{r})), \tag{15.59}$$

where $\mu_{xc}(n) \equiv (d/dn)(n\epsilon_{xc}(n))$. If the Hartree–Fock approximation is used for $\epsilon_{xc}(n)$ (the exchange-only local-density approximation) then

$$\mu_{xc}(n) = \frac{4}{3}\epsilon_{xc}(n) \simeq \Sigma_{HF}(k_F) = \frac{-e^2 k_F}{\pi} = \frac{-e^2 (3\pi^2)^{1/3}}{\pi} \cdot n^{1/3}. \tag{15.60}$$

This approximation, in which an attractive interaction proportional to $[n(\vec{r})]^{1/3}$ is used to approximately correct the Hartree approximation, has a long history in condensed matter physics going back to early work by Slater.

The ground state energy of the uniform-density (jellium model) electron system is not known exactly. There has been a long history of work aimed at calculating this quantity more accurately, using perturbative, variational, and non-perturbative quantum Monte Carlo methods which are beyond the scope of this book. The last approach is arguably the most accurate. While $\epsilon_{xc}(n)$ is still not known exactly, it is clear that in almost every instance the accuracy of calculations applying the local-density approximation is limited not by the relatively small uncertainty in this function, but by the local-density approximation itself. In many systems, some improvement to the local-density approximation can be achieved by invoking gradient corrections, usually in a partially phenomenological way.

---

[14] Note that $\epsilon_{xc}(n(\vec{r}))$ is a *function* of the density. The argument of the function is the local value of the density, $n(\vec{r})$. $\epsilon_{xc}(n(\vec{r}))$ is not a *functional* of the full *function* $n(\vec{r})$. A simple illustrative example of a non-local *functional* of the density is $f[n] = \int d^3r\, f_0(\vec{r})n(\vec{r}) + \int d^3r\, d^3r'\, f_1(\vec{r}, \vec{r}')n(\vec{r})n(\vec{r}')$.

With an explicit (but approximate) expression for the exchange-correlation functional in hand, one can calculate the ground state energy and density by minimizing the electron energy functional. Such calculations starting from microscopic Hamiltonians like Eq. (7.26) are often referred to as ***ab initio*** **or first-principles calculations**. A considerable amount of important information is accessible from such calculations, and we discuss some of them below.

• Structural determination. Given a set of atoms, what kind of crystalline structure will they form? To answer this question, we need to calculate the total energy of the system that includes the electronic terms in Eq. (7.26) *and* the Coulomb repulsion between nuclei, as a function of the nuclear positions:

$$E_{\text{tot}}(\{\vec{R}_I\}) = E(\{\vec{R}_I\}) + \sum_{I<J} \frac{Z_I Z_J e^2}{|\vec{R}_I - \vec{R}_J|}, \tag{15.61}$$

where $\vec{R}_I$ is the position of the Ith nucleus, Z_I is its charge, and $E(\{\vec{R}_I\})$ is the ground state energy of the electronic Hamiltonian (7.26) obtained by minimizing the energy functional $E[n]$. The crystalline structure (at zero temperature) is determined by minimizing $E_{\text{tot}}(\{\vec{R}_I\})$ with respect to $\{\vec{R}_I\}$. The crystalline structure can be experimentally measured using elastic X-ray scattering, as we discussed in Chapter 3.

• Cohesive energy. The minimum obtained above is the total energy of the crystal. It is usually measured in terms of the cohesive energy of the crystal, namely the energy difference between the crystal energy and the energy of isolated atoms:

$$E_{\text{coh}} = E_{\text{crystal}} - \sum_{\text{atom}} E_{\text{atom}}. \tag{15.62}$$

Negative cohesive energy means that the system lowers its energy by forming a crystal from individual atoms, and its absolute value is the net energy reduction due to crystallization.[15]

• Elastic constants and phonon spectra. The set of nuclear positions $\{\vec{R}_I\}$ that minimizes E_{tot} gives rise to equilibrium positions of the nuclei. Taking the second derivative of $E_{\text{tot}}(\{\vec{R}_I\})$ yields the elastic constants of the crystal introduced in Chapter 5, as well as the elastic tensor needed to calculate phonon spectra (including both acoustic and optical phonons).

• Phase diagram under pressure. The calculations described above assume the crystal lives in vacuum. In reality it is under atmospheric pressure, which generally has negligible effect on its structure and other properties. However, we sometimes study crystals under high pressure for the purpose of changing their overall density, structure, and electronic properties.[16] In the presence of pressure P and at finite temperature T, what needs to be minimized is the Gibbs free energy

$$G = E_{\text{tot}}(T) + PV - TS, \tag{15.63}$$

where S is entropy[17] and V is volume. At sufficiently low T we may neglect the last term as well as the difference between $E_{\text{tot}}(T)$ and the ground state energy $E_{\text{tot}}(\{\vec{R}_I\})$. Since the latter is accessible from DFT for all $\{\vec{R}_I\}$ (and thus any volume V), one can determine the crystalline structure under any pressure, including locations of structural phase transitions induced by pressure.

It is clear from the discussions above that a large amount of physical information is accessible from DFT-based *ab initio* calculations. We now spell out a few important caveats. The Kohn–Sham

---

[15] Some authors use the reverse sign convention in defining the cohesive energy.

[16] Experiments carried out with tiny samples in diamond anvils can reach enormous pressures of interest, for example, to geophysicists studying the earth's interior.

[17] Computation of the entropy involves the entropy of the phonons (which may need to be computed quantum mechanically) and the entropy of any low-energy electronic excitations.

scheme for applying the density functional formalism is built upon the strategy of treating the one-particle kinetic-energy term as accurately as possible, and doing the best one can with interaction terms. The resultant single-particle orbitals and corresponding eigenvalues are commonly understood (in the case of crystalline solids) as Bloch bands and band dispersions. However, just as in the Hartree–Fock approximation, these eigenvalues are Lagrange multipliers, whose interpretation as energies needs to be taken with a grain of salt. Furthermore, unlike in the case of the Hartree–Fock approximation, here the single-particle orbitals are *auxiliary* variables used to build up the density, which is the only fundamental quantity in DFT. As a result the physical meaning of the single-particle orbitals is not immediately clear either. Nevertheless, their interpretation as Bloch bands successfully explains the physical properties of many condensed matter systems. It is important to realize, however, that there are electronic systems where interactions play such a dominant role that this scheme fails completely. An especially important class of materials where the Kohn–Sham approach fails is the Mott insulator class. These materials would be predicted to be metals by Kohn–Sham calculations, but in fact are insulators (because the Coulomb repulsion among the electrons is so strong that they cannot move freely). We will discuss Mott insulators later, in Chapter 17, in connection with magnetism.

We close our discussion on DFT-based *ab initio* or first-principles calculations by stating that they apply to *all* electronic systems, not just crystalline solids. In fact, they played such a big role in determining structures of molecules (made of a handful instead of $\sim 10^{23}$ atoms) and studies of their properties, that Walter Kohn was awarded the Nobel prize in chemistry in 1998. In condensed matter physics, DFT is widely used not only to determine crystalline structures from first principles, but also to determine complex structures involving surfaces, interfaces, defects, and clusters etc. It has also been generalized to relativistic electrons, and systems with other forms of interactions.

15.7 Density–Density Response Function and Static Screening

We mentioned previously that the Hartree–Fock approximation for the electron gas (in the jellium model) has some special features, notably zero density of states at the Fermi energy, because of the long-range of the Coulomb interaction. These features of the Hartree–Fock approximation are not correct. For most purposes the long-range Coulomb interaction should be replaced by a screened interaction, which is defined as the interaction between two electrons together with the distortions of the charge density in the vicinity of each (due to the response of the other electrons to the presence of the two electrons in question). Below we discuss the screening of the interaction between one electron and another and between electrons and external charges. We start, though, by considering the screening of any external potential. We couch our discussion of screening in terms of DFT, which is perfectly suited for this purpose because it is formulated in terms of density – and it is the density which determines the Coulomb interactions. For simplicity, we consider the jellium model of the electron gas for this section and (most of) the rest of this chapter, where all eigenstates are labeled by a momentum quantum number defined in the *entire* momentum space (not just within the first Brillouin zone), and all response functions can be expressed as functions of a single wave vector and a frequency.

Consider an external potential $v_{\text{ext}}(\vec{r})$, for example the potential from an external charge $-Z$ located at the origin in an otherwise uniform electron gas. We limit our attention to the case where the external potential is weak enough that we can calculate the induced density in leading order, where it is proportional to the strength of the external potential, i.e. we will derive the linear response. When the charge-density change is small, we can expand the density around its uniform value. This gives rise to linear screening theory.

For the jellium model in the presence of a perturbing external potential $v_{ext}(\vec{r})$ we have

$$E[n] = E_{int}[n] + \int d\vec{r}\, n(\vec{r})v_{ext}(\vec{r}); \tag{15.64}$$

$$E_{int}[n] = E_{int}[n_0] + \frac{1}{2}\int d\vec{r}\int d\vec{r}'\,\delta n(\vec{r})\left.\frac{\delta^2 E_{int}[n]}{\delta n(\vec{r})\delta n(\vec{r}')}\right|_{n_0}\delta n(\vec{r}') + \cdots. \tag{15.65}$$

The linear term in the Taylor-series expansion of the internal energy functional can be dropped because the uniform density function $n(\vec{r}) = n_0$ must be an extremum of the internal energy functional.[18] Minimizing with respect to $n(\vec{r})$ gives[19]

$$\frac{\delta E[n]}{\delta n(\vec{r})} = \int d\vec{r}'\,\left.\frac{\delta^2 E_{int}[n]}{\delta n(\vec{r})\delta n(\vec{r}')}\right|_{n_0}\delta n(\vec{r}') + v_{ext}(\vec{r}) = 0 \tag{15.66}$$

$$\Rightarrow \quad \delta n(\vec{r}) = \int d\vec{r}'\,\chi(\vec{r},\vec{r}')v_{ext}(\vec{r}'), \tag{15.67}$$

with

$$\chi(\vec{r},\vec{r}') = -\left[\left.\frac{\delta^2 E_{int}[n]}{\delta n(\vec{r})\delta n(\vec{r}')}\right|_{n_0}\right]^{-1} \tag{15.68}$$

being the static density–density response function. Note that the inverse above is a (functional version of) matrix inversion defined so that

$$\int d\vec{r}\,\chi(\vec{x},\vec{r})\chi^{-1}(\vec{r},\vec{x}') = \delta(\vec{x} - \vec{x}'). \tag{15.69}$$

In a translationally invariant uniform electron gas these quantities can actually depend only on $(\vec{r} - \vec{r}')$. Hence we define the momentum-space density response function

$$\chi(\vec{q}) \equiv \int d\vec{r}\,\chi(\vec{r})e^{-i\vec{q}\cdot\vec{r}} \Rightarrow \chi(\vec{r}) = \frac{1}{\Omega}\sum_{\vec{p}}e^{i\vec{p}\cdot\vec{r}}\chi(\vec{p}), \tag{15.70}$$

where Ω is the sample volume, such that

$$\delta n(\vec{q}) = \int d\vec{r}\,e^{-i\vec{q}\cdot\vec{r}}\,\delta n(\vec{r}) = \int d\vec{r}\int d\vec{r}'\,e^{-i\vec{q}\cdot\vec{r}}\chi(\vec{r},\vec{r}')v_{ext}(\vec{r}')$$

$$= \chi(\vec{q})v_{ext}(\vec{q}). \tag{15.71}$$

Also from Eq. (15.69) we have

$$\chi^{-1}(\vec{p}) = 1/\chi(\vec{p}). \tag{15.72}$$

These relations tell us that, in the presence of translation invariance, physical quantities with different wave vectors decouple, and matrix relations simplify to arithmetic relations. Note that the wave-vector dependence of the susceptibility tells us that the density response is non-local in real space. In the presence of translation invariance, the easiest way to perform the functional inverse in Eq. (15.69) is to go to reciprocal space, use Eq. (15.72), and then transform back to real space.

---

[18] We use the fact that, since the system has translation symmetry in the absence of any external potential, the ground state electron density must be uniform.

[19] Note the rather direct analogy here to the mechanical response of a spring to an external force. The internal energy of the spring is quadratic in the displacement, while the external force contributes an energy that is linear in the displacement. Minimizing the total energy gives a displacement that is linear in the force and inversely proportional to the spring constant.

Equation (15.68) is formally exact. To make progress, however, we need explicit expressions for $E_{\text{int}}[n]$, which will inevitably involve approximations. To this end, we use Eq. (15.53) and write

$$E_{\text{int}}[n] = \frac{1}{2} \int d\vec{r} \int d\vec{r}\,' \, n(\vec{r}) \frac{e^2}{|\vec{r} - \vec{r}\,'|} n(\vec{r}\,') + T_S[n] + E_{\text{xc}}[n]. \tag{15.73}$$

For non-interacting electrons only $T_S[n]$ would be present, and we have

$$\left. \frac{\delta^2 T_S[n]}{\delta n(\vec{r})\delta n(\vec{r}\,')} \right|_{n_0} = -\chi_0^{-1}(\vec{r}, \vec{r}\,') = -\frac{1}{\Omega} \sum_{\vec{p}} \chi_0^{-1}(\vec{p}) e^{i\vec{p}\cdot(\vec{r}-\vec{r}\,')}, \tag{15.74}$$

where χ_0 is the static density–density response function of the *non-interacting* electron gas, which we can readily calculate from first principles. Combining this with the fact that the Fourier transform of $1/r$ is $4\pi/q^2$ in 3D, we obtain from the Fourier transform of (the inverse of) Eq. (15.68)

$$-\chi^{-1}(\vec{q}) = \frac{4\pi e^2}{q^2} - \chi_0^{-1}(q) + F_{\text{xc}}(q) \equiv +\frac{4\pi e^2}{q^2} - \Pi^{-1}(q), \tag{15.75}$$

where $F_{\text{xc}}(q)$ comes from the functional derivative of $E_{\text{xc}}[n]$, and $\Pi(q)$ is called the **polarization function** and is defined by the above equation. In most approximations we simply replace $\Pi(q)$ by $\chi_0(q)$, namely we ignore the exchange-correlation contribution and take into account only the effect of Coulomb interaction at Hartree level. But before doing that we first discuss the physical consequences of this density response.

Let us assume that v_{ext} is induced by some *external charge* brought in from the outside. Now we use an *infinitesimal test charge* to detect this potential. The interaction between this test charge and the external charge *plus* the induced charge of the electron system results in a *screened* potential:

$$v_{\text{sc}}(\vec{q}) = v_{\text{ext}}(\vec{q}) + \frac{4\pi e^2}{q^2} \chi(\vec{q}) v_{\text{ext}}(\vec{q}) \tag{15.76}$$

$$= v_{\text{ext}}(\vec{q}) \left(1 + \frac{4\pi e^2}{q^2} \chi(\vec{q}) \right) \tag{15.77}$$

$$= v_{\text{ext}}(\vec{q}) \left(1 - \frac{4\pi e^2/q^2}{(4\pi e^2/q^2) - \Pi^{-1}(q)} \right) \tag{15.78}$$

$$= v_{\text{ext}}(\vec{q}) \left(\frac{1}{1 - (4\pi e^2/q^2)\Pi(q)} \right) \equiv \frac{v_{\text{ext}}(\vec{q})}{\epsilon(q)}. \tag{15.79}$$

This equation defines the **static dielectric function**

$$\epsilon(q) = 1 - \frac{4\pi e^2}{q^2} \Pi(q), \tag{15.80}$$

which dictates how external charge is screened by an electric medium. For the case where the external potential comes from a charge $-Z$ at the origin we have

$$v_{\text{sc}}(\vec{q}) = \frac{4\pi Z e^2}{q^2 - 4\pi e^2 \Pi(q)} \approx \frac{4\pi Z e^2}{q^2 + q_{\text{sc}}^2}, \tag{15.81}$$

where in the last step we approximated $\Pi(q)$ by $\Pi(q = 0)$, and introduced

$$q_{\text{sc}}^2 \equiv -4\pi e^2 \Pi(q = 0). \tag{15.82}$$

Fourier transforming back to real space yields

$$v_{\text{sc}}(r) = \frac{Z e^2 e^{-q_{\text{sc}} r}}{r}. \tag{15.83}$$

We see that the original $1/r$ Coulomb potential is suppressed by an exponential factor due to screening, with a screening length of $1/q_{\text{sc}}$.

15.7.1 Thomas–Fermi Approximation

We now make the approximation $\Pi(q) \approx \chi_0(q)$, and take the long-wavelength limit $q \to 0$. This results in the so-called **Thomas–Fermi approximation**:

$$\chi(q) \approx \chi_{\mathrm{TF}}(q) = \frac{\chi_0(q=0)}{1 - (4\pi e^2/q^2)\chi_0(q=0)} \tag{15.84}$$

and

$$\epsilon(q) \approx \epsilon_{\mathrm{TF}}(q) = 1 - \frac{4\pi e^2}{q^2}\chi_0(q=0) = 1 + \frac{q_{\mathrm{TF}}^2}{q^2}, \tag{15.85}$$

where $q_{\mathrm{TF}}^2 \equiv -4\pi e^2 \chi_0(q=0) = 4k_{\mathrm{F}}/\pi a_{\mathrm{B}}$ and a_{B} is the Bohr radius.

In this case we are assuming that $v_{\mathrm{ext}}(\vec{r})$ is a slowly varying function, and can be taken as a *constant* locally. Its effect is simply changing the *local* chemical potential from the global chemical potential μ to $\mu - v_{\mathrm{ext}}(\vec{r})$. We thus find

$$n(\vec{r}) = n_0(\mu - v_{\mathrm{ext}}(\vec{r})), \tag{15.86}$$

where $n_0(\mu) = (2m\mu/\hbar^2)^{3/2}/3\pi^2$ is the free-electron density in 3D, and we find

$$\chi_0(q=0) = -\frac{\partial n_0}{\partial \mu} = -\frac{mk_{\mathrm{F}}}{\pi^2\hbar^2}. \tag{15.87}$$

Thus, within the Thomas–Fermi approximation, we have for the screening wave vector

$$q_{\mathrm{sc}}^2 \approx q_{\mathrm{TF}}^2. \tag{15.88}$$

We note that one can actually go beyond the linear-response case within the Thomas–Fermi approximation, because Eq. (15.86) assumes only that $v_{\mathrm{ext}}(\vec{r})$ is *smooth*, not that it has to be *small*.

15.7.2 Lindhard Approximation

In the **Lindhard approximation** we again use the approximation $\Pi(q) \approx \chi_0(q)$ but keep its full q dependence in Eqs. (15.75) and (15.80), which yields

$$\chi(q) \approx \chi_{\mathrm{L}}(q) = \frac{\chi_0(q)}{1 - (4\pi e^2/q^2)\chi_0(q)} \tag{15.89}$$

and

$$\epsilon(q) \approx \epsilon_{\mathrm{L}}(q) = 1 - \frac{4\pi e^2}{q^2}\chi_0(q). \tag{15.90}$$

The Lindhard approximation has a very simple interpretation. Here we have

$$\delta n(\vec{q}) = \frac{\chi_0(q)v_{\mathrm{ext}}(\vec{q})}{1 - (4\pi e^2/q^2)\chi_0(q)} = \chi_0(q)v_{\mathrm{ext}}(q)/\epsilon(q) = \chi_0(q)v_{\mathrm{sc}}(\vec{q})$$
$$= \chi_0(q)\left[v_{\mathrm{ext}}(\vec{q}) + \frac{4\pi e^2}{q^2}\delta n(\vec{q})\right], \tag{15.91}$$

namely we treat the system as if it were behaving like free electrons in its density response, but it is responding not only to the external potential, but also to the potential produced by the density response itself! The combination of the external potential plus the potential due to the density response is the screened potential. This is thus essentially a (self-consistent) Hartree approximation for the effect of the interactions on the static density–density response function.

To implement the Lindhard approximation we need the free-electron static density–density response function $\chi_0(q)$. A straightforward application of the (static) linear response theory (see Appendix A) yields (in d dimensions)

$$\chi_0(q) = 4 \int \frac{d^d \vec{k}}{(2\pi)^d} \frac{f_{\vec{k}}^\circ}{(\hbar^2/2m)[\vec{k}^2 - (\vec{k}+\vec{q})^2]}$$
$$= -2 \int \frac{d^d \vec{k}}{(2\pi)^d} \frac{f_{\vec{k}-\vec{q}/2}^\circ - f_{\vec{k}+\vec{q}/2}^\circ}{(\hbar^2/m)(\vec{k}\cdot\vec{q})}, \tag{15.92}$$

where $f_{\vec{k}}^\circ$ is the electron occupation number for momentum $\hbar\vec{k}$. At zero temperature we have $f_{\vec{k}}^\circ = \theta(k_{\rm F} - k)$, in which case we can evaluate the integral above to obtain in 3D

$$\chi_0(q) = -\left(\frac{\partial n_0}{\partial \mu}\right) L\left(\frac{q}{2k_{\rm F}}\right), \tag{15.93}$$

where

$$L(x) = \frac{1}{2} + \frac{1 - x^2}{4x} \ln\left|\frac{1+x}{1-x}\right| \tag{15.94}$$

is known as the **Lindhard function**. It is easy to check that $L(x \to 0) = 1$, thus Eq. (15.93) reduces to the Thomas–Fermi result of Eq. (15.87) in the limit $q \to 0$.

It is interesting to ask what the total amount of screening charge surrounding a charged impurity is. An impurity of charge $+Ze$ produces a perturbation $v_{\rm ext} = -4\pi Ze^2/q^2$ that diverges for $q \to 0$. However, the divergence of the dielectric function at small q causes the screened potential to be non-singular. From Eq. (15.91) we see that we must have

$$\lim_{q\to 0} \delta n(q) = +Z, \tag{15.95}$$

and thus the net screening charge exactly neutralizes the impurity charge. In this sense, the screening is perfect (when viewed at large length scales or small wave vectors). This sum rule for the screening charge has important consequences for the scattering phase shifts of the electrons that are (self-consistently) scattering from the screened potential. These will be described in Section 15.14.

The other thing to notice is the singularity of $L(x)$ at $x = 1$ or $q = 2k_{\rm F}$, which originates from the discontinuity of $f_{\vec{k}}^\circ$ at the Fermi surface. One can see from Fig. 15.1, shown later in this chapter, that $2k_{\rm F}$ is special because it is the largest momentum that a particle–hole pair of zero energy can have.[20] Fourier transforming the corresponding dielectric function to real space leads us to the so-called **Friedel oscillations** of the screened potential at long distance

$$v_{\rm sc}(r) \sim \frac{1}{r^3} \cos(2k_{\rm F} r), \tag{15.96}$$

which are missed in the Thomas–Fermi approximation. On the other hand, finite temperature and disorder (when present) smooth out the discontinuity[21] of $f_{\vec{k}}^\circ$ at $k_{\rm F}$, and suppress the Friedel oscillations at sufficiently large r.

---

[20] The Fermi gas is particularly susceptible to perturbations with wave vector $|\vec{q}| = 2k_{\rm F}$ because such perturbations can scatter electrons from one side of the Fermi sphere (near $\vec{k} \sim -\vec{q}/2$) to the diametrically opposite side ($\vec{k}+\vec{q} \sim +\vec{q}/2$). It turns out that there is a large phase space for such quasi-degenerate transitions between nearly parallel faces of the Fermi surface.

[21] Finite temperature smooths out the discontinuity in occupancy that occurs at the Fermi energy. Disorder, on the other hand, smears out the momentum, but not the discontinuity in occupancy as a function of energy. In the presence of disorder, energy eigenstates have indefinite momentum, and plane-wave states are eigenstates of momentum but have indefinite energy.

> **Exercise 15.7.** Use the general results of Appendix A to verify Eq. (15.92).

> **Exercise 15.8.** Calculate the Lindhard functions (as defined in Eq. (15.93)) in 1D and 2D, and show that the singularity at $q = 2k_F$ is stronger than that of the 3D case in Eq. (15.94) in both cases. In particular, the logarithmic divergence in 1D is closely related to the (logarithmic) Peierls instability we encountered in Chapter 7.

15.8 Dynamical Screening and Random-Phase Approximation

Obviously the Lindhard or self-consistent Hartree approximation can be applied not only to the static density response, but also to *dynamical* (or frequency-dependent) density–density response functions. This leads to the dynamical Lindhard approximation, more commonly referred to as the **random-phase approximation (RPA)** in the literature, where χ, χ_0, and ϵ in Eqs. (15.89) and (15.90) all become functions of both q and frequency ω, which describe the system's response to a time-dependent external potential with wave vector q and frequency ω:

$$\chi_{\mathrm{RPA}}(q, \omega) = \frac{\chi_0(q, \omega)}{1 - (4\pi e^2/q^2)\chi_0(q, \omega)}, \tag{15.97}$$

$$\epsilon_{\mathrm{RPA}}(q, \omega) = 1 - \frac{4\pi e^2}{q^2}\chi_0(q, \omega). \tag{15.98}$$

Here we need the dynamical density response function of free electrons, which can be obtained using linear-response theory (see Appendix A):

$$\chi_0(q, \omega) = 2 \int \frac{d^d \vec{k}}{(2\pi)^d} \frac{f^{\circ}_{\vec{k}-\vec{q}/2} - f^{\circ}_{\vec{k}+\vec{q}/2}}{(\hbar^2/m)(\vec{k} \cdot \vec{q}) + \hbar\omega + i\eta}. \tag{15.99}$$

Setting $\omega = 0$ we recover the free-electron static density–density response function, Eq. (15.92). It is interesting to inspect what happens at finite frequency in the long-wavelength limit. By taking the $q \to 0$ limit at finite ω one finds (for any dimension)

$$\chi_0(q \to 0, \omega) = \frac{n_0 q^2}{m\omega^2}. \tag{15.100}$$

We thus find in 3D[22]

$$\epsilon_{\mathrm{RPA}}(q \to 0, \omega) = 1 - \frac{4\pi e^2 n_0}{m_e \omega^2} = 1 - \frac{\Omega_{\mathrm{pl}}^2}{\omega^2} \tag{15.101}$$

where

$$\Omega_{\mathrm{pl}} = \sqrt{\frac{4\pi e^2 n_0}{m_e}} \tag{15.102}$$

is known as the plasma oscillation frequency or simply the **plasma frequency**.

---

[22] It is interesting to note that, depending on which one goes to zero first, the dielectric function in the zero-wave-vector and -frequency limit can go to either $+\infty$ or $-\infty$. Such sensitivity of response functions to the order of limits is a hallmark of condensed matter systems that support gapless excitations. The characteristic "phase velocity" of the perturbation ω/q goes to zero in one limit and infinity in the other, and helps explain the differences in the response in the two limits.

Exercise 15.9. Use the general results of Appendix A to verify Eq. (15.99), and then Eq. (15.100).

15.9 Plasma Oscillation and Plasmon Dispersion

As we saw in Section 15.7, the static density response of electrons gives rise to screening, which *weakens* the effects of an external potential, especially at long distance; this is reflected in the static dielectric function $\epsilon(q) > 1$. Screening also renders the *effective* electron–electron interaction short-ranged (since $\epsilon(q \to 0) \to \infty$), and removes the singularities encountered in Section 15.3 associated with the long-range nature of the Coulomb interaction. The situation, however, can be very different at finite frequency. For $\omega > \Omega_{pl}$ the density response of the electrons may *enhance* the effects of an external potential, resulting in a dielectric function $\epsilon(q, \omega) < 1$. In the extreme case, we can have $\chi(q \to 0, \omega \to \Omega_{pl}) \to \infty$ and $\epsilon(q \to 0, \omega = \Omega_{pl}) = 0$. At the plasma frequency the response of the system to an external perturbation *diverges*, because the perturbation is in resonance with an intrinsic oscillation frequency of the system. Hence this phenomenon is also referred to as the plasma resonance in the electron gas. Physically, this implies that the system can support density oscillation at Ω_{pl} *without* any external perturbation (or drive), which means there is an excitation of the system at precisely this frequency (or corresponding energy $\hbar\Omega_{pl}$). Since this excitation involves motion of all electrons in the system, it is an example of a collective mode.[23] In this section we discuss the physics of the plasma oscillation and its finite-wave-vector version known as the **plasmon mode**, from several different perspectives.

Another highly counter-intuitive aspect of dynamical screening is that, for $\omega < \Omega_{pl}$, we can have $\epsilon < 0$, thus changing the *sign* of the screened interaction and turning it *attractive*! This is essentially the physics behind the phonon-mediated attractive interaction between electrons that is responsible for superconductivity in the so-called conventional superconductors. The main difference is that the role of the plasma frequency is replaced by the Debye frequency, as we will discuss in Chapter 20.

15.9.1 Plasma Frequency and Plasmon Dispersion from the RPA

As we learned in Section 15.8, the dielectric function (at zero wave vector) is

$$\epsilon(\omega) = 1 - \frac{\Omega_{pl}^2}{\omega^2}. \tag{15.103}$$

While we obtained the results above by performing a fairly sophisticated quantum-mechanical calculation, miraculously $\hbar$ disappears in the expression above, suggesting that these results are actually of *classical* origin. That is indeed the case, as we will discuss in the following subsection.

To obtain the plasmon dispersion, we need to expand Eq. (15.99) for small q (but keeping ω finite) to obtain $\epsilon_{RPA}(q, \omega)$ in the same regime. We will not do this explicitly here but the reader is encouraged to attempt it and show that within the RPA we have the plasmon dispersion

$$\omega_{pl}^2(q) = \Omega_{pl}^2 \left[1 + \frac{3\mu q^2}{10\pi n e^2} + O(q^4) \right] = \Omega_{pl}^2 \left[1 + \frac{9q^2}{5q_{TF}^2} + O(q^4) \right]. \tag{15.104}$$

We note that the plasmon dispersion does involve $\hbar$, which enters the relation between the chemical potential μ (or the Fermi energy) and the electron density n, and is thus a quantum-mechanical effect.

---

[23] In the same sense that phonons are collective modes as they involve motions of all atoms that make up the crystalline lattice.

15.9.2 Plasma Frequency from Classical Dynamics

In this subsection we will obtain Eq. (15.102) by treating electrons as classical particles governed by Newtonian dynamics. Suppose that there is a density fluctuation $\delta n_{\vec{q}}$. This produces an electrostatic potential

$$\varphi_{\vec{q}} = -e\frac{4\pi}{q^2}\,\delta n_{\vec{q}}. \tag{15.105}$$

The fact that this diverges for small q is a result of the long range of the Coulomb force and explains why the plasma frequency remains finite at small q rather than dispersing linearly with q as an ordinary acoustic mode would. The corresponding electric field is

$$\vec{E}_q = -i\vec{q}\,\varphi_q. \tag{15.106}$$

According to Newton's laws, this electric field produces an acceleration

$$\dot{\vec{v}} = -\frac{e}{m_e}\vec{E}, \tag{15.107}$$

which gives a current density

$$\dot{\vec{J}} = -n_0 e\dot{\vec{v}} = \frac{n_0 e^2}{m_e}\vec{E}. \tag{15.108}$$

In writing this we have used the mean density n_0 rather than the instantaneous density since we are linearizing the equation of motion and $\vec{E}$ is already of first order in δn. Using the continuity equation

$$\nabla \cdot \vec{J} - e\frac{\partial n}{\partial t} = 0 \tag{15.109}$$

and the Maxwell (Poisson) equation

$$\vec{\nabla} \cdot \vec{E} = 4\pi(-en) \tag{15.110}$$

in combination with the above results yields

$$\ddot{n}_{\vec{q}} = -\Omega_{\mathrm{pl}}^2 n_{\vec{q}}, \tag{15.111}$$

with the (plasma) oscillation frequency Ω_{pl} given in Eq. (15.102). Thus the collective density oscillation mode is dispersionless here. Clearly, however, this derivation is valid only for $q \ll n_0^{1/3}$ since we treat the system as an elastic medium without worrying about the positions of individual particles. That is, we have implicitly assumed that Eq. (15.111) represents the equation of motion for a generalized harmonic oscillator coordinate, namely $n_{\vec{q}}$ can be used as a degree of freedom instead of the particle position. As we have seen earlier, the dynamics of $n_{\vec{q}}$ does involve quantum mechanics (especially the Fermi statistics of electrons), which gives rise to the plasmon dispersion at finite wave vectors. The essential point is that, at finite wave vectors, modulation of the density requires changing the wave vector of individual electrons, which costs additional kinetic energy.

Exercise 15.10. Plasma oscillations are longitudinal (the current vector is parallel to the propagation wave vector). It turns out that the plasma frequency is also relevant for transverse electromagnetic waves. In particular, for frequencies below the plasma frequency, electromagnetic waves cannot propagate through a plasma. Show that, for very low frequencies far below the plasma frequency, Maxwell's equations reduce to

$$\vec{\nabla} \times (\vec{\nabla} \times \vec{E}) \approx -\lambda^{-2}\vec{E}, \tag{15.112}$$

where the penetration depth is given by $\lambda = c/\Omega_{\mathrm{pl}}$. Use Eq. (15.108) to show that Maxwell's equations yield an exponentially damped solution with damping length λ.

> The penetration depth for aluminum is approximately 50 nm. Use this to estimate the plasma frequency and find the corresponding photon energy (in electron volts) at that frequency. The actual plasma frequency ($\hbar\Omega_{pl} \sim 15$ eV) is lower than this because of the contribution of the atomic core electrons, which produce a background dielectric constant.

15.9.3 Plasma Frequency and Plasmon Dispersion from the Single-Mode Approximation

In Chapter 4 we introduced the single-mode approximation (SMA) to calculate (approximately) the collective mode energy of the system. There we took the ratio between the first (frequency) moment of the dynamical structure factor $S(q, \omega)$ (which is known from the f-sum rule) and the static structure factor (or the zeroth moment of $S(q, \omega)$). The SMA becomes exact when there is a single excitation that contributes to $S(q, \omega)$, or, equivalently, the f-sum rule is exhausted by a single mode. In that case we can, of course, combine any two moments of $S(q, \omega)$ to obtain the mode energy. Defining the mth moment of $S(q, \omega)$ to be

$$S_m(q) = \int_{-\infty}^{\infty} d\omega[\omega^m S(q, \omega)], \qquad (15.113)$$

we have

$$[\Omega(q)]^{m-n} = S_m(q)/S_n(q) \qquad (15.114)$$

when

$$S(q, \omega) \propto \delta[\omega - \Omega(q)], \qquad (15.115)$$

namely a single mode with frequency $\Omega(q)$ contributes to $S(q, \omega)$. Equation (15.114) leads to other forms of the SMA. For example, another frequently used form of the SMA combines the first moment and the -1th (or first inverse) moment:

$$\Omega^2(q) \approx \frac{S_1(q)}{S_{-1}(q)} = \frac{\int_{-\infty}^{\infty} d\omega[\omega S(q, \omega)]}{\int_{-\infty}^{\infty} d\omega[S(q, \omega)/\omega]}. \qquad (15.116)$$

The advantage of this form is that the first inverse moment of $S(q, \omega)$ is related to the static density–density response function (see Appendix A). On combining that with the f-sum rule, the above becomes

$$\Omega^2(q) \approx -\frac{nq^2}{m\chi(q)}. \qquad (15.117)$$

Using the Thomas–Fermi approximation for the static density–density response function (15.84), we obtain the plasmon dispersion within the SMA:

$$\omega_{pl}^2(q) \approx \Omega_{pl}^2 \left[1 + \frac{\mu q^2}{6\pi n e^2} + O(q^4)\right]. \qquad (15.118)$$

A few comments are now in order. (i) The *exact* plasma frequency is obtained in the long-wavelength ($q \to 0$) limit; this is a consequence of the fact that the plasmon mode exhausts the f-sum rule in that limit. (ii) Unlike in the RPA, here we are able to obtain the plasmon dispersion *without* calculating any finite-ω response function (i.e. using only $\chi(q, \omega = 0)$). (iii) Compared with the RPA, the SMA *underestimates* the plasmon dispersion. While (i) and (ii) demonstrate the power of the SMA, (iii) reveals its limitation, namely it actually gives a weighted average of energies from all excitations at a given q that contribute to $S(q, \omega)$. In the present case, in addition to the plasmon mode, there are also particle–hole excitations at much lower energies (see Fig. 15.1(b)) that contribute to $S(q, \omega)$; their presence drags down the energy obtained from the SMA.

> **Exercise 15.11.** Think about how Eq. (15.117) resembles the expression for the frequency of a harmonic oscillator $\Omega^2 = K/m$ in terms of the ratio of the spring constant to the mass. Recall that the static suspectibility of the harmonic oscillator is inversely proportional to the spring constant.

> **Exercise 15.12.** Use the RPA and the SMA to obtain the plasmon dispersion of a 2D electron gas. Note that the 2D Fourier transformation of $1/r$ is $2\pi/q$.

15.10 Dielectric Function and Optical Properties

We have seen that electron–electron Coulomb interactions have very important effects on the dielectric function, which describes the response of the electrons to an external electric potential. For the jellium model described in previous sections, the most important response comes from the plasma mode. In more complicated condensed matter systems, many other collective degrees of freedom (including phonons) also respond to the external electric potential and contribute to the dielectric function $\epsilon(\vec{q}, \omega)$. Thus measuring $\epsilon(\vec{q}, \omega)$ reveals important physical properties of the system. This is the topic of the present section.

15.10.1 Dielectric Function and AC Conductivity

Since vacuum has $\epsilon = 1$, $\epsilon(\vec{q}, \omega) - 1$ describes the response of the charge density to the external electric potential. We have also learned earlier that the electrical conductivity $\sigma(\vec{q}, \omega)$, which in general is also a function of both the wave vector $\vec{q}$ and the frequency ω, describes the response of the charge current to the electric field.[24] We thus expect $\epsilon(\vec{q}, \omega)$ and $\sigma(\vec{q}, \omega)$ to be closely related, and establish their relation below.

Current and density are related by the continuity equation (15.109), which, upon Fourier transformation, becomes

$$i\omega e n(\vec{q}, \omega) + i\vec{q} \cdot \vec{J}(\vec{q}, \omega) = 0, \tag{15.119}$$

allowing us to write the charge density

$$-en(\vec{q}, \omega) = (1/\omega)[\vec{q} \cdot \vec{J}(\vec{q}, \omega)] = (1/\omega)\sigma(\vec{q}, \omega)[\vec{q} \cdot \vec{E}(\vec{q}, \omega)]. \tag{15.120}$$

On the other hand, $\epsilon(\vec{q}, \omega)$ relates the displacement field $\vec{D}$ to the electric field $\vec{E}$:

$$\vec{D}(\vec{q}, \omega) = \epsilon(\vec{q}, \omega)\vec{E}(\vec{q}, \omega) = \vec{E}(\vec{q}, \omega) + 4\pi \vec{P}(\vec{q}, \omega), \tag{15.121}$$

where $\vec{P}$ is the polarization. Using the fact that $\vec{\nabla} \cdot \vec{P} = 4\pi e n$, we find

$$en(\vec{q}, \omega) = \frac{\epsilon(\vec{q}, \omega) - 1}{4\pi}[i\vec{q} \cdot \vec{E}(\vec{q}, \omega)]. \tag{15.122}$$

Combining this with Eq. (15.120), we find

$$\epsilon(\vec{q}, \omega) = 1 + \frac{4\pi i}{\omega}\sigma(\vec{q}, \omega). \tag{15.123}$$

---

[24] As we have learned, in general the conductivity is a rank-2 tensor. Here, for simplicity, we assume it is a scalar, with no Hall component.

In Exercise 7.1 we found the ac conductivity of the Drude model (at $q = 0$) to be

$$\sigma(\omega) = \frac{n_0 e^2 \tau}{m_e} \frac{1}{1 - i\omega\tau}, \tag{15.124}$$

which results in

$$\epsilon(\vec{q} = 0, \omega) = 1 + \frac{4\pi i}{\omega} \frac{n_0 e^2 \tau}{m_e} \frac{1}{1 - i\omega\tau}. \tag{15.125}$$

For $\omega\tau \gg 1$, it reduces to Eq. (15.101). This is expected since, in this limit, the oscillation frequency is much higher than the scattering rate, so the effect of the latter is negligible, and we are effectively dealing with the jellium model with no impurity scattering.

As we learned earlier, $\epsilon(\vec{q}, \omega) - 1$ is a response function. As discussed in Appendix A for general situations, its real and imaginary parts describe the non-dissipative and dissipative responses, respectively. This is consistent with the fact that they are proportional to the imaginary and real parts of the conductivity, respectively. Real conductivity causes dissipation because the current is in phase with the electric field/voltage, while purely imaginary conductivity causes no dissipation because the current is out of phase with the electric field/voltage (examples include capacitors and inductors).[25]

15.10.2 Optical Measurements of Dielectric Function

Equation (15.123) suggests that one can measure $\epsilon(\vec{q}, \omega)$ by measuring the ac conductivity, at least in the long-wavelength limit. In reality, transport measurements are generally limited to low frequencies (GHz or microwave frequencies and below, thus low compared with the typical energy/frequency scales of many interesting physical processes in condensed matter, many of which were discussed in Section 9.3). As a result, the most widely used methods to determine $\epsilon(\vec{q}, \omega)$ involve optical measurements, using electromagnetic waves whose frequency is not too far from that of visible light (often loosely referred to as light). The corresponding wavelength is much larger than the lattice spacing, and we are thus effectively in the $q \to 0$ limit for most purposes. As a result we will drop $\vec{q}$ and focus on the frequency dependence of various physical quantities in this subsection. Because of Eq. (15.123), measurements of $\epsilon(\omega)$ are also referred to as optical conductivity measurements.

The (frequency-dependent) index of refraction of an optical medium is

$$N(\omega) = \sqrt{\epsilon(\omega)\mu(\omega)}. \tag{15.126}$$

For a non-magnetic medium the permeability is typically close to unity, $\mu(\omega) \approx 1$, and thus the dielectric function $\epsilon(\omega)$ completely determines the optical properties of the medium. In general $\epsilon(\omega)$ contains both real and imaginary parts, and so does the index of reflection:

$$N(\omega) = n(\omega) + i\kappa(\omega). \tag{15.127}$$

As a result, the wave vector of light in the medium $q = \omega/N(\omega)$ contains an imaginary part, indicating that light decays as it propagates in the medium. This, of course, indicates that the light is (or, equivalently, photons are) being absorbed in the dissipative medium.

One of the most widely used methods to determine $N(\omega)$ is to measure reflection of light at the interface between vacuum and the medium of interest. In the simplest case of normal incidence

---

[25] Conductivity is a somewhat unusual response function because the current is most naturally viewed as a response to the vector potential $\vec{A}$ (a good example of this is the persistent current we encountered before, and there will be more on this in Chapter 19), while the electric field $\vec{E}$ can be viewed as the time derivative of the vector potential. Thus $\vec{E} \sim i\omega\vec{A}$, and the two differ by an *imaginary* factor. As a result the conductivity, which is defined to be the current response to an electric field, has the roles of real and imaginary parts reversed when compared with more standard response functions as discussed in Appendix A.

(where for an isotropic medium there is no dependence on the polarization of the light), the reflection coefficient is[26]

$$R(\omega) = \left| \frac{1 - N(\omega)}{1 + N(\omega)} \right|^2 = \frac{[n(\omega) - 1]^2 + \kappa^2(\omega)}{[n(\omega) + 1]^2 + \kappa^2(\omega)}. \tag{15.128}$$

This equation alone is not enough to determine $N(\omega)$, as there are two unknowns (the real and imaginary parts). One can supplement this equation with the **Kramers–Kronig relations**, whose general form is discussed in Appendix A. Applying these relations to $\epsilon(\omega) - 1$ and using the fact that its real and imaginary parts are even and odd functions of ω, respectively (the reader is encouraged to verify that this follows from the corresponding property of the density–density response function), we have

$$\mathrm{Re}\,\epsilon(\omega) = 1 + \frac{2}{\pi} \int_0^\infty d\omega'\, \mathrm{Im}\,\epsilon(\omega')\mathcal{P}\frac{\omega'}{\omega'^2 - \omega^2}, \tag{15.129}$$

$$\mathrm{Im}\,\epsilon(\omega) = \frac{2\omega}{\pi} \int_0^\infty d\omega'\, \mathcal{P}\frac{\mathrm{Re}\,\epsilon(\omega') - 1}{\omega'^2 - \omega^2}. \tag{15.130}$$

It should be noted that using the Kramers–Kronig relations requires knowledge of physical quantities (such as the reflection coefficient) for *all* frequencies. In reality these properties are known only for limited ranges of ω. It is thus important that these ranges cover all important physical processes.

15.11 Landau's Fermi-Liquid Theory

It turns out that the (dimensionless) number r_s is not small in typical metals, indicating that the electron–electron Coulomb interaction is actually quite strong. Yet our theoretical description of metals that is based on *non-interacting* electrons, with proper mean-field-like treatment of Coulomb interactions using Hartree–Fock or density functional theory, appears to be quite successful in many cases. In this section we discuss the origin of this success, and explain it with Landau's Fermi-liquid theory, which is based on adiabaticity when interactions are turned on "slowly." We begin by a closer inspection of the free Fermi gas, which is the starting point of Fermi-liquid theory (hence its name).

15.11.1 Elementary Excitations of a Free Fermi Gas

For reasons that will become clear shortly, it is convenient in the following discussion to use the grand canonical (instead of canonical) ensemble, and allow the particle number to change (and be determined by the chemical potential μ) in the Hilbert space.[27] In this case we work with a μ-dependent grand Hamiltonian:

$$H_{\mathrm{G}} = H - \mu\hat{N}, \tag{15.131}$$

where $\hat{N}$ is the total particle number operator. Since $[H, \hat{N}] = [H_{\mathrm{G}}, \hat{N}] = 0$, eigenstates of H_{G} are also eigenstates of $\hat{N}$; it will thus *not* cause confusion if we replace $\hat{N}$ by its eigenvalue N. It is important to emphasize, however, that the actual value of N (or more precisely its mean value) in

---

[26] This result can be found in Jackson [18] and Zangwill [19], as well as in many textbooks on optics.

[27] The Hilbert space parameterized in terms of the occupancy of different single-particle orbitals is often called Fock space in such cases.

a particular state, say the ground state, depends on, and is ultimately controlled by, μ.[28] Now let us consider non-interacting electrons, with $H = T$ including the kinetic energy only:

$$T = \sum_{\vec{k},\sigma} \frac{\hbar^2 k^2}{2m} \hat{n}_{\vec{k},\sigma}, \qquad (15.132)$$

where $\hat{n}_{\vec{k},\sigma}$ is the number operator for the state with momentum $\vec{k}$ and spin σ. Owing to the Pauli exclusion principle, the only possible eigenvalues of $\hat{n}_{\vec{k},\sigma}$ are 0 or 1. For simplicity we consider electrons in free space; for Bloch electrons we simply replace $\hbar^2 k^2/2m$ by the conduction band energy and restrict $\vec{k}$ to the first Brillouin zone.

Using $\hat{N} = \sum_{\vec{k},\sigma} \hat{n}_{\vec{k},\sigma}$, we have

$$H_G = \sum_{\vec{k},\sigma} \left(\frac{\hbar^2 k^2}{2m} - \mu \right) \hat{n}_{\vec{k},\sigma} = \sum_{\vec{k},\sigma} \epsilon_{\vec{k}} \hat{n}_{\vec{k},\sigma}, \qquad (15.133)$$

where

$$\epsilon_{\vec{k}} = \frac{\hbar^2 k^2}{2m} - \mu \qquad (15.134)$$

is the kinetic energy measured from the chemical potential. Thus the ground state of H_G has $n_{\vec{k},\sigma}^{(0)} = 1$ for $\hbar^2 k^2/2m < \mu$ and $n_{\vec{k},\sigma}^{(0)} = 0$ for $\hbar^2 k^2/2m > \mu$. This is, of course, the familiar filled Fermi-sea state, once we identify[29] $\mu = \epsilon_F = \hbar^2 k_F^2/2m$.

We now turn the discussion to excitations. Clearly, an excited state corresponds to having

$$\delta n_{\vec{k},\sigma} = n_{\vec{k},\sigma} - n_{\vec{k},\sigma}^{(0)} \neq 0 \qquad (15.135)$$

for at least one state, but possibly more states. Owing to the Pauli principle, $\delta n_{\vec{k},\sigma}$ is constrained to be

$$\delta n_{\vec{k},\sigma} = \begin{cases} 0, +1 & (k > k_F); \\ 0, -1 & (k < k_F). \end{cases} \qquad (15.136)$$

Now let us focus on the simplest possible cases, where $\delta n = 0$ for all but one particular state $(\vec{k}, \sigma)$. For $k > k_F$ this corresponds to an added electron outside the Fermi surface, and we call that a single-electron excitation. For $k < k_F$ this corresponds to *removing* an electron inside the Fermi surface, and we call that a single-*hole* excitation. In both cases the excitation energy is

$$|\epsilon_{\vec{k}}| = \left| \frac{\hbar^2 k^2}{2m} - \mu \right| = \frac{\hbar^2 |k^2 - k_F^2|}{2m} \geq 0, \qquad (15.137)$$

as illustrated in Fig. 15.1(a). It is clear that we increase the energy of the system by adding an electron. Perhaps less obvious is the fact we also *increase* energy by *removing* an electron (or creating a hole).

---

[28] When first encountering the concept of the chemical potential μ while studying thermodynamics or statistical mechanics, students often get confused by its meaning. Likewise they are much less comfortable with the grand canonical ensemble, in which N is not fixed but μ is, than they are with the canonical ensemble. From Eq. (15.131) we see that μ can be understood as an incentive for a system to accommodate more particles; the bigger μ is, the more energy the system gains by increasing N. We thus control (without *a priori* fixing) N by controlling μ (the incentive). More on this point later.

[29] In this simple example, we see explicitly how the ground state particle number is determined by the competition of the incentive μ and the cost of accommodating the last particle, namely its kinetic energy ϵ_F (through k_F). While a simple relation between μ and k_F (or particle density) is obtained easily for non-interacting electrons, in general their relation is much more complicated in the presence of interactions. For non-interacting particles at finite temperature, the chemical potential can differ from the Fermi energy in cases where the band density of states is not symmetric about the Fermi energy.

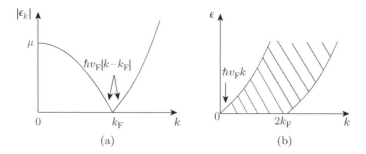

Figure 15.1 Energy–momentum relations of (a) a single electron or single hole, and (b) an electron–hole pair in a free Fermi gas. The energy of a particle–hole pair depends on the individual momenta of both the particle and the hole and not simply the net momentum. The hatched region represents the allowed range of energies for a given (net) momentum, $\hbar k$. $2\hbar k_F$ is the largest momentum change connecting two points on the Fermi surface. Beyond this limit the energy cost of making a particle–hole pair must be non-zero.

This is because in the grand Hamiltonian H_G, decreasing electron number by one comes with an energy cost of μ. (We lift the electron from inside the Fermi sea up to the chemical potential and move it into the reservoir.) It is also clear from Eq. (15.137) and Fig. 15.1(a) that the low-energy electron or hole excitations are those with k near k_F, where we can linearize their spectra:

$$|\epsilon_{\vec{k}}| \approx \frac{\hbar^2 k_F |k - k_F|}{m} = \hbar v_F |k - k_F|. \tag{15.138}$$

This is quite different from more typical cases (e.g. acoustic phonons), where low-energy excitations are those with (nearly) zero momentum.

The single-electron/hole excitations are "elementary" excitations of a free Fermi gas in the following sense. (i) As indicated in Eq. (15.137) and Fig. 15.1(a), they have a well-defined dispersion relation, namely their momenta *uniquely* determine their energies. (ii) They are building blocks of *all* excited states of the system, namely any excited state can be viewed as a *collection* of single-electron and/or -hole excitations, and the excitation energy is simply the sum of those of individual electron and hole excitations:

$$\delta E = \sum_{\vec{k},\sigma} \epsilon_{\vec{k}} \, \delta n_{\vec{k},\sigma} = \sum_{\vec{k},\sigma} |\epsilon_{\vec{k}} \, \delta n_{\vec{k},\sigma}|. \tag{15.139}$$

At this point it becomes clear why the grand canonical ensemble is advantageous over the canonical ensemble. In the latter we fix the number of electrons (as that of the ground state). As a result, the simplest excitation is a single electron *plus* a single hole. But such electron–hole-pair excitations are *not* elementary in the sense discussed above. First of all they do not have well-defined dispersion, as indicated in Fig. 15.1(b). (In particular, moving an electron from just below the Fermi surface to just above costs essentially zero energy, yet the momentum change can range from 0 to $2\hbar k_F$.) Secondly, there is no unique way to decompose multiple electrons and holes into individual pairs. We close by noting that the choice of the grand canonical ensemble is not just for its conceptual simplicity as illustrated already; in practice one often encounters physical processes in which the electron number of a system changes (as in tunneling or photoemission experiments), and one needs to compare energies between states with *different* electron numbers.

15.11.2 Adiabaticity and Elementary Excitations of an Interacting Fermi Gas

Trivial as it may be, the solvability of the free-electron Hamiltonian is tied to the fact that not only is N a good quantum number, but also $n_{\vec{k},\sigma}$ is a good quantum number for *every* $(\vec{k}, \sigma)$, because

$[T, \hat{n}_{\vec{k},\sigma}] = 0$. Related to this is the fact that T has infinitely many independent U(1) symmetries, one for every $(\vec{k}, \sigma)$, with $\hat{n}_{\vec{k},\sigma}$ being its generator (see Exercise 15.13).

Exercise 15.13.
(a) Find the unitary transformation which transforms the fermion destruction operators according to

$$\hat{U}_{\vec{k},\sigma} c_{\vec{k},\sigma} \hat{U}^\dagger_{\vec{k},\sigma} = e^{i\theta_{\vec{k}}} c_{\vec{k},\sigma}. \qquad (15.140)$$

You may need to consult Appendix J.
(b) Show that the free-electron Hamiltonian $\hat{T}$ is invariant under this transformation, regardless of the values of the $\theta_{\vec{k}}$.

Such a huge amount of symmetry is, of course, not the generic situation; once the two-body interaction V has been turned on,[30] electrons can scatter off each other, and $[H_G, \hat{n}_{\vec{k},\sigma}] = [V, \hat{n}_{\vec{k},\sigma}] \neq 0$. The only U(1) symmetry that survives is that associated with the conservation of total particle number N, and H_G is no longer exactly solvable except for some very special cases.[31]

Now imagine turning V on from zero "adiabatically" (even though there is no excitation gap to permit rigorous adiabaticity). Theoretically this can be achieved by introducing a family of grand Hamiltonians,

$$H_G(\lambda) = T + \lambda V - \mu \hat{N}, \qquad (15.141)$$

with λ increasing slowly from 0 to 1.[32] We note that $H_G(\lambda = 0)$ has a *non-degenerate* ground state (Fermi-sea state); as long as this remains the case for all $0 < \lambda \leq 1$, we know that the Fermi-sea state is adiabatically connected to the interacting ground state at $\lambda = 1$, with the same set of quantum numbers. We can thus label the interacting ground state with $\{\delta n_{\vec{k},\sigma} = 0\}$.[33] This adiabaticity breaks down if a quantum phase transition occurs at some value of λ (thereby scrambling the levels). The Fermi-sea state is then no longer adiabatically connected to the interacting ground state. We will discuss a number of examples of this in the next section; here we *assume* adiabaticity does hold for the ground state. If this is the case, we say that the interacting Fermi gas is a Fermi liquid.

What about excited states? To this end, let us consider the simplest type, namely a single electron with $|\vec{k}_0| > k_F$. If adiabaticity holds, this state could be labeled by $\delta n_{\vec{k},\sigma} = +1$. We find that it is *always* possible to find other excited states with *degenerate* energy, and the *same* good quantum numbers like total momentum and spin. One such possibility is a two-electron and one-hole excited state (illustrated in Fig. 15.2), with momenta $|\vec{k}_1|, |\vec{k}_3| > k_F$ and $|\vec{k}_2| < k_F$; the only constraints are

$$\vec{k}_0 = \vec{k}_1 + \vec{k}_3 - \vec{k}_2;$$
$$k_0^2 = k_1^2 + k_3^2 - k_2^2. \qquad (15.142)$$

It is easy to convince oneself that it is always possible to find solutions of the equations above, as well as corresponding ones for single-hole excitations. As a result, as soon as V is turned on, such

---

[30] Since we do not consider an external potential in this and the following sections, from now on we will use V to represent the two-body interaction.

[31] Most exactly solvable models of interacting Hamiltonians are 1D models. These are also special models with large numbers of *additional* conservation laws, although not as extensive as the free-fermion (or boson) models.

[32] It is understood implicitly that μ may need to be adjusted accordingly to keep the particle density a constant with λ increasing.

[33] One should be cautious in the presence of V because $\delta n_{\vec{k},\sigma}$ is not a good quantum number any more; this is simply to emphasize the fact we can use the same set of labels for interacting and non-interacting ground states when adiabaticity holds.

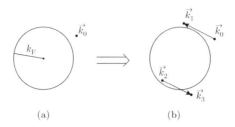

Figure 15.2 (a) A single-electron excitation at momentum $\vec{k}_0$. (b) A composite excitation of the same energy made of two electrons (at momenta $\vec{k}_1$ and $\vec{k}_3$) and a hole (at momentum $\vec{k}_2$). A scattering process between electrons at momenta $\vec{k}_0$ and $\vec{k}_2$ leads to decay of (a) to (b).

degenerate excitations will be (coherently) mixed in, and thus adiabaticity does not hold strictly for *any* excited states. In other words, one *cannot* follow excited states of $H_G(\lambda = 0)$ adiabatically to reach *exact* excited states of $H_G(\lambda = 1)$. Thus the situation seems quite hopeless. Let us emphasize that experimentally one *never* directly probes the ground state, as we are never at zero temperature. We are thus *always* probing excitations of the system, and it is the knowledge of them that physics seeks.

It took Landau's deep insight to rescue the possibility of describing an interacting Fermi gas starting from the non-interacting one. Landau observed that even though *strict* adiabaticity does not hold for excited states, *approximate* adiabaticity *does* hold for *low-energy* excited states. To this end let us consider the decay rate of the single-electron excitation state $|\vec{k}_0\rangle$, due to scattering processes like that illustrated in Fig. 15.2. Kinematic constraints in Eqs. (15.142) dictate that $\epsilon_{k_1}, |\epsilon_{k_2}|, \epsilon_{k_3} < \epsilon_{k_0}$. For a *low-energy* electron excitation with $\delta k_0 = k_0 - k_F \ll k_F$, we have

$$\epsilon_{k_0} \approx \hbar v_F \, \delta k_0, \tag{15.143}$$

thus $k_1 - k_F < \delta k_0$ and $k_3 - k_F < \delta k_0$. In other words, $\vec{k}_1$ and $\vec{k}_3$ are restricted to a thin shell of width δk_0 from the Fermi surface. Given this, combined with the constraints of Eqs. (15.142) and $k_2 < k_F$, one finds that the available phase space for scattering processes of Fig. 15.2 is proportional to δk_0^2. From Fermi's golden rule we conclude that the scattering rate of the electron obeys

$$1/\tau_{k_0} \propto \delta k_0^2. \tag{15.144}$$

From the uncertainty principle we thus find that there is an energy uncertainty of

$$\delta \epsilon_{k_0} \propto \delta k_0^2. \tag{15.145}$$

Comparing this with Eq. (15.143), we find that, for sufficiently small δk_0, we have

$$\delta \epsilon_{k_0} \ll \epsilon_{k_0}, \tag{15.146}$$

namely the uncertainty of the energy of a single-electron state is much smaller than its average energy, in the presence of V.

What does this mean? It means that, as V is turned on slowly, the single-electron state $|\vec{k}_0\rangle$ will evolve into an *approximate* eigenstate of $H_G(1)$, with uncertainty of energy much smaller than the average energy (simply referred to as the energy from now on); the lower the energy, the closer the state will be to an exact eigenstate. There is thus (very nearly) a one-to-one correspondence between the *low-energy* excited states of $H_G(0)$ and $H_G(1)$, and this correspondence is asymptotically exact in the limit of zero excitation energy. We can thus use the *same* indices, namely electron and hole numbers $\{\delta n_{\vec{k},\sigma}\}$, to label the low-energy excitations of $H_G(1)$. These excitations of the interacting system are referred to as quasiparticles and quasiholes. Obviously the allowed values of their occupation numbers satisfy the same constraints given by Eq. (15.136). The quasiparticles and quasiholes are elementary excitations of a Fermi liquid, in the same sense that electrons and holes are elementary excitations of a free Fermi gas. Thus the nature of the low-energy excitations of a Fermi liquid is very similar to that of a free Fermi gas, despite the fact that interaction may actually be quite strong. That is why a Fermi liquid behaves very much like a free Fermi gas at low temperature (so that only a low density of low-energy excitations is present) and when probed at low energy. As a result, it can be described using approximations starting from the free Fermi gas. In the following we present a more quantitative, but phenomenological, description of a Fermi liquid.

15.11.3 Fermi-Liquid Parameters

The eigenenergy, $E[\{\delta n_{\vec{k},\sigma}\}]$, of the excited state $|\{\delta n_{\vec{k},\sigma}\}\rangle$ is a functional of the quasiparticle distribution $\{\delta n_{\vec{k},\sigma}\}$. Since at low energy we expect very few quasiparticles and quasiholes around, it is natural to expand E to second order in $\delta n_{\vec{k},\sigma}$:

$$E[\{\delta n_{\vec{k},\sigma}\}] = E_0 + \sum_{\vec{k},\sigma} \epsilon_{\vec{k}} \, \delta n_{\vec{k},\sigma} + \frac{1}{2} \sum_{\vec{k},\sigma,\vec{k}',\sigma'} f_{\vec{k},\sigma,\vec{k}',\sigma'} \, \delta n_{\vec{k},\sigma} \, \delta n_{\vec{k}',\sigma'}, \tag{15.147}$$

where E_0 is the ground state energy,

$$\epsilon_{\vec{k}} = \frac{\delta E[\{\delta n_{\vec{k},\sigma}\}]}{\delta n_{\vec{k},\sigma}} \approx v_{\mathrm{F}}^*(k - k_{\mathrm{F}}), \tag{15.148}$$

and

$$f_{\vec{k},\sigma,\vec{k}',\sigma'} = \frac{\delta^2 E[\{\delta n_{\vec{k},\sigma}\}]}{\delta n_{\vec{k},\sigma} \, \delta n_{\vec{k}',\sigma'}} \tag{15.149}$$

is known as Landau's f-function, reflecting the fact that quasiparticles and quasiholes do interact with each other. The expression above neglects the possibility that these interactions result in scattering of quasiparticles which causes them to change their quantum numbers. That is, for the expression above, the individual quasiparticle state occupancies are good quantum numbers. The quasiparticles are themselves bare electron states dressed by particle–hole pairs that result from underlying scatterings. We are not neglecting those interactions (which can be substantial), but rather neglecting the scattering of two such dressed particles off each other. We will see below in Section 15.13 that, in the case of weak repulsive interactions, the Fermi liquid is indeed stable because the scattering of quasiparticles off each other at low energies is (nearly) negligible. The interactions between quasiparticles that we do keep are diagonal in the quasiparticle basis. They result microscopically (in part) from some of the virtual particle–hole pairs associated with one quasiparticle being Pauli blocked by the presence of the other quasiparticle, but they do not result in the quasiparticles scattering off each other and changing states.

In Eq. (15.148) $|\epsilon_{\vec{k}}|$ is the energy of a single quasiparticle or quasihole. It should be emphasized, though, that in the presence of V it is *not* the same as the kinetic energy cost of the free Fermi gas in Eq. (15.137). That is, the Fermi velocity $v_{\mathrm{F}}^* \neq v_{\mathrm{F}}$ is renormalized by the interactions. Very often v_{F}^* is parameterized by an interaction-renormalized effective mass m^*:

$$v_{\mathrm{F}}^* = \frac{\hbar k_{\mathrm{F}}}{m^*} = v_{\mathrm{F}} \frac{m}{m^*}, \tag{15.150}$$

which should not be confused with the band effective mass. When applied to Bloch electrons where a band effective-mass (referred to as band mass from now on) approximation is appropriate, m should be understood as the band mass, which is also called the "bare mass" in the literature. The effective mass m^* and renormalized Fermi velocity v_{F}^* are examples of Fermi-liquid parameters that receive renormalizations from interactions.[34] The difference between the dimensionless ratio $v_F/v_F^* = m^*/m$ and 1 is a measure of the strength of such renormalization.

The f-functions, on the other hand, depend continuously on $\vec{k}$ and $\vec{k}'$. Symmetry and other considerations can nevertheless simplify things, and reduce them to a discrete set of parameters as well. First of all, since the Fermi-liquid theory deals with low-energy quasiparticles and quasi-holes, we have $|\vec{k}|, |\vec{k}'| \approx k_F$; this reduces the $\vec{k}$ and $\vec{k}'$ dependence to their *directions*. Rotational invariance around the Fermi surface further reduces this dependence to the angle between them: $f_{\vec{k},\sigma,\vec{k}',\sigma'} = f_{\sigma\sigma'}(\theta_{\vec{k}\vec{k}'}) = f_{\sigma\sigma'}(-\theta_{\vec{k}\vec{k}'})$; we can thus expand $f(\theta)$ using orthogonal basis functions – Legendre polynomials $P_l(\cos\theta)$ in 3D, and $\cos(l\theta)$ (or Fourier transformation) in 2D. Secondly, spin rotation symmetry dictates $f_{\uparrow\uparrow} = f_{\downarrow\downarrow}$; $f_{\uparrow\downarrow} = f_{\downarrow\uparrow}$. We thus need deal only with $f_{\uparrow\uparrow}$ and $f_{\uparrow\downarrow}$. It is convenient for later applications to introduce their symmetric and anti-symmetric combinations:

$$f^{s,a}(\theta) = [f_{\uparrow\uparrow}(\theta) \pm f_{\uparrow\downarrow}(\theta)]/2. \tag{15.151}$$

As we will see, f^s impacts charge properties of the system, while f^a impacts spin properties. Thirdly, simple dimensional and scaling analyses indicate that f has the dimensions of energy, and is inversely proportional to the system volume. It would thus be desirable to combine it with a quantity that has the dimensions of inverse energy, and is proportional to the system volume, to obtain a dimensionless measure of its strength. One natural choice for such a quantity is the combined total density of states at the Fermi level (including both spin species):[35]

$$2D(0) = \sum_{\vec{k},\sigma} \delta(\epsilon_{\vec{k}}) = 2 \sum_{\vec{k}} \delta(\epsilon_{\vec{k}}) = \frac{2\Omega_d \mathcal{V}}{(2\pi)^d} \frac{k_F^{d-1}}{\hbar v_F^*}, \tag{15.152}$$

which takes the same form as that of a free Fermi gas, as long as one replaces v_F by v_F^*. Here d is the space dimensionality, $\mathcal{V}$ is the volume of the system, and $\Omega_d = 2\pi \, (4\pi)$ is the total solid angle for $d = 2 \, (3)$. Combining the considerations above leads to the definition of the following (discrete and dimensionless) Landau parameters:

$$F_l^{s,a} = 2D(0) \int \frac{d\omega_d}{\Omega_d} f^{s,a}(\theta) \times \begin{cases} P_l(\cos\theta) & (d=3); \\ \cos(l\theta) & (d=2). \end{cases} \tag{15.153}$$

Here $d\omega_d$ is the differential solid angle ($d\omega_2 = d\theta$, $d\omega_3 = 2\pi \sin\theta \, d\theta$). In terms of the Fs we have

$$f^{s,a}(\theta) = \frac{1}{2D(0)} \sum_{l=0}^{\infty} \begin{cases} (2l+1)F_l^{s,a} P_l(\cos\theta) & (d=3); \\ (2-\delta_{l,0})F_l^{s,a} \cos(l\theta) & (d=2). \end{cases} \tag{15.154}$$

Usually the sum is dominated by a handful of terms with small l. In the next section we discuss how to relate the Fermi-liquid parameters m^* and $F_l^{s,a}$ to physical observables.

---

[34] Interactions renormalize all the parameters in a Fermi liquid from the corresponding values of the free Fermi gas, except for k_F, which depends on the electron density only. The latter fact is known as **Luttinger's theorem**. For Bloch electrons with anisotropic band dispersions, the shape of the Fermi surface is no longer a sphere and can change with interaction, but Luttinger's theorem states that the volume it encloses cannot change.

[35] Throughout this chapter we will define the total density of states per spin as $D(0) = \mathcal{V}\rho(0)$, where $\mathcal{V}$ is the sample volume and $\rho(0)$ is the single-spin density of states (per unit volume per spin) at the Fermi level. Thus $D(0)$ is an extensive quantity (proportional to volume $\mathcal{V}$) having the dimensionality of inverse energy, while $\rho(0)$ is an intensive quantity which is independent of $\mathcal{V}$. In Section 7.7 $\rho(0)$ is called $\rho(\mu)$ or $\rho(\mu_0)$. Here quasiparticle energies are measured from the chemical potential and vanish on the Fermi surface by definition, whence the notation $\rho(0)$ and $D(0)$.

15.12 Predictions of Fermi-Liquid Theory

In principle, one can use the energy functional in Eq. (15.147) to calculate the grand partition function as a function of the temperature T, chemical potential μ, and other variables such as the magnetic field when they are present. This approach is valid for $k_B T \ll \mu$, where there are few quasiparticles and quasiholes, and they are all near the Fermi surface. From the grand partition function one can obtain various equilibrium thermodynamic functions of the system. In the following we discuss a number of them, which can actually be obtained fairly easily (without necessarily resorting to an explicit calculation of the grand partition function).

15.12.1 Heat Capacity

The heat capacity[36] at fixed volume $\mathcal{V}$ and particle number N (which means the chemical potential μ must be T-dependent) is

$$C_{\mathcal{V},N}(T) = \left.\frac{\partial \langle E \rangle}{\partial T}\right|_{\mathcal{V},N}. \tag{15.155}$$

We start by neglecting the last term in Eq. (15.147). In this case we simply have the spectrum of non-interacting fermions, and, in the low-temperature limit, $C_{\mathcal{V},N}$ is proportional to the product of the temperature and the density of states at the Fermi level:

$$C_{\mathcal{V},N}(T) = \frac{2\pi^2}{3} D(0) k_B^2 T. \tag{15.156}$$

This form is identical to that of a Fermi gas; see Section 7.7. It should be noted, however, that the density of states at the Fermi level, $D(0)$, has received renormalization from the renormalization of the Fermi velocity by the interactions. One way to emphasize this fact is to re-write Eq. (15.156) in terms of the free-fermion heat capacity $C_{\mathcal{V},N}^{(0)}(T)$:

$$C_{\mathcal{V},N}(T) = C_{\mathcal{V},N}^{(0)}(T) \frac{m^*}{m}. \tag{15.157}$$

We now argue that the last term in Eq. (15.147) does not alter the results above, in the low-T limit. This is due to the constraint

$$\delta N = \sum_{\vec{k},\sigma} \delta n_{\vec{k},\sigma} = 0. \tag{15.158}$$

Let us start with a mean-field approximation to its contribution to $\langle E \rangle$:

$$\left\langle \sum_{\vec{k},\sigma,\vec{k}',\sigma'} f_{\vec{k},\sigma,\vec{k}',\sigma'} \delta n_{\vec{k},\sigma} \delta n_{\vec{k}',\sigma'} \right\rangle \approx \sum_{\vec{k},\sigma,\vec{k}',\sigma'} f_{\vec{k},\sigma,\vec{k}',\sigma'} \langle \delta n_{\vec{k},\sigma} \rangle \langle \delta n_{\vec{k}',\sigma'} \rangle$$

$$= 0. \tag{15.159}$$

Here we used the fact that, due to rotational symmetry, $\langle \delta n_{\vec{k},\sigma} \rangle$ depends only on the magnitude of $\vec{k}$, while $f_{\vec{k},\sigma,\vec{k}',\sigma'}$ depends only on the *directions* of $\vec{k}$ and $\vec{k}'$; combining these with the constraint (15.158) leads to the result above.

---

[36] Which is the specific heat times the volume. The reader should compare equations of this section with those of Section 7.7.1, and note that the main difference is that $\rho(\mu)$, the density of states *per unit volume* at the Fermi energy, is replaced by $D(0)$ here.

The correction to the mean-field approximation is

$$\sum_{\vec{k},\sigma,\vec{k}',\sigma'} f_{\vec{k},\sigma,\vec{k}',\sigma'} \langle (\delta n_{\vec{k},\sigma} - \langle \delta n_{\vec{k},\sigma} \rangle)(\delta n_{\vec{k}',\sigma'} - \langle \delta n_{\vec{k}',\sigma'} \rangle) \rangle, \tag{15.160}$$

where the quantities in $(\cdot)$ represent *fluctuations*. Thus the correction is due to *correlations* in these fluctuations (cf. the density–density correlation function discussed in Chapter 2, although we are in momentum space here), which, in general, is not zero. This is the reason why mean-field theories are not exact. Here the key is that this correlation function is convoluted with a *slowly* varying function $f_{\vec{k},\sigma,\vec{k}',\sigma'}$ (on the scale of the distance between neighboring $\vec{k}$s, which is zero in the thermodynamic limit). This is like taking a Fourier transformation in the zero-wave-vector limit for this correlation function, and that results in a null outcome.[37] As a result, the last term of Eq. (15.147) makes no contribution to $\langle E \rangle$.

We thus conclude that Eqs. (15.156) and (15.157) give the correct low-T behavior of the heat capacity, which has the same linear-T dependence as a free Fermi gas, with a renormalization factor of m^*/m.

15.12.2 Compressibility

The isothermal compressibility of a system is defined by the change in volume with pressure at fixed temperature and particle number,

$$\kappa = -\frac{1}{\mathcal{V}} \frac{\partial \mathcal{V}}{\partial P}\bigg|_{T,N}. \tag{15.161}$$

Just like the heat capacity, we expect the isothermal compressibility to have the same qualitative behavior as that of the free Fermi gas. Using the **Gibbs–Duhem relation** from thermodynamics, the compressibility can be re-expressed in the more convenient form

$$\kappa = \frac{1}{n^2} \frac{\partial n}{\partial \mu}, \tag{15.162}$$

where n is the particle density.

Exercise 15.14. Derive Eq. (15.162) from Eq. (15.161).

For the free Fermi gas we have

$$\kappa^{(0)} = \frac{2D^{(0)}(0)}{n^2} \tag{15.163}$$

at $T = 0$ (and with very weak T dependence), and $D^{(0)}(0)$ is the density of states at the Fermi level for the free Fermi gas. Obviously we expect the latter to be replaced by $D(0)$ in a Fermi liquid, which results in a renormalization factor of m^*/m as for the heat capacity. But there is more; as we demonstrate below, due to the fact that κ measures the *change* of n in response to a change of μ, it also receives renormalization from the Landau parameters.

An increase in electron density, δn, results in added quasiparticles near the Fermi surface, and an increase of k_F; see Fig. 15.3(a). An increase in chemical potential, $\delta \mu$, therefore must compensate for the energy cost of adding one quasiparticle at the new Fermi surface at $k_F + \delta k_F$, such that its cost is zero within the new grand Hamiltonian (with chemical potential $\mu + \delta \mu$). Thus

---

[37] Another way to say this is that every $\vec{k}$ is coupled with infinitely many other $\vec{k}$s in the thermodynamic limit, or the coordination number is infinite. Mean-field approximations become exact in such limits.

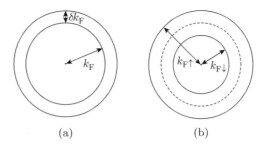

Figure 15.3 (a) A uniform expansion of the Fermi surface that results in a density increase. (b) Change of k_F in opposite directions for up- and down-spin quasiparticles, resulting in a net magnetization with no change in the total density. The dashed line represents the original Fermi surface.

$$\delta\mu = \hbar v_F^* \delta k_F + \sum_{k_F < k' < k_F + \delta k_F, \sigma'} f_{\vec{k},\sigma,\vec{k}',\sigma'} \, \delta n_{\vec{k}',\sigma'} = \hbar v_F^* \delta k_F (1 + F_0^s). \tag{15.164}$$

Comparing that with the case of free fermions, where $F_0^s = 0$ and $v_F^* = v_F$, we can immediately conclude that

$$\kappa = \kappa^{(0)} \frac{m^*/m}{1 + F_0^s} = \frac{2D(0)}{n^2(1 + F_0^s)}. \tag{15.165}$$

The reason why κ receives renormalization from F_0^s is that such a density change results in $\delta n_{\vec{k}',\sigma'}$ that is *independent* of $\theta_{\vec{k}'}$, and thus only the $l = 0$ component of $f_{\vec{k},\sigma,\vec{k}',\sigma'} = f_{\sigma,\sigma'}(\theta)$ contributes to the summation in Eq. (15.164). Also, because the change δn is the same for up- and down-spin quasiparticles, only the symmetric combination f^s enters the expression (see Fig. 15.3(a)).

15.12.3 Spin Susceptibility

The calculation of the spin susceptibility χ_s is very similar to that of the compressibility κ. A Zeeman field B that couples to the $\hat{z}$ component of spin,

$$H_Z = -\frac{g\mu_B B}{2\hbar} \sum_{i=1}^{N} S_i^z, \tag{15.166}$$

tends to split the Fermi surfaces of up- and down-spin electrons, by introducing quasiparticles of up-spin and quasiholes of down-spin without changing the total density (see Fig. 15.3(b)). A calculation similar to that for κ yields

$$\chi_s = \chi_s^{(0)} \frac{m^*/m}{1 + F_0^a}, \tag{15.167}$$

where $\chi_s^{(0)}$ is the free-Fermi-gas spin susceptibility. Comparing Eqs. (15.167) and (15.165), we find F_0^s is replaced by F_0^a here, due to the fact that $\delta n_{\vec{k}',\sigma'}$ is *opposite* for opposite σ', resulting in the anti-symmetric combination F_0^a.

Exercise 15.15. Derive Eq. (15.167).

15.12.4 Collective Modes, Dynamical and Transport Properties

With appropriate and straightforward extensions, Fermi-liquid theory can also describe the dynamical and transport properties of a Fermi liquid at sufficiently low temperature, low frequency, and long wavelength. The basic assumption (which is valid in the regime specified) is that quasiparticles and quasiholes behave as non-interacting entities[38] very much like electrons and holes in a free Fermi gas. As a result, the theoretical methods developed in earlier chapters for the latter can be applied to them, as long as appropriate interaction-renormalized Fermi-liquid parameters are used. This is also the reason why the dynamical and transport properties of a metal that is dominated by *interacting* electrons are very similar to those obtained using theoretical formalisms based on non-interacting electrons. In addition to quasiparticle and quasihole excitations, Fermi liquids also support collective modes, and the plasma mode is an example of that. These collective modes can also be described using Fermi-liquid theory.[39] Fermi-liquid theory also plays an important role in the study of liquid ^{3}He, where the atoms are strongly interacting but there are no long-range Coulomb forces. Here the collective "plasma" mode becomes a gapless **"zero-sound" mode** with linear dispersion at small wave vectors. Readers interested in these topics can consult numerous books and review articles (see, e.g., *Landau Fermi-Liquid Theory: Concepts and Applications* by Gordon Baym and Christopher Pethick, Wiley-VCH, 1991) on Fermi-liquid theory.

15.13 Instabilities of Fermi Liquids

As discussed in the previous sections, Landau Fermi-liquid theory is based on the assumption that, as the interaction V is turned on slowly, the free-fermion ground state evolves adiabatically to the interacting ground state. It can occur, however, that interactions make the Fermi-liquid ground state unstable, and the system goes through a quantum phase transition into a new phase (often with spontaneously broken symmetry). There are several different types of **Fermi-liquid instabilities**. Some of them can be described (and are in a certain sense predicted) by Fermi-liquid theory itself, and in such cases Fermi-liquid theory with proper extensions can also describe the new phase. Others involve physics beyond Fermi-liquid theory. Some instabilities are of a weak-coupling nature (i.e. occur even at infinitesimal V), while others are "strong-coupling" phenomena and occur only at finite V. We will discuss examples of all these below.

15.13.1 Ferromagnetic Instability

Equation (15.167) suggests that the spin susceptibility χ_s *diverges* as $F_0^a \to -1$. This indicates the ferromagnetic instability, as the system responds infinitely strongly to a weak Zeeman field that couples to the electron spins. In fact, for $F_0^a < -1$, the Landau energy functional Eq. (15.147) indicates that the system can *lower* its energy by *spontaneously* developing a spin polarization in the absence of external field, in the manner illustrated in Fig. 15.3(b). The energy as a function of magnetization is illustrated in Fig. 15.4.

A negative F_0^a means that overall we have $f_{\uparrow\downarrow} > f_{\uparrow\uparrow}$. This is the case for repulsive interactions, due to the exchange effect that we discussed earlier in Section 15.2. Owing to the anti-symmetry requirement of fermion wave functions, parallel-spin electrons have less chance of coming close, and cost less interaction energy. It thus lowers the interaction energy to have spins polarized.[40] The reason

---

[38] Or more precisely, as entities whose interactions are accurately described by mean-field theory. That is, the interactions do not cause strong scattering of the particles into new states.

[39] The plasma frequency is typically very high relative to the Fermi energy, so one might worry that Fermi-liquid theory is inapplicable. However, the high frequency is due solely to the long-range collective nature of the Coulomb force. (Recall that the plasma frequency has a purely classical expression that is independent of $\hbar$ and independent of the quantum exchange statistics of the underlying particles.) For purposes of computing the plasmon dispersion at small wave vectors, Fermi-liquid theory is still applicable because only excitations near the Fermi surface are involved.

[40] This physics is illustrated in the simple calculation of Exercise 15.5.

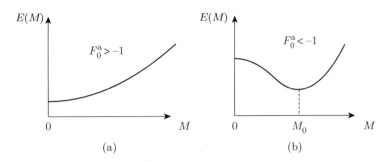

Figure 15.4 Energy as a function of magnetization for (a) $F_0^a > -1$ and (b) $F_0^a < -1$. The former is a Fermi liquid whose ground state is a spin singlet. The latter is a ferromagnet with a finite ground state magnetization M_0.

why the ferromagnetic instability occurs at *finite* interaction strength $F_0^a = -1$ (instead of 0^- or weak coupling) is that the gain of exchange energy needs to be sufficiently large to overcome the kinetic-energy cost of polarizing the fermions.

The ferromagnetic instability corresponds to a quantum phase transition in which the magnetization increases to a non-zero value (as illustrated in Fig. 15.4). This phase transition is triggered by a (true) level crossing between the singlet ground state and a magnetized state whose energy crosses below the singlet as F_0^a falls below -1.[41] Because of this, while *not* being adiabatically connected to the free-fermion ground state, a magnetized ground state *is* adiabatically connected to an excited state of the free Fermi gas, namely the Fermi-sea state with the same magnetization, as illustrated in Fig. 15.3(b). Because of this, Fermi-liquid theory not only "knows" about this instability (in terms of the critical Landau parameter), but can also describe the low-energy/temperature properties of the ferromagnetic phase as well. The most important difference here is that, because the ferromagnetic state breaks spin rotation symmetry spontaneously, the system supports gapless spin-wave excitations in addition to Landau quasiparticles and quasiholes. Thus extensions need to be made in Fermi-liquid theory to include these low-energy collective excitations which will be discussed in detail in Chapter 17.

15.13.2 Pomeranchuk Instabilities

The ferromagnetic instability is an example of a large class of Fermi-surface instabilities known as **Pomeranchuk instabilities**, in which the Fermi surface deforms spontaneously. Such instabilities occur when any one of the Landau F parameters becomes less than -1. In the ferromagnetic instability driven by F_0^a, the Fermi surfaces for each spin expand or shrink uniformly, and move in opposite directions for opposite spins. When a Pomeranchuk instability occurs in channel l, the Fermi surface distorts in a direction-dependent manner corresponding to the angle dependence of $P_l(\cos\theta)$ in 3D and $\cos(l\theta)$ in 2D. For example, when $F_2^s < -1$, the Fermi surface distorts from circular to elliptical in shape in the same way for both spins (see Fig. 15.5(a)). In such a **"nematic"** **Fermi liquid**, the orientation of the ellipse is selected spontaneously.

As another example, consider what would happen when $F_1^s < -1$. This would seem to lead to a uniform Fermi surface boost as illustrated in Fig. 15.5(b). However, in a system with Galilean invariance (i.e. isotropic quadratic band dispersion), such a center-of-mass boost always *costs* energy, and the possibility of a ground state with a spontaneous momentum boost does not occur. The cost

---

[41] This is a somewhat special case as magnetization (or total spin) is a conserved quantity, or good quantum number. Most (continuous) quantum phase transitions do not share this property and instead are characterized by a dense set of infinitesimally avoided level crossings. The difference between these two cases will be discussed in more detail in Chapter 17.

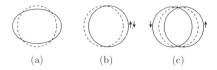

Figure 15.5 Fermi surfaces before (dashed lines) and after (solid lines) distortions due to Pomeranchuk instabilities. The details of the various distortions shown here are given in the text.

of a momentum boost in a system is solely in the kinetic energy, not the potential energy. This is because one can achieve the momentum boost by simply having the observer jump to a moving frame. In that frame the distances between the electrons remain the same and thus the interaction energy is unaffected. The only cost is the center-of-mass kinetic energy. In fact, by comparing the center-of-mass kinetic energy and the prediction of the Landau energy functional (15.147), one finds in 3D

$$\frac{m^*}{m} = 1 + \frac{F_1^s}{3}.$$ (15.168)

On the other hand, when $F_1^a < -1$, Fermi surfaces of opposite spins are boosted in opposite directions (see Fig. 15.5(c)), so the ground state is invariant under time-reversal transformation and carries zero net momentum and zero charge current. Galilean invariance does not exclude this possibility in which the ground state carries a spontaneous spin current.

All such ground states are adiabatically connected to excited states in the free Fermi gas with distorted Fermi surfaces. As a result Fermi-liquid theory can be used to describe the corresponding phases, once the gapless collective modes associated with corresponding spontaneously broken symmetries have been included. These gapless collective modes correspond to the amplitude of the Fermi-surface shape distortion oscillating slowly in space and time.

Exercise 15.16. Prove Eq. (15.168) in 3D and find the corresponding relation between m^* and F_1^s in 2D, when there is Galilean invariance.

Exercise 15.17. Use Eq. (15.147) to prove that the ground state energy can be lowered by distorting the circular Fermi surface when any of the Landau parameters becomes less than -1, and find the corresponding angular and spin dependences of the distortion.

15.13.3 Pairing Instability

In Fermi-liquid theory, the interaction between quasiparticles is described by the Landau f-function, which parameterizes the interaction contribution to the energy of a multi-quasiparticle state $|\{\delta n_{\vec{k},\sigma}\}\rangle$. Fermi-liquid theory is built on the argument that, for a single quasiparticle, the scattering rate into states containing additional particle–hole pairs is very small for quasiparticles near the Fermi surface. However, in the presence of multiple quasiparticles, scattering between them can become important in certain cases, and this is *not* described in Fermi-liquid theory. Would such scattering processes lead to any instability in a Fermi liquid? To address this question we examine here the simplest possible situation, namely the case with two quasiparticles. To maintain the highest possible symmetry, we consider the case that the total momentum of the two quasiparticles is zero, so we need only specify the momentum of one of the quasiparticles, which we will call $\vec{k}$; the other quasiparticle's momentum is thus $-\vec{k}$. To maximize simplicity, we will also assume that the scattering-matrix element is a constant:

$V_{\vec{k}\vec{k}'} = V_0.$[42] In real space this would correspond to a delta-function potential, thus the (fermionic) quasiparticles will be scattered only if they form a spin singlet, which we assume to be the case. The quantum number $\vec{k}$, which (without loss of generality) we identify as the momentum of the spin-up quasiparticle, uniquely determines the state of the *pair* $|\vec{k}\rangle = |\vec{k} \uparrow; -\vec{k} \downarrow\rangle$, and we have

$$\hat{H}_0|\vec{k}\rangle = 2\epsilon_{\vec{k}}|\vec{k}\rangle, \tag{15.169}$$

$$\langle\vec{k}'|\hat{V}|\vec{k}\rangle = V_{\vec{k}\vec{k}'} = V_0. \tag{15.170}$$

Here $\hat{H}_0$ is the "Fermi-liquid" Hamiltonian, which includes the interaction renormalizations of all Fermi-liquid parameters, but contains no scattering-matrix elements between quasiparticles. We want to solve the Hamiltonian $\hat{H}_0 + \hat{V}$ in the Hilbert subspace of two quasiparticles with zero total momentum. To have a well-defined problem, we need to specify constraints on $\vec{k}$. Clearly we have $k > k_F$. Since quasiparticles are well-defined near the Fermi surface, we also need to have an upper cutoff: $k < k_F + \Delta k_F$, with $\Delta k_F \ll k_F$. Equivalent to these is

$$0 < \epsilon_{\vec{k}} < E_c, \tag{15.171}$$

where the energy cutoff $E_c = \hbar v_F^* \Delta k_F$.

While this two-body problem is exactly solvable, and as we will see below the solution is *non-perturbative* in V_0, it is instructive to perform a *perturbative* renormalization-group analysis on the fate of V_0 when the cutoff is lowered toward 0, where all quasiparticles live right on the Fermi surface, which is where they are sharply defined in a Fermi liquid. The reader may wish to review Appendix I before proceeding with this section.

We introduce a varying energy cutoff Λ, whose initial value is E_c. The scattering-matrix element V should vary with Λ, such that the scattering *amplitude* for quasiparticles on the Fermi surface is invariant. To determine $V(\Lambda)$, we consider an infinitesimal change of Λ to $\Lambda - d\Lambda$. This eliminates some higher-order scattering processes that involve states with intermediate quasiparticle states with energy $\Lambda - d\Lambda < \epsilon < \Lambda$; see Fig. 15.6. V needs to be adjusted such that the *loss* of such processes is compensated for. At second order in V we have

$$V(\Lambda - d\Lambda) = V(\Lambda) + \sum_{\Lambda - d\Lambda < \epsilon_{\vec{k}''} < \Lambda} \frac{V^2(\Lambda)}{-2\Lambda}$$

$$= V(\Lambda) - \frac{D(0)V^2(\Lambda)}{2}\frac{d\Lambda}{\Lambda}, \tag{15.172}$$

where we used the fact that the energy denominator for the intermediate state $|\vec{k}''\rangle$ is -2Λ, and $D(0)$ is the total density of states per spin which comes from the summation over $\vec{k}''$.

Equation (15.172) can be recast in the form of a differential equation:

$$\frac{dV(\Lambda)}{d\ln(E_c/\Lambda)} = -\frac{D(0)V^2(\Lambda)}{2}. \tag{15.173}$$

This is another example of a renormalization-group flow equation, with the right-hand side being the β function, similar to that encountered in Chapter 11. From this equation we immediately see that the renormalization-group flow is qualitatively different for positive and negative V. As Λ decreases, a positive V decreases its magnitude and flows toward zero, justifying neglecting quasiparticle scattering in Fermi-liquid theory. In other words, a Fermi liquid is stable against positive V or repulsive

---

[42] It is a constant in the sense that it is independent of $\vec{k}$ or $\vec{k}'$. On the other hand, it scales inversely with the system volume $\mathcal{V}$, because that is how the probability that two particles come close together and scatter scales with $\mathcal{V}$. As we will see later, it is the combination $D(0)V_0$ that enters physical quantities, and this combination has no dependence on $\mathcal{V}$ because $D(0) \propto \mathcal{V}$.

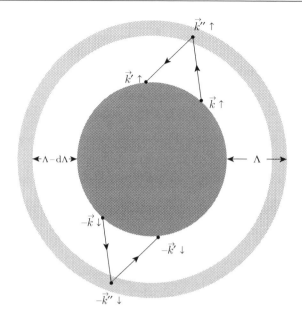

Figure 15.6 Illustration of renormalization of the quasiparticle-pair scattering-matrix element. When the energy cutoff Λ is lowered to $\Lambda - d\Lambda$, states in the lightly shaded shell are eliminated, and higher-order scattering processes involving these states are no longer allowed. The cutoff-dependent scattering-matrix element needs to be renormalized to compensate for this change.

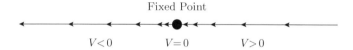

Figure 15.7 Renormalization-group flow diagram for the quasiparticle-pair scattering-matrix element V. V flows to zero if initially positive, and to $-\infty$ if initially negaive. $V = 0$ is a fixed point. Near $V = 0$ the rate of flow is reduced, and at $V = 0$ it vanishes.

interactions. A negative V, on the other hand, becomes *more* negative and flows toward $-\infty$. The flow diagram is illustrated in Fig. 15.7. It indicates that, for *attractive* interaction or negative V, quasiparticle scattering *cannot* be ignored at low energy, no matter how weak V is initially. Using renormalization-group jargon, V is irrelevant if it is positive, but *relevant* when it is negative.[43] The key point is that, for the special case where the momenta (and spins) of the two quasiparticles are oppositely directed (thereby being time reverses of each other), there is a large amount of phase space at low energies available for scattering. This case is special because momentum conservation guarantees that after the scattering the two momenta are still oppositely directed.

At what energy scale does such scattering dominate the physics? Integrating the flow equation (15.173), we obtain

$$V(\Lambda) = \frac{V_0}{1 + [D(0)V_0/2]\ln(E_c/\Lambda)}. \tag{15.174}$$

Consistently with earlier analyses, we find $\lim_{\Lambda \to 0} V(\Lambda) = 0$ if $V_0 > 0$. On the other hand, if $V_0 < 0$, $V(\Lambda)$ *diverges* when Λ reaches

---

[43] More precisely, V is *marginally* irrelevant/relevant when it is positive/negative, as there is no linear term in V in the β function.

$$\Lambda_c = E_c e^{-\frac{2}{D(0)|V_0|}}. \tag{15.175}$$

Clearly the perturbative renormalization-group analysis breaks down as Λ approaches Λ_c, and new physics must occur at this energy scale. A divergent attractive interaction suggests the formation of a bound state between the two quasiparticles, as we now demonstrate. Still restricting our attention to the subspace with only a single time-reversed pair of quasiparticles, we consider a bound state of the form

$$|\Psi_0\rangle = \sum_{\vec{k}} \psi_{\vec{k}} |\vec{k}\rangle \tag{15.176}$$

with eigenenergy $E < 0$:

$$(\hat{H}_0 + \hat{V})|\Psi_0\rangle = E|\Psi_0\rangle. \tag{15.177}$$

In terms of the wave function $\psi_{\vec{k}}$, we have

$$2\epsilon_{\vec{k}}\psi_{\vec{k}} + V_0 \sum_{\vec{k}'} \psi_{\vec{k}'} = E\psi_{\vec{k}}, \tag{15.178}$$

from which we obtain

$$\psi_{\vec{k}} = -\frac{V_0 C}{2\epsilon_{\vec{k}} - E}, \tag{15.179}$$

where

$$C = \sum_{\vec{k}'} \psi_{\vec{k}'}. \tag{15.180}$$

Substituting Eq. (15.179) into Eq. (15.180) yields the equation for E:

$$1 = -\sum_{\vec{k}} \frac{V_0}{2\epsilon_{\vec{k}} - E} = -D(0)V_0 \int_0^{E_c} \frac{d\epsilon}{2\epsilon - E}$$

$$= -\frac{D(0)V_0}{2} \ln\left(\frac{2E_c - E}{-E}\right). \tag{15.181}$$

There is *always* a solution for $V_0 < 0$, and, at weak (initial) coupling $D(0)|V_0| \ll 1$, we expect $|E| \ll E_c$; as a result we can replace $2E_c - E$ by $2E_c$, which leads to an explicit expression for E:

$$E = -2E_c e^{-\frac{2}{D(0)|V_0|}}. \tag{15.182}$$

We find the binding energy $|E|$ is indeed comparable to the energy scale Λ_c in Eq. (15.175), as expected.[44] This is a highly non-perturbative result, as the expression has an essential singularity at zero coupling strength, $V_0 = 0$. The non-perturbative nature of this bound state solution is already suggested by the *breakdown* of perturbative renormalization-group analysis earlier.

Since the pair wave function (15.179) has no angular dependence, the bound state has zero angular momentum or is an *s*-wave solution.[45] Such a wave function is symmetric under the exchange of the spatial coordinates of the pair. Owing to Fermi statistics the spin wave function of the pair must be anti-symmetric, corresponding to a singlet state.

As we have learned in quantum mechanics, an attractive potential/interaction always leads to bound states in 1D and 2D, while a *finite* strength of attractive potential/interaction is needed in 3D to support

---

[44] It turns out that a more correct treatment of the problem that includes Cooper pairs of holes near the lower cutoff finds that the binding energy is even larger. See Exercise 15.24.

[45] In this special model this is the only bound state solution. Solutions for general pairing interactions will be discussed in Section 20.8.

a bound state. The difference is due to the fact that the density of states is divergent/finite in 1D/2D at zero energy (which corresponds to zero momentum), whereas it goes to zero in 3D. The presence of a Fermi surface in the present case gives rise to a finite density of states $D(0)$ which cuts off sharply at the Fermi energy, regardless of dimensionality; so the situation is very much like in 2D in single-particle quantum mechanics.

The relevant nature of $V_0 < 0$, and in particular the bound state it supports, indicates that the Fermi liquid is *unstable* when it is present. This is because the system can lower its energy by exciting two quasiparticles out of the Fermi sea and letting them form a bound state to take advantage of the binding energy. This process does not stop at the two-quasiparticle level; more and more such bound pairs of quasiparticles (known as **Cooper pairs**) will form and condense to destroy the Fermi surface. For electrons, the resulting state is a superconductor whose macroscopic and microscopic descriptions will be presented in Chapters 19 and 20. For neutral fermions (e.g. ^{3}He atoms) this results in a superfluid state, which spontaneously breaks a "gauge" symmetry. Here we state only that this Cooper pairing instability is subtler than the ferromagnetic and Pomeranchuk instabilities, as Fermi-liquid theory does not "know" about it, in the sense that to reveal it requires physics beyond Fermi-liquid theory (quasiparticle scattering in this case). Another difference is that this is a *weak-coupling* instability, in that an infinitesimal negative V_0 immediately destabilizes the Fermi liquid. The perturbative renormalization group is ideally suited for revealing such instabilities since a perturbative calculation of the β function is reliable at weak coupling.

> **Exercise 15.18.** Derive Eq. (15.172) using the formalism of Appendix I.

> **Exercise 15.19.** In the text we considered the weak-coupling limit of $D(0)|V_0| \ll 1$. Now consider the opposite strong-coupling limit of $D(0)|V_0| \gg 1$, and find the expression for the bound state energy.

> **Exercise 15.20.** In the text we considered the bound state of a pair with zero total momentum. Now consider a pair with total momentum much smaller than the momentum cutoff ($q \ll \Delta k_\text{F}$), and find the expression for the bound state energy, in the weak-coupling limit.

15.13.4 Charge and Spin Density-Wave Instabilities

We have seen that a weak attractive interaction immediately leads to the Cooper-pairing instability, while a weak repulsive interaction is (marginally) irrelevant. On the other hand, a sufficiently strong repulsive interaction may lead to a ferromagnetic instability. Are there other possible instabilities associated with repulsive interaction? The answer is yes. We note that repulsive interaction between a pair of quasiparticles is equivalent to attractive interaction between a quasiparticle and a quasihole; it can thus lead to a pairing instability of quasiparticle–quasihole pairs. There are, however, two important differences between a quasiparticle–quasihole pairing instability and a quasiparticle–quasiparticle pairing instability. (i) Because the quasihole is below the Fermi surface and the quasiparticle is above the Fermi surface, the pairs necessarily carry finite momentum, and their condensation breaks translation symmetry spontaneously, leading to charge or spin density-wave states (depending on whether the pairs that condense carry spin or not). (ii) Owing to the difference in kinematic constraints related to (i), the quasiparticle–quasihole pairing instability does *not* occur at weak coupling, except for some very special cases.

Pairing is induced by scattering among the lowest-energy states of the pair, under the constraint of momentum conservation. For a Cooper pair, the lowest-energy states are those with both quasiparticles

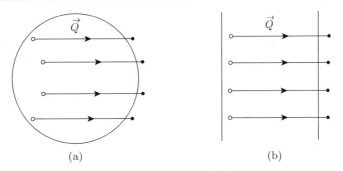

Figure 15.8 Quasiparticle–quasihole pairs with fixed momentum $\vec{Q}$. (a) For a generic Fermi surface such a pair has zero energy (or both the quasiparticle and the quasihole are on the Fermi surface) only at discrete points. (b) For a nested Fermi surface with parallel segments, quasiparticle–quasihole pairs with momentum $\vec{Q}$ that connects the parallel segments can be at zero energy on a finite area of the Fermi surface.

right outside the Fermi surface. Since the total momentum is zero, having one quasiparticle on the Fermi surface guarantees the same for the other quasiparticle, leading to a finite density of states *for the pair* at zero energy. A quasiparticle–quasihole pair carries finite momentum. Generically it has zero energy only at discrete points in momentum space, and as a result the quasiparticle–quasihole pair density of states approaches zero at zero energy (see Fig. 15.8(a) for illustration). Thus, generically, the density-wave instability requires a *finite* repulsive interaction, just like the ferromagnetic instability. Which one wins depends sensitively on details of the system, especially those of the Fermi surface. An exception to this is when the Fermi surface is "nested," namely it contains segments that are parallel to each other (see Fig. 15.8(b)). In such cases there is a finite quasiparticle–quasihole pair density of states at zero energy, and we have a weak-coupling instability toward density-wave formation, very much like the Cooper-pairing instability.

A specific type of charge density-wave state that electrons in the jellium model can form at large r_s is called the **Wigner crystal**, in which electrons (instead of atoms) form a close-packed crystalline state with one electron per unit cell. Obviously such a state minimizes the Coulomb interaction energy by having electrons staying away from each other, at the expense of a kinetic-energy cost associated with spatially localizing the electrons. Wigner crystals have been observed in 2D for electrons at low density trapped on the surface of liquid helium and in 2D electron gases in high magnetic fields where the Wigner crystal has to compete with fractional quantum Hall liquids, to be discussed in Chapter 16.

15.13.5 One Dimension

One dimension (1D) is very special in that the Fermi surface reduces to a pair of discrete Fermi points. They are thus always "parallel" and therefore nested. In fact, if we have a set of decoupled 1D wires arranged as a lattice in transverse directions, the Fermi surface is made of two parallel sheets perpendicular to the wires, similar to that in Fig. 15.8(b). As a result we have weak-coupling quasiparticle–quasihole pairing instability for repulsive interaction (as reflected by the divergence of the Lindhard function encountered in Exercise 15.8), and a quasiparticle–quasiparticle pairing instability for attractive interaction. We are thus led to the conclusion that the 1D Fermi liquid is unstable against *any* interaction; the only Fermi-liquid state is a free Fermi gas in 1D!

Another new feature in 1D is strong quantum fluctuations that do not allow spontaneously broken continuous symmetries. As a result of this, the pairing instabilities do *not* lead to ordered density-wave or superfluid/superconducting states. What we have instead is a critical phase with power-law decay of various correlation functions, known as a **Luttinger liquid**. We have, in fact, encountered an example of this in Chapter 6, where we showed that quantum fluctuations of phonons render the Debye–Waller factor zero for a 1D crystal at zero temperature. Thus a crystal (which is a special form of charge density wave) is unstable against quantum fluctuation in 1D. The resultant state is a 1D liquid state with power-law crystalline correlation, which is also an example of a Luttinger liquid.

Irrespective of whether we start with weak repulsive interactions in a 1D Fermi liquid or start with a Wigner crystal of strongly repelling electrons, proper treatment of the quantum fluctuations leads in both cases to a Luttinger liquid characterized by phonon-like collective density fluctuation modes. Remarkably, there is a one-to-one correspondence between excitations of a 1D free Fermi gas and free bosons in 1D. (A closely related correspondence appears in quantum Hall edge states, to be discussed in Section 16.9.) This can be used to attack the strongly interacting fermion problem by turning it into a weakly interacting boson problem via the technique of "**bosonization**." Readers interested in learning more about 1D correlated electron liquids are directed to Refs. [127, 128].

15.13.6 Two-Dimensional Electron Gas at High Magnetic Field

The starting point of Fermi-liquid theory is the free Fermi gas, with a unique Fermi-sea ground state that minimizes kinetic energy (within the limitation of the Pauli principle). In 2D and when a perpendicular magnetic field is present, the kinetic energy is quenched into a discrete set of Landau levels, each with macroscopic degeneracy. As we discussed in Chapter 12, when the Landau-level filling factor ν is an integer, there is a Landau-level (kinetic-energy) gap in the excitation spectrum, and the system exhibits the integer quantum Hall effect. The situation is completely different when ν is a fractional number. In this case, due to the Landau-level degeneracy, there is a massive ground state degeneracy for non-interacting electrons, associated with the different occupation configurations of the (partially filled) valence Landau level. In this case the ground state is completely determined by minimizing the Coulomb energy, and the kinetic energy plays no role (at least in the high-magnetic-field limit where the Landau-level spacing goes to infinity). Since the kinetic energy is completely degenerate, there is no scale against which to measure the potential energy, and even weak repulsion produces strong non-perturbative correlations in the ground state. A variety of strongly correlated non-Fermi-liquid phases (including the Wigner crystal) become possible. The most fascinating of these states, **fractional quantum Hall liquids**, will be discussed in Chapter 16.

15.14 Infrared Singularities in Fermi Liquids

We have seen that perturbation theory in an interacting Fermi gas is subtle. Landau's insight was that, despite the high density of nearly zero-energy excitations in Fermi gas, weak repulsive interactions generally do not severely perturb the system. A Landau Fermi liquid behaves much like a free-electron gas with various parameters (e.g. effective mass) renormalized by the interactions. On the other hand, we saw in the previous section that this picture can be unstable. For example, in the presence of arbitrarily weak *attractive* interactions, the Landau Fermi liquid is unstable with respect to the formation of Cooper pairs and, as we will see in Chapter 20, the opening of an excitation gap at the Fermi energy. In this section we will discuss some additional interesting features of the low-energy physics of the electron gas associated with the gapless spectrum of the Landau Fermi liquid and the finite density of states for low-energy excitations.

15.14.1 Perfect Screening and the Friedel Sum Rule

As a prelude to some of the singularities associated with the low-energy dynamics of the electron gas, we will return to the question of how an electron gas screens the Coulomb potential from a charged impurity. We saw in Eq. (15.95) that an impurity of charge $+Ze$ attracts a screening cloud of precisely $+Z$ electrons that exactly neutralizes the impurity and removes the leading $1/r$ term in the multi-pole expansion of the screened potential. **Jacques Friedel** showed that this leads to a very important **sum rule** for the phase shifts of the electrons viewed as non-interacting particles scattering from the self-consistently screened impurity potential:

$$2\sum_{\ell=0}^{\infty}(2\ell+1)\frac{\delta_\ell(k_F)}{\pi} = +Z, \tag{15.183}$$

where δ_ℓ is the scattering phase shift (evaluated at the Fermi energy) for particles of angular momentum ℓ scattering off the (spherically symmetric) screened potential. The factor $(2\ell+1)$ accounts for the multiplicity of the ℓth channel and the overall factor of 2 accounts for the spin multiplicity.

In order to define scattering phase shifts for individual particles, it is necessary to view them as non-interacting. One reasonable way to do this would be to use the Kohn–Sham method and some approximation to the density functional as described in Section 15.5. The Kohn–Sham orbitals (scattering states) derived will have phase shifts that obey the Friedel sum rule. It is not essential to have the exact density functional as long as the Hamiltonian contains the long-range part of the Coulomb interaction. This term is so important that the system will pay any finite energy price to screen it out at long distances. There may be small errors in the details of the screening charge distribution (and therefore in the individual scattering phase shifts), but not in its total charge content. If we did have access to the exact density functional, then of course we would be able to find the exact screening charge density. The Kohn–Sham orbitals would also be "exact," but these orbitals are not strictly physically meaningful (even though they give us the exact total density).

Another approach to this question is to assume that the electron–electron interactions are weak enough that the Landau Fermi-liquid picture applies. In this case we have (nearly) non-interacting quasiparticles which can have well-defined scattering phase shifts, possibly modified by the renormalization of the quasiparticle mass.

We can derive the Friedel sum rule by considering the subtle shift in the density of states caused by the presence of the scatterer [129]. At large distances r from the scattering center, the asymptotic solution of the radial Schrödinger equation for energy $\epsilon_k = \hbar^2 k^2/2m$ has the form

$$\psi_\ell(kr) \sim \frac{1}{kr}\sin\left(kr + \delta_\ell - \frac{\ell\pi}{2}\right). \tag{15.184}$$

In order to carefully count the density of states, we need to impose definite boundary conditions at some large but finite radius R. It does not particularly matter what boundary conditions we choose, but, for the sake of definiteness, we will require $\psi_\ell(kR) = 0$, which will in turn fix the allowed values of k. The nth allowed value is given by the implicit equation

$$k_n R + \delta_\ell(k_n) - \frac{\ell}{2}\pi = n\pi. \tag{15.185}$$

If we can solve this implicit equation we can determine the total number of states below energy ϵ_K in angular momentum channel ℓ from

$$N_\ell = \sum_{n=1}^{\infty}\theta(K - k_n)$$

$$\approx \int_0^K dk\,\frac{dn}{dk}, \tag{15.186}$$

where we have arrived at the second equation by treating n as a continuous variable. Using Eq. (15.185), we find

$$\frac{dn}{dk} = \frac{1}{\pi}\left(R + \frac{d\delta_\ell}{dk}\right). \tag{15.187}$$

The first term is simply the usual number of states in the absence of the scattering potential. The change in the number of states due to the scatterer is therefore

$$\Delta N_\ell = \int_0^K dk \, \frac{1}{\pi} \frac{d\delta_\ell}{dk} = \frac{1}{\pi}[\delta_\ell(K) - \delta_\ell(0)] = \frac{1}{\pi}\delta_\ell(k_F), \qquad (15.188)$$

where we used the fact that typically for short-ranged potential scattering, the phase shift vanishes at zero wave vector.[46] Since the bulk value of the Fermi energy is not shifted by the presence in the system of a single impurity, we can replace K by the Fermi wave vector.

With this result we have therefore derived the Friedel sum rule in Eq. (15.183). It is interesting to note that, while the total number of excess electrons in the screening cloud is an integer Z, individual angular momentum channels contribute fractions of an electron (except in the case where the potential is so short-ranged that only the s-wave scattering phase shift is non-zero).

Exercise 15.21. From the shift in the density of states due to the impurity, show that the total energy shift of the electron gas due to the presence of the impurity is given by

$$\Delta E = Z\epsilon_F - 2\sum_{\ell=0}^\infty (2\ell + 1) \int_0^{k_F} dk \, \frac{\hbar^2 k}{m_e} \frac{1}{\pi} \delta_\ell(k). \qquad (15.189)$$

N.B. Since this is a scattering problem, you do not need to think about the potential energy of the interaction with the scatterer. The energy of a given scattering state is given simply by its kinetic energy at large distances where it is not interacting with the scatterer.

15.14.2 Orthogonality Catastrophe

The perfect screening of an impurity of charge Ze requires Z electrons to flow in from infinity. It turns out that this means that the ground state of the electron gas in the presence of the impurity is (in the thermodynamic limit) orthogonal to the ground state in the absence of the impurity [129–131], a fact referred to as the **infrared or orthogonality "catastrophe."** To see why this is interesting, let us imagine the impurity potential is turned on suddenly. In the sudden approximation, the electron gas remains in its original state because the screening charge has not yet had time to flow in from infinity.[47] The state of the electron gas can be expressed as a superposition of the exact eigenstates of the new Hamiltonian. Because of the orthogonality catastrophe, the weight of the (new) ground state in this superposition is zero. That is, the sudden turning on of the potential inevitably creates excitations in the electron gas. In fact, one can think of the dynamics of the inflow of charge from infinity as the creation of an infinite number of infinitesimal low-energy excitations. The mean energy dumped into particle–hole pairs is finite, but the number produced is infinite and it is this "infrared" divergence which accounts for the orthogonality of the two ground states.

This phenomenon can be observed in the **X-ray photoemission** spectra of metals [129]. A high-energy X-ray can eject a core electron from one of the inner shells of an atom in a metal. To the valence electrons in the Fermi sea, the resulting **core hole** looks like a suddenly created charged impurity with $Z = +1$. The energy that goes into creating particle–hole pairs in the electron gas is removed from the kinetic energy of the outgoing photoelectron and can therefore be measured. Detailed calculations [129] show that the probability of energy loss ϵ to the particle–hole pairs has a power-law behavior for small energies:

---

[46] Assuming the potential has no bound states below zero energy, a possibility we will ignore for simplicity.

[47] Depending on the signs of the scattering phase shifts, some angular momentum channels may require particles to flow out to infinity.

$$P(\epsilon) \sim \theta(\epsilon)\epsilon^{\alpha}, \tag{15.190}$$

with the exponent given by

$$\alpha = 2\sum_{\ell=0}^{\infty}(2\ell + 1)\left[\frac{\delta_{\ell}(k_{\mathrm{F}})}{\pi}\right]^2. \tag{15.191}$$

The fact that $\alpha > 0$ implies that $P(0) = 0$, which in turn is a direct manifestation of the orthogonality catastrophe.

A related situation is found in the so-called **X-ray edge problem**. Here the X-ray energy is reduced until it has barely enough energy to excite the core electron up to the Fermi level. If we assume that the core electron is in a P shell and the dipole selection rule causes it to be placed in an S wave near the Fermi energy, the absorption cross section turns out to be a power law with a different exponent,

$$S(\epsilon) \sim \theta(\epsilon)\epsilon^{\alpha - 2\frac{\delta_0}{\pi}}, \tag{15.192}$$

where ϵ is the energy of the X-ray relative to the minimum required to excite the core electron to the ground state of the $(N + 1)$-particle electron gas. In general the exponent is negative and this leads to a divergence of the X-ray absorption cross section as the threshold is approached [129]. One can think of this power-law divergence as a kind of excitonic enhancement of the absorption cross section near threshold. In a semiconductor, the photoelectron can actually bind to the core hole, forming an exciton, and thereby contributing a sharp peak in the absorption spectrum (see Section 9.3). In a metal there is a continuous spectrum instead of a band gap at the Fermi level, preventing the formation of a true bound state (or exciton) there. As a result one has a (broadened power-law) resonance (instead of a sharp peak) just at the Fermi level, corresponding to a quasi-bound state.

We will not pursue detailed calculations here, but the qualitative ideas about infrared singularities introduced here will be relevant to our discussion below of the Kondo problem.

15.14.3 Magnetic Impurities in Metals: The Kondo Problem

Dilute magnetic impurities in metals (e.g. iron in gold) can have surprising effects on the electrical resistance and the magnetic susceptibility at low temperatures. For the case of antiferromagnetic coupling between the magnetic impurities and the conduction electrons, the impurities make an anomalously large contribution to the electrical resistance, which grows logarithmically as the temperature is lowered, before finally saturating. In the absence of any coupling of the impurity spin to the electron gas, the impurity would be a free spin and contribute a simple **Curie-law** term to the magnetic susceptibility that would diverge as T^{-1} as the temperature is lowered. For the case of antiferromagnetic coupling to the electron gas, the **local moment** of the impurity gets **"quenched"** by magnetic screening and its contribution to the susceptibility no longer diverges.

The **Kondo problem** is an idealization of the problem of a single magnetic impurity embedded in an electron gas, and its study played a central role in the development of the **renormalization-group method** for the analysis of many-body systems [132–134]. Interest in the problem arose in 1964 when Jun Kondo [132] noticed that the perturbation theory in powers of the magnetic coupling contained logarithmic divergences. The problem is defined by the deceptively simple-looking Hamiltonian

$$H = H_0 + \frac{J_{\perp}}{2}[S_+ s_- + S_- s_+] + J_z S_z s_z, \tag{15.193}$$

where H_0 is the kinetic energy of the free electron gas, $\vec{S}$ is the operator representing the impurity spin (with $S_{\pm} = S_x \pm i S_y$), and $\vec{s}$ is the conduction electron spin density at the location of the impurity (assumed to be the origin):

$$s_+ = \frac{1}{\mathcal{V}} \sum_{\vec{k},\vec{k}'} c^\dagger_{\vec{k}'\uparrow} c_{\vec{k}\downarrow}, \tag{15.194}$$

$$s_z = \frac{1}{\mathcal{V}} \sum_{\vec{k},\vec{k}'} \frac{1}{2} [c^\dagger_{\vec{k}'\uparrow} c_{\vec{k}\uparrow} - c^\dagger_{\vec{k}'\downarrow} c_{\vec{k}'\downarrow}], \tag{15.195}$$

with $\mathcal{V}$ being the sample volume and $s_- = (s_+)^\dagger$. In the remainder of this section we set $\hbar = 1$ for convenience.

The magnetic interaction is assumed (for simplicity) to be a delta function in space so the scattering-matrix element from state $\vec{k}$ to $\vec{k}'$ is independent of the momentum transfer.[48] In terms of the orbital angular momentum basis, the ultra-short range of the interaction means that only Fermi-sea electrons in the S wave ($\ell = 0$ channel) can interact with the impurity. The J_z term looks like an ordinary scalar impurity scattering term except that the sign of the interaction depends on the relative orientation of the impurity spin and the spin of the Fermi-sea electron. The $J_\perp$ term causes flip–flop scattering in which the impurity spin and the Fermi-sea spin mutually flip.

As we will learn in Chapter 17, magnetic interactions are primarily due to a combination of the Coulomb interaction and the Pauli exclusion principle, rather than direct magnetic effects which are extremely weak (e.g. magnetic dipole–dipole coupling). As a result, the interaction is often, to a very good approximation, spin-rotation invariant, which requires $J_\perp = J_z$. However, it is useful for present purposes to allow the possibility of anisotropic couplings, $J_\perp \neq J_z$. The case $J_\perp, J_z > 0$ corresponds to antiferromagnetic coupling, while ferromagnetic coupling corresponds to $J_\perp, J_z < 0$. As we will see below, the physics of these two cases is very different. For technical convenience later in the calculations, we note that the unitary transformation $U = e^{i\pi S_z}$ rotates the impurity spin quantization axes by 180° around the z direction. This takes S_x to $-S_x$ and S_y to $-S_y$, and thus changes the sign of $J_\perp$ but leaves J_z invariant. Hence, without loss of generality, we can always choose a basis in which $J_\perp$ is positive. In this case, $J_z > 0$ corresponds to antiferromagnetic coupling and $J_z < 0$ corresponds to ferromagnetic coupling.

If the Hamiltonian contained only the J_z term, there would be no spin-flip scattering and S_z would be a constant of the motion. Taking $S_z = +\frac{1}{2}$ (say), the spin-up electrons in the Fermi sea would see one sign of interaction with the impurity and the spin-down electrons would see the opposite, and each case could be solved separately and independently for the scattering phase shifts $\delta_{0\uparrow}$ and $\delta_{0\downarrow}$. From these we could (for example) straightforwardly solve for the contribution of the magnetic impurity to the electrical resistance or the magnetic susceptibility.

Exercise 15.22. In the absence of any coupling of the impurity spin to the electron gas, it would be a free spin and contribute a simple Curie-law term to the magnetic susceptibility that would be inversely proportional to the temperature: $\chi(T) = \chi_0/T$, where χ_0 is a T-independent coefficient. If the Hamiltonian contains only the J_z term, the Curie-law result is still valid but the prefactor χ_0 is modified. Use the result in Exercise 15.21 to determine the modification of χ_0 by taking into account the difference in the Fermi wave vector for spin up and spin down due to the spin susceptibility of the Fermi sea. Assume that the scattering phase shifts $\delta_{0\uparrow}$ and $\delta_{0\downarrow}$ for different spin configurations are known functions of energy.

The presence of the $J_\perp$ term renders the problem highly non-trivial. If the impurity spin suddenly flips, the scattering phase shifts seen by the Fermi sea electrons suddenly change. This leads to an infrared catastrophe reminiscent of the X-ray photoemission problem discussed above. In addition,

---

[48] But it is inversely proportional to the volume of the system, similarly to the scattering-matrix element in the Cooper-pairing problem studied in Section 15.13.3. Here this volume dependence is made explicit in Eqs. (15.194) and (15.195), as a result of which the Js in Eq. (15.193) have no volume dependence.

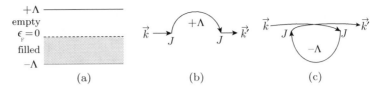

Figure 15.9 (a) High-energy cutoffs symmetrically placed above and below the Fermi level. (b) Virtual process with a particle near the upper cutoff, $+\Lambda$. (c) Virtual process with a hole near the lower cutoff, $-\Lambda$.

whenever the impurity spin flips one of the Fermi-sea electrons also flips. This is equivalent to suddenly removing one particle from the Fermi sea and replacing it with another of the opposite spin. This aspect of the problem makes it rather analogous to the X-ray edge singularity discussed above. Thus we expect interesting low-energy physics. But the problem is further complicated by the fact that repeated spin flips can occur, and they can interfere with each other. Consider two spin-up electrons in the Fermi sea. If the impurity is initially spin down, either *but not both* of the electrons could spin-flip scatter. Thus one must keep track of the entire spin history of the impurity because subsequent interactions with the Fermi sea are contingent upon it. This "memory" effect is the essence of the Kondo problem and must be treated with care. P. W. Anderson derived scaling laws for this problem that were based on a cutoff renormalization method [133]. Kenneth Wilson [134] extended these ideas to develop the modern version of the renormalization-group method and obtain an essentially exact numerical solution to the problem.

To begin our analysis, we first regularize the delta-function interaction defined in Eqs. (15.193)–(15.195) by introducing an "ultraviolet" cutoff that restricts the scattering to an energy range $\pm\Lambda$ on each side of the Fermi level as illustrated in Fig. 15.9(a). This means that no particles will be virtually excited higher than Λ above the Fermi energy and no virtual holes will be created deeper than $-\Lambda$. Our goal is to lower the cutoff from Λ to $\Lambda - \delta\Lambda$ and find a new Hamiltonian that has the same low-energy physics but a new high-energy cutoff. We can do this by using the general method developed in Appendix I to make a unitary transformation which (perturbatively) removes virtual excitations lying in the narrow energy range between the old and the new cutoffs. This unitary transformation will produce a new Hamiltonian similar in form to the old one but with modified coupling constants $J_\perp$ and J_z. As we shall see, this **"renormalization-group flow"** of the coupling constants will illuminate for us the essential physics of the Kondo problem.

To illustrate the method, let us begin with the simpler problem of a scalar impurity embedded in a spinless electron gas,

$$H = H_0 + \frac{J}{\mathcal{V}} \sum_{\vec{k},\vec{k}'} c_{\vec{k}'}^\dagger c_{\vec{k}}. \tag{15.196}$$

Using the methods of Appendix I, there are two types of virtual process that are eliminated when we lower the cutoff by an infinitesimal amount $\delta\Lambda$. The elimination of these high-energy virtual processes modifies the effective scattering from low-energy state $\vec{k}$ to low-energy state $\vec{k}'$ as illustrated in Fig. 15.9. The process shown in Fig. 15.9(b) produces

$$J(\Lambda - \delta\Lambda) = J(\Lambda) + \rho(0)\delta\Lambda \frac{J(\Lambda)^2}{-\Lambda}, \tag{15.197}$$

where we have approximated the energy denominator by $-\Lambda$ by assuming that $\epsilon_{\vec{k}}$, $\epsilon_{\vec{k}'}$, and $\delta\Lambda$ are negligibly small in comparison. Here $\rho(0) = D(0)/\mathcal{V}$ is the single-spin density of states (per unit energy per unit volume) and the number of high-energy states that have been eliminated is $D(0)\delta\Lambda = \mathcal{V}\rho(0)\delta\Lambda$.[49]

---

[49] We assume the density of states has very weak energy dependence and thus use the value at the Fermi level.

The process shown in Fig. 15.9(c) proceeds via virtual production of a hole near the lower cutoff, which is then filled by the incoming electron from state $\vec{k}$ scattering into it. This produces

$$J(\Lambda - \delta\Lambda) = J(\Lambda) + \rho(0)\delta\Lambda \, \frac{J(\Lambda)^2}{-\Lambda}(-1)^F, \tag{15.198}$$

where $F = +1$ because the final state differs from that of Fig. 15.9(a) by the exchange of one pair of fermions. In Fig. 15.9(b) the electron originally in state $\vec{k}$ ends up in state $\vec{k}'$. In Fig. 15.9(c) the electron originally in state $\vec{k}$ ends up deep inside the Fermi sea and an electron originally deep inside the sea ends up in $\vec{k}'$.

Because of the relative minus sign due to fermion exchange, the two processes cancel out, leading to the flow equation

$$\frac{dJ}{d\Lambda} = 0. \tag{15.199}$$

This tells us that a scalar potential is "marginal" – it neither increases nor decreases as the cutoff is lowered. This makes sense because the problem is quadratic in fermion operators and exactly soluble by simply expressing the Fermi-sea states in terms of the exact scattering states in the presence of the impurity. The associated phase shifts $\delta_\ell(k)$ describe the screening cloud around the impurity via the Friedel sum rule.[50] In this simple case, there is no physics which could cause the screening cloud to change as the cutoff is changed.

Exercise 15.23. For simple scalar-potential scattering we have non-interacting electrons, and one can obtain Eq. (15.199) by considering an analogous single-electron problem. The purpose of this exercise is to illustrate this point. In the absence of other electrons forming a Fermi sea, an electron with energy near zero is *not* in a low-energy state, but we can still apply the methods of Appendix I to it. (i) Show that, when one lowers the cutoff Λ, two types of virtual processes analogous (but not identical in the second case) to those illustrated in Figs. 15.9(b) and (c) get eliminated. (ii) Show that they produce renormalizations of J of the form

$$J(\Lambda - \delta\Lambda) = J(\Lambda) \mp \rho(0)\delta\Lambda \, \frac{J(\Lambda)^2}{\Lambda}, \tag{15.200}$$

respectively, as a result of which the net renormalization of J is zero due to their cancelation, resulting in the flow equation (15.199). (iii) Discuss why the second process, despite being physically quite different from the corresponding one with many electrons discussed in the text, produces the same renormalization. Note in particular that the virtual intermediate state has *lower* energy than the initial state in the present case, which is opposite to the many-electron case.

Exercise 15.24. From the text we learned that we cannot neglect the filled Fermi sea when doing the renormalization-group (RG) calculation, as the electrons deep in the Fermi sea also respond to perturbations like the scalar impurity potential. In this exercise we examine their contribution to the RG flow of the pairing interaction, which we studied in Section 15.13 but considered only the processes illustrated in Fig. 15.6 which are equivalent to those of Fig. 15.9(b) in the scalar impurity potential problem. (i) Find another class of processes in the pairing problem, that involve scattering of electrons deep in the Fermi sea and are analogous to those of Fig. 15.9(c) in the scalar impurity potential problem. (ii) Calculate its contribution to the RG flow, and show that,

---

[50] Because the impurity is not necessarily charged, the integer value Z in Eq. (15.183) might be zero. In fact, since the Kondo model completely neglects the Coulomb interaction, Z need not even be integer-valued.

instead of canceling out the term in Eq. (15.173), it doubles it. Discuss the origin of the sign difference between the present case and that of the scalar impurity potential. (iii) Use the corrected flow equation that includes the contributions from the deep-Fermi-sea electrons to determine the correct energy scale at which perturbative RG breaks down. This will turn out to be the correct energy scale for the superconducting gap, which we will learn about in Chapter 20. Compare it with Eq. (15.175) and discuss the origin and magnitude of their differences.

Let us turn now to the case of magnetic impurity scattering, and initially set $J_\perp = 0$. Similar arguments to those given above then lead to

$$\frac{dJ_z}{d\Lambda} = 0. \tag{15.201}$$

However, when we include both $J_\perp$ and J_z together, the cancelation no longer occurs because the corresponding terms in the Hamiltonian do not commute. Consider the diagrams shown in Fig. 15.10(a) and (b) which describe virtual processes that involve a spin-flip scattering and a non-spin-flip scattering, thus contributing to spin-flip scattering at low energy and renormalizing $J_\perp$. The process in Fig. 15.10(a) contributes

$$\frac{J_\perp(\Lambda - \delta\Lambda)}{2} = \frac{J_\perp(\Lambda)}{2} + \rho(0)\delta\Lambda \frac{[+\frac{1}{2}J_\perp(\Lambda)][-\frac{1}{4}J_z(\Lambda)]}{-\Lambda}, \tag{15.202}$$

while the process in Fig. 15.10(b) contributes

$$\frac{J_\perp(\Lambda - \delta\Lambda)}{2} = \frac{J_\perp(\Lambda)}{2} + \rho(0)\delta\Lambda \frac{[+\frac{1}{2}J_\perp(\Lambda)][+\frac{1}{4}J_z(\Lambda)]}{-\Lambda}(-1)^F. \tag{15.203}$$

The change in sign of the J_z matrix element cancels out the minus sign from the fermion exchange, so the two processes add. Taking into account the same diagrams but with the roles of J_z and $J_\perp$ interchanged gives an overall factor of 2. The diagrams in Fig. 15.10(c) and (d) involve two spin-flip scatterings, resulting in zero net spin flip, and these processes thus renormalize J_z. The calculation of the renormalization of J_z proceeds similarly, and the final result is the pair of flow equations

$$\frac{dJ_\perp}{d\Lambda} = -\frac{\rho(0)}{\Lambda} J_\perp J_z, \tag{15.204}$$

$$\frac{dJ_z}{d\Lambda} = -\frac{\rho(0)}{\Lambda} J_\perp^2. \tag{15.205}$$

We see immediately that, for antiferromagnetic coupling ($J_z > 0$), the minus sign in these equations causes the coupling to grow in strength as the cutoff is lowered, while the reverse is true for the ferromagnetic case. Multiplying the first equation by $J_\perp$ and the second by J_z, and taking their difference, gives

$$J_\perp \frac{dJ_\perp}{d\Lambda} - J_z \frac{dJ_z}{d\Lambda} = 0, \tag{15.206}$$

from which we obtain the following invariant for the flow:

$$J_\perp^2 - J_z^2 = C, \tag{15.207}$$

where C is a constant. This defines a set of hyperbolic scaling trajectories (parameterized by different values of C), shown in Fig. 15.11, along which the low-energy physics of the problem is invariant.

The flows in Fig. 15.11 show us, that for the ferromagnetic case, $J_\perp$ flows to zero. It is **"irrelevant" in the renormalization-group sense**, but "irrelevant" does not mean "zero." Once $J_\perp$ has flowed to zero, the Hamiltonian becomes exactly soluble and yields a Curie contribution to the susceptibiity

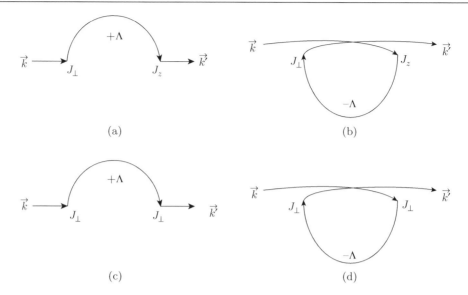

Figure 15.10 Panels (a) and (b) show virtual processes leading to renormalization of $J_\perp$. Panels (c) and (d) show virtual processes leading to the renormalization of J_z.

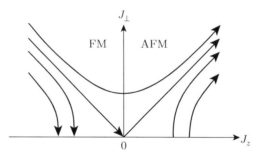

Figure 15.11 Hyperbolic renormalization-group flow lines in the parameter space for the Kondo problem. Spin-flip scattering controlled by $J_\perp$ is irrelevant in the ferromagnet ($J_z < 0$) and relevant in the antiferromagnet ($J_z > 0$). Each flow line obeys $J_\perp^2 - J_z^2 = $ constant. On the ferromagnetic side, the impurity spin yields a (renormalized) Curie-law contribution to the susceptibility at low temperatures. On the antiferromagentic side, the system flows to strong coupling and the impurity susceptibiity saturates and does not diverge at low temperature.

with a renormalized coefficient. On the antiferromagnetic side, $J_\perp$ grows and flows to strong coupling, meaning that it is **"relevant" in the renormalization-group sense**. At some point, the coupling becomes strong enough that our perturbative approximation to the flow equations becomes invalid. Nevertheless, the flow equations have taught us that the ferromagnetic and antiferromagnetic cases are very different. The essence of the Kondo effect in the antiferromagnetic case is that, no matter how weak the initial coupling is, it grows to such a strong value that the impurity spin binds into a singlet with its local screening cloud of Fermi-sea electrons, thereby effectively quenching the Curie susceptibility. This growth of the coupling also implies a strong contribution of the impurity to the electrical resistance of the metal.

For the SU(2)-invariant case the flow equations maintain this symmetry and reduce to

$$\frac{dJ}{d\Lambda} = -\frac{\rho(0)}{\Lambda} J^2, \tag{15.208}$$

which has the solution

$$\rho(0)J(\Lambda) = \frac{\rho(0)J_0}{1 + \rho(0)J_0 \ln(\Lambda/\Lambda_0)}, \tag{15.209}$$

where Λ_0 is the initial value of the cutoff and J_0 is the initial value of the coupling. These are very similar to the corresponding equations in the Cooper-pairing problem discussed in Section 15.13.3. The common feature is that (for a certain sign of interaction) the interaction is marginally relevant.

Throughout this discussion we have been implicitly assuming zero temperature. If the temperature is finite, we cannot neglect it indefinitely as we lower the cutoff. Roughly speaking, we should stop lowering the cutoff once it has reached the scale of the temperature. This leads to an effective coupling that scales with temperature as

$$\rho(0)J_{\text{eff}} = \frac{\rho(0)J_0}{1 + \rho(0)J_0 \ln(k_B T/\Lambda_0)}. \tag{15.210}$$

For the antiferromagnetic case, the logarithmic growth of the effective magnetic coupling as the temperature is lowered explains the logarithmic increase of the resistivity. Continued naive application of this equation to low temperatures leads to a divergence of the coupling at the characteristic Kondo temperature scale

$$k_B T_K = \Lambda_0 e^{-\frac{1}{\rho(0)J_0}}. \tag{15.211}$$

Because of the essential singularity at small values of the bare coupling J_0, this result cannot be obtained in ordinary perturbation theory in powers of the bare coupling J_0, since the radius of convergence vanishes. As noted above, the perturbative approximation to the renormalization-group flow equations breaks down and the coupling does not actually diverge. Rather, the expression for T_K correctly predicts the temperature scale at which the impurity spin is bound in a singlet with its local screening cloud and the physical properties stop changing as the temperature is lowered further (e.g. the Curie-law divergence of the magnetic susceptibility is cut off and the susceptibility saturates at a constant value).[51]

Kenneth Wilson's non-perturbative numerical treatment of the flow equations (including the flow of many new coupling constants associated with new terms generated in the Hamiltonian) gives quantitatively accurate results valid down to arbitrarily low temperatures [134]. Quite remarkably, Nathan Andrei [135, 136] and Paul Wiegman [137] were later able to provide exact analytic expressions for the many-body eigenstate wave functions using Bethe ansatz methods.

15.15 Summary and Outlook

Fermi-liquid theory, combined with band theory based on Bloch's theorem, provides a highly satisfactory description of the normal metallic phase of most metals that exist in nature. More exotic phases, some of which were briefly touched upon in this chapter, can be viewed as *descendants* of Fermi liquids, resulting from various types of relevant interactions that destabilize the Fermi-liquid phase. The remainder of this book is devoted to more detailed descriptions of some of these interaction-dominated phases.

---

[51] In this single-impurity problem this is a cross-over temperature scale where the physics changes, but there is no phase transition. In the Cooper-pairing problem (see in particular Exercise 15.24), there is actually a superconducting phase transition at the corresponding temperature scale. More on this in Chapter 20.

16 Fractional Quantum Hall Effect

We saw in our study of the integer quantum Hall effect that, as one moves to ever higher magnetic fields, the number of filled Landau levels grows smaller and eventually only a single Landau level is filled. At still higher fields this lowest Landau level becomes partially filled and, once the filling factor has fallen below $\nu \sim \frac{1}{2}$, one expects the Fermi level to lie below the percolation threshold. There will then be a mobility gap between the localized states at the Fermi level and the extended states in the middle of the disorder-broadened Landau level. The system would then be expected to be insulating, with both σ_{xx} and σ_{xy} vanishing. In the early history of the field, when sample disorder was moderately strong, this expectation was indeed fulfilled. With the invention of **modulation-doped** GaAs quantum wells (see Exercise 9.7), the disorder potential affecting the electron gas in the well was greatly reduced because the Si donors that supplied the electrons were located in a plane hundreds of ångström units away from the quantum well as illustrated in Fig. 9.15. The setback distance of the dopants greatly damps the high spatial Fourier components of the random disorder potential seen inside the quantum well, thereby reducing the backscattering of the electrons.

It was not really known what effect electron–electron interactions might have, but it was believed (correctly, it turns out) that, if the filling factor were sufficiently small and the disorder were sufficiently low, the electrons might freeze into a Wigner crystal, which is an electron lattice much like an ordinary solid of atoms. Because such a lattice would have gapless phonon-like excitations, it would presumably respond to, and be pinned by, any weak disorder potential, and the electron layer would be insulating. It came as a great shock when in 1982 Tsui, Störmer, and Gossard [82] discovered a quantized Hall plateau at filling factor $\nu = \frac{1}{3}$ with $\sigma_{xx} \to 0$ and $\sigma_{xy} = \frac{1}{3}e^2/h$. When the plateau appeared during the magnetic field sweep, Dan Tsui turned to his colleagues and jokingly said "Quarks!" Of course, the effect did not really involve quarks, but, equally incredibly, the electrons had in fact condensed into a new phase of matter that contained quasiparticles with **fractional charge** $q^* = \pm e/3$. Subsequent further improvements in sample quality have revealed many different quantized Hall plateaux at various rational fractional filling factors, as can be seen in Fig. 12.1. Each of these plateaux corresponds to a distinct phase of matter with novel and unexpected forms of electron correlation and excitations with fractional quantum numbers. Because $\rho_{xx} \longrightarrow 0$, the Hall current is dissipationless and hence there must be an excitation gap. This novel and quite unexpected physics is controlled by Coulomb repulsion between the electrons. It is best understood by first ignoring the disorder[1] and trying to discover the nature of the special correlated many-body ground state into which the electrons condense when the filling factor is a rational fraction. The novel ideas developed to understand these new phases of matter and their topological properties have had broad impact in theoretical physics beyond condensed matter, including even high-energy physics.

16.1 Landau Levels Revisited

In the limit of strong magnetic field, the spacing between Landau levels ($\hbar\omega_c \propto B$) is much larger than the characteristic Coulomb energy $e^2/\epsilon\ell \propto \sqrt{B}$ (where ℓ is magnetic length and ϵ is the dielectric constant of the host semiconductor); it is thus natural to neglect mixing between different Landau levels, and truncate the single-electron Hilbert space to that of a *single*, partially filled Landau level. For most of the observed fractional fillings, $\nu < 2$, thus we need only deal with the lowest Landau level. Without electron–electron interaction there would be a massive ground state degeneracy due to the many different ways to partially fill the degenerate lowest Landau level. Interactions lift this massive degeneracy and, as we will see, preferentially select certain highly correlated incompressible liquid ground states at certain filling factors. Because the kinetic energy is completely degenerate within the lowest Landau level, the effect of the Coulomb interaction cannot be treated perturbatively. The Coulomb scale is the only energy scale in the problem, and its effects are dramatic and highly non-trivial.

> **Exercise 16.1.** Using Eq. (12.38) and combinatorial analysis, find the numerical value of the degeneracy of a one-third-filled Landau level of non-interacting electrons at a magnetic field of 10 T for a sample with area 1 mm². (Use the Stirling approximation for the logarithm of the combinatorial factor and then exponentiate.) It would be difficult to carry out degenerate perturbation theory on a matrix of this dimension!

To study the effect of Coulomb interactions, we need to have a deeper understanding of the single-electron states in the lowest Landau level. In Section 12.5 we solved the Landau-level spectrum and wave functions using the Landau gauge. But physics is gauge-independent, and indeed in Section 12.8 we obtained the Landau-level spectrum *without* specifying a gauge. In the following we review this procedure, and in addition pay particular attention to the lowest-Landau-level wave functions. Consider the 2D single-electron Hamiltonian

$$H = \frac{1}{2m}\left(\vec{p} + \frac{e}{c}\vec{A}\right)^2 = \frac{1}{2m}\vec{\Pi}^2 = \frac{1}{2m}(\Pi_x^2 + \Pi_y^2), \qquad (16.1)$$

where

$$\vec{\Pi} = \vec{p} + \frac{e}{c}\vec{A}(\vec{r}) \qquad (16.2)$$

is the **mechanical momentum** of the electron. Unlike the **canonical momentum** $\vec{p}$, the different components of $\vec{\Pi}$ do *not* commute:

$$[\Pi_x, \Pi_y] = -\frac{i\hbar e}{c}(\partial_x A_y - \partial_y A_x) = \frac{i\hbar e B}{c} = \frac{i\hbar^2}{\ell^2}, \qquad (16.3)$$

---

[1] Of course, just as in the integer quantum Hall effect, if the disorder were truly zero, Galilean invariance would mean that $\rho_{xy} = B/nec$ would simply be linear in the magnetic field and not depend in any way on the states into which the electrons condense.

where for convenience we again choose $\vec{B} = -B\hat{z}$. We thus find the commutation relation between Π_x and Π_y is similar to that between x and p_x; thus Eq. (16.1) takes the form of a 1D harmonic oscillator Hamiltonian. We therefore introduce

$$a = \frac{\ell}{\sqrt{2}\hbar}(\Pi_x + i\Pi_y), \qquad a^\dagger = \frac{\ell}{\sqrt{2}\hbar}(\Pi_x - i\Pi_y), \tag{16.4}$$

yielding $[a, a^\dagger] = 1$ and

$$H = \hbar\omega_c(a^\dagger a + 1/2). \tag{16.5}$$

We thus immediately obtain the Landau-level spectrum

$$E_n = \hbar\omega_c(n + 1/2), \tag{16.6}$$

without specifying the gauge. Furthermore, a state in the lowest Landau level $|\psi\rangle$ must satisfy

$$a|\psi\rangle = 0, \tag{16.7}$$

a relation that is valid in *any* gauge.

How do we see that the Landau levels have (massive) degeneracy? To this end, we introduce another set of conjugate variables (similar to Π_x and Π_y), known as **guiding center coordinates**:

$$\vec{R} = (R_x, R_y) = \vec{r} - \ell^2(\hat{z} \times \vec{\Pi})/\hbar, \tag{16.8}$$

from which it is easy to show that

$$[R_x, R_y] = -i\ell^2. \tag{16.9}$$

We can thus introduce another set of harmonic oscillator operators,

$$b = \frac{1}{\sqrt{2}\ell}(R_x - iR_y), \qquad b^\dagger = \frac{1}{\sqrt{2}\ell}(R_x + iR_y), \tag{16.10}$$

with $[b, b^\dagger] = 1$. More importantly,

$$[R_i, \Pi_j] = [R_i, H] = [a, b] = [a, b^\dagger] = 0, \tag{16.11}$$

which means that R_i has only non-zero matrix elements between states *within the same Landau level*, and so do b and $b^\dagger$.[2] We thus have two independent harmonic oscillators, described by operators a and b, but the latter does not appear in the Hamiltonian in Eq. (16.5). That is, the frequency of the second oscillator is zero, and it is this which produces the massive degeneracy of the Landau levels.

Interestingly, $\vec{R}$ is in fact the position operator $\vec{r} \equiv (x, y)$ *projected onto a given Landau level*:

$$\vec{R} = P_n \vec{r} P_n, \tag{16.12}$$

where P_n is the projection operator onto the nth Landau level.[3] The degenerate states within the same Landau level can be generated using $\vec{R}$ or b and $b^\dagger$. Again, everything we said above is independent of the choice of gauge.

While all gauge choices are equivalent physically, certain gauges are more convenient to use than others. In Chapter 12 we chose the Landau gauge because the Hamiltonian is translationally invariant along one of the two directions, and the corresponding momentum is a good quantum

---

[2] Similarly, Π_i or a and $a^\dagger$ connect states in neighboring Landau levels *with the same intra-Landau-level quantum number*. See Exercise 16.3.

[3] The fact that R_x and R_y (which are Landau-level projected versions of x and y) no longer commute with each other (even though $[x, y] = 0$) has profound consequences on the dynamics of electrons confined to a given Landau level, on which we will have much more to say later. It is also the reason why the most localized wave packet that can be constructed in the lowest Landau level is a Gaussian of the form in Exercise 12.5.

number; it becomes particularly convenient when a confining potential with the same symmetry is present. In the next section we introduce another high-symmetry gauge choice, namely the **symmetric gauge**, whose rotational symmetry helps reveal the correlation physics of **fractional quantum Hall liquids**.

Exercise 16.2. Verify Eqs. (16.9) and (16.11).

Exercise 16.3. Construct the ladder operators a and $a^\dagger$ in the Landau gauge, and show that they have non-zero matrix elements only between states with the same (1D) momentum.

Exercise 16.4. Verify Eq. (16.12).

Exercise 16.5. We have learned in quantum mechanics that, for any single-particle wave function $\psi(x)$, the particle has uncertainty in its position x and momentum p, as summarized by the uncertainty relation $\Delta x \, \Delta p \geq \hbar/2$. Show that, for a wave function $\psi(x, y)$ in the nth Landau level, there is a similar uncertainty relation $\Delta x \, \Delta y \geq (n + 1)\ell^2$, where ℓ is the magnetic length. (Hint: it is useful to separate the electron coordinates x and y as combinations of the guiding center coordinates and components of mechanical momentum.)

16.2 One-Body Basis States in Symmetric Gauge

For reasons that will become clear later, it is convenient to choose the magnetic field to be $\vec{B} = -B\hat{z}$ and analyze the problem in a new gauge

$$\vec{A} = -\frac{1}{2}\vec{r} \times \vec{B} = -(Br/2)\hat{\varphi}, \tag{16.13}$$

known as the **symmetric gauge**. Here r and φ are polar coordinates. Unlike the Landau gauge, which preserves translation symmetry in one direction, the symmetric gauge preserves rotational symmetry about the origin (because $\vec{A}$ points in the azimuthal direction and its magnitude depends on r only). Hence we anticipate that the canonical angular momentum (rather than the y linear momentum) will be a good quantum number in this gauge.

For simplicity we will restrict our attention to the lowest Landau level only. With these restrictions, it is not hard to show that the solutions of the free-particle Schrödinger equation having definite angular momentum are (see Exercises 16.6 and 16.7)

$$\varphi_m(x, y) = \varphi_m(z, z^*) = \frac{1}{\sqrt{2\pi \ell^2 2^m m!}} z^m e^{-\frac{1}{4}|z|^2}, \tag{16.14}$$

where $z = (x + iy)/\ell$ is a dimensionless complex number representing the position vector, z^* is its complex conjugate, and $m \geq 0$ is an integer. (This convenient analytic form results from our choice of $\vec{B} = -B\hat{z}$.)

Exercise 16.6. Express the ladder operators a and b in terms of z and z^* and their derivatives, in the symmetric gauge.

Exercise 16.7. Verify that the basis functions in Eq. (16.14) do solve the Schrödinger equation in the absence of a potential and do lie in the lowest Landau level. There are many ways to do this. One is to re-write the kinetic energy in such a way that $\vec{p} \cdot \vec{A}$ becomes $\vec{B} \cdot \vec{L}$. Another is to look for a special state $\varphi_0(z, z^*)$ that is annihilated by both a and b, and then use $b^\dagger$ repeatedly to generate the other lowest-Landau-level states.

The (canonical as opposed to the mechanical) angular momentum of these basis states is of course $m\hbar$. If we restrict our attention to the lowest Landau level, then there exists only one state with any given canonical angular momentum, and only non-negative values of m are allowed. This "handedness" is a result of the chirality built into the problem by the magnetic field. The fact that the energy is independent of the canonical angular momentum is because the Hamiltonian involves the mechanical angular momentum. The reader may wish to review the distinction, which was discussed in Exercise 10.1.

It seems rather peculiar that in the Landau gauge we had a continuous 1D family of basis states for this 2D problem. Now we find that, in a different gauge, we have a discrete 1D label for the basis states! Nevertheless, we still end up with the correct density of states per unit area. To see this, note that the peak value of $|\varphi_m|^2$ occurs at a radius of $R_{\text{peak}} = \sqrt{2m\ell^2}$. The area $2\pi\ell^2 m$ of a circle of this radius contains m flux quanta. Hence we obtain the standard result of one state per Landau level per quantum of flux penetrating the sample.

Because all the basis states are degenerate, any linear combination of them is also an allowed solution of the Schrödinger equation. Hence any function of the form [138]

$$\Psi(x, y) = f(z)e^{-\frac{1}{4}|z|^2} \tag{16.15}$$

is allowed so long as f is *analytic* in its argument. In particular, arbitrary polynomials of any degree N,

$$f(z) = \prod_{j=1}^{N}(z - Z_j), \tag{16.16}$$

are allowed and are conveniently defined by the locations of their N zeros (or nodes) $\{Z_j; j = 1, 2, \ldots, N\}$. (Since the function must be analytic, it has no poles.)

Another way to see that Eq. (16.15) is a wave function in the lowest Landau level is to show that it satisfies Eq. (16.7). In the symmetric gauge we have

$$a = \frac{\ell}{\sqrt{2}\hbar}\left[p_x + ip_y - \frac{ieB}{2c}(x + iy)\right] = -\frac{i}{\sqrt{2}}\left(\frac{z}{2} + 2\frac{\partial}{\partial z^*}\right); \tag{16.17}$$

clearly we have

$$a\Psi(x, y) \propto f(z)\left(\frac{z}{2} + 2\frac{\partial}{\partial z^*}\right)e^{-\frac{1}{4}|z|^2} = 0 \tag{16.18}$$

as long as $f(z)$ is a single-valued function of z *only* (i.e. not a function of z^*). This also makes it clear why we do *not* have negative m states in the lowest Landau level; they would have to have negative powers of z, which makes the wave function non-normalizable due to divergence at the origin.

Another useful lowest-Landau-level solution is the so-called **coherent state** which is a particular exponential of z:

$$f_\lambda(z) \equiv \frac{1}{\sqrt{2\pi\ell^2}}e^{\frac{1}{2}\lambda^* z}e^{-\frac{1}{4}\lambda^* \lambda}. \tag{16.19}$$

The wave function using this exponential has the property that it is a narrow Gaussian wave packet centered at the position defined by the complex number λ. Completing the square shows that the probability density is given by

$$|\Psi_\lambda|^2 = |f_\lambda|^2 e^{-\frac{1}{2}|z|^2} = \frac{1}{2\pi\ell^2} e^{-\frac{1}{2}|z-\lambda|^2}. \qquad (16.20)$$

Because the guiding center coordinates do not commute with each other, this is (consistently with the uncertainty principle) the smallest wave packet that can be constructed from states within the lowest Landau level. Its square is effectively the expectation value of the projection of the delta-function density operator $\delta^2(\vec{r}-\vec{\lambda})$ onto the Hilbert space of the lowest Landau level. The reader will find it instructive to compare this Gaussian packet with the one constructed in the Landau gauge in Exercise 12.5.

A one-body potential $U(\vec{r})$ breaks Landau-level degeneracies. Just as the Landau gauge is convenient for studying translationally invariant potentials (along one direction; see Chapter 12), the symmetric gauge is convenient for studying rotationally invariant potentials $U(\vec{r}) = U(r)$, in which case m remains a good quantum number. In principle, one needs to solve the Schrödinger equation in the presence of $U(r)$ in this case, as it in general mixes different Landau levels. However, in the limit of strong B field such mixing is negligible, and $|\varphi_m\rangle$ of Eq. (16.14) remains an (essentially) exact eigenstate. The only effect of $U(r)$ is to give an m-dependent energy correction (thus lifting the Landau-level degeneracy):

$$u_m = \langle\varphi_m|U(r)|\varphi_m\rangle. \qquad (16.21)$$

16.3 Two-Body Problem and Haldane Pseudopotentials

Because the kinetic energy is completely degenerate, the effect of Coulomb interactions among the particles is non-trivial. To develop a feel for the problem, let us begin by solving the two-body problem. We note that the two-body Coulomb interaction $V(\vec{r}_1, \vec{r}_2) = V(|\vec{r}_1 - \vec{r}_2|)$ depends on the distance only, and thus respects rotation symmetry. The standard procedure to solve such a two-body problem in the absence of a magnetic field is to introduce center-of-mass and relative coordinates to reduce it to two decoupled one-body problems. Remarkably, such a decomposition is still possible in the presence of a *uniform* magnetic field, for the case of two particles with the *same* mass and charge. The potential energy is still rotationally invariant, but the kinetic energy is modified by the presence of the vector potential:

$$T = \frac{1}{2m_e}(\vec{\Pi}_1^2 + \vec{\Pi}_2^2) = \frac{1}{4m_e}\vec{\Pi}_{CM}^2 + \frac{1}{m_e}\vec{\Pi}_r^2, \qquad (16.22)$$

where m_e is the particle mass (we have the electron in mind), and the center-of-mass and relative mechanical momenta are

$$\vec{\Pi}_{CM} = \vec{\Pi}_1 + \vec{\Pi}_2; \qquad (16.23)$$

$$\vec{\Pi}_r = (\vec{\Pi}_1 - \vec{\Pi}_2)/2. \qquad (16.24)$$

The key to the decomposition is

$$[\Pi_{CM,\mu}, \Pi_{r,\nu}] = 0, \qquad (16.25)$$

which is true under the conditions stated above. It is easy to show that both the center-of-mass part and the relative part of the kinetic energy describe a charged particle moving in a uniform magnetic field, with the *same* cyclotron frequency and magnetic length ℓ; adding a two-body potential $V(r)$ is equivalent to adding a rotationally invariant potential to the relative part.

In the limit of strong magnetic field where Landau-level mixing (due to $V(r)$) can be neglected, the problem simplifies further, due to the analyticity properties of the wave functions in the lowest

Landau level. If we know the angular behavior of a wave function, analyticity uniquely defines the radial behavior. For example, for a single particle, knowing that the angular part of the wave function is $e^{im\theta}$, we know that the full wave function is guaranteed to uniquely be $r^m e^{im\theta} e^{-\frac{1}{4}|z|^2} = z^m e^{-\frac{1}{4}|z|^2}$.

Consider now the two-body problem for particles with relative angular momentum m and center-of-mass angular momentum M. The *unique* analytic wave function in the lowest Landau level is (ignoring normalization factors)

$$\Psi_{mM}(z_1, z_2) = (z_1 - z_2)^m (z_1 + z_2)^M e^{-\frac{1}{4}(|z_1|^2 + |z_2|^2)}. \tag{16.26}$$

If m and M are non-negative integers, then the prefactor of the exponential is simply a polynomial in the two arguments. Hence this state is made up of linear combinations of the degenerate one-body basis states φ_m given in Eq. (16.14), and the state therefore lies in the lowest Landau level. Note that, if the particles are spinless or spin-polarized fermions, then m must be odd to give the correct exchange symmetry. Remarkably, this is the exact (neglecting Landau-level mixing) solution for the Schrödinger equation for *any* central potential $V(|\vec{r}_1 - \vec{r}_2|)$ acting between the two particles. We do not need to solve any radial equation because of the powerful restrictions due to analyticity. There is only one state in the (lowest-Landau-level) Hilbert space with relative angular momentum m and center-of-mass angular momentum M. Hence (neglecting Landau-level mixing) it is an exact eigenstate of *any* central potential. Thus, assuming a rotationally invariant interaction, the restriction to the lowest Landau level uniquely determines the energy eigenfunction, independently of the Hamiltonian! (The eigenvalues do, of course, depend on the details of the Hamiltonian.)

The corresponding energy eigenvalue v_m is independent of M and is referred to as the mth **Haldane pseudopotential** [139],

$$v_m = \frac{\langle mM|V|mM\rangle}{\langle mM|mM\rangle}. \tag{16.27}$$

The Haldane pseudopotentials for the repulsive Coulomb potential are shown in Fig. 16.1.[4]

These discrete-energy eigenstates represent bound states of the *repulsive* potential. If there were no magnetic field present, a repulsive potential would, of course, have only a continuous spectrum, with no discrete bound states. However, in the presence of the magnetic field, there are effectively bound states because the kinetic energy has been quenched. Ordinarily two particles that have a lot of potential energy because of their repulsive interaction can fly apart, converting that potential energy into kinetic energy. Here, however (neglecting Landau-level mixing), the particles all have fixed kinetic energy. Hence particles that are repelling each other are stuck and cannot escape from each other. One can view this semiclassically as the two particles orbiting each other under the influence of $\vec{E} \times \vec{B}$ drift, with the Lorentz force preventing them from flying apart. In the presence of an attractive potential the eigenvalues change sign, but of course the eigenfunctions remain exactly the same (since they are unique)!

The fact that a repulsive potential has a discrete spectrum for a pair of particles is (as we will shortly see) the central feature of the physics underlying the existence of an excitation gap in the fractional quantum Hall effect. One might hope that, since we have found analyticity to uniquely determine the two-body eigenstates, we might be able to determine many-particle eigenstates exactly. The situation is complicated, however, by the fact that, for three or more particles, the various relative angular momenta L_{12}, L_{13}, L_{23}, etc. do not all commute. Thus we cannot write down general exact

---

[4] This is a good time to attempt Exercise 16.9. The reader has probably recognized that this calculation is identical to the calculation of the single-particle energy correction due to a rotationally invariant one-body potential in the strong-field limit, u_m, which was performed in the previous section.

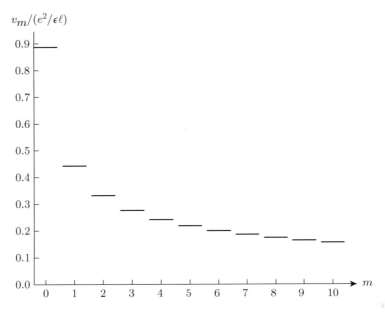

Figure 16.1 Haldane pseudopotentials for the Coulomb interaction, expressed in units of the characteristic Coulomb scale $e^2/\epsilon\ell$.

eigenstates.[5] We will, however, be able to use the analyticity to great advantage and make exact statements for certain special cases.

Exercise 16.8. Verify the claim that the Haldane pseudopotential v_m is independent of the center-of-mass angular momentum M.

Exercise 16.9. Evaluate the Haldane pseudopotentials for the Coulomb potential $e^2/\epsilon r$. Express your answer in units of $e^2/\epsilon\ell$. For the specific case of $\epsilon = 10$ and $B = 10\,\mathrm{T}$, express your answer in kelvins.

Exercise 16.10. Take into account the finite thickness of the quantum well by assuming that the one-particle basis states have the form

$$\psi_m(z, s) = \varphi_m(z)\Phi(s),$$

where s is the coordinate in the direction normal to the quantum well. Write down (but do not evaluate) the formal expression for the Haldane pseudopotentials in this case. Qualitatively describe the effect of finite thickness on the values of the different pseudopotentials for the case where the well thickness is approximately equal to the magnetic length.

---

[5] In fact we can still identify a special set of excited states and corresponding *exact* excitation energies for the N-electron system, associated with the center-of-mass degrees of freedom. Introducing the center-of-mass mechanical momentum $\vec{\Pi}_{\mathrm{CM}} = \sum_{j=1}^N \vec{\Pi}_j$ and the corresponding center-of-mass kinetic energy $T_{\mathrm{CM}} = \vec{\Pi}_{\mathrm{CM}}^2/(2Nm_e)$, we can always write the Hamiltonian as $H = T_{\mathrm{CM}} + H_{\mathrm{r}}$, where the relative part of the Hamiltonian, H_{r}, can be expressed as a function of $\vec{\Pi}_i - \vec{\Pi}_j$ and $\vec{r}_i - \vec{r}_j$. Thus $[T_{\mathrm{CM}}, H_{\mathrm{r}}] = [T_{\mathrm{CM}}, H] = 0$. As a result, the excitation energies of T_{CM}, $n\hbar\omega_{\mathrm{c}}$, are *exact* excitation energies of H.

16.4 The $\nu = 1$ Many-Body State and Plasma Analogy

So far we have found the one- and two-body states. Our next task is to write down the wave function for a fully filled Landau level. We need to find

$$\Psi_1[z] = f[z]e^{-\frac{1}{4}\sum_j |z_j|^2}, \tag{16.28}$$

where $[z]$ stands for $(z_1, z_2, \ldots, z_N)$ and f is a polynomial representing the Slater determinant with all states occupied; the subscript 1 in Ψ_1 stands for filling factor 1. Consider the simple example of two particles. We want one particle in the orbital φ_0 and one in φ_1. Thus (again ignoring normalization)

$$f[z] = \begin{vmatrix} (z_1)^0 & (z_2)^0 \\ (z_1)^1 & (z_2)^1 \end{vmatrix} = (z_1)^0(z_2)^1 - (z_2)^0(z_1)^1$$

$$= (z_2 - z_1). \tag{16.29}$$

This is the lowest-possible-order polynomial that is anti-symmetric. For the case of three particles we have

$$f[z] = \begin{vmatrix} (z_1)^0 & (z_2)^0 & (z_3)^0 \\ (z_1)^1 & (z_2)^1 & (z_3)^1 \\ (z_1)^2 & (z_2)^2 & (z_3)^2 \end{vmatrix}$$

$$= z_2 z_3^2 - z_3 z_2^2 - z_1^1 z_3^2 + z_3^1 z_1^2 + z_1 z_2^2 - z_2^1 z_1^2$$

$$= -(z_1 - z_2)(z_1 - z_3)(z_2 - z_3)$$

$$= -\prod_{i<j}^{3}(z_i - z_j). \tag{16.30}$$

This form for the Slater determinant is known as the **Vandermonde polynomial**. The overall minus sign is unimportant and we will drop it.

The single Slater determinant to fill the first N angular momentum states is a simple generalization of Eq. (16.30):

$$f_N[z] = \prod_{i<j}^{N}(z_i - z_j). \tag{16.31}$$

To prove that this is true for general N, note that the polynomial is fully anti-symmetric and the highest power of any z that appears is z^{N-1}. Thus the highest angular momentum state that is occupied is $m = N - 1$. But, since the anti-symmetry guarantees that no two particles can be in the same state, all N states from $m = 0$ to $m = N - 1$ must be occupied. This proves that we have the correct Slater determinant.

> **Exercise 16.11.** Show carefully that the Vandermonde polynomial for N particles is in fact totally anti-symmetric.

One can also use induction to show that the Vandermonde polynomial is the correct Slater determinant by writing

$$f_{N+1}(z) = f_N(z) \prod_{i=1}^{N}(z_i - z_{N+1}), \tag{16.32}$$

which can be shown to agree with the result of expanding the determinant of the $(N + 1) \times (N + 1)$ matrix in terms of the minors associated with the $(N + 1)$st row or column.

Note that, since the Vandermonde polynomial corresponds to the filled lowest Landau level (all states from $m = 0$ up to $m = N - 1$ are occupied), it is the *unique* state having the maximum density and minimum total angular momentum; hence it is an *exact* eigenstate for any form of (central) interaction among the particles (neglecting Landau-level mixing.).[6]

The (unnormalized) probability distribution for particles in the filled Landau-level state is

$$|\Psi_1[z]|^2 = \prod_{i<j}^N |z_i - z_j|^2 e^{-\frac{1}{2}\sum_{j=1}^N |z_j|^2}. \tag{16.33}$$

This seems like a rather complicated object about which it is hard to make any useful statements. It is clear that the polynomial term tries to keep the particles away from each other and gets larger as the particles spread out. It is also clear that the exponential term is small if the particles spread out too much. Such simple questions as "Is the density uniform?" seem hard to answer, however.

It turns out that there is a beautiful analogy to plasma physics, which was developed by R. B. Laughlin [83], which sheds a great deal of light on the nature of this many-particle probability distribution. To see how this works, we write it in the form of a Boltzmann weight in classical statistical mechanics:

$$|\Psi_m[z]|^2 = e^{-\beta U_{\text{class}}}, \tag{16.34}$$

where $\beta \equiv 2/m$ and

$$U_{\text{class}} \equiv m^2 \sum_{i<j} \left(-\ln|z_i - z_j|\right) + \frac{m}{4} \sum_k |z_k|^2. \tag{16.35}$$

In the present case $m = 1$, but we introduce m as a parameter for later convenience (because it turns out to be the inverse of the Landau-level filling factor). Also the inverse temperature $\beta \equiv 2/m$ simply fixes the normalization of U_{class} in a convenient way, but is otherwise arbitrary. It is perhaps not obvious at first glance that we have made tremendous progress, but we have. This is because U_{class} turns out to be the potential energy of a fake classical one-component plasma of particles of charge m in a uniform ("jellium") neutralizing background. Hence we can bring to bear well-developed intuition about classical plasma physics to study the properties of $|\Psi_1|^2$.

To understand this, let us first review the electrostatics of charges in 3D. For a charge-Q particle in 3D, the surface integral of the electric field on a sphere of radius R surrounding the charge obeys

$$\int d\vec{A} \cdot \vec{E} = 4\pi Q. \tag{16.36}$$

Since the area of the sphere is $A = 4\pi R^2$, we deduce

$$\vec{E}(\vec{r}) = Q\frac{\hat{r}}{r^2}, \tag{16.37}$$

$$\phi(\vec{r}) = \frac{Q}{r}, \tag{16.38}$$

and

$$\nabla \cdot \vec{E} = -\nabla^2 \phi = 4\pi Q \delta^3(\vec{r}), \tag{16.39}$$

---

[6] Another way to see that the Vandermonde polynomial corresponds to the maximum-density state in the lowest Landau level for spinless or spin-polarized electrons is the following. Owing to the analyticity requirement, any totally antisymmetric wave function must contain the Vandermonde polynomial as a factor; additional factors, if present, must be symmetric polynomials of the zs, which increases the total degree of the polynomial and spreads out the electron distribution, thus reducing the electron density.

where ϕ is the electrostatic potential. Now consider a 2D world where all the field lines are confined to a plane (or, equivalently, consider the electrostatics of infinitely long charged rods in 3D). The analogous equation for the line integral of the normal electric field on a *circle* of radius R is

$$\int d\vec{s} \cdot \vec{E} = 2\pi Q, \qquad (16.40)$$

where the 2π (instead of 4π) appears because the circumference of a circle is $2\pi R$ (and is analogous to $4\pi R^2$). Thus we find

$$\vec{E}(\vec{r}) = \frac{Q\hat{r}}{r}, \qquad (16.41)$$

$$\phi(\vec{r}) = Q\left(-\ln\left(\frac{r}{r_0}\right)\right), \qquad (16.42)$$

and the 2D version of Poisson's equation is

$$\nabla \cdot \vec{E} = -\nabla^2 \phi = 2\pi Q \delta^2(\vec{r}). \qquad (16.43)$$

Here r_0 is an arbitrary scale factor whose value is immaterial since it only shifts ϕ by a constant.

We now see why the potential energy of interaction among a group of objects with charge m is

$$U_0 = m^2 \sum_{i<j}(-\ln|z_i - z_j|). \qquad (16.44)$$

(Since $z = (x + iy)/\ell$ we are using $r_0 = \ell$.) This explains the first term in Eq. (16.35).

To understand the second term notice that

$$-\nabla^2\left[\frac{1}{4}|z|^2\right] = -\frac{1}{\ell^2} = 2\pi\rho_{\mathrm{B}}, \qquad (16.45)$$

where

$$\rho_{\mathrm{B}} \equiv -\frac{1}{2\pi\ell^2}. \qquad (16.46)$$

Poisson's equation tells us that $\frac{1}{4}|z|^2$ represents the electrostatic potential of a constant (negative) charge density ρ_{B}. Thus the second term in Eq. (16.35) is the energy of charge-m objects interacting with this negative background.

Notice that $2\pi\ell^2$ is precisely the area containing one quantum of flux. Thus the background charge density is precisely B/Φ_0, the density of flux in units of the flux quantum.

The very-long-range forces in this fake plasma cost huge (fake) "energy" unless the plasma is everywhere locally neutral (on length scales larger than the **Debye screening length**, which in this case is comparable to the particle spacing. See Box 16.1.) In order for the system to be neutral, the density n of particles must obey

$$nm + \rho_{\mathrm{B}} = 0, \qquad (16.47)$$

$$\Rightarrow \quad n = \frac{1}{m}\frac{1}{2\pi\ell^2}, \qquad (16.48)$$

since each particle carries (fake) charge m. For our filled Landau level with $m = 1$, this is of course the correct answer for the density, since every single-particle state is occupied and there is one state per quantum of flux.

We again emphasize that the energy of the fake plasma has nothing to do with the quantum Hamiltonian and the true energy. The plasma analogy is merely a statement about this particular choice of (variational) wave function. It says that the square of the wave function is very small (because U_{class} is large) for configurations in which the density deviates by even a small amount from $1/(2\pi\ell^2)$. The electrons can in principle be found anywhere, but the overwhelming probability is that they are found in a configuration which is locally random (liquid-like) but with negligible density fluctuations on long length scales. We will discuss the nature of the typical configurations again further below in connection with Fig. 16.2.

Box 16.1. Debye Screening

In a plasma or an electrolyte solution, the long-range Coulomb forces make it very energetically expensive to have large regions where the net charge density deviates significantly from zero. A test charge introduced into the system attracts a surrounding screening cloud of opposite charges which causes the electrostatic potential seen by another test charge to decay exponentially with the distance between the charges. The characteristic decay length for the screened potential is known as the Debye length (or, in electrolytes, the Debye–Hückel length). For cations and anions of charge $\pm Q$ each with number density n_0, the Debye–Hückel length is given by $\lambda_{\text{Debye}} = 1/\sqrt{2n_0 r_T}$, where r_T is the length scale at which the Coulomb interaction is equal to the thermal energy $4\pi Q^2/r_T = k_B T$. (If we include the effect of the dielectric constant associated with the polarizability of the ions and convert to SI units, we have instead $Q^2/\epsilon_0 \epsilon r_T = k_B T$.)

This result can be derived by noting that, in the absence of any external potential, the cation charge density is $\rho_+ = +Qn_0$ and the anion charge density is $\rho_- = -\rho_+$. In the presence of a self-consistently screened potential $V(\vec{r})$ the anion (cation) charge density is modified by the Boltzmann factor

$$\rho_\pm(\vec{r}) = \pm Qn_0 e^{-\beta(\pm Q)V(\vec{r})}. \tag{16.49}$$

Expanding to first order in the potential yields the linear-response screening charge density

$$\delta\rho(\vec{r}) = -2\beta n_0 Q^2 V(\vec{r}). \tag{16.50}$$

If we make the identification $\chi_0 = -2\beta n_0$ and plug this into the Thomas–Fermi screening expression in Eq. (15.85) we obtain the Debye–Hückel screening length shown above. The only real difference from Thomas–Fermi screening (besides the fact that both anions and cations participate in the screening) is the fact that the scale of potential change needed to significantly alter the charge density is set by the thermal energy scale rather than the scale of the Fermi energy.

16.4.1 Electron and Hole Excitations at $\nu = 1$

Just like in a free Fermi gas, excitations on top of the $\nu = 1$ ground state are (combinations of) electrons in the unoccupied single-particle orbitals and holes in occupied orbitals. The difference here is that, since all states in the lowest Landau level are occupied, the Fermi energy is inside the gap between Landau levels; as a result electron and hole excitation each have a finite gap (relative to the chemical potential), very much like in a band insulator. This gap is responsible for the incompressibility and quantum Hall effect. An electron excitation must reside in the next Landau level,[7] while a hole excitation resides in the lowest Landau level. For the rest of this subsection we focus on the holes as we can use the analyticity of the lowest-Landau-level wave functions to describe them very efficiently.

To create a single hole we need to remove an electron in the lowest Landau level. Among the lowest-Landau-level orbitals in the symmetric gauge, $|\varphi_{m=0}\rangle$ is the most localized (at the origin). Let

---

[7] If the (exchange-enhanced) Zeeman splitting is smaller than the Landau-level splitting, then the lowest-energy electron excitation is a spin-reversed electron still in the lowest Landau level.

us start by removing an electron from $|\varphi_0\rangle$, while leaving all other orbitals occupied. A straightforward evaluation of the single Slater determinant of this case results in the following single-hole wave function:

$$\Psi_{\text{hole}}[z] = \left(\prod_i z_i\right)\Psi_1[z]. \tag{16.51}$$

Clearly this state describes a single hole carrying charge $+e$, localized at the origin. The latter is reflected in the fact that this wave function vanishes when *any* of the electron coordinates approaches 0; thus the electron density is precisely zero at the origin. What if we now want to create the hole *not* at the origin, but at an arbitrary location Z? An obvious generalization of the wave function above is

$$\Psi_{\text{hole}}[Z, z] = \left(\prod_i (z_i - Z)\right)\Psi_1[z]. \tag{16.52}$$

Clearly now the electron density vanishes at Z. However, for $Z \neq 0$ this wave function no longer corresponds to a single Slater determinant made of $|\varphi_m\rangle$s; it is thus not immediately obvious that the hole carries unit charge. Fortunately this can be easily established using the plasma analogy again:

$$|\Psi_{\text{hole}}[Z; z]|^2 = e^{-\beta U'_{\text{class}}} \tag{16.53}$$

with

$$U'_{\text{class}} = \sum_{i<j}\left(-\ln|z_i - z_j|\right) + \frac{1}{4}\sum_k |z_k|^2 + \sum_k(-\ln|z_k - Z|). \tag{16.54}$$

Comparing with Eq. (16.35), we have the additional (last) term due to the first factor of the wave function (16.52). In the plasma analogy it corresponds to a fixed unit impurity charge located at Z. The perfect screening property associated with the long-range forces of the plasma guarantees that precisely one electron will be removed from its vicinity to maintain local charge neutrality. Thus Eq. (16.52) describes a single hole located at Z, again carrying unit charge.

It is straightforward to generalize Eq. (16.52) to multi-hole states with M holes located at $Z_1, Z_2, \ldots, Z_M$:

$$\Psi_{\text{holes}}[Z, z] = \left(\prod_{k=1}^{M}\prod_i (z_i - Z_k)\right)\Psi_1[z]. \tag{16.55}$$

16.5 Laughlin's Wave Function

When the fractional quantum Hall effect was discovered, Robert Laughlin realized that one could write down a many-body variational wave function at filling factor $\nu = 1/m$ by simply taking the mth power of the polynomial that describes the filled Landau level [83]:

$$f_N^m[z] = \prod_{i<j}^N (z_i - z_j)^m. \tag{16.56}$$

In order for this to remain analytic, m must be an integer. To preserve the anti-symmetry under particle-label exchange, m must be restricted to the odd integers. In the plasma analogy, the particles now have fake charge m (rather than unity), and so the density of electrons must be $n = 1/m2\pi\ell^2$ in order to maintain charge neutrality in the plasma. Thus the Landau level filling factor is $\nu = 1/m = \frac{1}{3}, \frac{1}{5}, \frac{1}{7}$, etc.

The **Laughlin wave function** naturally builds in good correlations among the electrons because each particle sees an m-fold zero at the positions of all the other particles. The wave function vanishes extremely rapidly if any two particles approach each other, and this helps minimize the expectation value of the Coulomb energy.

Since the kinetic energy is fixed we need only concern ourselves with the expectation value of the potential energy for this variational wave function. Despite the fact that there are no adjustable variational parameters (other than m, which controls the density),[8] the Laughlin wave functions have proven to be very nearly exact for almost any realistic form of repulsive interaction. To understand how this can be so, it is instructive to consider a model for which this wave function actually is the exact ground state. Notice that the form of the wave function guarantees that every pair of particles has relative angular momentum greater than or equal to m. One should not make the mistake of thinking that every pair has relative angular momentum precisely equal to m. This would require the spatial separation between particles to be very nearly the same for every pair, which is of course impossible.

Suppose that we write the Hamiltonian in terms of the Haldane pseudopotentials

$$V = \sum_{m'=0}^{\infty} \sum_{i<j} v_{m'} P_{m'}(ij), \tag{16.57}$$

where $P_m(ij)$ is the projection operator which selects out states in which particles i and j have relative angular momentum m. If $P_{m'}(ij)$ and $P_{m''}(jk)$ commuted with each other, things would be simple to solve, but this is not the case. However, if we consider the case of a "hard-core potential" defined by $v_{m'} = 0$ for $m' \geq m$, then clearly the mth Laughlin state is an exact, zero-energy eigenstate[139]:

$$V \Psi_m[z] = 0. \tag{16.58}$$

This follows from the fact that

$$P_{m'}(ij) \Psi_m = 0 \tag{16.59}$$

for any $m' < m$, since every pair has relative angular momentum of at least m.

Because the relative angular momentum of a pair can change only in discrete (even-integer) units, it turns out that this hard-core model has an excitation gap. For example, for $m = 3$, any excitation out of the Laughlin ground state necessarily weakens the nearly ideal correlations by forcing at least one pair of particles to have relative angular momentum 1 instead of 3 (or larger). This costs an excitation energy of order v_1.

This excitation gap is essential to the existence of dissipationless ($\rho_{xx} = 0$) current flow. In addition, this gap means that the Laughlin state is stable against perturbations. Thus the difference between the Haldane pseudopotentials v_m for the Coulomb interaction and the pseudopotentials for the hard-core model can be treated as a small perturbation (relative to the excitation gap). Numerical studies show that for realistic Coulomb pseudopotentials the overlap between the true ground state and the Laughlin state is extremely good.[9]

To get a better understanding of the correlations built into the Laughlin wave function it is useful to consider the snapshot in Fig. 16.2, which shows a typical configuration of particles in the Laughlin ground state (obtained from a Monte Carlo sampling of $|\Psi|^2$) compared with a random (Poisson)

---

[8] Strictly speaking, this is true for systems with rotational invariance. It was pointed out in Ref. [140] that there exists a one-parameter *family* of Laughlin states, with a continuous parameter that describes the degree of anisotropy of the state, and can be viewed as a variational parameter. The original Laughlin wave function is a very special member of this family, which may not give the best description of the ground state in the presence of an anisotropic effective mass and/or anisotropic interaction. The explicit wave functions of such anisotropic Laughlin states are constructed in Ref. [141].

[9] Of course, in the thermodynamic limit, the overlap between any approximate trial wave function and the exact ground state must go to zero. Numerical studies of finite-size systems show good overlap, which decays slowly with the system size. "Good overlap" means that the (square of the) overlap is much larger than $\mathcal{D}^{-1}$, the value achieved by a "typical" state randomly selected from a Hilbert space of dimension $\mathcal{D}$.

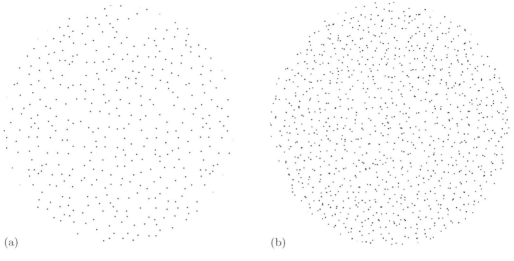

(a)   (b)

Figure 16.2 Comparison of typical configurations for a completely uncorrelated (Poisson) distribution of 1000 particles (a) to the distribution given by the Laughlin wave function for $m = 3$ (b). The latter is a snapshot taken during a Monte Carlo simulation of the distribution. The Monte Carlo procedure consists of proposing a random trial move of one of the particles to a new position. If this move increases the value of $|\Psi|^2$, it is always accepted. If the move decreases the value of $|\Psi|^2$ by a factor p, then the move is accepted with probability p. After equilibration of the plasma by a large number of such moves, one finds that the configurations generated are distributed according to $|\Psi|^2$. It is clear from inspection of the figure that the long-range forces in the effective plasma model produce a particle distribution which is much more uniform than the Poisson distribution in the left panel. Figure reprinted from R. B. Laughlin in [142] with permission from Springer. Copyright Springer-Verlag New York Inc. 1990.

distribution. Focusing first on the large-scale features, we see that density fluctuations at long wavelengths are severely suppressed in the Laughlin state. This is easily understood in terms of the plasma analogy and the desire for local neutrality. A simple estimate for the density fluctuations $\rho_{\vec{q}}$ at wave vector $\vec{q}$ can be obtained by noting that the fake plasma potential energy can be written (ignoring a constant associated with self-interactions being included)

$$U_{\text{class}} = \frac{1}{2L^2} \sum_{\vec{q} \neq 0} \frac{2\pi m^2}{q^2} \rho_{\vec{q}} \rho_{-\vec{q}}, \tag{16.60}$$

where L^2 is the area of the system and $2\pi/q^2$ is the Fourier transform of the logarithmic potential (which is easily derived from $\nabla^2(-\ln r) = 2\pi\delta^2(\vec{r})$). At long wavelengths ($q^2\ell^2 \ll 1$) it is legitimate to treat $\rho_{\vec{q}}$ as a collective coordinate of an elastic continuum. The distribution $e^{-\beta U_{\text{class}}}$ of these coordinates is a Gaussian and hence obeys (taking into account the fact that $\rho_{-\vec{q}} = (\rho_{\vec{q}})^*$)

$$\langle \rho_{\vec{q}} \rho_{-\vec{q}} \rangle = L^2 \frac{q^2}{4\pi m}. \tag{16.61}$$

We clearly see why the long-range (fake) forces in the (fake) plasma strongly suppress long-wavelength density fluctuations. We will return to this point later when we study collective density-wave excitations above the Laughlin ground state.

The density fluctuations on short length scales are best studied in real space. The pair distribution function $g(z)$ (see Section 2.5) is a convenient object to consider; $g(z)$ tells us the density at $r = |z|$ given that there is a particle at the origin:

$$g(z) = \frac{N(N-1)}{n^2 \mathcal{N}} \int d^2 z_3 \cdots \int d^2 z_N |\psi(0, z, z_3, \ldots, z_N)|^2, \tag{16.62}$$

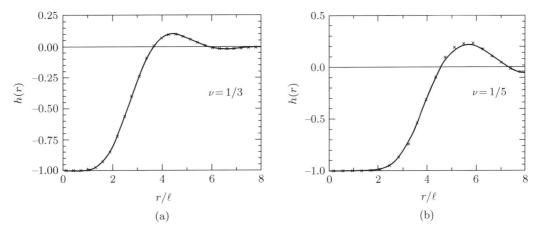

Figure 16.3 Plots of the two-point correlation function $h(r) \equiv 1 - g(r)$ for the Laughlin plasma with $\nu^{-1} = m = 3$ (a) and $m = 5$ (b). Notice that, unlike the result for $m = 1$ given in Eq. (16.63), $g(r)$ exhibits the oscillatory behavior characteristic of a strongly coupled plasma with short-range solid-like local order. Crosses are Monte Carlo data and solid lines are fits to the known general analytic form. Figure reprinted with permission from [143]. Copyright (1986) by the American Physical Society.

where $\mathcal{N} \equiv \langle \psi | \psi \rangle$, n is the density (assumed uniform) and the remaining factors account for all the different pairs of particles that could contribute. The factors of density are included in the denominator so that $\lim_{r \to \infty} g(r) = 1$.

Because the $m = 1$ state is a single Slater determinant, the plasma correlations needed to compute $g(z)$ can be obtained exactly analytically [83]:

$$g(z) = 1 - e^{-\frac{1}{2}|z|^2}. \tag{16.63}$$

Figure 16.3 presents numerical estimates of the pair correlation function $h = 1 - g$ for the cases $m = 3$ and 5. From the way in which the Laughlin wave function vanishes as two particles approach each other, one can see that, for the $\nu = 1/m$ state, $g(z) \sim |z|^{2m}$ for small distances. Because of the strong suppression of density fluctuations at long wavelengths, $g(z)$ converges exponentially rapidly to unity at large distances. For $m > 1$, g develops oscillations indicative of solid-like correlations, and the plasma actually freezes[10] at $m \approx 65$. The Coulomb interaction energy can be expressed in terms of $g(z)$ as[11]

$$\frac{\langle \psi | V | \psi \rangle}{\langle \psi | \psi \rangle} = \frac{nN}{2} \int d^2z \, \frac{e^2}{\epsilon |z|} \big[g(z) - 1 \big], \tag{16.64}$$

where the (-1) term accounts for the neutralizing background and ϵ is the dielectric constant of the host semiconductor. We can interpret $g(z) - 1$ as the density of the **"exchange-correlation hole"** surrounding each particle.

---

[10] That is, Monte Carlo simulation of $|\Psi|^2$ shows that the particles are most likely to be found in a crystalline configuration which breaks translation symmetry. Again we emphasize that this is a statement about the Laughlin variational wave function, not necessarily a statement about what the electrons actually do. It turns out that for $m \gtrsim 7$ the Laughlin wave function is no longer the best variational wave function. One can write down wave functions describing Wigner crystal states which have lower variational energy than the Laughlin liquid.

[11] This expression assumes a strictly zero-thickness electron gas. Otherwise one must replace $e^2/\epsilon |z|$ by $(e^2/\epsilon) \int_{-\infty}^{+\infty} ds \, |\Phi(s)|^2 / \sqrt{|z|^2 + s^2}$, where Φ is the wave-function factor describing the quantum-well bound state (see Exercise 16.10).

The correlation energies per particle for $m = 3$ and 5 are [144]

$$\frac{1}{N} \frac{\langle \Psi_3 | V | \psi_3 \rangle}{\langle \Psi_3 | \psi_3 \rangle} = -0.4100 \pm 0.0001 \tag{16.65}$$

and

$$\frac{1}{N} \frac{\langle \Psi_5 | V | \Psi_5 \rangle}{\langle \psi_5 | \psi_5 \rangle} = -0.3277 \pm 0.0002 \tag{16.66}$$

in units of $e^2 / \epsilon \ell$, which is ≈ 206 K for $\epsilon = 10$, $B = 10$ T. For the filled Landau level ($m = 1$) the exchange energy is $-\sqrt{\pi/8}$, as can be seen from Eqs. (16.63) and (16.64).

Exercise 16.12. Express the exact lowest-Landau-level two-body eigenstate

$$\Psi(z_1, z_2) = (z_1 - z_2)^3 e^{-\frac{1}{4}\left\{ |z_1|^2 + |z_2|^2 \right\}}$$

in terms of the basis of all possible two-body Slater determinants.

Exercise 16.13. Find the pair distribution function for a 1D spinless free-electron gas of density n by writing the ground state wave function as a single Slater determinant and then integrating out all but two of the coordinates. Use this first-quantization method even if you already know how to do this calculation using second quantization. Hint: take advantage of the following representation of the determinant of an $N \times N$ matrix M in terms of permutations P of N objects:

$$\text{Det } M = \sum_P (-1)^P \prod_{j=1}^{N} M_{jP_j}.$$

Exercise 16.14. Use the same method to derive Eq. (16.63).

16.6 Quasiparticle and Quasihole Excitations of Laughlin States

We can construct a "quasihole" wave function for the Laughlin states in a manner similar to Eq. (16.52):

$$\psi_Z^+[z] = \left[\prod_{j=1}^{N} (z_j - Z) \right] \Psi_m[z], \tag{16.67}$$

where Z is a complex number denoting the position of the hole and Ψ_m is the Laughlin wave function at filling factor $\nu = 1/m$. Clearly the electron density is zero at Z in this case as well. Unlike the $\nu = 1$ case, where an additional electron must go to a higher Landau level, here we can also construct a lowest Landau level "quasielectron" state:

$$\psi_Z^-[z] = \prod_{j=1}^{N} \left(2\frac{\partial}{\partial z_j} - Z^* \right) \Psi_m[z], \tag{16.68}$$

where the derivatives[12] act *only* on the polynomial part of Ψ_m. All these derivatives make ψ^- somewhat difficult to work with. We will therefore concentrate on the quasihole state ψ^+. The origin of the names quasihole and quasielectron will become clear shortly.

Just as in our study of the Laughlin wave function, it is very useful to see how the plasma analogy works for the quasihole state

$$|\psi_Z^+|^2 = e^{-\beta U_{\text{class}}} e^{-\beta V}, \tag{16.69}$$

where U_{class} is given by Eq. (16.35), $\beta = 2/m$ as before, and

$$V \equiv m \sum_{j=1}^{N} (-\ln|z_j - Z|). \tag{16.70}$$

Thus we have the classical statistical mechanics of a one-component plasma of (fake) charge-m objects seeing a neutralizing jellium background plus a new potential energy V representing the interaction of these objects with an "impurity" located at Z and having unit charge.

Recall that the chief desire of the plasma is to maintain charge neutrality. Hence the plasma particle will be repelled from Z. Because the plasma particles have fake charge m, the screening cloud will have to have a net reduction of $1/m$ particles to screen the impurity. But this means that the quasihole has **fractional electron number**! The (true) physical charge of the object is a fraction of the elementary charge,

$$e^* = \frac{e}{m}. \tag{16.71}$$

One might wonder whether the fractional charge calculated above is sensitive to the exact form of the variational wave functions (16.67) or (16.68) we used, which in turn depend on the choice of the Laughlin wave function as an *approximation* of the actual ground state. The answer is no. The fractional charge is actually a robust topological property of the fractional quantum Hall phase that the system is in, which does *not* change even when the Hamiltonian and its corresponding ground state wave function change, as long as the bulk excitation gap does not close. In fact the value of e^* is tied to the quantized value of σ_{xy}, as we demonstrate below through a physical process that creates a quasiparticle or a quasihole.

Consider piercing the 2DEG with an infinitesimally thin flux tube (say, a solenoid) at position Z as shown in Fig. 16.4, and slowly increasing the flux from zero to exactly one flux quantum, Φ_0. After this process, the Hamiltonian of the system is *equivalent* to the one we started with, since a unit flux quantum can be gauged away.[13] Owing to the presence of a gap in the bulk, the system stays in an eigenstate of the Hamiltonian during this process, *à la* the adiabatic theorem. However, it does *not* return to the ground state in this case; instead, an excitation has been created during the process.[14] Let us calculate the charge carried by this (localized) excitation. Consider a loop that encloses Z. Faraday's law tells us that there is an electric field that goes around this loop, which satisfies

$$\oint \vec{E} \cdot d\vec{l} = -\frac{1}{c} \frac{d\Phi(t)}{dt}. \tag{16.72}$$

---

[12] The derivative $\partial/\partial z_j$ appears because it is essentially the projection of z_j^* into the lowest Landau level. z_j^* lowers the angular momentum of particle j by one unit but violates the analyticity condition. The derivative also lowers the angular momentum by one unit but maintains the analyticity. This lowering of the angular momentum brings the particles closer together, and is to be contrasted with the increase of angular momentum created by the factors of z_j in the Laughlin quasihole wave function.

[13] The Aharonov–Bohm phase factor for an electron circling the flux tube is 2π, and hence the flux tube is effectively invisible when it contains an integer multiple of the flux quantum. A so-called **"singular gauge transformation"** can remove the 2π phase and hence the flux tube.

[14] The attentive reader can probably now see the similarity between the present reasoning and that of Section 13.5.3.

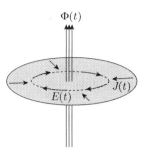

Figure 16.4 Construction of a Laughlin quasiparticle by adiabatically threading flux $\Phi(t)$ through a point in the sample. Faraday induction gives an azimuthal electric field $E(t)$, which in turn produces a radial current $J(t)$. For each quantum of flux added, charge νe flows into (or out of) the region due to the quantized Hall conductivity $\nu e^2/h$. A flux tube containing an integer number of flux quanta is invisible to the particles (since the Aharonov phase shift is an integer multiple of 2π) and hence can be removed by a singular gauge transformation. See Exercise 16.19.

Figure 16.5 Illustration of an example of "fake" charge fractionalization.

Owing to the non-zero Hall conductance, this electric field induces a net current toward Z, and after this process the amount of charge accumulated near Z is

$$Q = \sigma_{xy} \int_0^T dt \oint \vec{E} \cdot d\vec{l} = -\frac{\sigma_{xy} \Phi_0}{c} = -\frac{e}{m}, \tag{16.73}$$

which is precisely the expected quasiparticle charge. The value of the quasiparticle charge is directly related to the quantized Hall conductance. We see from this argument that the Laughlin quasiparticle or quasihole may be viewed as a composite of a fractional charge and a flux tube carrying one flux quantum, a point we will come back to when we discuss their statistics.

This is very strange! How can we possibly have an elementary excitation carrying fractional charge in a system made up entirely of electrons? To understand this, let us consider an example of another quantum system that seems to have fractional charge, but in reality does not. Imagine three protons arranged in an equilateral triangle as shown in Fig. 16.5. Let there be one electron in the system. In the spirit of the tight-binding model we consider only the 1S orbital on each of the three "lattice sites." The Bloch states are

$$\psi_k = \frac{1}{\sqrt{3}} \sum_{j=1}^3 e^{ikj} |j\rangle, \tag{16.74}$$

where $|j\rangle$ is the 1S orbital for the jth atom. The equilateral triangle is like a linear system of length 3 with periodic boundary conditions. Hence the allowed values of the wave vector are $\{k_\alpha = (2\pi/3)\alpha; \quad \alpha = -1, 0, +1\}$. The energy eigenvalues are

$$\epsilon_{k_\alpha} = -E_{1S} - 2J \cos k_\alpha, \tag{16.75}$$

where E_{1S} is the isolated atom energy and $-J$ is the hopping-matrix element related to the orbital overlap and is exponentially small for large separations of the atoms.

The projection operator that measures whether or not the particle is on site n is

$$P_n \equiv |n\rangle\langle n|. \tag{16.76}$$

Its expectation value in any of the three eigenstates is

$$\langle \psi_{k_\alpha} | P_n | \psi_{k_\alpha} \rangle = \frac{1}{3}. \tag{16.77}$$

This equation simply reflects the fact that as the particle tunnels from site to site it is equally likely to be found on any site. Hence it will, on average, be found on a particular site n only one-third of the time. The average electron number per site is thus 1/3. This, however, is a trivial example because the value of the measured charge is always an integer. Two-thirds of the time we measure zero and one-third of the time we measure unity. This means that the charge *fluctuates*. One measure of the fluctuations is

$$\langle P_n^2 \rangle - \langle P_n \rangle^2 = \frac{1}{3} - \frac{1}{9} = \frac{2}{9}. \tag{16.78}$$

This result is most easily obtained by noting that $P_n^2 = P_n$.

A characteristic feature of this "imposter" fractional charge e/m that guarantees that it fluctuates is the existence in the spectrum of the Hamiltonian of a set of m nearly degenerate states. (In our toy example here, $m = 3$.) The characteristic time scale for the charge fluctuations is $\tau \sim \hbar/\Delta\epsilon$, where $\Delta\epsilon$ is the energy splitting of the quasi-degenerate manifold of states. In our tight-binding example $\tau \sim \hbar/J$ is the characteristic time it takes an electron to tunnel from the 1S orbital on one site to the next. As the separation between the sites increases, this tunneling time grows exponentially large and the charge fluctuations become exponentially slow and thus easy to detect.

In a certain precise sense, the fractional charge of the Laughlin quasiparticles behaves very differently from this. An electron added at low energies to a $\nu = 1/3$ quantum Hall fluid breaks up into three charge-1/3 Laughlin quasiparticles. These quasiparticles can move arbitrarily far apart from each other and yet no quasi-degenerate manifold of states appears. The excitation gap to the first excited state remains finite. The only degeneracy is that associated with the positions of the quasiparticles. If we imagine that there are three impurity potentials that pin down the positions of the three quasiparticles, then the state of the system is *uniquely* specified. Because there is no quasidegeneracy, we do not have to specify any more information other than the positions of the quasiparticles.[15] Hence, in a deep sense, they are true *elementary particles* (or, perhaps more precisely, "emergent particles") whose fractional charge is a sharp quantum observable.

Of course, since the system is made up only of electrons, if we capture the charges in some region in a box, we will always get an integer number of electrons inside the box. However, in order to close the box we have to locally destroy the Laughlin state. This will cost (at a minimum) an energy comparable to the excitation gap. This may not seem important since the gap is small – only a few kelvins or so. But imagine that the gap were an MeV or a GeV. Then we would have to build a particle accelerator to "close the box" and probe the fluctuations in the charge. These fluctuations would be analogous to the ones seen in quantum electrodynamics at energies above $2m_e c^2$, where electron–positron pairs are produced during the measurement of charge form factors by means of a scattering experiment.

Put another way, the charge of the Laughlin quasiparticle fluctuates but only at high frequencies $\sim \Delta/\hbar$. If this frequency (which is $\sim 50\,\mathrm{GHz}$) is higher than the frequency response limit of our

---

[15] There also exist the so-called non-Abelian fractional quantum Hall states, whose quasiparticles are even more exotic than Laughlin quasiparticles. In their presence there *is* (quasi)degeneracy even when the positions are fixed. However, all the (quasi-)degenerate states have an identical charge distribution, and there is *no* charge fluctuation associated with such degeneracy. As a result, our discussion about charge fractionalization applies to this case as well. For further information see Section 16.13.

voltage probes, we will see no charge fluctuations. We can formalize this by writing a modified projection operator for the charge on some site n by

$$P_n^{(\Omega)} \equiv P^\Omega P_n P^\Omega, \tag{16.79}$$

where $P_n = |n\rangle\langle n|$ as before and

$$P^{(\Omega)} \equiv \theta(\Omega - H + E_0) \tag{16.80}$$

is the operator that projects onto the subset of eigenstates with excitation energies less than Ω. $P_n^{(\Omega)}$ thus represents a measurement with a high-frequency cutoff built in to represent the finite bandwidth of the detector. Returning to our tight-binding example, consider the situation where J is large enough that the excitation gap $\Delta = (1 - \cos(2\pi/3))J$ exceeds the cutoff Ω. Then

$$
\begin{aligned}
P_\alpha^{(\Omega)} &= \sum_{\alpha=-1}^{+1} |\psi_{k_\alpha}\rangle \theta(\Omega - \epsilon_{k_\alpha} + \epsilon_{k_0})\langle\psi_{k_\alpha}| \\
&= |\psi_{k_0}\rangle\langle\psi_{k_0}|
\end{aligned}
\tag{16.81}
$$

is simply a projector on the ground state. In this case

$$P_n^{(\Omega)} = |\psi_{k_0}\rangle \frac{1}{3} \langle\psi_{k_0}| \tag{16.82}$$

and

$$\left\langle \psi_{k_0} | [P_n^{(\Omega)}]^2 | \psi_{k_0} \right\rangle - \left\langle \psi_{k_0} | P_n^{(\Omega)} | \psi_{k_0} \right\rangle^2 = 0. \tag{16.83}$$

The charge fluctuations in the ground state are then zero (as measured by the finite-bandwidth detector).

The argument for the Laughlin quasiparticles is similar. We again emphasize that one cannot think of a single charge tunneling among three sites because the excitation gap remains finite no matter how far apart the quasiparticle sites are located. This is possible only because we are dealing with a correlated many-particle system.

To gain a better understanding of fractional charge it is useful to compare this situation with that in high-energy physics. In that field of study one knows the physics at low energies – this is just the phenomena of our everyday world. The goal is to study the high-energy (short-length-scale) limit to see where this low-energy physics comes from, or answer the question "What are the fundamental degrees of freedom and force laws that lead to our world?" By probing the proton with high-energy electrons, we can temporarily break it up into three fractionally charged quarks, for example.

Condensed matter physics in a sense does the reverse. We know the phenomena at "high" energies (i.e. room temperature) and the "fundamental" degrees of freedom (electrons in most cases), and we would like to see how the known dynamics (Coulomb's law and non-relativistic quantum mechanics) leads to unknown and surprising collective effects at low temperatures and long length scales. The analog of the particle accelerator is the dilution refrigerator.

To further understand Laughlin quasiparticles, consider the point of view of "flatland" physicists living in the cold, 2D world of a $\nu = 1/3$ quantum Hall sample. As far as the flatlanders are concerned, the "vacuum" (the Laughlin liquid) is completely inert and featureless. They discover, however, that the universe is not completely empty. There are a few elementary particles around, all having the same charge q. The flatland equivalent of Benjamin Franklin chooses a unit of charge which not only makes q negative but gives it the fractional value $-1/3$. For some reason the Flatlanders go along with this.

Flatland cosmologists theorize that these objects are "cosmic strings," topological defects left over from the "big cool down" that followed the creation of the universe. Flatland experimentalists call for the creation of a national accelerator facility which will reach the unprecedented energy scale of

10 K. With great effort and expense this energy scale is reached and the accelerator is used to smash together three charged particles. To the astonishment of the entire world, a new particle is temporarily created with the bizarre property of having integer charge![16]

Indication of fractionally charged objects has been seen in an ultrasensitive electrometer made from a quantum dot [145], and the reduced shot noise which such quantum dots produce when they carry current [64, 65]. The latter will be discussed in Section 16.9.1.

Because the Laughlin quasiparticles are discrete objects, they cost a finite energy to produce. Since they are charged, they can be thermally excited only in neutral pairs. The charge excitation gap is therefore

$$\Delta_c = \Delta_+ + \Delta_-, \tag{16.84}$$

where $\Delta_\pm$ is the quasielectron/quasihole excitation energy. In the presence of a transport current these thermally excited charges can move under the influence of the Hall electric field and dissipate energy. The resulting resistivity has the Arrhenius form

$$\rho_{xx} \sim \gamma \frac{h}{e^2} e^{-\beta \Delta_c / 2} \tag{16.85}$$

where γ is a dimensionless constant of order unity. Note that, just like in Chapter 9 (see Eq. (9.9)), the activation energy for the conductivity is $\Delta_c/2$ not Δ_c since the charges are excited in (particle–hole) pairs. Later, when we learn about superconductivity in Chapter 19 (see Section 19.10), we will see that there is a close analogy between the dissipation described here and flux-flow resistance caused by vortices in a superconducting film. The Laughlin quasihole is essentially a vortex because every electron has an extra unit of angular momentum in its presence. Correspondingly the Laughlin quasi-electron is essentially an anti-vortex because every electron has one fewer unit of angular momentum in its presence.

Theoretical estimates of Δ_c are in reasonably good agreement with experimental values determined from transport measurements in samples with ultra-low disorder. Typical values of Δ_c are only a few percent of $e^2/\epsilon\ell$ and hence no larger than a few kelvins. In a superfluid, time-reversal symmetry guarantees that vortices and anti-vortices have equal energies. The lack of time-reversal symmetry here means that Δ_+ and Δ_- can be quite different. Consider, for example, the hard-core model for which the Laughlin wave function Ψ_m is an exact zero-energy ground state as shown in Eq. (16.58). Equation (16.67) shows that the quasihole state contains Ψ_m as a factor and hence is also an exact zero-energy eigenstate for the hard-core interaction. Thus the quasihole costs zero energy. On the other hand, Eq. (16.68) tells us that the derivatives reduce the degree of homogeneity of the Laughlin polynomial and therefore the energy of the quasielectron *must* be non-zero in the hard-core model. At filling factor $\nu = 1/m$ this asymmetry has no particular significance, since the quasiparticles and quasiholes must be excited in pairs.

Recall that in order to have a quantized Hall plateau of finite width it is necessary to have disorder present (to remove Galilean invariance). For the integer case we found that disorder localizes the excess electrons allowing the transport coefficients not to change with the filling factor. Here it is the fractionally charged quasiparticles that are localized by the disorder. Just as in the integer case, the disorder may fill in the gap in the density of states, but the dc value of σ_{xx} can remain zero because of the localization. Thus the fractional quantum Hall plateaus can have finite width as the filling factor is varied.

---

[16] Whether this particle proves to be short- or long-lived depends on the details of the energetics. One could imagine a positive local potential that would make it energetically stable. The remarkable thing from our point of view as room-temperature beings is that the fractional charges are "deconfined" and can be separated from each other arbitrarily far without requiring infinite energy.

16.7 Fractional Statistics of Laughlin Quasiparticles

In addition to having fractional charge, Laughlin quasiparticles have the novel property of exhibiting **fractional statistics** that are intermediate between boson and fermion statistics. This unique feature is yet another manifestation of the important role of topology in the quantum Hall effect. Fractional statistics are possible only in 2D, and we will now explore what it means to have fractional statistics and how this situation comes about.

16.7.1 Possibility of Fractional Statistics in 2D

We have learned in quantum mechanics that there are two types of identical particles: bosons and fermions, obeying Bose and Fermi statistics, respectively. This conclusion follows from the following observation. The quantum state of a system of N identical particles is described by a wave function that depends on the particle positions and not on any past history of the positions (for simplicity we neglect internal degrees of freedom like spin, but their presence does not alter the discussion qualitatively.): $\psi(\vec{r}_1, \vec{r}_2, \ldots, \vec{r}_N)$. Now introduce the **exchange operator** between a pair of particles, say 1 and 2:

$$P_{12}\psi(\vec{r}_1, \vec{r}_2, \ldots, \vec{r}_N) = \psi(\vec{r}_2, \vec{r}_1, \ldots, \vec{r}_N) = \psi'(\vec{r}_1, \vec{r}_2, \ldots, \vec{r}_N). \tag{16.86}$$

The particles are identical and fundamentally indistinguishable. Hence permuting their positions, or, equivalently, permuting the arbitrary labels we have put on the particles, should not change the physics. Specifically, $(\vec{r}_2, \vec{r}_1, \ldots, \vec{r}_N)$ and $(\vec{r}_1, \vec{r}_2, \ldots, \vec{r}_N)$ actually correspond to the *same* configuration. As a result, we expect

$$\psi'(\vec{r}_1, \vec{r}_2, \ldots, \vec{r}_N) = p\psi(\vec{r}_1, \vec{r}_2, \ldots, \vec{r}_N), \tag{16.87}$$

where p is an eigenvalue of the operator P_{12}. Since we must have

$$|\psi'(\vec{r}_1, \vec{r}_2, \ldots, \vec{r}_N)|^2 = |\psi(\vec{r}_1, \vec{r}_2, \ldots, \vec{r}_N)|^2, \tag{16.88}$$

p can only be a phase factor (or $|p| = 1$). What values can p take? This is dictated by the fact that two exchanges brings the configuration back to $(\vec{r}_1, \vec{r}_2, \ldots, \vec{r}_N)$, or

$$P_{12}^2 = 1, \tag{16.89}$$

implying $p = \pm 1$. We understand that these two possible eigenvalues correspond to bosons and fermions, respectively, and the physics of a many-identical-particle system depends on p dramatically. For the simple case of non-interacting systems, bosons (with $p = 1$) all condense into the lowest-energy single-particle state, while fermions (with $p = -1$) are forced into high-energy states (up to the Fermi energy) due to the Pauli exclusion principle associated with the anti-symmetry of the wave function.

Simple as it is, the above is not the whole story about statistics. It turns out that the quantity that is crucial to the physics is not (just) p associated with the *mathematical* exchange, but rather the phase that arises from adiabatic transport of two particles along a path that actually gives a *physical* exchange of their positions. This phase factor depends both on the wave function *and* on the Hamiltonian. In some cases, it may also depend on the particular path taken by the particles as they trade places. However, it may also contain a contribution η which is purely topological. Since η does not necessarily equal p (although for bosons and fermions the two can easily be the same), we choose η to be the measure of statistics, and will justify this choice below. We start by carefully examining the configuration space of two identical particles, and contrast that with that of two distinguishable particles.

We can always decompose $(\vec{r}_1, \vec{r}_2)$ into a center-of-mass coordinate $\vec{R} = (\vec{r}_1 + \vec{r}_2)/2$ and a relative coordinate $\vec{r} = (\vec{r}_1 - \vec{r}_2)$; exchange involves (essentially) only the $\vec{r}$ space, so we ignore the $\vec{R}$ space

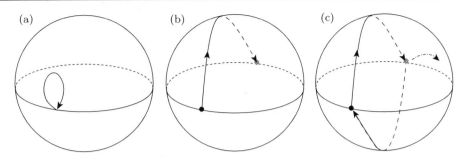

Figure 16.6 Paths showing the relative motion of two identical particles, by tracking their relative coordinate $\vec{r}$. (a) A path which does not involve an exchange. This path can be continuously deformed to a point. (b) A path involving a single exchange ($\vec{r} \to -\vec{r}$). This path is a closed loop since $\vec{r} \sim -\vec{r}$; however, it cannot be deformed to a point without cutting. (c) A path involving two consecutive exchanges. The dot–dash line shows how to deform this closed path to a point, by dragging it around the sphere. This is an illustration of the simple fact that "you cannot lasso a sphere."

in our discussion below. We further exclude the point $\vec{r} = 0$; physically this corresponds to a "hard-core" constraint that forbids two particles from being at the same spatial point. For a pair of identical particles, $\vec{r}$ and $-\vec{r}$ must be identified as the *same* point in their configuration space. It is the topology of this space (and its many-identical-particle generalization) that dictates the allowed values of η and possible types of statistics. We will see that its topology depends on spatial dimensionality in a crucial way.

We first take $d = 3$. Given the hard-core constraint ($r = |\vec{r}| > 0$), we can hold r fixed without loss of generality. We thus imagine two particles, moving in 3D relative to one another at a fixed separation (so that the locus of $\vec{r}$ is the surface of a sphere). The resulting manifold is a spherical surface, with opposite points identified (see Fig. 16.6). This is referred to as the (2D real) **projective sphere** (or plane)[17], or RP^2 [35, 43]. Let us now classify the kinds of closed paths in this configuration space. Figure 16.6(a) shows a path in the trivial class (no exchange). All paths in this class may be continuously shrunk to a point, and we therefore assign to these paths a (trivial) phase factor 1. In Fig. 16.6(b) we see another closed path (remember that $\vec{r} \sim -\vec{r}$, where $\sim$ stands for equivalence), which *cannot* be deformed to a point and thus represents a class distinct from the trivial class. This path represents a single exchange of the two particles, and its class receives the weight or phase factor η. Finally, in Fig. 16.6(c) we show a path which represents two exchanges (weight η^2). It is clear from the figure that this path may be continuously deformed (that is, without cutting) into the trivial path (Fig. 16.6(a)). It thus falls in the *same* **homotopy class** (see Appendix D). We thus find that, in 3D, two exchanges are topologically equivalent to zero exchanges. Thus the phase assigned to two exchanges must be that assigned to the case of no exchange. In other words, by a considerably more involved argument, for the 3D case we obtain the same result as Eq. (16.89): $\eta^2 = 1$. Hence there are only bosons and fermions. Formally, but also more precisely, the discussion above may be summarized by the equation $\pi_1(RP^2) = \mathbb{Z}_2$, namely the **fundamental group** (defined in Appendix D) of the projective sphere is the $\mathbb{Z}_2$ group [35, 43, 44]. This implies there are only two topologically inequivalent classes of closed paths in the configuration space of two identical particles in 3D. It can be shown that the same result is obtained in higher dimensions, namely $\pi_1(RP^{d-1}) = \mathbb{Z}_2$ for $d \geq 3$, although it is much harder to draw the configuration space in such cases.

The reward for our investment of effort in this rather formal language comes when we consider the 2D case. The appropriate manifold for this case is a circle with opposite points identified (see Fig. 16.7). Again it is clear that a single exchange cannot be deformed to the trivial path.

[17] For general dimensionality d, the corresponding manifold is RP^{d-1}.

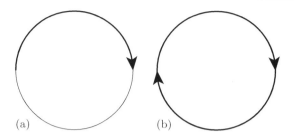

Figure 16.7 The 2D version of Fig. 16.6. (a) A closed path on the projective sphere involving a single exchange in 2D. (b) A closed path involving a double exchange. In contrast to the 3D case, in 2D neither of these paths can be continuously deformed to a point. Hence there is no requirement that the change in phase factor under particle exchange should square to unity, and fractional statistics is possible.

It is also clear, however, that neither can the two-exchange path made of two exchanges in the *same* direction be so deformed, nor in fact can the path representing n such exchanges, for any n. This means that the number of homotopy classes is infinite (namely $\pi_1(RP^1) = \mathbb{Z}$), rather than two. This also means that there is no constraint on the value of η^n, and thus on the value of η, for 2D identical particles.

From the above arguments we see why consideration of simple permutation symmetries of wave functions does not suffice in 2D. A given permutation has no information on the path by which the permutation is accomplished. We have shown that the topology of space for identical particles in 3D allows only two classes of exchange paths and thus only two values of the exchange phase η. These values, being ± 1, can be treated by a formalism which tracks only the net permutations of the particles. The 2D case, however, allows a continuum of values for η, because there is an infinite number of classes of exchange paths. For this case, one clearly needs more information than is contained in the simple two-valued permutation. We suggest, then, that it is an unfortunate historical accident that, on the one hand, there are only two kinds of wave functions which are eigenstates of permutation, while, on the other hand, our world is apparently (at least) 3D, so that we have been presented only with particles (bosons and fermions) whose exchange properties are path-independent. It was not until the discovery of the fractional quantum Hall effect in 2D electron gases that we had a realization of the richer possibilities for the exchange statistics (although the generic possibility had been anticipated theoretically [146, 147]).

There is another way of viewing the paths in Fig. 16.6 which is illuminating. The single-exchange path in Fig. 16.6(b) can be smoothly deformed into its time reverse by simple rotation about the "north pole" (Fig. 16.8). In quantum mechanics (for spinless particles), time-reversal transformation is equivalent to taking the complex conjugate of the amplitude. Since the path in Fig. 16.6(b) and its time reverse are equivalent, we have that $\eta^* = \eta$, which again gives $\eta = \pm 1$. For the 2D case, however, it is clear that there is no way to deform a path into its time reverse: these are paths with opposite winding numbers on a circle. This again tells us that η in 2D is not constrained to the values ± 1. We also see that 2D particles which are neither fermions nor bosons (known as "anyons") do not obey time-reversal symmetry.[18] This is in fact an outstanding signature of particles with non-standard statistics, which should be an aid both in thinking about their properties and in detecting their existence. The fractional quantum Hall effect, a phenomenon that exists in the presence of a strong magnetic field that breaks time-reversal symmetry, thus allows such exotic possibilities.

---

[18] There is one loophole to this claim. A time-reversal-symmetric system could have different flavors of anyons that occur in pairs with opposite values of the statistical angle θ. The two members of each flavor would transform into each other under time reversal.

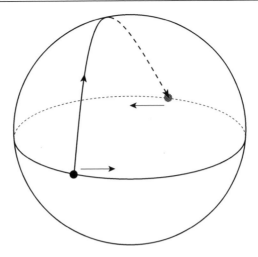

Figure 16.8 Illustration of how to deform the 3D exchange path (Fig. 16.6(b)) into its time reverse, without cutting, by simply rotating it about the "north pole" as shown – keeping the "equatorial" points diametrically opposite and leaving the polar point fixed.

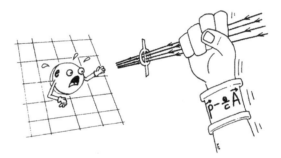

Figure 16.9 The piercing of a charge (which is confined to 2D) with a flux tube. The resulting composite object can have fractional statistics. To date, experimentalists have not succeeded in performing this operation; however, nature has been (as always) more clever. Figure reprinted with permission from [90]. Copyright 1989 World Scientific Publishing Co. Pte Ltd.

16.7.2 Physical Model of Anyons

Leinaas and Myrheim realized the mathematical possibility of fractional statistics in 2D [146]. Wilczek [147] constructed an explicit model of identical particles in 2D with $\eta \neq \pm 1$, which he termed **anyons** (since η can be an arbitrary phase factor). An anyon is modeled as a composite of a charge q and an infinitely thin flux tube piercing the charge (the construction of such an object is graphically illustrated in Fig. 16.9). In 2D, one can consistently allow the flux to take on any value, just as in the Aharonov–Bohm effect. Thus, the physical exchange of two identical 2D charge–flux composites by adiabatic transport can in principle give any phase $\eta = e^{i\theta}$, where θ is referred to as the "**statistical angle**."

The anyon picture allows us to present a simple model which again demonstrates the independence of η from p, and the fact that the physics is determined by the product ηp. We consider two bosons (meaning $p = +1$) with charge q bound to the ends of a rigid rod which is itself confined to a plane. Neglecting the center-of-mass motion and fixing the distance between the two particles, the problem has one degree of freedom: the relative angle ϕ. The two-body wave function ψ must be symmetric under exchange, since the particles are bosons; but exchange simply adds π to ϕ:

$$\psi(\phi + \pi) = +\psi(\phi).$$ (16.90)

The Hamiltonian, eigenfunctions, and eigenvalues are

$$H = \frac{L^2}{2I} = \frac{\hbar^2}{2I}\left(-i\frac{\partial}{\partial\phi}\right)^2,$$ (16.91)

$$\psi(\phi) = e^{im\phi}, \qquad E_m = \frac{(m\hbar)^2}{2I},$$ (16.92)

where L is the angular momentum and I is the moment of inertia. The symmetry requirement then restricts the boson spectrum to even values of m and the ground state energy is zero. Now we bind flux tubes to the bosons. Note that the particles never contact the flux, responding only to the vector potential. (We assume throughout that particles do not see their own flux.) The Hamiltonian in the presence of the flux tubes is (note the similarity to Exercise 10.1)

$$H' = \frac{\hbar^2}{2I}\left(-i\frac{\partial}{\partial\phi} + \frac{\theta}{\pi}\right)^2,$$ (16.93)

where

$$\theta = \pi\frac{q}{e}\frac{\Phi}{\Phi_0}$$ (16.94)

controls the phase under adiabatic exchange,

$$\eta = e^{i\theta}.$$ (16.95)

The new energy eigenvalues are

$$E'_m = \frac{\hbar^2(m + \theta/\pi)^2}{2I}.$$ (16.96)

For a suitable choice of flux, we can have $\theta/\pi = 1$, making the energy spectrum of bosons, whose wave function is symmetric (m even), precisely equivalent to that of fermions (without attached flux). One can in fact show that, for $\theta/\pi = 1$, every observable in the system has the value identical to what it would have been if the particles had been fermions (without flux tubes). For instance, the (mechanical) angular "Fermi" velocity in the ground state, $(\hbar/I)(\theta/\pi)$, is non-zero (even though $m = 0$) – again giving (for $\theta = \pm\pi$) the value appropriate to fermions. Thus if $p = +1$ ("bosons") but $\eta = -1$, the particles are effectively fermions. Conversely, if the permutation statistics is that of fermions ($p = -1$) and $\eta = -1$, the particles behave as effective bosons. In short, it is the product ηp that determines the observable physics.

For non-integral values of θ/π, the particles are effectively anyons: their statistics is "fractional." (Here we see that the term "fractional" statistics refers to the "statistics angle" θ being a fractional multiple of π.) This example, admittedly simple and artificial, nevertheless amply illustrates our claim: p (the permutation symmetry of ψ) does not necessarily equal η, the exchange phase, and it is the combination that determines the physics.

Exercise 16.15. Solve the toy model whose Hamiltonian is given in Eq. (16.93) by making a singular gauge transformation $U\psi = e^{-i\frac{\theta}{\pi}\varphi}\psi$. Show that this eliminates the vector potential from the Hamiltonian, at the expense of changing the boundary conditions that the wave function satisfies. Discuss how the exchange phase now follows from the new boundary conditions rather than from the Aharonov–Bohm phase associated with the vector potential.

16.7.3 Statistics Angle of Laughlin Quasiholes

As discussed in Section 16.6, a Laughlin quasiparticle (or quasihole) can be generated by piercing the fractional quantum Hall liquid at $v = 1/m$ with a unit flux quantum. It can thus be viewed as a composite of a (fractional) charge $q = e^*$ and a flux tube with $\Phi = \Phi_0$, which is precisely the model anyon! Applying Eq. (16.94), we immediately obtain the statistics angle of the Laughlin quasiparticle/quasihole to be

$$\theta = \pm \pi / m. \tag{16.97}$$

This result can also be obtained directly (and more rigorously) by calculating the Berry phase picked up by adiabatically exchanging two quasiholes, using the variational wave function (16.67), as we demonstrate below.

As a warm-up we first consider a *single* quasihole, and use the above-mentioned Berry-phase calculation to determine its charge. To this end, we go back to the single-quasihole wave function (16.67), but now need to pay close attention to its (Z-dependent) normalization, as Z is now a parameter that is varied in a Berry-phase calculation. We claim that, in order to have a Z-*independent* normalization, Eq. (16.67) needs to be modified to

$$\psi_Z[z] = e^{-\frac{|Z|^2}{4m}} \prod_{j=1}^{N} (z_j - Z) \Psi_m[z]; \tag{16.98}$$

note the (new) Z-dependent exponential factor. To show that this wave function has a Z-*independent* normalization, we note that $\langle \psi_Z | \psi_Z \rangle$ can be interpreted as the partition function of a classical one-component plasma in the presence of a unit impurity (fake) charge located at Z. Owing to the screening property of the plasma, this partition function (which is the exponential of the free energy) is independent of Z as long as the impurity does not get close to the edge of the plasma. Note that the added exponential factor in Eq. (16.98) translates into the interaction between the impurity charge and background potential in the plasma analogy, and without this term the analogy does *not* hold. We are now thus free to add a normalization *constant* to Eq. (16.98), and work with a properly normalized $|\psi_Z\rangle$.

We would like to calculate the Berry phase picked up by $|\psi_Z\rangle$ after we drag Z through a closed loop. This is a well-defined and (in principle) straightforward calculation, given the explicit form of the wave function (16.98). To simplify the calculation, however, we will now introduce some nice tricks, taking advantage of the (almost) holomorphic form of (16.98). First, instead of calculating the X and Y components of the Berry connection (see Eq. (13.11)), we work with the complex coordinates of a 2D plane $Z = X + iY$ and $Z^* = X - iY$, and the corresponding components of Berry connections:

$$A_Z = i \langle \psi_Z | \partial_Z \psi_Z \rangle, \qquad A_{Z^*} = i \langle \psi_Z | \partial_{Z^*} \psi_Z \rangle. \tag{16.99}$$

The only place in ψ_Z that Z^* appears is in the exponential factor, and so a straightforward calculation yields

$$A_{Z^*} = i \langle \psi_Z | \partial_{Z^*} \psi_Z \rangle = -i \frac{Z}{4m}. \tag{16.100}$$

The calculation for A_Z appears to be much more difficult because Z appears many times in ψ_Z. However, we can take advantage of the fact that $|\psi_Z\rangle$ is normalized:

$$i \, \partial_Z \langle \psi_Z | \psi_Z \rangle = 0 = A_Z + i(\partial_Z \langle \psi_Z |) | \psi_Z \rangle, \tag{16.101}$$

so that

$$A_Z = -i(\partial_Z \langle \psi_Z |) | \psi_Z \rangle = +i \frac{Z^*}{4m}, \tag{16.102}$$

where we have taken advantage of the fact that the only Z-dependent factor in $\langle \psi_Z |$ is the extra exponential factor in Eq. (16.98) (because, when we turn a ket into a bra, Z becomes Z^*).

From the above we can now reconstruct the X and Y components of the Berry connection,[19]

$$A_X = A_Z + A_{Z^*} = +\frac{Y}{2m}, \qquad A_Y = i(A_Z - A_{Z^*}) = -\frac{X}{2m}, \qquad (16.103)$$

which is nothing but the symmetric gauge of a uniform "statistical" magnetic field (Berry curvature), whose strength is $B_{\text{eff}} = -1/m$. Recalling that Z is a dimensionless coordinate, we can immediately find that the Berry phase picked up by dragging the quasihole through a loop is

$$\gamma = \oint \vec{A}(\vec{R}) \cdot d\vec{R} = -\frac{1}{m}\frac{\text{Area}}{\ell^2} = +\frac{2\pi}{m}\frac{\Phi}{\Phi_0}, \qquad (16.104)$$

where Φ is the (physical magnetic) flux enclosed in the loop.[20] This is nothing but the Aharonov–Bohm phase picked up by a (positive physical) charge $e^* = e/m$ moving in the original (physical) magnetic field, thus confirming our earlier calculation of the charge e^* using the plasma analogy.

We now generalize the above to the case of two quasiholes, by writing down the wave function with a constant norm:

$$\psi_{Z_1,Z_2}[z] = [(Z_1 - Z_2)^*(Z_1 - Z_2)]^{\frac{1}{2m}} e^{-\frac{|Z_1|^2 + |Z_2|^2}{4m}} \prod_{j=1}^{N} (z_j - Z_1)(z_j - Z_2) \, \Psi_m[z]. \qquad (16.105)$$

Note that the factor $[(Z_1 - Z_2)^*(Z_1 - Z_2)]^{\frac{1}{2m}}$ becomes the Coulomb interaction between the two impurities in the plasma analogy, which is what is needed to obtain a constant norm. The Berry connection and Berry-phase calculation proceeds (almost) identically as above; but, in addition to the contributions from two (fractionally) charged particles moving in the background magnetic field, we also obtain an *additional* phase factor when Z_1 is moved along a closed path that encircles Z_2. Without loss of generality, we can set $Z_2 = 0$ and $Z_1 = Z$. Let $a(Z)$ be the additional term in the Berry connection due to the quasihole at the origin. A straightforward calculation yields

$$a_Z = -\frac{i}{2m}\frac{1}{Z}, \qquad (16.106)$$

$$a_{Z^*} = +\frac{i}{2m}\frac{1}{Z^*}, \qquad (16.107)$$

from which we obtain

$$a_X = \frac{-1}{m}\frac{Y}{R^2}, \qquad a_Y = \frac{1}{m}\frac{X}{R^2}, \qquad (16.108)$$

$$\vec{a}(\vec{r}) = -\frac{\vec{R} \times \hat{z}}{mR^2} = +\frac{1}{m}\frac{1}{R}\hat{\varphi}, \qquad (16.109)$$

which is precisely the (symmetric-gauge) vector potential associated with a flux tube at the origin. The statistics angle is thus path-independent for any path which circles the origin (once) counterclockwise:

$$\theta \equiv \frac{1}{2} \oint d\vec{R} \cdot \vec{a}(\vec{r}) = +\frac{\pi}{m}. \qquad (16.110)$$

The factor of $1/2$ arises because one quasihole completely encircling the other is equivalent to two physical exchanges. (Circling half way around and then displacing the pair produces one exchange.)

---

[19] Note that, in such coordinate transformations, A_X transforms like $\partial_X = (\partial Z/\partial X)\partial_Z + (\partial Z^*/\partial X)\partial_{Z^*} = \partial_Z + \partial_{Z^*}$, and similarly $\partial_Y = i(\partial_Z - \partial_{Z^*})$, while A_Z and A_{Z^*} transform in the same way as ∂_Z and ∂_{Z^*}.

[20] Recall that we have taken $\vec{B} = -B\hat{z}$, so $\Phi < 0$ for a right-handed loop.

When a quasihole is moved around in a large loop encircling another quasihole, the total Berry phase can be found by combining the results from Eqs. (16.110) and (16.104):

$$\gamma_{\text{total}} = \gamma + 2\theta = +\frac{2\pi}{m}\left[-\frac{\text{Area}}{2\pi\ell^2} + 1\right]. \qquad (16.111)$$

Recall that the quasihole has positive electrical charge $e^* = e/m$. The first term in the square bracket is the number of flux quanta of the external field enclosed by the trajectory and is negative because the external field is pointing in the $-\hat{z}$ direction (and we have a positive charge traversing the loop in a right-handed sense, i.e. counterclockwise). The second term in the square brackets appears to indicate that there is a single additional quantum of flux associated with the creation of the quasihole (as depicted in Fig. 16.4), but the flux in this tube is oriented oppositely to that of the external field. Does this make sense? We know that uniformly increasing the strength of the external magnetic field while keeping the filling fraction fixed increases the number density of electrons, $\bar{n} = 1/m2\pi\ell^2$, thereby making the total charge inside the loop more *negative*. Thus it makes sense that the flux tube used to create the quasihole should be oriented in such a way as to *reduce* the total flux inside the loop so that a fraction of an electron is expelled from the loop, leaving the quasihole with positive charge.

We have been somewhat sloppy in interchangeably using the language of flux tubes and vortices in this discussion. On the one hand, we have the picture in Fig. 16.4, which involves actual insertion of a physical flux tube in order to create a quasihole. On the other hand, in Eq. (16.98) we simply wrote down the microscopic wave function living in the lowest Landau level (with uniform external magnetic field and no flux tube). This wave function has a "vortex" located at point Z because every electron has a phase winding of 2π about that point. If we had actually inserted a magnetic flux tube at point Z and adiabatically raised the flux from 0 to Φ_0, the kinetic-energy term of the Hamiltonian would be different, and the solution of the Schrödinger equation would have not been Eq. (16.98), but rather

$$\tilde{\psi}_Z[z] = e^{-\frac{|Z|^2}{4m}}\prod_{j=1}^{N}|(z_j - Z)|\Psi_m[z]. \qquad (16.112)$$

Having constructed the quasihole by adiabatically adding flux we can remove the (quantized) flux using a singular gauge transformation

$$\tilde{\psi}_Z(z) \rightarrow \prod_{j=1}^{N}e^{i\varphi_j}\tilde{\psi}_Z(z), \qquad (16.113)$$

where

$$\varphi_j = \text{Arg}(z_j - Z). \qquad (16.114)$$

This gauge change removes the quantized flux tube and brings us back to the solution in Eq. (16.98). This gauge change is singular because, while it leaves the wave function single-valued, it changes the phase winding of the wave function around the point Z. In a precise sense, this gauge change replaces the quantized magnetic flux tube in the Hamiltonian with a quantized vortex in the wave function. In principle one could do the Berry phase calculations in the gauge where the flux tubes are present, as we did in relating the Berry phase to the Aharonov–Bohm effect in Section 13.2. However, in the present case this would be quite complex.

Exercise 16.16. Using the plasma analogy, verify that Eq. (16.105) has a constant normalization independent of Z_1 and Z_2 as long as Z_1 and Z_2 are widely separated from each other and from the edges of the plasma. (Recall that the electrons carry fake charge m in the plasma and the impurities carry fake charge 1.)

Exercise 16.17.

(a) Derive Eq. (16.109) from Eq. (16.105).

(b) Instead of setting $Z_2 = 0$ and $Z_1 = Z$, define $Z = Z_1 - Z_2$. Moving Z halfway around the unit circle produces one exchange and moving Z all the way around the unit circle produces two exchanges. Re-derive the additional contribution to the Berry connection due to the factor $(Z^* Z)^{\frac{1}{2m}}$ and show that the statistics angle obeys

$$\theta = \frac{1}{2} \oint \left(dZ\, a_Z + dZ^*\, a_{Z^*} \right) = +\frac{\pi}{m}. \tag{16.115}$$

Exercise 16.18. Derive Eq. (16.112).

Exercise 16.19. Derive Eq. (16.113).

16.8 Collective Excitations

So far we have studied one particular variational wave function and found that it has good correlations built into it. To further bolster the case that this wave function captures the physics of the fractional Hall effect, we must now demonstrate that there is a finite energy cost to produce excitations above this ground state. We demonstrated that there is a gap for charged excitations in Section 16.6. In this section we will study the neutral collective excitations.

It turns out that the neutral excitations are phonon-like excitations similar to those in solids and in superfluid helium. We can therefore use a simple modification of Feynman's theory of the excitations in superfluid helium [143, 148], which we reviewed in Chapter 4. Unlike in Exercise 4.8, where we found that a free Fermi gas is not well described by the single-mode approximation (SMA),[21] we will find here that (a simple modification of) the SMA works extremely well for Landau-level electrons whose ground state is described by the Laughlin wave function.

As we mentioned previously, Feynman argued that in ^{4}He the Bose symmetry of the wave functions guarantees that, unlike in Fermi systems, there is only a single low-lying mode, namely the density mode. The paucity of low-energy single-particle excitations in boson systems is what helps make them superfluid – there are no dissipative channels for the current to decay into. Despite the fact that the quantum Hall system is made up of fermions, the behavior is also reminiscent of superfluidity, since the current flow is dissipationless. Let us therefore blindly make the single-mode approximation and see what happens.

From Eq. (16.61) we see that the static structure factor for the mth Laughlin state is (for small wave vectors only)

$$S(q) = \frac{L^2}{N} \frac{q^2}{4\pi m} = \frac{1}{2} q^2 \ell^2, \tag{16.116}$$

where we have used $L^2/N = 2\pi \ell^2 m$. The fact that this vanishes quadratically in q (rather than linearly as in superfluid helium) is a reflection of the long-range pseudo-Coulomb forces in the plasma analogy describing the Laughlin wave function. The wave function vanishes rapidly for configurations

---

[21] This is because, as can be seen in Fig. 15.1, there is a wide spread of particle–hole-pair energies for a given momentum transfer q.

in which the density of particles deviates at long length scales from that corresponding to Landau-level filling factor $\nu = 1/m$. As a result, the SMA Bijl–Feynman formula (4.64) then yields[22]

$$\Delta(q) = \frac{\hbar^2 q^2}{2 m_e} \frac{2}{q^2 \ell^2} = \hbar \omega_c. \qquad (16.117)$$

This predicts that there is an excitation gap that is independent of the wave vector (for small q) and equal to the cyclotron energy. It is in fact correct that at long wavelengths the oscillator strength is dominated by transitions in which a single particle is excited from the $n = 0$ to the $n = 1$ Landau level. This result was derived specifically for the Laughlin state, but it is actually quite general, applying to any translationally invariant liquid ground state.[23]

One might expect that the SMA will not work well in an ordinary Fermi gas due to the high density of excitations around the Fermi surface.[24] Here, however, the Fermi surface has been destroyed by the magnetic field and the continuum of excitations with different kinetic energies has been turned into a set of discrete inter-Landau-level excitations, the lowest of which dominates the oscillator strength.

For filling factor $\nu = 1$ the Pauli principle prevents any intra-Landau-level excitations, and the excitation gap is in fact $\hbar \omega_c$ as predicted by the SMA. However, for $\nu < 1$ there should exist intra-Landau-level excitations whose energy scale is set by the interaction scale $e^2/\epsilon \ell$ rather than the kinetic energy scale $\hbar \omega_c$. Indeed, we can formally think of taking the band mass to zero ($m_e \to 0$), which would send $\hbar \omega_c \to \infty$ while keeping $e^2/\epsilon \ell$ fixed. Clearly the SMA as it stands now is not very useful in this limit. What we need is a variational wave function that represents a density wave but is restricted to lie in the Hilbert space of the lowest Landau level. This can be formally accomplished by replacing Eq. (4.121) by

$$\psi_{\vec{k}} = \bar{\rho}_{\vec{k}}^0 \Psi_m, \qquad (16.118)$$

where the overbar and the superscript 0 indicate that the density operator has been projected into the lowest Landau level. The details of how this is accomplished are presented in Section 16.11.

The analog of Eq. (4.63) is

$$\Delta(k) = \frac{\bar{f}(k)}{\bar{s}(k)}, \qquad (16.119)$$

where $\bar{f}$ and $\bar{s}$ are the **projected oscillator strength** and **structure factor**, respectively. As shown in Section 16.11,

$$\bar{s}(k) \equiv \frac{1}{N} \langle \Psi_m | (\bar{\rho}_{\vec{k}}^0)^\dagger \bar{\rho}_{\vec{k}}^0 | \Psi_m \rangle = S(k) - \left[1 - e^{-\frac{1}{2}|k|^2 \ell^2} \right]. \qquad (16.120)$$

It turns out that the term in square brackets above is precisely the structure factor $S_{\nu=1}(k)$ of the filled Landau level [143]. Thus for $\nu = 1$ the projected structure factor vanishes exactly. This is simply because the Pauli principle forbids all intra-Landau-level excitations when the level is fully occupied. For the mth Laughlin state, Eq. (16.116) shows us that the leading term in $S(k)$ for small k is $\frac{1}{2} k^2 \ell^2$. Putting this into Eq. (16.120), we see that the quadratic terms cancel out. The long-range forces of the plasma analogy guarantee that the cubic term vanishes and thus the leading behavior for $\bar{s}(k)$ is quartic

$$\bar{s}(k) \sim a(k\ell)^4 + \cdots. \qquad (16.121)$$

---

[22] We will continue to use the symbol m_e here for the band mass of the electrons to avoid confusion with the inverse filling factor m.

[23] This is known as **Kohn's theorem**, stating that the oscillator strength is exhausted by the center-of-mass cyclotron mode at long wavelength.

[24] This expectation is only partly correct, however, as we saw when we studied collective plasma oscillations in systems with long-range Coulomb forces.

We cannot compute the coefficient a without finding the k^4 correction to Eq. (16.116). It turns out that there exists a **compressibility sum rule** for the fake plasma, from which we can obtain the exact result [143]

$$a = \frac{m - 1}{8}.$$

(16.122)

The projected oscillator strength is given by Eq. (4.53), with the density operators replaced by their projections. In the case of ^{4}He, only the kinetic-energy part of the Hamiltonian failed to commute with the density. It was for this reason that the oscillator strength came out to be a universal number related to the mass of the particles. Within the lowest Landau level, however, the kinetic energy is an irrelevant constant. Instead, after projection the density operators no longer commute with each other (see Section 16.11). It follows from these commutation relations that the projected oscillator strength is proportional to the strength of the interaction term. The leading small-k behavior is

$$\bar{f}(k) = b \, \frac{e^2}{\epsilon \ell} (k\ell)^4 + \cdots,$$

(16.123)

where b is a dimensionless constant. The intra-Landau-level excitation energy therefore has a finite gap at small k,

$$\Delta(k) = \frac{\bar{f}(k)}{\bar{s}(k)} = \frac{b}{a} \frac{e^2}{\epsilon \ell} + \mathcal{O}(k^2) + \cdots.$$

(16.124)

This is quite different from the case of superfluid ^{4}He in which the mode is gapless (or $\Delta(k \to 0) = 0$). However, like in the case of the superfluid, this **"magnetophonon"** mode has a **"magnetoroton"** minimum at finite k, as illustrated in Fig. 16.10.

Figure 16.10 also shows results from numerical exact diagonalization studies which demonstrate that the SMA is extremely accurate. Note that the magnetoroton minimum occurs close to the position of the smallest reciprocal lattice vector in the Wigner crystal of the same density. In the crystal the phonon frequency would go exactly to zero at this point. (Recall that in a crystal the phonon dispersion curves have the periodicity of the reciprocal lattice.)

Because the oscillator strength is (for small k) all in the cyclotron mode, the dipole-matrix element for coupling the magnetophonon/roton collective excitations to light is very small. They have, however, been observed both in Raman scattering [152, 153] and in optical absorption in the presence of

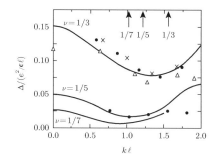

Figure 16.10 Comparison of the single-mode approximation (SMA) prediction of the collective mode energy for filling factors $\nu = 1/3, 1/5, 1/7$ (solid lines) with small-system numerical results for N particles. Crosses indicate the $N = 7$, $\nu = 1/3$ spherical system; triangles indicate the $N = 6$, $\nu = 1/3$ hexagonal unit cell system results of Haldane and Rezayi [149]. Solid dots are for the $N = 9$, $\nu = 1/3$ and $N = 7$, $\nu = 1/5$ spherical system calculations of Fano *et al.* [150]. The arrows at the top indicate the magnitude of the reciprocal lattice vector of the Wigner crystal at the corresponding filling factor. Notice that, unlike the phonon collective mode in superfluid helium shown in Fig. 4.4, the mode here is gapped. Figure reprinted from [151] by permission from Springer.

surface acoustic waves [154], and have been found to have an energy gap in excellent quantitative agreement with the SMA.

16.9 Bosonization and Fractional Quantum Hall Edge States

We learned in our study of the integer quantum Hall effect (Chapter 12) that gapless edge excitations exist even when the bulk has a large excitation gap. The chirality of these edge states is responsible for the dissipationless, quantized Hall transport. Owing to the similarity in transport properties between the integer and fractional quantum Hall states, we expect chiral edge states in the latter as well. We will begin our study of gapless edge states in the fractional Hall effect by first revisiting the integer case. Recall that the contour line of the electrostatic potential separating the occupied from the empty states could be viewed as a real-space analog of the Fermi surface (since position and momentum are equivalent in the Landau gauge). The charged excitations at the edge are simply ordinary electrons added or removed from the vicinity of the edge, and neutral excitations correspond to particle–hole pairs. In Chapter 12 we studied in some detail the case when the confining potential is translationally invariant, where the Landau gauge is convenient; physically that corresponds to a sample with Hall-bar geometry. Here we study the case with a rotationally invariant confining potential using the symmetric gauge, describing a disk (or circular)-shaped sample. One important difference here is that a disk has one edge (see Fig. 16.11(a)), while a Hall-bar has two edges (ignoring the small ends or else invoking periodic boundary conditions).

A confining potential gives rise to u_m monotonically increasing with m in Eq. (16.21). Thus, for N non-interacting electrons, all orbitals with $0 \leq m < N$ are occupied, forming a $\nu = 1$ integer quantum Hall liquid (see Fig. 16.12(a)). Now let us try to generate neutral excited states by moving electron(s) from occupied to unoccupied states. One thing we immediately notice is that this will always result in an *increase* of the total angular momentum M of the system, or $\Delta M > 0$. This

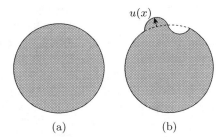

(a)   (b)

Figure 16.11 Illustration of a quantum Hall droplet (a) in the ground state and (b) in the presence of edge excitations, which can be understood as a distortion of its boundary (or edge wave) that conserves the area of the droplet.

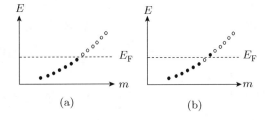

(a)   (b)

Figure 16.12 Edge state dispersion and electron occupation configurations for (a) the ground state and (b) the first (and only) excited state with $\Delta M = 1$.

reflects the *chirality* of the system. The smallest change is $\Delta M = 1$. There is one and only one such excited state, as illustrated in Fig. 16.12(b). It is fairly straightforward to show (by explicit calculation) that the corresponding wave function is

$$\psi_{\Delta M=1}([z]) = \left(\sum_{j=1}^{N} z_j \right) \Psi_1([z]). \tag{16.125}$$

But the easiest way to see this is by noticing that, due to the requirement of analyticity (or lowest-Landau-level constraint) and anti-symmetry due to Fermi statistics, any excited state wave function must take the form

$$\psi([z]) = P([z])\Psi_1([z]), \tag{16.126}$$

where $P([z])$ is a *totally symmetric* polynomial of $\{z_j\}$. For a state with fixed ΔM, $P([z])$ must also be a uniform polynomial with degree ΔM. Since the only totally symmetric and uniform polynomial with degree 1 is $\sum_{j=1}^{N} z_j$, Eq. (16.125) follows.

Equation (16.125) should be interpreted as saying that one of the electrons has had its angular momentum increased by unity, but we do not know which one (in order to preserve the overall anti-symmetry of the fermion wave function). This is to be contrasted with the symmetric polynomial factor that places a hole at the center of the droplet (see Eq. (16.51)),

$$P([z]) = \left(\prod_i z_i \right). \tag{16.127}$$

In this state the angular momentum of *every* particle is increased by unity, leaving a hole in the center and an extra particle at the edge. Thus we have $\Delta M = N$.

In general, we find that the number of excitations at a fixed value of ΔM, $D(\Delta M)$, is the same as the number of *linearly independent* uniform totally symmetric polynomials with degree ΔM. For example, $D(2) = 2$ because there are two such linearly independent polynomials: $\sum_{j=1}^{N} z_j^2$ and $\sum_{i<j}^{N} z_i z_j$. The reader is encouraged to verify by explicit construction of different particle occupation configurations that $D = 1, 2, 3, 5, 7$ for $\Delta M = 1, 2, 3, 4, 5$. As long as $\Delta M \ll N$, these all correspond to edge excitations because we are only rearranging electron occupation near the edge (or Fermi energy).[25]

We have thus found two different ways to determine $D(\Delta M)$. There is a third, more abstract way of doing this. It turns out that $D(\Delta M)$ is also equal to the number of different non-negative integer sets $\{n_l\}$ one can write down, subject to the constraint

$$\Delta M = \sum_{l=1}^{\infty} l \cdot n_l. \tag{16.128}$$

For example, for $\Delta M = 1$ the only possibility is $\{n_l\} = (1, 0, 0, \ldots)$, while for $\Delta M = 2$ we have $\{n_l\} = (2, 0, 0, \ldots)$ or $(0, 1, 0, \ldots)$. The reason why this method works in general is that (for sufficiently large N) there is a one-to-one correspondence between each fermion occupation configuration and a specific $\{n_l\}$, as we now demonstrate. Consider a fermion occupation configuration as illustrated in Fig. 16.13. We look for the first unoccupied orbital m_1, and define l_1 to be the total number of electrons in orbitals with higher $m > m_1$, and n_{l_1} to be the number of consecutive empty orbitals including m_1. We then look for the next unoccupied orbital m_2 that is *separated from* m_1 by at least one occupied orbital, and find the corresponding l_2 and n_{l_2}. We repeat the process until all occupied

---

[25] By taking linear combinations of excitations with different angular momenta ΔM we can construct localized excitations with well-defined angular coordinate like the one shown in Fig. 16.11(b).

$$n_j \qquad j \text{ electrons}$$

Figure 16.13 Relating the electron occupation configurations with a bosonic mode of angular momentum j and occupation number n_j.

orbtials have been exhausted. It is clear that we have $l_1 > l_2 > \cdots > 0$, and n_l can be any non-negative integer ($n_l = 0$ simply means l does not appear in a given configuration). It is also easy to show that Eq. (16.128) is satisfied with this construction.

It turns out, rather surprisingly, that the edge excitations in this Fermi system are in one-to-one correspondence with the excitations of a bosonic system. We may interpret n_l as the boson occupation number $b_l^\dagger b_l$ for bosons each carrying orbital angular momentum l. The number of different ways in which (chiral) fermion particle–hole pairs can carry total angular momentum ΔM is exactly the same as the total number of ways in which a set of bosons $\{b_l\}$ can carry the same total angular momentum. This identification is the mathematical basis of bosonization, a powerful method to study interacting electrons in 1D. Here the electrons live in 2D; but, as discussed earlier, the electronic states have a natural 1D labeling, and, physically, we are describing here edge excitations that propagate around a 1D boundary. Thus we can identify these as bosons living on the edge. Since the edge itself has no boundary,[26] we expect the bosons to obey periodic boundary conditions; this results in the quantization condition for its momentum $k = (2\pi/L)l$. Since the circumference $L = 2\pi R$, where R is the radius of the disk, we find its angular momentum to be $\hbar k R = l\hbar$, as expected. Thus this 1D interpretation is indeed valid. We also expect (for a smooth edge confining potential) the excitation energy to be proportional to $\Delta M \propto K$, where $K = \sum_k k \cdot n_k$ is the total momentum of these bosons, thus

$$E = \sum_k v\hbar k \cdot n_k, \qquad (16.129)$$

where v is the $\vec{E} \times \vec{B}$ drift velocity at the edge.

This looks amazingly similar to the 1D phonon spectrum! In fact we can formally introduce boson creation and annihilation operators in (angular) momentum space, and then use them to construct displacement fields in real space, in a manner similar to the phonon fields in Chapter 6 (see, e.g., Eq. (6.45)). There are two very important differences, however. First here k is constrained to be positive, thus these bosons propagate in only one direction[27] (that determined by the $\vec{E} \times \vec{B}$ drift), reflecting the chirality. Secondly, in (truly) 1D we can have longitudinal phonons only, meaning that the displacement is only along the propagation direction. In a magnetic field, as we learned in Chapter 12, the (1D) momentum actually encodes position (or displacement) in its perpendicular direction; thus the chiral edge bosons here are actually transverse! The physical meaning of the displacement they induce is the following. Because the bulk is incompressible, the only gapless neutral excitations must be area-preserving shape distortions such as those illustrated for a disk geometry in Fig. 16.11(b). The displacement field $u(x)$ (see Eq. (6.45), where j is replaced by a continuous 1D position x) is simply the local distortion that measures the change of the location of the boundary separating the incompressible liquid from vacuum. Because of the confining potential at the edges, these shape distortions have a characteristic velocity produced by the $\vec{E} \times \vec{B}$ drift.

In the case of a fractional quantum Hall state at $\nu = 1/m$ the bulk gap is caused by Coulomb correlations and is smaller but still finite. We can no longer use electron occupation to characterize either

---

[26] This is an example of a general mathematical result stating that the boundary of something has no boundary itself.

[27] Bosonization works only because negative angular momenta are forbidden here. An ordinary (non-chiral) 1D Fermi gas has two Fermi points, and you have to treat the excitations near the left- and right-moving Fermi surfaces separately. Here angular momentum plays the role of the Fermi momentum, and there is only a single Fermi point.

the ground state or edge excitations. However, the other two methods of labeling edge excitations, namely bosonization and in terms of totally symmetric polynomials, remain valid. Physically, this is because the only gapless excitations are again area-preserving shape distortions. More specifically, we expect the wave functions of such excitations to take the form

$$\psi([z]) = P([z])\Psi_m([z]). \tag{16.130}$$

This is a low-energy excitation because the good correlations built into the Laughlin wave function $\Psi_m([z])$ are maintained. These interaction-dominated 1D excitations (because they are confined to the edge) can propagate in the forward direction that is determined by the magnetic-field-induced chirality, and so they form a chiral Luttinger liquid [155].

Now let us consider charged excitations at the edge, which are also gapless. For integer edges these are simply electrons and holes near the Fermi energy. For fractional edges, however, the charge of each edge can be varied in units of e/m, just like in the bulk. To add a quasiparticle with charge $-e/m$ at the edge, we can simply create a quasihole with charge e/m at the origin, whose wave function is Eq. (16.67) with $Z = 0$. Since we do not change the electron number of the system, and the electron liquid is incompressible, the $1/m$ electron depleted from the origin must show up at the edge. We can create a large number $n \gg 1$ of quasiholes at the origin, resulting in an extension of the Laughlin wave function Ψ_m:

$$\psi_{mn}[z] = \left(\prod_{j=1}^{N} z_j^n\right) \Psi_m. \tag{16.131}$$

Following the plasma analogy, we see that this looks like a highly charged impurity at the origin which repels the plasma producing a fractional quantum Hall liquid living on an annulus with *two edges* as shown in Fig. 16.14. Each time we increase n by one unit, the annulus expands. We can view this expansion as increasing the electron number at the outer edge by $1/m$ and reducing it by $1/m$ at the inner edge. (Thereby keeping the total electron number integral as it must be.)

It is appropriate to view the Laughlin quasiparticles, which are gapped in the bulk, as being liberated at the edge. The gapless shape distortions in the Hall liquid are thus excitations in a "gas" of fractionally charged quasiparticles. This fact produces a profound alteration in the tunneling density of states to inject an electron into the system. An electron which is suddenly added to an edge (by tunneling through a barrier from an external electrode) will have relatively high energy (that is, poor correlation energy) unless it breaks up into m Laughlin quasiparticles (think about the decrease in Coulomb energy as m objects, each with charge $1/m$, move away from each other). This leads to an "orthogonality catastrophe," which simply means that the probability for this process is smaller and smaller for final states of lower and lower energy. As a result, the current–voltage characteristic for the **tunnel junction** becomes non-linear

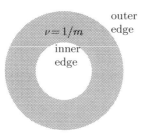

Figure 16.14 Illustration of an annulus-shaped sample with inner and outer edges. The shaded region is occupied by electron liquid.

$$I \sim V^m, \tag{16.132}$$

as we demonstrate below (in a heuristic manner).

The fractional quasiparticles are not fermions, but nevertheless they do have an approximately constant density of states at the edge just as ordinary electrons do at the $m = 1$ edge. If a bias voltage V is applied to a metallic contact, it raises the chemical potential by an amount eV above that of the Hall fluid. Electrons tunneling into the edge of the Hall fluid therefore arrive with an energy excess lying in the interval $[0, eV]$. Let us ignore matrix element effects and just count the total number of final states of m non-interacting quasiparticles that lie in this allowed interval:

$$N(V) = \frac{\rho^m}{m!} \int_0^\infty d\epsilon_1 \int_0^\infty d\epsilon_2 \cdots \int_0^\infty d\epsilon_m \, \theta \left(eV - \sum_{j=1}^m \epsilon_j \right), \tag{16.133}$$

where ρ is the quasiparticle density of states, θ is the step function, and the $m!$ accounts for the indistinguishibility of the quasiparticles. Simple dimensional analysis tells us that the integral is proportional to V^m. More precisely

$$N(V) = \left(\frac{1}{m!} \right)^2 (\rho eV)^m. \tag{16.134}$$

Thus we see that simple final state phase-space considerations can account for the non-linear increase in tunneling rate with voltage.

For the filled Landau level $m = 1$ the quasiparticles have charge $q = -e/m = -e$ and are ordinary electrons. Hence there is no orthogonality catastrophe and the I–V characteristic is linear, as expected for an ordinary metallic tunnel junction. The non-linear tunneling for the $m = 3$ state is shown in Fig. 16.15. This provides *indirect* evidence for fractional charged quasiparticles supported by the fractional quantum Hall liquid. A more direct measure of the quasiparticle charge is through shot noise in the current carried by edge states. As we discuss below, this shot noise is caused by tunneling of quasiparticles from one edge to another.

16.9.1 Shot-Noise Measurement of Fractional Quasiparticle Charge

As discussed in Section 10.5, shot noise is due to the discreteness of charge carriers, and can be used to measure the charge they carry. In this subsection we discuss how to use (quantum) shot noise to measure the (fractional) charge of Laughlin quasiparticles.

As discussed in Chapter 12, the absence of backscattering of electrons moving in opposite edges is crucial for the quantization of the Hall conductance. Similarly, the (quantized) Hall current is *noiseless* without backscattering. This can be seen from Eq. (10.73) by setting $T = 1$, corresponding to perfect transmission. The absence of backscattering is, of course, due to the spatial separation of edges. While the transport theory of fractional quantum Hall edges is more involved (since they cannot be described using free electrons), the conclusion on the absence of noise without backscattering still holds. One can, however, induce backscattering, and thus noise in the current, by bringing opposite edges together (usually through an electrostatic gating) to form a quantum point contact (see Fig. 16.16). At the point contact, quasiparticles can tunnel from one edge to the other, causing a backscattering current that is the *deviation* of the edge current from its quantized value in the absence of the tunneling:

$$I_{\text{back}} = I_{\text{qh}} - I = \sigma_{xy} V_{\text{H}} - I. \tag{16.135}$$

For fractional quantum Hall states I_{back} is dominated by the tunneling of fractionally charged quasiparticles through the gapped quantum Hall liquid separating the two edges in the region of the point contact. The same process also causes noise in the edge current. In the limit of low tunneling probabil-

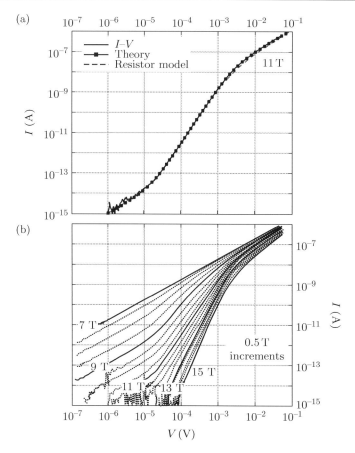

Figure 16.15 The non-linear current–voltage response for the tunneling of an electron into a fractional quantum Hall edge state. Because the electron must break up into m fractionally charged quasiparticles, there is an orthogonality catastrophe, leading to a power-law tunneling density of states, reflected in a non-linear I–V dependence over a wide range of current. The crossover to a linear dependence at low currents is due to the finite temperature. (a) Behavior at $\nu = 1/3$ that agrees well with theory [155, 156]; but the unexpectedly smooth variation of the exponent with magnetic field away from $\nu = 1/3$ shown in (b) is not yet fully understood. Figure reprinted with permission from [157]. Copyright (1998) by the American Physical Society.

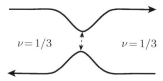

Figure 16.16 A fractional quantum Hall sample at filling factor $\nu = 1/3$ containing dissipationless edge modes. If the edges are pushed toward each other by electrostatic potentials applied to gate electrodes, a quantum point contact can form. Fractionally charged quasiparticles can tunnel from the left-moving edge to the right-moving edge (and vice-versa). These discrete "backscattering" events produce shot noise, whose intensity can be used to measure the quasiparticle charge.

ity (or $I_{\text{back}} \ll I_{\text{qh}}$), the *classical* shot-noise behavior Eq. (10.68) is expected, with the electron charge e replaced by the quasiparticle charge e^*, and the current I replaced by the backscattering current I_{back}:

$$S_I(\omega) = e^* \langle I_{\text{back}} \rangle. \tag{16.136}$$

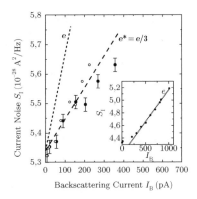

Figure 16.17 Shot noise data indicating that the elementary excitations tunneling between two edges of a FQHE sample at filling factor $\nu = 1/3$ carry fractional charge $e^* = e/3$. Figure reprinted with permission from [65]. Copyright (1997) by the American Physical Society. See also de Picciotto *et al.* [64].

The noise measurement using the setup of Fig. 16.16 indeed revealed the fractional charge of the 1/3 Laughlin states as well as other fractional quantum Hall states [64, 65]; representative data is shown in Fig. 16.17.

16.10 Composite Fermions and Hierarchy States

Our discussion so far has focused on the Laughlin states at filling factor $\nu = 1/m$, with $m = 3, 5, \ldots$ being an odd integer. It is clear from Fig. 12.1 that there are many other fractional quantum Hall plateaux corresponding to $\nu = q/p$, with p and q being mutually prime integers, and p is odd. These fractional quantum Hall states are generally referred to as **hierarchy states**, and are the focus of this section.

16.10.1 Another Take on Laughlin's Wave Function

Let us start with the Laughlin wave function at filling factor $\nu = 1/m$, with $m = 2p+1$ being odd, but look at it from a different perspective. Neglecting the common exponential factor for lowest-Landau-level wave functions, its polynomial part (16.56) can be written as

$$f_N^m[z] = \left[\prod_{i'<j'}^{N} (z_{i'} - z_{j'})^{2p} \right] \left[\prod_{i<j}^{N} (z_i - z_j) \right]. \tag{16.137}$$

The second factor is the familiar Vandermonde polynomial, and the wave function above would be describing the $\nu = 1$ integer quantum Hall state if the first factor were not there. Let us now scrutinize the first factor. It is very similar to the first factor of the multi-hole wave function (16.55), except that instead of creating a vortex with vorticity one at a fixed position, here the factor $\prod_{i'<j'}^{N} (z_{i'} - z_{j'})^{2p}$ puts a vortex with vorticity $2p$ on *every* electron! In other words, the first factor binds $2p$ vortices to the electrons, and these composite objects, termed **composite fermions** by Jain [158], form an integer quantum Hall state at $\nu = 1$!

As discussed in Section 16.6, such a vortex (with vorticity 1) can be created by adiabatically inserting a flux quantum Φ_0, in the direction *opposite* to the external magnetic field. Thus a composite fermion can also be viewed as an electron bound to $2p$ flux quanta, or flux $\phi = 2p\Phi_0$. Such charge–flux composites (see Fig. 16.9) are precisely the physical realization of the anyons discussed in Section 16.7.2. From Eq. (16.94) we obtain $\theta = 2p\pi$ and $\eta = e^{i\theta} = 1$, and thus there is *no* contribution to the statistics from the flux. As a result, the composite fermion has the *same* statistics as the electron and is fermionic, hence the name. The number of flux quanta seen by these composite fermions is

$$N'_\Phi = N_\Phi - 2pN = (2p+1)N - 2pN = N, \tag{16.138}$$

in which we subtracted the flux carried by the composite fermions from that of the external magnetic field. As a result, the ratio between the composite fermion number (which is the same as electron number) and the total flux quanta is one. While the flux carried by the composite fermions moves around with these novel particles, in a mean-field approximation one can approximate its effect by assuming it is actually static and partially cancels out the external magnetic flux. This naturally results in the understanding of the Laughlin states as the $\nu = 1$ integer quantum Hall state of composite fermions (carrying the appropriate amount of flux).

16.10.2 Jain Sequences

We have just made a nice connection between the Laughlin state for the fractional quantum Hall effect and the integer quantum Hall effect for composite fermions. More importantly, the composite-fermion concept leads to new *families* of fractional quantum Hall states, as we now demonstrate. Let us attach $\pm 2p$ flux quanta to electrons and turn them into composite fermions, where $\pm$ corresponds to the direction of flux being parallel/anti-parallel to the external flux. Similarly to Eq. (16.138), we have

$$N'_\Phi = N_\Phi \pm 2pN = N\left(\frac{1}{\nu} \pm 2p\right), \tag{16.139}$$

resulting in the composite-fermion filling factor

$$\nu_{CF} = \frac{N}{N'_\Phi} = \frac{1}{1/\nu \pm 2p}. \tag{16.140}$$

Putting the composite fermions in an integer quantum Hall state, $\nu_{CF} = q$ results in electron filling factor

$$|\nu| = \frac{q}{2pq \mp 1}. \tag{16.141}$$

We thus find that, for each integer p, we obtain two sequences of fractional quantum Hall states, corresponding to the $\pm$ signs, and $q = 1, 2, 3, \ldots$. These are known as Jain sequences. As an example, let us consider $p = 1$ and take the $+$ sign above. The corresponding sequence is $\nu = \frac{1}{3}, \frac{2}{5}, \frac{3}{7}, \ldots$. Taking the $-$ sign we obtain $\nu = \frac{2}{3}, \frac{3}{5}, \frac{4}{7}, \ldots$. Both sequences are prominently visible in Fig. 12.1. In fact the strongest hierarchy states (as measured by the transport gap) belong to the Jain sequences. Wave functions for these states have also been constructed [158].

16.11 General Formalism of Electron Dynamics Confined to a Single Landau Level

Thus far our discussion of the fractional quantum Hall effect has mostly been based on variational wave functions both for ground states and for excited states (including charged and neutral excitations, as well as edge states), whose forms are largely *independent* of the specific microscopic Hamiltonian (other than the qualitative features that electrons repel each other and the kinetic energy is quenched by a strong magnetic field). Such a variational approach is quite different from those discussed in earlier chapters, where the Hamiltonian is the starting point of any discussion, and we attempt to solve for its eigenvalues and eigenstates, with approximations if necessary. Naively one might think such a procedure can be carried out more straightforwardly in the fractional quantum Hall regime, because the quenching of kinetic energy leaves us with potential-energy terms of the Hamiltonian only, while in general the difficulty in solving a quantum mechanics problem (and one with many particles in particular) stems from the non-commutativity between kinetic- and potential-energy terms. As we are going to see in this section, this is *not* the case. Projection onto a single Landau level leads to non-zero commutators between electron density operators, which render even a Hamiltonian containing only potential-energy terms highly non-trivial.[28]

The potential-energy terms, including the single-particle potential and the electron–electron interactions, involve the position operator only, and can be expressed in terms of the density operator (or its Fourier transform). For simplicity we start with the single-particle case, and generalization to the many-electron case is straightforward. Consider a generic one-body potential term

$$U(\vec{r}) = \sum_{\vec{k}} U_{\vec{k}} \rho_{\vec{k}}, \tag{16.142}$$

with

$$U_{\vec{k}} = \frac{1}{A} \int d^2\vec{r}\,' e^{i\vec{k}\cdot\vec{r}'} U(\vec{r}\,'), \tag{16.143}$$

where A is the area of the system, and

$$\rho_{\vec{k}} = e^{-i\vec{k}\cdot\vec{r}}. \tag{16.144}$$

We now project $\rho_{\vec{k}}$ to the nth Landau level (with the aid of the guiding center coordinates defined in Eq. (16.8)):

$$\bar{\rho}_{\vec{k}}^{n} \equiv P_n \rho_{\vec{k}} P_n \tag{16.145}$$

$$= P_n e^{-i\vec{k}\cdot\vec{R}} e^{i\vec{k}\cdot(\vec{R}-\vec{r})} P_n = e^{-i\vec{k}\cdot\vec{R}} P_n e^{-i\vec{k}\cdot(\hat{z}\times\vec{\Pi})\ell^2/\hbar} P_n \tag{16.146}$$

$$= e^{-i\vec{k}\cdot\vec{R}} \langle n | e^{-i\vec{k}\cdot(\hat{z}\times\vec{\Pi})\ell^2/\hbar} | n \rangle = e^{-k^2\ell^2/4} L_n(k^2\ell^2/2) \bar{\rho}_{\vec{k}}, \tag{16.147}$$

where we used Eq. (4.110) in the last step, and

$$\bar{\rho}_{\vec{k}} = e^{-i\vec{k}\cdot\vec{R}} \tag{16.148}$$

is the *guiding center* density operator, whose form is *independent* of the Landau-level index n. The Landau-level dependence, on the other hand, is encoded in the Laguerre polynomial factor $L_n(k^2\ell^2/2)$; its presence leads to an n-dependent projected Hamiltonian from the *same* potential $U(\vec{r})$:

$$\hat{U}^n = \sum_{\vec{k}} U_{\vec{k}} e^{-k^2\ell^2/4} L_n(k^2\ell^2/2) \bar{\rho}_{\vec{k}}. \tag{16.149}$$

Using the commutator Eq. (16.9), it is easy to show that

$$\bar{\rho}_{\vec{k}} \bar{\rho}_{\vec{q}} = e^{\frac{i}{2}(\vec{k}\wedge\vec{q})\ell^2} \bar{\rho}_{\vec{k}+\vec{q}} \tag{16.150}$$

and

$$[\bar{\rho}_{\vec{k}}, \bar{\rho}_{\vec{q}}] = 2i \sin[(\vec{k}\wedge\vec{q})\ell^2/2] \bar{\rho}_{\vec{k}+\vec{q}}, \tag{16.151}$$

---

[28] The origin of all this lies in the fact that, as we saw in Section 16.1, the x and y components of the position operator no longer commute after projection onto a given Landau level.

where

$$\vec{k} \wedge \vec{q} = \hat{z} \cdot (\vec{k} \times \vec{q}). \tag{16.152}$$

Equation (16.151), which was first introduced into the physics literature in Ref. [143], is known to mathematicians as W_∞ algebra, but more commonly referred to as the Girvin–MacDonald–Platzman (GMP) algebra in the physics literature.

The attentive reader may have noticed that the algebra of Eqs. (16.150) and (16.151) is very similar to that of the magnetic translation operators we encountered in Chapter 12. This is *not* an accident; in fact up to a (gauge-dependent) phase factor, the magnetic translation operators and the guiding center density operators are the same thing! Thus

$$\bar{\rho}_{\vec{k}} \sim \tilde{T}_{(\hat{z} \times \vec{k})\ell^2}. \tag{16.153}$$

The reader is urged to verify this explicitly.

We see in Eq. (16.149) that the single-particle potential term of the Hamiltonian projected onto a Landau level can be written as a linear combination of $\bar{\rho}_{\vec{k}}$s. In fact this is true for *any* single-particle operator. To prove this statement, we need to consider a finite-size system with the shape of a parallelogram determined by its edges $\vec{L}_1$ and $\vec{L}_2$, satisfying magnetic periodic boundary conditions. Owing to the equivalence between $\bar{\rho}$ and $\tilde{T}$ mentioned above, we can formulate the boundary condition in terms of $\bar{\rho}$:

$$\bar{\rho}_{\vec{Q}_j} |\psi\rangle = |\psi\rangle, \tag{16.154}$$

with $j = 1$ or 2, and

$$\vec{Q}_j = (\hat{z} \times \vec{L}_j)/\ell^2. \tag{16.155}$$

The requirement

$$[\bar{\rho}_{\vec{Q}_1}, \bar{\rho}_{\vec{Q}_2}] = 0 \tag{16.156}$$

implies

$$\vec{L}_1 \wedge \vec{L}_2 = N_\Phi 2\pi \ell^2, \tag{16.157}$$

meaning that the system encloses N_Φ (an integer) flux quanta. The allowed values of $\vec{k}$ for which

$$[\bar{\rho}_{\vec{k}}, \bar{\rho}_{\vec{Q}_j}] = 0 \tag{16.158}$$

take the form

$$\vec{k} = [j_1 \vec{Q}_1 + j_2 \vec{Q}_2]/N_\Phi. \tag{16.159}$$

$\bar{\rho}_{\vec{k}}$ with $\vec{k}$ of the above form is a legitimate operator on the Hilbert space defined by the boundary conditions of Eq. (16.154), since acting with it on a state $|\psi\rangle$ does not alter the boundary conditions that the state satisfies. Clearly the number of *independent* such $\bar{\rho}_{\vec{k}}$s is N_Φ^2. On the other hand, N_Φ is the Landau-level degeneracy, which means that there are only N_Φ^2 linearly independent (single-particle) operators within a single Landau level. Thus any such operator can be written as a linear combination of the (allowed) $\bar{\rho}_{\vec{k}}$s.

Essentially everything above translates straightforwardly to the case of N identical particles (like spinless or spin-polarized electrons). In this case we have the (many-electron) guiding center density operator

$$\bar{\rho}_{\vec{k}} = \sum_{j=1}^{N} e^{-i\vec{k}\cdot\vec{R}_j}, \tag{16.160}$$

and Eq. (16.151) remains valid. All multi-particle operators can be expressed in terms of linear combinations of the $\bar{\rho}_{\vec{k}}$s, *and* their products (that is, polynomials of the density operators). If there is more than one species of particle present (e.g. electrons with both spin components), we need to introduce a guiding center density operator for each species. In the following we restrict our attention to a single species for simplicity.

The electron–electron interaction

$$\hat{V}_{ee} = \sum_{i<j} V(\vec{r}_i - \vec{r}_j), \tag{16.161}$$

after projection onto the nth Landau level, becomes

$$\hat{V}_{ee}^n = \frac{1}{A} \sum_{\vec{k}} \sum_{i<j} V_{\vec{k}} e^{-k^2\ell^2/2} [L_n(k^2\ell^2/2)]^2 e^{i\vec{k}\cdot(\vec{R}_i - \vec{R}_j)} \tag{16.162}$$

$$= \frac{1}{2A} \sum_{\vec{k}} V_{\vec{k}} e^{-k^2\ell^2/2} [L_n(k^2\ell^2/2)]^2 \bar{\rho}_{\vec{k}} \bar{\rho}_{-\vec{k}} + \text{constant}, \tag{16.163}$$

with

$$V_{\vec{k}} = \int d^2\vec{r}\,' \, e^{i\vec{k}\cdot\vec{r}'} \, V(\vec{r}\,'). \tag{16.164}$$

Just like the one-body Hamiltonian (16.149), the Landau-level dependence enters through the Laguerre polynomial factors (often referred to as Landau-level form factors), as a result of which electrons with the same (Coulomb) interaction can sometimes form *different* phases in different Landau levels at the same fractional (reduced) filling factor.

Now that we understand projection into a given Landau level, we are in a position to verify Eq. (16.120). The true structure factor evaluated in the nth Landau level is

$$S(k) = \frac{1}{N} \langle P_n \rho_{-\vec{k}} \rho_{\vec{k}} P_n \rangle, \tag{16.165}$$

whereas the structure factor evaluated using the projected density operators is

$$\bar{s}(k) = \frac{1}{N} \langle [P_n \rho_{-\vec{k}} P_n][P_n \rho_{\vec{k}} P_n] \rangle. \tag{16.166}$$

They differ because the former allows intermediate states outside the nth Landau level, whereas the latter does not. We have

$$S(k) = \frac{1}{N} \left\langle P_n \sum_{j,j'} e^{i\vec{k}\cdot(\vec{r}_j - \vec{r}_{j'})} P_n \right\rangle = 1 + \frac{1}{N} \left\langle P_n \sum_{j\neq j'} e^{i\vec{k}\cdot(\vec{r}_j - \vec{r}_{j'})} P_n \right\rangle$$

$$= 1 + \frac{1}{N} e^{-k^2\ell^2/2} [L_n(k^2\ell^2/2)]^2 \left\langle \sum_{j\neq j'} e^{i\vec{k}\cdot\vec{R}_j} e^{-i\vec{k}\cdot\vec{R}_{j'}} \right\rangle, \tag{16.167}$$

whereas

$$\bar{s}(k) = \frac{1}{N} \left\langle \sum_{j,j'} P_n e^{i\vec{k}\cdot\vec{r}_j} P_n e^{-i\vec{k}\cdot\vec{r}_{j'}} P_n \right\rangle \tag{16.168}$$

$$= \frac{1}{N} e^{-k^2\ell^2/2} [L_n(k^2\ell^2/2)]^2 \left\langle \sum_{j,j'} e^{i\vec{k}\cdot\vec{R}_j} e^{-i\vec{k}\cdot\vec{R}_{j'}} \right\rangle, \tag{16.169}$$

$$= S(k) - \left[1 - e^{-k^2\ell^2/2} [L_n(k^2\ell^2/2)]^2 \right]. \tag{16.170}$$

For the case of the lowest Landau level ($n = 0$) this reduces to Eq. (16.120).

The term in square brackets is precisely the structure factor for the nth filled Landau level. Specializing to the case of the lowest Landau level, this can be shown using the fact that the radial distribution function [83] is given exactly by Eq. (16.63) and the structure factor is related to this via

$$S(k) = 1 + \rho \int d^2r [g(r) - 1] + \rho (2\pi)^2 \delta^2(\vec{r}), \qquad (16.171)$$

where $\rho = 1/2\pi \ell^2$ is the mean density.

16.11.1 Finite-Size Geometries

Despite the absence of kinetic energy, the Hamiltonians projected to a given Landau level (Eq. (16.149) or Eq. (16.163), or their combinations) remain non-trivial due to the non-zero commutation relations between guiding center density operators, Eq. (16.151). There are, at the time of writing, no systematic theoretical methods to treat such Hamiltonians. As a result, microscopic studies of electronic properties of the 2DEG in the fractional quantum Hall regime have largely relied on finite-size numerical calculations (exact diagonalization in particular). To perform such studies, one first needs to choose a geometry for the system. Earlier we encountered two different geometries in our discussions of the quantum Hall effect: (i) the Hall-bar, which in our discussions is equivalent to a ribbon or cylinder, as we apply periodic boundary condition along the $\hat{y}$ direction; and (ii) the disk, in which we use the symmetric-gauge single-particle basis, and restrict the angular momenta of the electrons. In both cases we need (at least implicitly) a confining potential that prevents electrons from going to infinity due to their repulsion. Despite the common feature of *bulk* excitation gaps, there is a qualitative difference in the low-energy physics between (i) and (ii): (i) supports two sets of edge modes (one for each edge), whereas (ii) only supports one such set, as it contains only one edge. The sensitivity of the low-energy physics to the geometry and, in particular, topology of the system is a hallmark of topological phases of matter.[29]

There are two other geometries that are widely used in such finite-size studies, that do *not* contain edges: (iii) the torus and (iv) the sphere. In the following we discuss (iii) in some detail, and also briefly comment on (iv).

Imposing (magnetic) periodic boundary conditions in the form of Eq. (16.154) is equivalent to having a system with the geometry of a torus. In this case the system has no boundaries, and we do not have edge states. However, that does not mean that there are no low-energy states in the system. We will show below that, in the absence of a one-body potential ($U(\vec{r}) = 0$), there is a ground state degeneracy $D > 1$ except for the special case $N = N_\Phi$ (or a completely filled Landau level). We further argue that, when the system forms a fractional quantum Hall liquid, such a degeneracy is stable against a sufficiently weak disorder potential $U(\vec{r})$. To proceed, we construct the following center-of-mass magnetic translation operators along $\vec{L}_1$ and $\vec{L}_2$:

$$T_1 = \prod_{j=1}^{N} e^{i \vec{Q}_1 \cdot \vec{R}_j / N_\Phi}; \qquad T_2 = \prod_{j=1}^{N} e^{i \vec{Q}_2 \cdot \vec{R}_j / N_\Phi}. \qquad (16.172)$$

We learned about magnetic translation operators in Section 12.9. The operators above are essentially the same thing except that they translate the guiding center coordinates. We have

$$[T_\alpha, e^{i \vec{Q}_\beta \cdot \vec{R}_j}] = 0, \qquad (16.173)$$

---

[29] Clearly (i) and (ii) are topologically distinct. One can change the shape (geometry) of the cylinder and disk; as long as the topology is maintained, the number of edges will not change and neither will the number of branches of the edge modes.

$$[T_\alpha, H] = 0, \tag{16.174}$$

where α, β take the values 1 or 2, while $1 \le j \le N$ is the electron index. We also have

$$T_1 T_2 = T_2 T_1 e^{i2\pi\nu}, \tag{16.175}$$

where $\nu = N/N_\Phi$; we thus find that T_1 and T_2 do *not* commute with each other except for $\nu = 1$. Equations (16.174) and (16.175) imply that all eigenvalues of H come with degeneracies. For $\nu = q/p$ with p and q mutually prime integers, the degeneracy is an integer multiple of p (see Exercise 16.21). The minimum p-fold degeneracy is called the center-of-mass degeneracy because it is guaranteed by the algebra of the center-of-mass magnetic translation operators, Eq. (16.175). For $\nu = 1/m$ and the electrons forming a Laughlin-like fractional quantum Hall liquid, numerical studies [159] have shown convincingly that there are m exactly degenerate ground states ($D = m$), which are separated from all excited states by a finite gap.

This degeneracy is no longer exact when $U(\vec{r}) \neq 0$, in which case Eq. (16.174) no longer holds. However, as we show below, the lifting of the degeneracy due to $U(\vec{r})$ is exponentially small in the system size, thus the degeneracy is restored in the thermodynamic limit, and is *stable* against *any* symmetry-breaking perturbations (as long as the fractional quantum Hall phase is not destroyed by such perturbations). Let us return to the case of $U(\vec{r}) = 0$, and label the m degenerate ground states as $|l\rangle$, with $0 < l \le m$. The symmetry (16.174) allows us to choose the $|l\rangle$s to be eigenstates of T_1, while the algebra (16.175) dictates[30] that the corresponding eigenvalues are $e^{i2\pi l/m}$. Now consider the off-diagonal matrix elements of $U(\vec{r})$:

$$\langle l|\hat{U}^n|l'\rangle = \sum_{\vec{k}} U_{\vec{k}} e^{-k^2\ell^2/4} L_n(k^2\ell^2/2) \langle l|\bar{\rho}_{\vec{k}}|l'\rangle. \tag{16.176}$$

We now show that the matrix element $\langle l|\bar{\rho}_{\vec{k}}|l'\rangle = 0$ except for some special values of $\vec{k}$. From the (single-particle operator) algebra (16.150) we have

$$\bar{\rho}_{\vec{k}} T_1 = e^{i(\vec{k}\wedge\vec{Q}_1)\ell^2/N_\Phi} T_1 \bar{\rho}_{\vec{k}}, \tag{16.177}$$

$$\Rightarrow \langle l|\bar{\rho}_{\vec{k}} T_1|l'\rangle = e^{i(\vec{k}\wedge\vec{Q}_1)\ell^2/N_\Phi} \langle l|T_1\bar{\rho}_{\vec{k}}|l'\rangle, \tag{16.178}$$

$$\Rightarrow e^{i2\pi l/m} \langle l|\bar{\rho}_{\vec{k}}|l'\rangle = e^{i(\vec{k}\wedge\vec{Q}_1)\ell^2/N_\Phi} e^{i2\pi l'/m} \langle l|\bar{\rho}_{\vec{k}}|l'\rangle, \tag{16.179}$$

which implies $\langle l|\bar{\rho}_{\vec{k}}|l'\rangle = 0$ unless

$$(\vec{k}\wedge\vec{Q}_1)\ell^2/N_\Phi = 2\pi(l - l')/m \mod(2\pi). \tag{16.180}$$

This requires

$$k^2\ell^2 \sim O(N_\Phi), \tag{16.181}$$

leading to exponential suppression of the matrix element due to the exponential factor in Eq. (16.149), thus proving the robustness of ground state degeneracy against a weak random potential $U(\vec{r})$ [160], reflecting its topological nature. For more complicated fractional quantum Hall states, the ground state degeneracy D may not be equal to p and is thus *not* simply related to the center-of-mass degeneracy. In such cases the degeneracy becomes exact only in the thermodynamic limit, with or without $U(\vec{r})$. The torus degeneracy D is a widely used topological quantum number to characterize topological phases both in the fractional quantum Hall effect and in quantum spin systems.

---

[30] The only way in which T_1 and T_2 can both commute with H but not with each other is if T_2 maps different simultaneous eigenstates of T_1 and H onto each other. That is, T_2 is purely off-diagonal in the basis of eigenstates of T_1 and H.

As it turns out, it is also possible to put fractional quantum Hall liquids on a sphere [139]. In this case the ground state degeneracy is always one, and the low-energy Hilbert space is different from that of the torus. Again such sensitivity to the topology of the system is a hallmark of topological phases. More generally, the sphere and torus are two examples of compact 2D manifolds without a boundary. A topological invariant of such manifolds is their genus g, which (pictorially) is its number of holes or handles (see Fig. 13.5). The ground state degeneracy of a fractional quantum Hall liquid on a manifold of genus g can be expressed in terms of the **torus degeneracy** D as in [160]

$$\text{Degeneracy} = D^g. \tag{16.182}$$

The topological ground state degeneracy of Eq. (16.182) is a new feature *not* shared by other topological phases such as integer quantum Hall states and topological insulators. This indicates that fractional quantum Hall liquids are even more exotic than those (non-interacting) topological states. Using a term coined by Wen [15], fractional quantum Hall liquids are *topologically ordered*. The topological ground state degeneracy and anyonic excitations are consequences of this topological order, and are in fact closely related to each other (as we will discuss in Section 16.12). Integer quantum Hall states and topological insulators, on the other hand, despite having topologically protected edge states just like fractional quantum Hall liquids, are *not* topologically ordered. They do not have topological ground state degeneracy (that is, $D = 1$ in Eq. (16.182)), and their elementary excitations are simply electrons and holes, not fractionally charged anyons. We will encounter topologically ordered states again in our discussions of spin liquids (in Chapter 17) and superconductivity (in Chapters 19 and 20).

Exercise 16.20. Verify Eqs. (16.174) and (16.175).

Exercise 16.21. Show that, for $v = q/p$ with p and q mutually prime integers, Eqs. (16.174) and (16.175) imply that all eigenvalues of H come with degeneracies that are integer multiples of p. The key points are that both T_1 and T_2 commute with the Hamiltonian but not with each other. This is analogous to the situation of a particle moving in a central potential in which all three components of the angular momentum vector commute with the Hamiltonian but not with each other. This leads to the $(2L + 1)$-fold degeneracy among states with angular momentum L.

16.12 Relation between Fractional Statistics and Topological Degeneracy

We have now seen topological properties appear in many places in our discussion of the quantum Hall effect. We learned in Section 13.5 that the quantized Hall resistance in the IQHE is given by the Chern numbers of the occupied bands. In the FQHE we have seen that the Laughlin quasiparticles and quasiholes are vortex-like topological defects whose fractional charge can be derived both microscopically (Eq. (16.71)) and from a macroscopic argument based on the fractional quantized Hall conductance (Eq. (16.73)). We also found that the topological component of the Berry phase associated with moving quasiholes around each other shows that they obey fractional statistics (Section 16.7). Finally we found that FQHE states have non-trivial topological order associated with the fact that their ground state degeneracy depends on the genus of the manifold on which they live (Exercise 16.21). Our only remaining task now is to demonstrate that there is a deep connection between the fractional statistics of the topological defects and the topological order reflected in the ground state degeneracy. We will limit our discussion to the simple Laughlin states at filling factor $v = 1/m$, where the topological degeneracy is simply the center-of-mass degeneracy discussed above.

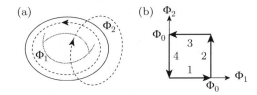

Figure 16.18 (a) An FQHE system with periodic boundary conditions represented here by a torus. Solenoids containing adjustable fluxes Φ_1 and Φ_2 wind through the two holes of the torus. (b) The path taken in time through parameter space by the two flux parameters as they vary from zero to one flux quantum Φ_0 and back to zero.

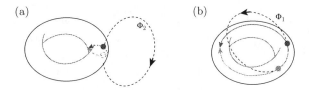

Figure 16.19 (a) Flux tube 2 containing a flux quantum is initially fully outside the torus. It is slowly moved into the torus, which produces a quasiparticle–quasihole pair (solid and dashed circles) at the points where it penetrates the surface of the torus. This causes the pair to move around the torus on a non-contractible loop and annihilate when the flux tube reaches the inner radius. (b) The analogous procedure for flux tube 1, which is gradually "laid down" on top of the torus and dropped into the interior. This causes a quasiparticle–quasihole pair to nucleate and travel on a non-contractible loop around the long dimension of the torus until the pair annihilates as the last segment of the flux tube drops into the hole.

Let us begin our discussion by considering a 2D square sample with N_e electrons subject to a constant perpendicular magnetic field B, and enclosing total flux $N_\Phi = BL^2/\Phi_0$ flux quanta. The filling factor is $\nu = N_e/N_\Phi = 1/m$ and the quantized Hall conductance is $\sigma_{xy} = \nu e^2/h$. Invoking quasi-periodic boundary conditions, the system becomes a torus with solenoids winding through each of the two holes in the torus as shown in Fig. 16.18(a). Let the flux be adiabatically varied along the path in parameter space shown in Fig. 16.18(b). Along the first leg of the square, Φ_2 is held constant at zero and Φ_1 increases from zero to one flux quantum. The resulting EMF induces a current parallel to the flux direction which displaces the center of mass of the electrons a distance $d = L/N_\Phi$.[31] Let us denote this translation operation by T_1.

The second leg displaces the center of mass by the same distance in the perpendicular direction. let us denote this second translation operation by T_2. The four legs of the path combine to take the center of mass along the sides of a square of area d^2 via the combined operation

$$T_2^{-1} T_1^{-1} T_2 T_1 = e^{-i2\pi\nu}, \tag{16.183}$$

where the last equality follows from the non-commutivity of the magnetic translation operators, Eq. (16.175). Equivalently, this phase results from the system acquiring an Aharonov–Bohm phase of $\theta = 2\pi N_e d^2 B/\Phi_0 = 2\pi\nu$. For $\nu = 1/m$, the mininum representation of the non-Abelian group of translations is m-dimensional, so the ground state is at least m-fold degenerate on the torus.

Let us now relate this topological degeneracy to the fractional statistics of the quasiparticles. We will do this by the following construction illustrated in Fig. 16.19(a). Here the operation T_2 is carried out not by varying the flux in the solenoid over time, but by moving the solenoid (containing a full

---

[31] This can be seen from the Landau gauge, where each of the N_Φ eigenstates moves to the position of its neighbor in momentum space. Recall from the discussion of Fig. 13.9 that threading a flux quantum through the hole in a cylinder with filling factor $\nu = 1$ moves exactly one electron from one edge to the other. This comes about precisely because the Landau level is filled and the flux tube translates each Landau gauge state into its neighbor.

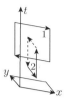

Figure 16.20 Space-time trajectories for two quasiparticles traversing non-contractible loops on the torus (here represented by a 2D square with periodic boundary conditions). The closed-loop trajectories correspond to inserting and removing the two flux tubes in Fig. 16.19 which in turn correspond to the sequence translations $T_2^{-1} T_1^{-1} T_2 T_1 = e^{-i2\pi\nu}$ in Fig. 16.18.

flux quantum) from outside the torus into the interior until it is linked with the torus just as before in Fig. 16.18(a). This nucleates a quasiparticle–quasihole pair, which then circles the small radius of the torus and annihilates when the solenoid completes its passage and enters the central large hole of the torus.[32] Going back to the picture in which the system is a planar square with quasi-periodic boundary conditions, we see that a charge $e^* = \nu e$ has passed a distance L from one edge of the sample to the other. This is precisely the same amount of current that flowed when the operation T_2 was carried out by the previous method, and indeed they are precisely equivalent operations that move the system from one degenerate ground state to another. Moving a quasiparticle or quasihole around a non-contractible loop produces a center-of-mass translation. As shown in Fig. 16.19(b), the operation T_1 is carried out similarly, and again the total charge displacement gives the same center-of-mass displacement as in the previous method.

We are now in a position to relate the non-commutativity of these two operations to the fractional statistics of the quasiparticles and quasiholes. It is easiest to visualize the space-time trajectory of the quasiparticles by mapping the torus back to the 2D square. The third dimension can then represent time, as illustrated in Fig. 16.20. Rather than drawing a quasiparticle moving halfway across the square and a quasihole moving backwards on the other half, we will take a quasiparticle to move all the way across the square as this is completely equivalent. The world lines have the same linking number as if one particle had circled the other, corresponding to a double exchange with topological phase $e^{i2\pi\nu}$. The non-universal parts of the phase (normally associated with the area swept out by the path) cancel out between the forward and backward parts of the loops, and as a result only this topological phase due to fractional statistics of the quasiparticles survives.

In summary, the fractional quantum Hall effect has a very rich topological structure, with many interconnections among the quantized Hall resistance, the fractional charge, and the fractional statistics of the quasiparticles.

16.13 Notes and Further Reading

The physics of fractional quantum Hall states, and more generally of electrons in a partially filled Landau level (often referred to as the fractional quantum Hall regime), is extremely rich. Interested readers can consult Refs. [142, 161–165] for more information. In this chapter we have only scratched its surface. In this section we briefly mention a few topics not covered in detail, and point the reader to relevant literature.

- Multi-component systems. In our discussions we have assumed that the electron spin has been fully polarized by the external magnetic field. In reality the Zeeman splitting (in GaAs quantum wells) is

---

[32] Recall that threading the 2DEG with a flux quantum generates a quasiparticle or quasihole (depending on the direction of the flux); see Fig. 16.4.

only of order 1 K for a magnetic field of order 10 T, which is significantly lower than the Coulomb energy scale $e^2/\epsilon\ell \sim 100$ K. As a result, the spin degree of freedom can be active in some circumstances. There are also important systems that support other internal degrees of freedom. One class of such systems is multi-layered 2DEGs, where two (or more) layers of 2DEG are placed in proximity, and coupled to each other via the inter-layer Coulomb interaction and/or tunneling. There is extremely rich physics associated with these internal degrees of freedom. For a review, see Ref. [166]. A particularly important class of multi-component quantum Hall states consists of those realized in single- and multi-layered graphene, due to the valley and other degeneracies. For a review, see Ref. [167].

- Hierarchy and composite-fermion (CF) states. As discussed in Section 16.10, the most prominent hierarchy states, namely the Jain sequences, can be understood as integer quantum Hall states of CFs. Many other hierarchy states can be understood by invoking interactions between CFs; interested readers can consult Ref. [158] for more details. Historically the hierarchy states were first understood as being due to Laughlin quasiparticle/quasiholes condensing into Laughlin-like correlated states formed by themselves; in this Haldane–Halperin hierarchy scheme [139, 168] fractional quantum Hall states are possible at any rational fractional filling $\nu = q/p$, as long as the denominator p is odd. We note that the Jain sequences end at filling factor $\nu = 1/2p$, where the CFs carrying $2p$ flux quanta see *zero* net flux. It was predicted [169] that this leads to a metallic state with a CF Fermi surface, which has received remarkable experimental support for the case of filling factor $\nu = 1/2$, a system of considerable interest and activity. This is just one example showing that there is extremely interesting strong-correlation physics in the fractional quantum Hall regime that goes beyond the "usual" fractional quantum Hall effect.

- $\nu = 5/2$ and non-Abelian fractional quantum Hall states. In 1987 a new fractional quantum Hall state was discovered at $\nu = 5/2$ [170]. It is different from those in Fig. 12.1 in two respects. (i) At this filling the lowest Landau level ($n = 0$) is completely filled (by both up- and down-spin electrons) and is effectively inert. Therefore this fractional quantum Hall state results from the $n = 1$ Landau-level electrons being at filling fraction $\nu = 1/2$. (ii) This is the first even-denominator fractional quantum Hall state, which cannot be understood using any of the hierarchy schemes. At the time of writing, the strongest contender is the Moore–Read state [171], which can be viewed as a Cooper-paired condensate formed by the CFs. In an ordinary metal, pairing and condensation of Cooper pairs leads to superconductivity, which will be discussed extensively in Chapters 19 and 20. In a strong magnetic field such pairing leads to fractional quantum Hall states. More specifically, the Moore–Read state corresponds to a weakly paired p-wave state (see Sections 20.7 and 20.8), which is topologically non-trivial. This in turn leads to more exotic properties. In addition to having fractional charge, quasiparticles/holes of the Moore–Read state obey so-called **non-Abelian statistics**, in the following sense. In the presence of these quasiparticles/holes, the state of the system is not fixed even when their positions are; the degeneracy of the resultant Hilbert space grows exponentially with the number of quasiparticles/holes. Exchanging these quasiparticles/holes results in a non-trivial unitary transformation within this degenerate Hilbert space, and different exchange operations in general do *not* commute with each other, hence the name non-Abelian statistics. For comparison, exchanging the Laughlin quasiparticles/holes results only in a phase change of the state, and subsequent exchanges result in additions of these phases, which do not depend on the order of the exchange; in this sense the fractional statistics obeyed by Laughlin quasiparticles/holes is Abelian. As it turns out, the Moore–Read state is a member of a sequence of non-Abelian fractional quantum Hall states known as Read–Rezayi states [172], which descend from the (Abelian) Laughlin state. Non-Abelian fractional quantum Hall states have potential applications in topologically protected quantum information processing and computation [173].

17 Magnetism

Magnetism is one of those wondrous effects that can startle you as a child and still fascinate you as a practicing physicist. It has had great practical importance in human history through the compass, the electric motor, non-volatile computer memory, and a host of other applications.

Classically, magnetic fields arise through electrical currents. Quantum mechanically, we have, in addition to this orbital motion effect, the remarkable concept of intrinsic spin and its associated magnetic moment.

Magnetism occurs both in metals such as iron as well as in certain types of insulators. We studied some magnetic effects in metals within the Landau Fermi-liquid theory in Chapter 15. In this chapter we will also study novel many-body effects in insulating magnets.

17.1 Basics

Magnetization is defined thermodynamically as

$$M^\mu(\vec{h}, T) = -\frac{\partial f}{\partial h^\mu}, \tag{17.1}$$

where f is the free-energy density and $\vec{h}$ is the magnetic field. The **magnetic susceptibility tensor** is defined by

$$\chi^{\alpha\beta} = \frac{\partial M^\alpha}{\partial h^\beta}. \tag{17.2}$$

In an isotropic material this is simply

$$\chi^{\alpha\beta} = \chi \delta_{\alpha\beta}, \tag{17.3}$$

a scalar times the unit tensor. Usually one is interested in the linear response so that implicit in Eq. (17.2) is a limit of $h \to 0$. Except for ferromagnets, most materials respond weakly to applied fields, and the magnetization is very nearly linear in h.

There are various different types of magnetic response. A **paramagnet** has $\chi > 0$, while a **diamagnet** has $\chi < 0$ and hence has $\partial^2 f / \partial h^2 > 0$. Thus a diamagnet is repelled from regions of high magnetic field. A **ferromagnet** has a finite magnetization even in the absence of an applied field. The magnetization is odd under time reversal (all velocities, currents, and spins change sign) and so a ferromagnet spontaneously breaks time-reversal symmetry. That is,

$$\lim_{h \to 0^+} M(h, T) \neq \lim_{h \to 0^-} M(h, T) \tag{17.4}$$

if $T < T_c$, where T_c is the critical temperature or Curie temperature (named for Marie's husband, Pierre).

An **antiferromagnet** has electronic spins that are oppositely aligned on two sublattices so that there is no overall ferromagnetic moment. A **ferrimagnet** has a similar staggered arrangement, but the magnetic moment on one sublattice is larger than that on the other so that there is a net ferromagnetic moment. Finally, some systems exhibit **spin-density-wave (SDW)** order in which the spins tumble or spiral with a period incommensurate with the crystalline lattice.

17.2 Classical Theory of Magnetism

In classical physics there are no intrinsic spins, so we need only concern ourselves with orbital magnetic effects. The Lorentz force clearly modifies the motion of charged particles, so one might expect there to be a change in the free energy when we apply a field and hence a finite magnetization. The free energy obeys

$$e^{-\beta F} = \prod_j \int \frac{d^3 \vec{p}_j}{(2\pi\hbar)^3} \int d^3 \vec{r}_j \, e^{-\beta H}, \tag{17.5}$$

where (for electrons)

$$H = \frac{1}{2m_e} \sum_j \left[\vec{p}_j + \frac{e}{c} \vec{A}(\vec{r}_j) \right]^2 + V. \tag{17.6}$$

Defining a new integration variable,

$$\vec{p}_j' \equiv \vec{p}_j + \frac{e}{c} \vec{A}(\vec{r}_j), \tag{17.7}$$

we can easily carry out the momentum integrations first to obtain a result which is totally independent of the vector potential $\vec{A}$. Hence the free energy is constant and the magnetization is strictly zero classically! This result is known as the **Bohr–van Leeuwen theorem**.

To understand this paradox, first note that this is *not* in conflict with the fact that the microscopic equations of motion are changed by the Lorentz force. The *dynamics* of a plasma is indeed modified by an applied field. What the theorem tells us is something different – namely that the static equilibrium thermodynamics is unaffected by the applied field. An immediate implication is that plasma physicists can never use magnetic traps to confine a plasma in equilibrium. Trapping is a purely dynamic non-equilibrium (and generally temporary) effect.

This theorem also tells us that the diamagnetic response which allows one to levitate a frog in a strong field gradient is a purely quantum-mechanical room-temperature effect [174]. Hence we need a quantum theory of magnetism.

17.3 Quantum Theory of Magnetism of Individual Atoms

Let us begin our investigation by focusing on the quantum treatment of the orbital motion in a single atom and continuing to ignore the intrinsic spin.[1] As we have already seen in Exercise 16.7, the Hamiltonian in Eq. (17.6) can be re-written in the symmetric gauge as

---

[1] Our discussion, of course, also applies to an individual ion.

$$H = \frac{1}{2m_e} \sum_{j=1}^{Z} \left\{ p_j^2 + \frac{e^2 B^2}{4c^2} r_{\perp j}^2 + \frac{e}{c} \vec{B} \cdot \vec{L}_j \right\} + V, \tag{17.8}$$

where

$$r_\perp^2 \equiv \left[\hat{B} \times \vec{r} \right]^2 = x^2 + y^2 \tag{17.9}$$

and Z is the number of electrons. Note that for the 2D case we considered previously it was not necessary to distinguish between $r_\perp$ and r.

Recall that the classical radius of the electron is given by

$$r_c \equiv \frac{e^2}{m_e c^2} = \alpha^2 a_B \tag{17.10}$$

and the Bohr magneton is defined by

$$\mu_B = \frac{e\hbar}{2m_e c}. \tag{17.11}$$

Using these, the Hamiltonian becomes

$$H = \sum_{j=1}^{Z} \frac{\vec{p}_j^2}{2m_e} + V + \frac{\mu_B}{\hbar} \vec{B} \cdot \vec{L} + \sum_j \frac{B^2}{8\pi} \pi r_{\perp j}^2 r_c, \tag{17.12}$$

where $\vec{L}$ is the total angular momentum. We can interpret the last term as the **magnetic pressure** (energy density of the magnetic field) multiplied by the volume of a disk of radius $r_{\perp j}$ and thickness r_c.

17.3.1 Quantum Diamagnetism

For simplicity let us first consider the case of an atom like Xe or Ar which contains only filled shells and has zero total angular momentum. Within first-order perturbation theory the energy shift due to the magnetic field is then simply[2]

$$\Delta E = \frac{B^2}{8\pi} \pi r_c \sum_j \langle r_{\perp j}^2 \rangle. \tag{17.13}$$

In SI units the magnetic pressure $B^2/8\pi$ becomes

$$\frac{B^2}{2\mu_0} = \frac{B^2}{8\pi \times 10^{-7}} \frac{\text{Pa}}{\text{T}^2}. \tag{17.14}$$

Thus 1 T yields a magnetic pressure of[3]

$$p \approx 3.98 \times 10^5 \text{ Pa} \approx 4 \text{ atm}. \tag{17.15}$$

Using

$$\pi r_c a_B^2 = \pi \alpha^2 a_B^3 \approx 2.48 \times 10^{-35} \text{ m}^3, \tag{17.16}$$

we have

$$\Delta E \approx 6.15 \times 10^{-11} \text{ eV} \left(\frac{B}{\text{T}} \right)^2 \sum_j \frac{\langle r_{\perp j}^2 \rangle}{a_B^2}. \tag{17.17}$$

---

[2] Because this term is quadratic in B, we properly should treat the $\vec{B} \cdot \vec{L}$ term up to second order in perturbation theory in order to be consistent. However, for an individual atom, $\vec{B} \cdot \vec{L}$ is a conserved quantity, and thus does *not* induce any change to the ground state energy or wave function, unless a level crossing is induced by a strong field.

[3] Think about what this means for the mechanical-strength requirements of a 30 T magnet!

Clearly this is a small energy compared with the 1 eV scale of typical atomic excitation energies and so our use of first-order perturbation theory is justified.

Because the energy rises with B, the response of a closed-shell atom is always diamagnetic. From Eqs. (17.1) and (17.13) we find that the magnetization is

$$M = -B\frac{r_c}{4}\sum_{j=1}^{Z}\langle r_{\perp j}^2\rangle n, \tag{17.18}$$

and the susceptibility is a dimensionless number

$$\chi = -\frac{n}{4}r_c\sum_{j=1}^{Z}\langle r_{\perp j}^2\rangle, \tag{17.19}$$

where n is the density of the atoms. Here we are treating each atom independently and measuring $\vec{r}_{\perp j}$ relative to the position of the appropriate nucleus. It is not hard to see that, since $r_c \propto \alpha^2$, $\chi \sim -10^{-5}$ for typical diamagnetic materials like water.

A diamagnetic object such as a drop of water or a small frog can be levitated in a field gradient due to the diamagnetic force

$$\vec{F} = \frac{\chi}{2}\int d^3\vec{r}\,\nabla(B^2). \tag{17.20}$$

If $\chi \sim -10^{-5}$

$$\vec{F} \sim -50\,\text{Pa}\int d^3\vec{r}\,\nabla\left(\frac{B}{\text{T}}\right)^2. \tag{17.21}$$

Notice that this diamagnetism is a purely quantum-mechanical effect and that it is manifest at room temperature despite the fact that the magnetic energy per atom is vastly smaller than the typical thermal energy. How can this be? Recall that normally we think of a system as being classical if thermal energies vastly exceed quantum energies. The resolution of this paradox is that there is another quantum energy scale which is much larger than the temperature, namely the electronic excitation gap in a typical diamagnetic atom. Because this gap is so large, the electrons are frozen in their s-wave ground state and the $\vec{B}\cdot\vec{L}$ term vanishes. The fact that the energy associated with the B^2 term is tiny does not matter since it always has the same sign for every thermal configuration of the atoms.

Unlike ordinary external forces which act on a body, diamagnetic forces act separately on each atom in the sample. Hence they can be used to largely compensate for the effects of gravity or to accelerate objects with little internal stress.

It is useful to ask what currents are flowing in a diamagnetic atom. These currents give rise to the magnetic moment which opposes the applied field. Since we are using first-order perturbation theory to calculate the energy change, we are using the *unperturbed* ground state. How then can it be possible to have any currents flowing? The answer to this paradox lies in the fact that the velocity operator changes from $\vec{p}/m_e$ to

$$\vec{v} = \frac{1}{m_e}\left(\vec{p} + \frac{e}{c}\vec{A}\right). \tag{17.22}$$

The wave function does not change but the operator does! The (canonical) momentum operator has zero expectation value just as it did before the field was applied.[4] However, there is a so-called diamagnetic current density

$$\vec{J}(\vec{r}) = -\frac{e^2}{c}\rho(\vec{r})\vec{A}(\vec{r}) = \frac{e^2}{2c}\rho(\vec{r})\vec{r}\times\vec{B}, \tag{17.23}$$

---

[4] This is not true for all gauge choices since the canonical momentum $\langle\vec{p}\rangle$ is gauge-dependent. Only the mechanical momentum $\langle\vec{p} + (e/c)\vec{A}\rangle$ is gauge-invariant.

where $\rho(\vec{r})$ is the electron number density. If we had chosen a different gauge so that $\nabla \cdot \vec{A}$ no longer vanished then the $\langle \vec{p} \rangle$ term would have changed to compensate for this, leaving the physical current gauge-invariant. The current from the $\langle \vec{p} \rangle$ term is usually referred to as the paramagnetic current. Note that, in a stationary quantum state, the continuity equation requires $\nabla \cdot \vec{J} = 0$. This is why Eq. (17.23) is valid only for gauge choices in which $\nabla \cdot \vec{A} = 0$.

Recall, from Exercise 10.1, that a nice illustration of all this can be obtained by considering an electron moving on the end of a rod that keeps it a fixed distance R from an axle. In the absence of a magnetic field, the eigenfunctions are

$$\psi_m(\varphi) = e^{im\varphi} \tag{17.24}$$

with eigenvalues

$$\epsilon_m = \frac{\hbar^2}{2I} m^2; \qquad m = 0, \pm 1, \pm 2, \pm 3, \ldots. \tag{17.25}$$

Now consider what happens when we add a uniform magnetic field $B\hat{z}$ parallel to the axis. The Hamiltonian becomes

$$H = \frac{\hbar^2}{2I} \left[-i\frac{\partial}{\partial\varphi} + \beta \right]^2, \tag{17.26}$$

where

$$\beta \equiv \left(\frac{eB\pi R^2}{c2\pi\hbar} \right) = \frac{\Phi}{\Phi_0}, \tag{17.27}$$

with $\Phi_0 \equiv hc/e$ being the flux quantum, and Φ is the flux enclosed by the circular orbit.

Notice the remarkable fact that the eigenfunctions are unchanged in the presence of the field and so are still given by Eq. (17.24). The energy eigenvalues are, however, changed, and become

$$\epsilon_m = \frac{\hbar^2}{2I}(m + \beta)^2, \tag{17.28}$$

illustrating the fact that the mechanical angular momentum

$$L_z^{\text{mech}} = \vec{r} \times \left(\vec{p} + \frac{e}{c} \vec{A} \right) \cdot \hat{z} \tag{17.29}$$

is offset by a constant β from the canonical angular momentum. The angular velocity is given by

$$\omega = \frac{\partial H}{\partial L_z} = \frac{\hbar}{I}(m + \beta). \tag{17.30}$$

Thus even for an s-wave ($m = 0$) state a diamagnetic current begins to flow as flux is added! Indeed, since the eigenfunction is totally unchanged, we can solve the time-dependent Schrödinger equation for any arbitrary time-dependent flux $\Phi(t)$. The solution is simply the old solution multiplied by a time-dependent phase

$$\psi(\varphi, t) = e^{im\varphi} e^{-if_m(t)}, \tag{17.31}$$

where $f_m(t) \equiv (\hbar/2I) \int_0^t dt' \left[m + \beta(t') \right]^2$ because

$$i\hbar \frac{\partial \psi(\varphi, t)}{\partial t} = \frac{\hbar^2}{2I} [m + \beta(t)]^2 \, \psi = H(t)\psi(\varphi, t). \tag{17.32}$$

The angular acceleration from Eq. (17.30) is

$$I\dot{\omega} = \hbar\dot{\beta} = \frac{e}{2\pi c}\dot{\Phi}$$
$$= -eE_\varphi R, \tag{17.33}$$

where E_φ is the azimuthal electric field caused by Faraday induction and so the RHS of the equation is just the torque.

So far we have considered only closed-shell atoms. We found that, despite their apparent inertness (due to the large excitation gap), these atoms do respond to an applied B field by generating diamagnetic currents. The large excitation gap prevents these currents from being due to changes in the quantum state of the atom. Rather they are due to changes in the current operator itself caused by the vector potential.

We have thus far assumed that we can treat individual atoms separately. This of course is not true for metals, where the valence electrons are distributed throughout the sample in extended plane-wave-like states. For the 2DEG studied in connection with the quantum Hall effect, the electrons occupy discrete Landau levels. In Section 17.9 we will revisit this and discuss their contribution to diamagnetism. Earlier in Section 8.5 we also discussed some other important physics associated with Landau-level formation (which was treated at semiclassical level there) in metals, namely quantum oscillations induced by Landau-level energies crossing the Fermi energy in a periodic fashion as B changes. The resulting oscillations in the magnetization are known as de Haas–van Alphen oscillations.

We turn now to the case of open-shell atoms, where the $\vec{B} \cdot \vec{L}$ term can induce a paramagnetic response.

17.3.2 Quantum Paramagnetism

We now turn to atoms with open shells so that the total orbital angular momentum quantum number $\ell > 0$. In this case we need to focus on the perturbation term which is linear in the magnetic field:

$$\Delta H = \frac{\mu_B}{\hbar} \vec{B} \cdot \vec{L} = \frac{1}{2}\hbar\omega_c m_z, \tag{17.34}$$

where ω_c is the cyclotron frequency and m_z is the azimuthal quantum number. That is, $\hbar m_z$ is the eigenvalue of the angular momentum operator component parallel to the field. In first-order perturbation theory the degeneracy is lifted, as illustrated by the "fan diagram" in Fig. 17.1.

At zero temperature the atom will sit in the lowest state and $\langle \Delta H \rangle$ will *decrease* with B. That is, the contribution of this term to the energy will be paramagnetic. Because this term is of first order in B, it will typically dominate over the diamagnetic term considered previously. Nevertheless, the energy scale is still quite small since (in SI units)

$$\mu_B = \frac{e\hbar}{2m_e} \approx 0.67 \text{ K/T}. \tag{17.35}$$

Thus at room temperature all energy levels of the multiplet will be nearly equally occupied, and the paramagnetic term in the free energy will be strongly suppressed. This is a simple manifestation of the Bohr–van Leeuwen theorem in the semiclassical limit $\hbar\omega_c \ll k_B T$. As a result of the thermal-fluctuation suppression of the paramagnetic contribution, the diamagnetic contribution can sometimes dominate (which is what allows the levitation of frogs in a magnetic field (gradient) despite the small amount of iron in their blood!).

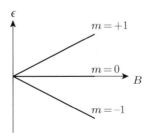

Figure 17.1 The eigenvalue spectrum for the case of angular momentum $\ell = 1$. The magnetic field splits the $2\ell + 1$ orbitals and causes them to "fan out" with increasing B.

Exercise 17.1. Derive the contribution of the paramagnetic term to the atomic susceptibility at finite temperature. Show that it is propotional to T^{-1}. This is known as the Curie law. This law is valid for independent, non-interacting magnetic moments. At temperatures below the scale of any interactions, the susceptibility typically saturates and the Curie law fails. The temperature scale below which the measured susceptibility begins to deviate from the Curie law is often used to characterize the scale of interactions.

17.3.3 Quantum Spin

Intrinsic spin is a purely quantum effect arising naturally only in relativistic quantum mechanics. The low-energy limit of the Dirac theory shows us that there exists a Zeeman coupling

$$H_Z = g\frac{\mu_B}{\hbar} \sum_j \vec{B} \cdot \vec{S}_j, \tag{17.36}$$

where $g = 2$ (neglecting QED corrections). Note that, for a spin-1/2 particle, the Zeeman splitting is precisely equal to $\hbar\omega_c$, the Landau-level splitting.[5] There is also a spin–orbit ($\vec{L} \cdot \vec{S}$) coupling which complicates things somewhat – it renders spin and orbital angular momentum no longer conserved quantities individually, but the total angular momentum $\vec{J} = \vec{L} + \vec{S}$ remains conserved. The effective g factor associated with $\vec{J}$ varies among different atoms as well as crystals, and may be controlled to some degree (e.g. via changing pressure) for the latter.

The Zeeman term is clearly paramagnetic. We have studied the Pauli paramagnetic response of non-interacting electrons and interacting electrons in the Fermi-liquid phase in Chapters 7 and 15, respectively. In the following sections we focus mainly on the collective magnetic states arising from the spin degrees of freedom.

17.4 The Hubbard Model and Mott Insulators

The **Hubbard model** is a toy model designed to capture the essence of magnetic effects due to the Coulomb interaction. It replaces the long-range Coulomb interaction by a simple on-site repulsion within the context of the tight-binding approximation. In second-quantized notation (see Appendix J for a brief overview of second quantization) the Hubbard Hamiltonian takes the form

$$H = -t \sum_{\langle ij \rangle, \sigma} (c_{i\sigma}^\dagger c_{j\sigma} + c_{j\sigma}^\dagger c_{i\sigma}) + U \sum_j n_{j\uparrow} n_{j\downarrow}. \tag{17.37}$$

The first term is the near-neighbor hopping of a tight-binding model (defined with orthonormal Wannier orbitals) and the second term costs a (two-body) potential energy U if a site is doubly occupied. Note that the Pauli exclusion principle allows double occupancy only by pairs of electrons of opposite spin. The hopping (t) term is often referred to as the "kinetic-energy" term, as it tends to delocalize the electrons through hopping and competes with the interaction (U) term. It should be kept in mind that t comes from the *combined* effects of the (original) electron kinetic energy and (one-body) periodic lattice potential, and characterizes the band in which electrons live.

---

[5] This fact means that the spin precession (Larmor) frequency and the orbital (cyclotron) frequency are identical if $g = 2$. Measurement of the small difference in these frequencies is a sensitive way to measure the quantum electrodynamic (QED) corrections to the value of $g - 2$.

Despite the fact that this model is an extreme oversimplification, it has resisted exact solution in spatial dimensions greater than one. Nevertheless, certain features of the physics of this model are understood.

The physics depends strongly both on the band filling, which can vary in the interval

$$0 \leq \langle n_{j\uparrow} + n_{j\downarrow} \rangle \leq 2, \tag{17.38}$$

and on the dimensionless ratio

$$-\infty < U/t < \infty \tag{17.39}$$

which controls the interaction strength. Let us begin our analysis with the case of half-filling:

$$\langle n_{j\uparrow} + n_{j\downarrow} \rangle = 1. \tag{17.40}$$

When $U = 0$ the ground state is a single Slater determinant made up from plane-wave states in the lower half of the band (indicated by the prime on the product below):

$$|\Psi\rangle = \prod_{\vec{k},\sigma=\uparrow,\downarrow}' \psi_{\vec{k}\sigma}^{\dagger} |0\rangle, \tag{17.41}$$

where the plane-wave creation operator is

$$\psi_{\vec{k}\sigma}^{\dagger} \equiv \frac{1}{\sqrt{N}} \sum_{j=1}^{N} e^{i\vec{k}\cdot\vec{R}_j} c_{j\sigma}^{\dagger}. \tag{17.42}$$

Each plane-wave state contains two electrons of opposite spin. Clearly this is an ordinary non-magnetic Fermi sea, representing a *metallic* ground state.

Consider now the case $U = \infty$ for which there can be no double occupancy. Here the states must be of the form

$$|\Psi\rangle = \left(\prod_{j=1}^{N} c_{j\sigma_j}^{\dagger} \right) |0\rangle. \tag{17.43}$$

There are 2^N different states according to the choice of spin orientation $\sigma_j = \{\uparrow, \downarrow\}$ on each lattice site. Since the particles cannot hop to any neighboring sites, both the (one-body) kinetic energy and the (two-body) potential energy vanish for every spin configuration. Hence the only energy eigenvalue is zero and it is 2^N-fold degenerate. Because of this degeneracy, no particular type of magnetic order is favored over any other, and the system is effectively a non-magnetic insulator with an infinite charge-excitation gap.

This is the third type of insulator we have encountered. Previously we have encountered band and Anderson insulators, both of which can be understood in terms of *non-interacting* electrons. Recall that in the Einstein relation we have the longitudinal conductivity

$$\sigma = e^2 D \frac{dn}{d\mu}. \tag{17.44}$$

In a band insulator we have the compressibility $\sim dn/d\mu = 0$ due to the band gap (the Fermi energy is inside the band gap and there are thus no states at the Fermi energy) resulting in $\sigma = 0$ at $T = 0$. Conversely, in an Anderson insulator $dn/d\mu > 0$, but the diffusion constant $D = 0$ due to disorder-induced localization. Here in this half-filled infinite-U Hubbard model we again have $dn/d\mu = 0$, but its origin lies in the (infinitely) strong electron repulsion, instead of an ordinary band gap. Such interaction-induced insulators are collectively known as **Mott insulators**.

It turns out the half-filled infinite-U Hubbard model is quite a special case. Consider what happens when we remove a *single* electron from this half-filled band. One can show that the ground state is given exactly by

$$|\Psi\rangle = \prod_{\vec{k}} \psi_{\vec{k}\uparrow}^{\dagger} |0\rangle, \tag{17.45}$$

where the product runs over all allowed $\vec{k}$s except for the single one corresponding to the highest kinetic energy. This state is therefore a fully ferromagnetic single Slater determinant with spin $S = \frac{1}{2}(N - 1)$ and a degeneracy of $2S + 1$.

It is clear that this is an exact energy eigenstate. What is less clear is that this is the ground state. However, a theorem by **Nagaoka** shows that this is the case. The ferromagnetic Nagaoka state shown above automatically obeys the no-double-occupancy rule due to the Pauli principle and the fact that all spins are aligned. Despite obeying this constraint, the kinetic energy is good because there is a "hole" at the top of the band. If the spins are not fully aligned then special care is needed to avoid double occupancy. Nagaoka showed that the "hole" is forced into states of less favorable kinetic energy by this constraint. This is related to the fact that, if all the spins are aligned, a hole can take any path between two points and the system ends up in the same final state in all cases (leading to quantum interference). On the other hand, if the spins are not all aligned, the spin configuration is "stirred up" in the wake of the hole motion and two different trajectories can lead to distinguishable final states.

Let us return now to the case of half-filling but with U finite. This case is believed to represent the insulating parent compounds of the high-temperature cuprate superconductors (to be discussed in Section 20.9). Just as in the $U = \infty$ case, large U gives insulating behavior. The main differences are that the charge-excitation gap is finite and different spin states are no longer precisely degenerate. It turns out that the ground state has antiferromagnetic **Néel order** for square and cubic lattices.

To see how this comes about, let us analyze a Hubbard model with only two electrons and two sites. The basic idea is that charge fluctuations leading to doubly occupied sites are energetically expensive and will occur with only small amplitude that can be computed perturbatively. We want to "integrate out" the high-energy charge fluctuations and be left with a low-energy effective Hamiltonian which (as it turns out) describes the dynamics of magnetic fluctuations and interactions. We present one method here, but the reader is also directed to Appendix I, where we provide a more general and powerful framework for deriving effective Hamiltonians for low-energy degrees of freedom and provide several examples (including a slightly different approach to the two-site Hubbard model) as illustrations.

We begin our analysis by noting that there are six possible states: $|\uparrow; \uparrow\rangle$, $|\downarrow; \downarrow\rangle$, $|\uparrow; \downarrow\rangle$, $|\downarrow; \uparrow\rangle$, $|\uparrow \downarrow; -\rangle$, $|-; \uparrow\downarrow\rangle$. Here the arrows represent spin orientation, "—" indicates that the site is empty, and the semicolon separates the left site from the right site. We will proceed by doing perturbation theory in a way which is backwards from the usual one. Since U is large, we will take the unperturbed Hamiltonian to be the potential energy,

$$V = U \sum_{j=1}^{2} n_{j\uparrow} n_{j\downarrow}, \tag{17.46}$$

and we will take the hopping kinetic energy to be the perturbation:

$$T = -t \sum_{\sigma=\uparrow,\downarrow} (c_{1\sigma}^{\dagger} c_{2\sigma} + c_{2\sigma}^{\dagger} c_{1\sigma}). \tag{17.47}$$

Since V commutes with the total spin operator $\vec{S} = \vec{S}_1 + \vec{S}_2$,[6] its eigenstates can be labeled by their total spin quantum number $S = 0, 1$, as the spectrum in Fig. 17.2 illustrates. There is a triplet

---

[6] This might not be so obvious due to the fact that a spin quantization axis has to be chosen in order to write down this U term. The reader is encouraged to verify that it is indeed invariant under spin rotation.

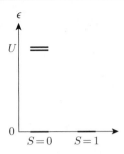

Figure 17.2 Eigenvalues for the two-site Hubbard model with two electrons in the absence of hopping. There are two singlets with energy U, one singlet with zero energy, and a triplet with zero energy.

$$|\uparrow;\uparrow\rangle, \quad \frac{1}{\sqrt{2}}\left(|\uparrow;\downarrow\rangle + |\downarrow;\uparrow\rangle\right), \quad |\downarrow;\downarrow\rangle$$

with energy $\epsilon = 0$, a singlet

$$|\psi_0\rangle = \frac{1}{\sqrt{2}}\left(|\uparrow;\downarrow\rangle - |\downarrow;\uparrow\rangle\right)$$

with energy $\epsilon = 0$, and two singlets with energy U because of double occupancy:

$$|\psi_{0L}\rangle = |\uparrow\downarrow;-\rangle \qquad \text{and} \qquad |\psi_{0R}\rangle = |-;\uparrow\downarrow\rangle.$$

These six states are all exact eigenstates of V, but we will mostly be interested in the low-energy states since we expect that the Hubbard interaction is generally much greater than typical thermal energies ($U \gg k_B T$).

Because the kinetic energy also commutes with $\vec{S}$,

$$[T, \vec{S}] = 0, \tag{17.48}$$

the perturbation can only mix the singlet states among themselves. The low-energy singlet and triplet states all obey

$$\langle\psi|T|\psi\rangle = 0. \tag{17.49}$$

(In fact, for the triplet state $T|\Psi\rangle = 0$ and it is an eigenstate of the full H.) Hence we have to go to second order in perturbation theory. Level repulsion will drive the low-energy singlet state downwards, lifting the spin degeneracy and leading to an effective antiferromagnetic low-energy spin Hamiltonian

$$H_{\text{eff}} = J\,\vec{S}_1 \cdot \vec{S}_2 + C, \tag{17.50}$$

with $J > 0$ and C being a constant. This effective Hamiltonian conveniently describes the low-energy spin physics, ignoring the high-energy charge-fluctuation physics at the scale U. This form is actually *uniquely* determined by the spin-rotation symmetry in the case of spin-1/2 particles.

To calculate J and C perturbatively in t/U note that

$$T|\psi_0\rangle = -2t\frac{1}{\sqrt{2}}\left(|\psi_{0L}\rangle - |\psi_{0R}\rangle\right), \tag{17.51}$$

so the second-order energy shift for $|\psi_0\rangle$ is

$$\Delta\epsilon = \frac{|\langle\psi_{0L}|T|\psi_0\rangle|^2}{-U} + \frac{|\langle\psi_{0R}|T|\psi_0\rangle|^2}{-U} = -\frac{4t^2}{U}. \tag{17.52}$$

Thus the energy of the triplet state remains zero, but the partial delocalization of the electrons in the singlet case reduces the kinetic energy from zero to $\epsilon = -4t^2/U$. These two results allow us to determine J and C using

$$H_{\text{eff}} = \frac{J}{2}\left[(\vec{S}_1 + \vec{S}_2)^2 - \frac{3}{2}\hbar^2\right] + C, \tag{17.53}$$

which yields

$$\epsilon_1 = \frac{J}{4} + C \tag{17.54}$$

for the triplet state, and

$$\epsilon_0 = -\frac{3}{4}J + C \tag{17.55}$$

for the singlet state. Thus

$$C = -\frac{t^2}{U} \tag{17.56}$$

and

$$J = \frac{4t^2}{\hbar^2 U}. \tag{17.57}$$

For the case of an infinite lattice one obtains (up to a constant which we will ignore) the so-called **Heisenberg spin model**

$$H = J \sum_{\langle ij \rangle} \vec{S}_i \cdot \vec{S}_j. \tag{17.58}$$

Because second-order perturbation theory couples only nearest-neighbor sites $\langle ij \rangle$, this result is independent of the spatial dimension and the details of the lattice structure. In the absence of spin–orbit coupling, the Coulomb interaction is invariant under spin rotations (choice of spin quantization axis), and the Heisenberg model captures this in the simplest possible way. For spin-1/2 particles with only near-neighbor interactions, Eq. (17.58) is the only possible interaction that is spin-rotationally invariant.

We note before closing this section that the antiferromagnetic spin exchange coupling, J, has its origin in the electron's desire to hop and lower its *kinetic energy*. Indeed, by allowing some amplitude for double occupancy, the potential energy actually goes *up* (though not by as much as the kinetic energy goes down). In some insulating materials (e.g. MnO and the insulating parent compounds of the high-temperature cuprate superconductors) the exchange between two magnetic cations occurs via a non-magnetic anion which acts as an intermediary sitting between them. This is referred to as **superexchange**. In these more complex models, the resulting effective J is usually antiferromagnetic (as in the Hubbard model we just discussed) but, depending on details, can be ferromagnetic. The qualitative rules developed by Goodenough [175] and Kanamori [176], which take into account such details as the spatial symmetry of the orbitals, are useful in predicting whether a given system will be ferromagnetic or antiferromagnetic.

The situation is very different for ferromagnetism in metals, which originates from lowering the repulsive Coulomb interaction by developing spontaneous magnetization (as discussed in Chapter 15). This Coulomb energy reduction via Pauli exclusion is often referred to as **direct exchange**. In this case the kinetic energy goes *up* rather than down.

In general, antiferromagnetism is often associated with Mott insulators (as the kinetic energy is frustrated by the strong Coulomb repulsion and thus the desire to release it is strong), while ferromagnetism most often arises in metals. As noted above, there are, however, exceptions. Some Mott insulators have a Heisenberg low-energy effective spin Hamiltonian with ferromagnetic exchange coupling $J < 0$, due to the coexistence of direct-exchange and superexchange processes.

> **Exercise 17.2.** Show that any rotationally invariant Hamiltonian for a system of two spin-1/2 particles can always be written in the form of Eq. (17.50).

17.5 Magnetically Ordered States and Spin-Wave Excitations

The ground states of Heisenberg-like spin Hamiltonians often have magnetic order and thus break spin-rotation symmetry (at least in spatial dimensions greater than one). The important collective excitations are **spin waves**, which can be viewed as Goldstone bosons associated with the broken symmetry. These are easiest to analyze in the *ferromagnetic* ($J < 0$) case, so we will begin there.

17.5.1 Ferromagnets

The ferromagnet is particularly simple because we can easily write down (one of) the exact ground states:

$$|S, S\rangle = |\uparrow\uparrow\uparrow\uparrow\uparrow \cdots \uparrow\rangle. \tag{17.59}$$

This fully polarized state has $S = N/2$, $S^z = N/2$. To see that it is an energy eigenstate, let us re-write the Hamiltonian as

$$H = J \sum_{\langle ij \rangle} \left[S_i^z S_j^z + \frac{1}{2}(S_i^+ S_j^- + S_i^- S_j^+) \right]. \tag{17.60}$$

Since any raising operator annihilates this state, only the Ising term contributes, and

$$H|S, S\rangle = E_0|S, S\rangle, \tag{17.61}$$

$$E_0 = JNz/8 = -|J|Nz/8, \tag{17.62}$$

where for convenience we have set $\hbar = 1$, z is the lattice coordination number, and thus $Nz/2$ is the number of near-neighbor bonds. To see that the state (17.59) is a ground state of H, simply notice that it is a ground state of *every single bond*, namely $J\vec{S}_i \cdot \vec{S}_j$.

To generate the other states in the ground state manifold, one simply applies the total-spin-lowering operator

$$S^- \equiv \sum_{j=1}^{N} S_j^-. \tag{17.63}$$

This yields the (unnormalized) state

$$|S, S - 1\rangle = |\downarrow\uparrow\uparrow\uparrow\uparrow \cdots \uparrow\rangle + |\uparrow\downarrow\uparrow\uparrow\uparrow \cdots \uparrow\rangle + |\uparrow\uparrow\downarrow\uparrow\uparrow \cdots \uparrow\rangle + \cdots, \tag{17.64}$$

which is a symmetric linear combination of single spin flips. Successive application of S^- generates a manifold of $2S + 1$ states. In the absence of a Zeeman coupling to an external field we have

$$[H, S^-] = 0, \tag{17.65}$$

and this manifold is degenerate in energy.

It turns out that a simple modification of the lowering operator can be used to generate exact spin-wave excited states:

$$S_{\vec{q}}^- = \frac{1}{\sqrt{N}} \sum_{j=1}^{N} e^{-i\vec{q}\cdot\vec{R}_j} S_j^-, \tag{17.66}$$

where $\vec{q}$ is a wave vector in the first Brillouin zone and we now have included the appropriate normalization factor. The spin-wave state

$$|\vec{q}\rangle = S_{\vec{q}}^{-} |S, S\rangle \tag{17.67}$$

is closely analogous to the single-mode approximation state we studied earlier in connection with superfluid helium and the quantum Hall effect. Unlike in those cases, however, the state here is an exact energy eigenstate.

A little algebra shows that

$$(H - E_0)S_{\vec{q}}^{-}|S, S\rangle = [H, S_{\vec{q}}^{-}]|S, S\rangle = \epsilon_{\vec{q}} S_{\vec{q}}^{-}|S, S\rangle, \tag{17.68}$$

where

$$\epsilon_{\vec{q}} \equiv +\frac{\hbar^2 z}{2}|J|(1 - \gamma_{\vec{q}}), \tag{17.69}$$

and

$$\gamma_{\vec{q}} \equiv \frac{1}{z}\sum_{\vec{\delta}} e^{-i\vec{q}\cdot\vec{\delta}} \tag{17.70}$$

is the same near-neighbor form factor that first appeared in our analysis of the band structure of the tight-binding model in Chapter 7. This is not a coincidence. One can view the fully aligned ground state as the vacuum and view the single flipped spin in the excited state as a "particle" (or a "magnon") that can hop from site to site using the $S^- S^+$ terms in the Hamiltonian. The Ising terms yield a constant which is just the right size to cause the excitation energy

$$\epsilon_{\vec{q}} \equiv E_{\vec{q}} - E_0 \tag{17.71}$$

to vanish in the limit of $\vec{q} \to \vec{0}$, as it must since $S_{\vec{q}}^{-}$ becomes S^- in that limit. Notice that the dispersion is quadratic,

$$\epsilon_{\vec{q}} \sim q^2, \tag{17.72}$$

for small wave vectors.

For a spin-1/2 ferromagnet we can view these particles as hard-core bosons. They are bosons because the creation operators

$$b_j^{\dagger} \equiv S_j^{-} \tag{17.73}$$

commute on different sites. They are not free bosons, however, because of the hard-core constraint (valid for the spin-1/2 case)

$$S_j^{-} S_j^{-} = 0. \tag{17.74}$$

Thus unfortunately we cannot generate two-magnon exact excited states by simply writing

$$S_{\vec{q}_1}^{-} S_{\vec{q}_2}^{-} |S, S\rangle. \tag{17.75}$$

At finite temperatures there is a finite magnon density and the constraint effectively turns this into an interacting-boson problem which cannot be solved exactly (except for $d = 1$).

If we work at very low temperatures we may be able to estimate the magnetization using the free-boson approximation, *if* the density of these (thermally excited) bosons is low so that their interactions have negligible effects. Calling the magnetization $M = S = N/2$ for the ground state, we have

$$M = N/2 - \sum_j b_j^{\dagger} b_j, \tag{17.76}$$

since the magnetization is lowered by one unit for each flipped spin present. In the free-boson approximation

$$M = N/2 - \sum_{\vec{q}} b_{\vec{q}}^{\dagger} b_{\vec{q}} \tag{17.77}$$

$$= N/2 - \sum_{\vec{q}} n_{\mathrm{B}}(\epsilon_{\vec{q}}), \tag{17.78}$$

where the boson operators

$$b_{\vec{q}}^{\dagger} = \frac{1}{\sqrt{N}} \sum_{j=1}^{N} e^{-i\vec{q} \cdot \vec{R}_j} b_j^{\dagger}, \tag{17.79}$$

$$b_{\vec{q}} = \frac{1}{\sqrt{N}} \sum_{j=1}^{N} e^{+i\vec{q} \cdot \vec{R}_j} b_j \tag{17.80}$$

satisfy the usual bosonic commutation relations (since we have relaxed the hard-core constraints):

$$[b_{\vec{q}}^{\dagger}, b_{\vec{q}'}^{\dagger}] = [b_{\vec{q}}, b_{\vec{q}'}] = 0; \tag{17.81}$$

$$[b_{\vec{q}}, b_{\vec{q}'}^{\dagger}] = \delta_{\vec{q}, \vec{q}'}. \tag{17.82}$$

At low temperatures, only long-wavelength magnons are excited, and we can make the quadratic approximation $\epsilon_{\vec{q}} \approx Aq^2$, where A is a constant. Thus

$$M \approx \frac{N}{2} - \frac{V}{(2\pi)^d} \int d^d\vec{q} \, \frac{1}{e^{\beta A q^2} - 1}. \tag{17.83}$$

For $d \leq 2$ there is an infrared divergence in the integral signaling the fact that the magnetization is destroyed at any non-zero temperature, and the rotation symmetry is restored. Thus $d = 2$ is the so-called "**lower critical dimension**" for the Heisenberg ferromagnet.

Mathematically this divergence is very similar to that encountered in the calculation of the Debye–Waller factor in Chapter 6, where we found that the Debye–Waller factor *vanishes* in $d = 1, 2$ at finite T, signaling that the spontaneously broken translation symmetry in the crystalline state is destroyed (that is, the full translation symmetry is restored) by thermal fluctuations. They also have very similar physical origins: given spontaneously broken *continuous* symmetries, gapless modes (long-wavelength spin waves and acoustic phonons) appear in the spectrum; it is their thermal fluctuations that destroy the broken symmetry or long-range order at non-zero temperatures. While the magnetization calculation here and the calculation of the Debye–Waller factor in Chapter 6 are approximate, there is a theorem due to Hohenberg, Mermin, and Wagner which establishes rigorously that spontaneously broken continuous symmetry is not possible for $d = 1, 2$ at finite T, for a wide range of models. The Heisenberg model is one example of this.

For $d = 3$ we have

$$M = \frac{N}{2} - \frac{V}{(2\pi)^3} 4\pi \int_0^{\infty} dq q^2 \, \frac{1}{e^{\beta A q^2} - 1}, \tag{17.84}$$

where we have moved the upper limit of the integration to infinity without serious error since the integrand is exponentially small at large q. Simple dimensional analysis shows that the result is

$$M = \frac{N}{2} \left[1 - \left(\frac{T}{T_0} \right)^{3/2} \right], \tag{17.85}$$

where $T_0 \sim J/k_{\mathrm{B}}$ is a temperature scale. This result tells us that thermal fluctuations do not immediately destroy the magnetization in $d = 3$, but rather do so gradually with a characteristic 3/2 power law as the temperature is raised.

Before proceeding to a discussion of antiferromagnets, it is useful at this point to back up a bit and re-derive the magnon dispersion relation via an equation-of-motion method. We will do this for the case of general spin s, but the manipulations are formally valid only in the limit of large s – that is, they constitute a semiclassical approximation. Nevertheless, we will see that they yield the exact dispersion relation even for $s = 1/2$.

Let us start with a single spin in a magnetic field:

$$H = -\lambda S^z, \tag{17.86}$$

and consider the Heisenberg equation of motion

$$\dot{S}^+ = \frac{i}{\hbar}[H, S^+] = -\frac{i}{\hbar}\lambda[S^z, S^+] = -i\lambda S^+, \tag{17.87}$$

which has the solution

$$S^x(t) + iS^y(t) = e^{-i\lambda t}\left[S^x(0) + iS^y(0)\right] \tag{17.88}$$

corresponding to simple gyroscopic precession of the spin about the z axis at a frequency of

$$\omega = \lambda. \tag{17.89}$$

Thus the quantum energy-level spacing

$$\hbar\omega = \hbar\lambda \tag{17.90}$$

is uniform as in a simple harmonic oscillator. This is the exact answer for Zeeman-split spin levels, although of course there are only $2s + 1$ such levels, rather than an infinite number as in a harmonic oscillator.

We now follow the same procedure for the full Heisenberg model to obtain

$$\dot{S}_i^+ = \frac{i}{\hbar}[H, S_i^+] = -\frac{i}{\hbar}|J|\sum_{\vec{\delta}}\left\{\frac{1}{2}S_{i+\delta}^+[S_i^-, S_i^+] + S_{i+\delta}^z[S_i^z, S_i^+]\right\}$$
$$= -i|J|\sum_{\vec{\delta}}\left\{-S_i^z S_{i+\delta}^+ + S_{i+\delta}^z S_i^+\right\}. \tag{17.91}$$

The RHS is not as simple as in the single-spin example. However, if we assume that s is large and that the spins are nearly perfectly aligned in the $\hat{z}$ direction, we can work to lowest order in the S^x and S^y fluctuations by making the replacement

$$S_i^z \rightarrow \hbar s, \tag{17.92}$$

to obtain

$$\dot{S}_i^+ = -i|J|\hbar s\sum_{\vec{\delta}}\left\{S_i^+ - S_{i+\delta}^+\right\}. \tag{17.93}$$

This can be solved by going to Fourier space,

$$S_{\vec{q}}^+(t) = e^{-i\omega_{\vec{q}}t}S_q^+(0), \tag{17.94}$$

with the dispersion relation

$$\hbar\omega_{\vec{q}} = |J|z\hbar^2 s(1 - \gamma_{\vec{q}}). \tag{17.95}$$

Even though this was derived for large s, it correctly reproduces the exact result we obtained in Eq. (17.69) for the case $s = 1/2$. By using Eq. (17.92) we have effectively dropped the hard-core constraint on the magnon bosons. This constraint becomes less and less important as s increases, since there can be as many as $2s$ bosons on a single site.

Exercise 17.3. Consider a set of three half spins arranged on an equilateral triangle and having ferromagnetic near-neighbor Heisenberg couplings. (It is useful to view this as a 1D chain of length three having periodic boundary conditions.)

Find all eight exact eigenstates and eigenvalues.

Hint: we know that the fully aligned state $|\uparrow\uparrow\uparrow\rangle$ is an eigenstate. Using Eq. (17.67), we know the exact single-spin-wave states. The periodic boundary conditions tell us that only three discrete values of q are allowed. One is $q = 0$. What are the other two?

Now you have four states. The remaining four are easily found from time-reversal symmetry by reversing all spins and complex conjugating all amplitudes.

The time reverse of a one-magnon state contains two flipped spins and is thus in a sense a two-magnon state. This trick gets us around the problem of not being able to create two independent magnons by the naive method discussed in connection with Eq. (17.75).

Exercise 17.4. Evaluate the constant T_0 in Eq. (17.85) in terms of J for the simple cubic lattice.

17.5.2 Antiferromagnets

We turn now to an analysis of the antiferromagnet ($J > 0$). Note that the Hamiltonian for the antiferromagnet is *identical* to that of the ferromagnet except for an overall sign change. Thus the eigenstates for the two are *exactly the same*. The fully aligned state in Eq. (17.59) is thus an exact eigenstate for the antiferromagnet. Unfortunately it has the highest possible energy rather than the lowest, since the signs of all the eigenvalues are reversed. Thus it is not very important to the thermodynamics, and we are forced to start from scratch in analyzing the low-energy physics of the antiferromagnet.

Another way to view this is to note that, classically, the free energies of the ferromagnet and antiferromagnet are identical, if we have a bipartite lattice (which is assumed in this subsection). This is because the two cases can be mapped into each other via the transformation $\vec{S} \to -\vec{S}$ on one of the sublattices. This transformation changes the sign of J but has no effect on the classical sum over spin orientations in the partition function. This transformation is *not* allowed in the quantum case because it destroys the canonical commutation relations

$$[S^\mu, S^\nu] = i\hbar\epsilon^{\mu\nu\gamma} S^\gamma. \tag{17.96}$$

These relations determine the dynamics of the spin waves. Classically this is unimportant, but quantum mechanically the spin-wave frequency $\omega_{\vec{q}}$ determines its energy $\hbar\omega_{\vec{q}}$ and hence its contribution to the thermodynamics. The existence of Planck's constant connects *dynamics* and *thermodynamics*.

On a bipartite lattice, the Ising part of the energy is minimized by having opposite ferromagnetic alignment on the two sublattices in the so-called **Néel state**:

$$|0\rangle = |\uparrow\downarrow\uparrow\downarrow\uparrow\downarrow\uparrow\downarrow\cdots\rangle. \tag{17.97}$$

It must be emphasized that this choice of ordering the z components of the spins is entirely arbitrary. The Heisenberg Hamiltonian is fully spin-rotation-invariant, and the sublattice magnetization in the Néel state can point in any direction. In essence we are defining the $\hat{z}$ direction using this.

While the Néel state captures the essence of the symmetry breaking in the ground state, it is unfortunately not an exact eigenstate when the transverse coupling terms $S_i^+ S_j^-$ are included. The transverse terms cause "zero-point motion," producing virtual spin flips,

$$\uparrow\downarrow \longrightarrow \downarrow\uparrow,$$

which reduce the staggered spin-order parameter defined by[7]

$$\hbar \vec{M}_s = \sum_{j \in A} \langle \vec{S}_j \rangle - \sum_{j \in B} \langle \vec{S}_j \rangle, \tag{17.98}$$

where the sums are over the two sublattices. The magnetization is reduced but remains finite (at $T = 0$) for $d > 1$. In 1D there are infrared divergences which destroy the magnetization even at zero temperature, as we will see shortly.

To analyze these fluctuations, we will again make a free-boson approximation that ignores the hard-core constraint. On the A sublattice we let $|\uparrow\rangle$ represent the vacuum and $|\downarrow\rangle$ represent one boson on the site. Thus

$$S_j^- \to \hbar b_j^{\dagger}; \qquad j \in A. \tag{17.99}$$

On the B sublattice it is the other way around: $|\downarrow\rangle$ is the vacuum and $|\uparrow\rangle$ is the one-boson state. Thus

$$S_j^+ \to \hbar b_j^{\dagger}; \qquad j \in B. \tag{17.100}$$

This transformation changes the Hamiltonian to the form

$$H = \hbar^2 J \sum_{\langle ij \rangle} \left[\left(\frac{1}{2} - b_i^{\dagger} b_i \right) \left(-\frac{1}{2} + b_j^{\dagger} b_j \right) + \frac{1}{2} \left(b_i^{\dagger} b_j^{\dagger} + b_i b_j \right) \right]. \tag{17.101}$$

Clearly the total number of bosons is no longer conserved due to terms like $b_i^{\dagger} b_j^{\dagger}$. This signals the fact that the staggered magnetization order parameter for the antiferromagnet is not a conserved quantity and is subject to zero-point quantum fluctuations as discussed earlier. This complication can, however, be dealt with by making a **Bogoliubov transformation** as long as only quadratic terms in the creation operators are retained. To this end, we make the linear spin-wave approximation and neglect quartic terms in Eq. (17.101) of the form $b_i^{\dagger} b_i b_j^{\dagger} b_j$:

$$H_{\text{LSW}} = \frac{z \hbar^2 J}{2} \sum_i b_i^{\dagger} b_i + \frac{\hbar^2 J}{2} \sum_{\langle ij \rangle} \left(b_i^{\dagger} b_j^{\dagger} + b_i b_j \right) + \text{constant}, \tag{17.102}$$

where z is the coordination number. From now on we will also treat the bs as regular instead of hard-core bosons, an approximation that is justifiable as long as the boson density remains small. To diagonalize H_{LSW}, we take advantage of its translation invariance and Fourier transform to wave-vector space (that is similar to what we did for ferromagnetic spin waves) and neglect the constant:

$$H_{\text{LSW}} = \frac{z \hbar^2 J}{2} \sum_{\vec{q}} \left[b_{\vec{q}}^{\dagger} b_{\vec{q}} + \frac{\gamma_{\vec{q}}}{2} \left(b_{\vec{q}}^{\dagger} b_{-\vec{q}}^{\dagger} + b_{\vec{q}} b_{-\vec{q}} \right) \right]. \tag{17.103}$$

This "almost" diagonalizes H_{LSW}, but not quite, due to the second term that mixes modes at $\vec{q}$ and $-\vec{q}$. (This is allowed by momentum conservation, and *lack* of particle number conservation!) Recall that in the ferromagnetic case there were no such terms, and Fourier transformation completely diagonalizes the spin-wave Hamiltonian (in the subspace of a single spin wave). To diagonalize (17.103) (which describes essentially decoupled sets of two coupled harmonic oscillators), we introduce the Bogoliubov transformation:

$$b_{\vec{q}} = \cosh \theta_{\vec{q}} \, \beta_{\vec{q}} + \sinh \theta_{\vec{q}} \, \beta_{-\vec{q}}^{\dagger}; \tag{17.104}$$

$$\beta_{\vec{q}} = \cosh \theta_{\vec{q}} \, b_{\vec{q}} - \sinh \theta_{\vec{q}} \, b_{-\vec{q}}^{\dagger}. \tag{17.105}$$

---

[7] Note that, unlike in the case of a ferromagnet, where the order parameter is simply the total magnetization and commutes with the Heisenberg Hamiltonian, the staggered magnetization does *not* commute with the Hamiltonian and is not a constant of the motion.

In the above $\theta_{\vec{q}} = \theta_{-\vec{q}}$ is real, and the carefully chosen coefficients (satisfying the mathematical identity $\cosh^2 \theta_{\vec{q}} - \sinh^2 \theta_{\vec{q}} = 1$) of the transformation guarantee that the βs satisfy the bosonic commutation relations just like the bs (the reader should verify these expressions):

$$[\beta_{\vec{q}}^{\dagger}, \beta_{\vec{q}'}^{\dagger}] = [\beta_{\vec{q}}, \beta_{\vec{q}'}] = 0; \tag{17.106}$$

$$[\beta_{\vec{q}}, \beta_{\vec{q}'}^{\dagger}] = \delta_{\vec{q},\vec{q}'}. \tag{17.107}$$

It is important to note from Eq. (17.105) that $\beta_{\vec{q}}$ carries the same momentum as $b_{\vec{q}}$, since it either destroys a boson with momentum $\vec{q}$, or creates a boson with momentum $-\vec{q}$. Substituting Eq. (17.104) into Eq. (17.103), and choosing $\theta_{\vec{q}}$ such that coefficients of off-diagonal terms like $\beta_{\vec{q}} \beta_{-\vec{q}}$ vanish, we obtain the diagonalized linear spin-wave Hamiltonian

$$H_{\text{LSW}} = \sum_{\vec{q}} E_{\vec{q}} \beta_{\vec{q}}^{\dagger} \beta_{\vec{q}}, \tag{17.108}$$

where

$$E_{\vec{q}} = \frac{z \hbar^2 J}{2} \sqrt{1 - \gamma_{\vec{q}}^2}, \tag{17.109}$$

and the proper choice of $\theta_{\vec{q}}$ that brings H_{LSW} to the form of Eq. (17.108) satisfies

$$\tanh(2\theta_{\vec{q}}) = -\gamma_{\vec{q}}. \tag{17.110}$$

Unlike the case of the ferromagnet, antiferromagnetic spin waves have a "relativistic" (i.e. linear rather than quadratic) dispersion which is gapless both at the Brillouin zone center ($q = 0$) and at the zone corner (for square, cubic, or hypercubic lattices, that corresponds to $\vec{q} = (\pi/a, \pi/a, \ldots)$, where a is the lattice constant). Furthermore, the spectrum is *linear* in δq for small deviations from both points:

$$\omega = c\,\delta q, \tag{17.111}$$

with the spin-wave velocity given by

$$c = \alpha \hbar^2 J z s a, \tag{17.112}$$

where α is a dimensionless constant of order unity whose precise value depends on the details of the lattice structure. While we have $s = 1/2$ in the above, it is actually valid for general s (see Eq. (17.122)). The existence of *two* branches of linear spin-wave excitations is what one expects from the Goldstone theorem (as applied to relativistic or Lorentz-invariant quantum field theories), since the Néel state breaks the spin-rotation symmetry, and two of the three generators of rotation (S^x and S^y) transform it non-trivially. While the Heisenberg model comes from non-relativistic quantum mechanics, in this case we have an *emergent* Lorentz invariance in the long-wavelength and low-energy limit, with the speed of light replaced by the spin-wave velocity (that is why we use c to represent both). Note that this happens in antiferromagnets, but not in ferromagnets.[8]

A simpler procedure to obtain the spin-wave spectra, which applies to general spin s, is to solve the Heisenberg equations of motion as we did for the ferromagnet. The analog of Eq. (17.91) is

$$\dot{S}_k^+ = +iJ \sum_{\delta} \left\{ -S_k^z S_{k+\delta}^+ + S_{k+\delta}^z S_k^+ \right\}. \tag{17.113}$$

---

[8] The same statement about the generators applies to the ferromagnet, but there the structure of the time derivatives in the Lagrangian is different and the theory is non-relativistic. In fact, the two non-trivial generators of rotation become canonically conjugate and describe the position and momentum of a single degree of freedom rather than being independent coordinates of two separate degrees of freedom.

Now, since there are two oppositely oriented sublattices, we make the following substitutions:

$$S_k^z \to +\hbar s, \qquad k \in A; \tag{17.114}$$

$$S_k^z \to -\hbar s, \qquad k \in B. \tag{17.115}$$

For $k \in A$ we obtain

$$\dot{S}_k^+ = -iJz\hbar s \left\{ S_k^+ + \frac{1}{z} \sum_\delta S_{k+\delta}^+ \right\} \tag{17.116}$$

and for $k \in B$ we have

$$\dot{S}_k^+ = +iJz\hbar s \left\{ S_k^+ + \frac{1}{z} \sum_\delta S_{k+\delta}^+ \right\}. \tag{17.117}$$

It is useful to define separate Fourier transforms for each sublattice:

$$S_{A(B)\vec{q}}^+ \equiv \sqrt{\frac{2}{N}} \sum_{k \in A(B)} e^{-i\vec{q}\cdot\vec{R}_k} S_k^+, \tag{17.118}$$

where $N/2$ is the number of sites in each sublattice and $\vec{q}$ is an element of the *magnetic Brillouin zone*, that is the Brillouin zone corresponding to a real-space lattice consisting of only one of the sublattices. Since the unit cell is doubled in real space, the magnetic Brillouin zone has only half of the volume of the usual structural Brillouin zone.

Recognizing that if $k \in A$ then $k + \delta \in B$ allows us to write the Fourier-transformed equations of motion as

$$\dot{S}_{A\vec{q}}^+ = -iJz\hbar s \left\{ S_{A\vec{q}}^+ + \gamma_{\vec{q}} S_{B\vec{q}}^+ \right\}, \tag{17.119}$$

$$\dot{S}_{B\vec{q}}^+ = iJz\hbar s \left\{ S_{B\vec{q}}^+ + \gamma_{\vec{q}} S_{A\vec{q}}^+ \right\}. \tag{17.120}$$

The dispersion relation is then found in the usual way from

$$\det \begin{pmatrix} (-i\omega_{\vec{q}} + iJz\hbar s) & iJz\hbar s\gamma_{\vec{q}} \\ -iJz\hbar s\gamma_{\vec{q}} & (-i\omega_{\vec{q}} - iJz\hbar s) \end{pmatrix} = 0, \tag{17.121}$$

which yields

$$\omega_q = Jz\hbar s \sqrt{1 - (\gamma_{\vec{q}})^2}, \tag{17.122}$$

in agreement with Eq. (17.109) for the case of $s = 1/2$. Note that here we used the *reduced*-zone scheme, with two lattice sites per unit cell, that reflects the fact that the Néel state breaks translation symmetry (in addition to spin-rotation symmetry). As a result, there are two modes for every $\vec{q}$ in the reduced zone, equivalent to one mode for every $\vec{q}$ in the extended-zone scheme used in the Bogoliubov transformation. The total number of modes is, of course, the total number of spins N in both schemes.

A disadvantage of the equation-of-motion approach is that it yields only the excitation spectrum, not the wave functions for either the ground state or the excited states. We do need these wave functions in order to calculate physical quantities like the sublattice magnetization, which suffers from quantum fluctuations. We thus return to the Bogoliubov transformation method for such calculations. Within the linear spin-wave approximation, the *reduction* of sublattice magnetization is

$$\Delta M_s = \sum_j \langle b_j^\dagger b_j \rangle = \sum_{\vec{q}} \langle b_{\vec{q}}^\dagger b_{\vec{q}} \rangle \tag{17.123}$$

$$= \sum_{\vec{q}} \left(\cosh^2 \theta_{\vec{q}} \langle \beta_{\vec{q}}^\dagger \beta_{\vec{q}} \rangle + \sinh^2 \theta_{\vec{q}} \langle \beta_{\vec{q}} \beta_{\vec{q}}^\dagger \rangle \right). \tag{17.124}$$

At $T = 0$ only the second term above contributes, and we have

$$\Delta M_s = \sum_{\vec{q}} \sinh^2 \theta_{\vec{q}} = \frac{L^d}{2} \int_{1BZ} \frac{d^d \vec{q}}{(2\pi)^d} \left(\frac{1}{\sqrt{1 - (\gamma_{\vec{q}})^2}} - 1 \right). \qquad (17.125)$$

For $d \geq 2$ the integral above is convergent and yields a relatively small correction, justifying the linear spin-wave theory (see Exercise 17.6). However, for $d = 1$ the integral diverges logarithmically at the zone center and zone boundary (where the spin-wave energy vanishes and quantum fluctuations are strongest). This signals the breakdown of linear spin-wave theory, and indicates that long-range Néel order is destroyed by quantum fluctuations in 1D. This is very similar to the logarithmic divergence that leads to vanishing of the Debye–Waller factor in 1D at $T = 0$, as we saw in Chapter 6. These are examples of a general result stating that divergent quantum fluctuations always preclude the possibility of spontaneously broken continuous symmetries in 1D, unless the order parameter is a conserved quantity (like that of a ferromagnet, which does *not* suffer from quantum fluctuations), or there are long-range interactions.[9]

> **Exercise 17.5.** Evaluate the constant α in Eq. (17.112) for
> (a) the 2D square lattice,
> (b) the 3D simple cubic lattice, and
> (c) the 3D FCC lattice.

> **Exercise 17.6.** Evaluate ΔM_s for 2D square and 3D cubic lattices at zero temperature using Eq. (17.125).

> **Exercise 17.7.** Show that, within the linear spin-wave approximation, ΔM_s diverges in 2D at finite temperature, indicating that thermal fluctuations restore spin-rotation symmetry. This is another manifestation of the Hohenberg–Mermin–Wagner theorem.

17.6 One Dimension

In the case of 1D Heisenberg spin chains, the physics is especially interesting. The exact results obtained previously for the ferromagnet apply in $d = 1$, but the physics of the antiferromagnet is not at all described by the semiclassical picture developed above. The problem is that the zero-point motion of the spins is large enough to disorder the Néel state at zero temperature. The ground state is thus **quantum disordered**.[10] Associated with this, there develop some quite surprising differences between the cases of integer and half-integer spins.

---

[9] This result is closely related to the Hohenberg–Mermin–Wagner theorem in classical statistical mechanics mentioned earlier. It can be shown using path-integral methods that, quite generally, a d-dimensional quantum system at $T = 0$ can be mapped onto a $(d + 1)$-dimensional classical system at finite T, with the extra dimension being "imaginary time." This is particularly true for systems with emergent Lorentz invariance (antiferromagnets and crystalline lattices being examples), where the resultant "space-time" is "isotropic" (because frequencies and wave vectors are linearly related). Thus such a 1D quantum system at $T = 0$ is equivalent to a 2D classical system at finite T, which cannot order. (In this context all systems behave classically at finite T and sufficiently low frequencies $\hbar\omega \ll k_B T$.)

[10] As we will see later, in some cases the system is quantum critical, with power-law decaying spin–spin correlations.

Let us begin with the idea of "wave–particle" duality for quantum spins. Quantum spins are strange objects, which behave in some ways like ordinary classical angular momentum vectors which can point in arbitrary directions. Yet, if we measure their projection along any given direction, we find that it can take on only a discrete set of $2s + 1$ values. The discreteness suggests that a collection of quantum spins might exhibit discrete excitations and a gap. The continuous-orientation picture suggests a continuous spectrum of gapless spin waves. To be specific, let us consider the case of spin 1/2. The two possible discrete states can be represented by the spinors $|\uparrow\rangle$ and $|\downarrow\rangle$. However, quantum mechanics allows for the possibility of a coherent linear superposition of these two states

$$|\psi\rangle = \cos(\theta/2)e^{-i\varphi/2}|\uparrow\rangle + \sin(\theta/2)e^{i\varphi/2}|\downarrow\rangle. \tag{17.126}$$

This continuous family of states is parameterized by the two polar and azimuthal angles θ and φ, determining the orientation of the expectation value of the spin vector. A quantum-mechanical process, of course, can be viewed (from a path-integral viewpoint) as a coherent superposition of *all* possible classical processes involving such continuous spin configurations; the *discrete* nature of quantum spin is a consequence of the interference effects of such superpositions. To this end, a fundamental difference between integer and half-integer spins shows up in the phase generated by a 2π rotation (in this section we set $\hbar = 1$ to simplify notation):

$$e^{i2\pi S^z}|sm\rangle = (-1)^{2m}|sm\rangle = (-1)^{2s}|sm\rangle, \tag{17.127}$$

indicating that such a rotation induces a sign change to the state ket for half-integer spin-s states, while no change occurs for integer spin-s states. We note that, since Eq. (17.127) applies for all m with fixed s, it actually implies an *operator identity*

$$e^{i2\pi S^z} = (-1)^{2s}. \tag{17.128}$$

One way to understand why this feature is particularly important in 1D is to think about how the spin-wave picture breaks down. Owing to the large quantum fluctuations in 1D, we cannot study a linearized theory of small quantum fluctuations around the classical ground state. Indeed, the quantum fluctuations become so large that the spins can rotate all the way through 2π and hence become sensitive to the phase that appears in Eq. (17.127). This phase can also be viewed as a Berry phase for a corresponding adiabatic spin rotation. Quantum interference (between trajectories with different winding numbers describing the spin rotations) is sensitive to such phase changes; in particular, the additional phase for half-integer spins tends to cause destructive interference and thus *suppresses* quantum fluctuations in 1D, effectively making these spins more "classical."

Spin 1/2 is the least classical possible spin value, and yet the continuous spin-wave picture works extremely well in dimensions greater than one. In the antiferromagnet the **staggered magnetization**

$$\vec{M}_s \equiv \sum_j (-1)^j \langle \vec{S}_j \rangle \tag{17.129}$$

is non-zero in the ground state (for $d > 1$). There exists a continuous family of spin-wave excited states in which the angles θ and φ precess at a frequency linearly proportional to the wave vector (at long wavelengths). The same holds true for all possible other spin values $s = 1, 3/2, 2, \ldots$. They are all effectively equivalent and the discreteness of the individual spin orientations is irrelevant to the low-energy physics.

The situation is very different in 1D spin chains, where we have seen that quantum fluctuations destroy the long-range staggered magnetic order even at zero temperature. It turns out that half-integer spins $s = 1/2, 3/2, 5/2, \ldots$ still have gapless excitations, but they are *not* spin waves. On the other hand, for integer spin values $s = 1, 2, 3, \ldots$, the discreteness becomes relevant and the spin chain exhibits the so-called **Haldane excitation gap**. The origin of such a major qualitative difference can be traced back to Eq. (17.127), and is clearly revealed by a theorem proved in the next subsection.

17.6.1 Lieb–Schultz–Mattis Theorem

Consider a spin-s chain of N spins, with nearest-neighbor antiferromagnetic Heisenberg coupling:

$$H = J \sum_{j=1}^{N-1} \vec{S}_j \cdot \vec{S}_{j+1} + J \vec{S}_N \cdot \vec{S}_1. \tag{17.130}$$

The last term makes the chain periodic (or turns it into a ring), and the Hamiltonian thus possesses lattice-translation symmetry in addition to spin-rotation symmetry. For the following we consider even N. In this case the **Lieb–Schultz–Mattis theorem** states that, for half-integer s, the separation between the ground and first excited state energies, $\Delta E = E_1 - E_0$, vanishes in the limit $N \to \infty$; more precisely, for sufficiently large N,

$$\Delta E < \alpha J / N, \tag{17.131}$$

where α is an N-independent finite number. Thus the excitation gap is *at most* $O(J/N)$.

The Lieb–Schultz–Mattis theorem is proved using the variational principle. Assume $|0\rangle$ is the exact ground state. If we can construct a state $|1'\rangle$ that is orthogonal to $|0\rangle$, then $E_{1'} = \langle 1'|H|1'\rangle$ provides an upper bound of E_1, thus

$$\Delta E \leq \Delta E' = E_{1'} - E_0. \tag{17.132}$$

The key step is constructing $|1'\rangle$ with desired properties, and this is achieved for half-integer spins through the so-called twist operator O_1, which is unitary (and thus does not change the norm of a state):

$$|1'\rangle = O_1|0\rangle, \tag{17.133}$$

$$O_1 = \exp\left[\frac{i2\pi}{N} \sum_{j=1}^{N} \left(j S_j^z \right) \right]. \tag{17.134}$$

Physically O_1 rotates each spin (around the z axis) by an amount that grows linearly from 0 to 2π along the chain, thus the name twist operator. To show that $\langle 1'|0\rangle = 0$, we inspect how O_1 transforms under lattice translation by one unit:

$$T \vec{S}_j T^\dagger = \vec{S}_{j+1}; \tag{17.135}$$

$$T O_1 T^\dagger = \exp\left[\frac{i2\pi}{N} \sum_{j=1}^{N} \left(j S_{j+1}^z \right) \right] \tag{17.136}$$

$$= O_1 \exp\left[-\frac{i2\pi}{N} \sum_{j=1}^{N} S_j^z \right] e^{i2\pi S_1^z}. \tag{17.137}$$

In the above we used the identification $\vec{S}_{N+1} = \vec{S}_1$. The last factor above is $(-1)^{2s}$ (see Eq. (17.128)), and for antiferromagnetic interactions we expect the ground state to be a singlet for even N, thus $\sum_{j=1}^{N} S_j^z |0\rangle = 0$. We also expect $|0\rangle$ to be an eigenstate of T (with some eigenvalue η), as guaranteed by the translation symmetry of H:

$$T|0\rangle = \eta|0\rangle. \tag{17.138}$$

As a result, we have

$$T|1'\rangle = T O_1 T^\dagger T|0\rangle = (-1)^{2s} \eta|1'\rangle, \tag{17.139}$$

namely, for half-integer s, $|1'\rangle$ and $|0\rangle$ are both eigenstates of T with *different* eigenvalues, and thus must be orthogonal to each other. We emphasize that this conclusion does not hold for integer s, or if translation symmetry is broken in H, in which case eigenstates of H are no longer eigenstates of T.

We now turn to the calculation of $\Delta E'$, which is a variational upper bound of ΔE for half-integer s:

$$\Delta E' = \langle 1'|H|1'\rangle - E_0 = \langle 0|O_1^\dagger H O_1 - H|0\rangle. \tag{17.140}$$

Using the fact O_1 is a combination of rotation operators, and the fact that every term in H is invariant under a *uniform* rotation, it is easy to show that, for a typical term in H, we have

$$O_1^\dagger \vec{S}_j \cdot \vec{S}_{j+1} O_1 - \vec{S}_j \cdot \vec{S}_{j+1} = (S_j^x S_{j+1}^x + S_j^y S_{j+1}^y)[\cos(2\pi/N) - 1]$$
$$- (S_j^y S_{j-1}^x - S_j^x S_{j-1}^y)\sin(2\pi/N). \tag{17.141}$$

The small but non-zero change caused by O_1 is due to the slow twist in the rotation. We note that the second term on the RHS above is *odd* under a uniform global $\pi/2$ rotation around the $\hat{z}$ axis (which turns $S^x \to S^y$ and $S^y \to -S^x$), followed by an inversion (which reverses the order of j and $j+1$). Since $|0\rangle$ is invariant (up to a phase) under such transformations, the second term must have zero expectation value. The expectation value of the first term is of order $O(1/N^2)$ for large N. Since there are N such terms in Eq. (17.140), we find

$$\Delta E < \Delta E' \sim O(J/N). \tag{17.142}$$

The Lieb–Schultz–Mattis theorem is thus proven. In the following subsection we explore its physical consequences in some detail.

> **Exercise 17.8.** The Lieb–Schultz–Mattis theorem remains valid in the presence of more-distant-neighbor spin–spin couplings, as long as such couplings are sufficiently short-ranged. Specifically, for the case of couplings that decay as a power law of distance, $J_{ij} \propto 1/|i-j|^\beta$, show that the theorem remains valid for $\beta > 3$.

17.6.2 Spin-1/2 Chains

For the nearest-neighbor Heisenberg antiferromagnetic spin-1/2 chain, there are indeed *gapless* excitations whose excitation energy goes to zero as $O(J/N)$, thus satisfying the Lieb–Schultz–Mattis theorem. In fact, the $O(J/N)$ scaling is what one expects from the linear antiferromagnetic spin-wave spectrum discussed in the last section, as the minimum wave vector is $O(1/N)$. On the other hand, we have also shown that there is *no* antiferromagnetic long-range order in 1D, yet such gapless excitation still exists. In fact, for this model, the ground state spin–spin correlation function can be shown to decay as a power law (with a logarithmic correction):

$$\langle 0|\vec{S}_i \cdot \vec{S}_j|0\rangle \sim \frac{(-1)^{i-j}\log^{1/2}|i-j|}{|i-j|^2}. \tag{17.143}$$

This is the typical situation for *quantum criticality*, with power-law (or quasi-) long-range order and gapless excitations. Owing to the slow decay of the spin–spin correlation function, if one were to take a snap shot of a *local* spin configuration, it would look like a Néel state (with spins alternately aligned and anti-aligned to some completely arbitrary quantization axis).

What is the nature of the low-energy excitations here? It turns out that the spin wave, which would normally represent a $\Delta S = 1$ excitation above the singlet ground state (because it is created by a spin-flip operator), splits up (or "fractionalizes") into two independent objects known as "**spinons**."[11]

---

[11] For a system with a fixed number of spins, the total spin quantum number of any Hamiltonian eigenstates can only differ from that of the ground state by an integer, even when the underlying spins are half-integer spins. Because of

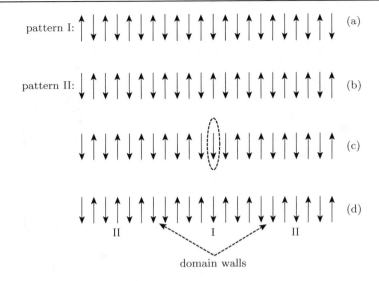

Figure 17.3 Two opposite-stagger ordered states in 1D, patterns I (a) and II (b). (c) Excitation caused by flipping a single spin in the center of the chain. (d) Fractionalization of the single flipped spin into two $s = 1/2$ domain walls separating patterns I and II of ordering (denoted by dashed arrows). The system moves from state (c) to state (d) by mutual spin flip of the pairs of spins on each side of the central spin. For the Heisenberg model the above picture should be viewed as a rough caricature of the physics. It can be rigorously justified for anisotropic "XXZ" models, where the XX and YY spin couplings are equal but the ZZ coupling is much larger. (The spinons are gapped excitations in this case.)

These objects are actually domain walls separating two possible stagger-ordered states as shown in Fig. 17.3. The reader has probably recognized that the mechanism of fractionalization here is very similar to that in the Su–Schrieffer–Heeger model discussed in Chapter 7.

This process is readily detectable using neutron scattering with spin-polarized neutrons in experiments that focus on the *spin-flip* scattering events. Such an experiment measures the dynamical spin structure factor of the system, as a neutron spin flip is always accompanied by a corresponding electron spin flip in the system (see Exercise 4.2). Owing to the kinematic constraints of energy, momentum, and total-spin conservation laws, an ordinary spin-wave excitation (or a magnon excitation with spin 1) would show up as a single sharply defined energy associated with each wavelength, very much like the single-phonon excitation discussed in Chapter 4. A pair of spinons, on the other hand, has much more phase space available to it (since only the sum of their individual momenta is fixed), and the spectral density is a convolution of the spectral densities of the individual spinons, similar to the two- or multi-phonon contribution to neutron scattering. See Fig. 17.4 for an example.

One can suppress the power-law Néel order by introducing frustration into the Hamiltonian, and the simplest way to do that is to introduce an antiferromagnetic second-nearest-neighbor coupling $J' > 0$, which penalizes parallel spins on the *same* sublattice:

$$H = J \sum_j \vec{S}_j \cdot \vec{S}_{j+1} + J' \sum_j \vec{S}_j \cdot \vec{S}_{j+2}. \tag{17.144}$$

this we may expect elementary excitations to carry an integer-spin quantum number. This is indeed the case both for ferromagnetic and for antiferromagnetic spin waves, whose total spin quantum numbers differ from those of the ground states by 1. The appearance of isolated or "deconfined" spin-1/2 spinons thus represents *fractionalization*. Of course, such spinons can only be created in pairs as long as we do not change the number of spins in the system, as a result of which the total spin of the system can change only by an integer.

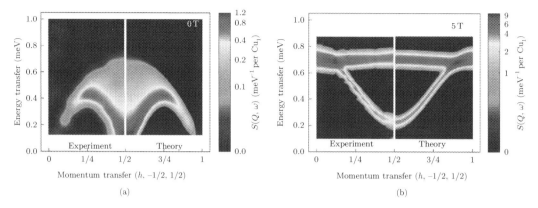

Figure 17.4 Excitation spectra of an antiferromagnetic spin-1/2 chain measured from inelastic neutron scattering. (a) The spin dynamical structure factor at zero magnetic field, where one sees a broad continuum indicating that a spin flip created by the neutron decays into two (or more) spinons, thus not having a fixed energy even when the total momentum is fixed. This should be contrasted with (b), where a strong enough magnetic field is applied to fully polarize the spins. In this case a spin flip is a sharp excitation (basically a ferromagnetic spin wave gapped by the magnetic field), which shows up as a sharp dispersion curve. Figure adapted from [177] by permission from Springer Nature. Copyright (2013).

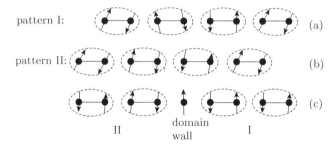

Figure 17.5 Two possible dimerization patterns of the ground states of the Majumdar–Ghosh model (a) and (b), and a spin-1/2 spinon excitation located at the domain wall between these two patterns (c). Each instance of two spins circled together indicates a singlet pair.

For the special case of $J' = J/2$, known as the **Majumdar–Ghosh model**, the ground states can be obtained exactly. In this case the Hamiltonian can also be written as

$$H = \frac{J}{2} \sum_j \left(\vec{S}_j \cdot \vec{S}_{j+1} + \vec{S}_j \cdot \vec{S}_{j-1} + \vec{S}_{j-1} \cdot \vec{S}_{j+1} \right)$$

$$= \frac{J}{4} \sum_j \left(\vec{S}_{j-1} + \vec{S}_j + \vec{S}_{j+1} \right)^2 + \text{constant}. \tag{17.145}$$

The two *exact* ground states are made of nearest-neighbor pairs of spins forming singlets, as illustrated in Figs. 17.5(a) and (b). The reason why they are exact ground states is the following. Equation (17.145) is a sum of three-spin (in sequence) cluster terms, and the ground state of each term should have total spin 1/2 for the corresponding cluster. The two states of Fig. 17.5 have the property that, among each cluster of three neighboring spins, two of them form a singlet, guaranteeing that the total spin of the cluster is 1/2; they are thus the ground state of every term in Eq. (17.145). It can be shown that there are no other ground states. These singlet bonds formed by nearest-neighbor spins are often called **valence bonds** (in analogy to the chemical bonds that are often referred to by the same name in chemistry), and such spontaneously dimerized states are also called

valence-bond solid (VBS) states, as the valence bonds form a periodic structure which spontaneously breaks the lattice translation symmetry of the original Hamiltonian.

In the Majumdar–Ghosh model the Lieb–Schultz–Mattis theorem is satisfied in a very unusual way: the ground state has double degeneracy, thus $\Delta E = 0$. The elementary excitations in this case are domain walls separating the two types of dimer patterns, as illustrated in Fig. 17.5(c), which again carry spin-1/2 and are thus (fractionalized) spinons. There is now a finite gap separating the spinon excitations from the ground states, as it is clear that the three-spin cluster with the spinon at the center is not in the ground state of the corresponding term in Eq. (17.145).

The Majumdar–Ghosh ground states represent a class of spontaneously dimerized (VBS) ground states, with short-range spin–spin correlations and a finite energy gap Δ separating excitations from the ground state(s). In fact, in this special case, the spin–spin correlation is identically zero beyond the nearest neighbors. Furthermore, lattice translation symmetry is broken spontaneously, which is responsible for the two-fold ground state degeneracy. The Majumdar–Ghosh model is a member of a class of models with qualitatively the same ground state properties, which are thus in the same *phase*. This phase is stable for $J' > J'_c \approx 0.2J$. A phase diagram of this J–J' model is illustrated in Fig. 17.6. For a *generic* member of this phase, we have a finite spin–spin correlation length ξ:

$$\langle S^z_j S^z_{j+r} \rangle \sim e^{-r/\xi}, \tag{17.146}$$

and the ground state degeneracy is *not* exact in a finite-size system, but rather the splitting is exponentially small:

$$\Delta E \sim e^{-N/\xi}. \tag{17.147}$$

For the very special Majumdar–Ghosh model, $\xi = 0$.

In this class of models the elementary excitations are spin-1/2 spinons associated with the domain walls (or solitons) separating the two patterns of dimerization. Again the mechanism of fractionalization is very similar to that in the Su–Schrieffer–Heeger model discussed in Chapter 7.[12]

One way to get around the Lieb–Schultz–Mattis theorem, but staying with half-integer spins (say, spin-1/2) is to break lattice translational symmetry in the *Hamiltonian*. The simplest example is a dimerized $S = 1/2$ chain. Figure 17.7 shows a spin chain in which the exchange coupling alternates between two different values, J_1 and J_2, with $J_1 > J_2$. In order to capture the essence of the ground state, let us consider the special case $J_2 = 0$.[13] Here we know that the exact ground state consists of independent singlets on the solid bonds illustrated in Fig. 17.7, which is identical to one of the Majumdar–Ghosh dimerized (or VBS) ground states. But in this case the ground state is *unique*, as

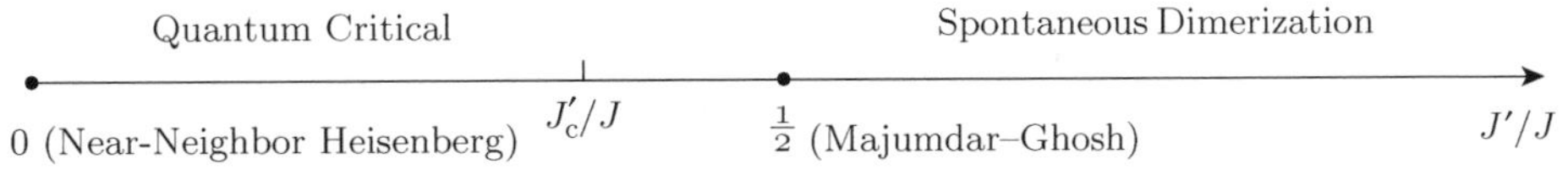

Figure 17.6 Phase diagram of the J–J' model, with nearest (J) and next-nearest (J') couplings of a spin-1/2 chain. For $J' > J'_c$, the ground state of the model exhibits spontaneous dimerization. The Majumdar–Ghosh (MG) model is a special point within the phase with spontaneous dimerization.

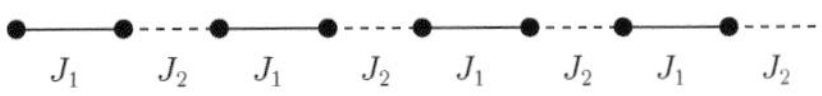

Figure 17.7 Dimerized $S = 1/2$ chain with alternating couplings $J_1 > J_2$.

---

[12] In fact the analogy goes further here, as spin dimerization is often accompanied by lattice dimerization. If this happens, the corresponding state is called a spin Peierls state.

[13] The reader should compare this with the extremely dimerized limit of the Su–Schrieffer–Heeger model discussed in Chapter 7.

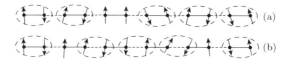

Figure 17.8 (a) A localized magnon carrying spin 1 of the J_1–J_2 model. Each instance of two spins circled together indicates a singlet pair. (b) A magnon separated into two spinons.

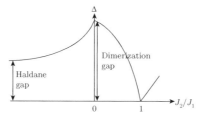

Figure 17.9 The excitation gap Δ for a dimerized spin-1/2 chain as a function of the bond strength ratio J_2/J_1. The spin-1 chain with a Haldane gap corresponds to $J_2 \longrightarrow -\infty$.

the dimerization pattern is not spontaneous but rather enforced by the (lower symmetry of the) Hamiltonian. There is a highly degenerate set of first excited states in which (any) one of the singlets is converted into a triplet, which can be viewed as a localized magnon carrying spin-1 (see Fig. 17.8(a)). It costs a finite energy to break the singlet bond and so there is an excitation gap for these magnons. If we now turn the second coupling J_2 back on, the valence bonds begin to fluctuate, but the essential character of the VBS ground state remains unchanged. This is because (for small enough J_2) perturbation theory in J_2 converges due to the finite excitation gap above the ground state. The singlets (or valence bonds) are more likely to be found on the J_1 links, and the excitation gap is stable against the J_2 perturbation. It turns out that the gap remains open for any value of $J_2 < J_1$, as illustrated in Fig. 17.9.

Since the ground state is dimerized as in the Majumdar–Ghosh model, one might also expect to have (deconfined) spin-1/2 spinon excitations as well. That is not the case, as illustrated in Fig. 17.8(b). In the region between the two spinons, singlet pairs are formed on the weaker (and thus "wrong") J_2 bonds, resulting in an energy cost that grows linearly with the spinon separation. As a result, the spinons are confined (like quarks in QCD) and thus *not* elementary excitations of the system, and the magnons are stable. The magnons show up as *sharp* peaks in inelastic neutron scattering, like phonons. This is in contrast to the broad spectrum seen in Fig. 17.4(a) that results from the magnon fractionalizing into two independent (deconfined) spinons, each of which carries part of the total momentum and total energy. See Fig. 17.10 for an indication of sharp magnon excitation in the closely related spin-1 chain, to which we now turn.

Exercise 17.9. Show that the Lieb–Schultz–Mattis theorem does not apply to the J_1–J_2 model discussed in here and illustrated in Fig. 17.7.

17.6.3 Spin-1 Chains, Haldane Gap, and String Order

Figure 17.9 also shows that the dimerization gap remains open even as the value of J_2 passes through zero and becomes negative. In the limit $J_2 \longrightarrow -\infty$, the pairs of spins on the J_2 links are forced to become parallel and their resulting triplet becomes an effective *spin-1* degree of freedom, as shown in Fig. 17.11. In this limit the J_1–J_2 model reduces to the spin-1 nearest-neighbor antiferromagnetic

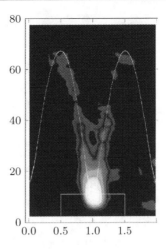

Figure 17.10 The spin-dynamical structure factor of an antiferromagnetic spin-1 chain measured from inelastic neutron scattering. The vertical axis is energy in units of meV, while the horizontal axis is the wave vector in units of π/a, where a is the lattice constant. A 9 meV Haldane gap is clearly visible at the Brillouin-zone boundary. In addition, the excitation energy distribution is much narrower than that of the spin-1/2 chain (compare with Fig. 17.4(a)), indicating sharp spin-1 magnon excitations (instead of deconfined spinon excitations). Figure adapted with permission from [178].

Figure 17.11 Reduction of the spin-1/2 J_1–J_2 model to a spin-1 chain in the limit $J_2 \to -\infty$. Two $S = 1/2$ spins coupled by J_2 (represented by a thick dot) are forced to form a triplet or effective spin-1, which is then coupled to its neighbors through the J_1 couplings (solid lines). This figure also illustrates the ideal valence-bond solid state which is the exact ground state of the Affleck–Kennedy–Lieb–Tasaki (AKLT) model: a spin 1 on a lattice site (thick dot) is viewed as two spin 1/2s forming a triplet. The AKLT ground state is constructed by having each spin 1/2 form a singlet pair with one of the spin 1/2s on the neighboring sites, and solid lines represent such singlet bonds. This state is uniquely defined due to the constraint that the two spin 1/2s on the same site must form a triplet (and thus be a symmetric state under interchange).

Heisenberg model, and the dimerization gap has now become the gap named after Haldane, who proposed in 1983 that an excitation gap exists not only for spin 1, but also for all antiferromagnetic integer spin chains [179, 180]. Figure 17.10 shows neutron-scattering results that indicate the presence of a sharp branch of magnon excitations with a **Haldane gap** for a spin-1 chain compound.

As in the case of the dimerized spin-1/2 chain, the ground state spin–spin correlation function of the spin-1 chain decays exponentially:

$$\langle S_j^z S_{j+r}^z \rangle \sim e^{-r/\xi}. \tag{17.148}$$

This apparent featurelessness obscures an underlying hidden order which is characteristic of the VBS state. To reveal this hidden order more clearly, we examine a variation of the Heisenberg model for the spin-1 chain:

$$H = J \sum_j [\vec{S}_j \cdot \vec{S}_{j+1} + \alpha(\vec{S}_j \cdot \vec{S}_{j+1})^2]. \tag{17.149}$$

The very special case of $\alpha = 1/3$ is known as the **Affleck–Kennedy–Lieb–Tasaki (AKLT) model**. The special property of the AKLT model lies in its spectrum of a *single* bond, as illustrated in Fig. 17.12 and compared with that of Heisenberg coupling. We see from this spectrum that the AKLT

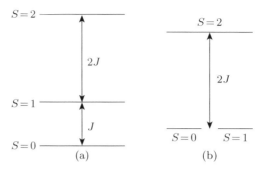

Figure 17.12 Spectra of two spin 1s coupled through antiferromagnetic Heisenberg (a) and AKLT (b) couplings. The singlet ($S = 0$) and triplet ($S = 1$) states are degenerate ground states for the AKLT coupling.

Hamiltonian can be written in the form of a sum of projectors onto the $S = 2$ state of each neighboring pair of spins, after removing an irrelevant constant:

$$H = 2J \sum_j P_{j,j+1}^{(S=2)}. \tag{17.150}$$

This allows one to construct an "ideal" valence-bond solid state, which can also be understood from Fig. 17.11 (see its caption). The reason why the AKLT state is the ground state of the AKLT model is quite similar to that of the spontaneously dimerized states being ground states of the Majumdar–Ghosh model. Let us examine a single pair of nearest-neighbor spin 1s. In general the total spin S can be 0, 1, or 2. Within the AKLT construction, it is the sum of the four spin 1/2s. Since two of them form a singlet, the possibility of $S = 2$ is eliminated, and S reduces to the sum of the remaining two spin 1/2s, thus $S = 0$ or 1. Since, for each term in the AKLT model both $S = 0$ and $S = 1$ are ground states (see Fig. 17.12), the AKLT state is a ground state of the entire AKLT Hamiltonian. It can be shown that this is the unique ground state in the presence of periodic boundary conditions.

Both the AKLT model and the Heisenberg model are members of the set of Hamiltonians whose (unique, non-degenerate) ground states are in the Haldane phase, exhibit a finite spin gap, and have short-range spin–spin correlation. Within the one-parameter family of Hamiltonians in the form of Eq. (17.149), the Haldane phase is stable for $-1 < \alpha < 1$. The AKLT model is a very special member of this family, playing a role very similar to that of the Majumdar–Ghosh model in the spontaneously dimerized phase for a spin-1/2 chain. It turns out that, if one examines (in the S^z basis) a typical configuration of the spins in the AKLT ground state, it looks like this:

$$+ \, 0 \, 0 - + 0 \, 0 \, 0 \, 0 - + - 0 + 0 \, 0 - + - 0 \, 0 \, 0 + 0 - . \tag{17.151}$$

Notice that, if we ignore the spins with $S^z = 0$, the remaining spins have perfect antiferromagnet order,

$$+ - + - + - + - + - + - + - . \tag{17.152}$$

Because of the random number of $S^z = 0$ sites inserted between the $S^z = \pm 1$ sites, the long-range antiferromagnetic order is invisible to the ordinary spin–spin correlation function which, as noted above, decays exponentially. This "hidden" order is, however, manifest in the non-local Rommelse and den Nijs **"string" correlation function** [181–183] defined by

$$\mathcal{O}_s^z(r) \equiv \left\langle S_j^z \exp\left\{ i\pi \sum_{k=j+1}^{j+r-1} S_k^z \right\} S_{j+r}^z \right\rangle. \tag{17.153}$$

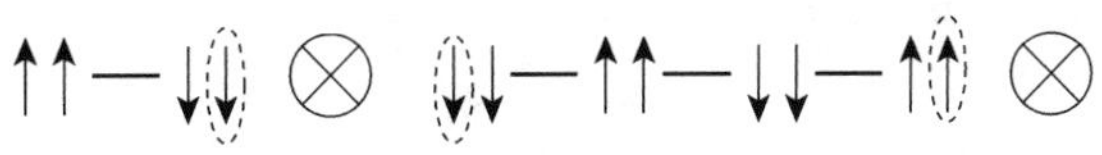

Figure 17.13 Valence-bond solid interrupted by non-magnetic impurities represented by the circles with crosses. The missing valence bonds liberate a *fractional* spin ($s = 1/2$) degree of freedom (indicated by the dashed ellipses) at the ends of each broken chain segment.

This object exhibits long-range order in the entire Haldane gapped phase:

$$\lim_{r \longrightarrow \infty} \mathcal{O}_s^z(r) \neq 0. \tag{17.154}$$

Breaking this (non-local) order to create an excitation costs a finite energy gap.

Because of the excitation gap, the spin degrees of freedom disappear at low temperature and the magnetic susceptibility becomes exponentially small. One of the remarkable and paradoxical features of the Haldane phase is that the introduction of *non-magnetic* impurities actually liberates spin degrees of freedom. These are not, as one might have naively expected, $\Delta S = 1$ excitations, but rather pairs of *fractional-spin* $\Delta S = 1/2$ excitations. The process is illustrated in Fig. 17.13. Recall that the spin 1 on each site is viewed as a pair of spin-1/2 objects bound ferromagnetically into a triplet. As noted earlier, the valence-bond solid ground state (in the absence of disorder) has each of the $s = 1/2$ objects pairing into a singlet bond with one of the spin-1/2 objects on the neighboring sites. (This is done in a symmetric way so that each site has total spin precisely equal to unity.) If one of the neighbors is a non-magnetic impurity, one of the singlet bonds is broken and *one-half* of the spin on the site is liberated. This is yet another example of fractionalization, perhaps even more striking than in spin-1/2 chains since here the fundamental degrees of freedom are spin-1 objects. As should have become clear by now, another place such half spins appear is at the end of a spin-1 chain (with open boundary conditions), where the last spin has only one neighbor, and only one of the spin 1/2s pairs up with another half spin on the neighboring site, and the remaining spin 1/2 is free.

These weakly interacting fractional spins are detectable experimentally. They remain free down to low temperatures and produce a Curie-like power-law tail in the magnetic susceptibility. This is an example of a quantum Griffiths singularity in which the Rommelse and den Nijs order (may) still be present but the gap is not [184, 185].

Exercise 17.10. Show that, up to a constant, all rotationally invariant couplings between two spin-s objects can be expressed in the form of

$$\sum_{k=1}^{2s} J_k (\vec{S}_1 \cdot \vec{S}_2)^k.$$

Exercise 17.11. Construct translationally invariant ideal VBS states for general integer-spin chains, in a manner similar to the spin-1 AKLT ground state.

Exercise 17.12. Construct Hamiltonians with rotationally invariant nearest-neighbor couplings that make the states constructed in Exercise 17.11 exact ground states. Are such Hamiltonians unique?

> **Exercise 17.13.** Show that the AKLT state does not have amplitude in spin configurations like $\cdots + + \cdots$ or $\cdots + 000 + \cdots$, namely a $+$ spin follows another $+$ spin, possibly with 0s separating them (but no $-$ spin).

17.6.4 Matrix Product and Tensor Network States

In the previous subsection we constructed the VBS ground state of the AKLT model, by introducing two (auxiliary) half spins to represent the original spin 1. We would like to write down the AKLT state *explicitly* in terms of the *original* $s = 1$ spins, say in the S_z basis:

$$|\psi_{\text{AKLT}}\rangle = \sum_{m_1 m_2 \cdots m_N} \psi_{m_1 m_2 \cdots m_N} |m_1 m_2 \cdots m_N\rangle, \tag{17.155}$$

where $m_j = 0, \pm 1$ is the eigenvalue of S_j^z. It appears that in order to accomplish this we need to specify a very large number ($\sim 3^N$) of amplitudes $\psi_{m_1 m_2 \cdots m_N}$. As we will see below, these amplitudes can be parameterized by a handful of matrices, and expressed elegantly in terms of matrix products.

To start, we first write a singlet pair between two half spins (a and b) in a matrix form:

$$\chi_{00}(a, b) = \frac{1}{\sqrt{2}}(|+1/2\rangle_a \otimes |-1/2\rangle_b - |-1/2\rangle_a \otimes |+1/2\rangle_b)$$

$$= (|+1/2\rangle_a, |-1/2\rangle_a) \begin{pmatrix} 0 & 1/\sqrt{2} \\ -1/\sqrt{2} & 0 \end{pmatrix} \begin{pmatrix} |+1/2\rangle_b \\ |-1/2\rangle_b \end{pmatrix}$$

$$= (|+1/2\rangle_a, |-1/2\rangle_a) S(|+1/2\rangle_b, |-1/2\rangle_b)^{\text{T}}, \tag{17.156}$$

where

$$S = \begin{pmatrix} 0 & 1/\sqrt{2} \\ -1/\sqrt{2} & 0 \end{pmatrix}. \tag{17.157}$$

To construct the AKLT state, we introduce two half spins, l_i and r_i (left and right) for each site i, and let r_i form a singlet with l_{i+1}. The resultant state is the tensor product of N singlet pairs:

$$|\psi\rangle = \prod_i (|+1/2\rangle_{r_i}, |-1/2\rangle_{r_i}) \begin{pmatrix} 0 & 1/\sqrt{2} \\ -1/\sqrt{2} & 0 \end{pmatrix} \begin{pmatrix} |+1/2\rangle_{l_{i+1}} \\ |-1/2\rangle_{l_{i+1}} \end{pmatrix} \tag{17.158}$$

$$= \sum_{\vec{a}, \vec{b}} S_{b_1 a_2} S_{b_2 a_3} \cdots S_{b_N a_1} |\vec{a}, \vec{b}\rangle, \tag{17.159}$$

where $a_i, b_i = \pm 1/2$ are the eigenvalues of $S_{l,i}^z$ and $S_{r,i}^z$, respectively, and $\vec{a} = (a_1, a_2, \ldots, a_N)$, $\vec{b} = (b_1, b_2, \ldots, b_N)$ are short-hand notations for collections of indices. To construct the AKLT state, we need to project $|\psi\rangle$ onto the triplet space for each pair of spins l_i and r_i, and in the S_i^z basis the projection operator takes the form

$$P = \sum_{m=-1}^{1} P^m, \tag{17.160}$$

where

$$P^{+1} = \begin{pmatrix} 1 & 0 \\ 0 & 0 \end{pmatrix}, \quad P^{-1} = \begin{pmatrix} 0 & 0 \\ 0 & 1 \end{pmatrix}, \quad P^0 = \begin{pmatrix} 0 & 1/\sqrt{2} \\ 1/\sqrt{2} & 0 \end{pmatrix}. \tag{17.161}$$

The reader is encouraged to verify these. We thus have

$$|\psi_{\text{AKLT}}\rangle \propto \left(\prod_i P_i\right)|\psi\rangle = \sum_{\vec{m}=\vec{a}+\vec{b}} P^{m_1}_{a_1 b_1} S_{b_1 a_2} P^{m_2}_{a_2 b_2} S_{b_2 a_3} \cdots P^{m_N}_{a_N b_N} S_{b_N a_1} |\vec{a}, \vec{b}\rangle$$

$$= \sum_{\vec{m}} \text{Tr}(P^{m_1} S P^{m_2} S \cdots P^{m_N} S)|\vec{m}\rangle$$

$$= \sum_{\vec{m}} \text{Tr}(A^{m_1} A^{m_2} \cdots A^{m_N})|\vec{m}\rangle, \tag{17.162}$$

where

$$A^m = P^m S \tag{17.163}$$

are 2×2 matrices. Comparing Eq. (17.162) with Eq. (17.155), we have

$$\psi_{m_1 m_2 \cdots m_N} = \text{Tr}(\tilde{A}^{m_1} \tilde{A}^{m_2} \cdots \tilde{A}^{m_N}), \tag{17.164}$$

where $\tilde{A}^m \propto A^m$ are re-scaled matrices so that the resultant wave function has the desired normalization. Also, because the trace is invariant under similarity transformations, these matrices can be transformed without changing the result. A popular choice for the $\tilde{A}^m$ is

$$\tilde{A}^{+1} = \begin{pmatrix} 0 & \sqrt{\frac{2}{3}} \\ 0 & 0 \end{pmatrix}, \quad \tilde{A}^{-1} = \begin{pmatrix} 0 & 0 \\ -\sqrt{\frac{2}{3}} & 0 \end{pmatrix}, \quad \tilde{A}^{0} = \begin{pmatrix} \sqrt{\frac{1}{3}} & 0 \\ 0 & -\sqrt{\frac{1}{3}} \end{pmatrix}. \tag{17.165}$$

We have thus successfully encoded the AKLT wave function (17.155) using three (very small) matrices!

The AKLT state is the simplest example of a so-called **matrix product state (MPS)**, which takes the form

$$|\psi_{\text{MPS}}\rangle = \sum_{\vec{m}} \text{Tr}(M_1^{m_1} M_2^{m_2} \cdots M_N^{m_N})|\vec{m}\rangle, \tag{17.166}$$

where m_i is the local basis state index, and $\{M_j^m\}$ is a set of matrices, not necessarily square (but that appropriately mesh with the neighboring ones in size so that matrix products are well-defined). For a translationally invariant MPS, the matrices depend only on m, *not* on the site index i. In this case they must be square matrices. An MPS has a very suggestive graphical representation, illustrated in Fig. 17.14.

The AKLT-type of VBS state can be generalized to spatial dimensions $d > 1$, and we discuss a specific example here. Consider a square lattice with a spin with $s = 2$ on each site. We can split it into four half spins, with each of them forming a singlet with one of the half spins of one of the four nearest neighbors. We then totally symmetrize the four half spins to ensure that they add up to

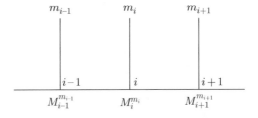

Figure 17.14 Graphical representation of a matrix product state (MPS). There is a matrix associated with a lattice site i, for a given local quantum number m_i. The two indices of the matrix must fuse with those of neighboring matrices and get contracted, representing matrix products.

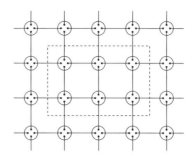

Figure 17.15 Graphical representation of an AKLT-type VBS state on a square lattice. Each site has a spin with $s = 2$, represented by four half spins (dots). The lines between dots on neighboring sites represent a singlet pair formed by two half spins. The circle around the four dots represents the fact that the four half spins are symmetrized to add up to total spin 2. The dashed line represents the boundary separating the whole system into two subsystems. It is clear that entanglement between the two subsystems arises from the singlet pairs formed across the boundary, and the entanglement entropy is proportional to the boundary length. See more discussion on this in Appendix F.

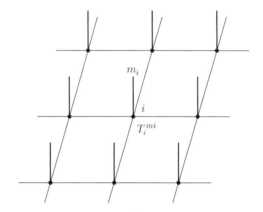

Figure 17.16 Graphical representation of a projected entangled pair state (PEPS). There is a rank-z tensor $T_i^{m_i}$ associated with a lattice site i, for a given local quantum number m_i. The z indices of the tensor must fuse with those of neighboring tensors and get contracted, representing tensor products.

total spin $s = 2$. Such an AKLT-type VBS state is illustrated in Fig. 17.15. These types of states are also called **projected entangled pair states (PEPS)**. A PEPS wave function can be written in a form that is a natural generalization of a MPS, namely each matrix is replaced by a rank-z (i.e. having z indices) tensor, where z is the coordination number of the lattice, and the wave function is constructed by taking the trace of tensor products. The need for such a generalization is quite obvious – a matrix has two indices, corresponding to the two neighbors of a lattice site in a 1D lattice as illustrated in Fig. 17.14. We need an index for each neighbor and thus, on a lattice with coordination number z, the matrix needs to be replaced by a rank-z tensor. The way in which these tensors are contracted to generate the wave function is illustrated in Fig. 17.16.

MPS and PEPS are examples of states constructed from networks of matrices and tensors, which are of great interest at the time of writing because they have the desired entanglement properties expected from ground states of "reasonable" Hamiltonians, namely those with sufficiently short-range couplings. It can be shown that ground states of very large classes of such Hamiltonians can be approximated by such states with arbitrary accuracy, as long as we are allowed to increase the sizes of the matrices or tensors, and powerful numerical methods based on this fact are being developed. More discussions of these points can be found in Appendix F.

The AKLT-type VBS state in any dimension respects lattice translational symmetry, and requires the size of the spin to match the lattice coordination number ($2s = z$). Most VBS states do break lattice translational symmetry, as in the case of spin-1/2 chains discussed earlier. We now turn our discussion to more general VBS states above 1D.

Exercise 17.14. As discussed in the text, one can construct an AKLT-type VBS state which has a valence bond on every link of the lattice. For a lattice with coordination number z, this analog of the AKLT state requires $s = z/2$ (and thus $s = 3/2$ on the honeycomb lattice and $s = 2$ spins on the square lattice). Each $s = z/2$ spin can then be viewed as a symmetrized combination of z $s = 1/2$ objects. By analogy with Eq. (17.150), the Hamiltonian that makes it the exact ground state can be written as a sum of spin projectors over near-neighbor pairs of sites:

$$H = \sum_{\langle ij \rangle} \vec{P}_{ij}^{S=z} = \sum_{\langle ij \rangle} \left[\sum_{k=1}^{z} J_k (\vec{S}_i \cdot \vec{S}_j)^k \right]. \qquad (17.167)$$

(a) Find the values of the parameters J_k for $s = 3/2$ spins on a honeycomb lattice (where each site has $z = 3$ neighbors).

(b) Do the same for $s = 2$ spins on the 2D square lattice.

17.7 Valence-Bond-Solid and Spin-Liquid States in 2D and Higher Dimensions

So far we have discussed only the case of antiferromagnetic coupling on bipartite lattices. On such lattices effects of quantum fluctuations are relatively small in 2D and 3D (see Exercise 17.6), and antiferromagnetic long-range spin order survives. If the lattice is a triangular, kagome, pyrochlore or other non-bipartite lattice, the situation is much more complex since the spins are *frustrated*. In such cases it is not possible to minimize the energies of all bonds even for *classical* spins when the basic plaquette of the lattice is a triangle (see Fig. 17.17(a)). Frustration can also be introduced in bipartite lattices by introducing more-distant-neighbor couplings; when antiferromagnetic coupling exists between spins on the same sublattice, antiferromagnetic long-range order gives rise to *positive* energies on these bonds, thus causing frustration. The 1D J–J' model of Eq. (17.144) is an example of that; see Fig. 17.17(c) for another example on square lattice.

The actual quantum ground states can have a variety of types of order or no order at all. For the triangular lattice with nearest-neighbor antiferromagnetic couplings only, spins do order in the pattern illustrated in Fig. 17.17(b). On the other hand, stronger frustration can suppress antiferromagnetic long-range order completely, and a very interesting question is what kind of states the system can

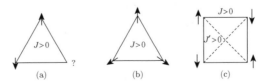

Figure 17.17 Frustration in triangle-based lattices. (a) When two spins are anti-parallel to minimize the antiferromagnetic bond that couples them, the third spin does not know in which direction to point. (b) A possible spin configuration that minimizes the total energy of the triangle, but not individual bonds. Quantum fluctuations on top of such compromised classical spin configurations are enhanced. (c) A square lattice with next-nearest-neighbor antiferromagnetic interactions is also frustrated.

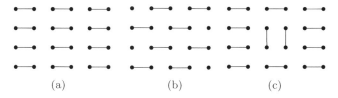

(a)   (b)   (c)

Figure 17.18 Two valence-bond solid (VBS) states on the square lattice. (a) Column VBS. (b) Staggered VBS. (c) In the column VBS, the valence bonds can easily resonate and mix with the valence-bond configuration shown here. Such resonance is not possible in the staggered VBS.

form. This section is devoted to discussing possible answers to this question. For simplicity we will restrict our discussion to spin-1/2 models with finite-range antiferromagnetic Heisenberg couplings that respect all the symmetries of the underlying lattice in this section.

An obvious possibility is a VBS state, which would be similar to what happens in a 1D spin-1/2 chain when a frustrating next-nearest-neighbor coupling J' is cranked up. In higher dimensions there are many possible VBS patterns; they all break lattice symmetries spontaneously, and a couple of examples on the square lattice are illustrated in Fig. 17.18. However, unlike the 1D case, here we do *not* have simple models like the Majumdar–Ghosh model whose exact ground states are these VBS states. As a result, quantum fluctuations of the valence-bond configurations are always present. We now argue that such fluctuations lift the degeneracy of the different VBS states, even though they have the same *variational* energy for simple models involving only finite-range Heisenberg couplings.

Let us start the discussion with a single bond connecting two spins, with the Hamiltonian $J\vec{S}_1 \cdot \vec{S}_2$. The Néel state $|\uparrow \downarrow\rangle$ has a variational energy of $-J/4$. The exact ground state is, as discussed earlier, actually the singlet state

$$\frac{1}{\sqrt{2}}(|\uparrow\downarrow\rangle - |\downarrow\uparrow\rangle), \tag{17.168}$$

whose energy is $-3J/4$, *significantly* lower than that of the Néel state. The origin of the factor of 3 difference lies in the fact that, while the Néel state minimizes only the $S_1^z S_2^z$ term in the Hamiltonian, the singlet state minimizes all three such terms (as guaranteed by its rotational invariance). This significant difference in energy is the driving force that stabilizes the valence-bond-based states discussed here and in the previous section.

It is worthwhile to understand from another perspective the difference between the Néel state and the singlet state formed by two spins. One would normally say that in the Néel state the two spins are anti-parallel, as one is up and the other is down. In fact this is true only for the z components of the spins; their x and y components are totally uncertain, and thus cannot be parallel or anti-parallel to each other. For the singlet state (17.168), the two spins are always anti-parallel to each other, no matter which component or quantization direction you choose to look at, as guaranteed by its rotational invariance. Paradoxically, the direction of each of the individual spins is actually *completely random*, which is again guaranteed by the rotational invariance of the state. The resolution of the paradox lies in the fact that the two spins are highly *entangled* in the singlet state (17.168); namely, if we trace out spin 1 to obtain the reduced density matrix of spin 2, we will obtain the identity matrix divided by its rank (2 in this case), indicating complete randomness (see Appendix F). Thus the fundamental difference between the Néel and singlet states is that the former is a product state (with no entanglement) and the latter is the maximally entangled state (a Bell state in quantum information terminology). As we will see later, entanglement is very useful in classifying various phases.[14]

---

[14] The differences between the singlet and Néel states discussed here are representative of those between entangled (which are intrinsically quantum) and unentangled (or product, which are essentially classical) states. In classical

The fact that each component of the spin for each particle is zero on average and yet the two spins are perfectly anti-parallel on any axis that we choose to measure is something that is impossible classically. The spins are mysteriously more strongly correlated than is possible classically (see Exercise 17.15). The strength of the correlations violates a bound discovered by John Bell for classical correlations. We say that in this situation the spins exhibit "non-classical correlations" resulting from entanglement. The canonical experimental signature for such non-classical correlations is that they violate the Bell inequality.

Exercise 17.15. Consider the situation of "classical" correlations of two spins described by an incoherent statistical mixture of $|\psi_{\uparrow\downarrow}\rangle = |\uparrow\downarrow\rangle$ and $|\psi_{\downarrow\uparrow}\rangle = |\downarrow\uparrow\rangle$ described by the density matrix

$$\rho_c = \frac{1}{2}|\psi_{\uparrow\downarrow}\rangle\langle\psi_{\uparrow\downarrow}| + \frac{1}{2}|\psi_{\downarrow\uparrow}\rangle\langle\psi_{\downarrow\uparrow}|. \tag{17.169}$$

(i) Show that all three components of each of the two spins have zero average. This is (superficially) similar to the singlet state (17.168).

(ii) Show that the z components of the two spins are maximally (anti-)correlated while the x and y components are completely uncorrelated.

(iii) Compare your result with the case of "quantum correlations" described by the pure-state density matrix

$$\rho_q = \frac{1}{2}\{|\psi_{\uparrow\downarrow}\rangle - |\psi_{\downarrow\uparrow}\rangle\}\{\langle\psi_{\uparrow\downarrow}| - \langle\psi_{\downarrow\uparrow}|\}. \tag{17.170}$$

Now let us move on to a single square with four spins, and nearest-neighbor couplings only. The corresponding Néel state (as illustrated in Fig. 17.17(c)) has energy $-J$. In this case we can have only two nearest-neighbor singlet bonds, resulting in energy $-3J/2$, still lower than that of the Néel state, although by a smaller factor. The reason for this reduction of the (energy-gain) factor is that a half spin can only form a singlet with another half spin, and only such valence bonds enjoy the $-3J/4$ energy gain; the energies of all other bonds are exactly zero. On the other hand, *every* bond gains the $-J/4$ energy in the Néel state. This is why Néel order eventually dominates in 2D and higher dimensions. Extending the same analysis to the entire square lattice shows that the Néel state is lower in energy than the VBS state by a factor of 4/3 if only nearest-neighbor coupling is present; this factor increases with increasing dimensionsality/coordination number. Thus some finite amount of frustration is needed to suppress Néel order.

The competition between forming singlets and forming Néel order is an illustration of the concept known as "**entanglement monogamy**." The more entangled spin A is with spin B, the less it can be entangled with any other spin C. Complete entanglement (e.g. perfect singlet or other Bell state formation) can occur between spin A and only *one* other spin at a time. As we saw above, complete entanglement implies non-classical correlations that are stronger than those of the Néel state. On the other hand, the spins in the Néel state are classically correlated with all their neighbors, and, if the

---

physics knowledge about the whole system is based on knowledge about its constituents. In particular, complete knowledge (namely, no uncertainty) about the whole system requires complete knowledge about *all* of its constituents. In a quantum world, as we have seen already, one may know *everything* about the whole system (via knowledge of the wave function), yet in the extreme example of a singlet state, we know *nothing* about the state of a single spin, as it does *not* have a definitive wave function, and its reduced density matrix (which is the best description available to us) carries maximum entropy! As Charles Bennett put it vividly, "A classical house is at least as dirty as its dirtiest room, but a quantum house can be dirty in every room, yet still perfectly clean overall." This highly counter-intuitive behavior is rooted in entanglement.

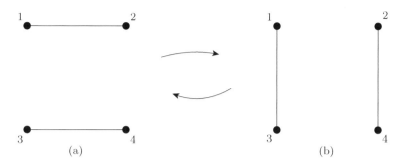

Figure 17.19 The two nearest-neighbor singlet pair configurations of a single square. Two spins connected by a solid line form a singlet pair. The two configurations (a) and (b) can mix (or resonate, as indicated by the arrows) with each other to lower the total energy of this four-spin cluster.

coordination number is high enough (and the lattice is not frustrated), this wins the competition for lowest variational energy.

We turn now to the possibility that spins can be in superposition states of different arrangements of valence bonds. Coming back to a single square, the new feature to notice, however, is that here there are two ways to form such nearest-neighbor singlet bonds or valence bonds, as illustrated in Fig. 17.19; they are labeled as $|(12), (34)\rangle$ and $|(13), (24)\rangle$, respectively.[15] Individually they are *not* eigenstates of the Hamiltonian that have couplings on all four nearest-neighbor bonds. This means that the Hamiltonian has terms that couple these two singlet states (of four spins), and the true ground state is an appropriate linear combination of them:

$$|\Psi_0\rangle \propto |(12), (34)\rangle + |(13), (24)\rangle. \tag{17.171}$$

The situation here is quite analogous to the tight-binding Hamiltonian we encountered in Chapter 7, where we have localized atomic orbitals coupled to each other by (off-diagonal terms of) the Hamiltonian, and eigenstates are their proper linear combinations.[16] States of the form (17.171) are referred to (again using chemistry terminology) as **resonating valence-bond (RVB) states**, since the singlet (or valence) bonds are jumping back and forth (or resonating) between different configurations to lower the total energy, in a manner similar to electrons jumping among atomic orbitals in a tight-binding model. By "jumping back and forth" we mean that the ground state is a coherent superposition of these different configurations.

Exercise 17.16. Prove that the two valence-bond states illustrated in Fig. 17.19 are linearly independent, but not orthogonal. Compute their overlap.

We are now in a position to discuss the difference in the energetics of different VBS states. As illustrated in Fig. 17.18(c), it is easy for the valence bonds to resonate and lower their energy (as in the single-square case discussed above) in the column VBS, but this is much more difficult in the staggered VBS. We thus expect the column VBS to have lower energy.

---

[15] Since a singlet state formed by two half spins is anti-symmetric under exchange, a sign convention is necessary. Here we adopt the convention that $|(ij)\rangle = (1/\sqrt{2})(|\uparrow_i \downarrow_j\rangle - |\downarrow_i \uparrow_j\rangle)$. Note that the anti-symmetry mentioned above means that $|(ij)\rangle = -|(ji)\rangle$.

[16] In the present case the analogy is actually quite complete in that, just like the atomic orbital, the valence-bond states are linearly independent but *not* orthogonal. Also, for four spins there are only two linearly independent singlet states, thus the two valence-bond states of Fig. 17.19 form a complete basis of the singlet subspace.

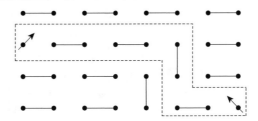

Figure 17.20 Two spinons in a column VBS background. There is a string (enclosed within dashed lines) connecting them, along which the energetically favorable column VBS pattern is disrupted. The path of the string is not unique, and the actual state will be a superposition of different possibilities.

As we saw in the previous section, elementary excitations of the 1D VBS state (as represented by the Majumdar–Ghosh model) are deconfined spin-1/2 spinons; we thus have the phenomenon of (spin) fractionalization in that case. Does fractionalization occur in VBS states of two and higher dimensions as well? The answer is *no*. To illustrate this point, let us create two separate spinons in the column VBS background, as illustrated in Fig. 17.20. We see that there is a path (or "string") connecting the two spinons, along which valence bonds form a staggered instead of a column pattern with neighboring valence bonds. This prevents the valence bonds along the string from resonating, and gives rise to an energy cost that grows linearly with the string length (and separation between the two spinons).[17] We thus find that the spinons are confined by a linear confining potential in this case; as a result they are *not* elementary excitations on top of a VBS ground state. The actual elementary excitations are (gapped) spin-1 magnons, which can be pictorially understood as *triplet bonds* that propagate through the lattice. They can be viewed as bound states of the spinons. The situation here is somewhat similar to a dimerized spin-1/2 chain (the J_1–J_2 model of Fig. 17.7), but the actual mechanism of confinement is fundamentally different – there the Hamiltonian *explicitly* breaks lattice translation symmetry, whereas here the symmetry breaking is *spontaneous* as in the Majumdar–Ghosh model.

We have now seen several examples of fractionalization in 1D (where this phenomenon is quite common). In all these examples the fractionalized excitations (the spinons discussed in the previous section and the solitons in the Su–Schrieffer–Heeger model discussed in Chapter 7) are domain walls separating topologically distinct ground states. This mechanism does *not* work in higher dimensions, as domain walls are extended (instead of point-like as they are in 1D) objects there, thus giving rise to a confining force that binds the fractionalized excitations (much like quarks are confined). The mechanisms for fractionalization in higher dimensions are much more intricate, and in general involve topological order and long-range entanglement.[18] The only example we have encountered so far is the Laughlin quasiparticle in a fractional quantum Hall liquid. We are now going to discuss our second example, namely (deconfined) spinons in a spin liquid.

Except for very special Hamiltonians such as the Majumdar–Ghosh model, there are in general quantum fluctuations in the VBS configurations. Such fluctuations are actually the source of the resonance energy. It is therefore entirely possible that such quantum fluctuations melt the VBS order

---

[17] The choice of the path of the string is not unique, and in fact the string will resonate ("vibrate") among different configurations. This will somewhat lower the energy ("string tension") but in general it will remain positive, meaning that the spinons remain confined, so long as the quantum fluctuations are not so large that the valence bond solid "melts" into a liquid. This possibility will be described further below.

[18] At the time of writing there is no widely accepted precise definition of long-range entanglement. Loosely speaking, it characterizes a highly entangled many-particle state that cannot be "smoothly" connected to a product state of local degrees of freedom. We will content ourselves in this book with discussing examples of such long-range entangled states in this and the following section.

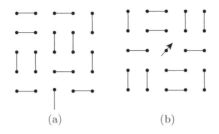

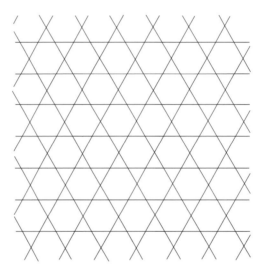

Figure 17.21 (a) A representative nearest-neighbor valence-bond configuration on a square lattice. The resonating valence-bond (RVB) state can be viewed as an equal-weight linear superposition of all such configurations. (b) A spinon in an RVB background. It is an unpaired spin living in a background of resonating valence bonds. Shown in the figure is a typical valence-bond configuration with a single unpaired spin.

Figure 17.22 The kagome lattice consisting of corner-sharing triangles.

to yield a "liquid" state of valence bonds. A representative wave function would be an equal-weight superposition of all nearest-neighbor valence-bond configurations (see Fig. 17.21(a) for a representative example on a square lattice), with proper signs that maximize the resonance energy gain. It can be viewed as a generalization of Eq. (17.171) to an infinite lattice. Such a state which does not break any spin rotation or lattice symmetry was termed a **resonating valence-bond (RVB) spin-liquid state** by P. W. Anderson in 1973, as a possible description of the ground state of the nearest-neighbor Heisenberg model on a triangular lattice. While it turns out that Néel order survives on the triangular lattice, there is evidence that the kagome lattice (illustrated in Fig. 17.22) with nearest-neighbor antiferromagnetic couplings supports some sort of spin-liquid ground state. This is because the kagome lattice consists of corner-sharing triangles and so is highly frustrated with respect to Néel order. The triangular lattice consists of *edge*-sharing triangles and has coordination number $z = 6$, whereas the kagome lattice has lower coordination $z = 4$, another fact that works against the Néel order.

Now let us consider a spinon in an RVB spin liquid. In this case it is an unpaired spin surrounded by resonating valence bonds; see Fig. 17.21(b) for a "snap-shot" of the spin/valence-bond configuration. Now we see a big difference with a spinon in a VBS background (see Fig. 17.20): there the spinon disrupts (or changes) the VBS pattern, which is the source of the linear confining potential that binds two spinons, but here the valence bonds are in a liquid state, so there is nothing to be disrupted! We may view this as saying that the valence-bond liquid "screens" the effect of an isolated spinon.[19] As a result, spinons are *deconfined*, and survive as elementary excitations in the RVB spin liquid.

There are many different types of spin liquids. They all support (fractionalized) spinon excitations; some of them are gapped, while others are gapless. The RVB spin liquid we have described is the simplest among the gapped spin liquids, which are topologically ordered in a sense similar to fractional quantum Hall liquids. In Section 17.7.1 we will discuss the nature of its topological order in some detail.

It is clear from our symmetry classification scheme that spin liquids are different from the Néel state (which breaks both spin-rotation and lattice translation symmetries) and VBS states (which respect spin-rotation but break lattice translation/rotation symmetries), as they break none of the symmetries of the Hamiltonian. The lack of symmetry breaking obscures the subtle nature of spin liquids. The more revealing way to distinguish between spin liquids and the other states is through entanglement. The (original) Néel state is a simple product state with no entanglement. The VBS state is "almost" a product state in the sense that spins are completely uncorrelated unless we happen to look at entanglement between two spins connected by a valence bond – and, furthermore, the valence bonds are fixed in space and not fluctuating. Another way to define "almost a product state" is that such states can be converted to true product states with unitary operations that consist of products of quasi-local unitaries, all of which commute with each other. In quantum information theory, such unitaries of "gate operations" are called **"low-depth" circuits** [186].

Quantum fluctuations around the Néel or VBS states do give rise to some enhancement of entanglement in the actual ground states, but such entanglement is "reducible" by low-depth circuits as described above. In spin liquids, on the other hand, we have "irreducible" entanglement that cannot be removed by low-depth circuits. This is because spin-liquid states consist of a huge superposition of many different (short-range) valence-bond configurations. Remember that it is superposition of different product states that is necessary (but not sufficient) for entanglement.

In classical statistical physics a liquid state is a "trivial" high-temperature phase of matter, with no order and (usually) much less interesting than the (lower-temperature) ordered phase; in some sense there is one, and only one, completely featureless classical liquid phase. This is *not* the case for quantum liquid phases at zero temperature, even though a quantum-mechanical problem can often be mapped onto a classical one with an extra (time-like) dimension via path integration. The difference lies in the fact that quantum mechanics is based on complex amplitudes, while classical statistical physics is based on (positive definite) real probabilities. Quantum fluctuations are usually weak in zero-temperature ordered phases (since fluctuations of the order parameter are restricted to the vicinity of its expectation values), making them very similar to their classical counterparts in most cases. The situation is quite different in quantum liquid phases. Here the dynamical variables have unrestricted fluctuations, making quantum interference effects very important. The Berry phase, a purely quantum effect, often has significant impact on the interference and long-distance/low-energy physics of the liquid states. The spin liquids discussed briefly here, and fractional quantum Hall liquids discussed in Chapter 16, are examples of such quantum liquids. We already see that they have rich and non-trivial internal structures, and "orders" that are *not* described by an order parameter associated with broken symmetries. Understanding the internal structures and orders of quantum liquids is a frontier of modern condensed matter physics.

17.7.1 $\mathbb{Z}_2$ Topological Order in Resonating Valence-Bond Spin Liquid

As discussed in Section 16.11, one indication of topological order is ground state degeneracy in topologically non-trivial manifolds. We thus consider a resonating valence-bond (RVB) spin liquid on

---

[19] Equivalently we may say that quantum fluctuations renormalize to zero the "string tension" between pairs of spinons.

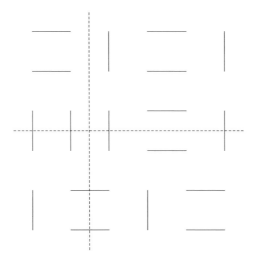

Figure 17.23 A valence-bond (solid segment) configuration on a 6×6 square lattice in which a cut along the y direction (vertical dashed line) always cuts through an even number of bonds. The same is true for a cut along the x direction (horizontal dashed line). The reader is encouraged to draw configurations in other topological sectors discussed in the text.

a torus. What this means is that we consider a finite-size system in the shape of an $L_x \times L_y$ rectangle, and impose periodic boundary conditions both in the x and in the y direction. In the following we assume L_x and L_y are both even for simplicity; the discussion can be generalized to other situations straightforwardly. Let us consider a generic valence-bond configuration as illustrated in Fig. 17.23, which seems featureless. To reveal the subtle pattern hidden in the state, consider "cutting" the torus along the y direction, and count the number of valence bonds it cuts through. This number of course depends on the precise location of the cut. But, because L_y is even, one can easily convince oneself that the *parity* of the numbers is the same for all the y-cuts, namely they are all even or all odd (the reader should ponder how the situation changes when L_y is odd). Exactly the same happens for cuts along the x direction. We can thus assign two independent $\mathbb{Z}_2$ quantum numbers to each valence-bond configuration, and divide them into four different sectors: even–even, even–odd, odd–even, and odd–odd. It is important to note that starting from a specific valence-bond configuration and making *local* changes through resonance processes depicted in Fig. 17.19 results in valence-bond configurations that have the *same* parity quantum numbers, belonging to the same sector (see the configurations of Figs. 17.18(a) and (c) for example). The difference between different sectors is thus *global*, or *topological* in nature. An RVB state is, of course, a linear combination of many different valence-bond configurations. However, they should be related to each other through the resonance processes depicted in Fig. 17.19, in order to take advantage of the resonance energy. As a result of this, we find that RVB states also decompose into four different topological sectors. We thus expect a four-fold ground state degeneracy for the RVB spin liquid on a torus [187], corresponding to $D = 4$ and $g = 1$ in Eq. (16.182). Because the different topological sectors of this particular type of RVB spin liquid are labeled by $\mathbb{Z}_2$ quantum numbers, it is also known as a $\mathbb{Z}_2$ spin liquid. Also similarly to the case of fractional quantum Hall liquids, such topological degeneracy cannot be lifted by disorder or any *local* perturbations in the thermodynamic limit, because the topologically distinct ground states are locally indistinguishable. This is a fundamental difference between topologically ordered states and (conventionally ordered) broken-symmetry states; for example, one can easily distinguish different valence-bond solid states like those illustrated in Figs. 17.18(a) and (b) by measuring local spin–spin correlation functions.

To gain some understanding of the topological order underlying valence-bond states [187, 188], consider the lattice shown in Fig. 17.24. The two sublattices have been indicated by solid circles (the

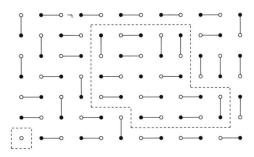

Figure 17.24 Illustration of topological order in a valence-bond state with two topological defects (unpaired spins). The existence of a topological defect can be deduced from the properties of a large loop enclosing it [187, 188]. Only in the case of an RVB liquid (as opposed to a valence-bond solid) will the two topological defects be unconfined from each other.

"black sublattice") and hollow circles (the "white sublattice"). All of the spins except two are paired up. Because we restrict ourselves to nearest-neighbor bonds, every bond connects sites on opposite sublattices. Thus the two topological defects are necessarily on opposite sublattices. The existence of a topological defect can be deduced from the properties of loops drawn around them at arbitrarily large distances. For simplicity consider a large loop chosen not to pass through any singlet bonds. Given the loop trajectory, we can compute the difference in the number of black and white points enclosed by the loop. This will be equal to the difference in the number of black and white topological defects enclosed by the loop. A related topological aspect can be seen if we allow the singlet bonds to resonate between configurations such as those illustrated in Fig. 17.19, namely the number of bonds cut by our trajectory will always be even. Similarly, if we choose a trajectory that passes through an odd number of bonds, quantum fluctuations will keep that number odd. That is, the number of bonds any trajectory passes through is conserved mod 2.

17.8 An Exactly Solvable Model of $\mathbb{Z}_2$ Spin Liquid: Kitaev's Toric Code

Our discussions in Section 17.7.1 have been qualitative, and perhaps mostly hand-waving. In this section we will present a concrete and exactly solvable model for the $\mathbb{Z}_2$ spin liquid due to Kitaev[189]. Because the different topological sectors of the model are robust against local perturbations, the states of this model have potential application as an error-correctable quantum memory for quantum information processing. In this context, the model and its associated states are known as the **Kitaev "toric code."** Similarly to the exactly solvable models we encountered in Section 17.6, to achieve solvability we need to modify the physically most natural nearest-neighbor Heisenberg model, except here the modification is much more significant.

Consider a square lattice, and put a spin 1/2 in the center of each bond of the lattice (instead of at each vertex or lattice site!). For an $L_x \times L_y$ rectangle, there are $2L_x L_y$ spins. We again impose periodic boundary conditions, thus forming a torus. We use s to label the "star" of four bonds emanating from a lattice site (or vertex), j to label an individual bond and its associated spin, and p to label a square (or plaquette). Both the star and the plaquette hold a total of four spins; see Fig. 17.25. The toric code Hamiltonian takes the form

$$H_{\text{tc}} = -\sum_s A_s - \sum_p B_p, \tag{17.172}$$

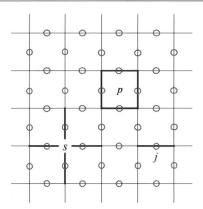

Figure 17.25 Illustration of the toric code model on a square lattice. Circles indicate the positions of spin-1/2 particles on the bonds of the lattice; j labels the spin located at the center of the jth lattice bond; p labels a plaquette containing a total of four spins; s labels a "star" also containing a total of four spins.

where

$$A_s = \prod_{j \in s} \sigma_j^z; \qquad B_p = \prod_{j \in p} \sigma_j^x. \qquad (17.173)$$

In the above $j \in s$ means the bond j has s as an end point, and $j \in p$ means j is one edge of the square p.

The Hamiltonian (17.172) looks quite strange and unphysical, as it involves four- instead of two-spin couplings, and the spin-rotation symmetry is badly broken. What is important, however, is that it is made of (albeit complicated) *local* couplings. As we will see, it captures all the universal topological properties of $\mathbb{Z}_2$ spin liquids. To reveal its solvability (which is by no means obvious), we note that all terms in (17.172) commute with each other:

$$[A_s, A_{s'}] = [B_p, B_{p'}] = [A_s, B_p] = 0. \qquad (17.174)$$

As a result, all eigenstates, including ground states, can be chosen to be eigenstates of all of the A_s and B_p. Since $A_s^2 = B_p^2 = 1$, they all have eigenvalues ± 1 and we expect (because of the minus signs in the Hamiltonian) a ground state to have

$$A_s = B_p = 1 \qquad (17.175)$$

for every s and p. Since we have $L_x L_y$ sites and the same number of squares, it appears that there are $2L_x L_y$ constraints in Eq. (17.175), which is the same as the number of spins in the system. If this were the case we would have a unique ground state. This is, however, not the case. In fact the A and B operators are not all independent; it is easy to show that

$$\prod_s A_s = \prod_p B_p = 1. \qquad (17.176)$$

As a consequence Eq. (17.175) represents only $2L_x L_y - 2$ independent constraints, two fewer than the number of spins. This results in a D-fold ground state degeneracy with $D = 2^2 = 4$. To see that this topological degeneracy is robust against disorder and does not rely on the translation symmetry of the Hamiltonian, let us introduce disorder into the Hamiltonian (17.172) by considering

$$H'_{\text{tc}} = -\sum_s J_s A_s - \sum_p K_p B_p, \qquad (17.177)$$

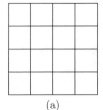 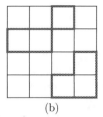 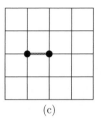 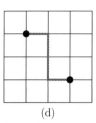

(a)   (b)   (c)   (d)

Figure 17.26 A square lattice in which the bonds with up spins are denoted by thin gray lines and those with down spins are denoted by thick black lines. (a) The configuration with all spins up. (b) For configurations satisfying $A_s = 1$ for all sites s, the thick bonds form closed loops. The no-loop configuration of (a) is a special case here. (c) Hitting one bond with σ^x in (a) creates a pair of charges (black dots) on neighboring sites connected by this bond. (d) Repeated applications of σ^x can move the two charges apart, with a string connecting them. Unlike the string in Fig. 17.20, however, the string here costs no energy and is invisible to local observables. The charges are thus deconfined and constitute well-defined quasiparticle excitations of the system. Since the string is *not* a closed loop, charges are topological defects of the ground state configurations depicted in (a).

where J_s and K_p are positive but otherwise random coupling constants. It is easy to see that the ground states still obey Eq. (17.176) and are thus unchanged, and so is the degeneracy.[20]

The discussion thus far might leave the reader with the impression that the Hamiltonian (17.172) is rather trivial. This is, of course, not the case. The highly non-trivial nature of the toric code can be appreciated by considering its ground state wave functions. Let us start with the constraint $A_s = 1$. To satisfy this constraint it is most natural to work in the σ^z basis for all the spins. In the ground state we must have an even number of up spins among the four. One way to understand this particular constraint is to view the up spins as background, and focus on the down-spin bonds; this constraint is equivalent to requiring that all such bonds form closed loops (see Fig. 17.26(b)).

There are, of course, many such loop configurations. For the system to be in one of the ground states, however, we must also satisfy the other constraint $B_p = 1$. One way to ensure this is to start with any loop configuration $|\text{loop}\rangle$ (which automatically satisfies $A_s = 1$ for any s) or arbitrary linear combinations of such loops, and then project it to the Hilbert subspace with $B_p = 1$:

$$|0\rangle = \prod_p P_p |\text{loop}\rangle, \qquad (17.178)$$

where

$$P_p = (1 + B_p)/2 \qquad (17.179)$$

is the projection operator that projects a state to one with $B_p = 1$, as guaranteed by the fact that the possible eigenvalues of B_p are ± 1. It is important to note that the operation of P_p does not change the eigenvalues of A_s because $[A_s, P_p] = 0$.

Let us consider in some detail what P_p does when it hits a particular loop configuration (or state). The constant term (1/2) of course does nothing. The other term $(B_p/2)$ is much more interesting. If none of the four bonds forming the square p is involved in any existing loops then it creates a small loop around p, because B_p turns all four up spins to down spins. For similar reasons, if such a small loop already exists, then B_p annihilates it. Most interestingly, if a subset of these four bonds consists of parts of other loop(s), then B_p changes their configurations (by inserting a "detour" into the path). We thus find that the operation $\prod_p P_p$ will generate a very complicated superposition of a huge number of loop configurations with arbitrarily large loops, even if the initial configuration in $|\text{loop}\rangle$ contains no loop at all (which is the case if we start with the state in which all spins are up, as in

---

[20] More generic local perturbations will lift this degeneracy and split the ground states, but the splitting vanishes exponentially with L_x or L_y, similarly to the situation for fractional quantum Hall liquids discussed in Section 16.11.

Fig. 17.26(a)). Such a massive superposition creates a long-range entangled and topologically ordered state, with properties identical to those of the $\mathbb{Z}_2$ RVB spin liquid state discussed in Section 17.7.1.

What are the possible outcomes of the projection operation in Eq. (17.178)? Naively, one might think this is sensitive to the initial loop configuration, which would result in many possible final states. However, our earlier analysis found that there are only four different ground states, indicating that many different initial loop configurations will result in the *same* ground state through the operation in (17.178). Then the natural question is how to distinguish the different ground states as well as the initial loop configurations that lead to them. The distinction lies in topologically non-trivial loops. As discussed before, a torus is *not* a singly connected manifold. As a result, there exist loops that wrap around the torus along either the x or the y direction, that cannot shrink to zero (or no loop). Pairs of such loops, on the other hand, can join through the operation of B_p and become a topologically trivial loop. However, this operation does not change the parity of such non-trivial loops. We thus find that we can assign two parity quantum numbers to each loop configuration, for the number of non-trivial loops wrapping around the x and y directions. Since the operation of B_p does not change these topological quantum numbers, we find that Eq. (17.178) generates four possible ground states, depending on the four possible combinations of these quantum numbers of the initial loop configuration. It should be clear by now that the four-fold ground state degeneracy of the toric code has the same topological origin as that of the RVB spin liquid discussed in Section 17.7.1, and such degeneracy takes the form of Eq. (16.182) in a genus-g lattice with $D = 4$. For a torus we have $g = 1$, but similar arguments make it clear that Eq. (16.182) applies to toric code on a genus-g lattice.

What about excited states of the Hamiltonian (17.172)? Owing to the property (17.174), they are simply states with some of the As or Bs equal to -1 instead of $+1$. The most elementary excitation is thus a lattice site with $A_s = -1$, or a square with $B_p = -1$. These are the two types of quasiparticles of the system. The first lives on a lattice site, and we will call this excitation "charge," while the second lives in a plaquette (which is also the smallest closed loop), and we will call it "flux," in obvious analogy with electromagnetism. The big difference here is that the charge and flux can only take values 0 or 1, and form (separate) $\mathbb{Z}_2$ sets under addition. (That is charge and flux are only conserved modulo 2.)

At this point it may seem that these elementary charge or flux excitations are very similar to a single (local) spin flip in a ferromagnet. Again this is not the case. In fact a *single* excitation cannot be created by any *local* operations. Let us consider a charge excitation first. To create one at site s, it appears that all one need do is flip the spin on one of the bonds connected to s using σ^x. It is easy to convince oneself that this operation actually creates a *pair* of charges on neighboring sites s and s', connected by the bond (see Fig. 17.26(c)). Additional operations may separate these two charges, but never eliminate one of them (see Fig. 17.26(d)). We thus find that charges can only be created in pairs. Exactly the same is true for flux excitations. Of course, mathematically this property is guaranteed by the constraints (17.176).

Physically, the somewhat strange properties discussed above reflect the fact that these are topological excitations (or defects in the topological order; see Fig. 17.26(d)) that are intrinsically *non-local*. Such non-locality is reflected by the fact that these excitations can feel the existence of each other even when they are very far apart. Naively there is no interaction among them, as the energy of the system is *independent* of their separations. However, as we demonstrate below, if one drags a charge around a flux, the system picks up a non-trivial Berry phase very much like that in the Aharonov–Bohm effect (hence the names charge and flux!). There is thus a "statistical interaction" between a charge and a flux, which is of topological nature.[21]

---

[21] Such statistical interaction is identical to that between anyons discussed in Section 16.7.2, except that here charge and flux are distinctive instead of identical particles. Distinctive particles with statistical interactions are often referred to as **mutual anyons**.

Consider a charge on site s, and a loop $\mathcal{L}$ that starts and ends at s. We can drag the charge around this loop using the operator $\prod_{j \in \mathcal{L}} \sigma_j^x$ (where $j \in \mathcal{L}$ means that the bond j is part of the loop $\mathcal{L}$), resulting in the final state

$$|\text{final}\rangle = \prod_{j \in \mathcal{L}} \sigma_j^x |\text{initial}\rangle. \tag{17.180}$$

Using the fact that

$$\prod_{j \in \mathcal{L}} \sigma_j^x = \prod_{p \in \mathcal{A}} B_p, \tag{17.181}$$

where $\mathcal{A}$ is the region enclosed by $\mathcal{L}$, we find

$$|\text{final}\rangle = |\text{initial}\rangle \tag{17.182}$$

when there is an even number of flux excitations (which are equivalent to zero due to the $\mathbb{Z}_2$ nature of the flux) inside the loop, while

$$|\text{final}\rangle = -|\text{initial}\rangle \tag{17.183}$$

when there is an odd number of flux excitations inside the loop, meaning that the charge picks up a -1 Berry phase when circling around the flux (or vice versa).

Making contact with the RVB spin liquid discussed in Section 17.7.1, we can identify the charge excitation of the toric code as the spinon of the RVB liquid. There also exist flux-like excitations in the RVB spin liquid, which are more complicated to visualize in terms of the spin degrees of freedom, and we will not discuss them further.

The ground state(s) of the Kitaev toric code model have all the defining universal properties of a $\mathbb{Z}_2$ spin liquid. However, one could argue that the original RVB physics is not easily recognizable in the toric code Hamiltonian (17.172) or its ground state (17.178): there are no fluctuating singlet bonds, because the toric code Hamiltonian is far from being rotationally invariant. Recall that, originally, the RVB spin liquid was proposed to describe antiferromagnets with (approximate) rotational invariance in spin space, and can be pictured as an equal-amplitude superposition of all nearest-neighbor valence-bond configurations. Indeed, while Anderson's variational RVB wave function was historically the first quantum many-body wave function with the kind of long-range entanglement that is also present in the Kitaev toric code, it long remained an open question whether it can actually appear as the ground state of a local rotationally invariant Hamiltonian. By now, a number of such local model Hamiltonians are known, which tend to be more complex than the Kitaev toric code. Of particular relevance here is a local parent Hamiltonian for the RVB state on the kagome lattice of Fig. 17.22 [190], which is expected to belong to the same $\mathbb{Z}_2$ spin liquid phase as the Kitaev toric code, and many of its defining properties have been demonstrated explicitly.[22]

17.8.1 Toric Code as Quantum Memory

The four degenerate ground states of the Kitaev model on the torus mean that the toric code can be used to store two **quantum bits** of information. That is, the set of quantum superpositions of these four ground states spans the same Hilbert space as superposition states of two spin-1/2 objects (**qubits**), $\vec{\sigma}_1, \vec{\sigma}_2$, which we will refer to as the "**logical qubits**" created out of the collection of N "**physical qubits**." If we have a square array of N physical qubits and manage to realize the toric code

---

[22] It turns out having a frustrated lattice is crucial for the (nearest-neighbor) RVB state to represent a gapped topological phase. The RVB state on non-frustrated bipartite lattices, on the other hand, has been shown to represent a critical point instead of a stable phase.

Hamiltonian and its periodic boundary conditions (a non-trivial task[23]), we can use the toric-code logical qubits as a quantum memory.

The basic idea behind this is that the two logical qubits are "topologically protected." That is, small deviations from the ideal Hamiltonian will lead to virtual production of pairs of "charges" or "fluxes" which will quickly recombine because of the energy gap. These will have little effect, since moving the state of the system from one logical quantum ground state to another would require one of the topological defects to travel on a non-contractible loop winding around the torus before returning to annihilate with its partner.[24] This requires tunneling a distance $\sim \sqrt{N}$ and is exponentially small. That is, if we do perturbation theory, the small parameter λ is the strength of the deviations from the ideal Hamiltonian divided by the excitation gap. Processes in which a (virtual) topological defect winds around the torus appear at order $\lambda^{\sqrt{N}}$.

The good news then is that the qubit states are topologically protected. That is, the (non-local) logical states and the distinction among them is invisible to local perturbations that flip or dephase individual physical qubits. The bad news is that the logical states are sufficiently well protected that it is difficult for us to change them when we want to carry out a single-qubit operation (e.g. rotation on the Bloch sphere). It is relatively easy to carry out a **Clifford group operation** on one of the logical qubits (i.e. bit flip σ_1^x, phase flip σ_1^z or both, which yields $i\sigma_1^y$). We do this by performing repeated Clifford operations on a line of physical spins along a non-contractible loop of the torus as described above. But what if we want to rotate the qubit through an angle θ around the x axis on the Bloch sphere? This requires

$$U_\theta = e^{i\frac{\theta}{2}\sigma_1^x} = \cos\left(\frac{\theta}{2}\right)\sigma_1^0 + i\sin\left(\frac{\theta}{2}\right)\sigma_1^x, \tag{17.184}$$

where σ_1^0 is the identity operation. Thus we have to carry out a quantum superposition of doing nothing and flipping a large number ($\sqrt{N}$) of physical spins. It appears to be difficult to achieve this superposition of macroscopically distinct operations, though proposals to get around this difficulty exist and are reviewed in Ref. [191].

A second problem is created by the fact that the energy gap for excitations is finite.[25] Even if the physical qubits are in good thermal equilibrium (something difficult to achieve inside an operating quantum computer) at low temperature, there will be a finite density of topological defects. The density will be exponentially small in the gap divided by the temperature, but it will be finite. Small deviations from the ideal Hamiltonian will allow these topological defects to travel (either coherently or diffusively) around the sample, producing logical errors.

The only way to beat these thermally induced errors is to introduce a "Maxwell demon" to "algorithmically cool" the system by rapidly and repeatedly measuring all the constraint operators ("error syndromes") A_s and B_p to determine when a thermal fluctuation has produced a local pair of defects. The demon must then quickly compute what local operations on the physical qubits it needs to carry out to cause the topological defects to recombine before they can change one of the topological winding numbers. This version of "**quantum error correction**" will be extremely challenging to achieve,

---

[23] The four-qubit interaction terms of the Hamiltonian are difficult to achieve directly (since interactions are typically two-body) and one must resort to an *effective* Hamiltonian obtained by having each of the four qubits in a star or plaquette virtually excite an ancilla qubit that has a strongly different transition frequency. It is possible to avoid the difficulty of realizing periodic boundary conditions by using the closely related "surface code" [191, 192] which has open boundary conditions.

[24] Recall the discussion in Section 16.12. There to go from one (topologically degenerate) ground state to another requires moving a Laughlin quasiparticle (a topological defect) around the torus, which is a non-contractible loop.

[25] There exist models in four spatial dimensions in which the topological defects are confined by a finite string tension (much like quark confinement in 3+1 space-time dimensions). These could be realized in three spatial dimensions via non-local Hamiltonians but this would be extremely challenging to realize experimentally.

especially when one considers the fact that no Maxwell demon will be perfect – it will introduce errors of its own when it reads out the error syndromes and tries to correct them. A very important concept in quantum error correction is that of the "break-even" point – that is, the point at which the error rate of the physical qubits is low enough and the Maxwell demon works well enough and quickly enough that a code using N physical qubits would yield collective logical qubits with coherence times exceeding that of the best of the individual physical qubits [193]. At the time of writing, it appears to be very challenging [191, 192] for the surface code to achieve or exceed the break-even point without considerable technical advances, but substantial efforts are being made.

Exercise 17.17. Prove Eqs. (17.174) and (17.176).

Exercise 17.18. Show that the state (17.178) satisfies (17.175).

Exercise 17.19. Prove Eq. (17.181).

Exercise 17.20. The toric code has a finite excitation gap. Find a lower bound for this gap for the Hamiltonian given in Eq. (17.177).

Exercise 17.21. Consider how the constraints in Eq. (17.176) need to be modified when the couplings J_s and K_p in Eq. (17.177) are allowed to take negative values, and show that the ground-state degeneracy remains exactly four.

Exercise 17.22. Consider the perturbing effect on the ground state(s) of the Hamiltonian given in Eq. (17.177) of a transverse magnetic field applied to a single site k via $V = \lambda \sigma_k^x$. Find the energy denominator that controls the strength of the perturbation and show that the effects of the perturbation do not propagate throughout the lattice but instead remain local. Use this fact to find the *exact* ground-state energy of the perturbed Hamiltonian.

Exercise 17.23. The locality of the response to a local perturbation provides a hint that helps explain why generic perturbations on the exactly soluble model have exponentially small effects in lifting the degeneracy of the ground state manifold. Suppose that the perturbation Hamiltonian $V = \lambda \sum_j \sigma_j^x$ acts on all sites of an $L_x \times L_y$ lattice. At what order in perturbation theory is the ground state degeneracy first lifted?

Exercise 17.24. Show that, as a flux is dragged around a charge, the system picks up a -1 Berry phase.

Exercise 17.25. Generalize the toric code to the 3D cubic lattice and discuss how the charge and flux excitations differ from the 2D case. What is the ground state degeneracy when periodic boundary conditions are imposed in all three directions?

17.9 Landau Diamagnetism

Our discussion of magnetism in condensed matter systems has thus far focused on Mott insulators, where the electrons' spin degrees of freedom dominate the physics. In an insulator, the electron kinetic energy is quenched, and orbital coupling between electron motion and the magnetic field (through the vector potential $\vec{A}$) is very weak, and can be described at the level of individual atoms or ions along the lines of Section 17.3. The situation is quite different in metals, where the orbital motion of itinerant electrons couples much more strongly with the field. We will show in this section that such coupling again gives rise to a diamagnetic response, known as **Landau diamagnetism**.

To reveal the physics behind Landau diamagnetism in the simplest possible setting, let us consider non-interacting and free electrons moving in 2D, with a magnetic field applied in the normal direction of the 2D plane. As we learned in Chapter 12, in this case the kinetic energy of the individual electrons is quantized into discrete Landau levels

$$\epsilon_n = \left(n + \frac{1}{2}\right)\hbar\omega_c, \tag{17.185}$$

with $n = 0, 1, \ldots$, and Landau-level degeneracy $N_\Phi = A/(2\pi\ell^2)$.[26] It is illuminating to view this result in terms of the electron density of states (DOS), as shown in Fig. 17.27. We know in 2D that the DOS is a constant when $H = 0$. The magnetic field "squeezes" states in the energy range $n\hbar\omega_c < \epsilon \leq (n+1)\hbar\omega_c$ into a single energy $(n+\frac{1}{2})\hbar\omega_c$, giving rise to a sequence of delta functions for the DOS.

Exercise 17.26. Show that $D\hbar\omega_c = N_\Phi$, where D is the DOS per spin when $H = 0$. This demonstrates that the number of states of each Landau level is the same as that in the energy range $n\hbar\omega_c < \epsilon \leq (n+1)\hbar\omega_c$ when $H = 0$. Show that the Landau-level filling factor is given by $\nu = \epsilon_F/(\hbar\omega_c)$, where ϵ_F is the electron Fermi energy at $H = 0$. Here we neglect electron spin.

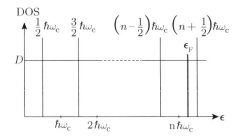

Figure 17.27 Density of states (DOS) per spin of a 2D electron gas. Without a magnetic field it is a constant D. With a magnetic field it is made of a sequence of delta functions located at $\epsilon_n = (n + \frac{1}{2})\hbar\omega_c$.

---

[26] Here we assume $|M| \ll |H|$ so that $B \approx H$. This needs to be checked for self-consistency in the end. For the case of the low-density 2D electron gases typically found in semiconductor quantum wells, this is an excellent approximation.

We are now in a position to show that this re-organization of electron energy levels in general *increases* the electron kinetic energy, thus giving rise to diamagnetism. The exercise above shows that, if the states in the energy range $l\hbar\omega_c < \epsilon \le (l+1)\hbar\omega_c$ are all occupied by electrons when $H = 0$, these electrons will fully occupy the lth Landau level, and their *average* energy does *not* change in the presence of the field. In particular, if the Landau-level filling factor ν is an integer, the energy difference between the cases with and without field is zero. On the other hand, if $\nu = n + \delta\nu$ is not an integer, the energy increases with field because electrons occupying the nth Landau level have energy $(n + \frac{1}{2})\hbar\omega_c$, which is higher than their average energy, $(n\hbar\omega_c + \epsilon_F)/2$ when $H = 0$ (see Fig. 17.27). This energy increase is the origin of Landau diamagnetism. At this point we already see that qualitatively Landau diamagnetism is a weak effect at weak field, because only electrons near the Fermi energy contribute.

We now calculate this effect quantitatively:

$$\Delta E(H) = E(H) - E(H=0) = 2D \int_0^{\Delta\epsilon_F} \left(\frac{1}{2}\hbar\omega_c - \epsilon'\right) d\epsilon' \tag{17.186}$$

$$= D \,\Delta\epsilon_F(\hbar\omega_c - \Delta\epsilon_F), \tag{17.187}$$

where $\Delta\epsilon_F = \epsilon_F - n\hbar\omega_c = \delta\nu\,\hbar\omega_c$, and the factor of 2 upfront takes into account both spin species. Obviously the energy change oscillates with ν with unit period. In terms of $1/H$ it oscillates with period $2e/(k_F^2\hbar c)$, which is another manifestation of quantum oscillations. We thus average $\Delta E(H)$ over $\delta\nu$[27] to obtain a smooth result:

$$\langle\Delta E(H)\rangle = \frac{1}{6}D(\hbar\omega_c)^2 = \frac{2}{3}D(\mu_B H)^2, \tag{17.188}$$

resulting in a diamagnetic susceptibility

$$\langle\chi_D\rangle = -\frac{4}{3}D\mu_B^2 = -\frac{2}{3}\chi_{\text{Pauli}}, \tag{17.189}$$

where

$$\chi_{\text{Pauli}} = 2D\mu_B^2 \tag{17.190}$$

is the **Pauli paramagnetic susceptibility** due to the electron spin (see Section 7.7.2). Our zero-temperature result (17.189) should be contrasted with Landau's original calculation performed at finite T with $\hbar\omega_c \ll k_B T \ll \epsilon_F$ (see, e.g., Ref. [194]):

$$\chi_{\text{Landau}} = -\chi_{\text{Pauli}}/3 = \chi_D/2. \tag{17.191}$$

Our result is twice as large as this. The discrepancy is not due to the effect of finite temperature but due to the fact that our simple calculation using a spatially uniform magnetic field has missed a subtle boundary effect which reduces the diamagnetic response by a factor of 2. If we carry out the calculation with a magnetic field that varies spatially with small but non-zero wave vector $\vec{q}$ this subtlety is avoided. We present this calculation in the context of the magnetic response of superconductors in Section 20.4.

In the above we used the fact that for free electrons in vacuum we have $\hbar\omega_c = 2\mu_B H$. For Bloch electrons in a crystal that can be described using an effective mass, $\hbar\omega_c$ needs to be properly re-scaled, and the (orbital) diamagnetic response may overwhelm the (spin) Pauli paramagnetic response. It is even possible to disentangle the two, especially for a (quasi-)2D system, as spin responds to the total field while the orbital response is (to a good approximation) sensitive only to the perpendicular

---

[27] Which is the appropriate thing to do in the presence of a smooth random potential where the local ϵ_F and ν varies smoothly.

component of the magnetic field, so changing the direction of the field will modify these two effects in different ways.

Now let us imagine what would happen if electrons were bosons. In this case they would all condense into the $\vec{k} = 0$ state when $H = 0$, and move to the lowest Landau level when the field is turned on. As a result, *all* (not just a small fraction near ϵ_F as in the fermionic case) of their energies go up:

$$\Delta E(H) = N\hbar\omega_c/2 \propto H, \tag{17.192}$$

resulting in a much stronger diamagnetic response:

$$M = -\frac{1}{V}\frac{\partial E(H)}{\partial H} = -N\mu_B/V, \tag{17.193}$$

namely we have a *finite* diamagnetic magnetization for even infinitesimal H. This is, of course, unphysical, as it would imply having $\vec{B}$ in the opposite direction to that of $\vec{H}$. What would actually happen is that M saturates at $-H/(4\pi)$ such that $B = 0$, namely the bosons would develop enough diamagnetic response that the magnetic flux is completely screened out. This is known as the **Meissner effect** or Meissner diamagnetism. This is indeed what happens in superconductors, where electrons form Cooper pairs that behave as bosons. We cover this in greater detail in our discussion of superconductivity.

18 Bose–Einstein Condensation and Superfluidity

So far in this book we have mostly been studying electrons, which are fermions. Owing to the Pauli exclusion principle, the ground state of non-interacting fermions is a Fermi sea. As we learned in Chapter 15 this state is *stable* against sufficiently weak repulsive interactions, namely both the ground state and the low-energy excitations of the interacting system are quasi-adiabatically connected to those of the non-interacting system. As a result, the thermodynamic and other properties of interacting fermions are often very similar to those of the free Fermi gas. This is, of course, the essence of Landau's Fermi-liquid theory.

In condensed matter physics we often encounter systems made of many bosons as well. In Chapter 4 we had a brief encounter with superfluid ^{4}He, a prototypical bosonic system. We have also discussed extensively phonons and photons, which are also bosons.[1] As mentioned in Chapter 7 and elsewhere, there is fascinating many-body physics in trapped-cold atom/molecule systems that are of interest to condensed matter physicists. Many of these systems are made of bosonic atoms or molecules. Coming back to electronic condensed matter physics, superconductors can (under certain circumstances) be (usefully but rather approximately) viewed as systems made of bosonic Cooper pairs. Similarly, excitons are charge-neutral analogs of Cooper pairs in which an electron pairs with a hole. As was discussed in Chapter 9 these bosonic excitations can hybridize with photons to form polaritons. As we will describe in this chapter, all these different bosonic objects can undergo Bose–Einstein condensation and exhibit signatures of superfluidity (even though their numbers are not strictly conserved). For these reasons we devote this chapter to bosons whose numbers are conserved (or can be approximated as being conserved), both for the topic's intrinsic importance and as preparation of our discussion of superconductivity in later chapters. As we will see, due to the difference in statistics, bosons behave differently from fermions in fundamental ways, starting even from the non-interacting case.

18.1 Non-interacting Bosons and Bose–Einstein Condensation

We start with non-interacting, spinless bosons, and as in Section 15.11 we use the grand canonical ensemble with the grand Hamiltonian

---

[1] An important difference for phonons and photons is that their numbers are not conserved. This leads to some fundamental differences from bosons whose numbers are conserved. This is the main focus of this chapter.

$$H_G = \sum_{\vec{k}} \left(\frac{\hbar^2 k^2}{2m} - \mu \right) \hat{n}_{\vec{k}}, \tag{18.1}$$

where μ is the chemical potential and $\hat{n}_{\vec{k}} = b_{\vec{k}}^\dagger b_{\vec{k}}$ is the number operator for wave vector $\vec{k}$, where $b_{\vec{k}}^\dagger$ and $b_{\vec{k}}$ are the corresponding boson creation and annihilation operators, which satisfy the usual harmonic oscillator commutation relations:

$$[b_{\vec{k}}, b_{\vec{k}'}^\dagger] = \delta_{\vec{k}, \vec{k}'}; \qquad [b_{\vec{k}}, b_{\vec{k}'}] = [b_{\vec{k}}^\dagger, b_{\vec{k}'}^\dagger] = 0. \tag{18.2}$$

Just as in the free-fermion case, for $\mu < 0$ the ground state is simply the vacuum with no particles. However, due to the *absence* of the Pauli exclusion principle, H_G has *no* ground state for $\mu > 0$, as one can lower the energy indefinitely by piling more and more bosons into any $\vec{k}$ state with $\hbar^2 k^2 / 2m < \mu$. As a result, if bosons are present at zero temperature, the *only* possible value for the chemical potential is $\mu = 0$. This is very different from the free-fermion case, in which all positive values of μ are allowed, and there is a one-to-one correspondence between μ and the fermion density (as determined by the Fermi wave vector k_F). Here, however, states with *any* number N of bosons occupying the $\vec{k} = 0$ (single-particle) state

$$|N\rangle = \frac{1}{\sqrt{N!}} (b_0^\dagger)^N |0\rangle \tag{18.3}$$

are exactly degenerate. Thus in this case N (or the density $n = N/V$) is totally undetermined by μ.

As we have learned in statistical mechanics (see, e.g., Ref. [28]), the phenomenon of a macroscopically large number of bosons occupying a single state is known as **Bose–Einstein condensation (BEC)**. In this chapter we will revisit BEC from the perspective of condensed matter physics, and focus on the importance of boson–boson interactions and the relation of BEC to superfluidity. As we will see, the pathology mentioned above (that μ can only be zero at $T = 0$ and it is unable to determine n) is cured by (repulsive) interactions. In this section, however, we will try to gain a deeper understanding of BEC for non-interacting bosons in preparation for later developments.

Clearly *any* superposition of the set of states (18.3) is also a ground state of H_G (18.1), and the set (18.3) is an orthonormal basis set of the ground state subspace, which has an enormous degeneracy. This basis is identical to the energy eigenstate basis of a simple harmonic oscillator, also labeled by a non-negative integer occupation number.[2] Another commonly used (*overcomplete* but *not* orthogonal) basis, which we carry over here with no modification, is the coherent-state basis, whose states are parameterized by a complex amplitude α,

$$|\alpha\rangle = e^{-|\alpha|^2/2} e^{\alpha b_0^\dagger} |0\rangle = e^{-|\alpha|^2/2} \sum_{N=0}^{\infty} \frac{\alpha^N}{\sqrt{N!}} |N\rangle, \tag{18.4}$$

and are eigenstates of b_0:

$$b_0 |\alpha\rangle = \alpha |\alpha\rangle. \tag{18.5}$$

The coherent states (18.4) are simply linear combinations of degenerate ground states and hence are themselves ground states. Nevertheless, they are highly unusual in the present context in that they do *not* have fixed particle number N, despite the fact that N is a conserved quantity for the Hamiltonian (18.1). Instead they are *coherent* superpositions of states with *different* N, with fixed relative phases among them. This is thus a classic example of a spontaneously broken symmetry. The symmetry (which is responsible for the conservation of total boson number) that is broken here is a U(1) symmetry, namely the U(1) transformation $b_{\vec{k}} \rightarrow e^{i\theta} b_{\vec{k}}$ and $b_{\vec{k}}^\dagger \rightarrow e^{-i\theta} b_{\vec{k}}^\dagger$, which leaves H_G of (18.1) invariant. On the other hand, under this transformation we find that $\alpha \rightarrow e^{i\theta}\alpha$ is *not* invariant.

---

[2] The reader should not be confused by the fact that in the harmonic oscillator case there is a unique ground state, and the energy eigenstate basis spans the entire Hilbert space!

The reader may feel uncomfortable thinking about states in which particle numbers fluctuate, despite the fact that the relative fluctuation is small in the thermodynamic limit (see Exercise 18.4). To help understand the symmetry-breaking nature of the non-interacting boson ground states we compare them with those of Heisenberg ferromagnets, that we studied in Section 17.5.1. There we have $2S + 1$ degenerate ground states, where S is the total spin, a thermodynamically large number, so the ground state degeneracy is huge. These states are labeled by a conserved quantum number S^z, which plays the role of the boson number N here. The conservation of S^z is again the consequence of a U(1) symmetry, corresponding to rotation around the $\hat{z}$ axis. The set of orthonormal ground states $\{|S, S^z\rangle\}$ of a ferromagnet corresponds to the set $\{|N\rangle\}$ here. The expectation value of the total spin operator $\vec{S}$ (or magnetization), which is the ferromagnetic order parameter, is non-zero only for its $\hat{z}$ component in these states. On the other hand, we also understand that the magnetization can point in any direction. A ground state with non-zero magnetization in the xy plane must be a linear superposition of states with different S^z. Thus S^z is no longer a good quantum number in such a state, and the U(1) rotation symmetry along the $\hat{z}$ direction is spontaneously broken. These states are analogous to the coherent ground states (18.4) here. Also the spin raising and lowering operators, $S^\pm = S^x \pm i S^y$, play roles very similar to $b_0^\dagger$ and b_0. Since their expectation values are linear combinations of the $\hat{x}$ and $\hat{y}$ components of the ferromagnetic order parameter, by this analogy we can identify the expectation value of b_0 as the order parameter of the BEC.

In fact, the analogy between non-interacting bosons and Heisenberg ferromagnets extends to excitations as well. The simplest excitations of H_G in (18.1) are states with $n_{\vec{k}} = 1$ for one of the non-zero $\vec{k}$s, namely a single boson outside the condensate. The dispersion relation of such an elementary excitation, which can be created using $b_{\vec{k}}^\dagger$, is obviously

$$\epsilon_{\vec{k}} = \frac{\hbar^2 k^2}{2m}. \tag{18.6}$$

It is analogous to a single spin-wave in the ferromagnet, which can be created using either $S_{\vec{k}}^-$ or $S_{\vec{k}}^+$ (depending on which ground state it acts on), and also has a *quadratic* dispersion.

Exercise 18.1. Prove Eq. (18.5).

Exercise 18.2. Prove that

$$|\alpha\rangle = U_\alpha |0\rangle, \tag{18.7}$$

where U_α is the unitary displacement operator

$$U_\alpha \equiv e^{\alpha b_0^\dagger - \alpha^* b_0}. \tag{18.8}$$

Exercise 18.3. Prove that the set $\{|\alpha\rangle\}$ defined in Eq. (18.4) spans the many-boson ground state subspace. (Hint: show that $|N\rangle$ can be expressed as an appropriate linear superposition of $\{|\alpha\rangle\}$ for any $N \geq 0$.)

Exercise 18.4. Calculate $\overline{N} = \langle\alpha|n_0|\alpha\rangle$ and $\overline{N^2} = \langle\alpha|n_0^2|\alpha\rangle$, and show that $\Delta N = \sqrt{\overline{N^2} - (\overline{N})^2} \ll \overline{N}$ for $\overline{N} \gg 1$.

18.1.1 Off-Diagonal Long-Range Order

The discussion above should have made it clear that the free-boson ground state, in which *all* bosons condense into a single state, can spontaneously break a U(1) symmetry. A natural question that follows is, then, if we did not have this insight, but were instead given a *specific* ground state in this enormous ground state subspace, how could we detect the broken symmetry? If the state handed to us is of the type (18.4), we can calculate the expectation value of b_0, and find that it is non-zero, which signals the broken symmetry. On the other hand, if what we are given is a state with fixed particle number as in (18.3), this method is ineffective. Here again our insight from magnetism (especially antiferromagnetism) comes in handy, namely we can detect order and broken symmetry by inspecting correlation functions of *local* operators, especially the large-distance behavior. We know that the ground state of the ferromagnet has the property that all the spins are parallel. This long-distance correlation property is more fundamental than the particular local orientation of the spins, which is completely arbitrary for an isotropic ferromagnet.

Obviously the analog of the local spin operator in the present case is the local boson field (see Appendix J)

$$\Psi(\vec{r}) = \frac{1}{\sqrt{V}} \sum_{\vec{k}} e^{i\vec{k}\cdot\vec{r}} b_{\vec{k}}, \tag{18.9}$$

where V is the system volume. Let us therefore examine the correlation function of $\Psi(\vec{r})$ with its Hermitian conjugate at a separate position:

$$\langle\langle \Psi^{\dagger}(\vec{r})\Psi(\vec{r}\,') \rangle\rangle = \frac{1}{V} \sum_{\vec{k}\vec{k}'} e^{-i\vec{k}\cdot\vec{r}} e^{i\vec{k}'\cdot\vec{r}\,'} \langle\langle b_{\vec{k}}^{\dagger} b_{\vec{k}'} \rangle\rangle = \sum_{\vec{k}} e^{i\vec{k}\cdot(\vec{r}\,'-\vec{r})} \frac{\langle\langle n_{\vec{k}} \rangle\rangle}{V}, \tag{18.10}$$

which is analogous to the spin–spin correlation function in a magnet. In the above $\langle\langle \cdot \rangle\rangle$ stands for the quantum-mechanical and thermal average, if the system is at finite temperature. For $T = 0$ we have $\langle\langle n_{\vec{k}} \rangle\rangle = N\delta_{\vec{k},0}$ and

$$\langle\langle \Psi^{\dagger}(\vec{r})\Psi(\vec{r}\,') \rangle\rangle = \frac{N}{V}, \tag{18.11}$$

which is independent of $(\vec{r}\,' - \vec{r})$, clearly indicating the presence of long-range order. It should also be clear that, even at finite T, as long as the macroscopic occupation of the $\vec{k} = \vec{0}$ state survives ($\langle\langle b_0^{\dagger}b_0 \rangle\rangle/V > 0$) in the thermodynamic limit $V \to \infty$ (which is the definition of BEC), such long-range order is present. We thus find BEC implies long-range order, and is therefore a symmetry-breaking state.

The correlation function (18.10) can, of course, also be calculated using first quantization; in fact it is the matrix element of the reduced density matrix (see Appendix F) of a single particle (chosen to be particle 1 without loss of generality since all particles are identical):

$$\rho(\vec{r}, \vec{r}\,') = \langle \Psi^{\dagger}(\vec{r})\Psi(\vec{r}\,') \rangle$$
$$= N \int d\vec{r}_2 \cdots d\vec{r}_N \, \psi^*(\vec{r}, \vec{r}_2, \ldots, \vec{r}_N)\psi(\vec{r}\,', \vec{r}_2, \ldots, \vec{r}_N). \tag{18.12}$$

In the above we assumed for simplicity that the system is in a pure state whose wave function is $\psi(\vec{r}_1, \vec{r}_2, \ldots, \vec{r}_N)$. The integrations over $\vec{r}_2, \ldots, \vec{r}_N$ trace these particles out, and the factor N in front of the integral is the proper normalization ensuring that the total particle number is N. Generalization to the case of a thermal ensemble is straightforward. We find that the long-range order exhibited by a BEC shows up in the off-diagonal matrix elements of the single-particle reduced density matrix, hence the name **off-diagonal long-range order (ODLRO)**.[3] This should be contrasted with more familiar types of order such as density-wave or crystalline order, which are periodic patterns in the *diagonal* matrix element of the single-particle reduced density matrix. This is simply spatial modulation of the average particle density.

One can use the single-particle reduced density matrix $\rho(\vec{r}, \vec{r}\,')$ defined in (18.12) to identify condensate(s) in the absence of translation symmetry (so single-particle eigenstates are no longer plane waves), and/or in the presence of interactions. Since $\rho(\vec{r}, \vec{r}\,') = \rho^*(\vec{r}\,', \vec{r})$ is a Hermitian matrix, we can find its orthonormal eigenfunctions with real eignevalues:

$$\int d\vec{r}\,' \, \rho(\vec{r}, \vec{r}\,') \chi_i(\vec{r}\,') = n_i \chi_i(\vec{r}), \tag{18.13}$$

in terms of which we have

$$\rho(\vec{r}, \vec{r}\,') = \sum_i n_i \chi_i(\vec{r}) \chi_i^*(\vec{r}\,'). \tag{18.14}$$

The eigenvalue n_i is the occupation number of the eigenfunction χ_i. If this is an *extensive* quantity so that we have $n_i \sim O(N)$, then we have a condensate, with χ_i being the condensate wave function. If the density matrix exhibits both ODLRO and diagonal density modulations (spontaneously broken translation symmetry) the system is referred to as a "**supersolid**." There is tantalizing experimental evidence of supersolids in cold-atom systems [195, 196].

Exercise 18.5. Consider the thermal distribution function for free bosons

$$\langle\langle \hat{n}_{\vec{k}} \rangle\rangle = \frac{1}{e^{(\epsilon_{\vec{k}} - \mu)/k_{\mathrm{B}}T} - 1} \tag{18.15}$$

with $\mu < 0$, so there is no BEC. Show that $\langle\langle \Psi^\dagger(\vec{r})\Psi(\vec{r}\,') \rangle\rangle$ decays exponentially and find the decay length.

18.1.2 Finite Temperature and Effects of Trapping Potential

The question of how raising the temperature eventually destroys the macroscopic occupation of the $\vec{k} = \vec{0}$ state deserves further discussion. Let us assume that there is no macroscopic occupation of any single state. At finite temperature the maximum density of bosons that we can have is then given by taking the chemical potential to approach zero from below, $\mu \to 0^-$. From Eq. (18.10) we see that the maximum possible total density is

$$\rho = \sum_{\vec{k}} \frac{\langle\langle n_{\vec{k}} \rangle\rangle}{V} = \int \frac{d^3 k}{(2\pi)^3} \frac{1}{e^{\beta \epsilon_k} - 1}, \tag{18.16}$$

where $\beta = 1/k_{\mathrm{B}}T$ and the conversion from the sum to the integral presumes that there is no macroscopic occupation of the $\vec{k} = \vec{0}$ state. Defining a dimensionless wave vector $x = \lambda k$, where the thermal de Broglie wavelength is defined via[4]

$$\lambda^2 = \frac{\beta \hbar^2}{2m}, \tag{18.17}$$

---

[3] The concept of ODLRO can be generalized to multi-particle reduced density matrices. For example, condensation of Cooper pairs leads to two-particle (rather than single-particle) ODLRO.

[4] Some authors define the thermal de Broglie wavelength with an additional factor of 2π on the RHS of Eq. (18.17).

we have

$$\rho = \frac{\lambda^{-3}}{2\pi^2} \int_0^\infty \frac{x^2\, dx}{e^{x^2} - 1} = \lambda^{-3} \frac{\zeta(3/2)}{8\pi^{3/2}} \sim 0.0586\lambda^{-3}, \qquad (18.18)$$

where ζ is the Riemann zeta function. Because we have set the chemical potential to zero, this is the largest possible density that can be achieved without macroscopic occupation of the ground state. Thus, if we can achieve a density greater than this, we will achieve BEC. For a given fixed physical density this means lowering the temperature below the critical temperature

$$T_c = \left(\frac{\rho}{\zeta(3/2)}\right)^{2/3} \frac{2\pi\hbar^2}{k_B m}. \qquad (18.19)$$

The physical interpretation of the above is that the onset of BEC occurs when the de Broglie wave packets (which have size $\sim \lambda$) of the individual bosons begin to overlap. This overlap can be forced by either decreasing the distance between the particles (i.e. increasing the density) or lowering the temperature to increase the de Broglie wavelength.

For the case of liquid ^{4}He (which is not very close to the ideal Bose gas case considered so far), we can simply use a refrigerator to achieve the necessary low temperature. Such refrigerators use adiabatic expansion of a working gas to produce cooling. For the case of cold atomic gases, adiabatic expansion lowers the temperature but unfortunately preserves the number of bosons in each mode and hence cannot increase the phase-space density $\rho\lambda^3$ or decrease the entropy. Instead one uses **evaporative cooling**, in which atoms with large kinetic energy are allowed to escape from a magneto-optical trap as the barrier to escape is gradually lowered. The density decreases during this process, but the temperature and the entropy fall more rapidly so that the net phase-space density increases, leading eventually to BEC. We thus find that a trap is not only necessary to confine the gas, but can also be manipulated for the purpose of cooling.

The calculations above were all carried out assuming that the free bosons live in an infinitely large box in 3D. In 1D and 2D the integral in Eq. (18.16) diverges. The critical density for BEC is thus infinite. Hence, for any finite density, the chemical potential is always negative and no BEC is possible. For the case of cold atoms in an isotropic 2D harmonic potential, the free-particle (or in the case of interactions, the mean-field) energy eigenvalues are discrete,

$$\epsilon(n_x, n_y) = \hbar\omega(n_x + n_y + 1), \qquad (18.20)$$

where ω is the trap frequency. Even in the limit of a weak trap ($\hbar\omega \ll k_B T$), the finite (but small) level spacing modifies the situation by introducing an infrared cutoff that (more or less) restores the BEC transition. For a large number of particles N, there is a rapid crossover (but not a true phase transition) in which N_0, the number of particles in the lowest energy state, dramatically increases. This rapid crossover (which we identify as the BEC "transition") occurs at a critical temperature

$$k_B T_c = \left(\frac{6N}{\pi^2}\right)^{1/2} \hbar\omega. \qquad (18.21)$$

This scale is much larger than the level spacing, indicating that the crossover is not simply a "freezing out" of excitations due to the finite gap to the first excited state.[5]

18.1.3 Experimental Observation of Bose–Einstein Condensation

The only stable bosonic system that remains a liquid (instead of solidifying into a crystal) in the zero-temperature limit is ^{4}He. In fact, below about 2.2 K it becomes a superfluid. As we discuss in the next

---

[5] Further details of the somewhat complex calculations for both the 1D and the 2D case can be found in [197].

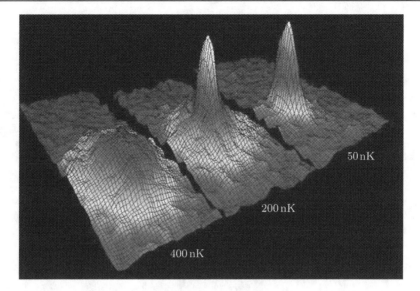

Figure 18.1 Images of the velocity distributions of ultra-cold rubidium atoms. Left frame: temperature just above the Bose–Einstein condensation temperature; center frame, just below the condensation temperature with the appearance of a condensate; right frame, further lowering of temperature, resulting in almost all atoms being in the condensate. Figure reprinted with permission from [198]. Copyright (2002) by the American Physical Society. Photo credit M. Matthews JILA.

section, the superfluidity is a consequence of, and thus an *indirect* indication of, BEC. However, due to the high density and resultant strong interaction in ^{4}He, one cannot readily probe the momentum distribution of the condensate directly.[6]

In 1995 BEC was directly observed in the momentum distribution of dilute gases made of alkali atoms, with a typical density of 10^{13}–10^{15} cm^{-3}. At such a low density the average distance between atoms is ~ 1000 Å or larger, which is much bigger than the size of the atoms, and their interactions can thus be quite small.[7] Such dilute gases are trapped using laser and/or magnetic fields, and cooled to astonishingly low temperatures (down to ~ 100 nK and in some cases even lower[8]) using sophisticated evaporative cooling methods that take advantage of the specific optical and magnetic properties of the atoms. Such temperatures are low enough for BEC to occur even at the low density of the atomic gas mentioned above. The momentum distribution of the atoms can be probed by removing the trap and letting the atoms fly out, and directly imaging the expanding atomic cloud.[9] Results of one of the original experiments are presented in Fig. 18.1, where the onset of BEC and the presence of a condensate are signaled by the appearance of a central peak formed by a large number of atoms that are *stationary*.

---

[6] The distribution can be measured with neutron-scattering techniques which show that only a small fraction of the particles resides in the zero-momentum condensate. The rest of the particles are virtually excited to higher states via particle–particle interactions. Equivalently we may say that the strong variation in the amplitude of the many-body wave function as any one particle moves amongst its fixed neighbors (and tries to avoid them) means that the overlap of that particle's state with the zero-momentum single-particle eigenstate is reduced.

[7] This is not always true. While the range of atomic interaction is of order the atom size of 1 Å, the relevant length scale is actually the scattering length a (defined in Eq. (4.33)) which can be much longer. In particular a diverges at the so-called **Feshbach resonance** where a new two-body bound state forms. There is fascinating many-body physics for strongly interacting bosons near Feshbach resonances, which we will not get into here.

[8] To comprehend just how cold 100 nK is, consider the fact that the thermal energy at this temperature corresponds to Planck's constant multiplied by an audio frequency $f \sim 2.1$ kHz!

[9] Once the cloud has expanded to many times its original radius, the particle positions are determined largely by the initial momenta of the particles rather than by their initial positions.

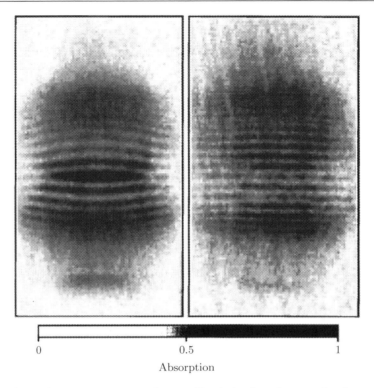

0   0.5   1

Absorption

Figure 18.2 Interference between two atom condensates. Two isolated condensates of sodium atoms are prepared separately and then allowed to expand and overlap. The interference fringes observed in the overlapping region clearly indicate the phase-coherent nature of the condensates. Because the condensates are not in communication prior to the expansion, the phase of the interference pattern varies randomly from shot to shot of the experiment. (See Exercise 18.6 for further discussion.) The two panels correspond to two different experimental settings. Figure reprinted with permission from [199]. Copyright 1997 by the American Association for the Advancement of Science; all rights reserved.

Since all atoms in a condensate are in the same state, a BEC provides a *coherent* source of a matter wave much like a laser. The phase-coherent nature of the condensate is revealed by preparing two *separate* condensates, and then letting them expand and overlap. The resultant interference pattern in the overlapping region as shown in Fig. 18.2 clearly indicates the coherence. Interestingly, because the condensates have not previously been in contact, they do not have a well-defined relative phase, and the interference pattern has a random phase from shot to shot of the experiment. See Exercise 18.6 for further discussion.

It should be emphasized that the atomic gases used in these experiments are in *metastable* states, since, at such low temperatures, the true stable states of alkali atoms are much denser crystalline solids. However, these metastable states have very long lifetimes (of order seconds), since destabilizing processes require collisions that involve at least three atoms,[10] which are extremely rare at such low density. The lifetime of these gases is not only long enough for the atoms to equilibrate via two-body collisions so that the temperature is well-defined and the system obeys the laws of thermodynamics and statistical mechanics, but also for experimentalists to perform the measurements mentioned above. The key experimental challenge with these systems is to keep the density low enough that three-body collisions do not make the lifetime too short, yet high enough that the BEC temperature

---

[10] Owing to conservation of energy and momentum, a two-body collision cannot lead to bound-state formation, and the atoms must therefore stay in the gaseous state. If two particles bind, a third body is required to carry off the binding energy.

can be reached and high enough that two-body collisions occur rapidly enough for the system to reach quasi-equilibrium. For many atoms the optimal density is roughly 10^{13} cm^{-3}.

Exercise 18.6. Consider two ideal BECs, each containing N atoms initially held in small traps located at positions $\vec{R}_\pm = \pm(d/2)\hat{x}$. The traps are released and the atom clouds expand far beyond their original diameters. The particles that reach the overlap region near $x = 0$ at a certain time t then have a relatively well-defined momentum $\vec{k}_\pm = \pm(md/2t)\hat{x}$. A (rather crude) model of the boson state in second quantization (see Appendix J) is

$$|\Psi\rangle = \left(b_{\vec{k}_+}^\dagger\right)^M \left(b_{\vec{k}_-}^\dagger\right)^M |0\rangle, \tag{18.22}$$

where $M \sim \mathcal{O}(N)$ represents the number of particles from each cloud in the overlap region.
- Show that the mean density $\langle n(\vec{r})\rangle$ is uniform and not modulated.
- Show that the density correlation function exhibits spatial modulation $\langle n(\vec{r})n(\vec{r}\,')\rangle - \langle n(\vec{r})\rangle\langle n(\vec{r}\,')\rangle \sim \cos[(2k_+(x - x')]$.
- Explain how to relate the above results to the fact that interference fringes are seen in each shot but the phase of the fringes is random from shot to shot.

18.2 Weakly Interacting Bosons and Bogoliubov Theory

We now consider a system with interacting bosons whose Hamiltonian is

$$H = \hat{T} + \hat{V} = \sum_i \frac{p_i^2}{2m} + \sum_{i<j} v(\vec{r}_i - \vec{r}_j). \tag{18.23}$$

In second quantization (see Appendix J) the Hamiltonian takes the form

$$H = \sum_{\vec{k}} \epsilon_{\vec{k}} b_{\vec{k}}^\dagger b_{\vec{k}} + \frac{1}{2V} \sum_{\vec{k}_1 \vec{k}_2 \vec{k}_3} v_{\vec{k}_1 - \vec{k}_3} b_{\vec{k}_1}^\dagger b_{\vec{k}_2}^\dagger b_{\vec{k}_3} b_{\vec{k}_1 + \vec{k}_2 - \vec{k}_3}, \tag{18.24}$$

where

$$v_{\vec{k}} = \int d\vec{r}\, e^{-i\vec{k}\cdot\vec{r}} v(\vec{r}). \tag{18.25}$$

Very often the interaction is sufficiently short-ranged that we can use an ultra-local approximation to the interaction $v(\vec{r}) = v_0 \delta(\vec{r})$, in which case $v_{\vec{k}} = v_0$ is a constant. We will consider this case for the rest of this chapter unless noted otherwise.

The Hamiltonian (18.24) is not exactly solvable due to the presence of the interaction term $\hat{V}$. To make progress we need to make approximations. The key insight is that, if the interaction is sufficiently weak, most of the bosons will remain in the $\vec{k} = 0$ state. As a result, the most important terms in the second (interaction $\hat{V}$) term of (18.24) are those that involve b_0 and/or $b_0^\dagger$. The first approximation is to keep only terms in $\hat{V}$ that contain four or two such operators (the reader should verify that there is no term in $\hat{V}$ that contains three of them):

$$\hat{V} \approx \frac{v_0}{2V} \left[b_0^\dagger b_0^\dagger b_0 b_0 + 4 b_0^\dagger b_0 \sum_{\vec{k}\neq 0} b_{\vec{k}}^\dagger b_{\vec{k}} + \sum_{\vec{k}\neq 0} (b_0^\dagger b_0^\dagger b_{\vec{k}} b_{-\vec{k}} + b_0 b_0 b_{\vec{k}}^\dagger b_{-\vec{k}}^\dagger) \right]$$

$$\approx \frac{N(N-1)v_0}{2V} + n v_0 \sum_{\vec{k}\neq 0} \left[b_{\vec{k}}^\dagger b_{\vec{k}} + \frac{1}{2N}(b_0 b_0 b_{\vec{k}}^\dagger b_{-\vec{k}}^\dagger + b_0^\dagger b_0^\dagger b_{\vec{k}} b_{-\vec{k}}) \right], \tag{18.26}$$

where we have used the fact that $N = b_0^\dagger b_0 + \sum_{\vec{k} \neq 0} b_{\vec{k}}^\dagger b_{\vec{k}}$, and in the second line we keep only terms proportional to N^2 and N.[11] The first term above is nothing but the Hartree–Fock energy of the state (18.3). The remaining term (in square brackets) consists of three parts. The first simply represents the change in energy of a particle outside the condensate due to its mean-field interaction with the background of other particles (most of which are in the condensate at weak coupling). The second represents the collision of two particles in the condensate and resultant scattering out of the condensate into equal and opposite momenta. The third term is simply the reverse of the second.

Now comes the crucial next approximation. For sufficiently weak interaction we expect

$$N_0 = \langle b_0^\dagger b_0 \rangle \approx N, \tag{18.27}$$

$$N_0 + 1 = \langle b_0 b_0^\dagger \rangle \approx N, \tag{18.28}$$

namely almost all bosons stay in the $\vec{k} = 0$ state. For such macroscopic occupation, the operators $b_0^\dagger$ and b_0 behave essentially as classical variables, and we can thus replace them by a pure number $\sqrt{N}$, which would be consistent with Eq. (18.27) if they were indeed c-numbers (instead of operators). Essentially we are saying that the commutator $[b_0, b_0^\dagger] = 1$ is negligible with respect to $\sqrt{N}$. With this approximation we arrive at a *quadratic* Hamiltonian

$$H_B = \frac{N(N-1)v_0}{2V} + \sum_{\vec{k} \neq 0} \left[(\epsilon_{\vec{k}} + nv_0) b_{\vec{k}}^\dagger b_{\vec{k}} + \frac{nv_0}{2} (b_{\vec{k}} b_{-\vec{k}} + b_{\vec{k}}^\dagger b_{-\vec{k}}^\dagger) \right], \tag{18.29}$$

which is of the same form as the linear spin-wave Hamiltonian (17.103) that we encountered in our study of Heisenberg antiferromagnets. Using the Bogoliubov transformation (Eq. (17.105), which was historically first introduced when studying the present problem by Bogoliubov), the Hamiltonian (18.29) can be brought to diagonal form

$$H_B = \sum_{\vec{k}} E_{\vec{k}} \beta_{\vec{k}}^\dagger \beta_{\vec{k}}, \tag{18.30}$$

with bosonic excitation spectrum

$$E_{\vec{k}} = \sqrt{\epsilon_{\vec{k}}^2 + 2nv_0\epsilon_{\vec{k}}}. \tag{18.31}$$

For large k (such that $\epsilon_{\vec{k}} \gg nv_0$) we have $E_{\vec{k}} \approx \epsilon_{\vec{k}}$, signaling that the excitation is essentially a single boson at momentum $\vec{k}$. In the opposite limit of long-wavelengths we find that

$$E_{\vec{k}} \to \sqrt{2nv_0\epsilon_{\vec{k}}} = \hbar v_s k \tag{18.32}$$

is a linear spectrum with a velocity

$$v_s = \sqrt{nv_0/m}. \tag{18.33}$$

This linear dispersion is qualitatively different from the non-interacting boson case, signaling that the long-wavelength excitations are collective modes (instead of being single-particle-like).

The change of the long-wavelength dispersion from being quadratic to linear can also be understood from a more macroscopic perspective. As we mentioned for free bosons the chemical potential μ is pinned to zero at $T = 0$, resulting in an infinite compressibility (see Eq. (15.162)):

$$\kappa = \frac{1}{n^2} \frac{\partial n}{\partial \mu} = \infty. \tag{18.34}$$

---

[11] Which is *not* the same as keeping terms proportional to N_0^2 and N_0, with $N_0 = \langle b_0^\dagger b_0 \rangle$ being the condensate number!

As a result, the sound velocity vanishes:

$$v_s = \sqrt{1/\kappa\rho} = 0, \tag{18.35}$$

where $\rho = nm$ is the mass density. In the presence of interactions, the ground state energy of the Bose liquid is no longer zero, and at the Hartree–Fock level it is (see Exercise 18.7)

$$E = \frac{N(N-1)v_0}{2V}. \tag{18.36}$$

We thus find

$$\mu = \left.\frac{\partial E}{\partial N}\right|_V = \frac{(N-1/2)v_0}{V} \to nv_0, \tag{18.37}$$

thus

$$\kappa = 1/(n^2 v_0) \tag{18.38}$$

is finite and $v_s = \sqrt{nv_0/m}$ in agreement with Eq. (18.33). We thus find that the long-wavelength collective mode is a sound wave, or phonon mode. This is consistent with the long-wavelength collective-mode spectrum for superfluid ^{4}He shown in Fig. 4.3.[12] The linearly dispersing phonon mode can also be understood as the Goldstone mode associated with the spontaneous breaking of the U(1) symmetry associated with the conservation of particle number, in the same sense that the linearly dispersing spin-wave excitations are the Goldstone modes associated with the spontaneous breaking of rotation symmetry in an antiferromagnet.

Perhaps the best example of a weakly interacting Bose-gas collective-mode spectrum is found in cold-atom systems. Figure 18.3 illustrates the linearly dispersing collective phonon mode at small wave vectors in ^{87}Rb. At larger wave vectors, the mode crosses over from linear to the quadratic dispersion expected for bare single-particle excitations, just as expected in the Bogoliubov theory. There is no hint of any roton minimum associated with density correlations (i.e. a peak in the static structure factor) such as occurs in a strongly interacting system like ^{4}He.

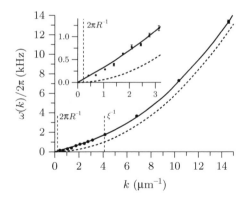

Figure 18.3 Collective-mode dispersion in a ^{87}Rb BEC. The inset shows the linear dispersion at small wave vectors. This crosses over to the quadratic free-particle spectrum at large wave vectors. The interactions are sufficiently weak that there is no roton minimum as in ^{4}He. The measured values (points) obtained from optical Bragg scattering agree very well with Bogoliubov theory derived from the local density approximation. The dashed line is the free-particle spectrum; R is the relevant system size and ξ is the so-called healing length which defines the wave-vector scale at which the dispersion crosses over from linear to quadratic. Figure adapted with permission from [200]. Copyright (2002) by the American Physical Society.

---

[12] However, unlike the Feynman single-mode theory discussed in Section 4.4, neither this analysis nor the Bogoliubov theory is sophisticated enough to yield the roton minimum that occurs in the spectrum at shorter wavelengths when the interaction is strong.

The microscopic Bogoliubov theory, of course, contains much more information. For example it also allows us to calculate the number of particles not in the $\vec{k} = \vec{0}$ state, or those *outside* the condensate (which therefore deplete the condensate):

$$\Delta N = \sum_{\vec{q} \neq 0} \langle b_{\vec{q}}^{\dagger} b_{\vec{q}} \rangle = \sum_{\vec{q} \neq 0} \left(\cosh^2 \theta_{\vec{q}} \langle \beta_{\vec{q}}^{\dagger} \beta_{\vec{q}} \rangle + \sinh^2 \theta_{\vec{q}} \langle \beta_{\vec{q}} \beta_{\vec{q}}^{\dagger} \rangle \right) \tag{18.39}$$

$$= \sum_{\vec{q} \neq 0} \sinh^2 \theta_{\vec{q}} = V \int \frac{d^d \vec{q}}{(2\pi)^d} \sinh^2 \theta_{\vec{q}}. \tag{18.40}$$

We find that ΔN is finite at $T = 0$ for $d > 1$ but diverges for $d = 1$, signaling that the condensate is depleted completely and the BEC is unstable against interaction (which results in quantum fluctuations) in 1D. Generalizing the calculation to finite T, one finds that ΔN diverges for $d = 2$ as well, signaling that the BEC is unstable against thermal fluctuations in 2D.[13] We thus find that, while non-interacting bosons behave very much like Heisenberg ferromagnets, interacting bosons behave like Heisenberg antiferromagnets, in terms of the long-wavelength collective-mode spectrum and the effect of quantum fluctuations on the condensate that gives rise to the order parameter.

Exercise 18.7. Prove $\langle N | H | N \rangle = N(N-1)v_0/2V$, where $|N\rangle$ is defined in Eq. (18.3).

Exercise 18.8. Find the parameter $\theta_{\vec{q}}$ in the Bogoliubov transformation Eq. (17.105) that is appropriate for the Bogoliubov Hamiltonian (18.29), and verify Eq. (18.31).

18.3 Stability of Condensate and Superfluidity

As we have just learned, interactions deplete the condensate somewhat for $d > 1$, but such depletion destroys the condensate completely in 1D. One might thus conclude that interactions are "bad" for BEC. Contrary to this naive expectation, we will in fact show in this section that repulsive interactions *protect* the condensate from fragmentation, namely having macroscopic numbers of bosons occupying multiple (single-particle) states. By the same mechanism they also protect the current of a moving condensate from decaying, thus turning the Bose liquid into a superfluid.

As we learned in Chapter 15, the exchange effect due to the anti-symmetry requirement of the wave function tends to keep fermions of the same spin (and other internal indices) apart, thus reducing the (repulsive) interaction energy. This leads, among other things, to ferromagnetism when the interaction is sufficiently strong. For bosons the wave function must be *symmetric* under exchange; we thus expect an opposite effect, namely they tend to get closer together, and thus *increase* the interaction energy. This is indeed true in most cases, as we can see by examining the wave function of two identical bosons occupying two orthonormal states ϕ and χ:

$$\psi(\vec{r}_1, \vec{r}_2) = \frac{1}{\sqrt{2}} [\phi(\vec{r}_1)\chi(\vec{r}_2) + \phi(\vec{r}_2)\chi(\vec{r}_1)]. \tag{18.41}$$

The probability density of finding the two particles at the same position $\vec{r}_1 = \vec{r}_2 = \vec{r}$ is $2|\phi(\vec{r})\chi(\vec{r})|^2$, which is *twice* that of the case when the two particles are distinguishable, for which we simply have $\psi(\vec{r}_1, \vec{r}_2) = \phi(\vec{r}_1)\chi(\vec{r}_2)$. This factor-of-2 enhancement is consistent with our intuition.[14]

---

[13] We are here ignoring the effect of harmonic traps discussed earlier.

[14] In quantum optics this factor-of-2 enhancement is known as photon bunching and is characteristic of thermal sources as opposed to coherent laser sources (where the photons are "condensed" into the same state) corresponding to BEC, to be discussed below.

There is, however, a very important exception to this enhancement, namely when the two bosons are in the *same* state ϕ. In this case we simply have a product state $\psi(\vec{r}_1, \vec{r}_2) = \phi(\vec{r}_1)\phi(\vec{r}_2)$, which is the same as that for distinguishable particles. Thus two identical bosons can lower their (repulsive) interaction energy by occupying the same state. This conclusion immediately generalizes to many-boson situations.

Such "attraction" in momentum space due to repulsion in real space obviously has a protecting effect on the condensate. There are many situations in which this is important for the stability of the condensate. While it is true that for non-interacting bosons there is a unique ground state with all bosons in the $\vec{k} = 0$ state, having some fraction of the bosons occupying another state (which results in "fragmentation") can cost arbitrarily low energy as long as the magnitude of the other state's momentum is sufficiently small. We can also have situations in which the boson dispersion has multiple degenerate minima, due to band structure or other effects. For non-interacting bosons there is a huge ground state degeneracy (even with *fixed* particle number) in this case, corresponding to the many different ways to distribute the bosons into these minima, which also results in fragmentation. With repulsive interaction there is a macroscopically large energy cost to such fragmentation, forcing the bosons to choose a *single* state to condense into.

Perhaps the most important effect of the repulsive interaction is protecting a moving condensate, or a *supercurrent*. As we learned in Section 8.8, scattering against impurities, phonons, etc. degrades the flow of electrons, as a result of which a voltage gradient is needed in order to maintain an electrical current in a metal. One would expect impurity and other scattering sources to have a similar effect on boson flows. This is indeed the case for *non-interacting* bosons, as can be seen easily from the following analysis.

Consider a *moving* condensate

$$|N\rangle_{\vec{k}} = \frac{1}{\sqrt{N!}}(b_{\vec{k}}^\dagger)^N |0\rangle. \tag{18.42}$$

An inelastic impurity or phonon process can scatter one of the bosons in the condensate to a different state $\vec{q}$, including $\vec{q} = 0$, which is energetically the most favorable (and thus most likely) process. Thus the state (18.42) which carries a current can relax back to the ground state (18.3), by scattering the bosons to the $\vec{q} = 0$ state one-by-one. Now consider the same scattering process in the presence of repulsive interaction, $v_0 > 0$. We find that the energy change of the system when one particle gets scattered to the $\vec{q} = 0$ state is

$$\Delta E_1 = -\frac{\hbar^2 k^2}{2m} + \frac{(N-1)v_0}{V} \approx -\frac{\hbar^2 k^2}{2m} + n v_0. \tag{18.43}$$

The first term above is the reduction of the kinetic energy of the scattered particle, but in addition there is an interaction energy *cost* due to the fact that this particle now has a *different* wave function from all other particles. Overall this results in an energy *barrier* for this scattering process if

$$\hbar k < \sqrt{2mn v_0}, \tag{18.44}$$

or

$$v < v_c = \sqrt{2n v_0/m} = \sqrt{2}v_s, \tag{18.45}$$

where $v = \hbar k/m$ is the condensate velocity, and v_c is the critical velocity at which the energy barrier vanishes. Now continue the scattering process such that the second, third, ... particles get scattered to the $\vec{q} = 0$ state. These result in *additional* energy costs for each scattering

$$\Delta E_1 \approx \Delta E_2 \approx \Delta E_3 \approx \cdots, \tag{18.46}$$

so the energy cost to relax the current carried by the state (18.42) keeps increasing until a finite fraction of the particles are in the $\vec{q} = 0$ state. At this point the total energy cost

$$\Delta E_{\mathrm{tot}} = \Delta E_1 + \Delta E_2 + \Delta E_3 + \cdots \tag{18.47}$$

is proportional to N. This thermodynamically large energy barrier which protects the current from relaxing is essentially the same as the energy cost for condensate fragmentation since, in order for the condensate current to drop, the condensate must fragment. In other words, the state (18.42) is *metastable* as long as Eq. (18.45) is satisfied. As a result of this, the flow cannot decay when the bosons form a condensate, resulting in superfluidity.

We should note that Eq. (18.45) is an overestimate of the critical velocity of the superflow, not only because Eqs. (18.42) and (18.3) are no longer exact in the presence of interactions (for a more correct calculation, see Exercise 18.10), but more importantly because there are energetically less costly processes (that may involve vortices) than those considered here, which can degrade the supercurrent. These were briefly mentioned in Section 4.4, and closely related processes that degrade a supercurrent in a superconductor will be discussed in Section 19.7.

As the reader must have appreciated by now, solving a quantum-mechanical problem that involves many particles is hard, especially when interactions are present. Even if one is handed a wave function, it is not always easy for one to figure out what kind of state/phase it actually describes. A good example of this is the Laughlin wave function we encountered in Chapter 16; without crucial insight from the plasma analogy, one would not know whether it describes an electron liquid or solid (Wigner crystal). The situation is much better when we have a condensate. This is especially true in the extreme case when *all* particles are in the condensate as in Eqs. (18.3) and (18.42), where the (single-particle) condensate wave function completely determines the many-particle wave function. Even in the more generic case where there is depletion due to either interaction or finite temperature, as long as a condensate is present, many important physical properties of the system are determined by the condensate wave function. It is thus possible to develop *macroscopic* descriptions of the system in terms of the condensate wave function, which is not only much simpler than the *microscopic* descriptions in terms of the many-particle wave function, but physically more transparent as one can rely on one's intuition from *single-particle* quantum mechanics. The **Ginzburg–Landau theory** for superconductivity, which will be developed in Chapter 19, is precisely based on the presence of a condensate wave function for Cooper pairs.[15] As a warm-up, let us compare the condensate wave functions of (18.3) and (18.42), which are (note the normalization is to N instead of 1, so that $|\psi(\vec{r})|^2$ is the local density)

$$\psi_0(\vec{r}) = \sqrt{N/V} \tag{18.48}$$

and

$$\psi_{\vec{k}}(\vec{r}) = \sqrt{N/V}\, e^{i\vec{k}\cdot\vec{r}}. \tag{18.49}$$

They have the same magnitude but differ in the $\vec{r}$-dependence of the phase (the overall phase is arbitrary as in quantum mechanics). It is the difference in the ($\vec{r}$-dependence of the) phase that dictates the difference in their physical properties; for example the local current density

$$\vec{J}(\vec{r}) = \frac{\hbar}{2m}\{\psi^*(\vec{r})[-i\,\nabla\psi(\vec{r})] + \psi(\vec{r})[i\,\nabla\psi^*(\vec{r})]\} = \frac{\hbar}{m}|\psi(\vec{r})|^2[\nabla\varphi(\vec{r})] \tag{18.50}$$

is zero for ψ_0 but finite for $\psi_{\vec{k}}$. In Eq. (18.50) $\varphi(\vec{r})$ is the phase of $\psi(\vec{r})$. We will encounter its generalization to non-uniform situations and, in particular, when the condensed particles carry charge in Chapter 19, where $|\psi(\vec{r})|^2$ corresponds to the local superfluid density and $(\hbar/m)\nabla\varphi(\vec{r})$ is its velocity.

---

[15] The reader may find the comments above similar to the motivation behind density functional theory, which also attempts to describe the system in terms of a single-variable function. A crucial difference is that the density functional theory applies only to the ground state, while the description based on the condensate wave function also applies to excited and non-equilibrium states.

Exercise 18.9. Verify Eqs. (18.43) and (18.46), and find the maximum energy barrier in Eq. (18.47).

Exercise 18.10. This exercise illustrates how to determine the critical superfluid velocity from the excitation spectrum, without using approximate states like those of Eqs. (18.42) and (18.3). Consider a generic many-particle system moving with (average) velocity $\vec{v}$, and write the Hamiltonian in the comoving reference frame in which the system has zero (average) velocity:

$$H_{\text{moving}} = \sum_i \frac{(\vec{p}_i + m\vec{v})^2}{2m} + \sum_{i<j} v(\vec{r}_i - \vec{r}_j) \qquad (18.51)$$

$$= H_{\text{lab}} + \vec{P}_{\text{tot}} \cdot \vec{v} + \text{constant}, \qquad (18.52)$$

where H_{lab} is the original Hamiltonian of Eq. (18.23) in the rest (or lab) frame, and $\vec{P}_{\text{tot}} = \sum_i \vec{p}_i$ is the total momentum of the system in the comoving frame, which is a conserved quantity in the absence of any external potential, or in the presence of Galilean invariance. (i) Show that the eigenstates of H_{moving} and H_{lab} are identical, and calculate the difference of the corresponding eigenenergies. (ii) Assume a single scattering process creates only a single elementary excitation in a superfluid whose spectrum is Eq. (18.31), and show that the critical superflow velocity v_c is v_s. (iii) Now consider the complicated spectrum that includes a roton minimum as illustrated in Fig. 4.3. Find v_c in this case. It should be noted that Galilean invariance (which is only weakly broken by scattering processes) plays a crucial role in the analysis here (originally due to Landau). In real systems, Galilean invariance is strongly broken by, e.g., the container that confines the fluid. As a consequence Landau's result provides only an upper bound of v_c.

18.4 Bose–Einstein Condensation of Exciton-Polaritons: Quantum Fluids of Light

As we learned in our study of semiconductors in Chapter 9, excitons are charge-neutral bound states of an electron in the conduction band and a hole in the valence band. Because the effective mass of the exciton is much less than that of ordinary atoms (and the de Broglie wavelength therefore much larger), one might imagine that excitons would be ideal candidates for observing Bose–Einstein condensation at elevated temperatures. It turns out, however, that they are rather sensitive to the presence of disorder and tend to become trapped at defect sites where they often annihilate through non-radiative recombination. These effects make it difficult to see any clear signatures of exciton BEC.

However, as we also learned in Chapter 9, excitons can hybridize with photons to form coherent composite excitations known as polaritons. These collective excitations arise because excitons can spontaneously annihilate (the electron and hole recombine) to emit a photon, but the photon can be reabsorbed to create an exciton. This coherent see-sawing back and forth between photon and exciton is illustrated schematically in Fig. 9.6(a). The net effect is that, in a uniform bulk sample, polaritons can move extremely rapidly to the edge of the sample and escape into free space as photons. This is of course a problem for trying to observe BEC. On the other hand, the hybridization of the exciton with the photon does turn it into a more delocalized object which is less likely to be sensitive to local impurity potentials. What we need is some way to maintain the exciton-photon hybridization but still keep the polariton from escaping the sample. Also, we need some way to drastically change the dispersion relation in Fig. 9.6(b) so that the collective mode has a well-defined (and very small) effective mass. It turns out that there is a beautiful way to do all of this at the same time, simply

QW

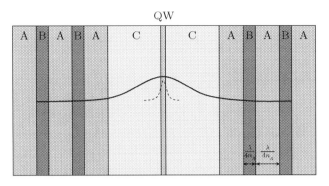

Figure 18.4 Simplified schematic illustration of a quantum well (QW) for confining exciton modes (dotted line) to 2D surrounded by spacer layers (C) and two distributed Bragg reflector (DBR) reflectors defining a $\lambda/2$ optical microcavity to confine the photon modes (solid line) to the same 2D region. The DBRs consist of alternating layers with different indices of refraction, n_A, n_B. The physical thicknesses of the A and B layers differ, but each has optical thickness equal to one-quarter of the relevant optical wavelength. Note that the actual electromagnetic mode structure is more complex and contains rapid oscillations not captured by the smooth envelope function depicted here. See [201] for a more detailed discussion.

by clever sample design. Experimentalists can not only create a "quantum fluid of light," but also detect its superfluid flow around obstacles and observe collective density oscillations ("sound waves of light"). See [201, 202] for reviews of this topic.

Figure 18.4 illustrates an edge view of a 2D quantum well (B) similar to those used to create 2DEGs for the quantum Hall effect, which has a smaller band gap than the surrounding A layers. This well therefore can confine optical excitons to the center of the structure, but leaves them free to move within the 2D plane. The differing band gaps in the A and B layers also causes their indices of refraction, n_A, n_B, to differ. Recall from our discussion of photonic crystals in Section 7.9 that one can take advantage of this to create a distributed Bragg reflector (DBR) which generates an optical cavity which confines the relevant photon modes to the 2D region surroundng the exciton layer. This dramatically enhances the exciton-photon hybridization and drastically reduces the ability of the resulting polaritons to escape (as photons) in the perpendicular direction. At the same time, this perpendicular confinement drastically alters the photon dispersion as we shall now see.

The confinement of the photon mode by the DBRs is not literally equivalent to hard-wall (i.e. vanishing) boundary conditions on the electromagnetic fields. However, to an excellent approximation we can replace the actual boundary conditions by the requirement that the photon wave vector has the form $\vec{k} = (k_x, k_y, k_\perp)$, where

$$k_\perp = \frac{\pi}{W}. \tag{18.53}$$

Here W is a phenomenological parameter representing an effective distance between the mirrors as if they imposed vanishing boundary conditions and were separated by half of a wavelength at the fundamental resonance frequency of the cavity formed by the DBRs. Loosely speaking, W represents the fact that the photon mode penetrates some distance into the DBRs as an evanescent wave. The photon dispersion then has the form

$$\omega = \frac{c}{\bar{n}}|\vec{k}| = \frac{c}{\bar{n}}\sqrt{k_x^2 + k_y^2 + k_\perp^2}, \tag{18.54}$$

where $\bar{n}$ is some effective mean index of refraction for the layered medium. For fixed $k_\perp$ this looks like the energy of a 2D massive relativistic particle. Expanding for small $\vec{k}_\parallel \equiv (k_x, k_y)$ gives for the energy in the non-relativistic limit

$$\hbar\omega = \hbar\omega_0 + \frac{\hbar^2 k_\parallel^2}{2M^*}, \tag{18.55}$$

where

$$\omega_0 = \frac{c}{n} k_\perp,$$

(18.56)

and the effective mass is given by

$$M^* = \frac{\bar{n}\hbar k_\perp}{c} = \frac{\bar{n}^2}{c^2} \hbar \omega_0.$$

(18.57)

Thus the confinement of the photon in the perpendicular direction gives it a "rest mass" and allows for the possibility that it propagates slowly (or not at all) in the other two directions.

How big is the rest mass of the photon in this geometry? It is convenient to compare it with the rest mass m_e of a (bare) electron:

$$\frac{M^*}{m_e} = \bar{n}^2 \frac{\hbar \omega_0}{m_e c^2}.$$

(18.58)

For later purposes we will want to choose the cavity resonance energy $\hbar \omega_0$ to be near the exciton energy and thus near (but below) the semiconductor band gap. This yields

$$\frac{M^*}{m_e} \sim 10^2 \frac{1.5\,\text{eV}}{0.5 \times 10^6\,\text{eV}} \sim 3 \times 10^{-4}.$$

(18.59)

Thus the effective mass is extremely low, which is a huge advantage in seeking to observe BEC.

Photons by themselves are essentially a non-interacting Bose gas. Strictly speaking, an ideal-gas BEC does not occur in 2D. However, what we really want to observe is superfluidity, which can occur in 2D for a repulsively interacting Bose fluid. We can achieve the desired repulsive interactions by hybridizing the photons with excitons to form polaritons. Excitons interact with each other via a combination of Coulomb interactions and Pauli exclusion, leading (at least at short distances) to an effective repulsive interaction which the polaritons will partially inherit through the hybridization. Because of the presence of the rest mass of the photons caused by the spatial confinement between the DBRs, the spectrum and the associated hybridization are very different than those depicted in Fig. 9.6(b). Figure 18.5 depicts the spectrum and hybridization when the photon rest energy is taken to be equal to the exciton energy. In contrast to the dispersion curves for polaritons moving freely in 3D shown in Fig. 9.6(b), in this case both the **upper and the lower polaritons** have a well-defined minimum in their energy at $k_\parallel = 0$ and the curvature about the minimum corresponds to an effective

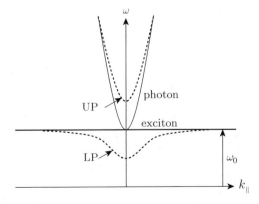

Figure 18.5 The energy spectrum for a confined photon moving in 2D with wave vector $k_\parallel$ and having rest energy tuned to be equal to the exciton energy. The confinement creates a photon rest mass that is orders of magnitude smaller than the exciton mass, and so the exciton dispersion curve looks flat on this scale. The photon and exciton modes hybridize, which lifts their $k_\parallel = 0$ degeneracy. The resulting upper (UP) and lower (LP) polaritons are weakly repulsively interacting bosons with masses orders of magnitude smaller than the exciton mass.

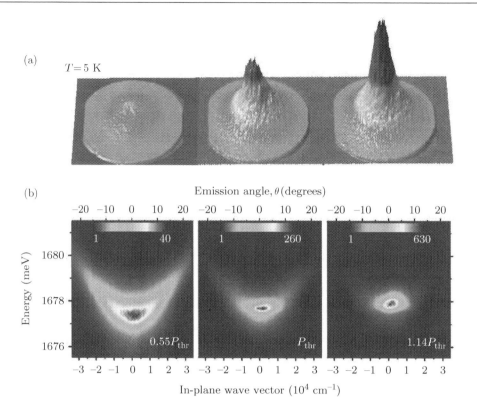

Figure 18.6 Experimental evidence for Bose–Einstein condensation of polaritons at temperature $T \approx 5$ K. (a) The quasi-equilibrium momentum distribution of polaritons at three different densities. The momentum vector $\vec{k}_\parallel = (k_x, k_y)$ is determined by the angle of emission of photons, which slowly escape from the condensate through the DBR, whose reflectivity is slightly less than unity. Note the remarkable similarity to the corresponding data for a cold-atom system at $T \sim 100$ nK shown in Fig. 18.1. (b) Direct display of the polariton population at different points on the dispersion curve. The energy is determined by the frequency of photons emitted from the condensate through the DBR. The momentum is determined from the angle of emission relative to the normal to the 2D system. Figure reprinted from [203] by permission from Springer Nature. Copyright (2006).

mass orders of magnitude smaller than the exciton mass (which is given by the reduced mass of the combined band mass of the electron and hole).

Experiments are generally performed on the lower polariton. While the number of polaritons is not strictly conserved, the DBR cavity makes the polariton lifetime sufficiently long that they can equilibrate. Figure 18.6 presents strong evidence that the lower polariton fluid can be maintained in steady-state quasi-equilibrium by laser pumping to create excitations in the upper polariton state which can relax via phonon emission into the lower polariton mode, which then thermalizes due to repulsive interactions. The data shows very strong evidence for BEC, with momentum distributions of the particles strikingly similar to those for cold atoms in Fig. 18.1. The fact that this can be achieved at $T \approx 5$ K instead of at sub-microkelvin scales is testament to the benefits of the extremely small polariton mass under 2D confinement. It is also possible to coherently pump the lower polariton branch, and, if the pumping laser beam is tilted away from the surface normal of the sample, a BEC with finite momentum can be created, allowing studies of dissipationless superfluid flow around impurity defects. This "superfluid of light" exhibits collective **Bogoliubov density modes** that travel at the "**speed of sound of light**," a novel concept if there ever was one! Readers interested in further details should consult reviews on the subject [201, 202].

19 Superconductivity: Basic Phenomena and Phenomenological Theories

Superconductivity was discovered in 1911 in the laboratory of H. Kamerlingh Onnes in Leiden, where it was found that the electrical resistivity of mercury completely disappears below a critical temperature $T_c \approx 4.19$ K. Kamerlingh Onnes won the Nobel Prize in 1913 for this discovery. Like the various quantum Hall effects and superfluidity, superconductivity is a remarkable many-body phenomenon in which quantum mechanics plays a fundamental role even at macroscopic scales.

Among the many (potential) applications of materials with zero electrical resistance are superconducting magnets which operate without dissipative energy loss. It was many decades, however, before such devices could be constructed, because of the peculiar nature of superconductivity in the presence of magnetic fields. Even though there is no Joule heating, it still costs energy to maintain the magnet at cryogenic temperatures. The discovery in 1987 of high-temperature superconductors which operate above the temperature of liquid nitrogen offers hope for much reduced refrigeration costs, but these are difficult materials with which to work and do not behave well in the presence of high magnetic fields.

In this chapter we discuss the basic phenomena exhibited by superconductors, and develop phenomenological (or effective) theories that capture the essence of superconductivity at the macroscopic level. A discussion of the microscopic theory of superconductivity is deferred to Chapter 20.

19.1 Thermodynamics

In 1933 Meissner and Ochsenfeld discovered that weak magnetic fields are expelled by superconductors. This **perfect diamagnetism**, now known as the **Meissner effect**, is intimately related to the perfect conductivity, as can be seen from the Maxwell equation

$$\nabla \times \vec{E} = -\frac{1}{c}\frac{\partial \vec{B}}{\partial t}. \tag{19.1}$$

In a perfect conductor, the electric field must vanish (at least in the limit of zero frequency where the inertia of the carriers can be neglected) since otherwise infinite current would flow. But this means that the magnetic flux density cannot change with time. Thus, if a perfect conductor at $B = 0$ is subjected to an applied field, the flux is unable to penetrate into the sample. Conversely, if a perfect conductor already contains a finite flux density, it cannot be removed.

It is indeed observed that a superconductor initially at $B = 0$ does not allow the penetration of a weak applied field. However, superconductivity is much more than simply perfect conductivity. A sample in the normal state at high temperatures and having an applied field does something very strange when it is cooled below the critical temperature for the onset of superconductivity. The flux does not stay trapped inside the sample as would be expected for the case of perfect conductivity. Instead, the flux is actually expelled from the sample. Thus the system arrives at the same final state irrespective of whether the field is applied before or after the temperature is lowered. This tells us that a superconductor with the flux expelled is in an equilibrium thermodynamic phase. It costs free energy to expel the magnetic flux against the applied magnetic pressure. Nevertheless, the system is willing to pay this price because it can gain the energy of condensation into the superconducting state. As we will shortly see, magnetic flux tends to frustrate this condensation, and that is why the system finds it favorable to expel the flux. The thermodynamic phase in which the electrons are superconducting and the magnetic flux has been expelled is known as the **Meissner phase**.

To put all this another way, a superconductor in equilibrium in a weak applied field H has a flux density $B(H, T)$ which is a *thermodynamic state function*. That is, B depends only on the current values of H and T, and is independent of the past history of the sample. In a perfect conductor the flux density depends on the past history and so is not a state function. The perfect conductor can be considered to be not in thermodynamic equilibrium because the relaxation time of the carriers is infinite. It turns out that the superconductor has finite conductivity while it is in the process of expelling the flux and does not achieve perfect conductivity until the flux has been fully expelled. This allows the flux to change with time and hence be expelled.[1]

The way that the sample excludes the magnetic flux is to generate currents at the surface which produce an opposing magnetic flux. The flux density does not fall to zero discontinuously inside the superconductor since this would require infinite current density on the surface. Rather the magnetic flux density goes to zero exponentially with a characteristic length λ known as the **penetration depth**. In typical superconductors this length lies in the regime 500–2000 Å at zero temperature and diverges as the temperature approaches T_c. Hence the diamagnetism is perfect only for macroscopic samples much larger than the penetration depth in size. These issues will be discussed in great detail in later sections.

19.1.1 Type-I Superconductors

If a sufficiently strong magnetic field is applied, the superconductivity is partially or completely destroyed. In so-called **type-I superconductors**, fields exceeding the thermodynamic critical field $H_c(T)$ completely destroy the superconductivity,[2] as illustrated in the phase diagram shown in Fig. 19.1(b). At the critical field the Meissner and normal phases are in equilibrium with each other. Hence the free-energy difference between the two phases can be found from[3]

$$F_S(T) + \frac{H_c(T)^2}{8\pi} = F_N(T), \tag{19.2}$$

---

[1] Because the flux expulsion is a first-order phase transition in which the magnetization changes discontinuously, hysteresis is sometimes observed, in which some of the flux gets trapped inside and fails to escape. It is also possible in small-size samples for supercooling to occur, in which the transition to the superconducting state does not occur until the sample has been cooled well below the equilibrium critical temperature. In both of these cases the superconductor is not in true equilibrium.

[2] The statements we are making here are in the context of mean-field theory, which is generally quite accurate in the low-temperature superconductors. However, in the high-temperature materials, fluctuation corrections to mean-field theory can be quite important and lead to a rich variety of new phenomena.

[3] See Eq. (19.66) and surrounding discussions for this.

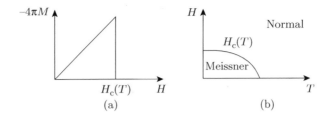

Figure 19.1 (a) Magnetization vs. applied field. (b) The phase diagram for a type-I superconductor.

where $F_{S(N)}$ is the free-energy density of the superconducting (normal) state in the absence of the applied field, and $H_c^2(T)/8\pi$ is the energy density cost to expel the field from the superconductor against the applied magnetic pressure.[4] Because the entropy is given by the (negative of the) temperature derivative of the free energy, the entropy change across the phase boundary is given by

$$S_S - S_N = \frac{H_c(T)}{4\pi} \frac{d H_c(T)}{dT}. \tag{19.3}$$

This is the analog of the **Clausius–Clapeyron equation** familiar from the theory of liquid–vapor phase transitions. As can be seen in Fig. 19.1(b), the temperature derivative of $H_c(T)$ is negative, implying that the entropy of the superconducting state is lower than that of the normal state. This is consistent with the idea that the electrons condense into a more ordered quantum state in the superconducting phase.

The entropy of the electrons in a normal metal is associated with the high density of electron–hole pair excitations around the Fermi surface. Well before the microscopic theory of superconductivity was developed there was considerable experimental evidence that the condensation of the electrons into the superconducting state was associated with the existence of a finite excitation gap. The presence of this gap means that the density of thermally excited (quasi)particle–hole pairs is much lower. It is this effect which accounts for the reduced entropy of the superconducting state. Experimental signatures of this gap can be obtained from the much reduced specific heat at low temperatures, the greatly reduced attenuation of ultrasonic waves transmitted through a superconductor, and the existence of a minimum frequency threshold for absorption of far-infrared radiation. It was also found that the thermopower of superconductors vanishes, strongly suggesting that the entropy transported by supercurrents vanishes.[5]

As shown in Fig. 19.1(b), the critical field starts out at zero at $T = T_c$, and so the superconducting phase transition at $H = 0$ is a continuous transition with no latent heat, while the transition at $H > 0$ is of first order. Using

$$B = H + 4\pi M, \tag{19.4}$$

we see that the magnetization of a type-I superconductor is given by

$$M(H, T) = -\frac{H}{4\pi}\theta[H_c(T) - H], \tag{19.5}$$

as illustrated in Fig. 19.1(a). Here θ is the step function. Note that the resulting discontinuity in M when H is finite is consistent with the first-order nature of the transition.

---

[4] We assume here that the free energy of the normal state is independent of H.

[5] See Exercise 8.9. This indicates that it is the (entropyless) condensate that carries the current, rather than the thermally excited quasiparticles as would be the case in an insulator or semiconductor (which also has a gap). There (see Section 9.1) the thermopower actually *diverges* in the low-temperature limit because the entropy carried by an *individual* quasiparticle/hole is huge, despite the fact the total entropy of the system (like that in a superconductor) is much lower than that in the normal metal due to the gap.

In many materials the empirical form

$$H_c(T) = H_c(0)\left[1 - \left(\frac{T}{T_c}\right)^2\right] \tag{19.6}$$

is approximately correct. Typical values for $H_c(0)$ are less than 10^3 Oe. Thus type-I superconductors do not make good candidates for magnet wire.

Exercise 19.1. Use the Clausius–Clapeyron result to find the discontinuity in the specific heat at the zero-field phase transition point in terms of the temperature derivative of the thermodynamic critical field.

19.1.2 Type-II Superconductors

Type-II superconductors behave somewhat differently. They also exhibit perfect diamagnetism up to a critical field $H_{c_1}(T)$, known as the lower critical field. Above this field the magnetization decreases continuously rather than suddenly. Magnetic flux begins to penetrate the sample (producing what is known as the "mixed state"), but does not destroy the superconducting state until the so-called upper critical field $H_{c_2}(T)$ has been reached, as illustrated in Fig. 19.2(a). This field can be as large as 10^5–10^6 Oe. The generic (mean-field-approximation) phase diagram for a type-II material is shown in Fig. 19.2(b). The transitions at both $H_{c_1}(T)$ and $H_{c_2}(T)$ are continuous or of second order, as indicated by the continuous change of magnetization there.

Before such materials can be used to construct magnets, however, a new problem must be overcome. In the mixed state the magnetic flux penetrates the sample in discrete bundles containing one superconducting flux quantum[6]

$$\Phi_S \equiv \frac{hc}{2e} \approx 2.068 \times 10^{-7} \text{ gauss cm}^2. \tag{19.7}$$

It is very important to note that, unlike the flux quantum Φ_0 defined previously in our study of the Aharaonov–Bohm effect (see Eq. (10.2)), the superconducting flux quantum is defined with a factor of 2 in front of the electron charge because the superconducting state involves a condensation of pairs of electrons. As a result[7]

$$\Phi_S = \Phi_0/2. \tag{19.8}$$

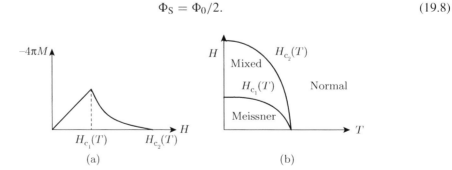

Figure 19.2 (a) Magnetization vs. applied field and (b) the phase diagram for a type-II superconductor.

---

[6] In SI units $\Phi_S = h/2e \approx 2.068 \times 10^{-15}$ T m^2, which is equivalent since one tesla is 10^4 gauss.

[7] In the remainder of the book we often simply refer to Φ_S as the flux quantum, as is often done in the literature. The reader should be careful in distinguishing between the (superconducting) flux quantum Φ_S and the flux quantum Φ_0 depending on the context.

These flux bundles tend to move and dissipate energy when a current is flowing. Thus the mixed state is not truly superconducting. Paradoxically, if the material is made less ideal by alloying or mechanical work hardening, the resulting defects can pin the flux bundles in place, thereby eliminating the flux flow dissipation (for current densities less than some critical value J_c) and restoring the superconductivity. It took many decades of research before this physics was understood and high-field magnet wire could be developed. The high-temperature superconductors have extremely high upper critical fields which might allow the construction of very powerful magnets. Unfortunately, the same physics that increases the upper critical field is also responsible for very low critical current densities, for reasons that will be made clear later.

Our first task is to develop a phenomenological model of the unusual electrodynamic properties of superconductors. We will then consider the microscopic origins of this phenomenology in the subsequent chapter.

19.2 Electrodynamics

In 1935 the brothers F. and H. London examined the electrodynamics of superconductors in an attempt to find some hints about the underlying physics. They assumed a two-fluid model in which supercurrents

$$\vec{J} = -en_s\vec{v}_s \tag{19.9}$$

are carried by a fraction of the electrons with density $n_s(T)$ moving with velocity $\vec{v}_s$. These electrons undergo free undamped acceleration in the presence of an electric field:

$$\frac{d}{dt}\vec{J} = \frac{n_s e^2}{m_e}\vec{E}. \tag{19.10}$$

Combining this with the Maxwell equation

$$\nabla \times \vec{E} = -\frac{1}{c}\frac{\partial \vec{B}}{\partial t}, \tag{19.11}$$

we obtain

$$\frac{\partial}{\partial t}\left(\nabla \times \vec{J} + \frac{n_s e^2}{m_e c}\vec{B}\right) = \vec{0}. \tag{19.12}$$

This equation is often used in plasma physics to describe the dynamics of mutual motion of matter and magnetic fields in highly conducting plasmas. Note, however, that, in the static limit where both the current and the magnetic field are constant, this equation places no constraints on either quantity. The London brothers realized that the Meissner effect implies that superconductors are more than simply perfect conductors. They proposed that, for some unknown reason, superconductors obey a more restrictive equation[8]

$$\left(\nabla \times \vec{J} + \frac{n_s e^2}{m_e c}\vec{B}\right) = \vec{0}, \tag{19.13}$$

now known as the **London equation**. Combining the London equation with the (microscopic) Maxwell equation

$$\nabla \times \vec{B} = \frac{4\pi}{c}\vec{J} + \frac{1}{c}\frac{\partial \vec{E}}{\partial t} \tag{19.14}$$

---

[8] This form of the equation gives a purely local relation between the current and the magnetic field. We will later generalize this to allow for a non-local response.

yields (in the static limit)

$$\nabla \times (\nabla \times \vec{B}) = \nabla(\nabla \cdot \vec{B}) - \nabla^2 \vec{B} = -\lambda^{-2} \vec{B}, \tag{19.15}$$

where

$$\lambda \equiv \left(\frac{4\pi n_s e^2}{m_e c^2} \right)^{-\frac{1}{2}} \tag{19.16}$$

is known as the **London penetration depth**. Using the fact that the magnetic field has no divergence, we see that the Meissner effect follows directly because the magnetic field is exponentially damped inside the superconductor. If, for example, we consider a superconductor filling the half space $x > 0$ and having a uniform field $B_0 \hat{z}$ outside, the solution of Eq. (19.15) inside the superconductor is

$$\vec{B}(\vec{r}) = B_0 \hat{z} e^{-x/\lambda}. \tag{19.17}$$

Thus the field penetrates only a small distance λ into the sample. Using Eq. (19.14), we see that the screening currents are similarly confined to the surface of the sample,

$$\vec{J} = \frac{c}{4\pi\lambda} B_0 \hat{y} e^{-x/\lambda}. \tag{19.18}$$

Returning to Eq. (19.13), we see that it can be re-written in terms of the vector potential as

$$\left(\nabla \times \vec{J} + \frac{n_s e^2}{m_e c} \nabla \times \vec{A} \right) = \vec{0}. \tag{19.19}$$

The solution of this equation gives the so-called **London equation** of the form

$$\vec{J} = -\frac{n_s e^2}{m_e c} \left(\vec{A} + \nabla\phi \right), \tag{19.20}$$

where ϕ is an as yet unknown scalar field. In the static limit, the continuity equation reduces to $\nabla \cdot \vec{J} = 0$. This then tells us that ϕ is determined by the solution of the equation

$$\nabla^2 \phi = -\nabla \cdot \vec{A}. \tag{19.21}$$

It is sometimes convenient to choose the so-called **London gauge** in which $\nabla \cdot \vec{A} = 0$, in which case we can choose[9] $\phi = 0$. In a general gauge, we simply have $\nabla\phi = \vec{A}_{\text{London}} - \vec{A}$.

F. London suggested that the London equation arises because of a "rigidity" of the superconducting state. Recall that the velocity operator in quantum mechanics is proportional to the mechanical momentum,[10]

$$\vec{v} = \frac{1}{m_e} \left(\vec{p} + \frac{e}{c} \vec{A}(\vec{r}) \right). \tag{19.22}$$

As discussed in Chapters 10 and 12, the canonical momentum is not a physical observable and is not gauge-invariant. Application of a longitudinal vector potential is simply a gauge change which changes the wave function and therefore the expectation value of the canonical momentum (see Box 19.1 for a discussion of this point). Hence the wave function cannot be "rigid" with respect to longitudinal gauge fields. On the other hand, if we assume that the wave function does not change when a weak transverse gauge field is applied, we can recover the London equation. In the ground state in the absence of a magnetic field (where we can make the gauge choice $\vec{A} = \vec{0}$) we expect to

---

[9] We have to be a little careful when we say "the" London gauge. The constraints that $\nabla \times \vec{A}$ is a specified function and $\nabla \cdot \vec{A} = 0$ do not uniquely specify $\vec{A}$. This is because we can add to $\vec{A}$ a term $\nabla\varphi$ subject to the constraint $\nabla^2 \varphi = 0$. This constraint does not enforce $\varphi = 0$. If, however, we put in a physical constraint that the current vanish at infinity, then $\vec{A}$ and hence $\nabla\varphi$ must vanish at infinity, and φ is uniquely determined up to an irrelevant overall constant.

[10] For simplicity we restrict our discussion and formulas here to one-body quantum mechanics.

have $\langle \vec{p} \rangle = \vec{0}$. If the state remains unchanged ("rigid") upon application of a weak transverse gauge field ($\nabla \cdot \vec{A} = 0$) then the velocity operator becomes

$$\vec{v} = \frac{e}{m_e c} \vec{A}(\vec{r}). \tag{19.23}$$

Combining this with Eq. (19.9) yields the London equation Eq. (19.20).

Box 19.1. Transverse vs. Longitudinal Vector Potential

In general a vector function $\vec{A}(\vec{r})$ can be decomposed into a sum over a transverse part $\vec{A}_{\text{trans}}(\vec{r})$ and a longitudinal part $\vec{A}_{\text{long}}(\vec{r})$:

$$\vec{A}(\vec{r}) = \vec{A}_{\text{trans}}(\vec{r}) + \vec{A}_{\text{long}}(\vec{r}),$$

with

$$\nabla \cdot \vec{A}_{\text{trans}}(\vec{r}) = 0; \qquad \vec{A}_{\text{long}}(\vec{r}) = \nabla \chi(\vec{r}).$$

The distinction between them is perhaps clearer in Fourier space:

$$\vec{q} \cdot \vec{A}_{\text{trans}}(\vec{q}) = 0, \qquad \vec{A}_{\text{long}}(\vec{q}) = i\vec{q}\chi(\vec{q}),$$

namely $\vec{A}_{\text{trans}}(\vec{q}) \perp \vec{q}$ (thus transverse) and $\vec{A}_{\text{long}}(\vec{q}) \parallel \vec{q}$ (thus longitudinal).

Coming back to the case $\vec{A}(\vec{r})$ being the vector potential, we find the corresponding magnetic field

$$\vec{B}(\vec{r}) = \nabla \times \vec{A}(\vec{r}) = \nabla \times \vec{A}_{\text{trans}}(\vec{r}),$$

which depends on $\vec{A}_{\text{trans}}(\vec{r})$ *only*. Thus $\vec{A}_{\text{long}}(\vec{r})$ has no physical content and is a pure gauge, whose effect can be canceled out by a proper choice of the wave-function phase. The London gauge is nothing but the transverse gauge in which $\vec{A}_{\text{long}} = \vec{0}$.

A somewhat special case is a constant vector potential $\vec{A}(\vec{r}) = \vec{A}_0$, which is both transverse and longitudinal. This ambiguity is related to the fact that it has no Fourier amplitudes at non-zero wave vectors. Since the corresponding magnetic field is zero (and thus has no physical effect), we normally view it as a pure gauge (or longitudinal) as well.

From the point of view of perturbation theory, one might imagine that it is possible that wave-function rigidity is equivalent to having a finite excitation gap in the energy spectrum. The argument is subtle, however, because we know that ordinary insulators are indeed weakly diamagnetic, but do not have the divergent diamagnetic susceptibility that we will see shortly in Eq. (19.37). The key point is that the superconductor may have a finite gap[11] for all single-particle excitations but not for a uniform collective current flow. The insulator has a gap that is produced (at least in part) by the periodic potential of the lattice and does not permit current flow.

Notice also that the Meissner effect cannot be explained by postulating that the electrons lock rigidly together into a crystalline state in which they all move together because the London equation tells us that the shear flow (the curl of $\vec{J}$) of the screening currents near the surface of the sample is finite. Thus the superconducting condensate must be a liquid capable of supporting shear.

Within the two-fluid picture, the superfluid behaves as if it has a rigid wave function while the normal fluid component does not. We will see later that the normal fluid component behaves much like the electrons in an ordinary metal, where it turns out that that paramagnetic current due to the non-rigidity of the wave function (that is, the part of the current related to $\langle -(e/m_e)\vec{p} \rangle$) precisely cancels

---

[11] In fact there exist so-called gapless superconductors in which there is *no* gap for single-particle excitations; see Chapter 20 for examples. But such gaplessness does not affect the statements here.

out the diamagnetic current $-(e^2/m_e c)\vec{A}$ at long wavelengths. It is this cancelation that prevents the Meissner effect in the normal state. Working through Exercises 10.1 and 19.2 is a good way to develop some physical intuition about paramagnetic and diamagnetic currents.

The reader has probably noticed that the London theory has a very different flavor from theories discussed earlier in this book. In earlier chapters we always start with a Hamiltonian that describes a specific system, solve it (with approximations if necessary), and deduce the physical properties of the system from the solutions. Here the procedures are reversed: we start from *known* physical properties (or experimental phenomena) of the system (the Meissner effect here), deduce appropriate equation(s) (the London equation in this case) that lead to such phenomena, and then try to deduce more properties from the phenomenological equations. The London theory is thus a typical example of such *phenomenological* (in contrast to *microscopic* or *ab initio*) theories. It is a *successful* phenomenological theory, as it gives more outputs than its input(s): the London equation not only reproduces the Meissner effect (which it used as an input), but also predicts the existence of the London penetration depth, which is an important measurable quantity for all superconductors. We will encounter other phenomenological theories later in this chapter and book, some of them even more successful.[12]

> **Exercise 19.2.** In our discussion of persistent currents in Exercise 10.1 and diamagnetism in Chapter 17, we learned how the wave function of a quantum rotor remains unchanged when magnetic flux is inserted along the rotor axis (when the symmetric gauge is used). This captures the essence of the "wave-function rigidity" in superconductivity. Show for such a quantum rotor that the paramagnetic current vanishes and that the current obeys an equation analogous to the London equation.

19.3 Meissner Kernel

The London equation assumed a local response of the supercurrent to the vector potential. In general, the current response to an applied vector potential is non-local because of the finite "size" of the Cooper pairs of electrons,[13] the finite mean free path, and other characteristic lengths which are finite. Ignoring the possibility that the response can also be non-local in time (since we are restricting our attention to very low frequencies), we have the general result (for a translationally invariant system)

$$J^{\mu}(\vec{r}, t) = -\frac{c}{4\pi} \int d^3\vec{r}\,' K^{\mu\nu}(\vec{r} - \vec{r}\,') A_{\nu}(\vec{r}\,', t), \tag{19.24}$$

where $K^{\mu\nu}$ is known as the **Meissner kernel**. Fourier transformation yields (dropping the time argument)

$$J^{\mu}(\vec{q}) = -\frac{c}{4\pi} K^{\mu\nu}(\vec{q}) A_{\nu}(\vec{q}). \tag{19.25}$$

Notice that we have not included a term analogous to the field ϕ in Eq. (19.20), and so one might worry that gauge invariance will be violated. Consider, however, the fact that a gauge change consists of adding a purely longitudinal contribution to the vector potential

$$A_{\nu}(\vec{q}) \longrightarrow A_{\nu}(\vec{q}) + i\, q_{\nu}\phi(\vec{q}). \tag{19.26}$$

---

[12] The reader will undoubtedly encounter less successful theories in the research literature, with *fewer* outputs than inputs.

[13] Cooper pairs were, of course, unknown to the London brothers in 1935.

Hence, if we impose the constraint

$$K^{\mu\nu}(\vec{q})q_\nu = 0, \tag{19.27}$$

gauge invariance will be automatically satisfied. Furthermore, the fact that in the static limit the current must be divergenceless implies that

$$q_\mu J^\mu(\vec{q}) = 0 = q_\mu K^{\mu\nu}(\vec{q}). \tag{19.28}$$

As discussed by Schrieffer [204], the symmetry property of the Meissner kernel $K^{\mu\nu}(\vec{q}) = \left[K^{\nu\mu}(-\vec{q})\right]^*$ can be used to show that gauge invariance implies local charge conservation (and vice versa).

For an isotropic system, these two constraints imply that the (static limit) Meissner kernel must be of the form[14]

$$K^{\mu\nu}(\vec{q}) = \left[\delta^{\mu\nu} - \frac{q^\mu q^\nu}{q^2}\right]K(q^2). \tag{19.29}$$

In order for the Meissner effect to occur, the Meissner kernel must have a finite value in the limit $q \rightarrow 0$. In fact, comparison with Eqs. (19.16) and (19.20) shows that the Meissner kernel is proportional to the inverse square of the penetration depth,[15]

$$\lim_{q\rightarrow 0}\lim_{\omega\rightarrow 0} K(q^2) = \lambda^{-2}. \tag{19.30}$$

Above the critical temperature, where the system is an ordinary metal, this quantity vanishes. If, however, we reverse the order of limits and keep the frequency small but finite, then there is a finite response due to the conductivity,[16]

$$\vec{J} = \sigma\vec{E} = \sigma\left(-\frac{1}{c}\frac{\partial\vec{A}}{\partial t}\right). \tag{19.31}$$

This tells us that the Meissner kernel in an ordinary metal obeys

$$K(q^2, \omega) = -\frac{4\pi i\omega\sigma(q, \omega)}{c^2}. \tag{19.32}$$

This correctly vanishes for $\omega \longrightarrow 0$, as it should since there is no Meissner effect. In a perfect metal with no disorder, the ($q = 0$) conductivity diverges at low frequencies as[17]

$$\sigma(\omega) = \frac{ne^2}{m(-i\omega)}. \tag{19.33}$$

Substituting this into Eq. (19.32) yields

$$\lim_{\omega\rightarrow 0}\lim_{q\rightarrow 0} K(q^2, \omega) = \frac{ne^2}{m_e c}. \tag{19.34}$$

If, however, we reverse the order of limits, and this is the key point, the normal metal exhibits a perfect cancelation between the paramagnetic and diamagnetic response and so obeys (as we will show in Section 20.4)

$$\lim_{q\rightarrow 0}\lim_{\omega\rightarrow 0} K(q^2, \omega) = 0. \tag{19.35}$$

---

[14] The rotation symmetry dictates that $K^{\mu\nu}$ must transform as an irreducible tensor under rotation. Its other symmetries (discussed above) dictate that it can only be a rank-2 spherical tensor which is traceless and symmetric, which takes the form shown. Note that the second term vanishes when acting on a transverse vector potential with $\vec{q} \cdot \vec{A}_{\vec{q}} = 0$, while the first cancels out when acting on a longitudinal vector potential with $\vec{A}_{\vec{q}} = i\vec{q}\phi(\vec{q})$.

[15] The $\omega \rightarrow 0$ limit must be taken first to reach the static limit. More on this later.

[16] We here choose the gauge in which the scalar potential vanishes so that the electric field is given by the time derivative of the vector potential (which may have both the longitudinal and transverse components), as we did in Section 8.2.2.

[17] See Exercise 7.1, and take the clean limit where the scattering time $\tau \rightarrow \infty$.

For small but finite q this describes the response to a long-wavelength transverse (static) vector potential, that is, to a long-wavelength static magnetic field whose strength is proportional to q^2. In a normal metal the response to such a field is finite and typically diamagnetic.[18] Thus we expect for small wave vectors

$$\lim_{\omega \to 0} K(q^2, \omega) \approx -4\pi q^2 \chi, \tag{19.36}$$

where χ is the "Landau diamagnetic" response of the metal that we discussed in Section 17.9. In a superconductor the perfect diamagnetism of the Meissner effect implies a divergent response at long wavelengths such that

$$\chi \approx -\frac{1}{4\pi (q\lambda)^2} \tag{19.37}$$

to yield a finite penetration depth.

19.4 The Free-Energy Functional

Let us now return to the thermodynamics and understand how the electrodynamics is related to it. As discussed in the previous section, the Meissner kernel tells us about the response to an externally applied vector potential. Let us consider the static limit and take advantage of the fact that the Meissner state and the normal state are both equilibrium phases. We can write the partition function (and hence the free energy) as a functional of the (static) externally applied vector potential,

$$Z[\vec{A}] = \mathrm{Tr}\left\{e^{-\beta H[\vec{A}]}\right\} \equiv e^{-\beta F[\vec{A}]}. \tag{19.38}$$

The *definition* of the equilibrium current density[19] is the functional derivative of the (matter) free energy with respect to the vector potential:

$$\left\langle J^{\mu}(\vec{r}) \right\rangle \equiv -c\, \frac{\delta F_{\mathrm{matter}}[\vec{A}]}{\delta A_{\mu}(\vec{r})}. \tag{19.39}$$

Without formal proof, one can see that this definition is sensible by using the Hellmann–Feynman theorem[20] to compute

$$-c\frac{\delta}{\delta A_{\mu}(\vec{r})} \left\langle \Psi \left| \frac{1}{2m} \left(\vec{p} + \frac{e}{c}\vec{A}\right)^2 + V \right| \Psi \right\rangle$$

$$= \frac{-e}{2m} \int d^3\vec{R}\, \Psi^*(\vec{R}) \left\{ \left(p^{\mu} + \frac{e}{c}A^{\mu}\right) \delta^3(\vec{r} - \vec{R}) \right.$$

$$\left. + \delta^3(\vec{r} - \vec{R}) \left(p^{\mu} + \frac{e}{c}A^{\mu}\right) \right\} \Psi(\vec{R}) \tag{19.40}$$

$$= \left\langle \Psi | J^{\mu}(\vec{r}) | \Psi \right\rangle \tag{19.41}$$

for a single-particle quantum system.

---

[18] Here we focus on the orbital effect of the magnetic field, and neglect Pauli paramagnetism coming from the spin response to the external magnetic field. See Chapter 17, in particular Section 17.9, and Section 20.4.

[19] Strictly speaking, we mean here the quantum/thermal expectation value of the current density.

[20] The Hellmann–Feynman theorem simply tells us that the energy shift is given by first-order perturbation theory and so we do not need to compute the change in the wave functions. Some care is required here to use infinitesimal functions $\delta A_{\mu}(\vec{R})$ which are purely transverse. Otherwise there is a first-order change in Ψ due to the gauge change.

The free energy can be expanded in a functional Taylor series in powers of the vector potential. Since there are no equilibrium currents in the absence of an externally applied vector potential, the linear term must vanish and the leading correction term must be quadratic:

$$F[\vec{A}] = F_0 + \int d^3\vec{r} \int d^3\vec{r}\,' A_\mu(\vec{r}) \Gamma^{\mu\nu}(\vec{r} - \vec{r}\,') A_\nu(\vec{r}\,') + \cdots. \qquad (19.42)$$

Because the free energy must be gauge-invariant, $\Gamma^{\mu\nu}$ must have the same tensor structure as the Meissner kernel $K^{\mu\nu}$. Indeed, it follows from Eqs. (19.39) and (19.24) that they must be linearly related:

$$F[\vec{A}] = F_0 + \frac{1}{8\pi} \int d^3\vec{r} \int d^3\vec{r}\,' A_\mu(\vec{r}) K^{\mu\nu}(\vec{r} - \vec{r}\,') A_\nu(\vec{r}\,') + \cdots. \qquad (19.43)$$

We can verify this result by substituting it into Eq. (19.39) and using the symmetry (valid in the static limit)

$$K^{\mu\nu}(\vec{r} - \vec{r}\,') = K^{\nu\mu}(\vec{r}\,' - \vec{r}). \qquad (19.44)$$

The result is precisely Eq. (19.24) as is needed.

We can now understand the divergence of the diamagnetic susceptibility in Eq. (19.37) from Eq. (19.43). A weak magnetic field $B_{\vec{q}} \hat{z} e^{i\vec{q}\cdot\vec{r}}$ at wave vector $\vec{q} = q\hat{x}$ can be described by a vector potential

$$\vec{A}_{\vec{q}} = \frac{B_{\vec{q}}}{iq} \hat{y} e^{i\vec{q}\cdot\vec{r}}. \qquad (19.45)$$

Thus a long-wavelength B field gives rise to a very large vector potential. If the Meissner kernel is finite, the response to the vector potential is finite and so the energy cost of the magnetic field is divergent. The system is willing to pay any energy cost to exclude the flux from the (bulk of the) system. This it does by generating surface currents which cost only a finite amount of energy.

To summarize, the defining characteristic of a superconductor is its response to a static transverse vector potential, which leads to the Meissner effect. A detailed discussion of the current correlations and response functions for metals, insulators, and superconductors can be found in Refs. [204, 205].

19.5 Ginzburg–Landau Theory

We saw in Eq. (19.38) that it is possible in principle to trace over ("integrate out") the electronic degrees of freedom to obtain the functional dependence of the free energy on the externally applied vector potential. From the requirement that this functional be consistent with the Meissner effect and the London equation we were able to deduce the leading terms in the functional Taylor-series expansion shown in Eq. (19.43). However, we have not yet obtained the microscopic theory of the Meissner kernel which enters this functional. This microscopic theory (to be described in the next chapter) was developed in 1957 by Bardeen, Cooper, and Schrieffer. Seven years prior to this, Ginzburg and Landau developed a phenomenological theory which, while not fully microscopic, did provide a physically appealing way to derive Eq. (19.43). The key feature of the advance made by Ginzburg and Landau was the identification of the **order parameter** for the superconducting phase transition.

Before proceeding, let us remind ourselves about what order parameters are. Landau developed the idea of the order parameter to describe the essential difference between ordered and disordered thermodynamic phases. Roughly speaking, an order parameter is something which is zero in the disordered phase and non-zero in the ordered phase. As we will see, it is often the case that the ordered phase spontaneously breaks a symmetry of the Hamiltonian. In this situation an important feature of the order parameter is that it captures the essence of this symmetry breaking.

Consider, for example, the Ising model of ferromagnetism in which each lattice site holds a classical spin $S = \pm 1$ which can point either up or down. The Hamiltonian is given by

$$H = -J \sum_{\langle ij \rangle} S_i S_j - h \sum_j S_j, \qquad (19.46)$$

where the sum is over near-neighbor pairs in the lattice. In a ferromagnet, the coupling J is positive, favoring locally parallel spin configurations. In the absence of an applied field h, we see that the Hamiltonian exhibits time-reversal symmetry and is invariant if all the spins are reversed:

$$H[S] = H[-S]. \qquad (19.47)$$

At high temperatures this symmetry guarantees that the net magnetization

$$m \equiv \frac{1}{N} \left\langle \sum_{j=1}^{N} S_j \right\rangle \qquad (19.48)$$

vanishes because there is equal probability for finding any given spin either up or down. In the high-temperature phase, entropy dominates over the energy preference for alignment, leading to a disordered state. At low temperatures, energy dominates over entropy, and the system orders by aligning its spins to produce a net magnetization. The Hamiltonian still exhibits the same time-reversal symmetry, which means that there is a precise degeneracy between states with positive and negative magnetization and they occur with equal probability. However, the system spontaneously chooses one or the other. The ergodic assumption that the system explores all possible states over time breaks down because, in the thermodynamic limit, there is an infinite barrier between the two possible magnetization states. The sign and magnitude of the magnetization give a precise characterization of this spontaneous symmetry breaking and hence constitute the order parameter for the transition. The precise mathematical definition of the spontaneous symmetry breaking for this system is

$$\lim_{h \to 0^+} \lim_{N \to \infty} m = - \lim_{h \to 0^-} \lim_{N \to \infty} m > 0. \qquad (19.49)$$

Note the order of limits in this expression. For any finite-size system, the partition function is analytic in h and the limits as h goes to zero from above and from below are equal to each other and both vanish. If, however, the thermodynamic limit is taken first, the two limits for h produce different results. This is the essence of spontaneous symmetry breaking.

To see that, below some critical temperature, the magnetized state is stable against thermal fluctuations, consider the situation at zero temperature in which the spins at the left edge of the sample are constrained to be up, while those at the right edge of the sample are forced to be down. Because the temperature is zero, the spins will be perfectly aligned parallel to the boundary spins, but somewhere in the middle of the sample there will have to be a domain wall separating the up and down domains. For the Ising model in d dimensions on a cubic lattice with L^d sites, the energy cost of this domain wall is

$$\Delta E = (2J) L^{d-1}. \qquad (19.50)$$

This diverges in the thermodynamic limit for any dimension greater than one. Hence we expect the ordered state to survive up to a finite critical temperature because nucleation of an infinite domain of reversed spins costs an infinite energy.[21] In 1D the density of domain walls $\sim e^{-2\beta J}$ is finite at any finite temperature, and so the system is always disordered.

---

[21] As the critical temperature is approached from below, thermal fluctuations in the domain-wall shape enhance the entropy associated with the domain wall, leading to a wall free energy which vanishes just at the critical point. This leads to the destruction of the long-range order.

Landau and his collaborators developed a general phenomenological theory of phase transitions and symmetry breaking. The central idea is to express the free energy as a function of the value of the order parameter. This can be done by defining the free-energy-like quantity as follows:

$$e^{-\beta N f(\varphi,T,h)} \equiv \text{Tr}_S \left\{ e^{-\beta H} \delta \left(\varphi - \frac{1}{N} \sum_j S_j \right) \right\}. \qquad (19.51)$$

The delta function picks out only those configurations where the magnetization is precisely φ. (We assume the thermodynamic limit so that φ is continuously distributed.) We see that by construction, the partition function is given exactly by

$$Z = \int_{-\infty}^{\infty} d\varphi \, e^{-\beta N f(\varphi,T,h)}. \qquad (19.52)$$

Because the argument of the exponential is an extensive quantity, the integral will be completely dominated by the value of the magnetization which minimizes f. Thus the equation

$$\left. \frac{\partial f(\varphi, T, h)}{\partial \varphi} \right|_{\varphi=m} = 0 \qquad (19.53)$$

defines an implicit function $m(T, h)$ for the magnetization.

It should be noted that we have been following the sloppy custom of referring to $f(\varphi, T, h)$ as the free-energy density. Strictly speaking, the (Gibbs) free-energy density is a function solely of T and h, and is given by

$$g(T, h) = f[m(T, h), T, h], \qquad (19.54)$$

where $m(T, h)$ is the solution of Eq. (19.53). The distinction between the Gibbs and Helmholtz potentials in magnetic systems will be discussed further below.

The procedure used in Eq. (19.51) is referred to as "integrating out" the spin degrees of freedom to obtain an effective free-energy function for the order parameter φ. Of course actually carrying out the trace over the spin degrees of freedom is an extremely difficult task which in general is impossible to perform exactly. By making a simple set of plausible assumptions, Ginzburg and Landau were brilliantly able to circumvent this problem.

The key ideas and assumptions are the following.

- $f[\varphi]$ is analytic in φ and hence admits a Taylor-series expansion in powers of φ.
- Near the critical temperature T_c, φ is small, implying that only leading terms in the Taylor-series expansion of f need to be kept.
- The symmetries of H can be used to eliminate terms in the expansion of f that violate these symmetries.
- The coefficients in the expansion are assumed to be analytic functions of temperature.

Thus for example, we might have the expansion

$$f[\varphi] = \alpha \varphi^2 + \frac{1}{2} \beta \varphi^4 - \gamma h \varphi + \cdots. \qquad (19.55)$$

Note that the thermodynamic relation $m = -\partial f / \partial h$ implies the coefficient $\gamma = 1$ in the expansion. The invariance of H under the transformation $\{S_j\} \to \{-S_j\}, h \to -h$ forbids the existence of a term linear in φ (unless it is also odd in h).

It turns out that the first assumption, that the free energy is analytic in the order parameter, is actually false. As we shall see below, using this assumption is equivalent to making the mean-field approximation. Fortunately, for the case of (low-temperature) superconductivity, the mean-field approximation will turn out to be extremely accurate. (This is related to the large size of Cooper pairs relative to the typical distance between electrons.)

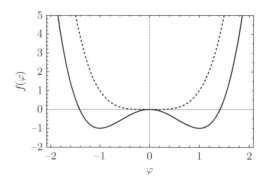

Figure 19.3 The Ising-model free energy as a function of the order parameter φ. Upper curve: $T > T_c$, and there is a single minimum at $\varphi = 0$. Lower curve: $T < T_c$, and there are two degenerate minima at $\varphi = \pm 1$.

Returning to the Ising model, unless we are able to explicitly carry out the trace over the spin degrees of freedom, we cannot explicitly evaluate the coefficients in the Taylor series (except to say which coefficients vanish by symmetry). A standard simplifying assumption is to take the coefficient β to be a constant and to take α to vary linearly through zero at the (mean-field) critical temperature:

$$\alpha(T) \propto T - T_c^{\mathrm{MF}}. \tag{19.56}$$

The free energy as a function of the order parameter is plotted in Fig. 19.3 for temperatures both above and below the critical temperature. We see that, below T_c^{MF}, the free energy has a symmetric pair of minima at finite values of the order parameter,

$$\varphi = m(T) = \pm \sqrt{\frac{-\alpha(T)}{\beta}} \propto \pm \left| \frac{T - T_c^{\mathrm{MF}}}{\beta} \right|^{\frac{1}{2}}. \tag{19.57}$$

The system spontaneously breaks time-reversal symmetry and chooses one of these two minima. The magnetization has a square root singularity as the critical temperature is approached from below. This of course is the usual mean-field result.

This simplest formulation of the theory is based on the idea that we are interested only in the symmetry breaking and the value of the order parameter. A more sophisticated version of the theory (which sometimes goes by the name of the **Landau–Ginzburg–Wilson theory**) allows for long-wavelength spatial fluctuations in the local value of the order parameter field $\varphi(\vec{r})$. The free energy is written as a functional of this field in the manner

$$e^{-\beta F[\varphi]} = \mathrm{Tr}_S\, P[\varphi, S] e^{-\beta H[S]}, \tag{19.58}$$

where P is the (functional) analog of the delta function in Eq. (19.51). It picks out only those microscopic spin configurations which are consistent with the specified spatial dependence of the coarse-grained magnetization $\varphi(\vec{r})$.

Again making a Taylor-series expansion (but this time a functional Taylor series) which includes gradient terms not forbidden by symmetry, we obtain the free-energy functional

$$F[\varphi] = \int d^d r \left\{ \frac{\rho}{2} |\nabla \varphi|^2 + \alpha \varphi^2 + \frac{1}{2} \beta \varphi^4 - h\varphi + \cdots \right\}. \tag{19.59}$$

The coefficient of the gradient term is positive, indicating that the spins prefer to order ferromagnetically, i.e. it costs energy to have a spatial variation of the magnetic alignment. Because for continuous phase transitions with diverging correlation lengths we are interested only in order-parameter fluctuations at very long wavelengths, it is usually legitimate to keep only the lowest-derivative terms in the gradient expansion. In developing the renormalization-group technique to analyze critical phenomena Wilson showed that the assumption of analyticity of this free-energy functional could be justified. The

Ginzburg–Landau free-energy function can be recovered from this by integrating out all of the components of φ with non-zero wave vectors. The non-analyticity of the Ginzburg–Landau free energy arises from infrared divergences which occur when integrating out the long-wavelength fluctuations.

Ginzburg and Landau understood that, in order to have the correct electromagnetic response, the free-energy functional for a superconductor must involve an order parameter that couples to the electromagnetic potential in a way that respects gauge invariance. We have already written down a gauge-invariant free energy in Eq. (19.43). While it is phenomenologically correct, it is unsatisfying in two respects. First, the Meissner kernel is highly non-local. For reasons of "naturalness" we would prefer to be able to write the free energy in a form analogous to Eq. (19.59) in which the tendency of the system to order is communicated from point to point by local derivative terms in the free-energy functional. Secondly, we have not identified the order parameter, and it does not appear in Eq. (19.43). Ginzburg and Landau realized that both of these difficulties could be simultaneously removed by postulating that the order parameter behaves like a quantum-mechanical wave function under gauge transformations. They had no way of guessing what such a complex order parameter field $\psi(\vec{r})$ might represent. Only later when the BCS microscopic quantum theory was developed would it come to be understood that $\psi(\vec{r})$ could be interpreted as the center-of-mass wave function of Cooper pairs of electrons. Ginzburg and Landau could, however, reasonably guess that $|\psi|^2$ somehow represents the local density of superfluid in the two-fluid model.

The Ginzburg–Landau free-energy functional is

$$F[\vec{A}, \psi] = \int d^3\vec{r} \ f[\vec{A}, \psi], \tag{19.60}$$

$$f[\vec{A}, \psi] = f_{\text{matter}}[\vec{A}, \psi] + \frac{|\vec{B}(\vec{r})|^2}{8\pi} \tag{19.61}$$

$$= \frac{1}{2m^*} \left| \left(\vec{p} + \frac{e^*}{c} \vec{A} \right) \psi \right|^2 + \alpha|\psi|^2 + \frac{\beta}{2}|\psi|^4 + \frac{|\vec{B}(\vec{r})|^2}{8\pi}, \tag{19.62}$$

where $\vec{B}(\vec{r}) = \nabla \times \vec{A}$ is the (microscopic) magnetic flux density. Because only the "covariant" derivative appears, this naturally has the same gauge invariance as ordinary quantum mechanics. For any longitudinal part added to $\vec{A}$, we can find a change of phase for ψ that gives back the original free energy. Because ψ describes pairs of electrons it turns out that $e^* = 2e$, a fact which was unknown to Ginzburg and Landau.[22] It is customary to choose $m^* = 2m_e$ but this is not essential since any other choice merely leads to a re-scaling of the field ψ (and the parameters α and β) to compensate.

It is remarkable that the London equation follows immediately from the Ginzburg–Landau free-energy functional in the ordered phase (with $|\psi| > 0$):

$$\vec{J}(\vec{r}) = -c \frac{\delta F_{\text{matter}}[\vec{A}, \psi]}{\delta \vec{A}(\vec{r})} = -c \frac{\partial f_{\text{matter}}[\vec{A}(\vec{r}), \psi]}{\partial \vec{A}(\vec{r})}$$

$$= -\frac{|\psi|^2 e^{*2}}{m^* c} \left(\vec{A}(\vec{r}) + \frac{c}{e^*} \nabla\varphi(\vec{r}) \right), \tag{19.63}$$

where $\varphi(\vec{r})$ is the phase of ψ. This is equivalent to Eq. (19.20), once we make the identification $n_s = 2|\psi|^2$.[23]

When studying superconductors in fixed externally applied magnetic fields, it is convenient to use the **Gibbs potential**

---

[22] Legend has it that Ginzburg suggested the possibility that $e^* \neq e$ to Landau, who "muttered darkly about gauge invariance and the subject was dropped" (A. A. Abrikosov, private communication).

[23] Note that $m^* = 2m_e$ and $e^* = 2e$. The way to understand these factors of two is to observe that $|\psi|^2$ is the Cooper-pair density, and each the Cooper pair is made of two electrons.

$$G(T, H) = F - \int d^3\vec{r} \; \vec{H} \cdot \vec{M}(\vec{r}), \tag{19.64}$$

where

$$\vec{M}(\vec{r}) \equiv \frac{1}{4\pi}\left(\vec{B}(\vec{r}) - \vec{H}\right) = \frac{1}{4\pi}(\nabla \times \vec{A} - \vec{H}) \tag{19.65}$$

is the magnetization. This Legendre transformation is convenient because G is canonically a function of H rather than M (see Box 19.2). The field energy term then becomes

$$\frac{1}{8\pi} \int d^3\vec{r} \left\{ H^2 + |\vec{H} - \nabla \times \vec{A}|^2 \right\}. \tag{19.66}$$

The first term is the field energy in the absence of the sample, and we can ignore it if H is held constant. The second term measures the magnetic work done in expelling flux from the sample.

Box 19.2. Magnetic Work

Let us begin with an analogy to a fluid system whose free energy is a function of temperature and volume: $dF = -S\,dT - P\,dV$, so that the pressure is

$$P = -\left(\frac{\partial F}{\partial V}\right)_T.$$

If we wish to study the case of constant external pressure, it is convenient to use the Gibbs potential $G(T, P) = F + PV$, which obeys

$$dG = -S\,dT + V\,dP \tag{19.67}$$

and hence is a canonical function of T and P. If the pressure is being supplied by a weight $W = PA$ placed on top of a piston of area A, then $U = Wh = PV$ is simply the gravitational potential energy of the weight. Hence G is the free energy of the combined fluid-plus-weight system.

If a magnetic system is subject to a time-independent external field $\vec{H}(\vec{r})$, then $-\int d^3\vec{r} \; \vec{H} \cdot \vec{M}$ represents the energy stored in the system that is maintaining the current supplied to the solenoid producing $\vec{H}$. Noting that $\delta\vec{M} = (1/4\pi)\delta\vec{B}$ we have (assuming no boundary terms when integrating by parts)

$$-\frac{1}{4\pi} \int d^3\vec{r} \; \vec{H} \cdot \frac{\delta\vec{B}}{\delta t} = +\frac{c}{4\pi} \int d^3\vec{r} \; \vec{H} \cdot (\nabla \times \vec{E})$$

$$= \frac{c}{4\pi} \int d^3\vec{r} \; \vec{E} \cdot (\nabla \times \vec{H})$$

$$= \int d^3\vec{r} \; \vec{J}_F \cdot \vec{E},$$

which is the power transmitted back into the current supply by the induced EMF when the magnetization in the sample changes. Thus $-\int d^3\vec{r} \; \vec{H} \cdot \vec{M}$ is the analog of the PV term in the fluid example and

$$G(T, H) = F(T, \vec{M}) - \int d^3\vec{r} \; \vec{H} \cdot \vec{M}.$$

In the normal state $\vec{B} = \vec{H}$ and this term vanishes. In the Meissner state $B = 0$ and $H^2/8\pi$ represents the magnetic pressure that must be overcome to expel the flux. As in the magnetic case studied previously, we will take $\alpha \sim T - T_c^{\text{MF}}$ and $\beta \sim$ constant. Then for $\alpha < 0$ we have the minimum free energy at

$$|\psi|^2 = \frac{|\alpha|}{\beta} \tag{19.68}$$

if the gradient term can be ignored. Let us suppose that this is the case and consider the situation where $\vec{A}$ is purely transverse. Because $\vec{p}\psi$ is purely longitudinal, these two terms cannot compensate for each other, and the energy rises. The minimum energy (assuming that $|\psi|$ is a constant) occurs for ψ constant, since then $\vec{p}\psi$ vanishes. The free-energy functional then becomes

$$f[\vec{A}, \psi] = \frac{1}{2m^*} \left(\frac{e^*}{c}\right)^2 |\vec{A}|^2 |\psi|^2 + \alpha|\psi|^2 + \frac{\beta}{2}|\psi|^4 + \frac{1}{8\pi}|\nabla \times \vec{A} - \vec{H}|^2. \tag{19.69}$$

At long wavelengths the $|\vec{A}|^2$ term will dominate as noted in Eq. (19.45). Comparison with Eq. (19.43) allows us to identify the Meissner kernel and hence the magnetic penetration depth

$$K = \lambda^{-2} = \frac{4\pi e^{*2}}{m^*c^2}|\psi|^2 = \frac{16\pi e^2}{m^*c^2}\frac{|\alpha|}{\beta}. \tag{19.70}$$

The assumption that ψ is a constant corresponds to the "wave function rigidity" postulated by London.

The assumption of wave-function rigidity is correct in the Meissner phase, at least in the interior of the sample, where the magnetic flux density is exponentially small. To prove that the Meissner phase always exists (for sufficiently weak H) we have to make less restrictive assumptions. Ignoring the small surface region, the bulk Gibbs free-energy density in the Meissner phase is optimized by a spatially constant order parameter (in the gauge where $\vec{A} = \vec{0}$ in the interior) of magnitude $|\psi|^2 = -\alpha/\beta$. The Gibbs free-energy density is thus

$$g_S = \frac{H^2}{8\pi} - \frac{\alpha^2}{2\beta}. \tag{19.71}$$

The corresponding free-energy density for the normal state with $\vec{B} = \vec{H}$ and $\psi = 0$ is $g_N = 0$. Hence the thermodynamic critical field (determined by $g_N = g_S$ is given by

$$H_c^2(T) = \frac{4\pi\alpha^2(T)}{\beta}, \tag{19.72}$$

yielding

$$H_c(T) = H_c(0)\left(1 - \frac{T}{T_c}\right) \tag{19.73}$$

if α obeys the standard assumption for its temperature dependence over the full temperature range (which of course it does not). This result does correctly predict that $H_c(T)$ vanishes linearly in $T - T_c$, in agreement with Eq. (19.6).

Exercise 19.3. Show that, with the magnetic field energy included in the free energy, extremizing the free energy with respect to the vector potential

$$\frac{\delta F}{\delta A_\mu(\vec{R})} = 0$$

correctly yields the (static) Maxwell equation

$$\nabla \times \vec{B} = \frac{4\pi}{c}\langle\vec{J}\rangle.$$

19.6 Type-II Superconductors

This derivation of the critical field for Meissner phase stability assumes that the only competing states have perfect flux exclusion or no flux exclusion at all (and zero order parameter). We also have to consider the possibility of type-II behavior in which flux (partially) penetrates and the order parameter remains finite, leading to the phase diagram in Fig. 19.2(b). In this case it is not appropriate to assume that ψ remains spatially constant. Optimizing the free energy with respect to $\psi(\vec{r})$,

$$\frac{\delta G}{\delta \psi^*(\vec{r})} = 0, \tag{19.74}$$

yields the non-linear Schrödinger equation

$$\left[\frac{1}{2m^*} \left(\vec{p} + \frac{e^*}{c} \vec{A} \right)^2 + \alpha \right] \psi + \beta |\psi|^2 \psi = 0. \tag{19.75}$$

If we make the simplifying assumption that $\vec{B} = B\hat{z}$ is spatially uniform, then we can use our knowledge of the quantum Hall effect to our advantage. The form of ψ which is variationally optimal for the gradient energy term is

$$\psi(x, y, z) = \psi_{\mathrm{LLL}}(x, y), \tag{19.76}$$

where ψ_{LLL} is one of the (many possible degenerate) lowest-Landau-level wave functions. The variational free energy will then have the form

$$G = \left(\frac{1}{2} \hbar \omega_{\mathrm{c}} + \alpha \right) \int d^3 \vec{r} |\psi_{\mathrm{LLL}}|^2 + \frac{\beta}{2} \int d^3 \vec{r} |\psi_{\mathrm{LLL}}|^4$$
$$+ \int d^3 \vec{r} \frac{(B - H)^2}{8\pi}, \tag{19.77}$$

where

$$\hbar \omega_{\mathrm{c}} \equiv \frac{\hbar e^* |B|}{m^* c} = \frac{\hbar e |B|}{m_{\mathrm{e}} c} \tag{19.78}$$

is the cyclotron energy. Now consider the case where $\alpha < 0$ and H and B are very small. The variationally optimal value of $\int d^3 \vec{r} |\psi_{\mathrm{LLL}}|^2$ will then be finite and the dependence of the free-energy density on B will have the form

$$g = \frac{\beta}{2} \langle |\psi_{\mathrm{LLL}}|^4 \rangle + \langle |\psi_{\mathrm{LLL}}|^2 \rangle \left(\alpha + \frac{\hbar e}{2m_{\mathrm{e}} c} |B| \right) + \frac{1}{8\pi} (B - H)^2, \tag{19.79}$$

where $\langle \cdot \rangle$ indicates the spatial average. For small enough H this will always have its minimum at $B = 0$, since the kinetic-energy term grows linearly with $|B|$. If we mix in higher-Landau-level solutions to further optimize the $|\psi|^4$ term, the coefficient of $|B|$ will only increase (because the kinetic energy increases when higher-Landau-level states get mixed in), further stabilizing the Meissner phase. Hence, for sufficiently weak H, total flux exclusion will always occur.

We can use the same method to find the upper critical field H_{c_2} where the order parameter goes to zero. From Eq. (19.77) we see that for $\hbar \omega_{\mathrm{c}} > 2|\alpha|$ the coefficient of the $|\psi|^2$ term is positive and the optimal value of the order parameter will be zero. Hence, using the fact that $B = H$ once the order parameter has vanished, we have

$$\frac{\hbar e H_{\mathrm{c}_2}}{m_{\mathrm{e}} c} = 2|\alpha|, \tag{19.80}$$

$$H_{\mathrm{c}_2} = \frac{2m_{\mathrm{e}} c}{\hbar e} |\alpha| \propto T_{\mathrm{c}} - T \tag{19.81}$$

for T less than but close to T_{c}.

A type-II superconductor is one in which H_{c_2} exceeds the thermodynamic field $H_c(T)$, defined by Eq. (19.2). Then, as the field is lowered through H_{c_2}, the system goes into a **mixed state** prior to entering the Meissner phase at $H_{c_1}(T)$ (which is smaller than $H_c(T)$; to be determined later).[24] If $H_{c_2} < H_c$ then the system is a type-I superconductor because it goes directly from the normal to the Meissner phases. Once the flux has been excluded there is no physical significance to the particular field value H_{c_2}. Comparison of Eqs. (19.72) and (19.81) shows that a system is a type-II superconductor if

$$\left(\frac{2m_e c}{\hbar e}\right)^2 > \frac{4\pi}{\beta}. \tag{19.82}$$

This can be cast into another form by introducing a new length scale ξ called the **Ginzburg–Landau coherence length**, which is the characteristic length over which the magnitude of ψ varies. Consider the situation $\alpha < 0$, $\vec{A} = \vec{H} = \vec{0}$ and take ψ to be real. Define the optimal value of ψ to be

$$\psi_\infty^2 = -\frac{\alpha}{\beta} \tag{19.83}$$

and define

$$\phi(\vec{r}) = \frac{\psi(\vec{r})}{\psi_\infty}. \tag{19.84}$$

The Gibbs free-energy density can then be written as

$$g = |\alpha|\psi_\infty^2\left\{-\phi^2 + \frac{1}{2}\phi^4 + \xi^2(T)|\nabla\phi|^2\right\}, \tag{19.85}$$

where

$$\xi^2(T) \equiv \frac{\hbar^2}{2m^*|\alpha|} \tag{19.86}$$

is the coherence length. Being the only length scale that enters the (properly normalized) free-energy density, it dictates how fast the order parameter can vary spatially.[25]

Using Eq. (19.86) in Eq. (19.81), we can write

$$H_{c_2} = \frac{\Phi_S}{2\pi\xi^2}, \tag{19.87}$$

where Φ_S is the superconducting flux quantum defined in Eq. (19.7). Recalling that the flux density is

$$B = \frac{\Phi_S}{2\pi\ell_S^2}, \tag{19.88}$$

where ℓ_S is the magnetic length (for particles with charge $2e$, as is appropriate for superconductors), we see that at the upper critical field the magnetic length is equal to the coherence length. Hence a short coherence length is needed for the system to be a type-II superconductor. The physics is that a short coherence length means that the amplitude variations of ψ present in the mixed state are not so energetically expensive.

---

[24] In a type-II superconductor there is no phase transition at $H_c(T)$, but it is still a useful quantity that serves as a reference value for magnetic field.

[25] One can verify this by, e.g., posing a boundary condition that $\phi = 0$ at the surface of a superconductor, and solve its spatial dependence to see how fast it recovers to the bulk value. ξ is basically the healing length of the order parameter in this case, as well as in vortices to be discussed later. For this reason the Ginzburg–Landau coherence length ξ is also referred to as the **healing length** in the literature.

A very important dimensionless number is the ratio of the penetration depth to the coherence length:

$$\kappa \equiv \frac{\lambda(T)}{\xi(T)}. \tag{19.89}$$

Combining Eq. (19.70) and Eq. (19.86) we have

$$\kappa^{-2} = \frac{16\pi e^2}{m^* c^2} \frac{\hbar^2}{2m^*} \frac{1}{\beta} = \frac{2\pi}{\beta}\left(\frac{e\hbar}{m_e c}\right)^2. \tag{19.90}$$

Comparison with Eq. (19.82) shows that the condition for a system to be a type-II superconductor is

$$\kappa > \frac{1}{\sqrt{2}}. \tag{19.91}$$

Roughly speaking, in a type-II material the order parameter varies more easily than the magnetic field. In a type-I material the reverse is true.

The Ginzburg–Landau theory is a highly successful phenomenological theory. With one input (the Ginzburg–Landau free-energy functional, in particular the existence of a wave-function-like order parameter that couples to electromagnetic potential in a gauge-invariant manner), the theory not only reproduced everything London theory has to offer, but also led to many new results. The most important among them is the *prediction* of the existence of type-II superconductors. In fact, most of superconducting phenomena (especially those associated with low-frequency/energy) are well described by the Ginzburg–Landau theory.

19.6.1 Abrikosov Vortex Lattice

After the development of the Ginzburg–Landau theory, Abrikosov realized that it is possible for there to be a mixed state in a type-II superconductor, which consists (at the mean-field level) of a regular lattice of vortices. To understand what this means, let us consider the regime in the phase diagram (Fig. 19.2(b)) where H is only slightly smaller than H_{c2} so that the order parameter is small. We have already seen that the optimal order parameter will be one of the many degenerate wave functions from within the lowest Landau level. This is a somewhat unusual circumstance in which many distinct modes simultaneously become unstable (that is, the operator in the quadratic part of the Ginzburg–Landau free energy simultaneously develops many negative eigenvalues). It is left to the quartic term to select out the mode which will be most favored. If the order parameter lies strictly in the lowest Landau level we can write it as

$$\psi(\vec{r}) = \chi \psi_{\text{LLL}}(x, y), \tag{19.92}$$

where χ is a scale factor and ψ_{LLL} is taken to be normalized:

$$\langle |\psi_{\text{LLL}}|^2 \rangle = L^{-3} \langle \psi_{\text{LLL}} | \psi_{\text{LLL}} \rangle = 1. \tag{19.93}$$

Then the Ginzburg–Landau free-energy density becomes

$$g = \left(\frac{1}{2}\hbar\omega_c + \alpha\right)\chi^2 + \frac{\beta}{2}\chi^4 \langle |\psi_{\text{LLL}}|^4 \rangle. \tag{19.94}$$

Clearly g will be minimized by finding the ψ_{LLL} that has the smallest possible value of $\langle |\psi_{\text{LLL}}|^4 \rangle$, subject to the constraint that the average density $\langle |\psi_{\text{LLL}}|^2 \rangle = 1$. That is, we want the wave function with the smallest possible variance in the density.

Recall from our quantum Hall discussion of the symmetric gauge that ψ_{LLL} is the product of a simple exponential and an analytic function defined by its zeros:

$$\psi_{\text{LLL}} = \prod_{j=1}^{N}(z - z_j)e^{-\frac{1}{4}|z|^2}. \tag{19.95}$$

Here $z = (x + iy)/\ell_S$ is the complex coordinate for the xy plane and ℓ_S is the (charge-$2e$) magnetic length. Clearly it is impossible to make $|\psi|^2$ perfectly constant because of all the zeros. However, invoking the Laughlin plasma analogy, we see that the density will be most nearly homogeneous if the average density of zeros is $1/2\pi\ell_S^2$ to maintain charge neutrality of the plasma,

$$|\psi_{\text{LLL}}(z)|^2 = e^{-\beta U(z)}, \tag{19.96}$$

where $\beta = 2$ and

$$U(z) \equiv -\sum_{j=1}^{N} \ln|z - z_j| + \frac{1}{4}|z|^2. \tag{19.97}$$

The electrostatic potential seen by the "particle" at z is due to a combination of a uniform background of charge density

$$\rho_{\text{B}} = -\frac{1}{2\pi\ell_S^2} \tag{19.98}$$

and a collection of fixed unit charges at the positions $\{z_j\}$. The closest we can come to having perfect cancelation of these two charge densities is to put the z_js on a triangular lattice of density $1/2\pi\ell_S^2$. Each of the analytic zeros gives ψ a phase winding of 2π and hence has a local circulating current, making it a vortex. This configuration is known as the **Abrikosov vortex lattice**, and each unit cell of this lattice encloses precisely one flux quantum as required by the charge neutrality of Laughlin plasma analogy.

Numerical analysis shows that[26]

$$\langle|\psi_{\text{LLL}}|^4\rangle \approx 1.159595\ldots \tag{19.99}$$

for the Abrikosov lattice solution. If the density were perfectly uniform, the result would have been unity. Hence the optimal configuration still yields a relatively uniform density despite the presence of the zeros required by the magnetic field.

19.6.2 Isolated Vortices

So far we have investigated the mixed state only near H_{c_2}, where the order parameter is very small. In that limit $B \approx H$ and the flux is very nearly uniform. This follows from the fact that

$$\nabla \times \vec{B} = \frac{4\pi}{c}\langle\vec{J}\rangle \tag{19.100}$$

and the screening currents are small because the order parameter is small.

We turn now to a very different limit. Consider the situation in a type-II material with an applied field H barely larger than H_{c_1}. In this case $\langle B\rangle \ll H$ and the vortices are very widely separated compared with both the coherence length ξ and the penetration depth λ. Because $H \ll H_{c_2}$, the magnitude of the order parameter is relatively large and the non-linearity of Eq. (19.75) cannot be ignored. To further complicate matters, the magnetic field is highly non-uniform and cannot make the lowest-Landau-level approximation.

To describe an isolated vortex running along the z axis and centered at $x = y = 0$ we make the ansatz

$$\psi(x, y, z) = \psi_\infty f(r_\perp)e^{i\varphi(x,y)}, \tag{19.101}$$

---

[26] See [206] for a straightforward method of calculating such averages, that uses the formalism of Section 16.11, but does *not* invoke the explicit form of the wave function ψ_{LLL}.

where $\psi_\infty^2 \equiv -\alpha/\beta, r_\perp^2 \equiv x^2 + y^2$, and φ is the azimuthal angle. This form describes a vortex because it has a phase winding of 2π around the z axis.[27] For small $r_\perp$, $f(r_\perp) \to 0$ because of the "centrifugal barrier." For large $r_\perp$, $f(r_\perp) \to 1$ in order to optimize the magnitude of the order parameter for the free energy in Eqs. (19.62) and (19.60).

Suppose as a first approximation that $\vec{B} = \vec{0}$ and choose the gauge $\vec{A} = \vec{0}$. In the region of large $r_\perp$ where $f(r_\perp)$ is a constant, the gradient of the order parameter is

$$\vec{p}\psi = \hbar(\nabla\varphi)\psi = \frac{\hbar\hat{\varphi}}{r_\perp}\psi. \tag{19.102}$$

The total gradient energy is therefore

$$T = \frac{\psi_\infty^2 \hbar^2 L_z}{2m^*} \int_0^{L_\perp} dr_\perp \, 2\pi r_\perp \frac{f^2(r_\perp)}{r_\perp^2}, \tag{19.103}$$

where L_z is the sample thickness and $L_\perp$ is the sample width. Since $f^2 \to 1$ at large distances, T diverges logarithmically with the sample size. Thus a vortex would seem to cost an infinite amount of energy if $B = 0$ as we have assumed. The system will clearly pay any finite energy cost to avoid this divergence. It can do this by adjusting the current distribution near the origin to produce just the right amount of flux so that the diamagnetic current cancels out the paramagnetic current at large distances:

$$\left(\vec{p} + \frac{e^*}{c}\vec{A}\right)\psi = \vec{0}. \tag{19.104}$$

From Eq. (19.102) we see that for large $r_\perp$ we must have

$$\vec{A} = -\frac{\Phi_S}{2\pi}\frac{\hat{\varphi}}{r_\perp}. \tag{19.105}$$

Since

$$\oint \vec{A} \cdot d\vec{r} = -\Phi_S \tag{19.106}$$

for any large contour enclosing the vortex, it follows from Stokes' theorem that the total magnetic flux must be quantized:

$$\int d^2 r_\perp \, \vec{B} \cdot \hat{z} = -\Phi_S. \tag{19.107}$$

So once again there is a one-to-one correspondence between a flux quantum and a vortex, which can be shown to be true in general. Note that an antivortex is described by the complex conjugate of Eq. (19.101) and carries the same flux with the opposite sign. These quantized fluxoids had historically been detected using magnetic decoration experiments that directly probe the magnetic flux, as illustrated, e.g., in Fig. 34.5 of Ref. [1]. Nowadays detections of vortex lattices and individual vortices rely more heavily on **scanning tunneling microscopes (STMs)** (see Appendix C), which measure the local electron tunneling density of states, a quantity which is very sensitive to the local structure of superconducting order parameter. See Fig. 19.4 for an example. This change of focus from magnetic field to order parameter is particularly important for strongly type-II superconductors that have large penetration depth (so the magnetic field tends to vary slowly and be uniform unless one is very close to H_{c_1}) and short coherence length (so the order parameter varies rapidly).

It is somewhat complicated to solve for the detailed spatial distribution of the magnetic flux density because of the non-linearity of the coupled set of differential equations obeyed by the flux density and

---

[27] Another way to understand Eq. (19.101) is to observe that it corresponds to a Cooper-pair wave function with angular momentum $l_z = \hbar$. Similarly, ψ^* has a phase winding of -2π and describes an anti-vortex.

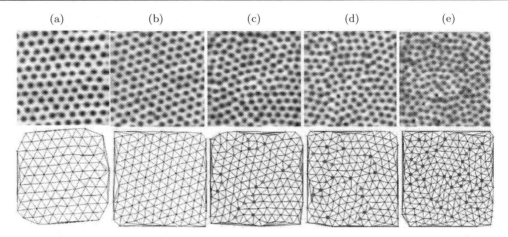

Figure 19.4 Vortex lattices of NbSe$_2$ under increasing magnetic fields: (a) 1.8, (b) 2.3, (c) 2.5, (d) 2.7, and (e) 3.3 T. The upper panels are images obtained using a scanning tunneling microscope (STM). The scanning area for all images is 375 nm × 375 nm. The lower panels provide the lattice structure. With increasing field the vortex density increases, along with fluctuation/distortion of the vortex lattice structure. At higher fields lattice defects are visible and labeled. Figure adapted with permission from [207]. Copyright (2008) by the American Physical Society. Credit: Argonne National Laboratory, managed and operated by UChicago Argonne, LLC, for the U.S. Department of Energy under Contract No. DE-AC02-06CH11357.

the order parameter. The details are presented in Tinkham [208], and we follow here his discussion of the limiting case $\kappa \gg 1$. In this extreme type-II limit ($\lambda \gg \xi$), one finds that $f(r_\perp)$ in Eq. (19.101) is close to unity everywhere except within a small distance ξ of the vortex line. We expect the diameter of the region containing the magnetic flux will be $\sim \lambda$ and hence much larger. Thus in the vast majority of the region of current flow the order parameter has fixed magnitude and the London equation (19.15) will be obeyed everywhere except in the vortex core region:

$$\lambda^2 \nabla \times (\nabla \times \vec{B}) + \vec{B} = \Phi_S \hat{z} \delta(x)\delta(y).$$ (19.108)

The RHS of this expression vanishes outside the core region to yield the London result. Its value in the core region has been approximated by a delta function whose strength guarantees that the flux quantization condition is satisfied since

$$\hat{z} \cdot \int d^2 \vec{r}_\perp \, \nabla \times (\nabla \times \vec{B}) = \oint (\nabla \times \vec{B}) \cdot d\vec{r} = 0,$$ (19.109)

where the latter integral is taken on a contour of large radius (where $\vec{B} = \vec{0}$) surrounding the vortex. Using $\nabla \cdot \vec{B} = 0$, Eq. (19.108) becomes

$$\nabla^2 \vec{B} - \lambda^{-2} \vec{B} = -\lambda^{-2} \Phi_S \hat{z} \delta(x)\delta(y),$$ (19.110)

from which we obtain the exact result

$$\vec{B}(r_\perp) = \hat{z} \frac{\Phi_S}{2\pi \lambda^2} K_0\left(\frac{r_\perp}{\lambda}\right),$$ (19.111)

where K_0 is the modified Bessel function which falls off exponentially at large distances:

$$\vec{B}(r_\perp) \sim \hat{z} \frac{\Phi_S}{2\pi \lambda^2} \left(\frac{\pi}{2} \frac{\lambda}{r_\perp}\right)^{1/2} e^{-r_\perp/\lambda}.$$ (19.112)

At short distances this approximate result has a logarithmic divergence which in reality is cutoff by the fact that ξ is finite. The magnetic field and order parameter configurations are illustrated in Fig. 19.5.

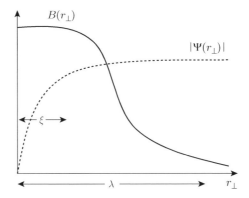

Figure 19.5 Magnetic-field and superconducting-order-parameter profiles of an isolated vortex.

The magnetic field energy and the kinetic energy of the currents contribute a certain energy per unit length of the vortex. This can be viewed as the "string tension" of an elastic string. Dimensional considerations suggest a field energy of order

$$t_{\text{field}} = \frac{1}{8\pi} \int d^2\vec{r}_\perp |\vec{B}|^2 \sim \frac{1}{8\pi} \left(\frac{\Phi_S}{\pi\lambda^2} \right)^2 \pi\lambda^2 = \frac{1}{8} \left(\frac{\Phi_S}{\pi\lambda} \right)^2 \tag{19.113}$$

per unit length. The kinetic-energy contribution turns out to be (see Exercise 19.5)

$$t_{\text{kinetic}} \approx \left(\frac{\Phi_S}{4\pi\lambda} \right)^2 \ln\kappa, \tag{19.114}$$

which overwhelms t_{field} for $\kappa \gg 1$. As a result, the total energy per unit length (or string tension) is

$$t \approx t_{\text{kinetic}} \approx \left(\frac{\Phi_S}{4\pi\lambda} \right)^2 \ln\kappa. \tag{19.115}$$

To convert to SI units it is convenient to use

$$\left(\frac{\Phi_S}{4\pi} \right)^2 = \frac{1}{8\alpha^2} a_B \text{ Ryd} = 1.96 \times 10^8 \text{ Å} \times k_B \text{ K}, \tag{19.116}$$

where α is the fine-structure constant, a_B is the Bohr radius, and Ryd is the Rydberg. For a typical elemental superconductor with a penetration depth at low temperatures of $\lambda \sim 500$ Å we find a very large string tension (ignoring the $\ln\kappa$ factor) of order

$$t \sim 10^3 \, k_B \text{ K/Å}. \tag{19.117}$$

This means that vortex lines tend to be extremely straight with negligible thermal fluctuations, even at higher temperatures where λ is increasing and the string tension is therefore decreasing. The situation is different in high-temperature superconductors (see Section 20.9) where λ can be considerably larger (~ 2000 Å), T_c is more than an order of magnitude larger and the coupling energy between the copper-oxide planes is quite weak. These effects combine to significantly enhance the thermal fluctuations of the vortices leading to very interesting new physics involving the statistical mechanics of fluctuating polymer-like vortex lines.

The string tension t characterizes the (free) energy cost of an isolated vortex in the absence of an external magnetic field $\vec{H}$. The external field, as we now show, *lowers* the energy cost, and makes it energetically favorable for the system to introduce vortices spontaneously once it reaches H_{c_1}. As discussed earlier, the field-energy term in the presence of $\vec{H}$ is (see Eq. (19.66))

$$\frac{1}{8\pi} \int d^3\vec{r} \left\{ H^2 + |\vec{H} - \vec{B}|^2 \right\} = \frac{1}{8\pi} \int d^3\vec{r} \left\{ 2H^2 + B^2 - 2\vec{H} \cdot \vec{B} \right\}. \tag{19.118}$$

The H^2 term above is independent of the state the superconductor is in, and the B^2 term has already been taken into account in the calculation of t. Thus the only new contribution to the vortex energy per unit length is due to the last term above:

$$\Delta t = -\frac{1}{4\pi} \int d^2 \vec{r}_\perp \, \vec{H} \cdot \vec{B} = -\frac{H \Phi_S}{4\pi}, \tag{19.119}$$

from which we obtain

$$H_{c_1} = \frac{4\pi t}{\Phi_S} \tag{19.120}$$

as the free-energy cost of a vortex becomes *negative* for $H > H_{c_1}$. In the extreme type-II limit of $\kappa \gg 1$, we find

$$H_{c_1} \approx \frac{\Phi_S}{4\pi \lambda^2} \ln \kappa = \frac{\ln \kappa}{\sqrt{2} \kappa} H_c \ll H_c. \tag{19.121}$$

We thus have the hierarchy $H_{c_1} \ll H_c \ll H_{c_2}$ in this limit.

> **Exercise 19.4.** We have shown that a vortex with vorticity 1 in the superconducting order parameter $\psi(x, y, z) = \psi_\infty f(r_\perp) e^{i\phi(x,y)}$ must carry with it exactly one flux quantum, $\Phi_S = hc/e^*$. Now consider a vortex with vorticity n, $\psi(x, y, z) = \psi_\infty f(r_\perp) e^{in\phi(x,y)}$, and show that it must carry with it exactly n flux quanta, in order for it to have finite energy per unit length.
>   Also show that the energy cost for a vortex with $n = 2$ exceeds the cost of two widely separated vortices with $n = 1$.

> **Exercise 19.5.**
> (i) Verify Eq. (19.114).
> (ii) Now consider $\lambda = \infty$, which corresponds to the case of a charge-neutral superfluid (to be discussed in Section 19.8). Show that in this case the vortex string tension diverges logarithmically with the system size.

19.7 Why Do Superconductors Superconduct?

We are finally in a position to address the most obvious and fundamental question about superconductors, namely why can current flow without resistance, even though in real materials there is inevitably disorder and thus electron-scattering processes? To address this question it is both conceptually and practically advantageous to consider a ring made of superconducting materials, similar to that in our discussion of persistent currents in normal metal rings in Chapter 10. In such a setup there is a closed path for the current to flow, and we avoid complications like contact resistance when we connect a superconductor to leads. In this setup the question posed in the title of the section is equivalent to the following: "Once a current flows in a superconducting ring, what prevents it from decaying with time?" As we saw in Chapter 11, such an initial current does decay due to electron scattering by the disorder potential in a normal metal.

Let us first set up a supercurrent flowing around the ring. For simplicity we assume that there is no external magnetic field, and the supercurrent is sufficiently weak that it does not generate appreciable magnetic field either. This is appropriate when the thickness of the ring is small compared with the

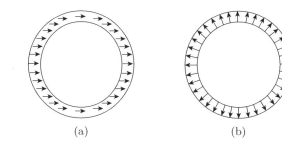

(a)   (b)

Figure 19.6 (a) Arrows indicating the phase of the superconducting order parameter in a ring with zero winding number and hence zero circulating current. (b) Phase winding +1, corresponding to a state with non-zero circulating current.

penetration depth.[28] As a result, we can choose a gauge such that $\vec{A} = \vec{0}$ everywhere. From Eq. (19.63) we find

$$\vec{J}(\vec{r}) = -\frac{|\psi|^2 e^*}{m^*} \nabla\varphi(\vec{r}), \qquad (19.122)$$

where $\varphi(\vec{r})$ is the phase of the order parameter ψ. Clearly $\varphi(\vec{r})$ is an angular variable that can be represented by the direction of an arrow confined to the plane of this page. First consider a zero-current state, as shown in Fig. 19.6(a). The orientation of the arrows in the figure represents the local phase angle $\varphi(\vec{r})$ at each point in space. The fact that the arrows are all parallel indicates that $\varphi(\vec{r})$ is a constant, and the ring is in a low-energy, zero-current state. Now consider the state of the system shown in Fig. 19.6(b), where the phase angle varies continuously, indicating that there is a current flow. The constraint that the order parameter be single-valued means that

$$\oint d\mathbf{r} \cdot \nabla\varphi = 2\pi n \qquad (19.123)$$

and hence the total circulation of current is quantized in discrete units. The integer n in Eq. (19.123) is invariant under smooth deformations of the orientation of the phase arrows and hence is a *topological invariant* known as the **winding number** (see Appendix D and Chapter 13).[29] An alternative representation of the winding of the phase is shown in Fig. 19.7, where we have $n = +3$. The superconducting ring is a circle, and the phase variable lives on a circle. The mapping from the position along the ring to the phase yields the trajectory along the surface of the torus as shown in the figure.

The fact that current-carrying states are topologically distinct from the zero-current state leads to the phenomenon of persistent current in a superconductor. The current-carrying states illustrated in Fig. 19.6(b) and in Fig. 19.7 are *not* the ground state[30] but are *metastable* because there is an energy barrier to be overcome to go from there to the ground state. Hence it is difficult for the current to decay. This can be seen in Fig. 19.8, which shows the singular deformation necessary to change the topological winding number of the phase. At some point along the ring the magnitude of the order parameter must be driven to zero so that the phase becomes ill-defined and hence changeable. This singularity is, of course, a vortex that we discussed extensively in Section 19.6. A vortex must move

---

[28] Taking into account the magnetic field generated by the current does not alter the discussions below qualitatively in this limit. We will consider the opposite limit later.

[29] The winding number is equivalent to the vorticity of a vortex as we encountered in Exercise 19.4. Some people would say there is a vortex with vorticity n that lives in the hole of the ring. We would like to avoid this language, and reserve the term vortex for a singularity in ψ in the interior of a superconductor as in the previous section, which we will be forced to encounter soon when we try to change the winding number of the ring.

[30] Unlike the persistent current we encountered in Chapter 10!

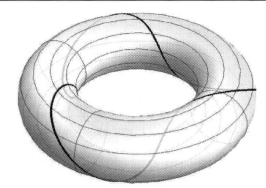

Figure 19.7 An alternative representation of the phase of the order parameter in a ring with topological winding number $n = +3$.

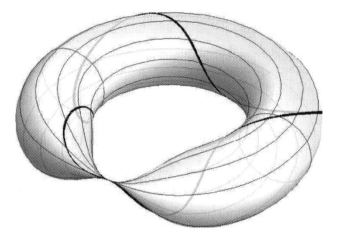

Figure 19.8 Illustration of the fact that the magnitude of the order parameter must go to zero at some point along the superconducting ring in order for the current to decay by changing the winding number of the phase from $n = 3$ to $n = 2$. There is a large energy cost when the order-parameter magnitude is forced to go to zero. This high energy barrier prevents the current from decaying even in the presence of thermal fluctuations.

across the ring at some point and because the free-energy density is minimized at a finite value of $|\psi|$, there is an energy barrier for this process that involves creating and moving a vortex inside the superconductor. At low enough temperature, this barrier is so large compared with $k_B T$ that the decay time of the current can easily exceed the age of the Universe. It is crucial to note that nothing in this argument relies on translation invariance. Thus disorder may reduce the size of the order parameter, but as long as the superfluid density $|\psi|^2$ is non-zero, current will persist.

To summarize the discussion above, here the supercurrent is protected by the *combined* effects of topology and energetics – topology dictates that the supercurrent can decay only through creating vortices and moving them across the superconductor. This vortex creation is heavily penalized by the (free) energetics of the superconductor.

Now let us come to the opposite limit, where the thickness of the ring is larger than the penetration depth. In this case the magnetic field generated by the supercurrent can no longer be neglected; in fact, for the same *energetic* reasons as illustrated in Exercise 19.4, the magnetic flux going through the ring (that is generated by the supercurrent) must be an integer multiple of the flux quantum. Here again the supercurrent cannot change continuously, but for *energetic* (instead of topological) reasons. The current has to change in discontinuous steps (that result in changes of integer flux quanta through the ring), again via creating and moving a vortex (now carrying quantized flux) inside the superconductor. The energetic penalty of such processes prevents the (metastable) current carrying state from decaying.

Exercise 19.6. When the superconducting wire is thick compared with the penetration depth, there is a path in the interior of the wire along which the current is exponentially close to zero everywhere around the ring. This means that the paramagnetic current and the diamagnetic current must cancel each other out at every point along that path. Show from this that the total magnetic flux through the hole in the ring is quantized in integer multiples of the superconducting flux quantum.

19.8 Comparison between Superconductivity and Superfluidity

Much of our discussion of superconductors in this chapter applies to superfluids as well. In particular, the Ginzburg–Landau theory can also describe superfluids, with one crucial difference: because superfluids are made of charge-neutral particles, the condensate charge $e^* = 0$, as a result of which there is *no* coupling between superfluids and electromagnetic fields.[31]

One can, nevertheless, simulate the effects of an external magnetic field in a superfluid in at least two different ways. The first is via rotation. In a rotating frame (namely, the frame provided by the rotating container that holds the fluid), the neutral particles experience a **Coriolis force**, whose form is very similar to the Lorentz force experienced by charged particles in a magnetic field. To see this, note that, if the particle has position $\vec{r}$ in the rotating frame, then the lab-frame velocity is

$$\vec{v}_\mathrm{L} = \dot{\vec{r}} + \vec{\Omega} \times \vec{r}, \tag{19.124}$$

where Ω is the angular velocity of the rotating frame as seen from the lab frame. The Lagrangian can be written in terms of the rotating-frame coordinates as

$$\mathcal{L} = \frac{1}{2}m|\dot{\vec{r}} + \vec{A}|^2 - V(\vec{r}), \tag{19.125}$$

where

$$\vec{A} \equiv \vec{\Omega} \times \vec{r}. \tag{19.126}$$

Expanding this, we have

$$\mathcal{L} = \frac{1}{2}m\dot{\vec{r}} \cdot \dot{\vec{r}} + m\dot{\vec{r}} \cdot \vec{A} + \frac{1}{2}m\vec{A} \cdot \vec{A} - V(\vec{r}). \tag{19.127}$$

The first two terms look like the kinetic energy of a particle of "charge" m moving in a vector potential $\vec{A}$ whose corresponding "magnetic field" is

$$\vec{B} = 2\vec{\Omega}. \tag{19.128}$$

The third term ruins the analogy but we can cancel it out by adding a parabolic potential

$$V_0(\vec{r}) = \frac{1}{2}m\Omega^2 r_\perp^2, \tag{19.129}$$

---

[31] Of course, the density fluctuations of a superfluid modulate its index of refraction, which does lead to coupling to optical fields and thus the possibility of Raman scattering.

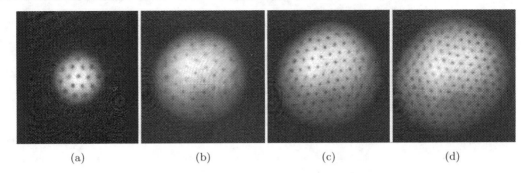

Figure 19.9 Triangular vortex lattice formed in rotating BECs with (a) 16, (b) 32, (c) 80, and (d) 130 vortices. Figure reprinted with permission from [209].

where $\vec{r}_\perp$ is the component of $\vec{r}$ normal to the rotation axis. This confining potential provides the centripetal acceleration needed to stabilize the circular motion. If we are able to tune the confining potential precisely, then the analogy to a free charged particle in a magnetic field becomes exact.

Exercise 19.7. If we interpret the neutral atoms in the BEC as having pseudocharge e matching that of the electron, answer the following questions.

(a) What is the magnetic field associated with a rotation rate of 100 Hz? (Express your answer in teslas.)

(b) What rotation rate would be needed to simulate a 10 T magnetic field?

As a result of this analogy, the equilibrium state of a rotating BEC above a threshold Ω is an Abrikosov vortex lattice, just like a superconductor in an external magnetic field H with $H_{c_1} < H < H_{c_2}$. Experimental observation of such a vortex lattice is unequivocal evidence of the presence of BEC; see Fig. 19.9. We should note, that since there is no way for a superfluid to "screen" the rotation, there is no notion of penetration depth here. Equivalently we can say that in a rotating superfluid $\lambda = \infty$, as a result of which $\kappa \to \infty$, so we are in the extreme type-II limit.

Another way to simulate the effect of a magnetic field on charged particles is to manipulate the interaction between (neutral) atoms and controlled laser beams to generate a so-called **synthetic gauge field** [210] in optical lattices. We will not delve into a detailed discussion of this subject here, but this method can be used to generate synthetic magnetic fields much higher than can be physically achieved with real magnets. The basic idea is to arrange the laser fields to modify the phase of the hopping-matrix elements for atoms in an optical lattice potential. By varying this phase in the appropriate manner from site to site, the particle can pick up a Berry phase when hopping around a closed loop in the lattice, as if it were a charged particle in a magnetic field. Similar techniques can be used to create **synthetic spin–orbit couplings** for neutral particles.

We now come to the most obvious connection between superconductivity and superfluidity – the (meta)stability of supercurrent. In Section 19.7 we discussed one mechanism for the supercurrent to decay – namely through nucleation and motion of vortices through the bulk, and the fact that there is a large (free) energy barrier for such processes that protects the supercurrent. This discussion applies to superfluids as well. In Section 18.3 we discussed another mechanism for the supercurrent to decay in a superfluid, namely fragmentation of the condensate via creating a macroscopically large number of elementary excitations. In principle such a mechanism exists in superconductors as well, although the way it works is more involved, as single-electron scattering tends to break a Cooper pair, resulting in an *additional* energy barrier for such processes. The critical current density is determined by the process with the lowest free-energy barrier, and is sensitive to details of the system/materials. We will not discuss this issue further other than briefly revisiting it in Section 19.10.

It should be very clear by now that the most important commonality between a superconductor and a superfluid is that they both have a condensate. As discussed in great detail in Chapter 18, the presence of a condensate implies long-range order and spontaneously broken symmetry. It turns out that the fact that in a superconductor the condensate is made of charged (instead of neutral) particles that couple to electromagnetic fields leads to a very subtle but fundamental difference between superconductivity and superfluidity. Such a coupling makes the condensate wave function ψ a *gauge-dependent* quantity, rendering its expectation value *not* a physical observable. As a result, the presence of this condensate does *not* indicate long-range order or spontaneous symmetry breaking in a superconductor.[32] An immediate consequence of this difference between superconductivity and superfluidity is that, unlike in the superfluid, there is *no* gapless Goldstone mode in superconductors. In fact, the coupling of the condensate also renders the photon, the quantum of electromagnetic field, massive. This is the famous **Anderson–Higgs mechanism** for gauge bosons to acquire mass. In a superconductor the presence of a mass for the electromagnetic field (the photon) is nothing but the Meissner effect. That is, the electromagnetic field cannot propagate freely in a superconductor, but instead decays with a characteristic length scale which defines the penetration depth.[33]

So what is the proper way to characterize the superconducting state, if it does not involve symmetry breaking? As it turns out, superconductivity is a topologically ordered state, in the same sense as in a fractional quantum Hall state and, in particular, as in the $\mathbb{Z}_2$ spin liquid discussed in Sections 17.7.1 and 17.8. To make the analogy closer, let us start by considering a 2D superconductor (or a superconducting thin film) on a torus; see Fig. 19.10. As we are going to see, there is a four-fold topological

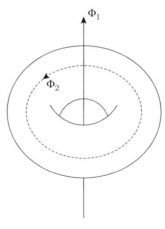

Figure 19.10 A 2D superconductor placed on a torus, with magnetic flux Φ_1 and Φ_2 going through the two holes but *not* into the superconductor itself.

---

[32] In fact, there are no gauge-invariant local observables associated with superconductivity that exhibit long-range order. For more detailed discussions, see Ref. [16].

[33] In a relativistic field theory a mass gap is related to a finite decay length via $\Delta = \hbar c/\lambda$. In a non-relativistic system the velocity has a different origin and in general is not the speed of light. It is also natural to discuss the fundamental difference between a superconductor and a normal metal in this context. As we discussed in Section 15.7, an external potential or electric field is screened in a metal, and the screening length plays a role very similar to that of the penetration depth in a superconductor. But a magnetic field is *not* screened in a normal metal. A superconductor, on the other hand, screens both electric and magnetic fields. Formally one can say that both longitudinal and transverse gauge potentials have acquired masses. If the system had Lorentz invariance these two masses would be the same, leading to the same screening length and penetration depth.

ground state degeneracy in this case. The four different ground states are distinguished by the amount of flux going around the two holes of the torus, Φ_1 and Φ_2. Their presence means we must have a non-zero vector potential $\vec{A}(\vec{r})$ on the surface of the torus, but as long as

$$\Phi_j = n_j \Phi_S, \qquad (19.130)$$

where $n_{j=1,2}$ are integers, they do not cost any (free) energy because we can always choose an appropriate phase of the condensate wave function $\psi(\vec{r})$ such that Eq. (19.104) is satisfied *everywhere*. One might think that in these cases we have only performed a gauge transformation from the case for $n_1 = n_2 = 0$, or no flux at all. But this is *not* the case. The crucial point here is the fact that the superconducting flux quantum Φ_S (corresponding to charge $e^* = 2e$ particles, or Cooper pairs) is only *one-half* that of the flux quantum for electrons, Φ_0 (see Eq. (19.8)). As a result, for n_j being an odd integer, the flux has a real physical effect, and *cannot* be gauged away for electrons that make up the superconductor. States with even and odd numbers of flux quanta therefore represent *different* physical states. The key point is that this degeneracy does *not* rely on any symmetry, and is thus of topological origin.[34] Using the same argument, if our 2D superconductor is put on a genus-g surface (see Fig. 13.5), the degeneracy is 4^g, corresponding to the different ways in which odd integer multiples of Φ_S can be placed in the holes, in agreement with Eq. (16.182) for $D = 4$.

What about 3D superconductors? In this case we can theoretically impose periodic boundary conditions for the electronic wave function along all three directions, which effectively puts the system on a 3D torus, $\mathcal{T}_3$. In this case we (effectively) have three holes in which the flux can reside *without* entering the superconductor itself. As a result, the topological degeneracy is eight in this case, which is the same as in the 3D toric code (see Exercise 17.25). Of course, in this case the "holes" are theoretical constructs, not physical holes that can accommodate real magnetic flux. In Chapter 20 we will discuss another consequence of this topological order, namely fractionalization of elementary excitations in superconductors.

19.9 Josephson Effect

In earlier sections we have argued that superconductivity results from the existence of a condensate, whose wave function is the order parameter of the Ginzburg–Landau theory. The Josephson effect is a remarkable manifestation of the existence of such a condensate. Instead of developing a fully microscopic theory, in this section we present only the minimal phenomenology needed to understand the Josephson effect.[35]

An ordinary (normal-state) tunnel junction consists of two metallic electrodes separated by a thin oxide barrier which allows electrons to tunnel quantum mechanically from one electrode to the other. Because, even in rather small (mesoscopic but not nanoscopic) grains, the size of the grain is much larger than the 1 Å scale of a typical Fermi wavelength, the "particle-in-a-box" quantum level spacing is extremely tiny and the density of states is essentially a continuum as shown in Fig. 19.11(a).

---

[34] Strictly speaking, the degeneracy is not quite exact, but rather the energy shift is exponentially small in the system size divided by the coherence length (the effective size of the Cooper pairs we will learn about in Chapter 20). One way to see this is to observe that the microscopic single-electron wave functions *do* change in the presence of an odd number of superconducting flux quanta through a hole in the torus, because effectively the periodic boundary conditions are replaced by antiperiodic boundary conditions. Provided that the superconducting **condensation energy** far exceeds the Thouless energy, this effect will be negligible. Another way to say it is that, if the electrons are tightly bound in Cooper pairs, they travel together as an effective charge-$2e$ object for which the superconducting flux quantum is invisible. There is an exponentially small probability that one member of a tightly bound pair will circle the torus and the other will not.

[35] In this section we follow the discussion in [211].

As a result, the tunneling of electrons is an incoherent process that is well described by the irreversible dynamics of Fermi's Golden Rule (see Appendix C for a more detailed discussion). Under voltage bias V the chemical potential is higher in one grain than it is in the other by an amount eV. Electrons in this energy interval are able to tunnel from the "cathode" to the "anode" without Pauli blocking due to the exclusion principle. As a result, the tunneling current is linear in the applied voltage (on voltage scales low compared with the tunnel barrier height) and the junction can be characterized as a simple resistor (shunted by the junction capacitance) as shown in Fig. 19.11(b).

Let us begin our discussion of the Josephson effect by considering a small isolated metallic electrode of some metallic superconductor. Because the electrode is isolated, the number of electrons in it is fixed and well-defined. For the moment, let us assume that the electron number has even parity. As discussed in Chapter 15, the essential physics of a superconductor is that an effective attractive interaction (e.g. resulting from virtual phonon exchange) leads to pairing of electrons into Cooper pairs. If the number of electrons in the electrode is even, then the quantum ground state has all of the electrons paired up into a special non-degenerate low-energy ground state. The system has a finite excitation gap, 2Δ, defined by the energy needed to break a Cooper pair into two separate quasiparticles. The scale of this gap is typically several kelvins. As illustrated in Fig. 19.12(a), the quasiparticle states form a near continuum *above* the gap. Only if the electrode is extremely tiny (on the few-nm scale) will the single-particle level spacing be appreciable. We will not consider this limit further.

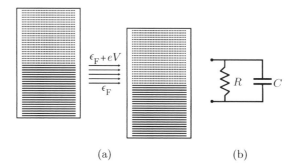

(a)　　　　　　　　(b)

Figure 19.11 (a) Normal metal tunnel junction with equilibrium Fermi energy ϵ_F and applied voltage bias V. The single-particle spectrum in each electrode is dense, and occupied (empty) states are indicated by solid (dashed) lines. (b) Neglecting certain many-body effects associated with the Coulomb interaction, the capacitance and tunnel conductance of the junction lead to the equivalent circuit shown. Figure reprinted with permission from [211].

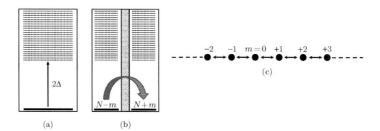

Figure 19.12 (a) The spectrum of a superconducting Cooper-pair box (CPB). For the case of an even number of electrons, there is a unique non-degenerate ground state separated from the excited states by a gap 2Δ. (b) A pair of CPBs connected by a tunnel barrier to form a Josephson junction. Ignoring the Coulomb energy, there is a large family of degenerate ground states labeled by an integer m representing the number of Cooper pairs transferred from one condensate to the other. (c) The "tight-binding" lattice along which the junction "moves" via Josephson tunneling between "sites" labeled by adjacent values of m. Figure reprinted with permission from [211].

Recalling that a temperature of 1 K corresponds to a frequency of approximately 21 GHz, we will be considering the limit of low temperatures and low frequencies relative to the gap: $k_B T$, $\hbar\omega \ll 2\Delta$. Hence to a good approximation we can say that the primary effect of superconductivity is to reduce the vast Hilbert space of the electrons in the electrode to a single quantum state, $|N\rangle$, labeled by the number of pairs which is a constant of the motion. All other states are separated from this unique quantum ground state by a large gap, and we will ignore them. Now consider a system with *two* metallic electrodes connected by a **tunnel junction** as shown in Fig. 19.12(b). We will again limit our attention to the quantum ground state of each electrode, assuming the electrons in each are fully paired up. Once again the total number of electron pairs in the system is fixed to some value $N = N_L + N_R$. Unlike in the case of the single electrode, the total number of pairs no longer uniquely specifies the quantum state. We must specify the number on each electrode. A useful notation for the set of low-energy states of this **Josephson junction** is

$$|m\rangle = |N_L - m, N_R + m\rangle, \qquad (19.131)$$

where m defines the number of pairs that have transferred through the tunnel junction from the left electrode to the right electrode starting from some reference state with pair number $N_{L(R)}$ on the left (right) electrode.

The two electrodes form a small capacitor, but for the moment we will ignore the Coulomb energy that builds up as Cooper pairs are transferred from one electrode to the other. In this case we then have a 1D family of states labeled by the integer m, which are degenerate in energy.[36] Remarkably, it turns out that the tunnel junction permits pairs of electrons to coherently tunnel together from one side to the other. This is captured by the phenomenological Hamiltonian

$$H_T = -\frac{1}{2} E_J \sum_m \{|m\rangle\langle m+1| + |m+1\rangle\langle m|\}. \qquad (19.132)$$

The parameter E_J is called the **Josephson coupling energy** and is a measure of the ability of Cooper pairs to coherently tunnel through the junction. We will provide a microscopic derivation of this Josephson coupling Hamiltonian in Chapter 20 using the BCS theory and the general method developed in Appendix I for derivation of low-energy effective Hamiltonians. For now, however, we will focus on understanding the rich and quite remarkable phenomenology of the Josephson junction.

We see that H_T causes m to either increase or decrease by unity, corresponding to the tunneling of a pair to the right or the left. We can gain some intuition by noticing that H_T is identical to that of a 1D tight-binding lattice model with near-neighbor hopping amplitude $E_J/2$, as illustrated in Fig. 19.12(c). The (unnormalized) eigenfunctions are plane-wave-like states labeled by a dimensionless "wave vector" $\varphi = ka$, where $a = 1$ is the "lattice constant":

$$|\varphi\rangle = \sum_{m=-\infty}^{+\infty} e^{im\varphi}|m\rangle. \qquad (19.133)$$

Recalling the cosine dispersion of the 1D tight-binding band with near-neighbor hopping (discussed in Section 7.4), we see that $|\varphi\rangle$ is an eigenstate of the Josephson tunneling Hamiltonian

$$H_T|\varphi\rangle = -E_J \cos\varphi|\varphi\rangle. \qquad (19.134)$$

Imagine a wave packet moving to the right on our tight-binding lattice. This corresponds to a net current of Cooper pairs coherently tunneling through the junction causing $\langle m \rangle$ to increase linearly

---

[36] Here we need to point out the crucial difference between the superconducting gap and the gap in a band insulator like silicon. The latter gap is tied to a particular density at which the electrons fill up all the valence band (bonding orbitals) and none of the conduction band. In a superconductor the particular density is not important – the gap follows the Fermi surface as it expands and contracts with density.

with time. The group velocity of the packet is given by the derivative of the energy with respect to the wave vector,

$$v_{\mathrm{g}}(\varphi) = \frac{1}{\hbar} \frac{\partial}{\partial \varphi} [-E_{\mathrm{J}} \cos \varphi], \tag{19.135}$$

so the net current flowing is given by the celebrated **(first) Josephson relation**

$$I(\varphi) = 2e v_{\mathrm{g}}(\varphi) = \frac{2e}{\hbar} E_{\mathrm{J}} \sin \varphi. \tag{19.136}$$

The maximum possible coherent (dissipationless) current occurs at $\varphi = \pi/2$ and is called the **critical current**,

$$I_{\mathrm{c}} = \frac{2e}{\hbar} E_{\mathrm{J}}. \tag{19.137}$$

If more current than this is forced through the junction, the voltage rises from zero to a high value above the excitation gap, and our low-energy effective model is no longer applicable.

As an alternative approach to the derivation of the Josephson relation for the current, let us define the operator $\hat{n}$ to be the number operator for the Cooper pairs transferred across the junction:

$$\hat{n} \equiv \sum_m |m\rangle m \langle m|. \tag{19.138}$$

Hamilton's equation of motion gives for the current operator

$$\hat{I} \equiv 2e \frac{d\hat{n}}{dt} = 2e \frac{i}{\hbar} [H_{\mathrm{T}}, \hat{n}] \tag{19.139}$$

$$= -i \frac{e}{\hbar} E_{\mathrm{J}} \sum_m \{|m\rangle\langle m+1| - |m+1\rangle\langle m|\}. \tag{19.140}$$

Next we simply note that the plane-wave energy eigenfunctions are also eigenfunctions of the current operator, obeying

$$\hat{I}|\varphi\rangle = I_{\mathrm{c}} \sin \varphi |\varphi\rangle, \tag{19.141}$$

which is of course equivalent to Eq. (19.136).

We continue for the moment to ignore the Coulomb interaction, but, as a first step toward including it, let us think about the situation where an external electric field is applied and maintained in such a way that there is a fixed voltage drop V across the tunnel junction. This adds to the Hamiltonian a term

$$U = -(2e)V\hat{n}. \tag{19.142}$$

Within the analogy to the tight-binding lattice being used here, this term corresponds to an additional linear potential (on top of the periodic lattice potential) produced by a constant electric field V/a.[37] Using the semiclassical equation of motion for Bloch electrons, Eq. (8.14) yields the equally celebrated **(second) Josephson relation**,

$$\hbar \, \partial_t \varphi = 2eV, \tag{19.143}$$

relating the time rate of change of the "momentum" $\hbar\varphi$ to the "force" $2eV$. Equivalently, the solution of the Schrödinger equation is

$$|\Psi(t)\rangle = e^{+\frac{i}{\hbar} E_{\mathrm{J}} \int_0^t d\tau \, \cos \varphi(\tau)} |\varphi(t)\rangle, \tag{19.144}$$

---

[37] Recall that a is the fake lattice constant, which has already been set to 1 in our notation; also $-2e$ is the charge of the Cooper pair.

where

$$\varphi(t) = \varphi(0) + \frac{2e}{\hbar} V t. \tag{19.145}$$

The overall phase factor in front of the wave function is not particularly important, but the time variation of the "wave vector" $\varphi(t)$ is extremely important because it leads to the **ac Josephson effect**, namely that dc voltage bias leads to an ac current

$$\langle \hat{I}(t) \rangle = I_c \sin[\varphi(0) + \omega t], \tag{19.146}$$

where the ac Josephson frequency is given by[38]

$$\omega = 2\pi \frac{2e}{h} V. \tag{19.147}$$

The inverse flux quantum in frequency units is[39]

$$\frac{2e}{h} \approx 483.5978525(30) \, \text{MHz}/\mu\text{V}. \tag{19.148}$$

Since frequency is the easiest physical quantity to measure with high accuracy, the ac Josephson effect finds great practical use in metrology to maintain (but not define) the SI volt.

We are now in a position to discuss the physical meaning of the (lattice) momentum-like parameter φ which determines the Josephson current through Eq. (19.136). Recall that in a bulk superconductor the local supercurrent density is proportional to the gauge-covariant derivative of the order parameter phase (see Eq. (19.63)); for simplicity let us assume that the vector potential $\vec{A} = \vec{0}$ for the moment, in which case Eq. (19.63) reduces to

$$\vec{J}(\vec{r}) \propto \nabla \varphi(\vec{r}). \tag{19.149}$$

Now consider applying the above to two (weakly) coupled superconductors, and that within each individual superconductor the order-parameter phase is a constant. In this case the derivative above should be replaced by the phase *difference* between the two superconductors when evaluating the supercurrent flowing between them (which is the Josephson current by definition):

$$I \propto \varphi_2 - \varphi_1, \tag{19.150}$$

when $\varphi_2 - \varphi_1$ is small. Comparing the above with Eq. (19.136), and expanding it to linear order in φ, we come to the conclusion that φ can be identified as the phase difference of the superconducting order parameters in the absence of $\vec{A}$:

$$\varphi = \varphi_2 - \varphi_1. \tag{19.151}$$

More generally, φ is the gauge-invariant phase difference:

$$\varphi = \varphi_2 - \varphi_1 + \frac{e^*}{c} \int_1^2 \vec{A}(\vec{r}) \cdot d\vec{r}. \tag{19.152}$$

Note that φ appearing in Eq. (19.136) as the argument of the sine function is not an accident; since φ_1, φ_2, and thus φ are defined modulo 2π, all physical observables must be periodic functions of φ with 2π periodicity.[40]

---

[38] The attentive reader may recognize that this oscillating Josephson current has the same mathematical origin as the Bloch oscillation of an electron in a tight-binding band, subject to a constant electric field.

[39] This is the 2014 CODATA recommended value, with the number in parentheses being the standard uncertainty in the last two digits.

[40] The attentive reader might be wondering why there cannot be a proportionality constant in Eq. (19.151). This periodicity requirement removes this ambiguity.

We are now in a position to promote φ to an operator $\hat{\varphi}$, whose eigenvalues and eigenstates are φ and $|\varphi\rangle$, respectively. Strictly speaking, since φ is only defined modulo 2π, $\hat{\varphi}$ is *not* well-defined; what *is* well-defined are periodic functions of $\hat{\varphi}$ with periodicity 2π, which can be expanded as summations of $e^{il\varphi}$, with l being an integer. Let us now figure out the explicit form of the operator $e^{il\varphi}$. From Eq. (19.136) it is clear that

$$\hat{I}(\hat{\varphi}) = \frac{2e}{\hbar} E_J \sin\hat{\varphi} = -\frac{ieE_J}{\hbar}(e^{i\hat{\varphi}} - e^{-i\hat{\varphi}}). \tag{19.153}$$

Comparing this with Eq. (19.140), we immediately find that

$$e^{i\hat{\varphi}} = \sum_m |m\rangle\langle m+1|; \qquad e^{il\hat{\varphi}} = \sum_m |m\rangle\langle m+l|. \tag{19.154}$$

We thus find that

$$[e^{i\hat{\varphi}}, \hat{n}] = e^{i\hat{\varphi}}, \tag{19.155}$$

or (somewhat sloppily ignoring the fact that $\hat{\varphi}$ is not actually well-defined)

$$[\hat{\varphi}, \hat{n}] = -i, \tag{19.156}$$

so that $\hat{\varphi}$ and $\hat{n}$ are canonical conjugates of each other. Unlike the more standard canonical conjugate $\hat{x}$ and $\hat{p}$, however, the 2π periodicity of $\hat{\varphi}$ forces the eigenvalues of $\hat{n}$ to be quantized integers, similarly to the quantization of the $\hat{z}$ component of orbital angular momentum as it is the canonical conjugate of an angle. It is the fact that Cooper pairs can be transferred only in integer units that leads to the Cooper-pair number being represented by the integer-valued angular momentum of a quantum rotor whose angular position is the phase variable $\hat{\varphi}$. Because $\hat{\varphi}$ is an angular variable, only periodic functions of this operator (such as $\cos\hat{\varphi}$) are actually well-defined operators that preserve the angular periodicity of the wave functions on which they operate.

Now the Hamiltonian of the junction with voltage bias V can be expressed in terms of $\hat{\varphi}$ and $\hat{n}$:

$$H = -E_J\cos\hat{\varphi} - 2eV\hat{n}, \tag{19.157}$$

and the pair of Hamilton equations

$$\frac{\partial\hat{n}}{\partial t} = \frac{\partial H}{\partial(\hbar\hat{\varphi})}, \tag{19.158}$$

$$\hbar\frac{\partial\hat{\varphi}}{\partial t} = -\frac{\partial H}{\partial\hat{n}} \tag{19.159}$$

corresponds to the pair of Josephson relations found in Eqs. (19.136) and (19.143).

In the discussions above the canonical pair $\hat{\varphi}$ and $\hat{n}$ correspond to the order-parameter phase difference and Cooper-pair number difference between two superconductors. It should be clear by now that a similar canonical pair can be introduced for each individual superconductor, so that (again in the absence of $\vec{A}$)

$$\hat{\varphi} = \hat{\varphi}_2 - \hat{\varphi}_1. \tag{19.160}$$

The canonical conjugates of $\hat{\varphi}_1$ and $\hat{\varphi}_2$ are the numbers of Cooper pairs in superconductors 1 and 2, respectively.

Exercise 19.8. Derive Eq. (19.141).

> **Exercise 19.9.** Verify for yourself that Eq. (19.144) does indeed solve the Schrödinger equation:
>
> $$i\hbar \, \partial_t |\Psi(t)\rangle = (H_T + U)|\Psi(t)\rangle. \tag{19.161}$$

> **Exercise 19.10.** Consider a single superconductor. As we argued above, it can be described by a canonical pair $\hat{\varphi}$ and $\hat{n}$, corresponding to its order-parameter phase and number of Cooper pairs, respectively; they satisfy the canonical commutation relation (19.156). (i) Find the wave functions of eigenstates of $\hat{n}$ in the $\hat{\varphi}$ representation, and show that the uncertainty of φ is maximum. (ii) Find the wave functions of eigenstates of $\hat{\varphi}$ in the $\hat{n}$ representation, and show that the uncertainty of n is maximum. These are reflections of uncertainty relations dictated by (19.156).

19.9.1 Superconducting Quantum Interference Devices (SQUIDS)

One of the most basic and useful superconducting circuits is the **superconducting quantum interference device (SQUID)** schematically illustrated in Fig. 19.13. The circuit consists of two Josephson junctions wired in parallel. Because there are two paths for Cooper pairs to take between the lower and upper parts of the circuit and because Josephson tunneling is completely coherent, we must add the resulting amplitudes coherently. The relative phase of the two tunnel amplitudes is controlled by the Aharonov–Bohm phase

$$\theta = 2\pi \frac{\Phi}{\Phi_S}, \tag{19.162}$$

which increases by 2π for every superconducting flux quantum added to the loop. The resulting periodic oscillations in the properties of the SQUID make it a very sensitive device for measuring magnetic flux. The oscillations also prove the relative phase coherence of Josephson tunnel events separated by macroscopic distances (mm to cm).

The flux through the loop is a combination of any externally applied flux plus flux generated by the currents flowing through the self-inductance of the loop. When we studied vortices in bulk type-II superconductors, we learned that, for energetic reasons, the current distribution around the vortex always adjusts itself to make the total flux an integer multiple of the superconducting flux quantum Φ_S. This makes the Aharonov–Bohm phase a multiple of 2π and makes the current vanish exponentially at large distances from the vortex core, thereby minimizing the energy. The contribution to the magnetic

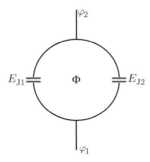

Figure 19.13 The circuit diagram of a SQUID consisting of a superconducting loop interrupted by two Josephson junctions and enclosing magnetic flux Φ. Cooper pairs can tunnel coherently through each of the Josephson junctions, and the interference of the associated amplitudes is controlled by the Aharonov–Bohm phase associated with magnetic flux enclosed by the loop.

flux from circulating currents in the SQUID depends on details of the circuit parameters (the loop self-inductance) but in many cases is quite small. For simplicity of presentation we will neglect the self-inductance of the loop and assume that Φ is simply the external flux which is continuously adjustable and is *not* quantized. This approximation is equivalent to assuming that the entire upper half of the circuit has the same fixed phase φ_2 and the entire lower half of the circuit has fixed phase φ_1. We saw in Eq. (19.63) that currents in superconductors are associated with gradients in the (gauge-invariant) phase. Our approximation here is equivalent to neglecting the gradient of the phase associated with the currents flowing in the wires of the circuit and assuming the entire (gauge-invariant) "phase drop" occurs across the Josephson tunnel junctions.[41]

Let us first consider the case of zero flux in the loop. Then the Josephson tunneling Hamiltonian is simply

$$H = -E_{J1} \cos(\varphi_2 - \varphi_1) - E_{J2} \cos(\varphi_2 - \varphi_1). \tag{19.163}$$

Because the tunneling amplitudes simply add, the circuit acts as if it is a single Josephson junction with effective Josephson coupling $E_{J1} + E_{J2}$. Now consider the case of magnetic flux Φ passing through the loop. The line integral of the vector potential around the loop will equal the enclosed flux. Let us choose a gauge in which the vector potential vanishes within the wires of the loop and has support only within the insulating barrier of the first tunnel junction. Then the Hamiltonian becomes

$$H = -E_{J1} \cos(\varphi_2 - \varphi_1 - \theta) - E_{J2} \cos(\varphi_2 - \varphi_1), \tag{19.164}$$

where θ is the Aharonov–Bohm phase in Eq. (19.162). Alternatively we can choose a more symmetric gauge with equal and opposite vector potential in each tunnel junction, which is convenient for the case where $E_{J1} = E_{J2} = E_J$, which we assume henceforth

$$H = -E_J \left[\cos\left(\varphi_2 - \varphi_1 - \frac{\theta}{2}\right) + \cos\left(\varphi_2 - \varphi_1 + \frac{\theta}{2}\right) \right].$$
$$= -2E_J \cos\left(\frac{\theta}{2}\right) \cos(\varphi_2 - \varphi_1). \tag{19.165}$$

We see from this result that the magnetic flux modulates the effective overall Josephson coupling and that the maximum dissipationless current that can be carried by the SQUID is given by

$$I_c = 2I_c^0 \left| \cos\left(\pi \frac{\Phi}{\Phi_S}\right) \right|, \tag{19.166}$$

where I_c^0 is the critical current of each individual junction. The fact that the critical current of the SQUID can be modulated all the way down to zero is unique to the symmetric case where the two junctions are identical and the quantum interference of the two paths is perfectly destructive.

> **Exercise 19.11.** Find the critical current as a function of the loop flux for an asymmetric SQUID with individual junction critical currents I_{c1} and I_{c2}.

Because the superconducting flux quantum is so small ($\Phi_S \approx 2.0678 \times 10^{-15}$ Wb), the SQUID is a very sensitive magnetometer. For a loop area of $A = 1\,\text{mm}^2$, the magnetic field that produces one superconducting flux quantum through the loop is $B_0 = \Phi_S/A = 2.0678 \times 10^{-9}$ T.

---

[41] This approximation is valid when the wires are big enough to carry currents without the need for large phase gradients and short enough that small phase gradients do not add up to large phase differences along the length. That is, we are assuming that both the kinetic inductance and the ordinary geometric (i.e. electromagnetic) inductance of the wires are negligible.

19.10 Flux-Flow Resistance in Superconductors

We can now combine the knowledge we have acquired in the sections above to understand the concept of **flux-flow resistance**. In order to build superconducting electromagnets, we need superconductors to continue to superconduct in the presence of high magnetic fields. We have already seen that, in type-I superconductors, an applied magnetic field is fully excluded by the Meissner effect, but suddenly destroys the superconductivity above a critical field. Type-II superconductors can tolerate much higher magnetic fields by allowing partial penetration of the magnetic field at the expense of having quantized vortices throughout the sample. Recall that each vortex has a phase winding of 2π around it. If the vortices move, the phase of the superconducting condensate therefore changes with time, and from the second Josephson relation we know that voltage will then appear in the superconductor, implying the existence of dissipation. Hence type-II superconductors transport supercurrents in the presence of a magnetic field only if the vortices are pinned by disorder in the materials. In fact the critical current generally *increases* with disorder up to a certain range of disorder strength.[42] This is reminiscent of the fact that dissipationless quantum Hall transport requires localization of the quasiparticles (and quasiholes) by disorder. Once again we see that perfection requires dirt![43]

19.11 Superconducting Quantum Bits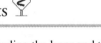

Many of the technologies of the twentieth century, including the laser and the transistor, were products of the revolution that began with the development of the quantum theory over the course of the first third of the century. A "second quantum revolution" now under way is based on a new understanding of how quantum devices can in principle be used to manipulate and communicate information in unexpected and powerful ways.

Information is physical and is encoded in the states of physical systems. A classical bit can be in two possible states, labeled 0 and 1. These states could represent a transistor being on or off, for example. A **quantum bit** (or a "qubit") is a physical system (e.g. an atom or an electron spin) that can be in two possible discrete energy states. The quantum state $|0\rangle$ could for example represent the ground state of the atom and the state $|1\rangle$ could represent the first excited state. (There are usually many other higher excited states, but we will assume we can ignore them.) The **superposition principle** tells us that there exist an infinite number of quantum states intermediate between these two states. These are superposition states of the form

$$|\psi\rangle = \cos(\theta/2)|0\rangle + e^{i\varphi}\sin(\theta/2)|1\rangle. \tag{19.167}$$

Every two-level quantum system is mathematically equivalent to a spin-1/2 and the angles θ and φ represent the orientation of the (pseudo)spin vector on the Bloch sphere.

One can view the superposition state as meaning that we are *uncertain* whether the qubit is in 0 or 1, which seems like a bad situation for the purposes of doing computations. On the other hand, one can also say that (in some sense) the qubit is both 0 and 1 at the same time. This leads to the idea of "quantum parallelism" in which a computer built of quantum bits might be able to carry out many different computations simultaneously in parallel using the superposition principle. The import of this possibility becomes clearer when we consider that a register of N classical bits can be in one of 2^N possible

---

[42] Because of this people often introduce disorder intentionally in the superconducting materials used to make superconducting electromagnets.

[43] Note that quantum Hall plateaux with perfectly quantized Hall resistance require disorder to stabilize, whose width often increases with disorder. In fact there is a very close analogy between the quantum Hall effect and superconductivity; see the appendix of Ref. [142].

states, whereas a register of N quantum bits can be in a superposition of all 2^N states at once (and, furthermore, there are many different such superpositions). One thus begins to see a hint of the possibility of exponential speedup of computations using quantum superposition. Quantum entanglement (see Appendix F) which produces non-classical correlations among different qubits is another resource that adds to the power of quantum hardware. The entire field of (theoretical) computer science is built on foundational theorems which tell us that the hardware implementation does not much matter, since any universal computer can be programmed to emulate any other. Quantum hardware is different, however, and certain tasks can be carried out exponentially faster than on classical hardware. For a simple primer on quantum information see [212] and for more detailed references on the subject see [213, 214].

There are many possible systems that could be used to store and manipulate quantum information, including atoms, ions, and superconducting electrical circuits. Here we will give a very brief introduction to quantum electrical circuits and the construction of "artificial atoms" that are based on Josephson junctions. The reader is directed to Appendix G for a brief primer on the quantization of the electromagnetic modes of simple linear circuits.

Let us consider the so-called **"Cooper-pair box,"** which consists of two small superconducting islands connected to each other via a Josephson junction.[44] Remarkably, this simple system acts as an "artificial atom" with quantized energy levels, and we can use the two lowest levels as the states of a superconducting quantum bit or "qubit." The transition frequencies between quantized energy levels can be engineered to lie in the microwave domain, and because of their large physical size (microns to millimeters) these artificial atoms can interact extremely strongly with microwave photons, leading to a novel new form of cavity quantum electrodynamics known as **"circuit QED"** [215]. The circuit QED architecture allows us to study the non-linear quantum optics of electrical circuits at the single-photon level and holds great promise as one of the leading technologies for the construction of quantum information processors [211, 215, 216].

In our discussion of the Josephson effect we have so far ignored the Coulomb energy of the electrons. As Cooper pairs tunnel from one island to the other they will charge the capacitor between the two islands. The Hamiltonian including both the Coulomb energy and the **Josephson energy** is given by (see also the discussion of the Hamiltonian for electrical circuits in Section G.3 of Appendix G)

$$H = \frac{Q^2}{2C} - E_{\mathrm{J}} \cos \hat{\varphi}. \tag{19.168}$$

The island charge built up via the tunneling process is given by

$$Q = 2e\hat{n}, \tag{19.169}$$

where $\hat{n}$ is the number of Cooper pairs that have tunneled through the junction. Defining the charging energy $E_{\mathrm{c}} = e^2/2C$, the Hamiltonian can be written

$$H = 4E_{\mathrm{c}}\hat{n}^2 - E_{\mathrm{J}} \cos \hat{\varphi}. \tag{19.170}$$

Returning to Eq. (19.157), we see that this is not quite the most general form of the charging energy. In the presence of an external bias voltage (intentionally applied or accidentally generated by stray charges trapped in the dielectric of the Josephson junction) there will be a term linear in $\hat{n}$. This is conventionally included by generalizing Eq. (19.170) to

$$H = 4E_{\mathrm{c}}(\hat{n} - n_{\mathrm{g}})^2 - E_{\mathrm{J}} \cos \hat{\varphi}, \tag{19.171}$$

---

[44] Equivalently, one could have a single small island connected to a large metallic ground. The physics will be the same.

where n_g is the so-called **"gate charge"** or **"offset charge"** induced on the capacitor by the external electric field. In ordinary electrical circuits we view the charge placed on a capacitor as a continuous variable, but here the integer eigenvalues of $\hat{n}$ are important. The number of Cooper pairs that have tunneled is an integer, and the offset charge can have a non-trivial effect on the spectrum of the Hamiltonian. Worse, because the offset charge can fluctuate in time (as stray charges move around), the energy levels of this system can fluctuate in time. Fortunately it turns out, however, that, in the limit $E_J \gg E_c$, the offset charge has an exponentially small effect on the spectrum. Cooper-pair boxes operated in this limit are known as **"transmon"** qubits [217], and at the time of writing hold the world record ($\sim 100\,\mu s$) for the phase coherence time (the time for which a quantum superposition of the ground and excited states maintains a well-defined phase, or, equivalently, there is no entanglement between the system and its environment). We will therefore ignore the offset charge from now on.

In the "position representation," where the wave function depends on the (angular) position variable $\Psi(\varphi)$, and consistently with Eq. (19.155), the number operator has the representation

$$\hat{n} = +i \frac{\partial}{\partial \varphi}. \tag{19.172}$$

The Hamiltonian in the position representation is thus

$$H = -4E_c \frac{\partial^2}{\partial \varphi^2} - E_J \cos \varphi. \tag{19.173}$$

As illustrated in Fig. 19.14, this has the same form as the Hamiltonian for a quantum rotor in the presence of a gravitational field which tries to keep the rotor near position $\varphi = 0$. The moment of inertia of the rotor is inversely proportional to E_c, and the gravitational field is proportional to E_J. For large enough gravitational field ($E_J \gg E_c$), the low-lying eigenstates will have support only in a limited region near $\varphi = 0$, and one can safely expand the cosine to obtain

$$H = H_0 + V, \tag{19.174}$$

where (ignoring a trivial constant in the Hamiltonian)

$$H_0 = -4E_c \frac{\partial^2}{\partial \varphi^2} + \frac{E_J}{2} \varphi^2 \tag{19.175}$$

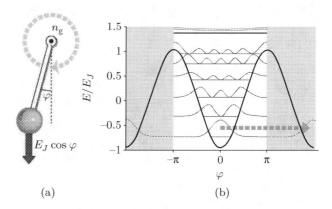

(a)   (b)

Figure 19.14 (a) The transmon qubit Hamiltonian is equivalent to that of a quantum rotor in a gravitational field. For large E_J/E_c (strong "gravity"), the effect of the offset charge n_g is exponentially small because it turns out to be related to the amplitude for the rotor undergoing a large excursion in which it rotates through a full 2π angle. (b) The square of the wave function for low-lying excited states in the cosine potential. Figure reprinted with permission from [217]. Copyright (2007) by the American Physical Society.

and

$$V \approx -\frac{E_J}{4!} \varphi^4. \tag{19.176}$$

Here we are effectively ignoring the fact that φ is an angular variable and treating it as non-compact (i.e. living on the real line). In that case, H_0 is simply a harmonic oscillator with "mass" $\hbar^2/(8E_c)$, "spring constant" E_J, and characteristic frequency $\Omega_J = \sqrt{8E_J E_c}/\hbar$, and V represents a negative anharmonicity resulting from the fact that the next term in the Taylor-series expansion of the cosine potential "softens" the potential.

The ground state wave function of H_0 is a Gaussian of the form

$$\Psi_0(\varphi) = \frac{1}{(2\pi \varphi_{ZPF}^2)^{\frac{1}{4}}} e^{-\frac{\varphi^2}{4\varphi_{ZPF}^2}}, \tag{19.177}$$

where

$$\varphi_{ZPF}^2 = \sqrt{\frac{2E_c}{E_J}} \tag{19.178}$$

is the variance of the phase (position) in the ground state due to zero-point fluctuations. Switching to second quantized notation in terms of raising and lowering operators, we can express the position operator as

$$\hat{\varphi} = \varphi_{ZPF}(a + a^{\dagger}), \tag{19.179}$$

and the conjugate charge as[45]

$$\hat{n} = +\frac{i}{2\varphi_{ZPF}}(a - a^{\dagger}). \tag{19.180}$$

The harmonic part of the Hamiltonian has the usual form

$$H_0 = \hbar\Omega_J\left(a^{\dagger}a + \frac{1}{2}\right). \tag{19.181}$$

The harmonic oscillator frequency Ω_J is known as the Josephson plasma frequency. Expressing the anharmonic term in second quantized form and making the rotating-wave approximation (i.e. neglecting off-diagonal terms in the number basis (the eigenstate basis of H_0) we have

$$V \approx -\frac{E_J}{4!}\varphi_{ZPF}^4(a + a^{\dagger})^4 \approx -\frac{E_c}{2}(a^{\dagger}a^{\dagger}aa + 2a^{\dagger}a), \tag{19.182}$$

and so the full Hamiltonian becomes (neglecting the constant-zero-point-energy term)

$$H \approx \hbar\omega_{01}a^{\dagger}a - \frac{E_c}{2}a^{\dagger}a^{\dagger}aa, \tag{19.183}$$

where the renormalized frequency which gives the first excitation energy is given by

$$\omega_{01} = \Omega_J - E_c/\hbar, \tag{19.184}$$

and the second transition frequency is given by

$$\omega_{12} = \omega_{01} - E_c/\hbar. \tag{19.185}$$

---

[45] Notice that, since we are no longer treating φ as an angular variable, $\hat{n}$ now has continuous eigenvalues, like ordinary linear momentum, rather than integer eigenvalues, like angular momentum.

The anharmonicity $\Delta f = (\omega_{12} - \omega_{01})/(2\pi) = -E_c/\hbar$ is for typical parameters in transmon qubits in the range of 100–300 MHz, while the transition frequency $\omega_{01}/(2\pi)$ is typically about 5 GHz. The anharmonicity is sufficiently large that a microwave pulse applied at frequency ω_{01} will not excite transitions to level 2 or higher so long as the envelope of the pulse is smooth and has a duration of more than a few nanoseconds. Thus the transmon qubit can be well approximated as a two-level system, i.e. a pseudo-spin-1/2 system whose superposition states are described by the orientation of the "spin" vector on the Bloch sphere. A much more complete pedagogical introduction to superconducting qubits and circuit QED can be found in [211].

Exercise 19.12. Derive Eqs. (19.177)–(19.185).

20 Microscopic Theory of Superconductivity

We learned in Chapter 15 that Fermi liquids are subject to a pairing instability if weak attractive interactions exist between the Landau quasiparticles. This pairing instability leads to superconductivity. In this chapter we will discuss: (i) the physical origin of the attractive interaction; (ii) the nature of the superconducting ground state, and its elementary excitations; (iii) prediction of physical properties of superconductors from the microscopic description; and (iv) classification of different superconducting phases. In particular, in (iii) we will not only re-derive the properties obtained in Chapter 19 from the London and Ginzburg–Landau theories and relate them to microscopic parameters, but also study those properties which are inaccessible from such effective theories.

The microscopic theory for superconductivity in what are now called conventional, low-transition-temperature (T_c) superconductors was advanced by **Bardeen, Cooper, and Schrieffer (BCS)** in 1957. This **BCS theory** is arguably the most successful theory in condensed matter physics. Even though the pairing mechanism is not entirely clear for high-T_c cuprate and some other unconventional superconductors at the time of writing, there is overwhelming evidence that pairing of electrons and condensation of such Cooper pairs, the most important and robust ingredients of BCS theory, are the origin of superconductivity in all these materials. The BCS and Ginzburg–Landau theories provide qualitatively correct, and often quantitatively accurate, descriptions of the superconducting phases of these materials.[1]

BCS theory not only describes superconductivity in electronic superconductors, but also fermionic superfluidity in ^{3}He, cold atomic gases, and high-density nuclear matter (in nuclei and in neutron stars) and possibly even quark matter. The corresponding energy scale, as measured, e.g., by T_c, ranges from nK to GeV! The applicability of the theory over dozens of orders of magnitudes in energy scale is testimony to its unparalleled robustness.

20.1 Origin of Attractive Interaction

The bare Coulomb interaction between two electrons is of course purely repulsive. What dictates the low-energy physics in metals, however, is the *screened* interaction between Landau quasiparticles. As

---

[1] Paradoxically, for high-T_c materials, it is often the *normal* state properties (above T_c or H_{c2}) that are hard to understand, and may not be captured by the Fermi-liquid theory. In such circumstances it is not clear whether the question of what the pairing "mechanism" is is even well-defined. What is clear is that the electrons are paired in the superconducting state.

we learned in Chapter 15, screening results from the response of the electron liquid to the presence and motion of an additional charged object, including a quasiparticle near the Fermi surface, which is the case of interest here. The screened interaction is significantly weaker than the bare Coulomb interaction, but remains purely repulsive within the Thomas–Fermi approximation.[2]

In Chapter 15 we considered screening due to other electrons, but did not take into account the possible screening effects due to the motion of nuclei and ions. As we will see below, this is justifiable for high (compared with typical phonon)-energy electronic processes, but at lower energy screening due to ion motion cannot be neglected, and leads to attractive interactions between quasiparticles due to its **over-screening effect**. We will also see that in this regime the frequency dependence (or equivalently the retardation in time) of the screened interaction is important.

Strictly speaking, the bare Coulomb interaction, while usually taken to be instantaneous in condensed matter physics, is mediated by exchanging (virtual) photons (which travel at the speed of light), and is thus retarded in time and has a frequency dependence. This retardation effect, however, has essentially no observable effect as electrons typically travel at speeds much lower than the speed of light.[3] The electronically screened interaction discussed in Chapter 15 *does* have frequency dependence, as described by the dielectric function $\epsilon(q, \omega)$. The characteristic frequency in $\epsilon(q, \omega)$ is the plasma frequency

$$\Omega_{pl} = \sqrt{\frac{4\pi e^2 n_e}{m_e}}, \tag{20.1}$$

which is typically of order $10\,\mathrm{eV}/\hbar$ or above in metals (here n_e and m_e are the electron density and mass, respectively). As a result, for electronic processes with $\omega \ll \Omega_{pl}$, we are in the static limit described by the Thomas–Fermi or Lindhard screening theory. On the other hand, if we were to go to high enough frequency with $\omega \lesssim \Omega_{pl}$, we would find (see Eq. (15.103)) $\epsilon(\omega) < 0$, resulting in *attractive* interaction! As we demonstrate below, the same phenomena occur due to screening from ion motion, except that this happens at a much lower (phonon) frequency. This is the essence of phonon-mediated attractive interaction between electrons (or more precisely, Landau quasiparticles).

To proceed, let us write down the full dielectric function that includes contributions from both electrons and ions:

$$\epsilon(q, \omega) = 1 - \frac{4\pi e^2}{q^2}[\Pi_e(q, \omega) + \Pi_{ion}(q, \omega)], \tag{20.2}$$

where $\Pi_e(q, \omega)$ and $\Pi_{ion}(q, \omega)$ are the polarization functions of electrons and ions, respectively. In Chapter 15 we included only the former. Now let us focus on the latter. Using arguments similar to those in Chapter 15, we expect

$$\frac{4\pi e^2}{q^2}\Pi_{ion}(q, \omega) \approx \frac{\Omega_p^2}{\omega^2} \tag{20.3}$$

for $\omega \sim \Omega_p$, where by analogy with Eq. (20.1)

$$\Omega_p = \sqrt{\frac{4\pi Z^2 e^2 n_{ion}}{M}} \tag{20.4}$$

---

[2] In more sophisticated descriptions of screening like Lindhard theory, one finds Friedel oscillations in the screened interaction at large distance, which already contains some *attractive components*; see Eq. (15.96). We will come back to this point later.

[3] The typical atomic scale for electron speeds is αc, where α is the fine-structure constant.

is the ion plasma frequency with M being the ionic mass, Ze being the ionic charge,[4] and $n_{\text{ion}} = n_e/Z$ being the ionic density. Because the ions are much more massive than electrons, we have $\Omega_p \ll \Omega_{\text{pl}}$. For typical electronic processes with energy scale much bigger than $\hbar\Omega_p$, the screening effect of the ions is negligible. But this is not the case for low-energy processes.

Identifying

$$\epsilon_{\text{el}}(q, \omega) = 1 - \frac{4\pi e^2}{q^2} \Pi_e(q, \omega) \tag{20.5}$$

as the dielectric function taking into account electron screening only (which is what we studied in Chapter 15), we have the screened interaction between electrons taking the form

$$V_{\text{sc}}(q, \omega) = \frac{4\pi e^2}{q^2} \frac{1}{\epsilon(q, \omega)} \approx \frac{4\pi e^2}{q^2} \frac{1}{\epsilon_{\text{el}}(q, \omega)} \frac{1}{1 - \Omega_p^2/[\omega^2 \epsilon_{\text{el}}^2(q, \omega)]} \tag{20.6}$$

$$\approx \frac{4\pi e^2}{q^2 + q_{\text{TF}}^2} \frac{\omega^2}{\omega^2 - \Omega_p^2(q)}, \tag{20.7}$$

where in the last step above we used the Thomas–Fermi approximation

$$\epsilon_{\text{el}}(q, \omega) \approx \frac{q^2 + q_{\text{TF}}^2}{q^2}, \tag{20.8}$$

and

$$\Omega_p(q) = \Omega_p/\sqrt{\epsilon_{\text{el}}(q, \omega)} \approx \Omega_p(q/q_{\text{TF}}) \propto q \tag{20.9}$$

at long wavelength. The physical interpretation of this linearly dispersing mode is that it represents an **acoustic phonon mode**, corresponding to collective motion of the ions. The reason why this collective-mode frequency is *not* Ω_p is because the charge redistribution due to ion motion is screened by the electrons, whose response time is $\sim 1/\Omega_{\text{pl}}$, much faster than any ion motion. Obviously the characteristic scale of Ω_p is the Debye frequency $\omega_D \sim v_s/a$, where v_s is the phonon velocity and a is the lattice spacing. We see that, for a *low-energy* electronic process with energy transfer less than $\hbar\omega_D = k_B\Theta_D$ (Θ_D is the Debye temperature), the effective interaction between electrons (or, more precisely, Landau quasiparticles) may actually be attractive!

Pictorially we imagine that the positive-ion background can distort in the presence of an electron. Because the ions are slow-moving, the excess positive background charge density remains in place after the electron has left. A second electron can then take advantage of this distortion to lower its electrostatic energy. As long as the electrons are separated in time, they will not repel each other directly, but rather effectively attract each other via the lattice distortion.

20.2 BCS Reduced Hamiltonian and Mean-Field Solution

To capture the important effect of phonon-mediated attractive interaction between quasiparticles at low energy, Bardeen, Cooper, and Schrieffer introduced the following simple Hamiltonian (known as the **BCS reduced Hamiltonian**) that captures the essence of the superconducting ground state and elementary excitations:

---

[4] This is the plasma frequency that the ions would have if the electrons were not mobile and only provided a neutralizing background charge density. In reality the frequency of ion motion is reduced to the (approximately linearly dispersing) acoustic phonon frequency due to screening from the electrons. The electrons are much less massive than the ions and can easily follow their motion and provide strong screening. More on this shortly.

$$H_{\text{BCS}} = \sum_{\vec{k},\sigma=\uparrow,\downarrow} \epsilon_{\vec{k}} n_{\vec{k}\sigma} + \sum_{\vec{k},\vec{k}'} V_{\vec{k},\vec{k}'} c^{\dagger}_{+\vec{k}\uparrow} c^{\dagger}_{-\vec{k}\downarrow} c_{-\vec{k}'\downarrow} c_{+\vec{k}'\uparrow} , \tag{20.10}$$

where $n_{\vec{k}\sigma} = c^{\dagger}_{\vec{k}\sigma} c_{\vec{k}\sigma}$ is a number operator, and $\epsilon_{\vec{k}} \approx v_{\text{F}}(k - k_{\text{F}})$ is the quasiparticle energy measured from the chemical potential (which vanishes on the Fermi surface),

$$V_{\vec{k},\vec{k}'} = \begin{cases} -V_0, & |\epsilon_{\vec{k}}| < \hbar\omega_{\text{D}} \text{ and } |\epsilon_{\vec{k}'}| < \hbar\omega_{\text{D}}; \\ 0, & \text{otherwise.} \end{cases} \tag{20.11}$$

Here $\hbar\omega_{\text{D}} = k_{\text{B}}\Theta_{\text{D}}$ is the Debye energy, the characteristic phonon energy scale. The justification for keeping the attractive interaction at low energy while neglecting repulsive interaction at higher energy (which may actually be stronger!) is that the former is (marginally) relevant, while the latter is (marginally) irrelevant under the renormalization group (see Chapter 15). In general, the interaction term would involve particle operators with four different wave vectors (which, because of momentum conservation, sum to zero if umklapp processes can be ignored). The BCS Hamiltonian is much more restrictive than this, and keeps only the interaction between pairs of quasiparticles with exactly opposite momenta and spins. This approximation is justified by using the fact that the pairing instability is the strongest in this case.[5] Recall from the discussion in Chapter 15 that, if we restrict our attention to a thin energy shell around the Fermi surface, then pairs of particles of exactly opposite momenta within the shell can scatter to any other momentum within the shell, while pairs of particles with non-zero total momentum must leave the thin shell and go to high-energy states when they scatter. This is why we can very effectively study the pairing stability of the normal state, as well as resultant superconducting ground state and quaisparticle excitations using the BCS reduced Hamiltonian (20.10).

For the non-interacting ($V_0 = 0$) ground state, the two single-particle states labeled by $(\vec{k}, \uparrow)$ and $(-\vec{k}, \downarrow)$ are either both occupied or both empty, because $\epsilon_{\vec{k}} = \epsilon_{-\vec{k}}$. This property remains valid in the presence of the (simplified) scattering (pairing) term in Eq. (20.10).[6] Thus, to study the ground state *only*, we can truncate the full Hilbert space to the smaller subspace in which $(\vec{k}, \uparrow)$ and $(-\vec{k}, \downarrow)$ are either both occupied or both empty, and map H_{BCS} to an effective spin-1/2 model. Introducing a pseudospin for each $\vec{k}$, we define

$$S^z_{\vec{k}} = (n_{\vec{k}\uparrow} + n_{-\vec{k}\downarrow} - 1)/2, \tag{20.12}$$

$$S^+_{\vec{k}} = c^{\dagger}_{\vec{k}\uparrow} c^{\dagger}_{-\vec{k}\downarrow}, \tag{20.13}$$

$$S^-_{\vec{k}} = c_{-\vec{k}\downarrow} c_{\vec{k}\uparrow}. \tag{20.14}$$

Because they are formed from pairs of fermion operators, it is easy to verify that they precisely satisfy the usual spin commutation relations. We will shortly make use of the fact that the pseudospin up and down states obey

$$|\downarrow\rangle_{\vec{k}} = |0\rangle, \tag{20.15}$$

$$|\uparrow\rangle_{\vec{k}} = c^{\dagger}_{\vec{k}\uparrow} c^{\dagger}_{-\vec{k}\downarrow} |0\rangle, \tag{20.16}$$

where $|0\rangle$ is the **vacuum state**.

---

[5] This is fine as along as we are studying ground state and elementary excitations on top of it. We do need to include attraction between pairs with finite total momentum when studying states carrying a supercurrent, as well as in many other circumstances.

[6] This breaks down once we allow pairs of quasiparticles with non-zero total momentum to scatter. Removing these processes is a major simplification of the BCS reduced Hamiltonian, rendering it exactly soluble [218]. As we will see shortly, the mean-field approximation is quite accurate for typical superconductors, and that is what we will pursue here.

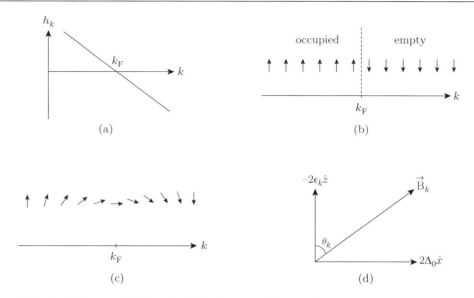

Figure 20.1 (a) The Zeeman field along the $\hat{z}$ direction provided by kinetic energy, in the pseudospin representation of the BCS reduced Hamiltonian (20.18). (b) The normal-state pseudospin configuration in the absence of the pairing interaction. (c) The pseudospin configuration in the BCS (superconducting) ground state. (d) Pseudo-magnetic-field coupling to a single pseudospin within the mean-field approximation.

Exercise 20.1. Verify that the pseudospins satisfy the usual spin commutation relations.

We can now write H_{BCS} in terms of these pseudospin operators:

$$H_{BCS} = \sum_{\vec{k}} \epsilon_{\vec{k}} (2S_{\vec{k}}^z + 1) + \frac{1}{2} \sum_{\vec{k},\vec{k}'} V_{\vec{k},\vec{k}'} (S_{\vec{k}}^+ S_{\vec{k}'}^- + S_{\vec{k}'}^+ S_{\vec{k}}^-) \tag{20.17}$$

$$= -2 \sum_{\vec{k}} h_{\vec{k}} S_{\vec{k}}^z + \sum_{\vec{k},\vec{k}'} V_{\vec{k},\vec{k}'} (S_{\vec{k}}^x S_{\vec{k}'}^x + S_{\vec{k}'}^y S_{\vec{k}}^y) + \text{constant.} \tag{20.18}$$

Here $h_{\vec{k}} = -\epsilon_{\vec{k}}$ plays the role of a Zeeman field in the $\hat{z}$ direction (see Fig. 20.1(a)). If $V_0 = 0$, this pseudo-Zeeman field aligns the pseudospins in the up direction for $k < k_F$, and in the down direction for $k > k_F$ (see Fig. 20.1(b)). The interaction term maps onto a bilinear coupling between the transverse components of the pseudospins. Importantly, for attractive interaction ($V_0 > 0$), this coupling is *ferromagnetic*, which, as we learned in Chapter 17, tends to align all the spins in the same direction.

There are several important differences between this pseudospin ferromagnet and that studied in Chapter 17. (i) The spins are labeled by momentum rather than position and the interaction is of infinite range in momentum space. That is, each spin interacts essentially equally with all the others with strength $-V_0$. (ii) Because the coupling exists only between the x and y components, pseudospin magnetization tends to develop in the xy plane (in the pseudospin space). We will assume it develops along the $\hat{x}$ direction, namely $\langle S_{\vec{k}}^x \rangle \neq 0$. Since the Hamiltonian (20.18) is invariant under rotation of pseudospin around the $\hat{z}$ axis, the appearance of non-zero $\langle S_{\vec{k}}^x \rangle$ indicates that this continuous symmetry is *spontaneously broken*. (That is, the ground state has lower symmetry than the Hamiltonian.) (iii) This ordering must overcome the effect of the kinetic-energy term, which tends to align pseudospins along the $\pm\hat{z}$ direction. The ordering field along the $\hat{z}$ direction gets stronger as one moves away from k_F, but vanishes right on the Fermi surface. We thus expect the pseudospin configuration to take the form shown in Fig. 20.1(c) even for infinitesimal attractive interaction.

As noted above, the spin–spin interaction is infinite-ranged in spin space. This strongly suggests that mean-field theory will be a good approximation because each spin couples in a sense to the average of all the others, and this average over many spins will thus not fluctuate much. To describe the symmetry-breaking ground state quantitatively, we therefore write $\vec{S}_{\vec{k}} = \langle \vec{S}_{\vec{k}} \rangle + \delta \vec{S}_{\vec{k}}$, substitute that into Eq. (20.18) and neglect terms that are quadratic in $\delta \vec{S}$.[7] This results in a mean-field Hamiltonian in which all the pseudospins decouple from each other:

$$H_{\text{MF}} = -\sum_{\vec{k}} \vec{B}_{\vec{k}} \cdot \vec{S}_{\vec{k}}; \tag{20.19}$$

$$\vec{B}_{\vec{k}} = -2\epsilon_{\vec{k}}\hat{z} - 2\sum_{\vec{k}'} V_{\vec{k},\vec{k}'} \langle S_{\vec{k}}^x \rangle \hat{x} = -2\epsilon_{\vec{k}}\hat{z} + 2\Delta_0\hat{x}, \tag{20.20}$$

where Δ_0 is an order parameter characterizing the symmetry breaking, and satisfies the self-consistent equation

$$\Delta_0 = V_0 \sum_{\vec{k}} \langle S_{\vec{k}}^x \rangle. \tag{20.21}$$

For later convenience we introduce

$$E_{\vec{k}} = \sqrt{\epsilon_{\vec{k}}^2 + \Delta_0^2} = |\vec{B}_{\vec{k}}|/2. \tag{20.22}$$

At zero temperature each pseudospin points in the direction of $\vec{B}_{\vec{k}}$, so the mean-field ground state wave function takes the form

$$|\psi_{\text{BCS}}\rangle = \prod_{\vec{k}} (u_{\vec{k}}|\downarrow\rangle_{\vec{k}} + v_{\vec{k}}|\uparrow\rangle_{\vec{k}}) = \prod_{\vec{k}} (u_{\vec{k}} + v_{\vec{k}} c_{\vec{k}\uparrow}^{\dagger} c_{-\vec{k}\downarrow}^{\dagger})|0\rangle, \tag{20.23}$$

where

$$u_{\vec{k}} = \sin\left(\frac{\theta_{\vec{k}}}{2}\right) = \sqrt{\frac{1}{2}\left(1 + \frac{\epsilon_{\vec{k}}}{E_{\vec{k}}}\right)}, \qquad v_{\vec{k}} = \cos\left(\frac{\theta_{\vec{k}}}{2}\right) = \sqrt{\frac{1}{2}\left(1 - \frac{\epsilon_{\vec{k}}}{E_{\vec{k}}}\right)}, \tag{20.24}$$

and $\theta_{\vec{k}}$ is the angle parameterizing the direction of $\vec{B}_{\vec{k}}$ (see Fig. 20.1(d)). Here $u_{\vec{k}}$ and $v_{\vec{k}}$ are also known as **BCS coherence factors**. We see that pseudospin ordering results in a state with coherent superposition between states whose total electron numbers differ by an even number, and is thus equivalent to the spontaneous breaking of the U(1) symmetry corresponding to particle number conservation. This important point will be further elaborated below.

The self-consistency equation (20.21) reduces to

$$\Delta_0 = \frac{V_0}{2} \sum_{\vec{k}} \sin\theta_{\vec{k}} = \frac{V_0}{2} {\sum_{\vec{k}}}' \frac{\Delta_0}{E_{\vec{k}}}, \tag{20.25}$$

where ${\sum_{\vec{k}}}'$ stands for summation over states satisfying $|\epsilon_{\vec{k}}| < \hbar\omega_{\text{D}}$. Canceling out Δ_0 on both sides and turning the summation into integration over the quasiparticle energy, the equation above reduces to

$$1 = D(0)V_0 \int_{-\hbar\omega_{\text{D}}}^{\hbar\omega_{\text{D}}} \frac{d\epsilon}{2\sqrt{\epsilon^2 + \Delta_0^2}} = D(0)V_0 \sinh^{-1}(\hbar\omega_{\text{D}}/\Delta_0), \tag{20.26}$$

resulting in

$$\Delta_0 = \hbar\omega_{\text{D}} \Big/ \sinh\left[\frac{1}{D(0)V_0}\right] \approx 2\hbar\omega_{\text{D}}e^{-\frac{1}{D(0)V_0}} \tag{20.27}$$

---

[7] This is the same procedure that we followed in deriving the Hartree equations in Chapter 15.

in the weak-coupling limit $D(0)V_0 \ll 1$. Here $D(0)$ is the normal-state quasiparticle density of states at the Fermi energy per spin species.

It is interesting to compare this expression for the superconducting gap with the expression in Eq. (15.182) for the Cooper-pair binding energy computed within the approximation that there is only a single Cooper pair outside the Fermi sea. The factor of 2 in the exponent means that the binding energy computed by Cooper *vastly* underestimates the true gap when the coupling is weak. This is because the Cooper problem neglects the fact that actually there are many paired electrons above the Fermi surface and many paired holes below the Fermi surface in the true (BCS mean-field) solution presented here. The reader may wish to revisit Exercise 15.24 to see how to correctly treat the pairing instability within the renormalization-group approach.

It is interesting to ask what happens for $V_0 < 0$, namely when we have repulsive interactions. Mathematically one finds that the only solution to Eq. (20.21) is $\Delta_0 = 0$, resulting in no pseudospin ordering and corresponding symmetry breaking. Physically this corresponds to no pairing when there is only repulsive interaction. We can gain additional insight by noting that, in the pseudospin mapping, repulsive interaction corresponds to *antiferromagnetic* coupling among the pseudospins. Because all of these pseudospins are coupled with equal strength, they are highly frustrated and this leads to no ordering, as we saw in Chapter 17.

As discussed earlier, a non-zero Δ_0 indicates spontaneous development of pseudospin magnetization in the xy plane that breaks the rotation symmetry along the $\hat{z}$ direction. Thus Δ_0 is the corresponding order parameter. In general this spontaneous magnetization can point in any direction in the xy plane, not just along the $\hat{x}$ direction as assumed thus far. We should thus generalize the definition of Δ_0 in Eq. (20.21) to

$$\Delta_0 = V_0 \sum_{\vec{k}} (\langle S_{\vec{k}}^x \rangle + i \langle S_{\vec{k}}^y \rangle), \tag{20.28}$$

where we find that Δ_0 is actually a *complex* order parameter, whose *phase* corresponds to the *direction* of pseudospin magnetization. It is thus a U(1) order parameter. It was shown by Gor'kov [219] that it is in fact proportional to the Ginzburg–Landau (complex) order parameter ψ that we encountered in Chapter 19 (more on this point later in Section 20.6).

This symmetry breaking is very easy to understand in the pseudospin language, by making the analogy to magnetism. However, in the original electron representation it leads to the BCS wave function (20.23) with *indefinite* charge! The particular phase of the coherent superposition of states with different Cooper-pair numbers is determined by the phase of Δ_0 (which determines the phase *difference* between the coherence factors $u_{\vec{k}}$ and $v_{\vec{k}}$). The indefiniteness of the charge is thus due to the appearance of the order parameter with a *definite* phase. This caused considerable confusion about gauge invariance and charge conservation when the BCS theory was first proposed. It turns out that when quantum fluctuations (beyond mean-field theory) in (the phase of) Δ_0 are treated correctly one does indeed find that the total charge is conserved and well-defined, but the local charge in one segment of a superconductor can be uncertain. We explored this concept earlier when we discussed the Josephson effect in Chapter 19, and will explore it further in Section 20.5.

20.2.1 Condensation Energy

Because of the pair coherence, the superconducting ground state (20.23) has higher kinetic energy but lower total energy than the normal state, which is simply a filled Fermi sea. The difference in total energy between the normal state and the superconducting state is known as the condensation energy, which we now calculate. We start with the normal state, whose kinetic energy is simply

$$\langle T \rangle_{\mathrm{N}} = \sum_{|\vec{k}| < k_{\mathrm{F}}} 2\epsilon_{\vec{k}} = \sum_{\vec{k}} 2\epsilon_{\vec{k}} \theta(-\epsilon_{\vec{k}}). \tag{20.29}$$

The interaction energy is[8]

$$\langle V \rangle_N = \sum_{\vec{k},\vec{k}'} V_{\vec{k},\vec{k}'} \langle S^x_{\vec{k}} S^x_{\vec{k}'} + S^y_{\vec{k}} S^y_{\vec{k}'} \rangle = \frac{1}{2} \sum_{\vec{k}} V_{\vec{k},\vec{k}}, \qquad (20.30)$$

where we used the fact that $\langle S^x_{\vec{k}} S^x_{\vec{k}'} + S^y_{\vec{k}} S^y_{\vec{k}'} \rangle = 0$ for $\vec{k} \neq \vec{k}'$ in the normal state. A similar calculation for the superconducting state yields

$$\langle T \rangle_S = \sum_{\vec{k}} 2\epsilon_{\vec{k}} |v_{\vec{k}}|^2 = \sum_{\vec{k}} \epsilon_{\vec{k}} \left(1 - \frac{\epsilon_{\vec{k}}}{E_{\vec{k}}} \right); \qquad (20.31)$$

$$\langle V \rangle_S = \langle V \rangle_N + \sum_{\vec{k} \neq \vec{k}'} V_{\vec{k},\vec{k}'} \langle S^x_{\vec{k}} \rangle \langle S^x_{\vec{k}'} \rangle = \langle V \rangle_N - \sum_{\vec{k}} \frac{|\Delta_0|^2}{2E_{\vec{k}}}. \qquad (20.32)$$

Putting everything together, we find the condensation energy

$$E_C = \langle T \rangle_N + \langle V \rangle_N - \langle T \rangle_S - \langle V \rangle_S \qquad (20.33)$$

$$= \sum_{\vec{k}}{}' \left[\frac{|\Delta_0|^2}{2E_{\vec{k}}} - |\epsilon_{\vec{k}}| \left(1 - \frac{|\epsilon_{\vec{k}}|}{E_{\vec{k}}} \right) \right] \qquad (20.34)$$

$$= |\Delta_0|^2 \sum_{\vec{k}}{}' \left(\frac{1}{|\epsilon_{\vec{k}}| + E_{\vec{k}}} - \frac{1}{2E_{\vec{k}}} \right) \qquad (20.35)$$

$$= 2D(0)|\Delta_0|^2 \int_0^\infty dx \left(\frac{1}{x + \sqrt{1+x^2}} - \frac{1}{2\sqrt{1+x^2}} \right) \qquad (20.36)$$

$$= \frac{1}{2} D(0)|\Delta_0|^2. \qquad (20.37)$$

Because the integral is convergent, in step (20.36) we allowed the integral's upper limit to go to ∞, even though the summation over $\vec{k}$ is restricted to $|\epsilon_{\vec{k}}| < \hbar\omega_D$. The error is negligible for weak-coupling superconductors satisfying $\Delta_0 \ll \hbar\omega_D$. As discussed in Chapter 19, the condensation energy is related to the critical field of a type-I superconductor:

$$E_C = \frac{L^3}{8\pi} H_c^2, \qquad (20.38)$$

where L^3 is the system volume. We see from this that microscopic theory allows us to make quantitative predictions of measurable quantities that are not possible in effective theories.

20.2.2 Elementary Excitations

Having understood the BCS superconducting ground state, we now study excitations on top of it. We can no longer use the pseudospin description because for excited states we must allow for fermion occupation configurations in which $\vec{k} \uparrow$ and $-\vec{k} \downarrow$ are *not* both occupied or both empty, and the pseudospin mapping breaks down. In other words, we now need to allow for *unpaired* electrons. We thus need to write the mean-field Hamiltonian in terms of the original electron operators (from now on we focus only on low-energy electronic states with $|\epsilon_{\vec{k}}| < \hbar\omega_D$):

$$H_{MF} = \sum_{\vec{k}}{}' \left[\epsilon_{\vec{k}} (c^\dagger_{+\vec{k}\uparrow} c_{+\vec{k}\uparrow} + c^\dagger_{-\vec{k}\downarrow} c_{-\vec{k}\downarrow}) - \Delta_0 (c^\dagger_{+\vec{k}\uparrow} c^\dagger_{-\vec{k}\downarrow} + c_{-\vec{k}\downarrow} c_{+\vec{k}\uparrow}) \right]. \qquad (20.39)$$

---

[8] It is often stated in the literature that the interaction energy is zero for the normal state. As we see here, this is not quite the case.

This Hamiltonian is very similar to Eq. (17.103), in that it is quadratic but includes terms made of two creation or two annihilation operators. The difference here is that we have fermionic instead of bosonic operators. Bogoliubov introduced a similar transformation (also known as a **Bogoliubov transformation**) to diagonalize such fermionic quadratic Hamiltonians:

$$\alpha_{\vec{k}} = u_{\vec{k}} c_{\vec{k}\uparrow} - v_{\vec{k}} c^{\dagger}_{-\vec{k}\downarrow}; \qquad (20.40)$$

$$\beta_{\vec{k}} = u_{\vec{k}} c_{\vec{k}\downarrow} + v_{\vec{k}} c^{\dagger}_{-\vec{k}\uparrow}. \qquad (20.41)$$

The α, β operators and their Hermitian conjugates satisfy fermion commutation relations, provided that $|u_{\vec{k}}|^2 + |v_{\vec{k}}|^2 = 1$, $u_{\vec{k}} = u_{-\vec{k}}$, and $v_{\vec{k}} = v_{-\vec{k}}$.

Exercise 20.2. Verify the statement above.

For H_{MF} of Eq. (20.39), the choice of $u_{\vec{k}}$ and $v_{\vec{k}}$ in Eq. (20.24) brings the mean-field Hamiltonian to diagonal form,

$$H_{\mathrm{MF}} = \sum_{\vec{k}}{}' E_{\vec{k}} (\alpha^{\dagger}_{\vec{k}} \alpha_{\vec{k}} + \beta^{\dagger}_{\vec{k}} \beta_{\vec{k}}), \qquad (20.42)$$

with $E_{\vec{k}}$ given by Eq. (20.22). We see that the physical meaning of $E_{\vec{k}}$ is simply the quasiparticle excitation energy of the superconductor.

Exercise 20.3. Verify Eq. (20.42).

The operators $\alpha^{\dagger}_{\vec{k}}$ and $\beta^{\dagger}_{\vec{k}}$ create fermionic quasiparticles with definite momentum ($\hbar\vec{k}$) and spin (up or down, respectively), but *not* definite charge, as they are made of coherent superpositions of electron creation and annihilation operators. Note that an electron of wave vector $\vec{k}$ and spin up carries the same momentum and spin as a hole at wave vector $-\vec{k}$ and spin down. This is why these coherent superposition quasiparticles carry definite spin and momentum, but indefinite charge. One can see from the pair creation and annihilation terms in Eq. (20.39) that an appropriate coherent superposition of zero and one Cooper pair at each wave vector will have lower energy than a state with a definite number of Cooper pairs. It is the pair creation/annihilation terms in the Hamiltonian that can scatter an electron into a hole and vice versa, since an electron and a hole differ in charge by one Cooper-pair charge.

These **Bogoliubov quasiparticles** are descendants of Landau quasiparticles in the normal state (a Fermi liquid), but their charge gets screened by the superconducting condensate.[9] In fact they become Landau quasiparticles for $k - k_{\mathrm{F}} \gg \Delta_0/v_{\mathrm{F}}$ ($|u_{\vec{k}}| \to 1$ while $|v_{\vec{k}}| \to 0$) and Landau quasiholes for $k_{\mathrm{F}} - k \gg \Delta_0/v_{\mathrm{F}}$ ($|v_{\vec{k}}| \to 1$ while $|u_{\vec{k}}| \to 0$), with the same spectrum (see Fig. 20.2(a)). But for $k = k_{\mathrm{F}}$ they are equal-weight combinations of an electron and a hole, with an excitation gap of Δ_0. This results in a gap in the quasiparticle excitation spectrum (see Fig. 20.2(a)) and in the density of states (Fig. 20.2(b)). The latter can be measured directly in electron tunneling spectroscopy between a normal metal and a superconductor; see Fig. 20.3(b) for an example.

It is useful to compare the Bogoliubov quasiparticle spectrum with that of Landau quasiparticles, which we first encountered in Fig. 15.1, and also included in Fig. 20.2(a). It is important to understand that in this representation the cost of creating an electron (Landau quasiparticle) with $k > k_{\mathrm{F}}$ is $\epsilon_k \approx v_{\mathrm{F}}(k - k_{\mathrm{F}})$. That is, it is the energy cost to move an electron from a reservoir at chemical potential μ to the excited state $\vec{k}$ above the Fermi surface. Correspondingly, the cost of creating a hole

---

[9] We can thus say that superconductors exhibit the phenomenon of spin–charge separation, in this particular sense.

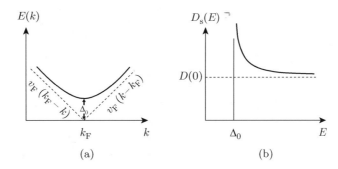

Figure 20.2 (a) The Bogoliubov quasiparticle spectrum (solid line) and corresponding Landau quasiparticle/quasihole spectrum (dashed lines). (b) The density of states (DOS) of Bogoliubov quasiparticles (solid line) and the corresponding DOS of Landau quasiparticles (dashed line).

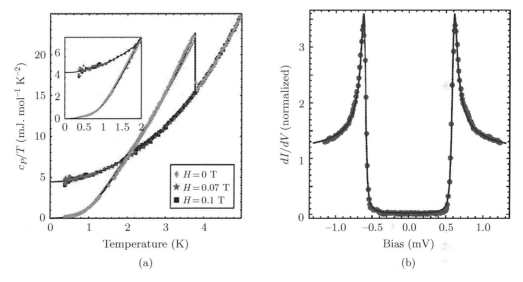

Figure 20.3 The specific heat and differential tunneling conductance of BiPd. (a) The specific heat divided by the temperature (C_P/T) as a function of temperature, for zero and finite magnetic fields. At zero field the material is superconducting for $T < T_c = 3.8$ K. The magnetic field suppresses superconductivity and reveals the normal-state property, with C_P/T approaching a constant as $T \to 0$, revealing a finite density of states at the Fermi energy. The increase of C_P/T with increasing T is due to the phonon contribution to C_P; C_P/T is much smaller in the superconducting state for $T \ll T_c$, due to the opening of the superconducting gap. (b) The differential tunneling conductance as a function of the bias voltage, at $T = 15$ mK. This is a direct measure of the density of states, from which a gap $\Delta \approx 0.6$ meV is clearly visible. The very small tunneling for bias voltage below the gap is due to thermal activation. $2\Delta/k_B T_c \approx 3.7$ is very close to the BCS prediction of Eq. (20.51). Figure adapted with permission from [220].

excitation in state $k < k_F$ is the cost of removing an electron from state $\vec{k}$ and moving it up into the reservoir at chemical potential μ. In the normal state, both of these costs are positive and go to zero as $|\vec{k}| \to k_F$. In the superconducting case, the energy cost remains finite and its minimum value is Δ_0.

One way to understand why inserting a single Bogoliubov quasiparticle costs a finite amount of energy is to recognize that it causes Pauli blocking, which interrupts the Cooper-pair coherence. The presence of a single (say, spin-up) fermion in state $\vec{k}$ blocks the Cooper pairs from scattering coherently into the pair state $(\vec{k}, \uparrow; -\vec{k}, \downarrow)$. This reduces the energy gain from the pair creation/annihilation terms in Eq. (20.39). In the language of the pseudospin mapping used earlier in our study of the ground state, such an unpaired single fermion makes the pseudospin mapping impossible for $\vec{k}$, as a result of which we lose one pseudospin and the corresponding (negative) pseudospin exchange energy.

Another type of elementary excitation in superconductors is the **vortices** that were discussed extensively in Chapter 19. We now demonstrate that they are analogs of the flux excitations in 2D and 3D Kitaev models discussed in Section 17.8 and Exercise 17.25, while Bogoliubov quasiparticles are like the charge excitations discussed there. The obvious similarity is that vortices carry magnetic (or physical) flux, and are point objects in 2D (and are thus particle-like excitations), and line or loop objects in 3D, just like the flux in 2D and 3D Kitaev models. Bogoliubov quasiparticles, on the other hand, are point objects in both cases, just like charge excitations in the toric code. The subtler but more fundamental similarity (which is again of topological origin), is the Berry phase a Bogoliubov quasiparticle picks up when circling a vortex due to the Aharonov–Bohm effect. One might think this phase is not well-defined because a Bogoliubov quasiparticle does not have a definitive charge. However, the key observation is that the quasiparticle is a linear superposition of an electron and a hole. Despite having opposite electric charge, the Aharonov–Bohm phase that they pick up,

$$e^{\pm ie\Phi_S/\hbar c} = -1, \tag{20.43}$$

is the *same*! We thus find that a Bogoliubov quasiparticle picks up a definitive -1 Berry phase from circling a vortex despite its indefinite charge, the same as that from a charge circling a flux in the toric code. This is another piece of evidence that a superconductor has the same topological order as a $\mathbb{Z}_2$ spin liquid, in addition to the same ground state degeneracy as was demonstrated in Section 19.8.

The Bogoliubov quasiparticle is yet another example of a fractionalized excitation in a condensed matter system, as it carries the spin but not the charge of an electron, similarly to a (neutral) soliton of the Su–Schrieffer–Heeger model. Here the fractionalization is due to the topological order, which is similar to what happens in fractional quantum Hall liquids and spin liquids.

Exercise 20.4. Calculate the Bogoliubov quasiparticle density of states (DOS) $D_s(E)$, and express it in terms of the normal-state DOS $D(0)$ (taken as a constant as we consider only states near k_F) and the gap Δ_0.

20.2.3 Finite-Temperature Properties

At finite temperature, thermally excited quasiparticles reduce the pairing order parameter Δ, in a manner similar to thermally excited spin waves reducing magnetic ordering that we studied in Chapter 17. In this case, Eq. (20.21) is replaced by

$$\Delta(T) = V_0 \sum_{\vec{k}} \langle c_{-\vec{k}\downarrow} c_{\vec{k}\uparrow} \rangle = V_0 {\sum_{\vec{k}}}' u_{\vec{k}} v_{\vec{k}} (1 - \langle \alpha_{\vec{k}}^{\dagger} \alpha_{\vec{k}} \rangle - \langle \beta_{\vec{k}}^{\dagger} \beta_{\vec{k}} \rangle) \tag{20.44}$$

$$= V_0 {\sum_{\vec{k}}}' \frac{\Delta(T)}{2E_{\vec{k}}} \tanh(\beta E_{\vec{k}}/2), \tag{20.45}$$

where $\beta = 1/k_B T$ is the inverse temperature and we used the fact that

$$u_{\vec{k}} v_{\vec{k}} = \frac{\Delta(T)}{2E_{\vec{k}}}. \tag{20.46}$$

At finite temperature we have

$$E_{\vec{k}} = \sqrt{\epsilon_{\vec{k}}^2 + \Delta^2(T)}, \tag{20.47}$$

reflecting the temperature dependence of the gap Δ.

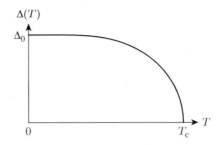

Figure 20.4 The temperature dependence of the order parameter/gap of an s-wave superconductor.

As the temperature T approaches the transition temperature T_c, we expect $\Delta(T \to T_c) \to 0$, and Eq. (20.44) becomes the self-consistency equation for T_c:

$$1 = V_0 \sum_{\vec{k}}{}' \frac{\tanh(\beta_c \epsilon_{\vec{k}}/2)}{2\epsilon_{\vec{k}}} = D(0)V_0 \int_{-\hbar\omega_D}^{\hbar\omega_D} \frac{\tanh(\beta_c \epsilon/2)}{2\epsilon}\, d\epsilon \tag{20.48}$$

$$= D(0)V_0 \ln(A\beta_c \hbar\omega_D), \tag{20.49}$$

resulting in

$$k_B T_c = 1/\beta_c = A\hbar\omega_D e^{-\frac{1}{D(0)V_0}}, \tag{20.50}$$

where $A \approx 1.13$ is a constant. Comparing this with Eq. (20.27), we find the **universal gap–T_c ratio** for weak-coupling s-wave superconductors:

$$\frac{2\Delta_0}{k_B T_c} \approx 3.5. \tag{20.51}$$

The temperature dependence of the gap (obtained from solving Eq. (20.44)) is illustrated schematically in Fig. 20.4, and knowing this allows one to calculate the thermodynamic properties of a superconductor straightforwardly. For example, Fig. 20.3(a) compares the specific heat of the superconducting and normal states of a particular superconductor. At low temperature ($k_B T \ll \Delta_0$), the specific heat is exponentially small due to the existence of a gap in the excitation spectrum. But as T approaches T_c it increases faster than, and eventually exceeds, the normal-state specific heat, which must be the case as there is no latent heat at the second-order phase transition between the two at $T = T_c$, so these two phases must have the same entropy.

20.3 Microscopic Derivation of Josephson Coupling

Equation (19.132) gave a phenomenological expression for the Hamiltonian H_T describing the tunneling of pairs of electrons through the barrier of a Josephson junction. We saw in Section 19.9 that this process is coherent and dissipationless since it leads to dc current flow with zero voltage drop across the junction (and ac current flow when the voltage drop is non-zero). The origin of this remarkable physics is the fact that a superconducting electrode containing an even number of electrons has a unique quantum ground state in which all the electrons are bound in pairs. It requires a finite energy to break a pair and so there is a gap to excited states above the ground state. Josephson tunneling therefore connects discrete quantum states which are degenerate (in the absence of Coulomb charging energy). The Josephson tunneling lifts this degeneracy, leading to the establishment of a well-defined phase difference across the junction and dissipationless current flow.

In sharp contrast, Eq. (C.15) gives the irreversible and dissipative tunneling of single electrons through a tunnel junction in the normal state. The derivation of this equation is based on Fermi's

Golden Rule since the single-particle tunneling is into a continuum of final states available above the Fermi level. Remarkably there is a deep connection between this expression and the Josephson coupling energy in the same junction when it is cooled into the superconducting state. To rather good accuracy, the value of the Josephson coupling E_J is given by the relation

$$E_J = \pi \Delta \frac{\hbar}{(2e)^2} G_N, \tag{20.52}$$

where $G_N = 1/R_N$ is the normal-state conductance. Using Eq. (19.137) relating the Josephson energy to the criticial current leads us to the **Ambegaokar–Baratoff relation** [221] connecting the superconducting **Josephson critical current** (at zero temperature) to the normal-state conductance

$$I_c = \frac{\pi}{2} \frac{\Delta}{e} G_N. \tag{20.53}$$

We will provide a microscopic derivation shortly, but one can understand Eq. (20.52) by noting that the microscopic tunneling Hamiltonian describes single-electron (not pair) tunneling across the junction. Thus coherent tunneling of a pair of electrons must be second order in the tunneling matrix element T_{lr}. In the intermediate state after the first tunneling, we have an unpaired electron in each island and so there is an energy of order 2Δ appearing in the denominator. Naively, this suggests that the Josephson coupling will scale inversely with the size of the superconducting gap. However, we note that, after the second electron has tunneled, we are left with a pair of holes in one island and an extra pair of electrons in the other. Taking into account the quantum amplitudes for both of these pairs to disappear into the condensate brings in two more factors of Δ (and two factors of the density of states to keep the dimensions correct), which explains why E_J is in fact linearly increasing with the superconducting gap. The normal-state conductance is computed using Fermi's Golden Rule for the tunneling rate, which is of course proportional to the square of the matrix element T_{lr}. Hence we see (qualitatively at least) the origin of Eq. (20.52).

One might wonder why E_J is not quartic in the tunneling matrix element, since the effective matrix element for the pair tunneling is quadratic, and perhaps we should square this as in Fermi's Golden Rule. The answer goes to the heart of the Josephson effect. Our effective Hamiltonian H_T connects discrete states (labeled by the number m of pairs that have tunneled), not a discrete state to a continuum. We will find not the irreversible incoherent dynamics of Fermi's Golden Rule (as we did in the normal metal case), but rather coherent tunneling which connects a set of degenerate states $\{|m\rangle\}$. We will in effect be doing degenerate perturbation theory (that is, finding a linear superposition of the degenerate states which diagonalizes H_T) and finding energy eigenvalues (and corresponding dynamics) which are first order in E_J, not second order.

To proceed with the microscopic derivation of Eq. (20.52), we note that the matrix element of the phenomenological Hamiltonian in Eq. (19.132) associated with the "unidirectional" hopping of a Cooper pair from the left electrode to the right electrode is $-E_J/2$. This is for the case where one has a definite number of Cooper pairs in each electrode. For the case where there is a definite phase relationship (and indefinite number difference) between the two electrodes, Eq. (19.134) tells us that the (unidirectional) Josephson tunneling matrix element is

$$M_{LR} = -\frac{E_J}{2} e^{i\varphi}. \tag{20.54}$$

On the other hand, the microscopic tunneling Hamiltonian in Eq. (C.11) sends an electron with spin σ from (say) left to right through the oxide tunnel barrier $T_{lr} c_{r\sigma}^{\dagger} c_{l\sigma}$, creating a broken pair and hence leaving a Bogoliubov quasiparticle in each electrode. We can apply the general methods developed in Appendix I to eliminate the coupling to this relatively high-energy intermediate state in order to obtain the effective low-energy Josephson Hamiltonian

$$M_{LR} = \sum_{r',l',\sigma'} \sum_{r,l,\sigma} \frac{T_{r'l'} T_{rl}}{-(E_l + E_r)} \langle \Psi | c_{l'\sigma'} c^{\dagger}_{r'\sigma'} c_{l\sigma} c^{\dagger}_{r\sigma} | \Psi \rangle, \tag{20.55}$$

where $E_{l,r}$ are the Bogoliubov quasiparticle energies in the intermediate state. Notice the important fact that, unlike the Fermi Golden Rule calculation, the second tunneling matrix element is not complex-conjugated. This is because both tunneling events carry an electron through the barrier in the same direction. It is this fact that will make the Josephson tunneling matrix element sensitive to the relative phase between the superconducting order parameters in the two electrodes.

In order to end up back in the ground state (despite the fact that two electrons have tunneled through the junction), the excess electrons on the right and the holes on the left need to pair up and disappear into the condensates. Since we are working perturbatively in powers of the hopping amplitude T, we can evaluate the matrix element in Eq. (20.55) to zeroth order in the hopping, in which case the left and right electrodes are uncorrelated,

$$\begin{aligned}
\langle \Psi | c_{l'\sigma'} c^{\dagger}_{r'\sigma'} c_{l\sigma} c^{\dagger}_{r\sigma} | \Psi \rangle &= \langle \Psi_L | c_{l'\sigma'} c_{l\sigma} | \Psi_L \rangle \langle \Psi_R | c^{\dagger}_{r\sigma} c^{\dagger}_{r'\sigma'} | \Psi_R \rangle \\
&= [u_l^* v_l]_L [u_r v_r^*]_R \, \delta_{l'\bar{l}} \delta_{r'\bar{r}} \delta_{\sigma'\bar{\sigma}} \\
&= e^{i\varphi} \left[\frac{\Delta}{2E_l} \right] \left[\frac{\Delta}{2E_r} \right] \delta_{l'\bar{l}} \delta_{r'\bar{r}} \delta_{\sigma'\bar{\sigma}}, \tag{20.56}
\end{aligned}$$

where we have used the BCS wave function in Eq. (20.23) and assumed it to be the same in each electrode except for the overall phase of the condensates. The phase difference φ between the two condensates is defined by

$$e^{i\varphi} = \frac{\Delta_L \Delta_R^*}{|\Delta|^2}. \tag{20.57}$$

The Kronecker delta functions enforce the fact that the pairs of electrons (and holes) have to be time reverses of each other if they are to disappear into the condensate.

Combining this result with Eq. (20.54), we obtain the following expression for the Josephson coupling energy:

$$\begin{aligned}
E_J &= T^2 D^2(0) \int_{-\infty}^{+\infty} d\epsilon \int_{-\infty}^{+\infty} d\epsilon' \frac{1}{E+E'} \frac{\Delta^2}{EE'} \\
&= 4T^2 D^2(0) \Delta^2 \int_{\Delta}^{+\infty} dE \int_{\Delta}^{+\infty} dE' \frac{1}{\sqrt{E^2 - \Delta^2}\sqrt{E'^2 - \Delta^2}[E+E']} \\
&= \pi^2 T^2 D^2(0) \Delta. \tag{20.58}
\end{aligned}$$

Using Eq. (C.15), it is straightforward to verify the claim in Eq. (20.52) that the Josephson energy is simply related to the normal-state conductance of the junction. The reader is encouraged to carefully verify all the steps in this derivation, paying particular attention to the minus signs associated with the ordering of the fermion creation and annihilation operators.

Exercise 20.5. As an alternative derivation, evaluate the matrix element in Eq. (20.56) by explicitly expressing the electron operators in terms of the two types of Bogoliubov quasiparticle operators and using the fact that the ground state is the quasiparticle vacuum.

In the above result, we have assumed T to be real but have deliberately written T^2 rather than $|T|^2$ to remind the reader that the Josephson coupling involves two electrons hopping in the same direction through the barrier, and so the matrix element actually depends on the choice of gauge.

20.4 Electromagnetic Response of Superconductors

As mentioned earlier, thermodynamic and many other properties of superconductors are easily cal-culable once the (T-dependent) quasiparticle spectrum is known. Many qualitative features are due to the gap in the spectrum, and in fact are similar to those of band insulators. The defining proper-ties of superconductors, however, lie in their response to electromagnetic fields, to which we turn in this section. We will focus on the Meissner effect, which, as discussed in Chapter 19, also implies zero resistivity. As we will see, the BCS coherence factors, which do not show up in thermodynamic properties, play crucial roles here.

In Section 19.2 we discussed London's remarkable insight that the Meissner effect originates from some kind of wave-function "rigidity" of the superconducting ground state, namely the wave function does *not* change when a (sufficiently weak) transverse vector potential is present. Here we discuss this point in more detail, and also from a slightly different viewpoint. We start the discussion from a single electron. The vector potential $\vec{A}(\vec{r})$ enters the kinetic-energy term of the Hamiltonian only:

$$T = \frac{1}{2m_e}\left[\vec{p} + \frac{e}{c}\vec{A}(\vec{r})\right]^2 \tag{20.59}$$

$$= \frac{p^2}{2m_e} + \frac{e}{2m_e c}\left[\vec{p}\cdot\vec{A}(\vec{r}) + \vec{A}(\vec{r})\cdot\vec{p}\right] + \frac{e^2}{2m_e c^2}|\vec{A}(\vec{r})|^2 \tag{20.60}$$

$$= T_0 + T_1 + T_2, \tag{20.61}$$

where $T_0 = p^2/2m_e$ is the original kinetic-energy term, while T_1 and T_2 correspond to the new terms that are linear and quadratic in $\vec{A}$, respectively. We will treat T_1 and T_2 as perturbations, and calculate their contributions to the ground state energy. For a ground state wave function $\psi_0(\vec{r})$ that is real and positive, we expect $\langle\psi_0|T_1|\psi_0\rangle = 0$, thus the only contribution at first order (in the perturbation series) to the ground state energy comes from T_2:

$$\Delta E_2 = \langle\psi_0|T_2|\psi_0\rangle = \frac{e^2}{2m_e c^2}\int d^3\vec{r}|\psi_0(\vec{r})|^2|\vec{A}(\vec{r})|^2. \tag{20.62}$$

This correction to the ground state energy, which is positive definite and thus corresponds to the diamagnetic response, takes a form consistent with that in Eq. (19.43), with an ultra-local "Meissner kernel"

$$K^{\mu\nu}(\vec{r} - \vec{r}') = \frac{4\pi e^2|\psi_0(\vec{r})|^2}{m_e c^2}\delta^{\mu\nu}\delta(\vec{r} - \vec{r}'). \tag{20.63}$$

In particular, by taking a functional derivative (see Appendix H) with respect to $\vec{A}(\vec{r})$, we obtain the current density

$$\vec{J}(\vec{r}) = -\frac{e^2|\psi_0(\vec{r})|^2}{m_e c}\vec{A}(\vec{r}), \tag{20.64}$$

which resembles the London equation with the superfluid density equal to the total electron density. In fact, this is how we understood the diamagnetism of an individual atom in Section 17.3.1, *by working in a specific gauge*.[10]

Unfortunately, we are not allowed to stop at this first-order perturbation treatment of T_2. The reason is that the second-order contribution of T_1 to the energy is also quadratic in $\vec{A}$, and thus of the same order as ΔE_2 above. One might think that, if there is a gap separating the ground state from all

---

[10] Another way to understand this result is, of course, to recognize that the mechanical momentum in the presence of a vector potential becomes $\vec{\Pi} = \vec{p} + (e/c)\vec{A}(\vec{r})$, thus Eq. (20.64) follows if the wave function does not change. This is how we understood persistent currents in Chapter 10.

excited states (which is the case for a superconductor) and the gap is large enough, this second-order correction due to T_1 would be negligible, since there will be a large energy denominator. But this argument is invalid, as can be seen from considering the special case in which $\vec{A}(\vec{r}) = \nabla \chi(\vec{r})$ is a pure gauge and cannot change the ground state energy E_0. But we *do* find a positive contribution from it in (20.62), which must be canceled out exactly by the contribution from T_1 (which is always negative and thus represents a paramagnetic contribution).[11] A related point is that the "Meissner kernel" of Eq. (20.63) is *not* gauge-invariant, and thus cannot be the complete response function.

Having understood the importance of treating the T_1 (paramagnetic) and T_2 (diamagnetic) terms on an equal footing, let us now consider the response of the system to a transverse vector potential along $\hat{y}$, with wave vector $\vec{q} = q\hat{x}$:

$$\vec{A}(\vec{r}) = A_0 \hat{y} \cos(qx), \tag{20.65}$$

describing a magnetic field

$$\vec{B}(\vec{r}) = -q A_0 \hat{z} \sin(qx). \tag{20.66}$$

London's wave-function rigidity would suggest that there is no response of the superconducting ground state wave function to this vector potential, and thus no correction to the ground state energy from the T_1 term. (Recall that the second-order energy correction comes from the first-order correction to the wave function.) We will see that this is indeed the case in the long-wavelength limit $q \to 0$. We will start, however, by first considering the normal state, and for simplicity we consider non-interacting electrons living in the 2D xy plane. In this case the unperturbed ground state is a filled Fermi sea:

$$|\psi_0\rangle = \prod_{|\vec{k}|<k_F,\sigma} c_{\vec{k}\sigma}^\dagger |0\rangle. \tag{20.67}$$

In second quantization, T_1 (for many-electron systems) becomes (see Appendix J)

$$T_1 = \frac{eA_0}{2m_ec} \sum_{\vec{k}\sigma} \hbar k_y \left(c_{\vec{k}+\vec{q}/2,\sigma}^\dagger c_{\vec{k}-\vec{q}/2,\sigma} + c_{\vec{k}-\vec{q}/2,\sigma}^\dagger c_{\vec{k}+\vec{q}/2,\sigma} \right). \tag{20.68}$$

A straightforward second-order perturbation calculation[12] yields its contribution to the ground state energy

$$\Delta E_1 = -\frac{2e^2 A_0^2}{m_ec^2} \sum_{|\vec{k}-\vec{q}/2|<k_F, |\vec{k}+\vec{q}/2|>k_F} \frac{k_y^2}{|\vec{k}+\vec{q}/2|^2 - |\vec{k}-\vec{q}/2|^2} \tag{20.69}$$

$$= -\frac{e^2 A_0^2}{m_ec^2} \sum_{|\vec{k}-\vec{q}/2|<k_F, |\vec{k}+\vec{q}/2|>k_F} \frac{k_y^2}{\vec{k}\cdot\vec{q}}. \tag{20.70}$$

The dimensionless sum above can be turned into an integral:

$$I(q) = \sum_{|\vec{k}-\vec{q}/2|<k_F, |\vec{k}+\vec{q}/2|>k_F} \frac{k_y^2}{\vec{k}\cdot\vec{q}}$$

$$= \frac{L^2}{(2\pi)^2} \frac{1}{q} \int_0^{2\pi} \left[\frac{\sin^2\theta}{\cos\theta} \int k^2\, dk \right] d\theta, \tag{20.71}$$

---

[11] This cancelation may be proven using the fact that $T_1 = -(i/\hbar c)[\chi(\vec{r}), T_0] = -(i/\hbar c)[\chi(\vec{r}), H_0]$, where H_0 is the unperturbed Hamiltonian, and following steps similar to those that led to the proof of the f-sum rule in Chapter 4.

[12] Every term in T_1 above creates a particle–hole pair with particle momentum $\vec{k} \pm \vec{q}/2$ and hole momentum $\vec{k} \mp \vec{q}/2$, which is an excited state of T_0, provided that the former is outside the Fermi sea and the latter is inside. Also important is the fact that every term in T_1 generates a *different* excited state. This calculation could also be done in first quantization.

where L^2 is the area of the system, and k and θ are the magnitude and angle of $\vec{k}$ measured from that of $\vec{q}$, respectively. The range of integration for k is determined by $|\vec{k} - \vec{q}/2| < k_F$, $|\vec{k} + \vec{q}/2| > k_F$. Obviously this range is proportional to q, thus $I(q \to 0)$ is a non-zero constant. A lengthy but elementary calculation finds

$$I(q) = \frac{k_F^2 L^2}{8\pi} \left[1 - \frac{1}{6} \frac{q^2}{k_F^2} + O(q^4) \right]. \tag{20.72}$$

> **Exercise 20.6.** Verify Eq. (20.72).

Using the fact that the total number of electrons is $N_e = k_F^2 L^2 / 2\pi$, and the single-electron density of states per spin is $D = mL^2/2\pi\hbar^2$, we have

$$\Delta E_1 = -\frac{N_e e^2 A_0^2}{4m_e c^2} + \frac{1}{6} D\mu_B^2 q^2 A_0^2 + O(q^4). \tag{20.73}$$

On the other hand, we have

$$\Delta E_2 = \frac{N_e e^2}{2m_e c^2} \overline{|\vec{A}|^2} = \frac{N_e e^2 A_0^2}{4m_e c^2}, \tag{20.74}$$

where the overbar stands for the spatial average. We thus find the correction to the ground state energy

$$\Delta E = \Delta E_1 + \Delta E_2 = \frac{1}{6} D\mu_B^2 q^2 A_0^2 + O(q^4) = \frac{1}{3} D\mu_B^2 \overline{B^2} + O(q^4), \tag{20.75}$$

namely the contributions to the Meissner kernel from T_1 and T_2 *cancel each other out* in the long-wavelength limit $q \to 0$. Obviously this cancelation occurs because the Fermi-sea ground state wave function of the free-electron gas is *not* rigid, and responds to the paramagnetic T_1 term substantially, by mixing in particle–hole pairs with momentum $\pm\hbar\vec{q}$. The remaining term $O(q^2 A_0^2) \propto B_q^2$, which is positive, corresponds to Landau diamagnetism. The result is consistent with what was obtained in Section 17.9, although here we do *not* need to consider edge-state contributions because we are dealing with a non-uniform magnetic field with zero average.[13]

We now consider the superconducting state's response to T_1. In this case the Fermi-sea ground state must be replaced by the BCS ground state (20.23), and the operator $c^\dagger_{\vec{k}+\vec{q}/2,\sigma} c_{\vec{k}-\vec{q}/2,\sigma}$ in T_1 creates a pair of Bogoliubov quasiparticles:

$$c^\dagger_{\vec{k}+\vec{q}/2,\uparrow} c_{\vec{k}-\vec{q}/2,\uparrow} |\psi_{BCS}\rangle = u_{\vec{k}+\vec{q}/2} v_{\vec{k}-\vec{q}/2} \alpha^\dagger_{\vec{k}+\vec{q}/2} \beta^\dagger_{-\vec{k}+\vec{q}/2} |\psi_{BCS}\rangle. \tag{20.76}$$

More importantly, there is another term in T_1 that generates the *same* excited state:

$$c^\dagger_{-\vec{k}+\vec{q}/2,\downarrow} c_{-\vec{k}-\vec{q}/2,\downarrow} |\psi_{BCS}\rangle = -u_{\vec{k}-\vec{q}/2} v_{\vec{k}+\vec{q}/2} \alpha^\dagger_{\vec{k}+\vec{q}/2} \beta^\dagger_{-\vec{k}+\vec{q}/2} |\psi_{BCS}\rangle. \tag{20.77}$$

We thus need to coherently add up their amplitude contributions to the matrix element connecting the ground state and the corresponding excited state (before squaring), resulting in

$$\Delta E_1 = -\frac{\hbar^2 e^2 A_0^2}{2m_e^2 c^2} \sum_{\vec{k}} \frac{k_y^2 |u_{\vec{k}+\vec{q}/2} v_{\vec{k}-\vec{q}/2} - u_{\vec{k}-\vec{q}/2} v_{\vec{k}+\vec{q}/2}|^2}{E_{\vec{k}+\vec{q}/2} + E_{\vec{k}-\vec{q}/2}}. \tag{20.78}$$

Without performing detailed calculations, we immediately see that ΔE_1 vanishes in the long-wavelength limit $q \to 0$, due to the interference between the two terms in the numerator, which causes it to go to zero, while the energy denominator is always finite due to the presence of a gap. We thus find

---

[13] As a result of which, it can be treated perturbatively. A uniform magnetic field, on the other hand, has singular effects and can never be treated perturbatively no matter how weak it is. One way to see this is to observe that for a constant B the vector potential always diverges at large distance.

that the paramagnetic T_1 term makes *no* contribution to the electromagnetic response, and the diamagnetic T_2 term gives rise to the Meissner effect, with the superfluid density equal to the electron density.

We note that the key to the vanishing of ΔE_1 in the superconducting state is the cancelation of the matrix element $u_{\vec{k}+\vec{q}/2}v_{\vec{k}-\vec{q}/2} - u_{\vec{k}-\vec{q}/2}v_{\vec{k}+\vec{q}/2}$ in the limit $q \to 0$. The existence of a gap in the excitation spectrum, while helpful, is *not* crucial to the this result. In fact, there exist unconventional superconductors whose gap vanishes at certain points on the Fermi surface, where Bogoliubov quasiparticles are actually gapless. (We will encounter examples of this later.) ΔE_1 still vanishes in the long-wavelength limit in these cases. We can, of course, also open up a band gap at the Fermi energy and turn the system into a band insulator. In this case there is a gap but no Meissner effect, and ΔE_1 is non-zero and still cancels out ΔE_2. The reason is that there are no coherence factors appearing in the matrix elements of T_1.

If one were given a *longitudinal* vector potential (or a pure gauge) $\vec{A}(\vec{r}) = \nabla\chi(\vec{r})$, and blindly followed the calculation above, one would again find $\Delta E_1 = 0$ in the long-wavelength limit, leading to an unphysical result $\Delta E = \Delta E_2 > 0$. The source of error here is that one needs to perform a related gauge transformation for the superconducting order parameter Δ, which *changes* the coherence factors as well as the quasiparticle spectrum. As a result, the cancelation of the matrix element mentioned above does not occur. We will discuss the gauge dependence of the superconducting order parameter in more detail in Section 20.6.

20.5 BCS–BEC Crossover

It is known that the **superfluidity** of a boson liquid (e.g. He^4) is due to Bose–Einstein condensation (BEC) of the bosons.[14] It is thus natural to think that superconductivity is due to BEC of Cooper pairs. In this section we examine how accurate this statement is. It turns out that, here and in many other circumstances, more insights can be gained by examining the BCS wave function and other aspect of superconductivity in real space.

To proceed, we note that the BCS wave function (20.23) may be written as

$$|\psi_{\text{BCS}}\rangle = \prod_{\vec{k}}(u_{\vec{k}} + v_{\vec{k}}c^{\dagger}_{\vec{k}\uparrow}c^{\dagger}_{-\vec{k}\downarrow})|0\rangle \propto \prod_{\vec{k}}(1 + g_{\vec{k}}c^{\dagger}_{\vec{k}\uparrow}c^{\dagger}_{-\vec{k}\downarrow})|0\rangle \qquad (20.79)$$

$$= \prod_{\vec{k}}e^{g_{\vec{k}}c^{\dagger}_{\vec{k}\uparrow}c^{\dagger}_{-\vec{k}\downarrow}}|0\rangle = e^{\sum_{\vec{k}}g_{\vec{k}}c^{\dagger}_{\vec{k}\uparrow}c^{\dagger}_{-\vec{k}\downarrow}}|0\rangle, \qquad (20.80)$$

where $g_{\vec{k}} = v_{\vec{k}}/u_{\vec{k}}$ and we used the fact that $(c^{\dagger}_{\vec{k}\uparrow}c^{\dagger}_{-\vec{k}\downarrow})^2 = (S^+_{\vec{k}})^2 = 0$ in the above. Defining

$$b^{\dagger}_{\text{pair}} = \sum_{\vec{k}}(g_{\vec{k}}/\gamma)c^{\dagger}_{\vec{k}\uparrow}c^{\dagger}_{-\vec{k}\downarrow} \qquad (20.81)$$

with

$$\gamma = \sqrt{\sum_{\vec{k}}|g_{\vec{k}}|^2}, \qquad (20.82)$$

we may write (20.80) as

$$|\psi_{\text{BCS}}\rangle \propto e^{\gamma b^{\dagger}_{\text{pair}}}|0\rangle. \qquad (20.83)$$

---

[14] As we saw in Section 17.9, non-interacting bosons have a much stronger diamagnetic response than fermions if they are charged. To properly understand superfluidity, however, (repulsive) interactions between the bosons must be taken into account, as was described in Chapter 18.

It is clear that $b_{\text{pair}}^\dagger |0\rangle$ is a normalized state for a *single* pair, whose total momentum is zero, and the relative wave function in real space (that is, the wave function describing the relative positions of the particles) is the Fourier transform of $g_{\vec{k}}/\gamma$.

If $b_{\text{pair}}^\dagger$ were an ordinary boson operator, Eq. (20.83) would describe a Bose condensate state, as in Eq. (18.4). Of course, another way to understand its BEC nature is if we project the state (20.83) to one with fixed boson number N. It then takes the form

$$|N\rangle \propto (b_{\text{pair}}^\dagger)^N |0\rangle, \tag{20.84}$$

namely every boson is in the same single-particle state. However, here $b_{\text{pair}}^\dagger$ is *not* a simple boson operator. Still, the discussion above suggests that we may interpret the BCS wave function as a condensed state of Cooper pairs created by $b_{\text{pair}}^\dagger$. If the pairs were sufficiently small, this analogy to BEC would become precise.

In fact, written in terms of electron operators, Eq. (20.84) describes the BCS superconducting ground state with a *fixed* number of N Cooper pairs and $2N$ electrons:

$$|\psi_{\text{BCS}}\rangle_{2N} \propto \left[\sum_{\vec{k}} g_{\vec{k}} c_{\vec{k}\uparrow}^\dagger c_{-\vec{k}\downarrow}^\dagger \right]^N |0\rangle. \tag{20.85}$$

We can obtain $|\psi_{\text{BCS}}\rangle_{2N}$ above from the original BCS wave function Eq. (20.23) (with indefinite particle number) by making a suitable linear superposition of the latter with different phases of Δ_0 (or $g_{\vec{k}} = v_{\vec{k}}/u_{\vec{k}}$):

$$|\psi_{\text{BCS}}\rangle_{2N} \propto \int_0^{2\pi} d\phi \, e^{-iN\phi} \prod_{\vec{k}} (u_{\vec{k}} + e^{i\phi} v_{\vec{k}} c_{\vec{k}\uparrow}^\dagger c_{-\vec{k}\downarrow}^\dagger)|0\rangle. \tag{20.86}$$

It is clear from the above that one can fix the particle/Cooper-pair number by taking a superposition of different order-parameter phases! Uncertain phase generates certain particle number (since they are conjugate to each other).

Exercise 20.7.   Verify Eq. (20.86).

One way to estimate the size of the Cooper pair (in the presence of many other Cooper pairs!) is to calculate the real-space pairing amplitude as a function of the distance separating the pair:

$$P(\vec{r}) = \langle \psi_{\text{BCS}} | \Psi_\uparrow^\dagger \left(\frac{\vec{r}}{2} \right) \Psi_\downarrow^\dagger \left(-\frac{\vec{r}}{2} \right) | \psi_{\text{BCS}} \rangle \tag{20.87}$$

$$= \frac{1}{L^3} \sum_{\vec{k}} e^{i\vec{k}\cdot\vec{r}} \langle \psi_{\text{BCS}} | c_{\vec{k}\uparrow}^\dagger c_{-\vec{k}\downarrow}^\dagger | \psi_{\text{BCS}} \rangle \tag{20.88}$$

$$= \frac{1}{L^3} \sum_{\vec{k}} u_{\vec{k}} v_{\vec{k}} e^{i\vec{k}\cdot\vec{r}} = \frac{1}{L^3} \sum_{\vec{k}} \frac{\Delta_0}{2E_{\vec{k}}} e^{i\vec{k}\cdot\vec{r}}, \tag{20.89}$$

where $\Psi_\uparrow^\dagger(\vec{r})$ is the up-spin electron creation (field) operator at $\vec{r}$ (see Appendix J). We thus find that $P(\vec{r})$'s Fourier transform is $\Delta_0/2E_{\vec{k}}$, which peaks at k_{F} and has a width of order $\Delta_0/\hbar v_{\text{F}}$. We thus expect $P(\vec{r})$ to oscillate with wave vector k_{F} and have a characteristic size

$$\xi_0 = \frac{\hbar v_{\text{F}}}{\pi \Delta_0}, \tag{20.90}$$

known as the **BCS coherence length**, that characterizes the Cooper-pair size. For a weak-coupling superconductor with $\Delta_0 \ll \hbar\omega_{\text{D}} \ll \epsilon_{\text{F}}$, we have

$$\xi_0 \gg \frac{\hbar v_{\mathrm{F}}}{\pi \epsilon_{\mathrm{F}}} \sim \frac{1}{k_{\mathrm{F}}}, \tag{20.91}$$

indicating that the Cooper-pair size is much *bigger* than the inter-electron spacing, thus Cooper pairs cannot be viewed as point-like bosons. In fact, it is because so many electrons live in the volume defined by a single Cooper pair that the mean-field BCS approximation works so well.

The fact that Cooper pairs have a finite size ξ_0 means that they respond to electromagnetic fields non-locally. More precisely, ξ_0 is the length scale that dictates how fast the Meissner kernel defined in Eq. (19.24) decays as $|\vec{r} - \vec{r}\,'|$ increases. In this context ξ_0 is also known as the **Pippard coherence length**, after Brian Pippard, who originally wrote

$$\xi_0 = a \frac{\hbar v_{\mathrm{F}}}{k_{\mathrm{B}} T_{\mathrm{c}}}, \tag{20.92}$$

where a is a number of order unity. Obviously, for weak-coupling superconductors Eqs. (20.90) and (20.92) can be made equivalent by properly choosing the value of a, because $k_{\mathrm{B}} T_{\mathrm{c}} \propto \Delta_0$. The Ginzburg–Landau coherence length $\xi(T)$ defined in Eq. (19.86), which is strongly dependent on the temperature T, is in principle a *different* length scale. In practice, however, we have $\xi(T \ll T_{\mathrm{c}}) \sim \xi_0$ for weak-coupling superconductors, as the Cooper-pair size limits how fast Δ can vary in space (see Section 20.6). However, in a strong-coupling superconductor in the BEC limit we can have $\xi(T \ll T_{\mathrm{c}}) \gg \xi_0$, and, for superfluids that result from condensation of point bosons, ξ_0 has no physical meaning (or should be understood as zero), but $\xi(T)$ remains well-defined and finite.

Electronic superconductors (in crystalline solids) are essentially always in this weak-coupling, or BCS, regime. On the other hand, in cold atomic gases it is possible to control the atom density (and thus Fermi energy) of fermionic atoms as well as their pairing interaction strength, such that one can tune the system from the weak-coupling/BCS regime to the strong-coupling/BEC regime, in which $|\Delta_0| \gg \epsilon_{\mathrm{F}}$. The latter corresponds to a Bose condensate of diatomic molecules. For s-wave pairing there is a smooth crossover between these two limits, during which the pairing gap remains open (the gap becomes half of the binding energy of the molecule in the BEC regime). As we will see later for non s-wave or unconventional pairing, there must be a quantum phase transition where the pairing gap vanishes.

One practical way to determine whether the system is in the BCS or BEC regime is to inspect the value of the chemical potential μ. For the weak-coupling (BCS regime), we expect μ to be very close to the Fermi energy ϵ_{F} for non-interacting fermions, which is positive. In the BEC regime the energy change of the system due to adding a *pair* of fermions is that of forming a molecule, which is the negative of the binding energy, $-2|\Delta_0|$. We thus expect $\mu \approx -|\Delta_0| < 0$. More precisely, the system is in the BEC limit if the chemical potential is below the bottom of the fermion band minimum.

20.6 Real-Space Formulation and the Bogoliubov–de Gennes Equation

In the earlier sections, our study of superconductivity has mostly been based on the BCS reduced Hamiltonian, which is formulated in momentum space, and relies heavily on translation symmetry. Translation symmetry is sometimes broken, for example by the presence of disorder, or (which is of particular interest) a magnetic field, and this renders the momentum-space formulation much less useful. In this section we develop a formalism (often associated with the names Bogoliubov and de Gennes, or BdG) that allows us to study pairing and superconductivity in real space. Although this formalism was originally formulated in the continuum (see Chapter 5 of [222]), here we focus on lattice models instead for their simplicity, and the continuum version can be reached by taking the continuum limit of a tight-binding lattice (see Section 7.4.3).

We consider the Hubbard model with *negative U*, which is probably the simplest model to describe electrons with an attractive interaction (one-band, only on-site attraction):

$$H = -t \sum_{\langle ij \rangle, \sigma} (c_{i\sigma}^{\dagger} c_{j\sigma} + c_{j\sigma}^{\dagger} c_{i\sigma}) - U \sum_{j} n_{j\uparrow} n_{j\downarrow} - \mu \sum_{j} (n_{j\uparrow} + n_{j\downarrow}), \tag{20.93}$$

where $U > 0$ characterizes the strength of attraction. Disorder can be included by adding an on-site single-electron potential of the form $\sum_{j} V_{j}(n_{j\uparrow} + n_{j\downarrow})$, and the effects of a magnetic field can be accommodated by changing the phase of the hopping-matrix elements (for the orbital effect) and having a different chemical potential for up- and down-spin electron (for the Zeeman term).

Other than being a lattice model formulated in real space, there are several (additional) important differences between the BdG model (20.93) and the BCS reduced Hamiltonian (20.10). (i) In (20.93) the attractive interaction is *not* restricted to electrons near the Fermi surface, and the high-energy cutoff previously provided by the Debye energy is replaced by the band width. The BdG model thus loses the physics associated with the fact that the attractive interaction is mediated by phonons (and is thus attractive only at low frequencies). (ii) In (20.93) the attractive interaction is (correctly) *not* restricted to pairs of electrons with total momentum zero. This allows for descriptions of states carrying net supercurrent, and is crucial for proper description of even the ground state when disorder and/or a magnetic field is present. Thus the BdG model contains some elements which are more realistic than the BCS pairing model and some which are less realistic.

To proceed with a mean-field solution, we write the interaction term as $-U \sum_{j} c_{j\uparrow}^{\dagger} c_{j\downarrow}^{\dagger} c_{j\downarrow} c_{j\uparrow}$, replacing the pair creation and annihilation operators by their expectation values, and arrive at the mean-field approximation of the interaction term:[15]

$$\hat{\Delta} = - \sum_{j} (\Delta_{j} c_{j\uparrow}^{\dagger} c_{j\downarrow}^{\dagger} + \Delta_{j}^{*} c_{j\downarrow} c_{j\uparrow}), \tag{20.94}$$

where the local pairing order parameter (which we allow to be spatially non-uniform) must satisfy the self-consistency equation

$$\Delta_{j} = U \langle c_{j\downarrow} c_{j\uparrow} \rangle. \tag{20.95}$$

Here Δ_{j}, or its continuum version $\Delta(\vec{R})$, is the microscopic counterpart of the Ginzburg–Landau order parameter introduced in Chapter 19. Crudely speaking it is the "Cooper-pair wave function." One way to understand the equivalence between them is to inspect how they transform under a gauge transformation $\vec{A} \to \vec{A} + \nabla \chi(\vec{R})$. Since the annihilation operator (which transforms in the same way as the electron function; see Appendix J) transforms as $c(\vec{R}) \to c(\vec{R}) e^{-\frac{ie\chi(\vec{R})}{\hbar c}}$, we find

$$\Delta(\vec{R}) \to \Delta(\vec{R}) e^{-\frac{2ie\chi(\vec{R})}{\hbar c}}. \tag{20.96}$$

Thus it carries effective charge $e^* = 2e$, just like the Ginzburg–Landau order parameter.

Equation (20.94) only allows pairs of electrons on the same site to be created or annihilated. More generally, one can have pair creation or annihilation on different sites:

$$\hat{\Delta} = - \sum_{ij} (\Delta_{ij} c_{i\uparrow}^{\dagger} c_{j\downarrow}^{\dagger} + \Delta_{ij}^{*} c_{j\downarrow} c_{i\uparrow}). \tag{20.97}$$

For $\Delta_{ij} = \Delta_{ji}$, the orbital wave function of the pair is symmetric, so their spin wave function must be anti-symmetric and thus a singlet. On the other hand, for $\Delta_{ij} = -\Delta_{ji}$, the orbital wave

---

[15] We have dropped an irrelevant constant term $-|\Delta_{j}|^2$ from the mean-field Hamiltonian, since it does not affect the mean-field eigenfunction.

function of the pair is anti-symmetric, thus the spin wave function is symmetric and corresponds to a triplet.[16]

In a translationally invariant system, we have

$$\Delta_{ij} = \Delta(\vec{R}_i - \vec{R}_j), \tag{20.98}$$

the Fourier transformation of which yields the pairing order parameter in momentum space:

$$\Delta(\vec{k}) = \frac{1}{N} \sum_j \Delta(\vec{R}_j) e^{-i\vec{k}\cdot\vec{R}_j}, \tag{20.99}$$

where N is the number of lattice sites. We see immediately that for on-site pairing with $\Delta_{ij} = \Delta\delta_{ij}$ we have that $\Delta(\vec{k}) = \Delta$ is a constant in momentum space (as in previous sections), but in general there is a $\vec{k}$ dependence.

The mean-field Hamiltonian in real space (also called the **Bogoliubov–de Gennes Hamiltonian**) is thus

$$\hat{H}_{\mathrm{BdG}} = \hat{H}_0 + \hat{\Delta}, \tag{20.100}$$

where

$$\hat{H}_0 = \sum_{ij,\sigma} H_0^{ij} c_{i\sigma}^\dagger c_{j\sigma} \tag{20.101}$$

includes all the hopping and on-site potential terms.

To diagonalize H_{BdG}, we note that it may be written as

$$\hat{H}_{\mathrm{BdG}} = (c_{1\uparrow}^\dagger, \ldots, c_{N\uparrow}^\dagger; c_{1\downarrow}, \ldots, c_{N\downarrow}) \begin{pmatrix} H_0 & -\Delta \\ -\Delta^\dagger & -H_0^* \end{pmatrix} \begin{pmatrix} c_{1\uparrow} \\ \cdot \\ \cdot \\ \cdot \\ c_{N\downarrow}^\dagger \end{pmatrix}$$

$$+ \text{ constant}, \tag{20.102}$$

where N is the number of sites, H_0 is the $N \times N$ matrix whose matrix elements are H_0^{ij}, H_0^* is its complex conjugate, Δ is the $N \times N$ matrix whose matrix elements are Δ_{ij}, and $\Delta^\dagger$ is its Hermitian conjugate. We have $\Delta^\dagger = \pm\Delta^*$ for singlet and triplet pairing, respectively. One way to motivate Eq. (20.102) is to perform a particle–hole transformation on the down-spin electrons, and view the pairing term $\hat{\Delta}$ as a term turning an up-spin electron into a down-spin hole.

We can diagonalize $\hat{H}_{\mathrm{BdG}}$ by first diagonalizing the $2N \times 2N$ matrix,

$$H_{\mathrm{BdG}} = \begin{pmatrix} H_0 & -\Delta \\ -\Delta^\dagger & -H_0^* \end{pmatrix}. \tag{20.103}$$

The structure of H_{BdG} guarantees particle–hole symmetry of its spectrum. It is easy to verify that if

$$\begin{pmatrix} H_0 & -\Delta \\ -\Delta^\dagger & -H_0^* \end{pmatrix} \begin{pmatrix} u \\ v \end{pmatrix} = E \begin{pmatrix} u \\ v \end{pmatrix}, \tag{20.104}$$

where u and v are N-component column vectors, then

$$\begin{pmatrix} H_0 & -\Delta \\ -\Delta^\dagger & -H_0^* \end{pmatrix} \begin{pmatrix} \mp v^* \\ u^* \end{pmatrix} = -E \begin{pmatrix} \mp v^* \\ u^* \end{pmatrix} \tag{20.105}$$

---

[16] For triplet pairing one could also allow for pairing terms between electrons with parallel spin like $\Delta_{ij} c_{i\uparrow}^\dagger c_{j\uparrow}^\dagger + \text{h.c.}$, where the anti-commutation relation between fermion creation operators guarantees $\Delta_{ij} = -\Delta_{ji}$. In the remainder of this section we consider only pairing between electrons with opposite spins for the sake of simplicity.

for singlet and triplet pairing, respectively. Thus the $2N$ eigenvalues of H_{BdG} come in $\pm E$ pairs, and we can use the N positive eigenvalues $E_l > 0$ with $l = 1, \ldots, N$, and their corresponding eigenvectors ψ^l (with $2N$ components) to construct quasiparticle operators:

$$\alpha_l^\dagger = \sum_{j=1}^N (\psi_j^l c_{j\uparrow}^\dagger + \psi_{N+j}^l c_{j\downarrow}), \tag{20.106}$$

$$\beta_l^\dagger = \sum_{j=1}^N (\psi_j^l c_{j\downarrow}^\dagger \mp \psi_{N+j}^l c_{j\uparrow}), \tag{20.107}$$

where $\mp$ corresponds to singlet and triplet pairing, respectively. $\hat{H}_{\text{BdG}}$ is brought to diagonal form in terms of these quasiparticle operators:

$$\hat{H}_{\text{BdG}} = \sum_{l=1}^N E_l(\alpha_l^\dagger \alpha_l + \beta_l^\dagger \beta_l). \tag{20.108}$$

The BdG approach is especially useful in situations where the superconductivity is spatially inhomogeneous. There, of course, one cannot use the plane-wave basis and must in general diagonalize the Hamiltonian numerically.

20.7 Kitaev's p-Wave Superconducting Chain and Topological Superconductors

We now apply the formalism developed in Section 20.6 to study what may appear to be the simplest model for superconductivity, namely *spinless* electrons with nearest-neighbor hopping and pairing on a 1D chain. As it turns out, this is the prototype of a **topological superconductor**, with non-trivial edge states described by **Majorana fermion** operators (see Appendix J and the pedagogical review by Beenakker [223]). This model was first studied by Kitaev, who pointed out that it supports a topologically trivial and a topologically non-trivial phase, and elucidated the nature of the latter.

Consider the **Kitaev Hamiltonian**

$$H_{\text{K}} = -\sum_j [t(c_j^\dagger c_{j+1} + c_{j+1}^\dagger c_j) + \mu c_j^\dagger c_j + \Delta(c_j^\dagger c_{j+1}^\dagger + c_{j+1} c_j)], \tag{20.109}$$

where t, μ, and Δ are all real parameters corresponding to hopping, the chemical potential, and the nearest-neighbor pairing amplitude. (Note that on-site pairing is not allowed for spinless fermions due to the Pauli principle. We will see shortly that we have p-wave-like rather than s-wave-like pairing in this case.) The lattice spacing is set to be unity. Going to momentum space, we have

$$\hat{H}_{\text{K}} = \sum_k [\epsilon_k c_k^\dagger c_k - \Delta(k) c_k^\dagger c_{-k}^\dagger - \Delta^*(k) c_{-k} c_k], \tag{20.110}$$

where

$$\epsilon_k = -2t \cos k - \mu, \tag{20.111}$$

$$\Delta(k) = i\Delta \sin k. \tag{20.112}$$

The resultant (single species, because there is no spin) quasiparticle spectrum is

$$E_k = \sqrt{\epsilon_k^2 + |2\Delta(k)|^2}. \tag{20.113}$$

We note that the superconducting order parameter in momentum space changes sign under $k \to -k$:

$$\Delta(k) = -\Delta(-k). \tag{20.114}$$

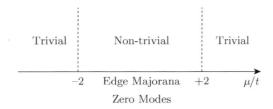

Figure 20.5 The phase diagram of the Kitaev Hamiltonian (20.109). The topologically non-trivial phase supports edge Majorana zero modes.

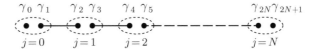

Figure 20.6 Majorana representation of Kitaev's p-wave chain. Each site j is represented by a pair of Majorana operators, γ_{2j} and γ_{2j+1} (circled together). For the special case of $\mu = 0$ and $\Delta = t$, couplings exist only between γ_{2j-1} and γ_{2j} (connected by a solid line), while γ_0 and γ_{2N+1} are not coupled to other Majorana operators.

This is, of course, due to the fact that the order parameter is odd under parity in real space: $\Delta_{j,j+1} = -\Delta_{j+1,j}$ (corresponding to the case of triplet pairing in Section 20.6). The odd parity of the order parameter is required in order to match the anti-commutation of the pair operators in Eq. (20.109): $c_{j+1}c_j = -c_j c_{j+1}$. This odd-parity pairing in 1D is also referred to as p-wave pairing, even though in 1D there is only the parity transformation and no (continuous) rotation. We will discuss classifying superconducting order parameters in terms of their transformation properties in more detail in Section 20.8.

An important consequence of (20.114) is that the superconducting pairing amplitude vanishes, $|\Delta(k)| = 0$, for $k = 0, \pi$. If $\epsilon_k = 0$ at one of these points, which is the case for $\mu = \pm 2t$, the quasiparticle spectrum E_k is *gapless*. As we will see, these values correspond to critical points separating topologically trivial and non-trivial phases of H_K.

To understand the phase diagram of H_K (Fig. 20.5), let us start with $\mu < -2|t|$, which, as discussed in Section 20.5, corresponds to the BEC regime in which the ground state may be viewed as a BEC of closely bound pairs of fermions. For large negative μ the fermions are expensive to create, and the pairing term creates virtual fermion pairs which do not live long enough for the kinetic energy to allow them to spread out. Each near-neighbor pair of sites is in a coherent superposition that is mostly vacuum, with a small amplitude for having a fermion pair present. Thus this phase connects smoothly to a low-density BEC of (nearly) point-like bosons, and is somewhat analogous to a topologically trivial band insulator that connects smoothly to a dilute collection of atoms. We thus identify the phase for $\mu < -2|t|$ to be (topologically) trivial. In particular, we do not expect interesting edge states to exist at the boundaries of the chain. The phase for $\mu > 2|t|$ is related by a particle–hole transformation, and thus of the same topological class.[17]

To reveal the topologically non-trivial nature of the phase with $|\mu| < 2|t|$, we consider a finite-length chain with $N + 1$ sites ($j = 0, 1, 2, \ldots, N$), and use the Majorana representation of fermion operators (see Appendix J). We introduce a pair of Majorana fermion operators, γ_{2j} and γ_{2j+1}, to represent the fermion operator c_j (see Fig. 20.6):

$$c_j^\dagger = \gamma_{2j} + i\gamma_{2j+1}, \qquad c_j = \gamma_{2j} - i\gamma_{2j+1}. \tag{20.115}$$

---

[17] Note that the BEC exists here even in 1D because the Kitaev Hamiltonian explicitly does not have the U(1) symmetry of particle number conservation. The Hamiltonian is of the BCS mean-field type, and does not suffer from the quantum fluctuations that in general prevent spontaneous symmetry breaking in 1D.

These Majorana operators are Hermitian and satisfy the anti-commutation relation

$$\{\gamma_j, \gamma_k\} = \frac{1}{2}\delta_{jk}. \tag{20.116}$$

In terms of these Majorana operators we have

$$c_j^\dagger c_j = -2i\gamma_{2j}\gamma_{2j+1}, \tag{20.117}$$

$$c_j^\dagger c_{j+1} + c_{j+1}^\dagger c_j = 2i(\gamma_{2j+1}\gamma_{2j+2} - \gamma_{2j}\gamma_{2j+3}), \tag{20.118}$$

$$c_j^\dagger c_{j+1}^\dagger + c_{j+1}c_j = 2i(\gamma_{2j+1}\gamma_{2j+2} + \gamma_{2j}\gamma_{2j+3}). \tag{20.119}$$

We find H_K simplifies tremendously in the special case of $\mu = 0$ and $\Delta = t$:

$$H_K = -4it(\gamma_1\gamma_2 + \gamma_3\gamma_4 + \cdots + \gamma_{2N-1}\gamma_{2N}) = -4it\sum_{j=1}^{N}\gamma_{2j-1}\gamma_{2j}. \tag{20.120}$$

In this case we see that H_K is made of pairs of Majorana operators γ_{2j-1} and γ_{2j} coupled to each other, but there is no coupling between different pairs.[18] More remarkably, γ_0 and γ_{2N+1} disappear from the Hamiltonian! The spectrum of H_K in this case is made of degenerate quasiparticle states with energy $|t|/2$, and an exactly zero-energy mode made of γ_0 and γ_{2N+1}. This zero mode has exactly half of its weight at site 0, and half at site N. Equivalently we can say that there is a **Majorana zero mode** at each end of the finite chain, keeping in mind that a Majorana mode is only half a real fermion mode. The appearance of such (fractionalized) gapless edge state(s) here bears some similarity to the appearance of half edge spins in the AKLT model studied in Section 17.6, and is characteristic of a topologically non-trivial phase. The AKLT representation of the spin 1 as a pair of spin-1/2 particles on each site is the natural language for seeing the spin fractionalization, much as the Majorana representation of the fermions is the natural language in which to see the fractionalization here.

It turns out that the appearance of edge Majorana zero modes is a generic property of the entire phase, with $|\mu| < 2|t|$ for H_K. Except for the special case discussed above, their energy is not exactly zero due to a small coupling between them (which decays exponentially with the chain length). This coupling moves the energy of the corresponding real fermion mode slightly away from zero. The situation is similar to a spin-1 chain in the Haldane phase that supports spin-1/2 degrees of freedom at the edge (generically with an exponentially small coupling between them), with the AKLT model being a very special example where this coupling is exactly zero.

This p-wave superconducting chain with $|\mu| < 2|t|$ is the simplest example of a topological superconductor that supports gapless Majorana fermion modes at its edges or boundaries. That superconductors can have non-trivial topological character can be understood from the fact that the Bogoliubov–de Gennes Hamiltonian (20.102) resembles a band Hamiltonian, and its solutions can be viewed as quasiparticle bands. As a result, topological superconductors can be classified in a manner similar to topological insulators (discussed in Chapter 14), but we will not delve into the details here. The interested reader may wish to attempt Exercise 20.8 to gain some sense of the topological nature of the Kitaev model.

Exercise 20.8. Use an appropriate version of the Anderson pseudospin representation of Eqs. (20.12)–(20.14) in the Kitaev model in Eq. (20.110) and study how the local "magnetic

---

[18] The reader may recognize the similarity between (20.120) and the extremely dimerized chain in Section 7.6, and the J_1–J_2 model of Section 17.6 with $J_2 = 0$.

> field" coupling to the pseudospins varies in magnitude and direction across the 1BZ. What is the difference between the topologically non-trivial regime $|\mu/t| < 2$ and the topologically trivial regime $|\mu/t| > 2$?

20.8 Unconventional Superconductors

The Kitaev model of the previous section is also the simplest example of an **unconventional superconductor**, whose superconducting order parameter Δ has a *lower* symmetry than that of the original lattice. In this particular example, the lattice has inversion (or parity) symmetry, but Δ changes sign (or is odd) under inversion, as can be seen either in real space, $\Delta_{i,i+1} = -\Delta_{i+1,i}$, or in momentum space, Eq. (20.114). Such behavior is quite different from that of the s-wave superconductor discussed in Section 20.2, where Δ is a constant in momentum space and *invariant* under *all* symmetry transformations, and is thus conventional.[19] In general, a superconductor is said to be unconventional if the order parameter Δ transforms non-trivially (i.e. is *not* invariant) under *some* symmetry transformation(s) of the system, including (lattice) translation. In the absence of an underlying crystalline lattice these transformations include combinations of translation, parity, (space and/or spin) rotation, and time-reversal. This section is devoted to general discussions of unconventional superconductors. We start by revisiting the Cooper problem.

20.8.1 General Solution of Cooper Problem

In Section 15.13.3 we solved the Cooper pairing problem with a constant pairing interaction in momentum space (see Eq. (15.170)). Here we revisit this problem for more general pairing interactions:

$$\langle \vec{k}|\hat{V}|\vec{k}'\rangle = V(|\vec{k} - \vec{k}'|) \approx V[2k_F \sin(\theta_{\vec{k}\vec{k}'}/2)] = \sum_l V_l e^{il(\theta - \theta')}, \qquad (20.121)$$

where $|\vec{k}\rangle$ is the state of a pair of electrons with opposite spins (which may form either a singlet or triplet) with momentum $\vec{k}$ and $-\vec{k}$, θ and k are the angle and magnitude of $\vec{k}$, $\theta_{\vec{k}\vec{k}'}$ is the angle between $\vec{k}$ and $\vec{k}'$, and

$$V_l = \frac{1}{2\pi} \int_0^{2\pi} d\theta \, e^{-il\theta} V[2k_F \sin(\theta/2)]. \qquad (20.122)$$

In the above we used the fact that $\vec{k}'$ and $\vec{k}$ are restricted to a thin shell close to the Fermi surface: $k_F < k, k' < k_F + \Delta k_F$, and in the last step of Eq. (20.121) we assumed a 2D system and performed a Fourier-series expansion.[20] The original Cooper model studied in Section 15.13.3 corresponds to the special case of $V_l = V_0 \delta_{0l}$.

Taking advantage of the rotation symmetry, we can choose to work with pair wave functions of fixed angular momentum l, namely

$$|\Psi_l\rangle = \sum_{\vec{k}} e^{il\theta} \psi_k |\vec{k}\rangle. \qquad (20.123)$$

---

[19] In general an s-wave or conventional superconductor may have a momentum-dependent Δ, as long as it is invariant under all the lattice symmetry transformations. In such cases it is also referred to as an extended s-wave superconductor.

[20] In 3D the expansion will be in terms of Legende polynomials, and the manipulations below can be carried out similarly. These expansions are very similar to those for the Landau functions in Section 15.11.3.

From the Schrödinger equation for the pair wave function,

$$(\hat{H}_0 + \hat{V})|\Psi_l\rangle = E_l|\Psi_l\rangle, \tag{20.124}$$

we obtain

$$E_l e^{il\theta} \psi_k = 2e^{il\theta} \epsilon_k \psi_k + \sum_{\vec{k}'} \langle \vec{k}|\hat{V}|\vec{k}'\rangle e^{il\theta'} \psi_{k'} \tag{20.125}$$

$$= 2e^{il\theta} \epsilon_k \psi_k + \frac{L^2}{(2\pi)^2} \sum_{l'} V_{l'} \int_0^{2\pi} d\theta' \, e^{il'(\theta-\theta')} e^{il\theta'} \int_{k_F}^{k_F+\Delta k_F} \psi_{k'} k' \, dk'$$

$$= 2e^{il\theta} \epsilon_k \psi_k + \frac{L^2}{2\pi} e^{il\theta} V_l \int_{k_F}^{k_F+\Delta k_F} \psi_{k'} k' \, dk', \tag{20.126}$$

thus

$$\psi_k = -\frac{V_l C_l}{2\epsilon_k - E_l}, \tag{20.127}$$

where

$$C_l = \frac{L^2}{2\pi} \int_{k_F}^{k_F+\Delta k_F} \psi_{k'} k' \, dk'. \tag{20.128}$$

Now, following steps similar to those in Section 15.13.3, we find that a bound state solution exists for any $V_l < 0$, with

$$E_l = -2E_c e^{-\frac{2}{D(0)|V_l|}} \tag{20.129}$$

at weak coupling, where $E_c = v_F \, \Delta k_F$.

A few comments are now in order. (i) The Fermi-liquid normal state suffers from a pairing instability, as long as *any* V_l is negative. This happens even if the interaction is predominantly repulsive, namely V_l is positive for most ls. It is perhaps interesting to note that, even if $V(\vec{k}' - \vec{k})$ is positive everywhere in momentum space, Eq. (20.122) may still yield negative V_l for some ls (we will analyze an example of this in Section 20.9)! When (weak) attractive interaction coexists with (strong) repulsive interaction, the instability tends to happen in channels with $|l| > 0$. Physically this is easy to understand: the centrifugal barrier that comes with the relative angular momentum keeps the pair apart to avoid the repulsion at short distance, while taking advantage of the attractive interaction at longer distance. (ii) For even l, the space part of the pair wave function is symmetric under exchange, thus the spin wave function must be anti-symmetric, resulting in a singlet. For odd l, the space part of the pair wave function is anti-symmetric, and the spin wave function must be symmetric, resulting in a triplet. (iii) Obviously the pairing instability of the normal state will be first triggered by the most negative V_l, as the temperature decreases. We thus expect the superconducting order parameter to have the symmetry corresponding to angular momentum l. (iv) Thus far we have assumed full (continuous) rotation symmetry to simplify the analysis. In the presence of a crystalline lattice, single-particle states with momentum $\vec{k}$ should be understood as Bloch states in the conduction band with lattice momentum $\vec{k}$. More importantly, the symmetry of the system is reduced to the corresponding point-group symmetry. In this case a similar analysis can be performed, with the Cooper-pair wave function and pairing order parameter classified according to how they transform under the lattice symmetry transformations, or, equivalently, in terms of the representations of the point group (just like different angular momentum states form different representations of the rotation group). In particular, if inversion symmetry is present, the pair wave function must be either even (symmetric) or odd (anti-symmetric) under inversion, which is equivalent to exchange of spatial coordinates of the pair, and the corresponding spin wave function must be anti-symmetric (singlet) or symmetric (triplet). (v) We have assumed that the electron spin is a good quantum number up to this point. This no longer holds in the presence of spin–orbit coupling. In this case the spin dependence of the pair wave function needs to be made explicit, namely $\psi_{\sigma_1\sigma_2}(\vec{k})$. As long as time-reversal symmetry is present, bound-state solutions made of pairs of single-particle states related to each other by time-reversal transformation (which are degenerate) can be found at weak coupling, provided that the density of states is finite at the Fermi energy.[21] On the other hand, broken time-reversal symmetry destroys the pairing instability, at least in certain channels. (vi) When time-reversal and inversion symmetries are both present, single-particle states come in degenerate pairs for every $\vec{k}$ (see Section 7.8). In this case a pseudospin quantum number can be introduced, and Cooper pairs can be classified into pseudospin singlets and triplets. However, in a non-centrosymmetric crystal (which has no inversion symmetry), the Cooper-pair wave function is in general a mixture of singlet and triplet.

20.8.2 General Structure of Pairing Order Parameter

In real space the most general form for a pairing term is

$$\hat{\Delta} = -\sum_{\sigma_1\sigma_2} \int d\vec{r}_1 \, d\vec{r}_2 [\Delta_{\sigma_1\sigma_2}(\vec{r}_1, \vec{r}_2) \Psi^\dagger_{\sigma_1}(\vec{r}_1) \Psi^\dagger_{\sigma_2}(\vec{r}_2) + \text{h.c.}]. \tag{20.130}$$

Clearly the pairing order parameter Δ is in general a function of the spins and coordinates of the pair of electrons (Cooper pair) it creates in Eq. (20.130). It is very useful to introduce the center-of-mass and relative coordinates of the Cooper pair:

$$\vec{R} = (\vec{r}_1 + \vec{r}_2)/2; \qquad \vec{r} = \vec{r}_2 - \vec{r}_1. \tag{20.131}$$

Performing a Fourier transformation with respect to $\vec{r}$, we obtain the pairing order parameter as a function of the center-of-mass coordinate $\vec{R}$ and *relative* momentum $\vec{k}$:

$$\Delta_{\sigma_1\sigma_2}(\vec{R}, \vec{k}) = \int d\vec{r} \, e^{-i\vec{k}\cdot\vec{r}} \Delta_{\sigma_1\sigma_2}(\vec{R} - \vec{r}/2, \vec{R} + \vec{r}/2). \tag{20.132}$$

With translation invariance Δ has no $\vec{R}$ dependence in the ground state,[22] and depends on $\vec{k}$ only. In this case Eq. (20.130) can be written as

$$\hat{\Delta} = -\sum_{\sigma_1\sigma_2\vec{k}} [\Delta_{\sigma_1\sigma_2}(\vec{k}) c^\dagger_{\sigma_1,+\vec{k}} c^\dagger_{\sigma_2,-\vec{k}} + \text{h.c.}], \tag{20.133}$$

as we have seen in earlier sections.[23] The resultant order parameter, $\Delta_{\sigma_1\sigma_2}(\vec{k})$, depends on the same variables as the Cooper-pair wave function $\psi_{\sigma_1\sigma_2}(\vec{k})$ discussed in the previous subsection (in fact Δ is often referred to as the Cooper-pair wave function itself!). It can thus be classified in the same way as ψ. We call a superconductor unconventional if Δ transforms non-trivially under certain symmetry transformation of the system.

The physical properties of unconventional superconductors are different from those of conventional superconductors in a number of important ways. We will discuss specific examples of these in Section 20.9.

---

[21] In other words a bound-state solution made of Kramers pairs of single-particle states is always possible. This is true even in the absence of translation symmetry, in which case single-particle eigenstates cannot be labeled by a momentum quantum number. Thus a disordered metal is still subject to Cooper-pairing instability, a fact known as **Anderson's theorem**.

[22] We will consider exceptions to this in the following subsection.

[23] In the presence of a crystalline lattice, lattice translation symmetry dictates that $\Delta_{\sigma_1\sigma_2}(\vec{r}_1, \vec{r}_2)$ is invariant under simultaneous lattice translations of $\vec{r}_1$ and $\vec{r}_2$. In this case $\vec{k}$ is restricted to the first Brillouin zone in Eq. (20.133), and $c^\dagger$ creates an electron in the conduction band with appropriate quantum numbers.

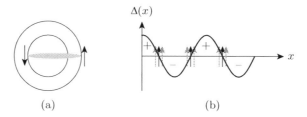

(a)   (b)

Figure 20.7 (a) Spin-split Fermi surfaces of the normal state. Pairing between up- and down-spin electrons on opposite sides of the Fermi surfaces results in a Cooper pair with finite momentum. (b) An inhomogeneous superconducting order parameter Δ that varies along the $\hat{x}$ direction. Unpaired spin-up electrons are localized along the domain walls where Δ goes through zero.

20.8.3 Fulde–Ferrell–Larkin–Ovchinnikov States

The superconducting order parameter Δ can also break (lattice) translation symmetry by spontaneously developing spatial dependence, namely having a dependence on the center-of-mass coordinate of the Cooper pair, $\vec{R}$. This possibility was first considered by Fulde and Ferrell [224], and Larkin and Ovchinnikov [225] in the 1960s, when they independently considered what kind of pairing instability exists when the normal state has spin-split Fermi surfaces due to either an external magnetic field,[24] as illustrated in Fig. 20.7(a), or an internal exchange field due to magnetic ordering. In this case it is not difficult to find that the lowest-energy pair to form is not that with zero total momentum as Cooper originally found, but that with net momentum $Q \approx k_{F\uparrow} - k_{F\downarrow}$. Fulde and Ferrell considered superconducting states in which the Cooper pairs all have the same momentum so that $\Delta(\vec{R}) \propto e^{i\vec{Q}\cdot\vec{R}}$, while Larkin and Ovchinnikov considered superconducting states in which $\Delta(\vec{R}) \propto \cos(\vec{Q}\cdot\vec{R})$, namely Cooper pairs occupy a superposition of states with momenta $\vec{Q}$ and $-\vec{Q}$.[25] More generally we can have

$$\Delta(\vec{R}) = \sum_{\vec{Q}} \Delta_{\vec{Q}} e^{i\vec{Q}\cdot\vec{R}}. \tag{20.134}$$

These types of superconducting states with non-uniform superconducting order parameter are collectively called **Fulde–Ferrell–Larkin–Ovchinnikov (FFLO)** states, or Cooper-pair density-wave states, as the squared magnitude of $\Delta(\vec{R})$, which can be viewed as the local density of Cooper pairs, is oscillatory in such states (except in the original Fulde–Ferrell version).

In the above we discussed why the Cooper pairs carry finite momentum from the perspective of the pairing instability of the normal state. Now we address the same question in the superconducting state. The difference between the FFLO and BCS superconducting states is that the former need to accommodate extra spin-up electrons which do not participate in (and in fact may block) pairing. If the superconducting order parameter were uniform, as in the BCS state, these extra electrons would have to be accommodated as spin-polarized Bogoliubov quasiparticles, each costing at least the gap energy Δ. By spontaneously developing an oscillatory superconducting order parameter that changes sign along certain nodal lines as illustrated in Fig. 20.7(b), it is much cheaper to accommodate the unpaired electrons along these nodal lines, where the quasiparticle gap *vanishes*. Also, by spatially localizing the spin-polarized electrons/quasiparticles in regions where the superconducting

---

[24] Usually the most important effect of a magnetic field is its orbital coupling to the superconducting order parameter as discussed in Chapter 19 and Section 20.4. In some circumstances, including a 2D superconductor subject to a magnetic field parallel to the 2D plane, the Zeeman effect could be the dominant one.

[25] In this subsection we focus on the dependence of Δ on the center-of-mass coordinate of the Cooper pair $\vec{R}$, and suppress the possible dependence on the relative momentum/coordinate.

order parameter is weakest, their damage to pairing is minimized. As a result, the FFLO state is energetically more favorable than the BCS state in the presence of spin polarization. The reader should compare this with the appearance of solitons (or domain walls) in the Su–Schrieffer–Heeger (SSH) model when extra charge carriers are present (see Section 7.6). Instead of simply putting the extra charge carriers in the conduction band, which is separated from the Fermi energy by a gap, the system pays the energy cost to introduce solitons with zero modes (i.e. the *local* band gap vanishes) where the charge carrier can be accommodated at zero energy. The net energy cost for deforming the order parameter (introducing the solitons) is less than the cost of placing the fermion above the band gap in the undistorted case. In FFLO states the nodal lines of the superconducting order parameter, which are domain walls separating regions of positive and negative Δ, are very similar to the solitons in the SSH model (separating regions of opposite dimerizations), which also support quasiparticle zero modes.

In the situation illustrated in Fig. 20.7(a), time-reversal symmetry is broken by the Zeeman splitting, as a result of which the weak-coupling instability is eliminated in all singlet (even angular-momentum l) channels; a *finite* pairing interaction strength is required to drive pairing and superconductivity in these channels.[26] However, weak-coupling pairing instabilities are still present in triplet (odd l) channels, as they involve pairing between parallel-spin electrons. Thus, if superconductivity is observed in a ferromagnetic metal or in a system close to ferromagnetic instability (in which, as we learned in Section 15.13.1, repulsive Fermi-liquid interactions are strong), in addition to FFLO states, one should also consider the possibility of triplet pairing, most likely in the p-wave channel. In fact the latter occurs in **superfluid** ^{3}**He**.

As discussed earlier, an external magnetic field has important effects in addition to Zeeman splitting. For this and other reasons, definitive evidence of the FFLO state in electronic superconductors does not yet exist at the time of writing. In trapped-cold-atom systems it is much easier to prepare spin-imbalanced fermions, with no other effects. Studies of fermion pairing in the presence of such an imbalance in 3D found phase separation between a BCS paired state and extra unpaired fermions [226, 227], instead of FFLO pairing. The situation turned out to be much more encouraging in 1D [228], where phase separation does not occur, and the paired state coexisting with imbalance is expected to be a 1D analog of the FFLO state, in which the fluctuating pairing order parameter has oscillatory correlations with a wave vector corresponding to the split of the Fermi wave vector [229].

The possibility of FFLO-type superconducting states has also been discussed for cases in which time-reversal symmetry is present, including in cuprate superconductors. In these contexts they are more often referred to as **Cooper-pair density-wave states**. For the possibility of realizing FFLO-type paired states in neutron stars and quark matter, where pairing comes from the QCD strong interaction, the reader is referred to Ref. [230].

20.9 High-Temperature Cuprate Superconductors

Ever since the discovery of superconductivity in mercury with a superconducting transition temperature temperature of $T_c \approx 4.19\,\mathrm{K}$, physicists have never stopped looking for superconductors with higher T_c, in the hope of finding room-temperature superconductors that could be used for lossless power transmission and many other applications. Progress along this road was very slow for the first 75 years after the 1911 discovery; before 1986 the highest T_c was about 23 K, which had been found in Nb_3Ge. A breakthrough came in 1986, when Bednorz and Müller found superconductivity in $La_{2-x}Ba_xCuO_4$ above 30 K. It was found shortly afterwards that this is just one member of a family

---

[26] Without time-reversal symmetry, the singlet pairing instability becomes similar to charge- and spin-density-wave instabilities (which, as discussed in Section 15.13, are particle–hole pairing instabilities), which require finite coupling stength to trigger.

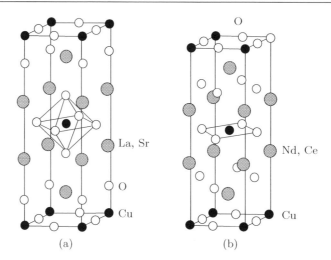

Figure 20.8 (a) A unit cell of $La_{2-x}Sr_xCuO_4$. (b) A unit cell of $Nd_{2-x}Ce_xCuO_4$. Note the CuO_2 planes at the top and bottom layers of both panels. Figure reprinted from [232] by permission from Springer Nature. Copyright Springer Nature (1989).

of copper-oxide compounds obtained by doping the same parent compound, $La_{2-x}M_xCuO_4$, with the dopant M being Ba, Sr, or Ca. See Fig. 20.8(a) for their structures. In the years following this discovery by Bednorz and Müller, more copper-oxide compounds were found, with ever-increasing T_c (and also greater complexity in their structures), including $YBa_2Cu_3O_7$ ($T_c \approx 92$ K), the YBCO family, $Bi_2Sr_2CaCu_2O_8$ ($T_c \approx 89$ K), and the BSCCO family, among others. The record (at the time of writing) at ambient pressure is held by $HgBa_2Ca_2Cu_3O_{8+\delta}$, with $T_c \approx 135$ K.[27] All of these compounds share a common feature, namely they all contain **copper-oxide (CuO_2) planes** (see Fig. 20.8), which are widely believed to be responsible for the superconductivity as well as many other important properties. These compounds are collectively known as **high-temperature cuprate superconductors**, or simply cuprates. In the following we analyze the (original) $La_{2-x}M_xCuO_4$ family (in particular $La_{2-x}Sr_xCuO_4$, which is the most studied compound in this family) in some detail due to its relatively simple structure, despite the modest $T_c \approx 39$ K for $x \approx 0.15$.

20.9.1 Antiferromagnetism in the Parent Compound

Let us start with the parent compound La_2CuO_4. Compared with other compounds we have encountered earlier in this book, it already has a fairly complicated structure (see Fig. 20.8(a)). We start our analysis with individual atoms (ions). Oxygen's outer-shell electron configuration is $2s^22p^4$. As in most cases, here its valence -2 leads to a closed-shell configuration $2s^22p^6$ for O^{2-}, which should be familiar from chemistry. Lanthanum's outer-shell electron configuration is $5d^16s^2$. It has valence $+3$, meaning that it loses all of its outer-shell electrons, again resulting in a closed-shell configuration for La^{3+}. Simple electron counting finds that copper must have valence $+2$ here. Removing two electrons from its $3d^{10}4s^1$ configuration results in $3d^9$ for Cu^{2+}. There is thus a single hole in the copper ion's $3d$ orbitals, and the system may be magnetic due to its net spin. As we will see shortly, this is indeed the case.

We now inspect the system from the band-structure viewpoint. Since we know (*a posteriori!*) that the CuO_2 planes dominate the electronic properties of the system, let us focus on them. Shown in Fig. 20.9(a) is a single plaquette (or unit cell) of a CuO_2 plane. The copper and oxygen ions form a

---

[27] T_c can be higher under pressure in cuprates and other superconductors. In fact, the highest record among all superconductors at the time of writing is held by H_2S, with $T_c = 203$ K under high pressure; see Ref. [231].

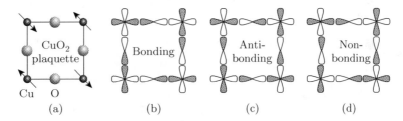

Figure 20.9 (a) A unit cell of the CuO_2 plane. The arrows represent the spins of the holes on Cu sites, which order antiferromagnetically at low temperature. (b)–(d) Illustrations of the copper $d_{x^2-y^2}$ orbital and oxygen p_x (on horizontal bonds) and p_y (on vertical bonds) orbitals. White lobes represent regions where the wave function is positive, and gray lobes represent regions where the wave function is negative. These orbitals hybridize when forming bands, which can be bonding, anti-bonding, or non-bonding, depending on the relative signs of the wave functions. The highest-energy band that hosts the hole corresponds to the anti-bonding hybridization. Figure reprinted with permission from [233]. Copyright (2003) by the American Physical Society.

square lattice with a three-point basis, with one copper and two oxygens in each unit cell. The relevant bands are, obviously, formed by the copper $3d$ and oxygen $2p$ orbitals. If the valence of copper were $+1$ we would have a $3d^{10}$ closed-shell configuration, namely all of these bands (or orbitals) would be fully occupied. Since we have $3d^9$ for Cu^{2+} instead, there is one hole for each unit cell. The question is, which band(s) do the holes go to? Or equivalently, which one of the bands formed by these orbitals has the highest energy? We know that the five $3d$ orbitals of Cu are in general split in a crystal environment (see Section 7.4.5). In the present case it is easy to see that the $d_{x^2-y^2}$ orbital (whose wave function is defined in Eq. (7.189)) should have the highest energy, since it stretches toward the neighboring oxygen ions, which are negatively charged (thus raising the electron potential energy). See Fig. 20.9 for illustration. As a result, the holes go to the band formed by the Cu $d_{x^2-y^2}$ orbital. Detailed band-structure calculations find that this band is indeed the closest to the Fermi energy, and thus the highest that is occupied by electrons consistent with our expectation. For many purposes we can capture the low-energy physics of cuprates by focusing on this band (or, equivalently, using a single-band model). In certain cases the presence of the oxygen p orbitals is important. The two most important ones are those that stretch toward Cu, as shown in Fig. 20.9, because they hybridize strongly with its $d_{x^2-y^2}$ orbital. When these orbitals are included, we have a three-band model.

Taking either viewpoint, we find that there is an *odd* number of holes (or electrons) in a unit cell, and simple band theory would thus predict that the system is a metal. In fact, La_2CuO_4, like *all* other undoped cuprates, is an antiferromagnetic insulator at low temperature. To understand the insulating nature and the antiferromagnetism, we need to take into account the electron–electron (or, equivalently, hole–hole) repulsive interaction. At the simplest level we just keep the single (copper $d_{x^2-y^2}$) band, and the interaction when two electrons (or holes) are on the *same* site. This results in the Hubbard model we studied in Section 17.4 on a square lattice. We already understood there that when the interaction U is large enough the system is a Mott insulator at half-filling, which clearly describes La_2CuO_4 and other undoped cuprates. Physically, this simply means that, due to Coulomb repulsion, there is a single hole *localized* on each Cu^{2+} ion, which is consistent with our earlier single-ion analysis. The spin carried by the hole is responsible for antiferromagnetism at low temperature.

As we also learned in Section 17.4, if the Hubbard U parameter is large compared with the band width, then at half-filling the Hubbard model reduces to an antiferromagnetic Heisenberg model. This further simplification is indeed appropriate for all undoped cuprates, and they all have similar antiferromagnetic Heisenberg exchange coupling $J \approx 10^3$ K, and **Néel ordering temperature** $T_N \approx$ 300 K.[28]

---

[28] As discussed in Section 17.5.2, $T_N = 0$ for the 2D Heisenberg model. The reason why $T_N > 0$ in cuprates is that there exists coupling between neighboring CuO_2 planes.

20.9.2 Effects of Doping

Physical properties of the compound change dramatically when a small fraction of La is replaced by Ba, Sr, or Ca. Let us analyse the effects of such doping. Ba, Sr, and Ca are all alkali earth elements with valence $+2$. They thus each donate one electron fewer than La does, and play a role very similar to that of acceptors in a semiconductor. As a result, each such dopant contributes an *additional* hole to the CuO_2 planes, and dopes them away from half-filling.[29] Most cuprates can be hole-doped, but there are a few that can be electron-doped. One example of such an electron-doped cuprate is $Nd_{2-x}Ce_xCuO_4$ (see Fig. 20.8(b)). Nd has valence $+3$; when it is replaced by the dopant Ce with valence $+4$, an additional electron goes to the CuO_2 planes and annihilates an existing hole. Thus Ce plays a role similar to that of a donor in a semiconductor.

The temperature-doping phase diagrams of $La_{2-x}Sr_xCuO_4$ and $Nd_{2-x}Ce_xCuO_4$ are shown together in Fig. 20.10 to illustrate their similarities as well as some differences. In both cases antiferromagnetism is suppressed by doping. Further increasing the doping level leads to superconductivity, which eventually disappears at a high doping level. Both of these features are common to all cuprates. For $La_{2-x}Sr_xCuO_4$ and other hole-doped cuprates there exists a large region in the phase diagram above the superconducting transition temperature where these systems behave anomalously, in the sense that they *cannot* be understood using Landau's Fermi-liquid theory, including evidence that electronic degrees of freedom are gapped in the *absence* of superconductivity. Such "pseudogap" and other anomalous behaviors remain mysterious at the time of writing, and we will not discuss them further in this book. For the remainder of this section we focus on the superconducting phase.

20.9.3 Nature of the Superconducting State

Owing to the unprecedentedly high T_c (at the time), there was initially some doubts on whether superconductivity in cuprates is due to pairing of electrons and condensation of the resultant Cooper pairs. Such doubts were quickly removed by the observation of flux quantization in units of

$$\Phi_S = \frac{hc}{2e} \tag{20.135}$$

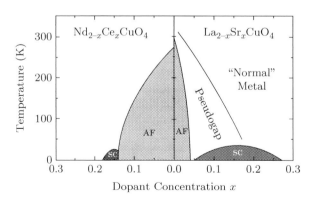

Figure 20.10 Phase diagrams of $La_{2-x}Sr_xCuO_4$ and $Nd_{2-x}Ce_xCuO_4$. SC stands for superconducting phase and AF stands for antiferromagnetic phase. Figure reprinted with permission from [233]. Copyright (2003) by the American Physical Society.

---

[29] The doped holes actually predominantly go to oxygen sites. They couple strongly to the holes on copper sites and form the so-called Zhang–Rice singlets [234]. As a result, one can still describe doped cuprates using a one-band model. For details see Ref. [234].

just like in conventional superconductors, thus indicating the charge of the superconducting condensate is $2e$. This leads to the following questions. (i) What is the internal structure (or symmetry) of the pairing order parameter in cuprates? (ii) What is the pairing mechanism?[30] As we will discuss below, experimentalists have succeeded in answering question (i), while at the moment there is no consensus on the answer to question (ii).

Early experiments indicated that, while the pairing is most likely to occur in a singlet channel in the cuprate superconductors, it is probably unconventional (or non-s-wave). Over the years evidence has been accumulating for a specific type of d-wave pairing, namely pairing in the $d_{x^2-y^2}$ channel. Let us first discuss what d-wave means in the presence of a crystal which does *not* have full rotation symmetry. The square lattice of the CuO_2 plane has a residual four-fold rotation symmetry, generated by rotations of 90°. A d-wave function changes sign under such a 90° rotation. A $d_{x^2-y^2}$ wave has its maxima in magnitude along the x and y directions, but with opposite sign, just like the copper $d_{x^2-y^2}$ orbital illustrated in Fig. 20.9. In its simplest form, the pairing order parameter resides on nearest-neighbor bonds of the square lattice *only*:

$$\Delta_{ij} = (\Delta/2)(\delta_{j,i\pm a\hat{x}} - \delta_{j,i\pm a\hat{y}}), \tag{20.136}$$

where a is the lattice constant. The d-wave nature is reflected in the opposite sign for the horizontal and vertical bonds. Fourier transforming to momentum space results in

$$\Delta_{\vec{k}} = \Delta[\cos(k_x a) - \cos(k_y a)]. \tag{20.137}$$

Note that $|\Delta_{\vec{k}}|$ (which is the quasiparticle gap) maximizes along the $\hat{x}$ and $\hat{y}$ directions, but *vanishes* (or has a node) along the diagonal $[(\pm\pi/a, \pm\pi/a)]$ directions of the first Brillouin zone. The latter is guaranteed by the $d_{x^2-y^2}$ symmetry, as long as the pairing order parameter is real[31] for all $\vec{k}$ (and thus does not break time-reversal symmetry).

There are numerous experimental results supporting the $d_{x^2-y^2}$ symmetry of the cuprate pairing order parameter. We discuss two of them in the following; one measures $|\Delta_{\vec{k}}|$ while the other probes the phase of $\Delta_{\vec{k}}$ directly.

As discussed in Section 9.3.1, angle-resolved photoemission spectroscopy (ARPES) is a very effective way to measure the electron (or quasiparticle) dispersion in a crystal. Applying it to a superconductor, ARPES measures the dispersion of Bogoliubov quasiparticles, which are, as discussed earlier, superpositions of electrons and holes. The magnitude of the superconducting gap can be extracted from such measurements. Figure 20.11 represents results of such a measurement of $|\Delta_{\vec{k}}|$, which is consistent with the existence of a gap node along the diagonal directions of the first Brillouin zone.

While the ARPES result clearly indicates a highly anisotropic gap (which is inconsistent with the simplest s-wave pairing with a constant gap) and suggests the presence of gap nodes, it cannot probe the sign change of $\Delta_{\vec{k}}$ upon going from the $\hat{x}$ direction to the $\hat{y}$ direction. Thus, to prove the $d_{x^2-y^2}$ symmetry of the pairing order parameter, we need a probe that is sensitive to the *phase* of $\Delta_{\vec{k}}$. As discussed in Section 19.9, the Josephson effect is very sensitive to the phase of superconducting order parameters. As a result, phase-sensitive probes of $\Delta_{\vec{k}}$ all involve (multiple) Josephson junctions. We discuss one of them below.

To probe the phase difference between different directions of $\Delta_{\vec{k}}$, we need to have Josephson junctions between crystals with different orientations. One configuration that involves three different orientations is illustrated in Fig. 20.12(a). The two crystals in the upper half of the panel are rotated by 30° and 60° from that of the crystal occupying the lower half, respectively. There are three junctions, and we will try to put all three of them in the ground state, which means that the

---

[30] A purist might argue that, in a strongly interacting system where mean-field theory may not apply, "the" mechanism is not a well-defined concept.

[31] Of course, there can be a global phase factor, which we will take to be real for simplicity.

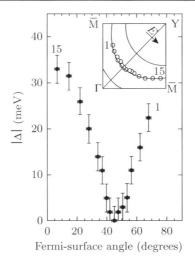

Figure 20.11 ARPES measurement of the direction dependence of the superconducting gap in $Bi_2Sr_2CaCu_2O_{8+x}$. There is clear indication of the gap vanishing along the diagonal direction (45°) of the first Brillouin zone, which is consistent with the $d_{x^2-y^2}$ symmetry of the pairing order parameter. Figure reprinted with permission from [235]. Copyright (1996) by the American Physical Society.

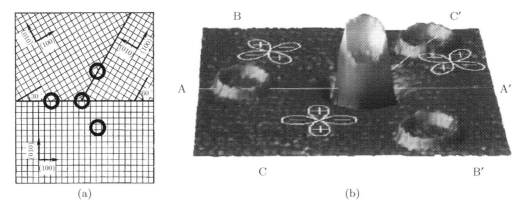

Figure 20.12 (a) Three crystals of $YBa_2Cu_3O_{7-\delta}$ with different orientations grown together to form three Josephson junctions. (b) Possible superconducting order-parameter configurations on the crystals, and flux equal to half of the superconducting flux quantum generated by the supercurrent to minimize the Josephson coupling energy on all three junctions. Figure reprinted with permission from [236]. Copyright (2000) by the American Physical Society.

gauge-invariant phase difference across each junction is zero. Let us first focus on the two vertical junctions. If we assume that the Cooper-pair tunneling from the lower crystal in the vertical direction is sensitive mostly to $\Delta_{\vec{k}}$ in the upper crystals, with directions of $\vec{k}$ close to the vertical direction, then the optimal order-parameter configuration of the two upper crystals would be indicated by those of the right panel of Fig. 20.12. However, such a configuration frustrates the junction between the two upper crystals, as the Cooper pairs crossing that junction see a π phase difference. Such a junction is known as a **π-junction**. It is easy to convince oneself that, no matter how we rearrange the order-parameter configuration among the three crystals, there is at least one junction that is frustrated, because the total phase difference for a loop that crosses all three junctions (and encloses the point where the three crystals touch) is π (modulo 2π).[32]

---

[32] The nature of such frustration is very similar to that felt by three spins on the vertices of a triangle that are coupled antiferromagnetically; see Fig. 17.17(a).

The way to release the frustration is to take advantage of the fact that the Josephson coupling energy is determined by the *gauge-invariant* phase difference across the junction, which also involves the line integral of the vector potential. Thus the system can spontaneously generate a supercurrent which induces a flux

$$\Phi = \frac{hc}{4e} = \frac{\Phi_S}{2} \tag{20.138}$$

around the joint. This flux contributes π to the total gauge-invariant phase difference around a loop enclosing the joint, canceling out that due to the sign change of $\Delta_{\vec{k}}$ along different directions. Another way to understand this is that, due to the existence of an odd number of π junction(s), the point where the three crystals meet can be viewed as a half vortex of the superconducting order parameter. For the same energetic reason as for why a vortex carries with it a flux quantum (see Section 19.6.2), this half vortex carries half of a a flux quantum. This half flux quantum was indeed observed experimentally (see Fig. 20.12(b)), providing convincing evidence for the sign change of $\Delta_{\vec{k}}$ due to d-wave pairing.

An extremely important consequence of $d_{x^2-y^2}$ pairing is the existence of gap nodes on the Fermi surface, around which the Bogoliubov quasiparticle dispersion is *gapless*. This is qualitatively different from the conventional s-wave superconductors, in which quasiparticles are *fully* gapped. Expanding the quasiparticle dispersion

$$E_{\vec{k}} = \sqrt{\epsilon_{\vec{k}}^2 + \Delta_{\vec{k}}^2} \tag{20.139}$$

around the nodal point $\vec{K}$ where $\epsilon_{\vec{K}} = \Delta_{\vec{K}} = 0$, we obtain a *linear* dispersion just like in graphene. As a result, the thermodynamic and some other properties of cuprate superconductors are very similar to those of graphene and other electronic systems with (massless) Dirac dispersion.

Exercise 20.9. Assume that the Fermi velocity at the node $\vec{K}$ is v_F and $\Delta_{\vec{K}+\delta\vec{k}} = v_g(\hat{e} \cdot \delta\vec{k})$, where $\hat{e}$ is the unit vector parallel to the Fermi surface at $\vec{K}$. Calculate the contribution to the low-temperature specific heat from this node.

20.9.4 Why d-Wave?

Before the discovery of superconductivity in cuprates, essentially all known superconductors were conventional (i.e. s-wave), with pairing mediated by phonons.[33] The situation is clearly different in the cuprates. In addition to the much higher T_c, as discussed in Section 20.9.3 there is overwhelming evidence that hole-doped cuprates are d-wave superconductors and thus unconventional. It is thus natural to suspect that the pairing in cuprates might *not* be mediated by phonons, but could rather be due to electron–electron interactions instead. This would naturally explain the unconventional nature of the pairing, since the electron–electron interaction is predominantly *repulsive*, while the centrifugal barrier due to the non-zero angular momentum of the Cooper pair prevents the electrons forming the pair from coming close, thus lowering the Coulomb energy (compared with s-wave pairing, where no such centrifugal barrier exists). Another consideration is that the energy scale associated with electron–electron interaction is much higher than that of phonons, and this could potentially lead to higher T_c.

At the time of writing there is no consensus on how electron–electron interactions lead to d-wave superconductivity in cuprates. The difficulty in theoretical study lies in the fact that the interaction is strong. Qualitatively, however, it seems reasonable to assume that the two most robust phases in

---

[33] The only possible exceptions are **heavy-fermion superconductors**, which we will not discuss in this book.

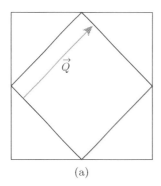
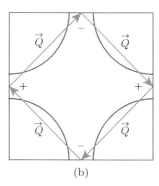

(a)   (b)

Figure 20.13 Fermi surfaces and the order-parameter configuration of cuprates as described by Hubbard-like models on a square lattice. The outermost square box is the first Brillouin zone, and the inner box in (a) and curved lines in (b) are Fermi surfaces. (a) The undoped case or half-filling. For the special case of only nearest-neighbor hopping the Fermi surface takes the shape of a diamond which is perfectly nested, with one of the nesting vectors $\vec{Q}$ shown. (b) The doped case. The pairing order parameter $\Delta_{\vec{k}}$ takes opposite signs in regions connected by $\vec{Q}$, resulting in a $d_{x^2-y^2}$ form for $\Delta_{\vec{k}}$.

cuprates, namely antiferromagnetism of the parent compound and d-wave superconductivity, are stabilized by the *same* type of repulsive electron–electron interaction. In the following we study their relation in a (perhaps overly) simplified model, in the weak-coupling limit.

As mentioned earlier, the simplest model to describe the copper–oxygen plane is the (positive-U) Hubbard model on a square lattice:

$$H = -t \sum_{\langle ij \rangle, \sigma} (c_{i\sigma}^{\dagger} c_{j\sigma} + c_{j\sigma}^{\dagger} c_{i\sigma}) + U \sum_{j} n_{j\uparrow} n_{j\downarrow}. \tag{20.140}$$

As we learned in Section 17.4, at half-filling and strong coupling ($U \gg t$), the system is a Mott insulator. Here we consider the opposite (weak-coupling) limit ($U \ll t$). In this case it is natural to first focus on the first term in Eq. (20.140), namely the hopping or band energy, which can be diagonalized by going to the momentum (Bloch-wave) basis:

$$\hat{T} = -t \sum_{\langle ij \rangle, \sigma} (c_{i\sigma}^{\dagger} c_{j\sigma} + c_{j\sigma}^{\dagger} c_{i\sigma}) = \sum_{\vec{k}\sigma} \epsilon_{\vec{k}} c_{\vec{k}\sigma}^{\dagger} c_{\vec{k}\sigma}, \tag{20.141}$$

where the band energy is

$$\epsilon_{\vec{k}} = -2t[\cos(k_x a) + \cos(k_y a)]. \tag{20.142}$$

A very special property of the dispersion above is

$$\epsilon_{\vec{k}} = -\epsilon_{\vec{k}+\vec{Q}}, \tag{20.143}$$

where $\vec{Q} = (\pm\pi/a, \pm\pi/a)$ is a so-called "nesting vector."[34] Related to this is the fact that at half-filling we have $\epsilon_F = 0$, and the Fermi surface takes the shape of a diamond (see Fig. 20.13(a)), with $\vec{Q}$ connecting opposite segments of the diamond. As we discussed in Section 15.13, such nested Fermi surfaces yield charge- and spin-density-wave (CDW and SDW) instabilities at weak coupling.

---

[34] Note that the choice of sign in front of π/a is arbitrary because different choices give rise to $\vec{Q}$s that differ by a reciprocal lattice vector of the square lattice.

To reveal this explicitly, we calculate the static charge-density susceptibility with $U = 0$ in Eq. (20.140), namely for non-interacting electrons, using the formalism of Appendix A:

$$\chi_0(\vec{q}, T) = 2 \sum_{\vec{k}} \frac{f_{\vec{k}} - f_{\vec{k}+\vec{q}}}{\epsilon_{\vec{k}} - \epsilon_{\vec{k}+\vec{q}}}, \tag{20.144}$$

where f is the Fermi distribution function for temperature T. Note the similarity to Eq. (15.92). It is easy to show that the spin-density susceptibility takes exactly the same form, and we do not distinguish between them.

Owing to the nesting property (20.143), we find that $\chi_0(\vec{Q}, T \to 0)$ diverges logarithmically (see Exercise 20.10). This is equivalent to the statement we made in Section 15.13 that a nested Fermi surface results in CDW or SDW instabilities at weak coupling.[35] It is unclear at this point, however, whether this weak-coupling instability leads to CDW or SDW order in the ground state. This depends on the interaction, to which we now turn.

We re-write the second (electron–electron interaction) term of (20.140) in terms of charge- and spin-density operators:[36]

$$\hat{V} = \frac{U}{4} \sum_j [(n_{j\uparrow} + n_{j\downarrow})^2 - (n_{j\uparrow} - n_{j\downarrow})^2] = \frac{U}{4} \sum_{\vec{q}} [\rho_{\vec{q}} \rho_{-\vec{q}} - S^z_{\vec{q}} S^z_{-\vec{q}}]. \tag{20.145}$$

This leads to the random-phase approximation (RPA; see Section 15.8) expressions for spin- and charge-density susceptibilities:

$$\chi^{\text{s,c}}_{\text{RPA}}(\vec{q}, T) = \frac{\chi_0(\vec{q}, T)}{1 \pm (U/4)\chi_0(\vec{q}, T)}. \tag{20.146}$$

Since $\chi_0(\vec{q}, T) < 0$, we find that $\chi^{\text{c}}_{\text{RPA}}(\vec{q}, T)$ is *no longer* divergent for $U > 0$, while $\chi^{\text{s}}_{\text{RPA}}(\vec{q}, T)$ now diverges at *finite* T, indicating that the ground state has SDW order. The energetic reason for why SDW order is favored over CDW order can be seen clearly in Fig. 20.14. In a CDW the probability of there being up- and down-spin electrons on the same site is enhanced, whereas in an SDW it is suppressed. It is also clear that the SDW state, which we studied here in the weak-coupling limit, is smoothly connected to the antiferromagnetic (or Néel-ordered) state in the strong-coupling (or Mott-insulator) limit studied in Section 17.5.2, described by the Heisenberg model.

CDW   SDW

Figure 20.14 Illustration of the electron configuration in a charge-density wave (CDW, left) and in a spin-density wave (SDW, right). Black dots represent lattice sites. Clearly, for CDW order the probability of finding two electrons on the same site is higher, costing more energy for $U > 0$.

Now consider the case with doping. In this case the Fermi surface is no longer perfectly nested, but it will be approximately so if the doping is not too high. As a result, $\chi_0(\vec{q})$ will not diverge as $T \to 0$, and neither will $\chi^{\text{s}}_{\text{RPA}}(\vec{q})$ as long as U is weak enough, so the ground state does *not* develop SDW or antiferromagnetic order. Instead, due to the approximate nesting, both $\chi_0(\vec{q})$ and $\chi^{\text{s}}_{\text{RPA}}(\vec{q})$ are

[35] Similarly, the weak-coupling pairing instability can also be revealed by an essentially identical logarithmic divergence in a properly defined pair susceptibility; see Exercise 20.11.

[36] We note that this decomposition is not unique. In addition, it does not maintain the explicit SU(2) spin rotation invariance of the model. The specific choice, however, does not affect the qualitative conclusions we will reach.

peaked (but remain finite) near $\vec{Q}$ at low T. This feature has a very important effect on the *screened* interaction, which we again use the RPA to evaluate:

$$\hat{V}_{sc} = \frac{U}{4} \sum_{\vec{q}} \left[\frac{\rho_{\vec{q}} \rho_{-\vec{q}}}{1 - (U/4)\chi_0(\vec{q})} - \frac{S_{\vec{q}}^z S_{-\vec{q}}^z}{1 + (U/4)\chi_0(\vec{q})} \right]. \tag{20.147}$$

We find the screened interaction between electrons with opposite spins, $V_{sc}^{\uparrow\downarrow}(\vec{q})$, is also peaked near $\vec{Q}$ (due to the contribution from the second term in the brackets, whose denominator is smaller than that for the first term), as illustrated in Fig. 20.15.

It appears that the screened interaction remains purely repulsive, since $V_{sc}^{\uparrow\downarrow}(\vec{q}) > 0$ everywhere in reciprocal space. In fact, however, it contains attractive interactions when viewed in real space. One way to see this is to consider the extreme case with $V_{sc}^{\uparrow\downarrow}(\vec{q}) \propto \delta_{\vec{q},\vec{Q}}$.[37] Fourier transforming to real space results in repulsion when the two electrons are on the same sublattice, but attraction when they are on opposite sublattices. The appearance of attraction is again due to *over-screening*, which is very much like attraction due to the electron–phonon interaction.

What kind of pairing state does such an interaction lead to? Recall that, in the pseudospin mapping introduced in Section 20.2, $V_{sc}^{\uparrow\downarrow}(\vec{q} = \vec{k} - \vec{k}')$ is the coupling between the transverse ($\hat{x}$ and $\hat{y}$) components of the pseudospins at $\vec{k}$ and $\vec{k}'$. Since $V_{sc}^{\uparrow\downarrow}(\vec{q}) > 0$, the couplings are antiferromagnetic, thus s-wave pairing corresponding to ferromagnetic ordering of the pseudospins is energetically unfavorable. To take advantage of the strongest antiferromagnetic coupling $V_{sc}^{\uparrow\downarrow}(\vec{q} \approx \vec{Q})$, the pseudospins, and thus the pairing order parameter $\Delta_{\vec{k}}$, should point in *opposite* directions between regions of momentum space that are connected by $\vec{Q}$. This naturally leads to a $d_{x^2-y^2}$ pairing order parameter, as illustrated in Fig. 20.13(b). It is interesting to note that this d-wave, as well as other unconventional superconducting states, corresponds to a pseudospin density-wave state in *momentum* space.

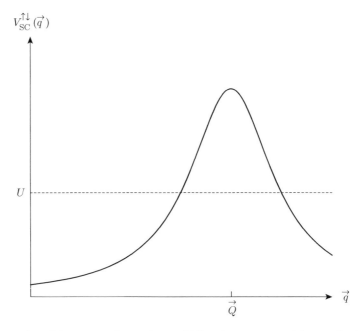

Figure 20.15 Illustration of the screened interaction (solid line) between up- and down-spin electrons of Eq. (20.147) in momentum space. For comparison, the bare interaction (dashed line) from the Hubbard U is a constant in momentum space.

---

[37] The reader may think that one needs to add $\delta_{\vec{q},-\vec{Q}}$ to the RHS of this expression. However, for $\vec{Q} = (\pi, \pi)$, $\vec{Q}$ and $-\vec{Q}$ are effectively the same wave vector.

While the above calculations based on the RPA are perturbative and do not necessarily apply in the regime of strong interactions, they do provide strong motivation for the idea that repulsive interactions combined with strongly wave-vector-dependent screening in a system with nearly nested Fermi surface segments naturally lead to the unconventional $d_{x^2-y^2}$ pairing observed in the hole-doped cuprates.

Exercise 20.10. Calculate $\chi_0(\vec{Q}, T)$ of Eq. (20.144) in the limit $T \to 0$ with the band dispersion (20.142), and identify the leading divergent term.

Exercise 20.11.

(i) Consider the response of a non-interacting electron system with a generic spin-degenerate Fermi surface to a pairing perturbation of the form $H' = -\Delta \sum_{\vec{k}}' (c_{\vec{k}\uparrow}^\dagger c_{-\vec{k}\downarrow}^\dagger + c_{-\vec{k}\downarrow} c_{\vec{k}\uparrow})$ (the prime in summation indicates it is restricted to $\vec{k}$s within a certain distance of the Fermi surface), which describes the metal coupled to a nearby (s-wave) superconductor via Josephson tunneling of Cooper pairs. As a result of this, there is a pairing induced in the metal. This is known as the **proximity effect**. Calculate the pairing response in the metal $\sum_{\vec{k}}' \langle c_{\vec{k}\uparrow}^\dagger c_{-\vec{k}\downarrow}^\dagger \rangle$, and show that the properly defined susceptibility diverges logarithmically as $T \to 0$. Discuss the similarity between this divergence and that in Exercise 20.10.

(ii) Now consider the case where there is a Zeeman splitting of the Fermi surfaces, as illustrated in Fig. 20.7. Show that the divergence is removed, and find the wave vector where this pairing susceptibility peaks, by considering a more general perturbation $H' = -\Delta_{\vec{q}} \sum_{\vec{k}}' (c_{\vec{k}+\vec{q}\uparrow}^\dagger c_{-\vec{k}\downarrow}^\dagger + c_{-\vec{k}\downarrow} c_{\vec{k}+\vec{q}\uparrow})$. This illustrates the fact that, in the presence of Zeeman splitting, the leading pairing instability is of FFLO instead of BCS type.

Appendix A

Linear-Response Theory

It is often the case that we are interested in the response of some physical observable in a quantum system to a weak perturbation acting on that system. A familiar example is electrical conductivity. In typical materials, the electrical current that flows is linear in the applied electric field. Another example is the magnetic susceptibility – the magnetization density of many materials is, to a good approximation, linear in the applied magnetic field. In this appendix we provide a general formulation of **linear-response theory** that is based on first-order perturbation theory. The expression for the linear response coefficient relating the observable to the perturbation is often called a **Kubo formula**, after R. Kubo who first derived the general expression for electrical conductivity.

A.1 Static Response

Consider a quantum-mechanical system with (unperturbed) Hamiltonian H_0, whose complete set of (normalized) eigenstates is $\{|\psi_j\rangle\}$ with corresponding eigenenergies $\{E_j\}$. Let us focus on one of the eigenstates, $|\psi_i\rangle$ (assumed to be non-degenerate for simplicity), and an observable $O = O^\dagger$. Without losing generality we can assume that $\langle \psi_i | O | \psi_i \rangle = 0$ (otherwise we can consider the observable to be $O' = O - \langle \psi_i | O | \psi_i \rangle$ instead).

Now perturb the system with a time-independent perturbation $H' = \lambda O$, where λ is a small parameter.[1] Straightforward first-order perturbation theory tells us that $|\psi_i\rangle$ is modified to

$$|\psi_i'\rangle = |\psi_i\rangle + |\delta\psi_i\rangle, \tag{A.1}$$

with

$$|\delta\psi_i\rangle = \lambda \sum_{j \neq i} \frac{|\psi_j\rangle \langle \psi_j | O | \psi_i \rangle}{E_i - E_j} + \mathcal{O}(\lambda^2). \tag{A.2}$$

One may worry that $|\psi_i'\rangle$ is no longer normalized; however, it is easy to see that the correction to its normalization is $\mathcal{O}(\lambda^2)$, while in linear response we are interested only in effects to order $\mathcal{O}(\lambda)$, and can thus ignore the change of normalization.

---

[1] The more general case where the perturbation couples to a different operator, rather than to the observable itself, will be treated further below.

We now consider the change of the expectation value of O in the state i, induced by the perturbation H':

$$\langle O \rangle_i = \langle \psi'_i | O | \psi'_i \rangle = 2\lambda \sum_{j \neq i} \frac{\langle \psi_i | O | \psi_j \rangle \langle \psi_j | O | \psi_i \rangle}{E_i - E_j} + \mathcal{O}(\lambda^2) \qquad (A.3)$$

$$= \chi_i \lambda + \mathcal{O}(\lambda^2), \qquad (A.4)$$

where

$$\chi_i = 2 \sum_{j \neq i} \frac{\langle \psi_i | O | \psi_j \rangle \langle \psi_j | O | \psi_i \rangle}{E_i - E_j} = 2 \sum_{j \neq i} \frac{|\langle \psi_i | O | \psi_j \rangle|^2}{E_i - E_j} \qquad (A.5)$$

is the corresponding (static) response function.

As an illustration, consider a simple harmonic oscillator of frequency ω_0 and spring constant $k = m\omega_0^2$, and let the perturbation be a force F coupling to the position, $H' = \lambda x = \lambda x_{ZPF}(a + a^\dagger)$, where $x_{ZPF} = \sqrt{\hbar/2m\omega_0}$ is the zero-point uncertainty of the position in the quantum ground state and we need to pay attention to the minus sign in $\lambda = -F$. If the initial state is the ground state $|0\rangle$, then Eq. (A.5) yields

$$\chi_0 = 2x_{ZPF}^2 \frac{|\langle 0 | a | 1 \rangle|^2}{-\hbar\omega_0} = -\frac{1}{k}, \qquad (A.6)$$

where we have used the fact that one-half of the zero-point energy of the ground state is potential energy, which means that $\frac{1}{2}k x_{ZPF}^2 = \frac{1}{4}\hbar\omega_0$. Linear-response theory is of course exact for a spring obeying Hook's law, and the quantum result from perturbation theory,

$$\langle x \rangle = (-F)\left(-\frac{1}{k}\right) = \frac{F}{k}, \qquad (A.7)$$

is exact for arbitrary strength of the force F; and this expression also agrees with the classical result.

It is often useful to consider perturbations of the form

$$H' = \lambda(O + O^\dagger), \qquad (A.8)$$

where $O \neq O^\dagger$ is non-Hermitian. For example, consider a single particle subject to a sinusoidal potential:

$$H' = 2\lambda \cos(\vec{q} \cdot \vec{r}) = \lambda(e^{-i\vec{q}\cdot\vec{r}} + e^{i\vec{q}\cdot\vec{r}}); \qquad (A.9)$$

in this case it is natural to choose

$$O = e^{-i\vec{q}\cdot\vec{r}} = \rho_{\vec{q}} \qquad (A.10)$$

to be the (single-particle version of the) density operator. If we now calculate $\langle O + O^\dagger \rangle_i$ along the lines of Eq. (A.3), we will in principle get four terms. However, very often only the cross terms of the form $\langle \psi_i | O^\dagger | \psi_j \rangle \langle \psi_j | O | \psi_i \rangle = |\langle \psi_i | O^\dagger | \psi_j \rangle|^2$ are non-zero for symmetry reasons. Using our density operator example above, in the presence of translation symmetry we expect $|\psi_i\rangle$ and $|\psi_j\rangle$ to be eigenstates of momentum; the reader should verify that this implies

$$\langle \psi_i | O | \psi_j \rangle \langle \psi_j | O | \psi_i \rangle = \langle \psi_i | \rho_{\vec{q}} | \psi_j \rangle \langle \psi_j | \rho_{\vec{q}} | \psi_i \rangle = 0 \qquad (A.11)$$

(unless $\vec{q} = 0$). In such cases we find

$$\langle O + O^\dagger \rangle_i = \chi_i \lambda + \mathcal{O}(\lambda^2), \qquad (A.12)$$

where

$$\chi_i = \sum_{j \neq i} \frac{\langle \psi_i | O | \psi_j \rangle \langle \psi_j | O^\dagger | \psi_i \rangle + \langle \psi_i | O^\dagger | \psi_j \rangle \langle \psi_j | O | \psi_i \rangle}{E_i - E_j}. \qquad (A.13)$$

Obviously χ_i is the zero-temperature response function of the system if $|\psi_i\rangle$ is the ground state. In thermal equilibrium at finite temperature, we have

$$\chi = \frac{1}{Z}\sum_i e^{-\beta E_i}\chi_i, \tag{A.14}$$

where Z is the partition function.

A.2 Dynamical Response

In the spirit of the (classical) calculation of the optical polarizability discussed in Section 7.10.1, let us consider the following (quantum) time-dependent perturbation:

$$H'(t) = \lambda O[2\cos(\omega t)]e^{\eta t}, \tag{A.15}$$

where $O = O^\dagger$ is some observable and $\eta = 0^+$ is an infinitesimal positive number. (We include the factor of 2 in the definition simply for convenience to cancel out the factor of 2 in the denominator of the definition of the cosine.) The η factor above ensures that $H'(t \to -\infty) \to 0$, so that the perturbation is turned off in the distant past and turns on slowly (adiabatically), reaching full strength in a broad region centered on $t = 0$, the time region of interest.[2] This factor is often kept implicit when non-essential, but we will keep it explicitly in our discussion below. Note that a general time-dependent perturbation can be written as combinations of terms like those in Eq. (A.15) via Fourier transformation.

Assume that the system is in state $|\psi_i\rangle$ for $t \to -\infty$. Using standard time-dependent perturbation theory, we write the state at time t as

$$|\psi_i'(t)\rangle = \sum_j c_j e^{-i\omega_j t}|\psi_j\rangle = e^{-i\omega_i t}(|\psi_i\rangle + |\delta\psi_i(t)\rangle), \tag{A.16}$$

with

$$|\delta\psi_i(t)\rangle = \sum_j (c_j - \delta_{ij})e^{i\omega_{ij}t}|\psi_j\rangle. \tag{A.17}$$

In the above $\omega_j = E_j/\hbar$ and $\omega_{ij} = \omega_i - \omega_j$. To order $\mathcal{O}(\lambda)$, the Schrödinger equation gives

$$i\hbar\frac{dc_j}{dt} = \langle\psi_j|H'|\psi_i\rangle e^{-i\omega_{ij}t}. \tag{A.18}$$

Integrating the above from $t' = -\infty$ (when $c_j = \delta_{ij}$) to $t' = t$ yields

$$|\delta\psi_i(t)\rangle = \frac{\lambda}{\hbar}e^{\eta t}\sum_{j\neq i}|\psi_j\rangle\langle\psi_j|O|\psi_i\rangle\left[\frac{e^{i\omega t}}{\omega_i - \omega_j - \omega + i\eta} + \frac{e^{-i\omega t}}{\omega_i - \omega_j + \omega + i\eta}\right], \tag{A.19}$$

which is correct up to order $\mathcal{O}(\lambda)$. Note that the $+i\eta$ "convergence factors" in the denominators are *not* put in by hand; they come from the time integral!

Assuming as before that $\langle\psi_i|O|\psi_i\rangle = 0$, the expectation value of the observable in the perturbed state is

$$\langle O\rangle_i = \lambda e^{\eta t}[\chi_i(\omega)e^{-i\omega t} + \chi_i(-\omega)e^{+i\omega t}], \tag{A.20}$$

---

[2] The fact that this convergence factor slowly blows up in the future is irrelevant because of causality.

where the susceptibility of the system when it is in state i is

$$\chi_i(\omega) = \frac{1}{\hbar} \sum_{j \neq i} |\langle \psi_i | O | \psi_j \rangle|^2 \left\{ \frac{1}{\omega_i - \omega_j + \omega + i\eta} + \frac{1}{\omega_i - \omega_j - \omega - i\eta} \right\}. \tag{A.21}$$

Note that, for any initial state, this obeys the fundamental symmetry relation

$$\chi_i(-\omega) = \chi_i^*(\omega) \tag{A.22}$$

as is required in order for $\langle O \rangle$ to be real. Clearly, in the limit $\omega \to 0$, $\chi_i(\omega)$ reduces to the static response function χ_i in Eqs. (A.5) and (A.13). Similarly, the dynamical response function of the system in thermal equilibrium is

$$\chi(\omega) = \frac{1}{Z} \sum_i e^{-\beta E_i} \chi_i(\omega). \tag{A.23}$$

We have so far considered only the case of the response to a perturbation O of the observable O itself. It is straightforward to generalize the above results to the case of the susceptibility $\chi_{AB}(\omega)$ of an observable A to a perturbation B. It is interesting to ask whether there is a simple relationship between this susceptibility and the corresponding susceptibility of B to a perturbation A. There is indeed a simple connection, and it is straightforward to derive the so-called **Onsager relation** between the two quantities. For systems with time-reversal symmetry we have

$$\chi_{BA}(\omega) = \chi_{AB}^*(-\omega). \tag{A.24}$$

The story is a bit more complicated for the case of thermal transport coefficients discussed in Section 8.9, where non-equilibrium temperature gradients are involved. There, depending on how things are defined, factors of temperature can appear in the Onsager relations, as in Eq. (8.171). We will not discuss these complications here, but refer the reader to [129] for the details.

As a simple illustration of dynamical susceptibility, let us return to the example of the oscillator model of atomic polarizability discussed in Section 7.10.1. The resonance frequency is ω_0, the oscillator (in this case undamped) has charge q and mass m, and is driven by a time-dependent electric field

$$H'(t) = -q E_0 x \cos(\omega t) e^{\eta t}. \tag{A.25}$$

From Eq. (A.15) we identify the observable to be the dipole moment $O = q x_{\text{ZPF}}[a + a^\dagger]$ and $2\lambda = -E_0$. Letting $|\psi_i\rangle$ be an eigenstate of the oscillator containing n quanta, Eq. (A.21) becomes

$$\begin{aligned}
\chi_n(\omega) &= q^2 x_{\text{ZPF}}^2 |\langle n | a + a^\dagger | n + 1 \rangle|^2 \left\{ \frac{1}{-\hbar(+\omega_0 - \omega) + i\eta} + \frac{1}{-\hbar(+\omega_0 + \omega) - i\eta} \right\} \\
&\quad + q^2 x_{\text{ZPF}}^2 |\langle n | a + a^\dagger | n - 1 \rangle|^2 \left\{ \frac{1}{-\hbar(-\omega_0 - \omega) + i\eta} + \frac{1}{-\hbar(-\omega_0 + \omega) - i\eta} \right\} \\
&= -\frac{q^2}{m_e} \frac{1}{\omega_0^2 - (\omega + i\eta)^2}.
\end{aligned} \tag{A.26}$$

The fact that this has only a single resonance frequency ($\pm\omega_0$) and that the susceptibility is independent of the initial excitation number n is a special property of the harmonic oscillator and is a reflection of the perfect linearity of the response. Noting that $2\lambda = -E_0$, we find that the polarizability is

$$\alpha(\omega) = -\chi(\omega) = \frac{q^2}{m_e} \frac{1}{\omega_0^2 - (\omega + i\eta)^2}. \tag{A.27}$$

Notice that the factors of $\hbar$ have completely canceled out, and (in the limit of negligible damping) our result is in precise agreement with the classical calculation of Section 7.10.1. One of the novel features of a harmonic oscillator is that, if one can couple only linearly to it (for purposes of driving it or measuring it), then it is extremely difficult to tell whether it is quantum or classical. Only by being able to measure non-linear operators such as the excitation number $a^\dagger a$ can one see discrete values that tell you that the energy of the system is quantized.

As another example, let us consider a single-particle or many-body system subject to a moving sinusoidal potential that couples to the density ρ:

$$H' = 2\lambda e^{\eta t} \cos(\vec{q} \cdot \vec{r} - \omega t) = \lambda e^{\eta t} (\rho_{\vec{q}} \, e^{i\omega t} + \rho_{-\vec{q}} \, e^{-i\omega t}). \tag{A.28}$$

If the system has translation invariance so that momentum is a good quantum number, certain terms in the linear-response expressions will vanish because of momentum conservation as they did in Eq. (A.11).

In this case it is useful to consider the responses of $\langle \rho_{\vec{q}} \rangle_i$ and $\langle \rho_{\vec{q}}^\dagger \rangle_i$ separately. The reader may object that, since $\rho_{\vec{q}}$ is not Hermitian, it is not an observable. However, the real and imaginary parts of $\rho_{\vec{q}}$ are separately Hermitian (and commute with each other), so the complex quantity $\langle \rho_{\vec{q}} \rangle_i$ is perfectly observable. Using momentum conservation, we find (again to order $\mathcal{O}(\lambda)$)

$$\langle \rho_{+\vec{q}} \rangle_i = \frac{\lambda}{\hbar} e^{\eta t} \sum_{j \neq i} e^{-i\omega t} \left[\frac{\langle \psi_i | \rho_{+\vec{q}} | \psi_j \rangle \langle \psi_j | \rho_{-\vec{q}} | \psi_i \rangle}{\omega_i - \omega_j + \omega + i\eta} + \frac{\langle \psi_i | \rho_{-\vec{q}} | \psi_j \rangle \langle \psi_j | \rho_{+\vec{q}} | \psi_i \rangle}{\omega_i - \omega_j - \omega - i\eta} \right]. \tag{A.29}$$

From the time dependence it is clear that $\langle \rho_{+\vec{q}} \rangle$ responds *only* to the $\rho_{-\vec{q}}^\dagger e^{-i\omega t}$ term in H'. We can thus define a corresponding dynamical (or ω- and $\vec{q}$-dependent) response function

$$\chi_i(\vec{q}, \omega) = \frac{1}{\hbar} \sum_{j \neq i} \left[\frac{\langle \psi_i | \rho_{+\vec{q}} | \psi_j \rangle \langle \psi_j | \rho_{-\vec{q}} | \psi_i \rangle}{\omega_i - \omega_j + \omega + i\eta} + \frac{\langle \psi_i | \rho_{-\vec{q}} | \psi_j \rangle \langle \psi_j | \rho_{+\vec{q}} | \psi_i \rangle}{\omega_i - \omega_j - \omega - i\eta} \right]. \tag{A.30}$$

This important result can now be used to determine the full density response in thermal equilibrium from Eq. (A.23). In Exercise 4.6, one is asked to show that the susceptibility can be written in terms of the *retarded* density–density correlation function defined in Eq. (4.106). The retarded correlation explicitly displays the *causality* constraint on the response which is the focus of the next section.

A.3 Causality, Spectral Densities, and Kramers–Kronig Relations

What is the meaning of the fact that the susceptibility is frequency-dependent? From Eq. (A.21) we see that the susceptibility has strong peaks or resonances when the perturbation oscillates at a frequency near one of the transition frequencies between energy eigenstates of the system. By Fourier transforming from frequency space to the time domain we see that in general the linear response to a time-dependent perturbation with time-varying strength $\lambda(t)$ has the form[3]

$$\langle O(t) \rangle = \int_{-\infty}^{+\infty} dt' \, \chi(t - t') \lambda(t'). \tag{A.31}$$

---

[3] The fact that χ depends only on the difference of the two times follows from the (assumed) time-translation invariance of the properties of the unperturbed system.

From **causality** we expect that observations at time t cannot be influenced by future values of $\lambda(t')$. Hence we expect $\chi(t - t')$ should vanish for $t' > t$. How does this come about? To see this requires understanding the properties of the susceptibility for complex frequencies.

While physically the frequency ω is real, it is often mathematically very useful to view the response function Eq. (A.21) as a function of *complex* ω, and inspect its analytic properties. It is clear from Eq. (A.21) that $\chi(\omega)$ has simple poles right below the real axis, as a result of which it is analytic in the upper half plane. An important consequence of this is that, when Fourier transforming $\chi(\omega)$ to the time domain,

$$\chi(t, t') = \chi(t - t') = \int_{-\infty}^{+\infty} d\omega \, e^{-i\omega(t-t')} \chi(\omega), \tag{A.32}$$

we find that $\chi(t < t') = 0$, because we can choose to perform the integral by closing the contour in the upper half plane. This result is precisely what one expects from causality.

In discussing linear-response functions, it is useful to introduce the concept of spectral densities. Using the fact that

$$\frac{1}{x + i\eta} = \mathcal{P}\frac{1}{x} - i\pi\delta(x), \tag{A.33}$$

where $\mathcal{P}$ stands for the principal part (when performing an integral), we find

$$\operatorname{Im}\chi(\omega) = -\frac{1}{2\hbar}[S(\omega) - S(-\omega)], \tag{A.34}$$

where Im stands for the imaginary part, and

$$S(\omega) = \frac{2\pi}{Z}\sum_{i,j} e^{-\beta E_i}\{|\langle\psi_i|O|\psi_j\rangle|^2 \delta[\omega - (E_j - E_i)/\hbar] \tag{A.35}$$

is called the **spectral density** of O. On the other hand, we can also express the response function $\chi(\omega)$ in terms of the spectral density:

$$\chi(\omega) = \int \frac{d\omega'}{2\pi\hbar} S(\omega') \left[\frac{1}{\omega - \omega' + i\eta} - \frac{1}{\omega + \omega' + i\eta}\right]. \tag{A.36}$$

In particular, the static response function

$$\chi = \chi(\omega = 0) = -2\int_{-\infty}^{\infty} \frac{d\omega'}{2\pi\hbar}\frac{S(\omega')}{\omega'}, \tag{A.37}$$

namely $-\chi$ is the (-1)st moment of corresponding spectral density. It is easy to verify that the above is equivalent to Eq. (A.14).

The spectral representation (A.36) of $\chi(\omega)$ can also be used to establish relations between its real (Re) and imaginary parts:

$$\operatorname{Re}\chi(\omega) = \int_{-\infty}^{\infty} \frac{d\omega'}{\pi}\operatorname{Im}\chi(\omega')\mathcal{P}\frac{1}{\omega' - \omega}, \tag{A.38}$$

$$\operatorname{Im}\chi(\omega) = -\int_{-\infty}^{\infty} \frac{d\omega'}{\pi}\operatorname{Re}\chi(\omega')\mathcal{P}\frac{1}{\omega' - \omega}. \tag{A.39}$$

These are known as **Kramers–Kronig relations**. Equation (A.38) follows directly from (A.36) and (A.33). Equation (A.39) can be obtained by further manipulations based on the analytic properties of $\chi(\omega)$.

The spectral density defined in Eq. (A.35) has a very clear physical meaning. We see that the expression resembles Fermi's Golden Rule, and indeed $|\lambda|^2 S(\omega)$ is precisely the Golden Rule transition rate from an initial state i to a final state j caused by the time-dependent perturbation, averaged over i based on the Boltzmann distribution. More specifically, the spectral density (or, equivalently, the imaginary part of the susceptibility) for positive ω describes the rate at which the system absorbs energy from the probe, while the spectral density at negative ω describes the rate at which the system emits energy into the probe. It follows therefore from Eq. (A.34) that Im $\chi(\omega)$ describes the **dissipative** part of the system response, and the net rate of energy absorption from the probe (i.e. the power dissipation) is given by

$$P = \hbar\omega|\lambda|^2 \left[-\frac{2}{\hbar} \operatorname{Im} \chi \right], \tag{A.40}$$

where the factor of $\hbar\omega$ accounts for the energy emitted or absorbed by each transition. Similarly, the real part of χ corresponds to the *non-dissipative* (or in electrical engineering language, the "**reactive**") part of the response.

In thermal equilibrium the relative probability of occupation of two energy eigenstates obeys the Boltzmann distribution

$$\frac{p_i}{p_j} = e^{-\beta(E_i - E_j)}. \tag{A.41}$$

Without loss of generality, let us take $E_i > E_j$. The ability of the system to emit energy $\hbar\omega = E_i - E_j$ into a probe via a transition from i to j and to absorb energy $\hbar\omega$ from the probe via a transition from j to i is determined by the occupation probabilities of the two states. It is straightforward to show from Eq. (A.35) that, for a system in thermal equilibrium, the spectral density of fluctuations obeys the condition of **detailed balance**:

$$\frac{S(-\omega)}{S(+\omega)} = e^{-\beta\hbar\omega}, \tag{A.42}$$

In the classical electrical engineering literature, the so-called **"one-sided" spectral density** of noise is defined as

$$S_{\text{tot}} = S(\omega) + S(-\omega). \tag{A.43}$$

From detailed balance it follows that

$$\frac{S(\omega) - S(-\omega)}{S(\omega) + S(-\omega)} = \tanh(\beta\hbar\omega/2). \tag{A.44}$$

Hence, using Eq. (A.34), we arrive at the **fluctuation–dissipation theorem** relating the dissipative part of the response to the noise spectrum:[4]

$$S_{\text{tot}}(\omega) = -2\hbar \coth(\beta\hbar\omega/2)\operatorname{Im} \chi(\omega). \tag{A.45}$$

In the classical limit ($\hbar\omega \ll k_{\text{B}}T$), this reduces to

$$S_{\text{tot}}(\omega) = -4\frac{k_{\text{B}}T}{\omega} \operatorname{Im} \chi(\omega). \tag{A.46}$$

In the quantum limit, it is useful to distinguish between $S(+\omega)$ and $S(-\omega)$, since they are not equal as they are in the classical limit. For the quantum case it is straightforward to find for positive frequencies

$$S(+\omega) = [1 + n_{\text{B}}(\beta\hbar\omega)][-2\hbar \operatorname{Im} \chi(+\omega)], \tag{A.47}$$

where $n_{\text{B}}(\beta\hbar\omega)$ is the Bose–Einstein occupation factor. (N.B. This is always present no matter whether we are dealing with bosons, fermions, or spins). Similarly, for negative frequencies

---

[4] Note that electrical engineers use time dependence $e^{+j\omega t}$ instead of $e^{-i\omega t}$ as we do in quantum mechanics, so the signs of the imaginary parts of all quantities are reversed in the engineering notation.

$$S(-\omega) = [1 + n_B(-\beta\hbar\omega)][-2\hbar\,\mathrm{Im}\,\chi(-\omega)] \tag{A.48}$$

$$= [n_B(+\beta\hbar\omega)][-2\hbar\,\mathrm{Im}\,\chi(+\omega)], \tag{A.49}$$

where we have used the fact that $\mathrm{Im}\,\chi(\omega)$ is odd in frequency and

$$1 + n_B(-\beta\hbar\omega) = -n_B(+\beta\hbar\omega). \tag{A.50}$$

We see clearly that Eqs. (A.49) and (A.47) obey detailed balance. Furthermore, the stimulated-emission factor $(1 + n_B)$ in $S(+\omega)$ confirms our picture that the spectral density of quantum noise at positive frequencies describes the ability of the system to absorb energy from a probe (because of stimulated emission from the probe into the system). Similarly, the factor n_B in $S(-\omega)$ confirms the notion that the spectral density of noise at negative frequencies describes the ability of the excited states of the system to emit quanta of energy into the probe.

It may be useful for the reader to (re-)visit Exercise 8.6 in light of the discussions above. The reader is also directed to Appendix G for further discussion and examples of linear response and the fluctuation–dissipation theorem for the case of electrical circuits. There we will derive the spectral density of voltage fluctuations associated with the dissipation of an electrical resistor and see how it differs at positive and negative frequencies.

Appendix B
B The Poisson Summation Formula

Suppose we have a function f of a real variable and we wish to sum the values of f on the sites of a uniform lattice of lattice constant a:

$$g = \sum_{n=-\infty}^{\infty} f(na). \tag{B.1}$$

It is possible to relate this to a sum on the reciprocal lattice. Define a function

$$g(x) = \sum_{n=-\infty}^{\infty} f(x + na). \tag{B.2}$$

Clearly g has the periodicity property

$$g(x + ma) = g(x), \tag{B.3}$$

where m can be any integer. Hence $g(x)$ can be expressed as a Fourier series

$$g(x) = \sum_{l=-\infty}^{\infty} \tilde{g}_l e^{ilGx}, \tag{B.4}$$

where

$$G \equiv \frac{2\pi}{a} \tag{B.5}$$

is the primitive vector of the reciprocal lattice, and $\tilde{g}_l$ are the Fourier coefficients given by

$$\tilde{g}_l \equiv \frac{1}{a} \int_0^a dx\, e^{-ilGx} g(x). \tag{B.6}$$

Substitution of (B.2) into (B.6) yields

$$\tilde{g}_l = \frac{1}{a} \int_0^a dx\, e^{-ilGx} \sum_{n=-\infty}^{\infty} f(x + na) \tag{B.7}$$

$$= \frac{1}{a} \sum_{n=-\infty}^{\infty} \int_{na}^{(n+1)a} dy\, e^{-il(y-na)G} f(y). \tag{B.8}$$

Using $e^{-ilnaG} = 1$, we have

$$\tilde{g}_l = \frac{1}{a} \sum_{n=-\infty}^{\infty} \int_{na}^{(n+1)a} dy \, e^{-ilGy} f(y) \tag{B.9}$$

$$= \frac{1}{a} \int_{-\infty}^{\infty} dy \, e^{-ilGy} f(y) = \frac{1}{a} \tilde{f}(Gl), \tag{B.10}$$

where $\tilde{f}$ is the Fourier transform of f. Now, from (B.4),

$$g(0) = \sum_{l=-\infty}^{\infty} \tilde{g}_l = \sum_{l=-\infty}^{\infty} \tilde{f}(Gl). \tag{B.11}$$

Combining this with (B.2) gives the **Poisson summation formula**

$$\sum_{n=-\infty}^{\infty} f(na) = \frac{1}{a} \sum_{l=-\infty}^{\infty} \tilde{f}(Gl). \tag{B.12}$$

This result can be very useful both in manipulating formal expressions and in improving the convergence. For instance, functions which decay slowly to zero in real space typically have Fourier transforms which decay rapidly.

As a simple example, suppose we have

$$f(x) = e^{-\frac{x^2}{2\sigma^2 a^2}}, \tag{B.13}$$

where $\sigma \gg 1$. Then the sum

$$g = \sum_{n=-\infty}^{\infty} e^{-\frac{n^2}{2\sigma^2}} \tag{B.14}$$

converges very slowly. However,

$$\tilde{f}(k) = \int_{-\infty}^{\infty} dx \, e^{-ikx} e^{-\frac{x^2}{2\sigma^2 a^2}} = \sqrt{2\pi\sigma^2 a^2} e^{-\frac{\sigma^2 a^2}{2}k^2}, \tag{B.15}$$

so that

$$g = \sqrt{2\pi\sigma^2} \sum_{l=-\infty}^{\infty} e^{-2(\pi\sigma)^2 l^2} \tag{B.16}$$

converges very quickly.

Notice that if we take only the $l = 0$ term in Eq. (B.12) we have

$$\sum_{n=-\infty}^{\infty} f(na) \approx \frac{1}{a} \int_{-\infty}^{\infty} dx \, f(x), \tag{B.17}$$

which is simply replacing the Riemann sum by its limit as $a \to 0$, namely the integral.

All of the above results are readily extended to higher-dimensional lattices, the factor $1/a$ being replaced by the unit cell volume Ω:

$$\sum_j f(\vec{R}_j) = \frac{1}{\Omega} \sum_{\{\vec{G}\}} \tilde{f}(\vec{G}). \tag{B.18}$$

It is often very useful to relate real space lattice sums to reciprocal lattice sums in this manner.

Appendix C

Tunneling and Scanning Tunneling Microscopy

We have learned in quantum mechanics that particles can tunnel through potential barriers that are higher than the particle's energy and thus classically forbidden. As a result, unless the potential barrier is infinitely high, particles cannot be truly confined by it. In this appendix we review this important physics and introduce the notion of the **tunneling Hamiltonian** that will be used repeatedly in the book, and apply it to understand how **scanning-tunneling microscopy (STM)** works.

C.1 A Simple Example

We start by considering a simple example to illustrate the basic physics of tunneling. Consider two identical quantum wells separated by an infinitely high potential barrier, as illustrated in Fig. C.1(a). In this case the two wells are decoupled, and a particle in one well can never tunnel to the other. Each well has a separate ground state and excited states; in this case these states have identical energies and wave functions (other than a trivial translation) because the two wells are identical. Let us assume that the energy gaps separating the ground states from excited ones are sufficiently large that we can ignore the excited states and an electron will be found only in one of the two degenerate ground states, labeled $|L\rangle$ and $|R\rangle$. Without loss of generality, let us set the ground state energy to be zero.

Now consider a very high but finite barrier separating the two wells, as illustrated in Figs. C.1(b) and (c). In this case the two wells are coupled, and eigenstates of the system will have support (non-vanishing amplitude) in both of them, as well as *inside* the barrier. The ground state (see Fig. C.1(b)) is nodeless and has even parity (when the potential has an inversion center), while the first (low-lying) excited state (see Fig. C.1(c)) has one node at the inversion center and has odd parity. Let us define the small energy splitting between these two states to be $2T > 0$. As long as the barrier is sufficiently high, the amplitude inside the barrier is negligible, and to a very good approximation the ground state $|G\rangle$ and first excited state $|E\rangle$ can be expressed as symmetric and anti-symmetric combinations of $|L\rangle$ and $|R\rangle$:

$$|G, E\rangle = \frac{1}{\sqrt{2}}(|L\rangle \pm |R\rangle). \tag{C.1}$$

An efficient way to capture the physics described above is through a tunneling Hamiltonian defined in the ground state space spanned by $|L\rangle$ and $|R\rangle$:

$$H_{\mathrm{T}} = -T(|L\rangle\langle R| + |R\rangle\langle L|). \tag{C.2}$$

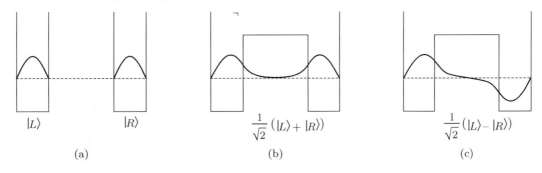

$|L\rangle$ $|R\rangle$ $\frac{1}{\sqrt{2}}\left(|L\rangle + |R\rangle\right)$ $\frac{1}{\sqrt{2}}\left(|L\rangle - |R\rangle\right)$

(a)   (b)   (c)

Figure C.1 Tunneling between quantum wells. (a) Two identical quantum wells separated by an infinitely high barrier. There are two degenerate ground states, $|L\rangle$ and $|R\rangle$, localized in the two wells, respectively. In (b) and (c) are shown two identical quantum wells separated by a barrier with finite height. The particle can now tunnel through the barrier, with the ground state (b) and first excited state (c) well approximated by symmetric and anti-symmetric combinations of $|L\rangle$ and $|R\rangle$.

It is easy to verify that the states (C.1) are eigenstates of (C.2) with eigenenergies $\mp T$. Equation (C.2) is the simplest example of a tunneling Hamiltonian. It is illuminating to write it in second-quantized form (see Appendix J for an introduction to second quantization):

$$H_{\mathrm{T}} = -T\left(c_{\mathrm{L}}^{\dagger} c_{\mathrm{R}} + c_{\mathrm{R}}^{\dagger} c_{\mathrm{L}}\right), \tag{C.3}$$

where $c_{\mathrm{L}}^{\dagger}$ creates a particle in $|L\rangle$ and c_{R} annihilates a particle in $|R\rangle$. It is clear that the first term in (C.2) or (C.3) moves a particle from $|R\rangle$ to $|L\rangle$, while the second term does the opposite.

Now consider a particle in a generic state

$$|\psi(t)\rangle = a(t)|L\rangle + b(t)|R\rangle, \tag{C.4}$$

with $|a|^2 + |b|^2 = 1$, and satisfying the Schrödinger equation

$$i\hbar\frac{d}{dt}|\psi(t)\rangle = H_{\mathrm{T}}|\psi(t)\rangle, \tag{C.5}$$

from which we find

$$i\hbar\frac{d}{dt}a(t) = -Tb(t), \tag{C.6}$$

$$i\hbar\frac{d}{dt}b(t) = -Ta(t). \tag{C.7}$$

This allows us to calculate the number current from left to right:

$$\langle I(t)\rangle = \frac{d}{dt}|b(t)|^2 = -\frac{d}{dt}|a(t)|^2$$
$$= \frac{T}{i\hbar}\left[a^*(t)b(t) - b^*(t)a(t)\right] = \langle\psi(t)|\hat{I}|\psi(t)\rangle, \tag{C.8}$$

where the current operator is

$$\hat{I} = \frac{T}{i\hbar}(|L\rangle\langle R| - |R\rangle\langle L|) = \frac{T}{i\hbar}\left(c_{\mathrm{L}}^{\dagger} c_{\mathrm{R}} - c_{\mathrm{R}}^{\dagger} c_{\mathrm{L}}\right). \tag{C.9}$$

C.2 Tunnel Junction

We now consider a general **tunnel junction** through which electrons can tunnel, described by the following Hamiltonian:

$$H = H_L + H_R + H_T, \tag{C.10}$$

where H_L and H_R are the Hamiltonians of the systems on the left and right sides of the junction that contain no coupling between them, $[H_L, H_R] = 0$, and the tunneling Hamiltonian is

$$H_T = -\sum_{l,r,\sigma} \left(T^*_{lr} c^\dagger_{l\sigma} c_{r\sigma} + T_{lr} c^\dagger_{r\sigma} c_{l\sigma} \right), \tag{C.11}$$

where l and r are labels of single-particle states on the left and right sides of the junction, respectively, and $\sigma = \uparrow, \downarrow$ is the spin label. We assume spin is not flipped in the tunneling process for simplicity, and the tunneling matrix element T_{lr} does not depend on spin. In reality the barrier in the junction is often made of an insulator with a large gap, connecting metal(s), semiconductor(s), or superconductor(s). But it can also simply be vacuum, whose energy is much higher than that of the relevant electronic states in a solid. The corresponding current operator is

$$\hat{I} = \frac{1}{i\hbar} \sum_{l,r,\sigma} \left(T^*_{lr} c^\dagger_{l\sigma} c_{r\sigma} - T_{lr} c^\dagger_{r\sigma} c_{l\sigma} \right). \tag{C.12}$$

In a tunneling experiment, one measures the current going through the junction (I) as a function of the voltage difference (V) between the two sides. The I–V curve tells us a lot about the electronic states of the two systems coupled by the junction. In particular, the differential conductance

$$G(eV) = \frac{dI}{dV} \tag{C.13}$$

directly probes the states on the two sides whose energies differ by eV, where $-e$ is the electron charge (see Fig. 19.11 for an illustration).

There are a number of subtleties in defining and using the tunneling Hamiltonian, but let us proceed to use it in a simple Fermi's Golden Rule calculation of the rate of electrons passing through a tunnel junction connecting two normal metals at zero temperature. The voltage drop across the tunnel junction is (by definition) the electrochemical potential difference between the two electrodes. Referring to Fig. 19.11, we will view this as a chemical potential difference of eV. The current flowing in the direction of the voltage drop is given by the Golden Rule expression

$$\begin{aligned} I &= e \frac{2\pi}{\hbar} \sum_{l,r,\sigma} |T_{lr}|^2 \theta(\epsilon_F + eV - \epsilon_l)\theta(\epsilon_r - \epsilon_F)\delta(\epsilon_r - \epsilon_l) \\ &= 2e \frac{2\pi}{\hbar} |T|^2 [eV D(0)] D(0), \end{aligned} \tag{C.14}$$

where $D(0)$ (see Eq. (15.152)) is the single-spin total density of states at the Fermi level (which we assume to be essentially independent of energy and equal on both sides of the junction). We have also ignored all the microscopic details of the matrix element T_{lr} and have simply replaced it by a constant[1] T. The step functions in this expression enforce the condition that the initial state must be occupied and the target final state must be initially empty. The overall factor of 2 accounts for the sum over the initial spin states. (The final spin state is not summed over because it is identical to the initial state.) We see that the current is linear in the voltage, and so this expression leads to the following result for the linear conductance of the tunnel junction:

$$G = 4\pi \frac{e^2}{\hbar} |T|^2 D(0)^2. \tag{C.15}$$

---

[1] The microscopic matrix element is impossible to compute from first principles because it depends exponentially on the thickness of the oxide tunnel barrier and varies strongly with disorder and local fluctuations in the barrier thickness. It is therefore best to simply replace its average magnitude squared by a constant which is fit to the junction conductance (which is easy to measure). We ignore here small Fermi-liquid correction factors associated with the fact that the Landau quasiparticles are electrons dressed with particle–hole pairs due to the interactions.

One might conclude from the above that tunneling directly probes the density of states (DOS) of the system. It turns out that this is *not* the case for non-interacting electrons, because the tunneling matrix element T_{lr} is also sensitive to the DOS, in a way that cancels out the DOS which comes from the sum over states [237]. This is very similar to the cancelation of the electron velocity (which is essentially the inverse DOS) in the calculation of the conductance of a quantum wire in Chapter 10, leading to the universality of the Landauer formula. The situation is more interesting in the presence of electron–electron interactions. In this case $G(eV)$ does probe the combined tunneling DOS of the system involved (per unit volume), which can be viewed as the DOS created by adding or subtracting an electron to or from the system (thereby converting a "bare" electron to a dressed quasiparticle):

$$\rho^{\mathrm{t}}(\epsilon) = \frac{1}{\mathcal{V}} \sum_l \rho_l^{\mathrm{t}}(\epsilon), \tag{C.16}$$

$$\rho_l^{\mathrm{t}}(\epsilon) = \frac{1}{Z} \sum_{i,f} e^{-\beta E_i} \left\{ |\langle f|c_{l\sigma}^\dagger|i\rangle|^2 \delta[\epsilon - (E_f - E_i)] \right.$$

$$\left. + |\langle f|c_{l\sigma}|i\rangle|^2 \, \delta[\epsilon + (E_f - E_i)] \right\}, \tag{C.17}$$

where Z is the partition function of the system, l is a single-particle state label, and i and f are labels of many-particle energy eigenstates of the system, with E_i and E_f being the corresponding energies. Note the similarity of the above to the definition of the dynamical structure factor in Eqs. (4.43) and (4.44). The reader should verify that $\rho^{\mathrm{t}}(\epsilon)$ reduces to the usual single-particle DOS (per spin per unit volume) for a non-interacting electron system $\rho(\epsilon)$, introduced in, e.g., Eq. (7.251). In the presence of interactions, they are not the same, and the cancelation mentioned above does not occur.

It should be noted that, in general, the **tunneling density of states** at the Fermi energy differs from, and is smaller than, the **thermodynamic density of states** $\frac{1}{2}dn/d\mu$ (in which we factor out the two-fold spin degeneracy), which is proportional to the inverse compressibility. This is particularly true at zero or low temperature, as the added particle (through tunneling) results in a state *locally different* from the ground or thermal state, and the system is not given a chance to relax. It is thus not uncommon that the system is compressible (i.e., $dn/d\mu > 0$), yet the tunneling DOS exhibits a pseudogap (or soft gap) at the Fermi energy. The Coulomb gap mentioned at the end of Section 11.9.2 and the power-law suppression of electron tunneling into fractional quantum Hall edges discussed in Section 16.9 are examples of these. All of these effects are associated with the correlated "dressed" states of the electron not being identical to those of the "bare" electron which suddenly tunnels into the interacting material.

C.3 Scanning Tunneling Microscopy

Scanning tunneling microscopy (STM), uses tunneling to study electronic states in real space with atomic-scale spatial resolution. The key idea is that the tunneling matrix element T_{lr} depends exponentially on the spatial separation between the electronic states l and r. If we can make the junction very small, then tunneling will probe electronic states in the immediate neighborhood of the junction, or the *local* tunneling density of states $\rho_l^{\mathrm{t}}(\epsilon)$, with l the local single-particle state that dominates the tunneling process. The basic idea is illustrated in Fig. C.2. In this case the tunneling Hamiltonian (C.11) is dominated by terms involving states of the apex atom of the tip and those right underneath it. By moving the tip around, one can then resolve individual atoms at the surface of the crystal being studied. See Fig. 3.16 for an example.

To measure **surface topography**, the STM can be operated with feedback control that adjusts the tip height to maintain a constant tunneling current. Because the current depends exponentially on the

Tunnel Tip

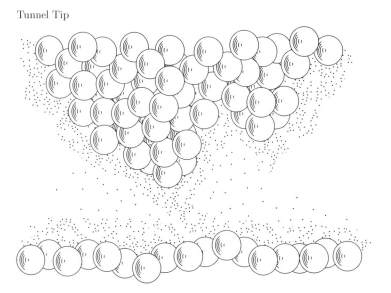

Figure C.2 Illustration of scanning tunneling microscopy (STM). A tunnel junction is made of a metallic tip (above) and a condensed matter system of interest (below), separated by vacuum, which provides the barrier. Circles represent atoms and black dots represent electrons. The tunneling current goes predominantly from the atom located at the apex of the tip into electronic states right underneath it. Figure reprinted with permission from [238]. Copyright (1987) by the American Physical Society.

height of the tip above the surface, there is a strong variation of tunneling current with height of the tip. As the tip is scanned across the surface, the feedback loop will adjust the height of the tip to maintain constant current. The feedback signal giving the height of the tip provides an extremely precise image of the surface topography with picometer resolution.

The STM can also be operated in "**spectroscopy mode**" in which the tip height is held fixed and the voltage applied to the tip is quickly scanned up and down.[2] The variation of the tunneling current with voltage gives an "I–V curve" which provides useful information about the electronic density of states under the STM tip. This information can in some cases be used to identify the atomic species underneath the tip location or to measure the variation of the superconducting gap with position on the surface of a superconductor.

> **Exercise C.1.** Show that, for non-interacting fermions, the local tunneling density of states in Eq. (C.17) reduces to the local density of states defined in Eq. (11.160).

---

[2] The servo loop that maintains the tunneling current must be shut off momentarily to allow the spectroscopic data to be taken. Equivalently, the spectroscopic bias voltage can be modulated at frequencies higher than the servo loop response can follow.

Appendix D
D Brief Primer on Topology

D.1 Introduction

Topology is the study of continuous functions and continuous deformations of objects [35, 43, 44]. Intuitively, a continuous function f smoothly maps closely neighboring points in its domain into closely neighboring points in the range of the function.[1] Of particular interest to us are **topological invariants**, quantities that do not change when a particular function or object is continuously deformed. The classic example that is often quoted is the fact that a donut (solid torus) can be continuously deformed into a coffee cup, and the genus (roughly speaking, the number of holes through the structure) is preserved in the process. It would require a discontinuous deformation to insert or remove a hole (or, in mathematical language, change the **genus**).

Before delving into invariants under continuous transformations, let us remind ourselves about plane Euclidean geometry. One possible definition is that this is the study of invariants under a much more restricted class of transformations, namely linear scale changes, rotations, and inversions. For example, each of the three interior angles of a triangle remains invariant if we shrink or enlarge the triangle, or rotate it.[2] Under more general continuous transformations, straight lines can become curved and triangles can become squares or circles. Because continuous transformations are so much more general than linear transformations, topological invariants will be much rarer than geometric invariants.[3] On the other hand, topological invariants are very deep and powerful precisely because they survive such a broad class of transformations. Consider, for example, a Hamiltonian characterized by a number of adjustable parameters. If a certain physical observable can be expressed as a topological invariant, we know that it will be robust against continuous deformations of the Hamiltonian

---

[1] The precise formal definition [35, 43] requires that the inverse mapping f^{-1} from any open set in the range always yields an open set back in the domain. As noted in [43], we cannot define continuity by the forward map f, since there exist discontinuous functions that map open sets in the domain to open sets in the range.

[2] This definition of plane geometry does not involve a discussion of absolute distances (we use a straight edge and a compass to draw geometric figures, but not a ruler). More generally, however, we can reasonably think of the geometry of an object as being defined by the distances between various points.

[3] As a toy example of a topological invariant, consider transforming one planar triangle into another planar triangle, which is a highly restricted continuous transformation, but more general than linear scale changes, rotations, and inversions. Under such a restricted continuous transformation the three interior angles will change, but their sum remains equal to π and is invariant. We thus find that, with more general transformations, we have fewer invariants.

parameters (unless a phase boundary is crossed) and therefore will be universal and independent of microscopic parameters that we may not otherwise be able to measure. The prime example of this is the quantized value of the Hall conductance in the integer quantum Hall effect which, as we discuss in Chapter 13, can be expressed in terms of the **Chern number** of the occupied electronic bands.

Instead of trying to be general and mathematically rigorous, our discussions in this appendix will be based on specific examples and intuitive (even hand-waving!).

D.2 Homeomorphism

A **homeomorphism** is a mapping from one space to another that is continuous and whose inverse exists and is also continuous. Two objects (mathematical spaces) that can be continuously mapped into each other by a homeomorphism are said to be homeomorphic or "topologically equivalent." Thus homeomorphism is an equivalence relation, namely if $A \sim B$ (where $\sim$ stands for equivalence) and $B \sim C$, then $A \sim C$. Such relations give rise to equivalence classes, in which all members of a given equivalence class are equivalent (in the present context, homeomorphic) to each other. We can thus use any member to represent the class, namely $[A] = [B] = [C] = \cdots$, where $[A]$ stands for the equivalence class that includes A. Topological invariants, namely quantities that do not change under homeomorphisms, can be used to categorize the equivalence class. These quantities are not *geometric*, because homeomorphisms distort geometries. Rather, topological invariants are much more general quantities like the genus. The genus is not the only topological invariant. Integrals of certain functions over the space can also be topological invariants. As we encountered in Chapter 13, the Hall conductance for a filled band can be expressed as an integral of the Berry curvature over the first Brillouin zone (a specific space). Remarkably, this integral takes on values that are only integers. Even more remarkably, these integer-valued integrals are topological invariants known as Chern numbers. Crudely speaking, it is impossible to smoothly deform one integer into another, and thus all spaces within an equivalence class are characterized by the same integer-valued topological invariant.

A popular example of homeomorphism mentioned above is that between a donut and a coffee cup – it maps a local region in the donut into a local region of the coffee cup. Imagine that the donut is made of clay, and we continuously push the clay around until we have a coffee cup. The homeomorphism tells us where in the donut a particular piece of clay was located when it was in the donut. More formally speaking, the homeomorphism tells us where the bits of clay ended up in the range (or where they came from in the domain), but does not tell us about all the pushes and pulls that got us from start to finish. All of the intermediate states in the continuous evolution between donut and coffee cup are in the same topological class. On the other hand, a donut and a solid ball are *not* homeomorphic to each other. In order to turn a ball made of clay into a donut, we have to punch a hole in the middle, which inevitably separates continuous regions in the ball, resulting in a *discontinuous* map between the two objects. In this simple example the topological distinction between a donut (or a torus) and a solid ball is reflected in their genus: $g = 1$ for a torus and $g = 0$ for a solid ball.

D.3 Homotopy

In the above we were looking at continuous and invertible maps between two spaces that are homeomorphic. We are also interested in maps that are between spaces that are not necessarily homeomorphic, and/or may not be invertible. The question is how to divide such maps into topologically equivalent classes.

As a simple example, consider a mapping of the unit circle (denoted as space S^1) onto another unit circle. The vector position on each circle is uniquely defined by its angular position,

$$\vec{r} = \vec{r}_0 + (\cos\theta, \sin\theta), \quad \vec{r}\,' = (\cos\theta\,', \sin\theta\,'), \tag{D.1}$$

and so we need only discuss mappings that connect the respective angles θ and $\theta\,'$. (We have offset the position of the first circle by an arbitrary displacement $\vec{r}_0$ to make it clear that the second circle is distinct from the first.) To specify the map, we require that the image start and end at the same point: $\vec{r}\,'(\theta = 0) = \vec{r}\,'(\theta = 2\pi) = (1, 0)$. Without loss of generality, we can take $\theta\,'(\theta = 0) = 0$.

Consider the following family of homeomorphisms labeled by a parameter t:

$$\theta\,'(\theta, t) = \theta + t \sin(k\theta/2), \tag{D.2}$$

where k is any non-zero integer. For $t = 0$ we have the identity map (if $\vec{r}_0 = 0$), and for $t \neq 0$ we have some other map. Notice that the term $\sin(k\theta/2)$ always vanishes for $\theta = 0$ and $\theta = 2\pi$. Thus for any t we still have a map from one circle to another circle, and any member of the entire family of maps for all t can be continuously deformed into any other member by varying t. We say that such maps are homotopic to each other, and call a continuously variable deformation like Eq. (D.2) a homotopy between the two maps. Obviously, **homotopy** is an equivalence relation among all maps that can be continuously deformed into each other.

Consider now a related family of maps defined by

$$\theta\,'(\theta, t) = m\theta + t \sin(k\theta/2), \tag{D.3}$$

where m is a non-zero integer. The previous case corresponds to $m = +1$, so let us now consider the case $m = -1$. This gives a map from one circle to the other that is very similar to the family studied above (and also provides a homeomorphism between the two unit circles), except that the path taken by the image $\theta\,'$ goes backwards (i.e. clockwise) around the circle as θ increases from 0 to 2π (i.e. travels counterclockwise around its circle). It is intuitively clear that these two maps are not homotopic. There is no way to continuously deform one into the other.

The integer-valued parameter m is a "**winding number**" that defines a **homotopy class** for these maps. All maps with the same value of m can be continuously deformed into each other, irrespective of the value of the parameters k and t. It is not possible to continuously deform one integer winding number into another (provided we exclude the possibility of zero radius for the circles). The existence of this infinite number of distinct homotopy classes labeled by the integer m tells us something important about the nature of the possible continuous mappings between these two spaces. For a general map $\theta\,' = f(\theta)$, we can compute the winding number via

$$m = \frac{1}{2\pi} \int_0^{2\pi} d\theta \, \frac{df}{d\theta}. \tag{D.4}$$

Intuitively we can interpret θ as "time" and $df/d\theta$ as "velocity" (not speed!), and so the total distance traveled is $2\pi m$.

To gain some intuition about the meaning of the winding number m, imagine the first unit circle is a rubber band with an arrow specifying its orientation, while the second unit circle is rigid, also with an arrow. Wrapping the rubber band around the rigid circle gives rise to a loop on the latter, and m is the number of times the rubber band wraps around the rigid circle. m is positive if the two arrows point in the same direction, and negative otherwise.

We note in passing that, except for $m = \pm 1$, the map (D.3) is not invertible, and thus does not provide a homeomorphism between the two unit circles. This is because (the reader should verify this) (D.3) is an $|m|$-to-one map.

In the above we considered maps from one unit circle to another unit circle. More generally we can consider maps from a unit circle (or S^1) to any space, the image of which is a loop in the target space. Loops are particularly important in the study of topological properties of spaces, as we now discuss.

D.4 Fundamental Group

A loop in space X can be viewed as a continuous map $f(\theta)$ from a unit circle (or S^1) to X. Let us focus on loops passing through a particular point x in X, and, without losing generality, we may assume $f(0) = f(2\pi) = x$. We say that such a loop is anchored at x. Now consider another such loop $g(\theta)$, with $g(0) = g(2\pi) = x$. We define the **composition** of the two loops f and g as

$$h(\theta) = f(\theta) \circ g(\theta) = \begin{cases} f(2\theta), & 0 \leq \theta \leq \pi; \\ g(2\theta - 2\pi), & \pi \leq \theta \leq 2\pi. \end{cases} \tag{D.5}$$

Clearly h is another loop anchored at x, which can be intuitively understood as the following: when one traverses the loop h, one first traverses the loop f, and then the loop g. Using $[f]$, $[g]$, and $[h]$ to represent the homotopy classes of loops represented by f, g, and h, respectively, we clearly have $[h] = [f \circ g] = [f] \circ [g]$.

We now apply this composition rule to the loops on a unit circle discussed in the previous section, and introduce the important concept of the **fundamental group** by viewing each of these equivalence classes as an element of a group. In the above example we have been able to label the different equivalence classes of maps F_m (i.e. maps equivalent under homotopy) according to their winding number m. The group operation (product) $F_m F_n$ is simply the composition of two maps, one from each equivalence class: $f_m \circ f_n$. Because the winding number for the composition of two maps is simply the sum of the winding numbers for each individual map, this new map is an element of the homotopy class F_{m+n}. Thus the group operation is closed (i.e. yields an element of the group) as required. The identity element of the group is therefore F_0, the equivalence class of maps with winding number zero. Finally, every element F_m has an inverse element, F_{-m} (and the identity is its own inverse as required). Thus the set $\pi_1 = \{F_m, m \in \mathbb{Z}\}$ and the associated operation satisfy all the requirements to be a group, and in fact this group is isomorphic to $\mathbb{Z}$, the group of integers under addition.

As noted above, the existence of different homotopy classes tells us something important about the types of mappings that exist between the two spaces, but now we see that the group structure defined by these maps tells us even more – hence the name *fundamental group*. To further illuminate the concept of the fundamental group, let us consider more examples.

The simplest case is perhaps when the target is the surface of a unit sphere, S^2 (see Fig. 13.3). It is intuitively clear that all loops on S^2 can be deformed into each other continuously. As a result, there is only one homotopy class of loops, and the corresponding fundamental group has a single (identity) element: $\pi_1(S^2) = 0$. Connected spaces with a trivial fundamental group are simply connected, because all loops can continuously shrink to a single point (namely all loops are contractible). Intuitively a simply connected space is a space that contains no hole.

Next consider the case where the target space is the surface of a torus, $\mathcal{T}$ (see Fig. 19.7). We can think of a torus as a square with coordinates (θ_1, θ_2) obeying periodic boundary conditions or, equivalently, as a circle with the property that each point on the circle defined by angular variable θ_1 is the center of another circle with another angular variable, θ_2. As a result, the torus has two different types of non-contractible loops, each with independent integer-valued winding numbers. Thus $\pi_1(\mathcal{T}) = \mathbb{Z} \times \mathbb{Z}$. The differences between their fundamental groups clearly indicate that S^2, $\mathcal{T}$, and the unit circle S^1 are topologically distinct spaces. Furthermore, the fundamental group gives us more quantitative information about the topology: the number of independent $\mathbb{Z}$ groups counts the number of holes in the space; the unit circle contains one hole, while the torus contains two holes.

In Section 16.7 we saw a subtler example. We learned that the configuration space describing a pair of indistinguishable (hard-core) particles is the **real-projective sphere**, RP^2 (which, because it is compact and non-orientable, is equivalent to the **projective plane** [43]). For the projective sphere, pairs of diametrically opposed points on an ordinary sphere are identified as a single point (which is why this is a useful construction for indistinguishable particles). "Straight lines" passing through a "point" on the projective sphere are great circles passing through the diametrically opposed pairs of points, just as in plane geometry there is an infinite number of "lines" passing through a single "point," but only one "line" that passes through two "points" (i.e. two pairs of diametrically opposed points). Maps from the unit circle S^1 to RP^2 have winding numbers that are only defined modulo 2. This is because, as illustrated in Fig. 16.6, the mapping (trajectory) corresponding to one exchange of a pair of particles is a non-contractible loop on RP^2, but paths corresponding to two exchanges can be continuously distorted to the identity. Thus the fundamental group $\pi_1(RP^2) = \mathbb{Z}_2$, corresponding to integers under addition modulo 2. Having a non-trivial fundamental group means that RP^2 is not simply connected (because it contains non-contractible loops), but, since the fundamental group does not contain a $\mathbb{Z}$ subgroup, the space contains no hole either, and thus $g = 0$.

D.5 Gauss–Bonnet Theorem

We have so far considered the properties of closed loops. Loops embedded in position space can be viewed as maps from the unit circle S^1 to (say) R^2 or R^3. By analogy with loops, a smooth surface (with genus 0) can be viewed as a mapping from the unit sphere into some embedding space, which for simplicity we will take to be R^3. Similarly, smooth surfaces with genus 1 can be viewed as mappings from some standard torus into the embedding space, and so forth for higher-genus surfaces. See Fig. 13.5 for a few examples. We will define the notion of local **Gaussian curvature**, and this will lead us to the beautiful **Gauss–Bonnet theorem**, which relates the integral of the local curvature C_G over the closed surface S to its genus g, which is the number of its handles:

$$\int_S C_G \, dA = 2\pi(2 - 2g), \tag{D.6}$$

which is invariant under smooth distortions of the surface since the genus is a topological invariant. To make sense out of Eq. (D.6), we first need to have a proper definition of the local curvature C_G.

We understand intuitively (and from rotational symmetry) that the surface of a sphere has constant curvature. A sphere with a very large radius (e.g. the earth approximated as a sphere) has barely perceptible local curvature, whereas a small sphere appears highly curved. It thus seems reasonable that the Gaussian curvature for a sphere should be monotone decreasing with radius. In order for the result in Eq. (D.6) to be dimensionless, the curvature must have units of inverse area. This suggests that the Gaussian curvature should therefore be taken to be simply $C_G = R^{-2}$. Plugging this into Eq. (D.6), we see that the radius of the sphere drops out and we obtain

$$\int_{-1}^{+1} R^2 \, d\cos\theta \int_0^{2\pi} d\varphi \, C_G = 4\pi, \tag{D.7}$$

which is the correct answer since the genus of a sphere is 0.

The steps required to generalize this result to arbitrary smooth closed surfaces will involve defining a local Cartesian coordinate system at each point on the manifold.[4] If we were being strict mathematicians we would do this with local tangent vectors and seek to describe the local Gaussian curvature in terms of its effect on the parallel transport of vectors around closed paths. The advantage of this approach is that one does not have to assume any particular embedding of the manifold in a higher-dimensional space. However, as physicists desirous of making connections to other related topics in this text, we are going to assume that the manifold is embedded in 3D Euclidean space so that we can define a unit vector normal to the manifold at each point.

---

[4] The ability to do this is guaranteed by the very definition of what a manifold is. In the present context, it is the "smoothness" of the smooth surfaces that makes them manifolds. For a more rigorous definition see Ref. [35].

Let us first practice on the sphere. Pick an arbitrary point on the sphere, and choose the unit vector $\hat{z}$ to be the local normal pointing in the outward direction. (We assume the surface is orientable.) Now consider the unique plane which is tangent to the sphere at that point and is therefore normal to $\hat{z}$. Now define a right-handed Cartesian coordinate system using orthogonal unit vectors in the plane obeying $\hat{x}_1 \times \hat{x}_2 = \hat{z}$. In the vicinity of the origin of this coordinate system, the surface is uniquely defined by its height relative to the plane, $z(x_1, x_2)$. By definition $z(0, 0) = \partial_{x_1} z(0, 0) = \partial_{x_2} z(0, 0) = 0$. Thus the leading-order terms in a Taylor-series expansion for z are

$$z(x_1, x_2) \approx \frac{1}{2} M_{\mu\nu} x_\mu x_\nu + \cdots, \tag{D.8}$$

where the Einstein convention for summation over repeated indices is implied and the matrix M is

$$M_{\mu\nu} = \partial_\mu \partial_\nu z\big|_{(0,0)}. \tag{D.9}$$

For a sphere $x_1^2 + x_2^2 + (z + R)^2 = R^2$, and thus $z \approx -(1/2R)(x_1^2 + x_2^2)$, so that the matrix M is given by

$$M = \begin{pmatrix} -1/R & 0 \\ 0 & -1/R \end{pmatrix}. \tag{D.10}$$

In general, M is a real symmetric matrix (with non-zero off-diagonal elements), and the Gaussian curvature is defined to be the determinant of M:

$$C_{\mathrm{G}} = \det M = M_1 M_2, \tag{D.11}$$

where M_1 and M_2 are the two eigenvalues of M. Thus, for the sphere $C_{\mathrm{G}} = 1/R^2$, as expected. For the more general case, we can always find a local coordinate system in which M in diagonal, and

$$z = \frac{1}{2}(M_1 x_1^2 + M_2 x_2^2). \tag{D.12}$$

For a flat surface, both eigenvalues are zero, whereas for a cylinder, only one of the eigenvalues is zero, but this is still enough to make the curvature vanish. This is consistent with the intuitive notion that a piece of paper can be wrapped around a cylinder without any folds or wrinkles. A sphere is the simplest surface where the curvature is positive everywhere. If we wrap a sheet of paper around the equator of a sphere, we have an excess of paper when we try to fold it to cover the poles. This is why Mercator-projection maps distort high latitudes so much. (The north pole is a single point on the globe, yet it is the entire upper edge of the rectangular Mercator map.) This is a reflection of the positive curvature of the surface. For a saddle, the two eigenvalues have opposite signs, so the Gaussian curvature is negative. Here we find that, if we round the paper into a partial cylinder to match one radius of curvature of the saddle, then, instead of an excess of paper, we have a shortage when we try to cover the rest of the saddle. The reader is encouraged to experiment with scissors and paper to verify these claims.

The construction of the matrix M is convenient for defining the Gaussian curvature at a point. However, it is not especially convenient for actually carrying out the integral over the entire closed surface. As mentioned previously, topologists prefer to talk about curvature in terms of properties related to parallel transport of tangent vectors on the manifold without having to discuss the embedding of the manifold in a higher-dimensional space. However, for our purposes as physicists, it turns out that there is another representation for the local curvature which is both more convenient and mathematically simpler, provided that one has an explicit embedding of the surface in R^3. In addition it gives a new way to visualize the global integral of the curvature by making use of the topological properties of a mapping between the closed surface of interest and the unit sphere. To see how this works, consider a closed surface in R^3 and let $\hat{n}(\vec{r})$ be the unique three-component unit vector pointing outward, normal to the surface at the point $\vec{r}$. The unit vector $\hat{n}(\vec{r})$ defines a point on the unit sphere and so the vector field representing the surface normals defines a mapping from the closed surface of interest to the unit sphere.

Now consider a small area element $dA = dx_1 \, dx_2$ on the smooth surface. If we move in a right-handed (counterclockwise when viewed from outside the surface) loop around the boundary of dA, the tip of the surface normal arrow will trace out a small closed loop on the unit sphere. This is an indication that the surface is curved at $\vec{r}$, as for a flat surface $\hat{n}(\vec{r})$ would be a constant. We expect the solid angle subtended by this small loop on the unit sphere to be proportional to the area dA:

$$d\Omega = K(\vec{r})dA. \tag{D.13}$$

Obviously $K(\vec{r})$ is a measure of how curved our surface is at $\vec{r}$, and can be defined as the curvature. Qualitatively, positive K means the loop traversed by $\hat{n}(\vec{r})$ has the same orientation as the loop around dA, while negative K means these two loops have opposite orientations. The reader is encouraged to verify that K is positive on a sphere but negative on a saddle. Quantitatively it is a straightforward exercise to show that

$$K(\vec{r}) = C_G(\vec{r}) = \hat{n}(\vec{r}) \cdot [\partial_{x_1}\hat{n}(\vec{r}) \times \partial_{x_2}\hat{n}(\vec{r})], \tag{D.14}$$

where x_1 and x_2 are (any) local Cartesian coordinates at $\vec{r}$. This immediately allows proof of Eq. (D.6) for any surface with $g = 0$:

$$\int_S C_G \, dA = \int_{S^2} d\Omega = 4\pi. \tag{D.15}$$

In the above we performed a change of integration variables from the surface of interest to those of a unit sphere (which is possible because of the diffeomorphism[5] between the two), and used the fact that $C_G = K$ is the Jacobian of this change of variables according to Eq. (D.13).

While we have yet to prove Eq. (D.6) for arbitrary g, it should already be clear by now that its LHS is a topological invariant, as two surfaces M_1 and M_2 that are diffeomorphic to each other must have $K_1(\vec{r}_1)dA_1 = K_2(\vec{r}_2)dA_2$ for any one-to-one map from M_1 to M_2 that results in $\vec{r}_1 \to \vec{r}_2$. This allows us to choose particular surfaces for a given g to prove Eq. (D.6). The reader is encouraged to carry out the integral of the LHS of Eq. (D.6) for a torus to show that it is zero, which is consistent with $g = 1$ on the RHS. In the following we offer a different approach that allows straightforward generalization to any g.

Consider a process illustrated in Fig. D.1 through which a genus-0 surface becomes a genus-1 surface. The genus-zero surface has two arms (Fig. D.1(a)) that join to form a tube in the genus-1 surface (Fig. D.1(c)), while no change occurs elsewhere in the surface. Imagine the special case in which each arm is made of a cylinder and a half sphere. Since the cylinder has zero curvature, the change of LHS of Eq. (D.6) is due to the disappearance of two half spheres or, equivalently, a sphere, which is -4π, consistent with the change of g by 1 in Eq. (D.6). Obviously this procedure can be generalized to construct surfaces with arbitrary g to prove Eq. (D.6) by induction. We note

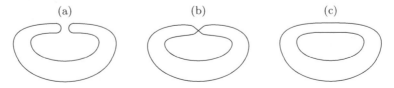

(a) (b) (c)

Figure D.1 Illustration of genus change of 2D surfaces. (a) A genus-0 surface with two arms. (b) The two arms (barely) join to form a neck. (c) A genus-1 surface forms after the two arms have joined.

[5] Diffeomorphism is basically homeomorphism between manifolds, with one-to-one mapping between the two that is not only continuous but also differentiable (or "smooth"). For a more rigorous definition see Ref. [35].

that, because of the change in topology, there is no continuous deformation that connects the surfaces in Figs. D.1(a) and (c). Instead, during the deformation one must encounter a singular situation like that illustrated in Fig. D.1(b), where the two arms join to form a neck. Obviously curvature is not well defined at the joint, just like for the degeneracy point of two bands discussed in Chapter 13 (see Fig. 13.8), where the local Berry curvature is not well defined for the two bands involved. In both cases it is through such singularities that the topology can change.

D.6 Topological Defects

Another place where topology plays an important role is in the mathematical description of defects in order-parameter fields in statistical mechanics [44]. Suppose we have a physical system in d dimensions (i.e. living in R^d), and suppose that at each point in space it is described by an order parameter. For a Bose condensate or a superconductor, the order parameter would be the complex-condensate wave function $\Psi(\vec{r})$. In the ordered phase, in order to minimize the free-energy functional (e.g. the Ginzburg–Landau free energy described in Section 19.5), it is typically the case that the magnitude of the order parameter is fixed (essentially) everywhere at a constant value (which, without loss of generality, we can take to be unity). Thus the complex order parameter is defined by its phase $\varphi(\vec{r})$, which defines a map from R^d to the unit circle S^1.[6] Such condensates can contain **topological defects** known as **vortices**. These are singularities around which the phase has a non-trivial winding. In 2D, such defects are points, while in 3D they are lines which either traverse the entire sample or form closed loops.[7] If we treat the vortex core as a forbidden region, then our physical space contains holes.

If we consider a closed loop in the physical space (punctured R^d), it will have some winding number (possibly 0) around the forbidden region. We can thus label all closed loops by this winding number that defines their homotopy class. Composition of such loops gives us the fundamental group $\pi_1(S^1) = \mathbb{Z}$. This is the first appearance that topology makes. Secondly, the value of the order parameter on the loop in R^d defines a mapping from a circle[8] to the order-parameter space, which in this case is also a circle. As we have already seen, this mapping also has fundamental group $\pi_1(S^1) = \mathbb{Z}$. If the path in (say, 2D) real space wraps counterclockwise once around a point vortex, the winding number for the phase along the path will be equal to the "**topological charge**" q of the vortex ($q = \pm 1$ for a single right-/left-handed vortex). If the loop in real space winds around a region n times, the total winding number for the map to the order parameter will be nQ, where Q is the total topological charge (the number of right-handed vortices minus the number of left-handed vortices) enclosed in the loop. Thus the topology of the loop (mapping from a circle to punctured R^2) and the mapping from the loop to the order-parameter space combine to give homotopy classes determined by the homotopy classes of the two mappings that are being composed here. Graphical representations of particular examples of this are given in Figs. 19.6 and 19.7. Figure 19.8 illustrates why it is necessary to drive the magnitude of the order parameter to zero at some point along a ring in order to continuously change the winding number of the phase. This is the same idea raised above, that the homotopy classes for the mapping from a circle to a circle are distinct and well-defined only if the target circle has non-zero radius.

---

[6] The unit circle is the space in which the reduced order parameter lives after fixing of its magnitude in this case. Such a reduction generalizes to other types of continuous order parameters, and we will encounter the example of magnetization soon. Let us call the reduced order-parameter space M. Free-energy functionals written in terms of order parameters with fixed magnitudes are also known as non-linear sigma models, to be distinguished from Ginzburg–Landau theories, in which the magnitude of the order parameter can vary and in particular can vanish.

[7] In practice the order-parameter field is not truly singular at these defects because the magnitude of the order parameter falls to zero very close to the defect to keep the gradient energy from diverging.

[8] This is because closed loops are continuous maps from a circle to the physical space.

The above discussion applies to point defects in 2D and line defects in 3D, where we can "surround" the defect with a simple loop (a mapping from the unit circle to the punctured R^d). Remarkably, we can detect the presence of the topological defect using a loop an arbitrary distance away, so long as it surrounds the defect (and only that defect). Thus there is a fundamental non-locality associated with the defect state even though the Hamiltonian or Ginzburg–Landau free-energy functional of the system may be completely local. This non-locality is due to the change of topology of the order-parameter configuration due to the presence of the defect.[9]

In addition to line defects it is also possible to have point defects in 3D. In order to surround a point defect in 3D we need not a loop as before, but rather a smooth closed surface (assumed to be topologically equivalent to a sphere). The closed surface can itself be viewed as a mapping from the unit sphere S^2 to the punctured R^3. The order parameter of the system (local phase of the condensate, local magnetic orientation, etc.) defines a map from R^3 to the order-parameter space. The point defects in the 3D order-parameter configurations are classified by the homotopy classes of the mapping from S^2 to the reduced order-parameter space, which form what is known as the **second homotopy group** $\pi_2(M)$[44], where M is the reduced order-parameter space.[10]

For example, consider a 3D magnet with local spin density described by a unit vector $\hat{m}(\vec{r})$. In this case $M = S^2$. If the spins are arranged in a "hedgehog" or monopole configuration (see Fig. 13.2) we might have (say)

$$\hat{m}(\vec{r}) = \lambda \frac{\vec{r}}{r} \tag{D.16}$$

everywhere outside a small core region at the origin where this mapping becomes singular and where the "charge" of the monopole is $\lambda = \pm 1$. Thus we are dealing with a punctured version of the position space R^3 which excludes the point defect. For simplicity, let us take the closed surface surrounding the point defect to be a unit sphere centered on the defect. The order parameter then provides a mapping from one unit sphere to another. The **"wrapping"** or **"engulfing" number** is a topological invariant for this mapping and thus

$$Q = \frac{1}{4\pi} \int_0^\pi d\theta \int_0^{2\pi} d\varphi [\vec{m} \cdot (\partial_\theta \vec{m} \times \partial_\varphi \vec{m})] = \lambda \tag{D.17}$$

gives the "topological charge" of the point defect (monopole). More complicated topological defects can have wrapping numbers that can be any integer, reflecting the fact $\pi_2(S^2) = \mathbb{Z}$. If the real-space sphere does not surround the point defect, then the wrapping number is zero.

A similar expression is useful in the analysis of the band structure of a two-band model. Suppose for simplicity we have a 2D tight-binding model with two orbitals per unit cell (or one orbital with two spin states). At each point $\vec{k} = (k_x, k_y)$ in the first Brillouin zone, the two Bloch band states are given by orthogonal superpositions of the two orbitals. We can think of these superpositions of two orbitals as defining the orientations of a pseudospin (or real spin) $\pm \hat{m}(\vec{k})$ for the upper and lower bands, respectively, that vary with the wave vector. According to Eqs. (13.51), (D.13), and (D.14), the Berry flux through a small area in 1BZ $dk_x\, dk_y$ is $\mp \frac{1}{2}[\vec{m} \cdot (\partial_{k_x} \vec{m} \times \partial_{k_y} \vec{m})]dk_x\, dk_y$. Since the first Brillouin zone is a torus, $\hat{m}(\vec{k})$ defines a mapping from the torus to the sphere that is characterized by the integer wrapping number

$$C = \mp \frac{1}{4\pi} \iint_{1\mathrm{BZ}} dk_x\, dk_y [\vec{m} \cdot (\partial_{k_x} \vec{m} \times \partial_{k_y} \vec{m})]. \tag{D.18}$$

---

[9] Using topology jargon, different order-parameter configurations form spaces known as fiber bundles, in which the physical space is the base while the order parameter at each space point is a fiber. For a more rigorous definition see Ref. [35]. The presence of topological defects changes the topology of the fiber bundle.

[10] In this context the fundamental group π_1 is also known as the first homotopy group. It should be clear now that there exists a sequence of homotopy groups, π_n, with n being a positive integer.

Because $\vec{m}$ is not the surface normal of the torus, the integral does not necessarily vanish as it did for the Gaussian curvature. Indeed, C can be any integer and is in fact the Chern number of the corresponding band. In Section 13.6 we discuss a specific two-band model whose bands have non-zero Chern numbers.

If we have time-reversal symmetry and if m represents a real spin (which is therefore odd under time reversal), then we have the symmetry constraint $m(\vec{k}) = -\vec{m}(-\vec{k})$ and the two bands form a degenerate Kramers doublet with total Chern number $C = 0$.[11] This physics is discussed in Section 14.2, where we describe the existence of a different and subtler $\mathbb{Z}_2$ topological quantum number.

The above remarks give only the barest introduction to the subject of topology and have left out many important topics which would require an entire advanced course to cover. However, they should assist the reader in understanding some of the basic topological ideas that appear in various chapters in this text.

Exercise D.1. Prove Eq. (D.14).

---

[11] Because the two bands related by time-reversal transformation must have degeneracies, their individual Chern numbers are not well-defined.

Appendix E
Scattering Matrices, Unitarity, and Reciprocity

Suppose that we have an arbitrary, possibly disordered, electronic (or optical) system with a time-independent Hamiltonian, with K ports which support incoming and outgoing electronic or electromagnetic waves. In the example illustrated in Fig. 10.7, if there are N ports on the left and M ports on the right, then $K = M + N$. Further suppose that the system is non-interacting (described by a quadratic Hamiltonian that conserves particle number). Under these conditions there is a linear relationship between the amplitudes of the incoming and outgoing waves in the K ports,

$$a^{\text{out}} = Sa^{\text{in}}, \tag{E.1}$$

where $a^{\text{in(out)}}$ is the vector of incoming (outgoing) amplitudes and S is a $K \times K$ complex matrix whose entries determine the amplitude and phase relations between the incoming and outgoing waves. In terms of Fig. 10.7, some of the terms in the S **matrix** describe transmission amplitudes t^{jk} from the left side to the right side, and some describe reflection amplitudes where the outgoing wave exits on the same side of the sample as it entered. (Thus t is an $M \times N$ rectangular matrix whose elements are a subset of the S matrix, and t is not itself unitary.)

Let us assume that the system is dissipationless so that all energy that flows into the system also flows out. The incoming power is proportional to the inner product

$$P_{\text{in}} = (a^{\text{in}})^{\dagger} a^{\text{in}}, \tag{E.2}$$

while the outgoing power is

$$P_{\text{out}} = (a^{\text{in}})^{\dagger} S^{\dagger} S a^{\text{in}}. \tag{E.3}$$

In order to guarantee that these are equal (energy is conversed) for all possible input amplitudes, we see that the S matrix must be unitary,

$$S^{\dagger} S = I, \tag{E.4}$$

where I is the identity matrix. Quantum mechanically, the squares of the amplitudes represent the probability that a photon or electron enters or leaves a particular port and the unitarity of the S matrix guarantees the conservation of probability. Ultimately, unitarity follows from the fact that, in quantum mechanics, time evolution under a Hermitian (self-adjoint) Hamiltonian is described by a unitary operator $U = \exp\{-(i/\hbar)Ht\}$ which preserves the norm of the state vector in Hilbert space.

Let us now consider the effect of time-reversal. For the case of electrons, the disordered region may contain a magnetic field which is odd under time-reversal. For photons, the scattering region may be "non-reciprocal" because it contains circulators (devices that also typically involve magnetic fields). Under time-reversal one should therefore make the replacement $S(+\vec{B}) \rightarrow S(-\vec{B})$ (since $\vec{B}$ changes sign under time-reversal). Since we are "watching the movie backwards," we should also interchange the input and output ports and, following the rules of quantum mechanics, complex conjugate all the amplitudes. As a result, we have

$$(a^{\text{in}})^* = S(-\vec{B})(a^{\text{out}})^*, \tag{E.5}$$

which leads to

$$(a^{\text{out}})^* = S^\dagger(-\vec{B})(a^{\text{in}})^*, \tag{E.6}$$

$$a^{\text{out}} = S^{\text{T}}(-\vec{B})a^{\text{in}}. \tag{E.7}$$

Comparison with Eq. (E.1) then yields

$$S^{\text{T}}(-\vec{B}) = S(+\vec{B}). \tag{E.8}$$

If the system is reciprocal ($\vec{B} = \vec{0}$), it follows that the S matrix is symmetric. In terms of Fig. 10.7, this means that the transmission amplitudes for a reciprocal system are symmetric, $t^{jk} = t^{kj}$, from which it follows that the transmission probabilities are symmetric $T^{jk} = T^{kj}$.

In the quantum Hall effect, edge-channel transport of electrons is chiral and thus highly non-reciprocal. See Chapters 12 and 16.

Appendix F

F Quantum Entanglement in Condensed Matter Physics

As we saw in Chapter 10, when the system being studied is entangled with its environment due to the ubiquitous coupling between them, it can no longer be described by a quantum-mechanical state or wave function. Instead we need to use its **(reduced) density matrix**.[1] This is because the system is in a mixed state (and thus *not* a pure state) when it is entangled, even if the Universe (system + bath) is in a pure state. The density matrix is also widely used in quantum statistical mechanics, where we deal with an *ensemble* of pure states that the system may be in, with appropriate probability. It turns out that a lot of insight into an isolated system (an idealization!) in a pure state (say the ground state) can be gained by artificially dividing the system into two or more parts, and studying how they are entangled with each other. We discuss how this is achieved, and start with some mathematical preparations.

F.1 Reduced Density Matrix

Let us assume the entire Hilbert space of the system is the tensor product of those of two subsystems, A and B:

$$\mathcal{H} = \mathcal{H}_A \otimes \mathcal{H}_B, \tag{F.1}$$

and the dimensionalities of $\mathcal{H}_A$ and $\mathcal{H}_B$ are N_A and N_B, respectively, so the dimensionality of $\mathcal{H}$ is $N = N_A N_B$. If $\{|i\rangle_A\}$ and $\{|j\rangle_B\}$ are orthonormal basis sets for $\mathcal{H}_A$ and $\mathcal{H}_B$, respectively, then their tensor product $\{|i\rangle_A \otimes |j\rangle_B\}$ forms an orthonormal basis set for $\mathcal{H}$. Any (pure) state in $\mathcal{H}$ can be expanded as

$$|\Psi\rangle = \sum_{i=1}^{N_A} \sum_{j=1}^{N_B} M_{ij} |i\rangle_A \otimes |j\rangle_B, \tag{F.2}$$

where the coefficients M_{ij} form an $N_A \times N_B$ (rectangular) matrix M.

To obtain the reduced density matrix for subsystem A, we trace out degrees of freedom in B from the total system density matrix $\rho = |\Psi\rangle\langle\Psi|$:

---

[1] Also referred to as the density operator, which should not be confused with the operator describing the density of particles. We will use the term density matrix to avoid possible confusion in this book.

$$\rho_A = \text{Tr}_B \rho = \sum_{j=1}^{N_B} \langle j|_B \Psi \rangle \langle \Psi|j \rangle_B = \sum_{i,i'=1}^{N_A} \sum_{j=1}^{N_B} M_{ij} M_{i'j}^* |i\rangle_A \langle i'|_A. \tag{F.3}$$

We thus find in the $\{|i\rangle_A\}$ basis

$$\rho_A = MM^\dagger. \tag{F.4}$$

Similarly,

$$\rho_B = M^\dagger M \tag{F.5}$$

in the $\{|i\rangle_B\}$ basis. Both are Hermitian and positive semi-definite as density matrices must be.[2] All entanglement properties between A and B are encoded in ρ_A and ρ_B. In particular, the entanglement entropy, defined to be the von Neumann entropy associated with ρ_A or ρ_B, quantifies the amount of entanglement between A and B:

$$S_E = -\text{Tr}[\rho_A \ln \rho_A] = -\text{Tr}[\rho_B \ln \rho_B]. \tag{F.6}$$

The last equality may not be immediately obvious, and will be made clear in the next section. We now discuss some simple examples.

Exercise F.1. (i) Show that $MM^\dagger$ and $M^\dagger M$ are Hermitian matrices. (ii) Show that their eigenvalues are non-negative. Hint: this does not require a complex calculation. Simply show that they satisfy the defining property of positive semi-definite matrices. (iii) Show that both $MM^\dagger$ and $M^\dagger M$ obey the requirement that density matrices have unit trace. Hint: use the fact that the state of the system plus bath is correctly normalized, $\langle \Psi|\Psi \rangle = 1$.

The simplest system that can be decomposed into two subsystems is made of two spin-1/2 particles. Let us consider one of the triplet states, with both spins up:

$$|\Psi\rangle = |\uparrow\rangle_A \otimes |\uparrow\rangle_B. \tag{F.7}$$

This is a product state with *no* entanglement between A and B. Going through the procedure above, we find that

$$\rho_A = \begin{pmatrix} 1 & 0 \\ 0 & 0 \end{pmatrix} \tag{F.8}$$

corresponds to a pure state since $\rho_A^2 = \rho_A$ is a projection operator. Another way to see that there is no entanglement is by calculating the entanglement entropy and finding it to be zero.

Now let us consider a more interesting case, when the two spins form a singlet:

$$|\Psi\rangle = \frac{1}{\sqrt{2}}[|\uparrow\rangle_A \otimes |\downarrow\rangle_B - |\downarrow\rangle_A \otimes |\uparrow\rangle_B]. \tag{F.9}$$

A straightforward calculation shows that in this case

$$\rho_A = \begin{pmatrix} 1/2 & 0 \\ 0 & 1/2 \end{pmatrix}. \tag{F.10}$$

We find $S_E = \ln 2$, and the two spins are clearly entangled. This is in fact the maximum entropy one can have in a two-level system.

---

[2] A positive semi-definite matrix ρ has the defining property that $\langle \psi|\rho|\psi \rangle \geq 0, \forall|\psi\rangle$. If ρ is Hermitian it follows that it has no negative eigenvalues (i.e. each eigenvalue is positive semi-definite). Density matrices must be positive semi-definite because, in the basis in which they are diagonal, the diagonal entries represent classical probabilities.

It is clear that entanglement must involve multiple terms in the decomposition (F.2), as is the case in the singlet example above, which contains two terms. But is it true that the more terms we have, the more entanglement we will get? The answer is no. Consider the following example:

$$|\Psi\rangle = \frac{1}{2}[|\uparrow\rangle_A \otimes |\uparrow\rangle_B + |\downarrow\rangle_A \otimes |\downarrow\rangle_B + |\uparrow\rangle_A \otimes |\downarrow\rangle_B + |\downarrow\rangle_A \otimes |\uparrow\rangle_B]. \tag{F.11}$$

It contains four terms, and has an even more complicated-looking reduced density matrix:

$$\rho_A = \begin{pmatrix} 1/2 & 1/2 \\ 1/2 & 1/2 \end{pmatrix}. \tag{F.12}$$

But it is easy to check $\rho_A^2 = \rho_A$ in this case, and there is thus no entanglement! A more straightforward way to see this is to recognize that we can write

$$|\Psi\rangle = \frac{1}{2}[|\uparrow\rangle_A + |\downarrow\rangle_A] \otimes [|\uparrow\rangle_B + |\downarrow\rangle_B], \tag{F.13}$$

indicating that this is actually a product state with both spins pointing in the $\hat{x}$ direction! The complication we encountered earlier is due to the fact that we have been working in the S_z basis, while the S_x basis would have simplified things considerably. We now develop a mathematical formalism that chooses the basis that reveals the entanglement most clearly.

F.2 Schmidt and Singular-Value Decompositions

We can diagonalize $MM^\dagger$ and $M^\dagger M$, and work in the basis in which ρ_A and ρ_B are both diagonal in subspaces A and B, respectively. It is clear that the double summation of Eq. (F.2) reduces to a *single* summation in this basis:

$$|\Psi\rangle = \sum_{l=1}^{r} s_l |l\rangle_A \otimes |l\rangle_B, \tag{F.14}$$

where $r \leq \min(N_A, N_B)$ is the rank of M, which is also the number of non-zero eigenvalues of ρ_A and ρ_B. In terms of their eigenvectors $|l\rangle_A$ and $|l\rangle_B$, we can write

$$\rho_A = \sum_{l=1}^{r} |s_l|^2 |l\rangle_A \langle l|_A; \qquad \rho_B = \sum_{l=1}^{r} |s_l|^2 |l\rangle_B \langle l|_B. \tag{F.15}$$

It is now clear that ρ_A and ρ_B share the same non-zero eigenvalues and must have the same entanglement entropy:

$$S_E = -\text{Tr}[\rho_A \ln \rho_A] = -\text{Tr}[\rho_B \ln \rho_B] = -\sum_{l=1}^{r} |s_l|^2 \ln |s_l|^2. \tag{F.16}$$

The way to write $|\Psi\rangle$ in the form of Eq. (F.14) is known as the **Schmidt decomposition**.

Exercise F.2. Show that the entanglement entropy in this case has an upper bound of $\ln r$.

Mathematically, the Schmidt decomposition can be viewed as an application of **singular-value decomposition (SVD)**, which states that any $N_A \times N_B$ rectangular matrix M can be brought to diagonal form by two unitary matrices:

$$M = UM'V^\dagger, \tag{F.17}$$

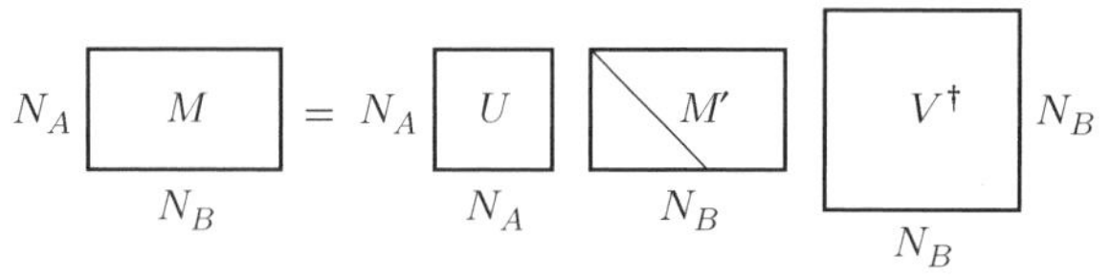

Figure F.1 Schematic illustration of singular-value decomposition. The diagonal line in the center box M' indicates that M' is a diagonal (but not necessarily square) matrix.

where U is an $N_A \times N_A$ unitary matrix that diagonalizes $MM^\dagger$, V is an $N_B \times N_B$ unitary matrix that diagonalizes $M^\dagger M$, and M' is an $N_A \times N_B$ rectangular matrix just like M, but with non-zero matrix elements only on the diagonal. The structure of the SVD is schematically illustrated in Fig. F.1.

F.3 Entanglement Entropy Scaling Laws

Entropy is one of the most important concepts in all of physics (and with powerful application to information theory). It is an extensive thermodynamic quantity, meaning that it is proportional to the volume of the system. The **entanglement entropy** S_E, defined in Eq. (F.6), does not necessarily follow this volume law (here the volume corresponds to that of the subsystem). In fact, for ground states of almost all local Hamiltonians (i.e. those including short-distance couplings only), it follows the so-called **area law**:

$$S_E \propto \Sigma \sim L_A^{d-1}, \tag{F.18}$$

where Σ is the area of the boundary between subsystems A and B, and L_A is the linear size of subsystem A (assumed to be the smaller one, and with sufficiently regular shape). Such scaling behavior can be understood heuristically by inspecting a projected entangled pair state (PEPS), as illustrated in Fig. 17.16. The key point is that in ground states entanglement comes from degrees of freedom near the boundary, that are entangled with each other across the boundary. The matrix product and tensor network states discussed in Section 17.6.4 capture this property and thus can be used to accurately approximate the ground states of a very large class of systems. The 2D $S = 2$ AKLT model, whose exactly soluble ground state is depicted in Fig. 17.15, provides a good illustration of the origin of the area law in a quantum spin liquid.

The area law for ground states of (reasonable) Hamiltonians is generic because of the notion of **monogamy of entanglement** (which is discussed in Section 17.7). Good variational states for local Hamiltonians will have strong non-classical correlations due to local (short-range) entanglement to optimize the energy. Entanglement monogamy implies that, if a particle is entangled strongly with a nearby particle, it cannot be strongly entangled with distant particles.

A random state in the Hilbert space $\mathcal{H}$, on the other hand, would have volume-law scaling (just like thermal entropy):

$$S_E \propto V_A \sim L_A^d, \tag{F.19}$$

where V_A is the volume of subsystem A. Roughly speaking, this is because in a typical state any particle in A could be entangled with any distant particle in B. Thus most of the states in $\mathcal{H}$ do not even "qualify" to be a ground state, and we should not waste our time on them. Instead we should focus on the tiny "corner" of $\mathcal{H}$, where states follow the area-law behavior (F.18). Most (if not all) of the states in this corner can be brought to some form of tensor network state, and that is the reason why the latter provide accurate approximations to the ground states of local Hamiltonians.

What about excited states? It is widely expected that highly excited states, namely those with a finite excitation energy *density* (or, equivalently, excitation energy proportional to the system's volume), would follow the volume law (F.19). This has been shown explicitly for free-fermion states.[3] We are used to the volume law from the fact that we expect the classical thermodynamic entropy to be extensive in the volume of a system. The same generally holds true for quantum systems but it has some peculiar implications. Suppose that a single excited energy eigenstate (with finite excitation energy density) has a volume law for its entanglement entropy. Now imagine freezing the positions of all the particles except one. We can think of the many-body wave function of the system as defining the single-particle wave function for that one particle. All the other particle coordinates can be thought of as parameters labeling that single-particle wave function. A volume law for the entropy implies that, if we pick another very distant particle and move it slightly, the wave function of the first particle *must* be significantly altered even though the two are highly separated. Suddenly the familiar concept of extensive entropy seems to have quite bizarre implications!

A known class of exceptions to the volume law mentioned above consists of highly excited states in **many-body localized systems** (see Section 11.11). Here they follow the area law instead, just like ground states. One way to understand this is to note that, in the basis where the local conserved observables $\{L_j\}$ are diagonal, all energy eigenstates are product states with no entanglement. When expressed in the original lattice basis, we expect entanglement to be dominated by the spreading of those L_js (see Eq. (11.169)) near the boundary, thus giving rise to area-law entanglement. Another way to understand this is to note that, for any energy eigenstate $|\{l_j\}\rangle$, one can always construct a *local* Hamiltonian $H = -\sum_j l_j L_j$ such that $|\{l_j\}\rangle$ is its *ground* state. So it must have area-law entanglement.

F.4 Other Measures of Entanglement

The reduced density matrices ρ_A and ρ_B encode all possible information about entanglement and hence are much richer than the single number represented by the entanglement entropy S_E. It is thus of interest to extract additional information from them by considering the full distribution of eigenvalues and the corresponding eigenfunctions of the density matrix. To this end, it is common to write

$$\rho_A = e^{-H_A}, \tag{F.20}$$

where H_A is a Hermitian matrix called the **entanglement Hamiltonian**. This form suggests a thermal density matrix with inverse temperature set to be unity.

> **Exercise F.3.** Show that the representation in Eq. (F.20) is possible for any positive semi-definite Hermitian matrix, which is the case here (see Exercise F.1).

The spectrum of H_A is called the entanglement spectrum. For quantum Hall and many other topological phases, it is found that the ground state entanglement spectrum resembles the corresponding edge-state spectrum [244], indicating that the ground state entanglement properties contain information about its excitation spectrum! It is hoped that this statement has much wider applicability, namely that a considerable amount of *dynamical* information is hidden in the ground state wave function, which can be decoded by inspecting its entanglement properties, *without* any knowledge of the system's Hamiltonian that actually controls its dynamics! (Compare this with what we learned within the

---

[3] See Ref. [239]. Simple as it may be, a free-fermion state (or single-Slater-determinant state) turns out to be highly non-trivial from the entanglement perspective. It is the first example of area-law violation above 1D for ground states [240, 241]. There are very few such examples currently known, all involving surfaces (instead of points) in momentum space that support gapless excitations, including the free Fermi gas [240, 241], Fermi liquids [242], and a class of coupled harmonic oscillators with phonon excitations vanishing on certain "Bose" surfaces [243].

single-mode approximation in Chapter 4 about excitations using only the ground state structure factor and one universal feature of the Hamiltonian, the oscillator-strength sum rule.)

For free-fermion (or single-Slater-determinant) states H_A takes the form of a free-fermion Hamiltonian. This allows significant simplification in the study of their entanglement properties.[4] Here it was found that the eigenstates of H_A contain very useful information as well [245, 246].

For highly excited states, it is expected that, for the so-called chaotic (in contrast to integrable) Hamiltonians, if subsystem A is sufficiently small compared with the whole system, ρ_A becomes the same as the thermal density matrix at the temperature where the system's internal energy density matches the energy density of the excited state being inspected. This property, known as thermalization, implies that the entanglement entropy becomes the same as the thermal entropy! Such thermalization has been demonstrated for typical excited states in free-fermion systems [239] (even though these are integrable systems). Again, many-body localized systems are exceptions to this, due to the very special integrability associated with the extensive local conservation laws discussed in Section 11.11. As a consequence, these systems fail to return to thermal equilibrium (or thermalize) once driven out of equilibrium, in the absence of coupling to external thermal baths.

F.5 Closing Remarks

We close this appendix by stating that entanglement, the most fundamental and counter-intuitive aspect of quantum mechanics, is a concept of central importance in quantum information science. It is finding increasingly wide application in condensed matter and other branches of physics at the time of writing. This is because it reveals fundamental properties of systems with many degrees of freedom, that are not easily accessible from correlation functions of local observables. Traditionally condensed matter physics has focused on such correlation functions, as they are closely related to response functions of the system being measured experimentally (see Appendix A). At the time of writing, attempts to directly measure the entanglement entropy in experiments have just begun [247].

---

[4] The situation is similar for ground states of coupled harmonic oscillator systems, whose wave functions are of Gaussian form. But it does not generalize to excited states in this case.

Appendix G

Linear Response and Noise in Electrical Circuits

In Appendix A, we derived linear-response theory and the fluctuation–dissipation theorem. To gain more insight into the concepts of linear response and the noise spectral density of thermal and quantum fluctuations, we will explore here examples involving linear dissipative electrical circuits. We will begin with an example from classical statistical mechanics and then return to the quantum case to make contact with the discussion in Appendix A. Along the way we will learn how to quantize electrical circuits, a concept relevant to the superconducting quantum bits discussed in Section 19.11.

G.1 Classical Thermal Noise in a Resistor

Electrical resistors cause dissipation. The fluctuation–dissipation theorem derived in Appendix A tells us that there will be noise associated with this dissipation and that, in the classical limit $\hbar\omega \ll k_B T$, the noise spectral density will be proportional to the temperature and the strength of the dissipation. To explore this concept in electrical-circuit language, let us model a resistor in thermal equilibrium at temperature T as a perfect resistor (no noise) plus a noise source which represents thermal fluctuations of voltages and currents in the circuit. Figure G.1 shows two completely equivalent representations of such noise sources. Figure G.1(a) shows an ideal resistor R (no noise) in series with a random voltage generator, which is also ideal (i.e. zero internal impedance), and represents the thermal fluctuations of the charges in the resistor. Figure G.1(b) show a completely equivalent representation in terms of an ideal current source (i.e. one with infinite internal impedance) in parallel with an ideal noiseless resistor. A little thought shows that the two representations will act on an arbitrary load attached to the terminals in exactly the same way, provided that the voltage noise source is related to the current noise source via

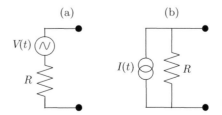

Figure G.1 A noisy resistor modeled as a perfect resistor R connected to an ideal noise source. (a) An ideal voltage noise source capable of producing voltage $V(t)$ irrespective of how much current is flowing. (b) An ideal current noise source capable of producing current $I(t)$ no matter what voltage is across it.

$$V(t) = RI(t). \tag{G.1}$$

For the purposes of the discussion in this section, we will ignore quantum effects altogether. The characteristic time scales for electronic motion are extremely short in comparison with anything that can be measured with a voltmeter, so it is appropriate (in the classical case) to assume a **"white-noise"** spectrum with **autocorrelation function**

$$\langle V(t)V(t')\rangle = \Gamma\delta(t - t') \tag{G.2}$$

for the voltage source in Fig. G.1(a). Here Γ is a constant to be determined from physical arguments about thermal equilibrium which we will give below. Equivalently, the currrent source in Fig. G.1(b) would obey

$$\langle I(t)I(t')\rangle = \frac{\Gamma}{R^2}\delta(t - t'). \tag{G.3}$$

We will focus for now on the voltage-source representation.

Consider the following definition of the Fourier transform:

$$V(\omega) \equiv \frac{1}{\sqrt{T_0}}\int_0^{T_0} dt\, e^{i\omega t}V(t). \tag{G.4}$$

Since $V(t)$ is random, the integral is a "random walk" (i.e. a "sum" of random numbers) whose value will be of order $\sim \sqrt{T_0}\times$ constant. Hence we divide by $\sqrt{T_0}$ to get a $V(\omega)$ whose statistical properties will be independent of T_0 (as long as T_0 greatly exceeds all characteristic times). Now consider the so-called **"power spectrum"** or "spectral density"

$$S_{VV}(\omega) \equiv \langle|V(\omega)|^2\rangle. \tag{G.5}$$

The units of $V(\omega)$ are $V/\sqrt{Hz}$ so the units of $S_{VV}(\omega)$ are V^2/Hz. We have

$$S_{VV}(\omega) = \frac{1}{T_0}\int_0^{T_0} dt \int_0^{T_0} dt'\, e^{i\omega(t-t')}\langle V(t)V(t')\rangle \tag{G.6}$$

$$= \frac{1}{T_0}\int_0^{T_0} dt' \int_{-t'}^{T_0-t'} dt''\, e^{i\omega t''}\langle V(t' + t'')V(t')\rangle. \tag{G.7}$$

Using time translation invariance of the correlation function yields

$$S_{VV}(\omega) = \frac{1}{T_0}\int_0^{T_0} dt' \int_{-t'}^{T_0-t'} dt''\, e^{i\omega t''}\langle V(t'')V(0)\rangle. \tag{G.8}$$

For T_0 very much greater than the autocorrelation time scale (see Eq. (G.20) for an example) we can safely extend the limits of integration:

$$S_{VV}(\omega) \approx \frac{1}{T_0}\int_0^{T_0} dt' \int_{-\infty}^{+\infty} dt''\, e^{i\omega t''}\langle V(t'')V(0)\rangle \tag{G.9}$$

$$= \int_{-\infty}^{+\infty} dt''\, e^{i\omega t''}\langle V(t'')V(0)\rangle = \langle|V(\omega)|^2\rangle. \tag{G.10}$$

This is the **Wiener–Khinchin theorem**: the spectral density and the autocorrelation function are Fourier transforms of each other. Thus

$$\langle V(t)V(0)\rangle = \int_{-\infty}^{+\infty} \frac{d\omega}{2\pi}S_{VV}(\omega)e^{-i\omega t}. \tag{G.11}$$

For the particular case of white noise, substitution of Eq. (G.2) into Eq. (G.10) gives

$$S_{VV}(\omega) = \Gamma. \tag{G.12}$$

Thus (classical) white noise has a "flat," frequency-independent spectral density.

To evaluate Γ, consider the circuit shown in Fig. G.1(a) and attach a load consisting of a simple capacitor whose capacitance is C. The voltage across the capacitor obeys

$$V_c = \frac{Q}{C}, \tag{G.13}$$

$$\dot{V}_c = \frac{\dot{Q}}{C} = \frac{I}{C}, \tag{G.14}$$

$$V_c(\omega) = I Z(\omega), \tag{G.15}$$

where Q is the charge on the capacitor, I is the current flowing around the circuit, and the impedance of the capacitor is[1]

$$Z(\omega) = \frac{1}{-i\omega C}. \tag{G.16}$$

The voltage across the capacitor is linearly related to the noise voltage via the impedance ratio

$$V_c(\omega) = V(\omega)\frac{Z(\omega)}{R + Z(\omega)} = \frac{V(\omega)}{1 - i\omega\tau}, \tag{G.17}$$

where

$$\tau = RC \tag{G.18}$$

is the characteristic time constant of the circuit.

The spectral density for the voltage across the capacitor is

$$\left\langle |V_c(\omega)|^2 \right\rangle = \frac{\left\langle |V(\omega)|^2 \right\rangle}{1 + \omega^2\tau^2} = \frac{\Gamma}{1 + \omega^2\tau^2}. \tag{G.19}$$

We see that the capacitor filters out the high-frequency noise because its impedance becomes small (relative to R) at high frequencies. Using Eq. (G.11), we have

$$\langle V_c(t) V_c(0) \rangle = \Gamma \int_{-\infty}^{+\infty} \frac{d\omega}{2\pi} \frac{1}{1 + \omega^2\tau^2} e^{-i\omega t}$$

$$= \Gamma \frac{e^{-|t|/\tau}}{2\tau}. \tag{G.20}$$

Thus we see that the voltage correlations fall off exponentially on a time scale given by the circuit's natural RC time constant.

The unknown constant Γ is readily obtained from the equipartition theorem. The equilibrium voltage distribution is

$$P(V) = \frac{1}{Z} e^{-\beta \frac{1}{2} C V^2}, \tag{G.21}$$

$$Z \equiv \int_{-\infty}^{+\infty} dV \, e^{-\beta \frac{1}{2} C V^2}, \tag{G.22}$$

where β is the inverse temperature and $\frac{1}{2}CV^2$ is the electrostatic energy stored in the capacitor. This yields

$$\langle V^2 \rangle = \langle V(0) V(0) \rangle = \frac{k_B T}{C}. \tag{G.23}$$

Thus, from Eq. (G.20), the resistor's thermal voltage noise spectral density is

$$S_{VV}(\omega) = \Gamma = 2\tau \frac{k_B T}{C} = 2R k_B T. \tag{G.24}$$

---

[1] Electrical engineers would replace $-i$ by $+j$ in this expression.

The equivalent current noise spectral density for the circuit model illustrated in Fig. G.1(b) is

$$S_{II}(\omega) = \frac{2}{R} k_B T. \tag{G.25}$$

There are various alternative expressions in the literature, which can be confusing. Sometimes the positive and negative frequencies are combined so that[2]

$$J_+(\omega) \equiv S_{VV}(\omega) + S_{VV}(-\omega) = 4R k_B T \tag{G.26}$$

and one uses frequency integrals only over positive frequencies,

$$\int_0^{+\infty} \frac{d\omega}{2\pi} = \int_0^{+\infty} df. \tag{G.27}$$

Sometimes the 2π in the integration measure is lumped into the spectral density,

$$\tilde{J}_+(\omega) = \frac{2}{\pi} R k_B T, \tag{G.28}$$

and the integration is of the form $\int_0^{+\infty} d\omega$. You have to be careful to check what convention each author is using. One sometimes sees $S_{VV}(\omega)$ described as the **"two-sided" spectral density** and $J_+(\omega)$ described as the **"one-sided" spectral density**.

Equation (G.24) is a classic example of the fluctuation–dissipation theorem which states that the thermal fluctuations in a particular variable are proportional to the dissipation felt by that variable. A similar relationship connects the fluctuating molecular forces acting on a particle in a fluid to the viscosity of the fluid. Large viscosity makes it more difficult for the particle to maintain the mean kinetic energy required by the equipartition theorem, and so the fluctuating forces must increase with the dissipation. For a more detailed discussion of the fluctuation–dissipation theorem, see Ref. [248].

G.2 Linear Response of Electrical Circuits

Suppose that we are given a black box containing a linear electrical circuit and having one port through which it can be probed. As illustrated in Fig. G.2(a), one can probe the system by injecting current at frequency ω and measuring the response of the voltage. The linear response coefficient is of course the (complex) impedance,

$$V(\omega) = Z(\omega)I(\omega). \tag{G.29}$$

This standard electrical response function is not in the form of the susceptibilty $\chi(\omega)$ that we defined in Eq. (A.21), but, as we will see when we develop the quantum theory of such circuits, it is very closely related. In particular, poles in $Z(\omega)$ occur at frequencies corresponding to the normal modes (i.e. resonances) of the circuit, just as poles in $\chi(\omega)$ do.

As a simple example, suppose that the black box contains a parallel LC resonator as illustrated in Fig. G.3(a). We further suppose that the oscillator is weakly damped by a small (real) admittance Y_0 (large resistance $R_0 = 1/Y_0$, added in parallel to the inductor and the capacitor). The impedance of this parallel combination is (in the electrical-engineering sign convention)

$$Z(\omega) = \frac{1}{j\omega C + 1/j\omega L + Y_0} = -j Z_0 \frac{\omega \omega_0}{\omega^2 - \omega_0^2 - j\omega\kappa}, \tag{G.30}$$

---

[2] Recall from Eq. (A.44) that $S(+\omega)$ and $S(-\omega)$ are related by detailed balance and that, in the classical limit ($\hbar\omega \ll k_B T$), they are in fact equal.

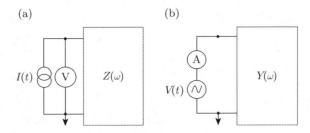

Figure G.2 (a) Experimental setup of a black box containing a linear electrical circuit and a single port (plus a ground port) through which it can be probed to measure its frequency-dependent impedance $Z(\omega)$ using an ac current drive and measuring the voltage response (a) or its admittance $Y(\omega)$ using an ac voltage drive and an ammeter (b). The current source and voltmeter have infinite internal impedance. The voltage source and ammeter have zero internal impedance.

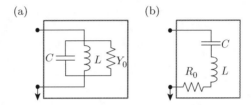

Figure G.3 (a) A parallel LC resonator weakly damped by a small (real) admittance Y_0 (large resistance $1/Y_0$) in parallel with the inductor and capacitor. This has a pole in the *impedance* in the complex frequency plane near the undamped resonance frequency $1/\sqrt{LC}$. (b) A dual circuit containing a series LC resonator and weakly damped by a small (real) impedance R_0 (large admittance $1/R_0$) in series with the inductor and capacitor. The dual circuit has a pole in the *admittance* near $1/\sqrt{LC}$ in the complex frequency plane.

where $Z_0 \equiv \sqrt{L/C}$ is the so-called characteristic impedance of the resonance, $\kappa \equiv Y_0/C = 1/R_0 C$ is the characteristic energy damping rate of the oscillator, and the resonance frequency is $\omega_0 = 1/\sqrt{LC}$. This expression can be recast as

$$Z(\omega) = -j\omega\omega_0 Z_0 \frac{1}{(\omega - j\kappa/2)^2 - \tilde{\omega}_0^2}, \tag{G.31}$$

where $\tilde{\omega}^2 \equiv \omega_0^2 - (\kappa/2)^2$. Recalling that $j = -i$ and taking the limit of infinitesimal damping, this expression bears a striking resemblance to the mechanical susceptibility for a quantum oscillator given in Eq. (A.26). We will explore this connection in more detail shortly.

Recall that the current source which is used to drive the oscillator in the setup shown in Fig. G.2(a) has infinite internal impedance, as does the voltmeter used to measure the response. Hence the measurement apparatus effectively applies open-circuit boundary conditions on the port of the black box. Thus the poles of $Z(\omega)$ correspond to the collective-mode frequencies under open-circuit boundary conditions for the black box. It is clear from Fig. G.3(a) that the parallel LC resonance requires open-circuit boundary conditions.

Let us turn now to consideration of the admittance $Y(\omega)$ as the linear response of the current to an applied voltage:

$$I(\omega) = Y(\omega)V(\omega). \tag{G.32}$$

Clearly the complex admittance is simply the inverse of the complex impedance. Hence the poles of Y are given by the zeros of Z. In order to understand the physical meaning of the poles of Y, it is useful to note that the physical setup appropriate for the measurement of the admittance (see Fig. G.2(b)) is quite different from that used for the measurement of the impedance. The voltage source and the ammeter used in the admittance measurement both have zero internal impedance. Hence the poles of Y correspond to the collective-mode frequencies of the black box under short-circuit boundary conditions on the port. It is clear from Fig. G.3(b) that the series LC resonator will not be able to freely oscillate at its natural resonance frequency unless the input port is shorted to ground to complete the circuit.

We have so far considered black boxes containing only a single LC resonator having either a pole in the impedance (for the parallel resonator) or a pole in the admittance (for the series resonator). In general the black box can contain an arbitrarily complex circuit. Indeed, the black box might not even contain discrete circuit elements. It might, for example, be a microwave cavity with multiple resonances, and the port might couple to these through an antenna inside the box. From the outside we will simply observe a sequence of poles in $Z(\omega)$ as we vary the frequency. It turns out that in between each pole there will be a zero (i.e. a pole in the admittance). Which of these two sets of poles corresponds to the naturally occurring collective mode frequencies depends on whether the antenna port is left open or is shorted. Furthermore, when the damping is zero (or at least very weak), knowing the locations of the poles and the residue of each pole is enough to completely characterize each resonance and represent it in terms of fixed parameter values for L and C.

G.3 Hamiltonian Description of Electrical Circuits

Let us consider the circuit in Fig. G.3(a), which will be driven by a current source as in Fig. G.2(a). It will turn out to be convenient to define a flux variable

$$\Phi(t) = \int_{-\infty}^{t} d\tau \, V(\tau), \tag{G.33}$$

where $V(\tau)$ is the voltage at the input port. Φ is called a flux variable because it has (SI) units of magnetic flux and because of the resemblance (except for a minus sign) of the expression $V(t) = \dot{\Phi}(t)$ to Faraday's law for the electromotive force associated with a time-varying flux through a loop. It follows that the current flowing through the inductor obeys $I_L = \Phi(t)/L$.

Current conservation at the input port tells us that the input current is equal to the sum of the three branch currents in the circuit. Hence

$$C\ddot{\Phi} = -\frac{\Phi}{L} - Y_0\dot{\Phi} + I(t). \tag{G.34}$$

Not surprisingly, this equation is precisely equivalent to Newton's law of motion for the coordinate $\Phi(t)$ of a harmonic oscillator of mass C, spring constant $1/L$, damping rate $\kappa = Y_0/C$, and external driving force $I(t)$. Indeed, in the absence of damping, the circuit dynamics is described by the Lagrangian

$$\mathcal{L} = \frac{1}{2}C\dot{\Phi}^2 - \frac{1}{2L}\Phi^2 + I(t)\Phi. \tag{G.35}$$

The canonical momentum conjugate to Φ,

$$Q = \frac{\delta\mathcal{L}}{\delta\dot{\Phi}} = C\dot{\Phi} = CV, \tag{G.36}$$

is nothing more than the charge stored on the capacitor. Using this we readily obtain the Hamiltonian in terms of the canonical momentum and the coordinate

$$H(Q, \Phi) = Q\dot{\Phi} - \mathcal{L} = \frac{Q^2}{2C} + \frac{\Phi^2}{2L} - I(t)\Phi. \tag{G.37}$$

We see that the electrostatic potential energy on the capacitor is represented here as kinetic energy and the magnetic energy stored in the inductor is represented here as potential energy.

Having derived the Hamiltonian, we are now in a position to find the susceptibility of a linear circuit to the external perturbation $-I(t)\Phi$ defined by the relation

$$\Phi(\omega) = -\chi(\omega)I(\omega). \tag{G.38}$$

In Appendix A we derived the susceptibility for a single harmonic oscillator. In general a black box containing a linear circuit will have many harmonic modes (each associated with a pole in the impedance if open-circuit boundary conditions apply). We can, however, very easily deal with this additional complication by simply noting that, since $\dot{\Phi}$ is the voltage, the susceptibility is simply proportional to the complex impedance:

$$\chi(\omega) = -\frac{1}{j\omega}Z(\omega) \tag{G.39}$$

(where again we are using electrical-engineering notation). This explains why the expression for the impedance of an LC resonator in Eq. (G.31) so closely resembles the susceptibility in Eqs. (A.26) and (A.27). Because we are dealing with harmonic oscillators, this expression is valid both classically and quantum mechanically. (Recall the discussion following Eq. (A.27).)

Evaluating the impedance for the parallel LC resonator in Eq. (G.30) at its resonance frequency ω_0 we obtain $Z(\omega_0) = R$. Converting Eq. (G.39) back to the quantum notation via $j \Longrightarrow -i$ and using Eq. (A.47), we find the flux noise associated with the resistor to be

$$S_{\Phi\Phi}(\omega_0) = [1 + n_{\mathrm{B}}(\beta\hbar\omega_0)]\frac{2\hbar}{\omega_0}R. \tag{G.40}$$

On resonance only the resistor contributes to the noise. There is no filtering from the LC part of the circuit. Since we can vary the LC circuit parameters to tune the resonance frequency arbitrarily, this expression is valid for any frequency ω. Recalling that the voltage is the time derivative of the flux, we see that the voltage noise spectral density of the resistor is

$$S_{VV}(\omega) = \omega^2 S_{\Phi\Phi}(\omega) = [1 + n_{\mathrm{B}}(\beta\hbar\omega)]2\hbar\omega R. \tag{G.41}$$

The resistor noise spectral density is plotted in Fig. G.4. We see that it rises linearly with frequency for large positive frequencies, asymptotically obeying

$$S_{VV}(\omega) \sim 2\hbar\omega R. \tag{G.42}$$

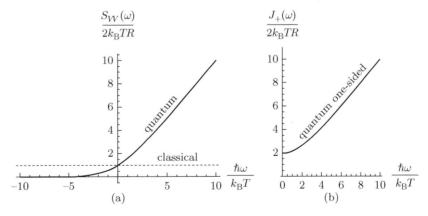

Figure G.4 (a) The two-sided spectral density $S_{VV}(\omega)$ of the voltage noise of a resistor in units of the classical (i.e. low-frequency) value $2k_{\mathrm{B}}TR$. (b) The one-sided spectral density $J_+(\omega) = S_{VV}(\omega) + S_{VV}(-\omega)$ in units of $2k_{\mathrm{B}}TR$.

The presence of Planck's constant tells us that this is quantum noise, or, in the parlance of quantum optics, "vacuum noise" associated with the zero-point energy of the circuit. The spectral density becomes exponentially small at large negative frequencies because the resistor is not able to emit high-energy quanta at low temperatures. In the classical region of low frequency relative to temperature, the spectral noise density reduces to

$$S_{VV}(\omega) = 2k_B T R, \tag{G.43}$$

in exact agreement with the result derived in Eq. (G.24) by purely classical arguments regarding a resistor in parallel with a capacitor (and with no inductor).

G.3.1 Hamiltonian for Josephson Junction Circuits

Since Φ is proportional to the current in the inductor, it is odd under time-reversal. This may make it seem more like a velocity and hence a peculiar choice for a coordinate. However, this (apparently arbitrary) choice is motivated by our desire to have a simple representation of non-linear circuits involving Josephson junctions. The energy stored in a Josephson junction is given by $-E_J \cos \varphi$, where φ is the gauge-invariant phase difference of the superconducting order parameter across the junction. However, using the second Josephson relation in Eq. (19.143), we see that the order parameter phase difference is simply a dimensionless version of the flux variable

$$\varphi = 2\pi \frac{\Phi}{\Phi_S}. \tag{G.44}$$

Each time the flux variable increases by one (superconducting) flux quantum, the phase difference winds by 2π. Thus, if we connect a Josephson junction to the input port of the black box in Fig. G.3(a), the Hamiltonian in Eq. (G.37) becomes

$$H' = H(Q, \Phi) - E_J \cos\left(2\pi \frac{\Phi}{\Phi_S}\right). \tag{G.45}$$

If we neglect the drive-current term and the inductor L, then we precisely recover the Hamiltonian for a capacitively shunted Josephson junction (Cooper-pair box/transmon qubit) originally derived in Eq. (19.168). If we retain the inductor then we have the Hamiltonian for an artificial atom (transmon qubit) coupled to a (single mode of a) microwave resonator. This is the circuit QED analog [211] of cavity quantum electrodynamics in quantum optics.

Appendix H

Functional Differentiation

In ordinary calculus, differentiation examines the effect of a small change in the argument of a function

$$\frac{df}{dx} \equiv \lim_{\epsilon \to 0} \frac{f(x + \epsilon) - f(x)}{\epsilon}. \tag{H.1}$$

Here we consider differentiation not of functions but rather of *functionals*. A functional is a mapping whose domain (argument) is a *function* and whose range (value) is a number. Thus, for example,

$$I[f] \equiv \int_{-\infty}^{+\infty} dx \, x^2 f(x) \tag{H.2}$$

is a functional whose argument is the function f. The **functional derivative** of I, written

$$\frac{\delta I}{\delta f(x_0)}, \tag{H.3}$$

is a measure of how sensitive the value of the functional is to small changes in the value of the function f at the point x_0. We make this small change by writing

$$f \to f + \epsilon g_{x_0}, \tag{H.4}$$

where now the change in the argument of the functional is not a number but rather a function whose value at the point x is

$$\epsilon g_{x_0}(x) = \epsilon \delta(x - x_0). \tag{H.5}$$

The functional derivative is then defined by

$$\frac{\delta I}{\delta f(x_0)} = \lim_{\epsilon \to 0} \frac{I[f + \epsilon g_{x_0}] - I[f]}{\epsilon}. \tag{H.6}$$

Applying this definition to the functional defined in Eq. (H.2), we obtain

$$\frac{\delta I}{\delta f(x_0)} = \lim_{\epsilon \to 0} \frac{1}{\epsilon} \left\{ \int_{-\infty}^{+\infty} dx \, x^2 \left\{ [f(x) + \epsilon \delta(x - x_0)] - [f(x)] \right\} \right\} = x_0^2. \tag{H.7}$$

The sensitivity of the value of the functional to changes in the value of the function f at the point x_0 is large when x_0^2 is large.

A familiar example from classical mechanics is the problem of finding the classical path q (as a function of time) that extremizes the action (which is a functional of the path)

$$S = \int_0^T dt' \, \mathcal{L}(q(t'), \dot{q}(t')). \tag{H.8}$$

The functional derivative must vanish,

$$\frac{\delta S}{\delta q(t)} = 0, \tag{H.9}$$

for all t. We can evaluate this as follows:

$$\frac{\delta S}{\delta q(t)} = \lim_{\epsilon \to 0} \frac{1}{\epsilon} \int_0^T dt' \left\{ \mathcal{L}\left(q(t') + \epsilon \delta(t' - t), \dot{q}(t') + \epsilon \dot{\delta}(t' - t)\right) - \mathcal{L}\left(q(t'), \dot{q}(t')\right) \right\}. \tag{H.10}$$

Keeping only the lowest-order terms in ϵ gives

$$\frac{\delta S}{\delta q(t)} = \lim_{\epsilon \to 0} \frac{1}{\epsilon} \int_0^T dt' \left\{ \frac{\partial \mathcal{L}}{\partial q} \epsilon \delta(t' - t) + \frac{\partial \mathcal{L}}{\partial \dot{q}} \epsilon \dot{\delta}(t' - t) \right\}. \tag{H.11}$$

Upon integrating the second term by parts, we recover the familiar Euler–Lagrange equation of motion,

$$\frac{\delta S}{\delta q(t)} = \frac{\partial \mathcal{L}}{\partial q(t)} - \frac{d}{dt} \frac{\partial \mathcal{L}}{\partial \dot{q}(t)} = 0. \tag{H.12}$$

Appendix I
Low-Energy Effective Hamiltonians

Many-body systems generally have Hamiltonian eigenstates spread over a wide range of energies. It is often the case that we are only interested low-energy states. In such cases it is useful to derive effective low-energy Hamiltonians by applying a unitary transformation that perturbatively eliminates terms that virtually excite particles into high-lying states. Standard examples include the effective spin Hamiltonian of the half-filled Hubbard model studied in Section 17.4, the Kondo problem studied in Section 15.14.3, and the BCS pairing instability studied in Chapter 20. Another example is Josephson tunneling of Cooper pairs studied in Section 19.9, where the first electron that tunnels has to break away from its pairing partner leading to a high-energy intermediate state. The tunneling of the second electron and the re-pairing on the other side of the junction leads to a low-energy final state.

I.1 Effective Tunneling Hamiltonian

As a simple example, consider fermions hopping on a lattice with three sites labeled 0,1 and 2 and let the middle site have a very large site energy $\Delta > 0$ as shown in Fig. I.1:

$$H = H_0 + V_+ + V_-, \tag{I.1}$$

$$H_0 = \Delta c_1^\dagger c_1, \tag{I.2}$$

$$V_+ = -t c_1^\dagger (c_0 + c_2), \tag{I.3}$$

$$V_- = V_+^\dagger = -t(c_0^\dagger + c_2^\dagger) c_1, \tag{I.4}$$

where $c^\dagger$ and c are fermion creation and annihilation operators.

If the near-neighbor hopping strength t were zero, the Hamiltonian would have spectrum $\{0, 0, \Delta\}$. In the limit of weak hopping $t \ll \Delta$, we expect that the low-energy states will have very little amplitude to have a particle occupying site 1. It is therefore useful to think about performing a unitary transformation that removes (to lowest order in perturbation theory) the terms in the Hamiltonian that connect the low- and high-energy sectors of H_0. This will decouple the two sectors and leave us with a low-energy Hamiltonian in which the site energies for sites 0 and 2 are slightly renormalized and there is an effective hopping term that directly connects sites 0 and 2. This effective hopping represents "tunneling" through the barrier represented by the middle site. To achieve this, consider the following unitary transformation:

$$W = e^{\lambda[V_+ - V_-]}, \tag{I.5}$$

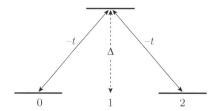

Figure I.1 Three-site tight-binding lattice with near-neighbor hopping strength $-t$ and the middle site having a large on-site energy Δ. A unitary transformation to eliminate coupling to the middle site yields an effective low-energy Hamiltonian for sites 0 and 2.

where λ is a (real) parameter to be determined. We note that V_+ and V_- are adjoints of each other, so W is indeed unitary. We also note that V_+ takes the system from the low-energy sector of H_0 to the high-energy sector and V_- does the reverse. Expanding W in powers of λ and recognizing that $V_+^2 = V_-^2 = 0$ and that $V_+ H_0 = H_0 V_- = 0$ gives

$$W H_0 W^\dagger = H_0 - \lambda \Delta [V_+ + V_-] - \lambda^2 \Delta [V_+ V_- - V_- V_+] + \cdots \tag{I.6}$$

and

$$W[V_+ + V_-]W^\dagger = [V_+ + V_-] + 2\lambda [V_+ V_- - V_- V_+] + \cdots . \tag{I.7}$$

Choosing $\lambda = 1/\Delta$ allows us to eliminate the terms which are linear in t that connect the high- and low-energy sectors, leaving (to second order in t)

$$
\begin{aligned}
\tilde{H} \equiv W H W^\dagger &= \left[H_0 + \frac{1}{\Delta}[V_+ V_- - V_- V_+] + \cdots \right] \\
&= \Delta c_1^\dagger c_1 + \frac{t^2}{\Delta}[c_1^\dagger c_1 (c_0 + c_2)(c_0^\dagger + c_2^\dagger) - c_1 c_1^\dagger (c_0^\dagger + c_2^\dagger)(c_0 + c_2)] \\
&= \tilde{H}_+ + \tilde{H}_-,
\end{aligned}
\tag{I.8}
$$

where

$$
\begin{aligned}
\tilde{H}_+ &= c_1^\dagger c_1 \left\{ \Delta + \frac{t^2}{\Delta}[(c_0 + c_2)(c_0^\dagger + c_2^\dagger) + (c_0^\dagger + c_2^\dagger)(c_0 + c_2)] \right\} \\
&= c_1^\dagger c_1 \left[\Delta + 2\frac{t^2}{\Delta} \right],
\end{aligned}
\tag{I.9}
$$

$$\tilde{H}_- = -\frac{t^2}{\Delta}[(c_0^\dagger + c_2^\dagger)(c_0 + c_2)]. \tag{I.10}$$

Notice that the transformed Hamiltonian $\tilde{H}$ commutes with $c_1^\dagger c_1$. The case $c_1^\dagger c_1 = 1$ corresponds to the high-energy sector and the case $c_1^\dagger c_1 = 0$ corresponds to the low-energy sector. Restricting our attention to the low-energy sector yields the effective Hamiltonian

$$\tilde{H}_- = -\frac{t^2}{\Delta} \left\{ [c_0^\dagger c_0 + c_2^\dagger c_2] + [c_2^\dagger c_0 + c_0^\dagger c_2] \right\}. \tag{I.11}$$

The first term in square brackets in Eq. (I.11) renormalizes the energies for sites 0 and 2 downward, which results from level repulsion associated with virtual processes in which a particle on one of those sites makes a virtual hop onto site 1 and then returns to its starting point. The second term in the equation corresponds to virtual processes in which a particle on site 0 virtually hops on to site 1 and then continues to site 2 (or vice versa), yielding an effective tunneling between the two sites on either side of the barrier.

While we arrived at the low-energy effective Hamiltonian (I.11) by considering fermions, the result actually applies to a bosonic system as well. The reader is encouraged to verify this using the original Hamiltonian (I.1), but with *bosonic* operators instead. The reason why this also works for bosons is that this is a non-interacting Hamiltonian, and the procedures followed above simply lead to an appropriate single-particle basis in which the high-energy site decouples from the low-energy ones. As we will see in the next example, the situation is very different in the presence of interactions, where the exchange statistics of the particles makes a huge difference.

I.2 Antiferromagnetism in the Hubbard Model

As a second example, let us consider a two-site version of the Hubbard model that was introduced in Eq. (17.37):

$$H = T + V, \tag{I.12}$$

$$T = -t \sum_{\sigma} (c_{1\sigma}^{\dagger} c_{2\sigma} + c_{2\sigma}^{\dagger} c_{1\sigma}), \tag{I.13}$$

$$V = +U \sum_{j=1}^{2} n_{j\uparrow} n_{j\downarrow}, \tag{I.14}$$

where $\sigma = \{\uparrow, \downarrow\}$ and $U \gg |t|$ represents strong on-site repulsion between the fermions. We would like to find a unitary transformation which is perturbative in the hopping t and removes the transitions that take the system from the low-energy sector (where no site is doubly occupied) to the high-energy sector (where one of the sites is doubly occupied). As discussed in Section 17.4, the resulting effective low-energy Hamiltonian will represent magnetism induced in the system not by magnetic interactions, but rather by a combination of Hubbard interactions and fermionic exchange statistics.

Let us divide the hopping term into three parts,

$$T = T_0 + T_+ + T_-, \tag{I.15}$$

where T_m corresponds to hops which change the number of doubly occupied sites by m. We would like to find a unitary transformation which eliminates T_+ and T_-. Noting that $T_+^{\dagger} = T_-$ and $T_0^{\dagger} = T_0$, and working by analogy with the previous example, we will use the unitary transformation

$$W = e^S; \quad S \equiv \frac{1}{U}(T_+ - T_-). \tag{I.16}$$

The transformed Hamiltonian is given by the expansion

$$\tilde{H} = WHW^{\dagger} = H + [S, H] + \frac{1}{2!}[S, [S, H]] + \cdots. \tag{I.17}$$

Using the simple commutation relation

$$[V, T_m] = mUT_m, \tag{I.18}$$

(where $m = +, -, 0$) we find that, to leading order in $1/U$, the transformed Hamiltonian is

$$\tilde{H} = WHW^{\dagger}$$
$$= V + T_0 + \frac{1}{U}\{[T_+, T_-] + [T_0, T_-] + [T_+, T_0]\} + \cdots. \tag{I.19}$$

Just as before, we have now removed (to low order in perturbation theory) terms that cause transitions between the low- and high-energy sectors. The number of doubly occupied sites is now a good quantum number ($[\tilde{H}, V] = 0$). For the special case of half-filling (two particles in total on two sites), there are no allowed hops that preserve the number of doubly occupied sites. Therefore T_0 annihilates every state and we can drop it. Further recognizing that in the low-energy sector there are no doubly occupied sites, T_- and V cannot act. We are therefore left with a very simple result for the low-energy sector:

$$\tilde{H} = -\frac{1}{U} T_- T_+. \tag{I.20}$$

In the original basis, this can be viewed as representing the effect of virtual transitions into and immediately out of the high-energy sector. In the new basis this is simply a term describing the low-energy dynamics of particle exchange. Since there are no charge fluctuations allowed in the new basis, only low-energy spin fluctuations remain. Let us therefore see how this can be expressed as a purely spin Hamiltonian.

If the two electrons have the same spin orientation, the Pauli principle forbids hopping and the Hamiltonian vanishes:

$$\tilde{H}|\uparrow\uparrow\rangle = \tilde{H}|\downarrow\downarrow\rangle = 0. \tag{I.21}$$

If, on the other hand, the two electrons have opposite spin, then the above Hamiltonian has the following matrix elements:

$$\langle\uparrow\downarrow|\tilde{H}|\uparrow\downarrow\rangle = \langle\downarrow\uparrow|\tilde{H}|\downarrow\uparrow\rangle = +2\frac{t^2}{U}, \tag{I.22}$$

$$\langle\uparrow\downarrow|\tilde{H}|\downarrow\uparrow\rangle = \langle\downarrow\uparrow|\tilde{H}|\uparrow\downarrow\rangle = -2\frac{t^2}{U}. \tag{I.23}$$

The factor of 2 arises because the virtual hopping can be from sites $1 \to 2 \to 1$ or $2 \to 1 \to 2$. That is, the intermediate doubly occupied state can be on either site 1 or site 2. The different signs result from the anti-commutation properties of the fermionic creation and annihilation operators and require some care to get right. In particular, it is essential to have a carefully defined convention for defining the states. One convention would be to always have the creation operators ordered with the site 1 operator to the left of the site 2 operator:

$$|\uparrow\downarrow\rangle = c_{1\uparrow}^\dagger c_{2\downarrow}^\dagger |0\rangle, \tag{I.24}$$

$$\langle\uparrow\downarrow| = \langle 0|c_{2\downarrow} c_{1\uparrow}. \tag{I.25}$$

Choosing the other order for the creation operators introduces a minus sign in the definition of the state. This is a perfectly acceptable "gauge choice" but must be used consistently throughout.

Equations (I.21)–(I.23) define matrix elements which correspond to the antiferromagnetic Heisenberg Hamiltonian

$$\tilde{H} = +2\frac{t^2}{U}\left\{\sigma_1^+\sigma_2^- + \sigma_1^-\sigma_2^+ + \frac{1}{2}(\sigma_1^z\sigma_2^z - 1)\right\} \tag{I.26}$$

$$= \frac{t^2}{U}\{\vec{\sigma}_1 \cdot \vec{\sigma}_2 - 1\} = J\vec{S}_1 \cdot \vec{S}_2 - C, \tag{I.27}$$

where $J = 4t^2/\hbar^2 U$ and $C = -t^2/U$, in agreement with Eq. (17.58). To arrive at the second line we have made use of the definition $\sigma^\pm = \frac{1}{2}(\sigma^x \pm i\sigma^y)$.

Exercise I.1. Repeat the analysis of this section for two spin-1/2 bosons, and show that the resultant spin exchange is ferromagnetic. Find the exchange coupling. Note that two bosons are allowed to occupy the same site even when they have the same spin, and the energy cost for such double occupancy is also U.

I.3 Summary

The above examples illustrate the general principle underlying the derivation of effective low-energy Hamiltonians based on elimination of virtual transitions into and out of high-energy states. This technique finds wide application in renormalization-group calculations. We lower a high-energy cutoff by performing a unitary transformation which (perturbatively) eliminates virtual excitations into states that lie between the old and new energy cutoff. This will lead to renormalization of the remaining coupling terms in the Hamiltonian. Whether the coupling constants increase or decrease as the cutoff is lowered tells us a great deal about the physics. If the coupling constants increase, the perturbative calculation will eventually break down as the cutoff is lowered, but we learn a great deal about the physics by knowing which of the effective couplings is most important.[1] This is the essence, for example, of the BCS effective Hamiltonian discussed in Chapter 20. This Hamiltonian focuses solely on pairing of electrons in time-reversed states (opposite spin and opposite momenta). This is because it is precisely these terms in the full Hamiltonian which are (marginally) "relevant" because they grow as the cutoff is lowered (provided that the original interaction is attractive).

---

[1] Very often in renormalization-group calculations it is necessary to perform an energy re-scaling such that the lowered cutoff is restored to the same value as before the higher-energy modes were eliminated. In the examples we encounter in this book, namely the Cooper/BCS and Kondo problems, it makes no difference to flow equations whether this procedure is performed or not, because it is always the dimensionless product of the coupling constant and density of states at the Fermi level that enters physical quantities. On the other hand, there are many situations in which the coupling constants need to be combined with the cutoff to form dimensionless combinations that enter physical quantities. In these cases it is the flow of these combinations that determines the importance of these couplings at low energies, and the most convenient way to obtain the correct flow equation is to perform the energy re-scaling mentioned above.

Appendix J

Introduction to Second Quantization

We saw in Section 15.2 that doing calculations with wave functions of many identical particles is quite cumbersome, even for a single Slater determinant, which is the simplest possible many-fermion wave function! The reason is, of course, that the wave function $\Psi(\vec{r}_1, \ldots, \vec{r}_N)$ depends on N variables (we suppress spin here for simplicity), where N is the particle number that is typically very large. The formalism of second quantization, which we introduce in this appendix, simplifies the bookkeeping by eliminating reference to the (first-quantized) wave function Ψ. In fact, it is not even necessary to specify the particle number N in second quantization, while in first quantization one does not even have a well-defined Hilbert space without N. This is in fact the most fundamental advantage of second quantization over first quantization, namely it allows description of states in which N is *not* fixed, and physical processes in which N changes. We note that second quantization is introduced in any textbook on many-body theory (see, e.g., [249]), and some textbooks on quantum mechanics (see, e.g., [250]), with varying degrees of mathematical rigor. Our exposition here will be heuristic, presented in the context of discussions of the main text. We will focus mainly on fermions, and make a comparison with the bosonic case occasionally.

J.1 Second Quantization

As usual, the first step is setting up an orthonormal set of basis states for the (entire) Hilbert space. In second quantization the entire Hilbert space includes states with all possible number of particles, and is often called the Fock space. It can be constructed in the following manner. We start from an orthonormal set of basis states for the single-particle Hilbert space, $\{|r\rangle\}$, where r is an index assumed to be discrete for simplicity (we will briefly consider the case of basis with continuous label later). An orthonormal set of basis states for the Fock space can be constructed by filling these single-particle states with either zero or one fermion, and we label them by the occupation numbers $\{n_r\}$. Obviously the total particle number is

$$N = \sum_r n_r. \tag{J.1}$$

The corresponding state $|\{n_r\}\rangle$ is the single Slater determinant introduced in Section 15.2 formed by those occupied single-particle states (or those with $n_r = 1$), with a sign convention to be specified shortly.

Among all these basis states there is a very special one for which all the occupation numbers are zero, which implies that $N = 0$ and thus there are *no* particles. We call this the **vacuum state**, and label it $|0\rangle$. The most crucial step in formulating second quantization is the introduction of a set of creation operators $\{c_r^\dagger\}$, that relates $|\{n_r\}\rangle$ to $|0\rangle$:

$$|\{n_r\}\rangle = \prod_r (c_r^\dagger)^{n_r} |0\rangle. \tag{J.2}$$

The equation above uniquely determines the algebraic properties of $\{c_r^\dagger\}$ and their Hermitian conjugates $\{c_r\}$, and sets a sign convention for $|\{n_r\}\rangle$. First of all we must have *all* creation operators anti-commuting with each other:

$$\{c_r^\dagger, c_s^\dagger\} = c_r^\dagger c_s^\dagger + c_s^\dagger c_r^\dagger = 0 = \{c_r, c_s\}. \tag{J.3}$$

The last equation follows from taking the Hermitian conjugate of the one before that. The special case of $r = s$ implies

$$(c_r^\dagger)^2 = 0 = (c_r)^2, \tag{J.4}$$

as required by the Pauli exclusion principle stating that there cannot be more than one fermion in the same single-particle state. For $r \neq s$ we also have $\{c_r^\dagger, c_s^\dagger\} = 0$ because changing the order of r and s in Eq. (J.2) is equivalent to switching rows of r and s in the Slater determinant (15.14), which results in an additional minus sign.

To complete the algebraic relations for the creation operators and their Hermitian conjugates (called annihilation operators), we also need the anti-commutators for c_r and $c_s^\dagger$:

$$\{c_r, c_s^\dagger\} = \delta_{rs}, \tag{J.5}$$

which follows from the orthonormal condition for $|\{n_r\}\rangle$:

$$\langle\{n_r'\}|\{n_r\}\rangle = \prod_r \delta_{n_r', n_r}, \tag{J.6}$$

in combination with Eq. (J.4).[1] The anti-commutator (J.5) also implies that the occupation number for state r takes the form

$$\hat{n}_r = c_r^\dagger c_r, \tag{J.7}$$

which is an *operator* whose possible eigenvalues are 0 or 1. It is easy to verify that

$$[\hat{n}_r, c_r] = -c_r, \qquad [\hat{n}_r, c_r^\dagger] = c_r^\dagger, \tag{J.8}$$

which are the same as the familiar commutation relations between the creation/annihilation operator and the number operator in a harmonic oscillator, thus the same names are used. In particular,

$$c_r|0\rangle = 0. \tag{J.9}$$

Having set up the orthonormal basis set for the Fock space, the next order of business is finding the representation of operators in second quantization. We will state without proof the following results, and provide comments and consistency checks afterwards.

- One-body operators:

$$\hat{O}^{(1)} = \sum_{j=1}^{N} \hat{o}^{(1)}(\vec{r}_j) \Rightarrow \sum_{rs} \langle r|\hat{o}^{(1)}|s\rangle c_r^\dagger c_s. \tag{J.10}$$

- Two-body operators:

---

[1] Consider the case where there is only one single-particle state. Then the Fock space is two-dimensional. It is a good exercise to check explicitly in this case that Eq. (J.5) follows from earlier equations.

$$\hat{O}^{(2)} = \sum_{j<k}^{N} \hat{o}^{(2)}(\vec{r}_j, \vec{r}_k) \Rightarrow \frac{1}{2} \sum_{r,s,r',s'} \langle r's'|\hat{o}^{(2)}|rs\rangle c_{r'}^{\dagger} c_{s'}^{\dagger} c_s c_r. \tag{J.11}$$

Generalizations to multi-particle operators are straightforward.

A few comments are now in order. (i) In second quantization, *all* operators are expressed as combinations of creation and annihilation operators. First-quantization operators (like $\hat{o}^{(1)}(\vec{r})$) are traded for their matrix elements in a specific basis. (ii) In first quantization summation is over particles, while in second quantization summation is over indices of single-particle basis states, and the particle index disappears. This has to be the case because in principle there is no need to index a particle as all particles are *identical*! Removing the particle label is the origin of the simplification brought by second quantization. (iii) In second quantization every index of an annihilation operator shows up as a ket index in the matrix element, and every index of a creation operator shows up as a bra index in the matrix element. (iv) Note that in Eq. (J.11) the order of ket indices is *opposite* to that of annihilation operator indices.

We can check the validity of Eqs. (J.10) and (J.11) for some special cases.

For the case $N = 1$, we are back to the familiar single-particle quantum mechanics. We thus have

$$\hat{O}^{(1)} = \hat{o}^{(1)}(\vec{r}) = \sum_{rs} |r\rangle\langle r|\hat{o}^{(1)}|s\rangle\langle s| = \sum_{rs} \langle r|\hat{o}^{(1)}|s\rangle |r\rangle\langle s|, \tag{J.12}$$

where we used the resolution of identity $\sum_r |r\rangle\langle r| = 1$. It should be obvious that the above is consistent with Eq. (J.10), as the operator $|r\rangle\langle s|$ takes the particle originally in state $|s\rangle$ to state $|r\rangle$, and thus is equivalent to $c_r^{\dagger} c_s$. This manipulation also reveals the origin of point (iii) above.

We now perform a less trivial check by calculating the expectation value of the two-body interaction term for a single Slater determinant of the form (J.2), and compare it with the first-quantization calculation performed in Section 15.2. We have in second quantization

$$\langle\Psi|\hat{U}|\Psi\rangle = \frac{1}{2} \sum_{rsr's'} \langle r's'|u|rs\rangle \langle\Psi|c_{r'}^{\dagger} c_{s'}^{\dagger} c_s c_r|\Psi\rangle. \tag{J.13}$$

We now make a few simple observations. (i) In order for $\langle\Psi|c_{r'}^{\dagger} c_{s'}^{\dagger} c_s c_r|\Psi\rangle$ to be non-zero, r and s must both be occupied states, and (ii) either $r' = r$, $s' = s$; or $r' = s$, $s' = r$. (iii) There is a relative minus sign for the two cases in (ii), due to the anti-commutation relation (J.3). With these observations it is easy to see that

$$\langle\Psi|\hat{U}|\Psi\rangle = \frac{1}{2} \sum_{rsr's'} \langle r's'|u|rs\rangle \left(\langle c_{r'}^{\dagger} c_r\rangle\langle c_{s'}^{\dagger} c_s\rangle - \langle c_{r'}^{\dagger} c_s\rangle\langle c_{s'}^{\dagger} c_r\rangle \right) \tag{J.14}$$

$$= \frac{1}{2} \sum_{rs} n_r n_s (\langle rs|u|rs\rangle - \langle rs|u|sr\rangle), \tag{J.15}$$

which is equivalent to Eq. (15.20). We note that having the annihilation operators in Eq. (J.11) properly ordered is crucial to obtain the correct result.

It is sometimes useful to consider a single-particle basis with a continuous label. The most widely used is that of position basis $\{|\vec{r}\rangle\}$, with normalization

$$\langle\vec{r}\,'|\vec{r}\rangle = \delta(\vec{r}\,' - \vec{r}). \tag{J.16}$$

The creation/annihilation operators in the position basis are also known as quantum field operators:

$$\Psi(\vec{r}) = \sum_s \psi_s(\vec{r}) c_s, \tag{J.17}$$

where $\psi_s(\vec{r}) = \langle \vec{r} | s \rangle$ is the wave function of $|s\rangle$ in the basis $\{|\vec{r}\rangle\}$. It is a good exercise to show that

$$\{\Psi(\vec{r}), \Psi^\dagger(\vec{r}\,')\} = \delta(\vec{r}\,' - \vec{r}). \qquad (\text{J.18})$$

Much of what we have discussed thus far carries over to the second-quantization description of many identical boson systems, with relatively little modification. For example we can introduce a set of boson creation operators $\{b_s^\dagger\}$ for each single-particle state, and construct a complete basis set for the Fock space in a way similar to Eq. (J.2):

$$|\{n_r\}\rangle = \prod_r \frac{1}{\sqrt{n_r!}} (b_r^\dagger)^{n_r} |0\rangle. \qquad (\text{J.19})$$

The two main differences are as follows: (i) n_r is no longer restricted to 0 or 1; and (ii) additional normalization factors are now needed. Also the anti-commutation relations among creation/annihilation operators for fermions need to be replaced by commutation relations that are identical to those of harmonic oscillators:

$$[b_r, b_s] = [b_r^\dagger, b_s^\dagger] = 0; \qquad [b_r, b_s^\dagger] = \delta_{rs}. \qquad (\text{J.20})$$

The way to construct operators in second quantization turns out to be identical to that for fermions.

We now discuss a subtle but profound difference between bosons and fermions. It lies in an innocent-looking mathematical difference between a commutator and an anti-commutator, namely that, if one switches the order between the operators, the commutator changes sign but the anti-commutator is *invariant*:

$$[A, B] = -[B, A]; \qquad \{A, B\} = \{B, A\}. \qquad (\text{J.21})$$

This means that one cannot tell which one is the creation operator and which one is the annihilation operator by inspecting the anti-commutator (J.5)! This allows for the possibility of particle–hole transformation for fermions, in which one exchanges the roles of creation and annihilation operators $c_r \Leftrightarrow c_r^\dagger$ for some of the single-particle states. This also leads to a transformation of the vacuum, if we use Eq. (J.9) as its definition. A well-known example of this is the Dirac theory of electrons, in which one defines vacuum to be the state in which all negative-energy states are already filled by electrons. This is the fundamental reason why spin-1/2 particles described by Dirac theory must be fermions, since such a transformation would not be possible for bosons!

J.2 Majorana Representation of Fermion Operators

We end this appendix with a very useful representation of fermion operators, in terms of **Majorana fermion** operators (see the review by Beenakker [223] for a discussion of Majorana fermions in condensed matter physics). To motivate this representation, let us recall how the creation and annihilation operators for bosons were first introduced in the study of a single harmonic oscillator. Up to numerical factors, we have $b = x + ip$ and $b^\dagger = x - ip$, namely the Hermitian operators x and p are the real and imaginary parts of the non-Hermitian operator b.[2]

We can perform a similar decomposition for fermion operators by writing

$$c^\dagger = \gamma_1 + i\gamma_2, \qquad c = \gamma_1 - i\gamma_2, \qquad (\text{J.22})$$

with

$$\gamma_1 = (c^\dagger + c)/2, \qquad \gamma_2 = (c^\dagger - c)/(2i) \qquad (\text{J.23})$$

---

[2] Remember that any operator O can be written as $O = O_R + iO_I$, with $O_R = (O + O^\dagger)/2 = O_R^\dagger$ and $O_I = (O - O^\dagger)/(2i) = O_I^\dagger$, which is the operator version of decomposing a complex number into real and imaginary parts.

being Hermitian operators playing roles similar to x and p in the construction of boson or harmonic oscillator creation/annihilation oscillators. These are known as Majorana or real (referring to the fact they are Hermitian) fermion operators. The anti-commutation relations (J.3) and (J.5) lead to those for the Majorana fermion operators:

$$\{\gamma_j, \gamma_k\} = \delta_{jk}/2. \tag{J.24}$$

For the simplest possible case in which there is only a single fermion mode with corresponding annihilation operator c, the Hilbert space is two-dimensional, and the only possible form of a Hamiltonian is $H = \epsilon \hat{n} = \epsilon c^\dagger c$, which, using (J.22), becomes

$$H = -2i\epsilon \gamma_1 \gamma_2. \tag{J.25}$$

Terms linear in γ (or c and $c^\dagger$) are not allowed in H even if they are Hermitian, because they change the parity of the fermion number and thus violate a fundamental conservation law of nature.

In Section 20.7, we discuss an interesting use of the Majorana representation in the Kitaev chain model.

References

[1] N. W. Ashcroft and N. D. Mermin (1976). *Solid State Physics* (Holt, Rinehart and Winston, New York).

[2] P. M. Chaikin and T. C. Lubensky (1995). *Principles of Condensed Matter Physics* (Cambridge University Press, New York).

[3] M. P. Marder (2010). *Condensed Matter Physics*, 2nd edn (Wiley, Hoboken).

[4] M. L. Cohen and S. G. Louie (2016). *Fundamentals of Condensed Matter Physics* (Cambridge University Press, New York).

[5] G. D. Mahan (2010). *Condensed Matter in a Nutshell* (Princeton University Press, Princeton).

[6] C. Kittel (2004). *Introduction to Solid State Physics,* 8th edn (Wiley, Hoboken).

[7] A. Altland and B. D. Simons (2010). *Condensed Matter Field Theory*, 2nd edn (Cambridge University Press, New York).

[8] A. Auerbach (1998). *Interacting Electrons and Quantum Magnetism* (Springer, Berlin).

[9] P. Coleman (2015). *Introduction to Many-Body Physics* (Cambridge University Press, New York).

[10] E. Fradkin (2013). *Field Theory in Condensed Matter Physics*, 2nd edn (Cambridge University Press, New York).

[11] N. Nagaosa (1999). *Quantum Field Theory in Condensed Matter Physics* (Springer, Berlin).

[12] S. Sachdev (2011). *Quantum Phase Transitions*, 2nd edn (Cambridge University Press, New York).

[13] R. Shankar (2017). *Quantum Field Theory and Condensed Matter: An Introduction* (Cambridge University Press, New York).

[14] A. M. Tsvelik (2003). *Quantum Field Theory in Condensed Matter Physics*, 2nd edn (Cambridge University Press, New York).

[15] X.-G. Wen (2003). *Quantum Field Theory of Many-Body Systems: From the Origin of Sound to an Origin of Light and Electrons* (Oxford University Press, Oxford).

[16] T. H. Hansson, V. Oganesyan, and S. L. Sondhi (2004). *Ann. Phys.* **313**, 497.

[17] A. Guinier (1963). *X-ray Diffraction* (W. H. Freeman, San Francisco).

[18] J. D. Jackson (1998). *Classical Electrodynamics,* 3rd edn (Wiley, New York).

[19] A. Zangwill (2013). *Modern Electrodynamics* (Cambridge University Press, Cambridge).

[20] L. I. Schiff (1968). *Quantum Mechanics,* 3rd edn (McGraw-Hill, New York).

[21] E. Merzbacher (1998). *Quantum Mechanics,* 3rd edn (John Wiley and Sons, New York).

[22] D. Shechtman, I. Blech, D. Gratias, and J. W. Cahn (1984). *Phys. Rev. Lett.* **53**, 1951.

[23] Y. L. Wang, H. J. Gao, H. M. Guo, H. W. Liu, I. G. Batyrev, W. E. McMahon, and S. B. Zhang (2004). *Phys. Rev.* **B70**, 073312.

[24] S. W. Lovesey (1984). *The Theory of Neutron Scattering from Condensed Matter* (Clarendon, Oxford).

[25] A. A. Aczel, G. E. Granroth, G. J. MacDougall, W. J. L. Buyers, D. L. Abernathy, G. D. Samolyuk, G. M. Sotcks, and S. E. Nagler (2012). Springer Nature: *Nature Commun.*, **3**, 1124.

[26] A. Clerk M. H. Devoret, S. M. Girvin, F. Marquardt, and R. J. Schoelkopf (2010). *Rev. Mod. Phys.* **82**, 1155.

[27] B. Fåk, T. Keller, M. E. Zhitomirsky, and A. L. Chernyshev (2012). *Phys. Rev. Lett.* **109**, 155305.

[28] K. Huang (1987). *Statistical Mechanics*, 2nd edn (John Wiley and Sons, New York).

[29] R. P. Feynman (1954). *Phys. Rev.* **94**, 262.

[30] W. S. Corak, M. P. Garfunkel, C. B. Satterthwaite, and A. Wexler (1955). *Phys. Rev.* **98**, 1699.

[31] R. V. Pound and G. A. Rebka (1960). *Phys. Rev. Lett.* **4**, 337.

[32] A. Zee (2016). *Group Theory in a Nutshell for Physicists* (Princeton University Press, Princeton).

[33] V. L. Moruzzi, A. R. Williams, and J. F. Janak (1977). *Phys. Rev.* **B15**, 2854.

[34] T. -C. Chung, F. Moraes, J. D. Flodd, and A. J. Heeger (1984). *Phys. Rev.* **B29**, 2341(R).

[35] M. Nakahara (2003). *Geometry, Topology and Physics* (IOP Publishing, Bristol).

[36] J. J. Sakurai and J. J. Napolitano (2010). *Modern Quantum Mechanics*, 2nd edn (Pearson India, Delhi).

[37] J. D. Joannopoulos, S. G. Johnson, J. N Winn, and R. D. Meade (2008). *Photonic Crystals: Molding the Flow of Light*, 2nd edn (Princeton University Press, Princeton).

[38] S. Assefa, G. S. Petrich, L. A. Kolodziejski, M. K. Mondoi, and H. I. Smith (2004). *Journal of Vacuum Science & Technology B: Microelectronics and Nanometer Structures Processing, Measurement, and Phenomena* **22**, 3363; doi: 10.1116/1.1821573.

[39] I. Bloch (2008). Springer Nature: *Nature*, **453**, 1016.

[40] J. Feldmann, K. Leo, J. Shah, D. A. B. Miller, J. E. Cunningham, T. Meier, G. von Plessen, A. Schulze, P. Thomas, and S. Schmitt-Rink (1992). *Phys. Rev.* **B46**, 7252.

[41] M. B. Dahan, E. Peik, J. Reichel, Y. Y. Castin, and C. Salomon (1996). *Phys. Rev. Lett.* **76**, 053903.

[42] H. Trompeter, W. Krolikowski, D. N. Neshev, A. S. Desyatnikov, A. A. Sukhorukov, Y. S. Kivshar, T. Pertsch, V. Peschel, and F. Leaderer (2006). *Phys. Rev. Lett.* **96**, 053903.

[43] C. Nash and S. Sen (1983). *Topology and Geometry for Physicists* (Academic Press, New York).

[44] N. D. Mermin (1979). *Rev. Mod. Phys.* **51**, 591.

[45] B. Vignolle, A. Carrington, R. A. Cooper, M. M. J. French, A. Mackenzie, C. Jaudet, D. Vignolles, C. Proust, and N. E. Hussey (2008). Springer Nature: *Nature*, **455**, 952.

[46] D. Schoenberg (1984). *Magnetic Oscillations in Metals* (Cambridge University Press, New York).

[47] Z. Wang, M. P. A. Fisher, S. M, Girvin, and J. T. Chalker (2000). *Phys. Rev.* **B61**, 8326.

[48] M. L. Cohen and J. R. Chelikowsky (1989). *Electronic Structure and Optical Properties of Semiconductors* (Springer, New York).

[49] R. N. Dexter, H. J. Zeiger, and B. Lax (1956). *Phys. Rev.* **104**, 637.

[50] D. N. Basov, M. M. Fogler, A. Lanzara, F. Wang, and Y. Zhang (2014). *Rev. Mod. Phys.* **86**, 959.

[51] S. M. Sze and K. K. Ng (2006). *Physics of Semiconductor Devices*, 3rd edn (Wiley, New York).

[52] H. van Houten and C. Beenakker (1996). *Physics Today* **49** (7), 22; doi: 10.1063/1.881503.

[53] M. A. Reed (1993). "Quantum Dots", *Scientific American* **268** (1), 118.

[54] R. A. Webb, S. Washburn, C. P. Umbach, and R. B. Laibowitz (1985). *Phys. Rev. Lett.* **54**, 2696.

[55] B. J. van Wees, L. P. Kouwenhoven, E. M. M. Williems, C. J. P. M. Harmans, J. E. Mooij, H. van Houten, C. W. J. Beenakker, J. G. Williamson, and C. T. Foxon (1991). *Phys. Rev.* **B43**, 12431.

[56] Y. Imry (2002). *Introduction to Mesoscopic Physics*, 2nd edn (Oxford University Press, New York).

[57] S. Feng, C. Kane, P. A. Lee, and A. D. Stone (1988). *Phys. Rev. Lett.* **61**, 834.

[58] C. W. J. Beenakker (1997). *Rev. Mod. Phys.* **69**, 731.

[59] Ya. M. Blanter and M. Büttiker (2000). *Phys. Rep.* **336**, 1–166.

[60] S. Feng, P. A. Lee, and A. D. Stone (1986). *Phys. Rev. Lett.* **56**, 1960.

[61] P. A. Lee, A. D. Stone, and H. Fukuyama (1987). *Phys. Rev.* **B35**, 1039.

[62] W. J. Skocpol, P. M. Mankiewich, R. E. Howard, L. D. Jackel, D. M. Tennant, and A. D. Stone (1986). *Phys. Rev. Lett.* **56**, 2865.

[63] S. B. Kaplan (1988). *Phys. Rev.* **B38**, 7558.

[64] R. de-Picciotto, M. Reznikov, M. Heiblum, V. Umansky, G. Bunin, and D. Mahalu (1997). Springer Nature: *Nature* **389**, 162.

[65] L. Saminadayar, D. C. Glattli, Y. Jin, and B. Etienne (1997). *Phys. Rev. Lett.* **79**, 2526.

[66] L. S. Levitov and G. B. Lesovik (1992). *JETP Lett.* **55**, 555.

[67] L. S. Levitov and G. B. Lesovik (1993). *JETP Lett.* **58**, 230.

[68] L. S. Levitov, H. Lee, and G. B. Lesovik (1996). *J. Math. Phys.* **37**, 4845.

[69] P. W. Anderson (1958). *Phys. Rev.* **109**, 1492.

[70] G. Bergmann (1984). *Phys. Rep.* **107**, 1–58.

[71] A. G. Aronov and Yu. V. Sharvin (1987). *Rev. Mod. Phys.* **59**, 755.

[72] H. F. Hess, L. K. DeConde, T. F. Rosenbaum, and G. A Thomas (1982). *Phys. Rev.* **B25**, 5578.

[73] A. C. Bleszynski-Jayich, W. E. Shanks, B. Peaudecerf, E. Ginossar, F. von Oppen, L. Glazman, and J. G. Harris (2009). *Science* **326**, 272.

[74] E. Abrahams, P. W. Anderson, D. C. Licciardello, and T. V. Ramakrishnan (1979). *Phys. Rev. Lett.* **42**, 673.

[75] F. J. Wegner (1976). *Z. Phys.* **25**, 327.

[76] A. L. Efros and B. I. Shklovskii (1975). *J. Phys. C: Solid State Phys.* **8**, L49.

[77] D. A. Huse, R. Nandkishore, and V. Oganesyan (2014). *Phys. Rev.* **B90**, 174202.

[78] M. Serbyn, Z. Papic, and D. A. Abanin (2013). *Phys. Rev. Lett.* **111**, 127201.

[79] H. L. Störmer (1992). *Physica B: Condensed Matter*: **177**, 1–4, pages 401–408.

[80] K. von Klitzing, G. Dorda, and M. Pepper (1980). *Phys. Rev. Lett.* **45**, 494.

[81] F. Schopfer and W. Poirier (2013). *J. Appl. Phys.* **114**, 064508.

[82] D. C. Tsui, H. L. Störmer, and A. C. Gossard (1982). *Phys. Rev. Lett.* **48**, 1559.

[83] R. B. Laughlin (1983). *Phys. Rev. Lett.* **50**, 1395.

[84] B. Huckestein (1995). *Rev. Mod. Phys.* **67**, 357.

[85] J. T. Chalker and P. D. Coddington (1988). *J. Phys. C: Solid State Phys.* **21**, 2665.

[86] S. L. Sondhi, S. M. Girvin, J. C. Carini, and D. Shahar (1997). *Rev. Mod. Phys.* **69**, 315.

[87] K. S. Novoselov, A. K. Geim, S. V. Morozov, D. Jiang, M. I. Katsnelson, I. V. Grigorieva, S. V. Dubonos, and A. A. Firsov (2005). Springer Nature: *Nature*, **438**, 197.

[88] D. J. Thouless, M. Kohmoto, M. P. Nightingale, and M. den Nijs (1982). *Phys. Rev. Lett.* **49**, 405.

[89] M. V. Berry (1984). *Proc. R. Soc. Lond.* **A392**, 45.

[90] F. Wilczek and A. Shapere Eds. (1989). *Geometric Phases in Physics* (World Scientific Publishing Co Pte Ltd., Singapore).

[91] D. Xiao, M.-C. Chang, and Q. Niu, (2010). *Rev. Mod. Phys.* **82**, 1959.

[92] D. Suter, K. T. Mueller, and A. Pines (1988). *Phys. Rev. Lett.* **60**, 1218.

[93] P. J. Leek, J. M. Fink, A. Blais, R. Bianchetti, Göppl, M., J. M. Gambetta, D. I. Schuster, L. Frunzio, R. J. Schoelkopf, and A. Wallraff (2007). *Science* **318**, 1889.

[94] N. Nagaosa, J. Sinova, S. Onoda, A. H. MacDonald, and N. P. Ong (2010). *Rev. Mod. Phys.* **82**, 1539.

[95] D. J. Thouless (1984). *J. Phys. C: Solid State Phys.* **17**, L325.

[96] R. B. Laughlin (1981). *Phys. Rev.* **B23**, 5632.

[97] B. I. Halperin (1982). *Phys. Rev.* **B25**, 2185.

[98] F. D. M. Haldane (1988). *Phys. Rev. Lett.* **61**, 2015.

[99] D. J. Thouless (1983). *Phys. Rev.* **B27**, 6083.

[100] Q. Niu (1990). *Phys. Rev. Lett.* **64**, 1812.

[101] L. J. Geerligs, V. F. Anderegg, P. A. M. Holweg, J. E. Mooij, H. Pothier, D. Esteve, C. Urbina, and M. H. Devoret (1990). *Phys. Rev. Lett.* **64**, 2691.

[102] H. Pothier, P. Lafarge, C. Urbina, D. Esteve, and M. H. Devoret (1992). *Europhys. Lett.* **17**, 249.

[103] R. L. Kautz, M. W. Keller, and J. M. Martinis (1999). *Phys. Rev.* **B60**, 8199.

[104] M. Switkes, C. M. Marcus, K. Campman, A. C. Gossard (1999). *Science* **283**, 1905.

[105] S. Nakajima, T. Tomita, S. Taie, T. Ichinose, H. Ozawa, L. Wang, M. Troyer and Y. Takahashi (2016). Springer Nature: *Nature Phys.* **12**, 296.

[106] M. Lohse, C. Schweizer, O. Zilberberg, M. Aidelsburger, and I. Bloch (2016). Springer Nature: *Nature Phys.* **12**, 350.

[107] C. L. Kane and E. J. Mele (2005). *Phys. Rev. Lett.* **95**, 226801.

[108] C. L. Kane and E. J. Mele (2005). *Phys. Rev. Lett.* **95**, 146802.

[109] B. A. Bernevig, T. L. Hughes, and S.-C. Zhang (2006). *Science* **314**, 1757.

[110] M. König, S. Weidman, C. Brüne, A. Roth, H. Bushmann, L. W. Molenkamp, X. L. Qi, and S. C. Zhang (2007). *Science* **318** (5851), 766–770, DOI: 10.1126/science.1148047

[111] L. Fu and C. L. Kane (2007). *Phys. Rev.* **B76**, 045302.

[112] J. D. Bjorken and S. D. Drell (1964). *Relativistic Quantum Mechanics* (McGraw-Hill, New York).

[113] D. Hsieh, D. Qian, L. Wray, Y. Xia, Y. S. Hor, R. J. Cava, and M. Z. Hasan (2008). Springer Nature: *Nature* **452**, 970–974.

[114] M. Z. Hasan and C. L. Kane (2010). *Rev. Mod. Phys.* **82**, 3045.

[115] C.-Z. Chang, J. Zhang, X. Feng *et al.* (2013). *Science* **340**, 167.

[116] S.-Y. Xu, I. Belopolski, N. Alidoust *et al.* (2015). *Science* **349**, 613.

[117] B. Q. Lv, H. M. Weng, B. B. Fu *et al.* (2015). *Phys. Rev.* **X5**, 031013.

[118] X. Wan, A. Turner, A. Vishwanath, and S. Y. Savrasov (2011). *Phys. Rev.* **B83**, 205101.

[119] H. Weng, C. Fang, Z. Fang, A. Bernevig, and X. Dai (2015). *Phys. Rev.* **X5**, 011029.

[120] S.-Ming Huang, S.-Y. Xu, I. Belopolski *et al.* (2015). Springer Nature: *Nature Commun.* **6** 7373.

[121] D. T. Son and B. Z. Spivak (2013). *Phys. Rev.* **B88**, 104412.

[122] A. A. Burkov (2015). *Phys. Rev.* **B91**, 245157.

[123] L. Fu (2011). *Phys. Rev. Lett.* **106**, 106802.

[124] A. Messiah (1964). *Quantum Mechanics* (Wiley, New York).

[125] P. Hohenberg and W. Kohn (1964). *Phys. Rev.* **136**, B864.

[126] W. Kohn and L. J. Sham (1965). *Phys. Rev.* **140**, A1133.

[127] T. Giamarchi (2004). *Quantum Physics in One Dimension* (Oxford University Press, New York).

[128] A. Imambekov, T. L. Schmidt, and L. I. Glazman (2012). *Rev. Mod. Phys.* **84**, 1253.

[129] G. D. Mahan (2000). *Many-Particle Physics*, 3rd edn (Kluwer Academic/Plenum Publishers, New York).

[130] J. J. Hopfield (unpublished, 1967); (1969). *Comments Solid State Phys.* **11**, 40.

[131] P. W. Anderson (1967). *Phys. Rev. Lett.* **18**, 1049.

[132] J. Kondo (1964). *Prog. Theor. Phys.* **32**, 37.

[133] P. W. Anderson (1970). *J. Phys. C: Solid State Phys.* **3**, 2436.

[134] K. G. Wilson (1975). *Rev. Mod. Phys.* **47**, 773.

[135] N. Andrei (1980). *Phys. Rev. Lett.* **45**, 379.

[136] N. Andrei, K. Furuya, and J. H. Lowenstein (1983). *Rev. Mod. Phys.* **55**, 331.

[137] P. B. Wiegman (1980). *JETP Lett.* **31**, 364.

[138] S. M. Girvin and T. Jach (1984). *Phys. Rev.* **B29**, 5617.

[139] F. D. M. Haldane (1983). *Phys. Rev. Lett.* **51**, 605.

[140] F. D. M. Haldane (2011). *Phys. Rev. Lett.* **107**, 116801.

[141] R.-Z. Qiu, F. D. M. Haldane, X. Wan, K. Yang, and S. Yi (2012). *Phys. Rev.* **B85**, 115308.

[142] R. E. Prange and S. M. Girvin, eds. (1990). *The Quantum Hall Effect*, 2nd edn (Springer-Verlag, New York).

[143] S. M. Girvin, A. H. MacDonald, and P. M. Platzman (1986). *Phys. Rev.* **B33**, 2481.

[144] D. Levesque, J. J. Weiss, and A. H. MacDonald (1984). *Phys. Rev.* **B30**, 1056.

[145] V. J. Goldman and B. Su (1995). *Science* **267** 1010.

[146] J. M. Leinaas and J. Myrheim (1977). *Nuovo Cimento* **B37**, 1–23.

[147] F. Wilczek (1982). *Phys. Rev. Lett.* **49**, 957–959.

[148] R. P. Feynman (1972). *Statistical Mechanics* (Benjamin, Reading).

[149] F. D. M. Haldane and E. H. Rezayi (1985). *Phys. Rev. Lett.* **54**, 237.

[150] G. Fano, F. Ortolani, and E. Colombo (1986). *Phys. Rev.* **B34**, 2670.

[151] S. M. Girvin (2000). "The Quantum Hall Effect: Novel Excitations and Broken Symmetries," in *Topological Aspects of Low Dimensional Systems*, ed. A. Comtet, T. Jolicoeur, S. Ouvry, and F. David. (Springer-Verlag, Berlin and Les Editions de Physique, Les Ulis).

[152] A. Pinczuk, B. S. Dennis, L. N. Pfeiffer, and K. West (1993). *Phys. Rev. Lett.* **70**, 3983.

[153] M. Kang, A. Pinczuk, B. S. Dennis, L. N. Pfeiffer, and K. W. West (2001). *Phys. Rev. Lett.* **86**, 2637.

[154] I. V. Kukushkin, J. H. Smet, V. W. Scarola, V. Umansky, and K. von Klitzing (2009). *Science* **324**, 1044.

[155] X.-G. Wen (1992). *Int. J. Mod. Phys.* **B6**, 1711.

[156] C. de C. Chamon and E. Fradkin (1997). *Phys. Rev.* **B56**, 2012.

[157] M. Grayson, D. C. Tsui, L. N. Pfeiffer, K. W. West, and A. M. Chang (1998). *Phys. Rev. Lett.* **80**, 1062.

[158] J. K. Jain (2007). *Composite Fermions* (Cambridge University Press, Cambridge).

[159] D. Yoshioka, B. I. Halperin, and P. A. Lee (1983). *Phys. Rev. Lett.* **50**, 1219.

[160] X. G. Wen and Q. Niu (1990). *Phys. Rev.* **B41**, 9377.

[161] M. Stone (1992). *Quantum Hall Effect* (World Scientific, Singapore).

[162] T. Chakraborty and P. Pietiläinen (1995). *The Quantum Hall Effects: Integral and Fractional* (Springer-Verlag, New York).

[163] S. Das Sarma and A. Pinczuk, eds. (1996). *Perspectives in Quantum Hall Effects: Novel Quantum Liquids in Low-Dimensional Semiconductor Structures* (Wiley-VCH, Weinheim).

[164] D. Yoshioka (2002). *The Quantum Hall Effect* (Springer-Verlag, New York).

[165] Z.-F. Ezawa (2013). *Quantum Hall Effects: Recent Theoretical and Experimental Developments*, 3rd edn (World Scientific, Singapore).

[166] S. M. Girvin and A. H. MacDonald, "Multi-Component Quantum Hall Systems: The Sum of their Parts and More," in Ref. [163].

[167] Y. Barlas, K. Yang, and A. H. MacDonald (2012). *Nanotechnology* **23**, 052001.

[168] B. I. Halperin (1984). *Phys. Rev. Lett.* **52**, 1583.

[169] B. I. Halperin, P. A. Lee, and N. Read (1993). *Phys. Rev.* **B47**, 7312.

[170] R. Willett, J. P. Eisenstein, H. L. Störmer, D. C. Tsui, A. C. Gossard, and J. H. English (1987). *Phys. Rev. Lett.* **59**, 1776.

[171] G. Moore and N. Read (1991). *Nucl. Phys.* **B360**, 362.

[172] N. Read and E. Rezayi (1999). *Phys. Rev.* **B59**, 8084.

[173] C. Nayak, S. H. Simon, A. Stern, M. Freedman, and S. Das Sarma (2008). *Rev. Mod. Phys.* **80**, 1083.

[174] A. Geim (1998). *Physics Today*, September, pp. 36–39.

[175] J. B. Goodenough (1955). *Phys. Rev.* **100**, 564; (1958). *J. Phys. Chem. Solids* **6**, 287.

[176] J. Kanamori (1959). *J. Phys. Chem. Solids* **10**, 87.

[177] M. Mourigal, M. Enderle, A. Klöpperpieper, J.-S. Caux, A. Stunault, and H. M. Rønnow (2013). Springer Nature: *Nature Phys.* **9**, 435.

[178] G. Xu, G. Aeppli, and M. E. Bisher (2007). *Science* **317**, 1049–1052, DOI: 10.1126/science.1143831

[179] F. D. M. Haldane (1983). *Phys. Lett.* **A93**, 464.

[180] F. D. M. Haldane (1983). *Phys. Rev. Lett.* **50**, 1153.

[181] K. Rommelse and M. den Nijs (1987). *Phys. Rev. Lett.* **59**, 2578.

[182] S. M. Girvin and D. P. Arovas (1989). *Physica Scripta* **T27**, 156.

[183] M. Kohmoto and H. Tasaki (1992). *Phys. Rev.* **B46**, 3486.

[184] R. A. Hyman, K. Yang, R. N. Bhatt, and S. M. Girvin (1996). *Phys. Rev. Lett.* **76**, 839.

[185] R. A. Hyman and K. Yang (1997). *Phys. Rev. Lett.* **78**, 1783.

[186] B. Zeng, X. Chen, D.-L. Zhou, and X.-G. Wen, Quantum information meets quantum matter, (https://arxiv.org/abs/1508.02595).

[187] S. A. Kivelson, D. S. Rokhsar, and J. P. Sethna (1987). *Phys. Rev.* **B35**, 8865.

[188] D. S. Rokhsar and S. A. Kivelson (1988). *Phys. Rev. Lett.* **61**, 2376.

[189] A. Yu. Kitaev (2003). *Ann. Phys.* **303**, 2.

[190] A. Seidel (2009). *Phys. Rev.* **B80**, 165131.

[191] A. G. Fowler, M. Mariantoni, J. M. Martinis, and A. N. Cleland (2012). *Phys. Rev.* **A86**, 032324.

[192] T. E. O'Brien, B. Tarasinski, and L. DiCarlo, Density-matrix simulation of small surface codes under current and projected experimental noise (https://arxiv.org/abs/1703.04136).

[193] N. Ofek, A. Petrenko, R. Heeres *et al.* (2016). Springer Nature: *Nature* **536**, 441–445.

[194] R. K. Pathria and P. D. Beale (2011). *Statistical Mechanics*, 3rd edn (Oxford, Academic Press).

[195] J. Leonard, A. Morales, P. Zupancic, T. Esslinger, and T. Donner (2017). Springer Nature: *Nature* **543**, 87.

[196] J.-R. Li, J. Lee, W. Huang, S. Burchesky, B. Shteynas, F. Çağrı. Top, A. O. Jamison, and W. Ketterle (2007). Springer Nature: *Nature* **543**, 91.

[197] V. Bagnato and D. Kleppner (1991). *Phys. Rev.* **A44**, 7439.

[198] E. A. Cornell and C. E. Wieman (2002). *Rev. Mod. Phys.* **74**, 875.

[199] M. R. Andrews, C. G. Townsend, H.-J. Miesner, D. S. Durfee, D. M. Kurn, and W. Ketterle (1997). *Science* **275**, 637.

[200] J. Steinhauer, R. Ozeri, N. Katz, and N. Davidson (2002). *Phys. Rev. Lett.* **88**, 120407.

[201] H. Deng, H. Haug, and Y. Yamamoto (2010). *Rev. Mod. Phys.* **82**, 1489.

[202] I. Carusotto and C. Ciuti (2013). *Rev. Mod. Phys.* **85**, 299.

[203] J. Kasprzak, M. Richard, S. Kundermann, A. Baas, and P. Jeambrun (2006). Springer Nature: *Nature* **443**, 409.

[204] J. R. Schrieffer (1988). *Theory of Superconductivity* (Addison Wesley, Reading).

[205] D. J. Scalapino S. R. White, and S. Zhang (1993). *Phys. Rev.* **B47**, 7995.

[206] K. Yang and A. H. MacDonald (2004). *Phys. Rev.* **B70**, 094512.

[207] M. Iavarone, R. Di Capua, G. Karapetrov, A. E. Koshelev, D. Rosenmann, H. Claus, C. D. Malliakas, M. G. Kanatzidis, T. Nishizaki, and N. Kobayashi (2008). *Phys. Rev.* **B78**, 174518.

[208] M. Tinkham (1996). *Introduction to Superconductivity*, 2nd edn (McGraw-Hill, New York).

[209] J. R. Abo-Shaeer, C. Raman, J. M. Vogels, and W. Ketterle (2001). *Science* **292**, 476–479, DOI: 10.1126/science.1060182.

[210] Y.-J. Lin, R. L. Compton, K. J. Garcia, J. V. Porto, and I. B. Spielman (2009). Springer Nature: *Nature* **462**, 628.

[211] S.M. Girvin (2014). "Circuit QED: Superconducting Qubits Coupled to Microwave Photons," in *Quantum Machines: Measurement and Control of Engineered Quantum Systems*: Lecture Notes of the Les Houches Summer School: Volume **96**, July 2011 (Oxford University Press, Oxford).

[212] S. M. Girvin (2014). "Basic Concepts in Quantum Information," in *Strong Light–Matter Coupling: From Atoms to Solid-State Systems*, ed. A. Auffves, D. Gerace, M. Richard, S. Portolan, M. Frana Santos, L. C. Kwek, and C. Miniatura (World Scientific, Singapore).

[213] M. A. Nielsen and I. L Chuang (2010). *Quantum Computation and Quantum Information* (Cambridge University Press, Cambridge).

[214] N. D. Mermin (2007). *Quantum Computer Science: An Introduction* (Cambridge University Press, Cambridge).

[215] R. J. Schoelkopf and S. M. Girvin (2008). Springer Nature: *Nature* **451**, 664.

[216] M. H. Devoret and R. J. Schoelkopf (2013). *Science* **339**, 1169.

[217] J. Koch, T. M. Yu, J. Gambetta, A. A. Houck, D. I. Schuster, J. Majer, A. Blais, M. Devoret, S. M. Girvin, and R. J. Schoelkopf (2007). *Phys. Rev.* **A76**, 042319.

[218] R. W. Richardson (1963). *Phys. Lett.* **3**, 277–279; (1964). *Nucl. Phys.* **52**, 221–238.

[219] L. P. Gor'kov (1959). *Sov. Phys. JETP* **36**, 1364.

[220] Z. Sun, M. Enayat, A. Maldonado, C. Lithgow, E. Yelland, D. C. Peets, A. Yaresko, A. P. Schnyder, and P. Wahl (2015). Springer Nature: *Nature Commun.* **6**, 6633.

[221] V. Ambegaokar and A. Baratoff (1963). *Phys. Rev. Lett.* **10**, 486; Erratum: *Phys. Rev. Lett.* **11**, 104.

[222] P. G. de Gennes (1966). *Superconductivity of Metals and Alloys* (W. A. Benjamin, New York).

[223] C. W. J. Beenakker (2013). *Annu. Rev. Condens. Matter Phys.* **4**, 113.

[224] P. Fulde and R. A. Ferrell (1964). *Phys. Rev.* **A135**, 550.

[225] A. I. Larkin and Y. N. Ovchinnikov (1965). *Sov. Phys. JETP* **20**, 762.

[226] M. W. Zwierlein, A. Schirotzek, C. H. Schunck, and W. Ketterle (2006). *Science* **311**, 492.

[227] G. B. Partridge, W. Li, R. I. Kamar, Y. Liao, and R. G. Hulet (2006). *Science* **311**, 503.

[228] Y.-A. Liao, A. S. C. Rittner, T. Paprotta, W. Li, G. B. Partridge, R. G. Hulet, S. K. Baur, E. J. Mueller (2010). Springer Nature: *Nature*, **467**, 567.

[229] K. Yang (2001). *Phys. Rev.* **B63**, 140511.

[230] R. Casalbuoni and G. Nardulli (2004). *Rev. Mod. Phys.* **76**, 263.

[231] A. P. Drozdov, M. I. Eremets, I. A. Troyan, V. Ksenofontov, and S. I. Shylin (2015). Springer Nature: *Nature* **525**, 73.

[232] Y. Tokura, H. Takagi, and S. Uchida (1989). Springer Nature: *Nature* **337**, 345.

[233] A. Damascelli, Z. Hussain, and Z.-X. Shen (2003). *Rev. Mod. Phys.* **75**, 473.

[234] F. C. Zhang and T. M. Rice (1988). *Phys. Rev.* **B37**, 3759.

[235] H. Ding, M. R. Norman, J. C. Campuzano, M. Randeria, A. F. Bellman, T. Yokoya, T. Takahashi, T. Mickiku, and K. Kadowaki (1996). *Phys. Rev.* **B54**, R9678(R).

[236] C. C. Tsuei and J. R. Kirtley (2000). *Rev. Mod. Phys.* **72**, 969.

[237] W. A. Harrison (1961). *Phys. Rev.* **123**, 85.

[238] G. Binnig and H. Rohrer (1987). *Rev. Mod. Phys.* **59**, 615.

[239] H.-H. Lai and K. Yang (2015). *Phys. Rev.* **B91**, 081110.

[240] M. Wolf (2006). *Phys. Rev. Lett.* **96**, 010404.

[241] D. Gioev and I. Klich (2006). *Phys. Rev. Lett.* **96**, 100503.

[242] W. Ding, A. Seidel, and K. Yang, *Phys. Rev.* **X2**, 011012 (2012).

[243] H.-H. Lai, K. Yang, and N. E. Bonesteel (2013). *Phys. Rev. Lett.* **111**, 210402.

[244] H. Li and F. D. M. Haldane (2008). *Phys. Rev. Lett.* **101**, 010504.

[245] M. Pouranvari and K. Yang (2013). *Phys. Rev.* **B88**, 075123.

[246] M. Pouranvari and K. Yang (2014). *Phys. Rev.* **B89**, 115104.

[247] R. Islam, R. Ma, P. M. Preiss, M. E. Tai, A. Lukin, M. Rispoli, and M. Greiner (2015). Springer Nature: *Nature* **528**, 77.

[248] F. Reif (1965). *Fundamentals of Statistical and Thermal Physics* (McGraw-Hill, New York)

[249] A. L. Fetter and J. D. Walecka (1971). *Quantum Theory of Many-Particle Systems* (McGraw-Hill, New York).

[250] E. Merzbacher (1998). *Quantum Mechanics*, 3rd edn (John Wiley and Sons, New York).

Index

^{3}He, p-wave superfluidity in, 621
r_s, of electron gas, 384
2DEG, *see* two-dimensional electron gas

Abrikosov vortex lattice, 569, 577
activation energy, 202
adiabatic
    theorem, 332, 447
    transport, 336
Affleck–Kennedy–Lieb–Tasaki (AKLT) model, 507
Aharonov–Bohm
    effect, 103, 222, 273, 336
    phase, 275, 326, 337
Ambegaokar–Baratoff relation, 604
Anderson
    –Higgs mechanism, 578
    localization, 198, 236, 253, 304
        in QHE, 305
        length, 255
        length, exponent, 321
        scaling theory, 283
        strong, 276
        suppression by spin–orbit coupling, 343
        weak, 254, 273
        weak anti-, 344
    theorem, 619
anomaly, chiral, 373
antiferromagnet, 4, 23, 481
anyon, 455
Arrhenius law, *see* activation energy
autocorrelation function, 666

backscattering, 317
    "glory" effect, 270
    quantum enhancement of, 270
band
    "twisted", 367

Chern, 351
conduction, 104
crossing, 352
gamma point, 201
index, 108
inversion, 200
Möbius, 365
magnetic, 329, 349
parity of, 367
rubber, 365
    $\mathbb{Z}_2$ topological classification of, 366
valence, 104
valley, 201
velocity
    anomalous, 358
width, 148
Bardeen–Cooper–Schrieffer (BCS), *see* BCS
BCS
    coherence factors, 597
    coherence length, 610
    reduced Hamiltonian, 594
    theory, 592
Berry
    connection, 333, 336, 457
        gauge choice for, 338, 340
    curvature, 166, 334, 346, 348, 648
        "radiated" by degeneracy, 341
        Dirac monopole, 341
        pseudovector nature of, 334, 348
    flux, 335
    phase, 274, 332, 334, 500
        geometric gates in quantum information processing, 341
        spin-s, 340
        spin-1/2, 339, 340
        spin–orbit coupling, 343

Bijl–Feynman formula (SMA), *see* Feynman
Bloch
    band
        magnetic, 344
        velocity, anomalous, 166, 346, 347
    oscillations, 168
    theorem, 105, 107, 164
Bogoliubov
    –de Gennes (BdG) Hamiltonian, 613
    –de Gennes (BdG) equation, 611
    modes, 548
    quasiparticle, 600, 604, 605, 627
    theory, 541, 542
    transformation, 496, 498, 540, 600
Bohr
    –Sommerfeld quantization, 176
    –van Leeuwen theorem, 481, 485
    magneton, 151, 482
    radius, 79
Boltzmann equation, 183, 265
    collision term, 183
        relaxation time approximation, 183, 185
    Drude conductivity, 190
    linearized, 187
    transport lifetime, 188
bond
    resonating, 122
    valence, 504
        solid (VBS), 505
Born approximation, 10, 16, 34, 47, 49, 98, 184, 254
Born–Oppenheimer approximation, 68, 70
Bose–Einstein
    condensation (BEC), 158, 209, 531, 545, 609

factor, 80, 90, 638
Bose–Einstein condensation (BEC), 532
bosonization, 83, 420, 465
boundary conditions
    periodic
        generalized, 327
Bragg
    diffraction, 44
    distributed reflector (DBR), 158, 546
    peaks, 36, 42, 89, 92
    reflection, 167
Brillouin zone, 31, 81
    boundary, 37, 74, 76, 109, 110, 128, 129, 167, 201, 499

causality, 59, 637
centrosymmetric, 29, 156, 619
charge, fractional, 430
Chern
    insulator, 355, 364
    number, 304, 349, 352, 362, 648
circuit QED, 588, 672
Clausius–Clapeyron equation, 551
Clifford group operation, 526
coherent state, 434, 532
cohesive energy, 131, 390
completeness relation, 47, 207
composite fermions, 469
composition, of loops, 650
compressibility, 66, 441, 487, 540, 645
    Gibbs–Duhem relation, 410
    isothermal, 252
    sum rule, 462
Compton effect, 11
condensation energy, superconductor, 579, 598
conductance, mesoscopic, 222–224, 233, 254, 284, 287
conductivity
    Drude, see Drude
    tensor, 103, 301
continuity equation, 257, 360, 398, 484, 554
Cooper pair, 417, 418, 429, 479, 530, 531, 563, 592, 604, 610, 619
    box, 588
    density wave states, 621
coordination number, 25, 120, 410, 491
core hole, 422
Coriolis force, 576
correlation function
    retarded, 59, 636
creation operator
    electron, 487

fermion, 613, 681
    phonon, 84
    photon, 14
critical exponent, 321
    dynamical, 93, 289, 291
cross section
    absorption, 423
    differential, 259
    scattering, 11
crystal
    field, 131
    liquid, 1, 19, 41
Curie
    law, 423, 486
    temperature, 481
current, persistent, see persistent current
curvature, Berry, see Berry
curvature, Gaussian, 341, 349, 651, 656
cyclotron
    frequency, 176, 307
    motion, circular, 307
    resonance, 176, 180, 210

de Haas–van Alphen oscillations, 178, 485
Debye
    –Waller factor, 40, 53, 88, 91, 95, 419, 493, 499
    screening, 440, 441
    temperature, 80–82, 103, 594
degeneracy
    Landau level, see Landau
    valley, 201
density
    N-representable, 387
    V-representable, 387
density functional theory, 378, 385, 544
density matrix, reduced, 250, 534, 659
density of states, 112, 115, 148
    in wave-vector space, 48, 227
    thermodynamic, 150, 645
    tunneling, 290, 645
        pseudogap in, 290
dephasing, see phase coherence
detailed balance, 186, 638
diamagnet, 4, 480
diamagnetism, perfect, 549
dielectric function, 278, 593
    Lindhard, 394
    optical, 401
    static, 393
diffusion
    as random walk, 257, 258
    constant, 190, 252

equation, 257
    quantum, 254, 265
        corrections, 268
Dirac
    electrons
        Hall effect for, 322
        Landau levels, 323
    equation, 125, 321, 368, 486
    fermion, 136, 177
        chirality, 353, 356
        mass, 362
    Hamiltonian, 343, 353, 362, 368
        four-component, 373
        massless surface states of, 370
        Weyl representation, 368
    mass, 137, 353
    monopole, 341
    negative energy solutions, 137
    point, 135, 136, 356, 369
    sea, 138
    string, 342
disorder
    annealed, 262
    quenched, 262
displacement field, elastic continuum, 64
distributed Bragg reflector (DBR), see Bragg
distribution
    Fermi–Dirac, 149, 247
    function
        pair, 17, 444
        radial, 17
        two-point, 17
    Poisson, 104, 245
drift velocity, 99
    $c\vec{E} \times \vec{B}/B^2$, 314
Drude
    conductivity, 100
        ac, 100
    model, 99, 190
        semiclassical, 164

edge mode
    chiral, 354, 365
    counter-propagating, 320
    gapless, chiral, 355
    helical, 363
edge states
    topological, 146
effective mass approximation, 120, 166
Ehrenfest theorem, 165, 167
Einstein
    A, B coefficients, 186
    relation, 190, 192, 252
elasticity, 64
    tensor, 70, 390

electric polarization, 360
electrochemical potential, 192, 228
electromagnetic field, quantization
    of, 13
electron
    classical radius of, 11
    tunneling spectroscopy, 600
electronic structure calculations, *ab*
        initio, 390
energy functional, 142, 386
    elastic, 66
entanglement, 588
    entropy, 251, 662
        area law of, 662
    Hamiltonian, 663
    monogamy, 515, 662
    with environment, 248, 250
envelope function, 125
evaporative cooling, 536
exchange
    -correlation hole, 445
    direct, 490
    energy, 380, 384
    hole, 384
    non-local, 380
    operator, 452

Fabry–Pérot cavity, 158, 234
Fermi
    arc, 372
        Golden Rule, 637
        wavelength, 225
fermionization, 83
ferrimagnet, 481
ferromagnet, 4, 480
Feshbach resonance, 537
Feynman
    Bijl–Feynman formula (SMA), 52,
        461
    Hellman–Feynman theorem, 116
    path integral, 236, 266, 337
Fick's law, 257
fine-structure constant, 100, 303
Floquet theorem, 113
fluctuation–dissipation theorem, 638,
        668
fluctuations
    rare, 242
flux
    -flow resistance, 451, 587
    quantum, 223, 304, 309, 311, 327,
        337, 342, 440
    tube, magnetic, 336, 447, 455
    variable, in electrical circuits, 670
form factor, 16
    atomic, 12
    crystal, 16
    multi-atom, 37

four-terminal impedance
        measurement, 233
fractional
    charge, 244, 447
    quantum Hall effect (FQHE), 301
    quantum Hall liquid, 420
    quantum number, 303
    statistics, 303, 452, 455, 456
Friedel
    oscillations, 395
    sum rule, 421
frog, levitation of, 481
Fu–Kane formula, 367
Fulde–Ferrell–Larkin–Ovchinnikov
        (FFLO) states, 620
functional derivative, 673
fundamental group, 174, 453, 650

Galilean invariance, 305, 413, 414
gamma point, *see* band
gate charge, *see* offset charge
gauge
    field, synthetic, 577
    Landau, 309
    symmetric, 309, 433
    transformation
        singular, 447
    transformation, large, 359, 361
Gauss–Bonnet theorem, 349, 651
Gaussian
    correlators, *see* Wick's theorem
    curvature, *see* curvature
genus, 349, 647
geometric phase, *see* Berry phase
geometry
    of Hilbert space, 331
    plane, 331
Gibbs potential, 563
Ginzburg–Landau
    coherence length, 567
    theory, 544, 559, 568
Girvin–MacDonald–Platzman
        algebra, 472
Goldstone
    mode, 7, 76, 541, 578
    theorem, 7, 71
graphene, 27, 134
    band sturcture, 362
Green's function, 262
guiding center, 310, 313, 432

Haldane
    gap, 500, 507, 509
    model, 356, 363
        mass, 357, 362
    pseudopotential, 436
Hall
    state, quantum anomalous, 370

Hall effect
    Anderson localization in, 318
    anomalous in ferromagnets, 348
    coefficient, 103, 173, 179, 181
    edge
        currents, 315
        states, chiral, 317
    fractional quantum (FQHE), 303,
        433
    integer quantum (IQHE), 303, 364
    Landauer transport picture, 315
    longitudinal resistance, 303
    percolation
        fractal dimension, 321
        picture, 318
        transition, 320
    plateau, 302
    quantum spin, 363
    resistance, 302
    resistivity, classical, 301
handle, *see* genus
harmonic approximation, 40, 69
Hartree
    approximation, 377
Hartree–Fock
    approximation, 378
    equations, 380
    self-energy, 383
heat capacity, *see* specific heat
Heisenberg spin model, 490
Hellman–Feynman theorem, *see*
        Feynman
HEMT, 219
hierarchy states, 469
homeomorphism, 648
homotopy, 649
    class, 453, 649
    fundamental group, *see*
            fundamental group
    group, second, 655
Hubbard model, 163, 486

independent-electron approximation,
        376
index theorem
    Atiyah–Singer, 144
insulator
    Anderson, 252
    band, 252
    Mott, 4, 252, 487
interband transitions, 168, 169

jellium model, 382, 419, 439
Josephson
    effect, 604
        (ac), 583
    energy, 581, 588, 605
    junction, 581, 672

π junction, 626
   critical current, 582, 604
   relation (first), 582
   relation (second), 582, 672
   relations, 584

Kane–Mele
   model, 363
kinetic inductance, 100, 586
Kitaev
   model, 614
   toric code, 521, 525
Kohn's theorem, 461
Kohn–Sham equations, 388, 389
Kondo problem, 423, 675
Koopmans' theorem, 382
Kramers degeneracy, 154, 363, 369
Kramers–Kronig relations, 402, 637
Kubo formula, 632

Lagrangian
   vector potential in, 337
Lamé coefficient, 65
Landau
   diamagnetism, 528, 558, 608
   Fermi liquid
      instabilities of, 412, 618
      nematic, 413
      Pomeranchuk instability of, 413
   Fermi liquid theory, 376, 385, 402
   gauge, see gauge
   level, 176, 177, 311
      degeneracy of, 327, 374, 435,
         472, 528
      filling factor, 304
      tilted, 314
Landau–Ginzburg–Wilson theory,
      562
Landauer–Büttiker
   "noise is the signal", 242
   formula, 229
   multi-terminal formula, 231
   multichannel transport, 276
   transmission eigenvalues, 239
   transport, 317
   voltage probe, 232
laser speckle, 233, 238, 267
lattice
   Bravais, 24
   optical, 159, 359
   reciprocal, 24, 30
   vector, 24
      primitive, 24
Laue diffraction, 36
Laughlin
   quasielectron state, 446
   quasihole state, 446
   wave function, 442

Lieb–Schultz–Mattis theorem, 501
Lindhard function, 395
linear response, 59, 632
Liouville's theorem, 182
local moment, 423
local-density approximation (LDA),
      389
localization
   Anderson, see Anderson
   many-body (MBL), 663
London
   equation, 553, 554
   gauge, 554
   penetration length, 550, 554
long-range order, 4, 19, 493, 502
   off-diagonal (ODLRO), 534, 535
   positional, 18
Lorentz
   covariance, 306
   force, 10, 301, 336
   number, 98
   transformation, 306
Lorentz force, 10
low-depth circuit, 519
lower critical dimension, 493
Luttinger liquid, 419
   chiral, 466
Luttinger's theorem, 408

Mössbauer effect, 15, 93
magnetic
   length, 309
   monopole, 342, 371
   pressure, 482
   susceptibility, 141, 480
magnetism, role of exchange in, 380
magnetization, 152
   definiton of, 480
   staggered, 500
magnetophonon mode, 462
magnetoresistance, 103, 181
   negative, due to weak localization,
      273
   Sharvin–Sharvin experiment, 275
   weak localization, 344
magnetoroton, 462
magnon, 45
Majorana
   fermion, 614, 615, 683
   zero-mode, 616
Majumdar–Ghosh model, 504
many-body localization (MBL), 297
mass action, law of, 202
matrix product state (MPS), 511
Matthiessen's rule, 99
Meissner
   effect, 530, 549, 606
   kernel, 556

phase, 550
mesoscopic
   conductance, non-self-averaging
      of, 233
   devices, 223
mid-gap state, 144
Miller indices, 32
mobility, 203
modulation doping, 219, 430
modulus
   bulk, 66
   shear, 66
momentum
   canonical, 307, 308, 431
   mechanical, 307, 308, 431
monopole, see Dirac, 353
MOSFET, 217
Mott insulator, see insulator, Mott

Néel
   order, 488
   state, 495
   temperature, 623
Nagaoka theorem, 488
noise
   $1/f$, 240, 243
   autocorrelation time, 243
   current, 242
   Johnson–Nyquist, 243
   nonequilibrium, 244
   partition, 246
      maximal, 246
   random telegraph, 245
   shot, 238, 239, 244
      fractional charge, 244, 467
      quantum, 245
   spectrum, 242
   white, 666

offset charge, 589
Onsager relation, 195, 635
orbit
   closed, 174
   electron, 174
   hole, 174
   open, 174
order parameter, 5, 22
   superconducting, 559
orthogonality catastrophe, 422, 466
oscillator strength, 50
   projected, 461
   sum rule, 52, 96
over screening, 630
   by phonons, 593

pair distribution function, see
      distribution function
paramagnet, 4, 480

participation ratio, inverse, 255
Pauli paramagnetic susceptibility, 529
Peierls instability, 140
persistent current, in mesoscopic rings, 224, 282
perturbation theory
  degenerate, 110
phase coherence, 223, 224, 235
  dephasing, 224
    inelastic scattering, 224, 248
    rate, 249
    role of entanglement, 248
  length, 224, 233, 287
phase transition
  continuous, 277
  metal–insulator, 253, 277
phonon, 45
  acoustic, 73, 76, 594
  mode of superfluid, 541
  optical, 67, 73
photoemission
  angle-resolved (ARPES), 210, 211, 362, 372, 625
Pippard coherence length, 611
plasma frequency, 396
Poisson
  distribution, *see* distribution, Poisson
  summation formula, 641
polariton, 86
  exciton, 86, 209
  phonon, 86
  upper, lower, 547
polarizability, 159
  atomic, 160
polarization
  electric, 360
  function, electron gas, 393
  spin, 152
polyacetylene, 141
power spectrum
  see spectral density, 666
projected entangled pair states (PEPS), 512
projective sphere, 453, 650
proximity effect, 631
pseudospin, 156, 298, 322, 343, 344, 352, 595, 630
pump, electron, 359

quantum
  bit (qubit), 525, 587
    logical, 525
    physical, 525
    transmon, 589
  critical point, 256, 288
  disordered, 499

dot, 221, 359
  error correction, 526
  point contact, 221, 467
  well, 189, 218
  wire, open channels of, 227
quasicrystal, 19, 41

random phase approximation (RPA), 396
random walk, *see* diffusion
renormalization group, 7, 277, 415, 417, 423, 562, 679
  flow, 425, 427, 595
resistance
  quantum of, 231
resistivity tensor, 102
resolvent, 295
  operator, 261
resonating valence bond (RVB) states, 516
Rydberg, 2, 79, 205, 208, 572

S-matrix, 230, 657
scanning tunneling microscope (STM), 43, 570, 642
  spectroscopy mode, 646
  surface topography mode, 645
scattering
  amplitude, 12, 184
  angle, 12
  cross section, 49, 101, 259
    differential, 48, 183, 260
  elastic, 12
  length, 49
  partial wave expansion, 189
  phase shift, 260
  Thomson, 10
  X-ray, 10
Schmidt decomposition, 661
semiconductor
  p–n junction, 212
  acceptor, 206
  band bending, 213
  depletion layer, 213
  diode, 212
  direct gap, 200
  donor, 204
  doped, 204
  heterostructure, 217
  indirect gap, 201
  inhomogeneous, 212
  intrinsic vs. extrinsic, 204
  inversion layer, 217
    electric subbands, 217
  quantum well, *see* quantum well
semimetal, 112
  graphene, 114, 136
  Weyl, *see* Weyl

shot noise, *see* noise
Shubnikov–de Haas oscillations, 178
single-mode approximation (SMA), 60, 399, 463
  Bijl–Feynman formula, *see* Feynman
singular-value decomposition (SVD), 238, 661
soliton, 141
sound, speed of, 66, 72, 81, 208
  of light, 548
specific heat
  classical, 79
  Debye model, 81
  Einstein model, 79
  electronic, 83, 150
spectral density, 637
  one-sided, 638, 668
  two-sided, 668
spin
  density wave (SDW), 481
  rotation symmetry, 154
  susceptibility, *see* susceptibility
  waves, 491
spin–charge separation, 148
spin-liquid state, 518
spin–orbit coupling, 151
  synthetic, 577
spinon, 502
SQUID, superconducting quantum interference device, 585
statistics
  angle, 455
  Fermi–Dirac, 103
  Laughlin quasiparticle, 457
  non-Abelian, 479
Stokes theorem, 334, 570
strain tensor, 65
string order, 508
structure factor
  dynamical, 48, 50, 89
    simple harmonic oscillator, 57
  projected, 461
  static, 16
Su–Schrieffer–Heeger (SSH) model, 140, 503, 517, 621
subband index, 219, 226, 229
superconducting gap-T_c ratio, 603
superconductor
  heavy fermion, 627
  high-temperature, 622
    copper-oxide (CuO_2) planes in, 622
  topological, 614
  Type-I, 550
  Type-II, 552
    mixed state of, 567
  unconventional, 617

superexchange, 490
superfluid, 609
superlattice, 158
   semiconductor, 219
superposition principle, 587
supersolid, 535
surface reconstruction, 43
susceptibility
   dissipative part of, 638
   reactive part of, 638
   spin, 152, 411, 424
symmetry
   chiral, 145
   inversion, 21, 348, 366
   particle–hole, 145, 196
   point, 26, 618
   rotation, 29
   spontaneously broken, 7
   time-reversal, 348, 363, 635
   translation, 5, 64, 85, 106, 171,
      254, 310, 383, 611, 633
      magnetic, 344
      spontaneously broken, 18, 20,
        22, 493
synchrotron radiation, 10

tensor, conductivity, *see* conductivity
thermal conductivity
   classical, 101
   of electrons, 195
thermoelectric effect
   Mott formula, 196
   phonon drag, 196
   thermopower, 196
Thomas–Fermi screening, 278, 394
Thouless
   charge pump, 358
   energy, 280
tight-binding model, 118
topological
   charge, 654
   defect, 654
   insulator, 200, 343, 364, 365, 616
      $\mathbb{Z}_2$ classification, 365

massless surface states of, 370
   strong, 3D, 368
   weak, 3D, 368
   invariant, 331, 647
   order, 303
   phase, 352
   phase, symmetry-protected (SPT),
      370
   protection, 331
   superconductor, 616
topology, 331
toric code, *see* Kitaev
transistor
   bipolar, 214
   field effect (FET), 217
translation
   magnetic, 326, 327, 329
   operator, 106
transport
   lifetime, *see* Boltzmann equation
   non-local, 224
   quantum, 222
   scattering theory of, 223
tunnel junction, 466, 581, 643
tunneling Hamiltonian, 642
turnstile, electron, 359
two-dimensional electron gas
      (2DEG), 217, 225, 301
two-level fluctuators, 240

umklapp process, 90, 595
unit cell
   conventional, 28
   magnetic, 327, 329
   primitive, 25
      Wigner–Seitz, 25
universal conductance fluctuations
      (UCF), 224, 233

vacuum state, 595, 680
valence bond solid (VBS), *see* bond
van Hove singularity, 115, 176
Vandermonde polynomial, 438

voltage, *see* electrochemical potential
vortex, 654
   lines, persistence length of, 572
   superconductor, 602

Wannier function, 123, 486
   non-local in topological bands,
      123, 351
wave packet
   Gaussian, 164
Weyl
   fermions, 371
   Hamiltonian, 371
   point, 371
   semimetal, 371
which path
   detector, 249
   information erasure, 250
Wick's theorem, 237
Wiedemann–Franz law, 98, 195
Wiener–Khinchin theorem, 243, 666
Wigner crystal, 419, 430
Wigner–Seitz cell, *see* unit cell,
      primitive
Wilson ratio, 152
winding number, 275, 336, 337, 574,
      649
wrapping (or engulfing) number, 655

X-ray
   edge problem, 423
   photoemission, 422

Zeeman energy, 151
zero mode, 144, 369
   boundary, 148
zero-point
   energy, 87
   fluctuations, 92, 94
zero-sound mode, ^{3}He, 412
zone scheme, 167
   extended, 74
   reduced, 74

PERIODIC TABLE

Image: David Freund / Photodisc / Getty.

Fundamental Constants (2014 CODATA Recommended values)

Quantity	Symbol	Value
Speed of light in vacuum	c	299 792 458 m/s (exact)
Planck constant	h	6.626 070 040(81) × 10^{-34} J s
Elementary charge	e	1.602 176 6208(98) × 10^{-19} C
Electron mass	m_e	9.109 383 56(11) × 10^{-31} kg
(energy units)	$m_e c^2$	0.510 998 9461(31) MeV
Inverse fine structure constant	α^{-1}	137.035 999 139(31)
Bohr radius	a_0	0.529 177 210 67(12) × 10^{-10} m
Rydberg	R_∞ ($R_\infty c$)	3.289 841 960 355(19) × 10^{15} Hz
(frequency units)	R_∞/h	13.605 693 009(84) eV
Bohr magneton	μ_B	
(frequency units)	μ_B/h	13.996 245 042(86) × 10^9 Hz/T
(temperature units)	μ_B/k_B	0.671 714 05(39) K/T
Boltzmann constant	k_B/h	2.083 6612(12) × 10^{10} Hz/K
(frequency units)	k_B	1.380 648 52(79) × 10^{-23} J/K
Superconducting flux quantum	$\Phi_S = \frac{2\pi}{2e} = 2\pi s$	2.067 833 831(13) × 10^{-15} Wb
Josephson constant	$\frac{2e}{2\pi} = \frac{1}{\Phi_0}$	483 597.8525(30) × 10^9 Hz/V
von Klitzing constant	$R_K = \frac{h}{e^2}$	25 812.807 4555(59) Ohms

Figures in parentheses indicate the uncertainty in the last two digits. T is Tesla, K is Kelvin, Wb is Weber. The ordinary flux quantum is $\Phi_0 = 2\Phi_S = \frac{h}{2e}$ in SI units and $\frac{hc}{2e}$ in cgs units as used in the text. N.B. At the time of this writing, a major revision of the SI system of units is under consideration in which exact numerical values for the Planck constant, the elementary charge, the Boltzmann constant, and the Avogadro constant will be defined.

Legend

- Atomic Number: 58
- Symbol: Ce
- Name: Cerium
- Atomic Weight†: 140.116
- Ground-state Configuration: [Xe]4f5d6s²
- Ground-state Level: ¹G°₄
- Ionization Energy (eV): 5.5387

†Based upon ¹²C. () indicates the mass number of the most stable isotope.

Period / Group data

Period 1
- 1 H Hydrogen 1.00794 1s ²S₁/₂ 13.5984
- 2 He Helium 4.002602 1s² ¹S₀ 24.5874

Period 2
- 3 Li Lithium 6.941 1s²2s ²S₁/₂ 5.3917
- 4 Be Beryllium 9.012182 1s²2s² ¹S₀ 9.3227
- 5 B Boron 10.811 1s²2s²2p ²P°₁/₂ 8.2980
- 6 C Carbon 12.0107 1s²2s²2p² ³P₀ 11.2603
- 7 N Nitrogen 14.0067 1s²2s²2p³ ⁴S°₃/₂ 14.5341
- 8 O Oxygen 15.9994 1s²2s²2p⁴ ³P₂ 13.6181
- 9 F Fluorine 18.9984032 1s²2s²2p⁵ ²P°₃/₂ 17.4228
- 10 Ne Neon 20.1797 1s²2s²2p⁶ ¹S₀ 21.5645

Period 3
- 11 Na Sodium 22.989770 [Ne]3s ²S₁/₂ 5.1391
- 12 Mg Magnesium 24.3050 [Ne]3s² ¹S₀ 7.6462
- 13 Al Aluminum 26.981538 [Ne]3s²3p ²P°₁/₂ 5.9858
- 14 Si Silicon 28.0855 [Ne]3s²3p² ³P₀ 8.1517
- 15 P Phosphorus 30.973761 [Ne]3s²3p³ ⁴S°₃/₂ 10.4867
- 16 S Sulfur 32.065 [Ne]3s²3p⁴ ³P₂ 10.3600
- 17 Cl Chlorine 35.453 [Ne]3s²3p⁵ ²P°₃/₂ 12.9676
- 18 Ar Argon 39.948 [Ne]3s²3p⁶ ¹S₀ 15.7596

Period 4
- 19 K Potassium 39.0983 [Ar]4s ²S₁/₂ 4.3407
- 20 Ca Calcium 40.078 [Ar]4s² ¹S₀ 6.1132
- 21 Sc Scandium 44.955910 [Ar]3d4s² ²D₃/₂ 6.5615
- 22 Ti Titanium 47.867 [Ar]3d²4s² ³F₂ 6.8281
- 23 V Vanadium 50.9415 [Ar]3d³4s² ⁴F₃/₂ 6.7462
- 24 Cr Chromium 51.9961 [Ar]3d⁵4s ⁷S₃ 6.7665
- 25 Mn Manganese 54.938049 [Ar]3d⁵4s² ⁶S₅/₂ 7.4340
- 26 Fe Iron 55.845 [Ar]3d⁶4s² ⁵D₄ 7.9024
- 27 Co Cobalt 58.933200 [Ar]3d⁷4s² ⁴F₉/₂ 7.8810
- 28 Ni Nickel 58.6934 [Ar]3d⁸4s² ³F₄ 7.6398
- 29 Cu Copper 63.546 [Ar]3d¹⁰4s ²S₁/₂ 7.7264
- 30 Zn Zinc 65.409 [Ar]3d¹⁰4s² ¹S₀ 9.3942
- 31 Ga Gallium 69.723 [Ar]3d¹⁰4s²4p ²P°₁/₂ 5.9993
- 32 Ge Germanium 72.64 [Ar]3d¹⁰4s²4p² ³P₀ 7.8994
- 33 As Arsenic 74.92160 [Ar]3d¹⁰4s²4p³ ⁴S°₃/₂ 9.7886
- 34 Se Selenium 78.96 [Ar]3d¹⁰4s²4p⁴ ³P₂ 9.7524
- 35 Br Bromine 79.904 [Ar]3d¹⁰4s²4p⁵ ²P°₃/₂ 11.8138
- 36 Kr Krypton 83.798 [Ar]3d¹⁰4s²4p⁶ ¹S₀ 13.9996

Period 5
- 37 Rb Rubidium 85.4678 [Kr]5s ²S₁/₂ 4.1771
- 38 Sr Strontium 87.62 [Kr]5s² ¹S₀ 5.6949
- 39 Y Yttrium 88.90585 [Kr]4d5s² ²D₃/₂ 6.2173
- 40 Zr Zirconium 91.224 [Kr]4d²5s² ³F₂ 6.6339
- 41 Nb Niobium 92.90638 [Kr]4d⁴5s ⁶D₁/₂ 6.7589
- 42 Mo Molybdenum 95.94 [Kr]4d⁵5s ⁷S₃ 7.0924
- 43 Tc Technetium (98) [Kr]4d⁵5s² ⁶S₅/₂ 7.28
- 44 Ru Ruthenium 101.07 [Kr]4d⁷5s ⁵F₅ 7.3605
- 45 Rh Rhodium 102.90550 [Kr]4d⁸5s ⁴F₉/₂ 7.4589
- 46 Pd Palladium 106.42 [Kr]4d¹⁰ ¹S₀ 8.3369
- 47 Ag Silver 107.8682 [Kr]4d¹⁰5s ²S₁/₂ 7.5762
- 48 Cd Cadmium 112.411 [Kr]4d¹⁰5s² ¹S₀ 8.9938
- 49 In Indium 114.818 [Kr]4d¹⁰5s²5p ²P°₁/₂ 5.7864
- 50 Sn Tin 118.710 [Kr]4d¹⁰5s²5p² ³P₀ 7.3439
- 51 Sb Antimony 121.760 [Kr]4d¹⁰5s²5p³ ⁴S°₃/₂ 8.6084
- 52 Te Tellurium 127.60 [Kr]4d¹⁰5s²5p⁴ ³P₂ 9.0096
- 53 I Iodine 126.90447 [Kr]4d¹⁰5s²5p⁵ ²P°₃/₂ 10.4513
- 54 Xe Xenon 131.293 [Kr]4d¹⁰5s²5p⁶ ¹S₀ 12.1298

Period 6
- 55 Cs Cesium 132.90545 [Xe]6s ²S₁/₂ 3.8939
- 56 Ba Barium 137.327 [Xe]6s² ¹S₀ 5.2117
- 57 La Lanthanum 138.9055 [Xe]5d6s² ²D₃/₂ 5.17
- 72 Hf Hafnium 178.49 [Xe]4f¹⁴5d²6s² ³F₂ 6.8251
- 73 Ta Tantalum 180.9479 [Xe]4f¹⁴5d³6s² ⁴F₃/₂ 7.5496
- 74 W Tungsten 183.84 [Xe]4f¹⁴5d⁴6s² ⁵D₀ 7.8640
- 75 Re Rhenium 186.207 [Xe]4f¹⁴5d⁵6s² ⁶S₅/₂ 7.8335
- 76 Os Osmium 190.23 [Xe]4f¹⁴5d⁶6s² ⁵D₄ 8.4382
- 77 Ir Iridium 192.217 [Xe]4f¹⁴5d⁷6s² ⁴F₉/₂ 8.9670
- 78 Pt Platinum 195.078 [Xe]4f¹⁴5d⁹6s ³D₃ 8.9588
- 79 Au Gold 196.96655 [Xe]4f¹⁴5d¹⁰6s ²S₁/₂ 9.2255
- 80 Hg Mercury 200.59 [Xe]4f¹⁴5d¹⁰6s² ¹S₀ 10.4375
- 81 Tl Thallium 204.3833 [Hg]6p ²P°₁/₂ 6.1082
- 82 Pb Lead 207.2 [Hg]6p² ³P₀ 7.4167
- 83 Bi Bismuth 208.98038 [Hg]6p³ ⁴S°₃/₂ 7.2855
- 84 Po Polonium (209) [Hg]6p⁴ ³P₂ 8.414
- 85 At Astatine (210) [Hg]6p⁵ ²P°₃/₂
- 86 Rn Radon (222) [Hg]6p⁶ ¹S₀ 10.7485

Period 7
- 87 Fr Francium (223) [Rn]7s ²S₁/₂ 4.0727
- 88 Ra Radium (226) [Rn]7s² ¹S₀ 5.2784
- 89 Ac Actinium (227) [Rn]6d7s² ²D₃/₂ 5.17
- 104 Rf Rutherfordium (261) [Rn]5f¹⁴6d²7s² ³F₂? 6.0?
- 105 Db Dubnium (262)
- 106 Sg Seaborgium (266)
- 107 Bh Bohrium (264)
- 108 Hs Hassium (277)
- 109 Mt Meitnerium (268)
- 110 Uun Ununnilium (281)
- 111 Uuu Unununium (272)
- 112 Uub Ununbium (285)
- 114 Uuq Ununquadium (289)
- 116 Uuh Ununhexium (292)

Lanthanides
- 58 Ce Cerium 140.116 [Xe]4f5d6s² ¹G°₄ 5.5387
- 59 Pr Praseodymium 140.90765 [Xe]4f³6s² ⁴I°₉/₂ 5.473
- 60 Nd Neodymium 144.24 [Xe]4f⁴6s² ⁵I₄ 5.5250
- 61 Pm Promethium (145) [Xe]4f⁵6s² ⁶H°₅/₂ 5.582
- 62 Sm Samarium 150.36 [Xe]4f⁶6s² ⁷F₀ 5.6437
- 63 Eu Europium 151.964 [Xe]4f⁷6s² ⁸S°₇/₂ 5.6704
- 64 Gd Gadolinium 157.25 [Xe]4f⁷5d6s² ⁹D°₂ 6.1498
- 65 Tb Terbium 158.92534 [Xe]4f⁹6s² ⁶H°₁₅/₂ 5.8638
- 66 Dy Dysprosium 162.500 [Xe]4f¹⁰6s² ⁵I₈ 5.9389
- 67 Ho Holmium 164.93032 [Xe]4f¹¹6s² ⁴I°₁₅/₂ 6.0215
- 68 Er Erbium 167.259 [Xe]4f¹²6s² ³H₆ 6.1077
- 69 Tm Thulium 168.93421 [Xe]4f¹³6s² ²F°₇/₂ 6.1843
- 70 Yb Ytterbium 173.04 [Xe]4f¹⁴6s² ¹S₀ 6.2542
- 71 Lu Lutetium 174.967 [Xe]4f¹⁴5d6s² ²D₃/₂ 5.4259

Actinides
- 90 Th Thorium 232.0381 [Rn]6d²7s² ³F₂ 6.3067
- 91 Pa Protactinium 231.03588 [Rn]5f²6d7s² ⁴K₁₁/₂ 5.89
- 92 U Uranium 238.02891 [Rn]5f³6d7s² ⁵L°₆ 6.1941
- 93 Np Neptunium (237) [Rn]5f⁴6d7s² ⁶L₁₁/₂ 6.2657
- 94 Pu Plutonium (244) [Rn]5f⁶7s² ⁷F₀ 6.026
- 95 Am Americium (243) [Rn]5f⁷7s² ⁸S°₇/₂ 5.9738
- 96 Cm Curium (247) [Rn]5f⁷6d7s² ⁹D°₂ 5.9914
- 97 Bk Berkelium (247) [Rn]5f⁹7s² ⁶H°₁₅/₂ 6.1979
- 98 Cf Californium (251) [Rn]5f¹⁰7s² ⁵I₈ 6.2817
- 99 Es Einsteinium (252) [Rn]5f¹¹7s² ⁴I°₁₅/₂ 6.42
- 100 Fm Fermium (257) [Rn]5f¹²7s² ³H₆ 6.50
- 101 Md Mendelevium (258) [Rn]5f¹³7s² ²F°₇/₂ 6.58
- 102 No Nobelium (259) [Rn]5f¹⁴7s² ¹S₀ 6.65
- 103 Lr Lawrencium (262) [Rn]5f¹⁴7s²7p? ²P°₁/₂? 4.9?

Printed in the United States
by Baker & Taylor Publisher Services